Materials Horizons: From Nature to Nanomaterials

Series Editor

Vijay Kumar Thakur, School of Aerospace, Transport and Manufacturing,
Cranfield University,
Cranfield, UK

Materials are an indispensable part of human civilization since the inception of life on earth. With the passage of time, innumerable new materials have been explored as well as developed and the search for new innovative materials continues briskly. Keeping in mind the immense perspectives of various classes of materials, this series aims at providing a comprehensive collection of works across the breadth of materials research at cutting-edge interface of materials science with physics, chemistry, biology and engineering.

This series covers a galaxy of materials ranging from natural materials to nanomaterials. Some of the topics include but not limited to: biological materials, biomimetic materials, ceramics, composites, coatings, functional materials, glasses, inorganic materials, inorganic-organic hybrids, metals, membranes, magnetic materials, manufacturing of materials, nanomaterials, organic materials and pigments to name a few. The series provides most timely and comprehensive information on advanced synthesis, processing, characterization, manufacturing and applications in a broad range of interdisciplinary fields in science, engineering and technology.

This series accepts both authored and edited works, including textbooks, monographs, reference works, and professional books. The books in this series will provide a deep insight into the state-of-art of Materials Horizons and serve students, academic, government and industrial scientists involved in all aspects of materials research.

Review Process

The proposal for each volume is reviewed by the following:

1. Responsible (in-house) editor
2. One external subject expert
3. One of the editorial board members.

The chapters in each volume are individually reviewed single blind by expert reviewers and the volume editor.

Abhay Kumar Singh

2D Transition-Metal Dichalcogenides (TMDs): Fundamentals and Application

Abhay Kumar Singh
Faculty of Engineering & the Built
Environment Dean's Office
University of Johannesburg
Johannesburg, South Africa

ISSN 2524-5384 ISSN 2524-5392 (electronic)
Materials Horizons: From Nature to Nanomaterials
ISBN 978-981-96-0246-9 ISBN 978-981-96-0247-6 (eBook)
https://doi.org/10.1007/978-981-96-0247-6

This Springer imprint is published by the registered company Springer Nature Singapore Pte Ltd.
The registered company address is: 152 Beach Road, #21-01/04 Gateway East, Singapore 189721, Singapore

Preface

This book is intended for students, researchers, and scientists working in the area of Transition Metal Dichalcogenides (TMDs) materials science and technology from a broad subject such as physics, chemistry and nanotechnology, materials science and technology. This fully packaged research book predominately would be helpful to undergraduate and postgraduate students, doctoral students, researchers, scientists, and technologists, who wish to build a career in the area of Transition Metal Dichalcogenides (TMDs) (or chalcogenide or beyond the graphene two-dimensional (2D) materials) based materials science and technology including nanoscience and nanotechnology.

The prime goal of this book is to provide a sound/or clear idea about beyond the graphene two-dimensional Transition Metal Dichalcogenides (2D TMDs) materials by providing their sound background with interpretations of the inside physiochemical mechanism as well as technological applications in various areas such as nanoelectronics, optoelectronics, topological insulators, biomedical, etc.

Herein the effort is made to present the book topics sequentially from their innovations to modern age developments in theoretical approaches and experimental interpretations including technical applications in a very simple way. This current novel research and technology book based on 2D TMDs contains 9 chapters, those topics are broadly dealing with the beginning introduction of 2D TMDs to their potential applications including the interpretation of their deep inside physiochemical mechanism at the nanoscale level.

The sound feature of the current book:

(1) Review the scenarios of 2D TMDs materials science for their wide range applications in distinct filed of nanotechnology devices and developments.
(2) Illustrate the scenarios of 2D materials science for the TMDs and vast potential utilities in the various technical areas for the device fabrications and developments.
(3) The detailed physio-chemical interpretations are addressed for various kinds of TMDs systems including preparation methods for the fabrication of the

advanced nanomaterials functionalization ability to fabricate various kinds of compositions.

(4) Enlightens the consequence of 2D TMDs materials for their prospective applications by providing a details description of novel fundamental properties of this kind of materials.

(5) Enlightens the consequence of 2D TMDs on their structure and defects-induced properties that impacts their wide range of applications in various fields.

(6) Enlightens the interesting spin-valley coupling that is considered the origin of the novel extraordinary physiochemical characteristics in TMDs and a play crucial role in their wide range of technical applications.

(7) Enlightens the TMD heterostructures formation with the kinds of materials that are useful to fabricate various high performances devices for distinct utility.

(8) Enlightens the emerging topological insulating behavior of TMDs in detail from their beginning to current age developments because of experimental and theoretical interpretations as well as their kinds of technical applicability.

(9) Enlightens an overview of emerging TMDs based devices for their kinds of applications in different areas and discussed in detail some of the important device's fabrication and their physical characteristic properties in terms of common uses.

(10) Special attention was also paid to one of the most emerging techniques "Atomic Layer Deposition (ALD)" and their applicability to fabricate TMDs based ultra-thin high-performance devices for these kinds of uses.

(11) Discusses the imminent projections and tasks to demonstrate the emerging novel TMDs and their potential nanoscale devices for the safety of society's health and advanced technology.

Chapter 1

Work deals with the emerging two-dimensional Transition Metal Chalcogenides (2D TMDs) materials starting from their beginning to modern age developments to be considered as the best alternative of burning 2D materials graphene, as the 2D TMDs offer kinds of superiority over the other 2D materials like graphene etc. Considering these facts in the running era leading investigators have paid attention to one the potential material "Transition Metal Chalcogenides" (TMDs) and recognized that they can possess two-dimensional configurations beyond the most efficient graphene. Their initial theoretical and experimental innovations predicted they can be considered one of the key materials for future potential applications. Although, very limited (or a few) books are available on these recently invented TMDs materials, therefore, it is desired to interpret TMDs developments and fundamental concepts sequentially with easy scientific interpretations for the common reader. Therefore, this chapter is predominantly devoted to chronological historical developments by providing a comprehensive overview and fundamentals of the TMDs two-dimensional materials with their classifications, including chalcogenide materials with their basic properties like semiconducting lone-pairs and alternation of the valence electron pairs and bonds hybridization property, etc., to distinguish the TMDs from the usual chalcogenide materials with the proper scientific descriptions based on the materials chemistry of the transition metals, such as the theory of the valence bonds, ligand-field approach, crystal-field approach, and band structure.

Chapter 2

Exploring the 2D TMDs and their synthetic process is also an important aspect of the compositions of the materials regard to get desirable structural and physical properties for potential uses in various technical areas. TMDs can be synthesized inform of monolayers, bilayers, and trilayers depending on requirements of structural and physical properties for the specific applications. Therefore, it is highly desirable to have a good understanding of the TMDs synthesis process. Considering the emergence this chapter deals with the typical material fabrication processes with their crucial parameters for the different kinds of scientific and technological uses. This work has provided descriptive knowledge about the synthetic approaches of kind of materials those helpful to develop desired structural and physical properties for the targeted technological purpose. Since different synthetic approaches to two-dimensional nanomaterials are largely affect their potential physical properties. Therefore, the process-dependent structural and physical properties of the TMDs have been addressed, by demonstrating the different synthetic approaches such as top-down; including liquid exfoliation, mechanical exfoliation, electrochemical exfoliation, and bottom-up; including growth mechanism, metal chalcogenisation via sulfurization/selenization, thermolysis from thiosalts, transition metal—chalcogen precursors via vapor pressure reaction, TMDs alloys growth, epitaxial growth of TMDs, atomic layer deposition (ALD) and layer transfer. Information about the different synthetic approaches is also significant to develop or building emerging high performances device applications in distinct areas. Considering their merits and demerits one can be selected accordingly to employ for the device fabrication based on TMDs materials.

Chapter 3

To be continued to explore the different aspects of TMDs materials detailed knowledge about the structural properties and their consequence on physical properties is also significantly important for the potential uses because TMDs have a unique structural property compared to other 2D materials that boosted the possibilities for their potential uses in the wide range of technical areas. Therefore, it is customary to explore TMDs' distinct structural and physical properties including the impact of the defects on these properties. Because TMDs' structural properties are one of the key parameters to define their possible utility for the specific purpose, modifications in structural configurations also make them more specific or general for their potential applications. Considering the view this work is more focused on the structural properties of different forms of the TMDs materials. Like bulk TMDs structure, individual triple layers structure, polymorphs structure, distorted structures, and a single layer or few layers structures including a description of the different phases and phase transition, stability, patterning, and tuneability of TMDs materials. Additionally, concrete knowledge about possible significant types of defect formations and their consequence on the structural properties of TMDs is also addressed. Taking into consideration, kinds of defects that govern structural formations have been also addressed, like point defects, doping, grain boundary, etc. Moreover, the impact of impurity on the TMD's electronic transport and metallicity, magnetic characteristics, and optical properties have also provided to get a better understanding of changes in monolayer and bilayer TMDs' physical characteristics. Hence overall this chapter work has provided a detailed understanding of the structures of the TMD materials and alternation due to impurity addition along with the variations in their physical properties.

Chapter 4

A piece of comprehensive knowledge of the physical properties of the TMDs materials is also useful to define their most suitable applicability in distinct technical areas. More precisely knowledge of physical properties is defined by their potential applications with their merit and demerits for specific uses. Therefore, it is essential to have a good understating of the various physical properties of TMDs that makes them different from the other kinds of existing materials. Considering the view this chapter work more emphasis on TMDs' different physical properties by defining the mechanical properties including strain engineering, thermal properties including thermal conductivity, thermal expansion, thermoelectric, optical properties including excitons; binding Energy, non-Rydberg excitonic, excitonic states in dark, excitonic collapse, Intra transition, dynamics, interactive exciton, mirror exciton as well as heterostructure excitons, excitons, and trions. Moreover, TMD material's electrical properties have also gained much attention and extensively explored due to their possible uses in high-processing electronics and optoelectronics device applications. Therefore, attention has also been paid to the electrical properties of the TMDs, thereby, a comprehensive discussion on electronic structure, ballistic transport, and mechanisms of scattering are also incorporated. Additionally, the two-dimensional TMDs are also considered to be prominent candidates that have an impassive magnetic property compared to other existing materials. It is extremely useful for memory, spintronics, computing, and in many ultra-high-processing demanding devices or technologies, therefore this key physical property of TMD materials has been also addressed by providing a discussion on the magnetic properties of TMD materials including the magnetism via grain boundaries, magnetism via vacancy and doping-induced magnetism.

Chapter 5

Emerging TMDs materials have adequate physical properties due to the existence of spin-valley coupling that makes them enables to offer distinct physical properties compared to other existing materials, including structural and physical superiority over the significant 2D material graphene. Therefore, the wide range of optical energy band gap and improved structural properties of TMDs with the spin-valley coupling makes them different from the other 2D materials. So it is important to address the TMDs' spin-valley coupling properties in detail for a better understanding of these materials. Significantly this work has addressed the spin-valley coupling in TMD materials and their impact on physical properties by providing a brief overview from before current era information. Including theoretical concepts and experimental findings descriptions on the degree of freedom and electronic structure, optical, electrical valley polarization, and valley coherence have been incorporated. Moreover, the valleytronics properties of TMD materials are one of the significant factors for their applications, therefore, this important property has been defined for the TMD materials with the discussion of the valley-spin concept to monolayers and bilayers. Moreover, also discussed other significant valleytronics properties, like selection rules for the valley and spin, valley Zeeman Effect and magnetic moment, valley Hall Effect and Berry phase effect, and nonlinear valley - spin currents through trigonal twisting for the monolayers and Bi-layers in detail. Additionally, the key inversion-symmetry induced also played a crucial role to define the physical properties of TMD materials, therefore, a separate discussion on inversion-symmetry-induced spin polarization and optics-induced Stark effect has been also included.

Chapter 6

Since it has been noticed that the pure TMD materials-based device have technical performances issue in actual applications compared to heterostructures, because due to mirror facts that the TMDs are also not completely from the physical deficiencies. Therefore, the attention moves toward the TMD heterostructures to get improved physical properties. Considering the emergence this work is more focused on the remarkable TMDs heterostructures by providing a discussion of the various aspects to build a deep understanding of such materials. Significantly this work has covered the past to present significant developments in these materials, including important Anderson's rule for the band alignment theoretical description. A significant description of TMDs materials heterostructures construction for their potential technical uses is also addressed. The strategies to construct 2D TMD heterostructures are also an important aspect for their different forms of possible combinations such as 0D–2D, 1D–2D, and 2D–2D heterostructures, thereby, details descriptions on each of them have incorporated that would be helpful for the specific utility. The band alignment in TMD heterostructures is also a crucial thing for their applications; therefore, it is also addressed to build a better understanding of the band formation behavior in this kind of material. Significantly vertical heterostructures can have scientific interest due to their structural and physical properties for the device's constructions and performances, therefore important vertical heterostructures have been also addressed by discussing key properties, like matching lattice heterostructures, strain minimize approach, individual (or single) layer heterostructures, Janus monolayer 2D TMDs, layer re-stacked heterostructures and vertically aligned heterostructures based on the experimental interpretations. Moreover, technically more demanding lateral TMD heterostructures have been also addressed including their significant properties, such as edge epitaxy, lithographic pattering approach, band gap engineering of LHs, and composition-dependent band alignment in LHs. Including TMD heterostructure's key governing physical parameters that can provide sound knowledge on their various aspects that may be decisive in the selection for the specific potential applications. Additionally, a useful discussion on mixed lateral and vertical hybrid structures has been also provided. That could define the structural and other physical parameter changes in hybrid structures, therefore having, a significant impact on overall technical performance based on such TMD materials devices. Hence this chapter work has provided depth and up-to-date knowledge on TMD heterostructures that would be useful for their potential technical applications in various areas.

Chapter 7

TMDs one of the most emerging and demanding properties is topological insulating that has been recognized to them as future prospective materials for ultra-high-processing devices in various areas. Specifically, the TMDs based topological insulators aspect of bulk to surface properties is the great interest for the fabrication of future next-generation ultra-high-performance devices. Therefore, the work emphasizes to TMDs' topological insulating properties of different kinds of materials by providing a sound background discussion including bulk TMD TIs aspects. Additionally, also defined the Berry curvature including chern invariance, chern of the insulator, topological protection of transport properties, TIs as a Dirac system, Hamiltonians, and kinetics. Moreover, the most significant topological properties in terms of surface-to-bulk conductivity have also been addressed with a concrete discussion on a monolayer, binary, and ternary composition's topological insulating properties. Crystal structures of the topological insulating materials are also playing a crucial role to define their topological phase, therefore, a discussion on the crystal of structures of TIs has also included providing a theoretical aspect on monolayer TIs crystal structure and experimental evidence to concepts of the wrinkled folded and scrolled TMDs. Since the band structures of topological insulating materials are quite different from the conventional materials, to provide a clear view of it a concrete general discussion on TIs band structure in a normal state has been also addressed. In addition to this, to discriminate the different kinds of superconductivity in topological phase mono layers a discussion on Ising superconductivity in monolayer TIs has been also provided by discussing Type-I, and Type-II including a theoretical aspect. Since the different kinds of topological phases occur in TIs, therefore, the significant nodal topological phase and aging effects including edge states transport of 2D TIs and their impact on topological edge states have been also addressed. Moreover, a general discussion on various significant properties of the TIs has been also incorporated such as cracking, carrier density, structural quality of thin films, choice of substrate and heterostructures, and superlattices.

Chapter 8

Emerging TMD materials can have a wide range of applications in various areas due to their adequate superior structural and physical properties than other existing materials. Therefore it is worth to address TMDs based possible technical device applications in distinct areas of uses. Therefore, this chapter's work is addressed various technical applications starting from the burning TMDs topological insulating property for the high-processing device applications based on the two-dimensional, three-dimensional, and new generation topological insulators (TIs) materials. By describing the TMDs as field effect transistors (FETs) devices that can serve superior to usual transistor with novel properties like flexible top-gated FETs, logic inverters, large-scale circuits, and analog and RF devices. Moreover, TMDs are most extensively used as light-emitting devices for cutting age high-processing optoelectronics devices. Therefore, TMDs based devices have been also addressed in detail by describing the light-emitting diodes (LEDs), LEDs in vertical stacking, single photon emitter, organic light-emitting diodes (OLEDs), and quantum dots (QDs) based light-emitting diodes. TMDs are also offered lasing properties through the population invasion of charges due to their unique structural property therefore a brief discussion has been also devoted to TMDs based lasers. Moreover, TMDs based devices are used as kinds of photodetectors, therefore a brief description is devoted to photo-detecting devices. In addition to this, TMDs are also considered to be potential candidates for photovoltaic device applications that can provide high-efficiency renewal energy throughput at a cheaper cost. TMDs mainly offers two category photovoltaic devices based on their physical properties namely traditional and transparent TMDs based solar cells that have been also addressed in this work. Moreover, TMDs relatively novel properties can also boost renewable energy technology such as solar water splitting due to the mechanism of electrocatalytic and photocatalytic types of water splitting. Therefore, the emerging highly demanding TMDs based water-splitting devices have also discussed by providing descriptions of the distinct types of potential solar cell devices. The layered TMDs are also become hot materials due to their biocompatibility and applications in biomedical science/technology as a variety of biosensors. Therefore, the kinds of TMDs based biosensors' working functionality and their specific uses have been also discussed, like fluorophore-labeled, label-free, absorption, electrochemical, and FET biosensors. Moreover, TMDs' adequate structural properties are also offered to integrate the TMD-TMD-based heterostructured devices by the layer stacking of either the same or distinct TMD materials; therefore, a discussion on this kind of devices has also incorporated, like fabrication and interpretation of graphene/TMD heterostructures. Additionally, considering the vast area of TMDs based devices applications a miscellaneous section has been also devoted to some important utilities, like, thermoelectric devices, TMDs for drug delivery, TMDs in photothermal therapy and biomedical imaging. Hence TMDs materials are the thrust materials that can have applications in diverse areas to fulfill the current/next-era high-processing technological demands.

Chapter 9

In recent years great attention has been paid to fabricating TMDs based next-generation high possessive ultra-thin devices with improved working efficiencies by Atomic Layer Deposition (ALD) technique. Now it has been recognized ALD is one most emerging techniques to fabricate 2D TMDs based devices for the next generation. Considering the emergence of the ALD technique utility this chapter work is devoted to this topic. By providing brief historical developments in the ALD technique with few significant prior fabricated TMD-based device's performance interpretations. Starting from the description of types of ALD precursors, growth characteristics, surface roughness, conformality, effect of deposition temperature and mechanics of thermal ALD with the initial surface reactions, reaction pathway process, and precursor chemisorptions. In addition to these also discussed ALD application on TMD materials to understand the basic surface properties of 2D TMDs and the physisorption impact on ALD. Moreover, to understand the formation of the uniform film 2D TMDs from ALD, descriptions have been also provided on the ALD on 2D TMDs without any surface treatment and uniform deposition on 2D TMD by surface functionalization. Moreover, to realize the actual synthetic application of the ALD process with the TMD materials some significant examples have been also discussed by providing descriptions of the MoS_2, $MoSe_2$, WS_2, WSe_2, HfS_2, and ReS_2. Moreover, a brief miscellaneous section also provided to cover those materials that have been previously fabricated by the ALD technique for various uses. Moreover, to realize the actual device fabrication and their performances some important TMD-based ALD process device fabrications have been also interpreted, specifically key physical characteristics and significant advancement in these devices in terms of existing and future applicability. In this order, ALD fabricated TMD-based photodetector and Field Effect Transistor (FET) (HfS_2 FETs and WS_2 FETs) devices and their critical working performances including significant physical characteristics have been also addressed.

Thus the applicability of the ALD technique for TMD-based materials is widely accepted due to the formation of high-quality thin film devices for modern and future optoelectronics and other kinds of device fabrications with reduced sizes and cost-effectiveness. Therefore ALD technique can be considered a potential technique for future high-performing TMDs based device fabrications.

Johannesburg, South Africa Abhay Kumar Singh

Acknowledgments The author sincerely acknowledged their collagenous, friend, and faculty members of the diverse departments for their valuable suggestions and encouragement during the preparation of the whole book contents. Additionally, the author is also heartily grateful to his all family members including Anshika Singh (wife) and lovely child Nirthik Singh for their consistent moral support during the course's massive work. The author is also gratefully thankful to the publisher of the book and journals cited in the bibliography in respective chapters.

Contents

Chapter 1
Basics of TMDs

1.1 Introduction

Dimensionality of nanomaterials is greatly influenced the electrochemical behavior and compositional atom arrangements. As intense, zero-dimensional structures, like nanoparticles and quantum dots possess the largest surface area and lowest surface energy with excellent activity and stability. Likewise, one-dimensional (1D) structures, like nanotubes and nanowires are widely used in the construction of micro-devices, such as fiber-shaped supercapacitors or microbial fuel cells [1]. Moreover, the two-dimensional (2D) structures offer more surface area, whereas the three-dimensional (3D) structures are a combination of 1D and 2D construction, which enable superior electron transfer efficiency via spatial construction. It is significant to note, the 2D materials provides superficial structures and advantages, such as promising effective active sites for catalytic reactions. They also form structures with a larger active surface area and open pore structures through stacking, which is helpful to enhance mass adsorption and electron transfer on the surface. Owing to these exceptional superiorities the 2D structure of nanomaterials offers extensive applications in various technological areas, like promising electrocatalytic energy conversion, etc. [2]. Such as, the 2D layered hetero-doped graphene and transition metal dichalcogenides (TMDs) nanosheets can serve as active and sustainable electrocatalysts for various optoelectronic devices [3, 4]. Therefore, it is desired to have an understanding of the structural characteristics, architectural features, and properties of 2D nanomaterials.

Graphene has attended great interest worldwide due to its specific monolayer sp^2 bonding of carbon atoms network [5]. Likewise, the layered TMDs have also recently gained great attention and scientists jumped into them due to their excellent electrical conductivity, heat conductivity, strong mechanical strength, ultra-high carrier mobility, and large surface area [5–7]. TMDs are also described from the relationship MX_2, their 2D morphologies boosted the attention owing to versatile physicochemical properties and potential applications. Like, their appropriate semiconducting

A. K. Singh, *2D Transition-Metal Dichalcogenides (TMDs): Fundamentals and Application*, Materials Horizons: From Nature to Nanomaterials,
https://doi.org/10.1007/978-981-96-0247-6_1

properties with sizable bandgaps are useful for a large quantity of applicability [8, 9]. However, 2D layered graphene can be also considered as an electrode material because it bears the brunt ability with high conductivity and large surface area [10, 11].

A wide range of layered materials has atomic layers weakly bonded together by van der Waals interactions, wherein they are readily isolated into single or few-layer nanosheets through mechanical exfoliation or liquid exfoliation processes [12–14]. The atomically thin 2D nanosheets can be derived from layered materials that possessed many interesting characteristics like graphene, such as excellent electronic properties, exceptional mechanical flexibility, and partial optical transparency. The 2D nanosheets typically have a well-defined crystalline structure with few surface dangling bonds that traditionally plague most semiconducting nanostructures. Therefore, they have excellent electronic properties that cannot be readily achieved in other kinds of nanostructured semiconductors. Beyond the typical semiconducting properties, the layered TMDs also offer a wide range of physical properties, including superconducting (e.g., $NbSe_2$), magnetic (e.g., $CrSe_2$), insulating (e.g., BN), topologically insulating (e.g., Bi_2 Te_3), and thermoelectric (e.g., Bi_2Te_3) properties. Moreover, the different TMD are also combined to create heterostructures or superlattices for further complexity and get entirely new functions.

A distinct feature of TMD, the van der Waals interactions in-between the neighboring layers allows much more flexible integration for the different materials without limitation of the lattice matching. So, they offer a great opening up of possibilities for the arbitrary combination to control different properties at the atomic scale [15]. Therefore, new physics and unique functions may emerge upon a combination of different layers, in which each layer has entirely different physical properties. Such layered materials may also function as excellent hosting materials for ion insertion and ion transport that can further modulate the electronic or topological nature of energy bands. Hence, the combination of TMDs could represent a new class of nearly ideal 2D material systems to explore the fundamental chemistry and physics at the limit of single-atom thickness, with the potential to open up exciting new technological opportunities beyond the reach of existing materials. It enables them to transform advances across diverse areas ranging from traditional electronics, optoelectronics, energy storage, flexible electronics, spintronic and quantum computing.

The layered TMD has a large family of materials, predominately monolayer TMDs are interesting due to their direct energy bandgaps and non-centrosymmetric lattice structure [16–18]. Like, MoS_2 can have tunable bandgaps that undergo a transition from an indirect bandgap in bulk crystals to a direct bandgap in monolayer nanosheets [16, 18]. Therefore, the diverse TMDs emerged as an exciting class of atomically thin semiconductors with tunable electronic structures. The layered TMDs' electronic structures also possessed special features except for the general characteristics of common semiconductors. Such as 2D crystals electrons have a honeycomb lattice structure with a pair of inequivalent valleys in the k-space electronic structure with an additional valley degree of freedom [19]. The additional degree of freedom of charge carriers leads to novel physics termed "valleytronics", which is manifested by polarized photoluminescence in monolayer MoS_2 and the valley Hall effect [17,

20–25]. Hence the exciting potential of atomically thin materials are directly beneficial for thin transistors [26], vertical tunneling transistors that promises unprecedented switching speed [27], vertical field-effect transistors (VFETs) to enable the 3D electronic integration [28], including other types of optoelectronic devices such as tunable photovoltaic devices and light emitting devices [29–34].

1.1.1 2D Materials

Two-dimensional (2D) materials are usually referred to as single-layer materials that became an interesting topic of worldwide research since the exfoliation of graphene in 2004 [35]. With the unambiguous contrast to their bulk counterpart, to know what's more fascinating with 2D materials under ultra-high specific surface areas and their diverse energy band structures that are sensitive to external perturbations and matter. The distinct surface behavior of 2D materials makes them competitive for devices as stated by Herbert Kroemer "The interface is the device" [36]. Research-based on 2D materials devices has not only contributed to the deeper understanding of the novel layered materials physics but also provided a significant platform for the potential opportunities in various fields ranging from electronics, and optoelectronics to energy and sensing applications.

Worldwide scientists started thinking toward 2D materials when in 1959, Richard Feynman gave an inspiring and influential lecture on the topic "There's Plenty of Room at the Bottom" [37]. In his famous talk, Feynman envisioned a scientific breakthrough in the field of physics with his questions "What could we do with layered structures with just the right layers? What would the properties of materials be if we could arrange the atoms the way we want them?". These Feynman questions are challenging for world scientists to manipulate and control things on the atomic scale. All these Feynman's questions were almost unresolved until 2004, a significant answer became 45 years later when the University of Manchester physicists Andre K. Geim, Konstantin S. Novoselov, and collaborators experimentally exfoliated and identified graphene, which was the 2D single atomic layer of carbon [35]. However, the history of graphene research can be traced back much further. Such as in 1947, Philip Wallace calculated the band structure of the one-atom-thick crystal [38], and Hanns-Peter Boehm synthesized graphene flakes through reductions of graphene oxide (GO) dispersions 15 years later [39]. Afterward, a large number of material scientists tried to produce kinds of one-layer graphite structures by employing exfoliation methods or thin-film growth technologies [40–42]. Thus the "graphene" or "graphene layer" was officially defined to introduce such a single atomic carbon layer of graphite structure by the International Union of Pure and Applied Chemistry in 1995 [43]. Based on the previous studies on ultrathin graphite and the tremendous rising interest of the scientific community in other carbon nanomaterials (such as fullerene and carbon nanotubes) [44–47]. More specifically scientists and engineers have paid devotion and interest in the energies of the new fascinating material to discover its remarkable science and potential for practical applications.

It is well known that most 2D layered materials exist in a bulk state. These materials have layered structures with weak interlayer van der Waals force to hold layers together. The key advantage of a layered structure allows for achieving monolayer or a few layers of 2D materials from mechanical exfoliation. While in-plane atoms are generally connected with strong covalent bonds [48, 49]. The monolayer 2D materials are possessed two-dimensional features in lateral x–y direction and their quantum confinement in the third dimension. This makes them unique materials to their bulk counterparts. The 2D material's high carrier mobility feature also fulfills the essential requirement of high-speed transistors. To achieve a high-performance transistor, a good ohmic contact, higher carrier mobility, and an appropriate band gap (~1 eV) are the basic requirements [50]. Although graphene is the most extensively studied 2D material with very high carrier mobility (~2×10^5 cm^2 V^{-1} s^{-1}) at the temperature of 5 K [51]. Nevertheless, the problem with this it is not an appropriate parameter for the logic application due to the graphene zero band gap that usually results very low on/off ratio (< 10) at the ambient temperature. To improve the on/off ratio scientists have proposed the opening of the bandgap of graphene. Therefore, the bandgap of the graphene can be also engineered up to ~ 400 meV, with a decreased mobility of less than 200 cm^2 V^{-1} s^{-1}. Similarly, graphene-based p-type devices (such as nanoribbons) are also offered a bandgap opening due to the various edge (armchair or zigzag) structures. It leads to a very high on/off ratio of ~ 10^6 with an extremely low charge carrier mobility of ~ 100–200 cm^2 V^{-1} s^{-1} compared to other graphene-based devices. Besides these devices' sub-threshold order of ~ 210 mV per decade, usually it is none desirable and ideal, it should be ~ 60 mV per decade at room temperature [52].

On other hand, TMDs, such as MoS_2 (1.8 eV), WS_2 (2.1 eV), $MoTe_2$(1.1 eV), and WSe_2 (1.7 eV) have desirable bandgaps [50, 53, 54]. Specifically, MoS_2 has been recognized as one of the most promising materials for logic devices, like metal–oxide–semiconductor field-effect transistors (MOSFETs). Its tunable bandgap and high on/off ratio at relatively cheaper prices also make them attractive. The monolayer MoS_2 indirect bandgap is usually around 1.8 eV, whereas its direct band gap is around 2.5 eV, while its bulk counterpart has an indirect band gap of 1.2 eV [16, 18, 55, 56]. Experimentally MoS_2 relatively large bandgap has achieved, but it has a low carrier mobility ~ 1 cm^2 V^{-1} s^{-1} without high-k dielectric gate material [12], however, its mobility can reach ~ 150 cm^2 V^{-1} s^{-1} at 300 K with HfO_2 as the top-gate layer [26, 57–59]. Though, a theoretical calculation based on density functional theory has predicted that the mobility of MoS_2 could be up to 400 cm^2 V^{-1} s^{-1} at room temperature [60]. Likely, another 2D material black phosphorus has been also considered a worth candidate due to its hole mobility up 10,000–26,000 cm^2V^{-1} s^{-1} [61].

Hence, in recent years various 2D materials beyond graphene have been explored and the research community paid much attention to them, including normal insulators and transition metal oxides, hexagonal boron nitride (h-BN), topological insulators (Bi_2Te_3), semiconductors (MoS_2, WSe_2, and black phosphorus), metals (TiS_2), superconductors ($NbSe_2$) and charge density waves (1T-TaS_2 at low temperatures), as depicted in Fig. 1.1. These materials can offer rich choices and high tunability in

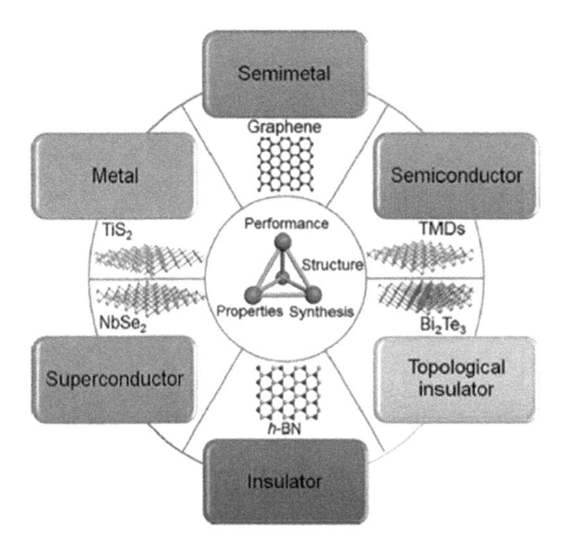

Fig. 1.1 Different forms of 2D materials (adopted from Li et al. [47])

form of 2D materials for the next generation devices with specific functions. Especially, their weak van der Waals interactions (compared to the strong covalent bond interaction inside covalently bonded material structures) allow the construction of promising building blocks for future electronics and optoelectronics by stacking 2D materials with multi-dimensional materials to form van der Waals heterostructures [15, 62–66].

1.1.2 Classification of 2D Materials

Usually, 2D materials family are categorized into five types: (1) graphene with single-layer having atomic arrangements in hexagonal honeycomb lattices, such as borophene, silicene, germanene, stanene, h-BN, and BP; (2) 2D Metal chalcogenides, like three-atom-thick transitional metal dichalcogenide (TMD) with a general stoichiometry formula MX_2 (M-represents the transition metals such as Mo, W, Ti, Nb, Re, Pt, etc., while, X represents chalcogen elements S, Se, or Te). There combinations with the III-VI and IV-VI groups families such as GaSe, InSe, GeSe, SnS, SnSe, SnS_2, $SnSe_2$, etc., including others like Bi_2Te_3, etc.; (3) 2D transition metal carbides and/or nitrides (M Xenes) has a general stoichiometric formula $M_{n+1}X_n$ (M-represents the transition metals Mo, Ti, V, Cr, Nb, etc., while, X represents C and/or N, and n is 1, 2, or 3), and their surfaces are terminated by O, OH, or F atoms; (4) 2D oxides or hydroxides, like, titania nanosheets; (5)2D organic materials, such as pentacene [47].

Thus, 2D materials can be categorized in several ways in terms of their electrical, mechanical, and transport properties, because they have possessed excellent mechanical, thermal, optical, and electronic properties [67–70]. The family members of 2D

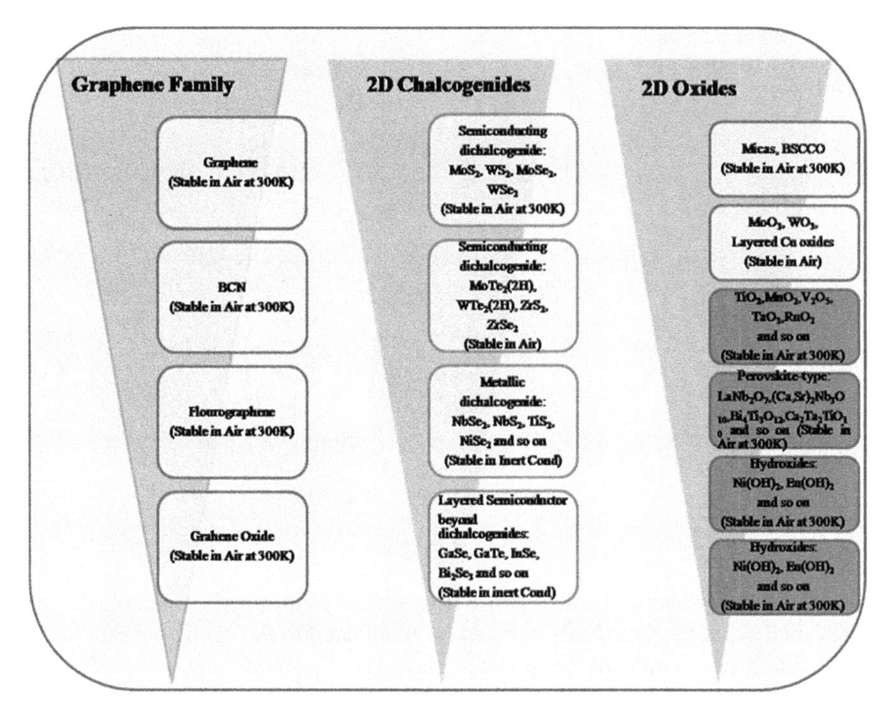

Fig. 1.2 Schematic for the family members of graphene, 2D chalcogenides, and 2D oxides

graphene, TMDs, and oxide materials are depicted in Fig. 1.2. Here dark boxes are showing monolayer fabrication by exfoliation process.

1.1.2.1 Layered Solids

The layered solids have consisted stacked array of two-dimensional layers with a high aspect ratio above to other and form three-dimensional macromolecular structures. A class of materials those bonding among the atoms of the same layer is much stronger compared to bonding among the atoms of adjacent layers called layered solids [71–73]. This kind of solid's chemical bonding anisotropy reveals interesting properties such as intercalation chemistry, ion exchange, and delamination to form colloidal dispersion of monolayers in solvents [71–75]. A schematic representation of a typical layered solid is shown in Fig. 1.3.

Hence, beyond the graphene 2D layered materials also reveal an interesting and exciting property. Materials such as hexagonal boron nitride (h-BN), dichalcogenides, tertiary compounds of carbo-nitrides, etc. [76, 77]. The rich variety of properties 2D layered systems offer to potential engineering on-demand, and create exciting prospects for device and technological applications, such as in electronics,

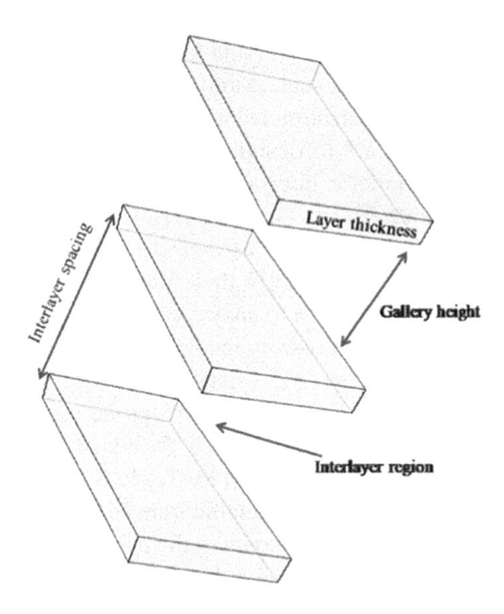

Fig. 1.3 Schematic for the typical layered solid

sensing, photonics, flexible electronics, energy harvesting and storage, thermal management, mechanical structures, catalysis, bio-engineering, gas adsorption, etc.

The family of 2D layered materials (like transition metal dichalcogenides) coordination and oxidation state of the metal atoms are also classified whether the transition metal dichalcogenide metallic, semimetallic (or semiconducting). Additionally, the superconductivity and charge density wave effect also exists in some transition metal dichalcogenides. Additionally, the chalcogenides of group III (GaSe, GaTe, InSe), group IV (GeS, GeSe, SnS, SnSe, etc.), and group V (Bi_2Se_3, Bi_2Te_3) may also possess graphite-like layered structures which offer a promising platform for the electronics, photonics and energy harvesting.

The most common type of 2D material is layered van der Waals solid which has a strong in-plane covalent or ionic bond and weak interlayer van der Waals bonding. The weak out-of-plane bonding allows the extraction of mono or a few layers of 2D materials from their bulk counterpart through mechanical or liquid exfoliation [78, 79]. These materials' dimensions in a lateral direction are up to a few micrometers with less than 1 nm in thickness. Likely well-studied MoS_2, $MoSe_2$, and WS_2 TMDs. Thus the quantum confinement effect is responsible turn the 2D layered materials (such as MoS_2, WS_2, WSe_2, $MoSe_2$, and $MoTe_2$), that accounted to indirect-bandgap semiconductors transformed into direct bandgap semiconductors from the bulk to monolayer [80, 81]. As an instance, MoS_2 has an indirect band gap (~ 1.2 eV) in the bulk form that transforms gradually into a direct bandgap (~ 1.8 eV) in form of the monolayer. That is analogous to pristine graphene with a band gap of ~ 0 eV and a few-layered h-BN with a band gap of ~ 5.5 eV. The intrinsic bandgaps of the layered materials imply possible their applications in electronics and photonics, as the compliments of the graphene and h-BN. Their inherent bandgap properties are coupled with the atomically smooth pristine interface that makes them potentially enable the low-power and low-dissipation devices to replace conventional semiconductors at ultra-scale. Similarly, layered structures such as hexagonal boron nitride (h-BN) have also attracted much due to their ability to form atomically thin dielectrics. It also offers to design kinds of device architectures possessing metal/dielectric/semiconductor interfaces. Like heterostructures field, effect transistors (FETs) or tunneling-FET devices are formed by stacking 2D semiconductors as the channel, 2D insulating layers as dielectric, and 2D metallic layers (graphene, $TaSe_2$, etc.) as gates. Subsequently, their interconnections are enabled energy-efficient transistor devices for digital to analog circuit applications.

Moreover, possibly indirect-to-direct bandgap transitions are also formed from a verity of bilayer MX_2 materials, in which M is a transition metal and X is a group VI element (S, Se, Te). That can be tuned through the application of an external electric field. As an instance, the density functional theory calculations of the bilayer MoS_2, WS_2, $MoSe_2$, and $MoTe_2$ have predicted their bandgaps to undergo a gradual semiconducting-to-metallic transition through the application of an external electric field oriented orthogonal to the plane of the layers. Moreover, quantum scale effects could also account for the superconductivity and charge density wave effects. It manifested at low temperatures in some 2D layered materials like $NbSe_2$, TaS_2, and Cu-intercalated $TiSe_2$. Thus, the transition metal in form of TMD is usually occupy trigonal prismatic or octahedral coordinates and forms a hexagonal structure [82]. Apart from TMDs, some other members of layered van der Waals solids also behave likewise, such as Sb_2Te_3 [83], vanadium oxide [84], and h-BN [85], etc.

Intercalation in Layered Solids

The inter-lamellar space between the successive layers are generally accommodate ions/molecules. Therefore, the intercalation process is generally described from the reversible and topotactic properties. This means, during the insertion or de-insertion of the guest species the overall structure of the layers remains intact. The held layers from the weak Van der Waal's forces can be separated, but it depends on the size of the ions/molecules. The process of intercalation/de-intercalation is archived from chemical or electrochemical methods. So, the electrochemical process also exploited for energy storage devices. For the instance, in the fabrication of rechargeable lithium-ion batteries, graphite or layered metal oxides are used as electrodes. That undergoes insertion/de-insertion of lithium ions; hence, it also contributes to the capacity delivering of the cell. This directly correlates to inserted ions to be exchanged by other ions when they are excessively present. This process is also known as ion-exchange reactions.

Delamination/Exfoliation

The layered solids delamination process is one of the most extensively studied phenomena [75]. The delamination process occurs when layers are separated, where each layer behaves as an independent particle. Almost all classes of layered solids are delaminated to monolayers or a few layers. Typical delamination are achieved by employing the thermal or chemical process. Such as graphite consists of neutral hydrophobic layers that can be delaminated thermally by heating at high temperatures in an inert atmosphere. While in the chemical process, the layers are separated by modification of interlayer region with hydrophilic or hydrophobic guest species and soaking the modified layered solid in a suitable solvent [75]. Complete delamination of layers is usually achieved from colloidal dispersion of monolayers by employing vigorous stirring or sonication processes. In this process layers held by weak Van der Waal forces the solvent molecules to penetrate the interlayer gallery resulting in the separation of layers. The delamination process is generally controlled by the solvation of the interlayer region, the lattice energy, and the layer charge density. Materials containing higher charge density require additional interlayer modifications to obtain a stable dispersion.

1.1.2.2 Ionic Solids

In this type of solid most of the divalent metal hydroxides and their derivatives crystallizes into layer form, whereas their structures are derived from the mineral brucite, $Mg(OH)_2$ [73, 86]. Divalent metal α-hydroxides, hydroxysalts, and hydrotalcite-like compounds belong to the anionic class clay [87, 88]. In such 2D material, the charged polyhedral layer exists between two layers of halide or hydroxide layers that are held together via electrostatic force between them. The ion exchange liquid exfoliation or ion intercalation is usually used to exfoliate mono or a few-layered 2D materials. Hence the typical layered ionic materials are possibly exfoliated from ion exchange methods, like $KCa_2Nb_3O_{10}$, $RbLnTa_2O_7$, $K_2Ln_2Ti_3O_{10}$, $La_{0.9}Eu_{0.05}Nb_2O_7$, etc. [89, 90].

α-Hydroxides of Nickel and Cobalt

Nickel and cobalt hydroxides exhibit polymorphism. The β-form crystallized structure is like mineral brucite that has an interlayer spacing of 4.6 Å [86]. The β-form consists of hexagonal packing of the hydroxyl ions with M_2^+ ions that occupy alternate layers of octahedral sites [86]. As a consequence stacking of charge-neutral hydroxide layers are the composition of $M(OH)_2$. That is held together by van der Waal's interaction, as the typical schematic structures of the α and β hydroxides are shown in Fig. 1.4.

Hydroxides of nickel and cobalt also exist in another hydrated form which is known as α-hydroxide. It has a similar structure to brucite with enhanced basal spacing. The partial protonation of the hydroxyl ions positively charged layers of the hydrated hydroxide generates the following form.

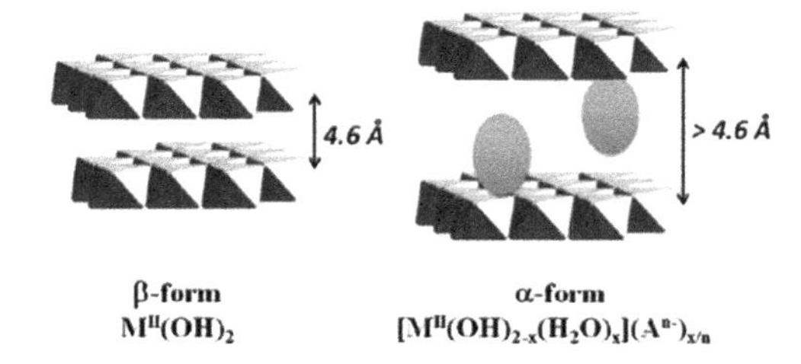

Fig. 1.4 Structures of the α and β-hydroxides

$$M^{II}(\mathrm{OH})_2 + XH^+ \leftrightharpoons \left[M^{II}(\mathrm{OH})_{2-x}(\mathrm{H_2O})_x\right]^{x+} \quad (1.1)$$

Such layers can be also formed intercalate anions yield compounds like $\left[M^{II}(\mathrm{OH})_{2-x}(\mathrm{H_2O})_x\right]^{x+}\left(A^{n-}\right)_{x/n}.m\mathrm{H_2O}$ [91–93]. Therefore, a variety of anions are possible, such as Cl^-, NO_3^-, and CO_3^{2-} that associates a long chain of alky carboxylates. Hence the basal spacing of α-hydroxides is much larger than 4.6 Å, it depends on the size and interlayer orientation of the intercalated anion, as depicted in structures of β and α-hydroxides in Fig. 1.4.

Layered Double Hydroxides

The layered double hydroxide stacking consists of an array of positively charged layers of mixed metal hydroxides having exchangeable anions charge balance in between the interlayer. This class's most common natural mineral is hydrotalcite with the typical composition $Mg_6A_{12}(OH)_{16}CO_3 \cdot 4H_2O$ [88]. Such layered materials usually represent by the following formula;

$$M^{II}_{1-x}M^{III}_x(\mathrm{OH})_2\left(A^{n-}_{x/n}\right).m\mathrm{H_2O} \quad (1.2)$$

whereas M^{II} = Mg, Co, Ni, Cu, Zn, Ca; M^{III} = Al, Fe, Cr, Ga, and A^{n-} represents the anion, like CO_3^{2-}, NO_3^-, SO_4^{2-} etc. While the empirical value of the "x" laying in between 0.25 and 0.33. The layered double hydroxide structures are derived from the mineral brucite $Mg(OH)_2$, whereas the Mg^{2+} ions are octahedrally surrounded by 6 OH^- ions. To form octahedral share edges of the infinite sheets [86, 87]. In case a fraction of Mg^{2+} amorphous substitution of other similar radii higher valiant metal ions [88]. Their layers acquire positive charges and anions intercalated between the layers to restore charge neutrality along with water molecules. Such metal hydroxide sheets are possibly stacked one above with other formed three-dimensional layered structures. These layers are usually held together by weak van der Waal's forces [88]. A schematic structure of the layered double hydroxides is illustrated in Fig. 1.5.

Fig. 1.5 Structure for the typical layered double hydroxide

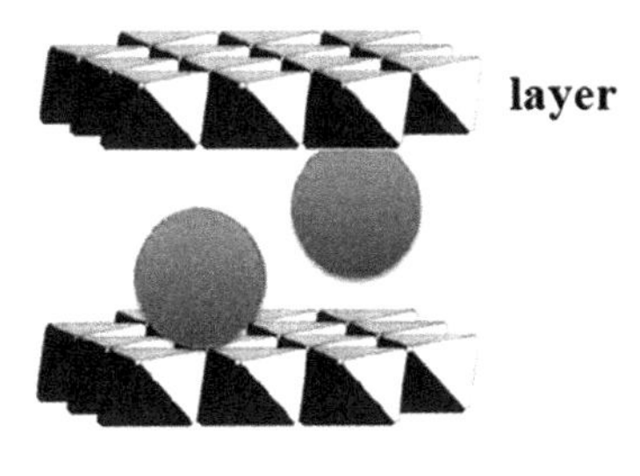

Layered Metal Oxides

The layered metal oxides such as $Cs_{0.7}Ti_{1.825}O_4$, $KCa_2Nb_3O_{10}$, $K_{0.45}MnO_2$, $K_4Nb_6O_{17}$, $RbTaO_3$, $KTiNbO_5$, and $Cs_{6+x}W_{11}O_{36}$ are commonly fabricated by stacking of negatively charged slabs that built from either corner or edge-shared MO_6 (M = Ti, Nb, Mn, Ta, W) octahedral units [15], whereas, the alkali metal cations (K^+, Rb^+, Cs^+) are occupied the interlayer space. The common feature in these layered oxides is cation-exchange through the interlayer alkali metal ions. Their ion-exchange and intercalation properties allow the chemical modifications of composition interlayer spacing at ambient temperature, by retaining the host slab units. Moreover, several layered oxides are also employed for photocatalytic water splitting, where some of them are capable to allow reversible redox reactions. These kinds of compositions could useful for energy storage applications.

Manganese Layered Oxide–Birnessite

Manganese oxide layered (or birnessite) material has been also recognized as a potential supercapacitor electrode [95]. Specifically, the heterovalent manganese cations (i.e., Mn^{3+} and Mn^{4+}) reduced the layers due to their edge sharing with the negative MnO_6 octahedra, as a schematic is depicted in Fig. 1.6. In which exchangeable solvated cations are accommodated in the interlayer to neutralize the negatively charged layers. Chemically birnessite is described from the formula $A_xMnO_2{\cdot}yH_2O$, where A represents H^+ or a metal cation like Na^+, K^+, or Ca^{2+}. Similarly, other birnessite layered solids are also feasibly exfoliated into monolayers [96]. In this process, exfoliated monolayers are usually assembled as functional composites [97–100]. Pioneering work has been also demonstrated by describing the pseudo-capacitative behavior of manganese dioxide [101, 102]. Thus, manganese dioxide has attended a lot of attention to fabricating MnO_2-based composite electrodes for the supercapacitor application. In which predominately a redox reaction in between III and IV oxidation states of Mn ions could involve forming pseudo capacitative (Faradic process) at the surface. On the other hand, in the bulk, the electrode's major charge storage mechanisms are due to manganese oxides. But due to poor conductivity

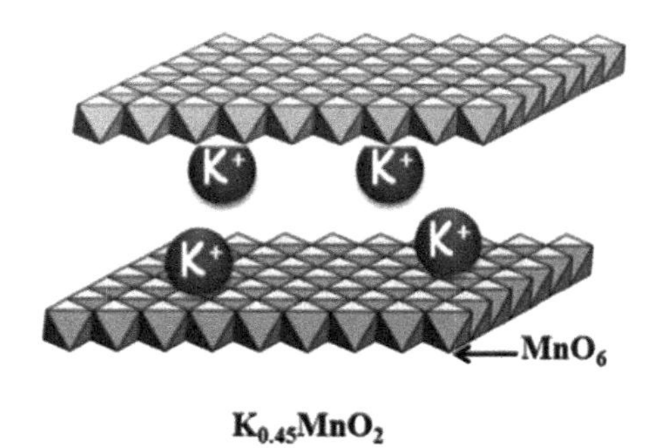

Fig. 1.6 Birnessite structure

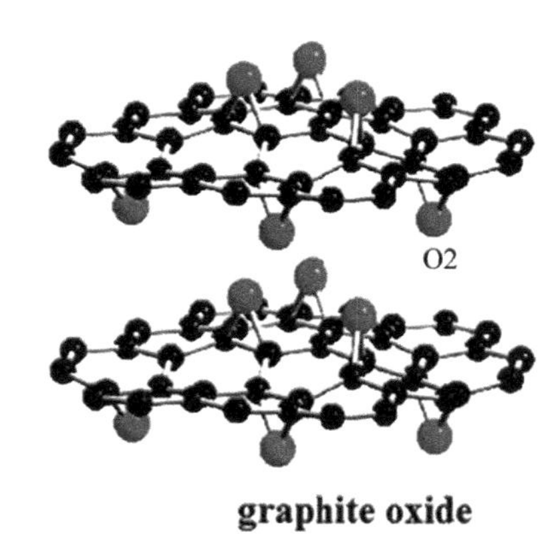

Fig. 1.7 Structures of graphite oxide

and structural instability, the hydrated manganese oxide possessed a low specific capacitance (100–200 Fg^{-1}).

Graphite Oxide

Graphite oxide (GO) [103] is a layered material that consists of hydrophilic oxygenated graphene oxide sheets (see Fig. 1.7) bearing covalently attached oxygen functional groups on their basal planes and edges [104]. There are various methods to synthesize GO, such as Brodie, Staudenmaier, Clauss, Boehm and Hofmann, Hummers, and Offeman [103, 105–107]. Depending on their extent of oxidation, inter-lamellar water contents, and interlayer distance. According to Lerf–Klinowski model, the GO basal planes are usually decorated from the epoxide and hydroxyl groups, while the carbonyl and carboxyl groups are located presumably at the edges [108–110].

The GO layers oxygen functionalities render it hydrophilic (lyophilic). However, its polar molecules readily intercalated into the interlayer and solvates, and as the consequence, complete delamination occurs [104]. Thus, the GO is considered to be an excellent host (or interlayer accommodation) for the long-chain aliphatic nanocomposite compounds; such as amines, ammonium ions [104, 111], transition metal ions [112], and metal oxide nanoparticles [113], hydrophilic molecules and polymers [114].

The delamination of graphite oxide occurs in a dilute alkaline aqueous solution. Additionally, the interlayer modifications of GO are also delaminated in organic solvents [115, 116]. Moreover, the two-dimensional sheets of GO are also formed through delamination while using them as precursors [117, 118]. GO could also use as a precursor of graphene sheets to make various graphene-based nanocomposites for high-performance device applications; such as battery electrodes, sensors, catalysts, potential hydrogen storage matrices, transistors, resonators, etc. [119–121].

1.1.2.3 Non-layered Solids–Surface-Assisted

2D nanostructured materials are commonly synthesized by making the layers artificially stacked on a substrate with arbitrary angles [122]. Predominately, these kinds of materials are fabricated by using epitaxial growth and chemical vapor deposition. As the typical example is silicone. Although such kinds of material's instability under ambient conditions is the real challenge to making them feasible in electronic applications [123, 124]. The typical examples of surface-assisted non-layered solids are Ge/Ag (100) and Ag(111) [125, 126], TiO_2 [127] and MgO/Mo (001) [128], and Al_2O_3/SiO_2 [129].

1.2 Chalcogenide Materials

Chalcogenides are materials **which associates** chalcogen **constituents**. The **word** 'chalcogen' **mentions** to group 16th group elements in the periodic table, sometimes it is known as the oxygen family. Sulfur (S), Selenium (Se), and Tellurium (Te) are the key elements of chalcogens, while the elements Oxygen (O) and radioactive polonium (Po) are also considered in this group. The term chalcogen was first introduced by Wilhelm Bilts's group around 1930 at the University of Honover, where it was proposed by Werner Fischer [130]. They stated that the heaviest elements, S, Se, and Te collectively can be considered as main "chalcogens", elements, usually term chalcogen is to be addressed only to these elements. The term "Chalcogens" reflects the Greek word "Chalcos", the literal meaning of this word is "Ore forming". Chalcogens (such as "S, Se, and Te") reacts with at least one of the electropositive element to form a stable chemical compound, this process formed compounds called chalcogenide. Hence chalcogenide has been recognized as an ore-forming element such that Sulfide, Selenide, and Telluride [131–133]. Whereas, in the chalcogenide compounds usually, chalcogens are basic elements. In the early few decades, chalcogen chemistry was exclusively centered on S, Se, and Te. It was noted that many factors contributed to their increasing importance in the chemistry of S, Se, and Te. Such as the development of synthetic methodologies to avoid the use of obnoxious reagents for a variety of technical applications [134].

Commonly chalcogens have oxidation states 2^-, 2^+, 4^+, and 6^+ with variable crystal structure features. They have six electrons in the outermost shell along with the usual electronic configuration ns^2 np^4. Their atomic volume, density, and atomic radius increase with the increasing atomic number. In practice, oxygen and sulfur are categorized as non-metals while selenium, tellurium, and polonium fall into the semiconductor or metalloid category. In the chalcogen group except for oxygen, all other elements are in a solid state. Chalcogen group elements usually exhibit allotropy characteristics. Sulfur, oxygen, selenium, polonium, and tellurium have more than twenty allotropes, nine, five, two, and one allotropes. This group material also shows high electronegativity behavior. Almost every chalcogen can play role in biological functions, such as either as a nutrient or a toxin. Therefore, it is worth discussing the common features of the three most widely used chalcogen elements;

Sulfur

Element sulfur was known since the old times of humanity. According to available information, it was known around four thousand years ago when Homer mentioned this element and later Bible described its common features in power. In the past centuries, sulfur was used extensively for many purposes such as painting, whitening linen, cosmetics, and medicine (about 1600 years B. C.). As the chemical element first Berzelius scientists discovered it in 1817. Subsequently, Mitcherlich discovered the allotropy of sulfur in 1822. Sulfur has been identified as a yellow solid with an atomic number ($Z = 16$) and an atomic mass of 32.06. Sulfur is usually appeared both as a native element and in the compound form [135]. The most significant sulfur allotropic forms have different structures such as octahedral or α-sulfur, prismatic or β-sulfur, γ-sulfur, δ- sulfur, and χ-sulfur. Element sulfur has a strong catenation tendency, like, polysulfide ${S^{2-}}_n$, and sulfones S_n. Generally, sulfur exists abundantly in a combined state of galena (PbS), cinnabar (HgS), copper pyrites ($CuFeS_2$), iron pyrites (FeS_2), zincblende (ZnS), gypsum ($CaSO_4.2H_2O$), heavy spar ($BaSO_4$) and celestite ($SrSO_4$). Sulfur could be oxidized into sulfur trioxide (SO_3), and sulfur dioxide (SO_2) with the ability to react moisture of the air. As the result, it produces sulfuric acid (H_2SO_4) and sulfurous acid (H_2SO_3), respectively.

Sulfur is also used commercially in the manufacturing of phosphate and ammonium fertilizers as the end-products. Sulfur has been also widely used in the synthesis of various chemicals and products such as detergents, gunpowder, matches, insecticides, agrochemicals, dyestuffs, fungicides, and petroleum refining including in the metallurgical applications. Sulfur can be also used directly in the commercial vulcanization of rubber. It also has ability to directly react with methane to produce carbon disulfide. It is used for the manufacture of cellophane and rayon. In chemical conversion, around 85% of elemental sulfur converts into sulfuric acid. This kind of conversion process is usually adopted in the extraction of phosphate ores for fertilizer manufacturing. Sulfuric acid applicability is not limited to discussing cases but also could be many more, such as sulfuric acid also used in oil refining, wastewater processing, and mineral extraction. Its other forms like sodium and ammonium thiosulfate are widely used as fixing agents in silver-based photography. In a similar fashion sulfites derived from burning sulfur are also extensively used for bleach paper. Moreover, the Leads Sulfur preservatives are also used to dry fruits, sugar, etc.

Selenium

Element selenium was discovered by J. J. Berzelius and J. G. Gahn in 1817, when they were sediment taken from the lead chamber of a sulfuric acid plant, in Gripsholm Sweden. Selenium has an atomic number ($Z = 34$) in the periodic table and an atomic mass of 78.96. Commonly, the element selenium is found in sulfide minerals (more than 40 minerals) [135]. Selenium has outermost orbit oxidation states 2^-, 0, 2^+, 4^+, and 6^+. Element selenium is less stable compared to sulfur; it is also a poor conductor of heat and electricity. However, the most of physical properties are similar to sulfur. The element selenium has more metallic characters therefore; it is classified as a nonmetal in the chalcogen group. Selenium is also used for different applications such as photoelectric cells, photocopies, solar cells, and light-sensitive devices owing to it conducts electricity much more efficiently in the presence of light and unlike in darkness. Thus, the most significant physical feature of selenium is its electrical properties. The various allotropic forms of selenium acts as a semiconductor. As we know that semiconductors are substances that can conduct an electric current better than nonconductors, but not better than conductors. Naturally, selenium is a reactive element that combined easily with hydrogen, fluorine, chlorine, and bromine. Moreover, it can also strongly react with nitric and sulfuric acids. It also combines with several metals to form compounds that are usually called selenides, like the compound cadmium selenide (CdSe). A large amount of consumption of selenium salts is toxic, while, trace amounts are necessary for cellular function in many organisms for all kinds of animals as well as humans. Particularly it's an ingredient in many multivitamins and other dietary supplements. Selenium usually exists in several allotropic forms; however, only three more stable allotropic forms are generally accepted. The amorphous selenium is generally in red, black, or glassy form. The most stable form of the element selenium has a crystalline hexagonal structure and it is usually found in a metallic gray, while the crystalline monoclinic selenium occurs in a deep red colure.

Tellurium

In the period of the Habsburg Empire, tellurium was discovered around 1782 when Franz Joseph Muller von Reichenstein found a mineral containing tellurium and gold. This newly investigated element was named the Martin Heinrich Klaproth in 1798, the meaning of the word *tellus* in Latin is earth. Tellurium has atomic number $Z = 52$ and the atomic mass is 127.60. Tellurium is the heaviest and non-radioactive metalloid member of the chalcogen family, it is a silvery white color having a similar metallic cluster to tin at room temperature. For traditional and industrial purposes tellurium has been extensively used in metallurgy, to fabricate various kinds of alloys, like steel or copper alloys [134]. In the twenty-first century, the significant importance of tellurium compounds has been recognized in both inorganic and organic chemistry including in materials science. A verity of potential applications of tellurium compounds in electronic industries makes them a hot material for semiconducting device manufactures. The semiconducting metal tellurides can have a wide range of potential utilities such as thermoelectric material for cooling devices, solar panels, etc. [136]. Naturally, tellurium occurs in the form of different isotopes possessing masses 120, 122, 126, 128, and 130. Tellurium's most stable isotope masses are 128 and 130. The elemental tellurium exists in a lower abundancy in the Earth's crust compared to other valuable materials, such as gold and platinum. Hence the significant uses of tellurium can be expressed in terms of the metallurgical additive, chemical industries, as a vulcanizing agent, accelerator in the processing of rubber, and as a component of the catalyst for synthetic fiber production. Particularly the highly pure tellurium is preferred use for electronics applications. It has attended tremendous attention from the scientific and technological industries due to its high applicability in devices, like thermal imaging, thermoelectric, phase change memory, and photoelectric devices.

Tellurium has been also used efficiently to fabricate solar cells devices/modules, such as the instance cadmium telluride (CdTe). National Renewable Energy Laboratory (NREL) has extensively fabricated and tested tellurium-based solar cell performances. They found these photovoltaic devices' efficiencies are impressive compared to other efficient potential materials solar cell devices. Therefore, CdTe solar has found great attention in recent years for their commercial production, and as the consequence, the demand for the element tellurium increases rapidly in recent years. Element tellurium and its compounds are mildly toxic and need to be handled very carefully, but acute poisoning is rare. Herein, the three prominent elements of the chalcogenide group elements; sulfur, selenium, and tellurium important physical properties are also summarized in Table 1.1.

Table 1.1 Key physical properties of the chalcogen elements sulfur, selenium, and tellurium

Properties	Sulfur (S)	Selenium (Se)	Tellurium (Te)
Atomic number	16	**34**	**52**
Atomic mass	32.07	78.96	127.60
Electronic structure	$[Ne]3s^1\ 3p^4$	$[Ar]\ 3d^{10}\ 4s^2\ 4p^4$	$[Kr]\ 4d^{10}\ 5s^2\ 5p^4$
M. P. (°C)	119 (S_β)	220.5	449.8
B. P. (°C)	444.6	684.8	989.8
Atomic radius (10^{-10} m)	1.00	1.40	1.37
Density (g/cm^3)	2.06 (S_α)	4.82	6.25
Covalent radius (10^{-10} m)	1.03	1.16	1.37
Electronegativity (Allerd-Rochow; Pouling)	2.44; 2.58	2.48; 2.55	2.01; 2.10
1st ionization potential (eV)	10.38	9.75	9.01

1.2.1 Semiconductor Lone Pairs

Chalcogen's outermost orbital shells (valence) electronic configurations are usually expressed as ns^2p^4 or $ns^2p^1{}_x p^1{}_y p^2{}_z$. Their valence electronic subshells are distributed into two electrons, the first one in the atomic *s*-orbital and the other dual electrons in three *p*-orbitals when electrons are uncoupled, whereas the remaining (third) solitary is engaged from a couple of electrons. The last paired electrons are generally called lone-pair (LP) electrons. These lone-paired electrons have a usual (not always) tendency and do not contribute to the formation of covalent bonds. On the other hand, usually *s*- subshell electrons also have chemically inactive behavior (but not always). Because of these key features a few significant electronic configurations of the chalcogens are discussed.

In the case of chalcogen atoms involve in chemical bonding formation, and then the change in atomic electronic arrangement depends on supplementary chemical constituents that exist within the complex. In this perfect process, every chalcogen atom (sulfur, selenium, and tellurium) forms double covalent bonds to their adjacent. Typically sulfur procedures rings, while the selenium gives us rings and chains, whereas the tellurium creates chains. All three key chalcogen group elements have solitary dual uncoupled *p*-electrons that actively involves in the construction of covalent bonds, whereas the subshell *s*-paired and LP *p*-orbital electrons still inactive. Though *p*-orbitals are in orthogonal position and their bonding angle is subtended by the chalcogen atoms both somewhat near to 90° from the LP electrons. And their orbital direction is vertical to a newly formed plane from the p_x and p_y orbitals. Specifically, the electronic configurations structure of selenium has followed the zeroth approximation which is nearly cubic, as the schematic is depicted in Fig. 1.8. More recently metal halide perovskites generic composition ABX_3 has also gained significant attention to generate favorable optical and electronic properties under the hybrid organic–inorganic structures. The metal halide perovskites are usually composed of an organic cation at A-site, a metal at B-site, and halides at the X-sites. In the perovskites configuration Se atoms are formed either extended SeN chains (−, −, −, −, −, etc. and +, +, +, +, +, +, +, +, etc.) or eight-membered rings (for +, −, +, −, +, −, +, −), their sign sequence is depending on the dihedral angle rotation, as a schematic is illustrated in Fig. 1.9. The first case could be correlated to the trigonal (or hexagonal) structure of the selenium, while the latter one is usually related to the monoclinic structure of selenium. However, in both cases attachment inside a chain/ ring is covalent. Additionally, bond formation among chains and rings is abundant feebler. Such bonding between them have be van der Waals (vdW) behavior, however, it is also defined as resonant [137, 138]. Even though the molecular interactions are significantly weaker, the weak interactions are mainly accountable to the long-range ordering in the crystalline form of selenium. Moreover, a significant interest has been also made in sulfur, selenium, and tellurium atoms owing to their similar electronic configurations of outer shells. Because of lighter sulfur atoms approaches to procedure rings whereas the weightier tellurium atoms favor to form chains. The proportion of the nearby distance among the chalcogenide atoms of the similar behavior chains

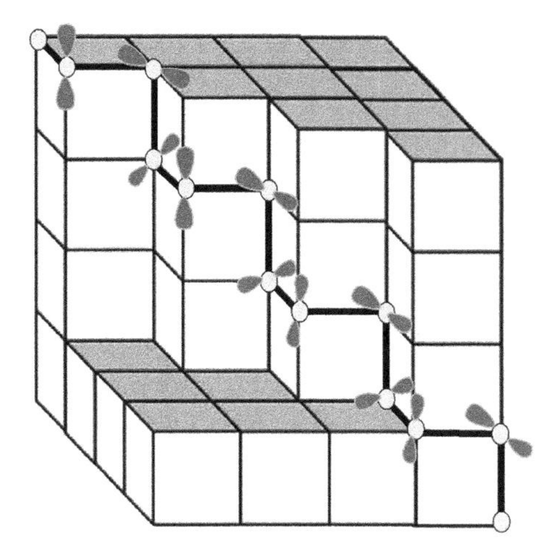

Fig. 1.8 Schematics of selenium zig-zag chain in a simple cubic lattice for a real structure in terms of intrachain and interchain atoms with the distances and bond angle

has followed a decreasing order (chain distance) pattern from S–Se–Te. It is directed to the presence of a stronger covalent component in the latter cases.

Moreover, the significant role of the LP in chalcogenide materials was first pointed out by Kastner [139], when he demonstrated the tetrahedrally bonded semiconductors. As Si or Ge are hybridized from sp^3 orbitals and further it splits into bonding (σ) and antibonding (σ^*) molecular states. That successively extended in the valence and conduction bands in a solid. Specifically, in terms of chalcogens (S, Se, Te), the *s* states exist sufficiently lower the *p* states, which means no reason to consider it. Because only two of the three *p* orbitals can contribute to the bond formation. Such process-formed chalcogens have twofold coordination and are left behindhand single nonbonding electron couple (LP). Normally in the solid, the unshared (or LP) electrons of an LP band are close to the native *p*-state energy. Whereas the bonding (σ) and antibonding (σ^*) bands split equally concerning their reference energy, and both σ and LP bands have occupancy. It leads bonding band cannot prevail longer at the top of the valence band, this reflects the important role of the LP band. As the typical schematic of the band diagram is illustrated in Fig. 1.10. Considering above discussed special case under the specific circumstances Kastner suggested the word 'LP semiconductors' for the chalcogenides [139, 140].

Likely, the discrete Se chain, somewhere the incorporation of atoms in terms of balls and covalent bonds sticks, charge density difference (CDD) isosurfaces properties can be also explored [141, 142], as the typical structure is illustrated in Fig. 1.11. The CDD implies the difference in electron density between the structure of the under-examined and isolated quasi-atoms [141, 142]. The appearance of a CDD cloud in the middle among dual atoms is usually correlated to the signature of a covalent bond. For an instance, a selenium chain overhead and toward side the CDD clouds is situated midway among the atoms which considered to be signature of the covalent bonds. Moreover, the covalent bonding signature may also appear in CDD clouds relative to non-bonding *p*-orbitals (LP electrons). Because LP electrons

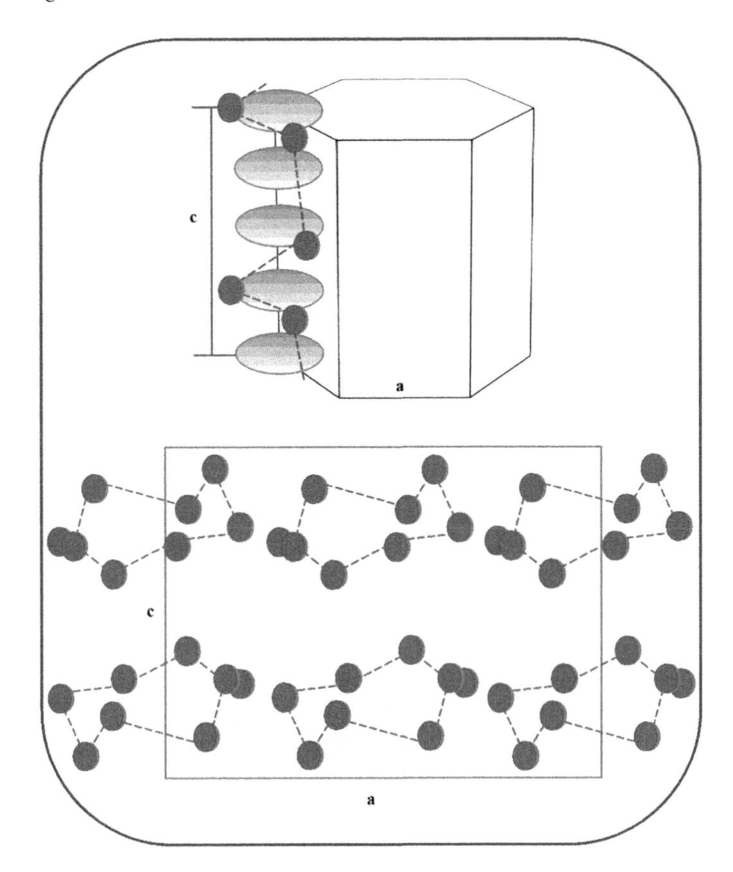

Fig. 1.9 Schematic for the selenium trigonal (top) and monoclinic (bottom) phases

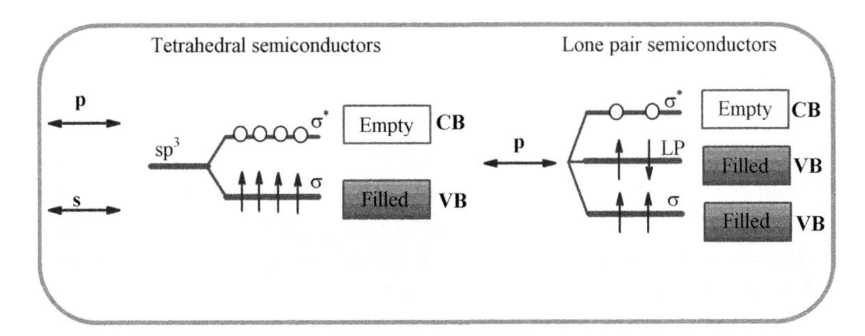

Fig. 1.10 Bonds for the tetrahedrally bonded and LP semiconductors

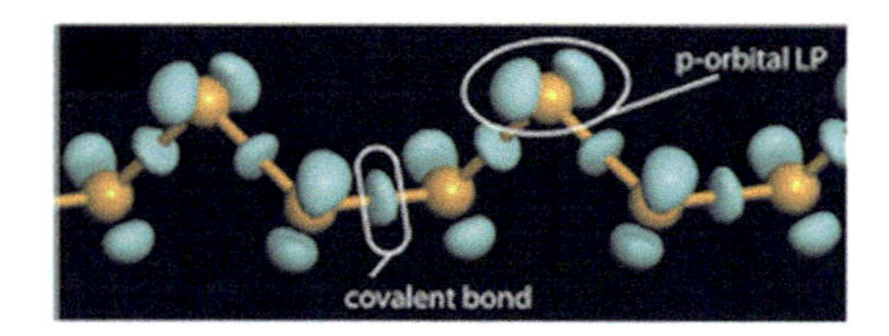

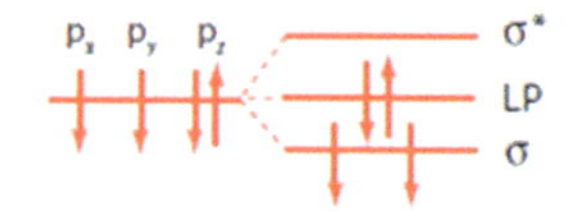

Fig. 1.11 Charge density difference (CDD) isosurfaces to a Se chain for both covalent bonds and lone-pair electrons (adopted from Kolobov et al. [142])

are also sometimes referred to for the two *s*- electrons, therefore, commonly the term LP is used for both non-bonding *p* as well as *s* orbital electrons.

Typically, materials wherein chalcogen atoms have this kind of electronic configurations, like chalcogenide glasses, $A^{V}{}_{2}$ $B^{VI}{}_{3,}$ and $A^{IV}B^{VI}{}_{2}$ crystals ($As_2S_{3,}$ GeS_2, GeTe, etc.). Usually, satisfy their valence requirements and the covalent coordination number of constituent elements following the so-called $8 - N$ rule. The valence is generally defined as to mean the number of the single (σ-type) covalent bond requirements of the atom to complete through the shared electrons comprising the bonds and its outer shell of *s* and *p* electrons. Since a complete shell contains 8 electrons, therefore, such atoms usually obey the "$8 - N$" rule for the condition $N > 4$. Where N is the number of valence electrons and their coordination number is defined by the $8 - N$ rule. For an instance, in the composition $As_2Se_{3,}$ all arsenic atoms are threefold coordinated and selenium atoms have twofold coordination. The $8 - N$ rule was suggested for chalcogenide glasses to justification for the observational insensitivity against foreign doping. The $8 - N$ rule is also often called as Mott rule [143].

More recently the halide perovskites have been also classified as the foundation of an emerging class of materials for a broad range of applications in various areas including renewable and sustainable applications [144]. These kinds of low-dimensional metal halides focused intensively owing to the success of hybrid lead halide perovskites as optoelectronic materials. Specifically, the light emission of low-dimensional halides based on $5s^2$ cations Sn^{2+} and Sb^{3+} is useful in a variety of applications that are complementary to three-dimensional halide perovskites with their unusual properties, such as broad band character and highly temperature-dependent lifetime. Thus exceptional properties can be derived from the chemistry of the $5s^2$ lone pair. Moreover, the $5s^2$ lone pairs also have exceptionally rich chemistry due to the combination of stereo activity and structural flexibility compared to neighboring metals. Additionally, it also followed the inert pair effect. Generally, the oxidative stability of ns^2 cations increases to moving down in the periodic table with $6s^2$ cations, such as Pb^{2+} and Bi^{3+} chemically inert lone pairs [145, 146] This kind of lone pair electrons also have strong stereochemical effects on the metal–ligand coordination environment. Though, these lone pair expressions usually distinct from the chemical stability of the "inert" lone pairs. The same heavier $6s^2$ cations also offer more space for the lone pair and reduced surface charge density on the cation than the lighter neighbors, thereby, increasing the polarizability and coordination number.

Collectively all these features allows to structurally inert lone pairs possess highly symmetric structures, like PbI_2 and BiI_3 [147].

Emission of the isolated center manifests itself in many different forms, such as a dopant in a host matrix or pure material, or under different conditions, like cryogenic or room temperature. The different fields of spectroscopy and materials science contributed to explaining the isolated center emission. Various theoretical models have been also demonstrated for their circulations. Among them, one of the important theoretical concepts based on a unified framework approach is widely accepted. Based on this approach there are two models widely used, first one is treating 0D centers as isolated ions addressed to the halide ligand field and the second one is the presentation of the limiting case to complete electronic and structural isolation in the framework of an extended solid. Both approaches lead to the same theoretical picture of the isolated MXz centers, as illustrated in Fig. 1.12a. Thus, the unification of the two models provides a framework that helpful to explain and predict many properties of emissive 0D metal halides. This model's first theoretical description was provided with the simple two-electron, it is so called Seitz model, as depicted in Fig. 1.12b. According to this model, a free ion with the ns^2 configuration has a singlet electronic ground state, it is usually denoted with atomic term symbol $^1S_0$. The splitting of the *ns np* excited states into ^1P (singlet) and ^3P (triplet) states is possibly due to coulomb (F) and exchange (G) interactions. Moreover, the 3P states also splits into non-degenerate states $^3P_2$, $^3P_1$, and $^3P_0$ due to the influence of the spin–orbit coupling interaction. The overall energy of all states could non-uniformly altered through the ligand field (in the case of the coordinating halides).

This kind of interaction with the ligand field of ns^2 ions may include a large degree of hybridization between the atomic orbitals of the central metal and ligands. This description has been verified with the help of molecular orbital (MO) theory, as the schematic illustrated in Fig. 1.12c. As per this model, the ground and excited states are localized on both central ions and ligands. Their atomic term symbols lead to $^1S_0$ which becomes like a_{1g}. However, their relative energy latter depends on the molecular geometry, such as octahedral, square-pyramidal, and disphenoidal. Their energies are summarized by Vogler and Nikolin in terms of various possible molecular geometries [148]. Though, it is very common to use atomic term symbols even for systems with a high degree of hybridization and denoted their resulting molecular states as $^1P_1$ and $^3P_0$-derived, etc. [149–151]. This model mainly serves to purpose: to explain the transitions that occur between the states in terms of their allowed or forbidden nature. It is also experimentally reflected in the measurement of emission decay time (longer radiative lifetimes for partially forbidden transitions and shorter for allowed ones) or the term of the corresponding excitation band intensity [152, 153]. As an example, the dramatic reduction of the radiative lifetime of Cs_2NaSbC_{16} from 1.4 ms at 5 K to 0.15 μs at higher temperatures, and no change in emission intensity until ~ 200 K. It entirely derived by the changes in the character of the radiative excited state from the forbidden $^3P_0 \rightarrow {}^1S_0$ transition at low temperatures and $^3P_1 \rightarrow {}^1S_0$ transition around at ~ 25 K [154]. However, the MO approach is more acceptable for comparing the optical properties of the different geometries and accurate prediction of ground and excited state geometries. Hence it is crucial for the

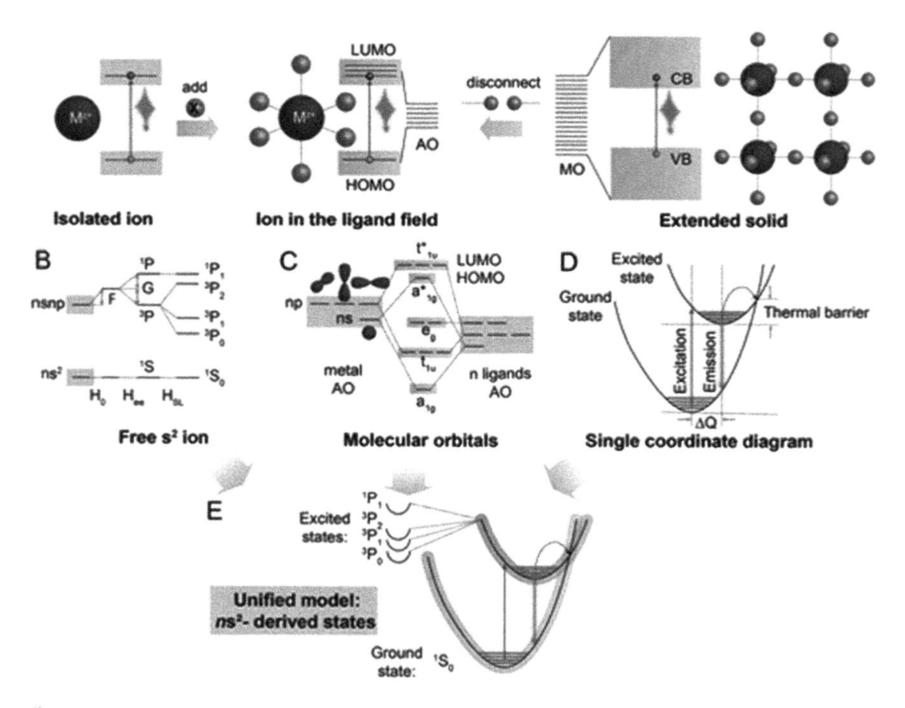

Fig. 1.12 Schematic to define a relationship between the atomic orbital model, molecular orbital model, and the semiconductor model of an extended solid. (B) Energy band diagram associated with the free ns^2 ion. (C) Energy band diagram of the metal–halide molecular orbitals. (D) Configurational coordinate diagram of the simplified STE model in 0D $5s^2$ metal halides. (E) A unified model with the configurational coordinate diagram, in which the ground and excited states are described using their atomic character as derived from the active ns^2 metal ion (adopted from McCall et al. [169])

0D systems models in which the halide bonds are not shared. Thus, the metal halide anion and its lone pairs are free to distort in the excited state, and their excited-state geometry is dramatically different than the ground-state geometry.

Although the Seitz model has been widely used to interpret the optical properties of ns^2 ion impurities in binary alkali metal halides including other oxide and halide hosts [153–155]. But the self-trapped excitations model gets more attention to interpret the optical properties of 0D metal halides, therefore, research focuses on ns^2 metal halide semiconductors. On the other hand, the dominancy of free excitons in prototypical 3D AMX_3 semiconductors has been described in terms of the structural versatility of metal halide yields where the self-trapped excitations exist [156]. The typical theoretical description of phonon-assisted self-trapped excitations has been provided in terms of recombination involvement in a single configuration coordinate diagram. Whereas the abscissa represents a change in molecular geometry of the excitation (Q) and the ordinate represents a change in energy, as illustrated in Fig. 1.12d [157]. As per this model the ground and excited states are the parabolic potential energy functions; it means their first approximation is harmonic. Even though in the single configuration coordinated diagram, the free exciton excited state can be parabola centered for the ground state having higher energy, free excitons not appears for the 0D $5s^2$ metal halides. Therefore, the localized self-trapped exciton model is useful for only a stable excited state. Moreover, the consistence with 1D or lower-dimensional systems with unstable free exciton states which probably localized upon any perturbation, it has to be considered [158]. The self-trapped exciton is considered to be a more relevant state for the 0D $5s^2$ metal halides. The simplification of the self-trapped excitations model of the single configuration coordinate diagram is almost identical to the generalized vibronic transitions as described by the Franck–Condon principle. Later works on it defined the self-trapped excitations model that's also consistent with the higher dimension metal halides.

Thus, the minima of each parabola reveal the equilibrium geometry of the ground or excited states, while the horizontal difference between these geometries (ΔQ) could be correlated to the structural change of the metal halide center that undergoes upon excitation. The difference between the excited and ground-state geometries also directly correlates to the large Stokes shift of 0D $5s^2$ metal halide emission. It reveals both absorption and emission transitions are raised from the different vibronic states (as denoted flat lines in the parabola), and their yields are analogous to individual transitions with emission probabilities under the Poisson distribution. The convolution in these transitions reveals a broadened emission peak, often with a tail toward lower energy [159]. However, the magnitude of the electron–phonon coupling is generally described from the Huang–Rhys parameter, that's usually larger for the stronger coupling. Hence it defines the degree of broadening associated with self-trapped excitations emission [157, 159].

Thus, the self-trapped excitations model is a configurationally coordinated diagram under a simplified projection to represent a complex energy surface from a two-dimensional plot, but it has limitations. Like in complex materials that have multiple emissive centers, therefore, it would be useful to characterize the self-trapped excitations emitters with the Toyozawa model's important features, such as the Huang–Rhys parameter and effective phonon energy [158–161]. Significantly this model can also rationalize the key property of self-trapped excitations emitters, such as the temperature-dependent emission quenching [152, 158]. Additionally, at the crossing of ground and excited states parabolas provides a nonradiative relaxation pathway from the excited-state minima through phonon excitation, it is also known as hot nonradiative recombination [158]. Moreover, the energy difference (height) between the intercept and the excited-states minima is also useful to define the energy barrier. Typically it defines the required order of the quench for the radiative process (see Fig. 1.12d).

Hence based on these theoretical descriptions one can say that the Seitz model is mainly toward the nature of the electronic states with the MO excitations associated with their useful degeneracy in the electric field of the ligand. While the single configuration coordinate model is useful to describe the phonon–exciton interactions. Moreover, interpreting the optical properties of 0D materials requires including all three models' theoretical concepts (i.e., unified model). A unified model schematic is depicted in Fig. 1.12e. This model is also based on the single configuration coordinate diagram, but the excited state is commonly split into several nondegenerate states. The

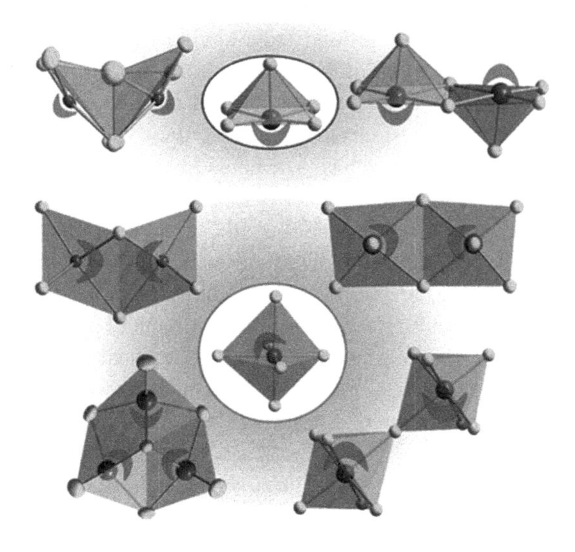

Fig. 1.13 Atomic environments of polynuclear 0D $5s^2$ metal halides from square pyramids and octahedra, and the lone pairs visualized with semicircular marks (adopted from McCall et al. [169])

splitting of each state reflects the nature of the contributing atomic states. Although in this model splitting of degeneracy, states vary with the coordination geometry, it also offers enough flexibility to describe the optical features.

Beyond these described isolated 0D units, an excess of polynuclear 0D cluster compounds based on $5s^2$ metal cations also exists [162–165], but their optical potential is still to be thoroughly explored. Although, the room temperature PL property has been noticed to $[Sb_2I_9]^{3-}$ dimmers for the composition $Cs_3Sb_2I_9$ with the weak emission intensity [166]. This could be due to the asymmetry in the polynuclear system. It provides a stable lone pair off-center in coupled units to enhance the nonradiative recombination, as illustrated in Fig. 1.13. However, the discovery of $Rb_7Sb_3Cl_{16}$ with its emissive distorted $[Sb_2Cl_{10}]^{4-}$and $[Pb_3Cl_{11}]^{5-}$dimers opens up the path for more optically active centers [167–169].

1.2.2 Alternation of the Valence Electron Pairs

Topological imperfections in a regular arbitrary linkage of the glasses are rises due to deviations of the 8 – N rule. More straight forward kinds of topological defects are due to unsatisfied atoms or dangling bonds. Therefore, the open-up S_8 ring generates two dangling bond defects. Moreover, the extended chain construction of selenium may also own dual dangling bonds at the terminating point. In which every dangling bond associates single electrically neutral electron. Specifically, in an amorphous network this kind of defect exists in large quantities because the Fermi level in chalcogenide glasses is held close to mid of the gap [170]. Typically the amorphous chalcogenides a dark electron spin resonance (ESR) are realized but signal has un-appeared, however, a greater amounts of defects states can be recognized as established from the many experimental methods. The discrepancy between the scientific approaches and experimental results leads that dangling bonds being informal to visualize, but they do not procedure key type of topological imperfections in chalcogenides.

To resolve this disagreement, P. W. Anderson suggested as, "owing to a strong electron–phonon connection, electron combination becomes energetically favorable in spite of endothermic cost in energy desired to keep dual electrons over a solitary location" [171]. These types of imperfections are known as negative correlation energy or negative-U, centers. The verification of this approach was provided by Street and Mott with the description of a circumstance instance of an elemental chalcogen [172]. They argued that the most of imperfections in chalcogens are couples of positively and negatively charged dangling bonds:

$$2D^0 = D^+ + D^- \tag{1.3}$$

here D represents the dangling bond and the superscript describes the charge. The absence of an ESR single leads to the system's diamagnetic behavior because in this situation all the electrons should be paired.

Further, this model work was extended by Kastner et al. [173] and presented the general form, the more generalized form is recognized as the *valence-alternation pair* (VAP) model or Kastner–Adler–Fritzsche approach. According to the generalized approach creation of over-coordinated imperfections due to the participation of LP electrons. The hypothesis of this model is based on an empty orbital of a positively charged dangling bond interaction together the LP electrons of a neighboring chain to form a threefold coordinated flaw, therefore, overall dropping the energy of the configuration:

$$C_1^+ + C_2^0 \rightarrow C_3^+ \tag{1.4}$$

here C represents chalcogen and subscripts correspond to coordination, while the superscripts reflect the charge. A typical VAP model bonds formation schematic is depicted in Fig. 1.14.

The overall gain in energy is due to the formation of an additional bond that derived the force to compensate the energy to create a doubly occupied site at the negatively charged dangling bond. Therefore various defect configurations energy levels exist in a-Se, as schematically depicted in Fig. 1.15.

This approach also leads that the most stable defect pairs is consist of the triply coordinated positive sites and singly coordinated negative sites, such as C^{+3} C^{-1} pairs. The energy of the VAP defects is approximately the identical as the energy of dual atoms inside a chain ($2C0_2$). In the case of two atoms the immediate proximity to each other, in this situation, the flaw couple is so-called an intimate valence-alternation couple. Whereas, in the case of nonaligned defects, a triply coordinated atom ($C0_3$) could be most stable under certain circumstances where chalcogen atoms coordination number three. Although the experimental evidence is more favorable to neutral dangling bond stability [174, 175].

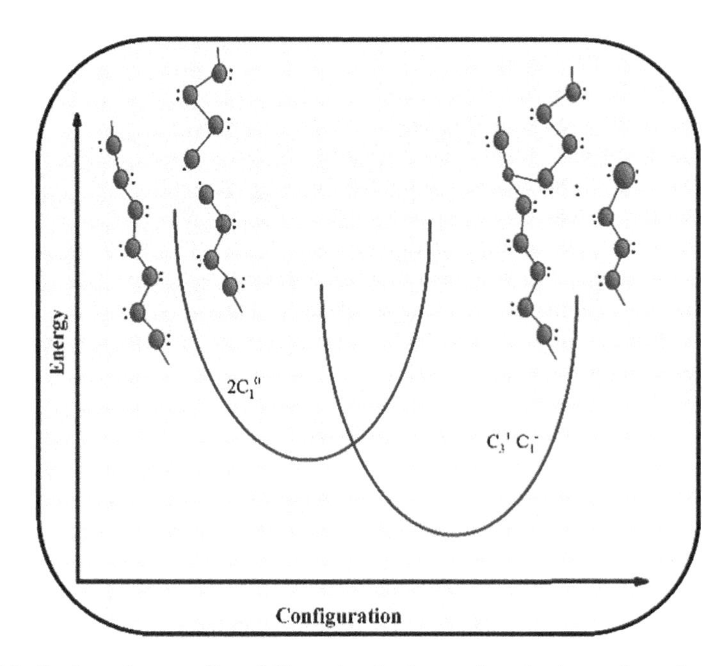

Fig. 1.14 Configuration coordinated illustration for the creation of valence-alternation couples in chalcogenide glasses, whereas the dots represents LP electrons

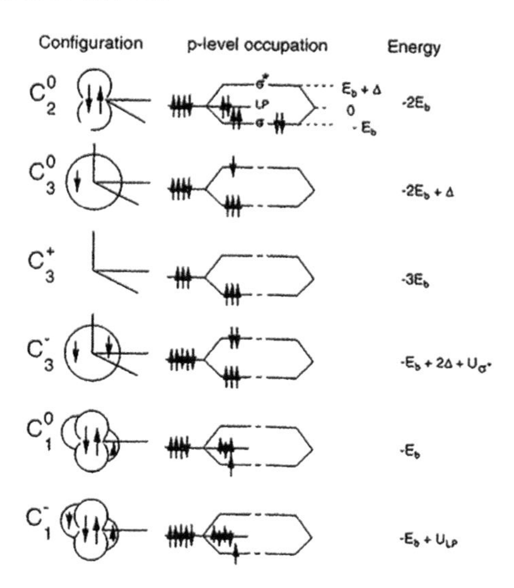

Fig. 1.15 Configurations and energy levels for the coordination defects in selenium. In configurations, straight lines represent the bonding (σ) orbitals, lobes represent the LP orbitals and large circles represent the antibonding (σ^*) orbitals. Arrows correspond to the electrons (adopted from Kastner et al. [173])

1.2.3 Donor–Acceptor Bonds

Almost every kind of chalcogenide (including some tellurides classified as "phase-change memory materials (PCMs)) has unique material properties, such as ultrafast crystallization rates and large (optoelectronics) properties contrasts between amorphous and crystalline phases [137, 138, 176–181]. The simultaneous occurrence of both amorphous and crystalline phases in these materials is rather counter intuitive. Therefore, the structural difference occurs between amorphous and crystalline phases, while the opposite trend could be also possible. Thereby, it is customary to consider electron delocalization in such materials. Hence, a single Lewis structure with the concept of bonding and nonbonding electron pairs cannot fully describe the electron distributions of both phases [137, 179]. Considering this issue conundrum, chemical-bonding models have been proposed with the assumption of a drastic change like chemical bonding between the amorphous and crystalline phases [138, 142].

Therefore, the useful existing models have been classified based on their length scale of delocalization. The key resonant bonding model [137, 138] is based on a limited number of Lewis-conforming resonant structures. It is suitable to describe simple crystals that possess minimal disorder, such as the nature, of structurally disordered amorphous PCMs beyond the validity of the resonant model. However, crystals possessing significant amounts of the structural disorder are difficult to describe from this model. Similar to PCM systems other materials also have metastability, such as vacancy-containing rocksalt-type crystalline GST or more general compositions along the pseudo-binary tie line of GeTe–Sb_2Te_3 [177, 182], Ag–In–Sb–Te and I–V–VI_2-type compounds (crystalline GeTe with vacancies) [183, 184]. Although the corresponding spatial scale cannot define very impressively, because the measurement of electron delocalization has a substantial metallic character and is designated as meta-valent bonding [185]. Moreover, a concept based on atomic orbital similarity was also demonstrated [186]. Whereas, the molecular-orbital (MO) approach that involves notable three-centers, four-electron (3c/4e) interactions, or the valence-bond (VB) theory of hyper bonding emphasis on linear triatomic bonding geometries that successfully described the hypervalent molecules chemistry [187, 188]. Later on, the formation of linear triatomic bonding geometries was also recognized in a solid state with exclusive evidence of amorphous GST and crystalline GeTe–Sb_2Te_3 alloys [141, 179]. The signature of electrons delocalization within the bonding configuration has been also noticed for the amorphous GST [179]. The key advantage of the hyper bonding-interaction model over other models, it characteristically could be linked to the microscopic properties of the materials.

Moreover, the concept of multicenter hyper bonding interactions was introduced and completed the picture of chemical bonding in chalcogenides [189]. According to this approach, the chemical-bonding interactions of amorphous/crystalline PCMs are associated with material properties. The newly proposed model significantly explains it without invoking any other new type of bonding interaction property contrasts between amorphous and crystalline phases of PCMs, in terms of their natural difference extents of hyper bonding. Thus, the concept of chemical-bonding interactions has a dramatic contrast with the existing theory of PCMs. Hence, the theories and concepts of donor–acceptor bonds have been innovated from time to time [189].

Similarly, to versatile PCMs the donor–acceptor model for the photovoltaic has been also demonstrated for one of the burning 2D materials, i.e., TMD [190]. A typical schematic for the TMD material represents in Fig. 1.16a–c. Material like the WSe structure with a bandgap of 1.87 eV behaves as a donor in this system, as illustrated in Fig. 1.16a. Whereas material WSe_2 with the bandgap 1.67 eV act as an acceptor, because its lower bandgap S could favor the band alignment. Since the donor consists of two layers of WSeS arrangements on top of each other. So, a strong dipole–dipole coupling occurs between the two donor layers due to the large permanent dipole moment in each layer. Consequently, the conduction band splits into bright and dark bands with an 18 meV energy gap. Moreover, the induced polarization in the acceptor layer can be calculated with the electric field from the donor layer. While the donor–acceptor coupling energy (γ_C) is obtained from the dipole-induced dipole interaction energy between the WSeS layers and the WSe_2 layer. In the equilibrium donor–acceptor, typical spacing is around 3 Å [191], with the corresponding coupling energy 515 μeV. The coupling energy may decrease with increasing donor–acceptor separation up to around 50 μeV for 10 Å separation. The typical radiative lifetime excitons range (5–17 ps) for the TMDs [192, 193, 194], with the corresponding recombination rate (γ_R) in the range of 40–130 μeV.

The energy band diagrams of the photovoltaic model are given in Fig. 1.16b, c. According to this representation, the standard donor–acceptor possible without dark state protection (Fig. 1.16c) corresponds to the improved approach with dark state protection. Both approaches have revealed the incoming photons generate excitations in the donor with the γ_R process, then transfer to the acceptor with the γ_C process and convert it into the current with the Γ process path. Although the γ_R the process is reversible but leads the radiative recombination loss. According to the photovoltaic model of TMDs, the formation of some new optically excitable states through strong excitonic coupling between the donor layers is also possible, as schematically illustrated in Fig. 1.16c. This kind of splitting of the conduction band in the donor favors the formation of a bright state "b" and a dark state "d". Though virtually all photon absorption and emission take place via the bright state. Moreover, the initial absorption of photons leads to the excitation of the bright state. The corresponding thermal relaxation process γ_{bd} gives us the excitations from the bright state "b" into the dark state "d", where the radiative recombination may forbidden. Since radiative recombination cannot occur through "d". Therefore, the photon re-emission suppress with fewer excitons that lost through the recombination before being transferred to the acceptor. Hence, the excitation transfers from d to α. The corresponding photocurrent is determined from the rate of Γ and α to β. Since the dark state protection from recombination also increases the number of excitons available in the acceptor, thereby, it probably gives a higher photocurrent in solar cells [190].

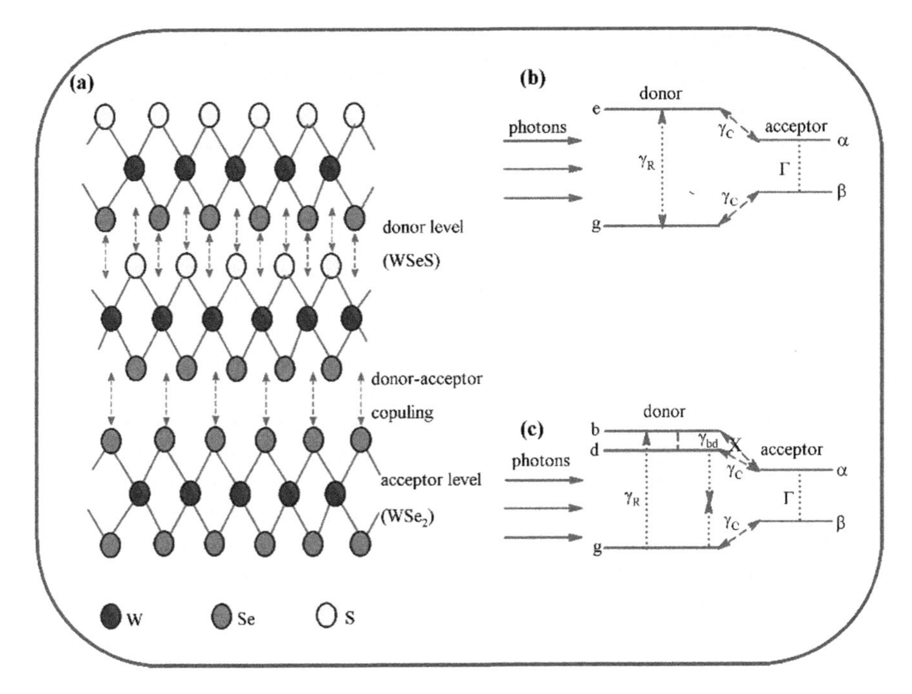

Fig. 1.16 Donor–acceptor model; **a** Structural model, **b** Slandered photon–electron conversion with donor and accepter, **c** Novel dark and bright states due to strong excitonic coupling under the dipole interactions

1.2.4 Hybridization of Bonds

Chalcogen generally has non-covalent interaction between the electrophilic region of an atom (S, Se, and Te) when a Lewis base is in the same intramolecular bonds or with other intermolecular bonds. Similarly, other interactions could also involve the main block elements, such as halogen bonding and pnicogen bonding due to secondary bond interaction. Chalcogen bonding interactions with halogen, hydrogen and pnicogen bonding are well described with the help of several theoretical approaches [194]. Though secondary bond interactions mechanism also has many common features. Like all the interactions involve an electrostatic and a covalent part. But specifically electrostatic part arises due to a Coulomb attraction between the negative electrostatic potential at the lone pairs (or π-bond of a Lewis base) and a region of positive electrostatic potential collinear to the covalent bond formed between a pnicogen, chalcogen, or halogen bonds due to electronegative substituents, this is the so-called σ-hole region [195–197]. The corresponding covalent portion could be due to charge transfer from the lone pair orbital of the Lewis base into the σ^* pnicogen, chalcogen, halogen orbitals, and electronegative substituent.

It gives to a 2e-delocalization and stabilization of lone pair orbitals, as the schematic is depicted in Fig. 1.17. The magnitude of 2e-delocalization is directly

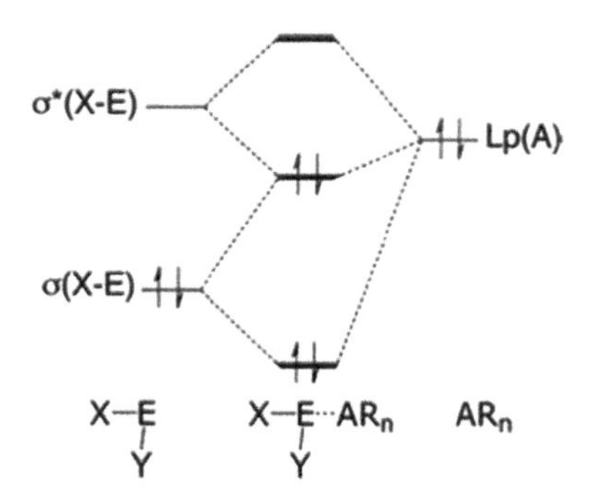

Fig. 1.17 2e-delocalization of an electron lone pair at the chalcogen acceptor (A) into the σ^*(XE) orbital of the chalcogen donor (adopted from Oliveira [194])

proportional to lone pair and σ^* orbital overlaps. Therefore, the energy gap $\Delta\epsilon(2e)$ is inversely proportional to the distance between the lone pair and σ^* orbitals.

As an intense, the delocalized $n \rightarrow \sigma^*$ orbitals due to the further decomposition of the molecular balance fragments constituents (such as thio-formamide and thiophene), as depicted in Fig. 1.18a, b. Thus the decomposition analysis has stabilized delocalized orbitals (see Fig. 1.18c, d), thereby, the hybridization orbitals provide a high-energy. Additionally it also leads the lone-pair orbitals occupied at even higher energy (see Fig. 1.18a) with unoccupied antibonding molecular orbitals (see Fig. 1.18b). Such decomposition possess the same molecular orbitals (see Fig. 1.18a, b) collectively to stabilize the form of amide lone pairs, irrespective of the α/β-connectivity, or the orientation of the thiophene ring [198]. Thus, sp^2 and sp^3 exist in the chalcogenide materials [194, 198, 199].

1.2.5 *Multiple Cantered Bonds*

Lone pair electrons (LPs) of the chalcogens can also serve to form multicenter bonds. Therefore, the hyper bonding interactions exist due to chemical bonding in chalcogens. As a typical example, the chemical interactions of a/c (amorphous/crystalline) PCMs. More specifically, Ge–Sb–Te (GST) memory material a/c crystallization behavior has revealed, a-GST charge density, the electron-localization function is negative to local energy density (or negative of the integrated crystal-orbital Hamilton population) with the monotonic increment for the decreasing inter atomic distance [200–203]. This kind of trend is predominately associated with covalent bonding. On the other hand, a broad distribution of inter-atomic distances also exists in metastable rocksalt-type c-GST (like a-GST) due to structural disorder [178, 204]. Hence, the bonding characteristics are the same for both a-GST and c-GST. This leads to surprising similarities in interatomic interactions of c-GST that are indistinguishable in a-GST. In more precise words, the chemical-bonding interactions in c-GST associates to subgroup interactions with the existing a-GST, but their interactions are not a different type to those present in a-GST. Although, several

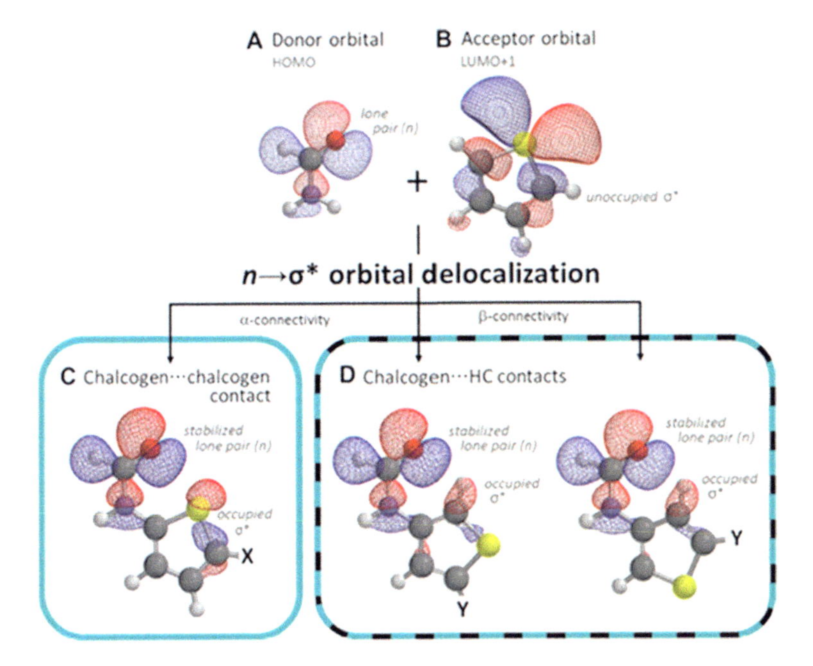

Fig. 1.18 Orbital decomposition analysis illustration of the hypothetical combination for the molecular fragments **a** + **b** in the three orientations, **a** HOMO containing the formyl oxygen lone pair (n) to stabilized with the (**c**) and (**d**) by the same set of antibonding σ^* orbitals of the (seleno/thio) phenefragment, irrespective of the orientation of the connected ring and the specific intramolecular contacts (adopted from Pascoe et al. [198])

meaningful differences could also exist between them, like inter atomic distance distributions of c-GST is longer than a-GST.

The conventionally a-GST structural network is considered to consist of ordinary two-centers via two-electron (2c/2e) covalent bonds, as depicted in Fig. 1.19a–i [138, 178]. Since the Te LPs also interact with the neighboring antibonding orbitals (through hyper bonding interaction), therefore, the LPs' delocalization-induced interactions form the antibonding orbitals, as illustrated in Fig. 1.19a-ii. According to quantum mechanical theory, the interactions of two-electron stabilization with the energy (ΔE_S) of an LP (n_A) via the 3c/4e interaction can be inversely proportional to LP antibonding orbital energy difference (ΔE_{A-BC}), as depicted in Fig. 1.19b [205]. That is approximately proportional to the extent of orbital overlaps between the LP (n_A) and the interacting antibonding orbital. For the sufficiently strong 3c/4e interactions, the lone pair's almost identical and collinear bonds are also formed that often denotes as "hyper bonds" or "ω bonds", as illustrated in Fig. 1.19a-iii [206]. Hence, the hyper blond pairs are considered to correspond to the axial bonds in a-GST, and their bonding characteristics perhaps distinguishable from the ordinary 2c/2e covalent bonds [179]. Their distinct features like (i) a (nearly) linear geometry of three bond atoms are effectively maximize the overlaps between LPs and antibonding orbitals; (ii) longer (weaker) bonds; (iii) higher polar covalency (i.e., ionicity) to positions Wannier functions (MLWFs), and the centers shifts toward Te atoms; iv) the stronger bonding-electron are usually delocalized. The well-established concept of "hyper bond" emphasizes the distinctive characteristics of 3c/4e bonds relative to ordinary 2c/2e covalent bonds [206]. Additionally, it also contains distinct characteristics of 3c/4e bonds relative to ordinary 2c/2e covalent bonds [206]. The bonding interactions of a-GST are strong in two different types; the first one ordinary 2c/2e covalent bonds (see Fig. 1.19a–i) and the second one through 3c/4e hyper bonds (see Fig. 1.19a-iii) that are associated with the weaker interactions of 3c/4e (see Fig. 1.19a-ii). Their multi-center interaction characteristics allow atomic species to accommodate the four s and p atomic orbitals (AOs), up to six bonds, therefore, the total LPs may exceed to ordinary Lewis maximum limit of four. Moreover, it has also addressed, how coordination number in a-GST (and in c-GST) approaches for all six bonds correspond to their three p orbitals that belong to perfect octahedral coordination. Including the formations of the total number of bonds and LPs to five or six hypervalent Ge or Sb atoms in a-GST [179]. The approach also leads to the initial bond lengths of 2c/2e bonds are increased or decreased with the formation of 3c/4e. Thus, the antibonding level of the B–C bond are occupied via interaction (see Fig. 1.19b). Moreover, the dependence of the hyper bonding tendency on the size of the bandgaps of materials are usually correlated to a strong hyper bonding tendency, specifically associated with smaller bandgaps of the materials. Further, the involvement of Te LPs in the formation of hyper bonds has also supported to enhance of the hyper bonds centers of Ge (or Sb) atoms, such as one, two, or threefold coordinated Te atoms as ligands, where LP electrons could donate in 3c/4e interactions (see Fig. 1.19a, b).

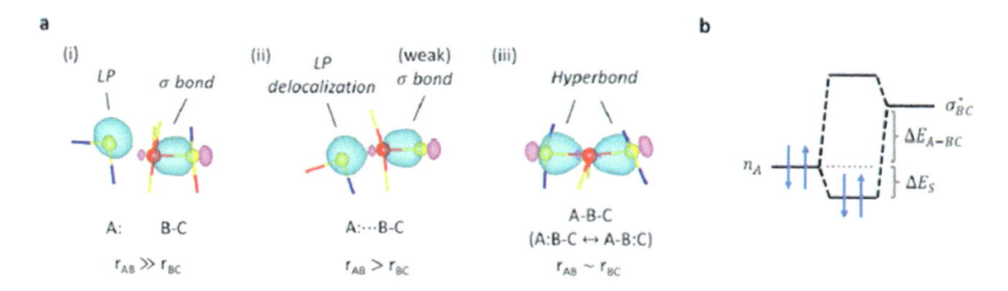

Fig. 1.19 Mechanism of 3c/4e hyper bonding interactions in a-GST models. **a** 3c/4e bonding configurations for (i), (iii) two limiting cases, and (ii) an intermediate interaction. (i) and (iii) correspond to an insignificant or strong 3c/4e interaction, resulting in a covalent σ-bond or hyper bond pair formation, respectively. The LP (nA) and bonding (σ_{BC}) orbitals are represented by isosurface plots of the maximally localized Wannier functions with light blue (positive) and pink (negative) contours. The antibonding orbital ($\sigma_{BC}{}^*$) is concentrated outside of the bonding region, thereby overlapping more with the LP than does σ_{BC}. The interaction (ii) reveals a nontrivial interaction between an LP orbital of A and the antibonding orbital of a B–C bonding pair. A: denotes an LP on an A atom, and ··· indicates a weak LP delocalization, while A–B–C denotes the formation of hyper bond pairs among A, B, and C atoms, which can be represented by the two ionic resonant structures, A: B–C and A–B: C. **b** A schematic interaction-energy diagram for an LP (nA) stabilization interaction with a nearby antibonding orbital ($\sigma_{BC}{}^*$) (adopted from Lee et al. [189])

Similarly, c-GST also possessed a significant difference with a-GST, owing to crystalline symmetry requiring Ge and Sb atoms to reside in (distorted) octahedral-ligand environments to allow the formation of the maximum of three perpendicular hyper bond pairs (i.e., six bonds). It directly correlate to be metastable c-GST crystalline structure that inherently provides a favorable condition for strong hyper bonding with near-linear alignments of p orbitals. Additionally, it also provides a sharp increase in hyper bond formation in c-GST, compared to a-GST. Such enhancement in hyper bond formation in c-GST is also directly correlated to the crystal-structure amplification effect. While the remaining contribution of the non-hyper bonds corresponds to ordinary 2c/2ecovalent bonds or weaker 3c/4e bonds as depicted in Figs. 1.19a-i, a-ii. These findings have established the high polar covalency of hyper bonds that is manifested in a higher degree of ionicity in c-GST compared to a-GST, due to charge transfer from Ge and Sb atoms to Te atoms [189].

Hence, the concepts of hyper bonding have revealed an elementary three-body structural motif in chalcogenides, which consist of either two equal bonds or a pair of short and long bonds with a perfect or close-to-linear bonding geometry. They can have a difference between the three configurations to originate different strengths of hyper bonding interactions. Their variations are generally continuous and differentiable between the configurations [179]. To avoid the experimental observation difficulties first and second spectral peaks are a rigorous justification for the disordered material systems, and the notion of a "Peierls distortion" for liquid, supercooled liquid, or glassy chalcogenides is often used [207]. To describe such linear bonding geometries, it has been also rephrased as a "weak hyper bonding interaction", therefore, the material-dependent abundance of such geometries in chalcogenide liquids has been rationalized by the theory of hyper bonding [189]. Although, it is an interesting aspect related to former hyper bonding models, the formation of both weak and strong hyper bonds due to the formation of energy levels near the top of the valence band with substantial antibonding character. Such antibonding-character states could be responsible for the amorphous and crystalline chalcogenides in which substantial linear triatomic motifs are present [179, 208]. Therefore, the overall decrease in antibonding character with vacancy formation leads to a stabilization of crystalline cubic GST that naturally accounted for the hyper bonding in terms of a reduced total number of hyper bonds with vacancy formation in the cubic crystalline structure.

1.3 Transition Metal Dichalcogenides (TMDs)

Two-dimensional transition metal dichalcogenides (TMDs) are a class of material that can serve as building blocks for next-generation technology [209]. The layered transition metal dichalcogenides (TMDs) in two-dimensional (2D) form (including few-layer and monolayer) have distinct properties from the bulk counterparts owing to the quantum confinement effect and broken inversion symmetry. Usually, TMDs defined from the formula MX_2, whereas M stands for a (sixfold coordinated) transition metal and X represents chalcogens, the latter threefold coordinated (also in the case of the IV-VI crystals), the typical structures according to the structural formula are depicted in Fig. 1.20a, b. However, in transition metal compounds bonding is commonly deliberated among unfilled orbitals of the metal and LPs of ligands. While in TMDs, the metal atoms provided four electrons to occupied the bonding states, the transition metal and chalcogens are in-general attributed from a recognized charge + 4 and − 2 [210]. In which the LP electrons of chalcogen atoms are positioned over sp^3-hybridized orbitals and terminated at the surfaces. Thereby, the coordination everywhere chalcogenides is disproportionate, that provided a noticeable cleavage features vertical to the hexagonal/trigonal equilibrium axis [191]. Hence, the nonappearance of dangling bonds makes the surfaces extremely steady and non-reactive. Considering various technically advanced features the increasing research efforts devoted to this field and the fantastic properties of 2D TMDs are continuously being revealed. Depending on the filling states of the d band of the metal elements. The 2D TMD semiconductors have ability to vary band structures, like semimetals, true metals, and superconductors for the distinct types of electronic devices. Upto-known layered TMDs are outlined in the periodic table as given with the shadow background in Fig. 1.21. A majority of 2D counterparts have been theoretically predicted as well as experimentally synthesized [209, 211, 212].

Typically, TMDs (e.g., MoS_2, WS_2, WSe_2, etc.) are consisting a sandwich structure of a transition metal layer (e.g., Mo, W, Nb) between two chalcogen layers

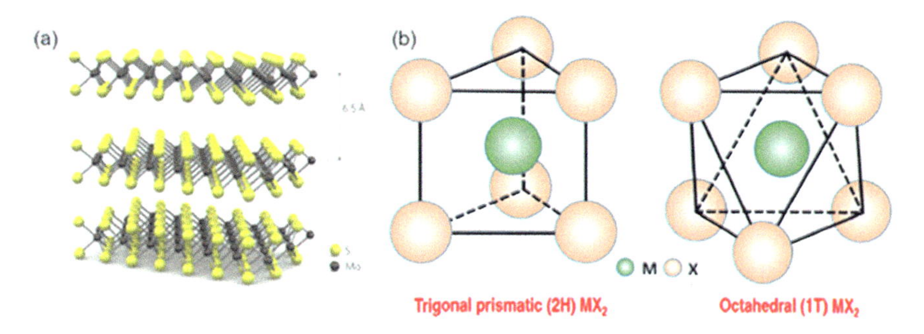

Fig. 1.20 Crystal structures of TMDs with a typical formula of MX_2; **a** Represents a three-dimensional model for the MoS_2 crystal structure; **b** Unit cell structures of the 2H-MX_2 and 1T-MX_2 (adopted from Lv et al. [220])

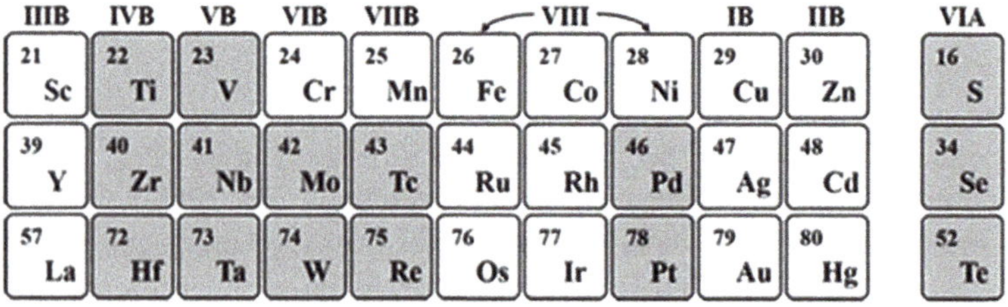

IIIB	IVB	VB	VIB	VIIB	VIII	VIII	VIII	IB	IIB	VIA
21 Sc	22 Ti	23 V	24 Cr	25 Mn	26 Fe	27 Co	28 Ni	29 Cu	30 Zn	16 S
39 Y	40 Zr	41 Nb	42 Mo	43 Tc	44 Ru	45 Rh	46 Pd	47 Ag	48 Cd	34 Se
57 La	72 Hf	73 Ta	74 W	75 Re	76 Os	77 Ir	78 Pt	79 Au	80 Hg	52 Te

Fig. 1.21 Well-known layered TMDs in the periodic table that highlighted with shadow

(e.g., S, Se, Te). TMDs are also characterized from the weak non-covalent bonding between layers and strong in-plane covalent bonding. The bulk TMDs are typically exfoliate into single or few-layered structures by physical or chemical approaches, such as adhesive tape exfoliation, solvent-assisted exfoliation, and chemical exfoliation via lithium intercalation [213–217]. Due to quantum confinement and surface effects, the monolayer and few-layered TMDs also allows a variety of exciting properties that have not usually appeared in their bulk counterparts. For the instance, the bulk semiconducting trigonal prismatic TMD possessed an indirect bandgap, but its single layers can have a direct electronic and optical bandgap [18], thereby, the enhanced photoluminescence (PL) has achieved [218]. Significant valley polarization has been also observed in monolayers of MoS_2. This effect is considered to be crucial for engineering valley-based electronic and optoelectronic devices [17]. Along with the mono- and few-layered MoS_2 and many other 2D layered TMDs are considered to be suitable for the exploration, including semiconductors (WS_2), semimetals (WTe_2, $TiSe_2$), true metals (NbS_2, VSe_2), and superconductors ($NbSe_2$, TaS_2)[210]. Significantly, the number of layers in TMD remarkably may also affect their properties [18]. As an intense second-order nonlinear optical response is strongly dependent on the number of layers. Whereas, the centrosymmetry of bulk MoS_2 crystals prohibits second-order nonlinear optical processes. Thus, in a monolayer or few-layer MoS_2 film with an odd number of layers, the inversion center almost disappeared, this allows a strong second-order nonlinear optical response [219]. Similarly, other semiconducting TMDs (e.g., WS_2, WSe_2, $MoSe_2$, etc.) also have an identical trigonal prismatic structure like MoS_2 with the direct bandgap as single molecular layers and indirect gap as bi- or multilayers [77]. Hence the structural control in TMD nanostructures is an important parameter to get desired electronic, optical, biological, chemical, and catalytic properties of the materials for specific uses [220–224]. Therefore, the distinct nanostructures of TMD materials can be achieved from the various synthesis processes that have been broadly categorized as the synthetic routes top-down (mechanical cleavage, chemical exfoliation) and bottom-up (direct wet chemical synthesis and chemical vapor deposition) approaches [220].

1.3.1 Materials Chemistry of the Transition Metals

Transition metals either individual or various chemical elements possessed valence electrons, these electrons participates in the formation of chemical bonds in form of two shells instead of only one. However, only term transition has no specific chemical significance, but it is a convenient name to distinguish the similarity of the atomic structures and resulting properties of the elements. Transition metals are generally occupy the middle portions of the long periods of the periodic table of elements between the groups on the left-hand side and the groups on the right. Particularly, they form Groups 3 to 12, as illustrated in Fig. 1.22.

Many of the transition metal elements are technologically important such as titanium, iron, nickel, and copper structures which are used for electrical applications.

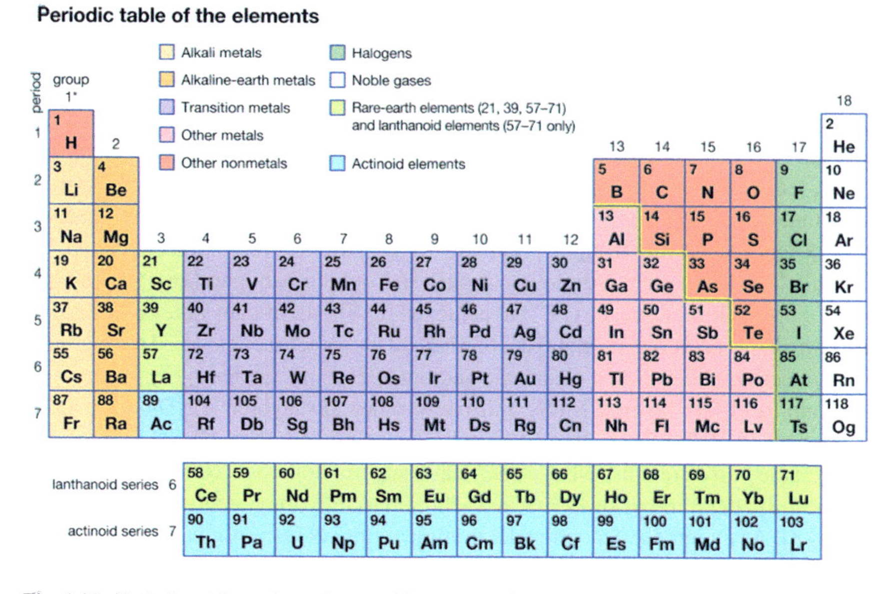

Fig. 1.22 Periodic table to show the transition metal with group elements

Moreover, transition metals are also formed many useful alloys to each other's as well as other group metallic elements. These elements also dissolve in mineral acids, although a few like platinum, silver, and gold are called "noble", this means they are unaffected by simple (nonoxidizing) acids. So, overall transition elements are subdivided based on their electronic structures into three main transition series, called the first, second, and third transitions series, and two inner transition series that is called lanthanoids and actinoids.

The main transition series begins with either scandium (symbol Sc, atomic number 21) or titanium (symbol Ti, atomic number 22) and it ends with zinc (symbol Zn, atomic number 30). The second series is started ends from the element yttrium (symbol Y, atomic number 39) to cadmium (symbol Cd, atomic number 48). The third series is from lanthanum (symbol La, atomic number 57) to mercury (symbol Hg, atomic number 80). These three main transition series contain 30 elements and they are called d-block transition metals. However, transition metals scandium, yttrium, and lanthanum do not form compounds analogous to others due to their chemistry being quite homologous to lanthanoids. Likely, zinc, cadmium, and mercury also exhibit few similar characteristics of the other transition metals, therefore, they are treated separately (see zinc group element). The remaining d-block transition metals with their few characteristics properties are listed in Table 1.2. It gives the first inner transition series includes the elements from cerium (symbol Ce, atomic number 58) to lutetium (symbol Lu, atomic number 71). These elements are also called lanthanoids (or lanthanides) due to their close chemistry with the lanthanum. Moreover, the actinoid series is consisting 15 elements from actinium (symbol A_c, atomic number 89) to lawrencium (symbol L_r, atomic number 103). All these inner transition series are covered under rare-earth elements and actinoid elements, however, more than 104 are considered transuranium elements.

Specifically, a transition metal is ‘an element their atoms have a somewhat occupied d sub-shell, or it gives upsurge to cations together anion complete d sub-shell.

Table 1.2 Important properties of the transition elements (https://www.britannica.com/science/transition-metal)

	Element with symbol	Atomic number	Atomic mass	Density (grams per cubic centimeter, 20 °C)	Melting point (°C)	Boilingpoint (°C)
1st main series	Titanium (Ti)	22	47.867	4.54	1668	3287
	Vanadium (V)	23	50.942	6.11	1910	3407
	Chromium (Cr)	24	51.996	7.14	1907	2672
	Manganese (Mn)	25	54.938	7.21–7.44	1246	2061
	Iron (Fe)	26	55.845	7.87	1538	2861
	Cobalt (Co)	27	58.933	8.9	1495	2927
	Nickel (Ni)	28	58.693	8.9	1455	2913
	Copper (Cu)	29	63.546	8.92	1085	2927
	Zirconium (Zr)	40	91.224	6.51	1855	4409
	Niobium (Nb)	41	92.906	8.57	2477	4744
	Molybdenum (Mo)	42	95.94	10.22	2623	4639
2nd main series	Technetium (Tc)	43	98	11.5	2157	4265
	Ruthenium (Ru)	44	101.07	12.41	2334	4150
	Rhodium (Rh)	45	102.906	12.41	1964	3695
	Palladium (Pd)	46	106.42	12.02	1555	2963
	Silver (Ag)	47	107.868	10.49	962	2162
	Hafnium (Hf)	72	178.49	13.31	2233	4603
	Tantalum (Ta)	73	180.948	16.65	3017	5458
	Tungsten (W)	74	183.84	19.3	3422	5555
3rd main series	Rhenium (Re)	75	186.207	21.02	3186	5596
	Osmium (Os)	76	190.23	22.57	3033	5012
	Iridium (Ir)	77	192.217	22.56	2446	4428
	Platinum (Pt)	78	195.084	21.45	1768	3825
	Gold (Au)	79	196.967	~ 19.3	1064	2856

Transition metals usually follow the electronic configuration rule (n − 1)$d^{1-10}ns^2$. It could correlate to their solitary goes across the row from left to right in the periodic table, thereby, electrons are usually included to (n − 1) d shell which occupied following to the Aufbau principle. By following Hund's rule "the states of electrons occupy the lowermost accessible energy levels formerly occupying the higher levels". Consequently, the (n − 1) d and ns shells are relatively nearby in energy, and half-occupied /or fully occupied shells are characterized from the enhanced constancy, sometimes an electron are also transported from s-shell to d-shell, like the atomic formation of copper is $3d^{10}4s^1$ rather $3d^9 4s^2$. Therefore, d-block transition metals possess s, p, and d orbitals and their n electrons in the d-orbital scan are termed ions together a d^n configuration, as the instance, Ti^{3+} has a d^1 ion, and Co^{3+} is a d^6 ion.

The greatest significant property of metallic elements is to behave as Lewis acids which procedure multiplexes together various kinds of Lewis bases. Where a metal compound made of a central metal atom or ion that's attached to single or additional neighbors, this is usually called a ligand (in Latin meaning of "ligature" is "bind'). The ligands ions or molecules may contain one or more pairs of electrons to share with the metal electrons. The metal complexes are also neutral, like Co $(NH_3)_3Cl_3$, positively charged, and [Nd $(H_2O)_9]^{3+}$ negatively charged.

Transition metals coordination number is also obtained by analyzing the size of the central metal ion because of the numeral of d electrons (and/or steric effects) raised to the ligands. Their multiplexes together coordination numbers are intermediate 2 and 9, it is well recognized. Specifically, 4–6 coordinations are classified utmost steady electronically and geometrically and their multiplexes together the coordination figures are the greatest abundant. A reason beyond that the transition metal's sixfold coordination is encountered in TMDs, here predominantly addressed only this issue.

In the six ligands coordination configuration the central metal octahedral (O_h) stoichiometry has the most stable geometry, therefore, the majority of such complexes can be assumed in such construction. The octahedral construction may also possess tetragonal (D_{4h}), rhombic (D_{2h}), or trigonal (D_{3h}) misrepresentations owing to electronic or steric impacts. Moreover, the six ligating atoms also assume to trigonal prismatic coordination. Besides, such coordination also possesses some multiplexes including a few metal compounds associating their coordination assembly; therefore, the octahedral coordination would be satirically fewer strained. Similarly, it is also well described that the bonding type of chalcogen atoms surrounding a metal is frequently trigonal prismatic in solid-state TMD, for instance, MoS_2 and WS_2 structures are depicted in Fig. 1.23. The stability of the trigonal prismatic configuration in TMDs are determined with the help of their electronic band structures. The determination of complex coordinates gives the chemical behavior of the ligands including information about their three-dimensional arrangements around the transition metal, such as in $Co(NH_3)_3Cl_3$, the Co metal is coordinated from the dual types of ligands. Regarding bonding of the complex transition metals, they are marginally dissimilar to the normal covalently bonded solids. The bonding area growths of the complex transition metals are usually determined by using the density functional theory.

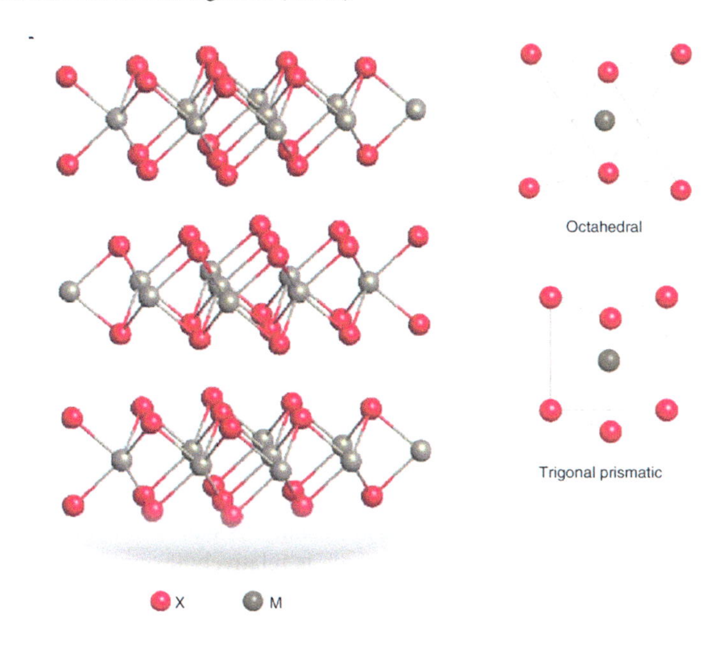

Fig. 1.23 Three-dimensional representation of a typical layered MX_2 structure, whereas the metal atoms are represented in gray and the chalcogen atoms in red. Two types of local coordination of the metal species depicted in the right panel, typically octahedral (top) and trigonal prismatic (bottom) (adopted from Jing et al. [151])

Among all these, it would be customary to mention a brief on dissuasion on second and third series elements for their better understanding. The second and third series transition metal elements have possessed almost similar chemical properties to the first series, but they have specific differences from the lighter elements of the group, which follow as:

(1) Although first series element cobalt forms a considerable number of tetrahedral and octahedral complexes with the + 2 oxidation state, this state characteristic to be consider for ordinary aqueous chemistry, but the + 2 states of rhodium (second series) and iridium (third series) are rare and relatively unimportant.
(2) The element manganese ion Mn is very stable with great significance in chemistry, while with the oxidation state + 2 technetium and rhenium are comparatively less impressive in terms of their practical utility.
(3) The element chromium has + 3 states and forms a great number of complexes, whereas molybdenum and tungsten have also possessed + 3 states, but they are not particularly in stable states, and form only a few complexes.
(4) Concerning oxo anions of first-row elements with higher oxidation states (like chromate, chromium in + 6 state) and permanganate (manganese, + 7 state) that usually form powerful oxidizing agents essentially restrict the function. Whereas their stoichiometric analogs, such as molybdate (molybdenum, + 6

state), tungstate (tungsten, + 6 state), pertechnetate (technetium, + 7 state) and perrhenate (rhenium, + 7 state) are quite stable and have an extensive diverse chemistry.

(5) Though in some cases it is also valid and useful analogies to build chemistry with the lighter element between the two heavier elements of the group. For the instance, the chemistry of complexes of rhodium in the + 3 state, is usually quite like complexes of cobalt in the + 3 state. Thus, the differences are more consistent than the similarities between the first lighter element and the heavier elements of each group.

(6) The heavier transition metals with a higher oxidation state are usually more stable than the elements in the first transition series. This is not only true for the oxo anions but also in the case of the higher halides as well. Thus, the heavier elements formed compounds such as ruthenium oxide, RuO(+8 state); tungsten chloride, WCl(+6 state); platinum fluoride, PtF(+6 state), etc. They have no analogs among the first-row elements, whereas the chemistry of aqua ions of the lower oxidation states, especially + 2 and + 3 states a dominant part of the chemistry of the lighter elements, which is relatively insignificant in terms of heavier ones.

(7) Moreover, the lanthanoid series of elements comes between the second and third regular transition series and they are formed by the filling of the 4f orbitals. Their 5d and 6 s orbitals act as the valence shells for the third transition series elements, that is imperfectly shielded by the 4f electrons to increase the nuclear charges. Thus, a steady contraction in the size of these orbitals through the lanthanoid series elements, their net result as the atomic and ionic radii of hafnium. Thus, the lanthanoid series is almost identical to the corresponding radii of the zirconium atom that lies just above it in Group 4. Hence zirconium and hafnium atoms have almost identical sizes and analogous electron configurations in all of their oxidation states. Indeed, it is very difficult to separate the two elements because of the great similarity in properties of their compounds. Concerning the third transition series, there is a steady but slow divergence in their properties to the second series elements in each group. So, a considerable difference appears to palladium, platinum, silver, and gold, but their differences are not as great, however, the lanthanoid contraction may not intervened to preclude greater disparity in the orbital sizes. Therefore, niobium and tantalum are not quite as similar to zirconium and hafnium, but the differences between them are slight. Similarly, molybdenum and tungsten, technetium and rhenium, ruthenium and osmium, and rhodium and iridium also show remarkable similarities in their chemistries.

1.3.2 Theory of the Valence Bond

The usual theory responsible for the creation of covalent bonds assumed that covalent bonds shaped while the atomic orbitals overlay. The forte of a covalent bond is

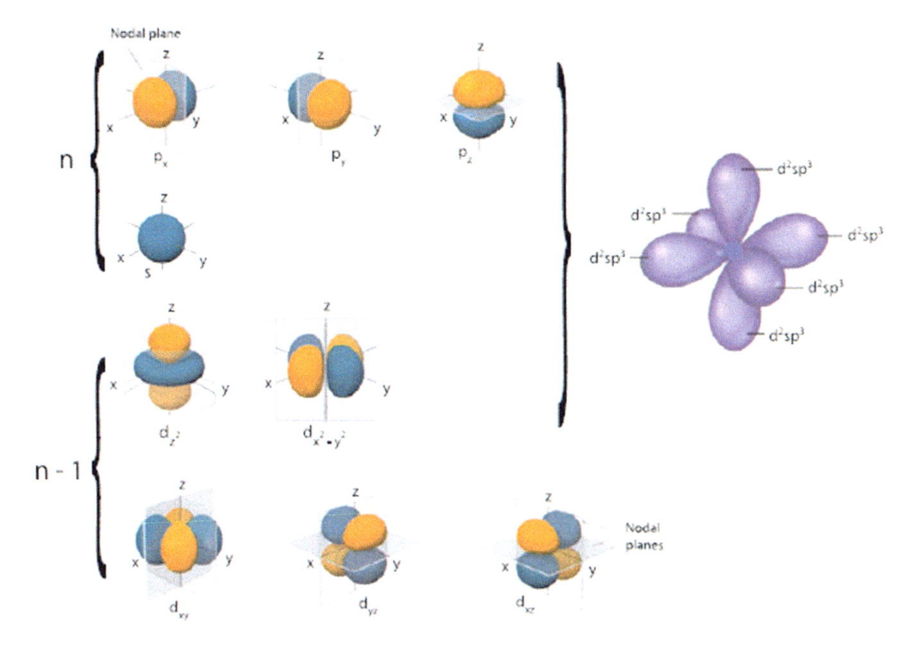

Fig. 1.24 The s, p, and d atomic orbitals of a transition metal (left) and the **respective** set of d^2sp^3-hybridized orbitals (right) with an octahedral bonding geometry of the metal (adopted from Averill et al., http://www.saylor.org/books)

proportionate to the integer of overlays. Subsequently also assumed that the atomic orbitals hybridized, and it maximized when they overlaps with adjacent atoms. Such as sp^3 hybridization of Si or Ge crystals, in which every atom is fourfold covalently coordinated. On the other hand, in transition metal, a set of six *s, p,* and *d*-orbitals are hybridized likely to sp^3 hybridization and generates six equivalent orbitals and directed to vertices of an octahedron, as the orbitals schematic is illustrated in Fig. 1.24.

However, it is informal to visualize the s-orbitals spheres and p-orbitals dumbells, while a representation of d-orbitals can be supplementary multifaceted. Especially when considering bonding geometry has five d-orbitals, referred to as $d_z{}^2$, $d_x{}^2{}_{-y}{}^2$, d_{xy}, $d_{yz,}$ and d_{xz}. Therefore, in an octahedral bonding geometry, the d_{xy} lobes lay among the x and y axes, d_{xz} lobes exist among the x and z axes, d_{yz} lobes lay among the y and z axes, $d_x{}^2{}_{-y}{}^2$ lobes occur over the x and y axes and final the $d_z{}^2$ (it also denotes $\Delta 3z^2 - r^2$) from the double lobes toward the z axes, they are a donut-shape ring exist over the xy plane surrounding the other duel lobes.

Similarly, for the formation of sixfold coordinated geometry materials, the bond hybridization requirements are either d^2sp^3 or sp^3d^2. There is a significant difference in-between these two types of bond formations are; whether the *d*-electrons are characterized from the similar quantum number n as the *s*- and *p*-electrons (sp^3d^2) or by the preceding number ($n - 1$). The latter case is usually related to d^2sp^3, the corresponding schematic is illustrated in Fig. 1.25. Their complexes are also

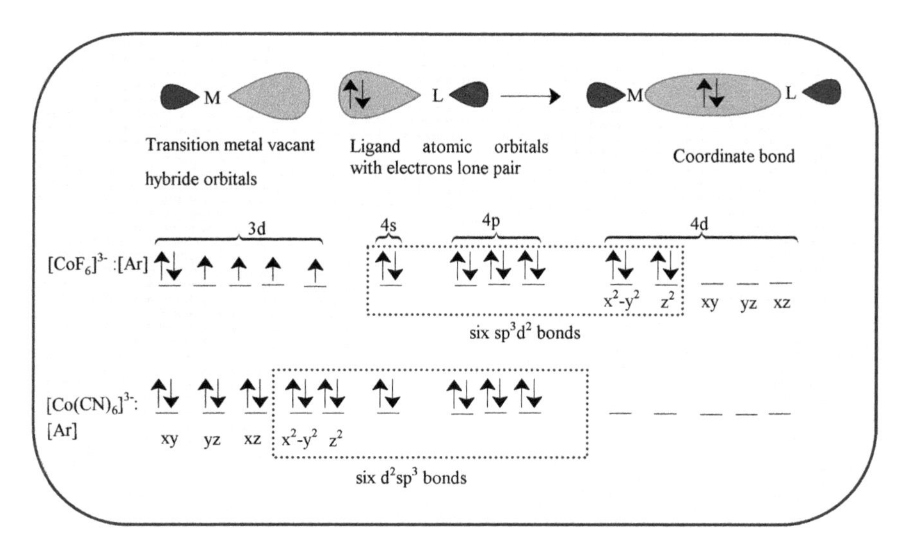

Fig. 1.25 Schematic of the formation of a dative (coordination) bond between a vacant hybridized orbital of a transition metal and LP electrons of ligands (top). At the bottom schematic for the involved electrons in the formation of octahedrally coordinated complexes $[CoF_6]_3^-$ and $[Co(CN)_6]_3^-$, using sp^3d^2 and d^2sp^3 hybridizations. The electrons represent the arrows and the opposite arrows for the opposite spins. The ligand electrons are represented by a relatively faint black color

occasionally called outer-orbital multiplexes and used d^2sp^3 orbitals are called inner-orbital complexes.

Since covalent bonds are normally shaped among couples of atoms when everyone contributes single electron per bond. Regarding bonding of transition metal multiplexes, they generally formed the bonds among vacant *d-s-p*-mixed (or hybridized) orbitals of the metal and LP (lone pair) electrons of ligands, as depicted in Fig. 1.25 (upper panel). This kind of formation is increasing to bonding and antibonding orbitals together σ symmetry nearby the metal–ligand bond axis. As a result, covalent bonds have an inductive manner, it is also called coordinate covalent bonds. Therefore, the transition metal complexes are also often called coordination complexes.

Additionally, with the help of the valence bond theory, one can also predict useful complex physio-chemical properties. As an example, to explain the cobalt magnetic properties, because they formed the outer complex, in this case, the 3d electrons remain undisturbed whereas the four electrons are still uncoupled; thereby the $CoF^{3-}{}_6$ should be paramagnetic (see Fig. 1.25). Another example is if the formed complex is inner then the redistributions of d-electrons could deliver double unoccupied orbitals for the coordination bond to ligands construction. Therefore, all Co d-electrons are paired and complex compound $[Co(CN)_6]^{3-}$, the overall it diamagnetic, it has been also verified experimentally [225]. Hence, the valence bond theory provides dual accurate options for the integer of unpaired electrons in a simplified manner, but it fails to provide the choice of the electron in between them [225].

1.3.3 Ligand-Field Approach

Electronic properties of the TMDs predominantly governs by the d-like bands and their filling [210]. Therefore, the ligands split the d-electrons levels due to a combination of different effects. Like the ligand field splitting results from hybridization with ligands' orbitals are dominant in the covalent systems [226]. As per well-established field theory, an octahedral (1T polymorph) d-shell splits into a lower-energy triplet (t_{2g}) and a high-energy doublet (e_g). Whereas a trigonal prismatic geometry (1H) has a low-energy triplet that can further split into a doublet and a singlet, it's generally lower in energy [226].

However, in the previous studies, the ligand field arguments are often given as simple intuitive starting points to understand various properties of the TMD materials. Specifically, a long-standing problem in the field of dichalcogenides when considering the relative stability between the two polymorphs [227, 228]. That is predominately controlled by the column of the transition metal M in the periodic table, which means the electron filling of the d-like bands. The most abundant d^0 TMDs are found in nature with the 1T polymorph. TMDs with usual occupation d^1-d^2 are more stable with the 1H polymorph, while the 1T polymorph is metastable in a distorted form [229]. However, TMDs with $n = 3$ are categorized as the most stable in a strongly distorted 1T phase with a 2×2 periodicity [191, 227, 230, 231]. But 1H polymorph lower energy than undistorted 1T prediction is considered to be a major issue. Moreover, TMD with the d^4–d^6 range has lower energy in the 1T phase compared to the 1H. Though, some materials unrelated to pyrite structures in certain cases may possess the most stable phases [232]. A natural explanation of this trend is followed for $n < 2$ d electrons. Therefore, the 1H phase becomes more stable concerning the 1T phase due to the filling of low-energy singlet increases. Whereas in the case of $n > 2$, the 1H polymorph becomes less favorable due to the increasing number of electrons, consequently, the higher-energy doublet gets filled [233, 234].

Moreover, the ligand field has a small charge-transfer energy that favors stronger hybridization. The total ligand field can be calculated from the 5-band d- model, it is the sum of the bare electrostatic crystal field and the contribution of the hybridization with various ligand states [226]. The splitting of the t_{2g}–e_g in 5-band is generally controlled by the competition between opposite trends. However, in both 4d and 5d orbitals splitting the hybridization trends dominated, so the total splitting may increase for the later-column materials [226].

Overall, the ligand field contribution to the splitting is almost constant, due to the effect of small charge-transfer energy in $3d^1$ materials (the most localized 3d electrons state). But inter orbital effects could be larger for the 5 orbital materials [226]. Hence based on the available evidence one can say, according to well-established ligand-field theory, the nine metal orbitals joints together the six ligand orbitals to create six bonding orbitals and six antibonding orbitals. These d_{xy}, d_{yz}, and d_{xz} orbitals have a dissimilar equilibrium those do not associate to σ-like ligand orbitals. The orbitals do not alter energy, typically they are known as non-bonding. The electrons couples of such orbitals are usually not participated in the bonding of the metal and ligands [225].

1.3.4 Crystal-Field Approach

A better understanding on transition metal complexes can be made by the interpretation of crystal field theory. This theory initially deals with five d-orbitals in a degenerated position. Further, the d orbitals may remain degenerate when all six negative charges are consistently dispersed onto the surface of a sphere, but their energy higher owing to repulsive electrostatic interactions between the negative charge spherical shell and d-orbitals electrons. Subsequently, if all the six negative charges associated with lone pair (LP) electrons of the ligands, they may bring near the metal atom (ion) in an octahedral array along the three principal axes, whereas the average energy of the d-orbitals remains unchanged and the overall degeneracy is lifted, as illustrated in Fig. 1.26. Specifically, the $d_z{}^2$ and $d_x{}^2{}_{-}y^2$ orbitals could be intensely impacted due to the fact they pointed directly into the charges. Therefore, electrons of these dual orbitals have greater energies to the d_{xy}, d_{yz}, and d_{xz} orbitals electrons. Thus, the initial five degenerated d-orbitals splits in binary energy levels that is unglued from a crystal-field splitting energy (Δ_o), as the schematic is depicted in Fig. 1.26. Generally, the lower and higher levels are designated from the t_{2g} and e_g. These names were given owing to the involvement of the group theory as well as the associated equilibrium of the states. The value of the energy levels splitting depending on the charge of the metal ion and metal location in the periodic table and the characteristic of the ligands. Since during the splitting of the d orbitals the crystal field is still unchanged, therefore, the total energy of the five d-orbitals are fall into dual e_g orbitals and their energy increases to the amount 0.6 Δ_o. On the other hand, the three t_{2g} orbital's energy has go down around 0.4 Δ_o. Similarly, interactions among the positively charged metal ion and the ligands from the total stabilization (lowering of the energy) of the system, therefore, a decrease in energy of all five d orbitals and their splitting unaffected (see Fig. 2.12, right).

Determining the different properties of a complex crystal field approximation is considered to be a crucial approach. The approximation theory demonstrates, if the energy of the splitting (Δ_o) is greater to the correlation energy of the binary electrons then they should be with opposite spins over the identical orbital (u_p). The metal ions electrons holds the configuration dn ($n \geq 4$), it could be more stable than the electrons occupied the three lesser orbitals. On the other hand, if the splitting energy (Δ_o) is lesser to the correlation energy ($\Delta_o < u_p$), then a few electrons possibly may paired off. Therefore, the total relative magnitude of splitting energy (Δ_o) and u_p is usually determined by summing the whole spin of the system. Typically, the previous case is correlated to a high-spin complex and the second one to a low-spin complex, as depicted in Fig. 1.27. Moreover, it has been correlated to the robust electrostatic field formation from the ligand giving a larger splitting. Regarding a sequence of complexes of the metals that belongs to the identical group in the periodic table

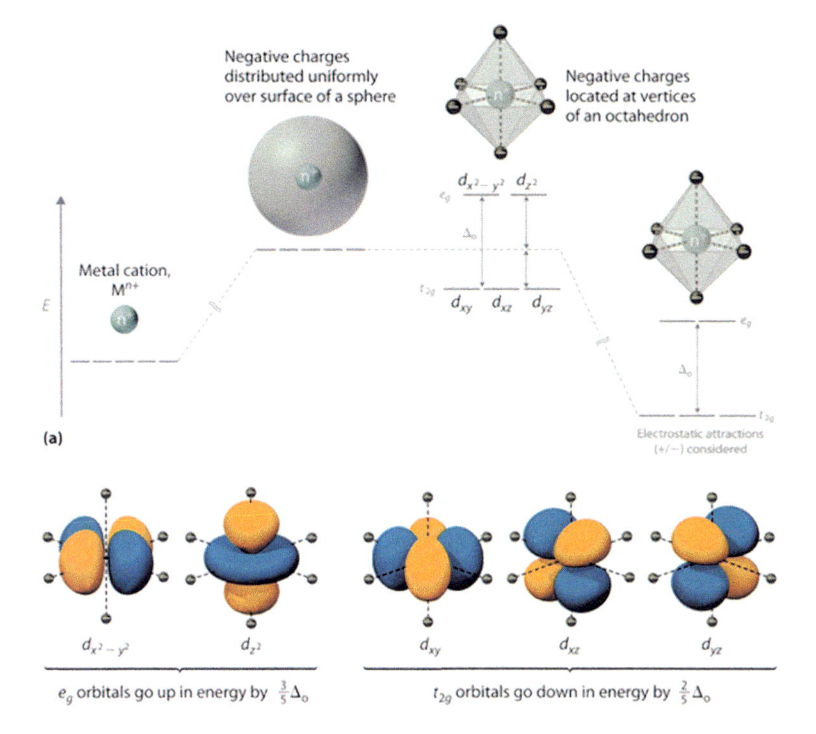

Fig. 1.26 Crystal field splitting of *d*-orbitals (adopted from https://chem.libretexts.org/Sandboxes/evlisitsynaualredu/Elena's_Book)

having the equal charge and ligands, their value of the splitting energy is usually in increasing order with the increasing principal quantum number: $\Delta_o(3d) < \Delta_o(4d) < \Delta_o(5d)$. Whereas in the case of the series possessed similar chemical legends, the total magnitude of the splitting energy (Δ_o) are found to be in decreasing order with the increasing size of the donor atoms.

Moreover, metal ions have an odd number of electronic configurations in the orbitals e_g, their solitary electron (or the third electron) may inhabit either single of dual degenerate e_g orbitals, this means they finds a degenerate ground state. However, according to the Jahn–Teller theorem [235], this kind of system cannot be considered stable. Therefore, system may experience a distortion with the decreasing regularity and split the degenerate states owing to orbital pushed up in an empty energy state. Thereby, the net result of the splitting to decrease the energy of the system [24, 237, 242–246]. Though, to describe the band structure of monolayer TMDs, the spin–orbit coupling (SOC) effect approach has been found unimpressive, due to a lack of inversion symmetry in monolayer TMDs [17, 247, 248].In this order, *Z*, *Y*. Zhu et al. have shown that the spin–orbit interaction induces splitting of ~ 0.15 eV at the topmost valence band in monolayer H-MoS_2. The SOC splitting can be larger for the monolayer TMDs, particularly for the heavy transition metal elements, therefore, the SOC strength increases rapidly with the increasing atomic numbers of constituent

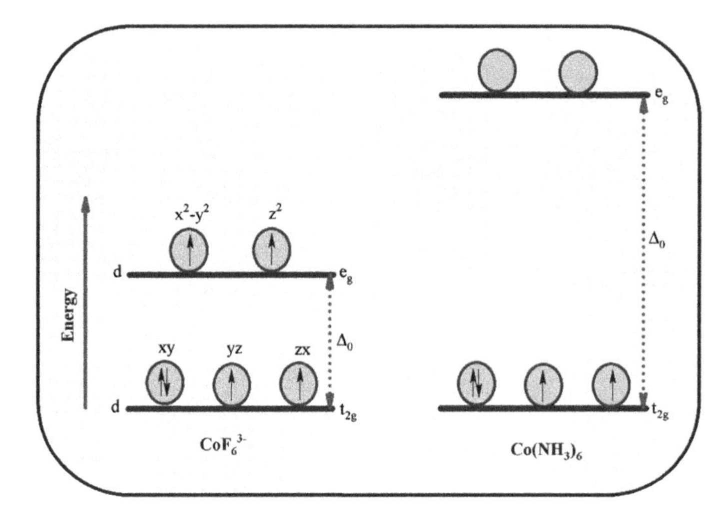

Fig. 1.27 Schematics for the strong (left) and weak (right) ligands to produce high-spin and low-spin complexes

eledegeneracy schematic is illustrated in Fig. 1.28. Hence, the collective distortion and decline in energy is typically denoted as the Jahn–Teller effect. However, the systems together an even number of electrons are usually energetically unfavorable due to Jahn–Teller distortion. Thus, it should be noted that the crystal field theory bonding in the complexes in terms of entirely electronic interactions with the consequence of legend charges on *d*-orbitals metal ion energies excluding the covalent bonds.

1.3.5 Band Structure

Owing to a great interest in metal dichalcogenides, systematic studies of the structural and electronic properties of the TMD family have demonstrated their devices. In the 1980s, Grasso et al. compressively reviewed the electronic properties of a series of bulk-layered materials with an augmented plane wave (APW) method [236]. Subsequently, efforts have been also made on understanding the basic electronic properties of monolayer TMDs including electronic band gaps, etc. [24, 237–241]. Additionally, several simulation methods have been introduced to measurements [249]. Moreover, Zhang et al. have presented detailed simulation results to band aliments of the TMD materials based on the density functional theory (DFT) calculations [250]. They have used the Vienna ab initio simulation package (VASP) by employing the projector-augmented wave (PAW) [251, 252]. They have used both local density approximation (LDA) and generalized gradient approximation (GGA) of Perdew-Burke-Ernzerhof (PBE) to describe the exchange–correlation potential [253, 254].

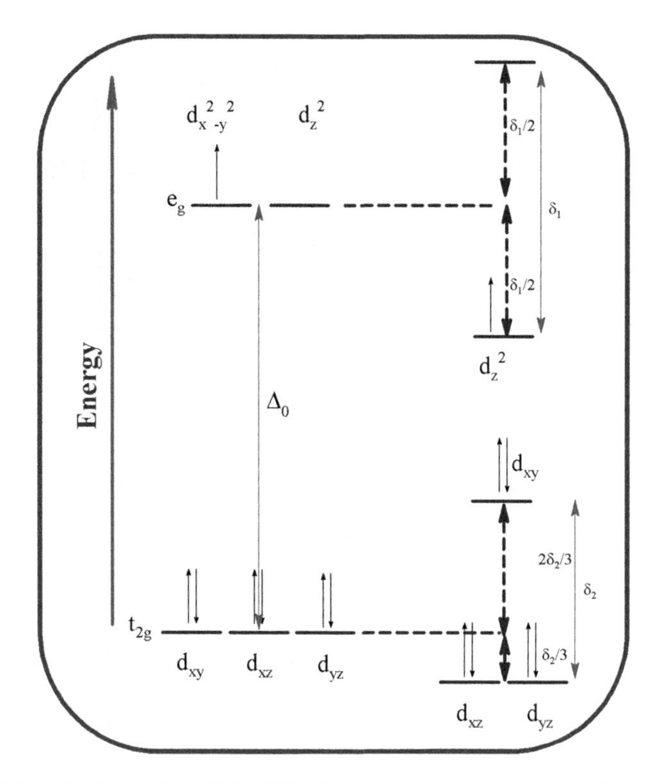

Fig. 1.28 Schematic for the Jahn–Teller Effect by showing the state moved up in energy **owing** to splitting, and the total energy of the system is decreased

In the beginning, the electronic band structure calculation of TMDs was interpreted with the help of the augmented spherical wave (ASW) method [255, 256]. This approach results were found to be in good agreement by using angle-resolved photoelectron spectroscopy. As per the findings, it was concluded that MoS_2, $MoSe_2$, and WSe_2 can have an indirect gap with the top of the valence band at Γ and the bottom of the conduction band halfway between Γ and K-points. Later this experimental interpretation also consisted of theoretical simulation studies. Moreover, a relationship between the ligand-field levels and the bands is also obtained from band structure calculations. Considering the significant point, the ligand field levels correspond to the centers of gravity of the appropriate d sub-bands in the periodic crystal and not to the band energies at $k = 0$ [257], as the schematic illustrated in Fig. 1.29.

Whereas the augment of the plane-wave (APW)-LCAO (linear combination of atomic orbitals) results for the tantalum 5d bands in 1T-TaS_2 has shown alongside the ligand-field levels related to the density-of-states curves corresponding to d sub-bands (here neglecting inter band hybridization). Likely APW-LCAO also results, the d bands of 2H-MoS_2 (as represented in Fig. 1.29) represent the histograms joining the various ligand-field levels with the density-of-states curves for the individual

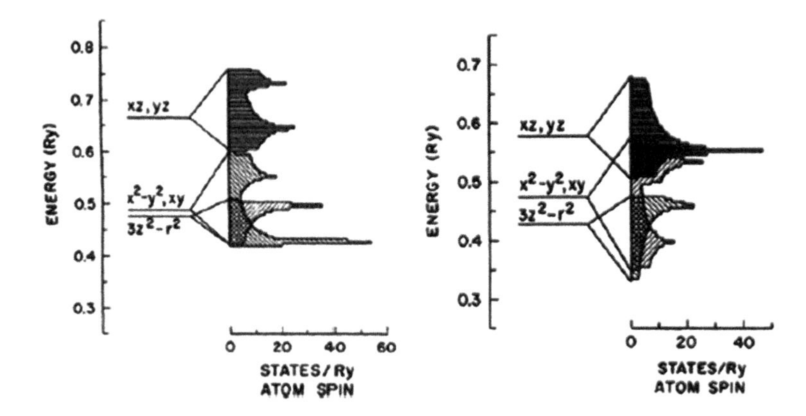

Fig. 1.29 Ligand-field levels and LCAO density-of-states curves for unhybridized d bands in 1T-TaS_2 (left) and 2H-MoS_2 (right) (adapted from Mattheiss, L. 1973 Phys. Rev. B 8, 3719)

molybdenum 4d sub-bands, by neglecting inter band hybridization and interlayer interactions [225].

In this order, Zhang et al.[250] have presented detailed theoretical results by demonstrating the density functional theory (DFT) calculations for the relative phase stability between trigonal prismatic structure (noted as H), octahedral structure (noted as T), and distorted octahedral structure (T′) in 24 types of 2D TMDs, by emphasizing combinations of group IV-VI transition metals and chalcogen species (S, Se, and Te). Their calculation showed the most stable phases of these TMDs. Moreover, the calculated band structures with spin–orbit coupling (SOC) and band alignments of monolayer TMDs concerning vacuum by using the Vienna ab initio simulation package (VASP) were also summarized [252]. Where the projector-augmented wave (PAW) [251] pseudo potentials were adopted for the both local density approximation (LDA) [253] and generalized gradient approximation (GGA) of Perdew-Burke-Ernzerhof (PBE) [254], to describe the exchange–correlation potential. Whereas the ground-level calculations are defined using the G_0W_0 approach to get more accurate band alignments of around 14 semiconducting monolayer TMDs. For the quasi-particle G_0W_0 calculations, Γ centered 12 × 12 × 1 Monkhorst–Pack K-point mesh has been used for Brillouin zone integration. Because the band gap center (BGC) is insensitive to the different exchange–correlation functionals [258]. To resolve this issue Toroker et al. have determined the band edge positions [259]. Therefore, the more appropriate conduction band minima (CBM) and valence band maxima (VBM) can be expressed as;

$$E_{\mathrm{CBM/VBM}} = E_{\mathrm{BGC}} \pm \frac{1}{2} E_g^{\mathrm{OP}} \tag{1.5}$$

1.3.5.1 Monolayer TMDs Band Alignment

Similarly, the band alignment of isolated monolayer TMDs have also described in terms of their phase stability and electronic structures. Usually, bandgaps of TMDs are obtained by defining the CBM, and VBM by employing the GGA-PBE and G_0W_0 methods, the different TMDs band gaps are tabulated in Table 1.3. Since the SOC effect is also impact their band gaps [244]. Especially in W-based H-monolayer TMDs owing to a large SOC effect of ~ 0.4 eV at the VBE [260]. Therefore, the work function in some monolayer TMDs deviates from their well-defined values. To provide a clear picture Zhang et al. have presented a details study by interpreting band alignments of various monolayer TMDs through the VBM and CBM plot, as illustrated in Fig. 1.30. They have divided the whole plot into three parts; the left part related to the band alignment of semiconducting monolayer TMDs, whereas the middle and right parts, are associated to metallic and semimetallic monolayer TMDs. Further, concerning H-monolayer VI-TMDs, with the increasing atomic indexes of chalcogen species S to Te, the VBE undergo a conspicuous energy increase accompanied by a relatively smaller energy increase of CBE, therefore, they give a decreasing energy gap. However, the larger atomic radius and decreased reactivity induces the weakened interaction between transition metal atoms and chalcogen atoms, correspondingly a larger lattice constant, it could be the reason to decrease in the band gaps. As an intense, for the same chalcogen, Mo is more reactive than W due to the intrinsic higher reactivity of 3d-electrons than 4d-electrons. Therefore, the overall energy levels of Mo-dichalcogenides may lower than those of W-dichalcogenides. Whereas the six outermost valence electrons of metal atoms bonded to six chalcogen atoms through trigonal prismatic coordination. This leads to the situation of a fully saturated valence band with empty band conduction, associating the semiconducting nature of H-monolayer VI-TMDs [250].

The band structures of the monolayer TMDs are also analyzed by considering the real space charge distributions corresponding to high symmetry K-points at band edges projection based on non-SOC calculations, as illustrated in Fig. 1.31. This analysis is provided insights effect of strain engineering on the bandgap. The H-monolayer of MoS_2, where the bonding states at Γ(V) and K(V) are predominately composed of of-plane $d_z{}^2$-orbitals and in-plane $d(x^2 - y^2) + d_{xy}$ orbitals. Therefore, the in-plane tensile strain weakened the inter-atomic $d_z{}^2$-$d_z{}^2$ bonding interaction more than the $d(x^2 - y^2) - d(x^2 - y^2)$ and $d_{xy} - d_{xy}$ inter-atomic bonding interactions. Their associated energy level shifts up to out-of-plane $d_z^2 - d_z^2$ bonding state, therefore, Γ(V) it becomes more than the in-plane bonding state K(V). However, the anti-bonding states ΓK(C) and K(C) of the valence band, whereas the K(C) is composed with a higher weight out-of-plane oriented orbital interaction compared to ΓK(C). Thus, the in-plane tensile strain shifts the energy position of K(C) down in a larger magnitude compared to ΓK(C). Their biaxial tensile strain is also inferred to produce a reduced indirect bandgap between Γ(V) and K(C). Similarly, the biaxial compressive strain are also possible with an increased indirect bandgap between K(V) and ΓK(C) [250]. Moreover, around 15% composition of *p*-orbitals at the valence band edge K(V) and ~ 88% composition of *d*-orbitals at the conduction band edge

Table 1.3 VBM (E_v), CBM (E_c), and band gaps (E_g) of different monolayer TMDs are listed

TMDs	DFT			G_0W_0			WF (eV)	Method
	E_g (eV)	E_v (e)	E_c (eV)	E_g (eV)	E_v (eV)	E_c (eV)		
H-MoS_2	1.59 (D)	− 5.86	− 4.27	2.71	− 6.42	− 3.71	5.07	PBE
	1.68 (D)			2.36	− 6.28	− 3.92	5.10	PBE
	1.58 (D)	− 6.13	− 4.55	2.48	− 6.32	− 3.84		LDA
H-$MoSe_2$	1.32 (D)	− 5.23	− 3.88	2.37	− 5.75	− 3.38	4.57	PBE
	1.45 (D)			2.04	− 5.62	− 3.58	4.60	PBE
	1.32 (D)	− 5.50	− 4.18	2.18	− 5.63	− 3.46		LDA
H-$MoTe_2$	0.94 (D)	− 4.75	− 3.78	1.89	− 5.24	− 3.35	4.29	PBE
	1.08 (D)			1.54	− 5.12	− 3.58	4.35	PBE
	0.93 (D)	− 5.04	− 4.11	1.71	− 5.11	− 3.40		LDA
H-WS_2	1.54 (D)	− 5.50	− 3.93	2.91	− 6.19	− 3.28	4.73	PBE
	1.82 (D)			2.64	− 6.11	− 3.47	4.79	PBE
	1.51 (D)	− 5.75	− 4.24	2.43	− 6.28	− 3.85		LDA
H-WSe_2	1.32 (D)	− 4.86	− 3.55	2.57	− 5.49	− 2.92	4.21	PBE
	1.55 (D)			2.26	− 5.46	− 3.20	4.33	PBE
	1.22 (D)	− 5.13	− 3.91	2.08	− 5.61	− 3.53		LDA
H-WTe_2	0.74 (D)	− 4.44	− 3.69	1.93	− 4.20	− 2.55	4.06	PBE
	1.07 (D)			1.62	− 4.97	− 3.35	4.16	PBE
T-ZrS_2	1.08 (I)	− 6.38	− 5.30	2.70	− 7.19	− 4.49	5.84	PBE
	1.19 (I)			2.56	− 7.14	− 4.58	5.86	PBE
	1.03 (I)	− 6.58	− 5.55	2.88	− 5.61	− 3.53		LDA
T-$ZrSe_2$	0.29 (I)	− 5.44	− 5.15	1.78	− 6.19	− 4.41	5.30	PBE
	0.50 (I)			1.54	− 6.14	− 4.60	5.37	PBE
	0.25 (I)	− 5.66	− 5.41	1.85	− 6.53	− 4.68		LDA
T-HfS_2	1.23 (I)	− 5.73	− 5.71	2.89	− 7.16	− 4.27	5.71	PBE
	1.27 (I)			2.45	− 6.98	− 4.53	5.75	PBE
	1.06 (I)	− 6.48	− 5.42	2.98	− 7.92	− 4.63		LDA
T-$HfSe_2$	0.45 (I)	− 4.91	− 5.37	1.95	− 6.14	− 4.19	5.17	PBE
	0.61 (I)			1.39	− 5.95	− 4.56	5.25	PBE
	0.30 (I)	− 5.57	− 5.26	1.96	− 6.53	− 4.58		LDA
T-SnS_2	1.59 (I)	− 7.05	− 5.46	3.04	− 7.77	− 4.74	6.55	PBE
	1.40 (I)	− 6.98	− 5.58	3.07	− 7.98	− 4.91		LDA
T-$SnSe_2$	0.78 (I)	− 6.00	− 5.22	1.84	− 6.57	− 4.74	5.79	PBE
	0.62 (I)	− 6.19	− 5.58	1.91	− 6.96	5.05		LDA
H-VSe_2	0.18(I)	− 5.51	− 5.33	M	M	M	5.42	PBE
H-VTe_2	0.17(I)	− 5.00	− 4.83	M	M	M	4.92	PBE
H-VS_2	SM	− 5.92	SM	M	M	M	5.92	PBE

(continued)

Table 1.3 (continued)

TMDs	DFT			G_0W_0			WF (eV)	Method
	E_g (eV)	E_v (e)	E_c (eV)	E_g (eV)	E_v (eV)	E_c (eV)		
T-VS_2	M	– 5.53	M	M	M	M	5.51	PBE
T-VSe_2	M	– 5.00	M	M	M	M	5.55	PBE
T-VTe_2	M	– 4.60	M	M	M	M	4.60	PBE
H-NbS_2	M	– 6.07	M	M	M	M	6.09	PBE
T-NbS2	M	– 5.33	M	M	M	M	5.32	PBE
H-$NbSe_2$	M	– 5.54	M	M	M	M	5.52	PBE
T-$NbSe_2$	M	– 4.91	M	M	M	M	4.90	PBE
H-$NbTe_2$	M	– 5.02	M	M	M	M	5.06	PBE
T-$NbTe_2$	M	– 5.14	M	M	M	M	4.62	PBE
H-TaS_2	M	– 5.91	M	M	M	M	5.93	PBE
T-TaS_2	M	– 5.09	M	M	M	M	5.06	PBE
H-$TaSe_2$	M	– 5.38	M	M	M	M	5.40	PBE
T-$TaSe_2$	M	– 4.68	M	M	M	M	4.66	PBE
H-$TaTe_2$	M	– 4.82	M	M	M	M	4.90	PBE
T-$TaTe_2$	M	– 4.83	M	M	M	M	4.40	PBE
T-TiS_2	SM	– 5.74	SM	SM	SM	SM	5.72	PBE
T-$TiSe_2$	SM	– 5.32	SM	SM	SM	SM	5.32	PBE
T-$TiTe_2$	SM	– 4.81	SM	SM	SM	SM	4.86	PBE
T-$ZrTe_2$	SM	– 4.81	SM	SM	SM	SM	4.56	PBE
T-$HfTe_2$	SM	– 4.70	SM	SM	SM	SM	4.72	PBE

Terms “D” and “I” in the parenthesis of E_g(D/I) indicate “direct” and “indirect” bandgaps, and “M” and “SM” denote the metallic and semimetallic nature of monolayer TMDs, respectively. Moreover, the term “WF” represents the work functions of semiconducting monolayer TMDs that are calculated by $(E_c + E_v)/2$, whereas the work functions of semimetallic and metallic monolayer TMDs calculated by the difference between VBM of DFT corresponding to vacuum level [250]

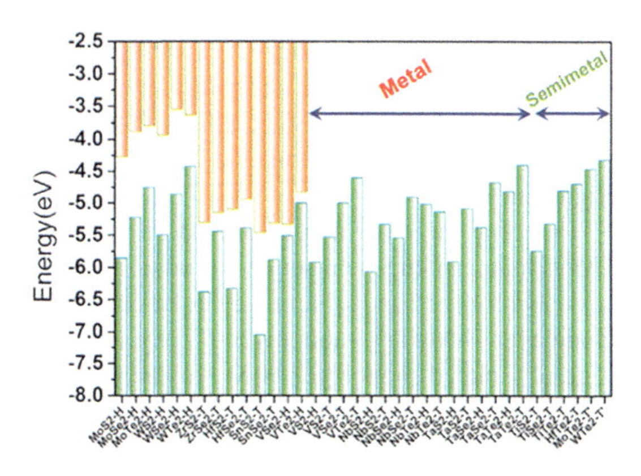

Fig. 1.30 Band alignment of the monolayer TMDs, by indicating CBM and VBM lower and upper columns respectively (adapted from Zhang, C. et al. 2017 2D Mater. 4, 015,026)

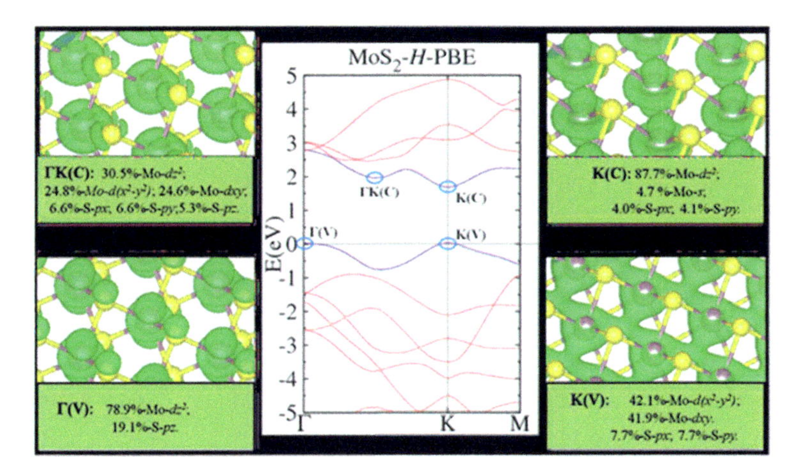

Fig. 1.31 Electronic structures of *H*-monolayer MoS_2 with projected charge distribution at band edges (Colored online). Here yellow and violet spheres represent the sulfur and molybdenum atoms. K(C) and K(V) are the conduction band edge and valence band edge at K-point, Γ(V) is the valence band edge at Γ point, and ΓK(C) is the conduction band edge between Γ and K-points (adopted from Zhang et al. [250])

K(C) satisfies the dipolar selection rule of $\Delta l = \pm 1$ for photon absorption-induced electron transitions. Since the partial population of *p*-orbitals in the valance bands restricts the photoluminescence peak intensity. It has also consisted of the experimental observation of the absorption peaks at ~ 1.8 eV in monolayer MoS_2. Where the two neighboring peaks are induced by the SOC splitting of the valence band edge, as represented in Fig. 1.31 [250].

1.3.5.2 Bilayer Heterostructures TMDs Band Alignment

Diverse promising heterostructures are stacked and their resultant band structure is not a simple combination of two monolayer TMDs. This kind of combination are generally described by Mott-Schottky or Anderson's rule [261, 262]. By including the additional description of the effects that occur at the interfaces under the charge transfer, charge redistribution, and interfacial dipoles leading to a shift of the band alignment [263, 264].

To define the band alignment before and after the stacking of two monolayer TMDs, one can choose a set of monolayer TMDs with negligible mismatch lattice constants for the band structure calculation of hetero stacks. Like band structures of monolayer H-$MoTe_2$, monolayer T-ZrS_2, and their hetero-stacks, the valence band edge of isolated monolayer H-$MoTe_2$ is higher than the conduction band edge of isolated monolayer T-ZrS_2. Therefore, the combination of the H-$MoTe_2$ monolayer with the monolayer T-ZrS_2 gives the band structure of H-$MoTe_2$-T-ZrS_2 bilayer hetero-stack, however, generally, it does not form with their band structures. This result demonstrates an obvious charge transfer from the valence band of monolayer H-$MoTe_2$ to the conduction band of monolayer T-ZrS_2. Their charge transfer are also correlated with the partially occupied valence band and conduction band around the Fermi level in the H-$MoTe_2$-T-ZrS_2 bilayer hetero-stack, as depicted in Fig. 1.31. By comparing the band structure of monolayer H-$MoTe_2$ and the projected band structure of H-$MoTe_2$ it has been noticed that a similar amount of lowering in the conduction bands and the valence bands, means they may possess almost the same band gap [250]. On the other hand, the T-ZrS_2 conduction band shifts down and valence shifted up. The inconsistency in band shifts between H-$MoTe_2$ and T-ZrS_2 could be due to a complex interfacial effect including dipole formation, charge redistribution, interlayer coupling, etc.

Moreover, the extraction of VBM, CBM, and Fermi levels from the band structures before and after stacking, the band alignment of H-$MoTe_2$-T-ZrS_2, H-$MoTe_2$-H-$NbSe_2$, and H-$MoTe_2$- T′-$MoTe_2$ bilayer heterostructures are used due to three types of hetero-stacks; semiconductor–semiconductor, semiconductor–metal and metal-semimetal. The band alignment variations of these three combinations before and after stacking are illustrated in Fig. 1.32a–f. In the first case, the H-$MoTe_2$-T-ZrS_2 semiconductor–semiconductor stacking reveals the VBM of H-$MoTe_2$ has a higher energy than the CBM of T-ZrS_2 (see Fig. 1.32a–c), and a charge transfer from H-$MoTe_2$ to T-ZrS_2 as depicted in Fig. 1.32d. Whereas the highest occupied state of the hetero-stack is aligned with the VBM of H-$MoTe_2$ after stacking, as the charge transfer and interlayer coupling are well-defined [263, 265, 266]. On the other hand, in the case of the H-$MoTe_2$-T-ZrS_2 stacking, the valence band offset (VBO) and the conduction band offset (CBO) between the H-$MoTe_2$ layer and T-ZrS_2 layer is around 0.72 and 1.11 eV. Hence, the bandgap of H-$MoTe_2$ is almost unchanged before and after stacking while the bandgap of T-ZrS_2 slightly reduced due to the complex interlayer coupling and the interfacial charges.

In the second case, the H-$MoTe_2$-H-$NbSe_2$ semiconductor–metal stacking leads the average energies of electrons in H-$MoTe_2$ higher than the H-$NbSe_2$, and their charge transfer occurs through the VBM of H-$MoTe_2$ to H-$NbSe_2$, as depicted in Fig. 1.32e. This representation leads the valence band of H-$MoTe_2$ to shift down to align with the VBM of the stacking, while the VBM of H-$NbSe_2$ is slightly lowered owing to charge injection, thereby, the Fermi level cannot change. Due to the existence of a large number of empty states around the highest occupied state of H-$NbSe_2$.

In the third case, H-$MoTe_2$-T′-$MoTe_2$-semiconductor-semimetal stacking revealed that the H-$MoTe_2$-H-$NbSe_2$ stacking charge transfer through T′-$MoTe_2$ to H-$MoTe_2$ is considered to be negligible, due to tiny upshift in VBM and CBM of H-$MoTe_2$ after stacking. It can be understood from the representation Fig. 1.32f compared to Fig. 1.32d, e, in which more transferred charges have accumulated at the interface rather than in the charge-accepting layer. Like metals, the shift of VBE positions in T′-$MoTe_2$ is very small because of its semimetallic behavior. It is due to small amounts of charge transfer through the small difference of chemical potential (or Fermi level) between the H-$MoTe_2$ and T′-$MoTe_2$. Whereas the Schottky barrier between H-$MoTe_2$ and T′-$MoTe_2$ for the 0.18 eV, is close to ohmic contact.

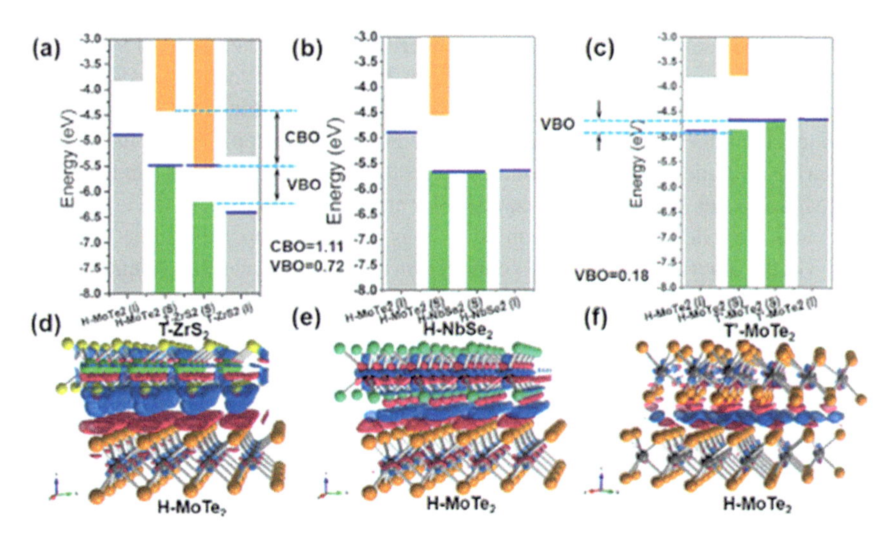

Fig. 1.32 Band alignments before and after the stacking; **a** H-$MoTe_2$ and T-ZrS_2, **b** H-$MoTe_2$ and H-$NbSe_2$; **c** H-$MoTe_2$ and T′-$MoTe_2$. The light gray columns indicate the VBM and CBM before the stacking. The blue horizontal bars denote the highest occupied states. The green and orange columns represent the VBM and the CBM after stacking, respectively. The valence band offset (VBO) and conduction band offset (CBO) are marked with black arrows. The charge density difference between the heterostructures, **d** H-$MoTe_2$-T-ZrS_2, **e** H-$MoTe_2$-H-$NbSe_2$, **f** H-$MoTe_2$-T-$MoTe_2$ and their two isolated TMD layer ($\Delta\rho = \rho$heterostructure-ρA-ρB) is calculated to show the charge redistribution after the stacking. The blue and pink clouds represent electron accumulation and depletion regions respectively (adapted from Zhang, C. et al. 2017 2D Mater. 4, 015,026)

Apart from these typical three kinds of hetero-stacks, the various TMDs bilayer hetero-stacks have been investigated [250, 267–269]. Thus, it is worth saying, the semiconductor–metal and semiconductor-semimetal stacking bands realignments reduced the Schottky barrier heights with a significant reduction in the contact resistance and an improvement in electrical performance for the TMD heterostructure-based devices. Theoretical studies have also demonstrated the Schottky barrier can be tuned by choosing various metallic TMDs as contact electrodes. Owing to the existence of the Van der Waals contact the 2D metal have advantages including facilitating carrier transport, weakened Fermi level pinning, and reduced electron-holes recombination [270].

1.4 Conclusions

In the conclusive remarks in this chapter, the author has discussed the basics of the TMDs, including an introductory note on such materials with the 2D materials. A discussion on classifications of 2D materials including layered solids and their types, ionic solids and types, and non-layered solids have also discussed. Since

TMDs are mainly chalcogen family materials, therefore, worth having a description of these materials, therefore, it has been discussed including a separate discussion on key group elements Sulfur, Selenium, and Tellurium. A separate section on the topic Transition Metal Dichalcogenides (TMDs) has also included and discussed including materials chemistry of the transition metals, theory of the valance bonds, ligand field approach, crystal filed approach and band structure (monolayer TMDs and bilayer heterostructures TMDs). An effort has been also made to provide key basic descriptions of the TMDs materials that would be helpful to make a better understanding of the subject as well as easy to build the scientific/ technical approach for the technologies based on these materials.

References

1. Yu M, Cheng X, Zeng Y, Wang Z et al (2016) Dual-doped molybdenum trioxide nanowires: a bifunctional anode for fiber-shaped asymmetric supercapacitors and microbial fuel cells. Angew Chem Int Ed 55:6762–6766
2. Cheng X, Sun M, Xie H, Li J (2018) Chapter 9—Transition metal dichalcogenides in energy applications. In 2D materials characterization, production and applications. CRC Press, Boca Raton
3. Ratha S, Rout CS (2013) Supercapacitor electrodes based on layered tungsten disulfide-reduced graphene oxide hybrids synthesized by a facile hydrothermal method. ACS Appl Mater Interfaces 5:11427–11433
4. Zhang G, Liu H, Qu J, Li J (2016) Two-dimensional layered MoS2: rational design, properties and electrochemical applications. Energy Environ Sci 9:1190–1209
5. Geim AK, Novoselov KS (2007) The rise of graphene. Nat Mater 6:183–191
6. Chen D, Tang L, Li J (2010) Graphene-based materials in electrochemistry. Chem Soc Rev 39:3157–3180
7. Huang X, Zeng Z, Fan Z, Liu J et al (2012) Graphene-based electrodes. Adv Mater 24:5979–6004
8. Huang X, Zeng Z, Zhang H (2013) Metal dichalcogenide nanosheets: preparation, properties and applications. Chem Soc Rev 42:1934–1946
9. Xu M, Liang T, Shi M, Chen H (2013) Graphene-like two-dimensional materials. Chem Rev 113:3766–3798
10. Roy-Mayhew JD, Aksay IA (2014) Graphene materials and their use in dye-sensitized solar cells. Chem Rev 114:6323–6348
11. Wang H, Hu YH (2012) Graphene as a counter electrode material for dye-sensitized solar cells. Energy Environ Sci 5:8182–8188
12. Novoselov KS, Jiang D, Schedin F, Booth TJ et al (2005) Two-dimensional atomic crystals. Proc Natl Acad Sci U S A 102:10451–10453
13. Halim U, Zheng CR, Chen Y, Lin Z et al (2013) A rational design of cosolvent exfoliation of layered materials by directly probing liquid–solid interaction. Nat Commun 4:2213
14. Coleman JN, Lotya M, O'Neill A, Bergin SD et al (2011) Two-dimensional nanosheets produced by liquid exfoliation of layered material. Science 331:568–571
15. Geim AK, Grigorieva IV (2013) Van der Waals heterostructures. Nature 499:419–425
16. Splendiani A, Sun L, Zhang Y, Li T et al (2010) Emerging photoluminescence in monolayer MoS_2. Nano Lett 10:1271–1275
17. Mak KF, He K, Shan J, Heinz TF (2012) Control of valley polarization in monolayer MoS2 by optical helicity. Nat Nanotechnol 7:494–498

18. Mak KF, Lee C, Hone J, Shan J et al (2010) Atomically thin MoS2: a new direct-gap semiconductor. Phys Rev Lett 105:136805
19. Wu SF, Ross JS, Liu GB, Aivazian G et al (2013) Electrical tuning of valley magnetic moment through symmetry control in bilayer MoS_2. Nat Phys 9:149–153
20. Mak KF, McGill KL, Park J, McEuen PL (2014) The valley Hall effect in MoS_2 transistors. Science 344:1489–1492
21. Cao T, Wang G, Han W, Ye H et al (2012) Valley-selective circular dichroism of monolayer molybdenum disulphide. Nat Commun 3:887
22. Xiao D, Liu GB, Feng WX, Xu XD et al (2012) Coupled spin and valley physics in monolayers of MoS2 and other group-VI dichalcogenides. Phys Rev Lett 108:196802
23. Zeng HL, Dai JF, Yao W, Xiao D et al (2012) Valley polarization in MoS_2 monolayers by optical pumping. Nat Nanotechnol 7:490–493
24. Kang J, Tongay S, Zhou J, Li JB et al (2013) Band offsets and heterostructures of two-dimensional semiconductors. Appl Phys Lett 102:012111
25. Bayliss SC, Liang WY (1982) Symmetry dependence of optical transitions in group 4b transition metal dichalcogenides. J Phys C: Solid State Phys 15:1283–1296
26. Radisavljevic B, Radenovic A, Brivio J, Giacometti V et al (2011) Single-layer MoS_2 transistors. Nat Nanotechnol 6:147–150
27. Britnell L, Gorbachev RV, Jalil R, Belle BD et al (2012) Field-effect tunneling transistor based on vertical graphene heterostructures. Science 335:947–950
28. Yu WJ, Li Z, Zhou H, Chen Y et al (2013) Vertically stacked multi-heterostructures of layered materials for logic transistors and complementary inverters. Nat Mater 12:246–252
29. Lopez-Sanchez O, Alarcon Llado E, Koman V, Fontcuberta A et al (2014) Light generation and harvesting in a van der Waals heterostructure. ACS Nano 8:3042–3048
30. Lee CH, Lee GH, van der Zande AM, Chen W et al (2014) Atomically thin p–n junctions with van der Waals heterointerfaces. Nat Nanotechnol 9:676–681
31. Cheng R, Li D, Zhou H, Wang C et al (2014) Electroluminescence and photocurrent generation from atomically sharp WSe_2/MoS_2 heterojunction *p–n* diodes. Nano Lett 14:5590–5597
32. Baugher BW, Churchill HO, Yang Y, Jarillo-Herrero P (2014) Optoelectronic devices based on electrically tunable p–n diodes in a monolayer dichalcogenide. Nat Nanotechnol 9:262–267
33. Pospischil A, Furchi MM, Mueller T (2014) Solar-energy conversion and light emission in an atomic monolayer p–n diode. Nat Nanotechnol 9:257–261
34. Duan X, Wang C, Pan A, Yu R et al (2015) Two-dimensional transition metal dichalcogenides as atomically thin semiconductors: opportunities and challenges. Chem Soc Rev 44:8859–8876
35. Novoselov KS, Geim AK, Morozov SV, Jiang D et al (2004) Electric field effect in atomically thin carbon films. Science 306:666
36. Kroemer H (2001) Nobel lecture: quasielectric fields and band offsets: teaching electrons new tricks. Rev Mod Phys 73:783
37. Feynman RP (1960) There's plenty of room at the bottom. Eng Sci 23:22
38. Wallace PR (1947) The band theory of graphite. Phys Rev 71:622
39. Boehm HP, Clauss A, Fischer GO, Hofmann U (1962) Dünnste Kohlenstoff-Folien. Z Naturforsch B 17:150
40. Shelton JC, Patil HR, Blakely JM (1974) Equilibrium segregation of carbon to a nickel (111) surface: A surface phase transition. Surf Sci 43:493
41. Van Bommel AJ, Crombeen JE, Van Tooren A (1975) The electrostatic potential in the surface region of an ionic crystal. Surf Sci 48:463
42. Lu X, Yu M, Huang H, Ruoff RS (1999) Tailoring graphite with the goal of achieving single sheets. Nanotechnology 10:269
43. Fitzer E, Kochling KH, Boehm HP, Marsh H (1995) Pure Appl Chem 67:473
44. Avouris P, Chen Z, Perebeinos V (2007) Carbon-based electronics. Nat Nanotechnol 2:605
45. Jariwala D, Sangwan VK, Lauhon LJ, Marks TJ et al (2013) Carbon nanomaterials for electronics, optoelectronics, photovoltaics, and sensing. Chem Soc Rev 42:2824

46. Li XM, Lv Z, Zhu HW (2015) Solar cells: carbon/silicon heterojunction solar cells: state of the art and prospects. Adv Mater 27:6549
47. Li X, Tao L, Chen Z, Fang H, Li X et al (2017) Graphene and related two-dimensional materials: Structure-property relationships for electronics and optoelectronics. Appl Phys Rev 4:021306–021331
48. Wang QH, Kalantar-Zadeh K, Kis A, Coleman JN et al (2012) Strano, electronics and optoelectronics of two-dimensional transition metal dichalcogenides. Nat Nanotechnol 7(11):699–712
49. Xu M, Liang T, Shi M, Chen H (2013) Graphene-like two-dimensional materials. Chem Rev 113(5):3766–3798
50. Wang XR, Shi Y, Zhang R (2013) Field-effect transistors based on two-dimensional materials for logic applications. Chin Phys B22(9):098505
51. Bolotin KI, Sikes KJ, Jiang Z, Klima M et al (2008) Ultrahigh electron mobility in suspended graphene. Solid State Commun 146(9):351–355
52. Wang X, Ouyang Y, Li X, Wang H et al (2008) Roomtemperatureall-semiconducting sub-10-nm graphene nanoribbonfield-effect transistors. Phys Rev Lett 100(20):206803
53. Ding Y, Wang Y, Ni J, Shi L et al (2011) First principles study of structural, vibrational and electronic properties of graphene-like MX2 (M = Mo, Nb, W, Ta; X = S, Se, Te) monolayers Phys B Condens Matter 406(11):2254–2260
54. Ataca C, Sahin H, Ciraci S (2012) Stable, single-layer MX_2 transition-metal oxides and dichalcogenides in a honeycomb-likestructure. J Phys Chem C 116(16):8983–8999
55. Klots AR, Newaz AKM, Wang B, Prasai D et al (2014) Probing excitonic states in suspended two-dimensional semiconductors by photocurrent spectroscopy. Sci Rep 4:6608
56. Rasmussen FA, Thygesen KS (2015) Computational 2D materials database: electronic structure of transition-metal dichalcogenides and oxides. J Phys Chem C 119(23):13169–13183
57. Fuhrer MS, Hone J (2013) Measurement of mobility in dual-gatedMoS_2 transistors. Nat Nanotechnol 8(3):146–147
58. Radisavljevic B, Kis A (2013) Reply to measurement of mobility in dual-gated MoS2 transistors. Nat Nanotechnol 8(3):147–148
59. Yu Z, Ong ZY, Pan Y, Cui Y et al (2016) Realization of room-temperature phonon-limited carrier transport in monolayerMoS_2 by dielectric and carrier screening. Adv Mater 28(3):547–552
60. Li X, Mullen JT, Jin Z, Borysenko KM et al (2013) Intrinsic electrical transport properties of monolayer silicone and MoS_2 from first principles. Phys Rev B 87(11):115418
61. Qiao J, Kong X, Hu ZX, Yang F, Ji W (2014) High-mobility transport anisotropy and linear dichroism in few-layer black phosphorus. Nat Commun 5:1–7
62. Liu Y, Weiss NO, Duan X, Cheng H et al (2016) Van der Waals heterostructures and devices. Nat Rev Mater 1:16042
63. Jariwala D, Marks TJ, Hersam MC (2017) Mixed-dimensional van der Waals heterostructures. Nat Mater 16:170–181
64. Novoselov KS, Mishchenko A, Carvalho A, Neto AHC (2016) 2D materials and van der Waals heterostructures. Science 353:aac9439
65. Ajayan P, Kim P, Banerjee K (2016) Two-dimensional van der Waals materials. Phys Today 69:38–44
66. Li X, Zhu H (2016) The graphene–semiconductor Schottky junction. Phys Today 69:46
67. Eda G, Fujita T, Yamaguchi H, Voiry D, Chen M et al (2012) Coherent atomic and electronic heterostructures ofsingle-layer MoS_2. ACS Nano 6:7311–7317
68. Li H, Lu G, Yin Z, He Q et al (2012) Optical identification of single- and few-layer MoS_2 sheets. Small 8:682–686
69. Tongay S, Zhou J, Ataca C, Lo K et al (2012) Thermally driven crossover from indirect toward direct bandgap in 2D semiconductors: $MoSe_2$ versusMoS_2. Nano Lett 12:5576–5580
70. Mak KF, Shan J (2016) Photonics and optoelectronics of 2D semiconductor transition metal dichalcogenides. Nat Photon 10:216–226

71. Alberti G, Costantino U (1996) Alberti G, Bein T (eds) Comprehensive supramolecular chemistry, layered solids and their intercalation chemistry. Elsevier, Oxford, vol 7
72. McCabe RW (1992) *Inorganic materials*. Wiley, New York, p 295
73. Ostwald HR, Asper R (1977) In: Leith RMA, Reidel D (eds) Physics and chemistry of materials with layered structures, vol 1, p 73
74. Rouxel J (1996) In: Alberti G, Bein T (eds) Comprehensive supramolecular chemistry, layered metal chalcogenides and their intercalation compounds. Elsevier, Oxford, vol 7
75. Jacobson AB (1996) Alberti G, Bein T (eds) Comprehensive supramolecular chemistry: colloidal dispersion of compounds with layer and chain structures. Elsevier, Oxford, vol 7, p 315
76. Final report of NSF/AFOSR Workshop on 2D Materials and Devices Beyond Graphene, May 2012—http://nsf2dworkshop.rice.edu/home
77. Wang QH, Kalantar-Zadeh K, Kis A, Coleman JN et al (2012) Electronics and optoelectronics of two-dimensional transition metal dichalcogenides. Nat Nano 7:699–712
78. Ahmed S, Yi J (2017) Two-dimensional transition metal dichalcogenides and their charge carrier mobilities in field-effect transistors. Nano-Micro Lett 9(1–23):50
79. Kaul AB (2015) Van der Waals solids: properties and device applications. Proc SPIE 9467 Micro Nanotechnol Sens Syst Appl VII:94670N(1–5). https://doi.org/10.1117/12.2178002
80. Zhao WJ, Ghorannevis Z, Chu LQ, Toh ML et al (2013) Evolution of electronic structure in atomically thin sheets of WS_2 and WSe_2. ACS Nano 7:791–797
81. Ma YD, Dai Y, Guo M, Niu CW et al (2011) Electronic and magnetic properties of perfect, vacancy-doped, and nonmetal adsorbed $MoSe_2$, $MoTe_2$ and WS_2 monolayers. Phys Chem Chem Phys 13:15546–15553
82. Enyashin AN, Yadgarov L, Houben L, Popov I et al (2011) New route for stabilization of 1T-WS_2 and MoS_2 phases. J Phys Chem C115:24586–24591
83. Kong D, Dang W, Cha JJ, Li H et al. Few-layer nanoplates of Bi_2Se_3 and Bi_2Te_3 with highly tunable chemical potential. Nano Lett 10:2245–2250
84. Yang S, Gong Y, Liu Z, Zhan L et al. Bottom-up approach toward single-crystalline VO_2-graphene ribbons as cathodes for ultrafast lithium storage. Nano Lett 13:1596–1601
85. Taha-Tijerina J, Narayanan TN, Gao G, Rohde M et al. Electrically insulating thermal nano-oils using 2D fillers. ACS Nano 6:1214–1220
86. Hooley JG (1977) Preparation and crystal growth of materials with layered structure
87. Reichle WT (1986) ChemTech 16:58
88. Trifiro F,Vaccari A (1996) In: Atwood JL, MacNicol DD, Davis JED, Vogtle F (eds) Comprehensive supramolecular chemistry. Pergamon Press, Oxford, vol 7 (Chapter 10)
89. Ebina Y, Sasaki T, Harada M, Watanabe M (2002) Restacked perovskite nanosheets and their Pt-loaded materials as photocatalysts. Chem Mater 14:4390–4395
90. Ozawa TC, Fukuda K, Akatsuka K, Ebina Y et al (2007) Preparation and characterization of the Eu3? doped perovskitenanosheet phosphor: $La_{0.90}Eu_{0.05}Nb_2O_7$. Chem Mater 19:6575–6580
91. Kamath PV, Therese GHA, Gopalakrishnan J (1997) On the existence of hydrotalcite-like phases in the absence of trivalent cations. J Solid State Chem 128:38–41
92. Ismail J, Ahmed MF, Kamath PV, Subbanna GN et al (1995) Organic additive-mediated synthesis of novel cobalt(II) hydroxides. J Solid State Chem 114:550–555
93. Rajamathi M, Thomas GS, Kamath PV (2001) The many ways of making anionic clays. Proc Indian Acad Sci Chem Sci 113:671–680
94. Ma R, Sasaki T (2010) Nanosheets of oxides and hydroxides: ultimate 2D charge-bearing functional crystallites. Adv Mater 22:5082–5104
95. Wei W, Cui X, Chena W, Ivey DG (2011) Manganese oxide-based materials as electrochemical supercapacitor electrodes. Chem Soc Rev 40:1697–1721
96. Post JE, Veblen DR (1990) Crystal structure determinations of synthetic sodium, magnesium, and potassium birnessite using TEM and the Rietveld method. Am Mineral 75:477–489
97. Liu ZH, Ooi K, Kanoh H, Tang WP, Tomida T (2000) Swelling and delamination behaviors of birnessite-type manganese oxide by intercalation of tetraalkylammonium ions. Langmuir 16:4154–4164

98. Omomo Y, Sasaki T, Wang LZ, Watanabe M (2003) Redoxable nanosheet crystallites of MnO_2 derived via delamination of a layered manganese oxide. J Am Chem Soc 125:3568–3575
99. Suzuki S, Takahashi S, Sato K, Miyayama M (2006) Key Eng Mater 320:223
100. Sasaki T (2007) J Ceram Soc Jpn 115:9
101. Lee HY, Goodenough JB (1999) J Solid State Chem 144:220
102. Lee HY, Manivannan V, Goodenough JB, Acad CR (1999) Sci Ser IIc: Chim 2:565
103. Brodie BC (1859) XIII. On the atomic weight of graphite. Philos Trans 249(1859):259
104. Croft RC (1960) Quartz. Rev 14(1):1
105. Staudenmaier L (1898) Process for the preparation of graphitic acid. Ber Dtsch Chem Ges 31:1481–1487
106. Clauss A, Boehm H (1957) Untersuchungen zur Struktur des Graphitoxyds. Z Anorg Chem 291:205–220
107. Hummers O (1958) Preparation of graphitic oxide. J Am Chem Soc 80:1339
108. Lerf A, He H, Foster M, Klinowski J (1998) Structure of graphite oxide revisited. J Phys Chem B 102:4477–4482
109. Lerf A, He H, Riedl T, Foster M et al (1997) ^{13}C and 1H MAS NMR studies of graphite oxide and its chemically modified derivatives. J Solid State Ionics 101–103:857–862
110. He H, Riedl T, Lerf A, Klinowski J (1996) Solid-state NMR studies of the structure of graphite oxide. J Phys Chem 100:19954–19958
111. Bourlinos AB, Gournis D, Petridis D, Szabo T et al (2003) Graphite oxide: chemical reduction to graphite and surface modification with primary aliphatic amines and amino acids. Langmuir 19:6050–6055
112. Kovtyukhova NI, Karpenko GA, Chuiko AA (1992) Bis(diarylphosphino)-1,1 binaphthyl (BINAP)-ruthenium(II) dicarboxylate complexes: new, highly efficient catalysts for asymmetric hydrogenations. J Inorg Chem 37:566–569
113. Cassagueau T, Fendler JH (1789) Preparation and layer-by-layer self-assembly of silver nanoparticles capped by graphite oxide nanosheets. J Phys Chem B 103:1789–1793
114. Compton OC, Nguyen ST (2010) Graphene oxide, highly reduced graphene oxide, and graphene: versatile building blocks for carbon-based materials. Small 6:711–23
115. Yang X, Makita Y, Liu Zh Ooi K (2003) Novel synthesis of layered graphite oxide−birnessite manganese oxide nanocomposite. Chem Mater 15:1228–1231
116. Nethravathi C, Rajamathi M (2006) Delamination, colloidal dispersion and reassembly of alkylamine intercalated graphite oxide in alcohols. Carbon 44:2635–2641
117. Du XS, Xiao M, Meng YZ, Hay AS (2004) Novel synthesis of conductive poly(arylene disulfide)/graphite nanocomposite. Synth Met 143:129–132
118. Nethravathi C, Rajamathi M (2008) Chemically modified graphene sheets produced by the solvothermal reduction of colloidal dispersions of graphite oxide. Carbon 46:1994–1998
119. Liang M, Zhi L (2009) Graphene-based electrode materials for rechargeable lithium batteries. J Mater Chem 19:5871–5878
120. Allen MJ, Tung VC, Kaner RB (2010) Honeycomb carbon: A review of graphene. Chem Rev 110:132–145
121. Tanenbaum DM, Parpia JM, Craighead HG, McEuen PL (2007) Electromechanical resonators from graphene sheets. Science 315:490–493
122. Hsu WT, Zhao ZA, Li LJ, Chen CH et al (2014) Second harmonic generation from artificially stacked transition metal dichalcogenide twisted bilayers. ACS Nano 8:2951–2958
123. Acun A, Poelsema B, Zandvliet HJW, van Gastel R (2013) Theinstability of silicene on Ag (111). Appl Phys Lett 103:263119
124. Tao L, Cinquanta E, Chiappe D, Grazianetti C et al (2015) Silicene field-effect transistorsoperating at room temperature. Nat Nanotechnol 10:227–231
125. Oughaddou H, Aufray B, Biberian JP, Hoarau JY (1999) Growth mode and dissolution kinetics of germanium thin films on Ag(001) surface: an AES–LEED investigation. Surf Sci 429:320–326
126. Golias E, Xenogiannopoulou E, Tsoutsou D, Tsipas P et al (2013) Surface electronic bands of submonolayerGe on Ag(111). Phys Rev B 88:075403

127. Hao B, Yan Y, Wang X, Chen G (2013) Synthesis of anatase TiO_2 nanosheets with enhanced pseudocapacitive contribution for fast Lithium storage. ACS Appl Mater Interfaces 5:6285–6291
128. Sterrer M, Risse T, Martinez Pozzoni U, Giordano L et al (2007) Control of the charge state of metal atoms on thin MgO films. Phys Rev Lett 98:096107
129. Jung H, Park J, Oh IK, Choi T et al (2014) Fabrication of transferable Al2O3 nanosheet by atomic layer deposition for graphene FET. ACS Appl Mater Interfaces 6:2764–2769
130. Fischer W (2001) A second note on the term chalcogen. J Chem Educ 78:1333
131. Bouroushian M (2010) Electrochemistry of metal chalcogenides. Springer, Heidelberg. ISBN: 978-3-642-03967-6
132. Todorov R, Tasseva J, Babeva T (2012) Thin chalcogenide films for photonic applications, photonic crystals innovative systems, lasers and waveguides. Dr. Alessandro Massaro (Ed.). ISBN: 978-953-51-0416-2
133. Gao MR, Xu YF, Jiang J, Yu SH (2013) Nanostructured metal chalcogenides: synthesis, modification, and applications in energy conversion and storage devices. Chem Soc Rev 42(2013):2986–3017
134. Chivers T, Laitinen RS (2015) Tellurium: a maverick among the chalcogens. Chem Soc Rev 44:1725–1739
135. Popescu MA (2002) Non-crystalline chalcogenides. National Institute of Materials Physics, Bucharest, Magurele, Romania. ISBN: 0-306-47129-9
136. Ibers J (2009) Tellurium in a twist. Nat Chem 1:508
137. Lucovsky G, White R (1973) Effects of resonance bonding on the properties of crystalline and amorphous semiconductors. Phys Rev B 8:660–667
138. Shportko K, Kremers S, Woda M, Lencer D et al (2008) Resonant bonding in crystalline phase-change materials. Nat Mater 7:653–658
139. Kastner M (1972) Bonding bands, lone-pair bands, and impurity states in chalcogenide semiconductors. Phys Rev Lett 28(6):355–357
140. Kastner M, Fritzsche H (1978) Defect chemistry of lone-pair semiconductors. Philos Mag B 37(2):199–215
141. Kolobov AV, Fons P, Tominaga J, Ovshinsky SR (2013) Vacancy-mediated three-center fourelectron bonds in GeTe-Sb_2Te_3 phase-change memory alloys. Phys Rev B 87:165206–9
142. Kolobov AV, Fons P, Tominaga J (2015) Understanding phase-change memory alloys from a chemical perspective. Sci Rep 5:13698–13711
143. Mott NF (1967) Electrons in disordered structures. Adv Phys 16:49–144
144. Remsing RC, Klein ML (2020) A new perspective on lone pair dynamics in halide perovskites. APL Mater 8:050902-1–50906
145. Sidgwick NV (1933) The covalent link in chemistry. CornellUniversity Press, Ithaca
146. Pyykko P (1988) Relativistic effects in structural chemistry. Chem Rev 88:563–594
147. Mercier N, Louvain N, Bi W (2009) Structural diversity and retrocrystal engineering analysis of iodometalate hybrids. Cryst Eng Comm 11:720–734
148. Vogler A, Nikol H (1993) The structures of s^2 metal complexes in the ground and sp excited states. Comments Inorg Chem 14:245–261
149. Li Z, Li Y, Liang P, Zhou T et al (2019) Dual-band luminescent lead-free antimony chloride halides with near-unity photoluminescence quantum efficiency. Chem Mater 31:9363–9371
150. Xu J, Li S, Qin C, Feng Z et al (2020) Identification of singlet self-trapped excitons in a new family of white-light-emitting zero-dimensional compounds. J Phys Chem C 124:11625–11630
151. Jing Y, Liu Y, Jiang X, Molokeev MS et al (2020) Sb^{3+} dopant and halogen substitution triggered highly efficient and tunable emission in lead-free metal halide single crystals. Chem Mater 32:5327–5334
152. Blasse G, Grabmaier BC (1994) Luminescent materials. Springer, Berlin
153. Blasse G (1990) Interaction between optical centers and their surroundings: an inorganic chemist's approach. Adv Inorg Chem 35:319–402

154. Oomen EWJL, Smit WMA, Blasse G (1987) The luminescenceof $Cs_2NaSbCl_6$ and $Cs_2NaSbBr_6$: a transition from a localized to adelocalized excited state. Chem Phys Lett 138:23–28
155. Jacobs PWM (1991) Alkali halide crystals containing impurity ionswith the ns^2 ground-state electronic configuration. J Phys Chem Solids 52:35–67
156. Jacobs PWM (2019) Alkali halide crystals containing impurity ions with the ns2 ground-state electronic configuration. J Phys Chem Solids 52:35–67
157. Pelant I, Valenta J (2012) Luminescence spectroscopy of semiconductors. Oxford University Press, New York
158. Ueta M, Kanzaki H, Kobayashi K, Toyozawa Y et al (1986) Theory of excitons in phonon fields. In: Excitonic processes in solids. Springer, Berlin, pp 203−284
159. McCall KM, Stoumpos CC, Kostina SS, Kanatzidis MG et al (2017) Strong electron−phonon coupling and self-trapped excitons in the defect halide perovskites A3M2I9 (A = Cs, Rb; M = Bi, Sb). Chem Mater 29:4129–4145
160. Luo J, Wang X, Li S, Liu J et al (2018) Efficient and stable emission of warm-white light from lead-freehalide double perovskites. Nature 563:541−545
161. Li S, Luo J, Liu J, Tang J (2019) Self-trapped excitons in all-inorganic halide perovskites: fundamentals, status, and potential applications. J Phys Chem Lett 10:1999–2007
162. Fisher GA, Norman NC (1994) The structures of the group 15 element (III) halides and halogenoanions. Adv Inorg Chem 41:233–271
163. Stöwe K, Beck HP (1994) Low temperature polymorphs of the compound In_3SnI_5. Z Kristallogr Cryst Mater 209:36–42
164. Krebs B, Ahlers FP (1990) Developments in chalcogen−halide chemistry. Adv Inorg Chem 35:235–317
165. Chabot B, Parthe E (1978) $Cs_3Sb_2I_9$ and $Cs_3Bi_2I_9$ with the hexagonal$Cs_3Cr_2Cl_9$ structure type. Acta Crystallogr Sect B: Struct Crystallogr Cryst Chem 34:645−648
166. McCall KM, Stoumpos CC, Kostina SS, Kanatzidis MG et al (2017) Strong electron−phonon coupling and self-trapped excitons in the defect halide perovskites $A_3M_2I_9$ (A = Cs, Rb; M = Bi,Sb). Chem Mater 29:4129−4145
167. Benin BM, McCall KM, Woerle M, Morad V et al (2020) The $Rb_7Bi_{3-3x}Sb_{3x}Cl_{16}$ family: a fully inorganic solid solution with room-temperature luminescent members. Angew Chem Int Ed 59:14490−14497
168. Zhou C, Lin H, Neu J, Zhou Y et al (2019) Green emitting single-crystalline bulk assembly of metal halide clusters with near-unity photoluminescence quantum efficiency. ACS Energy Lett 4:1579−1583
169. McCall KM, Morad V, Benin BM, Kovalenko MV (2020) Efficient lone-pair-driven luminescence: structure−property relationships in emissive $5s^2$ metal halides. ACS Mater Lett 2:1218−1232
170. Mott NF, Davis EA (1979) Electronic processes in non-crystalline materials, 2nd edn. Clarendon Press, Oxford
171. Anderson P (1975) Model for the electronic structure of amorphous semiconductors. Phys Rev Lett 34:953–955
172. Street RA, Mott NF (1975) States in the gap in glassy semiconductors. Phys Rev Lett 35:1293–1296
173. Kastner M, Adler D, Fritzsche H (1976) Valence-alternation model for localized gap states in lone pair semiconductors. Phys Rev Lett 37:1504–1507
174. Kolobov A, Kondo M, Oyanagi H, Durny R et al (1997) Experimental evidence for negative correlation energy and valence alternation in amorphous selenium. Phys Rev B56:R485–R488
175. Kolobov A, Kondo M, Oyanagi H, Matsuda A et al (1998) Negative correlation energy and valence alternation in amorphous selenium: an in situ optically induced ESR study. Phys Rev B 58:12004–12010
176. Ovshinsky SR (1968) Reversible electrical switching phenomena in disordered structures. Phys Rev Lett 21:1450–1453

177. Yamada N, Ohno E, Nishiuchi K, Akahira N (1991) Rapid-phase transitions of GeTe-Sb2Te3 pseudobinary amorphous thin films for an optical disk memory. J Appl Phys 69:2849
178. Kolobov AV, Fons P, Frenkel AI, Ankudinov AL et al (2004) Understanding the phase-change mechanism of rewritable optical media. Nat Mater 3:703–708
179. Lee TH, Elliott SR (2017) The relation between chemical bonding and ultrafast crystal growth. Adv Mater 29:1700814–8
180. Li XB, Chen NK, Wang XP, Sun HB (2018) Phase-change superlattice materials toward low power consumption and high density data storage: microscopic picture, working principles, and optimization. Adv Funct Mater 28:1803380
181. Lotnyk A, Behrens M, Rauschenbach B (2019) Phase change thin films for non-volatile memory applications. Nanoscale Adv 1:3836–3857
182. Yamada N, Matsunaga T (2000) Structure of laser-crystallized $Ge_2Sb_{2+x}Te_5$ sputtered thin films for use in optical memory. J Appl Phys 88:7020
183. Hoang K, Mahanti SD (2016) Atomic and electronic structures of I-V-VI2 ternary chalcogenides. J Sci: Adv Mater Dev 1:51–56
184. Edwards AH, Pineda AC, Schultz PA, Martin MG et al (2006) Electronic structure of intrinsic defects in crystalline germanium telluride. Phys Rev B 73:045210–045213
185. Wuttig M, Deringer VL, Gonze X, Bichara C et al (2018) Incipient metals: functional materials with a unique bonding mechanism. Adv Mater 30:1803777
186. Jones RO (2020) Phase change memory materials: rationalizing the dominance of Ge/Sb/Te alloys. Phys Rev B 101:024103
187. Rundle RE (1962) Negative ion–molecule reactions of SF4. Rec Chem Prog 23:195
188. Coulson CA (1964) The nature of the bonding in xenon fluorides and related molecules. J Chem Soc 1964:1442–1454
189. Lee TH, Elliott SR (2020) Chemical bonding in chalcogenides: the concept of multicenter hyperbonding. Adv Mater 32:2000340–2000349
190. Roy S, Hu Z, Kais S, Bermel P (2019) Enhancement of photovoltaic current generation through dark states in donor-acceptor pairs of tungsten-based transition metal di-chalcogenides (TMDCs). https://www.researchgate.net/publication/337241678
191. Wilson J, Yoffe A (1969) The transition metal dichalcogenides discussion and interpretation of the observed optical, electrical and structural properties. Adv Phys 18:193–335
192. Palummo M, Bernardi M, Grossman JC (2015) Exciton radiative lifetimes in two-dimensional transition metal dichalcogenides. Nano Lett 15:2794–2800
193. Moody G, Dass CK, Hao K, Chen CH et al (2015) Intrinsic homogeneous linewidth and broadening mechanisms of excitons in monolayer transition metal dichalcogenides. Nat Commun 6:8315–8316
194. Oliveira V, Cremer D, Kraka E (2017) The many facets of chalcogen bonding: described by vibrational spectroscopy. J Phys Chem A 121:6845–6862
195. Politzer P, Murray JS, Clark T (2010) Halogen bonding: an electrostatically-driven highly directional noncovalent interaction. Phys Chem Chem Phys 12:7748–7757
196. Clark T, Hennemann M, Murray J, Politzer P (2007) Halogenbonding: the σ-hole. J Mol Model 13:291–296
197. Politzer P, Murray J, Concha M (2007) Halogen bonding and the design of new materials: organic bromides, chlorides and perhaps even fluorides as donors. J Mol Model 13:643–650
198. Pascoe DJ, Ling KB, Cockroft SL (2017) The origin of chalcogen-bonding interactions. J Am Chem Soc 139:15160–15167
199. Ibrahim MAA, Safy MEA (2019) A new insight for chalcogen bonding based on point-of-charge approach. In: Phosphorus, sulfur, and silicon and the related elements, vol 194, pp 444–454
200. Bader RFW (1990) Atoms in molecules. Oxford University Press, New York
201. Savin A, Nesper R, Wengert S, Fassler TF (1997) ELF: the electron localization function angew. Chem Int Ed Engl 36:1808–1832
202. Cremer D, Kraka E (1984) Chemical bonds without bonding electron density—does the difference electron-density analysis suffice for a description of the chemical bond? Angew Chem Int Ed Engl 23:627–628

203. Dronskowski R, Bloechl PE (1993) Crystal orbital Hamilton populations (COHP): energy-resolved visualization of chemical bonding in solids based on density functional calculations. J Phys Chem 97:8617–8624
204. Matsunaga T, Yamada N, Kojima R, Shamoto S et al (2011) Phase-change materials: vibrational softening upon crystallization and its impact on thermal properties. Adv Funct Mater 21:2232–2239
205. Reed AE, Curtiss LA, Weinhold F (1988) Intermolecular interactions from a natural bond orbital, donor-acceptor viewpoint. Chem Rev 88:899–926
206. Weinhold F, Landis CR (2005) Valency and bonding: a natural bond orbital donor–acceptor perspective. Cambridge University Press, London
207. Gaspard JP (2016) Structure of covalently bonded materials: from the peierls distortion to phase-change materials. C R Phys 17:389–405
208. Wuttig M, Lusebrink D, Wamwangi D, Welnic W et al (2007) The role of vacancies and local distortions in the design of new phase-change materials. Nat Mater 6:122–128
209. Jung Y, Ji E, Capasso A, Lee GH (2019) Recent progresses in the growth of two-dimensional transition metal dichalcogenides. J Korean Ceram Soc 56:24–36
210. Chhowalla M, Shin HS, Eda G, Li LJ et al (2013) The chemistry of two dimensional layered transition metal dichalcogenide nanosheets. Nat Chem 5:263–275
211. Arul NS, Nithya VD (2019) Two dimensional transition metal dichalcogenides synthesis, properties, and applications. Springer Nature, Singapore
212. Choi W, Choudhary N, Han GH, Park J (2017) Recent development of two-dimensional transition metal dichalcogenides and their applications. Mater Today 20:116–130
213. Frindt RF (1966) Single crystals of MoS2 several molecular layers. Thick J Appl Phys 37:1928–1929
214. Joensen P, Frindt RF, Morrison SR (1986) Single-Layer MoS_2. Mater Res Bull 21:457–461
215. Nicolosi V, Chhowalla M, Kanatzidis MG, Strano MS et al (2013) Liquid exfoliation of layered materials. Science 340:1420
216. Zeng ZY, Yin ZY, Huang X, Li H et al (2011) Single-layer semiconducting nanosheets: high-yield preparation and device fabrication. Angew Chem Int Ed 50:11093–11097
217. Zeng Z, Sun T, Zhu J, Huang X et al (2012) An effective method for the fabrication of few-layer-thick inorganic nanosheets. Angew Chem Int Ed 51:9052−9056
218. Gutiérrez HR, Perea-López N, Elías AL, Berkdemir A et al (2013) Extraordinary room-temperature photoluminescence in triangular WS_2 monolayers. Nano Lett 13:3447–3454
219. Yin X, Ye Z, Chenet DA, Ye Y et al (2014) Edge nonlinear optics on a MoS2 atomic monolayer. Science 344:488–490
220. Lv R, Robinson JA, Schaak RE, Sun D et al (2015) Transition metal dichalcogenides and beyond: synthesis, properties, and applications of single- and few-layer nanosheets. Acc Chem Res 48:56–64
221. Hua H, Zavabeti A, Quana H, Zhua W et al (2019) Recent advances in two-dimensional transition metal dichalcogenides forbiological sensing. Biosens Bioelectron 142:111573–15
222. He J, Chai Y, Liao L (2018) Focus on 2D materials beyond graphene. Nanotechnology 29:010202–4
223. Huang HH, Fan X, Singh DJ, Zheng WT (2020) Recent progress of TMD nanomaterials: phase transitions and applications. Nanoscale 12:1247–1268
224. Sarkar AS, Stratakis E (2020) Recent advances in 2D metal monochalcogenides. Adv Sci 7:2001655–2001736
225. Kolobov AV, Tominaga J (2016) Two-dimensional transition-metal dichalcogenides. Springer International Publishing, Switzerland
226. Pasquier D, Yazyev OV (2019) Crystal field, ligand field, and interorbital effects in two-dimensional transition metal dichalcogenides across the periodic table. 2D Mater 6:025015
227. Kertesz M, Homann R (1984) Octahedral vs. trigonal-prismatic coordination and clustering in transition-metal dichalcogenides. J Am Chem Soc 106:3453–3460
228. Huisman R, De Jonge R, Haas C, Jellinek F (1971) Trigonal-prismatic coordination in solid compounds of transition metals. J Solid State Chem 3:56–66

229. Duerloo KAN, Li Y, Reed EJ (2014) Structural phase transitions in two-dimensional Mo- and W-dichalcogenide monolayers. Nat Commun 5:1–9
230. Whangbo MH, Canadell E (1992) Analogies between the concepts of molecular chemistry and solid-state physics concerning structural instabilities. Electronic origin of the structural modulations in layered transition metal dichalcogenides. J Am Chem Soc 114:9587–9600
231. Choi JH, Jhi SH (2018) Origin of distorted 1T-phase ReS2: first-principles study. J Phys Condens Matter 30:105403
232. Wang Y, Li Y, Chen Z (2015) Not your familiar two dimensional transition metal disulfide: structural and electronic properties of the PdS2 monolayer. J Mater Chem C 3:9603–9608
233. Yang H, Kim SW, Chhowalla M, Lee YH (2017) Structural and quantum-state phase transitions in van der Waals layered materials. Nat Phys 13:931–937
234. Santosh K, Zhang C, Hong S, Wallace RM et al (2015) Phase stability of transition metal dichalcogenide by competing ligand field stabilization and charge density wave. 2D Mater 2:035019
235. Englman R (1972) The Jahn-Teller effect in molecules and crystals. Wiley-Interscience, New York
236. Grasso V (1986) Electronic structure and electronic transitions in layered materials. Springer Science & Business Media, vol 7
237. Kang J, Tongay S, Li J, Wu J (2013) Monolayer semiconducting transition metal dichalcogenide alloys: stability and band bowing. J Appl Phys 113:143703
238. Gong C, Zhang H, Wang W, Colombo L et al (2013) Band alignment of two-dimensional transition metal dichalcogenides: application in tunnel field effect transistors. Appl Phys Lett 103:053513
239. Chang J, Register LF, Banerjee SK (2013) Atomistic full-band simulations of monolayer MoS_2 transistors. Appl Phys Lett 103:223509
240. Liu L, Kumar SB, Ouyang Y, Guo J (2011) Performance limits of monolayer transition metal dichalcogenide transistors. IEEE Trans Electron Dev 58:3042–3047
241. Ilatikhameneh H, Tan Y, Novakovic B, Klimeck G et al (2015) Tunnel field-effect transistor sin 2-D transition metal dichalcogenide materials. IEEE J Explor Solid-State Comput Devices Circuits 1:12–18
242. Shi H, Pan H, Zhang YW, Yakobson BI (2013) Quasiparticle band structures and optical properties of strained monolayer $MoS_2$and WS_2. Phys Rev B 87:155304–155308
243. Peelaers H, Van de Walle CG (2012) Effects of strain on band structure and effective masses in MoS_2. Phys Rev B 86:241401–241405
244. Yun WS, Han S, Hong SC, Kim IG et al (2012) Thickness and strain effects on electronic structures of transition metal dichalcogenides: 2H-M X_2 semiconductors (M = Mo, W; X = S, Se, Te). Phys Rev B 85:033305–033315
245. Han S, Kwon H, Kim SK, Ryu S et al (2011) Band-gap transition induced by interlayer van der Waals interaction in MoS_2. Phys Rev B 84:045409–045416
246. Cheiwchanchamnangij T, Lambrecht WR (2012) Quasiparticle band structure calculation of monolayer, bilayer, and bulk MoS_2. Phys Rev B 85:205302–205304
247. Roldán R, López-Sancho MP, Guinea F, Cappelluti E et al (2014) Momentum dependence of spin–orbit interaction effects in single-layer and multi-layer transition metal dichalcogenides. 2D Mater 1:034003
248. Cappelluti E, Roldán R, Silva-Guillén J, Ordejón P et al (2013) Tight-binding model and direct-gap/indirect-gap transition in single-layer and multilayer MoS_2. Phys Rev B 88:075409–075418
249. Zhu ZY, Cheng YC, Schwingenschlögl U (2011) Giant spin-orbit-induced spin splitting in two-dimensional transition-metal dichalcogenide semiconductors. Phys Rev B 84:153402–153405
250. Zhang C, Gong C, Nie Y, Ah Mi K (2017) Systematic study of electronic structure and band alignment of monolayer transition metal dichalcogenides in Van der Waals heterostructures. 2D Mater 4:015026
251. Blöchl PE (1994) Projector augmented-wave method. Phys Rev B 50:17953–17979

252. Kresse G, Furthmüller J (1996) Efficiency of ab-initio total energy calculations for metals and semiconductors using a plane-wave basis set. Comput Mater Sci 6:15–50
253. Ceperley DM, Alder B (1980) Ground state of the electron gas by a stochastic method. Phys Rev Lett 45:566–569
254. Perdew JP, Burke K, Ernzerhof M (1996) Generalized gradient approximation made simple. Phys Rev Lett 77:3865–3868
255. Coehoorn R, Haas C, De Groot R (1987) Electronic structure of $MoSe_2$, MoS_2, and WSe_2. II. The nature of the optical band gaps. Phys Rev B 35(12):6203–6206
256. Coehoorn R, Haas C, Dijkstra J, Flipse C et al (1987) Wold, electronic structure of $MoSe_2$, MoS_2, and WSe_2. I. band-structure calculations and photoelectron spectroscopy. Phys Rev B 35(12):6195–6202
257. Mattheiss L (1973) Band structures of transition-metal-dichalcogenide layer compounds. Phys Rev B 8(8):3719–3740
258. Zhuang HL, Hennig RG (2013) Single-layer group-III monochalcogenide photocatalysts for water splitting. Chem Mater 25:3232–3238
259. Toroker MC, Kanan DK, Alidoust N, Isseroff LY et al (2011) First principles scheme to evaluate band edge positions in potential transition metal oxide photocatalysts and photoelectrodes. Phys Chem Chem Phys 13:16644–16654
260. Yazyev OV, Kis A (2015) MoS_2 and semiconductors in the flatland. Mater Today 18:20–30
261. Nie Y, Hong S, Wallace RM, Cho K (2016) Theoretical demonstration of the ionic Barristor. Nano Lett 16(3):2090–2095
262. Bardeen J (1947) Surface states and rectification at a metal semi-conductor contact. Phys Rev 71:717–727
263. Zhang J, Wang J, Chen P, Sun Y et al (2016) Observation of strong interlayer coupling in MoS2/WS2 heterostructures. Adv Mater 28:1950–1956
264. Tongay S, Fan W, Kang J, Park J et al (2014) Tuning interlayer coupling in large-area heterostructures with CVD-grown MoS2 and WS2 monolayers. Nano Lett 14:3185–3190
265. Ceballos F, Bellus MZ, Chiu H-Y, Zhao H (2015) Probing charge transfer excitons in a MoSe2–WS2 van der Waals heterostructure. Nanoscale 7:17523–17528
266. Palummo M, Bernardi M, Grossman JC (2015) Exciton radiative lifetimes in two-dimensional transition metal dichalcogenides. Nano Lett 15:2794–2800
267. Chaves A, Azadani JG, Alsalman H, da Costa DR et al (2020) Bandgap engineering of two-dimensional semiconductor materials. 2D Mater Appl 4:1–29
268. Kang M, Kim B, Ryu SH, Jung SW (2017) Universal mechanism of bandgap engineering in transition-metal dichalcogenides. Nano Lett 17:1610–1615
269. Kumar A, Ahluwalia PK (2013) Semiconductor to metal transition in bilayer transition metals dichalcogenides MX_2 (M = Mo, W; X = S, Se, Te). Model Simul Mater Sci Eng 21:065015–065115
270. Liu Y, Stradins P, Wei S-H (2016) Van der Waals metal-semiconductor junction: weak fermi level pinning enables effective tuning of Schottky barrier. Sci Adv 2:1600069

Chapter 2
Fabrication Approaches of TMDs

2.1 Introduction

In recent years, much attention has been given to two-dimensional (2D) materials owing to their kinds of exciting physical and chemical properties. Despite of different kinds of 2D materials (like graphene), the vibrant research devoted on a new family of materials namely 2D transition metal dichalcogenides (TMDs) (typically these materials have been categorized beyond the graphene) to serve exciting, extraordinary physical and electronic properties that is useful in the kinds of potential applications based on these materials [1, 2]. Typically the layered TMDs materials are composed of two different groups of elements like transition metal (M) (groups 4–10) and a chalcogen (X) element such as sulfur (S), selenium (Se), or tellurium (Te).

TMDs are also exhibit different stimulating properties depending on their compositional variations and their natural behavior, such as semiconductors, metals, semimetals, and superconductors. Simulating properties of the TMDs are extensively dependent on compositional crystallographic orientations and stacking sequences within the formed crystals or thin films. Hence, the monitoring of the growth condition by controlling various processes and parameters, kinds of TMDs have been fabricated for the distinct uses, such as sensing, energy storage, energy conversion, catalysis, a memory device, etc. To make different kinds of TMDs various synthesis processes have been discussed in the current decade. Investigators not only paid emphasis on the synthesis process but also associated exciting properties due to their different electronic and physical structures. Several novel approaches have been demonstrated to fabricate 2D TMDs by describing exotic functionalities with the leading materials for their potential application in electronics and optoelectronics. Such as the binary materials are MoS_x, $MoSe_x$, WS_x, WSe_x, VS_x, and VSe_x, including ternary compositions of the TMDs [3, 4].

TMDs materials mainly involve the exfoliation process by the removal of layered bulk materials into a layer-by-layer stacking. Especially the involved forces facilitate the exfoliation either inside or outside, namely, intercalation-based exfoliation and or

A. K. Singh, *2D Transition-Metal Dichalcogenides (TMDs): Fundamentals and Application*, Materials Horizons: From Nature to Nanomaterials,
https://doi.org/10.1007/978-981-96-0247-6_2

direct exfoliation. Such liquid exfoliations significantly classified into two categories, typically first is classified as intercalation based on the exfoliation and second one is called direct exfoliation. First one is considered much easier than the second one owing to fact that the first process is derived by the force between two adjacent layers, whereas the second one is governed by exerted force on the basal plane or edges. In the exfoliation process predominately the thermodynamics and kinetics reactions mechanism dominate the kinetics. However, it is considered to be confusing due to uncertain reactions during the exfoliation processes.

Composition Li_xTS_2 has been also used as a common precursor for intercalation-based exfoliation. In which precursors Li_xTS_2 immersed into H_2O to react intercalated Li with H_2O and generates H_2, therefore, the reaction mechanism has been defined as Eq. 2.1, when the forces of the layers to go apart.

$$Li + H_2O \rightarrow LiOH + H_2O \tag{2.1}$$

Since Li, Na, and K vigorously react with H_2O, this means the larger atomic number substance reacts more vigorously than lighter one. To follow this general rule, it is obvious sodium and potassium-intercalated TMDs can be more easily exfoliates into monolayer or multilayer nanosheets. However, it has been also well-examined that the potassium-intercalated MoS_2 cannot be converted into a single layer or multiple layers after exfoliation. Therefore, only the reaction between the intercalated metal ions and H_2O is insufficient to explain the process. Hence, under these circumstances it would be better to reaction mechanism should be comprehensively defined with the available data evidence.

Like direct exfoliation, the sonication or shearing-assisted exfoliations are exhibited in Fig. 2.1. But such process findings have revealed there is no exact reaction mechanism to define interactions between the solvents and TMD. To overcome these shortcomings scientific community has made effort to focus on the thermodynamics of the solvents during exfoliation. By paying additional attention to the surface tension of the solvent and TMD match, to make exfoliation effective, but it is also not a universal rule for all cases. Additionally, the Hildebrand and Hansen's parameters have been also recognized as the additional factors to select the solvents. Collectively all these lead solvent and TMD match should have same cohesive energy. Due to fact that solvents are playing an important role in the direct exfoliation process, but still there is no general rule for the selection of the solvents.

In general are two key strategies frequently employed to fabricate TMD materials, namely the bottom-up approach and the top-down approach [3–5]. Particularly, in the top-down approach the MX_2 layers generally developed from their bulk materials via exfoliation. Typically, the top-down approach is applicable for the macroscopic scale when the initial material in a large dimension and processed externally to get nanostructure geometry from different technologies, like mechanical exfoliation, wet chemical exfoliation, ball milling, etc. On the other hand, the MX_2 layers on the targeted substrates from the elemental precursors can be developed from the bottom-up approach. It means, in the bottom-up approach miniaturization of materials up to their atomic level, and a large area uniform layer govern by the self-assembly

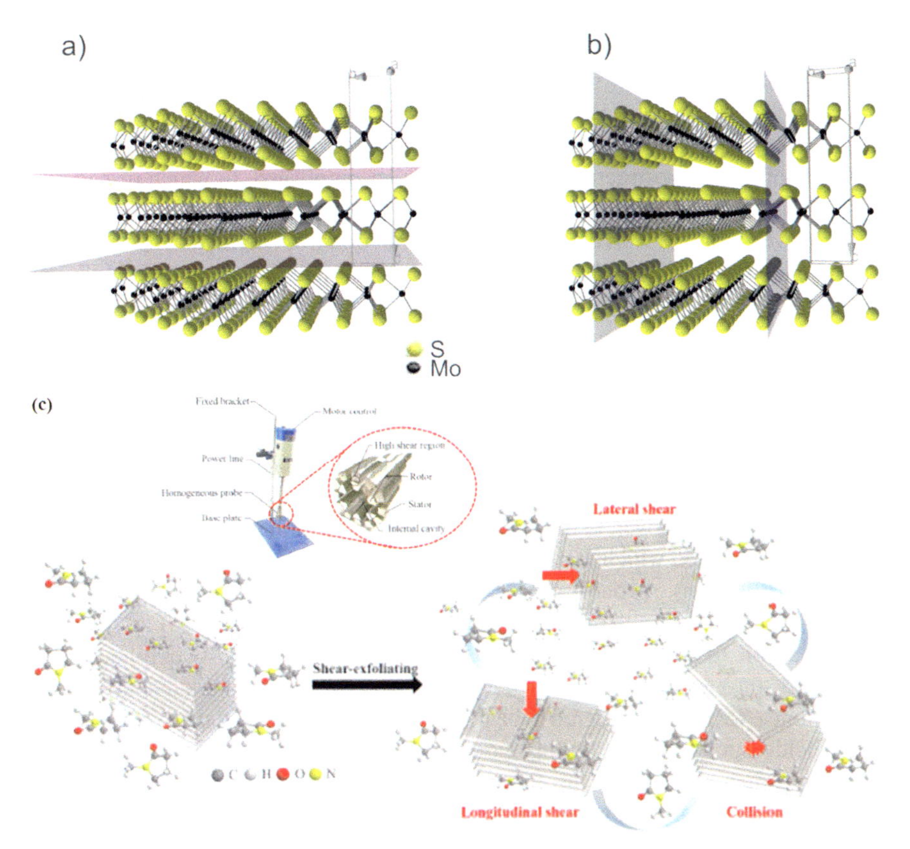

Fig. 2.1 Effects of sonication on the basal plane: **a, b** Mo–S–Mo bond during exfoliation in IPA (adopted from Lucia, S. C. et al. 2015 J. Phys. Chem. C 119, 3791–3801); **c** Schematic to shearing-assisted exfoliation of MoS_2 in NMP (adopted from Li, Y. et al. 2018 Ind. Eng. Chem. Res. 57, 2838–2846)

process through different synthetic techniques, like wet chemical synthesis, chemical vapor deposition (CVD), electrodeposition, etc. In general high-quality layered thin nanomaterials are produced by bottom-up technique for efficient applicability having great electronic properties and defect-free structure. Besides the versatile 2D material's successful developments, the bottom-up strategies are also not free from certain limitations in their large scale-production, due to their expensive synthesis process [3–5].

On the other hand, the key advantage of the top-down approach, it can easily handle, cheaper than the bottom-up approach, but it has a major obstacle to formation of poor material, it remains a great challenge for their industrial application. Therefore, it is worth to provide a discussion on both approaches of 2D TMDs preparation by demonstrating the various methods, like mechanical exfoliation, wet chemical exfoliation, chemical vapor deposition, electrodeposition, atomic layer deposition (ALD), etc. Discussion is not limited to these but also interpreted different process parameters that influence the structures and morphologies of the synthesized materials. By incorporating all these parameters of discussed two approaches a large number of TMDs monolayers have been synthesized, as illustrated in Fig. 2.2 [6, 7].

The history of the TMD materials synthesis started around 1985 when the synthesis of the heterostructures of TMDs was accomplished. In which researchers were studied the preliminary step of atomically thin van der Waals heterostructures [8]. However, around 1960s the mechanical exfoliation was employed to produce and investigated the properties of few-layer van der Waals materials, including MoS_2 and niobium diselenide ($NbSe_2$) [9–12]. By adopting the micromechanical exfoliation technique which is easiest and fastest process of individual crystal platelets for the different layered TMDs. Likewise, rubbing a fresh MoS_2 surface to another surface and its layered structure produced in form of monolayers. This process formed monolayer's monocrystallinity remains unaltered under ambient conditions without degradation in a long period. However, this process formed monolayer has a pronounced minority among thicker flakes; it is considered to be a major hurdle for large-scale production [13]. Moreover, to resolve such issues separation process of the colloidal solution was adopted which facilitates more efficient separation in terms of size (by mean centrifugation). As defined the weak van der Waals bonding (roughly ranging from 40 to 70 meV) and the surface tension (varying from 60 to 90 mJ m^{-2}).

In addition to all these the bulk parent crystals can be also exfoliated into nanolayers through the aforementioned approaches. The remarkable N-Methyl pyrrolidone (NMP) is considered to be most effective solvent. Besides NMP offering good yields but their high boiling point makes its removal difficult, so that the solvent residues remains in the final product [14, 15]. However, this exfoliation produced nanolayers possess high surface areas and aspect ratios. The aspect ratio is a crucial parameter for nanocomposites because it efficiently affects crack blunting, cracking, crack deflection, and microcracking [16–18]. Therefore, the variation of the aspect ratio could manipulate from the level of exfoliation and intercalation [19]. Commonly, a high degree of nanolayer exfoliation is revealed a higher aspect ratio and a greater number of nanolayers in a matrix [20]. Moreover, to get colloidal solution usually chemical modification/treatment and chemical intercalators are required for the interlayer repulsive forces that also stabilize the dispersion and distribution. Since intercalators also hinder the restacking of layers, therefore, overall repulsive forces are improved from the ultrasonication treatment. Additionally the power and duration of ultrasonication are also considered to be very crucial owing to breakage and fracture occur during the delamination of stacked layers [9, 21, 22]. To resolve this issue Coleman et al. [22, 23] have proposed direct liquid-based exfoliation. They showed the bulk TMDs like WS_2, MoS_2, molybdenum diselenide ($MoSe_2$), molybdenum ditelluride ($MoTe_2$), tantalum diselenides ($TaSe_2$), and $NbSe_2$ is well converted into nanolayers under the employed direct sonication-assisted exfoliation with the common solvents. Among them, NMP and isopropyl alcohol are also considered to be effective and favorable processes with the minimal decrease in the energy of exfoliation [24]. The significant advantages of such approaches they have not usually accompanied to any chemical steps. Moreover, these processes are also not air-sensitive and their yields impressive [25]. Similarly, Kalantar-Zadeh et al.

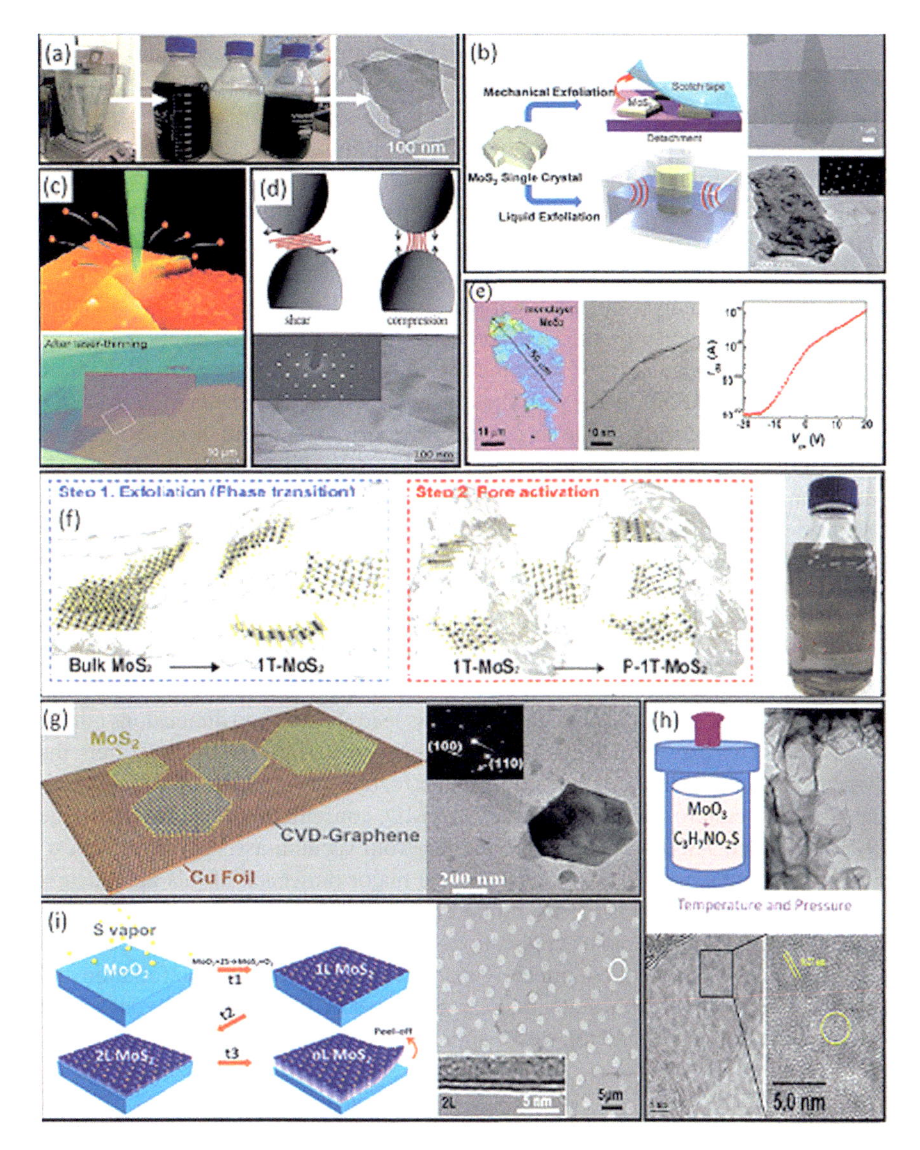

Fig. 2.2 Various conventional methods to synthesize TMDs: **a** Liquid exfoliation from the kitchen blender (adopted from Varrla, E. et al. 2015 Chem. Mater. 27, 1129–1139); **b** Mechanical exfoliation and liquid exfoliation from the scotch tape and ultrasonication process; **c** Thermal ablation by lasers (adopted from Castellanos-Gomez, A. et al. 2012 Nano Lett. 12, 3187–3192); **d** Exfoliation from the low-energy ball milling process; **e** Electrochemical exfoliation (adopted from Liu, N. et al. 2014 ACS Nano 8, 6902–6910); **f** Fluid dynamic exfoliation to ensure ultrahigh volumetric capacitance and capability (adopted from Jeong, J.M. et al. 2020 ACS Appl. Energy Mater. 3, 12078–12087); **g** van der Waals epitaxial growth (adopted from Shi, Y. et al. 2012 Nano Lett. 12, 2784–2791); **h** The hydrothermal method (adopted from Veeramalai, C. P. et al. 2016 Appl. Sur. Sci.389,1017–1022); **i** Chemical vapor deposition (adopted from Wang, X. et al. 2013 J. Am. Chem. Soc. 135, 5304–5307) (**b** and **d**, adopted from Ahmadi, M. et al. 2020 J. Mater. Chem. A 8, 845–883)

[15] have also used the grinding-assisted exfoliation approach in a liquid medium by adopting the Yao et al. approach [26], and produces mono- and few-layer thick flakes from MoS_2. According to their argument the selection of grinding solvent is also playing a crucial role in the grinding step to optimize the lateral size and thickness of the final products. Likewise, in situ growth the combination of aging, freeze-drying, sol–gel process, solvothermal reaction and high-temperature carbonization approaches are also recognized alternative methods to fabricate monolayer films by the chemical reaction of the surface. The advantage of these methods, the formed surface acts as a catalyst or just a template. Using these approaches, neat single layers has been achieved by the isolation of accompanied thicker layers. However, such processes are also not free from the certain weakness like they usually require high temperature. Additionally with them challenges remain whether the achieved layers exist independently on the surface significantly. Overall, all these challenging shortcomings makes them less applicable techniques, specifically for the large-scale production [27].

Considering several drawbacks of the Top-down approach to producing homogeneous TMD nanolayers in terms of shape and composition, bottom-up chemical-based methods have gained much attention in cutting-edge [14]. In this process, solution is initially heated up to the desired temperature to proceed the reaction at which the sources of growth decompose into monomers [28]. Ultimately, the monomers reach the supersaturation state, results the crystals start to grow spontaneously. Moreover use of the organic surfactants leads the desired dimensions (shape and size) TMDs. These kinds chemical synthesis process of TMDs also offers the advantage to control shape, composition, and size, by controlling the reaction conditions like temperature, choice of sources of TMD, concentration, and reaction time [29, 30]. So that, TMD monolayer production from vacuum-assisted vapor deposition on metal substrates is also available. The major drawback of this approach is the areas of the produced layer are limited to several nm^2, which is considered to be a key hurdle for the most of applications [21]. All these issues discussed based on the samples which are predominately made from the mechanical exfoliation and stacking [8, 31–34]. Apart from them, a large number of TMDs have also developed by using both bottom-up and top-down approaches [9, 21, 35–38].

2.1.1 Top-Down Approach

The top-down approach also describes an exfoliation method that essentially involves peeling off single or a few layers of materials from bulk TMDs. This process can occur in the air through mechanical exfoliation and chemical or electrochemical methods. The top-down method includes exfoliation, micromechanical, ball milling, liquid phase exfoliation, and electrochemical exfoliation. Generally, synthesis protocols follow the heating of the pure precursor of corresponding elements in a fused silica ampoule at an appropriate temperature and portion. This is time-consuming and has a slow cooling rate process. Moreover, other popular techniques like ultrathin thio/

selenophosphates from the micromechanical exfoliation to achieve crystal quality and features for specific applications [5, 39].

2.1.1.1 Liquid Exfoliation

Liquid-phase exfoliation is also a useful approach for the large-scale, low-cost mass production of 2D materials for various applications. This approach achieved 2D materials are also useful to deposit on a variety of substrates, owing to the advantage the whole process is usually performed at a low temperature. The liquid-phase exfoliation process is also useful for thin-film transistors, inkjet-printed electronics, conductive electrodes, and nanocomposites [40–43]. The liquid exfoliation is also useful to perform the direct ultra-sonication. By employing this technique can be fabricated single-layer and multilayer nanosheets of a few layered TMD materials, such as MoS_2, WS_2, $MoSe_2$, $NbSe_2$, $TaSe_2$, $MoTe_2$, $MoTe_2$, etc. Besides, the numerous advantages of the liquid-phase exfoliation approach they have also a few significant shortcomings, like small grain size, high defect density, high possibility of contamination due to chemical groups, and possible phase transitions of the exfoliated TMDs. Despite of few significant shortcomings several liquid-phase exfoliation methods have been used in the recent past, such as solvent-based and ion intercalation methods [44, 45].

In recent years it has been also identified that liquid exfoliation is useful for the large-scale production of single and few-layered metal phosphorous tri chalcogenides (MPX_3) through exfoliation of the bulk crystals [46–55]. Usually in this process, the bulk crystals are dispersed in the kinds of solvents (like ethanol, acetone, water, proposal, ethanol–water mixture, and dimethyl formamide) and vigorously ultrasonicated, which results in the separation of the un-exfoliated bulk products. Likewise the ultrasonic liquid exfoliation allows to convert stacked bulk into a single nanosheet, under the action of the Ar/O_2 plasma treatment for the oxygen doping, as the typical case of $FePSe_3$ [56, 57]. Using the same way, the $Ni_{1-x}Fe_xPS_3$ bulk has also exfoliated from the ultrasonication-assisted approach by using N, N-dimethyl formamide (DMF) [57]. Moreover, the $MnPS_3$flakes are also exfoliated from the bulk materials on polymethyl methacrylate (PMMA) and polyvinyl alcohol (PVA)-coated SiO_2/Si substrates, by the removing the PMMA layer from acetone and subsequent rinsing of isopropanol alcohol [39, 58].

Solvent-Based Exfoliation Method

Predominately the solvent-based exfoliation process consist immersion, insertion, exfoliation, and stabilization. A schematic diagram of the solvent-based exfoliation procedures is illustrated in Fig. 2.3. To achieve efficient exfoliation from this process the solvents must be fully immersed with 2D material during sonication. Moreover, solvents should exfoliate the material at high concentrations by keeping restacking of the exfoliated 2D material. Additionally, it must satisfy the conditions of surface

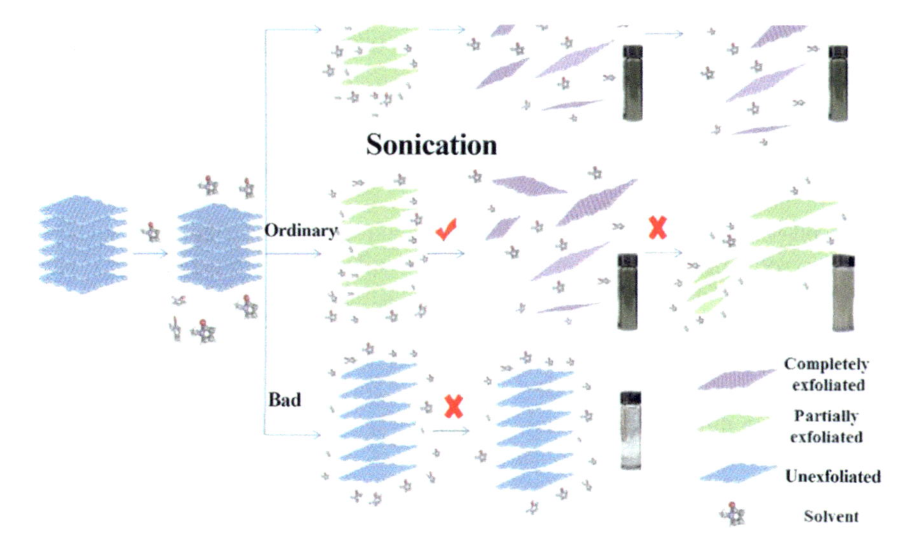

Fig. 2.3 Schematic for the solvent-based exfoliation (adopted from Shen et al. [44])

tension and Hildebrand and Hansen solubility parameters to select the proper solvent for a selected 2D material. Considering all these basic solvent requirements, IPA/water, acetone/water, and THF/water have been recognized best solvents. However, the optimal volumetric ratio of solvent and water depends on the characteristic of the 2D material. Such as, a 1:1 IPA/water ratio is favorable for the graphene, hBN, WS_2, $MoSe_2$, while 7:3 for the MoS_2 [44].

Ion Intercalation Method

The basic concept behind the ion intercalation process is the intercalation of the impurities in-between layers of bulk TMD crystal, to increase the interlayer spacing. The advantage of the increase in interlayer space, it can reduce the van der Waals force between the energy barrier and exfoliation. The intercalants could be alkali metals, organometallic, polymers, and atomic species. Especially, lithium-ion is considered to be a good material owing to its high reduction potential and high mobility ability. Intercalation in lithium, such as n-butyllithium (n-BuLi) solution in hexane has been frequently used for the TMDs. Where n-Bu acts as a transfer of an electron to TMD layers and Li^+ is responsible for ion intercalation of the charge balance. To improve the lithium intercalation efficiency usually, ultrasonication or microwaves is used. Whereas the lithium-ion intercalated TMD bulk crystal is generally exfoliated from the hydrolyzing and sonication processes. A typical lithium-intercalated exfoliation process schematic is illustrated in Fig. 2.4a, b [59]. The advanced Li-ion intercalation has been also achieved by using the electrochemical approach, as depicted in Fig. 2.4b. It has been establish that lithium intercalation from the electrochemical

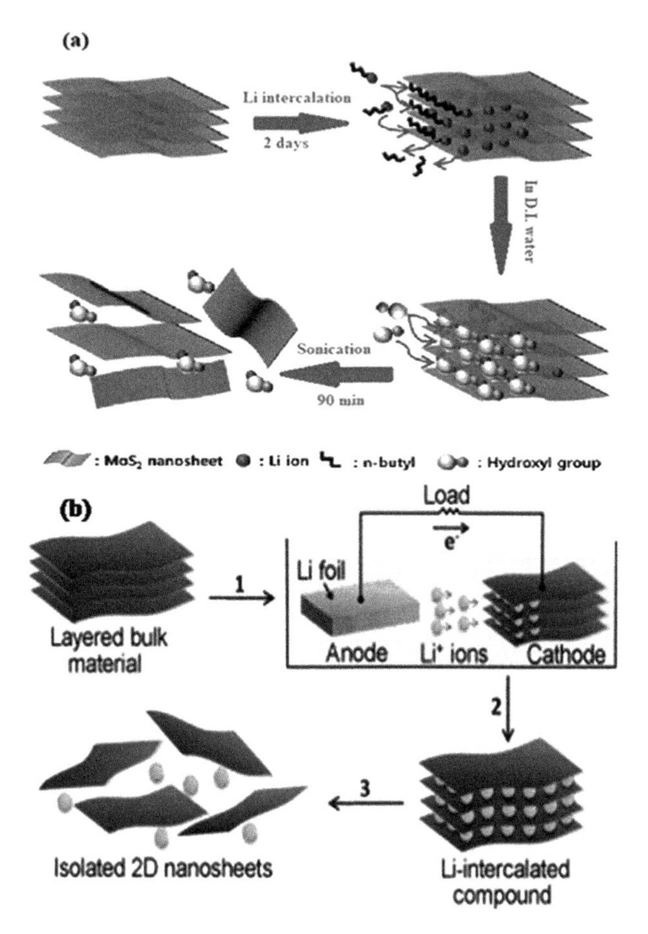

Fig. 2.4 Schematic illustration for the ion intercalation: **a** process of lithium intercalation and exfoliation (adopted from Kang, K. et al. 2014 Langmuir 30 98669873); **b** TMD material electrochemical lithium intercalation and exfoliation process (adopted from Ikram, M. et al. 2020. https://doi.org/10.5772/intechopen.94892)

approach is faster with the controllable manner [60]. To execute the lithium intercalation process, a voltage is required to apply between anodic lithium foil and cathodic bulk TMD in an electrolyte. Usually in this process, the Li^+ ions are placed between the TMD layers, and the resultant Li^+ ion-intercalated TMD bulk crystal generates the TMD nanosheets. Thus, the exfoliation process offers a structural deformation of the exfoliated material. Moreover, lithium-ion intercalation is also associated with charge transfer from n-BuLi to TMD crystal. Under influence of the charge transfer TMD structure is also changed from 2 H to 1 T phase [45]. The structural phase change is considered to be more favorable with the increasing lithium dosage. However, under the annealing process or infrared (IR) radiation exposure, the intercalation-induced phase transformation could be reversible [61, 62].

2.1.1.2 Mechanical Exfoliation

Mechanical exfoliation is one of the most efficient ways to produce the cleanest, highly crystalline, and atomically thin nanosheets of layered materials. First, this method was used to obtain the graphite monolayer; that is so-called graphene [13]. This process is very simple, cost-effective, and applicable to all kinds of van der Waals materials. The natural bulk crystal's high-quality monolayer has been achieved from this process [63]. Besides all these impressive features this process is time-consuming and non-scalable [64]. In the current era, various types of mechanical exfoliation methods have been used, like Scotch tape, ball milling, roll milling, gel-assisted exfoliation, metal-assisted exfoliation, and layer-resolved splitting (LRS) [65]. The key mechanical exfoliation methods; Scotch-tape, metal-assisted, and LRS methods are discussed in this section. Considering the fact, the Scotch-tape method is the representative method of mechanical exfoliation, LRS is the unique method for the wafer-scale monolayer from the mechanical exfoliation and the metal-assisted exfoliation is an intermediate process between mechanical and LRS approach.

Scotch-Tape Method

First-time Novoselov et al. got success to isolate a two-dimensional (2D) monolayer from the bulk crystal, by using the Scotch-tape process, as the typical procedure is illustrated in Fig. 2.5 [66, 67]. In this method, a bulk crystal of 2D material puts into the middle of the adhesive tape and it thins by the repetition of the folding and unfolding of the tape. The thinned 2D material layers/tape spread throughout the SiO_2 substrate. To increase the adhesion between the substrate and thinned 2D material layers a uniform pressure is applied. At the end of the process, the adhesive tape is usually gently remove and leaves the monolayers of 2D material on the substrate. This process fabricated TMD MoS_2 monolayer mobility has been measured it is in-between 0.5 and 3 cm^2/V s [68]. The introduction of this novel approach had attracted great attention to exploring different properties of the TMD materials, due to its easy, simple, and low fabrication cost. To explore the different properties of this process fabricated TMDs were characterized as distinct layers of the MoS_2 using the Raman spectroscopy technique. The findings showed that the Raman $E^1{}_{2g}$ mode spectral peak is in a decreasing order with the thickness, while the A_{1g} mode spectral peak is in an increasing order corresponding to thickness. The distinct frequencies between the Raman modes $E^1{}_{2g}$ and A_{1g} has been also achieved with varying thicknesses [69]. Subsequently, the successful fabrication of suspended MoS_2 monolayers has established the effective Young's modulus with an average breaking strength of 2,706,100 and 23 GPa [70]. Moreover, the direct bandgap of the MoS_2 has been defined as around 1.88 eV, whereas the improved mobility of the MoS_2 has demonstrated around 200 cm^2/Vs with a high on/off ratio of approximately 13,108. Additionally, their photo responsivity at a gate voltage of 50 V has measured 7.5 mA/W [65, 71, 72]. Indeed, innovations are not limited to only these outlined key developments, it is still ongoing.

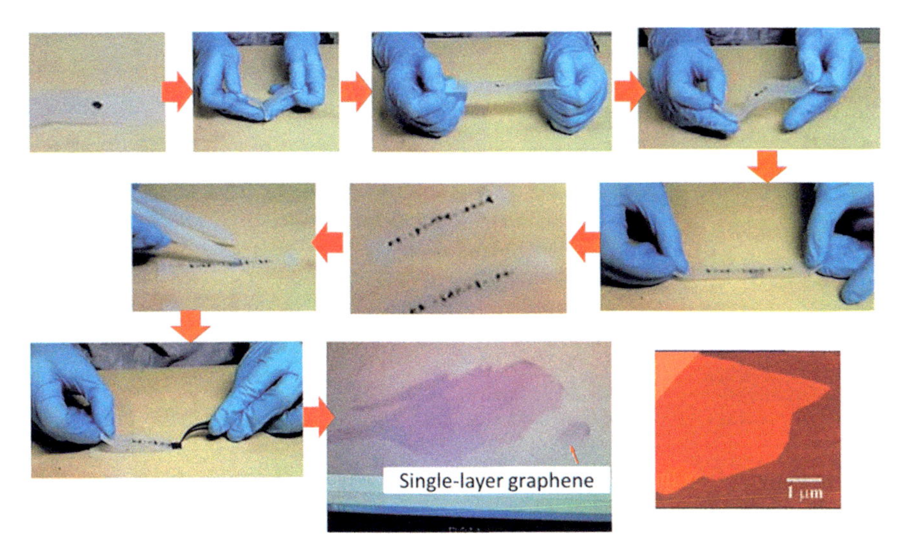

Fig. 2.5 Scotch-tape method procedure to produce a monolayer of 2D material (adopted from Qi et al. [67])

Metal-Assisted Approach

Besides the Scotch-tape method is very simple and promising approach for the production of monolayer TMD possessing distinct impressive features. It suffers to large area monolayer TMD production which is still challenging from this method. Therefore, the metal-assisted approach is considered to be a useful method to produce large flake-size monolayer TMDs. In the metal-assisted synthesis approach, adhesion between metal and TMD is a critical parameter. Considering several parameters gold has defined to be a suitable metal due to its strong semi-covalent interaction with top layer chalcogen atoms of the TMD. Hence, using this process the gold-assisted exfoliation is possible by the deposition of gold layer (100–150 nm), which ultimately combines with the thermal release tape, as the schematic is given in Fig. 2.6. In early investigations, it noted that the exfoliated monolayer MoS_2 is up to around 500μm^2. Moreover, this process gold exfoliated TMD surface powerful STEM imaging technique characterization has demonstrated that the distance between the gold surface and the attached top sulfur atom is around 3.5 Å for the MoS_2 [73]. Typically this investigation was established that the gap between the gold and sulfur surface atoms larger than their covalent bond of 2.2 Å. This outcome also revealed that the interaction between gold and sulfur atoms has a strong van der Waals bonding instead of the chemical bond interaction. Moreover, it has been also demonstrated that the binding energy between gold and MoS_2 is decreasing rapidly with the increasing distance. It reveals for the pristine gold layer a high-quality exfoliation is desired. Hence, these experimental findings are considered as evidence of exposed gold surface in the air for a long duration with a less transfer amount of MoS_2 monolayer. The missing gold

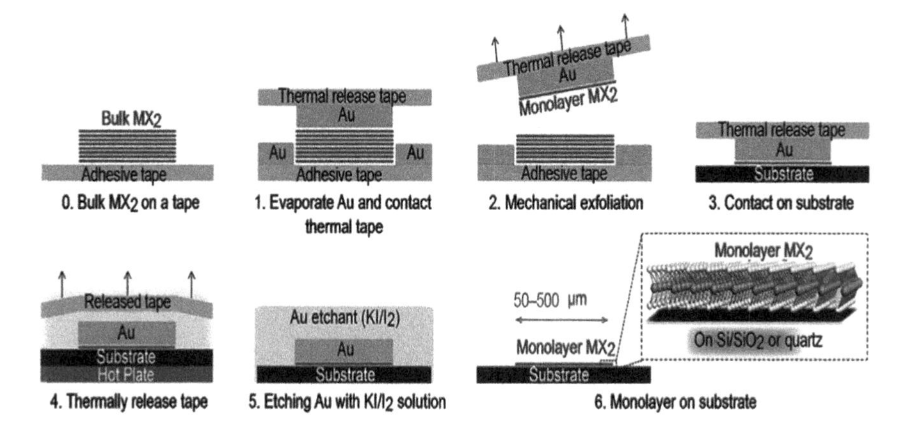

Fig. 2.6 Illustration for the gold-assisted exfoliation process (adopted from Zavabet, A. et al. 2020 Nano-Micro Lett. 12, 66)

atoms on the surface have been well described theoretically using the DFT simulation. These remarkable outcomes have also led the large vacancies in the gold surface to reduce the overall binding force between TMD and the assisting gold surface.

Layer-Resolved Splitting Approach

Since both Scotch-tape and metal-assisted exfoliation methods have difficulty with their scalability, therefore, investigators have paid attention to fulfilling the lack of the wafer-scale size exfoliation approach. This could answer with a new method, namely, it is called layer-resolved 2D material splitting technique [74]. This new approach is useful for the isolated 2-inch wafer-size WS_2 monolayer. Using this process performed the chemical vapor deposition (CVD) and grown WS_2 multilayer on a sapphire substrate for the exfoliation instead of the naturally synthesized bulk crystal. It has been found that the CVD-grown top layer is irregular and discontinuous, while the inner layers have uniform and continuous films. Moreover, the successive deposition of Ni thick film on top of the multilayer WS_2 WS_2 multilayer detached to the sapphire substrate. Subsequently, another thick Ni layer has also deposited on the bottom of the multilayer WS_2, to split the WS_2 monolayer one by one from the bottom. However, the layer-resolved splitting method developed film quality was found inferior to the Scotch tape method. This is because the used samples were grown by the CVD process instead of the naturally grown samples.

Therefore, an additional thinning method is desired to remove the layers of the produced thin film of TMD. To form few layers of the TMD flake on a substrate usually has been used the thermal energy or laser. The thermal energy process usually applied to sublimate TMD flake from the upper layer [75]. To perform the experiment a laser has also used to produce monolayer TMD by removal of the extra layers from exfoliated thick TMD. To achieve it Hu et al. have used 532 nm layer with 2.5 mW

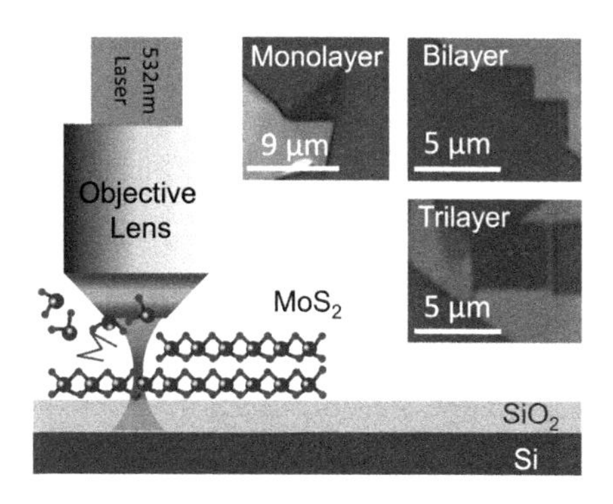

Fig. 2.7 Laser-assisted thinning process for the TMD. Typical device setup with the thinning outcomes for monolayer, bilayer, and trilayer (adopted from Hu et al. [77])

for the 0.2 s exposure time. Their findings showed that after 13 scans the original 10-layer MoS_2 reduced into monolayer MoS_2 with a size of around 10 μm [76]. They also defined the mechanism behind the laser thinning of the TMD multilayer as the thermal thinning in the case of the MoS_2 heated upto 603 K. They chose this heating temperature because it falls under the laser exposure as defined by the theoretical simulation (around 669 K). Under the control of the exposure time and power use of this approach is reduce a bulk MoS_2 upto a desired thickness. Moreover, this approach also enables to generate square pattern of MoS_2 trilayer, bilayer, and monolayer from bulk material, as depicted in Fig. 2.7. It has also confirmed that the thickness reduction of the MoS_2 recognized from the Raman technique. Specifically, in terms of the frequency interval of the $E^1{}_{2g}$ and A_{1g}modes Raman spectral peaks that's around 25.3 cm^{-1} and 19.4 cm^{-1} corresponding to bulk and monolayers [77]. The key advantage of this method, it can provide the position and structure control of monolayer TMD. Hence, this approach would be useful for the fabrication of potential novel devices with structural control.

2.1.1.3 Electrochemical Exfoliation

Electrochemical exfoliation or electrodeposition technique is the convenient process for the production of the metallic or semiconducting thin films on a base material or a substrate by the electrochemical reduction of metallicions from electrolytes. In the modern world, this process is considered an emerging and flexible technique for the fabrication of thin film with the reduce grain size. To initiate electrodeposition process generally a cathode (working electrode) and an anode (counter electrode) are desired to immerse into an electrolyte-containing cell/bath. The electrochemical exfoliation occurs owing to charge transfer reaction under the influence of the external potential across the substrate/electrolyte interface. The typical schematic diagram of an electrochemical exfoliation process setup consist working electrode/cathode, anode, electrolyte-containing bath, current source, and ampere/voltmeter is depicted in Fig. 2.8.

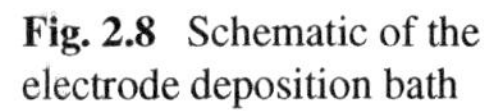

Fig. 2.8 Schematic of the electrode deposition bath

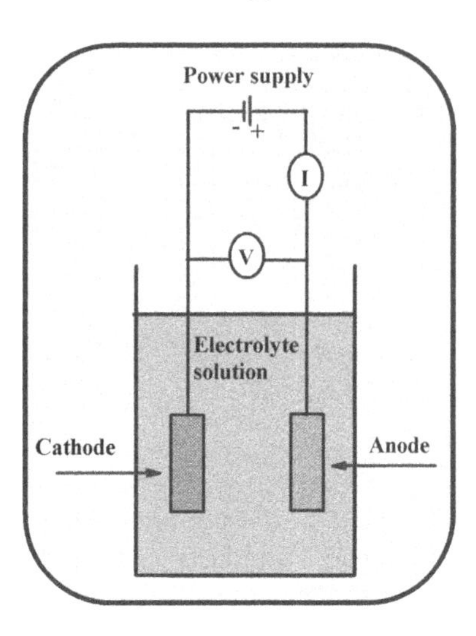

Moreover, the electrochemical exfoliation technique is an adequate, fast, simple, and inexpensive process to produce different kinds of nanostructured materials, such as thin films, nanoparticles, nanowires, nanocrystalline materials, and nanocomposites possessing controlled characteristics parameters, like composition, geometry, and morphology [78]. In this process deposition occurs through pure metal deposition or co-deposition of the alloy from electrolyte under the impact of the applied external current. By employing this approach 2D and 3D nanostructures are possibly deposit on a targeted substrate. The significant advantage of this deposition technique is the controlled synthesis process of nanostructured materials to get desired geometry, morphology, and crystallographic orientation by the direct growth on a substrate under controlled operating conditions and bath chemistry [79]. Specifically, this technique ability like highly controllable, repeatable and scalable nature of electrodeposition has attracted much attention for the fabrication of TMDs nanostructures accompanying a verity of exciting characteristics and efficient device functionality with low cost.

Such as the MoS_2 nanostructured has been prepared by employing different types of electrochemical techniques using the aqueous and non-aqueous cathodic electrodeposition process [4]. In a simple case, the DC bias has applied between MoS_2 and the Pt wire to achieved the electrochemical exfoliation, starting with a low positive bias to wet the bulk MoS_2 followed by a larger bias to exfoliate the crystal. As the consequence, several MoS_2 flakes dissociated from the bulk crystal and became suspended in the solution (see Fig. 2.9). Under the external positive bias application to the working electrode, the oxidation of water produces –OH and –O radicals that assembled around the bulk MoS_2 crystal. The radicals and/or $SO^{2-}{}_4$ anions insert themselves between the MoS_2 layers and weaken the vdW interactions between the layers. Whereas the oxidation of the radicals and/or anions releases the O_2 and/or SO_2, therefore, the MoS_2 interlayers greatly expand. Ultimately, MoS_2

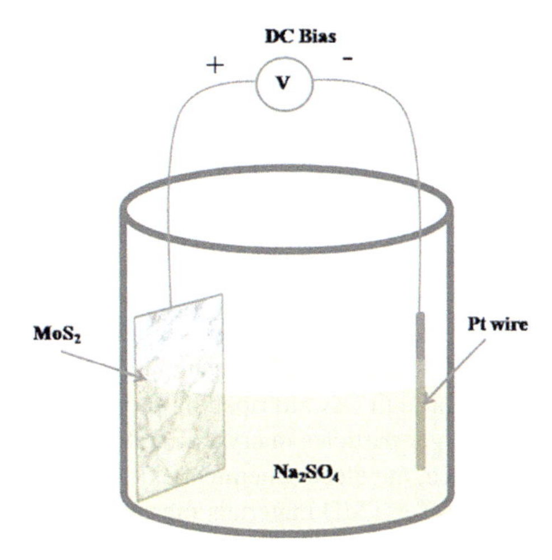

Fig. 2.9 Schematic setup to electrochemical exfoliation of bulk MoS_2 crystal

flakes detached from the bulk MoS_2 crystal by the erupting gas, and in the end, it is suspended in the solution [4].

A similar process is also used by Albu-Yaron et al., they have synthesized the MoS_2 through both aqueous (water) and non-aqueous (ethylene glycol) routes by using sodium tetrathiomolybdate ion at different temperatures on a conductive glass substrate [80]. The two-electron reduction of tetrathiomolybdate ions ($MoS_4{}^{2-}$) are produced MoS_2 by following the reaction mechanism.

$$MoS_4^{2-} + 2e^- + 4H^+ \rightarrow MoS_2 + 2H_2S \tag{2.2}$$

In this process potassium chloride and ammonium chloride are generally used as supporting electrolytes, whereas, as the proton donor the ethylene glycol containing electrolyte would be suitable. This innovation revealed the formation of an amorphous MoS_2 matrix surrounded to nanocrystallites and nanoclusters of MoS_2 under aqueous and non-aqueous electrolyte modes. Additionally, it also revealed the electrodeposition process are also widely influence the formation of a crystalline embryo under the varying temperature electrolytic medium, substrate, and operating voltage. Significantly, it has also remarked that after increasing the temperature structural quality might be deteriorated cause of the enhancement in various types of defects. Therefore, it is desired to control structural morphology under varying temperatures to achieve a high-quality defect-free structure with the sensational characteristics for prospective device applications.

Similarly, another remarkable study has shown the synthesis of MoS_2 using the electrolyzing aqueous solution of $(NH_4)_2[MoS_4]$, it lead the same amorphous MoS_2 formation is possible with the mixed phases of MoS_2 and MoS_3 [81]. Moreover, to avoid the formation of amorphous material and get a pure crystalline structure, a non-aqueous-bases electrochemical exfoliation technique has been considered suitable by using room temperature ionic liquids, specifically when precursors taken into

1:1 molar ratio. Typically such experiment performed on a glassy carbon substrate. Where the substrate dip into an electrochemical bath containing Pt as a reference electrode followed by potentiodynamic deposition (scan rate 100 mV/s, temperature range from room temperature to 100 °C) and chronoamperometric deposition (constant temperature, varying potential from 2.7 to 1 V). To continue this order, the temperature-dependent morphology growth at room temperature has been also examined for this process synthesized MoS_2 [82]. It showed that the initial amorphous nature and with the increasing temperature transformed into nanoparticles further agglomerated to form nanocluster material. Continue increase in temperature up to 100 °C the attached nanoparticles cluster forms a thin crystalline film of pure phase through Oswald ripening process, where the smaller nanoparticles merge and built larger particles in crystalline form.

Hence, the electrochemical exfoliation technique is one of the significantly used methods for TMD materials fabrication. Yet, a large number of TMDs are reported to produce using this approach. A few important TMDs materials which have developed from this approach using different fabrication techniques are summarized in Table 2.1.

Table 2.1 TMD preparation through CVD and their corresponding growth criteria with the types of structural growths

Materials	Precursors	Synthesis temperature (°C)	Substrate	Structural growth
MoS_2	Mo-coated SiO_2 and sulfur	750	Oxide-coated SiO2	Monolayer (hexagonal lattice)
	$(NH_4)_2MoS_4$	1000	Sapphire	Trilayer (polycrystalline)
	Ammonium thiomolybdate	400	Graphene-coated Cu foil	Single crystalline hexagonal flakes
WS_2	WO_x-coated Si/ SiO_2 substrate and sulfur	750–950	Oxide-coated Si/ SiO_2 substrate	Monolayer and multilayer stacked structure
VS_2	VCl_3 and sulfur	500–600	SiO_2/Si substrate	Nanosheets (single crystalline hexagonal 1T structure)
WSe_3	WO_3 and selenium powder	850	Sapphire	Monolayer flakes
$MoSe_2$	MoO_3 and selenium pellets	850	SiO_2/Si substrate	Monolayer in the single crystalline island of triangular geometry
VSe_2	VCl_3 and Se powder	500–600	Mica, sapphire	Single crystalline VSe2 nanosheets in triangular or hexagonal form

2.1.2 *Bottom-Up Approach*

Much progress has been made in the synthesis of TMDs from various techniques based on the bottom-up approach. Owing to fact that properties of as-prepared TMDs are frequently used for different synthetic processes. Like solution based wet chemicals getting are getting much attention to synthesize kinds of TMDs. However, most TMD materials have been synthesized using the chemical vapor deposition (CVD) method and it is considered an adequate technique for the synthesis [25]. Subsequently, other bottom-up deposition methods such as molecular beam epitaxy [26–28] magnetron sputtering [29–31] pulsed-laser deposition [32–34], and atomic layer deposition (ALD) [35–37], etc., have also employed for the direct synthesis of TMD layers [38]. Thus, in this section, our discussion is focused on these fabrication techniques of TMDs based on a bottom-up approach.

2.1.2.1 Wet Chemical Route

Since solution-phase reaction offer synthetic chemistry at low temperatures compared to the vapor-phase reaction route, resulting sophisticated nanostructures like quantum dots, nanorods, nanoparticles, etc. Their specific geometry can be controlled with help of the various process parameters and the final product may achieve predictably. Solution route synthesized high-quality nanostructured are also tailored more surprising properties at a low cost. Therefore, in recent years, much attention has been paid to the synthesis of TMDs through a solution-phase bottom-up approach. The properties of the as-prepared TMDs vary extensively with the use of different synthetic regimes, precursors, and stabilizing ligands, to giving rise TMDs with a wide range of morphologies, phases, and applications [83].

Hydrothermal and Solvothermal Process

Hydro/solvothermal methods refer to the synthesis of materials in closed vessels at high temperatures and pressure, it is usually carried out in the water (hydrothermal) or other non-aqueous solvents (solvothermal) [84, 85]. Hydro/solvothermal fabrication approaches are commonly used to synthesize a variety of nanomaterials. These methods are also known as the wet-chemical bottom-up tactic to synthesize TMD-based nanomaterials. Specifically solvothermal and hydrothermal approaches to synthesized nanoscale TMDs were introduced in the years 1998 and 2000. When Zhan et al. developed a solvothermal method for the production of nanocrystalline MoS_2 using pyridine as the solvent, and MoO_3 as the element sulfur precursor [86]. Their outcomes revealed the formation of thin platelets of semiconducting 2H MoS_2 haying an average diameter around 100 nm. Subsequently, Fan et al. followed this approach closely and successfully synthesized nanocrystallites of MoS_2 and $MoSe_2$ first time from the hydrothermal method [87]. They used Na_2MoO_4 as the Mo source and $Na_2S_2O_3$/Na_2SeSO_3 as the sulfur/selenium source. They confirmed the amorphous MoS_2 and $MoSe_2$ formation with on-average crystallite sizes of 4 and 7 nm. Further, they annealed at 350 °C for 9 h to get a semiconducting 2H phase for both MoS_2 and $MoSe_2$. In both cases, hydrazine was involved as the reducing agent in the reduction of Mo (VI) to Mo (IV). Therefore, the synthesis work paved the way to use of hydro/solvothermal methods for the bottom-up synthesis of ultrathin nanoscale TMDs.

The hydro/solvothermal methods of TMDs became much popular recently to produce nanostructures with different morphologies like nanosheets, nanotubes, nanorods, nanowires, and nanoflowers [88, 89]. Wang et al. investigated the self-assembled behavior of monolayer MoS_2 by using the humble solvothermal approach. They used octylamine as the solvent and ethanol as an additive [90]. They showed that the absence of ethanol results the MoS_2 nanosheets. While, the presence of ethanol leads hierarchical 3D tubular structures accompanied 2D nanosheets [83]. Moreover, It has also concluded that the formation of the tubular structure due to individual nanosheets, instead of a nanotube of restacked sheets. It could be due to blend factors, like, ethanol in amine is considered to be a poor solvent that may disrupt the interaction between nanosheets. However, layered nanomaterials normally formed nanotubes under the certain conditions. The use of octylamine is also helpful in capping ligands to prevent restacking of the sheets. Collectively these factors also gives raise the formation of tubular arrangements of nanosheets, rather than single crystalline MoS_2 nanotubes. Additionally, it has also demonstrated that the diameter and length of the tubular structures are possibly tuned by changing the nature of alcohol (for instance use of the butanol). Moreover, it has been also pointed out, this synthesis process is highly scalable, such as 2 g of MoS_2 hierarchical tubes being produced, hence this approach has advantages for potential commercial uses.

Hot-Injection Process

La Mer and Dinegar proposed a set of conditions in the 1950s that are considered necessary to synthesize the monodisperse colloidal solutions of nanoparticles. The defined conditions are allowing involvement of a rapid nucleation and a slower nuclei growth in a controlled manner to form larger particles [91]. These basic conditions motivate a different group of investigators for the development a new synthetic approach to colloidal QDs, like, cadmium sulfide, selenide, and telluride. All of them have the involvement of a rapid injection of organometallic precursors into vessels at high temperatures containing coordinating ligands and solvents [92]. Therefore, the term "hot-injection" has become common for the QDs synthesis. This is frequently used for a wide variety of QDs production, beyond the cadmium led QDs [93, 94]. Moreover, a numerous efforts have been also made to employ the hot-injection for the production of monodisperse TMDs. A small variation also impacts their stabilizing ligands, precursors, reaction time, and temperature, therefore, produced TMDs possess various phases, sizes, and morphologies.

2.1.2.2 Chemical Vapor Deposition (CVD) and Growth Mechanism

Chemical Vapor Deposition (CVD) synthetic approach is usually connected to bottom-up process for the preparation of vertical 2D heterostructures. Using this approach it is also possible to grow the planar multi-junction heterostructures. Such as an atomically sharp heterojunction and a clean interface in vertically stacked heterostructures has been fabricated. Hence this is useful technique for the production of scalable and controllable vertical and planar heterostructures, yet a variety of approaches has been explored for the growth of heterostructures using CVD process.

Typically, CVD is considered to be a useful method for the large-area deposition of atomically thin TMD on a comparatively matching substrate. Usually, TMDs growths are achieved from the CVD by following these two ways: sulfurization and selenization. In the both cases vapor-phase reactions occur by transition metal to sulfur and selenium. In general TMDs are grown with different morphologies and structural orientations such as monolayer, bilayer, and multilayer stacking possessing distinct crystallographic orientations and make single crystalline, polycrystalline, or pure crystalline forms.

In the CVD process, the vapor-phase reaction has occurred on a transition metal-coated substrate or bare substrate in a chalcogen (S or Se) atmosphere. This allows to reaction take place for the vapor phase of both transition metal and chalcogen. This technique's procedure parameters, such as deposition temperature, temperature ramp, carrier gas flow rate, and vacuum pressure are significantly influenced by the crystal structure, surface morphology, electronic property and orientation of the fabricated film. Thus, the key role of the CVD system is generation of vapor precursors of source material and it deliver to vapor form reactants through the reactor at the designated temperatures and pressure. Typical schematic of a three-zone CVD furnace is depicted in Fig. 2.10. Generally this system consist a reaction chamber which equipped with three zones pattern that useful for the placement of substrate and subsequent precursors, with the temperature controller heating system. Usually, the CVD process following the growth kinetics owing to adsorption of reactant species in a vapor phase which governs from the comparatively high decompose temperature of corresponding precursors of the substrate surface. The subsequent heterogeneous reaction gets initiated due to development of the freshly formed film on the substrate surface and by-product species. The formed by-product removes to the reactor chamber by following the diffusion process. Subsequently for the transportation of vaporized reactants on the targeted substrate in the reactor, the carrier gasses like H_2 and N_2 (reactive gases) or Ar (inert gases) are often employed followed by specific flow rate. Generally, sulfurization and selenization processes occur in the Ar and H_2 gases mixed atmosphere under a certain ratio which, respectively, participates in the nucleation and reduction process. This process also offers advantage of adjustable deposition rate; however, a low deposition rate is favorable to grow a high quality thin film. Using this technique well designed good quality epitaxial film with identical film thickness and decent surface morphology to be achieved.

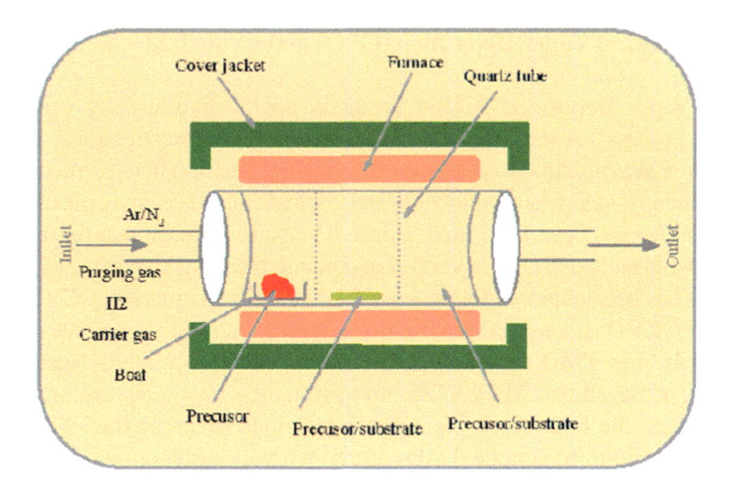

Fig. 2.10 Schematic for the typical CVD growth

One-Step CVD Approach

In this order Gong et al. [95] have successfully demonstrated the vertical and planar MoS_2/WS_2 heterostructures by using the single step CVD process, as the typical schematic is illustrated in Fig. 2.11a. According to their demonstration the in-plane epitaxial growth is possible at lower temperature (around 650 °C). Since, this energy was not enough for the nucleation of WS_2 for the MoS_2, therefore, the higher temperature (850 °C) is considered to be suitable for the formation of vertically stacked structure owing to existence of van der Waals heterostructure which is recognized as more thermodynamically stable.

Two-Step CVD Approach

The two-step CVD approach is the more commonly adopted technique. In this technique, as-grown layered crystals are used as a substrate for the second layer. This process has also adapted for both the vertically stacked and lateral heterostructures which are predominately governed by the rate of gas flow and process time. As an instance, Li et al. [96] have reported the growth of 2D $GaSe/MoSe_2$heterostructure by using a two-step CVD technique. They also stated that the MX/MX_2 vertical heterostructures can have commensurate superstructures because of the large lattice misfit between the two layers. Moreover, the fabrication of a stacked TMD/hBN heterostructure has also realized by using the Ni–Ga alloy and Mo foil as the substrate without any intermediate operations. Therefore, the Ni–Ga alloy promotes the formation of the hBN honeycomb lattice, while the Mo foil act as a source of Mo [97].

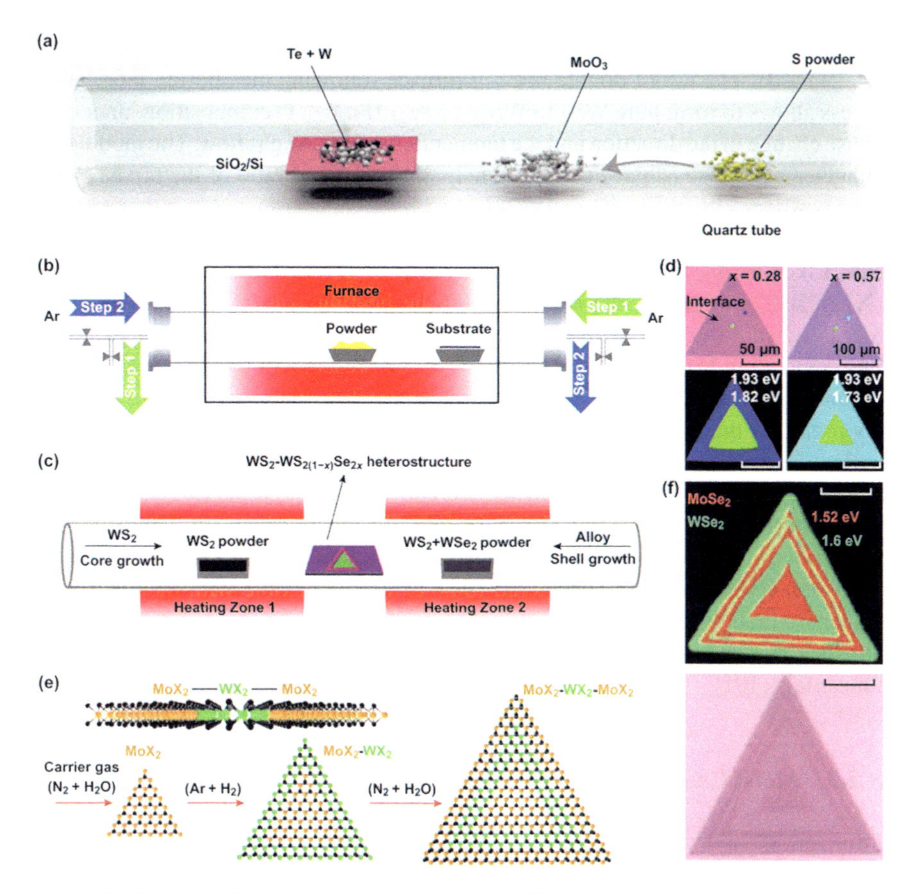

Fig. 2.11 Schematic for the chemical vapor deposit (CVD) growth: **a** Sketch for the synthetic procedure of MoS_2/WS_2 heterostructures by employing one-step CVD technique. Where the S powder is placed at the upstream, while the wafer containing mixed powder of W and Te is kept at a downstream; **b** Representation of multi-step CVD growth realization via directed switchable carrier gas flow and cooling procedure; **c** Process of modulable progression for the $WS_2 - WS_{2(1-x)}Se_{2x}$ $(0 < x \leq 1)$ monolayer lateral heterostructures under the twofold heating quartz tube; **d** Depicts the optical images and PL intensity mappings to heterostructures at the distinct x; **e** Reflect basics of one-pot fabrication approach; **f** Depicts to optical image and PL intensity mappings for the hetero-superlattice containing sharp interlines (adopted from Liu et al. [102])

Multi-step CVD Approach

The multi-step CVD method is developed by altering the route and modulation of gas flow components [98–100], the schematic of this process is depicted in Fig. 2.11b. Therefore, this approach allows construct a sharper heterojunctions boundary that is desired for the multi-junction heterostructures sequential growth**.** Using this approach, Biyuan Zheng et al. [98] have investigated an effective process for the fabrication of modulated $WS_2 - WS_2(1 - x)Se_{2x}(0 < x \leq 1)$ monolayers planar heterostructures associating tunable band alignments feature, as depicted in (see Fig. 2.11c, d). The use of double heating furnace following the altering fabrication process allows to form pure WS_2 to $WS_{2(1-x)}Se_{2x}(0 < x \leq 1)$ composition, under the precise change in the direction and the temperature of the Ar gas flow. The modulated parameter x has also achieved with the help of the defined ratio of the mixed WS_2/ WSe_2 powders.

This kind of controlled growth has also obtained by altering the gas flow components. Like one-pot fabrication approach by insertion of two precursors (MoX_2 and WX_2 mixed powders) in the same boat at the heating zone [99], usually in this process substrate is kept at the lower temperature. This mechanism has different steps depicted in Fig. 2.11e. The gas flows in the presence of the $N_2 + H_2O$ (g) to promote only MoX_2 growth. On the other hand gas flow switching to Ar + H_2 (5%) only promotes the growth of WX_2. To terminates the reaction process reversing the gas flow. Corresponding demonstrated optical pictures and Raman intensity mapping at 240 and 250 cm^{-1} have revealed the well-specified spatial distribution with the precise interface of the $MoSe_2$ and WSe_2 in the multi-junction heterostructure regime, as illustrated in Fig. 2.11f. Hence compared to one-step and two-step CVD mythology (multi-step CVD) has been considered to be extensive flexible with the controlled manner. So that, this approach supports the possibility to fabricate spatially selected optoelectronic devices owing to their separation of electrons and holes in the kinds of materials [101].

2.1.2.3 Metal Chalcogenisation Via Sulfurization/Selenization

Sulfurization

Using the sulfurization process Zhan et al. first reported the successful growth of MoS_2. The Mo thin film was coated on a SiO_2 substrate by placing it in the higher temperature zone, whereas the ceramic sample holder containing uncontaminated sulfur was kept in the lower temperature zone. In a well-designed growth circumstances the vapor-phase reaction began around 750 °C, at this temperature sulfur and Mo effectively deposited on a large area under a governed circumstance. Moreover, the MoS_2 polycrystalline layer on the SiO_2 substrate possessed grain size in the range of 10 to 30 nm has been also achieved. This investigation also revealed that MoS_2 thin film can have a fluctuating morphology with stacked alignment. It has also defined that the fabricated MoS_2 thin film order of two and three layers structures with interlayer spacing ~ 6.6 ± 0.2 Å. However, their HRTEM structural interpretation also revealed that CVD-grown MoS_2 film has also Moiré patterns. The scalability behavior of CVD-grown thin films also offered good uniformity under adequate size control. The thickness of the MoS_2 is also accessible by the CVD process which depends on the as-prepared Mo film and the area of the substrate used for the device fabrication. Later Liu et al. also demonstrated a similar result using the CVD method with different approaches with distinct precursors [102]. Afterward, a vast number of investigators conducted using the CVD approach for the sulfurization of the distinct materials deposited on a substrate to develop high-quality TMD materials thin films for the various efficient devices [3].

Selenization

The process of the selenization is the common technique for the formation of transition metal selenide using the CVD method. Selenization is usually performed via the straight construction of transition metal selenide or by the creation of the transition oxide on a suitable substrate before selenization. Owing to low chemical reactivity of selenium usually H_2 gas is desired to perform the selenization process under the robust reduction of transition metal oxide precursor; the chemical reaction can be expressed as:

$$WO_3 + 3Se + H_2 \rightarrow WSe_2 + H_2O + SeO_2 \quad (2.3)$$

Using this approach, the large area WSe_2 monolayer with high crystalline quality has been synthesized, moreover, using the CVD process selenization of WO_3 is also possible [103]. In this process, hydrogen introduces as a carrier gas during the whole reaction process. Typically optimized value of the carrier gasses Ar/H_2 are often used in the ratio 4:1 to get good synthesized materials. The significant demonstration outcomes showed a better crystallinity quality of the WSe_2 film, the corresponding materials hole and electron mobility were achieved ~ 90 and 7 cm^2/Vs. Typically these results are useful for the implication for the logic circuit combination. Moreover, precursor materials like WSe_2, WO_3, and selenium powder are used to grow thin films, these precursor materials usually positioned into the central temperature zone in the upper stream area of the tube furnace which heated up to 270–950 °C. To confirm substrate sustainability during the whole process sapphire was used as a substrate material which located in the lower side in the furnace. To begin the crystal growth kinetics Ar and H_2 were used as the carrier gasses with 80 and 20 sccm, which helpful for the transition of selenium and WO_3 vapor in the direction of targeted substrate. To continue this order Huang et al. have also contributed by demonstrating the temperature dependence morphology changes owing to changes in nucleation density, caused by the synthesis time. Their results revealed that the fabricated WSe_2 at 850 °C have a triangular shape crystalline structure possessing a lateral size around 10–50 μm. Moreover, it has been also directed that the lower fabrication temperature could be accountable for the higher density nucleation and superior regularity.

Similarly, by following the identical process prepared the monolayer WSe_2 flakes and explored their shapes with the varying (increasing) temperature [104]. This investigation revealed that the increase in growth time results a larger single crystalline material having a bigger domain extent under the triangular geometry. Innovations are not limited only the above discussed few cases, but also there are several reports on $MoSe_2$ fabrication using CVD to demonstrate monolayer and few layers with the distinct crystallinity phases [105–107]. Such as a high-quality uniform $MoSe_2$

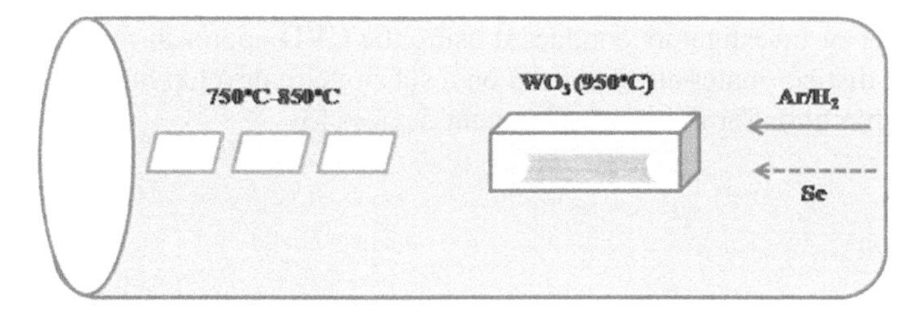

Fig. 2.12 Typical schematic for the growth of WSe_2 layers on sapphire substrates by the reaction of WO_3 and Se powders in a CVD furnace

growth generally evaluate by implication of MoO_3 and selenium pellets for the CVD at the atmospheric pressure. To grow the $MoSe_2$ temperature kept around 750 °C (heating ramp 50 °C/min) for the 20 min to complete the entire reaction process [109]. In the entire growth process, the Ar and H_2 (15% of Ar) carrier gasses mixture has been used at 50 sccm. Typical illustration of CVD setup for the growing $MoSe_2$ is depicted in Fig. 2.12. This fabrication process yielded within the desired pressure and reveal a triangular even $MoSe_2$ monolayer, as the consequence it produces a large single crystalline islands, approximately triangles size in the range of a few tens to 100 μm or more than 100 μm.

Moreover, typical vanadium diselenide (VSe_2) TMD-layered material are also synthesized by consisting vanadium atoms between the two selenium atoms layers which represents the CdI_2-type structure. Moreover, the VSe_2 growth is also impressively influenced when use van der Waals epitaxy approach to match substrate (mica, sapphire, etc.) and fabricated material. While earlier the VSe_2 growth from different approaches including usual TMDs by employing vapor-phase reaction of the relevant precursor unlike to metal oxide reduction process. Owing to the existence of strong sigma bonding in vanadium oxide precursor, however, it's difficult to performed oxide reduction via the selenization process. Therefore, growth kinetics from this mode should be distinctly controlled by using supplementary vanadium source materials such as VCl_3. Moreover, it has also been noticed that the growth of 1T VSe_2 monocrystalline nanosheets possible with high crystallinity and high electrical conductivity (10^6 S m^{-1}) possessing 4–10 times larger size compared of other TMDs. However, CVD fabricated VSe_2 has a uniform crystalline size with stoichiometric uniformity possessing distinct shapes of hexagonal and triangular configuration. The stoichiometric constancy of VSe_2 nanosheets has been also confirmed by demonstrating the color mapping.

The high-quality $NbSe_2$ monolayer growth is also possible by employing the CVD carrier gas action and metal oxide reduction process at 795 °C for 16 min [110]. Significantly it leads distinct types of TMD having variable crystal structures and morphology orientations with the control layer numbers, including growth parameter alternation.

2.1.2.4 Thermolysis from Thiosalts

The thermolysis process is also useful to deposit MoS_2 sheets, for instance, the dip-coating in ammonium thiomolibdates [$(NH_4)_2MoS_4$, it converted into MoS_2 by annealing at 500 °C under the Ar/H_2 flow to eliminate the residual solvents, whereas the NH_3 molecules and other by-products dissociate from the precursors. This process is also followed in the sulfurization process in sulfur vapor at 1000 °C [102]. Although the thermolysis of ammonium thiomolybdate also produces a better quality MoS_2 thin film, however, in case of large area TMDs fabrication by employing this process is the challenging task due to the technical limitations of uniform and ultra-thin ammoniumthiomolybdate film development, generally this process fabricated thin films belongs to polycrystalline structure.

A vast number of technical findings have demonstrated that MX_2 synthesis via sulfurization/decomposition of pre-deposited metal-based precursors is effective to prepare large-area MX_2 thin layers. But it has also various limitations, such as, it is difficult to control the thickness of pre-deposited metal oxide or metal thin film, thus their wafer-scale uniformity is limited. To get a better quality TMD with the desired number of layers, the thickness of metal oxides requires to be accurately controlled. Moreover, investigators have also made effort to improve the **fabrication** process via the deposition of metal oxide layers using atomic layer deposition [111]. Adopting this synthetic approach also gets atomically thin TMD nanosheets with good thickness controllability and uniformity.

2.1.2.5 Transition Metal-Chalcogen Precursors Via Vapor Pressure Reaction

Usually in the CVD process powder forms of transition metal oxide and chalcogen are frequently used as the precursors. In which precursors are evaporated at high temperatures and they adsorb to the substrate where the chalcogenization of transition metal oxide occurs and form TMD [112–114]. Using the technique of transition metals -chalcogenide precursors through the vapor pressure reaction several authors have reported their sound findings from time to time [115–117]. Typically transition metal oxide and chalcogen powder are located at different temperature zones owing to fact that the sublimation temperature difference. It reveals that the CVD furnace temperature has two independent heating zones for providing better control of evaporation of the precursors. Generally, the transition metal oxide is kept in a high temperature zone than chalcogen for the evaporation of both precursors at the same time. In which the gas phase transition metal oxide may be adsorbed on the substrate while the chalcogen gas deliver to the substrate surface with the help of inert carrier gas (Ar or N). As the consequence, the TMD monolayer forms on the substrate surface afterward chalcogenization of pre-adsorbed transition metal oxide. However, hydrogen gas is also occasionally used for the improvement in the reduction of transition metal oxide to get better chalcogenization [66]. For such a process typical growth temperatures between 750–950 °C would be considered suitable. The fundamental experimental demonstrations give the relationship between the ratio of transition metal (M) and chalcogen (X), the growth temperature, edge structure, and shape of the evolution of TMDs as depicted in Fig. 2.13a. Such as Wang et al. demonstrated that the crystal domain have an M zig-zag edge with a triangle shape if the M:X ratio is larger than 1:2. While, in case of M:X ratio is smaller than 1:2, the domain may have an X zig-zag edge and triangle shape. However, in the case of an equal M:X ratio (1:2) domain possess an alternate M and X zig-zag edge with a hexagonal or truncated triangle character [118]. In addition to this Yang et al. have also defined some additional feature, such as the growth temperature could be connected **to** M:X ratio and the domain shape, as illustrated in Fig. 2.13b. Moreover, they have also specified three remarkable growth conditions to decide the domain shapes which represents as the three-point star, triangle and hexagonal flakes [119].

Moreover, the completion of the nucleation allows the increment in TMD grain size by initiation of adjacent grains merging. In circumstances the grain growth stops overlapping and keeps stacking on top of each other. Thus, the large area of monolayer TMD is growth also possible by employing the thermal-CVD process. Additionally, other experimental findings are also demonstrated that the successful fabrication of wafer-scale single crystal WS_2 and MoS_2 monolayers. Likely, Lee et al. have fabricated the single-crystalline hexagonal boron nitride (SC-hBN) growth via the self-collimation of B and N edges inherently. To achieve this they used the sodium tungstate dihydrate ($Na_2WO_4.2H_2O$) to dissolve in acetylacetone as a W precursor, whereas the sodium molybdate dihydrate ($Na_2MoO_4.2H_2O$) dissolved in acetylacetone to get as a Mo precursor. While the liquid ammonium sulfide solution ($(NH_4)_2S$) has used as S precursor and carrier gases H_2 and Ar to complete the grain growth process. These findings showed the triangle shaped WS_2 and MoS_2 monolayers throughout the substrate with a line arrangement in a direction. Though, a longer growth time is desired for the wafer-size single-crystalline WS_2 and MoS_2 monolayers formation [120].

Transition Metal–Chalcogen Precursor Growth Via Metalorganic Chemical Vapor Deposition (MOCVD)

Large-scale production is a critical issue for TMDs for their commercialization. MOCVD is considered to be also a good alternative to scale up the growth of TMDs. This technique is investigated for the growth and development of TMDs. It can also use for the deposition of semiconducting thin films [121–123]. To grow the TMD materials form the MOCVD method usually organic molecules gasses containing transition metals (Mo or W) and chalcogens (S, Se) are flow over a substrate, further, it decompose by thermal energy and deposit TMD thin film on a substrate. The deposition parameters of the MOCVD are usually controlled through the partial pressure of the precursors. This method also offers a uniform deposition of the TMD thin films on a large-size substrate, the typical MOCVD schematic diagram is illustrated in Fig. 2.14 [124].

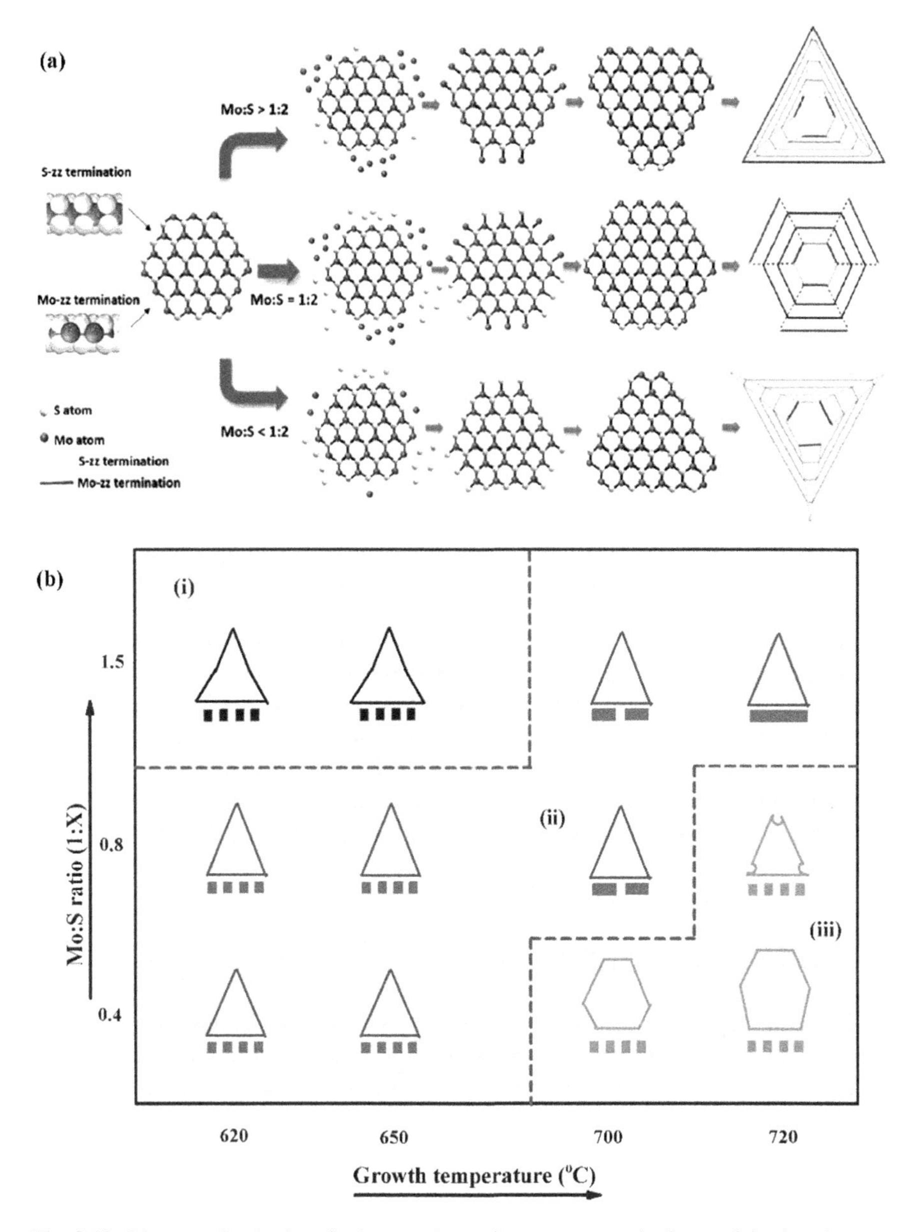

Fig. 2.13 Diagrams for the domain shape and growth parameters: **a** the figure of the domain as per M:X ratio (adopted from Shanshan Wang et al. [120]); **b** the domain profile corresponds to nominal M:X ratio and growth temperature

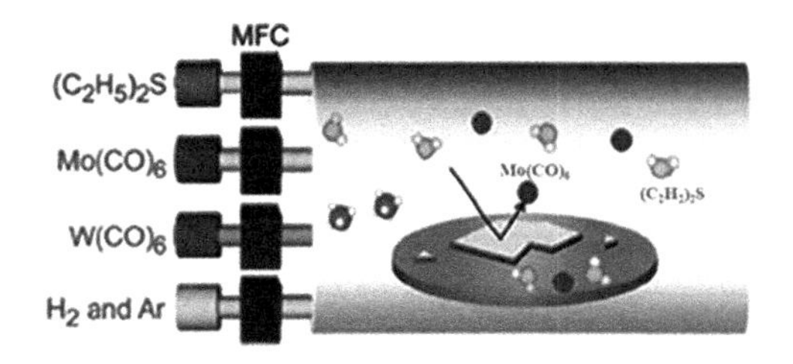

Fig. 2.14 The schematic diagram for the MOCVD setup (adopted from Mandyam, S. V. et al. 2020 J. Phys. Mater. 3, 024008)

Using this technique, the homogenous monolayer and a few layers of MoS_2 and WS_2 on the 4-inch substrate have been demonstrated [124]. For the deposition SiO_2 substrate it is generally used with the precursors molybdenum hexacarbonyl [$Mo(CO)_6$], tungsten hexacarbonyl [$W(CO)_6$],ethylene disulfide [$(C_2H_5)_2S$], Ar, and H_2 [124]. Whereas the gas Ar has usually used as a carrier gas and H_2 gas to improve the grain size and crystallinity quality. Similarly, another important study demonstrated that the WSe_2 thin film on different kinds of substrates, the results showed the alternation in temperature, pressure, ratio of transition metal and chalcogen elements and substrate impact also affects the morphologies of the fabricated films [125]. Moreover, a larger grain size has also achieved by applying the high pressure, temperature with the variable Se: W ratio. Additionally, this kind of work also extended with the higher growth rates of TMD in terms of a few layers of MoS_2 using pulsed MOCVD [126]. Hence, in the MOCVD process controls the number of layers and allowing to growth a homogenous film on a larger substrate, however, it has also some debatable issues like use of the toxic precursors and high-cost equipment. This technique also offers smaller grains size, under the suitable conditions, nonetheless, this technique has capability to produce wafer-scale consistency in a monolayer.

Chemical Vapor Transport (CVT) Process

CVT deposition process concept was innovated in the mid of the nineteenth century to grow single-crystal materials. In the beginning, Schafer performed a systematic investigation on CVT and described the process of migration. To continue this Fischer et al. contributed by employing the sealed ampoule for the CVT method first time. A typical setup of the CVT is illustrated in Fig. 2.15 [127]. In the CVT process usually, a powder precursor (AB(s)) is kept into the source zone (high temperature) with the transport agent (L(g)) gas. In further processing, the evaporated precursor decomposed and reacts to moving agent. Therefore, collectively gases approaches toward the lower temperature region, typically it is designated as a sink (deposition zone). A converse reaction can also take place in the deposition zone which results a single crystal structure cause of reformation. Using this technique there are two ways to find out the TMD monolayer. From the first way by exfoliation with the TMD bulk crystal, and the second one is by growing TMD monolayer to the substrate. Moreover, this process offers to avoid mechanical exfoliation to obtain monolayer

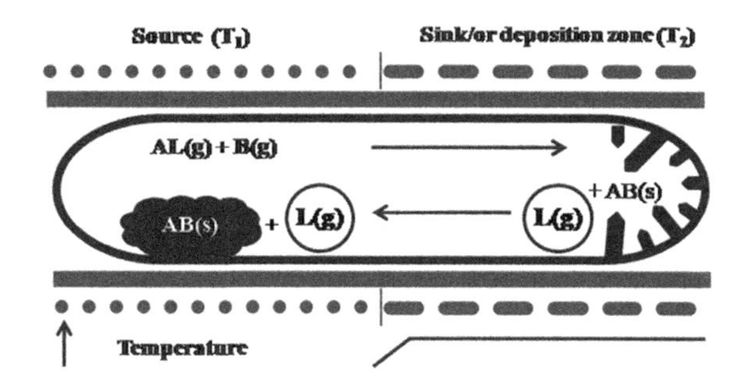

Fig. 2.15 Schematic illustration of the typical CVT system

MoS_2 after bulk MoS_2 crystal growth. Overall CVT method provides a better quality TMDs, but keeps in mind the procedure of experimental sample preparation is the difficult and laborious job [128, 129].

2.1.2.6 TMDs Alloys Growth

Yet, TMDs have been extensively explored owing to applications because of their intrinsic optical and electrical properties. TMD materials altering properties also makes them a suitable candidate for the specific applications with the improved devices performance. There is numerous fabrication techniques have been introduced for the manipulation of TMD properties, such as CVD synthesis, doping, etc. More specifically doping method are used to fabricate such kind of alloying materials. Predominately doping methods are sub-categorized into three categories; substitutional, interstitial and charge doping [130–132]. Significantly substitutional doping has also gotten much attention from investigators due to its scientific and technical reliability. Typically in the substitutional doping, the contaminated atoms replaced the host atoms and forms covalent bonds. Therefore, the formed alloys would be stable and their property degradation is lesser than other types of doping [133]. Since usually TMD consisting cation elements via transition metal (Mo and W) and anion elements via chalcogenide (S and Se), therefore, two distinct types of cation and anion substitutional doping are preferably considered.

Cation Substitutional Growth of TMDs

Cation substitutional doping growth method in which transition metal (TMD) host atoms are generally replaced from the impurity atoms. Like W doping in Mo-based TMDs (and vice versa), **it is** typically known as the cation substitution growth process that can influence the optical properties of the fabricated materials. The optical photoluminescence (PL) characterization of TMDs reflects their peak value, whereas the peak positions are shifted off the substitutional doped transition metal atoms and chalcogen atoms [65]. It has also well described the PL intensities of WS_2, MoS_2,

WSe_2, and $MoSe_2$ monolayers on SiO_2 substrate at room temperature. The PL peaks at 2.03, 1.88, 1.67, and 1.57 eV has been noticed for the WS_2, MoS_2, WSe_2, and $MoSe_2$, respectively [134, 135]. Moreover, the concentration-dependent W and Mo alloys $Mo_{12x}W_xSe_2$ monolayers PL peaks tuning has also favored it [136]. Furthermore, different kinds of metal atoms are also utilized as the cation substitutional doping to alter the physical properties of the TMDs. Like Rhenium (Re) also used as a donor to act as an n-type dopant and Niobium (Nb) has been used as an acceptor to act as a p-type dopant. It has been also well described that doped MoS_2 with Re shifts the Fermi level up about 0.5 eV, as a consequence it leads to degeneracy in n-type doping [137]. The I–V characteristics for the doped and undoped MoS_2 monolayers have been also measured [137]. A similar behavior electrical behavior has been also recognized for the Re-doped MoS_2 that is closer to metallic character rather than semiconductor. Similarly, the use of Nb as a p-type dopant to transit from inherently n-type MoS_2 to extrinsic p-type doped MoS_2 has been also studied. Typically the 0.5% Nb doping concentration provides an ohmic contact in-between the doped MoS_2 and Ti electrodes instead of the Schottky barrier. The vertical stacking of the Nb-doped and undoped MoS_2 p-n homojunction also provides a gate tunable current rectification [138].

Anion Substitutional Growth of TMDs

Anion substitutional dropped growth is the technique to replace host atoms of chalcogens in TMDs with nonmetal dopants. Like fabrication of Se-doped MoS_2 alloy monolayers ($MoS_{2x}Se_{2-2x}$) by varying the Se concentration [139]. This study has demonstrated that the PL peak position changes in-between 659 to 789 nm depending on compositional amounts of Se. Owing to substitutional doping of Se the transition is possible from WS_2 to WSe_2. In the whole process not only change the PL energy of WS_2 monolayer but also alter the semiconductor types (n- to p-type). Subsequently, it also found that the WS_2, MoS_2, and $MoSe_2$ can have inherent n-type semiconductor properties, whereas WSe_2 possessed p-type property. In this order also demonstrated the PL intensity and threshold voltages of $WS_{2x}Se_{2-2x}$ with the different values of x. It revealed the changes in PL peaks with p-type WSe_2 to n-type WS_2 [140]. Similarly, they doped a few layers of WS_2 and MoS_2 with chloride molecules (Cl) as a dopant reduces the contact resistance and Schottky barrier width [141]. Moreover, nitrogen is also used to replace the sulfur atoms of MoS_2. In this case, a positive shift in threshold voltage (Vth) has been noticed, which leads to the nitrogen-doped MoS_2 being a p-type material [142].

Hence tailoring the TMDs properties are important to get improved performances of the devices for the wide range of applications. Specifically in terms of the magnetic material's substitutional dopped growth of the ferromagnetic TMD alloys that generates a consistent magnetic phase. In recent years 2D dilute magnetic semiconductors are also attended a great attention because of their applications in spintronic and memory devices.

2.1.2.7 Epitaxial Growth of TMDs

TMD materials thin films homo and heteroepitaxial growth are one of the most important methods for the technology. To achieve high-quality over layers in hetero epitaxial one should overcome the problem of lattice matching between the substrate and the grown layer. The lattice-matching of the TMD is especially severe as tetrahedrally bonded covalent semiconducting materials like GaAs and Si. This happens because dangling bonds on the surface of a substrate which is connected only to the atoms of well-lattice-matched materials, whereas the length and angle of the covalent bonds cannot be changed easily.

Several techniques are also used to grow TMDs, including CVT, atomic layer deposition (ALD), and chemical vapor deposition (CVD) [125, 143]. In this order, molecular beam epitaxy (MBE) has been also introduced with various significant superiority to well established techniques [144]. Use of highly pure base materials (99.9999%) and ultra-high vacuum growth atmosphere (typically lower than 10^{-9} Torr) allows reduction in the contamination amounts within the material. Specifically, a low growth temperature is preferred in the MBE technique to grow the vertical heterostructures with negligible degradation or intermixing/interaction compare to use diverse materials. This technique has a slower growing rate (typically on the order of 1A ° min^{-1}), and it allows us to get an accurate thickness control. This is helpful to achieve the thickness dependence band structure of the TMDs. Moreover, due to solid source use in MBE it's also provided flexibility for the TMD growth. This kind of flexibility of the MBE technique allows fabricating alloys like the mixing of chalcogen (e.g., MoS_xSe_{2-x}, $W_xMo_{1-x}Se_2$), or other magnetic material intentionally doped TMDs [145–147]. The MBE versatility also offers to achieve graded doping profiles and compositions, while the other growth methods can achieve it only by the modification of relevant source fluxes throughout the growth process [148]. The MBE thin film fabrication conducts in the ultra-high vacuum atmosphere; therefore, it also offers several in-situ characteristics, unlike the low vacuum surroundings CVD or ALD techniques. Likely most commonly used technique is in situ reflection high energy electron diffraction (RHEED). From this technique crystal structure, growth morphology and growth rate are determined [149]. However, numerous optical approaches have been also used for the determination of alloying composition and their layer thickness, such as X-ray photoelectron spectroscopy (XPS) and scanning tunneling microscopy/spectroscopy (STM/S) [150, 151]. All these advanced features of MBE make it an ideal technique for the research to explore essential nucleation/growth phenomena and inherent property of TMDs.

The history of the epitaxial growth of TMDs started when Atsushi Koma (in 1984) first demonstrated growth by using MBE. At that time growth was termed as "van der Waals epitaxy" (VDWE) owing to absence of covalent bonds in-between the substrate and epilayer, in which individually vdW-like forces presents [152]. According to their interpretations, the instant layered growth of the TMD could be realized for the bulk lattice spacing with the lattice mismatch to the substrate. They have also demonstrated creation of the VDWE field, however active research was done in this area in 1994. Around 1994 Prof. Wolfram Jaegermann's and fellow researcher initiated research work in the area of TMD development by employing the metal–organic molecular beam epitaxy [153]. They reported a detail investigation on the growth kinetics, nucleation, and growth optimization. This group as well as other groups has also conducted research in this area from time to time, but at that time it was still a niche area [154, 155]. The great attention on the epitaxial growth of TMDs has been devoted around 2014 considering their potential application as the network materials for the innovative logic devices [156–160]. Yet, a huge number of TMDs as well as other layered materials have been investigated and interpreted, such as WTe_2, $HfSe_2$, $MoTe_2$, $HfTe_2$, $ZrSe_2$, $MoSe_2$, $SnSe_2$, WSe_2, etc. [158, 161–167].

Van Der Waals Nature of TMDs

Conventionally epitaxial growth of three-dimensional materials has been categorized from the Frank-van der Merwe (layer-by-layer), Volmer-Weber (island), or Stranski–Krastanov (layers1islands) modes [168]. The Frank-van der Merwe type of growth is usually related to strong interaction between the deposited atoms and substrate possessing a high diffusion rate, this mode would be more useful for 2D-dimensional growth. The Volmer-Weber growth offers weaker substrate interaction between the deposited atoms with the lower diffusion rate, therefore, it results epilayers in form of 3D-dimensional islands. Stranski–Krastanov type of growth is typically appearing for the stressed structures; by means in this type of growth a substantial lattice mismatch in-between the epilayer and substrate. This kind of growth essentially progresses via the layer-by-layer deposition mode, but their island growth due to relaxation of the strain.

Several investigators in their technical reports have made effort to categorize the TMDs VDWE by demonstrating the conventional growth modes [169, 170]. But the problem with all these growth modes is to provide description of the materials growth in case out-of-plane interaction deposition compared to in-plane interaction. Therefore, they allow accumulated strain by demonstrating an increase in epilayer thickness. But in actuality, the comparative in-plane and out-of-plane interactions are not hold for the TMDs and other 2D materials. However in such materials inert surfaces out-of-plane bonding in-between the epilayer and substrate (for initial growth) including the successive layers of the deposited TMD (for multilayer growth) shows the vdW behavior, by mean it is externally weak compare to strong in-plane covalent bonding. Hence the conventional growth approach is not very much useful for the epitaxial growth of TMDs.

Mostly basic models based on the equilibrium thermodynamics, but with them crucial kinetic issues like diffusion and nucleation cannot ignore. Several film studies in the recent years have demonstrated that in the most of growth conditions TMD deviates due to equilibrium like realization of fractal grain morphology [171]. Therefore, herein one has also discussed some important key factors that can influence the TMD films growth process.

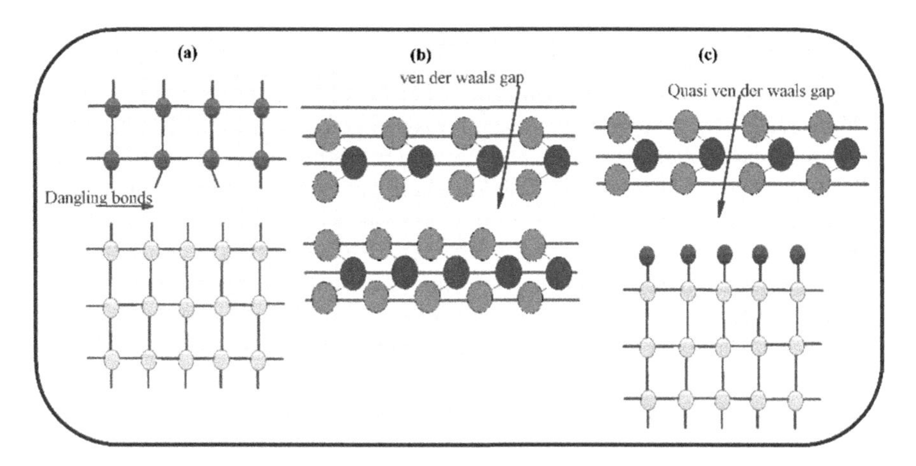

Fig. 2.16 **a** Schematic for the fundamental epitaxy of 3D materials, in which covalent bonding at the epilayer/substrate interface can induce strain in a lattice-mismatched system; **b** Schematic for the VDWE of a TMD on a 2D substrate where the weak vdW bonds at the interface may allow for unstrained growth even to significant lattice mismatch; **c** Schematic of VDWE for a TMD on a passivated 3D material to similarly unstrained growth

Substrates

Of course, substrates play a vital character in the epitaxial growth development. Usually the development of conventional 3D semiconductors is essentially restricted due to lattice mismatch between the epilayer and substrate. Specifically, such kinds of growth mechanisms govern the issue of the IIIV group materials and their heterostructures, owing to fact that they requires rigorous innovative work and improvement in adequate routes including implication of metamorphic buffer layers and wafer bonding [172–174]. Particularly, mismatch in the lattice consequence of the compressive or tensile strain within the epilayer in-plane bonding and begins the growth process, as depicted in Fig. 2.16a. While reaching the epilayer at the crucial thickness, the strain relaxation process begins via the defects formation, as the instance unfit dislocations at the interface. As a result, a defect containing epilayer growth is free of strain to its lattice constant.

Another side, the VDWE process includes the growth mechanism on a substrate material without (ideally) unsaturated bonds at the surface, so it leads the inert growth of epilayers, such as TMDs. The lacking of dangling bonds discloses that the none-existence of covalent bonds between the epilayer and substrate throughout the deposited area, so, it reflects they are connected to only weak vdW-like forces, as illustrated in Fig. 2.16b. It is very much similar to weak vdW bonding which existing in-between two successive layers in TMDs and other kinds of 2D materials, so this adequate feature make the process is also useful for the mechanical exfoliation [175]. Thus, vdW bonding is sufficiently weak but their lattice mismatch characteristics are not allowing to strain or defects at the interface [161].

Inert Type Substrates

The MBE grown TMD initially demonstrated to the Koma group by using the $NbSe_2$ on MoS_2 [152]. They were demonstrated the RHEED to interpret the lattice constant for the monolayer thickness TMD matches with the estimated lattice constant of the bulk $NbSe_2$. Their results showed the lattice discrepancy around 13% for the epilayer and substrate. Additionally, they also explored a streaky RHEED pattern and related it to uniform and smooth growth of the film. Also, their Auger electron spectroscopy interpretation established the material stoichiometry with analogous quality for the freshly exfoliated bulk $NbSe_2$. Moreover, the extensive work on growth of such materials with the non-TMD substrates and alloying compositions $NbSe_2$ and $MoSe_2$ on muscovite (a form of mica: $KAl_2(AlSi_3O_{10})(F, OH)_2$) also provided the identical abrupt interface. According to them a few textures of the thin film appeared owing to hexagonal TMD growth on a pseudohexagonal substrate [176, 177]. To conduct these experiments verities of the 2D substrates used either geological material or growing from the CVT that provides a fresh exfoliation for the production of a clean interface before loading into the MBE system. The significant step edges were created with the help of following exfoliation process, those acts as traps for the deposit adatoms, the growth rate of such edges substantially faster than the inert terraces [178, 179].

Non-van der Waals Type Substrate

There are a few successes to grow with the 3D-dimensional substrates, the initially fabricated substrate surface made inert by employing the chemical passivation process. This kind of growth is usually called "quasi" vdW, in which gap between the epilayer and substrate as depicted in Fig. 2.16c. A typical example is the $MoSe_2$ growth on F-terminated CaF_2 [170, 180]. To achieve it initially CaF_2 was cleaved and their successive annealing under the vacuum produces an F-terminated surface. It also noted that the $MoSe_2$ developed on CaF_2 has a somewhat low quality than films developed on graphene. Alike outcomes have also achieved with the sulfur and selenium terminated GaAs (111) [181–183] and sapphire substrates [184]. Whereas the $MoSe_2$ and $NbSe_2$ growth on both As and Ga terminated GaAs surface reconstruction being identical quality for the both surface terminations. Thus, the growing mechanism of 3D materials allowing a simple integration of these materials into the Si platform that's essentially required for the incorporation with CMOS process flows and fabrication of devices. Though, in the most of cases still favorable substrates are SiO_2 (or other dielectrics) or Si (100). Attention has been also paid to growing such materials with few promising outcomes, but more attention is required on this topic. Additionally, another important research related to this technology is selective area growth.

Heterostructures

Heterostructures are generally grown in situ by using the key growth technology MBE. This technology's slower growth rate permits the construction of immediate interfaces of the materials under their complex structures. Additionally, the non-existence of mismatch dependent strain and defects in VDWE also reveals a heterostructure configuration, which exclusively governs the estimated electronic and optical properties. Therefore, the MBE technique offers construction of structural parameters in extra exotic forms compare to TMD/TMD heterostructures. This allowing to inert materials could be used for the construction of the Heterostructures, such as graphene [158], topological insulators [185], and II-VI materials [186]. Moreover, the atomically thin behavior and their thickness govern properties additionally enhances the possibility of the combination of such materials. In this order, several investigators have reported their findings, such as Koma et al. have developed the MBE heterostructure first time using different kinds of TMDs, such as $NbSe_2$ on MoS_2 bulk crystal [152]. By continuing various pioneering work for the several vertical TMD/TMD heterostructures have been potentially constructed using MBE, like $MoSe_2/MoS_2$, TaS_2/MoS_2, $TaS_2/MoSe_2$, $TaS_2/MoTe_2$, $TaS_2/HfSe_2$, $TaS_2/NbSe_2$ [187–191], HfS_2/WSe_2 [155], $MoTe_2/MoS_2$ [171, 192] $HfSe_2/MoS_2$ [157, 162], WTe_2/MoS_2, $MoSe_2/MoTe_2$ [161, 163]. In this sequence other kinds of highly heterogeneous systems have been also fabricated, such as $NbSe_2$ pyrolytic graphite with the high orientation (HOPG) [193], $MoSe_2/SnS_2$ [194], TaS_2/HOPG [191], HfS_2/HOPG [154], $MoSe_2$/Bi2Se_3 [195], $HfSe_2$/HOPG [157, 162], WSe_2/HOPG [178, 196], $MoTe_2/Bi_2Te_3$, $MoSe_2/Bi_2Se_3$, $MoSe_2$/HOPG [163], WTe_2/HOPG, WTe_2/Bi_2Te_3 [161]. Yet a huge number of TMD materials reported having distinct kinds of electronic properties (band-gap, electron affinity, work function, etc.) showed a greater flexibility for the designing and fabrication of the electronics and nanoelectronics devices [197].

Nucleation and Growth

Typically TMD nucleation and growth from the MBE includes a multilayered interchange with the altering kinetic processes, like adsorption/desorption with the substrate diffusion, adatom connection, and edge diffusion. Collectively all such difficult concerns and commonly known problems of chalcogen integration likewise comparatively inferior thermal management prior to chalcogen desorption pointed out to get an optimal growth condition is challenging. Due to these issues, most of the MBE-grown TMD have been demonstrated in form of multilayer, islanded films with sub-100 nm grains. Reduction of nucleation is also important to allowing the extended grains growth of the material. Owing to fact that a film is developed by the successive deposition of smaller TMD grains that have a higher concentration at the grain boundaries. The higher concentrations of grains at the boundaries acts as scattering centers for the transporting carriers and ultimately it influence the mobility of the material [198]. Moreover, it has been also recognized that the experimentally achieved kinetic parameters could be integrate with the help of Monte-Carlo theoretical simulations. Initially this kind of integration from TMD thin film development groups operating wrongly with the distinct growth parameters [178]. To resolve the key issues, three stage approach of the nucleation and growth for the development of VDWE was adopted to restrict the metal–metal adatom interactions including substrate temperature keep high with a high chalcogen background pressure throughout the whole growth process, thereby an improved 2D growth is possible with a significant grain sizes.

Moreover, temperature of the substrate is also playing a crucial role in the construction of the nuclei, therefore it impacts to the grains shape and produced thin film quality. A vast number of investigators have made effort to grow TMDs by MBE to achieve a homogenous nucleation and successively follows a high temperature annealing to enhance the materials crystalline characteristics [199, 200]. But this strategy is viable for the pseudomorphic covalent systems; so that, the possibility of the single crystal growth is feasible. While, in case of the TMDs a negligible interaction is desired with the substrate, therefore, the high nucleation rates just reduce the grain size to encourage nucleation govern islands. Thus, TMDs nucleation mainly due to occurrence of the step edges and existence of defects in the dangling bonds. Though, in the absence of step edges and defects the nucleation can also occur due to metal–metal adatom interactions on the inert terraces during VDWE. Hence, for the reduction in nucleation and achieve the superior grains growth, the metal–metal interactions should be restricted or minimal. Therefore, high substrate temperatures should be preferred to promote desorption of the TMDs metal adatoms.

Since the sticking coefficient of the chalcogen atoms is considerably low compare to transition metals at the identical temperature [201]. Therefore, interactions in metal–metal adatoms are mainly governed by the nucleation procedure in VDWE. Under the maintained low transition metal fluxes with the combination of a high chalcogen flux substantially also reduce nucleation process and favored the 2D grains growth [158, 163, 166, 184, 185]. Several investigations it has demonstrated that the increasing chalcogen—metal ratio up to enough high level (~ 400:1) also increased the grain size with the limited island growths. The collective effect of the high substrate temperature, high chalcogen flux, and a low metal flux usually maximized the possibility of chalcogen atoms diffusion and interaction with the transition metal atom or pre-existing nucleus, therefore enables a lower nucleation concentrations of the 2D grains growth [178]. Moreover, altering the parameters also offers sufficiently large grains with the less islands formation.

Grown TMD Grain Morphology

In the growth of the TMDs, the substrate temperature and source fluxes also influence significantly grains shape and structural morphologies, because of small chalcogen—metal ratio it gives us fractal or dendritic growth (collectively it classified as branch growth) with the subsequent formation of the metallic clusters throughout the nucleation process, whereas the higher chalcogen-metal concentration ratio provides us symmetrical grains shape [158, 166, 179]. The high substrate temperatures has also favored to diffusion which controls the grain morphology. Moreover, in some simulation demonstrations, it has also predicted that the branch grain morphologies may formed during the low temperature growth. This could be due to the competition between adatoms attachment with the pre-existing grains and the diffusion of the adatoms along the grains edges. Because the low temperature grown adatoms does not have sufficient energy to diffuse through the grain edge for the formation of identical shapes. In addition to this, both growth rate and attachment rate are also generally enough high in the low temperature growth process. On the other hand, the high temperature growth offers edge diffusion, therefore formed grains have a symmetrical shape. Later, it also verified the experimentally for both $MoTe_2$ and WSe_2 compositions [178, 179]. Additionally, the diffusion and attachment rate can be also tuned with the enhancing amounts of the source flux, as the consequence it may increase the attachment rate and break the symmetrical state between edge diffusion and newly attached adatom.

Significant Issues with the Chalcogen Incorporation

A variety of high quality TMDs growth has been experimentally demonstrated using selenides and tellurides. But serious problem is the induction of chalcogen has been realized in a few materials, owing to a high sticking coefficient of the transition metals contrast to chalcogen elements [171]. At the beginning of the VDWE process, the growth of $NbSe_2$ was demonstrated with the initial growth (3 nm) in form of even single crystal [202]. It can be transformed into 3D-dimensional growth to fabricated thicker films. Similar characteristics have been also realized in the growth of $HfSe_2$ on HOPG [157]. The developed streaky RHEED patterns transformations in the Debye ring pattern associating spots is depicted in Fig. 2.17a, b. Such interpretation leads the morphological transition of thin film from homogenous to rough with their increasing thickness. These findings were concluded with the statements; the excess of transition metal adatom concentration results the excessive surface nucleation in a little surface migration time. Similarly, other TMDs' characteristics have also been shown, like WTe_2 (see Fig. 2.17c). In this remarkable study, beam interruption was used for both the $NbSe_2$ and WTe_2. The flat stoichiometric films over the growth surface are also achieved, as the typical streaky RHEED pattern is illustrated in Fig. 2.17d.

The incorporation of the chalcogen concentric amounts in TMD also influences the phase formation within the material. As the intense, the $MoTe_2$ developed thin film has revealed the phase formation within the material have a strong dependence on the growth rate and ratio of Te and Mo. In the case of high Te and Mo ratio, the formed film has semiconducting behavior with the trigonal prismatic (2H) phase development. Whereas in case of the lower Te: Mo ratio developed thin film few regions may have more metallic characteristics under the distorted octahedral (1T′) phase structure. The phase transformation from 1 T' has been also recognized in the annealed 2H $MoTe_2$ at 500 °C. The chalcogen and metal ratio has also impacted the quality

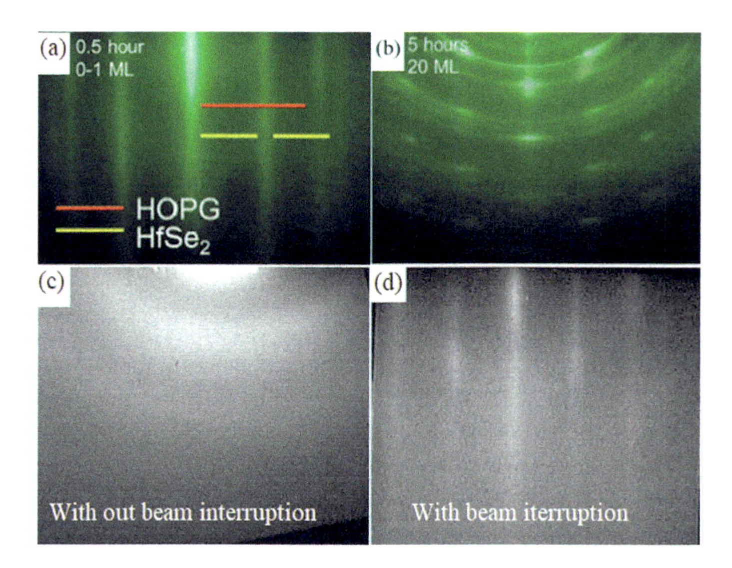

Fig. 2.17 RHEED pattern of $HfSe_2$ grown on HOPG: **a** Conventional growth of a (0–1) monolayer (ML) film with a streaky pattern to show a flat film; **b** Debye ring pattern with spots to show the rough polycrystalline film for the 20 MLs thicker film (**a** and **b** adopted from Yue et al. [158]); **c** A blurry Debye ring pattern for the WTe_2 system when the film grown without beam interruption; **d** RHEED pattern for the WTe_2 to show the stoichiometric, flat crystalline growth under the beam interruption (**c** and **d** adopted from Walsh, L. A. et al. 2017 2D Mater. 4, 025044)

of the developed material as recognized for the WSe_2-TMDs [178]. However, the lower ratio Se-W developed material has revealed a 3D dimensional island structure associating protrusions within the island centers. The bilayer WSe_2 also has also shown an identical lateral grain size without protrusions for a high Se: W ratio. The increasing Se flux removes the bump but suffers from a low Se-W ratio therefore it forms W-rich nuclei which have been confirmed by using XPS technique. Hence, the metal assembling with the inferior Se-passivation favors the vertical growth of WSe_2, whereas the higher Se and W ratio allows restrict the metal clustering, so, it leads the possibility of both vertical and lateral growths together.

2.1.2.8 Atomic Layer Deposition (ALD)

Atomic layer deposition (ALD) is one of the precisive chemical vapor deposition techniques by providing a sequential exposure of chemical species. It has been one of the most suitable thin films deposition approach to fabricate**d** nanoscale devices [203–205]. Specifically, ALD technique offers the self-limiting growth development features of the precursor molecules through the chemisorption on the surface reaction sites and the subsequent reactant molecule's reaction with the adsorbed precursor species and forms a sub-nanometer thick film within a cycle. Due to adequate deposition feature, it also offers numerous superiority including atomic-scale thickness control and a greater area homogeneity. Additionally, ALD also offers high conformal coating even on a 3D-complex structure like nano flakes [206–208]. Moreover, compared to usual 3D materials the 2D structured materials have a distinct surface chemistry, therefore, ALD grown 2D material features are generally possessing diverse aspect. Such as there is no dangling bond within the grown TMD which directly correlated to the formation of chemically inert surface on the basal plane [209]. It means no chemisorption of the chemical species on the surface. However, the fabricated 2D TMDs also have intrinsic defects while the synthesis process, due to unequal spreading of reaction sites throughout the TMD which creates non-homogenous deposition on the surface in a completed ALD process [210]. Therefore, it defines the crucial limitation of the ALD for the application 2D materials.

Significant TMDs Surface Properties for the Conventional Materials

Significant difference between usual materials and 2D TMDs is the occurrence of reaction sites over the surface. Since it is well known that the usual material has dangling bonds over the surface which responsible to reacts with distinct chemical species. In the similar way this concept can also applicable to ideally infinite 2D materials, but according to theoretical concepts there is no dangling bonds exist due to their peculiar crystal configuration, therefore, 2D materials are considered to be chemically inert [211]. Hence, the none occurrence of the dangling bonds over the basal plane is one of the significant benefit with the 2D TMDs, owing to this property they avoids electrical deterioration at the interface traps [212]. The ALD growth process is also free of dangling bonds that suppress material development throughout the surface. Moreover, when precursor (or reactant) exposed over a substrate which full of the dangling bonds then chemical reaction occur on the upper part of the surface through the reaction with the existing dangling bonds (i.e., surface reaction sites), this situation correlates to starting of the thin film growth. But nonexistence of the reaction sites over the surface in pristine 2D TMD, therefore, inhibition over the basal plane of 2D TMD thin film has appeared [210, 213].

In general 2D TMD consisting distinct two or more periodic table elements, typically one element from transition metal group and another one belongs to chalcogen group. Therefore, it is quite obvious they have various kinds of defects compared to other types of 2D materials such as graphene in which only a single carbon element exists. The 2D MoS_2 TMD also have various kinds of point defects, like mono sulfur vacancy, disulfur vacancy, complex vacancy in which Mo atom surrounded by three neighboring sulfur atoms, in the similar way a distinct complex vacancy where the Mo atom surrounded from three disulfur pairs, antisite defects etc. [214]. Likely to graphene the adsorption of chemical species on 2D TMDs is also different which depends to the chemical characteristics or types of defect sites. The various kinds of molecule adsorption on 2D MoS_2 have been investigated [215]. The estimated adsorption energies for the various types of molecules (CO_2, CO, O_2, H_2O, NO, N_2, and H_2) and basal plane (or defects such as S vacancy, S divacancy, Mo vacancy, one or two Mo atoms on S mono and divacancy and one or two S atoms on Mo mono and divacancy) have been defined for the 2D MoS_2 by employing the density functional theory (DFT) approach. The theoretical interpretations revealed that the all kinds of molecules have greater adsorption energy on defect sites compare to basal plane. Generally adsorption energy shows a variable behavior depending on the formed surface and the type of used chemicals. Additionally, the defects of the 2D TMDs also play a vital role as the high reactive and trap sites for the distinct molecules under exposure of the functional chemical group elements [216].

Classical ALD

According to the classical theory of ALD, a film can be formed on reaction sites of the surface by the chemical reaction between chemical species (i.e., precursor and reactant) [203]. Such a chemical reaction is typically known as dissociative chemisorption, which occurs due to the influence of external energy, ultimately forming a chemical bond between adsorbent (solid) and adsorbate (gas). Since in the whole ALD process, the chemical species only reacts with surface reaction sites of the substrate, while the residual molecules are not reacted, thereby, they finally purged out. This kind of self-limiting growth based on chemisorption is irreversible in general [217].

In contrast to this chemisorption-based reaction, another adsorption mechanism, physisorption becomes more significant in the modern ALD process, especially in case of the 2D materials. Unlike the chemisorption-based process, in this alternative method the chemical species approaches on top of the 2D surface and their weak intermolecular force (e.g., van der Waals force) physically attached on the surface without any kind of bond modifications among the adsorbent and adsorbate. Since physisorption is one of the fundamental bonding theories, this was frequently employed for the ALD process until carbon-based materials (i.e., carbon nanotube or graphene) came out to the world because most of the ALD reactions conducted on the substrate full of dangling bonds which enables a favorable chemical reaction on the surface. But graphene itself lacks the dangling bonds on the top of the surface, by means it is chemically inert on the surface, resulting the growth inhibition or abnormal nucleation growth on top of the graphene surface [218–220]. To overcome this problem, several investigators have tried to physically attach molecules for artificial adsorption sites for the reaction with subsequent chemical species.

Usually, to interpret this kind of ALD growth plot the Lennard–Jones potential model is used, as depicted in Fig. 2.18a, b. According to this molecular adsorption model concept, the depth of potential well directly correlates to the attractive force between solid (substrate) and gas (precursor or reactant), and the x-axis defines the distance between them. This model leads to the attractive force of physisorption is usually comparatively low ($\sim$ 40 meV) for a fairly large distance ($>$ 3 Å) contrast to the chemisorption ($\sim$ 1 eV) for a small distance ($<$ 3 Å) [222, 223]. Owing to its weak bond strength and relatively long distance, the physically attached molecules are easily detached from the external energy source, like thermal heating, long purging

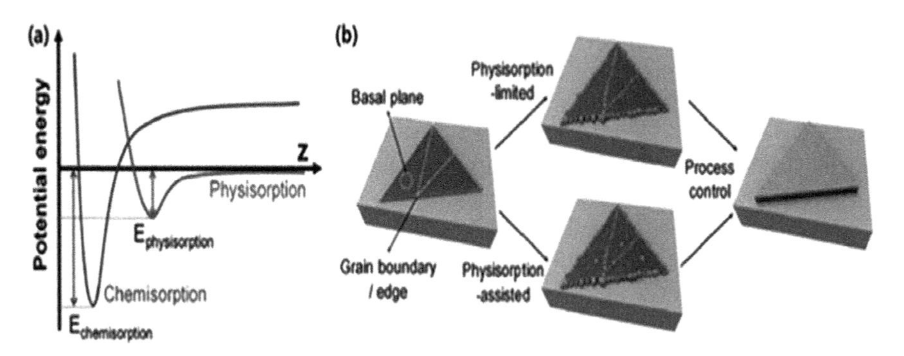

Fig. 2.18 **a** Lennard–Jones potential energy curve for the physisorption and chemisorptions; **b** Schematic for the ALD growth on the defect-free 2D TMD surface with and without physisorption (adopted from Nam et al. [259])

time, etc. [224, 225]. Since the physisorption process does not allow the bond alteration for the adsorbate and adsorbent, therefore it could be considered flexible. The simplest schematic of ALD for the 2D materials with and without physisorption is illustrated in Fig. 2.18b. Since the ALD process for the 2D substrate is advantageous to the defect or edge sites, not for the basal plane. As the intense, suppose few chemical species are physically attached to the basal plane of the 2D material, this kind of attachment creates the reaction sites for the used precursor, as the consequence nucleation occurs on the basal plane. In simple words, if molecules are physically attached throughout the basal plane, then it leads a homogenous thin film fabrication on dangling bonds free 2D TMD.

Uniform Film Deposition from ALD

In the modern world using the ALD technique highly uniform up to sub-10 nm thick 2D TMD thin films have been obtained, this is basic requirement for the integration of nanoscale devices to high performance. In more than a decade, various reports of ALD on 2D TMDs have been presented. Herein one has discussed the several ALD process 2D TMD thin film fabrication methods.

Uniform TMDs Thin Film Fabrication Without Surface Treatment from ALD Technique

ALD research with the 2D TMDs was begun from the exfoliation of TMD flakes in form single crystal TMD. Around the year 2011 the MoS_2 based top-gate transistors with or without high-k dielectrics different properties have been investigated and interpreted [71]. To fabricate the monolayer MoS_2 field effect transistor (FET) consisting no high-k dielectrics, a MoS_2 monolayer (~ 6.5 Å) was developed through the mechanical exfoliation on a 270-nm-thick SiO_2/homogeneously silicon doped substrate. Owing to combined effects of the mobility of 2D-material with high-k dielectrics results an enhanced dielectric screening [226–229]. Considering the remarkable achievement in this area several investigators have preferred to deposit ALD HfO_2 on a MoS_2 monolayer following no any action of the additional treatment. Specifically the selection of high-k dielectrics HfO_2 based on their high dielectric constant (k ~25) and a wide energy band gap (~5.7 eV) [230]. To deposit HfO_2, tetrakis (dimethylamino) hafnium (TDMAHf) was used where Hf played the role as the precursor and HO_2 act as the reactant, a 30 nm HfO_2 film was fabricated over the MoS_2 flake to make top-gate dielectric. This process fabricated FET electrical properties was extensively enhanced. However, the direct ALD deposition over to without treated MoS_2 surface may encourage the frequent nucleation thereby the inconsequential changes appeared in the surface structural morphology before and after ALD HfO_2. Moreover, the nucleation and growth in such ALD fabricated thin film preferably happening on the surface which consisting high concentration of hydroxyl (–OH)-terminated groups, that generally considered as the adsorption sites for the precursors [231]. Although, the TMD flakes exfoliated from the mechanical process have shown none existence of intrinsic OH species, therefore, it has produced a relatively flat thin film. This has been confirmed by the demonstration of thick HfO_2, in which the nucleation of HfO_2 started to connect eventually to form a somewhat even thin film [232].

Similarly, there is no particular difficulty in the deposition of comparatively thick dielectric (> 15 nm) over to dangling bonds free TMD substrates by employing the ALD. But keep in mind the thickness of the dielectric must be in decreasing manner to integrate the complementary metal oxide semiconductor (CMOS) [233]. Moreover, the electrical properties of TMD based FET has shown a significant enhancement with a small enough gate dielectric thickness, specifically in the sub-10 nm range. This range fabricated devices have provided a better control of the working channel with a greater drive current [229]. Thus, the deposition of ultrathin high-k dielectrics with excellent uniformity on 2D TMD layers is the requirement for high-performance device fabrication.

Initially, a smooth thin film (< 10 nm) on 2D TMDs was obtained under a control process temperature. In the year 2012 Al_2O_3 ALD performed over the exfoliated MoS_2 and boron-nitride (BN), and based on their experimental findings theoretical approach was interpreted [213]. According to this interpretation, ALD deposition of Al_2O_3 can be achieved over mechanical exfoliation developed 2D MoS_2, similarly BN films also possible by using the trimethylaluminum (TMA), and H_2O from the keeping difference in the process temperatures by providing description of temperature effect on the even deposition. It has also demonstrated that over the MoS_2 and BNs substrates afterward 111 cycles ALD deposition of Al_2O_3 it depends on processing temperature (200, 300, and 400 °C). Moreover, a minimal thickness of Al_2O_3 has been achieved with the deposition growth rate of Al_2O_3 on a 10 nm SiO_2 substrate. However, it also interpreted that there was no temperature dependent growth rates for the ALD Al_2O_3 over the SiO_2 in the temperature range 200–400 °C, while the chemical adsorption of TMA with the OH groups was also unaffected in this temperature range. Whereas the ALD deposited Al_2O_3 over the BN and MoS_2 found to be intensely influenced from the growth temperature. The ALD deposited Al_2O_3 over the BN and MoS_2 at 200 °C was also found with the superb homogeneity and free of the voids or pinholes. Moreover, with the increasing process temperature the deposition of Al_2O_3 was also in an increasing order, but it not free from the numerous voids and pinholes. While the processing temperature reaching at 400 °C, several Al_2O_3 islands formed throughout the surface. In order described several parameters with the help of the experimental findings the surface coverage of ALD deposited Al_2O_3 over the MoS_2 was also calculated which slightly higher than that on BN. So, it could directly correlate to the aspect of the film growth extensively depend on the process temperature which is typically connected to physisorption of the precursor over the MoS_2 and BN. This could be due to the precursor (or reactant) molecule physical adsorption throughout the dangling bonds free substrate to form a weak van der Waals interaction. With the increase in temperature a few physisorbed molecules could also desorbed through breaking the weak bonds and inhomogeneous reaction sites over the TMD. Whereas the physisorbed molecules over the basal plane slowly separated with the increasing temperature and the edge sites of TMDs remain acts as the reaction sites owing to existence of the dangling bonds [218, 219]. This finding lead that with the ALD process homogenous deposition of Al_2O_3 films over the BN and MoS_2 surfaces is not possible while the process temperature increase enough.

Moreover, investigations have also focused to explore interaction of precursors and 2D materials surface and define the mechanism of reaction for the primary ALD cycles over the 2D materials, adsorption energy for the TMA and MoS_2 or BN a DFT based interpretation introduced. As per the theoretical interpretation the physical adsorption energy of TMA over the BN surface could be up to 8.7 kcal/mols, typically it is greater than H_2O over the BN, while the energy of physisorption of TMA over the MoS_2 is around 21.9d kcal/mol, this is also greater than the H_2O above the MoS_2. Therefore, the DFT calculation established the TMA is more extensively physisorbed over the 2D materials compared to H_2O. Additionally, the TMA have greatest interaction energy to MoS_2 due to existence of positively charged Al that more efficiently interacts with the negatively charged N and S compared to negatively charged O. Additionally the polarizabilities of TMA and MoS_2 is also greater than the H_2O and BN. These calculations are provided the confirmation about the high coverage of Al_2O_3 over the MoS_2 and BN for the high temperature range (300–400 °C). In addition to this, an investigation has been also concluded that the temperature dependence growth of the ALD process Al_2O_3 onto the 2D substrate could predominantly control through the physisorption of the precursors in place of the chemisorption.

Uniform Deposition on 2D TMD with the Surface Treatment

Although lowering the process temperature is one of the potential approaches to get a homogeneous thin film over the 2D TMDs, nonetheless, it also suffers from the unwanted impacts as well. As an example, the low temperature ALD deposition gives us a poor film density with the numerous precursor contaminations (e.g., C, N, O, etc.) owing to inadequate activation energy of the reaction for the precursor and surface reaction sites [233, 234]. Therefore, it is obvious to get a homogenous thin film by employing the ALD without compromising the film quality by considering several factors.

Functionalization of the Molecule on 2D TMDs In 2013 report has become on ALD deposition of HfO_2 over a mechanically exfoliated MoS_2 substrate. TDMA Hf and H_2O have been used for HfO_2 deposition at 200 °C [210]. According to the interpretation, the HfO_2 30 nm thin layer can be deposited on MoS_2 to cover the entire layer, but the deposition was non-uniform [71, 213, 236]. The key demonstration was the presumption of the residue elimination procedure. To fabricate TMD based FET, performed the successive layer coating of the organic materials following the elimination from the suitable solvents carried out after the transferring or lithography event [236, 237]. Although generated residues during the processes it hardly removed even after UHV annealing, this means it remains as surface contaminants that acts as a reaction sites. Moreover, the effect of deliberately attached molecules (like organic or solvent contaminants) has been also explored and demonstrated that it promotes the nucleation of ALD HfO_2 on MoS_2 flakes. This goal could be achieved by soaking the acetone or N-methyl-2-pyrrolidone (NMP) or spin coating with poly (methyl methacrylate) (PMMA) prior to ALD deposition of HfO_2. Results revealed no noteworthy alteration afterward ALD deposition of the MoS_2 that immersed into acetone, however, a remarkable difference in phases has been noticed after PMMA coating or the NMP treatment. Subsequently, it was also demonstrated that the organic residue or solvent remains over to the MoS_2 substrate that acts as a nucleation promotion site to improve the smoothness of the ALD-deposited HfO_2. It also confirms the impact of the purge time of the ALD process on the physisorption of chemical species. Moreover, in case of the very lengthy purge time among the precursor and reactant exposure (over 6 h) disappear the HfO_2 associated characteristic peak, while the individual Hf4f and O1s peaks of the precursor and reactant purge time in between 5 and 10 s. It considered as an evidence of the physisorbed precursor molecules that detached during the longer purge time. Thus, the physisorption or molecule addition (functionalization) enhances the nucleation of ALD deposition with the lower amounts of reaction sites over the TMD surface.

Later, it was also successfully demonstrated that the Al_2O_3 deposition over to exfoliated WSe_2 with the organic functionalization process via the titanyl phthalocyanine (TiOPc) [238]. This work provided also a boost for the uniform ALD deposition on graphene using TiOPc [239]. As per the first principal calculation, the robust binding energy between dimethyl aluminum and the surface species generates through the detachment of TMA with the surface hydroxyl group and the elements of TiOPc(C, N, or O) over 1.5 eV. This leads TMA can add to the TiOPc coating that positively provides reaction sites for successive reactants. Afterward, intensive research was made with the WSe_2 to deposit onto the well-arranged pyrolytic graphite (HOPG) by employing molecular beam epitaxy (MBE), subsequently a monolayer of TiOPc has also fabricated by employing the MBE. Their highly ordered and defect-free morphologies indicate that the homogenously thin layer of TiOPc monolayer is possibly deposit from the MBE-grown WSe_2.

Plasma-Assists Treatment In the ALD fabrication process, plasma treatment is one of the most vibrantly used techniques in modern days, this readily primes to any surface better acceptance of the followed process [240, 241]. Due to high energetic species during the plasma ignition the functional groups generates on the substrate surface by the reaction in-between reactive ion/electron species and the surface group [242]. Therefore, the plasma assists surface treatment are widely used for carbon nanotubes (CNTs), and graphene [243, 244]. Materials that have no dangling bond on the surface typically it can improve the nucleation of the ALD process by creating the hydroxyl species throughout the surface. But there is a serious issue with them the pristine substrate which may damage afterward plasma ignition owing to highly energetic ion bombardment from the ion/electron species [245]. Regarding the thin 2D TMDs compare to bulk materials the effect of plasma damage is considerably more crucial. Hence, it is advantageous for the functionalization of the ultra-thin 2D TMD surface with the less destruction.

The impact of oxygen plasma pretreatment on ALD deposition growth of Al_2O_3 and HfO_2 over the 2D MoS_2 is generally achieved [246]. To minimize damage the plasma process preferred to use remote inductive coupled plasma. It means the plasma glowing and surface treatment belongs to distinct reactor area. The benefit of isolated plasma surface treatment like TMD surface cannot undergo an intensive ion and electron bombardment while the treatment process [247]. Using this technique, various ALD depositions on distinct TMDs at different temperatures have been conducted, such as growth features of the ALD deposited Al_2O_3 and HfO_2 over to pure MoS_2 at 250 °C as the thickness function.

Further, to improve the ALD film surface coverage over the MoS_2, usually the O_2 plasma treatment is performed before the thin film fabrication with the variable treatment time (10, 20, and 30 s). This lead the surface coverage of the ALD fabricated Al_2O_3 films remarkably enhance afterward 10 s of plasma pretreatment (93%), compare to plasma untreated (77%) sample. Almost a complete coverage of Al_2O_3 films has been achieved in 30 s plasma pretreatment. Therefore, the overall existing surface reaction sites increases as the plasma treatment time increasing. A likewise tendency of surface coverage enhancement through the plasma pretreatment to the ALD deposition of HfO_2 has been also recognized. As per the demonstration, the optimum plasma treatment time for the homogenous ALD deposition over to 2D MoS_2 has been achieved around 30 s. Moreover, interpretation for the both Al_2O_3 and HfO_2 deposition having distinct thicknesses afterward 30 s plasma treatment over the MoS_2 has been also explored. According to facts nearly desired deposition around 10-nm-thick Al_2O_3 and HfO_2 over to MoS_2 is achievable afterward the 30 s of O_2 plasma treatment. The root means square (RMS) surface roughness of ALD-deposited Al_2O_3 and HfO_2 has been also interpreted in terms of the various plasma pretreatment times. In both Al_2O_3 and HfO_2 cases, the RMS surface roughness slowly decreases with the increasing plasma pretreatment time. It could be directly correlated to generation of additional surface reaction sites while the increase in treatment time. Hence, the dielectrics would be more homogenously deposited in case of the increasing plasma-treatment time.

Advances in this area, in 2018 investigators have introduced the direct plasma treatment on the CVD-grown $MoSe_2$ layer and get a homogenous versatile class of the dielectric. To passivate the bulky $MoSe_2$ from unintended molecular adsorption that encourages the decline in FET performance [248]. Because $MoSe_2$ has a lesser energy bandgap (1.57 eV for monolayer, 1.09 for bulk) compare to MoS_2 (1.88 eV for monolayer, 1.23 for bulk), so, it provides superior electrical characteristics of the FET devices [134, 249].

Moreover, investigations are also made to use H_2O plasma treatment to get a uniform dielectric deposition on the MoS_2 substrate [250]. The use of H_2O plasma generates –OH species over the TMD surface. Since H_2O contains a hydrogen atom itself, which is dissimilar to O_3 or O_2 plasma, therefore, the hydroxyl radical (–OH*) generates under the plasma ignition [251, 252]. The created hydroxyl groups may adsorb through the MoS_2 substrate and act as the reaction site, therefore, a homogenous and adequate quality Al_2O_3 are achieved over to MoS_2 flake. Thus by employing the H_2O plasma treatment fruitfully has been achieved a smooth Al_2O_3 film over the exfoliated MoS_2 up to 1.5 nm. Typically it is called a state of art at a thickness homogeneous deposition over the TMD material. Additionally, the Raman and PL characterizations peak intensities are un-decreased afterward the H_2O plasma treatment instead of that these characteristics peaks increased due to carbonaceous impurities during the MoS_2 removal under the plasma action (self-cleaning effect). Thus, this technique is considered to be a promising approach for TMD surface treatment.

UV-O_3 Treatment to Surface Hydroxylation Although plasma treatment are also promotes the nucleation of ALD deposition over the TMD surface through the formation dangling bonds onto the surface. But, it also harms the ultra-thin TMD himself due to high sensitive species formation while the plasma explosion, as the consequence decline in its intrinsic and exceptional features. Therefore, the formation of the dangling bonds on the 2D TMDs surfaces without disadvantage is desired. To overcome this crucial issue researchers have made an effort to develop a facile way that generates reaction sites throughout the 2D TMD surface without itself destruction. Therefore, in 2014 introduced the room temperature ultraviolet-ozone (UV-O_3) treatment for the growth of ALD deposited Al_2O_3 on the 2D MoS_2 [253]. This is one of the usual treatment techniques to removal of the surface contamination through the generated highly reactive radicals by the UV and O_3 action [254]. In modern days, it becomes attractive method to develop a hydrophilic surface by formation of the hydroxyl species over the surface [255]. Using this surface treatment it has estimated that ALD deposition over the 2D TMD to provide a homogenous film with preserving its basic structure [256, 257]. Like nondestructive MoS_2 functionalization process without breaking sulfur-molybdenum bonds during the UV-O_3 exposure. Mostly for the removal of top layers to bulk MoS_2 investigators have preferred mechanical exfoliation using adhesive tape. Usually, afterward the exfoliation, samples are loaded into the ultrahigh vacuum (UHV) within 5 min and subsequently transferred into a compartment for UV-O_3 treatment. This process has been extensively used for the different TMDs surface treatments to fabricate high-performance devices.

Direct Plasma-Enhanced ALD (PE-ALD) on 2D TMDs As discussed in the previous section UV-O_3 treatment can provide enthusiastic outcomes for the generation of reaction sites onto the MoS_2 without harming the host material. Besides the enthusiastic outcomes, long exposure of UV over the TMDs might be induced photodegradation, therefore, it deteriorates the electrical characteristics of the TMDs [258, 259]. So, investigators paid attention to finding another way which covers whole reaction sites throughout the 2D TMD surface without employing UV light. Therefore, the ALD nucleation onto MoS_2 using O_3 as the reactant has been explored. As per this process, O_3 has used as a strong oxidant for the ALD, due to its high reactivity including numerous benefits, such as inferior contaminations and high density to use as the reactant [260]. Since O_3 itself cannot produce S–O surface functional groups, so that reaction sites created the subsequent precursors onto MoS_2 [210]. Therefore, to complete this process frequently used the two-step deposition method to the Al_2O_3. ALD deposition of Al_2O_3 at 200 °C using TMA and H_2O with 3 nm minimal thickness has lead inferior homogeneity over the MoS_2 flake. However, the homogeneity of the ALD deposited Al_2O_3 remarkably enriched when O_3 used as a reactant at the place of H_2O, but the formed film remain not free from huge numbers of pinholes. Therefore, to improve the film surface quality preferred to use five cycles ALD process of Al_2O_3 at 30 °C, to fix a seeding layer before the ALD deposition of Al_2O_3 at the temperature 200 °C. This process obtained Al_2O_3 showed even RMS roughness (0.23 nm) compared to pure MoS_2 (0.18 nm). Here the reactive O generates due to O_3 detachment that provides precursor to reaction sites by the formation of the weak S–O bonds. Such insincerely created surface reaction species may also desorb or diffuse onto the surface with increasing temperature; ultimately it leads a high surface roughness. To avoid undesirable effects like thermal desorption of reaction sites a low-temperature ALD process are preferred [261].

More advances in this area, move toward a novel way of homogenous dielectric deposition onto 2D TMDs using the PE-ALD technique. The PE-ALD high sensitive chemical species generates through the plasma reaction (e.g., O*, H*, etc.) [262–264]. The use of high reactive radicals PE-ALD not only permits to the development at a lower temperature but also makes a high-density film in comparison to thermal ALD [265, 266]. Therefore, efforts have been made to reduce the destruction of 2D TMDs owing to high sensitive plasma species, like the use of isolated plasma or comparatively mild reactant source. Moreover, deposition of homogenous Al_2O_3 and HfO_2 onto exfoliated multilayer MoS_2 (6–8 nm) from the remote PE-ALD has been also recognized [267]. To achieve the goal, TMA and TEMAHf as the precursors and H_2O and O_2 plasma as the reactants have been used. Both Al_2O_3 and HfO_2 PE-ALD grown films have shown the homogenous surface over several layers of MoS_2, and it irrespective to growth temperature, furthermore, in contrast to thermal ALD it shown frequent nucleations on top of the MoS_2, as depicted in Fig. 2.19. The use of PE-ALD enables the achievement of a uniform thin film of less than 3.5 nm with sub-nanometer scale roughness. The PE-ALD also induced the dangling bonds on the surface directly therefore, the physisorption factor does not affect the initial growth characteristics. Therefore, it provides an ultrathin dielectric thin film ($\sim$3.5 nm) over the MoS_2 with the sub-nanometer scale roughness. The fabricated and demonstrated

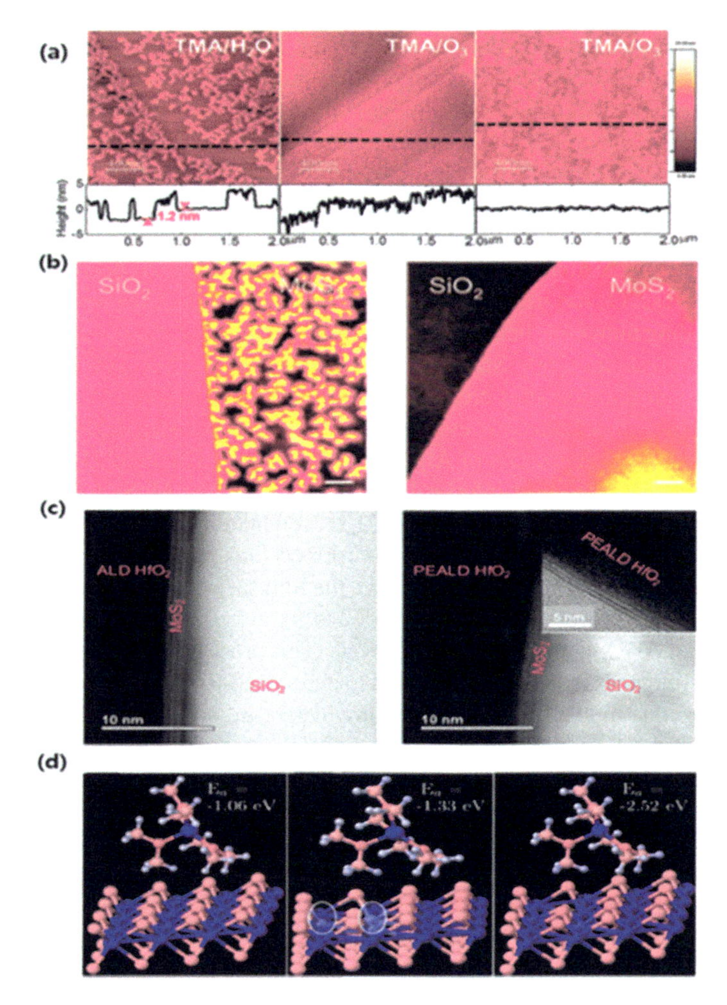

Fig. 2.19 **a** AFM images and height profiles for the ALD Al_2O_3 on MoS_2 using H_2O (30 cycles) (left), O_3 (30 cycles) at 200 °C (middle), and O_3 (five cycles of the seeding layer at 30 °C, followed by 45 cycles of ALD at 200 °C) (right) (adopted from Cheng et al. [232]); **b** AFM images for the ALD Al_2O_3 on MoS_2 using water (left) and O_2 plasma (right) (adopted from Price et al. [266]); **c** Cross-sectional TEM images of thermal ALD (left) and PE-ALD (right) of HfO_2; **d** Shows the estimated binding energy for the TDMAH corresponding to pristine MoS_2, sulfur vacancies and hydrogen-passivated sulfur vacancies (**c** and **d** adopted from Price et al. [267])

MoS_2 FET devices using such dielectrics have shown an alteration in electrical features that is fairly different. Like, PE-ALD Al_2O_3 deposited onto MoS_2 FET, the PE-ALD HfO_2 process un-degraded electrical characteristics of MoS_2 FET with the fairly enriched performance; this is the contradictory result as the ALD HfO_2 or PE-ALD Al_2O_3.

Though PE-ALD do not allowing deteriorate the electrical features of the MoS_2-based FET, additionally also remarkably improve the performance. It makes feasible the number of MoS_2 layers around 6–8 nm, corresponding to 9–12 layers of MoS_2. This method reveals a thick flake that has a negligible effect of oxidation or damage because only the topmost TMD layer affects with the plasma species. This remarkable investigation also revealed the effect of PE-ALD on comparatively thin, mono, bi, and trilayer of MoS_2 [268]. Whereas the CVD grow MoS_2 have various intrinsic defects on the surface compared to mechanically exfoliated films. Hence the electrical property degradation in monolayer MoS_2 FET, like V_{th}shift, I_{on}/I_{off} ratio decreased. Moreover, in case of the bi and trilayer MoS_2 FET a declination in the electrical characteristics. It usually correlates to TEM analysis in which the topmost layer of MoS_2 distracted afterward the PE-ALD action, as depicted in Fig. 2.19c. The TEM image of trilayer MoS_2 of thermal ALD deposition has different three layers in- between SiO_2 and HfO_2, it reveals there is no destruction afterward completing the thermal ALD deposition. In contrast to this, the fuzzy three layers of MoS_2 have also noticed, it generally correlated to plasma damage. Moreover, the high-resolution TEM image has also visualized the two MoS_2 layers toward the lowermost side with a fuzzy layer on the upmost MoS_2 flake. This is also evidence about the topmost layer of MoS_2 destructed while the PE-ALD action. Moreover, the first-principal calculation has also provided the estimated adsorption energy for the Hf precursor of MoS_2 which depends on diverse surface stoichiometry, such as pristine, sulfur vacancies, and H-passivated vacancies (see Fig. 2.19d). This calculation also provided the adsorption energy of the precursor 1.06 eV for pristine, 1.33 for the sulfur vacancies, and 2.52 eV for H-passivated vacancies in MoS_2. Moreover, the theoretical backbone also reveals the H-terminated vacancies generation in MoS_2 by the PE-ALD process, that's the most stable. Therefore, the PE-ALD promotes the nucleation and growth of the film under the ALD process by forming the species on the top most 2D TMD layer [260].

2.1.2.9 Layer Transfer

As one has discussed previously TMDs are a class of materials with a rich catalog of novel properties, and many of TMDs may go beyond graphene. Usually, TMDs are represented with the general formula MX_2, where M is a transition metal atom, and X is a chalcogen atom (usually S, Se, or Te). Owing to their wide range of possibilities of applicability in kinds of modern technologies, it is obvious to fabricate heterostructures from individual TMD films, to customize stacking order. This is the technical requirement to get desired a systematic methodology for transferring large-scale TMD films from their growth substrate onto a target substrate without compromising the intrinsic structural and physicochemical properties in form of 2D TMDs. Thus, any versatile TMD layer transfer approach should allow for a uniform separation of the film from its growth on a substrate, by maintaining the structural integrity of the film during the transfer steps [269, 270].

Since heterogeneous integration of 2D TMD layers is distinguishable from the tailored components as predicted to enable even more exotic functionalities impossible with conventional thin film semiconductor growth technologies [32, 271–280]. Generally, 2D TMD layers exert weak van der Waals (vdW) attraction to the underlying growth substrate. Because it allows individually assembling them in a layer-by-layer manner to achieve desired electronic structures that imply new venues for 2D heterojunction devices with tailored band offsets [270, 281]. Although the atom-thick semiconductor heterostructures technically challenging to integrate with conventional thin film growth because of their intrinsic lattice match constraints that impose crystallographic limitation for the choice of materials to integrate. Nevertheless, a few critical fundaments advantages inherent to 2D TMDs toward their exploration for novel technologies; (1) It requires to develop viable strategies to transfer 2D TMD layers from the originally grown substrate and integrate them on the secondary substrate with the desired functionalities (2) The intrinsic mechanical and electrical properties of the transferred 2D layers should not be compromised throughout the layer integration, that should also be uniformly preserved on a wafer scale. (3) The layer integration process should be universal for the kinds of 2D TMDs and different materials substrates without being limitation to specific kinds for technological versatility [270].

In the modern world, the most frequently used approach to transfer and integration of 2D TMD layers based on the chemical etching of the growth substrates involving protection polymers (e.g., polymethyl-methacrylate (PMMA)) and subsequent chemical lift-off [282–284]. However, this kind of strategy tends to result in the fragmentation of individual 2D layers as they employ solution-based chemicals to etch away both the protection layer and growth substrates (e.g., silicon dioxide (SiO2) or sapphire wafer) [283]. Though, they have scalability limitation issues in terms of heterogeneously stacking up 2D layers of multiple components in a controlled manner and difficult to apply for a variety of unconventional substrates. Additionally, the intrinsic material properties of 2D layers may compromise and damage by the used chemicals though their transfer and integration stages. Besides these advantages and disadvantages, chemical approaches herein are discussed along with a few important other approaches, however, there are several other TMDs layer transferring approaches being introduced based on different techniques, such as PMMA-assisted transfer, Polydimethylsiloxane (PDMS)-assisted transfer, Polystyrene (PS)-assisted transfer and other polymer-assisted transfer[269].

Layer Transfer from Water-Assisted 2D Layer Integration

Large area 2D TMDs layer integration on a substrate via a water-assisted 2D layer transferring technique is illustrated in Fig. 2.20. A complete process is usually achieved with these steps: (1) Transition metals deposition on the surface of growth substrates (SiO_2/Si) followed by their conversion to 2D TMD layers through chemical vapor deposition (CVD). (2) The grown 2D TMDs on SiO_2/Si substrates immersion

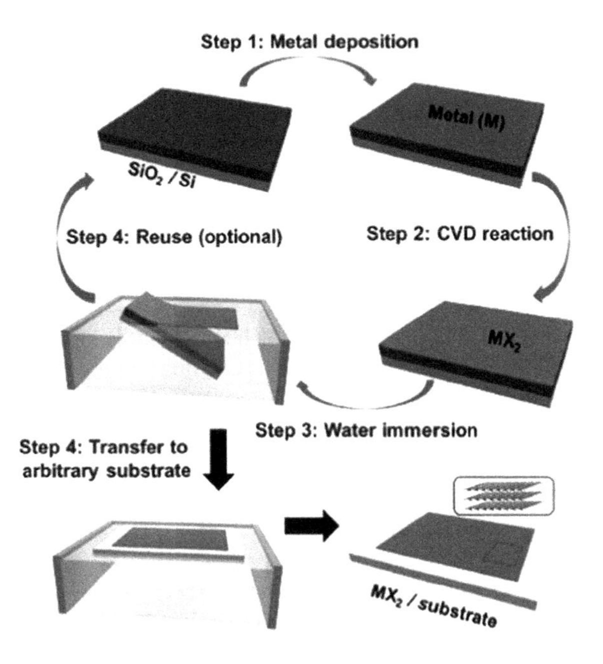

Fig. 2.20 Schematic for the illustration of water-assisted green integration to CVD-grown 2D TMD layers on arbitrary substrates (adopted from Watson et al. [268])

inside water followed by spontaneous 2D layer separation. (3) Transfer and integration of the delaminated 2D TMD layers onto secondary substrates inside water. (4) Originally grown substrate recycling to get additional 2D TMD growth (optional). The water-assisted 2D layer separation is generally carried out with the help of two slightly different ways. The first approach is based on a slow immersion of the entire 2D TMD-grown substrate inside water, while the second one is to deposit water droplets only on the sample surface to leverage its buoyancy. The simplest whole water process is without involving any kinds of additional chemicals to separate transfer and integrate the 2D TMD layers. Although generally also employed the combined use of polymeric protective materials (PMMA or Polyvinyl alcohol (PVA)) and chemical etchants for the removal of SiO_2 [283, 284]. However, in almost entire conventional approaches, the 2D TMDs' operational complexity inherent are susceptible to chemical degradation due to the corrosive nature of hydrogen fluoride (HF) or strong bases (sodium or potassium hydroxide (NaOH or KOH)) involvement in the processes [282, 284]. Therefore, the structural integrity is significantly altered by the solution-based chemicals (acetone) employed to rinse away the protective polymers that affect the polymer residuals. On the other hand, the water-assisted approach yields a completely clean and homogenous integration for the variety of 2D TMDs on a large centimeter scale. This approach also have an advantage to intrinsically free chemicals-associated structural damage, like the original SiO_2/Si substrates after 2D layer separation has reused for the subsequent growth of additional 2D TMD materials, as depicted in step-4 in Fig. 2.19 [270].

Metal-Assisted Transfer

Despite the various efforts to minimize the problem of polymer residues, it remains a critical issue to use polymer supports for transfer. To overcome this, alternative supports have been introduced that do not suffers from such drawbacks. Metal-assisted support has been recognized as suitable substitute due to their larger adhesion energy compared to polymers. This process reveals TMDs can have less prone to tearing. As a method outlined a Cu/TRT assembly has used to transfer CVD-grown MoS_2 from a SiO_2/Si substrate to a target [285], as the schematic is illustrated in Fig. 2.20. Though this approach successfully resolved the polymers residue issue, but it is not free from the cracks and holes in the transferred film. It is due to mainly mechanical strain incurred from peeling with thermal release tape (TRT). To advancement, another method has been also introduced by utilizing a Cu support layer but without TRT [286]. This approach is based on water intercalation to delaminate the MoS_2 from its growth substrate. The buoyancy force supplied by the water which key to preventing damage to the thin PDMS/PMMA/Cu/MoS_2 assembly during peeling, as the rigid Cu support.

Hence metal support methods are considered one of the more robust layers transferring approach, but it has also a similar drawback to polymer supports that requires removal via chemical etching in the last transferring step, therefore, it results damage in the films. Additionally, electron beam evaporation is a softer metal deposition method than sputtering, but it may also damage the film. The overall fabrication cost issue is also a crucial thing because this process is relatively expensive due to the used metal that restricts its use in industrial applications.

Impact of Layer Transfer Techniques on 2D TMDs Film Quality

By using different kinds of transfer approaches diverse TMDs have been successfully transferred onto a wide range of different substrates. In general for the 2D TMDs-based novel technological applications, it is essential to transferred films should have a high-quality fidelity to preserve the material's internal properties. This means the structural, chemical, and electronic properties of transferred TMD films should be carefully verify [282, 285, 287–293]. Since many transfer methodologies suffer from process-related weaknesses, such as trapped bubbles, polymer residues, cracks, or wrinkles. All these characteristic features may degrade the device's overall performance. Like, inhomogeneous or uncontrollable strain is detrimental to photoluminescence (PL) and optical applications [294–296]. Similarly, cracks, wrinkles, or polymer residue strongly affects the device resistivity and electron mobility [283, 297, 298]. Moreover, these distinct effects are different that significance depending on the application. Such as, polymer residue does not have a large influence over the PL signal of 2D TMDs [298]. Thus, in the case of optical applications, these issues may ignore. However, ideally, a perfect transfer entails the functional continuity of the 2D film before and after the transferring is only differentiate from the substrate. But in practice, it does not occur depends on the used method and modifications of the film results. Therefore, drawbacks associated with transferring CVD-grown TMD films are crucial things. Moreover, characterization techniques used to quantify these drawbacks and methods to improve the film quality post-transfer are also significant [269].

2.2 Conclusions

In the concluding view, one has discussed the typical material fabrication processes with their crucial parameters for the different kinds of scientific and technological uses. Therefore, this chapter work has provided descriptive knowledge about the synthetic approaches for the kinds of materials that is helpful to explore the desired properties for the specific technological purpose. More accurately one can say the different synthetic approaches of the two-dimensional nanomaterials largely affect their physical properties and potential applicability. Therefore, it is customary to know the process-dependent properties of the TMDs, thereby, one has also discussed the different synthetic approaches such as, top-down; including liquid exfoliation, mechanical exfoliation, electrochemical exfoliation, and bottom-up; including growth mechanism, metal chalcogenisation via sulfurization/selenization, thermolysis from thiosalts, transition metal—chalcogen precursors via vapor pressure reaction, TMDs alloys growth, epitaxial growth of TMDs, atomic layer deposition (ALD) and layer transfer. Thus the information about the different synthetic approaches helpful for their scientific interpretations and emerging applications in various areas. Based on their merits and demerits one can select suitable approach to employ for the targeted/general utility of the TMDs materials.

References

1. Chia X, Eng AYS, Ambrosi A, Tan SM et al (2015) Electrochemistry of nanostructured layeredtransition-metal dichalcogenides. Chem Rev 115:11941–11966
2. Choi W, Choudhary N, Han GH, Park J et al (2017) Recent developmentof two-dimensional transition metal dichalcogenides and their applications. Mater Today 20:116–130
3. Arul NS, Nithya VD (2019) Two dimensional transition metal dichalcogenides. Springer Nature, Singapore Pte Ltd.
4. Kolobov AV, Tominaga J (2016) Two-dimensional transition-metal dichalcogenides. Springer International Publishing, Switzerland
5. Dong R, Kuljanishvili I (2017) Review article: progress in fabrication of transition metal dichalcogenidesheterostructure systems. J Vac Sci Technol B 35:030803–030814
6. Ahmadi M, Zabihi O, Jeon S, Yoonessi M et al (2020) 2D transition metal dichalcogenide nanomaterials:advances, opportunities, and challenges in multifunctionalpolymer nanocomposites. J Mater Chem A 8:845–883
7. Zhang Q, Mei L, Cao X, Tang Y et al (2020) Intercalation and exfoliation chemistries oftransition metal dichalcogenides. J Mater Chem A 8:15417–15444
8. Spah R, Lux-Steiner M, Obergfell M, Bucher E et al (1985) n-$MoSe_2$/p-WSe_2 heterojunctions. Appl Phys Lett 47:871–873

9. Butler SZ, Hollen SM, Cao L, Cui Y et al (2013) Progress, challenges, and opportunities in two-dimensional materials beyond graphene. ACS Nano 7:2898–2926
10. Frindt R (1966) Single crystals of MoS2 several molecular layers thick. J Appl Phys 37:1928–1929
11. Frindt RF (1972) Superconductivity in ultrathin $NbSe_2$ layers. Phys Rev Lett 28:299–301
12. Consadori F, Fife AA, Frindt RF, Gygax S (1971) Construction and properties of weak-link detectors using superconducting layer structures. Appl Phys Lett 18:233–235
13. Novoselov KS, Geim AK, Morozov SV, Jiang D et al (2005) Two-dimensional gas of massless Dirac fermions in graphene. Nature 438:197–200
14. Halim U, Zheng CR, Chen Y, Lin Z et al (2013) A rational design of cosolvent exfoliation of layered materials by directly probing liquid–solid interaction. Nat Commun 4
15. Nguyen EP, Carey BJ, Daeneke T, Ou JZ et al (2015) Investigation of two-solvent grinding-assisted liquid phase exfoliation of layered MoS_2. Chem Mater 27:53–59
16. Boo WJ, Liu J, Sue HJ (2006) Fracture behaviour of nanoplatelet reinforced polymer nanocomposites. Mater Sci Technol 22:829–834
17. Kim J, Han NM, Kim J, Lee J et al (2018) Highly conductive and fracture-resistant epoxy composite based on non-oxidized graphene flake aerogel. ACS Appl Mater Interf 10:37507–37516
18. Naebe M, Abolhasani MM, Khayyam H, Amini A et al (2016) Crack damage in polymers and composites: a review. Polym Rev 56:31–69
19. Boo WJ, Sun L, Warren GL, Moghbelli E et al (2007) Effect of nanoplatelet aspect ratio on mechanical properties of epoxy nanocomposites. Polymer 48:1075–1082
20. Li B, Zhong W-H (2011) Review on polymer/graphite nanoplatelet nanocomposites. J Mater Sci 46:5595–5614
21. Mas-Balleste R, Gomez-Navarro C, Gomez-Herrero J, Zamora F (2011) 2D materials: to graphene and beyond. Nanoscale 3:20–30
22. Coleman JN, Lotya M, O'Neill A, Bergin SD et al (2011) Two-dimensional nanosheets produced by liquid exfoliation of layered materials. Science 2011(331):568–571
23. Nicolosi V, Chhowalla M, Kanatzidis MG, Strano MS et al (2013) Liquid exfoliation of layered materials. Science 340:1226419
24. Zhu S, Gong L, Xie J, Gu Z et al (2017) Design, synthesis, and surface modification of materials based on transition-metal dichalcogenides for biomedical applications. Small Methods 1:1700220
25. Tang Q, Zhou Z (2013) Graphene-analogous low-dimensional materials. Prog Mater Sci 58:1244–1315
26. Yao Y, Tolentino L, Yang Z, Song X et al (2013) High-concentration aqueous dispersions of MoS_2. Adv Funct Mater 23:3577–3583
27. Zhang Y, Zuo L, Huang Y, Zhang L et al (2015) In-situ growth of few-layered MoS2 nanosheets on highly porous carbon aerogel as advanced electrocatalysts for hydrogen evolution reaction. ACS Sustain Chem Eng 3:3140–3148
28. Peng X, Manna L, Yang W, Wickham J et al (2000) Shape control of CdSe nanocrystals. Nature 2000(404):59–61
29. Kongkanand A, Tvrdy K, Takechi K, Kuno M et al (2008) Quantum dot solar cells. Tuning photoresponse through size and shape control of $CdSe-TiO_2$ Architecture. J Am Chem Soc 130:4007–4015
30. Anand DK, Tiwari P, Kim Y, Lee H (2017) Bifunctional oxygen electrocatalysis through chemical bonding of transition metal chalcogenides on conductive carbons. Adv Energy Mater 7:1602217
31. Das S, Robinson JA, Dubey M, Terrones H et al (2015) Beyond graphene: progress in novel two-dimensional materials and van der Waals Solids. Annu Rev Mater Res 45:1–27
32. Geim AK, Grigorieva IV (2013) Van der Waals heterostructures. Nature 499:419–425
33. Britnell L, Gorbachev RV, Jalil R, Belle BD et al (2012) Field-effect tunneling transistor based on vertical graphene heterostructures. Science 335:947–950

34. Gao G, Gao W, Cannuccia E, Taha-Tijerina J et al (2012) Artificially stacked atomic layers: toward new van der Waals solids. Nano Lett 2012(12):3518–3525
35. Gupta A, Sakthivel T, Seal S (2015) Recent development in 2D materials beyond graphene. Prog Mater Sci 73:44–126
36. Balendhran S, Ou JZ, Bhaskaran M, Sriram S et al (2012) Atomically thin layers of MoS2via a two step thermal evaporation–exfoliation method. Nanoscale 2012(4):461–466
37. Joensen, P., Frindt, R. F., Morrison, S. R. 1986"Single-layer MoS_2" Mater. Res. Bull., 21, 457–461
38. Hoang Huy VP, Ahn YN, Hur J (2021) Recent advances in transition metal dichalcogenide cathode materials for aqueous rechargeable multivalent metal-ion batteries. Nanomaterials 11:1517
39. Samal R, Sanyal G, Chakraborty B, Rout CS (2021) Two-dimensional transition metal phosphoroustrichalcogenides (MPX3): a review on emergingtrends, current state and future perspectives. J Mater Chem A 9:2560–2591
40. Torrisi F, Hasan T, Wu W, Sunet Z et al (2012) Inkjet-printed graphene electronics. ACS Nano 6:2992–3006
41. Zeng X, Hirwa H, Metel S, Nicolosi V et al (2018) Solution processed thin film transistor from liquidphase exfoliated MoS_2 flakes. Solid State Electron 141:58–64
42. Blake P, Brimicombe PD, Nair RR, Booth TJ et al (2008) Graphene-based liquid crystal device. Nano Lett 8:1704–1708
43. Xiao J, Choi D, Cosimbescu L, Koech P et al (2010) Exfoliated MoS_2 nanocomposite as ananode material for lithium ion batteries. Chem Mater 22:4522–4524
44. Shen J, He Y, Wu J, Gao C et al (2015) Liquid phase exfoliation of two-dimensional materials by directly probing and matchingsurface tension components. Nano Lett 15:5449–5454
45. Zeng Z, Yin Z, Huang X, Li H et al (2011) Single-layer semiconducting nanosheets: high-yield preparation and device fabrication. Angew Chemie Int Ed 50:11093–11097
46. Mukherjee D, Sampath S (2017) Few-layer iron selenophosphate, FePSe3: efficient electrocatalyst toward water splitting and oxygen reduction reactions. ACS Appl Energy Mater 1:220–231
47. Mukherjee D, Austeria PM, Sampath S (2016) Two-dimensional, few-layer phosphochalcogenide, $FePS_3$: a new catalyst for electrochemical hydrogen evolution over wide pH range. ACS Energy Lett 1:367–372
48. Jenjeti RN, Kumar R, Austeria MP, Sampath S (2018) Field effect transistor based on layered $NiPS_3$. Sci Rep 8:1–9
49. Kumar R, Jenjeti RN, Austeria MP, Sampath S (2019) Bulk and few-layer MnPS3: a new candidate for field effect transistors and UV photodetectors. J Mater Chem C 7:324–329
50. Kumar R, Jenjeti RN, Sampath S (2019) Bulk and few-layer 2D, p-MnPS3 for sensitive and selective moisture sensing. Adv Mater Interfaces 6:1900666
51. Kumar R, Jenjeti RN, Sampath S (2020) Two-dimensional, few-layer MnPS3 for selective NO_2 gas sensing under ambient conditions. ACS Sens 5:404–411
52. Jenjeti RN, Kumar R, Sampath S (2019) Two-dimensional, few-layer NiPS3 for flexible humidity sensor with high selectivity. J Mater Chem A 7:14545–14551
53. Zhu W, Gan W, Muhammad Z, Wang C et al (2018) Exfoliation of ultrathin FePS3 layers as a promising electrocatalyst for the oxygen evolution reaction. Chem Commun 54:4481–4484
54. Silipigni L, Basile A, Barreca F, De Luca G et al (2017) Partial reduction of graphene oxide upon intercalation into exfoliated manganese thiophosphate. Philos Mag 97:2484–2495
55. Qing Y, Jin W, Xin-Yao S, Tao W et al (2019) Pulse generation of erbium-doped fiber laser based on liquid-exfoliated $FePS_3$. Chin Phys B 28:084208
56. Hao Y, Huang A, Han S, Huang H et al (2020) Plasma-treated ultrathin ternary FePSe3 nanosheets as a bifunctional electrocatalyst for efficient zinc–air batteries. ACS Appl Mater Interfaces 12:29393–29403
57. Huang H, Shang M, Zou Y, Song W et al (2019) Iron phosphorus trichalcogenide ultrathin nanosheets: enhanced photoelectrochemical activity under visible-light irradiation. Nanoscale 11:21188–21195

58. Song B, Li K, Yin Y, Wu T et al (2017) Tuning mixed nickel iron phosphosulfide nanosheet electrocatalysts for enhanced hydrogen and oxygen evolution. ACS Catal 7:8549–8557
59. Luong DH, Phan TL, Ghimire G, Duong DL et al (2019) Revealingantiferromagnetic transition of van der Waals MnPS3 via vertical tunneling electrical resistance measurement. APL Mater 7:081102
60. Lee JH, Jang WS, Han SW, Baik HK (2014) Efficient hydrogen evolution by mechanically strained MoS_2 nanosheets. Langmuir 30:9866–9873
61. Shi S, Sun Z, Hu YH (2018) Synthesis, stabilization and applications of 2-dimensional 1T metallic MoS_2. J Mater Chem A 6:23932–23977
62. Fan X, Xu P, Zhou D, Sun Y et al (2015) Fast and efficient preparation of exfoliated 2H MoS2 nanosheets by sonication-assisted lithiumintercalation and infrared laser-induced 1T to 2H phase reversion. Nano Lett 15:5956–5960
63. Eda G, Yamaguchi H, Voiry D, Fujita T et al (2011) Photoluminescence from chemically exfoliated MoS_2. Nano Lett 11:5111–5116
64. Li H, Wu J, Yin Z, Zhang H (2014) Preparation and applications of mechanically exfoliated single-layer andmultilayer MoS_2 and WSe_2 nanosheets. Acc Chem Res 47:1067–1075
65. Yuan H, Dubbink D, Besselink R, ten Elshof JE (2015) The rapid exfoliation and subsequent restacking oflayered titanates driven by an acid-base reaction. Angew Chemie Int Ed 54:9239–9243
66. Kang K, Chen S, Yang EH (2020) Synthesis, modeling, and characterization of 2D materials, and their heterostructures. Elsevier, pp 247–264, Chapter 12
67. Qi H, Wang L, Sun J, Long Y (2018) Production Methods of Van der Waals heterostructures based on transition metal dichalcogenides. Crystals 8:35
68. Yi M, Shen Z (2015) A review on mechanical exfoliation for the scalable production of graphene. J Mater Chem A 3:11700–11715
69. Novoselov KS, Jiang D, Schedin F, Booth TJ et al (2005) Two-dimensional atomic crystals. Proc Natl Acad Sci USA 102:10451–10453
70. Lee C, Yan H, Brus LE, Heinz TF et al (2010) Anomalous lattice vibrations of single- and fewlayer MoS_2. ACS Nano 4:2695–2700
71. Bertolazzi S, Brivio J, Kis A (2011) Stretching and breaking of ultrathin MoS_2. ACS Nano 5:9703–9709
72. Radisavljevic B, Radenovic A, Brivio J, Giacometti V et al (2011) Single-layer MoS_2 transistors. Nat Nanotechnol 6:147–150
73. Yin Z, Li H, Li H, Jiang L et al (2012) Single-layer MoS2 phototransistors. ACS Nano 6:74–80
74. Velický M, Donnelly GE, Hendren WR, McFarland S et al (2018) Mechanism of gold-assisted exfoliation of centimeter-sized transition-metal dichalcogenide monolayers. ACS Nano 12:10463–10472
75. Shim J, Bae SH, Kong W, Lee D et al (2018) Controlled crack propagation for atomic precision handling of wafer-scale two-dimensional materials. Science 362:665–670
76. Lu X, Utama MIB, Zhang J, Zhao Y et al (2013) Layer-by-layer thinning of MoS2 by thermal annealing. Nanoscale 5:8904–8908
77. Hu L, Shan X, Wu Y, Zhao J et al (2017) Laser thinning and patterning of MoS2 with layer-by-layer precision. Sci Rep 7:15538
78. Li H, Zhang Q, Yap CCR, Tay BK et al (2012) From bulk to monolayer MoS_2: evolution of Raman scattering. Adv Funct Mater 22:1385–1390
79. Mohanty US (2011) Electrodeposition: a versatile and inexpensive tool for the synthesisof nanoparticles, nanorods, nanowires, and nanoclusters of metals. J Appl Electrochem 41:257–270
80. Nasirpouri F (2017) Electrodeposition of nanostructuredmaterials. Springer International Publishing
81. Albu-Yaron A, Levy-Clement C, Hutchison JL (1999) A study on MoS_2 thin films electrochemically deposited in ethylene glycol at 165 C. Electrochem Solid-State Lett 2:627–630

82. Merki D, Fierro S, Vrubel H, Hu X (2011) Amorphous molybdenum sulfide films as catalystsfor electrochemical hydrogen production in water. Chem Sci 2:1262–1267
83. Murugesan S, Akkineni A, Chou BP, Glaz MS et al (2013) Roomtemperature electrodeposition of molybdenum sulfide for catalytic and photoluminescence Applications. ACS Nano 7:8199–8205
84. Coogan A, Gunko YK (2021) Solution-based bottom-up synthesis of group VI transition metal dichalcogenides and their applications. Mater Adv 2:146–164
85. Morey GW, Niggli P (1913) The hydrothermal formation of silicates: a review. J Am Chem Soc 35:1086–1130
86. Dubois T, Demazeau G (1994) Preparation of Fe_3O_4 fine particles through a solvothermal process. Mater Lett 19:38–47
87. Zhan JH, Zhang ZD, Qian XF, Wang C et al (1998) Solvothermal synthesis of nanocrystalline MoS2 from MoO3 and elemental sulfur. J Solid State Chem 141:270–273
88. Fan R, Chen X, Chen Z (2000) A novel route to obtain molybdenum dichalcogenides by hydrothermal reaction. Chem Lett 29:920–921
89. Chakravarty D, Late DJ (2015) Microwave and hydrothermal syntheses of WSe_2 micro/nanorods and their application in supercapacitors. RSC Adv 5:21700–21709
90. Liu Y, Zhang N, Kang H, Shang M et al (2015) WS_2 nanowires as a high-performance anode for sodium-ion batteries Chem Eur J 21:11878–11884
91. Wang PP, Sun H, Ji Y, Li W et al (2014) Three-dimensional assembly of single-layered MoS_2. Adv Mater 26:964–969
92. Lamer VK, Dinegar RH (1950) Theory, production and mechanism of formation of monodispersed hydrosols. J Am Chem Soc 72:4847–4854
93. Murray CB, Norris DJ, Bawendi MG (1993) Synthesis and characterization of nearly monodisperse CdE (E = sulfur, selenium, tellurium) semiconductor nanocrystallites. J Am Chem Soc 115:8706–8715
94. Bai X, Purcell-Milton F, Gunko Y (2019) Optical properties, synthesis, and potential applications of Cu-based ternary or quaternary anisotropic quantum dots, polytypic nanocrystals, and core/shell heterostructures. Nanomaterials 9:85
95. Murray CB, Kagan CR, Bawendi MG (2000) Synthesis and characterization of monodisperse nanocrystals and close-packed nanocrystal assemblies. Annu Rev Mater Sci 30:545–610
96. Gong YJ, Lin JH, Wang XL, Shi G et al (2014) Vertical and in-plane heterostructures from WS_2/ MoS_2 monolayers. Nat Mater 13:1135–1142
97. Li XF, Lin MW, Lin JH, Huang B et al (2016) Two-dimensional GaSe/$MoSe_2$ misfit bilayer heterojunctions by van der Waals epitaxy. Sci Adv 2:1501882
98. Fu L, Sun YY, Wu N, Mendes RG et al (2016) Directgrowth of MoS_2/h-BN heterostructures via a sulfide-resistant alloy. ACS Nano 10:2063–2070
99. Zheng BY, Ma C, Li D, Lan JY et al (2018) Bandalignment engineering in two-dimensional lateral heterostructures. J Am Chem Soc 140:11193–11197
100. Sahoo PK, Memaran S, Xin Y, Balicas L (2018) One-pot growth of two-dimensional lateral heterostructuresvia sequential edge-epitaxy. Nature 553:63–67
101. Zhang ZW, Chen P, Duan XD, Zang KT (2017) Duan, Robust epitaxial growth of two-dimensional heterostructures, multiheterostructures, and superlattices. Science 357:788–792
102. Liu Y, Zhang S, He J, Wang ZM et al (2019) Recent progress in the fabrication, properties, and devices of heterostructures based on 2D materials. Nano-Micro Lett 11(13):1–24
103. Liu KK, Zhang W, Lee YH, Lin YC et al (2012) Growth of large-area and highly crystalline MoS_2 thin layers on insulating substrates. Nano Lett 12:1538–1544
104. Huang JK, Pu J, Hsu CL, Chiu MH et al (2013) Large-area synthesis of highly crystalline WSe2 monolayers and device applications. ACS Nano 8:923–930
105. Liu B, Fathi M, Chen L, Abbas A et al (2015) Chemical vapor deposition growth of monolayer WSe2 with tunable device characteristics and growth mechanism study. ACS Nano 9:6119–6127
106. Chen J, Zhao X, Tan SJ, Xu H et al (2017) Chemical vapor deposition of large-sizemonolayer $MoSe_2$ crystals on molten glass. J Am Chem Soc 139:1073–1076

107. Zheng B, Chen Y (2017) Controllable growth of monolayer MoS_2 and $MoSe_2$ crystals using three-temperature-zone furnace. In: IOP conference series: materials science and engineering. IOP Publishing, pp 012085, vol 274
108. Shaw JC, Zhou H, Chen Y, Weiss NO et al (2014) Chemical vapordeposition growth of monolayer $MoSe_2$ nanosheets. Nano Res 7:511–517
109. Wang X, Gong Y, Shi G, Chow WL et al (2014) Chemical vapor deposition growth of crystalline monolayer $MoSe_2$. ACS Nano 8:5125–5131
110. Zhang Z, Niu J, Yang P, Gong Y et al (2017) Van der Waals epitaxial growth of 2D metallic vanadium diselenide single crystals and their extra-high electrical conductivity. AdvMater 29:1702359
111. Wang H, Huang X, Lin J, Cui J et al (2017) High-quality monolayer superconductor$NbSe_2$ grown by chemical vapour deposition. Nat Commun 8:394
112. Song JG, Park J, Lee W, Choi T et al (2013) Layer-controlled, wafer-scale, and conformal synthesis of tungsten disulphide nanosheets using atomic layer deposition. ACS Nano 7(12):11333–11340
113. You J, Hossain MD, Luo Z (2018) Synthesis of 2D transition metal dichalcogenides by chemical vapor deposition with controlled layer number and morphology. Nano Converg 5:26
114. Cain JD, Shi F, Wu J, Dravid VP (2016) Growth mechanism of transition metal dichalcogenide monolayers: the role of self-seeding fullerene nuclei. ACS Nano 10:5440–5445
115. Zhou D, Shu H, Hu C, Jiang L et al (2018) Unveiling the growth mechanism of MoS2 with chemical vapor deposition: from two-dimensional planar nucleation to self-seeding nucleation. Cryst Growth Des 18:1012–1019
116. Imanishi N, Kanamura K, Takehara Z (1992) Synthesis of MoS2 thin film by chemical vapor deposition method and discharge characteristics as a cathode of the lithium secondary battery. J Electrochem Soc 139:2082
117. Lee YH, Zhang XQ, Zhang W, Chang MT et al (2012) Synthesis of large-area MoS_2 atomic layers with chemical vapor deposition. Adv Mater 24:2320–2325
118. Zhang W, Li X, Jiang T, Song J et al (2015) CVD synthesis of Mo(12x)WxS2 and MoS2(12x) Se2x alloy monolayers aimed at tuning the bandgap of molybdenum disulphide. Nanoscale 7:13554–13560
119. Kang K, Godin K, Yang EH (2015) The growth scale and kinetics of WS2 monolayers under varying H_2 concentration. Sci Rep 5:13205
120. Wang S, Rong Y, Fan Y, Pacios M et al (2014) Shape evolution of monolayer MoS2 crystals grown by chemical vapor deposition. Chem Mater 26:6371–6379
121. Lee JS, Choi SH, Yun SJ, Kim YI et al (2018) Wafer-scale single-crystal hexagonal boron nitride film via self-collimated grain formation. Science 362:817–821
122. Nishizawa J, Kurabayashi T (1990) Mechanism of gallium arsenide MOCVD. 41:958–962
123. Nishizawa J (1983) On the reaction mechanism of GaAs MOCVD. J Electrochem Soc 130:413
124. Soga T, Hattori T, Sakai S, Takeyasu M et al (1984) MOCVD growth of GaAs on Si substrates with AlGaP and strained superlattice layers. Electron Lett 20:916–918
125. Kang K, Xie S, Huang L, Han Y et al (2015) High-mobility three-atom-thick semiconducting films with wafer-scale homogeneity. Nature 520:656–660
126. Eichfeld SM, Hossain L, Lin YC, Piasecki AF et al (2015) Highly scalable, atomically thin WSe2 grown via metalorganic chemical vapor deposition. ACS Nano 9:2080–2087
127. Kalanyan B, Kimes WA, Beams R, Stranick SJ et al (2017) Rapid wafer-scale growth of polycrystalline 2H-MoS2 by pulsed metalorganic chemical vapor deposition. Chem Mater 29:6279–6288
128. Binnewies M, Glaum R, Schmidt M, Schmidt P (2013) Chemical vapor transport reactions—a historical review. Zeitschrift für Anorg und Allg Chemie 639:219–229
129. Dave M, Vaidya R, Patel SG, Jani AR (2004) High pressure effect on MoS2 and $MoSe_2$ single crystals grownby CVT method. Bull Mater Sci 27:213–216
130. Ubaldini A, Jacimovic J, Ubrig N, Giannini E (2013) Chloride-driven chemical vapor transport method for crystal growth of transition metal dichalcogenides. Cryst Growth Des 13:4453–4459

131. Hallam T, Monaghan S, Gity F, Ansari L et al (2017) Rhenium-doped MoS_2 films. Appl Phys Lett 111:203101
132. Sasaki S, Kobayashi Y, Liu Z, Suenaga K et al (2016) Growth and optical properties of Nb-doped WS_2 monolayers. Appl Phys Express 9:071201
133. Chae WH, Cain JD, Hanson ED, Murthy AA et al (2017) Substrate-induced strain and charge dopingin CVD-grown monolayer MoS_2. Appl Phys Lett 111:143106
134. Kim Y, Bark H, Kang B, Lee C (2019) Wafer-scale substitutional doping of monolayer MoS2 films for high performance optoelectronic devices. ACS Appl Mater Interfaces 11:12613–12621
135. Gusakova J, Wang X, Shiau LL, Krivosheeva A et al (2017) Electronic properties of bulk and monolayer TMDs: theoretical study within DFTframework (GVJ-2e method). Phys Status Solidi 214:1700218
136. Cadiz F, Courtade E, Robert C, Wang G et al (2017) Excitonic linewidth approaching the homogeneous limit in MoS2-based van der Waalsheterostructures. Phys Rev X7:021026–12
137. Tongay S, Narang DS, Kang J, Fan W et al (2014) Two-dimensional semiconductor alloys: monolayer $Mo_{12x}W_xSe_2$. Appl Phys Lett 104:012101
138. Zhang K, Bersch BM, Joshi J, Addou R et al (2018) Tuning the electronic and photonic properties of monolayer MoS_2 via in situ rhenium substitutional doping. Adv Funct Mater 28:1706950
139. Suh J, Park TE, Lin DY, Fu D et al (2014) Doping against the native propensity of MoS_2: degenerate hole doping by cation substitution. Nano Lett 14:6976–6982
140. Li H, Duan X, Wu X, Zhuang X et al (2014) Growth of alloy MoS2xSe2(1x) nanosheets with fully tunable chemical compositions and optical properties. J Am Chem Soc 136:3756–3759
141. Duan X, Wang C, Fan Z, Hao G et al (2016) Synthesis of WS2xSe22x alloy nanosheets with composition-tunable electronic properties. Nano Lett 16:264–269
142. Yang L, Majumdar K, Liu H, Du Y et al (2014) Chloride molecular doping technique on 2D materials: WS_2 and MoS_2. Nano Lett 14:6275–6280
143. Azcatl A, Qin X, Prakash A, Zhang C et al (2016) Covalent nitrogen doping and compressive strain in MoS_2 by remote N_2 plasma exposure. Nano Lett 16:5437–5443
144. Jin Z, Shin S, Kwon DH, Han SJ et al (2014) Novel chemical route for atomic layer deposition of MoS2 thin film on SiO2/Sisubstrate. Nanoscale 6:14453–14458
145. Cho AY, Arthur JR (1975) Molecular beam epitaxy. Prog Solid State Chem 10:157–191
146. Li H, Duan X, Wu X, Zhuang X et al (2014) Growth of alloy $MoS_{2x}Se_{2(1-x)}$ nanosheets with fully tunable chemical compositions and optical properties. J Am Chem Soc 136:3756–3759
147. Defo RK, Fang S, Shirodkar SN, Tritsaris GA (2016) Strain dependence of band gaps and excitonenergies in pure and mixed transition-metal dichalcogenides. Phys Rev B 94:155310
148. Zhang KH, Feng SM, Wang JJ, Azcatl A et al (2015) Manganese doping of monolayer MoS_2: the substrateis critical. Nano Lett 15:6586–6591
149. Woodall JM, Freeouf JL, Pettit GD, Jackson TN et al (1981) Ohmic contacts to normal-GaAs using graded band-gaplayers of $Ga_{1-X}In_xAs$ grown by molecular-beam epitaxy. J Vac Sci Technol 19:626–627
150. Haeni JH, Theis CD, Schlom DG (2000) RHEED intensity oscillations for the stoichiometric growth of $SrTiO_3$ thin films by reactive molecular beam epitaxy. J Electroceram 4:385–391
151. Springthorpe AJ, Humphreys TP, Majeed A, Moore WT (1989) In-situ growth-rate measurements during molecular-beam epitaxyusing an optical-pyrometer. Appl Phys Lett 55:2138–2140
152. Zhou JJ, Li Y, Thompson P, Chu R et al (1997) Continuous in situ growth rate extraction using pyrometric interferometry and laser reflectance measurement during molecular beam epitaxy. J Electron Mater 26:1083–1089
153. Koma A, Sunouchi K, Miyajima T (1984) Fabrication and characterization of heterostructures with subnanometer thickness. Microelectron Eng 2:129–136
154. Tiefenbacher S, Sehnert H, Pettenkofer C, Jaegermann W (1994) Epitaxial-films of WS_2 by metal-organic Van-Der-Waals epitaxy (MO-VDWE). Surf Sci 318:L1161–L1164

155. Kreis C, Traving M, Adelung R, Kipp L et al (2000) Tracing the valence band maximum during epitaxial growth ofHfS_2 on WSe_2. Appl Surf Sci 166:17–22
156. Kreis C, Werth S, Adelung R, Kipp L et al (2002) Surface resonances at transition metal dichalcogenide heterostructures. Phys Rev B 65:153314
157. Zhang Y, Chang TR, Zhou B, Cui YT et al (2014) Direct observation of the transition from indirect to direct bandgap in atomically thin epitaxial $MoSe_2$. Nat Nanotechnol 9:111–115
158. Yue RY, Barton AT, Zhu H, Azcatl A et al (2015) $HfSe_2$ thin films: 2D transition metal dichalcogenides grown by molecular beam epitaxy. ACS Nano 9:474–480
159. Vishwanath S, Liu XY, Rouvimov S, Mende PC et al (2015) Comprehensive structural and optical characterization of MBE grown MoSe2 on graphite, CaF2 and graphene. 2D Mater 2:024007
160. Liu HJ, Jiao L, Yang F, Cai Y et al (2014) Dense network of one-dimensional midgap metallic modes inmonolayer $MoSe_2$ and their spatial undulations. Phys Rev Lett 113:066105
161. Ugeda MM, Bradley AJ, Shi SF, da Jornada FH et al (2014) Giant bandgap renormalization and excitonic effects in a monolayer transition metal dichalcogenide semiconductor. Nat Mater 13:1091–1095
162. Walsh LA, Hinkle CL (2017) Van der Waals epitaxy: 2D materials and topological insulators. Appl Mater Today 9:504–515
163. Vishwanath S, Liu XY, Rouvimov S, Basile L et al (2016) Controllable growth of layered selenide and tellurideheterostructures and superlattices using molecular beam epitaxy. J Mater Res 31:900–910
164. Aminalragia-Giamini S, Marquez-Velasco J, Tsipas P, Tsoutsou D (2017) Molecular beam epitaxy of thin $HfTe_2$ semimetal films. 2D Mater 4:015001
165. Tsipas P, Tsoutsou D, Marquez-Velasco J, Aretouli KE et al (2015) Epitaxial $ZrSe_2$/ $MoSe_2$ semiconductor v.d. Waals heterostructures on wide band gap AlN substrates. Microelectron Eng 147:269–272
166. Vishwanath S, Liu XY, Rouvimov S, Mende PC et al (2015) Comprehensive structural and opticalcharacterization of MBE grown$MoSe_2$ on graphite, CaF2 and graphene. 2D Mater 2:024007
167. Liu HJ, Jiao L, Xie L, Yang F et al (2015) Molecular-beam epitaxy of monolayer and bilayer WSe_2: a scanning tunnelling microscopy/spectroscopy study and deduction of exciton binding energy. 2D Mater 2:034004
168. Venables J (2000) Introduction to surface and thin film processes. Cambridge University Press, Cambrige, 372 pp
169. Saidi WA (2014) Van der Waals epitaxial growth of transition metal dichalcogenides on pristine and N-doped graphene. Cryst Growth Des 14:4920–4928
170. Tiefenbacher S, Pettenkofer C, Jaegermann W (2000) Moire patternin LEED obtained by van der Waals epitaxy of lattice mismatchedWS_2/$MoTe_2$(0001) heterointerfaces. Surf Sci 450:181–190
171. Diaz HC, Chaghi R, Ma YJ, Batzill M (2015) Molecular beamepitaxy of the van der Waals heterostructure $MoTe_2$ on MoS_2:phase, thermal, and chemical stability. 2D Mater 2:044010
172. Olsen GH (1975) Interfacial lattice mismatch effects in III-V compounds. J Cryst Growth 31(Dec):223–239
173. Hull R, Fischercolbrie A (1987) Nucleation of GaAs on Siexperimental evidence for a 3-dimensional critical transition. Appl Phys Lett 50:851–853
174. Krost A, Schnabel RF, Heinrichsdorff F, Rossow U (1994) Defect reduction in GaAs and InP grown on planarSi(111) and on patterned Si(001) substrates. J Cryst Growth 145(1–4):314–320
175. Novoselov KS, Geim AK, Morozov SV, Jiang D et al (2004) Electric field effect in atomically thin carbon films. Science 306:666–669
176. Saiki K, Ueno K, Shimada T, Koma A (1989) Application of Vander Waals epitaxy to highly heterogeneous systems. J Cryst Growth 95(1–4):603–606
177. Ueno K, Saiki K, Shimada T, Koma A (1990) Epitaxial-growth oftransition-metal dichalcogenides on cleaved faces of mica. J Vac Sci Technol A 8:68–72

178. Yue R, Nie Y, Walsh LA, Addou R et al (2017) Nucleation and growth of WSe_2: enabling large grain transition metal dichalcogenides. 2D Mater 4:045019
179. Jiao L, Liu HJ, Chen JL, Yi Y et al (2015) Molecular-beam epitaxy of monolayer $MoSe_2$: growth characteristics and domain boundary formation. New J Phys 17:053023
180. Koma A, Ueno K, Saiki K (1991) Heteroepitaxial growth by Vander Waals interaction in one-dimensional, 2-dimensional and 3-dimensional materials. J Cryst Growth 111:1029–1032
181. Ueno K, Shimada T, Saiki K, Koma A (1990) Heteroepitaxial growth of layered transition-metal dichalcogenides on sulfur terminated GaAs (111) surfaces. Appl Phys Lett 56:327–329
182. Nishikawa H, Shimada T, Koma A (1996) Epitaxial growth of $TiSe_2$ thin films on Se-terminated GaAs(111). B J Vac Sci Technol A 14:2893–2896
183. Dong SN, Liu XY, Li X, Kanzyuba V, Yoo T, Rouvimov S et al (2016) Room temperature weak ferromagnetism in $Sn_{1-x}Mn_xSe_2$ 2D films grown by molecular beam epitaxy. APL Mater 4:032601
184. Hotta T, Tokuda T, Zhao S, Watanabe K et al (2016) Molecular beam epitaxy growth of monolayer niobiumdiselenide flakes. Appl Phys Lett 109:133101
185. Aretouli KE, Tsoutsou D, Tsipas P, Marquez-Velasco J et al (2016) Epitaxial 2D$SnSe_2$/2D WSe_2 van der Waals heterostructures. ACS Appl Mater Interfaces 8:23222–23229
186. Ueno K, Takeda N, Sasaki K, Koma A (1997) Investigation of the growth mechanism of layered semiconductor GaSe. Appl Surf Sci 113:38–42
187. Koma A, Yoshimura K (1986) Ultra sharp interfaces grown with vanderwaals epitaxy. Surf Sci 174:556–560
188. Ohuchi FS, Shimada T, Parkinson BA, Ueno K et al (1991) Growth of $MoSe_2$ thin-films with Van der Waals epitaxy. J Cryst Growth 111:1033–1037
189. Parkinson BA, Ohuchi FS, Ueno K, Koma A (1991) Periodic lattice-distortions as a result of lattice mismatch in epitaxial-films of 2-dimensional materials. Appl Phys Lett 58:472–474
190. Shimada T, Ohuchi FS, Koma A (1993) Polytypes and charge density waves of ultrathin TaS2 films grown by Van-Der-Waalsepitaxy. Surf Sci 291:57–66
191. Murata H, Koma A (1999) Modulated STM images of ultrathin $MoSe_2$ films grown on MoS_2(0001) studied by STM/STS. Phys Rev B 59:10327–10334
192. Diaz HC, Ma YJ, Chaghi R, Batzill M (2016) High density of(pseudo) periodic twin-grain boundaries in molecular beamepitaxy-grown van der Waals heterostructure: $MoTe_2/MoS_2$. Appl Phys Lett 108:191606
193. Shimada T, Yamamoto H, Saiki K, Koma A (1990) RHEED intensity oscillation during epitaxial-growth of layered materials. Jpn J Appl Phys 29:L2096–L2098
194. Ohuchi FS, Parkinson BA, Ueno K, Koma A (1990) Van derWaals epitaxial-growth and characterization of MoSe2 thin-filmson SnS_2. J Appl Phys 68:2168–2175
195. Xenogiannopoulou E, Tsipas P, Aretouli KE, Tsoutsou D et al (2015) High-quality, large-area$MoSe_2$ and $MoSe_2/Bi_2Se_3$ heterostructures on AlN(0001)/Si(111)substrates by molecular beam epitaxy. Nanoscale 7:7896–7905
196. Park JH, Vishwanath S, Liu XY, Zhou HW et al (2016) Scanning tunneling microscopy and spectroscopy of air exposure effects on molecular beam epitaxy grownWSe_2 monolayers and bilayers. ACS Nano 10:4258–4267
197. Gong C, Zhang HJ, Wang W, Colombo L (2015) "Band alignment of two-dimensional transition metaldichalcogenides: application in tunnel field effect transistors. Appl Phys Lett 107:053513
198. Ago H, Fukamachi S, Endo H, Solis-Fernandez P et al (2016) Visualization of grain structure andboundaries of polycrystalline graphene and two-dimensionalmaterials by epitaxial growth of transition metal dichalcogenides. ACS Nano 10:3233–3240
199. Tsoutsou D, Aretouli KE, Tsipas P, Marquez-Velasco J et al (2016) Epitaxial 2D$MoSe_2$ ($HfSe_2$) semiconductor/2D $TaSe_2$ metal van der Waals heterostructures. ACS Appl Mater Interfaces 8:1836–1841
200. Roy A, Movva HCP, Satpati B, Kim K et al (2016) Structural and electrical properties of $MoTe_2$ and $MoSe_2$grown by molecular beam epitaxy. ACS Appl Mater Interfaces 8:7396–7402

201. Nie YF, Liang CP, Zhang KH, Zhao R et al (2016) First principles kinetic Monte Carlo study on the growth patterns of WSe_2 monolayer. 2D Mater 3:025029
202. Yamamoto H, Yoshii K, Saiki K, Koma A (1994) Improved heteroepitaxial growth of layered NbSe2 on GaAs (111). B J Vac Sci Technol A 12:125–129
203. George SM (2010) Atomic layer deposition: an overview. Chem Rev 110:111–131
204. Kim H, Lee HBRBR, Maeng WJJ (2009) Applications of atomic layer deposition to nanofabrication and emerging nanodevices. Thin Solid Films 517:2563–2580
205. Choi W, Choudhary N, Han GH, Park J et al (2017) Recent development of two-dimensionaltransition metal dichalcogenides andtheir applications. Mater Today 20:116–130
206. Zhu C, Xia X, Liu J, Fan Z et al (2014) TiO_2 nanotube @ SnO_2 nanoflake core–branch arrays for lithium-ion battery anode. Nano Energy 4:105–112
207. Ansari MZ, Parveen N, Nandi DK, Ramesh R et al (2019) Enhanced activity of highly conformal and layered tin sulfide (SnS x) prepared by atomic layer deposition (ALD) on 3D metal scaffold towards high performance supercapacitor electrode. Sci Rep 9:10225
208. Cremers V, Puurunen RL, Dendooven J (2019) Conformality in atomic layer deposition: Current status overview of analysis and modelling. Appl Phys Rev 6:021302
209. Chhowalla M, Shin HS, Eda G, Li LJ et al (2013) The chemistry of two-dimensional layered transition metal dichalcogenide nanosheets. Nat Chem 5:263–275
210. McDonnell S, Brennan B, Azcatl A, Lu N et al (2013) HfO_2 on MoS_2 by atomic layer deposition: adsorption mechanisms and thickness scalability. ACS Nano 7:10354–10361
211. Jena D (2013) Tunneling transistors based on graphene and 2-D crystals. Proc IEEE 101:1585–1602
212. Chuang HJ, Tan X, Ghimire NJ, Perera MM et al (2014) High mobility WSe2 p- and n-type field-effect transistors contacted by highly doped graphene for low-resistance contacts. Nano Lett 14:3594–3601
213. Liu H, Xu K, Zhang X, Ye PD (2012) The integration of high-k dielectric on two-dimensional crystals by atomic layer deposition. Appl Phys Lett 100:152115
214. Zhou W, Zou X, Najmaei S, Liu Z et al (2013) Intrinsic structural defects in monolayer molybdenum disulfide. Nano Lett 13:2615–2622
215. González C, Biel B, Dappe YJ (2017) Adsorption of small inorganic molecules on a defective MoS2 monolayer. Phys Chem Chem Phys 19:9485–9499
216. Hajati Y, Blom T, Jafri SHM, Haldar S et al (2012) Improved gas sensing activity in structurally defected bilayer graphene. Nanotechnology 23:505501
217. Puurunen RL (2005) Surface chemistry of atomic layer deposition: a case study for the trimethylaluminum/water process. J Appl Phys 97:121301
218. Wang X, Tabakman SM, Dai H (2008) Atomic layer deposition of metal oxides on pristine and functionalized graphene. J Am Chem Soc 130:8152–8153
219. Xuan Y, Wu YQ, Shen T, Qi M et al (2008) "Atomic-layer-deposited nanostructures for graphene-based nanoelectronics. Appl Phys Lett 92:013101
220. Nam T, Park YJ, Lee H, Oh IK et al (2017) A composite layer of atomic-layer-deposited Al_2O_3 and graphene for flexible moisture barrier. Carbon 116:553–561
221. Lazic P (2016) Physics of surface, interface and cluster catalysis. In: IOP, Bristol, England, pp 2.1–2.25
222. Lippert E (1960) The strengths of chemical bonds. Angew Chemie 72:602
223. Jeong SJ, Kim HW, Heo J, Lee MH et al (2016) Physisorbed-precursor-assisted atomic layer deposition of reliable ultrathin dielectric films on inert graphene surfaces for low-power electronics. 2D Mater 3:035027
224. Muneshwar T, Cadien K (2018) Surface reaction kinetics in atomic layer deposition: an analytical model and experiments. J Appl Phys 124:095302
225. Konar A, Fang T, Jena D (2010) Effect of high-κgate dielectrics on charge transport in graphene-based field effect transistors. Phys Rev B 82:115452–7
226. Newaz AKM, Puzyrev YS, Wang B, Pantelides ST et al (2012) Probing charge scattering mechanisms in suspended graphene by varying its dielectric environment. Nat Commun 3:734

227. Chen F, Xia J, Ferry DK, Tao N (2009) Dielectric screening enhanced performance in graphene FET. Nano Lett 9:2571–2574
228. Jena D, Konar A (2007) Enhancement of carrier mobility in semiconductor nanostructures by dielectric engineering. Phys Rev Lett 98:136805–4
229. Robertson J (2004) High dielectric constant oxides. Eur Phys J Appl Phys 28:265–291
230. Kobayashi NP, Donley CL, Wang SY, Williams RS (2007) Atomic layer deposition of aluminum oxide on hydrophobic and hydrophilic surfaces. J Cryst Growth 299:218–222
231. Son S, Yu S, Choi M, Kim D et al (2015) Improved high temperature integration of Al2O3 on MoS2 by using a metal oxide buffer layer. Appl Phys Lett 106:021601
232. Cheng L, Qin X, Lucero AT, Azcatl A et al (2014) Atomic layer deposition of a high-k dielectric on MoS2 using trimethyl aluminum and ozone. ACS Appl Mater Interfaces 6:11834–11838
233. Groner MD, Fabreguette FH, Elam JW, George SM (2004) Low-temperature Al_2O_3 atomic layer deposition. Chem Mater16:639–645
234. Swerts J, Peys N, Nyns L, Delabie A et al (2010) Atomic layer deposition of strontium titanate films using Sr(#2#1Cp)2 and $Ti(OMe)_4$. J Electrochem Soc 157:26
235. Fang H, Chuang S, Chang TC, Takei K et al (2012) High-performance single layered WSe2 p-FETs with chemically doped contacts. Nano Lett 12:3788–3792
236. Qiu H, Pan L, Yao Z, Li J et al (2012) Electrical characterization of back-gated bi-layer MoS2 field-effect transistors and the effect of ambient on their performances. Appl Phys Lett 100:123104
237. Park JH, Fathipour S, Kwak I, Sardashti K et al (2016) Atomic layer deposition of Al_2O_3 on WSe2 functionalized by titanyl phthalocyanine. ACS Nano 10:6888–6896
238. Park JH, Movva HCP, Chagarov E, Sardashti K et al (2015) In situ observation of initial stage in dielectric growth and deposition of ultrahigh nucleation density dielectric on two-dimensional surfaces. Nano Lett 15:6626–6633
239. Shul J, Pearton SLP (2001) Handbook of advanced plasma processing techniques. Plasma Phys Control Fusion 43:372
240. Steckelmacher W (2002) On minimizing the heat leak of current leads in cryogenic vacuum systems. Vacuum 42:779–785
241. Kumar N, Yanguas-Gil A, Daly SR, Girolami GS et al (2009) Remote plasma treatment of Si surfaces: Enhanced nucleation in low-temperature chemical vapor deposition. Appl Phys Lett 95:144107
242. Hsueh YC, Wang CC, Liu C, Kei CC et al (2012) Deposition of platinum on oxygen plasma treated carbon nanotubes by atomic layer deposition. Nanotechnology 23:405603
243. Vervuurt RHJ, Karasulu B, Verheijen MA, Kessels WMM et al (2017) Uniform atomic layer deposition of Al_2O_3 on graphene by reversible hydrogen plasma functionalization. Chem Mater 29:2090–2100
244. Nan H, Zhou R, Gu X, Xiao S et al (2019) Recent advances in plasma modification of 2D transition metal dichalcogenides. Nanoscale 11:19202–19213
245. Yang J, Kim S, Choi W, Park SH et al (2013) Improved growth behavior of atomic-layer-deposited high-k dielectrics on multilayer MoS2 by oxygen plasma pretreatment. ACS Appl Mater Interfaces 5:4739
246. Lucovsky G (1989) J Vac Sci Technol B Microelectron Nanom Struct 7:861
247. Hong S, Im H, Hong YK, Liu N et al (2018) Flexible electronics: mechanically tunable magnetic properties of flexible SrRuO3 epitaxial thin films on mica substrates. Adv Electron Mater 4:1800308
248. Kim S, Maassen J, Lee J, Kim SM et al (2018) Interstitial Mo-assisted photovoltaic effect in multilayer MoSe2 phototransistors. Adv Mater 30:1705542
249. Huang B, Zheng M, Zhao Y, Wu J et al (2019) Atomic layer deposition of high-quality Al_2O_3 thin films on MoS2 with water plasma treatment. ACS Appl Mater Interfaces 11:35438–35443
250. Gorbanev Y, O'Connell D, Chechik V (2016) Non-thermal plasma in contact with water: the origin of species. Chem Eur J 22:3496
251. Nguyen SVT, Foster JE, Gallimore AD (2009) Operating a radio-frequency plasma source on water vapor. Rev Sci Instrum 80:083503

252. Azcatl A, McDonnell S, Santosh KC, Peng X et al (2014) MoS_2 functionalization for ultra-thin atomic layer deposited dielectrics. Appl Phys Lett 104:111601
253. Vig JR (1985) UV/ozone cleaning of surfaces. J Vac Sci Technol A 3:1027
254. Clark T, Ruiz JD, Fan H, Brinker CJ et al (2000) A new application of UV−ozone treatment in the preparation of substrate-supported, mesoporous thin films. Chem Mater 12:3879–3884
255. Huh S, Park J, Kim YS, Kim KS et al (2011) UV/ozone-oxidized large-scale graphene platform with large chemical enhancement in surface-enhanced Raman scattering. ACS Nano 5:9799–9806
256. Sun H, Chen D, Wu Y, Yuan Q et al (2017) High quality graphene films with a clean surface prepared by an UV/ozone assisted transfer process. J Mater Chem C 5:1880–1884
257. Choi W, Cho MY, Konar A, Lee JH et al (2012) High-detectivity multilayer MoS(2) phototransistors with spectral response from ultraviolet to infrared. Adv Mater 24:5832–5836
258. Ahn S, Kim G, Nayak PK, Yoon SI et al (2016) Prevention of transition metal dichalcogenide photodegradation by encapsulation with h-BN layers. ACS Nano 10:8973–8979
259. Nam T, Seo S, Kim H (2020) Atomic layer deposition of a uniform thin film on two-dimensional transition metaldichalcogenides. J Vac Sci Technol A 38:030803
260. Luo YR (2007) Comprehensive handbook of chemical bond energies. CRC Press, Boca Raton
261. Kim H (2011) Characteristics and applications of plasma enhanced-atomic layer deposition. Thin Solid Films 519:6639–6644
262. Rossnagel SM, Sherman A, Turner F (2016) Plasma-enhanced atomic layer deposition of Ta and Ti for interconnect diffusion barriers. J Vac Sci Technol B18:2016
263. Kim H, Rossnagel SM (2002) Growth kinetics and initial stage growth during plasma-enhanced Ti atomic layer deposition. J Vac Sci Technol A20:802
264. Potts SE, Keuning W, Langereis E, Dingemans G et al (2010) Low temperature plasma-enhanced atomic layer deposition of metal oxide thin films. J Electrochem Soc 157:P66
265. Kim H, Chung I, Kim S, Shin S et al (2015) Improved film quality of plasma enhanced atomic layer deposition SiO_2 using plasma treatment cycle. J Vac Sci Technol A 33:01A146
266. Price KM, Schauble KE, McGuire FA, Farmer DB et al (2017) Uniform growth of sub-5-nanometer high-κ dielectrics on MoS2 using plasma-enhanced atomic layer deposition. ACS Appl Mater Interfaces 9:23072
267. Price KM, Najmaei S, Ekuma CE, Burke RA et al (2019) Ex-solution synthesis of sub-5-nm FeOx nanoparticles on mesoporous hollow N, O-doped carbon nanoshells for electrocatalytic oxygen reduction. ACS Appl Nano Mater 2:4085–6097
268. Watson AJ, Lu W, Guimaraes MHD, Stöhr M (2021) Transfer of large-scale two-dimensional semiconductors: challenges and developments. 2D Mater 8:032001
269. Kim JH, Ko TJ, Okogbue E, Han SS (2019) Centimeter-scale green integration of layer-by-layer 2D TMD vdW heterostructures on arbitrary substrates by water-assisted layer transfer. Sci Rep 9:1641
270. Liu Y, Weiss NO, Duan X, Cheng HC et al (2016) Van der Waals heterostructures and devices. Nat Rev Mater 1:16042
271. Novoselov KS, Mishchenko A, Carvalho A, Castro Neto AH (2016) 2D materials and van der Waals heterostructures. Science 353:461
272. Lin Z, McCreary A, Briggs N, Subramanian S et al (2016) 2D materials advances: from large scale synthesis and controlled heterostructures to improved characterization techniques, defects and applications. 2D Mater 3:042001
273. Choudhary N, Islam MR, Kang N, Tetard L et al (2016) Two-dimensional lateral heterojunction through bandgap engineering of MoS2 via oxygen plasma. J Phys Condens Matter 28:364002
274. Yazyev OV, Kis A (2015) MoS_2 and semiconductors in the flatland. Mater Today 18:20–30
275. Chiu MH, Zhang C, Shiu HW, Chuu CP et al (2015) Determination of band alignment in the single-layer MoS2/WSe2 heterojunction. Nat Commun 6:7666
276. Chiu MH, Tseng WH, Tang HL, Chang YH et al (2017) Band alignment of 2D transition metal dichalcogenide heterojunctions. Adv Funct Mater 27:1603756
277. Duesberg GS (2014) A perfect match. Nat Mater 13:1075–6

278. Li MY, Chen CH, Shi Y, Li LJ (2016) Heterostructures based on two-dimensional layered materials and their potential applications. Mater Today 19:322–335
279. Das S, Robinson JA, Dubey M, Terrones H, Terrones M (2015) Beyond graphene: progress in novel two-dimensional materials and van der Waals solids. Annu Rev Mater Res 45:1–27
280. Kang J, Tongay S, Zhou J, Li J et al (2013) Band offsets and heterostructures of two-dimensional semiconductors. Appl Phys Lett 102:012111
281. Gurarslan A, Yu Y, Su L, Yu Y et al (2014) Surface energy-assisted perfect transfer of centimeter-scale monolayer and few-layer MoS_2 films onto arbitrary substrates. ACS Nano 8:11522–11528
282. Lin YC, Zhang W, Huang JK, Liu KK et al (2012) Wafer-scale MoS2 thin layers prepared by MoO3 sulfurization. Nanoscale 4:6637–6641
283. Liu KK, Zhang W, Lee YH, Lin YC et al (2012) Growth of large-area and highly crystalline MoS2 thin layers on insulating substrates. Nano Lett 12:1538–1544
284. Lin Z, Zhao Y, Zhou C, Zhong R et al (2016) Controllable growth of large-size crystalline MoS2 and resist-free transfer assisted with a Cu thin film. Sci Rep 5:18596
285. Lai S, Jeon J, Song YJ, Lee S (2016) Water-penetration-assisted mechanical transfer of large-scale molybdenum disulfide onto arbitrary substrates. RSC Adv 6:57497–57501
286. Liu K, Yan Q, Chen M, Fan W et al (2014) Elastic properties of chemical-vapor-deposited monolayer MoS_2, WS_2, and their bilayer heterostructures. Nano Lett 14:5097–5103
287. Zhang T, Fujisawa K, Granzier-Nakajima T, Zhang F et al (2019) Clean transfer of 2D transition metal dichalcogenides using cellulose acetate for atomic resolution characterizations. ACS Appl Nano Mater 2:5320–5328
288. Yeh PC, Jin W, Zaki N, Zhang D et al (2014) Probing substrate-dependent long-range surface structure of single-layer and multilayer $MoS_2$by low-energy electron microscopy and microprobe diffraction. Phys Rev B 89:155408–155409
289. Tongay S, Fan W, Kang J, Park J et al (2014) Tuning interlayer coupling in large-area heterostructures with CVD-grown MoS2 and WS2 monolayers. Nano Lett 14:3185–3190
290. Shi J, Ma D, Han GF, Zhang Y et al (2014) Controllable growth and transfer of monolayer MoS2 on Au foils and its potential application in hydrogen evolution reaction. ACS Nano 8:10196–10204
291. Jain A, Bharadwaj P, Heeg S, Parzefall M et al (2018) Minimizing residues and strain in 2D materials transferred from PDMS. Nanotechnology 29:265203–265209
292. Pu J, Yomogida Y, Liu KK, Li LJ et al (2012) Highly flexible MoS2 thin-film transistors with ion gel dielectrics. Nano Lett 12:4013–4017
293. Xu ZQ, Zhang Y, Lin S, Zheng C et al (2015) Synthesis and transfer of large-area monolayer WS2 crystals: moving toward the recyclable use of sapphire substrates. ACS Nano 9:6178–6187
294. Cadiz F, Courtade E, Robert C, Wang G et al (2017) Excitonic linewidth approaching the homogeneous limit in MoS_2 based van der Waals heterostructures. Phys Rev X 7:021026–021112
295. Yin H, Zhang X, Lu J, Geng X et al (2020) Substrate effects on the CVD growth of MoS_2 and WS_2. J Mater Sci 55:990–996
296. Xu P, Neek-Amal M, Barber SD, Schoelz JK et al (2014) Unusual ultra-low-frequency fluctuations in freestanding graphene. Nat Commun 5:3720
297. Liu X, Huang K, Zhao M, Li F (2020) A modified wrinkle-free MoS2 film transfer method for large area high mobility field-effect transistor. Nanotechnology 31:055707
298. Wang X, Kang K, Godin K, Fu S et al (2019) Effects of solvents and polymer on photoluminescence of transferred WS2 monolayers. J Vac Sci Technol B 37:052902

Chapter 3
Structures and Defects of TMDs

3.1 Introduction: Bulk TMDs

Bulk TMDs are recognized as three-dimensional (3D) materials which have been known and used for common uses for a long time. The previous applications of bulk TMDs (such as solid lubricants) are based on their adequate mechanical properties which are characterized by the presence of van der Waals bonds between the successive layers. However, after a great innovation of graphene boosted the 2D materials research, great attention has been also paid to the TMDs to conduct forefront research in solid-state materials. The key interest in TMD materials research has been focused on their monolayer and few-layer structures, including interest in 3D TMDs for advanced research. Thus, the structural features of TMDs are useful for the specific targeted application. Therefore, in this chapter, we will discuss 3D TMDs' structural properties, and related significant issues to build an understanding of the 2D TMDs including leading research in this area.

Since any material structural properties also influenced largely by the existence of kinds of defects, owing to the fact the defects are also significantly useful to develop material for a specific application. Therefore, attention has been also paid to this critical issue of TMD materials and discussed the impact of the various defects on their structures. Moreover, detail has been also provided on impurities containing TMDs' physical properties.

3.1.1 *Individual Triple Layer Structure*

The typical layered structure of a TMD is illustrated in Fig. 3.1. As represented in the layered TMD structure every individual layer consist of three atomic planes along with a typical thickness of 6–7 Å. In which hexagonal planes of metal atoms are sandwiched between the two planes of the chalcogen atoms. However, some cases

A. K. Singh, *2D Transition-Metal Dichalcogenides (TMDs): Fundamentals and Application*, Materials Horizons: From Nature to Nanomaterials,
https://doi.org/10.1007/978-981-96-0247-6_3

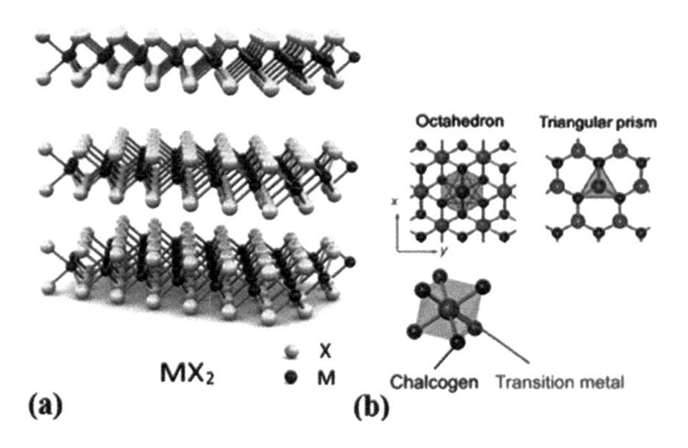

Fig. 3.1
a Three-dimensional representation of the structure of transition metal dichalcogenides (TMDs); **b** top view of monolayers constructed from octahedral and triangular prismatic coordination (both from Derakhshi, M. et al. 2022 J. Funct. Biomater. 13, 27)

deviated from this structure, owing to the hexagonal in-plane geometry of metal ions possibly being distorted; therefore, formed layers cannot be planar. This depends on the structural context as well as their applications; sometimes it refers to triple layers, but in the majority of cases, it is called monolayers (or single layers).

In the case of transition metal complexes, their bondings are usually described in terms of the empty orbitals of the metal and lone pairs ligands of TMDs. Where the metal atoms provides four electrons to complete the bonding states, including transition metal and chalcogens that are attributed with a formal charge + 4 and − 2, respectively [1]. As the consequence, coordination around the chalcogenides would be lopsided, thereby it leads the market cleavage properties perpendicular to the hexagonal/trigonal symmetry axis [2]. Moreover, absence of the dangling bonds make the surfaces extremely stable and unreactive. The typical valance electron distribution in a TMD schematic is depicted in Fig. 3.2a. Its more simplified picture which provides in terms of non-bonding electrons and considered to be located on the *s*-orbital with chalcogen pure *p*-orbitals to make metal chalcogen bonds. This theoretical DFT interpretation also leads to lone-pair electrons of chalcogen atoms that are arranged inform of sp^3-hybridized orbitals, as depicted in Fig. 3.2b.

Moreover, DFT analysis is also revealed that the metal atoms within a triple layer typically have six fold coordinated and their bonding geometry may either trigonal prismatic or octahedral, as illustrated in Fig. 3.1. Whereas in the trigonal prismatic arrangement two chalcogenide planes form a slab that directly stacked above to each other. On the other hand, the octahedral arrangement is staggered. Moreover, the preferred phase adoption from the TMD depends mainly on the *d*-electron count of transition metal, although their certain dependence on the relative size of the atoms. Like group 4 metals have octahedral structures, and most group 5 metals also possessed octahedral structures, while some of them have trigonal prismatic structures. Moreover, the reverse situation is also possible for group 6 metals. Further, group 7 TMDs analysis again gives the octahedral structures, but this result was in a distorted manner. Similarly, group 10 TMDs analysis has also demonstrated their octahedral structures.

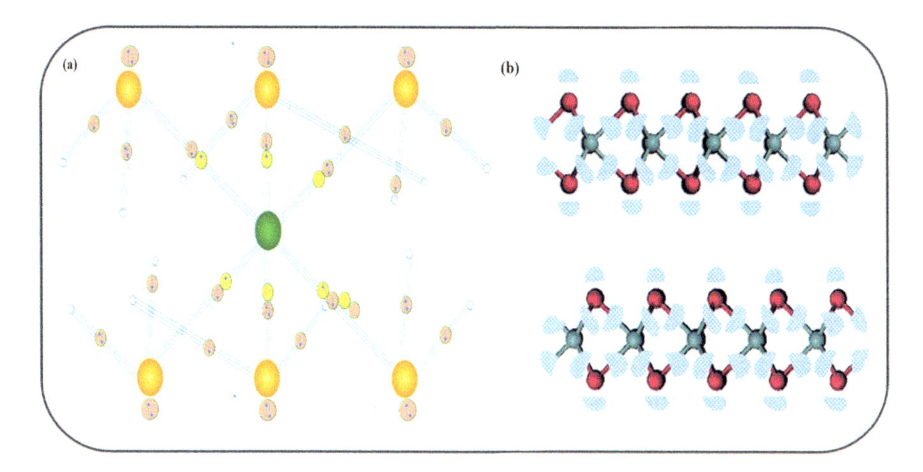

Fig. 3.2 **a** Schematic of a fragment of an octahedral TMD structure to demonstrate the distribution of valence electrons. The metal atoms are given in the dark and the chalcogen atoms are in the light. The electrons of each species are indicated by arrows by the following convention. The unpaired electrons of the covalent bonding are represented with a single arrow in a circle. Whereas the lone-pair electrons that formed dative (coordinate) bonds are represented with two arrows of opposite directions in a circle. The vacant orbitals are represented from the empty circles (as the transition metal atom provides four bonding electrons plus two vacant orbitals); **b** schematic to relaxed structure for the $MoTe_2$ and CDD isosurfaces charge accumulation in the vdW gap, associated with sp3-hybridization of chalcogen atoms

3.1.2 Polymorphs Structures

Depending on coordination environments and stacking orders of polyhedra (trigonal prisms or octahedra) predominantly in form of three polymorphs 1T, 2H, and 3R, the typical crystal structure of bulk MoS_2 is illustrated in Fig. 3.3. Monolayer of MoS_2 consists of two hexagonal planes of S atoms separated by one plane of Mo atoms, whereas the Mo coordination can be either trigonal prismatic or octahedral. Yet, five polymorphs including 1H, 1T, 1T′, 1T″, and 1T‴ have recognized for the monolayer (or bulk) structure of the TMDs.

Typically, in the polytypes crystal structure of TMDs consisting 1T, 2H, and 3R phases with several monolayers in their unit cell. Usually, letters T, H, and R are related to the structural symmetry, to represent the trigonal, hexagonal, and rhombohedral [1]. The 1T phase is usually related to Mo-S octahedral coordination, while the 2H and 3R phases correspond to the trigonal prismatic coordination phase. These coordination phases are belonging to different point groups, such as D_{6d} to 1T, D_{6h} to 2H, and C_{3V} corresponding to 3R. The thermodynamically stable phase 2H phase generally occurred naturally in existing MoS_2 crystals, like molybdenite. While both 1T and 3R phases are metastable in the state that is often found in the synthetic TMD. The bulk TMD 1T phase was first introduced by a German scientist, Robert Schçllhorn [3]. By demonstrating pioneering research on potassium extraction from ternary $KMoS_2$ to get 1T MoS_2. With the remarks the octahedral coordination in

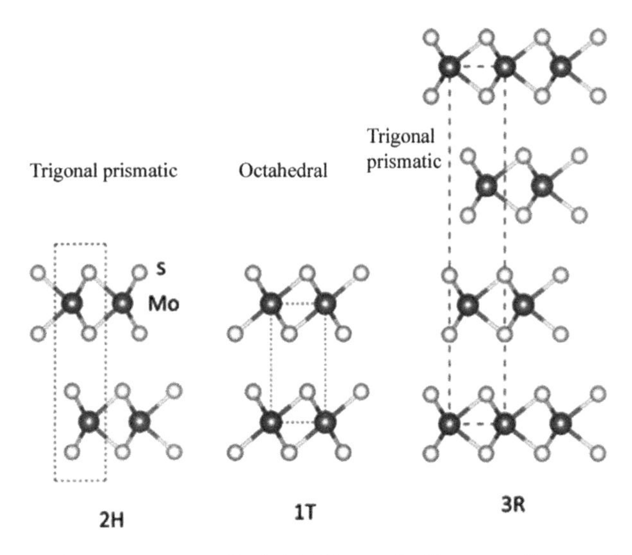

Fig. 3.3 Typical polytypes structure of bulk MoS2 crystals. The unit cells are enclosed from the dashed lines for different polytypes Mo-S coordination, such as 2H, 1T, and 3R phases (adopted from Saha, D. et al. 2020 J. Electrochem. Soc. 167, 126517)

1T MoS_2 could associate a distorted layer-type structure, their work also established the relation between the unit cells of 1T MoS_2 and2H MoS_2:$a_{1T} = \sqrt{3}a_{2H}$ and $c_{1T} = 1/2c_{2H}$. Additionally, they have also concluded that in these materials metallic conductivity is a new phase.

The experimental evidence has revealed the existence of two types of superstructures (a $\times$ 2a and $\sqrt{3}a \times \sqrt{3}a$), depending on the potassium extraction process of precursor $K_x(H_2O)_yMoS_2$ [4–6]. Therefore, the formation of different superstructures could be related to the degree of structural distortion, including Mo-Mo dimerization and even trimerization. Although different superstructures have been recognized as "1T-MoS_2", however, results are presented in a more clarified way from the other research groups, by introducing an advanced strategy for the synthesis of 1T MoS_2 using ternary $LiMoS_2$ crystal as a precursor [7]. Investigators were concluded that the individual S-Mo-S unit can consist of MoS_6 octahedra sharing edges without Mo-Mo clustering [8, 9].

Moreover, to provide a comprehensive description of the layered TMDs properties, the simulation results have been also discussed, such as the ab initio first-order principles-based investigation including stability, exfoliation energy, and electronic properties of TMDs. This study has revealed the formations of Ti-, V-, and Cr group transition metals and S, Se, and Te in three different polymorphic phases: 2H, 1T, 1T′ [9]. The most common structural phases of the TMDs are 2H, 1T, and $1T_d$, as illustrated in Fig. 3.4a–d [10]. The polycrystalline TMD structure's top and lateral views are depicted in Fig. 3.4a–c, whereas the polytype structure (see Fig. 3.4d) have related to supercells that are useful to access the magnetic ordering. The 2H structure (as depicted in Fig. 3.4a) contains a hexagonal lattice with 2 formula units (f.u.) per

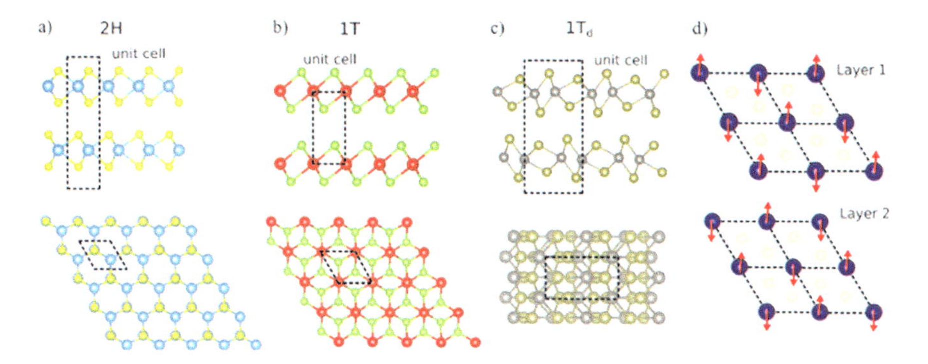

Fig. 3.4 TMDs polytypes crystal structures lateral and top view: **a** Trigonal prismatic (2H); **b** octahedral (1T); **c** distorted octahedral (1T$_d$), the corresponding unit cells are represented in dashed lines; **d** Schematic of the supercell to define the magnetic ordering, where the red arrows represent the initial magnetic moments in transition metal atoms (adopted from Bastos et al. [9])

unit cell, corresponding to atoms planes in AbABaB stacking sequence (here capital and lower-case letters representing the chalcogen and metal atoms planes, respectively), belonging to the $P6_3/mmc$ space group [11]. As depicted in Fig. 3.4 (b), the 1T structure has a hexagonal lattice with 1 f.u. per unit cell with the Ab-CABC stacking sequence, their corresponding space group is $P\bar{3}m1$ [11], each layer of the 1T$_d$ (as depicted in Fig. 3.4 c) could generate from a 1T monolayer by reconstructions of a 2×1 orthorhombic cell, it could originate the dimerized lines of metal atoms, thereby, distortion exists under the Peierls transition mechanism [12]. Moreover, the 1T$_d$ structure has possibly composed of an orthorhombic lattice with 4 f.u. in the unit cell, corresponds to the $Pnm2_1$ space group. Additionally, the bonding geometry symmetries are corresponding to D_{6h}, D_{3d}, and C_{2v} point groups for the 2H, 1T, and 1T$_d$. In addition to these, it is also customary to outline the possibilities about some TMDs can have polymorphs structural phases, as depicted in Fig. 3.5.

Thus form the ab initio structural demonstration has been addressed the materials composed of Ti-, V-, and Mo-group metals possess no fully occupied d-orbitals, which could lead in some of the TMDs' non-zero magnetic moment, as experimentally noticed in the case of the VS_2 and VSe_2 with the ferromagnetic ordering at low temperatures [11, 45]. Moreover, the intrinsic magnetism in bulk TMDs to be also addressed with the help of supercells containing eight f.u. model, as the typical antiferromagnetic configuration is illustrated in Fig. 3.4d. Thus, the concerning non-magnetic and ferromagnetic orderings of the unit cell would be employed [9, 13].

3.1.3 Distorted Structures

Since the single layer TMD phase depends extensively on transition metal d-orbital electron concentration. It means the occupation of d-orbital makes them useful for

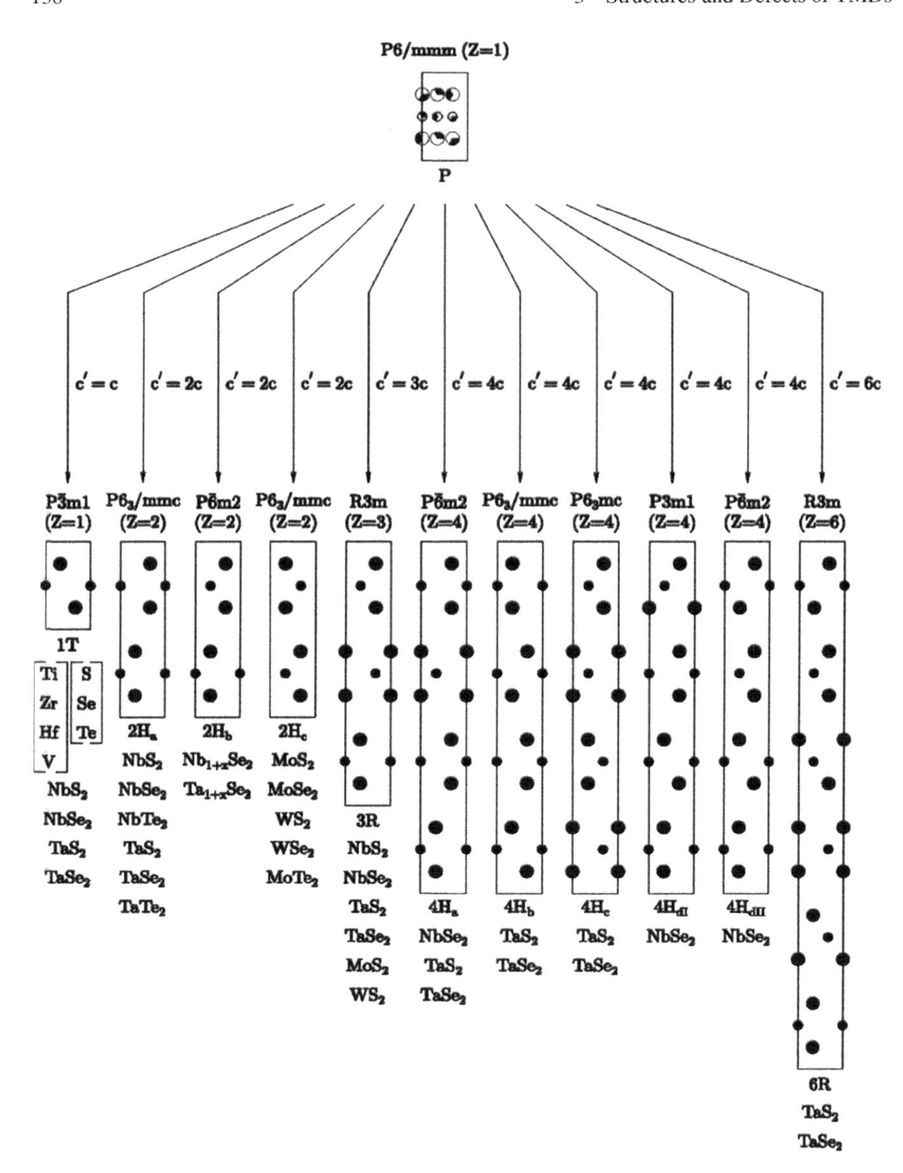

Fig. 3.5 Mechanisms of ordering for the formation of 11 polytype structures in TMDs. The projections of (11–20) polytypes exhibited hexagonal structures. Small circles represent the metal atoms and large circles the chalcogens (adopted from Katzke, H. et al. 2014 Phys. Rev. B 69, 134111)

the phase manufacturing. There are a number of instances of phase alteration through the chemical reactions, like periodic table 6th group TMDs, such as MoS_2, $MoSe_2$, WS_2 [14–22]. Likely, it has also demonstrated that the development of the metallic 1T phase in MoS_2 due to lithium intercalation, that induce a reduction in MoS_2 and increase the electron concentration within the d-orbitals [23]. Therefore, group 6 TMDs with the 1T phase become characteristically negatively charged [24]. Another similar example is the lithiation of TaS_2which also encourage the phase alteration from the semiconducting 1T phase to the metallic 2H phase [25]. Hence, in general, the phase alteration in formed single layer TMDs is uncompleted, because it also contains layer fractions of both 2Hand 1T phases [26]. Moreover, the group 6 TMDs 1T phase is also not allowing formation of a perfect (a × a) 1T phase of TiS_2 but it has more in a distorted 1Tphase [5, 8, 9, 16, 27, 28].

Usually, periodic table 6th group TMDs formed 1T phase is considered to be unstable, therefore, to improve the stability of the formed superlattice structure it is usually stabilized from the additional charges over to surface of the TMDs. Typically, four distinct types of constructions have been suggested to describe it: tetramerization (2a × 2a), trimerization ($\sqrt{3}$a × $\sqrt{3}$a), and zigzag chain (2a × a). As the typical atomic structures of the disordered phase in TMDs is illustrated in Fig. 3.6a, b. However, for the reorganization distinct type of distortion scientific community has a big debate over a long time. Such as restacked distortion in MoS_2 and WS_2 has been noticed using electron diffraction. This observation has revealed the distorted structure of restacked (T) ($MoS_2$and WS_2) that can form zigzag network likely to WTe_2 structure possessing a shorter M–M distances to W and Mo (0.27 nm and 0.29 nm) [24]. The distorted atomic distances could also substantially shorter than typical values of the 2H phase, such as 0.315 and 0.312 nm for the WS_2 and MoS_2. This experimental evidence was further verified in the various studies with good agreement [26, 27, 29, 30]. Moreover, single layer exfoliated MoS_2/WS_2 afterward lithium intercalation illustrates the crank chains (2a × a) from the deformed WTe_2-like (T) phase of MoS_2 and WS_2 collectively with the perfect (T) TiS_2 phase. Another concentrated MoS_2 distributed throughout the solution or inserted with the hydroxide molecules or metal atoms have also revealed the occurrence of a (2a × a) framework within the solution. The ($\sqrt{3}$a × $\sqrt{3}$a) structure building has been also noticed in the oxidized $K_x(H_2O)_yMoS_2$ (with no creation of Mo^{6+}) or restacked (T) MoS_2 when treated with Br_2 [5, 31]. However, likewise actions with WS_2 have not induced any significant alterations in the structural building of (T) WS_2, whereas the absence of crank deformation has been noticed in almost all cases [24].

Surprisingly few TMDs are obviously deformed like WTe_2 and ReS_2. In which every layer consists crank chains of the transition metal atoms. The natural disturbance occurrence in a bulk crystal of ReS_2 allows to each layer behave as a monolayer [32]. However, further decrement of ReS_2 increases the distortion and it forms Re_4 rhombus constellations [33].

Thus, the distorted structure of TMDs can be induced by modifications in the metal–metal bond distance that finally affect their electronic structure. As the intense, ReS_2 associating 3 electrons in d-orbital could have a metallic character. Their strong distortion between the layers allows formation of endless zigzag network of Re atoms

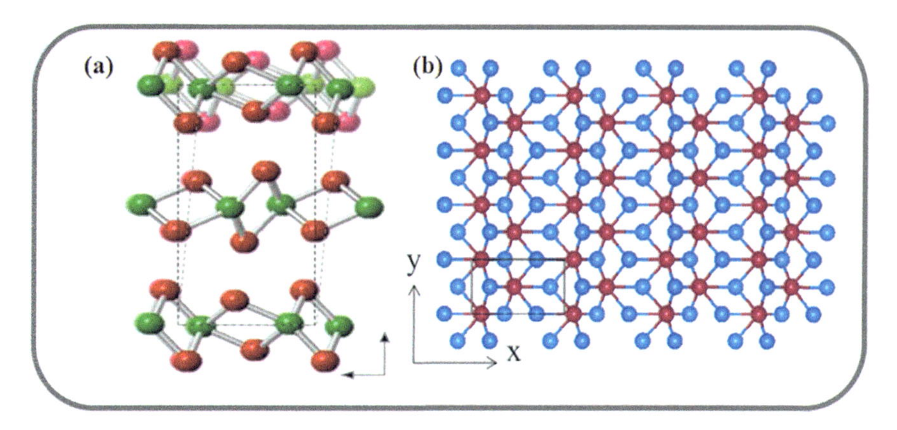

Fig. 3.6 **a, b** Distorted atomic structure of octahedral (T) and (T_d) phase (adopted from **a** Zhang, K. et al. 2016 Nat. Commun. 7, 13552 and **b** Santosh et al. [20])

to opening the energy gap in the band structure [34, 35]. Therefore, ReS_2 is a direct bandgap semiconductor with a bandgap of 1.35 eV, while its bulk and monolayer bandgap are 1.45 eV [32]. Additionally, their mono layers distortion in TMDs are also strained [36].

3.2 Single Layer or Few Layers of Structures

A monolayer sheet of transition metal dichalcogenide consists of a plane of transition metal sandwiched between two layers of chalcogens to form a trilayer structure. This kind of multiple-trilayers stacking that held together by weak out-of-plane van der Waals interactions to form a bulk material. Predominantly two types of distinct orientations are responsible to provide the unique structure of the TMDs: the first is "basal planes" and the second one is "edge planes".

The existence of these two different kinds of orientations in a monolayer sheet allows the anisotropic properties in the material, as the intense, surface inertness of the basal plane compared to the high surface energy of the edge plane. The higher anisotropic property of a material is usually directly related to crystal orientation. Moreover, significant differences in the properties may also appear depending on the transition metal coordination with the chalcogen and stacking sequence of multiple layers. Thus, a single (or monolayer) layer TMD has a trigonal prismatic or octahedral metal coordination phase which is usually referred to as 2H and 1T, the typical crystallographic structures are illustrated in Fig. 3.7 [37–39]. The TMD multiple structures or several different polymorphic structures also occur, as the consequence, TMD materials offer a variety of different properties.

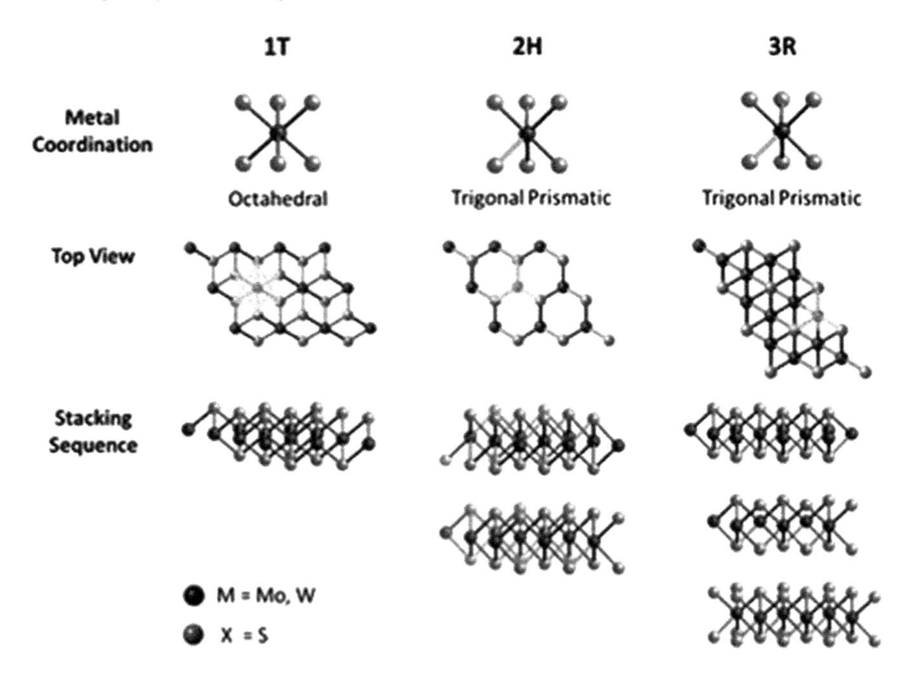

Fig. 3.7 Metal coordinations and stacking sequences of TMD structural unit cells: tetragonal symmetry (1T) corresponds to octahedral coordination stacking sequences and hexagonal symmetry (2H) and rhombohedral symmetry (3R) corresponds to dissimilar stacking sequences of trigonal prismatic single layers (adopted from Toh et al. [38])

Most commonly existing polymorphs are defined as 1T, 2H, and 3R, in this representation digit is related to the corresponding number of layers within the crystallographic unit cell, whereas the letters are representing the type of symmetry. The letter T stands for tetragonal (D_{3d} group), H for hexagonal (D_{3h} group), and R for rhombohedral (C_{3v5} group). The formation of 1T reflects metallic behavior, while others 2H and 3R are usually correlated with the semiconducting behavior of a TMD material. So that each TMD polymorph is possessing unique structural and electronic properties in which different catalytic properties are emerged.

Since transition metal coordination by chalcogens and the stacking sequences of multiple layers of TMD materials are usually elicit different electronic properties. Therefore, knowing the stacking sequencing in TMDs is an important parameter. Usually, the stacking sequence in TMDs is represented by the three letters which indicate the relative positions of the chalcogenide- metal atoms in each layer [39]. More precisely layers stacking sequence in TMDs are generally described with the help of the following points.

1. 1T phase single layer TMD has a stacking sequence like AbC. This kind of stacking sequence symmetry allows the octahedral phase in case of the multiple layers formation which could have an AbCAbCAbC possible sequence.

2. TMD 2H phase has a stacking sequence order AbABaB in which chalcogen atoms are overlapping along the structure formation axis accompanied by the metal atoms of the adjacent layer.
3. Formation of rhombohedral symmetry responsible for all the layer's trigonal prismatic phases, their position of metal and chalcogen atoms may shift, therefore, the achieved layers stacking sequence should be AbACaCBcB in each unit cell.

3.2.1 2H Phase

The 2H semiconducting phase has a trigonal prismatic arrangement of atomic layers under AbA stacking belonging to group symmetry D_{3h}, as depicted in Fig. 3.7. Group VI MX_2 monolayers with the 2H structural phase usually reflect semiconducting behavior with bandgaps in-between 1 and 2 eV [40–42]. Thus, 2H phase TMDs considered to be promising semiconductors for flexible electronics applications [40]. The 2H structural phase also gives rise to metallic edge states associated with the electrocatalytic activity [43]. In general, the primitive unit cell of the 2H phase is hexagonal.

3.2.2 1T Phase

2D-TMDs exist with the different structural phases due to configuration differences of the transition metal atoms component. The quasi-metallic1T′ phase also exists in 2D-TMDs with a unique in-plane atomic arrangement that comprises a distorted sandwich structure, where the transition metal atoms form a period-doubling 2×1 structure compared to an array of the1D zigzag chains [44–47]. The 1D zigzag feature of 1T′ phase 2D-TMD is significantly different from the quasi-1D structures to other classes of the 2D layered structures, typically it is known as transition metal tri chalcogenides. Thereby, they forms MX_3 structural configuration by incorporating a trigonal prismatic structure that is aligned to form a long needle-like crystal along a single axis [48–50]. The existence of a unique 1D zigzag periodic structure in 1T′ phase gives rise to 2D-TMDs' strong anisotropic properties which could have a significant impact on their electronic properties [51].

Similarly, 2D-TMDs also commonly form polymorphs of trigonal prismatic 1H phase and the octahedral 1T phase. Although various experimental and theoretical studies have suggested that the 1T phase structure of 2D TMDs is dynamically unstable under free-standing conditions [44, 51–53]. Moreover, like Peierls distortion, the 1T phase is relaxed and buckle spontaneously into a thermodynamically more stable distorted phase [54]. Moreover, their electronic structural calculations have revealed that the 1T phase possessed metallic character [55, 56]. Therefore, to distinguish the 1T and 1T′-phases, it is significant to note that the charge density wave phase should be typically recognized at low temperatures in 1T phase TMD systems (e.g., T_c ~ 120 K for $TaSe_2$ and T_c ~ 35 K for $NbSe_2$). Their charge density wave lattice distortion in the form of the periodic 1D zigzag chain structure. Thus the 1T has a unique structure to the 1T′-phase of 2D-TMD, which is also recognized at room temperature [26, 57, 58]. The typical 1T and 1T′-phases of the TMDs are illustrated in Fig. 3.8a, b.

Besides above discussed structural phases in TMDs, they may also have other additional structural phases with unique optical and electronic properties. Like $MoTe_2$ and WTe_2 are usually undergo a first-order phase transition from the monoclinic 1T′ phase to an orthogonal structure phase under the lowering the temperature, this is well-known $1T_d$ phase. Although structures of the 1T′ and $1T_d$ phases are similar, an apparent difference is the dislocations between the stacking layers, as depicted in Fig. 3.8c. This means the resultant symmetry changes between these two structural phases [59, 60]. Similarly, structural changes have been also noticed for the $1T_d$ phase TMDs that also possessed quasi-metallic electronic properties analogous to their 1T′ phase counterparts [61, 62].

Such inversions are also appeared in the formation of the 1T′ phase when brought structural distortion of its 1T phase counterpart in a monolayer. Similarly with the

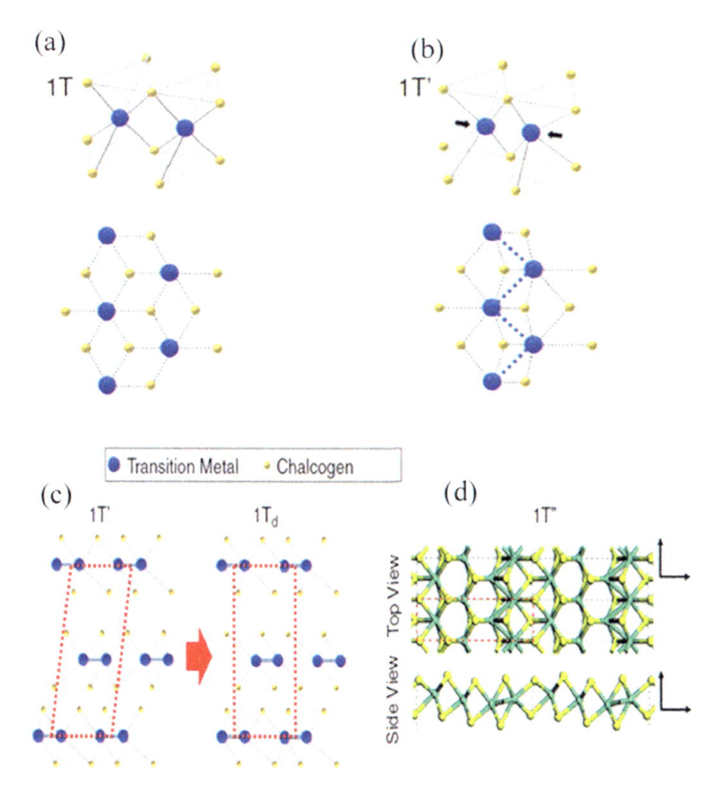

Fig. 3.8 **a** Octahedral (1T) phase; **b** distorted (1T′) phase; **c** Stacking orders to distinguish the 1T′ and $1T_d$ phases; **d** atomic structures of 1T″-phase monolayer TMD (adopted from Tang et al. [54])

other stable polymorphs it also forms 1T″ phase, their structural formation usually octahedrally coordinated, as depicted in Fig. 3.8d [63]. A strong association between the transition metal atoms to form M–M configurations is preferred, because it dimerized with the M atoms to form the 1T′ phase, whereas the trimerization process is responsible for the formation of the 1T″ phase. The 1T″ phase usually has a wide bandgap semiconductor unlike to quasi-metallic 1T′ phase, as the example 1T″ phase monolayer MoS_2 with an indirect bandgap 0.27 eV [64]. However, regarding the relative stability between each octahedral phase there is no consensus to monolayer MoS_2. Though a few individual studies have been demonstrated the different levels of stability between each structural phase [64–66].

3.2.3 Significance of Phase Transition

Properties of TMDs are generally determined with the help of their phases, such as the phase transition is triggered by the modulation of the electron density. However, there is the key issue to how precisely control the phase transition in TMDs is still challenging. Therefore, understanding the mechanism and the conditions under which the phase transition can be triggered for the better utilization of TMDs in the future. The knowledge of TMDs also helpful to improve their prospective applications, more precisely phenomena involved in phase transition processes [67–69]. As an example, alkali metal ion–intercalated MoS_2 pre-lithiation treated phase transition also improves the cycle stability and rate capability compared with the bulk or exfoliated–restacked MoS_2 (Fig. 3.9a) [70]. Moreover, considering the layered structure of MoS_2 the energy storage matching application is also possible because it requires electrode materials with a large specific surface area to store Li^+/Na^+ and good conductivity to provide electrons transport [71, 72]. However, the semiconducting 2H phase is comparatively not considered to be a good choice for an electrode, while the 2H–1T/1T′ transition via alkali metal ion intercalation makes the 1T/1T′ metallic phase is more desirable for an electrode as energy storage. In the 2H–1T′ phase transition process it has been also recognized the MoS_2 2H phase splits into three orientation variants. This kind of splitting provides a more homogeneous phase conversion with a uniform charge distribution compared to bulk or exfoliated–restacked MoS_2.

The advantage of metallic 1T/1T′ phase is also useful to fabricate a superior contact device [30]. As an intense (Fig. 3.9b) a thin layer of 1H $MoTe_2$ flake converts into 1T′/1H/1T′ configuration under the laser treatment. Their metallic 1T′ phase can have good contact with both the deposited metal electrode and 1H $MoTe_2$. Therefore, a significant improvement in carrier mobility ($\approx$ 50 times) would be achieved [73]. Similarly, the electronic device of the 1T′ ($MoTe_2$)–1H(MoS_2)–1T′ ($MoTe_2$) heterojunction also offers enhanced conductivity by employing chemical vapor deposition possess [74]. Another key study based on a device with 1T′ phase contact has also demonstrated a larger current than the 1H phase contact even if the latter channel is remarkably shorter than the former one.

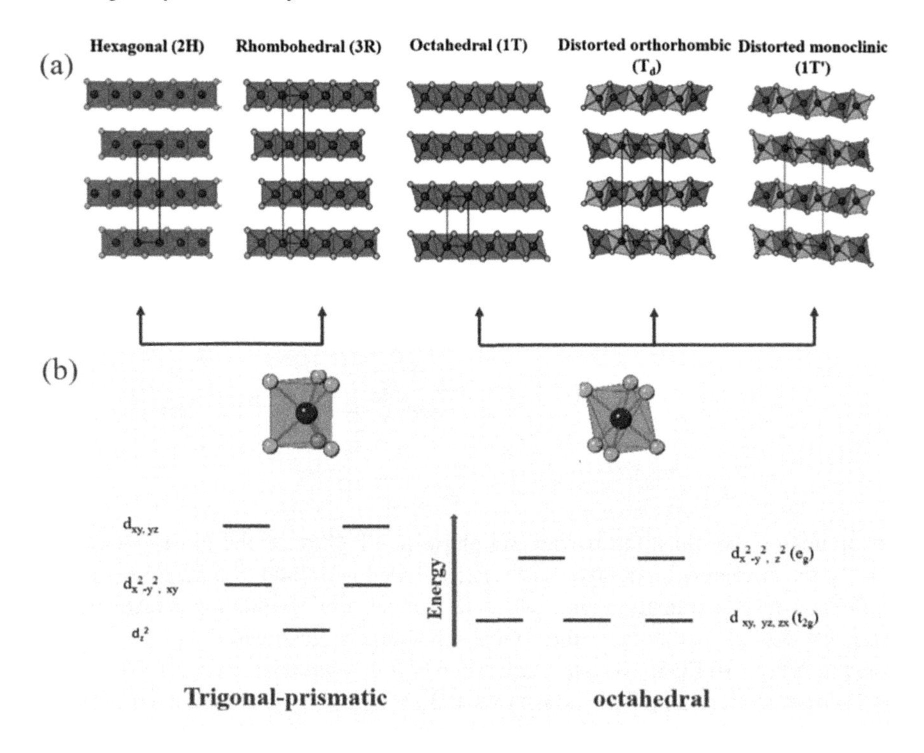

Fig. 3.9 Schematic for atomic structures of various phases in TMDs: **a** bonding and stacking modes in six kinds of phase structures (adopted from Sokolikova et al. [90]); **b** comparison of d-orbital spitting due to spin-orbital coupling in 1H and 1T phases

Similarly, the phase transition has been also utilized for many prompt novel applications, like, triggering the surface chemical reaction and surface Raman enhancement. Since in both stem electrons transferred from electron-rich metallic 1T/1T′ to MX_2 reactant or probe molecules. Therefore, the phase transition from 2H to 1T/1T′ also significantly increases the possibility of a charge transfer between MX_2 and the environment. It could be precisely controlled the locations or patterns of the reaction or Raman signal through MX_2 phase control, it is useful for their prospective applications. Hence, the phase transition not only provides the possibility TMDs to use with merit due to the formation of distinct kinds of phases but also have a wide platform for many novel applications concerned with the change in electron structures.

3.2.3.1 Strategies of Phase Transition for TMDs

Charge Doping Theoretical Approach

As per the theoretical approach, it has been predicted that the charge doping in TMD can effectively reduce the transition barrier between the 2H and 1T phases [66, 75].

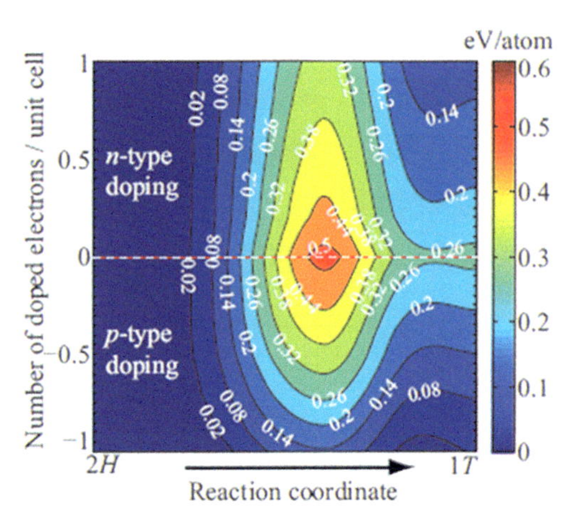

Fig. 3.10 Shows the charge doping requirement for the phase transition from 2H to 1T phase (adopted from Zhuang [66])

Theoretically, the transition barrier 2H phase to 1T phase is the most extensively studied for the MoS_2 monolayer and it is estimated at around ≈ 0.52 eV per atom, while their reverse transition under the zero charges doping situation is estimated at around ≈ 0.24 eV per atom, as the typical schematic is illustrated in Fig. 3.10. It has noted in 2H to 1T (1T to 2H) phase transition barrier significantly decreased from $\approx$ 0.52 eV per atom (0.24 eV per atom) to ≈ 0.3 eV per atom (≈ 0.2 eV per atom) and ≈ 0.1 eV per atom (≈ 0.08 eV per atom) with the increasing doping concentration from 0 to 1 per unit cell. Therefore, the 1T phase cannot be considered stable without an external stabilizing factor; thereby it should further undergo a transition to the 1T′ phase [30].

However other TMDs like $MoTe_2$ have an energy difference between the 2H and 1T′ phase, but MoS_2 is much larger than $MoTe_2$ [75]. A comparison of the energy difference between the 2H and 1T′ phases in monolayer MoS_2 and $MoTe_2$ and charge doping alternation provides a more clear view. Such as under a zero-charge doping condition, the energy difference in MoS_2 has been estimated $\approx$ at 0.5 eV per recipe with the constant stress and ≈ 0.6 eV per recipe under constant area. On the other hand, in the case of the $MoTe_2$, it has estimated around ≈ 0.03 and ≈ 0.07 eV per recipe. Thus, to trigger the phase transition, the surface charge density ≈ -0.29e or 0.35e per recipe is desired for the MoS_2 monolayer under constant stress, while comparatively a smaller value ≈ -0.04e or 0.09e per recipe is desired for the $MoTe_2$ monolayer, with the corresponding smaller charge density 10^{14}e cm^{-2}. This charge is usually accessed with the help of conventional experimental conditions, like in electrostatic gating [34].

Moreover, heteroatom intercalation or surface chemical modification has also been extensively explored by providing effective charge doping of the TMDs [76–79]. Predominately alkali ion intercalation of the TMDs has been targeted for both theoretical and experimental demonstrations. As the well-defined concepts of the alkali ion, it intercalate into the layer interval of TMDs to form A_xMX_2 (A for alkali ion). Generally, this process of charge doping governs from alkali ions to TMD.

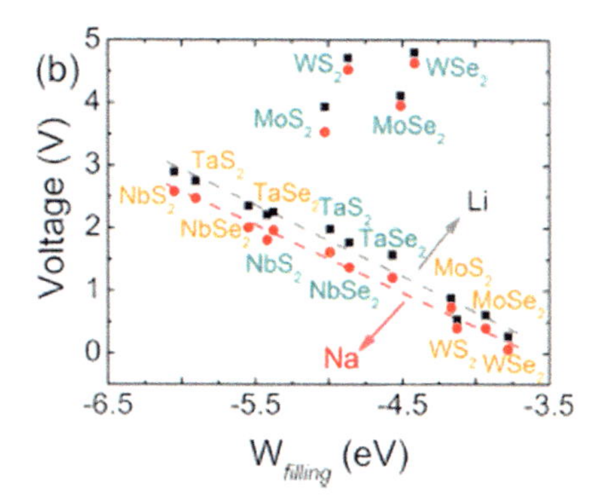

Fig. 3.11 The W filling for different MX_2; orange and green labels represent 2H and 1T phases, whereas red and black plots represent voltages of the Na and Li intercalation (adopted from Fan et al. [80])

The corresponding theoretical approach information is depicted in Fig. 3.11. Herein $W_{filling}$ represents the required energy to transfer an extra electron into an empty state above the Femi level of pristine MX_2. It directly correlates to the energy gain of TMDs during intercalation. A larger absolute value of $W_{filling}$ could provide a larger energy gain during the intercalation process [80]. As the intense, MoX_2 and WX_2 have a smaller absolute value of $W_{filling}$. By means it reveals a smaller energy with lower voltage gains for the MoX_2 and WX_2 compared to NbX_2. In more precise words, the alkali ion intercalation more easily occurs in MoX_2 and WX_2 than the NbX_2. In addition to this, surface hydrogen modification has been also a widely accepted approach for electronic structure modulation. By employing this approach transition of semimetallic graphene to semiconductor or insulator has been extensively explored in the past decades [81, 82]. Surface hydrogen modification on TMDs theoretical calculations has also predicted that it occurs with the chalcogen atoms. It cannot only trigger the phase transition between 2H and 1T but also significantly modify the magnetism and conductivity of TMDs [83].

Additionally, there are several other promising charges doping strategies have been also explored for the phase transitions of the TMDs, such as some of the significant approaches are given:

- Doping of electrostatic gating.
- Doping of alkali ion intercalation.
- Doping of 2D electrode.
- Doping of vacancies or substitutional heteroatoms.

Synthesis-Controlled Approach

The synthesis-controlled growth approaches of the TMDs are broadly categorized into two subgroups; in the first one growth is controlled by the conditions and the second one is connected to plenary growth. Although there are several methods have been employed to fabricate TMDs [34]. The CVD is one of the most extensively used methods for the fabrication of high-quality TMDs [84]. The stable phase

TMD formation widely depends on their fabrication technique, such as temperature, precursor type, the ratio of precursors, and the growth procedure. The smallest energy difference ($\approx$ 0.03 eV) between 2H and 1T′ phases in $MoTe_2$, is considered to be suitable to an ideal case for the modulation of phase transition via synthesis conditions [75]. The intense complete phase transition cycle of 1T′–2H–1T′ has been also demonstrated by controlling the growth temperature and the rate of tellurization (Fig. 3.12a) [85]. This demonstration leads to the 1T′ phase could be favorable under a low temperature and slow tellurization rate, or a high temperature with a fast tellurization rate, whereas the 2H phase is considered to be a dominant component in film for the moderate temperature and tellurization rate. This description has been established the growth temperature largely influenced from the amount of Te than Mo owing to the melting point of the Te source being much lower than the Mo source. Thus, it is convenient to say that the 2H phase forms in a narrow growth window with the proper Te to Mo ratio. The stichometry of the ratio of Te and Mo should be neither too low nor too high. By adjusting a fine-tuning of the tellurization rate, a novel $MoTe_2$ film can be precisely achieved through a two-step growth process, as the schematic is depicted in Fig. 3.12b [86]. Moreover, modulation of the tellurization rate with the heating temperature of the Te source gives a fast tellurization induce1T′ phase and a slow in 2H phase.

Controlling the tellurization rate is not only the parameter to get a high quality of TMD, but it also depends on the selection of a proper cooling down rate to form the final phase. The quenching of the system around $\approx$ 450, 350, and < 100 °C, may develop the 1T, 1T′, and 2H phases of the $MoTe_2$, as the schematic represents in Fig. 3.12c [87]. Particularly, the 1T phase formation is only realized in the presence of both CO_2 and H_2, this could be due to the interplay between CO_2 and H_2 along with the emergent $MoTe_2$ film. The corresponding theoretical calculations have also predicted that the 1T phase may possess the best thermodynamic stability at a high temperature, followed by the 1T′ and 2H phases with the temperature cools down. Thus the quenching of the system at a proper temperature under a specific gas environment gives desired specific phases of the TMDs.

Mostly studies are available on 2D $MoTe_2$ to interpret their 2H and 1T′ phases. However, in recent years the 3R phase $MoTe_2$ has been also identified in CVD process fabricated TMDs [88–90]. To achieve the key 3R phase during the synthesis process $MoCl_5$ precursor was heated with a separate heater. Since $MoCl_5$ has a much lower melting point than other Mo precursors, like MoO_x. Therefore, it offers modulation of the Mo/Te ratio in growth. Additionally, the bonding energy of Mo–Cl is also lower than the Mo–O, it is usually used in conventional MoOx as a precursor. Moreover, it also has advantages in the formation of Mo-Te at a low temperature. To avoid the phase transition of 1T′ at a high temperature. As the consequence, the ultrathin single crystalline flake with a thickness of less than 9 nm has also fabricated. The experimental examinations with the help of scanning transmission electron microscopy (STEM) have revealed the existence of 3R stacking mode with the hexagonal ring at the center atom for the non-monolayer location [34]. Moreover, the angle-resolved second harmonic generation (SHG) polarization pattern has been also recognized with the hexagonal symmetry for both the parallel and perpendicular

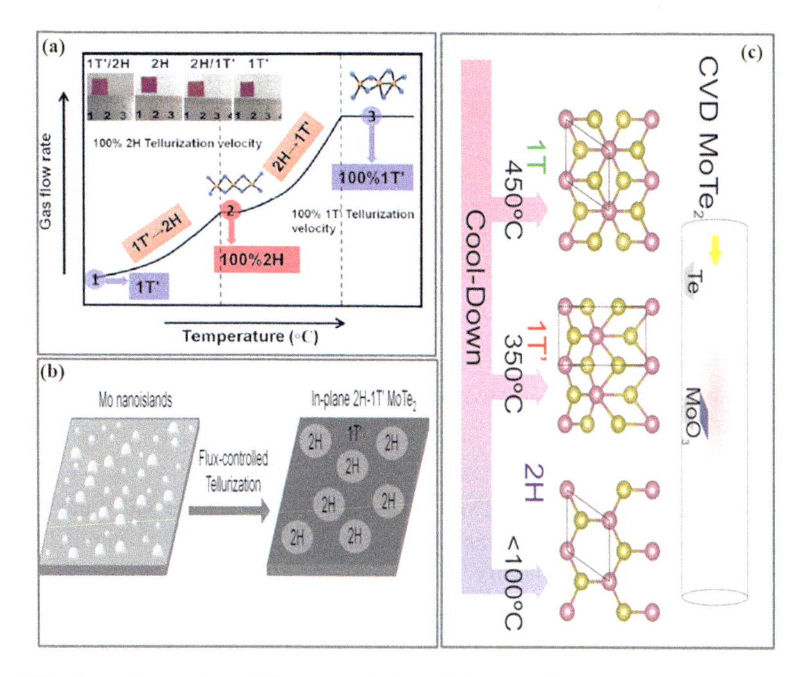

Fig. 3.12 Control growth conditions: **a** evolution of phases in $MoTe_2$ with growth temperature and gas flow rate (adopted from Yang et al. [85]; **b** schematic for the growth process to mixed phases $MoTe_2$ film and pattern strategy (adopted from Guo, J. et al. 2022 Nanomaterials 12, 110); **c** schematic the final phase of $MoTe_2$ film for the different quenching temperatures (adopted from Empante et al. [87])

configurations, this is also another direct evidence of similar symmetry of the 3R phase with the 2H phase.

Additionally, the medium-assisted growth also performed to get an induced phase transition in $MoTe_2$ [91]. The fabricated $MoTe_2$ 2H phase energy (≈ 0.6 eV) is found to be lower than the 1T′ phase of the pristine state. Although, the energy of the 1T′ phase has a distinct decrease and becomes ≈ 0.153 eV lower than the 2H phase in turn after adding one iodine (I) atom into the configuration. Moreover, with the I atom, the Mo-Te bond in the 1T′ phase elongates around $\approx 1\%$, whereas the 2H phase almost unchanged ($\approx 0.29\%$). Thus the lowers the energy in 1T′ phase facilitates the phase transition between 2H and 1T′.

Similarly, a verity of polynary TMDs has been also interpreted [92, 93]. Based on the superiority of similar structures to TMDs. In the TMDs structure, either a transition metal or chalcogenides are substituted by other atoms in the same group. A verity of TMDs alloys is also reported in form of quaternary ($Mo_xW_{1-x}S_{2x}Se_{2(1-x)}$) possessing a larger tunable bandgap range (1.60–2.03 eV) compared to their ternary alloys $Mo_{1-x}W_xS_2$ and $MoSe_{2(1-x)}S_{2x}$ [94]. The mixing energies in $Mo_xW_{1-x}S_{2x}Se_{2(1-x)}$ are so small, this indicates a significant local variability in electronic properties occurrence, which influenced the bandgaps on a large

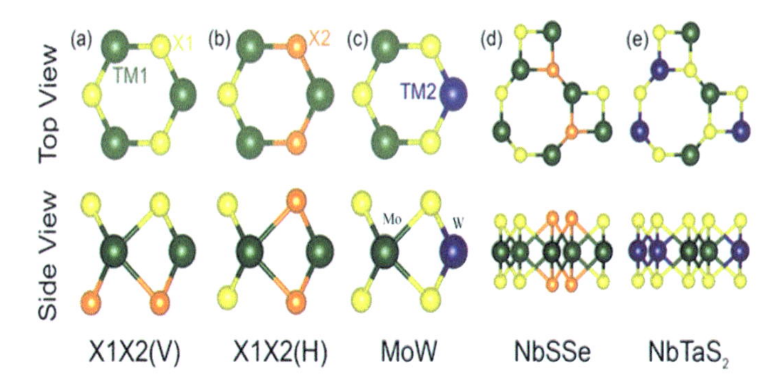

Fig. 3.13 Atomic structures of the TMDs for largest bandgap alloys (adopted from Lucking, M. C. et al. 2018 Sci. Rep. 810118)

scale. As the facts the quaternary TMD alloys ($MoW(SeS)_2$) have wider and smaller band gaps, these two different kinds of structures are illustrated in Fig. 3.13. The alloys components can also modify by adjusting the growth temperature, therefore, the corresponding photoluminescence (PL) position are also possibly tuned [34].

Thus polynary (or polymeric) TMDs are possessed tunable bandgaps in the range of their parent one. The kind of polynary TMDs alloyed band gaps have been tuned up to several hundreds of meV. Although, a much larger tuning range of bandgap would be achieved in polynary TMDs alloy when it is alloyed with metallic and semiconducting materials. As the intense, WSe_2 and WTe_2 are possessed stable phases of semiconducting 2H and metallic $1T_d$ (1T′) in a suitable environment. Under the fabrication process their alloys forms $WSe_{2(1-x)}Te_{2x}$, which undergo a phase transition from 2H to 1T′, whereas the value of x could vary between 0 to 1 [95]. However, as per past demonstrations, the 2H phase can be maintained even if x is lower than 0.4 and 1T′ phase that's favorable for the larger $\times$ 9(~ 0.7). Thus, the phase 2H or 1T′ phase is formed under the varying x value in the range of 0.5–0.6. Moreover, a different mild transition is also possible in the homophase TMDs alloys, while an abrupt phase transition process could be in a heterophase 2H–1T′ system, in such cases a continuously tunable band gap is not be possible [34, 79, 88–90].

3.2.3.2 Phase Stability

This is another important parameter to define the relative mechanism for the phase stability of the TMD materials. Structural transformation under eminent pressure, the high symmetry 1T structure of different group TMDs is unstable and makes lower-symmetry octahedral phase 1T′, that is due to distortion. Therefore, the ground state energy differences of the different phases could interpret in terms of phase stability of the TMD materials [10]. Such as the relative phase stability mechanism based on the energy difference calculation of various TMD materials, as depicted in Fig. 3.14. The illustrated diagram is useful to correlate the phase stability of T and H that arises due to the number of d electrons available in a particular group of transitional metals in the periodic table and their corresponding d-orbital energy level splitting under the presence of the ligand field that formed by surrounding chalcogen atoms [2, 30]. Similarly, in an ambient condition, almost all VI group TMD materials also have a layered bulk crystal structure composed of monolayers, whereas the symbol X represents atoms belonging to trigonal prismatic coordination around the M atoms. However, some exceptions are also possible like WTe_2. This TMD material has a stable state with distorted octahedral coordination (T_d). In general, VI group MX_2 structured monolayers TMDs are stable with the H structural phase having semi-conducting behavior and a direct bandgap in the range of 1–2 eV [2, 96, 97]. These materials show significant overlaps of transition metal d-orbitals with the p-orbital of chalcogens to form a strong p-d-orbital hybridization at the valance band. The circumstances of overlaps in metal d-orbitals with ligand orbitals offer a stabilizing factor for trigonal prismatic coordination. Therefore, H-phase hybridization in a compound splits the d band into sub-bands. As an intense, compound MoS_2 state dz^2 is filled and has a bandgap of 1.8 eV for the monolayer.

The DFT energy calculation has also demonstrated that the TMDs in the order of E (H-WSe_2) < E(T_d-WSe_2) < E (T-WSe_2) with the preferred stability order H-WSe2 > T_d-WSe_2 > T-WSe_2. The H-WSe_2 phase is possessing energy 0.25 eV which more stable than T_d and 0.74 eV, whereas the T_d phase is considered to be more stable than the T phase. In the similar way E (H-$MoTe_2$) < E (T_d-$MoTe_2$) < E (T-$MoTe_2$) also have stability order: H-$MoTe_2$ > $T_d$$MoTe_2$ > T-$MoTe_2$. The H-$MoTe_2$ phase has energy 0.03 eV with greater stability than T_d (0.49 eV), it reflects the T_d phase also more stable than the T phase. The DFT energy calculations have also revealed that the H-$MoTe_2$ and T_d-$MoTe_2$TMDs phases could attribute with a very small stabilization energy difference. On the other hand, the stability order of WTe_2 seems to be different than that of the above TMDs as expressed by their total energy ordering E (T_d-WTe_2) < E (H-WTe_2) < E (T-WTe_2). Therefore, the order of stability of these TMD materials is expressed in this order T_d-WTe_2 > HWTe_2 > T-WTe_2 (see Fig. 3.14) [20].

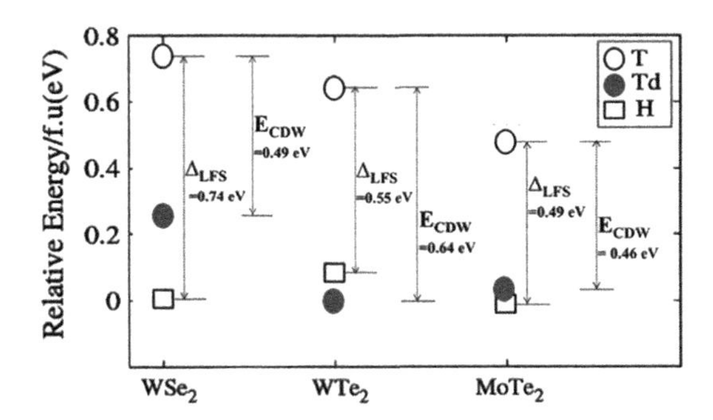

Fig. 3.14 Relative stability for the various phases of TMDs with trigonal prismatic (H), octahedral (T), and distorted octahedral (T_d) coordinated geometries. The open circle, open square, and filled circle refer to the octahedral (T), trigonal prismatic (H), and distorted octahedral (T_d) coordinated phases (adopted from Santosh et al. [20])

Further DFT calculation based on the thermodynamical stabilities of the WSe_2 and $MoTe_2$ has revealed that it can stabilize the H structure and WTe_2 stabilized the T_d structure. Predominately considering the conceptual problem, why a particular TMD is stabilized in the distorted octahedral (distorted T or T_d) structure? So, innovations were focused on both ligand field stabilization (LFS) energy and the charge density wave (CDW) [57, 98]. These stabilizations have been correlated to a macroscopic quantum state consisting of a periodic modulation of electronic charge density accompanied by a periodic lattice distortion in quasi-1D or quasi-2D layered structured materials. As depicted in Fig. 3.14 the ligand field stabilization energy (Δ_{LFS}) is 0.74 eV in-between the H and T phases, with a larger energy gain for the WSe_2 TMD. Whereas the charge density wave is accompanied by a periodic lattice distortion and CDW T_d phase stabilization energy (E_{CDW}) 0.49 eV. Similarly, for the composition $MoTe_2$, $\Delta_{LFS} \approx 0.49$ eV and $E_{CDW} \approx 0.46$ eV has also obtained, these are very close to each other in magnitude. Considering these promising characteristics such materials have been recognized for the further investigation. Particularly, the T_d phase of $MoTe_2$ at a high temperature [99]. The H-$MoTe_2$ has been identified as an excellent candidate for the phase change material under the application of 0.3 to 3% tensile strains [40]. Composition $MoTe_2$ has E_{CDW} larger than D_{LFS} compared to WTe_2 could be a reason behind the larger bonding distance of W-Te which reduced the ligand splitting energies of WTe_6 configuration with the corresponding D_{LFS}. Moreover, the larger W-Te distance of the periodic lattice distortion is also useful to match the electronically induced CDW, this leads to larger E_{CDW}. Thus, the energy of the CDW and the periodic lattice distortion is due to the charge density wave instability that drives distorted octahedral T_d geometry with large stabilization energy. As the energy diagrams of H and T_d structures are illustrated in Fig. 3.14. The crossing points of the W (Se, Te)$_2$ and (Mo, W) Te_2TMD materials the H and T_d phases have the same energy with bi-stable behavior. The typical values of Gibbs free energy for the (Mo, W) Te_2 are also defined at crossing energy points of $Mo_{0.67}W_{0.33}Te_2$ at room temperature [20]. Where the composition Mo-W, metallic $T_{d,}$ and semiconducting H structures are interchangeable by external perturbations/ physical parameters; such as temperature change, mechanical strain, or an external electric or magnetic field. The external changes may an abrupt corresponding sudden change in the conductivity of TMD. This mechanism is also considered to be a promising approach to developing steep for the electronic device to fulfill the rapid conductivity change requirement.

The orbital projected density of states is another useful approach to defining the phase stability of TMD materials. Typically, this approach is expressed by the density of states vs total energy plots, as depicted in Fig. 3.15. Where the T-WTe_2 has a high density of states (DOS) at E_F, and T_d with a reduced DOS due to CDW/periodic lattice distortion stabilization. Although the band gap opening in 1D Peierls instability (along the x-axis in Fig. 3.15c) has a frustrated form for the 2D system owing to other direction (y-axis in Fig. 3.15c) containing regular periodicity. Their corresponding metallic electronic structure leads to a reduced DOS at a Fermi level (E_F) rather than a gap opening. This is a direct consequence of H-phase stability derived from the different mechanisms of the ligand-field. Owing to rearrangement from the

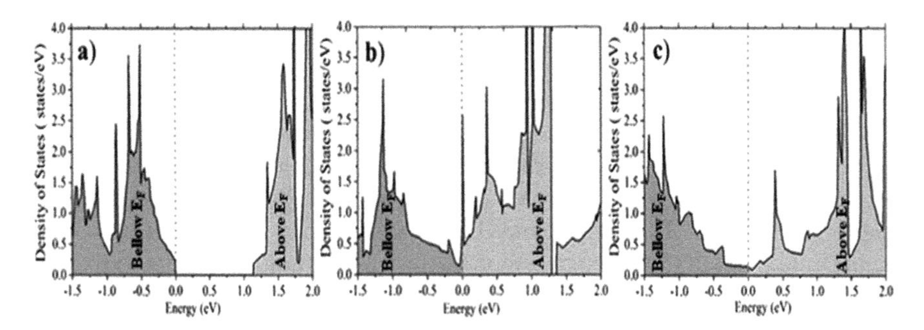

Fig. 3.15 The total d-orbital density of states corresponds to different coordination geometries: **a** trigonal prismatic (H); **b** octahedral (T); **c** distorted octahedral (T_d). The zero of the energy refers to the Fermi level, indicated by the dotted vertical lines. The filled states below the Fermi level represent the valance band, whereas the states above the Fermi level represent the conduction band (adopted from Santosh, K.C. et al. 2015 2D Mater. 2, 035019)

bonding configuration and further splitting of the d-orbital energy levels. Therefore, the H-phase gap opening due to the ligand field stabilization energy, however, the energy cost of ligand orbitals is also responsible for the re-hybridization structural changes from T (octahedralW-X6) to H (trigonal prism W-X6) coordination. Since the transition metals are usually considered in the octahedral crystal field formed by the representation X, thereby, the M d levels splits into t_{2g} (d_{xy}, d_{yz}, d_{zx}) and (d_{x2-y2}, d_{z2}) states. Where the d electrons mainly occupied the t_{2g} levels, leading to almost filled t_{2g} bands in the octahedral crystal structure. But there are a few hybridizations between M d- and X p-orbitals also exist. These states are near to Fermi level due to the consequence of the X p-orbitals and Md hybridization in the valence band and the Md-orbitals in the conduction band for the trigonal prismatic phase. Moreover, the further splitting of the energy levels is due to the reduction in the kinetic energy of the electronic system. Hence, the d-DOS splitting and the competing effects of D_{LFS} and E_{CDW}/periodic lattice distortion could be considered origin of the phase change in TMD. But note that the carrier density is significantly reduced after the phase transition to $T_{d,}$ it pointing out the density of states in the Fermi level is not enough to open a bandgap in T_d-WTe_2 [20]. Similarly in the presence of CDWs, the conductivity is also reducing. The energetic advantage of distortion to stabilized valence electrons is specifically large for the bulk TMDs, such as 1T-TaS_2 and 1T-$TaSe_2$, CDW formation at room temperature [100].

3.2.3.3 Local Phase Patterning

The local 2H-1T phase transition of TMD is called phase patterning, it can be achieved in a controllable way, as the typical process illustrated in Fig. 3.16a [10]. Cho et al. investigated the phase patterning in the composition $MoTe_2$ by using the laser [73]. They irradiated a high-energy laser beam on the $MoTe_2$ layer to distort the 2H phase into the 1T phase (WTe_2-like structure) via the phase transition process. The phase transition results were confirmed by interpreting the Raman spectroscopy analysis. Though, this technique does not avoid the thinning of the $MoTe_2$ layer owing to laser irradiation. Their interpretation of Raman spectroscopy outcomes was also not free from contamination, such as oxides and heterophase homojunction structures. However, it did not find any evidence about the reverse phase transition from 1T to 2H. The key advantage of this approach that used laser energy generated temperature was around 400 °C. This temperature limit was less than the previously defined temperature of 880 °C for the composition $MoTe_2$ phase transition. The reason behind that, the reaction occurs at a lower temperature than the Te sublimation temperature (~ 400 °C), therefore, the 2H phase region is reserved without Te vacancies [73].

Another Ar plasma approach has been also introduced for the phase patterning of the TMD materials (see Fig. 3.16b). In this approach around 40 s, Ar plasma-treated MoS_2 undergoes 2H to 1T phase transition [101]. The basic principle of this approach is like the laser phase patterning by creating the chalcogen atom vacancies. This process hitting area via plasma cannot control by the delivery instrument, therefore, it is costmary to undesired regions that should be protected from the Ar plasma. Additionally, if the treatment duration increases it leads to the increasing S defects elimination of the 1T phase [101]. Thus, the discussed these two approaches are very similar based on the creation of the irreversible chalcogen vacancies, which become a driving force for the phase transition from 2H to 1T. Since both approaches exposed only the specified area, therefore, they can be used for selective phase

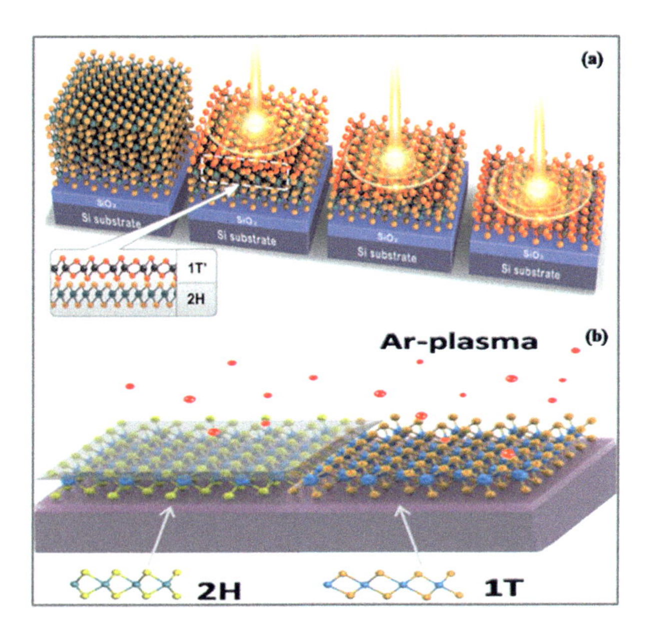

Fig. 3.16 **a** Schematic of the laser-irradiation process (adopted from Guo, J. et al. 2022 Nanomaterials 12, 110); **b** Schematic of the plasma-treated process for the formation of 1T Phase MoS_2 (adopted from Zhu et al. [101])

patterning or creating in-plane heterophase structures. But these approaches have a major drawback in terms of their destructive nature, by means they are not offering reverse changes from the 2H to 1T phase state [51].

Moreover, to achieve a more efficient process of local phase patterning in TMDs, several innovations have been deliberated in various studies [10, 51, 79, 90, 102–104]. Considering the key issue of chalcogen donor electrons to form the vacancies in TMDs. The chalcogen vacancies can play a crucial role to promote the 2H to 1T phase transformation following the twin mechanism process: by the creation of additional charges and flagging of the metal–chalcogen bonds that provides the structural alteration [105]. As the intense, MoS_2 energy dissimilarity in the creation of 2H and 1T phase is decrease up to almost zero while vacancy amounts extended up to 8 at%[106]. Additionally, theses vacancies also acts as point defects and facilitate as the nucleation sites for the 1T′ phase under the 2H matrix. To deduce the suitable technique other methods have been also introduced, such as strict electrochemical etching to get 2H to 1T structural transition [105]. By demonstration of 1T phase content in multi-layered MoS_2 has been altered either by increase in the etching capability or the reaction time, but findings revealed the three-dimensional spreading of the metastable phase in an uncontrolled manner [105]. The typical phase transformation patterns of the composition $NbSe_2$ are illustrated in Fig. 3.17 [107].

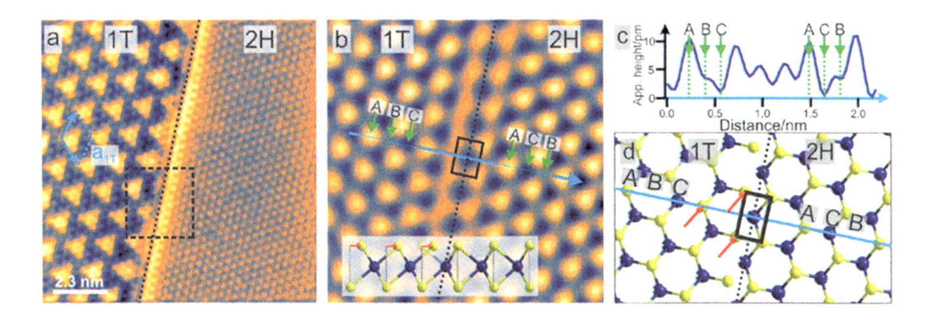

Fig. 3.17 Atomically sharp phase boundary between 1T- and 2H-$NbSe_2$ at 77 K (scan parameters: I = 1.7 nA, and U = −2 V): **a** The pristine surface (right) does not exhibit charge modulations, while a clear CDW order indicates 1T-$NbSe_2$ (left). The charge modulation of the 1T phase (1T-CDW) manifests itself as a commensurate ($\sqrt{13} \times \sqrt{13}$)R13.9° superstructure (blue arrows) with a lattice constant a_{1T} of 12.4 Å; **b** close-up of the dashed square in panel a represents an FFT-filtered image to remove the electronic CDW modulation. Whereas each bright dot corresponds to one selenium atom in the top Se layer. Across the boundary (black dotted line) and a rectangular alignment (black box), the distinct positions on the surface are highlighted by green arrows A–C. The bottom inset shows a side view of a model of one $NbSe_2$ layer and red arrows indicate the shifted Se top plane after the 2H–1T transformation; **c** Line profile along the blue line in panel b represented by green arrows indicates the corresponding lattice positions in panel b; **(d)** Schematic of a single $NbSe_2$ layer at the phase boundary, in which yellow spheres represent the Se top plane in STM image and blue spheres represent subsurface of Nb. Se atoms in the bottom layer of 1T-$NbSe_2$ are colored pale yellow. Red arrows indicate the top layer shift that leads to a rectangular Se atom arrangement at the boundary as highlighted in black (adopted from Bischoff et al. [107])

3.2.3.4 Phase Tunability

Since some TMDs also exist in either an octahedral or trigonal prismatic coordination, such as VI-B group TMDs. This implies a semiconducting to metallic transition due to a change in coordination from H-structure to T-structure, as depicted in the energy diagram in Fig. 3.18. The semiconducting 2H phase transition to metallic 1T phase occurs when material electrons doped, as the instance MoS_2 alkali atom intercalation. The same analogy can be also applied to monolayer materials to make metallic contacts to semiconducting TMDs by locally transform phase [30, 72]. This leads to the monolayer materials switched by external stimuli like an applied electric field in a field effect device [108]. Similarly, other TMD materials also have naturally different polymorphs and they exhibit structural transitions as a function of temperature. $MoTe_2$ has a high-temperature metallic 1T′ phase whereas their 2H phase lies in the ground state at lower temperatures. Moreover, the transition temperature can be also tuned by the stoichiometric variation and metallization of 2HeMoTe_2 flakes using the local laser heating to transform it into a 1T′ phase [73].

In a similar way, group V-B TMDs are also transform their phases with the tunability ability, like vanadium dichalcogenides. As per density functional theory (DFT) calculations its monolayer forms a 1H ground state, however, the bulk material generally forms a 1T phase. This is direct evidence of formation energies in these two very similar polymorphs. Although material monolayer grown from the MBE technique has revealed structural phase 1T but not 1H phase [102]. Besides the structural similarities in different TMDs, their electronic structures are widely different due to a distinct number of d-electrons, this property of TMDs makes them

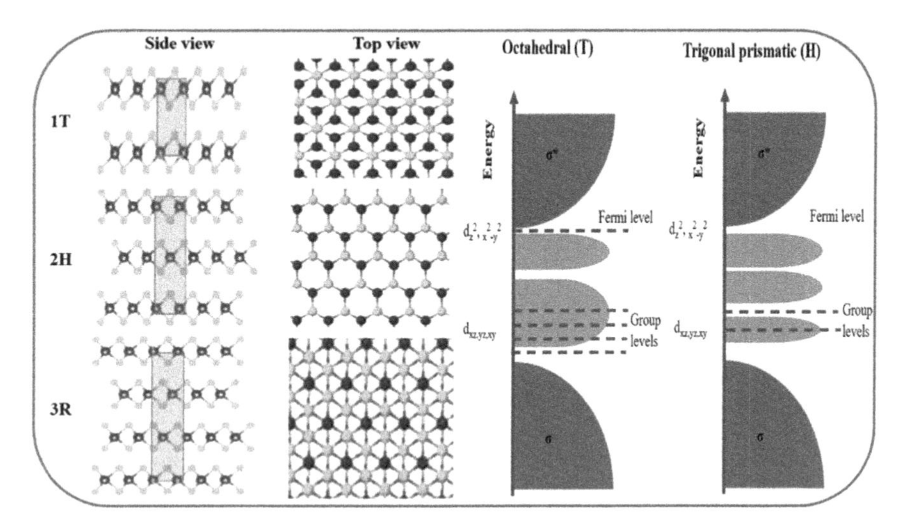

Fig. 3.18 Schematic for the 1T, 2H, and 3R TMDs, including electron energy diagram of the octahedral and trigonal prismatic coordinated TMs and the Fermi level position dependence on the number of d-electron

highly versatile. Hence synthesis of TMD materials truly single layer provides a better understanding to control the various quantum phenomena.

The lamellar effect (or stacking) and intrinsic features under exterior circumstances (like compression and electrical field) are generally used for the modulation of the electronic property. Like the high compression activities of MoS_2 up to 38.8 GPa, in which phase variations $2H_c$ to $2H_a$ has been expected at about 20 GPa [109]. Theoretically, it has been established by showing the unrestricted shared descending of the formed layers; therefore, the original $2H_c$ stacking changes into a 2H stacking [110]. In contrast to this, the $MoSe_2$ upholds the $2H_c$ stacking with no phase transformation under the applied higher compression [111]. Similarly, under pressure, the vibrant procedure of phase alteration for the MoS_2, $MoSe_2$, and $MoTe_2$ has been introduced by employing the first-principle approach [112]. In the higher compression, the phase alteration occur $2H_c$ to $2H_a$ for the MoS_2, whereas the vigorous obstacle increase as the compression increases owing to a reduction in layer distance. This theoretical interpretation has also consisted the experimental findings. While there is no phase alteration occurring in $MoSe_2$ and $MoTe_2$ owing to the robust race among the van der Waals force and Coulomb exchanges. Therefore, the electronic structure of MoS_2 evolves the transformation from semiconductor to metal under pressure [113]. This theoretical interpretation also reveals a transition from direct to indirect bandgap in the monolayer with the occurrence of metallization in multilayer structures. Moreover, it also describes the band splitting due to diverse orbital features at distinct k-points and alterations under applied compression. Under a pressure, similar changes are also appeared in $MoSe_2$ and $MoTe_2$ electronic structures. Additionally, their applied potential can be manipulated through the electronic structure via exterior electric fields, thereby, the phase transformation occurs due to order and disorder transformation [114, 115].

In a similar fashion group-VB, the layered dichalcogenides also have charge density wave (CDW) phases in TMDs, specifically, TaS_2, $TaSe_2$, and $NbSe_2$ [116, 117]. Likewise, 1T-TaS_2 also have a layered structure in which both the Mott state and the several abnormal CDW states exist. The CDW phase in 1T-TaS_2 due to the composite Fermi surface uncertainty like the electron–phonon interconnection and Coulombic exchanges. Therefore, the several phases occurs at ambient compression as the function of temperature, together with high-temperature metallic disproportionate CDW (ICCDW, $T < 550$ K), approximately equal CDW (NCCDW, $T < 350$ K), and commensurate CDW (CCDW, $T < 180$ K) [118]. On the other hand, the lower temperature state CCDW has a greatest characteristic structural building, which is usually categorized by the $\sqrt{13} \times \sqrt{13}$ super lattice star-of-David-like structures formed by the twelve Ta atoms clusters around the thirteenth Ta atom [119]. The typical pressure–temperature phase illustration is according to resistivity measurement as depicted in Fig. 3.19a. This result demonstrates, the superconductivity compression range 3–25 GPa belongs to NCCDW stage at a conversion temperature about 5 K. It reveals that the whole system converts into the metallic when the temperature beyond the superconductivity conversion temperature. Additionally, the CDW phases in 1T-TaS_2 with the thickness of the film have revealed that both CCDW/NCCDW and NCCDW/ICCDW phase transitions can be modulated, as illustrated in Fig. 3.19b [120]. With the thickness reduction the phase transition temperature is also decreased. Moreover, the phase transition is abruptly disappeared at a critical thickness (nearly 10 nm), therefore, in this material the insulating characteristics are emerged below 3 nm [104]. Continuing investigations have also revealed that the phase alteration among CDW and the metallic phase of the 1T-TaS_2 film can be performed with bias-voltage [121]. As the H TaS_2 has lone a single-phase transformation in a metallic phase to an disproportionate CDW phase, this is dissimilar to usual 1T phase transition [122]. The experimental evidences have been also revealed that together charge concentration wave and superconducting phase are possible under the monolayer limit [123, 124]. Moreover, an incommensurate CDW state also occurs in bulk 2H-$NbSe_2$ at the temperature lower the CDW conversion temperature (TCDW of 33.5 K), while the superconductivity exists below the transition temperature T_c (7.2 K). However, the superconductivity alteration temperature is decreased with the layer thickness reduction. A noteworthy increment in TCDW at 33 K for the bulk and 145 K for the monolayer has been also interpreted. This result has a good agreement with the theoretical predictions [125, 126]. Thus, an improvement of the electron–phonon complex element at $q \approx q_{CDW}$ plays a crucial part to form CDW phase in $NbSe_2$. The CDW phases of single layer 2H-$NbSe_2$ theoretical simulation interpretation for the various structures with different distortion patterns compatible with 3×3 CDW structure has been demonstrated [127, 128]. The existence of many CDW phases have been also established experimentally in 2H-$NbSe_2$ [129]. However, in isostructural material 2H-$TaSe_2$, the 3×3 CDW order still unaffected in bulk to single layer transition [130].To interpret it several experimental and theoretical results have been addressed based on physical mechanisms involvement of electron–phonon joining, Fermi surface nesting, and saddle-point peculiarities debateable concepts, but it is still challenging to further innovation.

The phase tunability in superconductive TMDs is also whichever inherent or due to chemical doping, applied pressure, and electrostatic doping. Such as, in pure TaS_2 and $TiSe_2$ superconductive from the nonmetallic lower temperature phase by evolvements of the NCCDW phase at the ambient compression [118]. The superconducting transition is also controllable from the chemical doping in Cu/Pd [131, 132]. Where the superconducting phase transpires as the CDW phase and it suppress from the doping in the area limited everywhere the crucial doping density [131]. The pressure encouraged superconductivity has been also recognized in 1T-$TiSe_2$ [133]. The 2H phase in bulk 2H-$NbSe_2$ superconductivity also coexists together CDW phase at low temperature at a crucial temperature $T_c = 7.2$ K. Its superconducting feature remains within the 2D limitation up to the conversion temperature ($T_c = 1.9$ K) [124]. Similarly, metallic H-TaS_2 and H-$NbSe_2$ is also possessed intrinsic superconducting behavior due to greater spin–orbit joining under the atomic layer restriction [134]. The archetypal band insulator (like MoS_2 thin flakes) superconductivity is also induced due to electrostatic carrier doping [135]. It's possible through the traveling ions in a voltage-biased ionic liquid located over the sample. However, the gate persuaded superconductivity of the MoS_2 multilayers via bulk to monolayer is well studied with the remarks a noticeable suppression of T_c from bilayers to monolayer. Because of the probable infinitesimal mechanisms of feebler superconductivity

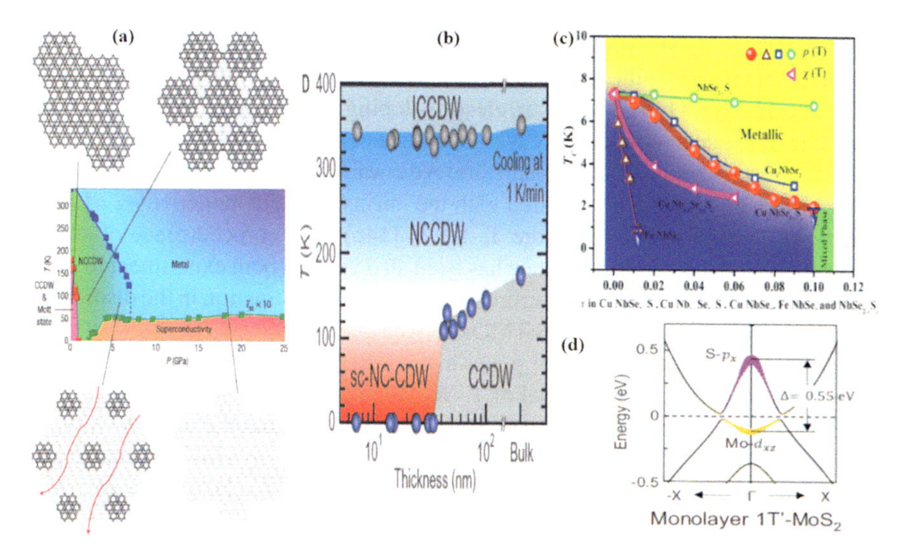

Fig. 3.19 **a** Temperature–pressure phase diagram for the 1T-TaS2 (adopted from Biesner, T. et al. 2020 Crystals 10, 4); **b** schematic pictures of a Ta atom network in the CCDW (left), hexagonal NCCDW (middle), and ICCDW (right) phases. The dark circles represent the Ta atoms displaced from their undistorted lattice coordinates (adopted from Yoshida et al. 2015 Sci. Adv. 1, e1500606); **c** superconducting polymorphism with the critical temperature in TMDs (adopted from Yan, D. et al. 2019 arXiv:1907.01776); **d** band structure of monolayer 1T′-MoS2 near the Γ point (adopted from Choe et al. [147])

in the monolayer [136, 137]. Thus the superconductivity regularly exists in a two-dimensional configuration [138]. As the identical superconductivity phenomena is derived from the gate in WS_2 and $MoSe_2$ films [139, 140]. Additionally, compression is also a key flexible physical constraint to modify the crystal's electronic structure. In the compression alteration from $2H_c$ to $2H_a$ in MoS_2 is associated change semiconductor to metal. Moreover, increase in the compression over 100 GPa, the $2H_a$-MoS_2 is also attributed superconductivity, the corresponding mechanism has been also elucidated by the DFT interpretation [141]. The co-occurrence of two Fermi pockets toward distinct high regularity lines gives change in both valence and conduction bands, therefore a rise in their overlapping energy can be connected to the occurrence of superconductivity. The electronic feature of WTe_2 has been also verified at high pressure [142]. Their structural alteration from T_d ($Pmn2_1$) to 1T′ ($P2_1/m$) is nearly at 4–5 GPa and the changeover compression is near to the emergence compression of the superconductivity. So, compression persuaded superconductivity in WTe_2 has been recognized. The pure 1T-MoS_2 also has superconducting properties at the superconducting transition temperature of 4 K [143]. The various superconducting TMD materials of group-IVB TMDs along with their crucial temperatures are concisely summarized in Fig. 3.19c. Due to near-spin degeneracy, superconductivity also exists in electron-doped TMDs. Therefore, splitting of the spin degeneracy in the momentum space to hole-doped TMDs has been suggested for the topological superconductors [144]. To verify it K-intercalated type-II Weyl semimetal T_d-WTe_2 has been synthesized by the liquid ammonia, their superconductivity is recognized at the T_c around 2.6 K [145]. Moreover, electron doping also plays a significant part for the tuning of the superconductivity with no lattice distraction and K-intercalation of the WTe_2 host topological nontrivial state. Afterward, a novel monoclinic phase of WS_2 (2 M-phase) has been identified with the intrinsic superconductivity 2 M WS_2 at the critical transition temperature T_c 8.8 K [146]. The co-occurrence of superconductivity and the topological state has been also verified both experimentally and theoretically. As per theoretical interpretation, the lattice distraction in the disordered octahedral phase (T′) in group-VIB TMDs derived due to inherent band reversal close to the Fermi level in chalcogenide p and metal d bands (for T′-MS_2 and T′-MSe_2) compared to d–d kind band reversal of T′-WTe_2 [147]. As the intense T′-MoS_2 (see Fig. 3.19d) band inversion also leads two Dirac cones without the consideration of spin–orbit coupling. Thus the spin–the orbit coupling opens-up a fundamental gap at the Dirac points. That can provide upswing to the quantum spin Hall phase or 2D Z_2 topologically insulating phase. Moreover, MoS_2 and additional 1T′-MX_2 also have Z_2 nontrivial band topology [148, 149]. Therefore, the external electric field turns the topological state with the on-and-off process [44].

Phase Control

In the real world, the phase control of the grown TMDs one of the key issues for specific applications. Considering this critical issue several approaches have been proposed by the investigators, such as the energy difference approach in phases H and T. For the instance, group VI-B TMDs have a considerable energy difference in their transformed 1T/1T′ phase and H-phase pristine in MoS_2, however, a phase change cannot be easily possible in the pristine materials. Similarly, other materials of this group like MoS_2 or WS_2 transformed the 2H phase into 1T or 1T′ due to the alkali atom intercalation energy difference [72, 150, 151]. Similarly, ammonium in an aqueous solution has also used to stabilize the 1T phase [152]. The Li-intercalation in MoS_2 single crystal and their associated phase transition to 1T phase is useful for the photoemission application. The binding energy shift of about 0.8 eV in the Mo-3d core level is considered to be a convenient differentiation and assessment of the concentration of the 1T and 2H phases [153, 154]. The chemically exfoliated monolayer of the MoS_2 and WS_2 also have heterogeneity that contains 1T, 1T′ and H-phases [14, 26, 29, 31, 155]. It is usually correlated to a consequence of charge transfer of the TMDs owing to organic compounds used for the exfoliation process, like butyl lithium. Although their 1T phase has retained when it dried, this revealed the adsorbed ions such as protons and remaining alkali ions that compensate for a negative charge on the 1T-MoS_2 [154, 156]. The 1T-WS_2 also reflects the stability without Li ions [29] due to their 1T phase formation in MoS_2 or WS_2. It is in a metastable state with a significant energy barrier of ~ 1 eV in the reformation of the H-phase [3, 29, 156, 157], however, annealing in a vacuum is useful to reform the stable H-phase. This is also a direct indication to control the local modification in

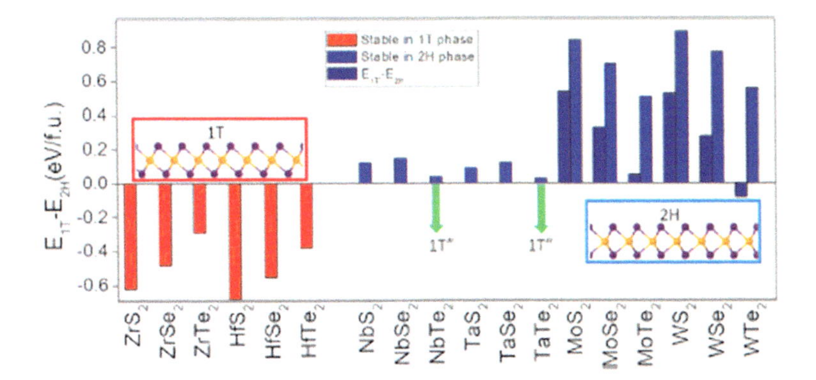

Fig. 3.20 Formation energies for the H and 1T (1T′) phases. 1T phase is preferred to Zr- and Hf-dichalcogenides. For Nb- and Ta-dichalcogenides the 2H phase is only slightly preferred over the 1T phase. The Nb- and Ta-Te_2 known bulk structure for the distorted 1T″ phase, this is not under consideration here. In the group VIB TMDs, only WTe_2 has the 1T′ phase as the lowest energy the other materials prefer the 2H phase with 1T′-$MoTe_2$ being very close in energy to its 2H phase (adopted from Lasek et al. [102])

MoS_2 into 1T (1T′) and H patterns, which can be exploited to make metallic contact pads for the semiconducting H-phase [158]. Moreover, concerning electrochemical or photocatalytic applications the 1T phase or a phase mixture are often considered to beneficial get a higher HER in 1T MoS_2 compared to their H-phase. The monolayer $MoTe_2$ phase transition is usually monitored from the STEM to define a transition which precedes via an intermediate phase between the 1T and H-phases. So, the transformation into the T-phase can be induced and controlled by the electron beam [159].

Similarly, group V-B TMDs phase formation is also control by employing this approach. As the $NbSe_2$ and $TaSe_2$ phase 1T to 1H formation depending on the growth temperature. Group V-B TMDs' different phases energies are very close according to DFT calculations, as depicted in Fig. 3.20 (here note that $TaTe_2$ and $NbTe_2$ do not form the 'simple' 1T structure, but they can form a distorted structure, which is referred as 1T″ structure). In the same fashion VSe_2 and VS_2 also forms a more stable H-phase rather than the 1T phase as per the DFT calculation predictions [102]. Although in the experimental investigations, only 1T phase has been achieved for both bulk and monolayer. This could be due possible CDW phase lowering the energy of the 1T phase that sufficiently stabilized it over the 1H phase, but the overall formation energy of these two phases is very close to each other.

As illustrated in Fig. 3.20, the 2HeTaS_2 has a lower energy configuration than its 1T phase but its thermal annealing in a vacuum gives a 1T crystal due to the surface layer transforming into a 2H phase [160]. It can be verified from the characteristics of the CDW structures. Likewise, the transformation of the 2HeTaS_2 surface layer on the 1T-TaS_2 bulk with the enhanced superconducting transition temperature 2.1 K compared to bulk 2HeTaS_2 crystals have also noticed including phase switching in $TaSe_2$ and $NbSe_2$, etc. [107, 161–163].

Moreover, group VIII-B (10) TMDs are also following the concept of the energy phase difference in low-energy argon sputtering, and their subsequent annealing transforms the $PdSe_2$ surface from the bulk truncation to a novel vdW nanoribbon structure [164]. The more conventional hexagonal group 10 TMDs' usual 1T structures can have most energetically stable phase, whereas the 1H phase has significantly higher formation energy [165]. Although, an unexpected 1T$PtSe_2$ monolayer on a Pt (111) crystal formation due to partial transformation in 1H phases [166]. Such transformation was triggered by the annealing-induced Se-loss. It has been interpreted as the Se-vacancies arrange at the 1T-1H phase boundary and stabilized the 1H phase. Therefore, it gives rise to triangular 1H domains within the 1T matrix and produced an ordered pattern. So, the Se-3d core level peak of the 1H phase shifted around 0.4 eV to higher binding energy concerning 1Tphase. Therefore, the relative intensities of the components of XPS are useful to measure the concentration of the transformed surface. This kind of transformation of the 1H phase could be reversible for the Se exposed surface. The importance of Se-deficiency in the transformation process also reflects in the 1H phase that can adsorb pentacene more strongly than the 1T phase. Thereby, the 1T/1H patterned surface is useful as an adsorption template. Hence the molecular adsorption on monolayer TMDs would be useful for the modification and control of the properties.

Distinction of Metastable Group VI TMD Polymorphs

The reliable distinction of the metastable polymorphs of group VI TMDs is also significantly important for the experimental exploration of the electronic features of various states. In a large number of studies, the key dissimilarity among 1T and 1T′ structural polymorphs was frequently ignored, however, attention has been also paid to the 1T′ phase, like 1 T or commonly mentioned as the metallic state [29, 167, 168]. The 1T and 1T′ phases are usually dominant and unambiguously distinguished from the high-resolution transmission electron microscopy (HR TEM) and scanning transmission electron microscopy (HR STEM) pictures of the atomically thin nanosheets. In which 1T′ phase feature belongs to zigzag chains of transition metal atoms, as depicted in Fig. 3.21a, b [26, 68]. Though the 1T and 2H phases have a threefold regularity in the ab plane, they nearly not distinguishable in the HR TEMpictures. However, the 1T and 2H domains of the group VI TMD monolayers is distinguishable from the high angle annular dark-field (HAADF STEM) pictures by reflecting the the intensity of chalcogen pillars, as depicted in Fig. 3.21c, d [26]. A discernible variation in signal intensity in trigonal prismatic 2H structure due to two chalcogen atoms overlying in the way of the upcoming beam. Such kinds of the upcoming beam make it dissimilar and nearly analogous to a metal location to produce a honeycomb lattice in STEMimage (see Fig. 3.21e). While in the octahedral 1T structure, a robust dissimilarity among the transition metal and chalcogen spots forms a hexagonal lattice (see Fig. 3.21f). Additionally, the stacking arrangements of the 1H monolayers in bi- and triple layer nanosheets have been also recognized. Extreme care should be taken in the interpretation of the microscopy

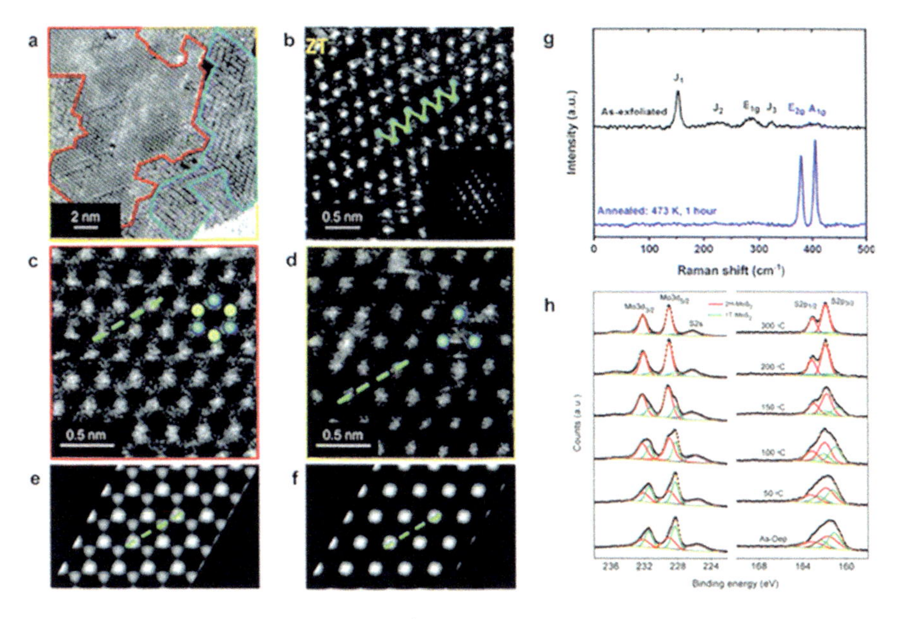

Fig. 3.21 The distinction of 1H, 1T, and 1T′ polymorphs; **a** STEM image of chemically exfoliated MoS_2 displaying co-existing 1H, 1T, and 1T′ phases, the corresponding areas enclosed by red, yellow, and blue curves; **b–d** represents the HAADF STEM images of the 1T′, 1H, and 1T MoS_2 lattice; **e, f** simulated HAADF STEM images of the ideal 1H and 1T MoS_2 monolayers; **g** Raman spectra for the chemically exfoliated 1T′ (top) and annealed 2H (bottom) MoS_2 flakes; **h** XPS spectra of the Mo 3d, S 2 s and S 2p core level electrons of mixed-phase 1T/2H MoS_2 nanosheets to demonstrate gradual recovery of the 2H phase during thermal annealing, the 1T and 2H phase contributions represent from green and red (adopted from Sokolikova et al. [90])

data, specially Fourier-filtered pictures because residual derivatives which provide upswing to quasi-atomic pillars [169]. Moreover, it has been also noticed that the 2H and 3R polytypes are possibly distinguishable by the intensity profiles of selected area electron diffraction (SAED) patterns along the [001] zone axis [170]. Usually, the [99] and [108] diffraction spots are near with identical intensities in the 2H phase, whereas in the 3R phase, the intensities are approximately 3-times different [170].

The spectroscopic techniques, such as Raman spectroscopy and X-ray photoemission spectroscopy (XPS) also offers a comparatively correct approach to detect co-occurring polymorphs in TMD nanostructures throughout micron-sized areas. As an instance, the chemically exfoliated MoS_2 octahedral 1T phase possess two Raman active modes, E_g and $A1_g$, at 287 and 408 cm^{-1}, while the $E^1{}_{2g}$ mode of 2H phase (at 383 cm^{-1}) absent [16]. Additionally, three unexplored modes J_1, J_1, and J_3, at 156, 226, and 333 cm^{-1} are also recognized (see Fig. 3.21g). The additional three modes are attributed due to 2a × a super structure in an octahedral phase. The wider set of Raman active modes of the 1T′ and T_d phases correlated to their inferior regularity compare to the 2H phase, which is typically connected to WTe_2 and $MoTe_2$ [171–174]. The 1T′ phase J_1, J_2, and J_3 modes signature can also achieve in MoS_2, WS_2, $MoSe_2$, and WSe_2 single crystals and few-layered nanosheets [175–179]. Here it should be keep in view that the J_1 mode constantly occurs with the utmost intensity in Raman spectra for a almost pure phase of 1T′, such as group VI disulfides [68, 178, 179]. Hence, inclusive Raman signal intensity is expressively inferior for the 1T(1T′) phase TMDs compared to the 2H phase for the identical samples with the equivalent thickness, like MoS_2 and WSe_2 nanosheets [158, 177, 180–182]. Similarly, another useful technique XPS spectra could also provide information about the phase distinction in 1T and 1T′ by demonstrating the peak's systematic shifting toward the lower binding energy side compared to the 2H phase, although, the XPS technique is not sensitive to the differences between the 1T and 1T′ phases (Fig. 3.21h) [36]. Moreover, in the chemically exfoliated TMDs, the swing toward a lesser binding energy side is characteristically due to the fractional decrease of transition metal centers throughout alkali metal intercalation [176]. Thus, using the various high-processing microscopic and spectroscopic characterizing tools TMDs' different phases can be distinguished.

3.3 Defects in TMDs

Though defects have been extensively explored in the bulk materials since several decades with the governing role to determine the key material properties. But it has found a great attention only after the successful growth of graphene using chemical vapor deposition (CVD), therefore, the kinds of defect behaviors in two-dimensional (2D) materials. In recent years researcher focus has shifted toward various burning 2D materials, such as graphene, monolayer insulator, and semiconducting materials, including h-BN, transition metal dichalcogenides (TMDs), and phosphorene, to underscore the importance of defect engineering. As the intense, MoS_2 is an n-type semiconductor having relatively low electron mobility. More significantly, the large-scale growth of various 2D films prerequisite requirements for their technological applications. To predictably introduced a variety of defects including dislocations, grain boundaries (GBs), and edges. The produced defects are in detrimental situation to their electronic applications. Therefore, to facilitate meaningful engineering or control of the physical properties, it is important to develop a deeper understanding of the defect formation, structures, properties, and their response to external environmental conditions. To examine the individual defects various powerful microscopy techniques have been employed, specifically to provide imaging (or characterization) of individual defects. Modern microscopic characterization tools can also serve as a route to create kinds of nanostructures, phase interfaces, GBs, and nanowires, in addition to conventional vacancies. The purposely generated vacancies can be also filled with foreign atoms to achieve desired doping and create kinds of nanostructures. To provide rich platforms for the designing of new devices and exploration of novel condensed-matter physics. Additionally, the shear openness of 2D crystals compared to bulk materials also offers a new degree of freedom with the significant advantage of defect structures engineered, their physical properties are generally controlled by the environment [183].

Thus, 2D materials signify a huge class of materials with diverse atomic arrangements and chemical configurations with an extensive range spectrum of electronic characteristics like insulators to superconductors. The family of TMD materials is much more diversified as depicted schematic in Fig. 3.22, their elemental mixtures are painted in the periodic table (more than 40 species) [1]. As we know the binary TMD monolayer materials typically have an MX_2 stoichiometry (where M is the cation that is frequently contained transition metal and X is a chalcogen, normally S, Se, or Te). And several monolayer TMD materials crystallize in 1H and 1T phases [36]. The 1H phase contains a layer of transition metal atoms crammed among double layers of chalcogen atoms in the trigonal prismatic construction. While the cation and anion in the 1T phase produces an octahedral coordination. Based on the misrepresentation of the cation location the 1T phase is further sub-divided into the inherent 1T, 1T′, and 1T″ phases [184].

TMDs also experience a layer-dependent changeover in which the band structure is converted an indirect to a direct bandgap semiconductor [96, 185]. Due to this adequate transition ability, TMDs become more attractive for electronics and optoelectronics [97, 186]. Although TMD-based devices are possessed n- or p-type behaviors, this contradicts to "one would imagine from a flawless crystal structure without unsaturated bonds" [187]. However, the experimentally obtained usual device mobilities lies below the theoretical predictions [187, 188]. So, the photoluminescence (PL) measurement can resolve the emission peak within the optical

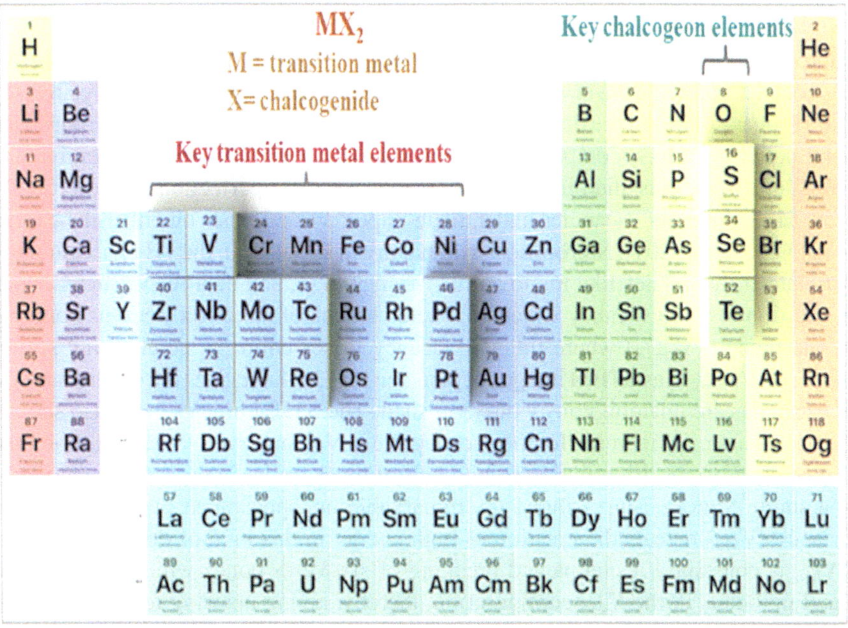

Fig. 3.22 Schematic of all possible TMDs with their elemental composition highlighted

bandgap, but it's again not reliable with the band structure governed from a perfect crystal configuration [189]. Raman spectroscopy also provides an intensity rise in LA (M) mode usually connected to the phonon vibrational mode, as a function of the crystal disordering [190]. All these evidences are indicated to a modest point; the structural defects in TMDs cannot be just avoided [188].

Thus, many different defects forms exist in 2D materials such as vacancies, antisite defects, interstitials (adatoms) and line defects, GBs (geometric defects), the typical structures are illustrated in Fig. 3.23a, all of them are induced different properties [41, 188, 191–195]. Thus chalcogen vacancies may incorporate the in-gap states and control the carrier transportation and optical features of 2D TMDs [96, 97], while the metal vacancies or metal antisite defects empower robust flaw spin splitting and generate confined magnetism [196, 197]. Similarly, the extended line defects also provide upswing to exciting electronic effects, like carrier-concentration waves [198–200]. Hence, broadly defects in 2D TMDs (or in 2D materials) are classified based on their dimensionality, like zero-dimensional (point defects, dopants, or 'non-hexagonal' rings), one-dimensional (grain boundaries, edges, and in-plane heterojunctions) and two-dimensional (layer stacking of different layers or vdW solids, wrinkling, folding, and scrolling) [188, 192]. The various kinds of significant defects in 2D materials are summarized in Fig. 3.23b.

3.3.1 Point Defects

Point defects are one of the humblest and utmost plentiful kinds of imperfection in 2D materials. Meanwhile the atomic structure of TMD materials is considerably diverse, like 1H phase of MoS_2 is a demonstrative monolayer material in the TMD group. The point defects of MoS_2 composition have been extensively explored. Predominately point defects are classified in the vacancy and antisite imperfections. As the CVD-grow MoS_2 inherent vacancies and antisite imperfections exists on either the cation or anion position [201]. Moreover, a trigonal prism structure in the 1H phase directly correlates to vacancies like mono sulfur vacancy (V_S), and disulfur vacancy (V_{S2}). In addition to a Mo vacancy which connected to lack of three sulfur atoms bonding in one plane (V_{Mo+3S}) where the Mo vacancy losses entire its neighboring sulfur supports (V_{Mo+6S}), as STEM images depicted in Fig. 3.24a. It gives the different types of vacancies are unambiguously distinguished. The STEM images have also shown there is no intensity increase for the vacancy site in V_{S2}, however, some weak contrast may leave at V_S position that is around half of the S_2 pillar in the lattice analogous to a single S atom. Surprisingly, the remarkable nonappearance of the single Mo vacancy (V_{Mo}) reveals the creation of high energy in this kind of defects. Because a V_{Mo} is constantly accompanied breaking of six Mo-S bonds which leaves six S dangling bonds in the V_{Mo} configuration, thereby, giving upswing to high creation energy [202]. Overall Mo vacancy in MoS_2 move toward to procedure imperfection multiplexes with the existing S vacancies. While an antisite imperfection in which a Mo atom occupied the place of an S_2 column (namely as MoS_2) and vice versa

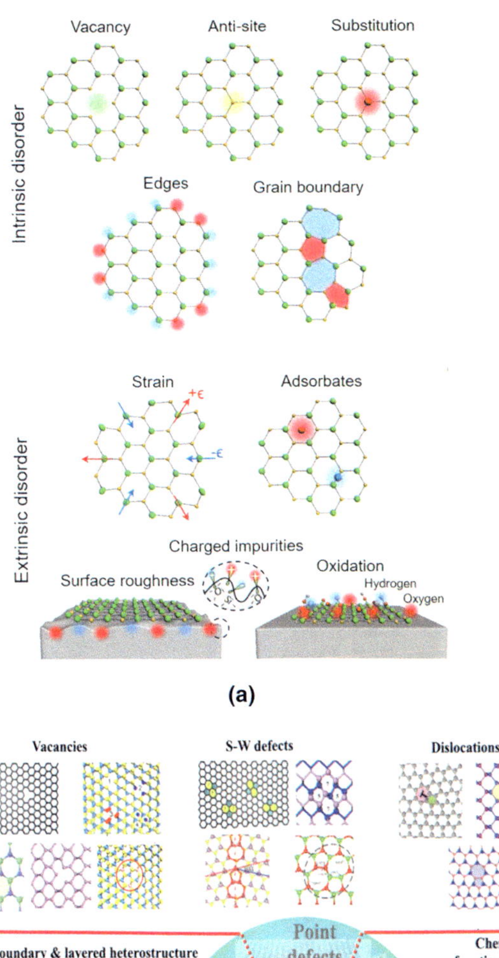
Vacancy
Anti-site
Substitution
Intrinsic disorder
Edges
Grain boundary
Strain
+ε
-ε
Adsorbates
Extrinsic disorder
Charged impurities
Surface roughness
Oxidation
Hydrogen
Oxygen

(a)

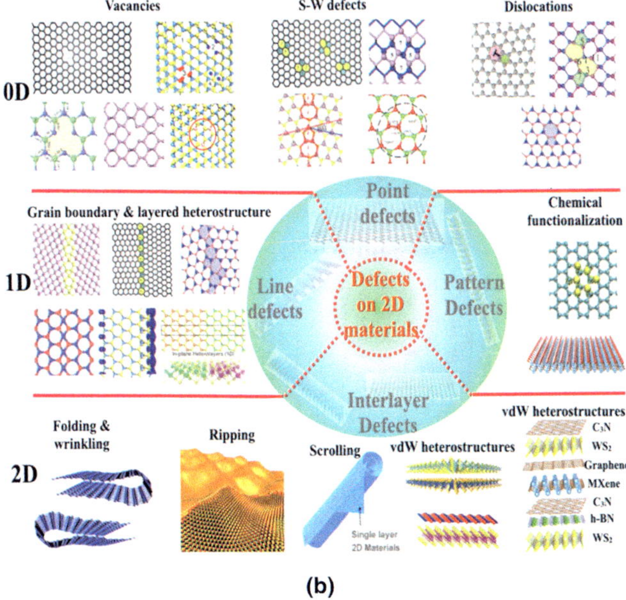
Vacancies
S-W defects
Dislocations
0D
Grain boundary & layered heterostructure
1D
Point defects
Line defects
Defects on 2D materials
Pattern Defects
Interlayer Defects
Chemical functionalization
vdW heterostructures
Folding & wrinkling
Ripping
Scrolling
2D
Single layer 2D Materials
vdW heterostructures
C_3N
WS_2
Graphene
MXene
C_3N
h-BN
WS_2

(b)

◂**Fig. 3.23** **a** Various kinds of defects in 2D materials: Types of disorder in 2D materials; Intrinsic disorder: vacancy, antisite, substitution, edges; Extrinsic disorder: strain, adsorbates, surface roughness, charged impurities in the substrate, and oxidation (adopted from Daniel Rhodes et al. 2019 Nature Materials, Nature Publishing Group, https://doi.org/10.1038/s41563-019-0366-8.hal-02335515), **b** Various kinds of defects in 2D materials. Analogous to macroscopic crystalline materials, structural defects in graphene and other 2D materials TMDs have different dimensionalities. Zero-dimensional (0D) point defects consist of vacancies, Stone–Wales (S–W) defects, adatoms, dislocations, and substitutions. One-dimensional (1D) linear defects arise in situations different from those of bulk crystals due to the reduced dimensionality. Not only edge dislocations, but also grain boundaries are 1D lines along which atoms are arranged abnormally. Also, 1D defects including edges and phase interfaces. Interstitials between layers can bridge adjacent layers, forming higher dimensional structures. Stacking failure is another typical defect in few-layer graphene and stacking it with other 2D layered materials. Particular 2D defects such as folding, wrinkled, displacement, ripple, and heterostructures stacked vertically (adopted from Khossossi et al. [192])

($S2_{Mo}$) also appeared in the CVD fabricated sample (see Fig. 3.24a). Both kinds of antisite defects experience an extra atomic shift due to the contravention of the native threefold equilibrium.

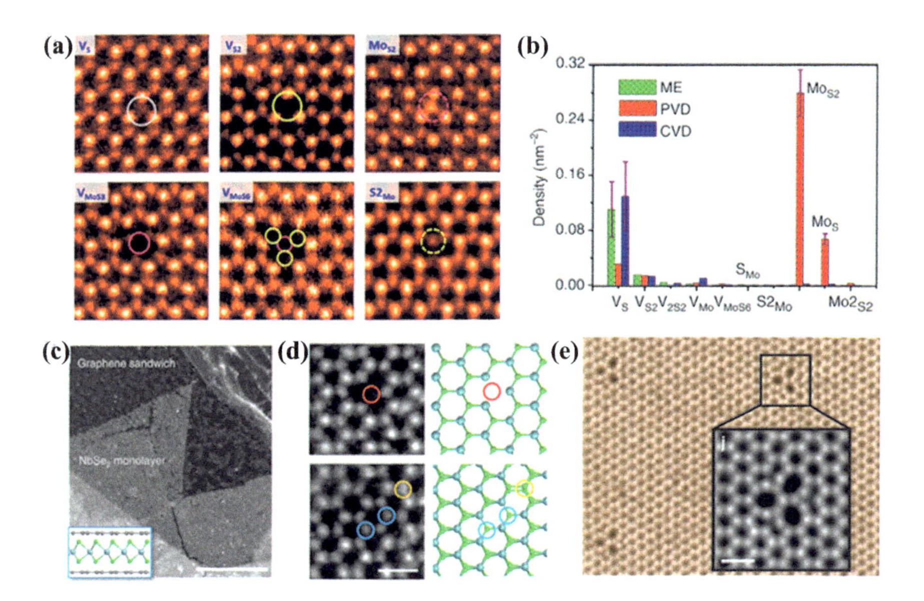

Fig. 3.24 Atomic structure of point defects in monolayer TMD; **a** atomic resolution STEM ADF images of different intrinsic point defects including single vacancy, vacancy cluster, and antisite defects for the CVD-grown monolayer MoS_2 (adopted from Zhou et al. [201]); **b** statistical histogram of the density for the various point defects in monolayer MoS_2 (**b** Hong et al. [208]); **c** low-magnified STEM image of graphene sandwiched air-sensitive monolayer $NbSe_2$; **d** atomic STEM image of Se vacancy and the antisite defects (C and D adopted from Wang et al. [211]); **e** STEM image for the threefold symmetry rotational defect in WSe_2 (adopted from Lin et al. [218])

In general, V_s is universal across almost every monolayer even in the mechanically exfoliated, owing to its lowermost creation energy (~ 2 eV) for entirely kinds of structural imperfections [201–204]. This is also speculated V_S types that participate to n-type doping in MoS_2 devices [205–207]. Thus, the construction of such vacancies depends on the development conditions and used technique. This statement could be verified with the help of the creation energy of the imperfections due to diverse chemical capabilities. As the intense, the construction energy of V_S in a sulfur lacking situation is usually much lower compared to sulfur-rich conditions [184]. Therefore, a high concentration V_S is achieved under a Mo-rich environment. By varying chemical potential it can also serve for two extreme conditions. These outcomes have recognized the development of unpackaged Mo and alpha-S corresponds to Mo-rich and S-rich environments. That directly correlates to the construction of the vacancies and antisite imperfections which alters with the growth thermodynamics factors. Additionally, some studies have also pointed out that the defects formation energy also affects by the charge state [202], however, yet there is no satisfactory explanation has been provided. Though a comparison of the same composition TMD materials growths by employing the different methods have been demonstrated based on the dependency of the imperfection inhabitants on the growing situations. The different kinds of imperfections in monolayer MoS_2 developed by the different methods (such as CVD, physical vapor deposition (PVD), and ME) have been demonstrated as the leading class of imperfections variation, like V_S in ME and CVD, and MoS_2 antisite defects in PVD growth, as depicted in Fig. 3.24b [208]. This could be owing to the variance in the used flux of precursor materials for the CVD and PVD methods.

Hence point imperfection constructions noticed in typical MoS_2 monolayer has been universalize for the kinds of TMD materials monolayer to 1H phase formation. Similarly, chalcogen vacancies also give us a T-phase for the compositions $MoTe_2$, ReS_2, $ReSe_2$, etc. [209, 210], however, these material's phases may easily damage under electron beam irradiation. Though several TMDs like MoS_2, $MoSe_2$, WS_2, WSe_2, $MoTe_2$ 1H phase, and ReS_2. $ReSe_2$ 1T$''$ phase is steady in suitable circumstances. In addition to this, many TMD monolayers are also disposed to oxidation while the fabrication earlier electron microscopy characterization. To prevent such damage effort has been made to sandwich the air-sensitive $NbSe_2$ monolayer, to crystallize a superconductor material in the 1H phase in the dual successive layers [211, 212], as depicted STEM in Fig. 3.24c. This investigation leads to the oxidation still occurring, whereas the pure triangular configuration preserved and their corresponding defective atomic structure are explored [211]. Thus, the experimental pieces of evidence have established that an inherent mono selenium vacancy, diselenium vacancy, and antisite defects exist in the CVD fabricated $NbSe_2$ monolayer (see Fig. 3.24d), like MoS_2. Additionally, it has been also noticed that the point flaws has not affected the superconductivity of the CVD developed $NbSe_2$ monolayer.

Chalcogen vacancy of TMD can also play a significant role in their structural reconstruction, interlayer fusion, and alteration of the native electronic construction due to their large aggregation ability [202, 204, 213–215]. This ability of the materials makes them enables catalytic applications [216]. The rotational point defects that are involved in the formation of a sequence bond revolution are likely to the construction of Stone–Wales flaws that are also evident in the TMD monolayer [217]. Moreover, the existence of trefoil-shaped defects in the WSe_2 monolayer is also realized [218]. This kind of point defect evolves three diselenium vacancies in the peripheral sublattice associating a sequence of following revolutions of the metal chalcogen bonds up to 60 degrees rotation. The advantage of this kind of rotational defect is to maintain the threefold regularity in the trigonal prismatic lattice that increases the p-type doping persuading the local magnetic moments. Furthermore, the rotational flaws are also expanding by the successive bonds revolutions; therefore, they forms five and eightfold member rings within the lattice (see Fig. 3.24e).

3.3.2 Doping

Doping is also one of the vibrant methods for altering the electrical and optical features of 2D materials. Dopants in material are also adding new characteristics, such as magnetism and spintronics features. But it remains a big task to use old-style ion implantation procedures for the doping 2D materials because the extremely thin behavior allows to easily destruct or form unwanted imperfections due to high-energy ion irradiation. Usually, approaches adopted for doping 2D materials including usual substitutional doping procedures for the unpackaged materials, while the new methods enable the layered and extremely thin nature of 2D materials, such as carrier transformation, intercalation, and electrostatic gating [219]. The variant selection of doping methods provides to 2D materials numerous benefits to accurately control their features with exciting novel features.

Therefore, dopants are desirable processes to foreign species that can adjust the Fermi level, to modify the electronic structure or incorporate additional charges into the electron system of 2D materials. Particularly controlled doping from the ion implantation has been widely used for bulk semiconductors. The dopants are usually classified into two categories namely substitutional and surface doping to produce useful defects in 2D TMD materials.

Substitutional doping can be achieved by exchanging the cationic or anionic atoms in the lattice with a heteroatom while the fabrication or prior- fabrication treatments, such as inserting or diffusion. To create this kind of doping defect the dopant elements possess the identical amount of valence electrons as a substituted atom which makes it easier to adjust the defects density and concentration of the material. While the elements with different valence electron numbers would be useful to produce n-type or p-type conductivity to act as donors or acceptors. In the substitutional doping process exchange of an atom generally persuades lattice distortions and restricts the kind of elements. In terms of charge transfer doping it is considered to be an emerging method that enables the van der Waals nature of 2D material's surfaces, and their larger surface region exploits the connections among the 2D materials, whereas the surface-adsorbed dopants include adatoms, ions, and molecules. It is also useful to avoid lattice distortion in 2D crystals; therefore, it is useful for the choice of numerous diverse kinds of dopants and effective doping to maintain a great mobility of the 2D material [220]. Thus substitutional doping is the method of incorporation of the doping component covalently into the lattice by exchanging the natural element. The incorporation of additional foreign elements generally reasons lattice distortions including perturbations that influenced the crystal arrangement and electronic features. As the intense, sulfurization/selenization at high temperatures or deposition of transition metals into the usual and fast post-synthesis approaches to doped chalcogens and metal atoms in TMDs [221]. The dopants are also produced defects or distorted structures in some other cases [221]. As a consequence overall material quality is degraded by persuading a novel structure that is useful to modify the electronic and catalytic features. So, in such a doping process the fundamental issues of the 2D materials should be addressed; (1) kinds of elements should be integrated into the TMD lattice, (2) dopant should stabilize the crystal structure or developed a novel phase, (3) how will dopant influence the electronic characteristics and functionalities. To achieve these objectives the theoretical guidance also narrows down the issue of the doped 2D materials. Like theoretically it has expected the thermodynamic constancy and electronic characteristics of numerous 2D TMD binary compounds with their desirable band gaps. Moreover, the lattice mismatch between the dopant and host elements should be fewer than 3%, and the corresponding change of the metal–chalcogen bonds lesser than 0.1 Å [222]. Additionally, it has been also directed way of the induction of metal dopants into 2D materials from the usual post-synthesis metal deposition method, for that a few best favorable metal dopants proposed to add magnetism in 2D TMDs [223]. This could also help to achieve experimental challenges for the steady 2D compounds associating novel functions, such as magnetism.

The isoelectronic substitution doping also considered to be an effective method to dope 2D semiconductors to tune their band gap with reduced dislocations and defects owing to the high tolerance of the complex compound. However, unlike electron-rich atoms, the isoelectronic dopants have not provided additional electrons to adjust the electronic structure, but they can be utilized as electronegativity to modify the charge concentration. Thus isoelectronic dopants normally induced non-degenerate doping with no significant modification in their basic electronic properties, thereby, it would be a suitable technique to create both n-type and p-type doping. The isoelectronic metal dopants in general induced a wider range of alternation in the band gap of 2D TMDs. It also allows the localization of the d-orbitals metals transition that mainly governs the electronic property of the materials. As the intense, the isoelectronic doping of W in $MoSe_2$ allowing tuning for the both band gap and valence band breaking owing to robust spin-orbital combination [224]. A theoretical study based on the calculations has also anticipated that the isoelectronic doping of W into $MoSe_2$ may suppress profound fault levels to maintain their essential electronic features. The profound levels in 2D $MoSe_2$ exit due to Se vacancies that alleviated to form 2D $Mo_{1-x}W_xSe_2$ permits with the lower W amounts. The addition of element W modulates the band structure and pushes the charges into the profound states close to the conduction band edges [225]. Therefore, under the controlled proportion of precursors the WO_3 and MoO_3, $W_{1-x}Mo_xSe_2$ $(0 < x < 0.18)$ TMD crystals generates

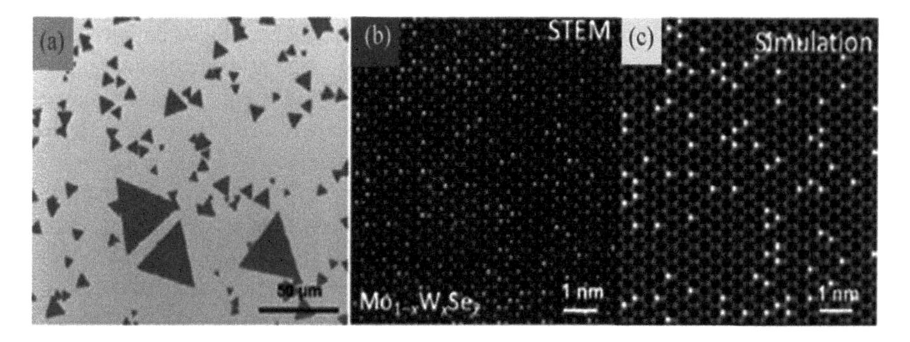

Fig. 3.25 Isoelectronic doping of W in ML $MoSe_2$ and induced properties: **a** SEM image for the ML W-doped $MoSe_2$ grown on SiO2 substrate; **b** STEM image to show formed random alloy; **c** simulated image for the formed random alloy (adopted from Xufan Li et al. 2016, https://web.ornl.gov/~geohegandb/Li%20AdvMat16%20Isoelect%20doping.pdf)

deep states charges by using the CVDtechnique. The different places doping concentration can be easily explored from the atomic resolution STEMimaging technique, it has not only ensure the evidence about the formation of the ultimate arbitrary compound in the lattice but also the reduction of Se vacancies by a factor of 50% in $Mo_{0.82}W_{0.18}Se_2$ related to $MoSe_2$, as the typical image represents in Fig. 3.25a–c. In addition to these, it has been also noticed that the corresponding PL intensity expressively improved in $Mo_{0.82}W_{0.18}Se_2$. Their fabricated device characteristics have demonstrated with the enhancing W doping levels in $MoSe_2$, their n-type characteristics feature is intimidated, while an enhancement in p-type transport behavior has also noticed.

The p-type $MoSe_2$ semiconductor from the tiny W doping levels and n-type pure $MoSe_2$ together are also construct an atomically thin p-n homojunction by stacking $Mo_{1-x}W_xSe_2$ and $MoSe_2$ layers through a wet transfer process. This kind of fabricated p-n homojunction device offers an exceptional gate-tunable performance afterward annealing [226], they can use for the application of light emitting diodes and photovoltaics to get effective current rectification and photovoltaic response, owing to their interface possessing unbroken band arrangement and a inferior amount of carrier trap sites than that of heterojunctions [227].

Chalcogen dopants are also intensely influence the electronic features of 2D TMDs, owing to the fact, the degree of d-state fraternization depending on the transition metal and the analogous chalcogen ligand atmosphere [95]. Usually, the post-synthesis doping is due to processing, like higher temperature sulfurization/selenization to produce arbitrary doping of chalcogens in the 2D TMDs [228, 229]. In such a process the vapor phase alteration generally desired a higher temperature (500–1000 °C) to fast-track the ion interchange rate (i.e., Se–S orTe–S exchange) to anneal the 2D TMD [221]. The most critical stage in post-synthesis doping is to relocate the ingredients of 2D TMD, such as first generating vacancies as the host for inward dopants. Their dislocation threshold energy is usually defined as the required smallest energy that one ingredient desires to left its lattice position to form a flaw;

typically it is 6–8 eV/atom for the chalcogen and 20–30 eV/atom for the transition metal in 2D TMDs [230]. This required threshold every (6–8 eV/atom) provides to atoms by the CVD furnace heating around the equivalent thermal temperature (it is typically above the 700 °C) to generate a great number of imperfections. Moreover, ion implantation is another competent approach to dope the materials. Traditionally higher energy ions (above the 100 eV/atom) are used in ion implantation technique to sputter the chalcogens and induct unwanted faults. Moreover, the laser plasma plumes usually generate pulsed laser deposition (PLD) to dislocate the ingredients in a controlled way; it also provides lower kinetic energy to induce dopants at the empty lattice locations [231]. This method also depend on the collision among inward dopants implanted in plasma clouds and the ingredients of 2D TMD, however, other traditional methods are based on the flow of the chemical vapors. Additionally, it also not requires higher temperature annealing. The PLD approach is also considered to be a useful technique to form a well-ordered TMD compound (e.g., the Janus structure), in which the top and bottom layers in a TMD consisting diverse chalcogens. The equilibrium is broken and the occurrence of stable dipole moments inherently offers various novel functions, such as piezoresponse, catalytic behavior, charge separation, and the Rashba effect [232, 233].

Composition MoSSe Janus structures is generally formed by employing the direct exposer of the $MoSe_2$ to S vapor. The high cohesive energy of MoS_2 than $MoSe_2$ [232, 234] allows for to removal top S layer in MoS_2 and Se from $MoSe_2$ successively under the hydrogen plasma [235]. In both processes the induced defects levels in the Janus structure owing to the higher processing temperature or higher plasma energy. To decrease the imperfection level significantly should restrict the kinetic energy of the inward chalcogen atoms. As demonstrated the Se species with less kinetic energy than the 10 eV/atom could be implanted into WS_2 ML by employing pulse laser plasmas [236]. Their background Ar pressure is generally controlled by tuning the kinetic energy for the choosy selenization of the topmost layer of WS_2 ML to get a better class WSSe Janus structure at 300 °C.

3.3.3 Line Defects

The one-dimensional (1D) defects are known as line defects, the typical line defect in monolayer $MoSe_2$ is depicted in Fig. 3.26a, b [213]. However, other (1D) defects are boundaries and edges also fall in the one-dimensional (1D) defects category. A single line imperfection contains of lone row of Se vacancies that are allied in a rectilinear direction. Their rebuilding where the hexagonal rings in the line imperfections are compacted in which their orientation vertical. This allows to metal atoms dragged in the direction of each other to rebalance the additional electrons persuaded by the absent Se atoms. On the other hand, the dual line flaw includes twofold rows of Se vacancies and experiences a considerable renovation. The Mo and Se atoms within the line imperfection alterations into an octahedral-like coordinate and advance contract

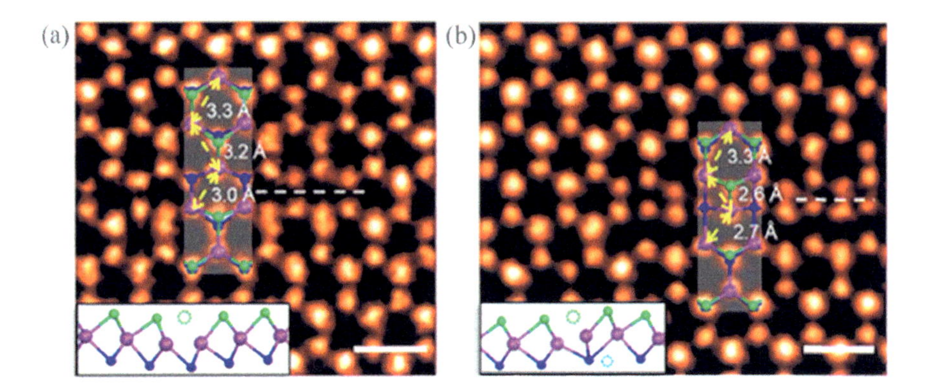

Fig. 3.26 Line defects in monolayer TMD: (A, B) STEM images of the line defects in monolayer $MoSe_2$ (adopted from Lin et al. [213])

the Mo-Mo distance within the dual line fault, such as a ductile strain vertical to the horizontal route of the imperfection.

The further reduction in Mo-Mo bonds induced a metallic state by creation of dual line faults 1D metallic networks within the 2D semiconducting lattice [213], alike line imperfections also exist in MoS_2 [214]. Here it should be noted these two kinds of line faults are not inherent imperfections in originally developed samples, but owing to the growth and accumulation of chalcogen vacancies that are governed by the electron irradiations. These line faults could also exists afterward thermodynamic action, such as higher temperature annealing, in which the thermal energy stimulates the chalcogen vacancies, likely to have a similar influence to electron beam irradiation [213]. Thus, the accumulation of chalcogen vacancies results the structural relaxation that gives to a low system energy as compared to the same amounts of vacancies arbitrarily dispersed within the lattice [214, 237]. Moreover, the remote vacancies primarily have an isotropic strain which is adjusted in the monolayer, however, due to electron beam (or thermal) excitation like an isotropic strain outline may disturb, therefore line defects forms along the zigzag ways contingent on the anisotropic strain outline persuaded by the relocation and accumulation of chalcogen vacancies [214]. Hence the way of line imperfection can be choosily aligned under an applied original strain for the free-standing flakes.

3.4 Grain Boundary

TMD materials grain boundaries are other significant structural fault that expressively impacts the electrical and optical features [205, 238]. Grain boundaries are formed due to misorientation or misalignments between two crystalline domains while the structural development. The mirror twin boundaries (so named 60 degrees grain boundaries) in which two domains have 60° of misorientation, therefore, they have mirror symmetry to each other toward the joining area. Usually, a mirror twin boundary comprises four and eight-member rings, as depicted in Fig. 3.27a, b [183, 201, 205, 213, 238]. Using the STEM technique overlaid atomic structures of dual distinct kinds of mirror twin boundaries can be recognized in the $MoSe_2$ monolayer. Type one (Fig. 3.27a) is known as point sharing 4|4P mirror twin boundary, in which binary domains linked at Se ended zigzag boundaries to form a chain of four-member rings through sharing the Se_2 columns. Another one (Fig. 3.27b) has also an alike construction but their domains move half of a unit cell concerning each other. A displacement in the Se_2 columns corresponding to separate Se atoms at the edge has also visualized. This kind of construction is called an edge-sharing 4|4E twin boundary [213]. The chemical composition of the border area is possibly also distinguished these twin boundaries in terms of Se-deficiency. Moreover, the MoS_2 monolayers are also produced a similar structure [201] through the alteration among two edge constructions underneath electron irradiation [213]. Such two distinct kinds of identical boundaries are appearing due to the creation of the chalcogen vacancy underneath electron irradiation. Because of their nucleation and growth of triangular reversal domains among the three metallic 4|4P or 4|4E twin boundaries, as illustrated in Fig. 3.27c [213].

The MBE-grown $MoSe_2$ has been also attributed to an inherent great attentiveness of triangular reversal domains associated with the 4|4P and 4|4E identical boundaries interlocking in the lattice, it could be due to the Se lacking growing situation [240,

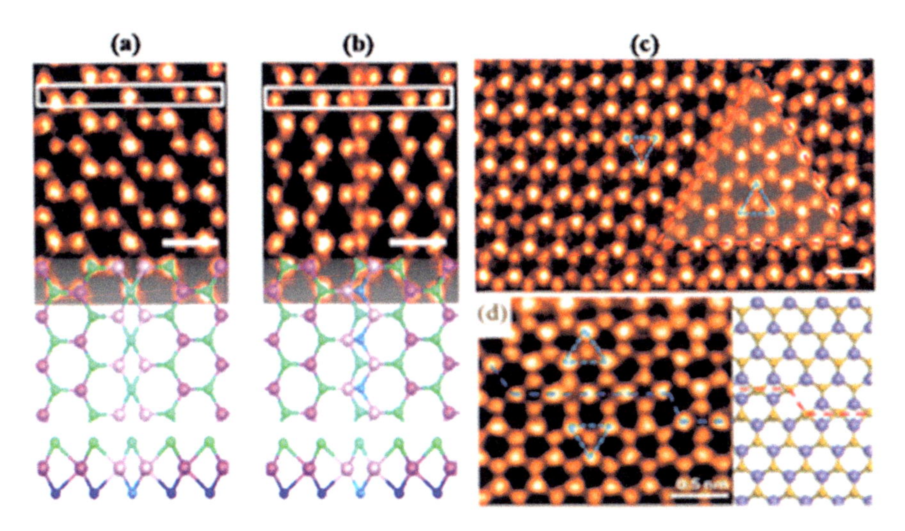

Fig. 3.27 Twin boundaries and inversion domains in TMD; **a, b** Atomic resolution STEM ADF images and the corresponding atomic models of mirror twin boundaries in monolayer $MoSe_2$ for the different configurations. The two inversion domains can be connected by Se_2 point sharing (4|4P, **a**) or edge-sharing (4|4E, **b**) structure; **c** STEM ADF image of a triangular inversion domain with three twin boundaries embedded in monolayer $MoSe_2$ (**a**, **b**, and **c** adopted from Lin et al. [213]); **d** STEM ADF image and atomic model of a twin boundary consisting of 4|8 rings in CVD developed monolayer $MoSe_2$ (adopted from Lu et al. [239])

241]. However, armchair edge like of twin boundaries have been also forecast theoretically [242], but experimentally it has been found only zigzag edge kinds, it is owing to the higher creation energy of the armchair kind borders. Additionally, these two different kinds of mirror twin boundaries are also exist in various CVD developed TMDs, the inherent eight-member rings finalized to four-member rings and create zigzag 4|8 boundaries, specifically, while the two domains have not straight edges [184]. The typical example of the formation of eight-member rings, it serves as kinks toward the twin boundary to accommodate the steps of two domains, as illustrated in Fig. 3.27d [239]. Moreover, such mirror twin boundaries also influences the optical and electrical characteristics of the monolayer material [238]. Likewise twin boundaries have been also recognized in WTe_2. Though, the WTe_2 crystallized in the 1T′ phase in which the point-sharing frontier remain tolerate at the interface with the Te columns that are somewhat displaced vertical to the basal plane of the lattice [243].

As per the theoretical description of the insight relationship between the grain boundary structure and their charge transport, the momentum conservation rules and first-principles quantum transport calculations are usually employed [244]. The different linear grain boundary structures have possessed both high transparency and perfect reflection from the boundaries that possibly produced the charge carriers within a large window of energy values. As per the quantum transport calculations interpretation, typically the extended line defects could be due to the occurrence of a periodic sequence of octagonal and pentagonal sp^2-hybridized rings [245–247]. Additionally, their electron density at the Fermi level may very low, therefore, any perturbation of the electronic structure have profound effects on transport in 2D materials. This is due to reduced carrier density at the domain boundaries [248]. Hence the Friedel oscillations near the boundary open an energy gap for electrons with wave vectors perpendicular to the interface. The size gap may different for the monolayer and bilayer sides due to their different chiralities. Since heterojunctions creates an asymmetric transport behavior in forward and reverse bias conditions. Therefore, asymmetry itself behaves as a potential drop with a magnitude depending on the direction of the current across the monolayer and bilayer boundaries. While the homojunction such as monolayer and bilayer boundaries do not have any asymmetry even though they still possess boundary resistance due to the influence of low carrier density and gap opening near the interface [193].

3.5 Dislocations

The tilt grain boundaries are formed whenever two crystalline domains revolve at a definite angle. Along with successively two lattices connected at the grain boundary to produce a smooth monolayer. Therefore, a minor angle tilt grain boundary considered to be as an array of dislocation cores in bulk materials. Likewise, the tilt grain boundaries have been also visualized as composed of the dislocations in 2D materials. The dislocation cores can also form from a variety of ring motifs such as

4|6, 4|8, 5|7, 6|8, as depicted in Fig. 3.28a–e [242]. The motifs 5| 7 consist of five and seven-member rings (Fig. 3.28b), it is the most common dislocation cores in tilt grain boundaries of 2D TMD monolayers. As per the basic principle, the tilt angle 5|7 dislocation core can be adopted by either a metal-rich or a chalcogen-rich dislocation core arrangement. By allocation either metal–metal or chalcogen-chalcogen homo-element bonds among the five and seven-member rings. It depends on the exact positioning and corresponding comparative displacement of the two domains [242]. However, the experimental interpretations have revealed that only 5|7 dislocation cores with metal–metal homo-element bonds that is usually visualized in CVD-grown TMDs [201, 239]. However, it has been also predicted that much inferior creation energy in a Mo-rich chemical situation for the 5|7 dislocation cores which associated with chalcogen-chalcogen homo-element bonds [242]. This contradiction is due to an actual grain border characteristically consisting of manifold kinds of dislocation cores and occasionally even comprises huge void. In addition to this, the atomic structures of tilt grain boundaries are also fluctuate considerably contingent on the misorientation angle. The experimentally obtained grain boundary in monolayer MoS_2 has been also associated with a tilt angle of 18.5 degrees, as depicted in Fig. 3.28a. Therefore, the most common 5|7 and 6|8 dislocation cores are created as their **comprehensive** atomic structures presented in Fig. 3.28 (b, c) [201]. Particularly, the Mo-Mo homo-element bond in the 5|7 dislocation core transfers the structure into a 6|8 dislocation core at the identical grain boundary owing to the addition of a sulfur column, however, it is energetically promising only under the S-rich environment. A grain boundary with a similar tilt angle (B17.5 degrees) is also another region of the same specimen having a different set of dislocation cores with four and sixfold rings structures (4|6 motifs), as depicted in Fig. 3.28d, e. The pristine 4|6 core transformation to 5|7 dislocation core owing to sulfur atoms S-orientation which forms by the removal of two sulfur atoms (see Fig. 3.28d). Similarly, the incorporation of Mo into the 4|6 core also provides an identical result (Fig. 3.28e). Hence, these two types of dislocation cores are considered to be energetically promising only in the Mo-rich environment [201]. A variety of grain boundary structures can be formed by employing the tilt angles concept based on the native variation of Mo and S basis attentions while the material growth. The exact atomic structures are generally controlled by supervisory the thermodynamic parameters. The strain may also accumulate in such a dislocation core that makes them movable underneath electron irradiation. However, the straight surveillance of dislocation relocation at the grain boundaries generally has a diverse manner owing to the trigonal prismatic structure of 1H-TMD [249].

3.6 Edges

Since 2D surfaces are terminated as a 3D bulk material, therefore, 1D edges symbolize a significant heterogeneity for 2D materials with an extensive diversity of characteristics, like dangling bonds and exclusive structures (such as armchairs or zigzag

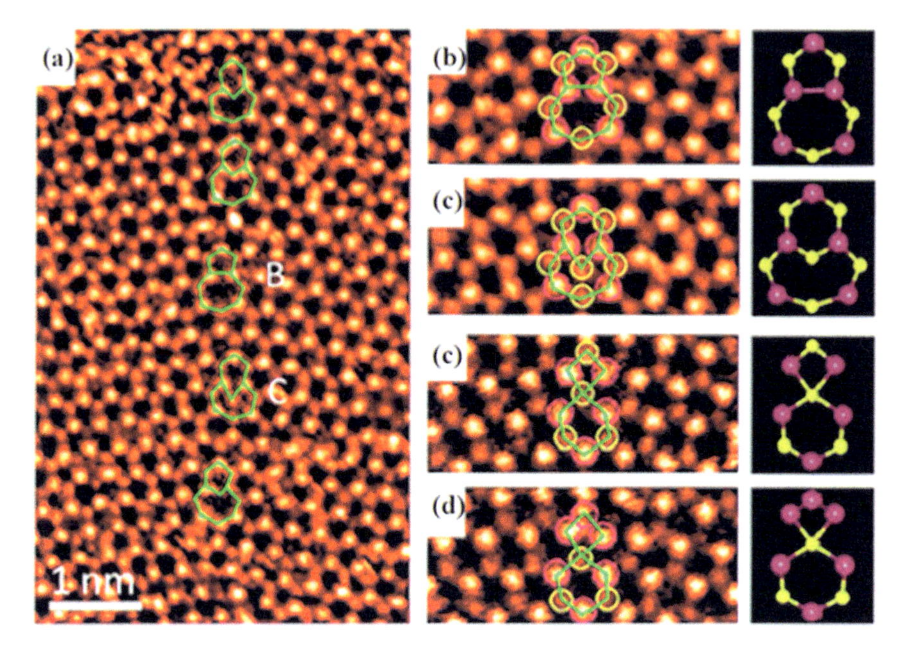

Fig. 3.28 Atomic structures for the tilt grain boundaries in monolayer TMD; **a** atomic resolution STEM ADF image of an 18.5 degrees grain boundary containing different types of dislocation cores; **b, c** the zoom-in images of different dislocation cores, as highlighted in **a**, the dislocation core contains either five- and sevenfold rings (5|7, **b**) or six- and eightfold rings (6|8, **c**); **d, e** dislocation core from a 17.5 degrees grain boundary to another type of structure that consists of four- and sixfold rings (4|6, **d**) and with Mo substitution (adopted from Zhou et al. [201])

edges), by breaking the translational symmetry of a 2D crystal. The edge features are also responsible for the novel functionalities which fairly diverse from those of the basal plane, like metallic states [250], ferromagnetism [251], topological states [252], enhanced second harmonic generation [253], PL [254], and active sites for catalysis [255]. Moreover, from the synthesis vision, the 2D materials' edges are also important to constructing types of edges which controlled by factors like chemical potentials of the diverse reactants, growth rate, and morphology of 2D flake [256]. Typically four kinds of edges are possible with a hexagonal structure of GaSe, as depicted in Fig. 3.29a [257]. In which Ga-terminated zigzag (ZZ) edge, Se-terminated ZZ edge, and armchair edge with a 19°-tilted edge. Their corresponding chemical potential difference, $\Delta\mu$, among Ga and Se also plays a crucial part to decide the creation energy of the diverse edges (Fig. 3.29b). The corresponding growing or etching rate for every edge of the concluding equilibrium profile of the 2D flake is illustrated in Fig. 3.29d. If $\Delta\mu$ closes to 0, the growth rate of the diverse edges usually vary in order of armchair > Ga-ZZ > Se-ZZ. While the morphology of the GaSe crystalline flake mainly reflects deliberate rising edges, however, it ultimately stabilized into a triangular shape with exposed Se-ZZ edges (Fig. 3.29c). On the other hand, ambiguous etching leads to edges with the fastest etching rate. Such edges dominates

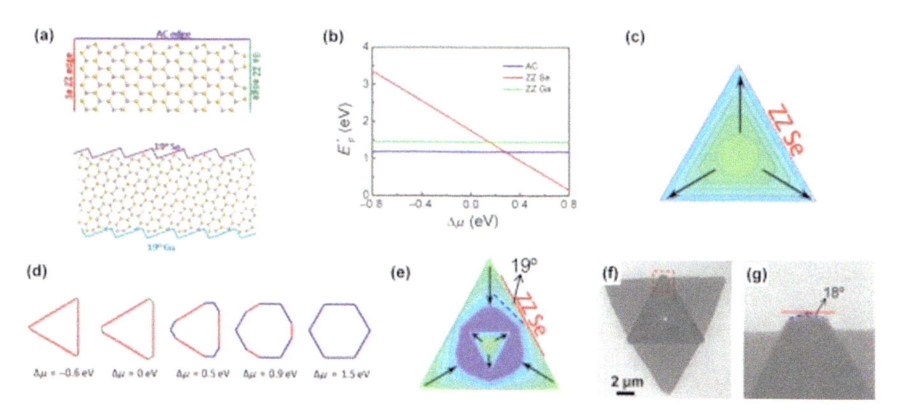

Fig. 3.29 Edge developed in the growth processes; **a** atomic structure of Ga-ZZ, Se-ZZ, armchair, and 19° tilted edges on ML GaSe; **b** formation energy for initiating the growth of ZZ and armchair edges on GaSe as a function of Δμ; **c** simulated growth of GaSe from a circular nucleus at Δμ = 0.06 eV; **d** equilibrium shapes of ML GaSe flakes at different Δμ defined from Wulff construction; **e** simulated etching process of as-grown GaSe domains at Δμ = 0.06 eV; **f, g** SEM images for the bilayer GaSe flake at the initial etching stage (adopted from Li et al. [257])

the flake's perimeter; therefore, the 19°-tilted edge (Fig. 3.29e, f, g) due to kinks on the edge that can act as dynamic sites for the atom detachment. Moreover, chemical potential variation is also play a crucial role in variation of the type of edge which offers an option to modify the kinds of unprotected edges and crystal morphology to get anticipated edge accompanying functionalities on demand.

The most explored Mo and S edges are terminated at ZZ in 2D MoS_2 crystals having a hexagonal phase. The variety of novel edge formations have been recognized by supervising the chemical potentials and other processing parameters. As an instance, with the high Mo chemical potential (μ Mo) the major edges in monolayer MoS_2 has been achieved by using the molecular beam epitaxy (MBE) technique. These new edges like distorted 1T edge, Mo-Klein edge, and Mo-ZZ edge of Mo are possibly terminated, this feature is also correlated to vigorous magnetism beyond room temperature and is useful for an adequate hydrogen growth reaction [258]. The vacuum annealing process of MBE and CVD is also an impressive way to generate new edges in 2D materials [259, 260]. The edge growth and transformation of W-doped $MoSe_2$ has been also realized at the atomic scale in real time using in situ heating in a STEM. The edge creation and rebuilding of monolayer $Mo_{0.95}W_{0.05}Se_2$ in STEM underneath diverse native chemical surroundings can be also tracked by supervising the quantity of carbon residue and heating time [259]. The manipulation of μMo leads to distinct edge structures while the in situ heating. With the decreasing quantity of carbon residue four separate steady edges are created including Se-ZZ terminated with -Mo_6Se_6 NW and Se, Mo-ZZ terminated with Mo_6Se_6 NW and Se, as depicted in Fig. 3.30(a, b). Whereas with the increasing heating period, the Se-ZZ-Se edges reconstructs the ZZSe-Mo-Mo_6Se_6 NW edges, as the process illustrated in Fig. 3.30(c–f). As per the prediction, all four steady edges can have different

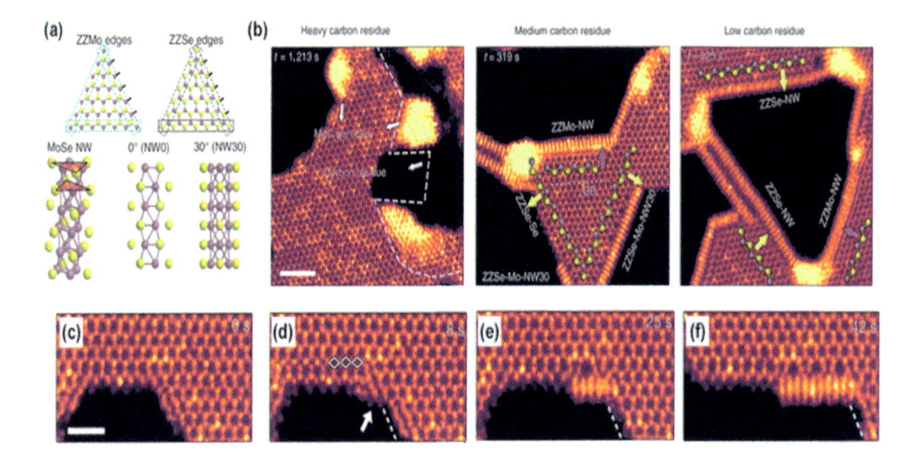

Fig. 3.30 Edge evolution and reconstruction of suspended ML for the W-doped $MoSe_2$ under in situ heating in STEM; **a** the atomic structure of Mo-oriented zigzag (ZZMo) edge, Se-oriented zigzag (ZZSe) edge, and Mo_6Se_6 NW formed on edge with two commonly observed projections at 0° and 30°; **b** atomic resolution STEM images of edges after in situ heating for times indication. Chemical potential is regionally different due to the reaction between Mo and C surface residue, leading to different types of reconstructed exposed edges, at the scale bar: 2 nm; **c–f** atomic resolution STEM images to the process of edge reconstruction, at the scale bar: 1 nm (adopted from Sang, X. et al. 2018 Nat. Commun. 9, 2051)

electronic characteristics and magnetic ordering. Like, two NW-terminated edges it could be metallic owing to the metallic nature of free-standing Mo_6Se_6 NWs [261].

However, surprisingly current STM and spectroscopy studies have also demonstrated the 1D metallic property of such edges' quantum-confined Tomonaga – Luttinger liquid (TLL) behavior [260]. The metallic Se-GB4-Se edge also offers numerous exciting electronic belongings, like stemming from the metallic GB4 mirror twin boundary that shows the charge density waves and 1D TLL behavior [62–64]. Although, theoretically also anticipated metallic properties of the Se-GB4-Se edge and ferromagnetism for the ZZSe-Mo-NW30 edge that are still unexplored experimentally. Hence, the edges are joined with other prolonged faults in 2D TMDs to deliver a perfect system by adding novel functions, like spin, magnetism, and edge conduction states. The approach for appreciating novel edge structures in 2D $MoSe_2$ could apply to an extensive variety of TMDs by providing an exceptional chance for the topological study of different types of edges including their electronic, magnetic, and catalytic properties.

3.7 Stacking Layer Structures

Since different stacking patterns and twist angles exist in 2D materials which induce important amendments in their features by altering the interlayer connections. For the instance, CVD developed MoS_2 bilayers have diverse twist angles with a constant direct band gap, while the indirect bandgap varies through the alteration in twist angle due to a modification in their interlayer electronic coupling [262]. Furthermore, the artificially stacked $MoSe_2$ bilayers with mutual rotational angles ($< 5°$) has been also recognized from the LF Raman spectroscopy [263] in form of Moire-like circular patches feature under high symmetry stacking arrangements (AA′, AB′, and A′B), likely to stacking near 60° (Fig. 3.31a). Usually their patch sizes continuously in decreasing order with the aberration from 60° and stacking becomes entirely unaligned near 30°. It means a change in twist angle the location and size of the Moire-like circular patches allowing to a precise arrangement in 2D materials. Additionally, regulation the interlayer coupler and Moire periodic potentials are also considered to be a crucial entity for scheming novel artificially stacked 2D materials with emergent functionalities.

Mounding or sewing two monolayers of diverse 2D materials is also useful to procedure a heterostructure. The heterostructures of 2D materials also offer different

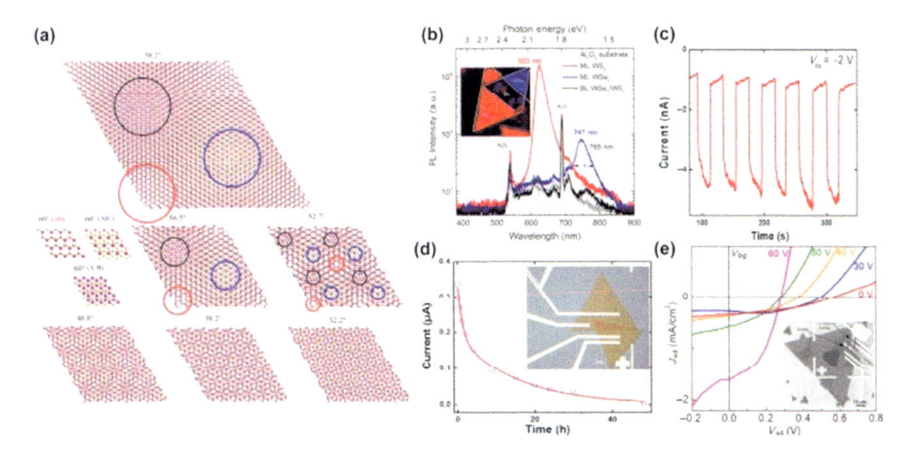

Fig. 3.31 Emerging properties from 2D bilayers and heterostructures; **a** commensurate bilayer $MoSe_2$ systems with twist angles above 30° illustrates how stacking evolves with twist angle. Red, blue, and black circles highlight the presence of 2H, AB′, and A′B stacking in a twist angle. The more the angle deviates from 60°, the smaller 2H, AB′, and A′B stacking areas appeared. For 32.2° the stacking a completely disordered (adopted from Puretzky et al. [263]); **b** PL spectra of WSe_2/WS_2 BL heterostructure with a twist angle of 25.3° to show a strong PL quenching (adopted from Wang et al. [264]); **c** device based on $MoSe_2/Mo_{1-x}W_xSe_2$ bilayer heterostructure fabricated by thermal annealing to show photodetector behavior; **d** device based on $MoSe_2/Mo_{1-x}W_xSe_2$ bilayer heterostructure without thermal annealing depicts persistent photocurrent (c and d adopted from Li, X. F. et al. 2016 J. Mater. Res. 31, 923–930); **e** gate-tunable photovoltaic behavior of the $MoSe_2$/GaSe bilayer heterostructure (adopted from Li, X. et al. 2016 Sci. Adv. 2, e1501882)

electronic properties of the coupled crystals with the induced additional functionalities. Such as 2D TMD heterostructures forms type-II junctions, to facilitate effective partition for the photoexcited electrons and holes with excessive potential applications, like photodetectors, photovoltaics, sensors, etc. Likewise, the WSe_2/WS_2 bilayer heterostructures with arbitrary twist angles using the wet transfer method is also fabricate [264, 265]. It has a robust interlayer electronic coupler persuaded considerable PL quenching, effective charge separation, fast charge transfer, and enhanced light absorption (Fig. 3.31b).

However, the PL spectrum of such 2D materials cannot confirm interlayer excitons owing to weakened interlayer exciton transition by the defects or residue introduced during the fabrication of the heterostructures. Similar behavior has been also interpreted for the $MoSe_2/Mo_{1-x}W_xSe_2$ heterostructures fabricated by wet process (Fig. 3.31c, d) [228]. The $MoSe_2/Mo_{1-x}W_xSe_2$ heterostructures also forms atomically thin p-n junctions with a clean interface, tunable gate, and photodetecting behavior by thermal annealing. However, without thermal annealing, $MoSe_2/Mo_{1-x}W_xSe_2$heterostructures have persistent photoconductivity owing to the photo population of trap states from adsorbates at the interface. Thus, formation a spotless and atomically sharp interface is significantly essential. The CVD developed bilayer heterostructures generally possessed much better interfaces, thereby this process fabricated bilayer heterostructures delivered superior optoelectronic properties. For instance, the definite interlayer locations and atomically sharp interfaces of $GaSe/MoSe_2$ heterostructures are formed a p–n junction with a type-II band alignment. That reflects a tunable gate photovoltaic response with effective charge transfer (Fig. 3.31e) [266]. Hence it is customary to demonstrate a universal method that allows the reliable production of precise bilayers and heterostructures with controlled stacking arrangements including spotless interfaces for the next-generation optical, electrical, and optoelectronic devices.

3.8 Ripples

As per theoretical predictions, the truly flat 2D materials are unbalanced because thermal unpredictability can abolish the 2D nature when the aspect ratio is extreme. The discovery of graphene has provided the option to people start realizing the real suspended 2D materials that cannot be completely flat, but they have ripples, by mean they have finite surface roughness and deformation, that is stabilized in their 2D nature [267]. Similarly, the morphology of ripples in as-synthesized MoS_2 monolayers is also realized by using atomic force microscopy [268] and electron diffraction [269]. The height of such ripples typically orders a nanometer. The periodic ripples are also intentionally generated in MoS_2 monolayers by using a laser beam during the growth process [270]. Over the macroscopic scale, ripples changes the bond length of the lattice gradually; therefore, it introduces the strain that would be useful to engineer the electronic properties of 2D materials [268, 271, 272]. Moreover, the ripples in other materials monolayer (such as h-BN) in principal should be the same order of magnitude as graphene due to the structural similarity.

3.9 Effect of Impurity

3.9.1 Electronic Transport and Metallicity

Dislocations and grain boundaries generally act as scattering centers or sinks for the charge carriers and reduce their mobility. The density of grain boundaries can be different types with increasing orientations in the case of a large specimen (~ μm size). Therefore, the electrical performance of the materials deteriorates progressively [273], which also consisted of the observations in conventional semiconductors. However, the development of nanotechnology in the modern world allows for a more accurate direct examination of the role of the individual grain boundary. At the nanoscale dimension, the distinct electrical behaviors have revealed two different types of grain boundaries, i.e., typical tilt and twin grain boundaries [238, 273]. The electrical property measurements of the FET devices built with a single GB possessed reduced mobility by half compared to that of pristine MoS_2 with the tilt grain boundaries, while the twin grain boundaries have even higher conductivity [238, 273]. Though, theoretical predictions based on first-principles calculations for a few selected grain boundary structures can have high conductivity along few-atom wide 60° (inversion) twin grain boundaries [274]. Later, the STM study of $MoSe_2$ revealed the mid-gap-state induced metallicity near 60° twin grain boundaries [275]. A systematic theoretical analysis has also demonstrated the metallicity of twin grain boundaries as an intrinsic property of 2D TMDs that arises due to polar discontinuity across the grain boundaries [276, 277]. Surprisingly, the transport properties of the grain boundaries also depends on prior electrical signals [278], in which the grain boundaries probably change their structures due to the migration of S vacancies. Therefore, usually it is correlated to the realization of the proof-of-concept non-volatile memory element—memristor—in 2D materials.

3.9.2 Magnetic Properties

The tilt grain boundaries can also act as recombination centers for electrons and holes, which harms the optical applications, such as the example chemical photoelectron and photovoltaic devices. Remarkably, if the grain boundary's localized states are partially occupied under a controlled gating or doping, their final grain boundaries become magnetic according to the Stoner model. The typical example of the tilt grain boundaries is the magnetization density of Mo- and S-rich 5|7 structures [279]. The increasing tilt angle allows the overlapping between the localized states, therefore, splitting exchange becomes greater. This can be directly correlated to the overall system turning into half-metallic. However, magnetism also exists in graphene due to the dangling bond imperfections, such as vacancies, adatoms, or edges, it can be easily annealed. Hence in contrast to grain boundaries, the topological defects are considered to be robust against local reconstructions, thereby, the overall magnetism preserved under the kind of disturbances and as a consequence, they are showing a considerable potential for spintronics applications.

3.9.3 Optical

The electrical transport of grain boundaries interpretation is also be useful to describe the optical properties of grain boundaries that vary from material to material. Predominately twin grain intensity has been noticed for twin grain boundaries. Although there is no accurate information on grain boundary structures available including experimental observations, however, it is very difficult to say exactly which kind of defects plays a crucial role. Though it is believed that the doping and strain effects contributes in appearance of PL variance, however, it needs further details remain to be unraveled.

3.9.4 Chemically Enhanced Plasticity

Mechanical properties of polycrystalline materials predominantly depend on the migration behaviors of dislocations under external stress or other stimuli. The glide of B- or N-5|7 in h-BN governs by SW rotation [280]. That is only possible under electron beam irradiation because of the high activation barrier [281, 282]. In contrast to this, the migration mechanisms and barriers of dislocations in TMDs also depends on sensitively specific core structures that vary. Like 4|6 and 6|8 cores migrates under a direct rebonding (RB) mechanism which mainly involves a single metal atom and other dislocations glide including generalize SW (SWg) rotation. However single-atom movement through the RB mechanism usually has low barriers in the range of 0.5–0.7 eV, whereas the SWg process is a concern to several-atoms movement that requires two to four times higher energy barriers around 1.1–2.4 eV. Moreover, the dislocation structures can be also controlled by chemical potential, it provides an interesting and unexpected phenomenon as most mobile dislocations (4|6 and 6|8) that thermodynamically favorable at the same time. However, 4|6 or 4|8 structures also have a dominant appearance at a specific concentration that directly correlates to their plastic deformation macroscopically. The fast movement of individual 6|8 s has been realized experimentally under electron beam irradiation [249]. It is considered to be a direct evidence of the whole grain boundary migration. A similar process has been also achieved for memristor devices in which the grain boundary moves with 3 μm in 1000 s [278].

3.10 Impurity Effect on Bilayer TMDs

The lack of stacking fault energy results in characteristic stacking boundaries in bilayer graphene. The existence of boundaries gives sharing of soliton-like structures between different stacking domains (like AB and AC stackings), as depicted in Fig. 3.32 [283–285]. The strain-induced accommodation profound the ripples in the bilayer graphene that has been noticed in TEM observation and theoretically demonstrated from the atomistic simulations (Fig. 3.32). The identical structure factors (rocking curves) of both stacking variants AB and AC are also following under the $11\bar{2}0$ reflections. The corresponding stacking faults are bordered therefore partials are not visible. In contrast to this, the $2\bar{2}00$ reflections show a local transition of stacking sequence from AB to AC (and vice versa) in the bilayer grapheme, it considered to be a direct evidence of a real intensity variation. These kinds of boundaries have been also predicted in host novel topological electronic states that are protected by symmetry [286, 287]. It could be related to the explanation of the puzzling subgap conductance in gated bilayer graphene.

3.11 Conclusions

In the conclusive view, one has continued the various aspects of TMD materials by exploring their different structural and physical properties including the impact of the defects on their properties. Owing to the fact materials' structural properties are one of the key parameters to define their possible utility for a specific purpose and modifications. The structural configurations are also makes them more specific or general for their potential applications. Considering this view one has discussed the structural properties of different forms of the TMDs materials, such as bulk TMDs structure, individual triple layers structure, polymorphs structure, distorted structures, and a single layer or few layers structures, including a description of different phases and phase transition, stability, patterning and tuneability of TMDs materials. Moreover, a descriptive description of possible significant types of defects has been also taken into consideration, such as point defects, doping, and grain boundary. Additionally, a brief discussion on the impact of impurity on TMDs' electronic transport and metallicity, magnetic characteristics, and optical properties have been also provided for an understanding of the changes in these physical characteristics. Moreover, a short discussion has been also included on the topic effect of impurity on bilayer TMDs. Thus overall in this chapter work, one has provided a detailed understanding of the structures of the TMD materials and alternation due to impurity addition along with the variations in their physical properties.

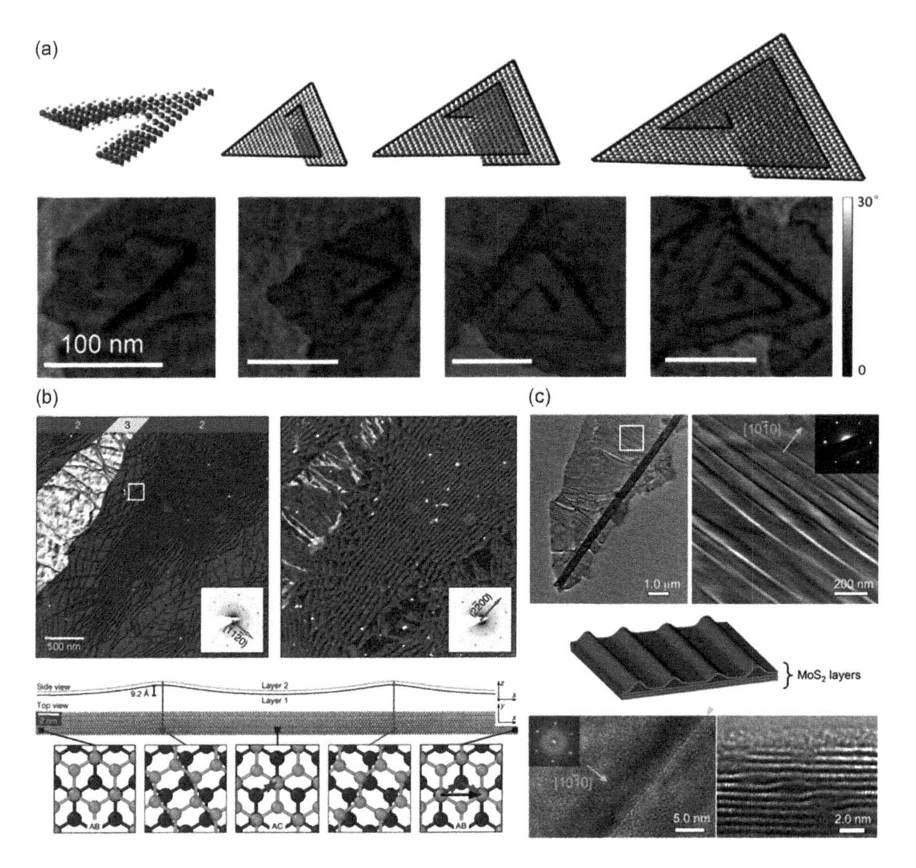

Fig. 3.32 **a** Schematic illustration (top row) and AFM image (bottom row panels) of screw dislocations in MoS_2 at different growth stages; **b** top panels: Dark-field TEM images of the same area of a multilayer graphene membrane reflections, the numbers in the top-left indicate the bilayer and the trilayer, the dark lines in the bilayer/trilayer regions correspond to an extended network of partials between the layers. The bottom is an atomic model for a pair of parallel 60 partials in the side and top view, with dislocation cores marked by vertical arrows. An enlarged view for the gradual change in the stacking sequence, AB–AC–AB, across the two partials, the short arrows are the Burger vectors of type (1/3); **c** top panel: left is a TEM image for the MoS_2 film, with a magnified view of the white squared region shown to the right, and inset is the electron diffraction pattern. Middle panel: schematic of surface "ripplocations" in MoS_2. Bottom panel: an HRTEM image of a ripplocation (left) and a TEM image of the cross-section view of multilayered MoS_2 containing buried ripplocations (adopted from Zou et al. [183])

References

1. Chhowalla M, Shin HS, Eda G, Li LJ et al (2013) The chemistry of two dimensional layered transition metal dichalcogenide nanosheets. Nat Chem 5:263–275
2. Wilson J, Yoffe A (1969) The transition metal dichalcogenides discussion and interpretation of the observed optical, electrical and structural properties. Adv Phys 18:193–335

3. Wypych F, Schollhorn R (1992) 1T-MoS2, a new metallic modification of molybdenum disulphide. J Chem Soc Chem Commun 1386–1388
4. Wypych F, Weber T, Prins R (1997) Scanning tunneling microscopic investigation of $K_x(H_2O)_yMoS_2$. Surf Sci 380:L474–L478
5. Wypych F, Weber T, Prins R (1998) Scanning tunneling microscopic investigation of 1T-MoS2. Chem Mater 10:723–727
6. Wypych F, Solenthaler C, Prins R, Weber T (1999) Electron diffraction study of intercalation compounds derived from 1T-MoS_2. J Solid State Chem 144:430–436
7. Zeng Z, Jin H, Rudolph M, Rominger F et al (2018) Gold(III)-catalyzed site-selective and divergent synthesis of 2-aminopyrroles and quinoline-based polyazaheterocycles. Angew Chem Int Ed 57:1246–1249
8. Zhao W, Pan J, Fang Y, Che X et al (2018) Metastable MoS_2: crystal structure, electronic band structure, synthetic approach and intriguing physical properties 24:15942–15954
9. Bastos CMO, Besse R, Da Silva JLF, Sipahi GM (2019) Ab initio investigation of structural stability and exfoliation energies in transition metal dichalcogenides based on Ti-, V-, and Mo-group elements. Phys Rev Mater 3:044002–044010
10. Kolobov AV, Tominaga J (2016) Two-dimensionaltransition-metal dichalcogenides. Springer International Publishing
11. Hulliger F, Lévy F (1977) Structural chemistry of layer-type phases, 1st edn. physics and chemistry of materials with layered structures 5. Springer Netherlands
12. Besse R, Caturello NAMS, Bastos CMO, Guedes-Sobrinho D (2018) Size-induced phase evolution of MoSe2 nanoflakes revealed by density functional theory. J Phys Chem C 122:20483
13. Gao D, Xue Q, Mao X, Wang W et al (2013) Ferromagnetism in ultrathin VS2nanosheets. J Mater Chem 1:5909–5916
14. Joensen P, Frindt RF, Morrison SR (1986) Single-layer MoS_2. Mater Res Bull 21:457–461
15. Miremadi BK, Morrison SR (1988) The intercalation and exfoliation of tungsten disulphide. J Appl Phys 63:4970–4974
16. Sandoval SJ, Yang D, Frindt RF, Irwin JC (1991) Raman study and lattice dynamics of single molecular layers of MoS_2. Phys Rev B: Condens Matter Mater Phys 44:3955–3962
17. Yang D, Sandoval SJ, Divigalpitiya WM, Irwin JC et al (1991) Structure of single-molecular-layer MoS_2. Phys Rev B: Condens Matter Mater Phys 43:12053–12056
18. Gordon RA, Yang D, Crozier ED, Jiang DT et al (2002) Structures of exfoliated single layers of WS_2, MoS_2, and $MoSe_2$ in aqueous suspension. Phys Rev B: Condens Matter Mater Phys 65:125407
19. Lee CH, Silva EC, Calderin L, Nguyen MAT (2015) Tungsten ditelluride: a layered semimetal. Sci Rep 5:10013
20. Santosh KC, Zhang C, Hong S, Wallace RM (2015) Phase stability of transition metal dichalcogenide by competing ligand field stabilization and charge density wave. 2D Mater 2:035019
21. Hu Z, Wu Z, Han C, He J (2018) Two-dimensional transition metal dichalcogenides: interface and defect engineering. Chem Soc Rev 47:3100–3128
22. Qian Z, Jiao L, Xie L (2020) Phase engineering of two-dimensional transition metal dichalcogenides. Chin J Chem 38:753–760
23. Py MA, Haering RR (1983) Structural destabilization induced by lithium intercalation in MoS2 and related compounds. Can J Phys 61:76–84
24. Heising J, Kanatzidis MG (1999) Exfoliated and restacked MoS2 and WS2: ionic or neutral species? Encapsulation and ordering of hard electropositive cations. J Am Chem Soc 121:11720–11732
25. Ganal P, Olberding W, Butz T, Ouvrard G (1993) Soft chemistry induced host metal coordination change from octahedral to trigonal prismatic in 1T-TaS_2. Solid StateIonics 59:313–319
26. Eda G, Fujita T, Yamaguchi H, Voiry D et al (2012) Coherent atomic and electronic heterostructures of single-layer MoS_2. ACS Nano 6:7311–7317

27. Qin XR, Yang D, Frindt RF, Irwin JC (1991) Real-space imaging of single-layer $MoS_2$by scanning tunneling microscopy. Phys Rev B: Condens Matter Mater Phys 44:3490–3493
28. Qin XR, Yang D, Frindt RF, Irwin JC (1992) Ultramicroscopy 42–44(Part 1):630–636
29. Voiry D, Yamaguchi H, Li J, Silva R et al (2013) Enhanced catalytic activity in strained chemically exfoliated WS2 nanosheets for hydrogen evolution. Nat Mater 12:850–855
30. Kappera R, Voiry D, Yalcin SE, Branch B, Gupta G, Mohite AD, Chhowalla M (2014) Phase-engineered low-resistance contacts for ultrathin MoS2 transistors. Nat Mater 13:1128–1134
31. Heising J, Kanatzidis MG (1999) Structure of restacked MoS_2 and WS_2 elucidated by electron crystallography. J Am Chem Soc 121:638–643
32. Tongay S, Sahin H, Ko C, Luce A et al (2014) Monolayer behaviour in bulk ReS2 due to electronic and vibrational decoupling. Nat Commun 5:3252
33. Fujita T, Ito Y, Tan Y, Yamaguchi H et al (2014) Chemically exfoliated ReS_2 nanosheets. Nanoscale 6:12458–12462
34. Wang R, Yu Y, Zhou S, Li H et al (2018) Strategies on phase control in transition metal dichalcogenides. Adv Funct Mater 28:1802473
35. Kertesz M, Hoffmann R (1984) Octahedral vs. trigonal-prismatic coordination and clustering in transition-metal dichalcogenides. J Am Chem Soc 106:3453–3460
36. Voiry D, Mohite A, Chhowalla M (2015) Phase engineering of transition metal dichalcogenides. Chem Soc Rev 44:2702–2712
37. Chia X, Eng AYS, Ambrosi A, Tan SM et al (2015) Electrochemistry of nanostructured layered transition-metal dichalcogenides. Chem Rev 115:11941–11966
38. Toh RJ, Sofer Z, Luxa J, Sedmidubsky D (2017) 3R phase of MoS2 and WS2 outperforms the corresponding 2H phase for hydrogen evolution. Chem Commun 53:3054–3057
39. Chauhan H, Deka S (2021) Fundamentals and supercapacitor applications of 2D materials. Chapter 6. Elsevier
40. Duerloo KAN, Li Y, Reed EJ (2014) Structural phase transitions in two-dimensional Mo- and W-dichalcogenide monolayers. Nat Commun 5:4214–9
41. Li X, Tao L, Chen Z, Fang H et al (2017) Graphene and related two-dimensional materials: structure-property relationships for electronics and optoelectronics. Appl Phys Rev 4:021306
42. Xu HM, Gu C, Zhang XL, Shi L et al (2021) Phase-controlled 1T transition-metal dichalcogenide-based multidimensional hybrid nanostructures. CCS Chem 3:58–68
43. Jaramillo TF, Jørgensen KP, Bonde J, Nielsen JH et al (2007) Identification of active edge sites for electrochemical H_2 evolution from MoS_2 nanocatalysts. Science 317:100–102
44. Qian X, Liu J, Fu L, Li J (2014) Quantum spin Hall effect in two-dimensional transition metal dichalcogenides. Science 346:1344–1347
45. Yin X, Wang Q, Cao L, Tang CS et al (2017) Tunable inverted gap in monolayer quasi-metallic MoS2 induced by strong charge-lattice coupling. Nat Commun 8:486
46. Yin X, Tang CS, Wu D, Kong W et al (2019) Unraveling high-yield phase-transition dynamics in transition metal dichalcogenides on metallic substrates. Adv Sci 6:1802093
47. Tang CS, Yin X, Yang M, Wu D et al (2020) Anisotropic collective charge excitations in quasimetallic 2D transition-metal dichalcogenides. Adv Sci 7:1902726
48. Stolyarov MA, Liu G, Bloodgood MA, Aytan E et al (2016) Breakdown current density in h-BN-capped quasi-1D TaSe3 metallic nanowires: prospects of interconnect applications. Nanoscale 8(34):15774–15782
49. Island JO, Molina-Mendoza AJ, Barawi M, Biele R et al (2017) Electronics and optoelectronics of quasi-1D layered transition metal trichalcogenides. 2D Mater 4:022003
50. Geremew A, Bloodgood MA, Aytan E, Woo BWK et al (2018) Current carrying capacity of quasi-1D ZrTe3 Van Der Waals nanoribbons. IEEE Electron Device Lett 39:735–738
51. Kim J, Lee Z (2018) Phase transformation of two-dimensional transition metal dichalcogenides. Appl Microsc 48:43–48
52. Jin Q, Liu N, Chen B, Mei D (2018) Mechanisms of semiconducting 2H to metallic 1T phase transition in two-dimensional MoS2 nanosheets. J Phys Chem C 122:28215–28224
53. Keum DH, Cho S, Kim JH, Choe DH et al (2015) Bandgap opening in few-layered monoclinic MoTe2. Nat Phys 11:482–486

54. Tang CS, Yin X, Wee ATS (2021) 1D chain structure in 1T′-phase 2D transition metal dichalcogenides and their anisotropic electronic structures. Appl Phys Rev 8:011313
55. Calandra M (2013) Chemically exfoliated single-layer MoS2: stability, lattice dynamics, and catalytic adsorption from first principles. Phys Rev B 88:245428
56. Kadantsev ES, Hawrylak P (2012) Electronic structure of a single MoS2 monolayer. Solid State Commun 152:909–913
57. Wilson JA, Di Salvo FJ, Mahajan S (1975) Charge-density waves and superlattices in the metallic layered transition metal dichalcogenides. Adv Phys 24:117–201
58. Moncton DE, Axe JD, DiSalvo FJ (1975) Study of superlattice formation in 2H-$NbSe_2$ and 2H-$TaSe_2$ by neutron scattering. Phys Rev Lett 34:734–737
59. Dawson WG, Bullett DW (1987) Electronic structure and crystallography of MoTe2 and WTe2. J Phys C Solid State Phys 20:6159
60. Sun Y, Wu SC, Ali MN, Felser C et al (2015) Prediction of Weyl semimetal in orthorhombic $MoTe_2$. Phys Rev B 92:161107
61. Ali MN, Xiong J, Flynn S, Tao J et al (2014) Large, non-saturating magnetoresistance in WTe_2. Nature 514:205–208
62. Yan X, Lv Y, Li L, Li X et al (2017) Artificial gravity field, astrophysical analogues, and topological phase transitions in strained topological semimetals. npj Quant Mater 2:31
63. Geng X, Sun W, Wu W, Chen B et al (2016) Pure and stable metallic phase molybdenum disulfide nanosheets for hydrogen evolution reaction. Nat Commun 7:10672
64. Ma F, Gao G, Jiao Y, Gu Y et al (2016) Predicting a new phase (T′′) of two-dimensional transition metal di-chalcogenides and strain-controlled topological phase transition. Nanoscale 8:4969–4975
65. Bruyer E, Di Sante D, Barone P, Stroppa A et al (2016) Possibility of combining ferroelectricity and Rashba-like spin splitting in monolayers of the 1T-type transition-metal dichalcogenides MX2(M=Mo, W;X=S, Se, Te). Phys Rev B 94:195402
66. Zhuang HL, Johannes MD, Singh AK, Hennig RG (2017) Doping-controlled phase transitions in single-layer MoS_2. Phys Rev B96:165305
67. Chhowalla M, Voiry D, Yang J, Shin HS et al (2015) Phase-engineered transition-metal dichalcogenides for energy and electronics. MRS Bull 40:585–591
68. Tan SJ, Abdelwahab I, Ding Z, Zhao X et al (2017) Chemical stabilization of 1T′ phase transition metal dichalcogenides with giant optical Kerr nonlinearity. J Am Chem Soc 139:25042511
69. Chen W, Gu J, Liu Q, Luo R et al (2018) Quantum dots of 1T phase transitional metal dichalcogenides generated via electrochemical Li intercalation. ACS Nano 12:308–316
70. Leng K, Chen Z, Zhao X, Tang W (2016) Phase restructuring in transition metal dichalcogenides for highly stable energy storage. ACS Nano 10:9208–9215
71. Zhu Y, Peng L, Fang Z, Yan C et al (2018) Structural engineering of 2D nanomaterials for energy storage and catalysis. Adv Mater 30:1706347
72. Acerce M, Voiry D, Chhowalla M (2015) Metallic 1T phase MoS2 nanosheets as supercapacitor electrode materials. Nat Nanotechnol 10:313–318
73. Cho S, Kim S, Kim JH, Zhao J et al (2015) Phase patterning for ohmic homojunction contact in $MoTe_2$. Science 349:625–628
74. Naylor CH, Parkin WM, Gao Z, Berry J et al (2017) Synthesis and physical properties of phase-engineered transition metal dichalcogenide monolayer heterostructures. ACS Nano 11:8619–8627
75. Li Y, Duerloo KA, Wauson K, Reed EJ (2016) Structural semiconductor-to-semimetal phase transition in two-dimensional materials induced by electrostatic gating. Nat Commun 7:10671–8
76. Wang X, Chen X, Zhou Y, Park C et al (2017) Pressure-induced iso-structural phase transition and metallization in WSe_2. Sci Rep 7:46694–46699
77. Yagmurcukardes M, Bacaksiz C, Senger RT, Sahin H (2017) Hydrogen-induced structural transition in single layer ReS_2. 2D Mater 4:035013

78. Iyikanat F, Kandemir A, Ozaydin HD, Senger RT et al (2017) Hydrogenation-driven phase transition in single-layer $TiSe_2$. Nanotechnology 28:495709
79. Li W, Qian X, Li J (2021) Phase transitions in 2D materials. Nat Rev Mater. https://doi.org/10.1038/s41578-021-00304-0
80. Fan S, Zou X, Du H, Gan L et al (2017) Theoretical investigation of the intercalation chemistry of lithium/sodium ions in transition metal dichalcogenides. J Phys Chem C 121:13599–13605
81. Elias DC, Nair RR, Mohiuddin TM, Morozov SV et al (2009) Control of graphene's properties by reversible hydrogenation: evidence for graphane. Science 323:610–613
82. Pumera M, Wong CH (2013) Graphane and hydrogenated graphene. Chem Soc Rev 42:5987–5995
83. Qu Y, Pan H, Kwok CT (2016) Hydrogenation-controlled phase transition on two-dimensional transition metal dichalcogenides and their unique physical and catalytic properties. Sci Rep 6:34186–34213
84. Shi Y, Li H, Li LJ (2015) Recent advances in controlled synthesis of two-dimensional transition metal dichalcogenides via vapour deposition techniques. Chem Soc Rev 44:2744–2756
85. Yang L, Zhang W, Li J, Cheng S et al (2017) Tellurization velocity-dependent metallic–semiconducting–metallic phase evolution in chemical vapor deposition growth of large-area, few-layer $MoTe_2$. ACS Nano 11:1964–1972
86. Yoo Y, DeGregorio ZP, Su Y, Koester SJ et al (2017) In-plane 2H–1T′ $MoTe_2$ homojunctions synthesized by flux-controlled phase engineering. Adv Mater 29:1605461
87. Empante TA, Zhou Y, Klee V, Nguyen AE et al (2017) Chemical vapor deposition growth of few-layer MoTe2 in the 2H, 1T′, and 1T phases: tunable properties of MoTe2 films. ACS Nano 11:900–905
88. You J, Hossain MD, Luo Z (2018) Synthesis of 2D transition metal dichalcogenides by chemical vapor deposition with controlled layer number and morphology. Nano Convergence 5:1–13
89. Yang D, Hu X, Zhuang M, Ding Y et al (2018) Inversion symmetry broken 2D 3R-$MoTe_2$. Adv Funct Mater 28:1800785
90. Sokolikova MS, Mattevi C (2020) Direct synthesis of metastable phases of 2D transition metal dichalcogenides. Chem Soc Rev 49:3952–3980
91. Zhang Q, Xiao Y, Zhang T, Weng Z et al (2017) Iodine-mediated chemical vapor deposition growth of metastable transition metal dichalcogenides. Chem Mater 29:4641–4644
92. Feng Q, Zhu Y, Hong J, Zhang M et al (2014) Growth of large-area 2D $MoS_2(_1$-x) Se_2x semiconductor alloys. Adv Mater 26:2648
93. Rhodes D, Chenet DA, Janicek BE, Nyby C et al (2017) Engineering the Structural and Electronic Phases of MoTe2 through W Substitution. Nano Lett 17:1616–1622
94. Susarla S, Kutana A, Hachtel JA, Kochat V et al (2017) Quaternary 2D transition metal dichalcogenides (TMDs) with tunable bandgap. Adv Mater 29:1702457
95. Yu P, Lin J, Sun L, Le QL et al (2017) Metal–semiconductor phase-transition in $WSe_{2(1-x)}Te_{2x}$ monolayer. Adv Mater 29:1603991
96. Mak KF, Lee C, Hone J, Shan J, Heinz TF (2010) Atomically thin MoS_2: a new direct-gap semiconductor. Phys Rev Lett 105:136805–136814
97. Wang QH, Kalantar-Zadeh K, Kis A, Coleman JN et al (2012) Electronics and optoelectronics of two-dimensional transition metal dichalcogenides. Nat Nanotechnol 7:699–712
98. Li G, Hu WZ, Qian D, Hsieh D et al (2007) Semimetal-to-semimetal charge density wave transition in 1T−$TiSe_2$. Phys Rev Lett 99:027404–027414
99. Brown BE (1966) The crystal structures of WTe_2 and high-temperature $MoTe_2$. 20:268–274
100. Darencet P, Millis AJ, Marianetti CA (2014) Three-dimensional metallic and two-dimensional insulating behavior in octahedral tantalum dichalcogenides. Phys Rev B 90:0451345–0451355
101. Zhu J, Wang Z, Yu H, Li N et al (2017) Argon plasma inducedphase transition in monolayer MoS_2. J Am Chem Soc 139:10216–10219
102. Lasek K, Li J, Kolekar S, Coelho PM et al (2021) Synthesis and characterization of 2D transition metal dichalcogenides: recent progress from a vacuum surface science perspective. Sur Sci Rep 76:100523–100552

103. Yang H, Kim SW, Chhowalla M, Lee YH (2017) Structural and quantum-state phase transitions in van der Waals layered materials. Nat Phys 13:931–937
104. Huang HH, Fan X, Singh DJ, Zheng WT (2020) Recent progress of TMD nanomaterials: phasetransitions and applications. Nanoscale 12:1247–1268
105. Gan X, Lee LYS, Wong K, Lo TW et al (2018) 2H/1T phase transition of multilayer MoS2by electrochemical incorporation of S vacancies. ACS Appl Energy Mater 1:4754–4765
106. Ding W, Hu L, Dai J, Tang X et al (2019) Highly ambient-stable 1T-MoS2 and1T-WS2 by hydrothermal synthesis under high magnetic fields. ACS Nano 13:1694–1702
107. Bischoff F, Auwarter W, Barth JV, Schiffrin A et al (2017) Nanoscale phase engineering of niobium diselenide. Chem Mater 29:9907–9914
108. Wang Y, Xiao J, Zhu H, Li Y et al (2017) Structural phase transition in monolayerMoTe2 driven by electrostatic doping. Nature 550:487–491
109. Aksoy R, Ma Y, Selvi E, Chyu MC et al (2006) X-ray diffraction study of molybdenum disulfide to 38.8 GPa. J Phys Chem Solids 67:1914–1917
110. Hromadova L, Martoňák R, Tosatti E (2013) Structure change, layer sliding, and metallization in high-pressure MoS_2. Phys Rev B: Condens Matter Mater Phys 87:144105–6
111. Zhao Z, Zhang H, Yuan H, Wang S et al (2015) Pressure induced metallization with absence of structural transition in layered molybdenum diselenide. Nat Commun 6:7312
112. Fan X, Singh DJ, Jiang Q, Zheng W (2016) Pressure evolution of the potential barriers of phase transition of MoS2, MoSe2 and MoTe2. Phys Chem Chem Phys 18:12080–12085
113. Fan X, Chang C-H, Zheng W, Kuo J-L (2015) The electronic properties of single-layer and multilayer MoS2 under high pressure. J Phys Chem C 119:10189–10196
114. Ramasubramaniam A, Naveh D, Towe E (2011) Tunable band gaps in bilayer transition-metal dichalcogenides. Phys Rev B: Condens Matter Mater Phys 84:205325–10
115. Tan W, Wei Z, Liu X, Liu J et al (2017) Ordered and Disordered Phases in Mo_{1-x} W_x S_2 monolayer. Sci Rep 7:15124
116. Wilson J, Di Salvo F, Mahajan S (1974) Charge-density waves in metallic, layered, transition-metal dichalcogenides. Phys Rev Lett 32:882–885
117. Rossnagel K (2011) On the origin of charge-density waves in select layered transition-metal dichalcogenides. J Phys Condens Matter 23:213001
118. Sipos B, Kusmartseva AF, Akrap A, Berger H et al (2008) From Mott state to superconductivity in 1T-TaS_2. Nat Mater 7:960–965
119. Fazekas P, Tosatti E (1979) Electrical, structural and magnetic properties of pure and doped 1T-TaS_2. Philos Mag B 39:229–244
120. Yu Y, Yang F, Lu XF, Yan YJ et al (2015) Gate-tunable phase transitions in thin flakes of 1T-TaS_2. Nat Nanotechnol 10:270–276
121. Geremew AK, Rumyantsev S, Kargar F, Debnath B et al (2019) Bias-voltage driven switching of the charge-density-wave and normal metallic phases in 1T-TaS_2 thin-film devices. ACS Nano 13:7231–7240
122. Sugai S (1985) Lattice vibrations in the charge-density-wave states of layered transition metal dichalcogenides. Phys Status Solidi B 129:13–39
123. Xi X, Zhao L, Wang Z, Berger H et al (2015) Strongly enhanced charge-density-wave order in monolayer $NbSe_2$. Nat Nanotechnol 10:765–769
124. Ugeda MM, Bradley AJ, Zhang Y, Onishi S et al (2016) Characterization of collective ground states in single-layer $NbSe_2$. Nat Phys 12:92–97
125. Calandra M, Mazin I, Mauri F (2009) Effect of dimensionality on the charge-density wave in few-layer 2H-$NbSe_2$. Phys Rev B Condens Matter Mater Phys 80:241108-4
126. Khestanova E, Birkbeck J, Zhu M, Cao Y et al (2018) Unusual suppression of the superconducting energy gap and critical temperature in atomically thin $NbSe_2$. Nano Lett 18:2623–2629
127. Silva-Guillén JÁ, Ordejón P, Guinea F, Canadell E (2016) Electronic structure of 2H-NbSe2 single-layers in the CDW state. 2D Mater 3:035028
128. Lian C-S, Si C, Duan W (2018) Unveiling charge-density wave, superconductivity, and their competitive nature in two-dimensional $NbSe_2$. Nano Lett 18:2924–2929

129. Guster B, Rubio-Verdú C, Robles R, Zaldívar J et al (2019) Coexistence of elastic modulations in the charge density wave state of 2H-$NbSe_2$. Nano Lett 19:3027–3032
130. Ryu H, Chen Y, Kim H, Tsai H-Z et al (2018) Persistent charge-density-wave order in single-layer $TaSe_2$. Nano Lett 18:689–694
131. Morosan E, Zandbergen HW, Dennis B, Bos et al (2006) Superconductivity in $CuxTiSe_2$. Nat Phys 2:544–550
132. Morosan E, Wagner KE, Zhao LL, Hor Y et al (2010) Multiple electronic transitions and superconductivity in Pd_xTiSe_2. Phys Rev B: Condens Matter Mater Phys 81:094524–5
133. Kusmartseva AF, Sipos B, Berger H, Forro L et al (2009) Pressure induced superconductivity in pristine $1T-TiSe_2$. Phys Rev Lett 103:236401–236404
134. Sergio C, Sinko MR, Gopalan DP, Sivadas N et al (2018) Tuning Ising superconductivity with layer and spin–orbit coupling in two-dimensional transition-metal dichalcogenides. Nat Commun 9:1427
135. Ye J, Zhang Y, Akashi R, Bahramy M et al (2012) Superconducting dome in a gate-tuned band insulator. Science 338:1193–1196
136. Costanzo D, Jo S, Berger H, Morpurgo AF (2016) Gate-induced superconductivity in atomically thin MoS2 crystals. Nat Nanotechnol 11:339–344
137. Piatti E, De Fazio D, Daghero D, Tamalampudi SR et al (2018) Multi-valley superconductivity in ion-gated MoS2 layers. Nano Lett 18:4821–4830
138. Costanzo D, Zhang H, Reddy BA, Berger H et al (2018) Tunnelling spectroscopy of gate-induced superconductivity in MoS_2. Nat Nanotechnol 13:483–488
139. Shi W, Ye J, Zhang Y, Suzuki R et al (2015) Superconductivity series in transition metal dichalcogenides by ionic gating. Sci Rep 5:12534
140. Jo S, Costanzo D, Berger H, Morpurgo AF (2015) Gate-induced superconductivity in atomically thin MoS_2 crystals. Nano Lett 15:1197–1202
141. Chi Z, Chen X, Yen F, Peng F et al (2018) Superconductivity in pristine $2Ha-MoS_2$ at ultrahigh pressure. Phys Rev Lett 120:037002–037006
142. Lu P, Kim J-S, Yang J, Gao H et al (2016) Origin of superconductivity in the Weyl semimetal WTe2 under pressure. Phys Rev B 94:224512
143. Fang Y, Pan J, He J, Luo R et al (2018) Structure re-determination and superconductivity observation of bulk 1T MoS_2. Angew Chem Int Ed 57:1232–1235
144. Hsu Y-T, Vaezi A, Fischer MH, Kim E-A (2017) Topological superconductivity in monolayer transition metal dichalcogenides. Nat Commun 8:14985
145. Zhu L, Li Q-Y, Lv Y-Y, Li S et al (2018) Superconductivity in potassium-intercalated T_d-WTe_2. Nano Lett 18:6585–6590
146. Fang Y, Pan J, Zhang D, Wang D et al (2018) Discovery of superconductivity in 2M WS2 with possible topological surface states. Adv Mater 26:1901942
147. Choe D-H, Sung H-J, Chang KJ (2016) Understanding topological phase transition in monolayer transition metal dichalcogenides. Phys Rev B 93:125109-7
148. Soluyanov AA, Gresch D, Wang Z, Wu Q et al (2015) Type-II Weyl semimetals. Nature 527:495–498
149. Ugeda MM, Pulkin A, Tang S, Ryu H et al (2018) Observation of topologically protected states at crystalline phase boundaries in single-layer WSe_2. Nat Commun 9:3401
150. Li Q, Yao Z, Wu J, Mitra S et al (2017) Intermediate phases in sodium intercalation into MoS2 nanosheets and their implications for sodium-ion batteries. Nano Energy 38:342–349
151. Liu L, Wu J, Wu L, Ye M et al (2018) Phase-selective synthesis of 1T0 MoS2 monolayers and heterophase bilayers. Nat Mater 17:1108–1114
152. Liu Q, Li X, He Q, Khalil A et al (2015) Gram-scale aqueous synthesis of stable few-layered 1T-MoS2: applications for visible light-driven photocatalytic hydrogen evolution. Small 11:5556–5564
153. Papageorgopoulos CA, Jaegermann W (1995) Li intercalation across and along the van der Waals surfaces of MoS2 (0001). Surf Sci 338:83–93
154. Eda G, Yamaguchi H, Voiry D, Fujita T et al (2011) Photoluminescence from chemically exfoliated MoS2. Nano Lett 11:5111–5116

155. Yaremko AM, Yukhymchuk VO, Romanyuk YA, Baran J et al (2017) Theoretical and experimental study of phonon spectra of bulk and nano-sized MoS2 layer crystals. Nanoscale Res Lett 12:82
156. Voiry D, Salehi M, Silva R, Fujita T et al (2013) Conducting MoS2 nanosheets as catalysts for hydrogen evolutionreaction. Nano Lett 13:6222–6227
157. Tsai HL, Heising J, Schindler JL, Kannewurf CR et al (1997) Exfoliated-restacked phase of WS2. Chem Mater 9:879–882
158. Kappera R, Voiry D, Yalcin SE, Jen W et al (2014) Metallic 1T phase source/drain electrodes for field effect transistors fromchemical vapor deposited MoS2. Apl Mater 2:92516
159. Lin YC, Dumcenco DO, Huang Y-S, Suenaga K (2014) Atomic mechanism of the semiconducting-to-metallic phase transition in single-layered MoS2. Nat Nanotechnol 9:391–396
160. Wang Z, Sun Y-Y, Abdelwahab I, Cao L et al (2018) Surface-limited superconducting phase transition on 1T-TaS2. ACS Nano 12:12619–12628
161. Zhang J, Liu J, Huang J, Kim P et al (1997) Creation of nanocrystals via tip induced solid-solid transformations. Mater Res Soc Symp Proc 466:89–94
162. Zhang J, Liu J, Huang JL, Kim P et al (1996) Creation of nanocrystals through a solid-solid phase transition induced by an STM tip. Science 274:757–760
163. Wang H, Lee J, Dreyer M, Barker BI (2009) A scanning tunneling microscopy study of a new superstructure around defects created by tip-sample interaction on 2H-NbSe2. J Phys Condens Matter 21:265005
164. Nguyen GD, Oyedele AD, Haglund A, Ko W et al (2020) Atomically precise PdSe2 pentagonal nanoribbons. ACS Nano 14:1951–1957
165. Weston A, Zou Y, Enaldiev V, Summerfield A et al (2020) Atomic reconstruction in twisted bilayers of transitionmetal dichalcogenides". Nat Nanotechnol 15:592–597
166. Lin X, Lu JC, Shao Y, Zhang YY et al (2017) Intrinsically patterned two-dimensional materials for selective adsorption of molecules and nanoclusters. Nat Mater 16:717–721
167. Mahler B, Hoepfner V, Liao K, Ozin GA (2014) Colloidalsynthesis of 1T-WS2 and 2H-WS2 nanosheets: applications for photocatalytic hydrogen evolution. J Am Chem Soc 136:14121–14127
168. Guo Y, Sun D, Ouyang B, Raja A, Song J et al (2015) Probing the dynamics of the metallic-to-semi conducting structural phase transformation in MoS2crystals. Nano Lett 15:5081–5088
169. Zhao X, Ning S, Fu W, Pennycook SJ et al (2018) Differentiating polymorphs in molybdenum disulfide viaelectron microscopy. Adv Mater 30:1802397
170. Shi J, Yu P, Liu F, He P et al (2017) 3R MoS2 with broken inversion symmetry: a promising ultrathin nonlinear optical device. Adv Mater 29:1701486
171. Chen S, Goldstein T, Venkataraman D, Ramasubramaniam A et al (2016) Activation of new Raman modes by inversion symmetry breaking in type II Weyl semimetal candidateT'-$MoTe_2$. Nano Lett 16:5852–5860
172. Ma X, Guo P, Yi C, Yu Q et al (2016) Raman scattering in the transition-metaldichalcogenides of 1T0-MoTe2, Td-MoTe2, and Td-WTe2. Phys Rev B 94:214105
173. Chen SY, Naylor CH, Goldstein T, Johnson ATC et al (2017) Intrinsic phonon bands in high-quality monolayer T0 molybdenum ditelluride. ACS Nano 11:814–820
174. Lee CH, Cruz Silva E, Calderin L, Nguyen MAT et al (2015) Tungsten ditelluride: a layered semimetal. Sci Rep 5:10013
175. Yu Y, Nam GH, He Q, Wu XJ et al (2018) High phase-purity 1T0-MoS2- and 1T0-MoSe2-layered crystals. Nat Chem 10:638–643
176. Tan SJR, Sarkar S, Zhao X, Luo X et al (2018) Temperature- and phase-dependent phonon renormalization in 1T0-MoS2. ACS Nano 12:5051–5058
177. Sokolikova MS, Sherrell PC, Palczynski P, Bemmer VL et al (2019) Direct solution-phase synthesis of 1T0 WSe2 nanosheets. Nat Commun 10:712
178. Pierucci D, Zribi J, Livache C, Gréboval C et al (2019) Evidence for a narrow band gap phase in1T0 WS2 nanosheet. Appl Phys Lett 115:032102

179. Guo C, Pan J, Li H, Lin T et al (2017) Observation of superconductivity in 1T0-MoS2 nanosheets. J Mater Chem C 5:10855–10860
180. Fan X, Xu P, Zhou D, Sun Y et al (2015) Fast and efficient preparation of exfoliated2H MoS2 nanosheets by sonication-assisted lithium intercalation and infrared laser-induced 1T to 2H phase reversion. Nano Lett 15:5956–5960
181. Mohite AD, Chhowalla M (2014) Phase-engineered low resistance contacts for ultrathin MoS2 transistors. Nat Mater 13:1128–1134
182. Ma Y, Liu B, Zhang A, Chen L et al (2015) Reversible semiconducting-to-metallic phase-transition in chemical vapor deposition grown monolayerWSe2 and applications for devices. ACS Nano 9:7383–7391
183. Zou X, Yakobson BI (2017) Defects in two-dimensional materials. https://doi.org/10.1017/9781316681619.021
184. Lin J, Zhou W (2018) Defect in 2D materials beyond graphene (Chapter 6). In: Defects in advanced electronic materials and novel low dimensional structures. Elsevier. https://doi.org/10.1016/B978-0-08-102053-1.00006-5
185. Splendiani A, Sun L, Zhang Y, Li T et al (2010) Emerging photoluminescence in monolayer MoS_2. Nano Lett 10:1271–1275
186. Jariwala D, Sangwan VK, Lauhon LJ, Marks TJ et al (2014) Emerging device applications for semiconducting two-dimensional transition metal dichalcogenides. ACS Nano 8:1102–1120
187. Schmidt H, Giustiniano F, Eda G (2015) Electronic transport properties of transition metal dichalcogenide field effect devices: surface and interface effects. Chem Soc Rev 44:7715–7736
188. Lin Z, Carvalho BR, Kahn E, Lv R (2016) Defect engineering of two-dimensional transition metaldichalcogenides. 2D Mater 3:022002
189. Tongay S, Suh J, Ataca C, Fan W et al (2013) Defects activated photoluminescence in two-dimensional semiconductors: interplay between bound, charged, and free excitons. Sci Rep 3:2657
190. Mignuzzi S, Pollard AJ, Bonini N, Brennan B et al (2015) Effect of disorder on Raman scattering of single-layer MoS_2. Phys Rev B 91:195411
191. Cai H, Yu Y, Lin YC, Puretzky AA (2021) Heterogeneities at multiple length scales in 2D layered materials: from localized defects and dopants to mesoscopic heterostructures. Nano Res 14:1625–1649
192. Khossossi N, Singh D, Ainane A, Ahuja R (2020) Recent progress of defect chemistry on 2D materials for advanced battery anodes. Chem Asian J 15:3390–3404
193. Hus SM, Li AP (2017) Spatially-resolved studies on the role of defects and boundarie sin electronic behavior of 2D materials. Prog Surf Sci 92:176–201
194. Mahjouri-Samani M, Liang LB, Oyedele A, Kim YS et al (2016) Tailoring vacancies far beyond intrinsic levels changes the carrier type and optical response in monolayer MoSe2−x crystals. Nano Lett 16:5213–5220
195. Carozo V, Wang YX, Fujisawa K, Carvalho BR et al (2017) Optical identification of sulfur vacancies: Bound excitons at the edges of monolayer tungsten disulfide. Sci Adv 3:1602813
196. Li WF, Fang CM, Van Huis MA (2016) Strong spin-orbit splitting and magnetism of point defect states in monolayer WS_2. Phys Rev B 94:195425
197. Avsar A, Ciarrocchi A, Pizzochero M, Unuchek D (2019) Defect induced, layer-modulated magnetism in ultrathin metallic $PtSe_2$. Nat Nanotechnol 14:674–678
198. Barja S, Wickenburg S, Liu ZF, Zhang Y et al (2016) Charge density wave order in 1D mirror twin boundaries of single-layer $MoSe_2$. Nat Phys 12:751–756
199. Koós AA, Vancsó P, Szendrő M, Dobrik G et al (2019) Influence of native defects on the electronic and magnetic properties of CVD grown $MoSe_2$ single layers. J Phys Chem C 123:24855–24864
200. Jolie W, Murray C, Weiß PS, Hall J et al (2019) Tomonaga-luttinger liquid in a box: Electrons confined within MoS_2 mirror-twin boundaries. Phys Rev 9:011055
201. Zhou W, Zou X, Najmaei S, Liu Z et al (2013) Intrinsic structural defects in monolayer molybdenum disulfide. Nano Lett 13:2615–2622

202. Noh JY, Kim H, Kim YS (2014) Stability and electronic structures of native defects in single layer MoS_2. Phys Rev B 89:205417
203. Komsa HP, Krasheninnikov AV (2015) Native defects in bulk and monolayer MoS_2 from first principles. Phys Rev B 91:125304
204. Liu D, Guo Y, Fang L, Robertson J (2013) Sulfur vacancies in monolayer MoS2 and its electrical contacts. Appl Phys Lett 103:183113
205. Najmaei S, Liu Z, Zhou W, Zou X et al (2013) Vapour phase growth and grain boundary structure of molybdenum disulphide atomic layers. Nat Mater 12:754–759
206. Lee YH, Zhang XQ, Zhang W, Chang MT et al (2012) Synthesis of Large-area MoS2 atomic layers with chemical vapor deposition. Adv Mater 24:2320–2325
207. Radisavljevic B, Radenovic A, Brivio J, Giacometti V et al (2011) Single-layer MoS_2 transistors. Nat Nanotechnol 6:147–150
208. Hong J, Hu Z, Probert M, Li K et al (2015) Exploring atomic defects in molybdenum disulphide monolayers. Nat Commun 6:6293
209. Keyshar K, Gong Y, Ye G, Brunetto G et al (2015) Chemical vapor deposition of monolayer rhenium disulfide (ReS2). Adv Mater 27:4640–4648
210. Lin YC, Komsa HP, Yeh CH, Bjorkman T et al (2015) Single-layer ReS_2: two-dimensional semiconductor with tunable in-plane anisotropy. ACS Nano 9:11249–57
211. Wang H, Huang X, Lin J, Cui J et al (2017) High-quality monolayer superconductor $NbSe_2$ grown by chemical vapour deposition. Nat Commun 8:394
212. Nguyen L, Komsa HP, Khestanova E, Kashtiban RJ et al (2017) Atomic defects and doping of monolayer $NbSe_2$. ACS Nano 11:2894–2904
213. Lin J, Pantelides ST, Zhou W (2015) Vacancy-induced formation and growth of inversion domains in transition-metal dichalcogenide monolayer. ACS Nano 9:5189–5197
214. Komsa HP, Kurasch S, Lehtinen O, Kaiser U (2013) From point to extended defects in two-dimensional MoS_2: evolution of atomic structure under electron Irradiation. Phys Rev B 88:35301
215. Lin J, Zuluaga S, Yu P, Liu Z (2017) Novel Pd_2Se_3 two dimensional phase driven by interlayer fusion in layered $PdSe_2$. Phys Rev Lett 119:16101
216. Li H, Tsai C, Koh AL, Cai L et al (2015) Activating and optimizing MoS_2 basal planes for hydrogen evolution through the formation of strained sulphur vacancies. Nat Mater 15:48
217. Stone AJ, Wales DJ (1986) Theoretical studies of icosahedral C60 and some related species. Chem Phys Lett 128:501–503
218. Lin YC, Bjorkman T, Komsa HP, Teng PY et al (2015) Three-fold rotational defects in two-dimensional transition metal dichalcogenides. Nat Commun 6:6736
219. Luo P, Zhuge FW, Zhang QF, Chen YQ et al (2019) Doping engineering and functionalization of two-dimensional metal chalcogenides. Nanoscale Horiz 4:26–51
220. Zhang XJ, Shao ZB, Zhang XH, He YY et al (2016) Surface charge transfer doping of low-dimensional nanostructures toward high-performance nanodevices. Adv Mater 28:10409–10442
221. Taghinejad H, Rehn DA, Muccianti C, Eftekhar AA et al. Defect-mediated alloying of monolayer transition-metal dichalcogenides. ACS Nano 12:12795–12804
222. Kutana A, Penev ES, Yakobson BI (2014) Engineering electronic properties of layered transition-metal dichalcogenide compounds through alloying. Nanoscale 6:5820–5825
223. Karthikeyan J, Komsa HP, Batzill M, Krasheninnikov AV (2019) Which transition metal atoms can be embedded into two-dimensional molybdenum dichalcogenides and add magnetism? Nano Lett 19:4581–4587
224. Li XF, Basile L, Yoon M, Ma C et al (2015) Revealing the preferred interlayer orientations and stackings of two-dimensional bilayer gallium selenide crystals. Angew Chem Int Ed 54:2712–2717
225. Huang B, Yoon M, Sumpter BG, Wei SH et al (2015) Alloyengineering of defect properties in semiconductors: suppression of deep levels in transition-metal dichalcogenides. Phys Rev Lett 115:126806

226. Li XF, Lin MW, Basile L, Hus SM et al (2016) Isoelectronic tungsten doping in monolayer $MoSe_2$ for carrier type modulation. Adv Mater 28:8240–8247
227. Jin Y, Keum DH, An SJ, Kim J et al (2015) A van der Waals homojunction: ideal p-n diode behavior in $MoSe_2$. Adv Mater 27:5534–5540
228. Duan XD, Wang C, Fan Z, Hao GL et al (2016) Synthesis of WS2xSe2–2x alloy nanosheets with composition-tunable electronic properties. Nano Lett 16:264–269
229. Li HL, Liu HJ, Zhou LW, Wu XP et al (2018) Strain-tuning atomic substitution in two-dimensional atomic crystals. ACS Nano 12:4853–4860
230. Komsa HP, Kotakoski J, Kurasch S, Lehtinen O et al (2012) Two-dimensional transition metal dichalcogenides under electron irradiation: defect production and doping. Phys Rev Lett 109:035503
231. Mahjouri-Samani M, Gresback R, Tian MK, Wang K et al (2014) Pulsed laser deposition of photoresponsive two-dimensional GaSe nanosheet networks. Adv Funct Mater 24:6365–6371
232. Zhang J, Jia S, Kholmanov I, Dong L et al (2017) Janus monolayer transition-metal dichalcogenides. ACS Nano 11:8192–8198
233. Li RP, Cheng YC, Huang W (2018) Recent progress of Janus 2D transition metal chalcogenides: from theory to experiments. Small 14:1802091
234. Kang J, Tongay S, Zhou J, Li JB (2013) Band offsets and heterostructures of two-dimensional semiconductors. Appl Phys Lett 102:012111
235. Lu AY, Zhu HY, Xiao J, Chuu CP et al (2017) Janus monolayers of transition metal dichalcogenides. Nat Nanotechnol 12:744–749
236. Lin YC, Liu CZ, Yu YL, Zarkadoula E et al (2020) Lowenergy implantation into transition-metal dichalcogenide monolayers to form Janus structures. ACS Nano 14:3896–3906
237. Han Y, Hu T, Li R, Zhou J (2015) Stabilities and electronic properties of monolayer MoS_2 with one or two sulfur line vacancy defects. Phys Chem Chem Phys 17:3813–3819
238. Van Der Zande AM, Huang PY, Chenet DA, Berkelbach TC et al (2013) Grains and grain boundaries in highly crystalline monolayer molybdenum disulphide. Nat Mater 12:554–561
239. Lu X, Utama MIB, Lin J, Gong X et al (2014) Large-area synthesis of monolayer and few-layer MoSe2 films on SiO2 substrates. Nano Lett 14:2419–2425
240. Lehtinen O, Komsa HP, Pulkin A, Whitwick MB et al (2015) Atomic scale microstructure and properties of se-deficient two-dimensional MoSe2. ACS Nano 9:3274–3283
241. Hong J, Wang C, Liu H, Ren X et al (2017) Inversion domain boundary induced stacking and band structure diversity in bilayer MoSe2. Nano Lett 17:6653–6660
242. Zou X, Liu Y, Yakobson BI (2013) Predicting dislocations and grain boundaries in two dimensional metal-disulfides from the first principles. Nano Lett 13:253–258
243. Zhou J, Liu F, Lin J, Huang X et al (2017) Large-area and high-quality 2D transition metal telluride. Adv Mater 29:1603471
244. Yazyev OV, Louie SG (2010) Electronic transport in polycrystalline graphene. Nat Mater 9:806–809
245. Lahiri J, Lin Y, Bozkurt P, Oleynik II et al (2010) An extended defect in graphene as a metallic wire. Nat Nanotechnol 5:326–329
246. Gunlycke D, White CT (2011) Graphene valley filter using a line defect. Phys Rev Lett 106:136806
247. Recher P, Trauzettel B (2011) Viewpoint: a defect controls transport in graphene. Physics 4:25
248. Clark KW, Zhang XG, Gu G, Park J et al (2014) Energy gap induced by Friedel oscillations manifested as transport asymmetry atmonolayer-bilayer graphene boundaries. Phys Rev X 4:011021
249. Azizi A, Zou X, Ercius P, Zhang Z et al (2014) Dislocation motion and grain boundary migration in two-dimensional tungsten disulphide. Nat Commun 5:4867
250. Lauritsen JV, Nyberg M, Vang RT, Bollinger MV et al (2003) Chemistry of one-dimensional metallic edge state sin MoS_2 nanoclusters. Nanotechnology 14:385–389
251. Li YF, Zhou Z, Zhang SB, Chen ZF (2008) MoS2 nanoribbons: high stability and unusual electronic and magnetic properties. J Am Chem Soc 130:16739–16744

252. Fei ZY, Palomaki T, Wu SF, Zhao WJ et al (2017) Edge conduction in monolayer WTe2. Nat Phys 13:677–682
253. Chakraborty C, Goodfellow KM, Dhara S, Yoshimura A et al (2017) Quantum-confined stark effect of individual defects in a van der Waals heterostructure. Nano Lett 17:2253–2258
254. Gutiérrez HR, Perea-López N, Elías AL, Berkdemir A et al (2013) Extraordinary room-temperature photoluminescence in triangular WS_2 monolayers. Nano Lett 13:3447–3454
255. Jaramillo TF, Jørgensen KP, Bonde J, Nielsen JH et al (2007) Identification of active edge sites for electrochemical H2 evolution from MoS2 nanocatalysts. Science 317:100–102
256. Ma T, Ren WC, Zhang XY, Liu ZB et al (2013) Edge-controlled growth and kinetics of single-crystal graphene domains by chemical vapor deposition. Proc Natl Acad Sci USA 110:20386–20391
257. Li XF, Dong JC, Idrobo JC, Puretzky AA (2017) Edge-controlled growth and etching of two-dimensional GaSe monolayers. J Am Chem Soc 139:482–491
258. Zhao XX, Fu DY, Ding ZJ, Zhang YY et al (2018) Mo-terminated edge reconstructions in nanoporous molybdenum disulfide film. Nano Lett 18:482–490
259. Sang XH, Li XF, Zhao W, Dong JC (2018) In situ edge engineering in two-dimensional transition metal dichalcogenides. Nat Commun 9:2051
260. Xia YP, Wang B, Zhang JQ, Jin YJ et al (2020) Quantum confined Tomonaga-Luttinger liquid in Mo_6Se_6 nanowires converted from an epitaxial $MoSe_2$ monolayer. Nano Lett 20:2094–2099
261. Venkataraman L, Hong YS, Kim P (2006) Electron transport in a multichannel one-dimensional conductor: molybdenum selenide nanowires. Phys Rev Lett 96:076601
262. Liu KH, Zhang LM, Cao T, Jin CH et al (2014) Evolution of interlayer coupling in twisted molybdenum disulfide bilayers. Nat Commun 5:4966
263. Puretzky AA, Liang LB, Li XF, Xiao K et al (2016) Twisted MoSe2 bilayers with variable local stacking and interlayer coupling revealed by low-frequency Raman spectroscopy. ACS Nano 10:2736–2744
264. Wang K, Huang B, Tian MK, Ceballos F et al (2016) Interlayer coupling in twisted WSe_2/WS_2 bilayer heterostructures revealed by optical spectroscopy. ACS Nano 10:6612–6622
265. Wang SS, Robertson A, Warner JH (2018) Atomic structure of defects and dopants in 2D layered transition metal dichalcogenides. Chem Soc Rev 47:6764–6794
266. Li XF, Lin MW, Lin JH, Huang B et al (2016) Two dimensional GaSe/MoSe2 misfit bilayer heterojunctions by van der Waals epitaxy. Sci Adv 2:1501882
267. Meyer JC, Geim AK, Katsnelson MI, Novoselov KS (2007) The structure of suspended graphene sheets. Nature 446:60
268. Luo S, Hao G, Fan Y, Kou L et al (2015) Formation of ripples in atomically thin MoS2 and local strain engineering of electrostatic properties. Nanotechnology 26:105705
269. Brivio J, Alexander DTL, Kis A (2011) Ripples and layers in ultrathin MoS2 membranes. Nano Lett 11:5148–5153
270. Liu H, Chi D (2015) Dispersive growth and laser-induced rippling of large-area single layer MoS2 nanosheets by CVD on c-plane sapphire substrate. Sci Rep 5:11756
271. Conley HJ, Wang B, Ziegler JI, Haglund RF et al (2013) Bandgap engineering of strained monolayer and bilayer MoS2. Nano Lett 13:3626–3630
272. Miro P, Ghorbani-Asl M, Heine T (2013) Spontaneous ripple formation in MoS2 monolayers: electronic structure and transport effects. Adv Mater 25:5473–5
273. Najmaei S, Amani M, Chin ML, Liu Z et al (2014) Electrical transport properties of polycrystalline monolayer molybdenum disulfide. ACS Nano 8:7930–7937
274. Zou XL, Liu YY, Yakobson BI (2013) Predicting dislocations and grain boundaries in two-dimensional metal-disulfides from the first principles. Nano Lett 13:253–258
275. Liu HJ, Jiao L, Yang F, Cai Y et al (2014) Dense network of one-dimensional midgap metallic modes in monolayer $MoSe_2$ and their spatial undulations. Phys Rev Lett 113:066105
276. Zou XL, Yakobson BI (2015) Metallic high-angle grain boundaries in monolayer polycrystalline WS2. Small 11:4503–4507
277. Gibertini M, Pizzi G, Marzari N (2014) Engineering polar discontinuities in honeycomb lattices. Nat Commun 5:5157

278. Sangwan VK, Jariwala D, Kim IS, Chen KS et al (2015) Gate-tunable memristive phenomena mediated by grain boundaries in single-layer MoS2. Nat Nanotech 10:403–406
279. Zhang ZH, Zou XL, Crespi VH, Yakobson BI (2013) Intrinsic magnetism of grain boundaries in two-dimensional metal dichalcogenides. ACS Nano 7:10475–10481
280. Zou XL, Liu M, Shi Z, Yakobson BI (2015) Environment-controlled dislocation migration and super plasticity in monolayer MoS2. Nano Lett 15:3495–3500
281. Gibb AL, Alem N, Chen JH, Erickson KJ et al (2013) Atomic resolution imaging of grain boundary defects in monolayer chemical vapor deposition-grown hexagonal boron nitride. J Am Chem Soc 135:6758–6761
282. Cretu O, Lin YC, Suenaga K (2014) Evidence for active atomic defects in monolayer hexagonal boron nitride: a new mechanism of plasticity in two-dimensional materials. Nano Lett 14:1064
283. Butz B, Dolle C, Niekiel F, Weber K et al (2014) Dislocations in bilayer graphene. Nature 505:533–537
284. Alden JS, Tsen AW, Huang PY, Hovden R et al (2013) Strain solitons and topological defects in bilayer graphene. Proc Natl Acad Sci USA 110:11256–11260
285. Lin JH, Fang WJ, Zhou W, Lupini AR et al (2013) AC/AB stacking boundaries in bilayer graphene. Nano Lett 13:3262–3268
286. Vaezi A, Liang YF, Ngai DH, Yang L et al (2013) Topological edge states at a tilt boundary in gated multilayer graphene. Phys Rev X 3:021018
287. Zhang F, MacDonald AH, Mele EJ (2013) Valley Chern numbers and boundary modes in gapped bilayer graphene. Proc Natl Acad Sci USA 110:10546

Chapter 4
2D TMDs Properties

4.1 Introduction

To deduce another alternative to 2D graphene scientific community has explored other ultrathin-layered materials with versatile properties. That could have a 2D atomic structure with potential novel properties, the transition metal dichalcogenides (TMDs) have been recognized as one of the potential 2D materials, which gained extensive attention due to their useful physicochemical properties and have a broad potential application [1]. Likewise, in graphite, the layered TMDs can be also thinned to single or few-layered nanosheets with distinct emergent physical characteristics different from those of their bulk counterparts. To date, a verity of methods has been executed to obtain 2D TMD films, including alkali metal intercalation, mechanical cleavage exfoliation, chemical vapor deposition, and wet-chemical synthetic method [2–6]. Therefore, herein one has discussed the various potential physical properties of the TMD materials.

4.1.1 *Mechanical Properties*

Mechanical properties are the key aspect that people pay attention to explore a new material. Predominantly, stone, bronze, and iron are the three milestone materials that appeared synchronically in ancient times. Stone is rigid, but brittle and difficult to processing, while bronze and iron are stronger, tougher, and easier to shape. Their utilizations have greatly changed human history. Since increasingly fast discoveries of new functional materials are made in modern times, understanding of such materials physical properties considered to be a significant parameter. There are significant criteria for seeking super-strong or super-rigid materials for applications ranging from people's daily life to space exploration [7–10]. Concerning semiconductors, are indispensable complements to the electrical and optical properties of materials [11,

A. K. Singh, *2D Transition-Metal Dichalcogenides (TMDs): Fundamentals and Application*, Materials Horizons: From Nature to Nanomaterials,
https://doi.org/10.1007/978-981-96-0247-6_4

12]. Mechanical properties are also play vital roles in the designs of flexible, stretchable, and epidermal electronics which potentially dominates the future electronics industry [13–15].

Two-dimensional (2D) materials electrons and phonons are limited in a planar dimension; therefore, their several properties may deviate from their 3D counterparts [16]. Like graphene and transition metal dichalcogenides (TMDs); such as molybdenum disulfide (MoS_2) in which monolayer limit the features of massless Dirac fermions and an intrinsic direct band gap [17–19], with the various differ optical and electrical properties [20]. But an important question remains, whether the mechanical properties of 2D materials become also different in the monolayer limit. Likely to 3D materials the elastic modulus E and Poisson's ratio can be also used to determine the elastic properties of 2D materials. Whereas the E is also called Young's modulus under the applied strain uniaxially and express in term of stress and strain. Poisson's ratio is defined as the negative ratio of transverse to axial strain under uniaxial stress. Additionally, the significant tensile strength parameter describes the maximum tension for a material. However, these key parameters have to be renormalized by the planar elastic energy that leads to units of J/m^2 or N/m. Although 2D modulus and strength are also considered to be more suitable to describe 2D materials: specifically for comparison between 2D and 3D materials. The 2D parameters can be also converted into 3D ones by dividing the 2D values to the thickness of the 2D materials. Therefore, using these concepts the conventional elastic theory can be readily applied to 2D systems. Though, several new aspects of mechanical properties in 2D systems different from 3D systems have been also demonstrated [20, 21].

Considering the various concepts of 2D materials a valuable theoretical DFT approximation of the mechanical properties of monolayer TMDs can be described in terms of mechanical bending [22]. The typical nanoribbons structure is depicted in Fig. 4.1a–c. The bending stiffness of several TMDs is tabulated in Table 4.1. According to this data, in general bending stiffness rises since S to Se to Te (see Table 4.1) reflecting a toughening of the nanoribbons since S to Se to Te. The d^0 complexes, particularly S and Se, together to $PdTe_2$ which has a inferior (< 3 eV) bending stiffness. The occurrence of such inferior fluctuating inelasticity in these complexes results enormous changes in the indigenous tension including the carrier concentration picture underneath machine-derived twisting. However, the 1H complexes have an advanced twisting rigidity with a sophisticated flexural inflexibility in contradiction of mechanical twisting. The projected twisting toughness value for MoS_2 (12.29 eV) consists of the reported experimental values $6.62-13.24$ eV and $10-16$ eV [23, 24]. Therefore, exploring the trend of mechanical strengths concerning transition metals filling of the d-band valence electrons property is important. The occupation of the d band upsurges to the transition metal group IV ($\sim$ sparsely-filled) to VI ($\sim$ half-filled) to X ($\sim$ filled) in the identical row in the periodic table. In which both numbers Y_{2D} and S_b are increasing with the increasing amount of valence d electrons up to the shell become approximately half-filled. As the in-plane toughness and twisting toughness of 1H-NbS_2 and 1H-TaS_2 of the group V (d^1) transition metals. The in-plane toughness of NbS_2 and TaS_2 have been achieved at 95.74 and 115.04 N/m, correspondingly. The corresponding twisting toughness is around

4.87 and 6.43 eV, correspondingly (for NbS_2 and TaS_2). Hence TMDs (TX_2) have an equal chalcogen atom possess a trend $d^0 < d^1 < d^2$ in their stiffness. On the other hand, a decrement trend in both Y_{2D} and Sb can be seen whereas moving to half-filled (d^2) with approximately filled (d^6) for the d-band transition metal. Furthermore, the greater twisting toughness of group VI complexes is decreasing due to changing the phase since 1H to distorted 1T phase, as the example 1H to 1T' alteration of $MoTe_2$.

Mathematically, it can be expressed from the following relationship

$$t_{\text{eff}} = \sqrt{12S_b/Y_{2D}} \tag{4.1}$$

and

$$Y_{3D} = Y_{2D}/t_{\text{eff}} \tag{4.2}$$

The effective thickness and 3D Young's modulus is usually estimated by the effective thickness combination $d_{X\text{-}X}$ distance and the whole active deterioration extent of electron concentration into the vacuum. However, experimentally it is challenging

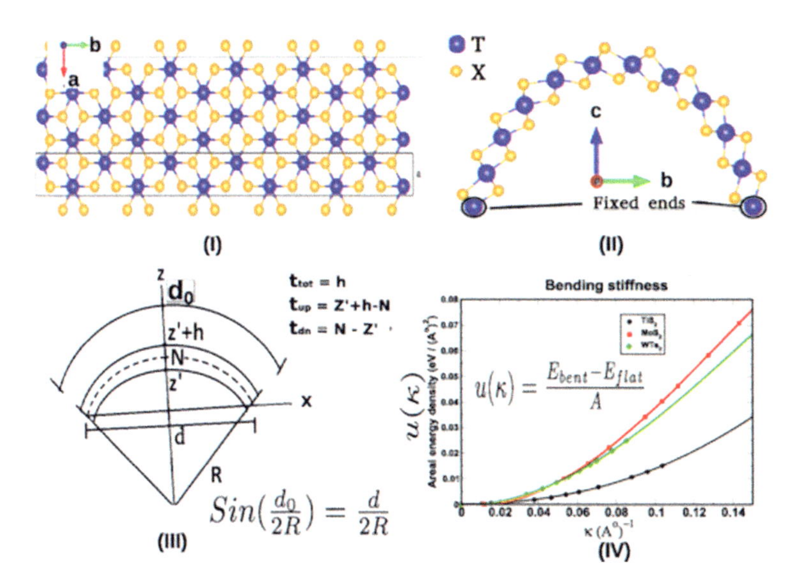

Fig. 4.1 **a** Illustrate a nanoribbon (enclosed by a rectangle) that is taken to simulate an extended sheet of 1T monolayer; **b** the lattice constant with the ribbon extended along the *a* axis and a vacuum to 20 Å inserted along *b* and *c* axes and bent it for 1T nanoribbon; **c** schematic for the bending a thin plate, where the d_0, d, and R are the lengths of a thin plate before bending, length after bending and radius of curvature. N is the neutral surface denoted by a dashed line, t_{tot}, t_{up}, and t_{dn} are the physical thicknesses of the bent nanoribbon, assuming to middle layer coincides with the neutral surface (*N*); **d** areal bending energy density vs bending curvature curve to determine the bending stiffness, where E_{bent}, E_{flat}, *A* are the total energy of bent nanoribbon, the total energy of flat nanoribbon and cross-sectional area of flat nanoribbon (length × width) (adopted from Nepal et al. [22])

Table 4.1 The ground state parameters for the TMD mono-layers possessing 1H or 1T phase: relaxed in-plane lattice constant, a; Metal-chalcogen and - chalcogen distance, d_{T-X}, and d_{X-X}, X-T-X angle, θ_{X-T-X}; cohesive energy per atom, E_c; in-plane (ν_{in}) and out-of-plane (ν_{out}) Poisson's ratios; 2D Young's modulus, Y_{2D}; Bending stiffness, S_b; effective thickness, t_{eff}

T^{4+}	TMDs	a (Å)	$d_{T\text{-}X}$ (Å)	$d_{X\text{-}X}$ (Å)	$\theta_{X\text{-}T\text{-}X}$ degree	E_c (eV/atom)	ν_{in}	ν_{out}	Y_{2D} (N/m)	S_b (eV)	t_{eff} (Å)	Y_{3D} (Y_{2D}/t_{eff}) (GPa)
d^0	TiS_2	3.42	2.42	2.80	90.16	6.80	0.17	0.42	85.20	2.25	2.25	378.67
	$TiSe_2$	3.55	2.55	3.04	91.76	6.17	0.23	0.43	59.74	2.86	3.03	197.72
	$TiTe_2$	3.76	2.77	3.44	94.55	5.41	0.24	0.38	44.46	3.29	3.77	117.93
	ZrS_2	3.67	2.57	2.87	88.14	7.35	0.19	0.52	83.76	2.13	2.21	379.00
	$ZrSe_2$	3.81	2.70	3.12	90.14	6.71	0.22	0.47	71.30	2.57	2.63	271.10
	$ZrTe_2$	4.01	2.91	3.53	92.94	5.89	0.18	0.44	43.16	3.01	3.66	117.92
	HfS_2	3.62	2.53	2.85	88.65	7.35	0.19	0.52	85.78	2.82	2.51	341.75
	$HfSe_2$	3.75	2.66	3.09	90.37	6.67	0.21	0.47	77.75	3.64	3.00	259.17
	$HfTe_2$	3.98	2.88	3.47	92.58	5.80	0.15	0.41	46.77	3.92	4.01	116.63
d^2	MoS_2	3.17	2.40	3.10	80.56	7.86	0.26	0.30	141.59	12.29	4.08	347.03
	$MoSe_2$	3.30	2.53	3.31	81.86	7.22	0.26	0.32	114.97	14.60	4.94	232.73
	MoTe1H	3.51	2.71	3.59	83.04	6.54	0.28	0.34	87.88	14.63	14.63	155.54
	$MoTe_2$-1T′	3.65	–	–	–	6.54	0.28	0.46	61.85	7.28	4.75	130.21
	WS_2	3.16	2.40	3.10	80.25	7.91	0.26	0.33	143.92	12.61	4.10	351.02
	WSe_2	3.29	2.53	3.32	82.16	7.20	0.33	0.35	130.03	14.48	4.62	281.45
	WTe_2-1T′	3.61	–	–	–	6.49	0.35	0.60	86.79	8.96	4.45	195.03
d^6	$PdTe_2$	3.96	2.67	2.73	83.91	4.07	0.32	0.64	61.82	2.78	2.94	210.27

(continued)

Table 4.1 (continued)

T^{4+}	TMDs	a (Å)	$d_{T\text{-}X}$ (Å)	$d_{X\text{-}X}$ (Å)	$\theta_{X\text{-}T\text{-}X}$ degree	E_c (eV/atom)	ν_{in}	ν_{out}	Y_{2D} (N/m)	S_b (eV)	t_{eff} (Å)	Y_{3D} (Y_{2D}/t_{eff}) (GPa)
	PtS_2	3.52	2.37	2.45	84.25	5.73	0.29	0.58	105.81	5.66	3.20	330.65
	$PtSe_2$	3.68	2.49	2.60	84.83	5.32	0.26	0.59	87.01	6.33	3.74	232.65
	$PtTe_2$	3.95	2.66	2.74	84.15	5.07	0.26	0.57	81.41	4.58	3.29	247.45

The representations of T^{4+} such as d^0, d^2 and d^6 [22]

to describe the entire active deterioration extent of the electronic carrier spreading. So, in practice taken a range from the $d_{X\text{-}X}$ distance to the interlayer metal-metal distance within the bulk structure for the real width, that can give us a range of the bending stiffness and in-plane stiffness [23, 24]. By using Eq. (4.1), one can evaluate a practical worth of the real width for a verity of TMDs. Though, the calculated real width t_{eff} for a few compositions TX_2($T = $ Ti, Zr, Hf; $X = $ S, Se) have been also achieved less than their d_{X-X} distance, as depicted in Fig. 4.2. It could be correlated to the statement the bending is much easier than stretching. A similar outcome has been also demonstrated for the real width of the carbon monolayer using the several approaches [25–28]. Moreover, using Eq. (4.1) investigators have also estimated the effective thickness of the carbon monolayer around 0.7–0.9 Å [25, 26, 29]. This is much less than the normal spacing (3.4 Å) between sheets of graphite. Thus such strong relevance indicates the possible breakdown of Eq. (4.1) to estimate the effective thickness in case of the atomically thin carbon layer [26]. Additionally, 3D Young's modulus also allows us to estimate comparative the strength of various 2D and 3D materials by using Eq. (4.2), like MoS_2 in contradiction of steel. Likely to 2D in-plane toughness, the 3D Young's modulus of TMD monolayers declines since S to Se to Te. The greater exploration of the real width has provided a vast over estimation of the 3D Young's modulus for the group IV complexes with sulfur considered to be chalcogen atom. As the MoS_2 and WS_2 have a large 3D Young's moduli 347.03 and 351.02 GPa, it agrees with the investigational value of 270 ± 100 GPa for the MoS_2 TMD [30].

Semiconductor industries have also focused attention on experiential research of TMDs and examined several potential compositions, such as MoS_2, WS_2, and WSe_2. Key attention to the TMDs owing to their bandgap in which crossover from indirect bandgap to direct bandgap for the bulk and monolayer [31]. Mechanical properties of TMD layers first were studied to deduce less defectiveness by exfoliating the flakes of MoS_2. In a similar study, AFM nanoindentation was used to measure the mechanical properties of exfoliated mono- and bilayer MoS_2 [30]. These findings lead to the in-plane 2D modulus of monolayer MoS_2 being 180 ± 60 N/m (270 ± 100 GPa) and their average breaking strength around 15 ± 3 N/m (23 GPa), this is several times lower than monolayer graphene but still much stronger than steel. Their prevention generally lies in-between 0.02 and 0.1 N/m. Moreover, the bilayer MoS_2 have a 2D modulus of 260 ± 70 N/m, corresponding to a lower 3D modulus of 200 ± 60 GPa, due to defects or interlayer sliding [30]. In this order, the elastic properties of suspended thick MoS_2 sheets with 5–25 layers have been also explored [32]. By considering the bending rigidity with the addition of one more linear term, (where t and r are the thickness and the radius of membrane), therefore following Eq. (4.3) can be expressed as,

$$F = \left(\sigma_0^{2D}\pi\right)\delta + \left(E^{2D}\frac{q^3}{r^2}\right)\delta^3 \tag{4.3}$$

Here F is the applied point force, δ is the indentation depth at the center, r is the hole radius, $q = 1/\left(1.05 - 0.15\nu - 0.16\nu^2\right)$ is a dimensionless constant determined by

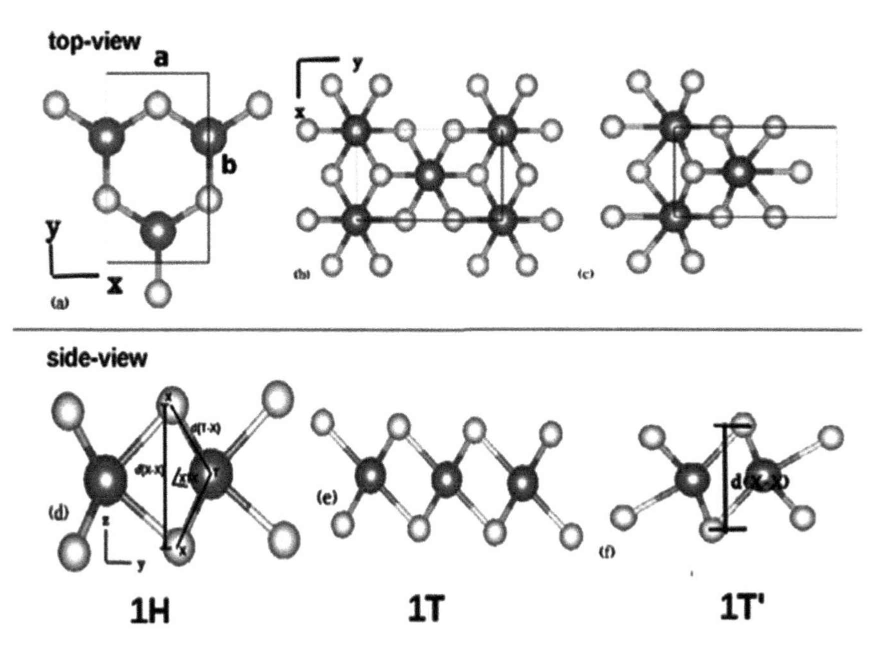

Fig. 4.2 Rectangular unit-cells of types 1H, 1T, and 1T of WTe_2; the first row of the top view (**a**–**c**) and second one (**d**–**f**) corresponds to the side view; $d(T - x)$ is metal chalcogen distance, $\angle xTx$ is an angle made by two $d(T - x)$ sides and $d(X - X)$ (or $d\ X - X$) is the distance between the outer and inner layer of flat monolayer bulk TMDs (adopted from Nepal et al. [22]).

the Poisson's ratio of the membrane, where E^{2D} and are the 2D modulus and the 2D pretension, the typical schematic of the mechanical properties probing 2D materials by AFM nanoindentation is illustrated in Fig. 4.3.

Besides the additional term, the modulus and the pretension can be similarly derived by fitting the experimental force-displacement curves. This leads to the mean Young's modulus is 330 ± 70 GPa., and prevention of 0.13 ± 0.10 N/m [32]. Both values have been found higher than the previously demonstrated outcomes [33].

Another key issue with the mechanical properties of TMDs whether fabrication techniques impact them, such as CVD samples having similar mechanical properties as exfoliated samples. Although CVD is the widely applied approach to grow large-area materials cost-effectively. Nonetheless CVD process-grown samples usually

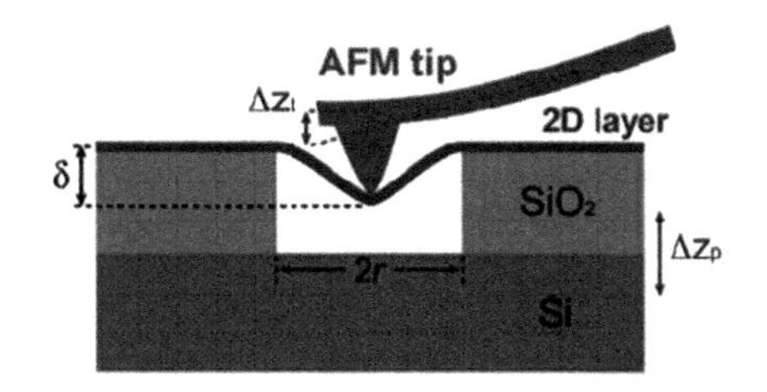

Fig. 4.3 Illustration for the probing mechanical properties of 2d materials using afm nanoindentation (adopted from Liu [20])

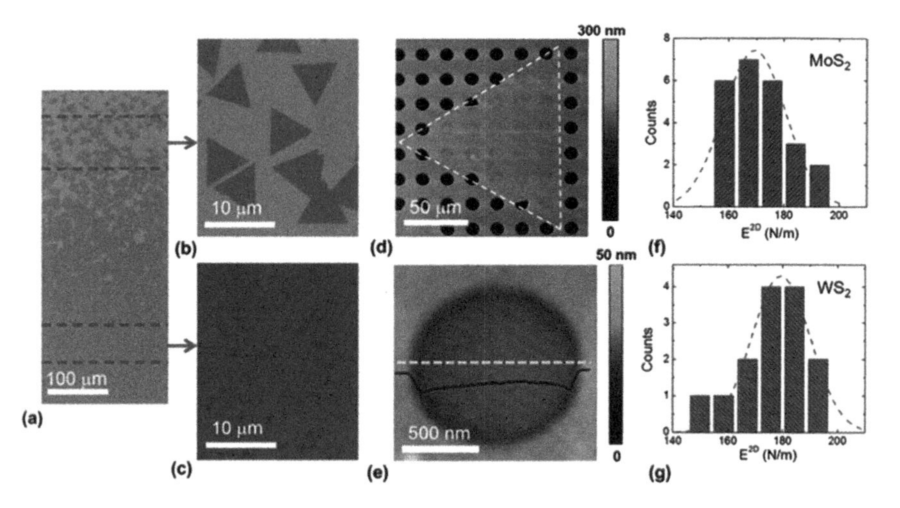

Fig. 4.4 Results for the CVD-grown MoS_2 and WS_2 monolayers: **a–c** as-grown CVD MoS_2 that consists of the isolated triangle and continuous film; **d** AFM image to entire triangle transferred onto a holey substrate; **e** AFM image of a suspended membrane over a hole; **f, g** histograms of E^{2D} for MoS_2 and WS_2 including corresponding Gaussian distribution (adopted from Liu [20])

have more defective compared to single crystals. Besides this drawback, people have developed a CVD process to synthesize large areas as well as wafer-scale, such as MoS_2 [4, 34–38]. The as-grown specimen can possess isolated triangles of monolayer crystals at different positions on the substrate slightly far away from the precursors, and it merged into continuous film closer to the precursors (Fig. 4.4a, c). Moreover, the transferred CVD MoS_2 and WS_2 with a similar PDMS stamping process onto holey substrates has been also achieved (see Fig. 4.4d, f) [39]. AFM nanoindentation probe of modulus monolayer CVD MoS_2 and WS_2 findings have revealed 171 ± 11 (264 GPa) and 177 ± 12 N/m (272 GPa), respectively (see Fig. 4.4f, g). Almost similar values can be also achieved for the MoS_2 and WS_2 that originates due to their similar lattice constants and bonding energies. Thus the mechanical properties of TMDs are an important parameter and it depends on various factors [20, 21, 38].

4.1.2 Strain Engineering

Ultra-strength materials stresses can be up to 1/10 of their ideal strength (the highest achievable stress of a defect-free crystal at zero temperature), by mean “Smaller is Stronger” [40]. 2D materials atomically thick behavior materials have non-hydrostatic stresses ability up to their ideal strength without inelastic relaxation. Like, a large elastic strain-of up to 30% could be realized in ideal single-crystal graphene [41, 42]. Similarly, MoS_2 has been also used as an ideal 2D semiconductor under the uniaxial or biaxial strain possessing a tunable bandgap [43], their phase change could excite under a sufficiently large strain [44]. Likewise, $MoTe_2$ only < 1% strain triggers phase transition from semiconducting to metallic phase [44, 45]. However, other TMDs significantly require a large biaxial or uniaxial strain to induce the phase transition. Similarly, change in lattice structure also leads to variation in other physical properties of 2D materials. Such as thermal conductivity is usually related to the phonon vibration, it is usually tuned via simply mechanical strain. In general, any kind of straining leading a decrease in thermal conductivity due to the softening of the phonon models [46]. Certain 2D materials are also possessed piezo-electricity properties (such as h-BN, MoS_2), these materials' polarization are closely linked to the lattice symmetry. Similarly, the chemical reactivity also influenced by the existing strain in 2D materials. As an intense, monolayer MoS_2 under strain is a promising catalyst for hydrogen evolution reaction (HER). Nevertheless as per experimental strain engineering at the low strain stage it is mostly under 5%, which is much smaller compared to the ideal maximum strain (~ 20–30%). Therefore, there is a large gap between the ideal strain and the achieved strain in 2D materials [47].

4.1.2.1 Theoretical Prediction

Since a vast number of the TMDs phase alteration among 2H and 1T′ that triggers underneath together the uniaxial and isotropic strains, excluding the WTe_2, because it is triggered only under compression [44]. Additionally, usually the phase transition process occurs suddenly, by mean it is an irregular jump among the *a*, and *b*-axes, the corresponding space is generally recognized while the changeover procedure, as illustrated in Fig 4.5a. Under precise control of the employed stress, a mix phase point of steady 2H and 1T′ is also obtainable. Such stress is produced by the approaches alike flexible substrate or atomic force microscopy (AFM) tips asserting on a suspended sample (see Fig. 4.5b). This approach directly provides a path about the possibility to construct a homojunction via in situ phase conversion d. Although numerous 2D TMDs tension needs to activate the phase conversion considerably bigger, such as the 15 Nm^{-1} for $2H - MoS_2$, this is nearby the breaking threshold of 2D TMDs. Compare to this, $MoTe_2$ requires very less stress (6.9 Nm^{-1}) owing to its narrow energy gap among the 2H and 1T′ phases, in which lonely a modest (< 3%) uniaxial stress is desired for the phase alteration. Moreover, the AFM tip has been also used to get a localized strain into 2D TMDs films. This allowing to location of the phase transition are precisely designed (see Fig. 4.5c) [48]. Since the thermodynamic issue is also one of the key factor in this kind of phase transition, therefore, it should be considered and estimate for the conversion since 2H to 1T′ phase around ≈50 s in $MoTe_2$, the 1T′ phase transits to 2H once more in numerous tens of minutes afterward eliminating the exterior strength. Surprisingly, the theoretical examination revealed that the time period can be smaller to 1T′–2H conversion under the existence of the contiguous 2H domain or applied 2H preferring stresses.

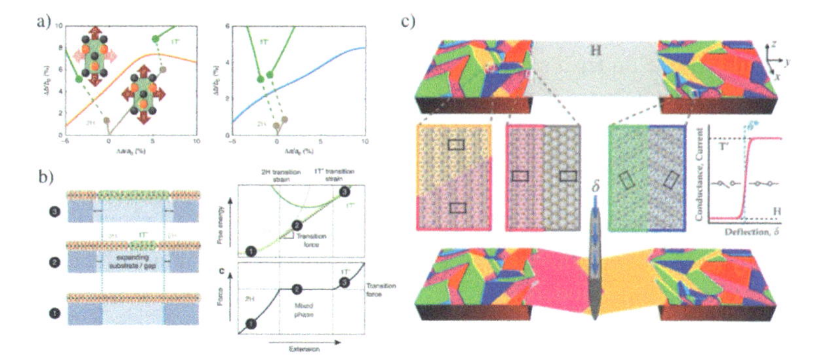

Fig. 4.5 Theoretical prediction of strain-induced phase transition; **a** Left panel was used for the Perdew–Burke–Ernzerhof functional to calculate the crystal energy, whereas the HSE06 functional used in the right panel from starting to their stress-free 2H equilibrium values a_0 and b_0, the lattice constants a and b evolve in response to progressive application of a uniaxial-load (F_y) or a "hydrostatic" isotropic tension (6); **b** tensile mechanical deformation induces phase transition (**a** and **b** adopted from Duerloo et al. [44]); **c** schematic for the localized reversible strain-induced transformation from H to T′ in a partially supported TMD monolayer. Pre-existing T′ domain patterns (top) are bridged by indentation-induced T′ domains (bottom). Examples of three interface structures determined from density functional theory (DFT) and a schematic of the expected conductance or current versus indenter deflection behavior at fixed voltage are also included (adopted from Berry et al. [48])

4.1.2.2 Strain by AFM Tips

Collectively through resonating substrate and AFM tip, a governable strain is usually indigenously applied on a 2D TMD film. Upto definite strain standards the different Raman signals by diverse phases of the $MoTe_2$ are willingly collected, as depicted in Fig. 4.6a. Therefore, the strain is required to trigger the phase transition changes with temperature. From the experimental point of view, the value of the desired strain monotonically decreased with the increasing temperature, this finding has been considered a direct evidence of the combined effect of strain and temperature [49]. Additionally, the temperature is also influenced other kinds of stimuli, such as electric field, chemical doping, and so on with a similar effect. A homogeneous and reversible phase transition could be also realized under room temperature at a tensile strain of 0.2%.

4.1.2.3 Strain by the Substrates

As discussed the AFM tip provides a precise and local strain in TMD film, but it is still an inefficient strategy and also has an inadequate utility in technical area. To get a governable strain at a greater scale the pre-patterned substrate is considered to be a superior option [50, 51]. Like, the construction designs of cones and pyramids are usually invented over a sapphire substrate for the advancement, to achieve

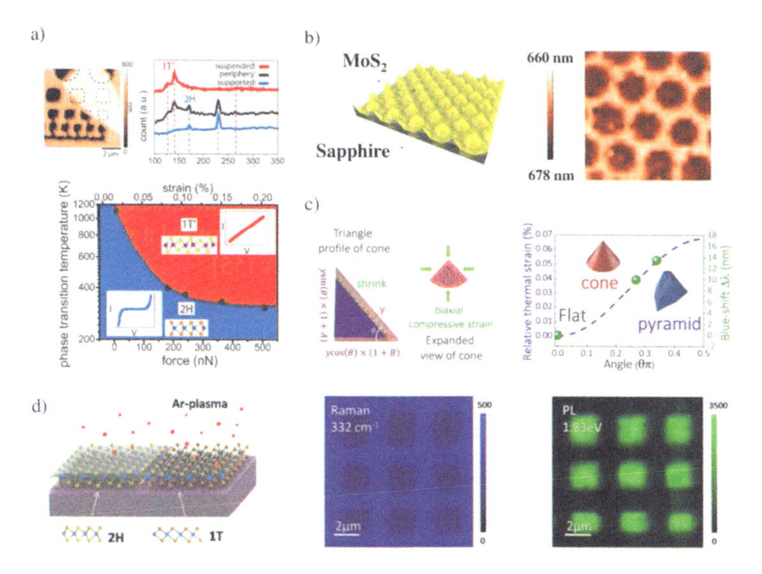

Fig. 4.6 Different methods to introduce strain: **a** AFM image for 2H $MoTe_2$ flake deposited on a hollow substrate and the corresponding Raman spectra at different sites. The bottom image corresponds to the relation between external force and phase transition temperature (adopted from Song et al. [49]); **b** Schematic of MoS_2 on pattern substrate and uniform PL intensity of MoS_2 on cones; **c** Model to explain thermal strain formed on different environmental temperatures and the relation between relative thermal strain and angle of cones (**b** and **c** adopted from Wang [50]); **d** Schematic of Ar plasma–induced phase transition from 2H to 1T in MoS_2 and Raman/PL of patterned mixed phase MoS_2 treated with Ar plasma (adopted from Zhu [54])

it 2 nm MoO_3 precursors deposited followed the subsequent sulfurization process to grow epitaxial bilayer MoS_2 on patterned substrates (see Fig. 4.6b). Raman and PL measurements revealed the uniform peak shifts on cones and pyramids due to a uniform induced strain in such types of substrates. Likewise, various simulation results it has also directed the strain style can be modified by ductile to compressive via a thermal expansion coefficient value of substrate, which would be defined to precise angle among the flat and hypotenuse flanks of the pattern surface (see Fig. 4.6c). In the different words, a precise strain is possible by distinct substrate designs.

4.1.2.4 Strain Due to Physical Collision

Besides the comparatively multifaceted substrate strategy, the physical crash can also reduce the target substantial strain to persuade phase changeover [52, 53]. Like, a superficial Ar plasma technique has been used to activate the phase alteration among 2H–1T′ in monolayer MoS_2 [54]. The key view of this approach is based on uses of low energy Ar plasma for the bombarding 2H-MoS_2, that impact the 2H phase in dual behavior: to induce the S vacancies that facilitate descending of the S layer. The additions of pattern mask the 2H–1T′ hybrid structures' specific patterns allowing to fabricate at a large scale (see Fig. 4.6c). Moreover, the benefit of the Ar plasma is none of heteroatom contamination incorporated while the phase conversion process. It has been also noted that at 40% of the 2H phase transformed into 1T′ phase by 40s small energy plasma action. However, the high energy plasma or prolong time action may reduce the metallic performance of the 1T′ phase. The transformed 1T′ phase is significantly steady in a suitable condition besides afterward a week exposure; it is owing to the steadying consequence of S vacancies [55].

4.2 Thermal Properties

2D heterostructure structure thermal transport is one of the fundamental issues for their physics and engineering applications. The heat dissipation due to charge carrier scattering in nanostructures-based electronic and photoelectric devices. The generation of waste Joule heat often leads to the formation of high-temperature hot spots in devices [56]. Ultimately it affects the stability and performance of devices. Therefore, the choice of 2D building blocks with high thermal conductivity alleviates Joule heat dissipation. Whereas, the low thermal conductivity of the materials is an essential requirement to fabricate thermoelectric devices to convert waste heat into electrical energy. The 2D heterostructures involvement causes an increase in the Seebeck coefficient and a lowering in thermal conductivity depending on carrier energy filtering or quantum confinement effects. The heat dissipation and thermoelectric generator comparison are also due to the design of functional thermal devices such as rectifiers, transistors, and logic gates to handle heat signals that are required in controllable thermal transport. In general, the structural diversities combined with structural defect and external stress fields realize rich diversity of thermal properties in 2D heterostructures, which makes them enable the different thermal fields. Further, the interfaces of heterostructures make the thermal transport mechanisms differ in single 2D materials and conventional bulk materials, either through inducing a contact thermal resistance or new phonon modes [57].

Therefore, thermal assessment is the significant factor in the strategy and utility of electronic devices. It is considered to be a restrictive feature for device fabrication [58–64]. An atomic width monolayer thermal assessment is extremely crucial in TMDs for the electronics. Whereas the sufficiently localized Joule heating in extremely-thin narrow space can effortlessly produce "hot spots" [65, 66]. Although previously it has been demonstrated experimentally and theoretically the thermal conductivity of monolayer TMDs can be 2–3 orders inferior to graphene. The discussed twice limits of the 2D materials causes of the critical blockage for the effective thermal assessment of TMD-based devices. Moreover, the electron scattering at an interface, phonon scattering at an interface are also playing a crucial part in performance of nanoscale devices, it is considered to be extra vital than the material himself [71–73]. Concerning monolayer TMDs inferior thermal conductivity and thermal conductance at the interface considerably impact the functionality and consistency of the devices.

4.2.1 Phonon Dispersion in TMDs

Thermal properties of TMDs are generally described in terms of atomic vibrations (i.e., phonon energy). Therefore, it is customary to define the lattice vibrational modes (phonons) of the materials. As the distinctive instance, MoS_2 is a layered dichalcogenide, in which every MoS_2 unit cell contains 3 (1 Mo and 2 S) atoms to form 3 acoustic and 6 optical phonon modes, as illustrated in Fig. 4.7a [74]. The three acoustic phonon modes are belonging to longitudinal acoustic mode, transverse acoustic mode, and flexural acoustic mode. The longitudinal acoustic mode can be connected to compressional waves and the corresponding atomic movements toward the wave propagation direction. The transverse acoustic mode also correlates to shear waves corresponding to the in-plane movements vertical to the propagation direction. While the flexural acoustic mode belongs to the out-of-plane atomic displacements. Thus a phonon dispersion has been described from a critical correlation among the phonon wave vector q and phonon energy E or frequency ω ($E = \hbar\omega$, where $\hbar$ is the Planck constant). The typical phonon dispersion and density of states of $MoSe_2$ and WS_2 are depicted in Fig. 4.7b, c. Where lower frequency acoustic phonon branches of MoS_2 up to 233 cm^{-1} due to $Mo(xy)$, $S(xy)$, and $Mo(z)$ vibrations of identical masses of the Mo and 2S. On the other hand, WS_2 up to 182.3 cm^{-1} due to $W(xy)$ and $W(z)$ vibrations cause W atoms abundant heavier masses [74].

The short wave vector q lies nearby the midpoint of the Brillouin zone, at which the frequency of the longitudinal acoustic and transverse acoustic modes have a rectilinear dispersion [75–78]. The corresponding longitudinal acoustic mode group velocity could be several times (around five times) lower than the transverse acoustic mode in graphene [79]. Whereas, the flexural acoustic mode has a nearly quadratic dispersion [80, 81]. The corresponding phonon group velocity is commonly expressed to the relationship $= d\omega/dq$, the typical plot between q and ω as depicted in Fig. 4.8. According to demonstration longitudinal acoustic and transverse acoustic modes group velocities are drastically decreasing with the increasing phonon frequency, whereas the flexural acoustic mode group velocity in growing order with the increasing frequency up to the maxima value at the central point of the plot and finally it decreased up to zero at the edge of the zone [79].

4.2.2 Phonon Relaxation and Mean Free Path

Phonon relaxation time is usually connected to their traveling through the materials because a kind of scattering process occurs, like phonon–phonon umklapp scattering,

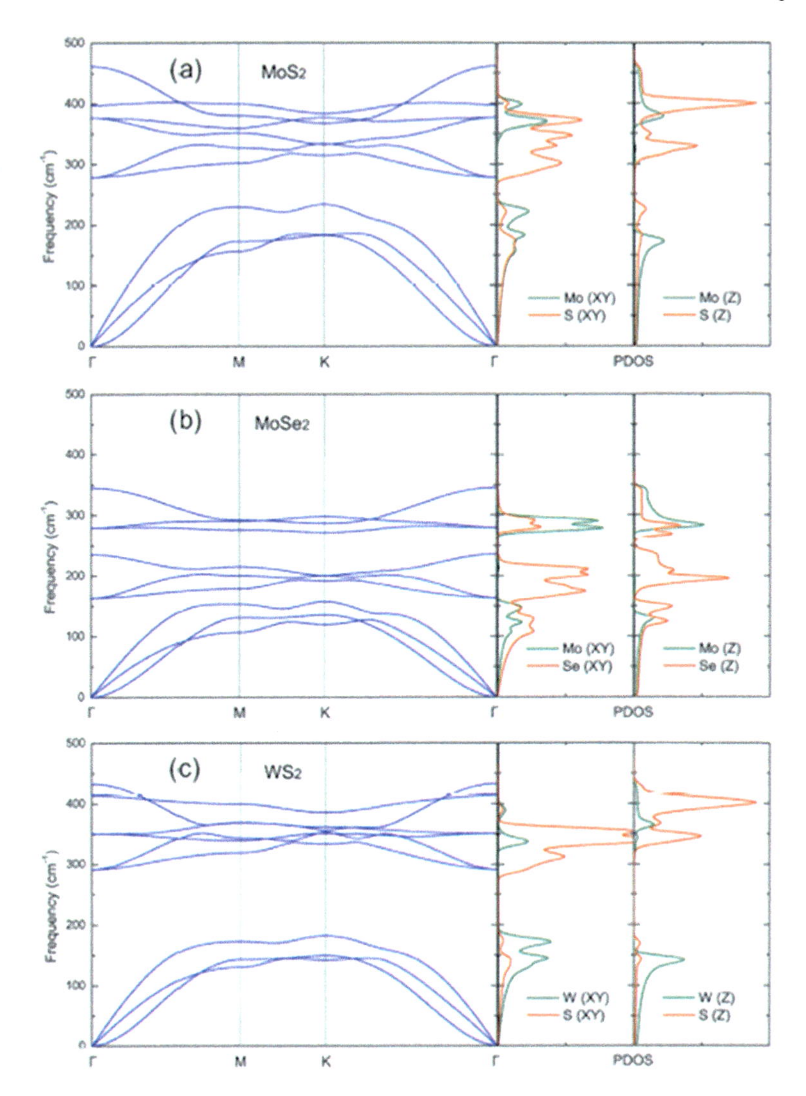

Fig. 4.7 Phonon dispersion and projected PDOS: **a** MoS_2; **b** $MoSe_2$; **c** WS_2 (adopted from Peng et al. [74])

boundary scattering, and imperfections scattering. The extrinsic periphery scattering and imperfections scatting issues are usually resolved by refining material quality, whereas the intrinsic phonon relaxation process/time phonon–phonon umklapp scattering is interpreted with the help of the well-known Klemens's time-dependent perturbation, the theoretical relationship is given as [82, 83].

$$\frac{1}{\tau_{qs}} = 2\gamma_{qs}^2 \frac{K_B T}{M v^2} \frac{\omega_{qs}^2}{\omega_m} \tag{4.4}$$

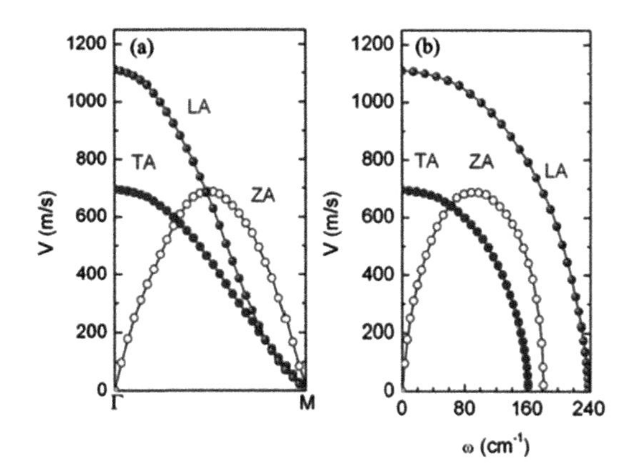

Fig. 4.8 Acoustic phonon modes phonon group velocity along G–M direction in MoS_2 with a function: **a** wave number; **b** frequency (adopted from Cai et al. [79])

Here τ_{qs} is the lifetime of diverse phonon branches, M is the atomic mass, ω_m is the Debye frequency, T is the temperature, K_B is the Boltzmann constant and v is the average sound speed, which expressed to relationship $2/v^2 = 1/v^2_{\text{longitudinal}} + 1/v^2_{\text{transverse}}$.

Phonon mean the free path is also a significant parameter to describe the size dependence thermal features. In 2D materials, umklapp scattering limits the phonon mean free path (λ), typically that is described from the help of the relaxation time τ and group velocity v. The phonon mean free path at the q point and s polarization is usually expressed from the relationship $\lambda_{qs} = v_{qs}\tau_{qs}$. As the intense TMD material, MoS_2 mean free path in decreasing order with the increasing frequency. Whereas the longitudinal acoustic mode has longest mean free path compared to transverse and flexural acoustic modes. This is owing to the collective belongings of the frequency depending relaxation time and group velocity [70].

4.2.3 *Thermal Conductivity*

Thermal conductivity k of TMDs is usually defined by Fourier's law [84];

$$J - k\nabla T \tag{4.5}$$

or

$$k = -\frac{J}{\nabla T} \tag{4.6}$$

where J, T, and κ are the heat current density, temperature, and heat conductivity, whereas the $\nabla T(\nabla T = dT/dz)$ represents temperature incline toward the heat flux direction. Since thermal conductivity is also connected to specific heat from the relationship $k \approx \Sigma C v \lambda$, here C is specific heat and v,λ are the average, group velocity and mean free path [85]. Commonly this relationship is useful for the diffusive transport circumstances when material dimensions are additionally greater than the phonon mean free path ($L \gg \lambda$) [81]. To calculate heat flux in monolayer TMDs, usually the thickness of 2D materials "h" chooses for the interlayer distance of the respective bulk materials, for instance typically, $h = 4.41\ ^{\circ}A$ in case of the MoS_2[86].

The low heat conduction (thermal conductivity) in TMDs has been studied extensively, and the various TMDs materials' thermal conductivities for the monolayers are summaries in Table 4.2. Theoretically, the thermal conductivity of monolayer MoS_2 at room temperature has been obtained ($103\ W\ m^{-1}\ K^{-1}$) using the Boltzmann transport equation [87]. However, by employing the non-equilibrium Green's function its thermal conductivity was calculated at around $23.3\ W\ m^{-1}\ K^{-1}$ [88]. Whereas, the experimentally thermal conductivity of the monolayer MoS_2 has been demonstrated at $34.5\ W\ m^{-1}\ K^{-1}$ [68]. The different values of thermal conductivity attribution for the same material by using a distinct approach with their advantages and disadvantages reveals the thermal conductivity of TMDs crucial to the imperfections or the substrates, it induces additional phonon scattering and reduces the inherent thermal conductivity. Like, in non-equilibrium Green's function harmonic force constants are considered to be ballistic or semi-ballistic phonon transportation and ignore the anharmonic phonon–phonon scatting in TMDs. Similarly, the Boltzmann transport equation approach limits the first-order anharmonicity. Other simulation approaches like molecular dynamics has been also used extensively to describe the low thermal conductivities of the TMDs based on the natural accounts of the lattice anharmonicity. Although, the accuracy of the molecular dynamic limits the quality of the empirical inter atomic potentials. Moreover, to explore the various TMD materials another most extensively explored monolayer is WSe_2, its thermal concavity around $3.935\ W\ m^{-1}\ K^{-1}$ [89]. This value is one order lower than the thermal conductivity of MoS due to ultralow Debye frequency and the occurrence of heavy atoms mass in WSe_2. Likely graphene thermal conductivity of TMDs also showing temperature dependence behavior with the increasing temperature decreasing conductivity [89, 90].

TMDs also show in-plane thermal conductivity in a decreasing order when coming into contact with a substrate owing to damping of the flexural acoustic phonon [61, 81, 91]. It has been recognized from the optothermal Raman technique. At room temperature, the thermal conductivities of suspended monolayers MoS_2 and $MoSe_2$ are 84 and $59\ W\ m^{-1}\ K^{-1}$. However, the thermal conductivities of TMDs monolayers may differ from the different substrates. Like thermal conductivities of the monolayers MoS_2 and $MoSe_2$ in a decreasing order 55 and $24\ W\ m^{-1}\ K^{-1}$, when SiO_2 is used as a substrate [91].

4.2.4 *Thermal Expansion*

The thermal expansion coefficient is also a vital thermal feature of TMDs. The thermal expansion coefficient (TEC) is due to unmatch among the substrate and

Table 4.2 Thermal conductivities for the different monolayer TMDs at room temperature [70]

TMDs	Process	Type	k (W m^{-1} K^{-1})
MoS_2	Exfoliated	Suspended	34.5
MoS_2	First principles calculation	Suspended	23.2
MoS_2	Optothermal Raman technique	Suspended	84
MoS_2	Boltzmann transport equation	Supported	103
MoS_2	Optothermal Raman technique	Supported on gold	55
$MoSe_2$	Boltzmann transport equation	Suspended	54
$MoSe_2$	First principles calculation	Suspended	17.6
$MoSe_2$	Optothermal Raman technique	Suspended	59
$MoSe_2$	Optothermal Raman technique	Supported on SiO_2	24
WSe_2	Boltzmann transport equation	Suspended	3.935
WS_2	Boltzmann transport equation	Suspended	142
WS_2	Raman spectroscopy	Suspended	32
WS_2	First principles calculation	Suspended	31.8

introduced added inner strain in the TMD material (such as MoS_2) layer. This directly results to change in the optical and electronic functionality in TMD devices owing to the thermal strain. It is also a key parameter to define the thermal constancy and inherent optical features of few-layer TMD-based devices [92].

Thus, the building block of nanodevices like monolayer/or a few layers (such as MoS_2) temperature differences between the substrate and TMD (MoS_2) layers results the TEC owing to the inclusion of thermal strain inside the TMD layer. The key reason occurrence of TEC we cannot avoid the self-heating effect during the operation of devices, subsequently, it also induced thermal strain by TEC mismatch, therefore, it directly impacts the electronic and transport properties of devices [93, 94]. Thermal strain also affects the performance of devices even resulting in cracking failure when it exceeds the bearing limit of the TMD layer (MoS_2) [95].

Although the direct measurement of TEC is most extensively explored for the monolayer MoS_2 that is hindered when attached to the substrate. Therefore, researchers have extensively explored the theoretical calculations [96–98]. The positive linear TEC of two-dimensional (2D) MoS_2 has been estimated by employing the first-principles based on quasi-harmonic approximation [96]. Density functional perturbation theory has been also used to calculate the phonon spectra of 2H-MoS_2 structures [99]. Whereas a negative-positive crossover in the TEC of monolayer MoS_2 at 20 K proposed which attributed to negative and positive modes [97].

To explore the TEC Raman spectroscopy is also recognized as a flexible instrument for 2D TMDs [100–102]. Using this technique temperature effects on the Raman modes of 2D TMDs have been widely investigated [103–108]. Although, the temperature behaviors of the Raman peaks of TMDs are still controversial. In some reports, the peak positions varied linearly with increasing temperatures [103–106]. Whereas, in some studies both monolayer and few-layer $MoSe_2$ and WSe_2 have

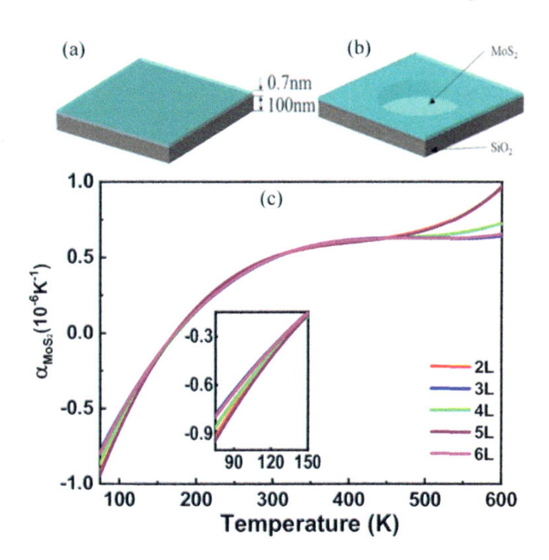

Fig. 4.9 Schematic for the finite element model of MoS_2; **a** supported; **b** suspended (**a** and **b** adopted from Yang et al. [115]); **c** calculated TECs of MoS_2 for the different numbers of layers, the inset figure shows a magnified view of the TECs in the temperature range of 75–150 K (adopted from Lin et al. [92])

also shown a linear temperature dependency [104]. The linear temperature dependence of Raman modes was also investigated in monolayer $Mo_{1-x}W_xS_2$ [106]. On the other hand, in some studies temperature dependency of the Raman peak locations in TMDs has also defined as a nonlinear function. The temperature-dependent Raman spectroscopy of the substrate bonding of MoS_2 and WS_2 can be expressed to temperature-dependence Raman modes by using a third-order polynomial function [107–109]. Indeed, the temperature dependence of Raman modes of 2D TMDs differs, the TEC mismatch between TMDs and substrates is extensively play an important role in the temperature evolution of Raman modes. Preferably, to eliminate the substrate effects, suspended TMDs have been used to study the intrinsic properties [110–114]. Around two to ten times improvements in the movement and on/off portion are also possible in suspended monolayer MoS_2 [110]. To describe the elastic coefficients collectively 2D elastic modulus and Young's modulus are used for the suspended multilayer WSe_2 [111]. Including intrinsic thermal conductivity for the monolayer and few-layer MoS_2 [68, 112]. Thus the TEC of the 2D materials is still debatable and under theoretical and experimental investigation in contrast to supported and suspended substrates [92, 115–118]. The typical finite element model for supported and suspended substrates schematics including calculated TECs for MoS_2 with different numbers of layers is illustrated in Fig. 4.9a–c.

4.2.5 Thermoelectric

Thermoelectric (TE) materials make it possible to drive electric currents by using temperature gradients, and conversely cooling of a system can be achieved by using a voltage difference, namely the Seebeck and Peltier effects. The actual performance of

TE conversion is quantified from the dimensionless figure of merit (ZT). This includes strongly interrelated electronic and thermal transport properties to the material. These kinds of inter-relation among Seebeck coefficient (S), electrical conductance (G), thermal conductance (κ), and significant enhancement in ZT are extremely complex issues [119]. Although efforts have been made to improve the low TE efficiency of 2D materials, the major hurdle is to exceed more than 1 typical ZT value for the bulk material [120]. To resolve this issue a quest has been made with the TMDs, specifically semiconducting TMDs have a wide range of bandgaps and lower lattice thermal conductivities that makes them favorable for their use as a TE material [119].

Moreover, thermoelectric (TE) materials can be also used to recover waste heat and for solid-state cooling, which important to alleviates the energy crisis [121–125]. A few bulk materials like group III nitrides also have a good TE performance at high temperatures, although their thin-layer films have a better TE conductivity [126–131]. Another key reason for the attraction of TMDs as TE materials is their density of states increases near the Fermi energy level, ultimately it also increases the Seebeck coefficient (S). Therefore, TE properties of low-dimensional TMDs have been systematically explored [132]. Additionally, the 2D layered TMDs are also interesting because of their unique structural, physical, and chemical properties. That can be easily achieved through various methods, such as mechanical exfoliation, physical vapor deposition, chemical vapor deposition, and hydrothermal synthesis [133]. TMDs are also widely used in the fabrication of high-performance nano-electronic devices due to their high carrier mobility and good flexibility [134–138]. As an instance, the TE performance of single-layer $MoSe_2$ can have $ZT < 0.2$, due to its high lattice thermal conductivity of $\sim$ 60 W/m K [132]. Similar results are also expected from the single-layer MoS_2 [139].

TMDs are also used for the minimization of the frequency gap to effectively reduce the lattice thermal conductivity [140]. Moreover, it also reduced by changing the stoichiometric ratio of TMDs [141, 142]. Other materials, like single-layer MoSSe, have been also studied which lays the foundation for experimental and theoretical studies of such materials [143]. More recently Janus Mo monolayers and single layers of PtSSe have been also recognized as useful photocatalysts for decomposing water with infrared light due to the small band gap and inherent electric field [133]. The 2D PtSSe has also provided a new platform for the overall water-splitting reaction with high experimental feasibility [144]. Other single-layer Janus-type platinum dichalcogenides have also revealed the structure of Janus crystals that can significantly improve the piezoelectric constants of PtX_2 crystals (X = S, Se, Te) and add a new piezoelectric degree of freedom in the out-of-plane direction [133]. A detailed experimental evolution of TE characteristic parameters for TMDs has been also explored [145]. The theoretical and experimental aspects of the various TMDs in terms of TE for advanced applications including biophysics have been also explored in recent years [145–147].

4.3 Optical Properties

Semiconducting TMDs, MX_2 (M = Mo, W, etc) have an indirect band gap in their bulk phase. Whereas the decreasing number of layers allows their existence as a point at which the band gap transform into an indirect to a direct gap. The transition of the band gap has been extensively studied both theoretically and experimentally for the various TMD materials, such as typical MoS_2 using DFT simulations as well as angle-resolved photoelectron spectroscopy (ARPES) [148–150]. However, earlier experimental evidence of the transition from an indirect to a direct gap was based on the observation of increasing photoluminescence yield, as the number of layers decreasing [151–156]. Moreover, the direct observation of changing the value of the band gap with the number of layers have also demonstrated with the help of the spatially resolved optical absorption spectroscopy for the MoS_2 flakes [157].

The monolayer, bilayer, trilayer, and bulk TMD (such as MoS_2, WSe_2) observed band structures by ARPES are illustrated in Fig. 4.10. The MoS_2 bulk-like layer shows a maximum valance band at $\overline{\Gamma}$ and conduction band minimum at K. Whereas, with the decreasing number of layers the states at $\overline{K}$ move upwards in energy toward the V_{BM} at $\overline{\Gamma}$, further, it reaches above the states at $\overline{\Gamma}$, in the case of the MoS_2 monolayer. Using this process, the band structure evolution can be explained with the help of the quantum confinement of electrons at $\overline{K}$ and $\overline{\Gamma}$. Where the electrons at K derived from in-plane Mo $d_{x^2-y^2}/d_{xy}$ orbitals that are unaffected by the confinement in the z-direction [149]. While the states at Γ are governed from both orbitals Mo d_{x^2} and Sp_z contributions [149]. Therefore, with the decreasing number of layers the interplanar contributions at $\overline{\Gamma}$ certainly reduced.

Similar behavior can also notice in various TMDs including $MoSe_2$, $MoTe_2$, $MoSe_2$, and tungsten dichalcogenides such as WSe_2, and WS_2 [148, 149, 158–163]. The tellurides also have similar behavior, like $MoTe_2$, WTe_2, Bi_2Te_3, and GaTe/$HfSe_2$ [164–167]. However, experimental verification of such behavior in WTe_2 could be more due to the existence of a thermodynamically preferred type II Weyl semimetallic phase [139, 168]. Owing to increasing band gap a reduction in the number of layers by the reduction in interactions between neighboring atomic layers. Thus this behavior is likely to be held in almost all kinds of semiconducting TMDs having a 2H crystal structure.

The photoluminescent behavior of monolayers, bilayers, and multilayers of WS_2 and WSe_2 are also significant and interesting things [169]. Because the TMD multilayer regime is usually showing a small feature due to photoluminescence across the direct gap, while a larger feature occurs due to the indirect gap. TMDs' photoluminescent indirect peak movement higher energy side with the reduced number of layers indicating the increasing band gap, whereas, in the case of the direct gap it does not change [169]. However, a direct photoluminescent peak becomes more intense with decreasing layer number [169], this could be related to photoluminescent quantum yield. Almost identical behavior has also recognized for other VI group TMDs [18, 19, 151, 154].

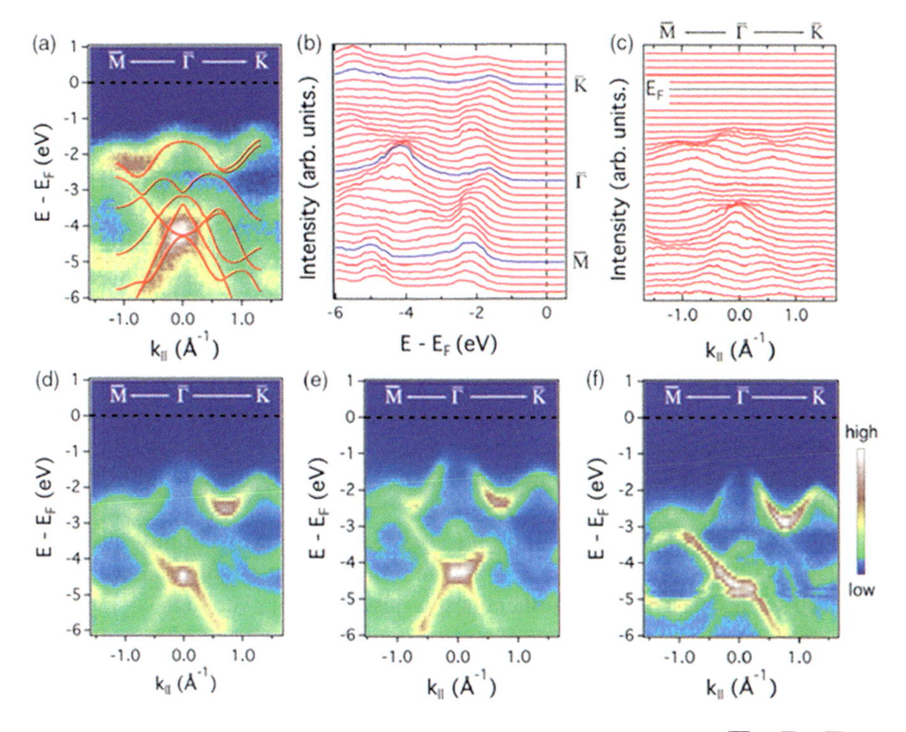

Fig. 4.10 **a** ARPES band map for the exfoliated monolayer MoS_2 along the $\overline{M} - \overline{\Gamma} - \overline{M}$ high symmetry lines; **b, c** corresponding EDCs and MDCs; **d–f** ARPES band maps for the exfoliated bilayer, trilayer, and bulk MoS_2 (adopted from Jin et al. [440])

Usually, the absorption spectrum of the 2D material in the infrared–visible range is dominated by a step function-like spectrum. This is due to the joint density of states and the matrix elements close to the band edges. TMDs also exhibit strong resonance features close to the absorption edge due to excitonic effects. The doped TMD monolayers also show trion quasiparticles [170–175]. Like, two-hole, one-electron, or one-hole, two-electron quasiparticles, analogous to H_2^+ and H^- ions, including bi-excitons (bound states of two holes, two electrons) [176–188]. The binding energy of TMD excitons approximately an order of magnitude larger [189–199], it can be accessed from the quantum well-type structures [200, 201], ultimately it is useful for the high-temperature excitonic devices. Schematics of the excitons, trions, and bi-excitons are illustrated in Fig. 4.11. Likewise, a tuneable excitonic LED has been also demonstrated for the monolayer WSe_2 p–n junctions [202].

Nonlinear optical properties of the semiconducting monolayer TMDs are also important, specifically, interest in naturally breaking inversion symmetry owing to a reduction in dimensionality [203]. Although, this property of TMDs is in beginning (or few studies) phase and reports are limited. Such as second-order nonlinear susceptibility in WS_2 approximately three orders of magnitude larger than other common nonlinear materials [204]. Additionally, the nonlinear refractive index of

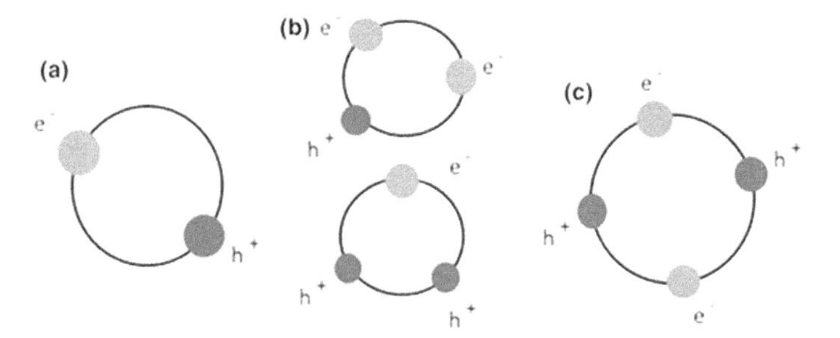

Fig. 4.11 Schematic diagram: **a** an exciton; **b** negatively and positively charged trions; **c** bi-excitons

TMDs can also tune [205]. Moreover, the nonlinear absorption (i.e., two-photon absorption processes, etc.) are also the size and layer number dependent [206–208]. This kind of flexibility of TMD materials is allowing to further alteration in their optical properties.

Thus TMDs are impressive for all-optical switches and other ultrafast photonics applications [209, 210]. However, to fabricate reliable devices with nonlinear properties and surface passivation a further quest is desired [211]. Although composition MoS_2 has been also used to produce nanoscale optical components, including microlenses and gratings to achieve high control of light with higher accuracy compared to traditional bulk materials [212]. Likely to group VI TMDs, group V TMDs ($M = V$, Nb, Ta) are not having semiconducting behavior, but they are metallic in most cases [213–215], therefore, their optical properties are relatively limited.

TMD materials (like TiS_2) laser interaction studies also revealed their semimetallic nature in bulk or 2D sheets, in which the band gap opens to a direct gap at 2.87 eV, as determined by the photoluminescence measurement [216]. An almost similar result would be achieved from the TiS_2 sheets [216]. TMD composition TiS_2 is also interesting for applications where optically transparent layers are required, like optoelectronics and photovoltaics. It can also use as the n-type layer in a perovskite n-i-p solar cell [217].

4.3.1 Excitons

Excitons are a tightly bound state of the electron and hole together under an attractive Coulomb interaction [218]. The occurrence of excitons resonances strongly enhances the transitions in the continuum of unbound electrons and holes under the light interaction. Excitons also depend on their radii, when radii are small, their properties remain to a large extent within the Wannier-Mott regime and preserve analogies to the electronic structure of the hydrogen atom. Such materials have almost ideal 2D confinement and their reduced dielectric screening from the environment favors the Coulomb attraction between the holes, the electrons one to two orders of magnitude stronger compared to traditional quasi-2D systems [190, 218]. These materials reflect a little dielectric constant, like ionic crystals, electrons and holes that are strongly bounded to every one inside the identical or adjacent neighbor unit cells. These kinds of excitons are known as Frenkel excitons, their usual binding energy is in the range of 0.1–1eV. Typically Frenkel type of excitons has been recognized as ionic crystals and organic molecular crystals by collection of aromatic molecules, whereas in the semiconductors the dielectric constant is usually big. So, electric field transmission by the valence electrons reduced the Coulomb interaction among the electrons and holes, and as a consequence, the exciton radii (great) is more than the lattice distance. Therefore, the influence of the lattice strength is also incorporated into the effective masses of the electron and hole [190, 218].

Coulomb electron-hole interaction can be separated into direct and exchange contributions, both of which included long-range and short-range interactions [219]. The long-range contribution is usually correlated to the Coulomb interaction that acted at inter-particle distances, in real space it larger than inter-atomic bond lengths (like, for small wave vectors in reciprocal space compared to the size of the Brillouin zone). Whereas the short-range contribution due to the overlap of the electron and hole wave functions at the scale of the order lattice constant ($a_0 \simeq 0.33\,\text{nm}$ in monolayer WSe_2), typically within one or several unit cells (like, large wave vectors in reciprocal space).

Generally, direct Coulomb interaction describes the interaction of positive and negative charge distributions of the electrons and holes. Where the long-range part of the direct interaction is mainly due to enveloping of the electron-hole pairs that are weakly sensitive to the particular form of the Bloch functions: such as valley and spin states. That depends on the dimensionality and dielectric properties of the system. Their electrostatic origin also contributes dominantly to the exciton binding energy. In contrast to this, the short-range part of the direct interaction is due to the Coulomb attraction of the electrons and the holes within the same or neighboring unit cells. This kind of interaction is sensitive to a particular form of the Bloch functions, thereby, as a rule, they are considered together with the corresponding part of the exchange interaction. Moreover, in the semi-classical depiction, the long-range direct interaction corresponds to attractive Coulomb forces between opposite charges. Therefore, an electron and a hole can form a bound state, as the neutral exciton, with strongly correlated relative positions of the two constituents in real space, as schematically illustrated in Fig. 4.12a. The concept of correlated electron-hole motion is depicted in Fig. 4.12b, where the modulus squared of the electron wave function relates to the position of the hole for ground state exciton in monolayer MoS_2. In monolayer TMD the exciton Bohr radius is usually order of one to a few nanometers and the correlation between an electron and a hole extends over several lattice periods. Thus, exciton has an intermediate nature between the so-called Wannier-Mott or large-radius semiconductors (such as GaAs and Cu_2O and Frenkel exciton), in which the corresponding charge transfer between nearest lattice sites. The majority of experimental demonstrations have revealed the Wannier-Mott effective mass approximation to be appropriate even for quantitative predictions.

The increasing amounts of the excitons in the four particle complexes like bi-excitons to be also shaped such configuration, as characterized in 2D TMDs [220,

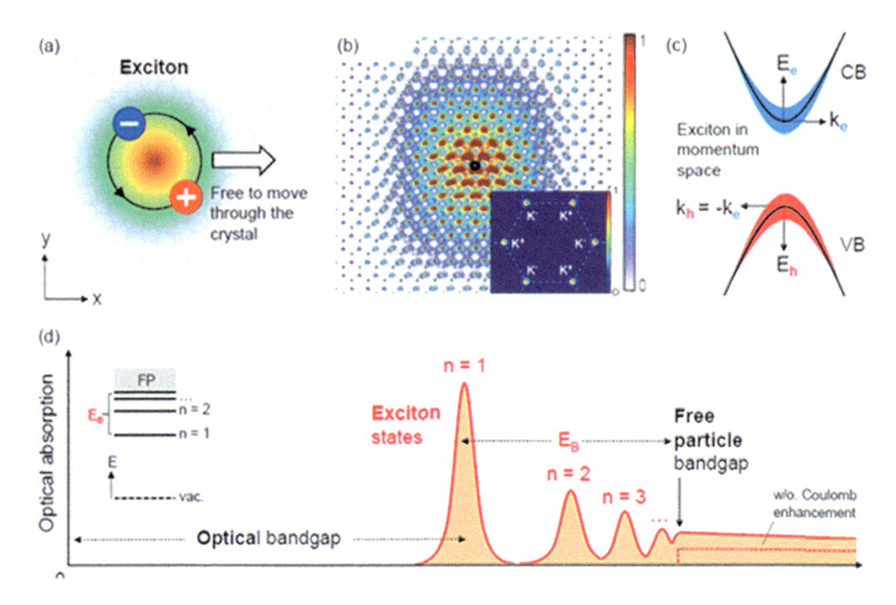

Fig. 4.12 **a** Schematic for real-space representation of the electron-hole pair bound in a Wannier-Mott exciton, by showing strong spatial correlation of the two constituents. The arrow indicates the center of the mass wave vector that is responsible for the motion of the exciton as a whole; **b** Illustration of a typical exciton wave function for the monolayer MoS_2. The modulus squared of the electron wave function is plotted in color scale (grayscale) for the hole position x_{ed} at the origin. The inset shows the corresponding wave function in momentum space across the Brillouin zone, including both K^+ and K^- exciton states; **c** Representation to exciton in reciprocal space, with the contributions of the electron and hole quasiparticles in the conduction (CB) and valence (VB) bands, schematically shown by the width of the shaded area; **d** Schematic for the illustration of optical absorption of an ideal 2D semiconductor including the series of bright exciton transitions below the renormalized quasiparticle band gap. The Coulomb interaction for the enhancement of the continuum absorption in the energy range exceeding EB, the exciton binding energy. The inset shows the atom-like energy level scheme of the exciton states, designated by their principal quantum number n, with the binding energy of the exciton ground state ($n = 1$) denoted by EB below the free particle bandgap (FP) (adopted from Wang et al. [218])

221]. Their binding energy is in general greater than the order of magnitude of the conventional quantum well structures. Additionally, the impartial quasiparticles and charged quasiparticles (trions) are possible, where the dual electrons with a hole create a negatively charge trion, as depicted in Fig. 4.13.

4.3.1.1 Binding Energy

The binding energy of excitons is intensely depending on dimensional parameter "α" in a structure. Usually, it represents from the αD space, typically exciton binding energy is described from the described expression [222]:

Fig. 4.13 Schematic for the quasiparticles, exciton (left) and trion (right), in the two-band model

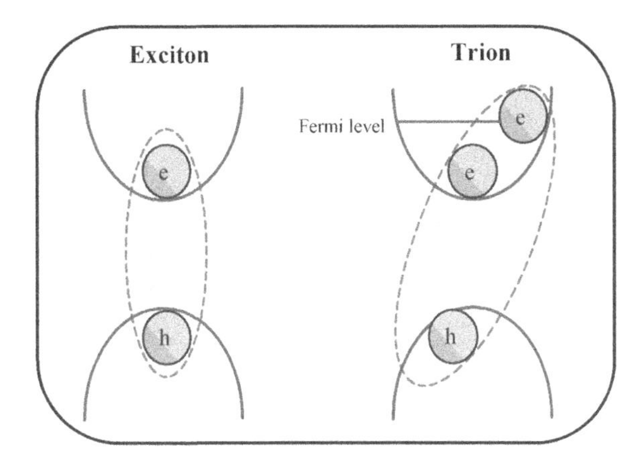

$$E_n = -\frac{E_0}{\left(n + \frac{\alpha-3}{2}\right)^2} \tag{4.7}$$

here n represents the principal quantum number and E_0 corresponds to the Rydberg exciton.

The yield of an exciton binding energy is generally larger by number 4 in 2D than in 3D due to the absence of a dielectric effect. However, within the boundary of atomically thin materials, the dielectric transmission may reduce due to electric field lines connecting the electron and the hole extending exterior to the sample, as depicted in Fig. 4.14a, it gives an uniform greater enrichment factor. However, experimental investigations of monolayer TMDs demonstrated the enhancement of the exciton binding energy that is quite distinct owing to a much diverse dielectric atmosphere exterior to the sample [223, 224], as witnessed in both monolayer and bulk WSe_2 in an optical investigation under the higher magnetic fields.

Fundamentally excitonic feature of a thin 2D layer distinct to the 3D bulk semiconductor of the identical material, similar to actual-space source manner of the TMDs schematic is given in Fig. 4.14a. In contrast bulk phase, the electrons and holes formation of exciton in a monolayer are largely restricted in a plane. In addition to this, they also experiences a reduction in screening due to the change in the dielectric environment. These effects have two major impacts on the electronic and excitonic properties of the material [225, 226]. As depicted in Fig. 4.14b, the quasiparticle band gap increases for the monolayer, therefore, it enhance the electron-hole interaction, ultimately leads an increase in the binding energy of the exciton. Owing to this reason, in a low-dimensional material like carbon nanotubes or 2D TMDs, excitons have together Wannier-Mott and Frenkel characters. More specifically, a large excitons can be characterized as Wannier type, while, strongly bounded would be related to Frenkel kind. Whereas the exciton Bohr radius is usually estimated from the GW-BSE approach, which an order of 1 nm [227–230]. Typically in MoS_2 the operational Bohr radii is 9.3 Å for the monolayer and 13.0 Å for the bilayer together operative binding energies of 0.224 and 0.106eV [231]. It also leads the ground state

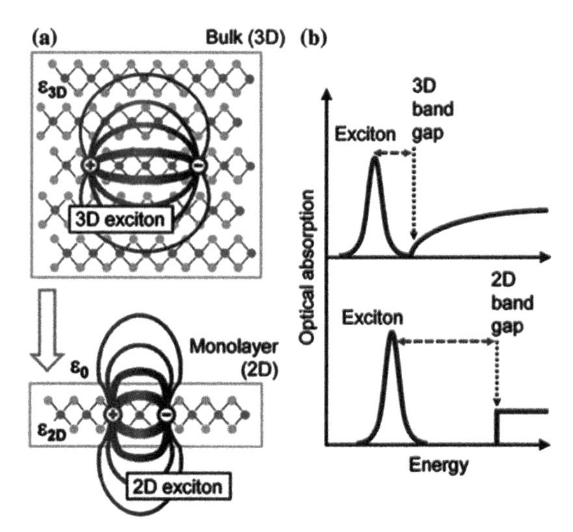

Fig. 4.14 **a** Real-space representation of electrons and holes bound into excitons for the three-dimensional bulk and a quasi-two-dimensional monolayer. The changes in the dielectric environment are indicated schematically by different dielectric constants ε_{3D} and ε_{2D} and by the vacuum permittivity ε_0; **b** represents the impact of the dimensionality on the electronic and excitonic properties by optical absorption. The transition from 3D to 2D is expected to lead to an increase of both the band gap and the exciton binding energy (indicated by the dashed) (adopted from Chernikov et al. [195])

binding energy 0.897eV for the monolayer and 0.424eV for the bilayer which is significantly larger than 65 Å [232, 233].

Such behavior is due to a large enough dielectric function that allows the three-dimensional extent of the wave function extends above a small to a number of nanometers. On the other hand, inferior transmission under a vacuum or in the dielectric surrounding is favorable for the greater binding energy. However, significantly excitonic bound states are also plays a crucial contribution in altering the optical features of the lower dimensional materials [234–236]. Specifically, the robust three-dimensional confinement and decreased in screening effects related to 3D solids allow good exciton wave function localization in the momentum space. Moreover, the 2D gap system dipoles allow inter-band transitions; therefore, the optical absorption spectrum belonging to non-interacting boundary (i.e., ignoring the Coulomb interaction) could behave alike a step function. As a consequence strong electron-hole interaction could be responsible for the redshifts owing to a large amount of spectral weight that ultimately gives a qualitatively different spectrum accompanied by a series of new excitonic levels below the quasiparticle bandgap, as illustrated in Fig. 4.14b.

2D quantum wells are usual semiconductors in which the electron-hole contact is somewhat feeble. So, the exciton binding energy accurate determination is a crucial task, like, the energy gap among the principal excitonic peak and the band edge absorption step. Because their energy locates around tens of meV and is vulnerable to environment screening and temperature is broadening. However, experimentally a single-layer TMD (MoS_2) cannot have an absorption step and indirect to direct band, as recognized in 2D $WS_{2(1-x)}Se_{2x}$ alloys [18, 19, 237]. The double absorption peaks reflects the spin-orbit splitting nearby the Kohn-Sham energy band gap which followed the density functional theory (DFT). This theoretical approximation leads that the peaks prior inferred as direct band edge conversions. In the extensive studies of first-principles calculations on MoS_2 monolayer using the GW method [238], investigators have predicted that a quasiparticle bandgap can be larger around one electron volt compare to initially experimentally demonstrated assessment [198, 230, 231]. Moreover, the pertinent estimation based on first-principles GW-BSE theory [239] also disclosed the energy gap inconsistency originates owing to robust excitonic influence. In this order, a comprehensive first-principles study of the various TMDs based on the electronic structure has been also interpreted in rather a diverse manner than previously [240, 241]. The Moiré superlattices have been also recognized in two-dimensional semiconductor hetero-bilayers, by using this approach the number of electrons per effective atom can be tuned [242]. Additionally, density function calculations for the thrust application such as photovoltaic and topological insulators have been also verified [241–243].

The interpretation of band gaps and excitonic effects in TMD monolayers from the first-principles calculations is considered to be a state of art. Using this art can also calculate quasiparticle band structure and optical feedback tighter with the electron-electron exchanges and excitonic assistances by employing the GW and Bethe-Salpeter expressions (GW-BSE). To estimate accurately excitonic states essentially required a fine k-space sampling because it correlates to the bounded states in actual space [230], with a greater energy cut-off. Additionally, a huge quantity of bands is desired to get a superior settlement to the experimental findings. As interpreted a $72 \times 72 \times 1$ k grid with a 476 eV energy cut-off and 6000 bands have a good settlement to the experimental outcomes [230]. According to the theoretical prediction, the optical spectrum among 2.2 and 2.8 eV is not unremarkable, because they comprises several bright and dark excitonic states, and therefore, they expanded due to electron-phonon interactions.

The theory of neutral and charged excitons in monolayer TMDs depending on an active mass model of excitons and trions has been demonstrated using ab-initio approximation; this study has also provided an appropriate management of transmission in twofold dimensions [227]. Using this approach estimated exciton binding energies have been found in a proper settlement to greater level many-body calculations depending on the Bethe-Salpeter expression. The analytical approach by the study of optical properties TMD (MoS_2) has been also interpreted [244–246].

The exciton band structure has been also interpreted in terms of perspective mirror threefold rotational and time-reversal separate equilibriums [247, 248]. Moreover, it has been also anticipated the promising occurrence of lower energy exciton states to nearby the Brillouin zone center and the Brillouin zone corners, whereas the excitons have electrons and holes in conflicting valleys (K and K′), or in other world excitons belongs to holes in Γ valley and electrons in the both K or K′ valley. The monolayer MoS_2 can have an indirect gap under the excitation spectral characterization. On the other hand, in the bulk TMDs indirect band transition corresponds to the valance band maximum (VBM) at the center of the hexagonal Brillouin zone (Γ point) and the conduction band minimum lie nearly halfway in the Γ-K direction [249, 250]. Thus compared to bulk TMDs monolayer TMDs have a lower symmetry point.

The optical spectra of the monolayer group VIB TMDs has also revealed the existence of dual lower energy exciton peaks related to perpendicular conversions at the K point of the Brillouin zone. Their spin-orbit-splits valence band has a dual degenerate conduction band. Such excitons are mainly located nearby 2D geometry with powerful bounding states owing to weak dielectric transmission in the monolayer. Experimentally splitting values are 146 meV in the monolayer and 174 meV in the bilayer TMDs. That is fairly good agreement containing the energy difference among the A and B exciton peaks in the recorded absorption spectrum [18].

To describe it the Mott-Wannier effective mass theory should be considered in a plane, by supposing the distance scale of the exciton Bohr radius. Therefore, together electron and hole would be localized at $z = 0$ in a plane. The Coulombic interplay in the plane defined by a dominant dielectric constant $k = \sqrt{\varepsilon_{II}\varepsilon_{I}}$, whereas ε_{II} and ε_{I} are the parallel and perpendicular dielectric tensor components corresponding to *the c-axis*. The typical magnitude of $\varepsilon_{II} = 2.8(4.2)$ and $\varepsilon_{I} = 4.2(6.5)$ have defined for monolayer and bilayer TMDs. The achieved monolayer/bilayer values are very much lesser than the bulk magnitudes. The difference in values are due to the used different approaches [251]. Although different methods calculations also agreed with a robust decrease of the dielectric screening in the monolayer in contrast to the bulk, such as ground and excited state of monolayer $MoSe_2$ [252].

The large excitonic effects in MoS_2, WS_2, WSe_2, and MoSSe have been also noticed [198, 253, 254]. Using the DFT approach photoemission has been described as the Kohn- Sham energies which do not properly resemble to quasiparticle dynamisms, thereby, needs to appropriately define the electron adding or elimination actions. To resolve the issue, the GW approximation can be utilized, whereas, exterior the mean field, the independent-particle DFT approach has been preferred. The DFT approach is precisely accounted to the many-body electron-electron interactions. Additionally, it also useful for the electron-hole interactions to resolve the complexity using the Bethe-Salpeter equation (BSE) to the dual element Green's function. Moreover, the spin orbit splitting has also considered crucial for the level of theory in the follow the trends $\Delta^{\mathrm{PBE}} < \Delta^{G_0W_0} < \Delta^{\mathrm{HSE}}$.

Typically the large excitonic effect absorption spectra have been also calculated by using the BSE approach. Several absorption spectra of monolayers have been studied by indicating the existence of dual powerful compact excitonic peaks that arise due to perpendicular alterations at the K point due to spin-orbit splitting valence band to a doubly degenerate conduction band. The different types of TMDs findings have also demonstrated that the exciton binding becomes weaker with the heavier chalcogen. This could be due to an increase in the dielectric screening through the additionally diffusive orbitals in the case of the weighty chalcogens. Whereas, the exciton splitting may larger in WX_2 compounds compared to their MoX_2 counterparts [198].

4.3.1.2 Non-Rydberg Excitonic

Experimental challenges in the determination of exciton binding energy in 2D TMDs due to linear optical methods that are commonly used for bulk semiconductors or conventional semiconductors based on the identification of the onset band-to-band transitions in the optical absorption or emission spectrum. Like onset band-to-band transitions cannot be observed in 2D TMDs accurately due to the transfer of oscillator strengths from the band-to-band transitions to the fundamental exciton states. Including lifetime broadening and potential overlap in energy with exciton states which originated from higher energy bands or different parts of the Brillouin zone. Therefore, it opens a path for investigators to find alternative approaches to define the exciton agitated states and estimate their binding energy owing to level gaps defined by the model. The simplest 2D hydrogenic model has been introduced based on the existence of an electron-hole pair under the 2D interactions through the Coulombic potential due to dielectric confinement and nonlocal screening [255, 256], the related energy spectrum is typically called Rydberg series [195]. The occurrence of such potential leads to an unconventionally spaced Rydberg series of excited excitons that generates an anomalous ordering of s, p, and d levels [257]. Typically, these excited states allow direct estimation of the free-particle band gap and binding energy of the ground state [257, 258]. Both key material parameters are usually difficult to measure in monolayer TMDs, but they are necessarily sensitive to the surrounding dielectric environment [259–261].

The measurements of the 1s and 2s/2p states permit to obtain the exciton binding energy by employing 2D hydrogenic model. Using this model exciton properties of the various TMDs have been experimentally assessed in term of the excitonic Rydberg series, considering the agitated states of the bounded electron-hole couples tagged to the resemblance of the hydrogen series as 2s, 3s, and so on. The corresponding resonance energy parting of a hydrogenic progression like Wannier excitons has been defined. Moreover, the light coupling of the excited states may reduce compare to the core changeover, it means their spectral weightiness reduction with an enhancing quantum number [195]. As the instance, the reflectance difference $\triangle R/R = \left(R_{\text{Sample}} - R_{\text{Substrate}}\right)/R_{\text{Substrate}}$ of the WS_2 monolayer sample at a temperature of 5 K, as illustrated in Fig. 4.15a [195]. Its reflectance contrast $\triangle R/R$ spectrum profile have several pronounced peaks belonging to A, B, and C excitons in WS_2. Additionally, a small feature toward the low energy side of the peak A relates to a charged exciton, or trion possessing binding energy order of 20–30meV. Moreover, A exciton could also relate to the fundamental band gap of the material [195]. By highlighting weak signatures of the high order excitonic alterations. The derivation of the reflectance difference $(\mathrm{d}/\mathrm{d}E)(\Delta R/R)$ plot toward the high-energy side of the exciton 1s ground state, whereas the manifold extra peaks are associated to 2s, 3s, 4s, and 5s states of A exciton. Corresponding peak locations are usually taken out by monitoring the particular points of modulation.

Therefore, to compute the exciton binding energy one should define the quasiparticle band gap respect to the energy of a detached electron-hole pair. Most simply when an electron-hole pair interacts through a Coulomb attractive central potential

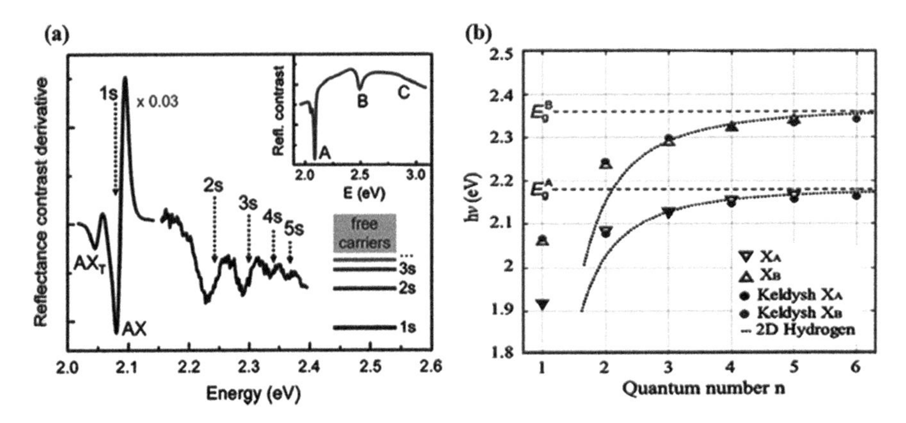

Fig. 4.15 **a** The derivative of the reflectance contrast spectrum $(d/dE)(\Delta R/R)$ for the WS_2 monolayer. The exciton ground state and the higher excited states are labeled by their respective quantum numbers (schematically shown at the bottom right). The spectral region around the 1s transition (AX) and the trion peak (AX_T) of the A exciton is scaled by a factor of 0.03 for clarity. The inset shows the as-measured reflectance contrast ($\Delta R/R$) for comparison, allowing for the identification of the A, B, and C transitions (adopted from Chernikov et al. [195]); **b** Experimental (empty triangles) and theoretical (filled circles, solid lines) transition energies for the exciton states. Theoretical values are obtained by fitting them to the effective mass model. The fits of the $n > 2$ states to a 2D hydrogen model in black dashed lines. Dashed horizontal lines indicate the theoretically calculated values for the quasiparticle bandgaps $E_g^A = 2:18\,\text{eV}$ and $E_g^B = 2:36\,\text{eV}$ (Vaquero et al. [245])

then it forms a series of excitonic, Rydberg-like states possessing a definite parity likely to the hydrogen model. In 2D materials, exciton energies are extremely weak with the quantum number n that preclude by a simple fitting of the data according to hydrogen model. In which $n = 3-5$ peaks predominately considered hydrogenic and their data points fits provide a quasiparticle band gap of Eg, using this value exciton binding energy can be extracted.

Rydberg series theoretically describes with the help of effective mass theory for the excitons in 2D-TMDs. If suppose the spectral location of the detected peaks and their related quantum number n, as illustrated in Fig. 4.15b. According to representation we can get unquestionably the spectral delegates of the peaks by the spectral locations fit [227]. The numerical solution of the effective mass Schrödinger equation for the radically symmetric exciton states "ns" possessing energies E_n, is expressed as [245];

$$H\psi_{ns}(r) = \left(E_n - E_g\right)\psi_{ns}(r) \tag{4.8}$$

whereas E_g defines the quasiparticle bandgap and H represents the Hamiltonian of the relevant coordinate $r = |r_e - r_h|$

$$H = -\left({\hbar^2}/{2\mu}\right){d^2}/{dr^2} + V(r) \tag{4.9}$$

here μ is the effective masse (or reduced mass) corresponding to A and B excitons. In thin layered semiconductors, electron-hole attractions $V(r)$ are not just expressed as a Coulombic potential owing to it greatly influenced from the nonlocal screening through the inserted medium. Therefore, the screened electron-hole attraction (or interaction) is more precisely defined from the Keldysh potential [255].

$$V(r) = -\left(\pi e^2/2r_0\right)[H_0(kr/r_0) - Y_0(kr/r_0)] \tag{4.10}$$

whereas H_0 and Y_0 are the zero-order Struve and Bessel functions and κ reflects the dielectric constant of the implanted medium. The parameter r_0 relates to the screening length owing to the 2D polarizability of the TMD (such as 1L-MoS_2). In these circumstances, it is worth using Keldysh potential method to interpret the Coulombic potential $V(r) = -e^2/kr$, for the larger distance ($r \gg r\_0/k$), but it may diverge logarithmically at the short distance ($r \ll r_0/k$).

A usual missing with three fitting parameters for every excitonic Rydberg series, explicitly E_g, κ and r_0. The principal assessment of the quasiparticle bandgap E_g can be assumed with the help of the spectral location of the extremely excited states ($n > 2$). The radii of the excitons lay in the expanse $V(r) = -e^2/kr$ and their corresponding energy levels are well-fitted in terms of the 2D hydrogenic Rydberg series by using the following relationship;

$$E_n = E_g - \frac{\mu e^4/2\hbar^2 k^2}{(n - 1/2)^2}. \tag{4.11}$$

Using this relationship various layered TMDs materials have been defined [195, 226, 245, 262, 263]. As the typical theoretical and experimental findings fit for the 1L-MoS_2 is represented in Fig. 4.15b [245].

4.3.1.3 Excitonic States in Dark

The resonant generated by photon absorption under the normal light incidence is typically connected to optical bright excitons. But their subsequent scattering event with other excitons, the electrons or phonons, and defects also induces the spin flips that considerably change the exciton momentum. Under such circumstances a more complex generation process occurs, typically it is known as non-resonant optical excitation or electrical injection, which forms a variety of exciton states. As the consequence an exciton not necessarily (or able) to recombine radiatively, as an instance, the optical transition is spin forbidden. This kind of exciton describes the optically dark condition and their corresponding transition states are known as dark excitonic states. The dark excitons also generates in an alternative way when a hole and an electron inject electrically or come together to form an exciton possessing total angular momentum $\neq 1$ or large center of- mass momentum K_{exc}. Therefore, may or not excitons directly interact with light either through the absorption or emission of a single photon, it depends on the center of mass wave vector K_{exc}, the relative motion wave function, valley, and spin states of the electron and hole. Based on experimental and theoretical studies a schematic is illustrated in Fig. 4.16a.

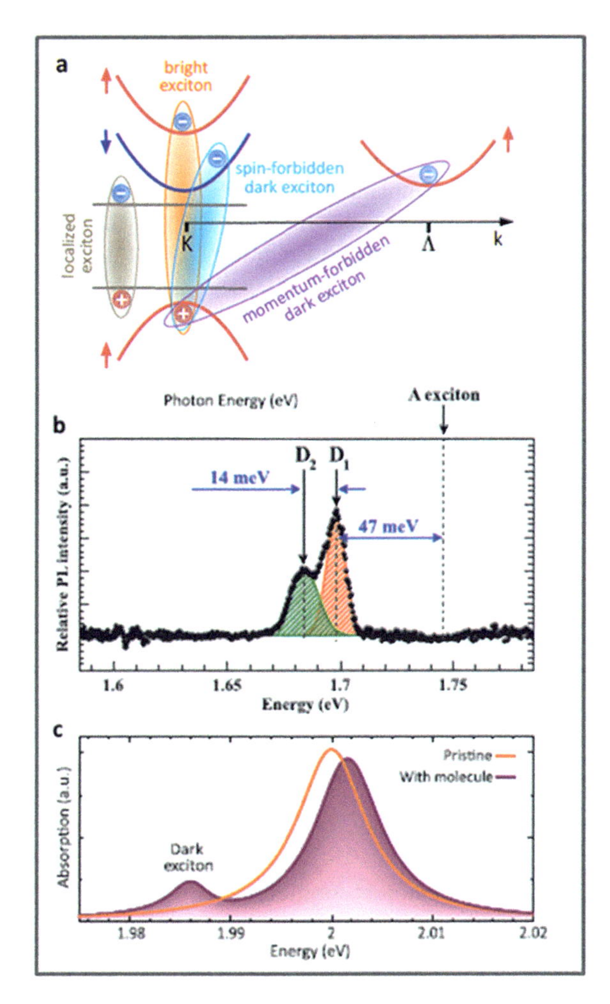

Fig. 4.16 Diverse exciton kinds in atomically thin nanomaterials and associated heterostructures: **a** excitons are Coulomb-bound electron-hole pairs (ovals in the picture). The momentum-forbidden dark excitons containing electrons and holes situated at distinct valleys in the momentum space. Spin-forbidden dark excitons contain electrons and holes with contradictory spin. These states are accessible by light owing to the absence of desired momentum relocation and spin-flip, so, the electrons and holes in localized excitons trapped into a contamination persuaded potential are illustrated; **b** Relevance PL spectra for the WS_2 monolayers, where the signal measured at the magnetic field $B = 1$ T subtracted from the signal at zero field. The twofold extra resonances (D_1, D_2) underneath the A exciton associated to spin-forbidden dark excitons; **c** picture of momentum-forbidden dark excitons in occurrence of molecules from a novel identifying technique (adopted from Mueller et al. [276])

Dark exciton usually contains two dominant processes namely spin-forbidden dark excitons and momentum-forbidden dark excitons. In the spin-forbidden dark excitons, spin-orbit attraction lifts the spin degeneracy of the conduction band with a diverse symbol for the tungsten and molybdenum dependent TMDs [264]. Although in MoS_2 and $MoSe_2$ spins are allowing bright excitons with the minimal energy, while in WS_2 and WSe_2 spin-forbidden states involving Coulombic bounded electrons and holes having a contradictory spin energetically lowest, as depicted in Fig. 4.16a.

This is an outstanding settlement to the temperature dependence PL yield experimental achievements [265]. Enhancing of such states has been mainly established from the lower temperature magneto-PL tests, in which robust in-plane magnetic fields employed [171, 266, 267]. A couple of the charge carrier spin via the Zeeman effects have been used for mixing the spin-split bands, thereby, softens, spin-selection rules are also described. The dark excitons a clear features that appears approximately at 50 meV, it is below bright exciton resonance in magneto-PL spectra of tungsten-based TMDs, as illustrated in Fig. 4.16b. This outcome showed a peak doublet structure to describe the splitting of the spin-forbidden states owing to the influence of the Coulombic interchange coupler. Moreover, the spin-forbidden dark excitonic states can also activate from an alternate method by combination of surface plasmon modes to the involvement of only WSe_2 monolayers on the patterned silver surfaces [268]. The dark excitons of WSe_2 also possessed a nonzero dipole moment perpendicular to the material plane that allowed them to twosome in surface plasmon modes.

On the other hand, intervalley the dark excitons with Coulomb-bound electrons and holes situated at diverse valleys in the momentum space revealed significant scattering networks for the exciton-phonon procedure. Particularly, the dark KΛ excitons and hole situated at the K and electron at the Λ valley are the pronounced anxiety to tungsten created TMDs, because it is supposed at this stage the energetically lowermost ground state, as schematically depicted in Fig. 4.16a. Additionally, the scanning tunneling spectroscopy interpretation has also revealed the Λ valley nearby 80 meV under the K valley in monolayer WSe_2 [269], the corresponding momentum forbidden KΛ excitons can be energetically inferior to the bright KK states. Moreover, the strain depending PL measurement has also demonstrated the comparative spectral variations among K and Λ valley having a conflicting characteristics for the MoS_2 and WSe_2 that implies the former one as an indirect gap material [270]. Theoretical and experimental joint investigations have also suggested the homogenous line width of excitonic resonances in the tungsten-based TMDs. Therefore, in such TMDs scattering of optical and acoustic phonons lies within the energetically lesser deceitful KΛ excitons. It provides to a substantial expansion even at low temperatures. Their excitonic resonance line width could also control with strain [271]. Whereas, the exciton associated with substantially narrow that come to be additional symmetric in selenium containing monolayer materials, however, there is no alteration is detected for WS_2, and an increasing line width may appear for the MoS_2 [272]. These variations are traced back to the strain persuaded variations in the exciton-phonon scattering owing to the amendment of the spectral space among bright and dark excitons. In this order, the phonon aided radiative recombination passageways to momentum-forbidden states are also recognized from the PL measurements [273]. Some theoretical works have also predicted the activation of such excitons in the existence of extraordinary dipole molecules [274]. The distribution in the translational symmetry of the TMD lattice provides the desired momentum to reach the KΛ excitons depends on their molecular dipole movement and molecular exposure. An effective exciton-molecule combination consequence to a novel exciton transition (Fig. 4.16c) that exploits to a dark exciton established recognition of the molecules. The existence of momentum-forbidden dark states can be instrumently confirmed through the probing the intra-excitonic 1s-2p changeover [275]. By differentiating the optical outcome soon afterward the optical excitation, subsequent thermalization allows the alteration of the bright exciton population into the energetically lowermost momentum-forbidden dark excitonic states in monolayer WSe_2 [276].

A schematic diagram of the WSe_2 monolayer illustrates the bright and dark transitions in Fig. 4.17. As per representation, the appearance of two distinct ns Rydberg series of near-band edge excitons in the WSe_2 monolayer. First is the routinely observed bright exciton series that is related to upper conduction band sub-band. While the dark exciton series is usually connected to lesser conduction band sub-band [195, 262, 263, 277]. The large n-limit energy difference between bright and dark exciton states have been verified from the measurement of the spin-orbit splitting conduction band $\Delta_c = E_g^B - E_g^D$ (here E_g^B and E_g^D are the corresponding states). The single-particle band gaps are related to higher and inferior conduction band sub-bands. The energy gap Δ_{DB} among the ground states of dark and bright excitons also depending on the corresponding E_b^D and E_b^B, their analogous energies that can be expressed from the term $\Delta_{DB} = \Delta_c + \left(E_b^D - E_b^B\right)$.

The experimental verification of encapsulated WSe_2 monolayer and hexagonal boron nitride (hBN) flakes that fabricated on a Si substrate, from the usual approaches of the mechanical exfoliation and deterministic dry transfer techniques. The experimental findings comprised of a low-temperature (4.2 K) magneto photoluminescence which performed in magnetic fields up to 30 T employed toward the monolayer plane, as illustrated in Fig. 4.18a, b. The low-range photoluminescence spectra of the WSe_2 monolayer has been established the set of multiple photoluminescence transitions by probable creations of numerous diverse excitonic states containing nontrivial indirect/intervalley excitons and excitonic multiplexes, like trions and bi-excitons [226]. Results also recognized the typical order of emission peaks associated to the Rydberg series of $1s^B$, $2s^B$, and $3s^B$, up to $4S^B$ states of bright excitons. Their PL spectrum at $B = 0$ also showed a weak changeover associated to the ground state, $1s^D$ which correlated to dark exciton [226]. There is no extra emission characteristics have been achieved among $1s^B$ and $2s^B$ states. Due to the lower energy state, it has mostly populous (under optical excitation) and emission from this state legitimated toward the monolayer plane. The weak doublet spectrum splitting ($\approx$ 0.7 meV 6, 8) could be ignored because of their small negligible energy.

The appearance of an effective magnetic brightening of the ground state of the dark exciton in the WSe_2 monolayer has been also established by the contemporary tests. Such as the intensity of the $1s^D$ emission increases gradually as a function of $B_{\parallel}$, finally controls the PL spectrum in the boundary of high magnetic fields

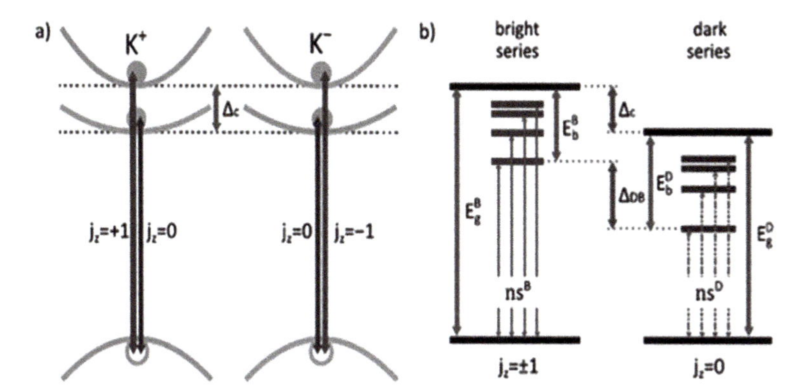

Fig. 4.17 Schematic of the bright and dark transitions in WSe_2 monolayer: **a** representation of the interband transitions in a single particle picture. The spin-up and spin-down states are shown as light-black parabolas. Optically active and inactive transitions at K^+ and K^- valleys are indicated with arrows. The light arrow represents the conduction band spin-orbit splitting (Δ_c); **b** energy diagram of excitonic transitions: Rydberg series of bright (ns^B) and dark (ns^D) resonances. The light-gray arrows represent the band gaps for bright and dark excitons $E_g^{B/D}$ and the binding energies of the ground states of bright and dark excitons $E_b^{B/D}$. The energy splitting between ground states of bright and dark excitons (Δ_{DB}) and the conduction band spin-orbit splitting (Δ_c) are represented by the arrows (adopted from Kapuściński et al. [226])

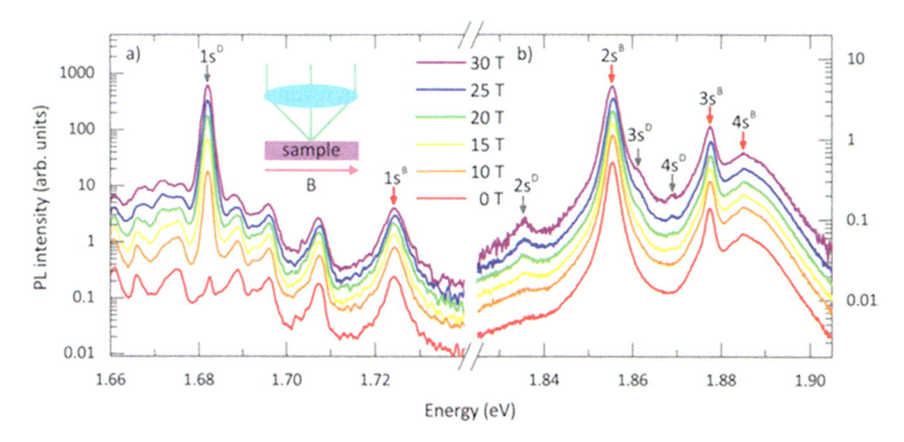

Fig. 4.18 Magnetic field-dependence of emissions of the bright and dark exciton Rydberg states. Low-temperature photoluminescence (PL) spectra in arbitrary (arb.) units at selected in-plane magnetic fields in the energy region: **a** ground state; **b** excited state. Spectral shifts normalized to the $1s^B$ feature in the lower energy region and the $2s^B$ feature in the higher energy region. Annotation the intensity scales differ for these two regions. The inset schematically represents displays the backscattering geometry used in the experiment, where the excitation and the collected light beams are quasi-perpendicular to the sample surface (adopted from Kapuściński et al. [226])

(see Fig. 4.18a). Significantly, these transitions have been governs by the employing in-plane magnetic field and it appeared in a high spectral range (see Fig. 4.18b), associated with the excited states of bright excitons. These additional conversion labels $2s^D$, $3s^{D,}$ and $4s^D$ are exhibited in Fig. 4.18 that accredited to excited Rydberg states of the dark exciton.

The magnetic brightening of dark exciton states due to an upsurge in comparative intensity $I_D(n)/I_B(n)$ of dark ns^D emission peak concerning its bright ns^B complement, approximately it following the $I_D(n)/I_B(n) = b_n B_{\|}^2$ rule. However, this rule is not useful for the ground state to the in-plane fields $B_{\|} > 20\ T$, because the comparative intensity may deviate to the quadratic dependency owing to shortened the dark exciton lifetime, therefore, it lead to a decrease in the dark exciton population. Thereby, the data points of $B_{\|} < 20\ T$ should be appropriate to estimate b. The coefficient b_n defines the relative oscillator strength $r_n = r_{ns}^D / r_{ns}^B$ and relative population $P_n = N_{ns}^D / N_{ns}^B$ (under optical excitation) of corresponding dark ns^D and bright ns^B states,$b_n = r_n.p_n$. It has been also recognized that dark and bright states cannot be far from being equally populated pairs of the exited states ($p_n \approx 2$ for $n = 2,\ 3$). Whereas, the population of the ground dark state ($1s^D$) has estimated to be five orders of magnitude higher than that of the $1s^B$ ground bright state ($p_n \approx 10^5$)[226].

Energy illustrations of both dark and bright exciton series are the most interesting to define the energy levels. The central positions of $1s^D$ and $1s^B$ emission peaks $\left(E_{1s}^{D} = 1.86\,\text{eV} \text{ and } E_{1s}^{B} = 1.725\,\text{eV}\right)$ is usually measured to the spectra which recorded with the sensible accuracy, while the corresponding dominant energies of alterations related with excited excitonic states extracts by replicating the higher energy portion of the spectra with manifold Lorentzian resonances. A typical diagram is depicted in Fig. 4.19, whereas both bright and dark series excitonic resonances are following well-recognized concepts and it can be defined from the semiempirical formula [277]:

$$E_n^{\text{D/B}} = E_g^{\text{D/B}} - \frac{Ry^{\text{D/B}}}{(n+\delta)^2} \tag{4.12}$$

Here, $E_g^{D/B}$ is the dark/bright bandgap related to the lower/upper conduction subbands $Ry^{D/B}$ belongs to the first approximation, that recognized the active Rydberg energy for the dark/bright exciton relationship $Ry^{D/B} = 13.6\,\text{eV}.\mu^{D/B}/\left(\varepsilon^2 m_0\right)$, here $\mu^{D/B}$ is the reduced mass valence band hole and electron for the upper/lower conduction sub-bands. Whereas ε is the dielectric constant of the surrounding material and m_0 is the mass of an electron. The δ is the parameter that accounted for the 2D screening length. In the case of hBN encapsulated WSe_2 dielectric screening is around $\delta = -0.09$. By extrapolating the Fig. 4.19 plots the Rytova–Keldysh potential is also obtained [262, 278]. The taken out values for the dark and bright bandgaps are around $E_g^D = 1.8802 \pm 0.0004\,\text{eV}$ and $E_g^B = 1.8940 \pm 0.0006\,\text{eV}$ and amplitude of the conduction band spin-orbit splitting $\Delta_c = E_g^B - E_g^D = 14 \pm 1\,\text{meV}$. Moreover, the binding energies of bright and dark excitons are around $E_b^B = 170.6 \pm 0.4\ \text{meV}$ and $E_b^D = 198.4 \pm 0.8 me$. Whereas the energy ratio $\rho_{\text{DB}} = E_b^D / E_b^B = 1.16$, and

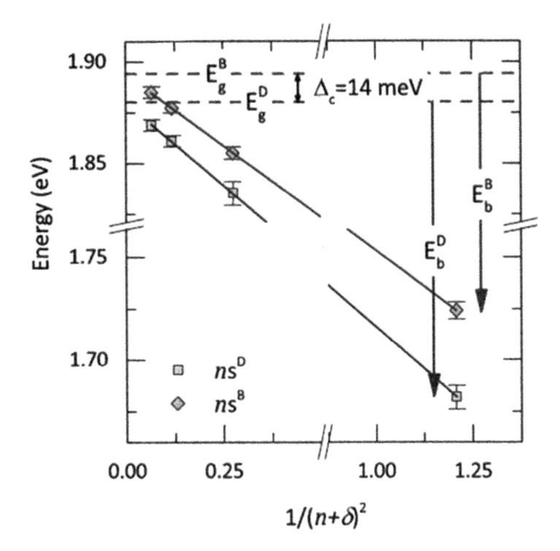

Fig. 4.19 Dark and bright excitonic Rydberg series; Experimentally defined transition energies for the bright (diamonds, ns^B) and dark (squares, ns^D) exciton s-series as a function of $1/(1+\delta)^2$ 1, where δ = − 0:09. The error bars represent the line widths of the observed emission lines. The solid and lines represent fits to the data with the model. The dashed lines denote the band gaps for dark $E_g^D = 1.880\,\text{eV}$ and bright $E_g^B = 1.894\,\text{eV}$ excitons, while the arrows correspond to binding energies of bright ($E_b^B = 171\,\text{meV}$) and dark $E_b^D = 198\,\text{meV}$ excitons. The conduction band splitting ($\Delta_c = 14$ meV) is also indicated (adopted from Kapuściński et al. [226])

$\mu^D/\mu^B = 1.18$ and dielectric constant $\varepsilon = 4.5$. It has also predicted the difference between E_b^D and E_b^B is possible because it not arises only owing to direct Coulombic term but also to the exchange term. As the bright and dark excitons consist of parallel or antiparallel spin configurations [279–281]. Hence the altercation interaction in energy dissimilarity among the bright and dark excitons binding energies in TMD monolayers also plays a crucial role.

4.3.1.4 Excitonic Collapse

Excitons are usually viewed inform of bosons by paired electrons and holes under a tightly bounded state in the Coulomb interactions. It may collapse into a phase coherent state known as Bose-Einstein condensation at low temperatures as detected from the photoluminescence spectra. Since the excitons lifetime and binding energy are crucial to observe exciton BEC experimentally in solids. In solids, the long exciton radiative lifetime allows excitons to build up a quasi-equilibrium state before recombination. Particularly excitons in solids with an indirect band gap are possessed long lifetimes. Therefore, exciton lifetime enhances from the three-dimensional parting of electrons and holes by the band alignment and external electric fields [282].

The analysis of the qualitative shape of exciton wave functions leads it diverse for robust (the energy goes to zero at μ, Lévy index = 1 signifying the exciton ionization) and weak screenings, as depicted in Fig. 4.20a, b. On which contrary exciton collapses as its energy tens to minus infinity. In both screening regimes, the wave functions are sensitive to the Lévy index μ. particularly, for weak screening μ diminished from 2 to 1, therefore, the wave functions become progressively less localized in k space. The corresponding amplitudes diminish like $\mu \rightarrow 1$. So, the wave functions become almost delocalized with zero amplitude. This situation can also correlate to "extra strong" localization in the coordinate space, at which wave function degenerates into (almost) Dirac δ—function, this is the condition of the exciton collapse, by mean "falling" of the electron to the hole. Whereas, the opposite situation may also occur in the strong screening, in which the momentum space of the wave function obtains Dirac δ function shape to $\mu \rightarrow 1$. Whereas the coordinate space corresponds to the nearly delocalized wave function, it belongs to the ionization of the exciton. Thus the interaction among screening effects and disorder leads whichever to exciton ionization or its collapse for the robust disorder. Moreover, it could also stated that, the interaction among screening and robust disorders such as substance amorphization, it is common at the semiconductor boundaries, whereas a higher mechanical stresses happen and it abolish by ionizing or collapses contingent on the relationship among screening radius, Lévy index, and orbital quantum number m of the excitons, however, it prevents the functional ability of the devices such as solar cells and/or light emitting diodes [283].

TMD substances like MoS_2, WS_2, and WSe_2, excitons have a higher binding energy that makes them utmost thermally steady [218]. But influences to the imperfections and kinds of the disorders of these excitons features (like their relaxation and decoherence mechanisms), it is remain debatable to get a superior extent. Typically, the exciton binding energy of the 2D dichalcogenide semiconductor WS_2 has

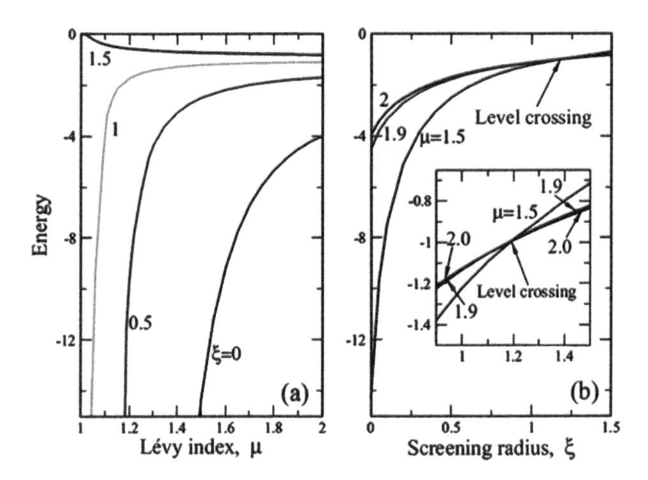

Fig. 4.20 Ground state energy as a function of Lévy index μ and dimensionless screening radius ξ: **a** near curves to define the screening radius; **b** Lévy index. The Inset of panel **b** shows details of the behavior of ground state energy at $\xi > 0.9$, where energy levels crossing occurs approximately at $\xi \approx 1.15$. The levels crossing point is the same for all $1 < \mu < 2$ and occurs for finite screening only (adopted from Kirichenko et al. [283])

been obtained everywhere 0.32 eV, it is enough higher than (around 0.05 eV) normal 3D materials [195]. This result has established the 2D TMD materials are decent member for the optoelectronic and photonic devices at the room temperature. Their in-plane exciton localization radii can be around 5 nm [195, 218]. The arithmetical estimations with the potential like Wannier-Mott-like exciton have also been sound realization of the atomically thin WS_2 and other monolayer TMDs with screening radius $r_0 \approx 75$ Å [195, 196].

A simpler correlation of the excitonic collapse can be interpreted as; above the excitonic states (those probed by linear processes to *s*-states, those probed by two-photon processes to *p*-states) it is normally established. The relativistic quantum mechanics' robust Coulombic coupler leads in extreme cases to an excitonic collapse escorted by the integration of the 1*s* states in the band structure continuum [284, 285]. According to the simple selection rules the *s*-like excitonic states are combined to the light field, as the consequence gives us famous excitonic Rydberg series with the lowermost optical active state to 1*s* exciton resonance depending on the selection rules. Though, the *s*-kind optical selection rules following the implicit supposition of the system excited from the non-interacting ground state. While, the intensely interrelating system ground state himself excitonic [285], such optical intra-excitonic transitions governs by basically diverse optical selection rules and show distinct peaks by making *p*-states bright and *s*-states dark. This prospect is considered to be a rather exotic need for further investigation. Keeping the view of the energy locations of the several witnessed features assigned by this methodology approved (within the Rydberg model) with the experimental and the usual assignment, as illustrated in Fig. 4.21. Moreover, it would be also accounted for a usual description of the resonant upsurge of second harmonic generation for energies consisting to 1s resonances excitons.

4.3.1.5 Intra Transition

Interband absorption reveals the ability to generate bound electron-hole pairs, while the intra-excitonic absorption probes existing excitons through transitions from the 1*s* ground state to higher relative-momentum states, such as 2*p* by absorption of mid-infrared phonons. A typical schematic of the selection rule for the 2D tungsten-based TMD monolayer and different possible types of transitions is illustrated in Fig. 4.22a–f. Fig. 4.22c–f, it represents the involvement of different kinds of exciton transitions in their usual symbols, such as the bright exciton (X_0), dark exciton (X_D), positive trion $\left(X^+\right)$, intervalley trion $\left(X_1^-\right)$, intravalley trion $\left(X_2^-\right)$, and the exciton–plasmon $\left(X^{'-}\right)$[286, 287]. Among these several additional emergent excitonic states, like charge-neutral biexciton (XX) and trion–exciton multiplexes $\left(XX^-\right)$, positive dark trion $\left(X_D^+\right)$ and negative dark trion $\left(X_D^-\right)$, intervalley exciton (X_i), and dark exciton phonon replication $\left(X_D^R\right)$ owing to the competent gate controllability may also appear.

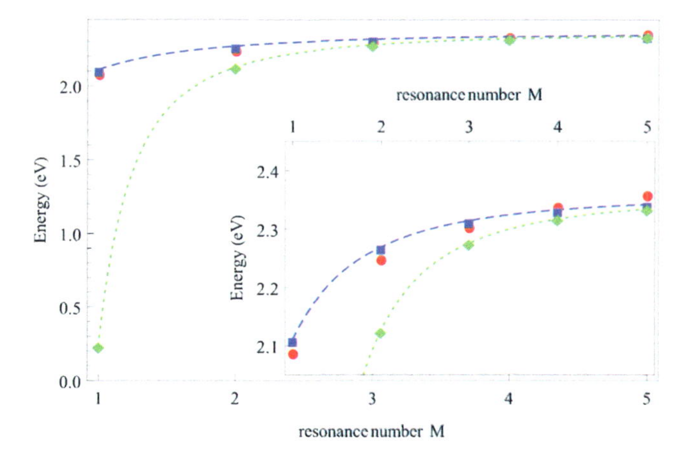

Fig. 4.21 The spectral position of the energetically lowest bright excitonic transitions with $j = \pm 3/2$ in WS_2. The squares are the computed results assuming bright *p*-type excitons, the dots show the experimental data and the diamonds indicate the theoretical predictions for the *s*-type bright states. The dashed and dotted lines represent energies. The *x*-axis label denotes the number *N* of the experimentally noticed bright exciton resonances that related to the principle quantum number via $n = N - 1$. The inset figure shows the results for the energetically higher states with a finer energy resolution (adopted from Stroucken, T. et al. 2014, arXiv:1404.4238)

Interesting intra (or inter-layer) valley transition is not only due to the spin-forbidden dark exciton to combine to the chiral phonon and produce the phonon replication with circular polarization, but it also depends on momentum-forbidden intervalley dark exciton interaction from the chiral phonon which located at K point, as the consequences circularly polarized light emission [288]. If we suppose circularly polarized excitation occurs in the K valley, then intervalley exciton consisting one electron in the K′ valley and one hole in the K valley. The electron transitions toward a simulated state in the K valley by releasing a chiral phonon afterward recombined to the hole in the valence band of the similar valley. The emitted photon can have a definite helicity, as the schematic is depicted in Fig. 4.23a. As presented in Fig. 4.23b, the peak X_i constantly occurs among the X_D^R and XX^-, and the energy dissimilarity among X_i and X_D around 16 meV. The magneto-PL spectra also demonstrated that X_i has a sharpest curve for all excitonic states associated to achieve g factor −12.5. This value also consists of the theoretic calculations of the spectrum g factor of the intervalley exciton [184]. The valley determined PL spectra have also revealed X_i possess a analogous valley polarity to bright exciton, 36.7% versus 39.5% [184]. Moreover, the lifetime of the X_i spectra of the 200 ps TRPL line trace have been also examined, this findings has revealed that it is equivalent to the dark exciton, but their magnitude order is greater to the bright exciton and intervalley trion. The lengthy existed intra-exciton with valley evidence is considered to be a promising route to realizing excitonic valleytronics and their coupling with the chiral phonon useful to exploiting chiral phonon for valley-spin manipulation [184, 276].

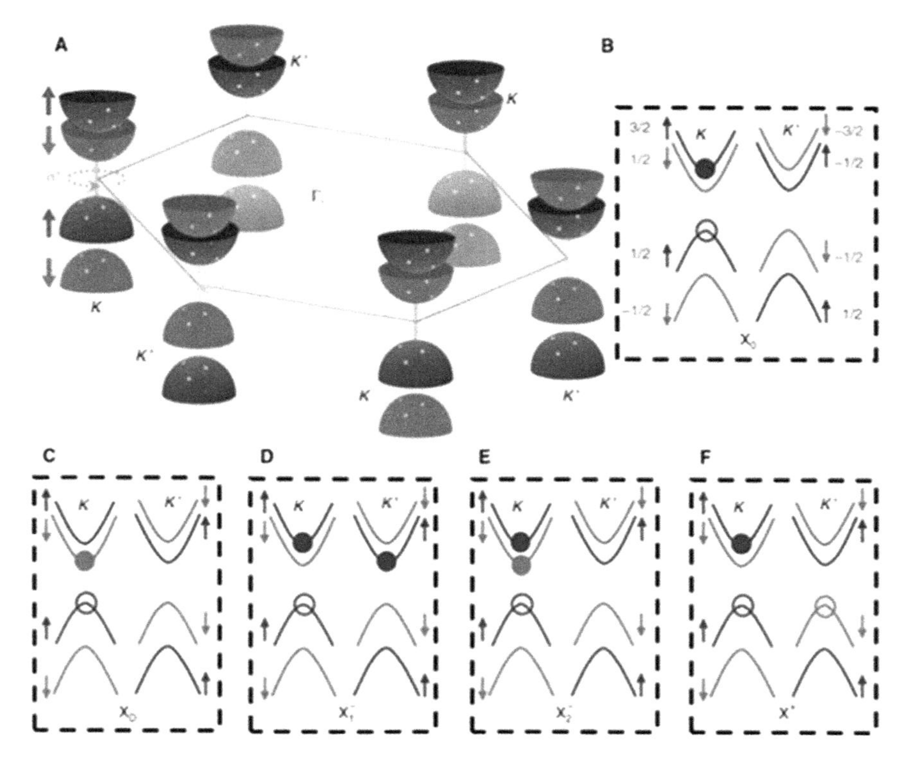

Fig. 4.22 Excitonic states in 2D tungsten-based TMD monolayer; **a** Valley-dependent optical selection rules for interband transitions in 2D monolayer tungsten-based TMDs. Both the conduction and valence bands are spin-split due to the strong SOC, resulting in the spin-valley locking. σ^+ polarized light only couples to the K valley, whereas σ^- polarized light only couples to the K′ valley. Light and dark black stand for a spin up and spin down; **b–f** Excitonic configurations of the bright exciton 0 (X_0), dark exciton (X_D), intervalley negative trion (X_1^-), intravalley negative trion (X_2^-), and positive trion (x^+). The z-component of the total angular momentum for each band, with the Berry phase contribution considered (adopted from Li et al. [184])

4.3.1.6 Dynamics

Carrier and exciton dynamics of monolayer TMDs have been extensively studied by the most common ultrafast pump-probe spectroscopy. This is an ideal tool to investigate the intrinsic ultrafast optical phenomena in the 2D regime. To interpret the ultrafast formation and recombination of excitons [289–291], the many-body correlations between excitonic complexes [188, 292, 293], valley/spin depolarization dynamics [294–297], giant bandgap renormalization and transient optical gain under the intense optical excitations have been used [298, 299], including ultrafast interface charge transfer in TMDs van der Waals heterostructures [300]. Besides, a comprehensive physical picture of the exciton dynamics and transient optical nonlinearities is usually absent, owing to the influence of carrier screening effect on the optical response of exciton resonance inconsistency, such as excitation wavelength

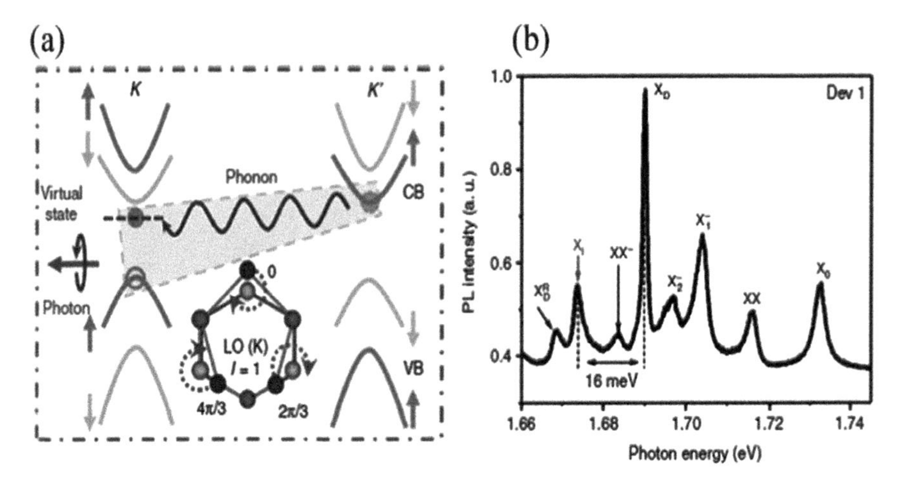

Fig. 4.23 Intervalley exciton in h-BN–encapsulated monolayer WSe_2: **a** Schematic of the intervalley exciton and chiral phonon coupling. The intervalley exciton consists of one electron and one hole in the different valleys. The electron can transit to a virtual state located at the hole-sitting valley by emitting a chiral phonon and then recombining with the hole, emitting a circular photon. Inset shows the schematic of the E′chiral phonon mode at the K point; **b** Photoluminescence spectrum of monolayer WSe_2. Along with the established excitonic states, a novel peak X_i locates between the phonon replica of the dark exciton X_D^R and charged biexciton XX^-, with the -16 meV lower than XD (adopted from Li et al. [184])

and temporal resolutions [299–302]. Due to controversy like exciton–exciton interactions such as Coulombic repulsion or atom-like attraction [302–305]. Moreover, under the intermediate and high regime excitation densities, the explicit assignment of Auger-type nonradiative mechanisms is also required to understand the associated nonlinear optical phenomena [291, 306–310]. Like the photoexcited electron-hole system of monolayer TMDs remains in an unexpected non-equilibrium state at the nanosecond (ns) timescale [291, 311]. And dynamics of other quasiparticles, such as trions, remain largely unexplored [312].

Therefore, the dynamical responses of TMDs (or all kinds of semiconductors) are broadly divided into coherent and incoherent regimes [313, 314]. The incoherent exciton dynamics of TMDs include processes like exciton thermalization, radiative/non-radiative recombination, intra and intervalley scattering, exciton dissociation, and exciton–exciton annihilation [315]. While the coherent exciton response is usually related to the dynamics of the exciton formation process, it is less explored due to the limited availability of sequential resolution (i.e., > 100 fs for pump-probe optical spectroscopy and > 1 ps to time-resolved photoluminescence) [299, 316, 317]. Therefore, our attention to discussing the dynamics of TMDs based on the ultrafast pump-probe approach, however, the valley dynamics will be discussed in the next chapter.

Recently, a study with the chronological resolution of ultrafast difference reflectivity (ΔR/R) spectroscopy of the fewer than 30 fs dynamics to exciton formation in

1LMoS has been demonstrated [315] By use of a pump-probe technique is generally evaluated the excited state population by monitoring the ΔR/R value. Specific attention has been also paid to in what way lengthy time takes in a preliminary inhabitant of higher energy photoexcited electrons and holes to liberate the least energy of the exciton states, as depicted 1s states for the A and B excitons. Moreover, the approach to how the formation time depends upon the energy of the initial state (see Fig. 4.24). The noted dynamics are to be found remarkably fast, by means the lowest energy excitons from the high-energy populations when a characteristic timescale as fast as 10 fs. Meanwhile, the increasing energy of the initial state of photo-injected carriers increases the pump photon energy, as consequence possibilities are; (i) the development period of the excitons is increasing gradually with the growing energy; (ii) the preliminary sub-100 fs fast degeneration constituent of the excitons would vanish; (ii) the dynamics of a leisurelier degeneration procedure (on the timescale of picoseconds) can be connected to liberation of a thermal inhabitation of excitons which does not considerably alternated.

Schematically ultrafast exciton dynamics can be summarized by the exciton construction procedure (see Fig. 4.24a) with the important optical excitonic changeovers (see Fig. 4.24b) of 1L-MoS_2. The lowermost energy A and B excitonic resonances arisen to the optical alteration among the twofold uppermost energy valence bands and the lowermost energy conduction bands at the K and K' points in the Brillouin zone that pinpointed at the energies 1.88 and 2.03 eV (see Fig. 4.24b). Moreover, the illustration of Fig. 4.24c is also revealed both A and B optical alterations via twin manifolds of Rydberg-like series of bound states, which combine to a continuum of unbound electron-hole states [318, 319]. The corresponding resonances belong to lowermost energy (or ground state) excitons for the A and B multifolds (A1s and B1s states). Such states can possessed the greatest oscillator strengths with the dominating optical outcome above the respective high energy unbound states. Moreover, it also gives the estimated exciton binding energy order of 350 meV, by measuring the energy dissimilarity among the 1s exciton and the lowermost energy state in the exciton continuum [315]. The pump photon energies width between the 2.29–2.75 eV (with bandwidths spanning from 70 to 140 meV), it is useful to get all energies photoexcitations as an early inhabitants of excitons at or sound into the exciton continuum for A excitons. Likewise, B excitons at energies overhead $\sim$ 2.4 eV has been also achieved. The subsequent formation dynamics and formation time ($\tau(E)$ (see Fig. 4.24c) of the probe energies also gives us the rise and decay dynamics of the ΔR/R signals for the 1s states of the A and B excitons.

Under the tunable pump pulses photo-inject electron-hole pairs excess energy increases concerning E_G. In which the time-delayed broadband probe pulses to be quantified the build-up and early decay dynamics of both A1s and B1s excitons. The experimental temporal resolution of the sub-30-fs equipment has revealed the $\Delta R/R(E_{\text{probe},}\tau)$ map as a function of the pump-probe delay τ and the probe photon energy E_{probe} for two different pump photon energies: 2.29 and 2.75 eV across and above the E_G, as illustrated in Fig. 4.25a, b. Whereas, to reflect the Si substrate the ΔR/R value should be equivalent to a double-pass differential transmission (ΔT/T). The experimental evidence showed a comparable qualitative response with the positive

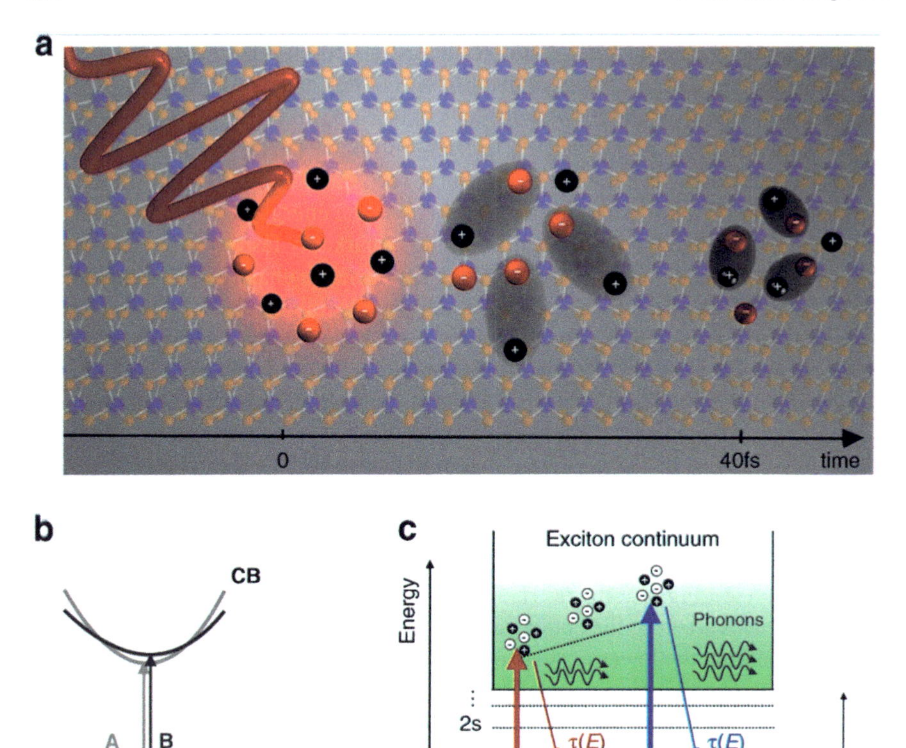

Fig. 4.24 The formation process of exciton in 1L-MoS_2: **a** cartoon of the exciton formation process after photo-injection of free electron-hole pairs; **b** Schematic for illustration of the single particle band structure of 1L-MoS_2 at the K/K′ points. The two arrows represent A/B excitonic transitions, split due to the strong spin-orbit interaction at the K/K′ points of the Brillouin zone; **c** Sketch of the pump-probe experiment, to show a few-optical-cycle laser pulse injects free electron/hole pairs at increasing energies above the exciton continuum (E_G). These quasiparticles lose their initial kinetic energy and scatter down via a cascade process mediated by phonons to lower-lying discrete excitonic states until they reach the 1s excitonic state. The timescale $\tau(E)$ of this relaxation process reflects the absorption change in a probe beam that tuned on resonance with the 1s state, due to the Pauli blocking effect (adopted from Trovatello et al. [315])

strong features associated with A_{1s} (centered at $E_{probe} = 1.88$ eV) and B_{1s} (centered at $E_{probe} = 2.03$ eV) exciton states. Thus $\Delta R/R$ spectrum can have a symmetric profile for each excitonic resonance without any shift in peak maxima, under the temporal window 100–200 fs. However, the feeble and wide adverse characteristics have been noticed at greater and inferior probe energy transitions owing to a photonic consequence caused by the interference of manifold reflections of the inward beam.

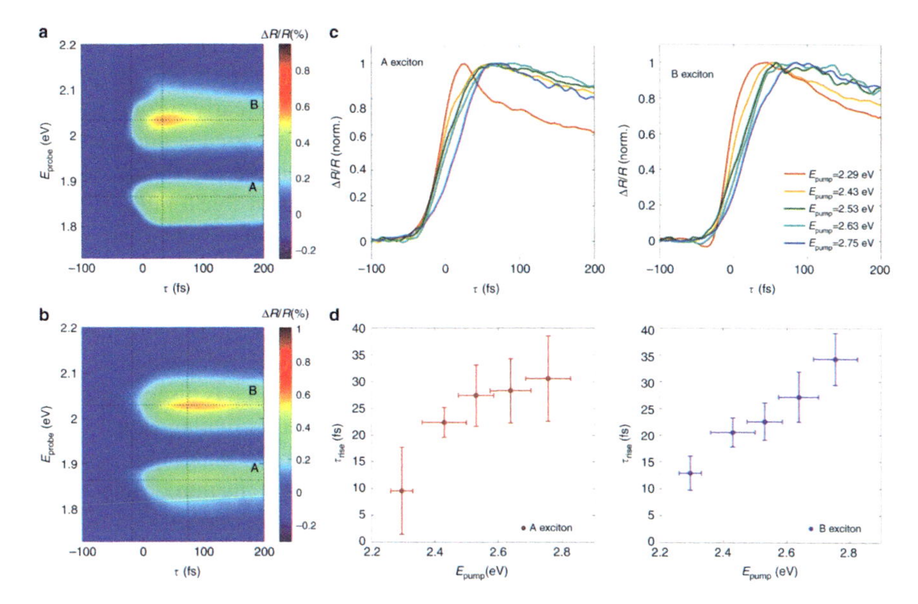

Fig. 4.25 The formation process for energy-dependent exciton: the $\Delta R/R$ maps measured on 1L-MoS_2 photoexcited **a** at 2.29 eV; **b** at 2.75 eV, at room temperature. The horizontal dashed lines pass through the maximum of the ΔR/R spectrum at the energies of the A_{1s}/B_{1s} exciton transitions, while the vertical lines mark the temporal range from 10 to 90% of the build-up signal. The excitation fluence is 5 μJ cm^{-2}. Pump and probe beams have parallel and linear polarizations. The time zero is defined, for each measurement, as the maximum of the cross-correlation signal between the pump and the probe pulses; **c** temporal cuts of the $\Delta R/R$ maps across the A_{1s} and B_{1s} excitonic resonances for increasing pump photon energy; **d** pump photon energy dependence of τ_{rise}. Horizontal error bars are determined by the bandwidth of the pump pulses and vertical error bars are defined from the fits of the time t_{races} (adopted from Trovatello et al. [315])

However, the temporal window explored ~200 fs gives a $\Delta R/R$ spectrum a symmetric outline in almost every excitonic resonance with a un-shift peak maxima.

Moreover, a closer inspection has also revealed the formation time τ_{rise} of the ΔR/R of the A_{1s} and B_{1s} excitons that could be longer for the higher pump photon energies. Therefore, the excitation energy becomes close to E_G and their signal is a quasi-instantaneous (by mean pulse width limited) build-up. While with the increase in pump photon energy τ_{rise} is also significantly increasing. The high temporal resolution has provided a better-quantified effect, as depicted in Fig. 4.25c, the temporal cuts for A_{1s} and B_{1s} exciton peak with the growing pump photon energies. The corresponding construction periods has been also assessed by the fits of the sequential traces in the progressive space among 100 and 200 fs by employing the following relationship;

$$f \propto \left(1 - e^{-t/\tau_{rise}}\right) * H^*\left(A_1 e^{-t/\tau_1} + A_2 e^{-t/\tau_2}\right) \tag{4.13}$$

where H is the heavy side function centered at $\tau = 0, \tau_1$ and $E(\omega)$ (they are decay constants), A_1 and A_2 are the amplitudes of every deterioration constituents. Their fit function can be a complex Gaussian profile respect to the instrumental feedback that is evaluated from the cross-connection profile among the pump and the probe, as depicted in Fig. 4.25d. The τ_{rise} gradually enhances with the early surplus energy of the photoinjected charge carriers. Remarkably, in case of the greater pump energy excitation (i.e., 3.75 eV) the τ_{rise} sensitively deviates with the increasing energy trend. Tentatively flattening in the formation dynamics is responsible for the different phonon-mediated scattering processes due to the involvement of the electronic states too away from the K/K' points. The feeble negative incline could be related to negative times in the B exciton. The excitation photon energy of 2.29 eV is attributed, it is well known pump-perturbed free-induction decay (PPFID)[320]. The procedure happens while the probe pulse precedes the pump, therefore, free-induction degeneration field emission is due to the specimen excited by the probe pulse which is disconcerted by the interplays with the pump pulse. It also gives increase in oscillating signals with the negative delays. Therefore, the PPFID impacts disregard to building the period of the specimen, causing, its amplitude additional one order of magnitude and lesser to the blanching of the excitonic peak. Additionally, this consequence also reveals the exceptionally fast deterioration for the low pump photon energy, it happens on a period analogous to that of the building and disappears too far as the pump adjusted for the high energy photons. This influence is more vibrant for the A exciton in ΔR/R trace, as depicted in Fig. 4.25c.

4.3.1.7 Interactive Exciton

Several studies on the exciton–exciton interactions, specifically the nonlinear optical response of various atomically thin two-dimensional layered materials have been examined [259, 321–323]. TMDs' inversion symmetry breaking and efficient light-matter interaction makes them ideal to investigate nonlinear optics [324, 325]. In various TMDs, such as MoS_2, $MoSe_2$, $MoTe_2$, and WSe_2. WS_2, ReS_2, etc. second and third-order nonlinear parametric mechanisms are usually generated from the monochromatic laser illumination [326–334]. However, a well-accepted view of the optical excitation in a semiconductor at the band gap typically results in luminescence at lower energy. A resonant laser excitation of the lowest-energy exciton results pronounced photoluminescence (PL) emission at higher energy compared to excitation laser of different TMD materials [335].

The efficiency of the exciton-exciton interaction is usually calculated from the excitation-induced dephasing (EID) of excitonic resonances under an incoherent limit. The existence of intravalley and intervalley as well as intralayer and interlayer exciton-exciton scattering channels is depicted in Fig. 4.26a–d. The temperature and screening dependence EID is evident from microscopic insights into the fundamental nature of scattering between intralayer and interlayer excitons. The mass asymmetry between electrons and holes, including the overlap of Bohr radii, are also considered key quantities to determine exciton-exciton efficiency. As interpreted the hBN encapsulated WSe_2 monolayer also provides the choice of substrate that can influence the Auger processes [309, 323, 336] due to dominant intravalley scattering channels within the K,∧, and K' valley, as illustrated Fig. 4.26b, d. The temperature and density dependence EID for the WSe_2 monolayer is depicted in Fig. 4.27a–c. Experimental finding revealed that the EID increases linearly with exciton density n, i.e., $\gamma^{KK} = \gamma_{x-x}n$ with the slope γ_{x-x}. However, the slope of the EID showed high temperature dependent with $\gamma_{x-x} = 2.9 \times 10^{-10}(4.9 \times 10^{-10})\ \mu\,\text{eV}\,\text{cm}^2$ for $T = 300\,(50\,\text{K})$ (see Fig. 4.27b). This behavior generally governs by the temperature dependence of the exciton distributions of the bright KK and momentum-dark K∧ and KK′ states (see Fig. 4.27c). At low temperatures ($T < 30$ K), KK′ excitons reveal the EID, because most of the excitons reside in the energetically lowest state, as depicted in Fig. 4.27b [335]. Moreover, the important dark excitonic states EID are most efficient for the tungsten-based TMDs. Similarly, in $MoSe_2$ the KK state is usually energetically to the lowest [337], therefore, bright excitons are expected to dominate EID at all temperatures. While the WSe_2 monolayer on a sapphire substrate have quantitatively larger values of EID [338].

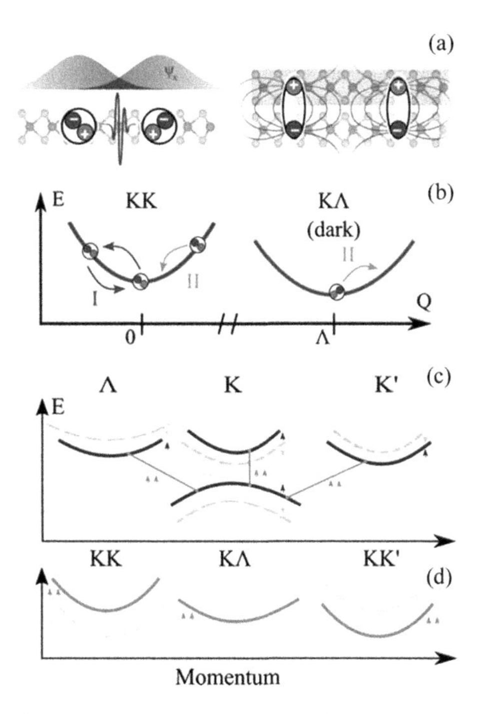

Fig. 4.26 **a** Schematic for the exciton-exciton scattering in a TMD monolayer (left) and a van der Waals heterostructure (right), while the intralayer interaction is determined from the wave function overlap, the interlayer coupling resembles a dipole-dipole interaction; **b** Scattering channels in monolayer WSe_2 including intravalley (I) and intervalley processes (II); **c** Schematic of the electronic band structure of monolayer WSe_2 in the vicinity of the high symmetry K, K′ and Λ points, indicated to spin-allowed transitions (lines); **d** Excitonic band structures for the energetic ordering of exciton states as a function of the center-of-mass momentum. Note that the momentum-dark KK′ and K Λ states are energetically lower than the bright KK exciton (adopted from Erkensten et al. [335])

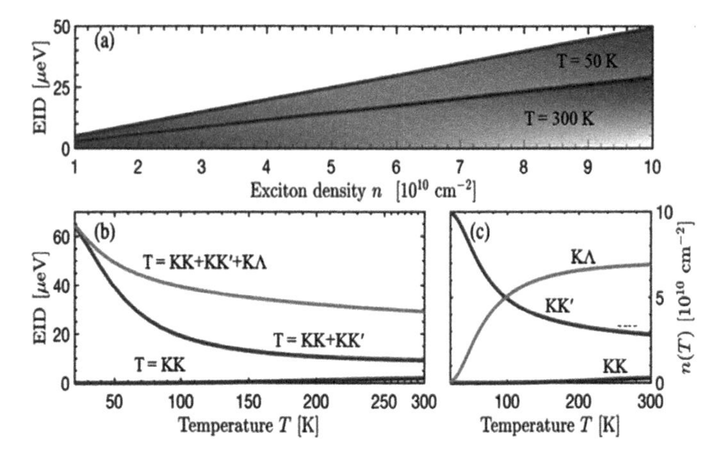

Fig. 4.27 Temperature and density dependence of excitation-induced dephasing in monolayer WSe_2: **a** EID as a function of exciton density "n" for two different temperatures; **b** EID as a function of temperature to show single scattering contributions including intravalley scattering (KK) as well as intervalley scattering (K Λ, KK$'$) at fixed exciton density $n = 10^{11}$ cm^{-2}: **c** Temperature-dependent density $n(T)$ for the KK, K Λ, and KK' excitons adopted from Erkensten et al. [335])

To understand the relationship of the semiconductors, it is crucial to consider the role of light-induced excitons (electron—hole pairs) in parametric processes and their effect on the *n*th-order nonlinear susceptibility $\chi^{(n)}(\omega)$. In 2D materials, it is more appropriate to use surface conductivity $\sigma(n)$ to describe the nonlinear optical response about sheet current $j^{(n)}(\omega)$ and electric field $E(\omega)$, where $\sigma(n)$ represents a $(n+1)$th-rank tensor which strongly depends on symmetry of the crystal structure of the medium and its composition. The contribution of light-induced excitons to the $\sigma(n)$ of TMDs particularly responsible for resonances between the laser frequency and excitonic states that favor certain multiphoton-based mechanisms. Such resonances and the selection rules for one-photon and two-photon PL have been demonstrated for the systems with reduced dimensionality like carbon nanotubes and TMD monolayers [259, 339]. In the case of TMD monolayers, the role of excitons are considered more relevant to their multilayer and bulk counterparts. Due to Coulomb interaction between electrons and holes enhances the quantum confinement and reduced the dielectric screening inherently in an atomically thin system [259]. The enhancement leads to an increase in radiative transition probabilities and the implication of the presence of excitons in the nonlinear response of monolayers. To explain this process a microscopic theory based on the role of wave functions symmetries of the excitons in nonlinear optical response has been also interpreted [340]. Later, investigators used this theory to describe the second-order susceptibility by explaining the second harmonic generation at exciton resonances in WSe_2 [259]. However, the generation of excitons via nondegenerated multiphoton absorption and its influence on the nonlinear response of TMD monolayers and the PL emission is still unresolved [252, 341, 342].

4.3.1.8 Mirror Exciton

A surprising observations has also made to optical properties of MoS_2 which strongly influenced when placed close to a gold-coated surface [343]. Compared to dark field image and scattering spectra from monolayers MoS_2 on quartz. The CCD-captured pictures have revealed remarkably different scattering colors of MoS_2 crystals with the distinct substrates. Such as scattering on quartz has a blue color, while the scattering on Au shown green. Usually, the scattered light intensity of MoS_2 on quartz is twofold robust to MoS_2 on Au. It has also noted that for both systems, the light scattered supplementary powerful to the edges of the monolayer than the midpoint, although their spectral shapes are almost unaltered. In the either cases, their twice K-excitonic states in energy levels splits owing to robust spin-orbit coupler. It has been also noted that the color of the material in the optical pictures is governed by robust (diverse) scattering at the greater energies compare to the K-excitonic resonances. While the smaller wavelength's resonant scattering amplitude is roughly four times robust as scattering to the K-excitons. Remarkably for a quartz substrate, only lone scattering peak has been recognized, whereas, for MoS_2 on a gold system, the center at 439 nm for light dispersed to the edge of the crystal, and a second robust scattering peak seemed at 502 nm showing green crystals. It corresponding peak has been accompanying to an exciton band nesting nearby the Γ -point, at which the valence and conduction band dispersals are parallel, therefore, the van Hove singularity with excitonic characteristics [230, 232].

Reports on the mirror are not limited to initial observations but come regularly to explore various TMDs' distinct properties, to describe investigators have made efforts efficiently by introducing the various experimental and theoretical studies [344]. Since two TMD monolayers are bound by the weak van der Waals force, therefore, a theoretical model based on decoupled bilayer and interlayer coupling as a perturbation has been introduced [345]. According to this approach under the vanishing interlayer coupling limit, the monolayer Bloch wave functions for the $\tau \boldsymbol{K}$ valley expressed as $\psi_{n,k}(r) = e^{i(\tau \mathbf{K}+k).r} u_{n,k}(r)$ (here symbols are in their usual meaning). So, the Bloch wave function $\psi_{n,0}$ constructs the local basis functions like;

$$\psi_{n,0}(r) = \frac{1}{\sqrt{N}} \sum_{R} e^{i\tau \boldsymbol{K}.\boldsymbol{R}} D_n(r - \boldsymbol{R}) \tag{4.14}$$

where N is the unit cell number of the respective monolayer, and $D_n(r - R)$ is the linear combination of the atomic orbitals localized near the metal position **R**, which depends on the valley index τ. If we consider the time reversal relation between the two valleys, D_n in the same band but opposite valleys related by a complex conjugate then the Bloch wave function is expressed as $\psi_{n,k}(r) \approx e^{i(\tau \mathbf{K}+k).r} u_{n,k}(r) = e^{ikr} \psi_{n,0}(r)$, the corresponding equation can be written as;

$$\psi_{n,k}(r) = \frac{1}{\sqrt{N}} \sum_{R} e^{i(\tau \boldsymbol{K}+\boldsymbol{k}).\boldsymbol{R}} D_n(r - \boldsymbol{R}) \tag{4.15}$$

where $\sum_R e^{i(\tau\boldsymbol{K}-\boldsymbol{k}).\boldsymbol{R}} D_n(r-\boldsymbol{R}) \approx D_n(r-\boldsymbol{R})$ for the low energy electrons and holes with small $|\boldsymbol{k}|$, and $D_n(r - R)$ which is well localization near **R**. In bilayer stacking the in-plane crystalline axes of the two layers along the same direction (R-type stacking), and the two metal atoms in different layers horizontally overlap at the in-plane (*xy*) coordinate origin. Likewise, any other stacking configuration can be also obtained using the configuration through a θ-angle rotation of the upper layer around the coordinate origin that also follow the translation $-r_0$ to the lower layer, as depicted in Fig. 4.28. Further addition of the interlayer coupling as a perturbation of the hopping integral between the two wave functions $\psi_{n'k'}$ and $\psi_{n,k}$ that are located in the upper and lower layer. So, the perturb wave function could be modified as illustrated in Fig. 4.28. Simplifying the expression gives us $\sum_{k'k}$ only a few terms k' and k small magnitude that greatly reduced the number of $\tau k - \tau' k'$. As depicted in Fig. 4.28c three groups of $k.\boldsymbol{K}$, $\hat{C}_3\boldsymbol{K}$ and $\hat{C}_3^2\boldsymbol{K}$ on the thickest circle that are closest to Γ and correspondingly most pronounced term $|\tau_{n'n}|$: $-2\boldsymbol{K}$, $-3\hat{C}\boldsymbol{K}$ and $-2\hat{C}_3^2K(k_{1,2}, \hat{C}_3k_{1,2}$ and $\hat{C}_3^2k_{1,2})$ which correlates to the second (third) closest to Γ, and their corresponding $|\tau_{n'n}|$ values connected to much weaker transitions [345].

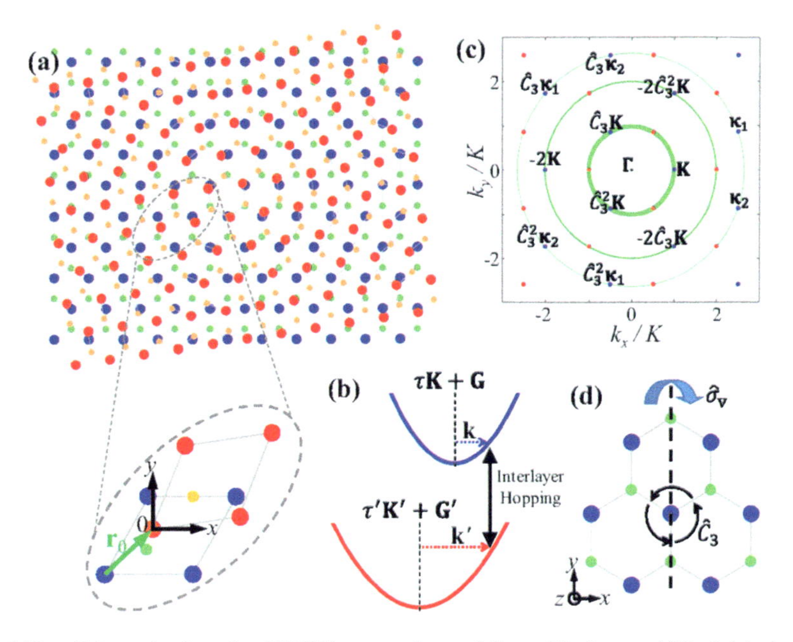

Fig. 4.28 **a** Schematic of a twisted TMD homo- or hetero-bilayer. The large red (blue) dots denote the metal atoms in the upper (lower) layer, and the small orange (green) dots denote the chalcogen atoms in the upper (lower) layer. The enlarged view shows two unit cells in the upper and lower layers. The in-plane (*xy*) coordinate origin is set on a metal atom in the upper layer; **b** two wave vectors in different layers show the overlap in momentum space to satisfy the momentum conservation of interlayer hopping; **c** the blue dots denote $\kappa \equiv \boldsymbol{K}+\boldsymbol{G}$ points, and the red dots are their time reversals. Thicker green circle means smaller $|\kappa|$ thus larger $|tnn'(\tau\kappa)|$; **d** Illustration of the $2\pi/3$-rotational ($\hat{C}_3$) symmetry and the in-plane mirror ($\hat{\sigma}_v$) symmetry of monolayer TMDs(adopted from Wang et al. [345])

Usually, the monolayer hexagonal lattice structure has both the $2\pi/3$-rotational $\hat{C}_3$ symmetry and the in-plane mirror $\hat{\sigma} v$ symmetry, as illustrated in Fig. 4.28d. The hopping terms $tnn(\mathbf{q})$ with the same $|\mathbf{q}|$ values but different $\mathbf{q}$ directions are also related to these symmetry operations. More specifically the mirror reflection is devoted to $\hat{\sigma} v$ for a real space vector $\mathbf{r} = (r_x, r_y)$ over the vertical yz plane, i.e., $\hat{\sigma}_v r = \left(-r_x, r_y\right)$, or on a wave vector $\boldsymbol{q} = \left(q_x, q_y\right)$ as $\sigma_v q = \left(-q_x, q_y\right)$. So, it is obvious $\hat{\sigma}_v \boldsymbol{K} = -\boldsymbol{K}$, under the mirror reflection $\psi_{\tau,0,n}(\hat{\sigma}_v r) = \psi_{\tau,0,n}(r) = \psi_{-\tau,0,n}(r) = \psi^*_{\tau,0,n}(r)$, here the last step represents the time reversal relation between two valleys. Both the upper and lower layer can have yz-plane mirror symmetry, corresponding to appropriate R stacking ($\theta = 0°$), or H stacking ($\theta = 60°$) [345].

Since in the homo-bilayer TMDs conduction and valence bands of the structures are possessed twofold degeneracy at $\tau \boldsymbol{K}$ point (without spin-orbit coupling), but the presence of interlayer coupling causes a finite energy level splitting which provides information on the hopping term of $\tau_{n'n}(q)$. As the modulated R type ($\theta = 0°$)/ or H type ($\theta = 60°$) TMD homobilayer structures with varying r_0 (thickness), the successive two layers are commensurate and their interlayer hopping between τK in the lower layer and $\tau' K'$ in the upper layer when $\tau' = \tau$ for R stacking, and $\tau' = -\tau$ for H stacking. The simplification of the terms gives most of the $\boldsymbol{r}_0$ values corresponding to R- or H-type bilayer structures that are unstable, this means they do not exist in nature. However, such structures can locally exist in anion commensurate bilayer with a large-scale moire superlattice pattern [346–348]. In a local region with a size much larger than a monolayer lattice constant but much smaller than a supercell. Their atomic registry between the two layers are locally indistinguishable from an R- or H-type commensurate bilayer [346–348]. As $\boldsymbol{r}_0$ varies from position to position in a supercell therefore **r**0-dependent conduction/valence band energy also shifts, this energy shift is usually responsible for the observation of position-dependent local band gap modulation [51].

In the modern era, specific attention has been paid to excitons control in an atomically thin film with a mirror [349]. Because of the radiative properties of the excitons that are drastically modified by placing microcavities in photons above a mirror [350, 351]. In such a system light emitted in the upward direction interferes with the light emitted downward and subsequently reflected by the mirror. It depends on the phase of the reflected light relative to the directly emitted light, usually, their interference is constructive to enhance the emission and a shortened radiative lifetime of the excitons. While in case of destructive interference prolongs the radiative lifetime. Similar effects can also be noticed in several individual dipolar optical emitters, such as atoms, ions, molecules, and quantum dots [349]. Near a mirror, the resultant change in radiative rates limits to a few tens of percent because only a small fraction of optical modes is reflected by the mirror onto the quantum emitter. It means the inplane momentum of delocalized excitons in a homogeneous, atomically thin membrane is conserved during radiative decay. Such light reflection by a planar mirror interacts deterministically with the same excitonic mode. Therefore, it is possible to get enhanced control over to radiative properties of the excitons using a mirror, including near-complete suppression of radiative decay when the reflection loss is small [350, 351].

Recently in a study of twisted angle t-$MoSe_2$/$MoSe_2$ device brooking the mirror symmetry by displaying black and gray areas related to interchanging domains with rhombohedral stacking regularity. The domain formations are due to rearrangements of atoms in every TMD layer to retain interlayer commensurability. In the case of the near-0° target twist angle a minimal twist with asymmetrical, micrometer-sized AB ($Mo_{top}Se_{bottom}$) and BA ($Se_{top}Mo_{bottom}$) domains has been observed that's big sufficient to be picture seeming optically. Such abnormality in the domains correlates to indigenously changing twist angles owing to strain inhomogeneity [352]. According to the interpretation material spatial dependent optical properties it could be understood based on the electronic band structure of AB/BA ($Mo_{top}Se_{bottom}$/$Se_{top}Mo_{bottom}$) domains in t-$MoSe_2$/$MoSe_2$. Contrasting usual 2H $MoSe_2$ bilayers, the crystal structures of the AB (BA)domains do not has mirror/inversion symmetry because the topmost and lowest layers are undoubtedly distinguishable from the crystal structure, electrons, and holes which specially exist in the upper or lower layers, therefore, it leads interlayer excitons with the favorite dipole positioning.

The fragmented mirror asymmetry in AB ($Mo_{top}Se_{bottom}$) domain appears with the features of the states at band edges. Density functional theory calculations of the AB-stacked $MoSe_2$/$MoSe_2$ bilayer have demonstrated that the valence band maxima lay at the Γ point and the conduction band minima at the Q or K point that depends on the computation constraints, as illustrated in Fig. 4.29a. Whereas the hole wave function at the Γ point is similarly dispersed throughout the layers, therefore, the electron wave function is additionally contained in the upper layer (as compared with the bottom) while the conduction band minimum at Q or K point, as depicted in Fig. 4.29b. The K-point band (100%) asymmetry is much stronger than Q-point (~ 60%). This indicates that electrons favor to exist in the topmost layer in AB-stacked $MoSe_2$/ $MoSe_2$ and their consequent momentum-indirect interlayer excitons are accountable for the peak transition that should have a descending dipole alignment. Since BA stacking is fairly a mirror picture of AB stacking to the horizontal plane, therefore, the dipole positioning of the interlayer exciton should in upwards to BA domains. The density functional theory calculations have also revealed the electron-hole partition of the Γ–K and Γ–Q interlayer excitons are around at 0.34 and 0.07 nm, while the experimental findings showed that the electron-hole separation around 0.26 nm. The corresponding AB-stacked $MoSe_2$/$MoSe_2$ binary monolayers band arrangement at the K (or K′) point, and the K–K intralayer excitons in the lowermost layer belong to inferior energy than those in the upper layer, as illustrated in Fig. 4.29c. Therefore, electrons and holes to the distinct layers can be formed momentum-direct, K–K interlayer excitons [353–356]. Thus K–K interlayer excitons are accountable for peaks that not showing reflection at zero energy, but they acquires oscillator strength and become resonant together intralayer excitons, as depicted in Fig. 4.28d.

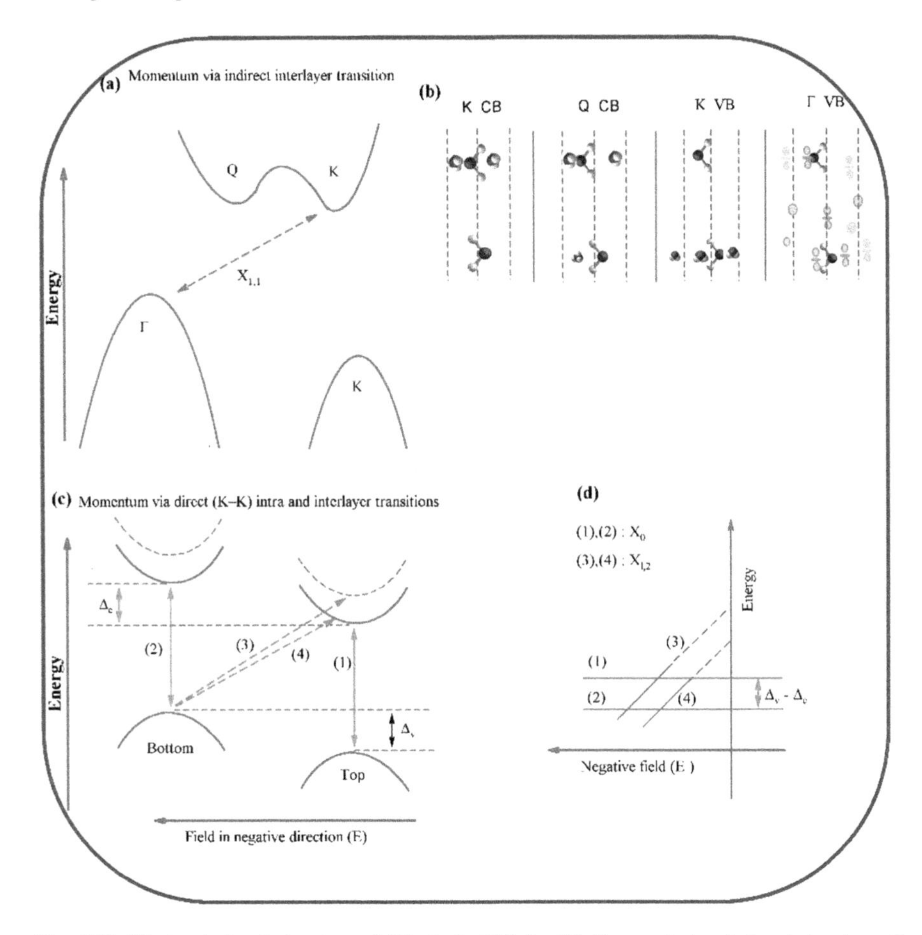

Fig. 4.29 Electronic band structure of AB-stacked $MoSe_2/MoSe_2$: **a** electronic band structure of an AB-stacked $MoSe_2/MoSe_2$ bilayer with the VBM at the Γ point and the CBM at either the K or Q points; **b** contour plots of the electron wave functions at the conduction band K and Q points and the hole wave functions at the valence band K and Γ points; **c** the electronic band structure of AB-stacked $MoSe_2/MoSe_2$ at the K valley. Electronic states localized in the top (bottom) are drawn in cyan (medium black) color, while the solid and dashed line represents spin up and down states. The optical transitions (1) and (2) correspond to momentum-direct intralayer transitions (X_0), while (3) and (4) represent the momentum-direct (K–K) interlayer transitions ($X_{I,2}$). The value of the energy splitting between the top and bottom layer K valley conduction band (Δ_c) defined from the DFT calculations is around 50 meV and the valence band (Δ_v) is 62 meV at $E_z = 0$. The black arrow at the bottom represents the negative direction of an out-of-plane electric field. The energy of the interlayer transition from the bottom layer VBM to the top layer CBM (3 and 4) is reduced under negative E_z; **d** energy of X_0 (1 and 2 in **c**) and $X_{I,2}$ (3 and 4 in **c**) transitions as a function of E_z. Due to the reduced binding energy of $X_{I,2}$, the energy of $X_{I,2}$ at $E_z = 0$ is higher than X_0 (Sung et al. [352])

4.3.2 Heterostructure Excitons

Van der Waals heterostructures composed of atomically thin layers of TMD offer an opportunity to realize artificial materials with designable properties [357]. Especially allows the realization of excitons with remarkably high binding energies [195, 257]. The interlayer excitons in TMD heterostructures are characterized by binding energies exceeding 100 meV to make them stable at room temperature and their lifetime one to two orders of magnitude larger than that of intra-layer excitons [358–360]. However, interlayer excitons has also observed at room temperature in TMD heterostructures [361]. The interlayer exciton high binding energy of TMD heterostructures makes them a material platform to explore high-temperature quantum Bose gases and also create realistic excitonic devices. Additionally, van der Waals heterostructures also offers the possibility of stacking different kinds of TMDs on top of each other [357]. Type-II band alignment stacking is possible in a wide range of TMD heterostructures [364]. Typically their optically excited states are known as interlayer excitons. Such excitons consist of an electron localized in one of the TMD layers and a hole localized in the other TMD layer that can play a crucial role in excitonic superfluidity [362–364].

Typically three kinds of band arrangements such as straddling/type-I, staggered/type-II, and broken/type-III constructs the semiconductor heterostructures [254]. Their formation is understood by Anderson's rule [365]. However, to interpret the van der Waals heterostructures in TMDs various theoretical estimations have been demonstrated based on the band structures and type-II arrangement, considering the conduction and valance bands residing in opposite monolayers [254, 364, 385]. Not only theoretical but also several experimental determinations of the band alignments have been also carried out to explore TMD heterostructures [366–368]. As the intense, a type-II arrangement together a valence band offset of 0.83 eV and a conduction band offset of 0.76 eV for the MoS_2/WSe_2 heterostructure by using micro-beam X-ray photoelectron spectroscopy and scanning tunneling microscopy/spectroscopy (STM/STS) [369]. Through employing higher resolution XPS and UV–Vis absorption spectroscopy type-II alignment in $MoS_2/MoTe_2$ heterostructure valance band offset 0.9 eV and conduction band offset 0.46 eV it has been also realized [370]. Using the submicrometer angle-resolved photoemission spectroscopy valance band offset of $MoSe_2/WSe_2$ heterostructures has been experimentally determined around 0.3 eV [371].

The interlayer hybridization is also crucial to interlayer coupler strength that (specifically for the interlayer expanse and twist angle among the component of monolayers) impacts on the band types of TMD heterostructures. Most of such heterostructures have followed the preservation at the K valleys in contradictory layers to procedure a direct band gap, through insignificant or feeble interlayer hybridization close to valleys [345, 372–376]. Moreover, the Γ and Λ (or Q) valley's important interlayer hybridization has also remarked in an interlayer linking possessing forte of numerous hundred meV, it is analogous to the band offset [345]. This kind of robust interlayer combination also leads to great energy changes (redshift) and allows to transfer of the conduction and valance bands of the heterostructures corresponding to the $\Gamma/\wedge$ (or Q) valleys, therefore, the formation of an indirect band gap is possible in some TMDs [377, 378], a similar process to form homobilayers and bulk TMDs [250, 371, 379].

The formation of bulky alterations in band offsets at several valleys has also influenced the modifications in the orbital features; it is reflected to be a significant factor for the subsequent abnormal interlayer hybridization in momentum space [371]. It has been noticed that the bands at K valleys usually characteristics an in-plane orbital, like at $\Gamma/\wedge$ (or Q) valleys have been featured by an out-of-plane orbital character [380]. So, interlayer distance and twist angle affects the interlayer connections and/or the interlayer combination forte that could considerably adjust the bands at the $\Gamma/\wedge$ (or Q) valleys. This directly correlates to the energies of the Γ and $\wedge$ (or Q) points depending on the interlayer expanse and stacking way [372, 375, 381, 382].

Type-II band arrangement of utmost TMD heterostructures carrier transformation has also attracted much attention [300, 383]. As the instance the extremely fast carrier transformation in MoS_2/WS_2 heterostructures which demonstrated with the help of the photoluminescence (PL) diagramming and femtosecond pump-probe spectroscopy [300]. The demonstrations lead to the holes in the MoS_2 monolayer proficiently transmitted in the WS_2 monolayer within 50 fs afterward photoexcitation. Later on, the electrons and holes conversion in contradictory directions on a sub-picosecond time scale and consequently form the $MoS_2/MoSe_2$ heterostructure [384]. The charge transference dynamics in TMD heterostructures have been also expansively examined, in which the interfacial carrier transformation is establish comprehensively ultrafast (mostly within 100 fs) and independent to twist angle [375, 385–389]. It revealed that the interlayer van der Waals coupler in heterostructures is usually too feebler to intralayer covalent bonding, therefore, interlayer charge carrier transferring is not so rapid compared to intralayer exciton dynamics, whereas the momentum incongruity $\pm$ K/$\pm$ K$'$ of the valleys unavoidable owing to lattice incongruity and twist-angle among the integral of the monolayers by indicating the opposite valleys of the same monolayer, as depicted in Fig. 4.30. Additional valleys like Q and Γ valleys are also followed a similar trend.

Numerous innovations have been also reported to describe the inherent mechanism for the effective charge transferring in heterostructures of TMDs [346, 390, 391]. Such as in TMD monolayers exciton binding energy (0.5–1 eV) is analogous to crucial Frenkel exciton [198, 231, 392, 393], while their wave function favors a Wannier–Mott type with the electron-hole parting extension above a number of tens of unit cells (as calculated exciton Bohr radius ~1–3 nm) [227, 230, 257]. Thus in TMD heterostructures constituent of monolayers in contrast to weak van der Wale coupler, their layer parting is usually fewer to 1 nm. Therefore, the layer detached electrons and holes may remain go through robust Coulombic exchanges to make bounded exciton states [300, 394]. The bounded exciton states are dynamically encouraging and they contest to the intralayer exciton states [300]. It leads the photoexcited electrons and holes to having an analogous prospect to produce layer detached bounded exciton states with the intralayer exciton states. They are also perhaps accounted for the proficient interlayer charge transformation in heterostructures [300].

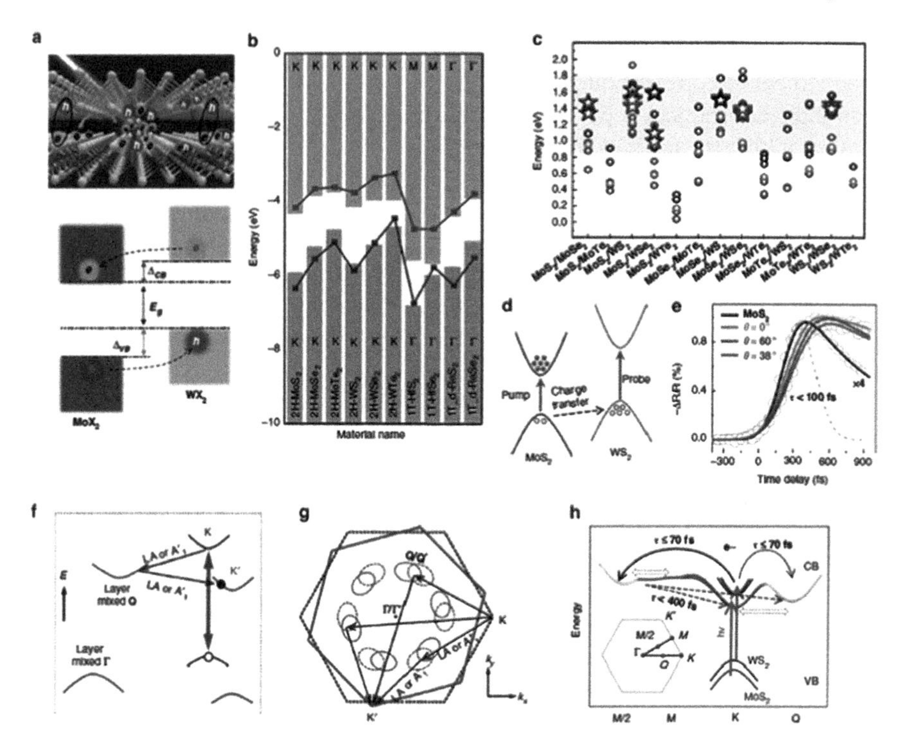

Fig. 4.30 Band alignment and ultrafast charge transfer in TMD vdW heterostructures: **a** Schematic of the side-view structure (top) and the type-II band alignment (bottom) of a TMD vdW hetero-bilayer. The e/h is electron/hole; Δ_{CB} is the conduction band offset of constituent monolayers; Δ_{VB} is the valence band offset of the constituent monolayers; Eg is the band gap of the hetero-bilayer; **b** Calculated band alignments of various TMD monolayers. The bar and line-point plots represent the CBM and VBM values obtained from PBE and HSE06. The positions of CBM and VBM on the Brillouin zone are also exhibited in the figure; **c** Plot of the calculated band gaps (E_g) and the experimental PL energies of the interlayer excitons versus various TMD vdW heterostructures with a type-II band alignment. The calculated E_g values are represented from the circles. Most of them overlap in the region from ~ 1.3 to ~ 1.5 eV; **d** Band alignment of a MoS_2/WS_2 hetero-bilayer to show the hole transfers from the VBM of MoS_2 to WS_2 after optically pumping the MoS_2 A-exciton; **e** Twist-angle independent charge transfer dynamics ($\tau < 100$ fs) defined from the selectively probing of the WS_2 A-exciton of the MoS_2/WS_2 hetero-bilayer; **f, g** Schematic for the phonon scattering-mediated interlayer electron transfer process in the energy and momentum. LA ($A_1{}'$) represents the longitudinal acoustic phonons; **h** Schematic for the ultrafast electron scattering from K to M, M/2, and Q valleys within 70 fs (arrows) in a MoS_2/WS_2 hetero-bilayer and the subsequent electron scattering from M/2 and M back to K′and Q′valleys in the other layer within 400 fs (dashed arrows) (adopted from Jiang et al. [254])

A similar mechanism also exists in phonon scattering combined interlayer hybridized circumvents to momentum incongruity of quick charge transformation [345]. The interlayer carrier transformation proficiently takes place in twofold consecutive processes: first one, the photoexcited electron scattered to the K valley toughly layer hybridized Q/Q' valley (Γ/Γ' valleys for a hole) by the release of an intralayer phonon, subsequently it relaxed from the Q′ valley (Γ′ valley for a hole) to K′ valley in the contradictory layer by producing additional phonon. This kind of interlayer carrier transformation (Q/Q' or Γ/ Γ′ valleys) is usually fast (<50 fs) owing to the robust interlayer combination (or hybridization) in the corresponding areas. Whereas the position of Γ cannot affect due to strong layer mixing under the interlayer twisting, meanwhile, the corresponding Q valleys areal ways on a circle area with the robust interlayer joining for every twist angle. In the either processes, the twist-angle is independent of the carrier transformation. This phenomena has been also verified from the MoS_2/WS_2 hetero-bilayer using time-resolved and angle-resolved photoemission spectroscopy. In which the extremely fast scattering of electrons to the K to M, M/2, and Q valleys within 70 fs [290], through the electron scattering from M/2 and M back to the K′ and Q′ valleys in another layer indoors 400 fs, as depicted in Fig. 4.30.

4.3.3 Trions

Neutral excitons tend to dominate optical properties in monolayer (ML) TMDs, whereas, the more complex exciton species also plays an important role. More specifically charged excitons or trions are the species in which an exciton binds together another electron (or hole) to form a negatively (or positively) charged in a three-particle state. Such as random doping in TMD layers forms n-type [395] that reflects the negative trions likely, in which adsorbates do not introduce additional significant changes to the doping level. Usually, trion binding energy in semiconductor nanostructures is around 10% of the exciton binding energy [396]. Their corresponding neutral exciton binding energy order of 500 meV, so, the yield has been estimated trion binding energy of several tens of meV.

In monolayer MoS_2 tightly bound negative trions with a binding energy of about 20 meV has been verified, as depicted in Fig. 4.31a [397]. That is one order of magnitude larger than the binding energy of the well-studied quasi-2D systems, like the first observed trions in II-VI quantum wells. Particularly at low temperatures well separated neutral and charged excitons in monolayer $MoSe_2$ possessed a trion binding energy of approximately 30 meV, with the tunable charge structures, as depicted in Fig. 4.31b [398]. Moreover, the full bipolar transition is also possible from the neutral exciton to either positively or negatively charged trions, depending on the sign of the applied gate voltage. The binding energies of these two kinds of trion species may similar, but they have minor differences in the effective masses of electrons and holes [399]. It has been also noticed that the energy separation of neutral excitons and trions is the sum of trion-binding energy (specifically in the zero-density case) and the term proportional to the Fermi energy of the free charge carriers [218, 225]. Trion signatures in PL at sufficiently large free carrier densities have been also noticed in absorption measurements [195, 298, 397, 400, 401]. Thus, in general, the electrical charge tuning of excitons has been verified in monolayer TMDs devices, including

WSe_2 and WS_2 [181, 312, 400, 402, 403]. Moreover, studies have also extended toward the biexcitons in addition to neutral and charged excitons [184, 225, 312].

The fundamental difference between the conventional quantum well structures and monolayer TMDs is the carriers with an additional degree of freedom, this is typically called the valley index. The existence of the valley index leads the several optically bright and dark configurations [404–407]. Therefore; the system gives rise to potentially complex recombination and polarization dynamics [408]. The tunable charge of monolayers has encapsulated hexagonal boron nitride to get narrow optical transitions with low-temperature line widths below 5 meV (typically), as depicted in Fig. 4.31c. As per the interpretations the fine structure of trions related to the occupation of the same or different valleys by the two electrons [218]. The comparison of the charge tuning in ML WSe_2, and ML $MoSe_2$ have been also studied [181, 184, 312, 315, 409]. Most studies have demonstrated that the highest-energy valence band and the lowest-energy conduction band belong to antiparallel and parallel spins in ML WSe_2 and ML $MoSe_2$. Thus the concept of trion can be interpreted as the three-particle complex that is useful at low carrier densities. Thus the elevated densities intriguing of new many-body effects are also possible as predicted by several groups [218].

4.3.4 *Electrical Properties*

4.3.4.1 Electronic Structure

Experimentally most extensively studied 2H MoS_2 lattice parameters can be defined from $a = b = 3, 16$ Å and $c = 12.29$ Å, as illustrated in Fig. 4.32 [410–412]. In theoretical verification in-plane lattice parameter values have shown good consistency with a few percent of discrepancy (3.13 Å) while the examined from the local density approximation (LDA), on the other hand, the theoretical discrepancy has been also found around 3.23 Å from the generalized-gradient approximation (GGA) [413, 414]. The discrepancy between experiment and theory is more considerable for the out-of-plane (c) parameter [415] because in-covalent interactions can play a crucial role in the bonding of the MoS_2 layers. However, both of them have been not well described from the approaches LDA and GGA calculations. Their diverse solution's possibility of non-covalent interactions gives the slightly different c parameter and layer-by-layer binding energy [416]. With the addition of a correction for the weak non-covalent interactions, a detailed theoretical approach has been also demonstrated in which reasonable values for the c parameter 11.89 and 13.07 Å for LDA and GGA are described [413].

Moreover, it has been also noticed that in the multilayer MoS_2 structure with different numbers of S–Mo–S layers, the monolayer to bulk form transformation cannot have a significant difference in term a and b lattice parameters corresponding to the bulk structure, including without any significant change in Mo–S bond lengths and S–S distances. Additionally, from the GGA approach energy band gap for the

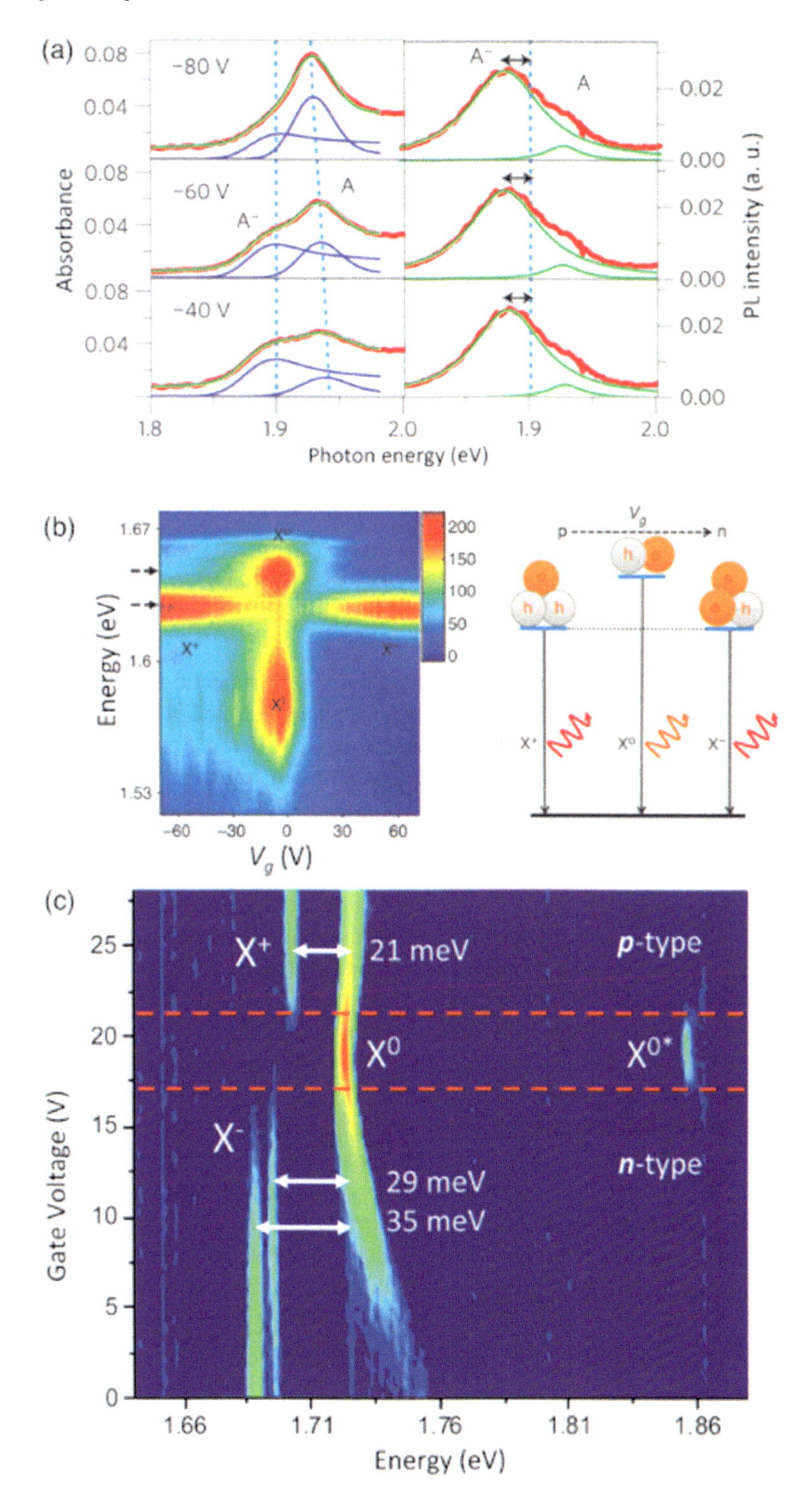
(a)
−80 V
−60 V
−40 V
A⁻
A
Absorbance
PL intensity (a. u.)
Photon energy (eV)
(b)
Energy (eV)
V_g (V)
X⁺
X⁰
X⁻
p
n
V_g
(c)
X⁺
21 meV
p-type
X⁰
X⁰*
X⁻
29 meV
35 meV
n-type
Gate Voltage (V)
Energy (eV)

◀**Fig. 4.31** **a** Absorbance and photoluminescence spectrum to show the signatures of neutral (A) and charged (A^-) excitons in a charge tunable MoS_2 monolayer; **b** Color contour plot of PL from an electrically gated $MoSe_2$ monolayer that can be tuned to show emission from positively charged (X^+) to negatively charged (X^-) trion species; **c** Contour plot of the first derivative of the differential reactivity in a charge tunable WSe_2 monolayer. The n and p-type regimes are manifested by the presence of X^+ and X^- transitions. Around charge neutrality, the neutral exciton X_0 and an excited state X^{0*} are visible (adopted from Wang et al. [218])

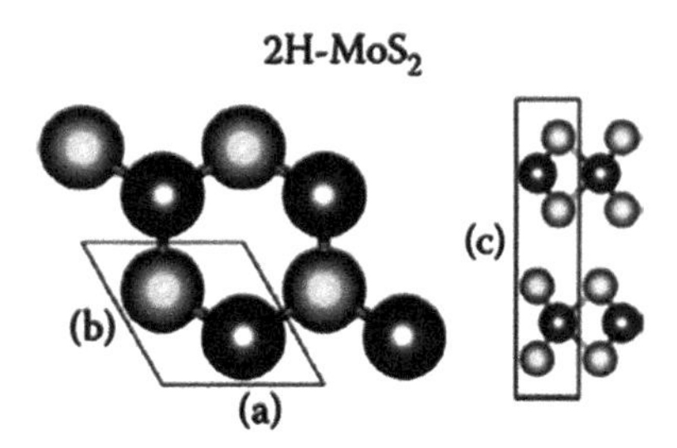

Fig. 4.32 Top and side view of the 2H-MoS_2 crystal structure; **a**, **b**, and **c** are the in-plane (out-of-plane) lattice parameters. S and Mo are represented in light and dark colors, respectively

MoS_2 structure above the 1.18 and 0.62 eV from the LDA calculations [413], as illustrated in Fig. 4.32. It is considered to be an excellent agreement with the experimental value of 1.23 eV [411, 417].

To continue this journey various theoretical and experimental findings based on TMDs' electronic structural properties have been demonstrated [418–423]. Considering most common TMDs properties in the band structures creating d orbitals in metal atoms [424]. The distinct d-electron sums of transition metal filling the nonbonding d bands to dissimilar degrees, consequential in an extensive variability in electronic features together with semiconductor, metal, superconductor, etc. [425].

Usually, semiconducting TMDs band gaps belong close infrared to the visible region is a encouraging property for the photonics and optoelectronics applicability [426–428, 202, 429–431]. However, corresponding bulk materials can have an indirect bandgap, whereas, at the bottom conduction band and the top valence band, the maximum is located at different high symmetry points of the Brillouin zone. While, in the case of the monolayers it changes into a larger direct-band gap semiconductor [231, 432–435]. Therefore, monolayers TMDs electronic properties are considerably tuned due to quantum confined effect in the out-of-plane direction with the decreased symmetry and variation in hybridization among p_z orbitals of chalcogen atoms and d orbitals of metal atoms nearby Γ point while the weakening depressed in thickness [19, 434, 436]. Although the exception is also possible when indirect-to-direct band gap transition and band gap size enlargement, in this situation few layer or monolayer 2D TMDs also exhibits robust excitonic effects owing to the feeble charge carrier screening and following stronger Coulombic interplays by reduction in thickness [195, 198, 227, 230, 257, 259, 397, 436, 437].

Typical MoS_2 simplified band diagram is illustrated in Fig. 4.33, it represents semiconducting TMD by exhibiting the lowermost conduction band c1 and the uppermost splitting in valence bands v1 and v2. The indirect band gap E_g' with the bulk feature minima of the conduction band locates between Γ and K points, while the maxima

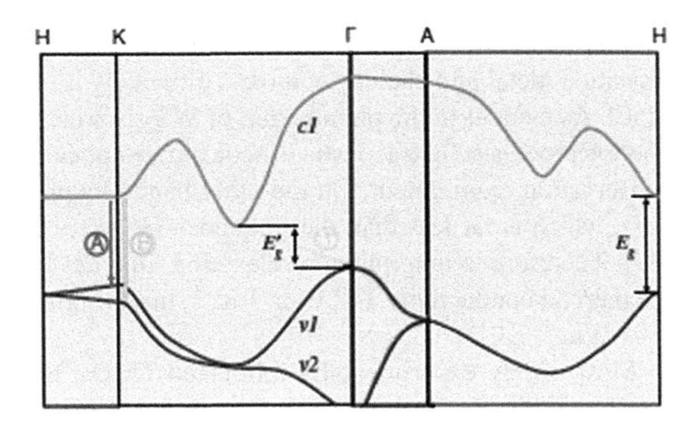

Fig. 4.33 Simplified band structure for the bulk MoS_2, by showing the lowest conduction band c_1 and the highest split valence bands v_1 and v_2. A and B are the direct-gap transitions and I is the indirect-gap transition. E'_g is the indirect gap for the bulk, and Eg is the direct gap for the monolayer (adopted from Mak et al. [18]).

of the valence band are situated at the Γ point, while the direct band gap E_g of the monolayer positioned at K point. The valence band everywhere the K point splits due to the influence of spin-orbit coupling [18, 158, 438, 439]. Under the direct-gap conversions among the maxima of the splitted valence bands and the minima of the conduction band, usually it is known as A and B excitons; this is responsible for the appearance of dual discrete lower energy resonance peaks in the optical spectra. On the other hand, in the case of indirect-gap transition intensity enhancement is appeared at an even low-energy resonance peak. In TMDs indirect-to-direct band gap changeover controllability and robust excitonic effect for the monolayer have been extensively explored through interpreting absorption spectra and photoluminescence [18, 19, 422, 440–443].

4.3.5 *Ballistic Transport*

Since the efficiency of heat dissipation in thermal conductivity is crucial for the TMD-based device performance and reliability. Therefore, it gains great attention for the theoretical and experimental investigators to explore the various TMDs [444, 445]. As the typical example, theoretically and experimentally demonstrated room temperature thermal conductivity of monolayer MoS_2 can be larger than 83 $Wm^{-1} K^{-1}$ for 1 μm. It usually correlates to a sample smaller size (much smaller than the mean free paths) that allows the concept of phonons to conduct heat without being scattered, this is known as ballistic property [446–448]. However, the larger sample's room temperature phonon mean free paths of TMDs hundreds of nanometers, therefore, the ballistic transport occurs in TMDs devices with lengths of tens of nanometers or less [87, 449]. Such a ballistic transport in MoS_2 may persist up to 30 nm at room temperature, even at larger lengths with lower temperatures [444].

Therefore, the ballistic transport property of TMDs has been widely studied by demonstrating the transition metal-chalcogen weak bonds for the special sandwich structure, such a special structure helps to raise the phonon properties. Additionally the adequate crystal structure atomic masses another factor that efficiently affects the thermal conductivities of TMDs. Because of a large mass difference between the transition metal and chalcogen atoms, ultimately it induces substantial phonon-gap [450]. As evident in the phonon gap of WS_2 is around 3.3 THz, likely under three-phonon processes (like acoustic + acoustic → optical), which strongly favors energy conservation requirement. On the other hand, the phonon gap in MoS_2 is only 1.4 THz, which is far less than the frequency range (0–6.9 THz) of acoustic branches [87]. Therefore, a long phonon relaxation time has been noticed for the WS_2 with the thermal conductivity 142 Wm^{-1} K^{-1}, this is significantly greater than the MoS_2 (103 Wm^{-1} K^{-1}).

Most widely experimentally fabricated TMDs, like MoS_2 and $MoSe_2$, thermal transport properties have been also demonstrated, by interpreting the thermal conductivity of MoS_2 around 26 Wm^{-1} K^{-1} at room temperature [451, 452]. Similarly, the thermal conductivity of $MoSe_2$ has been also demonstrated around 54 Wm^{-1} K^{-1} under the iterative solution [87]. Moreover, the relative contribution of different phonon polarization branches of thermal conductivity also noticed similar for silicene rather than graphene because MoS_2 cannot have an individual atomic plane. Work has also predicted that the room-temperature thermal conductivity is around 23.2 W Wm^{-1} K^{-1} for MoS_2, by the numerical simulation [79]. More simulations research has been also presented by the various investigators by showing the thermal conductivities of MoS_2 and $MoSe_2$ around 110.43 and 43.88 Wm^{-1} K^{-1} [453]. The 2D MXenes hexagonal transition metal carbides/nitrides structure is also highly attractive due to their promising technological applications, like energy storage, thermoelectric conversion, and flexible devices. According to demonstration performance of the TMDs devices strongly depends on the thermal transport properties of the constituent material. Moreover, low thermal conductivity and excellent conducting properties are desired to fabricate the thermoelectric devices, additionally, the micro-nano electronic devices also require a material with high thermal conductivity. Because of these key requirements, various studies have demonstrated that heat dissipation is a major challenge to MXene-based devices due to their large aspect ratios and heterogeneous structures [454–458]. In recent years TMDs' thermal conductivity experientially has been measured for various materials by varying their layers at room temperature, however, it has been noted that the increased thermal conductivity with a decreasing order temperature [57, 67, 69, 459–466].

Moreover, it is also significant to discuss the key ballistic transport property in TMDs namely thermoelectric property which directly useful for their device applications. Thermoelectric (TE) materials offer to drive electric currents using temperature gradients and conversely cooling a system just by using a voltage difference. Theoretical approaches to thermoelectric property demonstrate the logarithmic derivative of the electronic transmission gives us to Seebeck coefficient at low temperatures, it can be obtained by using the Mott formula [119].

$$S(T, \mu) \approx \frac{\pi^2 k_B^2 T}{3e} \frac{d \ln \tau(E)}{dE} \tag{4.16}$$

Hear symbols are in their usual meaning. The chemical potential (μ) chosen around the band edges near to maximum thermoelectric figure (ZT) value. The difference between μ at the valence band edge (conduction band edge) and at the μ where p-type ZT (n-type ZT) has a maximum, it is crucial to determine the optimal doping levels in the semiconductors. As the demonstrated facts most of the TMDs have relatively low ZT [119]. Various theoretical studies on TE properties of $MX_2(M = \text{Mo}, W;\ x = S,\ \text{Se})$ monolayers have been also carried out. The most of studies predicted that MoS_2 monolayer n-type ZT around 0.04, by using the Boltzmann equation and equilibrium molecular dynamics (EMD) simulations [139]. It has been noted that the different n-type ZT values (0.87/1.35) layer thickness dependence and their constant k_{ph} values in a diffusive regime. However, some investigators have also overestimated the ZT values compared to well-studied MoS_2, $MoSe_2$, WS_2, and WSe_2 in the frame of ballistic transport [467].

In addition to various investigated compounds, $ZrSe_2$, HfS_2, and $HfSe_2$ are considered to be dynamically stable in both the 2H and 1T phases. A substantial difference in their performances has been also demonstrated [119]. In this kind of TMDs, both n-type and p-type ZT values of the 2H phases are generally much larger than the 1T phase, and corresponding k_{ph} of the 1T phases is always slightly higher than the 2H phases. Due to this reason, their electronic band structures exist. The frontier bands in the 2H phase are less dispersive than the 1T phase. Whereas valance band maxima are almost flat, this leads to sharp changes in the density of state and the electronic transmission spectrum and resulting a rise to enhance S and ZT. Therefore, the p-type ZT values of the 2H phases is up to five to six times higher than the 1T phases. While the n-type ZT difference does not as dramatic as in the p-type ZT. But in the case of the larger 2H phase, the 1T phase of $ZrSe_2$, $HfSe_2$, and HfS_2 is due to an account of the phonon scatterings [468, 469]. In a study, investigators have also quantified the role of k_{ph} on ZT [119]. They demonstrated that at the higher temperature (above 500 K) band gap of TMD (like $ZrTe_2$) cannot sufficiently support an efficient TE response. Therefore, the hybrid function calculation provides a more appropriate band gap to describe abrupt (or ballistic) changes in the transmission spectra at the valence band edges of TMDs (such as $ZrSe_2$, HfS_2, and $HfSe_2$). Their outcome has also revealed TE response of TMDs enhanced by reducing k_{ph}, thereby it is useful to employ for phonon engineering. Hence, the Seebeck coefficient has reduced due to the simultaneous contribution of p- and n-type carriers, especially at higher temperatures [119, 147].

4.3.6 Mechanisms of Scattering

Particularly low frequency noise analysis are considered to be an essential tool to characterize electronic fluctuations caused by several scattering sources associated with crystalline imperfections, phonon scattering, and trapping/de-trapping of charge carriers at the interface [470–475]. More significantly to identify the main scattering sources using the model–dependent analysis and possibly develop a method of its suppression by providing feedback for device optimization. Most low-frequency noise studies have focused on the low-frequency noises through the channels without eliminating contact effects [473–476]. In TMD materials metal contacts are made by the formation of the Schottky barrier that is usually unavoidable (such as in Si–MOSFETs), this becomes another significant noise source. The kinds of newly formed sources are often so dominant that cannot be resolved unambiguously [477]. Therefore, attention has been also paid to the issue of Coulomb scattering of mono and multilayer TMDs which is generally related to Coulomb scattering impurity density at the interface [478].

Scattering mechanisms in TMDs can be discussed based on induced vacancy in states. The induced vacancy in the gap states acquires a dual character due to charging of the vacancy site, this is usually treated as combined short-range and long-range Coulomb (i.e., charged impurity) scatterers, typically MoS_2 vacancy S, as illustrated in Fig. 4.34 [479–482]. Such vacancy gives rise to three atoms in-gap states. It has been noted that the lowest state is doubly occupied in the undoped, charge-neutral material $(V_g = 0)$, therefore, it behaves as a donor state to trap the holes in p-doped materials. Whereas the upper states are almost empty and it behaves as deep single-electron acceptors, this traps the electrons in n-type doped $(V_g > 0)$ materials [479, 483, 484]. However, the doubly charged vacancy sites are prohibited by a large on-site Coulomb repulsion energy. Nevertheless, *S* vacancies can acquire a net charge in p- or n-type doped MoS_2 (positive and negative, respectively) owing to changes in their gap states [485]. Additionally, the induced vacancies in TMDs usually persuaded the unfilled in-gap states, the empty state inhabited state just overhead the valence band, in which edges are generally lacking [485, 486]. In the second situation, the vacancy may still impartial in the p-doped material.

Moreover, *T* matrix formation based approximation has been also introduced to demonstrate the scattering amplitude for the short-range component of the scattering potential, which is strongly renormalized concerning its value. Considering problem-charged vacancies' short-range potential is irrelevant in comparison to the Coulomb contribution to the scattering potential, therefore, the effective reduction

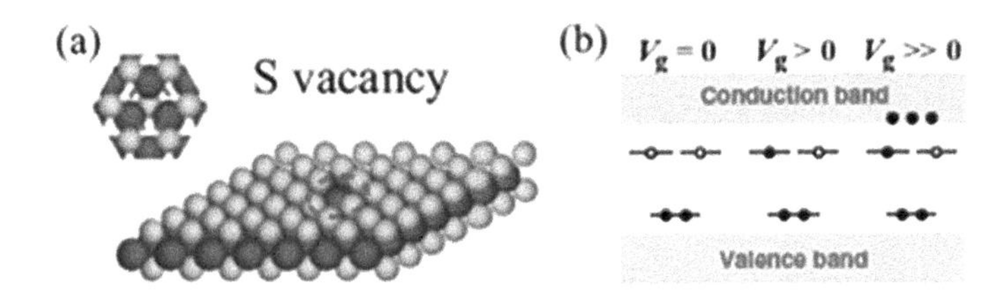

Fig. 4.34 Atomic vacancies in 2D TMDs: **a** Top and side views of the atomic structure of a sulfur vacancy in 2D MoS_2; **b** Sulfur-vacancy-induced in-gap states, in case of the undoped material $(V_g = 0)$, the state close to the valence band maximum is occupied with two electrons from the missing sulfur anion, while with an applied gate voltage, $V_g > 0$, one electron is trapped in the upper in-gap states before carriers are introduced into the conduction band $(V_g >> 0)$ (adopted from Kaasbjerg et al. [490])

in vacancies due to charged impurities. The dominance of Coulomb disorder scattering can give rise to a strong temperature and density dependence of the mobility, it directly relates to temperature-dependent screening of the Coulomb potential. In the case of the degenerate low-temperature ($T \ll T_F$) the Coulomb disorder like short-range disorder due to the efficient carrier screening in TMDs, and corresponding yields behavior like $\mu \sim n^0 T^0$. Whereas in the case of the nondegenerate ($T \geq T_F$), the screening efficiency has been strongly reduced due to the temperature dependence of the 2D screening function. Experimentally it has been also noticed that the metallic characteristics with the movement reductions with growing temperature and subsequently enhancing the charge density. This feature is intrinsic in the 2D semiconductors at the quantum-classical boundary ($T = T_F$) among the homogeneous and non-homogeneous states [487]. So, temperature and density intermission under the changeover is highly depending on the material constraints, such as spin and valley degeneracy, and effective mass. At the crucial carrier densities (10^{11}–10^{13} cm^{-2}) of the TMDs, the Fermi temperature usually belonging to the range $1\,\text{K} \lesssim T_F \lesssim 100\,\text{K}$, this reveals the quantum-classical crossover easily accessible in the temperature range for the weak phonon scattering [488, 489]. On the other hand, in TMDs, vacancies remain neutral with the p-type doped material, and their mobility is significantly higher with much weaker temperature dependence. Their qualitatively different density dependences on the mobility decrease with the carrier density. It has been also categorized as an inherent behavior due to atomic-vacancy-limited transport in TMDs with charge-neutral vacancies.

The low energy transport in TMD the K, K' valence and conduction band valleys are illustrated in Fig. 4.35. As per the low-energy Hamiltonian and Boltzmann transport theory the wave function is usually expressed as $\psi_{\sigma\tau k}(r) = \frac{1}{\sqrt{A}} e^{ikr} \chi_{\sigma\tau}$ where A is the sample area and the valence ($\sigma = v$) and conduction ($\sigma = c$) band eigenspinors $\chi_{\sigma\tau}$ that dominant near the band edges. The corresponding bands are parabolic with the effective energy $\varepsilon_{\sigma\tau k} = E_{\sigma\tau} \pm \hbar^2 k^2/2m_\sigma^*$, where the term $\varepsilon_{\sigma\tau k}$ represents band-edge position, as depicted for the spin dependence in Fig. 4.35. Typical conduction band splitting of MoS_2 has been achieved for the 2–3 meV with the neglecting spin degeneracy, whereas, DFT calculated effective mass values are $m_v^* = 0.59$ and $m_c^* = 0.48$ for the 2D MoS_2 [490].

Moreover, the scattering mechanism not only governs above discussed significant parameters but also depends on several parameters; such as conductivity which is limited from the elastic disorder scattering, like, the atomic point defects can act as strong short-range scatterers and produce multiple scattering of the same defect to infinite order in scattering potential, their relaxation time in a long-range Coulomb potential due to the localized nature of the in-gap states dielectric function and extrinsic and intrinsic screening, as demonstrated in various theoretical and experimental reports [490–493].

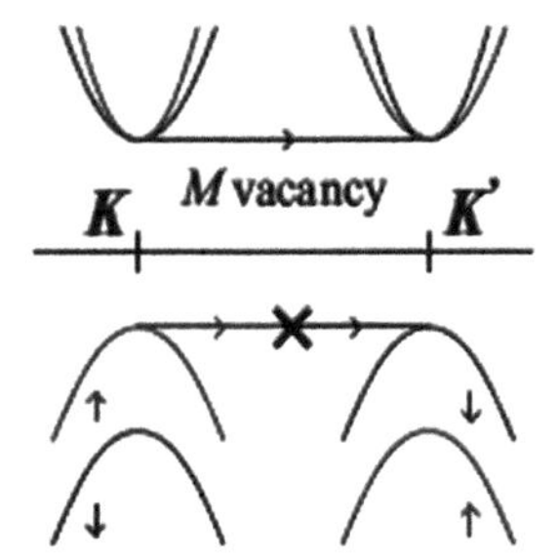

Fig. 4.35 Band structure and disorder scattering in the K, K′ valleys of 2D TMDs. The intervalley scattering is strongly suppressed by the large spin-orbit splitting in the valence band (for some TMDs also in the conduction band), and a symmetry-induced selection rule for defects with C_3 symmetry and centered at the M and X sites in both the valence and conduction bands (adopted from Kaasbjerg et al. [490])

4.4 Magnetic Properties

As earlier discussed TMDs has the formula MX_2 in which M is a transition metal and X is a chalcogen. TMDs can have a layered structure and crystallize in several polytypes, including 2H-, 1T, 1T′, and T_d-type lattices [494]. Much attention has been paid to M = Mo or W because the 2H forms of these compounds have semiconducting behavior, they are mechanically exfoliated to a monolayer. As the bulk $MoTe_2$ 2H-form is a semiconductor with an indirect band gap of 0.88 eV. Therefore, the unique structural properties of TMDs especially the monolayer form is promising for the device applications, such as magnetoresistance and spintronics, high on/off ratio transistors, optoelectronics, valley optoelectronics, superconductors, and hydrogen storage [495, 496]. Most of the interesting properties arise on account of the strong spin-orbit interaction present in these materials due to the occurrence of the heavy metal ion. Several studies have focused on the spin-orbit coupling to explore the interesting consequences of magnetism, such as M = Sc, Ti, V, Cr, Mn, Fe, Ni, Nb, W, Ta, Sn, and x = S, Se [495–497]. Other TMDs those have similar structure to MoS_2, such as CrS_2, $CrSe_2$, FeS_2, $FeSe_2$, MoS_2, $MoSe_2$, WS_2, WSe_2, TaS_2, $TaSe_2$ are energetically favors the 2H phase, whereas TiS_2, $TiSe_2$, MnS_2, $MnSe_2$, NbS_2, $SnSe_2$ and SnS_2 favorably to 1-T phase. Moreover, ScS_2, $ScSe_2$, VS_2, VSe_2, CoS_2, $CoSe_2$, NiS_2, $NiSe_2$and $NbSe_2$ are stable in both phases with competitive stability [497].

The 2D TMDs sandwich structures in the X-M-X sequence, where transition metal atoms are sandwiched between two layers of chalcogen atoms. In which each transition metal atom is surrounded by six chalcogen atoms. The five formerly degenerate d orbitals of 3d transition metal ion split into energy as it is bonded to the chalcogen ligands, and formed the structural phases trigonal prismatic H-phase, octahedral T-phase, distorted octahedral 1T'-type, and T_d-type. Under the crystal field with D_{3h} symmetry in the H phase, the five degenerate 3d orbitals split into a single state a_1 (d_{z2}) and two twofold degenerate states e_1 (d_{x2-y2}/d_{xy}) and e_2 (d_{xz}/d_{yz}). Whereas in the T phase, the triangle sublattice of transition metal atoms enhance the first-neighboring coordination number 6 and forms an octahedral crystal field. On the other hand, the d states split into t_{2g} and e_g manifolds. Moreover, further trigonal distortion, t_{2g} degeneracy lifted and forms higher-lying a_{1g} level and twofold degenerate e_g states in T' and T_d phases. Thus, the magnetic properties of TMDs due to the splitting and filling behavior of d orbitals of the transition metal ions under the different crystal fields. However, for the same transition metal ion, the electronegativity of a chalcogen atom also plays a crucial role, therefore, the lighter chalcogen atoms draw more electrons from the metal ions and affects their on-site magnetic moments.

Computationally it has been also confirmed that intrinsic long-range magnetic ordering realization on transition metal sites for the delocalized p states of S/Se/Te atoms in a variety of transition metal dichalcogenides. In the early stage, it has been extensively explored for the TMD monolayers stability based on DFT calculations and predicted that out of 88 combinations only 52 H or T structures free standing phase occurs. With these, it has also predicted that H-phase TMDs (such as M = Cr, Mo, V, Mn, Co, and W, and X = S, Se, and Te) to be ferromagnetic metals possessing net magnetic moments in the range of 0.2–3.0 lB for each composition (MX_2) [498]. Similarly, the magnetic properties of MTe_2 (M = Ti, V, Cr, Mn, Fe, Co, Ni) monolayers for both H and T phases have been also explored [499]. The outcomes indicated that H-VTe_2, T-$MnTe_2$, and H-$FeTe_2$ forms the ferromagnetic metals with magnetic moments 0.78, 2.80, and 1.48 l B, while the T-VTe_2 recognized as an indirect band gap semiconductor with the magnetic moment 1.0 lB. It has also estimated that the Curie temperatures for the H-VTe_2, T-VTe_2, T-$MnTe_2$, and H-$FeTe_2$ monolayers is around 301, 33, 88, and 229K. Additionally, non-collinear DFT calculations revealed that the HVTe_2, T-VTe_2, and H-$FeTe_2$ monolayers have in-plane easy magnetization direction with magnetocrystalline anisotropy energy values 0.51, 1.74, and 2.57meV for each configuration. Whereas the T-$MnTe_2$ possessed a perpendicular easy magnetization axis with MAE of 0.54me. Moreover, the DFT calculations within LDA+U approximation single-layer VS_2 1T phase strongly correlated to material, while the 2H structure of monolayer VS_2 falls into a ferromagnetic semiconductor category [500]. Their magnetic moments can be localized on V atoms and they couple ferromagnetically through superexchange interactions mediated by the S atoms. The calculations of magnetic anisotropy also revealed an easy plane magnetic moment in 2H VS_2. In this order the magnetic properties of 2D NbS_2 and ReS_2 nanosheets have been also examined, both of them have bipolar magnetic semiconductors with spin gaps 0.27 and 1.63 eV and corresponding Curie temperatures to be 141 and 157K predicted.

A large family of magnetic TMDs such as VSe_2, VTe_2, $MnSe_x$, $NbTe_2$, $NbSe_2$, and $CrTe_2$ has also drawn great attention to both experimental and theoretical aspects due to their room-temperature T_C. Specifically few materials associated T phase (as depicted in Fig. 4.36) have metallic nature. Therefore, the magnetic coupling mechanism stems from the competition between indirect superexchange and itinerant exchange interactions. In addition to this, metallic TMDs also belong to charge density wave (CDW) systems, thereby, the phase transition steers the change of

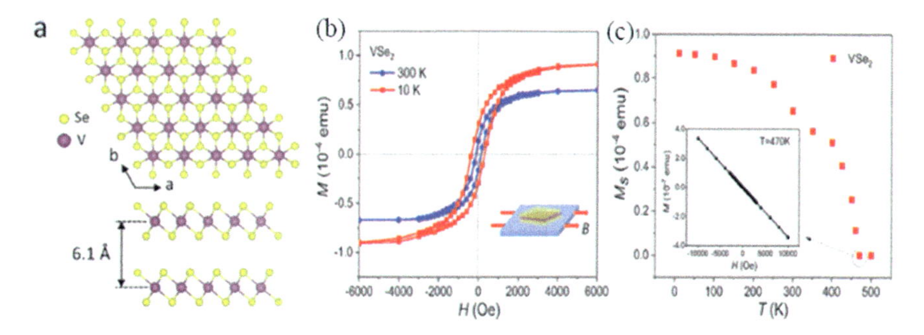

Fig. 4.36 **a** Top and side views of 1T TMD lattice; **b** M–H hysteresis loop of bare VSe_2 flakes on SiO_2 substrate under an in-plane magnetic field at 300 and 10 K; **c** temperature-dependent saturated magnetization (Ms) of bare VSe_2 from 10 to 500 K, inset M–H curve at 470 K indicating to loose of magnetization (adopted from Yu et al. 2019, http://lujionggroup.science.nus.edu.sg/wpcontent/uploads/2019/10/Yu_et_al-2019-Advanced_Materials.pdf)

electronic structures. Therefore, it is also customary to clarify the correlation effect and the possible competition between CDW ordering and magnetic ordering [501].

4.4.1 Magnetism Via Grain Boundaries

Grain boundaries (GBs) can also modify the TMD materials' properties, such as electrical transport, mechanical properties, and magnetic or optical response. Like enhanced electrical transport parallel to GBs and the absence of electronic transport perpendicular to GBs. Therefore, electronic transport property changes under the existence of GBs and their statistics collect in terms of grain boundary (GB) angle. Any angle of the GBs could also reduce the mobility and enhance the photoluminescence depending on the type of GB. Additionally, it also enhances biexciton emission from the GBs regions. However, it is still not much clear whether such an effect comes from grain boundary/sec or due to other changes within the material close to GBs, like strain, defects concentration, Fermi level position, etc [502].

Regarding TMD's magnetic property, the first-principle calculations study has demonstrated that the dislocation cores in MX_2 TMDs can possess substantial magnetism due to partial occupancy of spin-resolved localized electronic states, along with the significant spin-spin interactions, while the dislocations aligned to GB. Their magnetic energy depends on spin ordering and electronic properties of GBs and tilt angle, such as the kind of component dislocations and doping levels. The GBs **contains** 5|7 rings have been recognized as ferromagnetic half-metallic nanowires in the range of middle slope, whereas GBs contains 4|8 cores are antiferromagnetic semiconductors [503].

According to the first principle calculation the projection of the basal plane of the MX_2 sheets associated to the M atom in a sublattice and it superimposed X' to

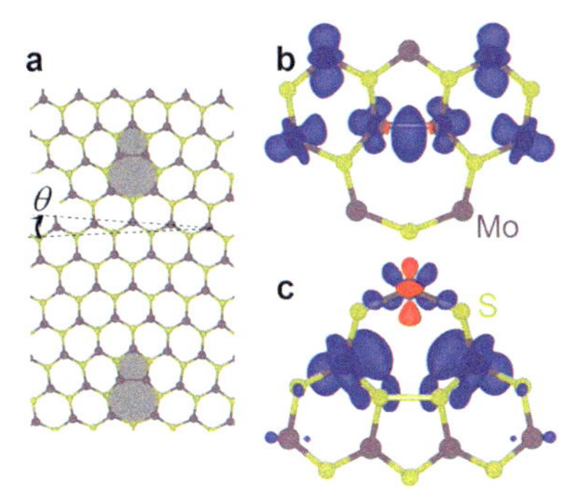

Fig. 4.37 Structure of grain boundary in single-layer MoS2: **a** a symmetric GB with a tilt angle (9°). Each repeat cell includes a Mo-rich ⊥ with a homoelemental Mo-Mo bond, tagged as Mo-rich ⊥. Reversing the grain tilt to -9° creates an S-rich ⊤ with two S-S bonds. Isosurface plots (2×10^{-3} e/A^3) show the magnetization densities; **b** Mo-rich ⊥; **c** S-rich ⊤. Blue and red colors denote positive and negative values of the magnetization density, respectively (adapted from Zhang et al. [503])

another, as illustrated in Fig. 4.37a. It results in a dislocation in the MX_2 structure by removing half of an armchair atomic line and reconnecting seamlessly all of the resultant dangling bonds yields 5|7core. The corresponding strain energy is usually proportional to the square of the Burges vector |b|2 (this is the lowest for the 5|7 core). Nonetheless, the bi-elemental characteristic of MX_2 has two types of 5|7 dislocations: first is an M-rich dislocation possessing M-M bond and $b = (1, 0)$ and it defines as ⊥the second one is X-rich dislocation which is designated as ⊤ with the X-X bond and $b = (0, 1)$. The unfavorable homoelemental energetic bonds are usually avoided by formation of a 4|8 core (created by removing two parallel zigzag atomic chains). Although this kind of dislocation possessed $\sqrt{3}$ times larger Burgers vector $b = (1,\ 1)$. A larger strain energy $\sim$ $|b|2$ forms due to an isolated 4|8 dislocation. However, for the compact alignment in a high-tilt GB they are favorable.

MoS_2 is the extensively experimentally studied TMD by demonstrating the presence of ground state magnetization densities of isolated Mo-rich ⊥ and S-rich ⊤ dislocations, as depicted in Fig. 4.37b, c. A low-tilt 9° GB system has a well-separated defect at low dislocation density. Their spin polarization can be extremely indigenous at the dislocations entrenched within a suitable 2D semiconductor host. Specifically, in case of the Mo-rich ⊥ situation, the whirl is predominantly due to the Mo-Mo bond and corresponding four nearest-neighbor Mo atoms. While in S-rich ⊤ the spin sits into three Mo atoms of the heptagon, and binary atoms head-to-head of S-S pair. In these kinds of dislocation, overall magnetic moment is $\mu = 1.0\,\mu_B$. The energy dissimilarity among the nonmagnetic and magnetic states is obtained at ~36 meV per dislocation. Therefore, the magnetism of dislocations is robust in defect-embedded 2D material that is distinctly different from magnetic edge states along bare zigzag edges [504–508].

Since dislocations are usual ingredients of the grain boundaries and it contributed to magnetism. As depicted in Fig. 4.37a, the symmetrical GB lines bisects the tilt angle θ among the neighboring grains have higher strain energy than a symmetric one. And their GBs magnetic moment per unit length M as a function of the tilt angle θ (below $\theta = 32^{\circ}$). The linear density of dislocations in low-angle GBs are usually obtained from the relationship $2\sin(\theta/2)/b \approx \theta/b$. It directly gives us to rectilinear movement density proportionate to the tilt angle $\approx \mu\theta/b$, here μ is the movement of every dislocation. In the MoS_2 system, the Mo-rich GBs tolerate the proportionation and stands for a bigger tilt angle to S-rich GBs. The indigenous movements of the S-rich T reduced as the S-S bonds become extra distorted at greater θ. Moreover, it has been also noticed that the above 32°total number of 5|7 dislocations declines and hexagons with homoelemental bonds have appeared (symmetric "s-hexagons"). GBs with s-hexagons further increases the magnetic moment up to about $\theta = 47^{\circ}$, while at higher angles the magnetism collapses because GB approaches a state of pure s-hexagons at $\theta = 60^{\circ}$. The obtained $M = 1.75\,\mu_B/\text{nm}$ for the Mo-rich GB and $M = 1.55\,\mu_B/\text{nm}$ to S-rich GB, at $\theta = 47^{\circ}$.

Similarly above 47° the 4|8, the core could be additionally promising than the 5|7 core, which has a superior Burgers vector without homoelement bonding. In the occurrence of 8-fold rings at the greater tilt angles, such as at $\theta = 60^{\circ}$ the termination of stress fields among the contiguous 4|8s lowering the GB energy by ~3.2 eV/nm from that of an arrangement of homoelement bonding (i.e., all s-hexagons). Therefore, the magnetic moment of the 4|8 GBs suddenly upsurges and reaches upto 60°, and $2.1\,\mu_B/\text{nm}$ in magnitude, (greater than the highest obtained by 5|7 boundaries), the corresponding spin-polarization energy is around 45 meV per dislocation, which is similar to ferro- and antiferromagnetic states. The 4|8s is also possible in antisymmetric zigzag GBs, such as invariant to mirror reflection with the component reversal. It also composed of in 8|4|4|8 units, with the corresponding $M = 1.2\,\mu_B/\text{nm}$ [42]. Although, the strong magnetic 4|8 and 8|4|4|8 structures surpasses the 5|7 dislocations at the high tilt. Moreover, the comparative energies of GBs with ferromagnetic or anti-ferromagnetic order among the contiguous dislocations give GBs contains 5|7s, ferromagnetic structure energetically favored to Mo-rich ounce periphery with $E_{\text{antifero}} - E_{\text{fero}}$ around ~8 meV at $\theta = 22^{\circ}$ and it increase upto ~14 meV for more closely-spaced dislocations at $\theta = 32^{\circ}$. Whereas the GBs with 4|8 cores favors the anti-ferromagnetic structure, at $\theta = 60^{\circ}$ the obtained energy $E_{\text{antifero}} - E_{\text{fero}} \approx -32\,\text{meV}$.

The Mapping of DFT calculated energies of the 1D closest neighbor according to the Ising model $= -J\sum_i S_i S_{i+1}$, (here J corresponds to spin-spin interaction, S_i represents the spin of the ith dislocation ($S = 1/2$)) revealed the GBs containing 5|7 dislocations connected to positive J that increase as the separation between dislocations and decreases for the larger tilt angle, while the GBs composed of 4|8s have a negative J, which is larger in magnitude for higher tilt angles. The most strongly-interacting local moments, at $\theta = 32^{\circ}$ the ferromagnetic ordering, and $\theta = 60^{\circ}$ anti-ferromagnetic ordering.

Hence dislocations and grain boundaries in MoS_2 are magnetic due to the spin-polarized density of states that projected onto just the atoms of the Mo-rich dislocation, compared to ideal defect-free MoS_2. In which two new localized states predominately Mo $4d_{x2-y2,xy}$ and $4d_{z2}$ have a small admixture of S_{3px} within the bulk band gap, it typically designated as δ and $\delta*$. Predominantly δ state revealed a bond attachment feature nearby the Mo-Mo bonds, whereas the $\delta*$ reflects the antibonding feature. The indigenous surplus of Mo deteriorates the crystal strength somewhat, therefore, such states energies marginally related to dislocations in case of the S-rich GB. With the suitable doping level such confined states are somewhat engaged, thereby the δ state crossover the Fermi level consequentially the density of states triggering spontaneously to spin splitting. So, the spin polarization of the Mo-rich $\perp$ reverberations the three-dimensional spreading of the δ state [503].

Similarly, the influence of grain boundary on the magnetic features of the monolayer WS_2 has been also demonstrated by interpreting experimental and theoretical structural parameters, as illustrated in Fig. 4.38a–f. As per the experimental findings two tilt GBs (as excited by the missiles) with 12° and 22° misorientated angles associating numerous 6|8 edge dislocations. For the greater tilt of 22° more dislocation density has been recognized toward the grain boundary which gives greater native strain. This kind level strain nearby the GBs provides a minor deformation of the ounces out of the WS_2 monolayer plane, this indicates the loss of sulfur atoms in the annular dark field (ADF) STEM images, it could be responsible for the local distortion. Both GBs and dominantly comprised of 6|8 edge dislocations are represented in Fig. 4.38d, e, which derived from 5|7 defects by the insertion of a two-atom column of S between the W–W bonds [509]. Particular 6|8 dislocations are considered to be additional favorable in the S-rich circumstances compare to 5|7 dislocations in TMDs [509]. Similar 6|8 dislocations have also appeared in MoS_2 [510]. The distance between dislocation cores along the 12°GB has been obtained at ~ 1 nm, close to the simulated structure of the 9°GB composed of 6|8s, as illustrated in Fig. 4.38c. Moreover, first-principles electronic structure calculations indicate that 6|8 dislocations induced a sequence of locally confined states in the band gap of the WS_2 monolayer, as depicted in Fig. 4.38f. That could serve as sinks for charge carriers to determine the electron transportation in WS_2. Significantly, nonmagnetic 6|8 dislocations contrast with magnetic 5|7 dislocations in TMDs [503]. It directly correlates to the incorporation of two sulfur atoms in the 5|7 dislocations to quench the radical electronic states, therefore, it providing electronic stabilization of 6|8 dislocations.

Strain field also impacts the GB of the WS_2, the map strain fields at the dislocation cores in the low- and high-angle BGs at 12° and 22°, and their e_{xy} map parameter is illustrated in Fig. 4.39a–j. This result showed a considerable amount of shear strain present at the dislocation cores. As Fig. 4.39b, g–j revealed e_{xx}, e_{xy}, e_{yy} strain, and rotation xy maps for the 12° and 22° grain boundaries. The GB has a shear strain (e_{xy}) in the range of 5–25% (corresponding to distances ranging from 0.50 to 0.035 nm from the dislocation core), while a considerable high strain in the range of 5–58% (corresponding to distances in the range 0.50–0.035 nm from the dislocation core) has been recognized for the 22° GB. Additionally, GB corresponding to 22° also has a larger e_{xx}, e_{yy} strain fields, and lattice rotation xy compared to the 12°

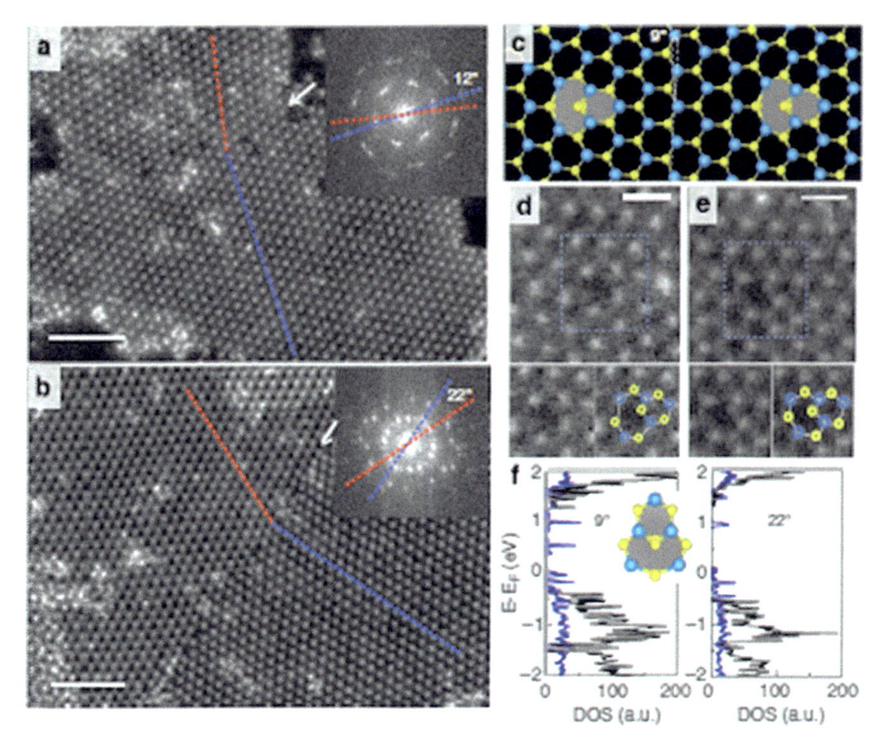

Fig. 4.38 Low- and high-angel grain boundary structure for the monolayer WS_2 with its electronic structure: **a** ADF-STEM image of a tilt grain boundary in a single-layer WS_2 sheet and the corresponding FFT, with a misorientation angle of 12° between two grains as marked by dashed lines. scale bar 2 nm; **b** ADF-STEM image and the corresponding FFT of a grain boundary with a 22° misorientation angle, scale bar 2 nm; **c** atomic structure model to calculate 9° grain boundary composed of 6|8 dislocations; **d, e** the magnified regions of a 6|8 dislocation core in 12° and 22° grain boundaries, respectively. The lower images highlight the position of Wand S atoms and the 6|8 dislocation structure, the scale bar is 5 A; **f** total electronic density of states (gray line) for the 9° (left) and 22° (right) grain boundaries composed of 6|8 structures, respectively. The blue line shows the states contributed by atoms forming the dislocations (adopted from Azizi et al. [511])

grain boundary. This kind of high-strain field associated to dislocations leads unique migration dynamics and reconstruction mechanisms [511].

Similarly, the migration of GBs are also interpreted for the significant $MoTe_2$ monolayer, as illustrated in Fig. 4.40a–i [512]. By showing mirror-symmetric GBs between two 1H grains at around 60° misorientation. Initially, the area around the boundary belongs to single-crystalline, therefore, misoriented grains only appeared. Moreover, molybdenum atoms at the grain boundary (typically) undergo correlated migration by half a lattice vector, as the arrow indicated in the figure. The density functional theory (DFT) simulations lead to the first migration of the Mo atom at the boundary with an energy barrier of 2.3 eV and it drops to 2.0 eV for the second one. The migration of the mirror-symmetric GBs in $MoTe_2$ starts with the displacement

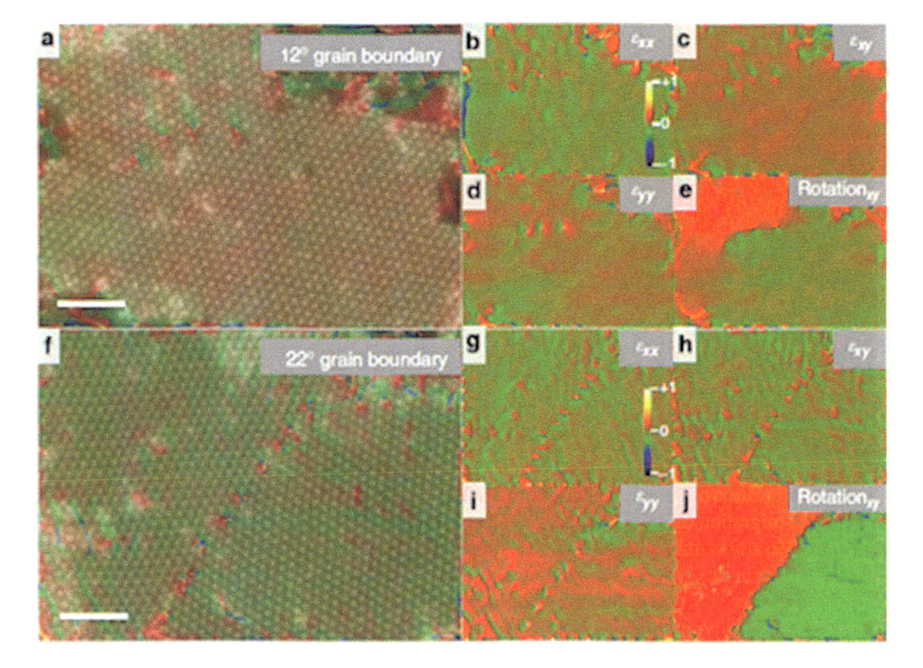

Fig. 4.39 Strain-field mapping of the low and high-angle grain boundaries of the monolayer WS_2: **a** a superimposed image of the 12° grain boundary and its E_{xy} strain map, scale bar 2 nm; **b** e_{xx}; **c** e_{xy}; **d** e_{yy}; **e** rotation xy maps for the 12° grain boundary; **f** superimposed image of the 22° grain boundary and its corresponding strain field e_{xy}, scale bar 2 nm; **g** e_{xx}; **h** e_{xy}; **i** e_{yy}; **j** rotation xy maps for the 22° grain boundary (adopted from Azizi et al. [511])

of Mo atoms by half a lattice vector, very similar to the grain boundary migration of $MoSe_2$ [513]. In both materials, chalcogen vacancies are formed in the lattice by electron irradiation before Mo atoms shift to stabilize the system. However, GBs in WSe_2 and WS_2 starts to migrate with the movements of chalcogen atoms around the dislocation core in contrast to $MoTe_2$ and $MoSe_2$ [511, 514–516]. Thus various studies have demonstrated to GBs' impacts on the magnetic properties of a variety of TMD materials, including above discussed significant TMDs results [517–519].

4.4.2 Magnetism Via Vacancy

It is difficult to avoid vacancy defects within the materials (including layered materials), from both top-down and bottom-up nanomaterials synthesis strategies. Since vacancy defects are to be useful for specific applications (such as defect-induced magnetism-based TMD devices). So, with modern techniques, various vacancies can be created with high precision [520–525]. As an instance, recently vacancy defect formation in monolayer $PdSe_2$ has been studied by adopting the well-known concept

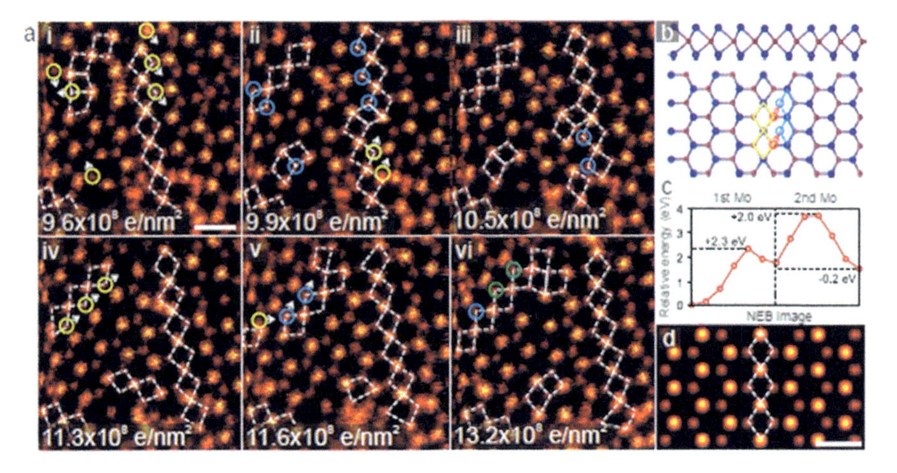

Fig. 4.40 Migration of grain boundaries in a $MoTe_2$ monolayer; **a** STEM-HAADF image sequence showing the migration of grain boundaries. The arrows on the images show the direction of movement of the Mo atoms marked with circles; **b** a simulated model of the grain boundary. The migration of the boundary atoms is schematically indicated with arrows, similar to the experimental images; **c** migration barriers for two of the Mo atoms at a four-unit model of the grain boundary calculated from DFT by employing nudged elastic band method; **d** image simulation of the grain boundary model (adopted from Elibol et al. [512])

of TMD prone to form chalcogen vacancies, the pentagonal structural features with high exfoliation energy give us more flexibility for the vacancy defect formation in $PdSe_2$compared to Mo dichalcogenides [526]. As monolayer $PdSe_2$ distinct defects are illustrated in Fig. 4.41. In which selenium (V_{Se}) and dual Pd and Se vacancy (V_{Pd+Se}), palladium monovacancy (V_{Pd}), Se vacancy (V_{Se+Se}) with consisting pentavacancy palladium and four nearest selenium (V_{Pd+4Se}). All the defected structures are generally relaxing in the symmetrical geometries including single vacancy placed in a $4 \times 4 \times 1$ 2D $PdSe_2$ supercell. The existence of vacancies is significantly elongated to the adjacent $P_{d\text{–}Se}$ bond by $\sim$ 4–7% [526].

Valance band (VBM) and conduction band (CBM) defective replicated STM images structure with the vacancy viewing of the topmost Se layer and their opposed sideview is depicted in Fig 4.41a. When a selenium monovacancy finds in the upper Se layer (top), the crater of electron density in both VBM and CBM states on the topmost view, whereas, the opposed view correlates to the bright atom in the VBM. The bright atoms are the adjacent foreigner to Se atom that is inhabited by 0.35e additional compare to other Se atoms owing to carrier rearrangement. While the identical atom is in dark CBM states STM picture directly correlates to the extra charge transferring to the occupying states. A alike picture also reveals the V_{Pd+Se} bi-vacancy in the STM, as depicted in Fig. 4.41b. The occurrence of a vacancy in the topmost layer persuades a decline in the electron density for the VBM and CBM states, and a bright atom appears in the VBM lowest layer. Whereas other kinds of vacancies are indentify symmetrical to the $PdSe_2$ sheet plane, but they are not much significant as classify in the topmost and lowermost views. However, a P_d

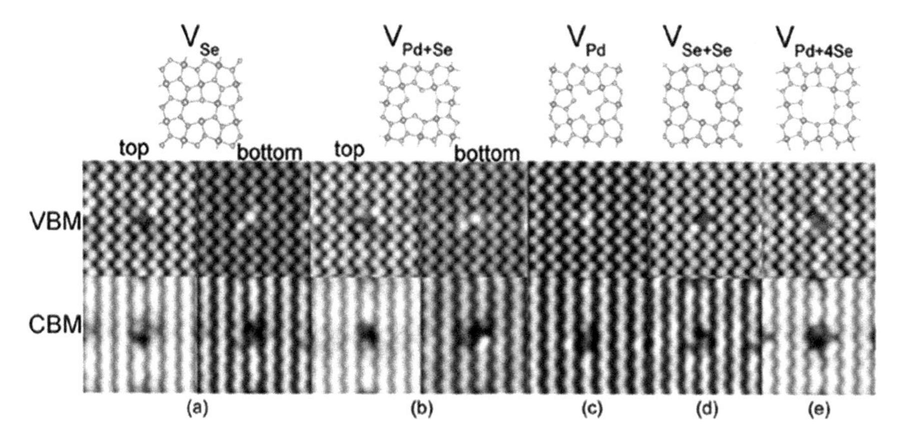

Fig. 4.41 Relaxed atomic structures of defect SL $PdSe_2$ and respective simulation STM images: **a** VSe; **b** V_{Pd+Se}; **c** V_{Pd}; **d** V_{Se+Se}; **e** V_{Pd+4Se}. "Top" and "bottom" correspond to the different views for the layer of vacancy location. VBM and CBM STM simulated for **a** $V_s = -0.5$ eV and $V_s = 1.8$ eV, **b** $V_s = -0.5$ eV and $V_s = 1.8$ eV, **c** $V_s = -0.5$ eV and $V_s = 1.9$ eV, **d** $V_s = -0.8$ eV and $V_s = 1.8$ eV, and **e** $V_s = -0.5$ eV and $V_s = 1.6$ eV, respect to the Fermi level (adopted from Kuklin et al. [526])

vacancy (V_{Pd}, see Fig. 4.41c) lie in the middle layer of $PdSe_2$ with the accessibility owing to the leading Se-Se surface charge. Therefore, vacancy persuaded charge redistributions give a straight visualization of a bright/dark atom of Se in the VBM/CBM states with a supplementary charge carrier. A similar outcome has been also accessed for a simple single-atomic vacancy from the experimental STM picture of bulk $PdSe_2$ [527]. Moreover, the fourth type of double-selenium vacancy (V_{Se+Se}, see Fig. 4.41d) is likely to single V_{Se}, but either sides of the monolayer may look identical STM picture. A more multifaceted V_{Pd+4Se} vacancy (see Fig. 4.41e) also provides the noticeable tetragonal dips in the STM due to non occurrence of the binary topmost selenium atoms. Therefore, selenium atoms adjacent neighbors have an excess charge of 0.2 e with the spin-polarized states [526].

To estimate the stabilities of the vacancies containing TMDs, the vacancy formation energy is the useful parameter, it can be obtained from the following relationship:

$$E_f = E_v - E_{df} + \sum_i n_i \mu_i \tag{4.17}$$

where E_v and E_{df} are the total energies of the imperfection and imperfection-free $PdSe_2$ monolayers, while the n_i is the quantity of atoms of kind I that can remove to the supercell to the defect, and μ_i is the corresponding chemical potential. The calculated μ_{Pd} and μ_{Se} under Pd-rich (Se-rich) conditions have been achieved − 1.45 (− 1.81) and − 1.94 (− 1.76) eV [526]. The fault creation energy calculation has revealed that V_{Pd} and V_{Se} are due to the mainly imperfections in the $PdSe_2$ monolayer. In the case of a Pd-rich situation Se vacancies desire the lowermost energy expenditures with the creation energy around 1.64 eV, this is the typical energy range for the transition-metal dichalcogenide family, like $MoSe_2$ and MoS_2 [516, 528]. While, in the case of Se-rich situations, the development energy of a Pd vacancy has the bottommost value around 1.48 eV, this leads Se-rich in poor condition compared to Pd. The vacancy formation of V_{Pd} to be considered too informal compare to V_{Mo} creation of $MoSe_2$ or MoS_2. The palladium ions cannot be strongly enclosed by the chalcogen atoms likely to Mo in $MoSe_2$ and MoS_2. Therefore, Pd is more accessible to remove from the layer. Also significant factor in defect formation has a robust binding energy among the $PdSe_2$ layers in its unpackaged assembly. This means it offers robust interlayer binding and comparatively lower vacancy construction energy with a great quantity of Se imperfections which produced while the mechanical separation of the layers, whereas the middle position of the Pd atoms is preserved to them from the imperfection construction throughout the mechanical cleavage. The vacancy creation energies of bi-vacancies also correlates to the development of V_{Pd+Se} which desired fewer energy to the construction of binary distinct monovacancies V_{Pd} and V_{Se}, The creation energy of selenium bi-vacancies is greater to the energy superposition of double Se vacancies. It demonstrates a monovacancy V_{Se} has a superior propensity to produce vacancy and fewer probabilities to be merged. Although the energy created in a multifaceted V_{Pd+4Se} vacancy is also uppermost, it can be produced by a stepwise mechanism [526].

The complex vacancy densities of states (DOS) have been considered intensively embedded states and it correspondingly narrows the band gap. Significantly the V_{Pd+Se} bi-vacancy band gap is decreased up to 0.84 eV possessing the maximum embedded states at −0.10 and 0.82 eV, as illustrated in Fig. 4.42. Whereas the Se bi-vacancy V_{Se+Se} reflects four embedded states (−0.44, −0.02, 0.40, 0.72 eV) band gaps in a monolayer with a reduced width of 0.34 eV. The four embedded states kind of V_{Pd+4Se} vacancy persuades the spin-polarized states owing to a huge quantity of dangling bonds. Their spin-up and spin-down band gaps are 0.59 and 0.53 eV. Thus twofold jointed states in the spin-up network have a highest at 0.53 and 0.98 eV and four utmost of 0.00, 0.55,0.86, and 1.01 eV analogous to spin-down states. Moreover, their closest-adjacent of selenium atoms takes a surplus charge of 0.2 e and reveals magnetic moments of $0.1-0.4\ \mu_B$ per atom with the overall native magnetic moment of $2\ \mu_B$ per flaw, as depicted in Fig. 4.42f. Hence the accurate regulation of vacancies delivers a platform for the persuaded magnetism in TMDs [526, 529].

4.4.3 Doping-Induced Magnetism

Magnetism-associated edges, grain boundaries, and vacancies opens interesting possibilities, but they also associated with a certain problems such as the stability of these entities. The induced magnetism is usually achieved in TMDs from a variety of methods. Including atomic doping [530–540], surface functionalization [540–543], depositing on a substrate with magnetic proximate effect [540, 544, 545], and building edge state [503, 540, 546]. Specifically, the doping approach is to be

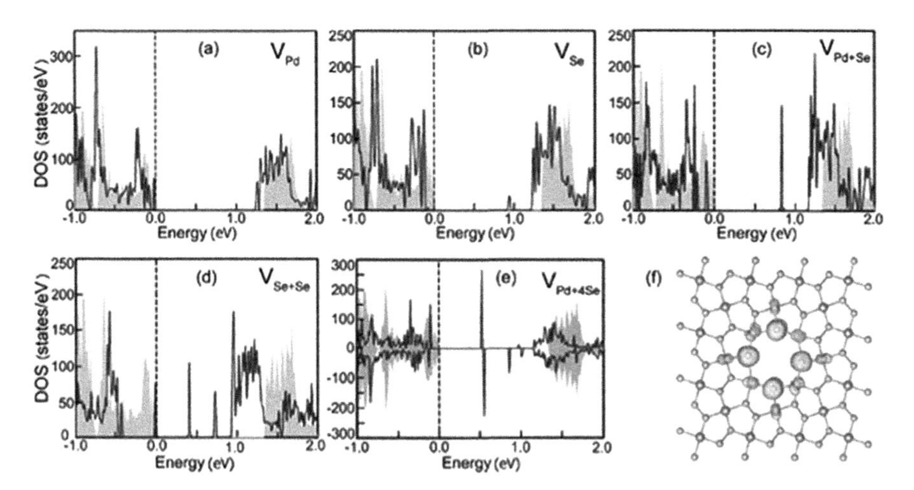

Fig. 4.42 TDOS defected in 2D $PdSe_2$: **a** V_{Pd}; **b** V_{Se}; **c** V_{Pd+Se}; **d** V_{Se+Se}; **e** V_{Pd+4Se} in lines to the defect-free $PdSe_2$ monolayer (filling), the Fermi level is set to 0 eV; **f** spatial distribution of spin density (0.001 e/Å$_3$) in the monolayer contained V_{Pd+4Se} vacancy (adopted from Kuklin et al. [526])

considered an effective way to tailor material electronic properties and implement new functionalities.

Magnetic doping is considered to be an effective tool for the functionalization of TMDs. This kind of defect doping induces magnetism in an atomic layered structure, in which the long-range magnetic order arises due to the presence of single-atom defects in combination with an intrinsic discriminating mechanism in sublattices [534, 540]. TMDs also offer a perfect platform with appropriate defective localization by various kinds of doping approaches, such as introducing vacancy, atomic substitution, and displacement. The TMDs doped by transition metal atoms (Fe, Co, Ni, Mn, etc.) with large magnetic moments have been recognized as diluted magnetic semiconductors that have promising applications in spintronics [534]. As per theoretical investigations on long-range ferromagnetic (FM) ordering in Mn-doped MoS_2, $MoSe_2$, $MoTe_2$, and WS_2 by substituting Mo or W sites to Mn, the first-principles calculations lead an anti-ferromagnetic (AFM) exchange between Mn d states and p states of the chalcogen atoms. It can be effectively regulated for the long-range FM exchange of Mn atoms [547]. Similarly, MoS_2 monolayer doped by various transition metals has been also demonstrated ferromagnetism by Mn, Zn, Cd, and Hg for the concentration of 6.25%, while the Co and Fe dopants AFM due to Jahn-Teller distortions [548]. Moreover, the electronic and magnetic properties of monolayer $MoTe_2$ are also due to the incorporation of Mo vacancy or substituting Mo in various 3d transition metals. This study demonstrates a stable magnetic moment for such cases, except for the Ni defect. Similarly, ferromagnetism in 1 T-$ZrSe_2$ monolayer by doping V, Cr, Mn, and Fe is also possible [549]. It has also demonstrated that the magnetic moments of dopants increase by including Hubbard potential U_{eff} according to density functional theory, by indicating a transition from the low to high spin state.

In addition to this, the diluted magnetic semiconductors have been also predicted in other 2-D materials including TMDs, through the magnetic dopants. In a theoretical demonstration the room-temperature ferromagnetism with Curie temperature of 629 K in Fe-doped SnS_2 with magnetic momentum of 2.0 μ_B per Fe atom is also predicted. It has also demonstrated the long-range ferromagnetism with intralayer sites occupied by the dopants, whereas it is paramagnetic with Fe fixed at interlayer sites. Another DFT study on the magnetic order of monolayer $MoSe_2$ with the different elements doped critical temperature has been also demonstrated, as depicted in Fig. 4.43a, b [539]. Moreover, investigations with the possible magnetically ordered state of Fe and Mn-doped $MoSe_2$ at an atomic substitution of 15% at a temperature of 5 K have been also carried out, as depicted in Fig. 4.44a, b. The magnetization in the out-of-plane direction $S_Z = S_Z/|S|$ (see Fig. 4.44a), the magnetically ordered state in V and Co-doped $MoSe_2$ can also FM along to out-of-plane informal alignment. While the Cr substitution magnetically ordered state also reflects AFM. Moreover, the magnetically ordered state of V-doped $MoSe_2$ saturates to a faultless FM state to the out-of-plane informal alignment. Whereas, Co dopants are bunches of FM-oriented to Co ions, in which long-range order may missing. Furthermore, in the case of Cr doping, the magnetically ordered state can have a randomized magnetic order with a random spins orientation. The randomization probably due to AFM magnetically ordered state that indicates to a magnetic hindrance for the clustering, resulting in a randomization in positioning of the magnetic moments [539].

On the other hand, the magnetically well-arranged state in Fe and Mn-doped $MoSe_2$ at an atomic substitution of 15% at a temperature of 5 K, owing to the magnetic relaxation-alignment is in-plane (see Fig. 4.44b). The plot of the in-plane angle (ϕ) for

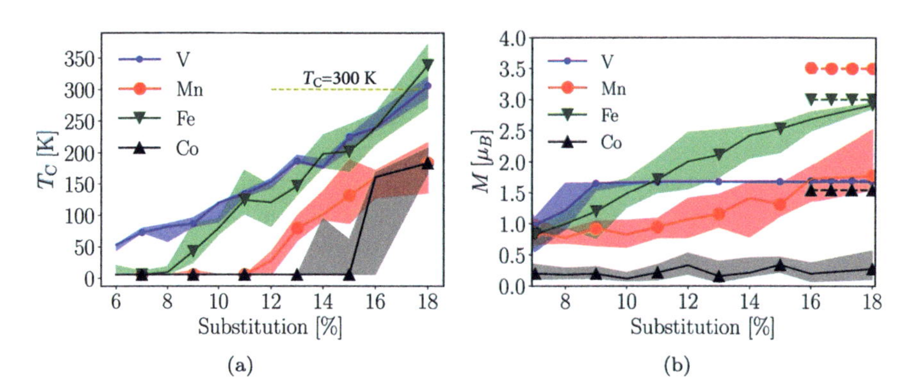

Fig. 4.43 Critical temperature and magnetization for the doped $MoSe_2$: **a** critical temperature (Curie temperature for V and Co and K-T transition temperature for Fe and Mn) of V, Mn, Fe, and Co-doped $MoSe_2$ as a function of atomic substitution. The solid lines show the median critical temperature obtained from the MC simulations, and the shading shows the variance between the 25th and the 75th percentile; **b** the average magnetization of V, Mn, Fe, and Co-doped $MoSe_2$ as a function of atomic substitution. The solid lines show the median magnetization from the MC simulations and the shading shows the variance between the 25th and the 75th percentile. The dotted lines show the starting saturation magnetization for each dopant, as per DFT (adopted from Tiwari et al. [539])

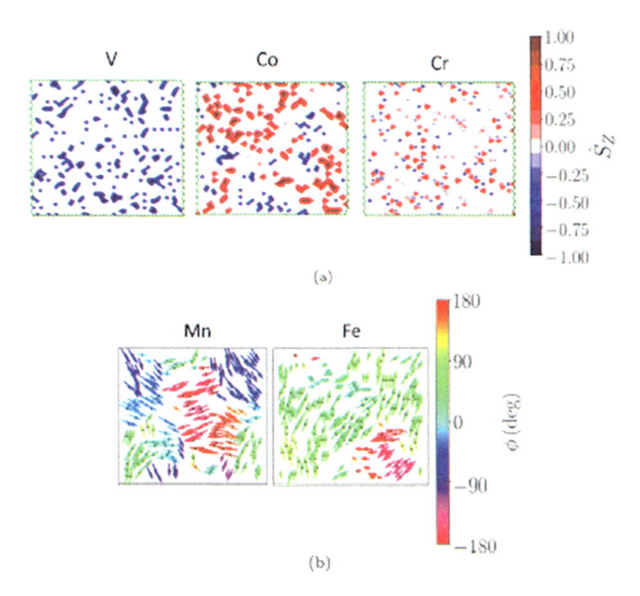

Fig. 4.44 Possible magnetically ordered states in doped MoSe2: **a** the magnetically ordered state of a V, Cr, and Co-doped sample of MoSe2 for the atomic substitution 15%, at a temperature of 5 K; **b** the magnetically ordered state of an Mn, Fe-doped sample of $MoSe_2$ for the atomic substitution 15%, at a temperature of 5 K. ϕ is the in-plane angle of the magnetic moment (adopted from Tiwari et al. [539])

every dopant atom:$\phi = \cos^{-1}\left(\frac{S_x}{S_\|}\right)$, where, $S_\|$ is in-plane magnetization. Both Mn and Fe orientations of the magnetic movement is remain arbitrary for a few short-range (< 8 Å) ordering. Therefore, two distinct effects have been interpreted for both Fe and Mn. First one is domain formation with FM clusters and the second one belongs to none perfect ferromagnetic clusters formation owing to their magnetic ordering disruptions at somewhat lengthier expanses, the small infringement of magnetic order at a lengthier expanse owing to Kosterlitz-Thouless changeover feature, when temperatures underneath the K-T conversion temperature, thereby, the long-range magnetic ordering does not occur [550].

Moreover, the sub-nanometer 1D substitution network in monolayer MoS_2 has also accomplished experimentally using a dislocation-catalyzed approach [551, 552]. It opens up a prospect to generate an extensive variety of spintronic devices with interesting promising features. A first principle calculation study on linear atomic doping of MoS_2 by substituting S atoms with B, C, N, or F, and Mo atoms with Mn, Fe, Co, or Ni has been studied [552]. The outcome of this investigation has revealed that the B, C, N, or Ni linear-doped MoS_2 are usually non-magnetic (NM), and the semiconducting- to-metallic changeover occurs when B, N, and Ni lined atomic doping is designed, whereas the semiconducting characters conserved with the replacement of C atoms in MoS_2. The NM, FM, and antiferromagnetic (AFM) ordering are deliberated for all magnetic structures to obtain the respective ground states. The supercells of lined-doped MoS_2 are shaped by extension of its unit cells sideways a zigzag direction. As the energies of diverse magnetic coupler ordering are summarized in Table 4.3, it indicates the ground states of F, Mn, Fe, and Co lined-doped monolayer MoS_2 in almost every ferromagnetic, whereas for the confirmation the structural constancy of magnetic monolayers, the establishment energy per dopant atom can be estimated with the help of the expression 4.17. The obtained formation energies have revealed the fabrication of these structures belong to exothermic (see Table 4.3). There is no noticeable imagined frequency to exists for the F, Mn, Fe, or Co-doped MoS_2, this indicates these monolayers are dynamically stable [552].

Typical spin-charge densities with the different concentrations are illustrated in Fig. 4.45a–l, in which spin polarization is mostly confined at transition metal atoms and neighboring to dopant atoms in the case of F substitution. The magnetic moments of transition metal atoms exchanged monolayers mostly due to the distribution of dopant atoms and the adjacent Mo atoms (see Fig. 4.45a–l and Table 4.4). The spin-charge density has revealed the magnetic moment is predominantly positioned at the Co and two neighboring Mo_1 and Mo_2 atoms. Whereas the most of the spin is largely contained at Fe atoms with the relatively feeble minority spin produced at Mo_2 atoms in Fe-doped MoS_2, while the mainstream of spin allocates at Mn atoms with minority spin resided in the twofold adjacent Mo_1 and Mo_2 atoms for Mn-doped MoS_2.

Therefore, the mechanism of 1D defect-induced magnetism in 2D MoS_2 monolayers transferred the electrons from the Mo atom to the F atom (see Table 4.4), owing to the smaller atomic radius of fluorine compared to sulfur. As the consequence positively charged Mo atom forms a shorter Mo–F bond to Mo–S. Compared to an S atom, an F dopant can move toward the Mo_1 atom slightly and result in a stronger interaction than the Mo_2 atom. Thus, a shorter Mo_1–F bond to Mo_2–F are generated. The five S atoms next to Mo_1 are generally accepted with 10/3 electrons, while the F atom has withdrawn most of the 2/3 electrons. Though, the Mo_2 is enclosed to four S and two F atoms. So, it is wondered to 8/3 electrons can relocate from Mo_2 to four S atoms. Hence a segment of the residual electrons of Mo_2 moves to the F atom owing to the relatively feeble interaction among Mo_2 and F atoms. Thus, a magnetic moment of 0.46 μ_B is persuaded on Mo_2 atoms with negligible spin polarization of F and Mo_1 atoms [552].

Table 4.3 Computed formation energies E_f (eV per dopant atom) and the energy differences of NM, FM, and AFM states (meV/dopant) of the four doped MoS_2 for the ferromagnetic ground state [552]

Dopants	*Ef*	NM-FM/	AFM-FM/
F	−1.39	36	52
Fe	−3.31	167	154
Mn	−3.91	97	72
Co	−3.80	209	194

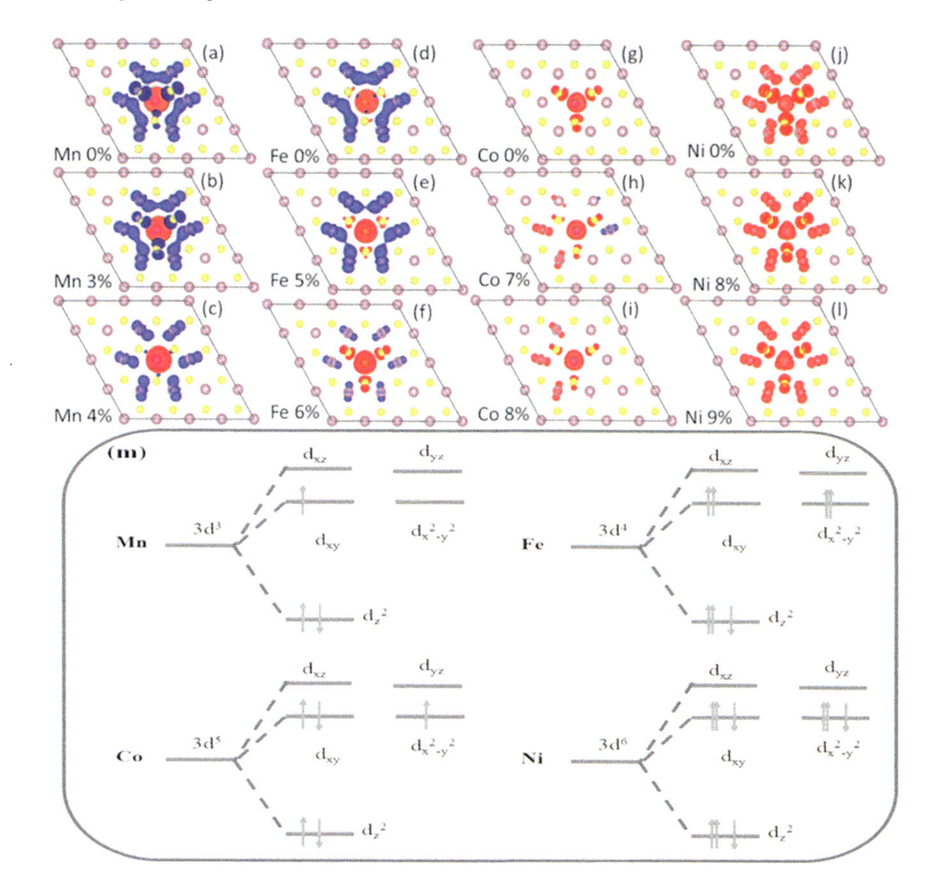

Fig. 4.45 **a–l** Isosurface plots of the spin-charge density for Mn, Fe, Co, and Ni-doped MoS_2 monolayer under different biaxial strains, here purple, yellow, and brown spheres are Mo, S, and TM ions, respectively, whereas red and blue represents spin-up and spin-down charge densities; **e** the electronic configuration of transition metal (adopted from Zhu, Y. et al. 2018 AIP Advances 8, 055917)

Table 4.4 Bader charge analysis with electron gain or loss on dopants is defined by the positive and negative values

Line defects	Charge transfer (e/atom)	Magnetic moments (μ_B)				Magnetic order	
		Mo_1	Mo_2	Dopant	Total	2X	3X
F@MoS2	0.69	0.03	0.46	0.05	0.58	NM	NM
Mn@MoS2	−1.24	−0.16	−0.10	1.43	1.06	FM	FM
Fe@MoS2	−0.73	0.03	−0.10	1.69	1.76	FM	FM
Co@MoS2	−0.76	0.45	0.15	0.34	0.91	FM	FM

The localized magnetic moments on the transition metal atoms and total magnetic moments per dopant atoms are summarized with the majority and minority spins expressed by the positive and negative values [552]

The dissimilar magnetic movements of transition metal-doped systems have been also interpreted from the crystal field theory. According to this theory, the metal ions are coordinated with D_{3h} symmetry, as the consequence we get double degenerate bands $\left(d_{xz}d_{yz}\right),\left(d_{xy}, d_{x^2-y^2}\right)$ and a non-degenerate band (d_{z^2}). Meanwhile the Mn, Fe, Co, and Ni atoms are positively charged with d^3, d^4, d^5, and d^6 arrangement, therefore, the Ni-doped MoS_2 is NM. Typical d-electron arrangements of dopants are illustrated in Fig. 4.45m. It has also been noticed that the Mo_1 and Mo_2 atoms possessed confined magnetic moments of 0.45 and 0.15 μ_B in the case of Co linear-doped MoS_2. However, the magnetic moment of Co composed with the twofold adjacent Mo atoms is equivalent to 0.94 μ_B. Hence the extended range FM combination is strengthened by the S atom intermediated charge transfer [540, 552].

4.5 Conclusions

The conclusive remakes, in continuation to the previous three chapters in which attention has been paid to TMDs materials' basic properties such as synthesis and various forms of structural descriptions including kinds of defects formations and their impact on properties. This chapter work has provided an understanding of the TMDs' various physical properties like mechanical properties including strain engineering, thermal properties including thermal conductivity, thermal expansion, and thermoelectric, optical properties including excitons; binding Energy, non-Rydberg excitonic, excitonic states in dark, excitonic collapse, Intra transition, dynamics, interactive exciton, mirror exciton including heterostructure excitons, excitons, and trions. Moreover, TMD materials electrical properties have been extensively explored owing to the possibility of their uses in high-processing optoelectronic devices for different applications. Attention has been also paid to the electrical properties of the TMDs by providing a comprehensive discussion on electronic structure, ballistic transport, and mechanisms of scattering. Additionally, two-dimensional TMDs can also have an impassive magnetic property compared to other existing materials. Therefore, it is worth discussing this key physical property of TMD materials, thereby, a discussion on the magnetic properties of TMD materials has been also included in this work by providing the magnetism via grain boundaries, magnetism via vacancy, and doping-induced magnetism.

References

1. Huang HH, Fan X, Singh DJ, Zheng WT (2020) Recent progress of TMD nanomaterials: phase transitions and applications. Nanoscale 12:1247–1268
2. Py M, Haering R (1983) Structural destabilization induced by lithium intercalation in MoS_2 and related compounds. Can J Phys 61:76–84
3. Li H, Lu G, Wang Y, Yin Z et al (2013) Mechanical exfoliation and characterization of single- and few-layer nanosheets of WSe2, TaS2, and TaSe2. Small 9:1974–1981

4. Lee YH, Zhang XQ, Zhang W, Chang MT et al (2012) Synthesis of large-area MoS_2 atomic layers with chemical vapor deposition. Adv Mater 24:2320–2325
5. Mahler B, Hoepfner V, Liao K, Ozin GA (2014) Colloidal synthesis of 1T-WS2 and 2H-WS2 nanosheets: applications for photocatalytic hydrogen evolution. J Am Chem Soc 136:14121–14127
6. Sun Y, Fujisawa K, Lin Z, Lei Y et al (2017) Low-temperature solution synthesis of transition metal dichalcogenide alloys with tunable optical properties. J Am Chem Soc 139:11096–11105
7. Treacy MMJ, Ebbesen TW, Gibson JM (1996) Exceptionally high young's modulus observed for individual carbon nanotubes. Nature 381:678–680
8. Wong EW, Sheehan PE, Lieber CM (1997) Nanobeam mechanics: elasticity, strength, and toughness of nanorods and nanotubes. Science 277:1971–1975
9. Lee C, Wei XD, Kysar JW, Hone J (2008) Measurement of the elastic properties and intrinsic strength of monolayer graphene. Science 321:385–388
10. Tian YJ, Xu B, Yu DL, Ma YM et al (2013) Ultrahard nanotwinned cubic boron nitride. Nature 493:385–388
11. Ozgur U, Alivov YI, Liu C, Teke A et al (2005) A comprehensive review of zno materials and devices. J Appl Phys 98:103
12. Wu JQ (2009) When group-iii nitrides go infrared: new properties and perspectives. J Appl Phys 106:011101
13. Sun YG, Rogers JA (2007) Inorganic semiconductors for flexible electronics. Adv Mater 19:1897–1916
14. Rogers JA, Someya T, Huang YG (2010) Materials and mechanics for stretchable electronics. Science 327:1603–1607
15. Kim DH, Lu NS, Ma R, Kim YS et al (2011) Epidermal electronics. Science 333:838–843
16. Novoselov KS, Jiang D, Schedin F, Booth TJ et al (2005) Two-dimensional atomic crystals. Proc Natl Acad Sci U S A 102:10451
17. Novoselov KS, Geim AK, Morozov SV, Jiang D et al (2005) Two-dimensional gas of massless dirac fermions in graphene. Nature 438:197–200
18. Mak KF, Lee C, Hone J, Shan J et al (2010) Atomically thin MoS_2: a new direct-gap semiconductor. Phys Rev Lett 105:136805–136814
19. Splendiani A, Sun L, Zhang YB, Li TS et al (2010) Emerging photoluminescence in monolayer MoS_2. Nano Lett 10:1271–1275
20. Liu K, Wu J (2016) Mechanical properties of two-dimensional materials and heterostructures. J Mater Res 31:832–844
21. Yengejeh SI, Wen W, Wang Y (2021) Mechanical properties of lateral transition metal dichalcogenide heterostructures. Front Phys 16:13502–13507
22. Nepal NK, Yu L, Yan Q, Ruzsinszky A (2019) First-principles study of mechanical and electronic properties of bent monolayer transition metal dichalcogenides. Phy Rev Mater 3:073601
23. Casillas G, Santiago U, Barrón H, Alducin DA et al (2014) Elasticity of MoS_2 sheets by mechanical deformation observed by in situ electron microscopy. J Phys Chem C 119:710–715
24. Zhao J, Deng Q, Ly TH, Han GH et al (2015) 2015 "Two-dimensional membrane as elastic shell with proof on the folds revealed by three-dimensional atomic mapping. Nat Commun 6:8935
25. Yakobson BI, Smalley RE (1997) Fullerene nanotubes: C 1,000,000 and beyond: Some unusual new molecules—long, hollow fibers with tantalizing electronic and mechanical properties—have joined. Am Sci 85:324–337
26. Wang Q (2004) Effective in plane stiffness and bending rigidity of armchair and zigzag carbon nanotubes. Int J Solids Struct 41:5451–5461
27. Kudin KN, Scuseria GE, Yakobson BI (2001) C_2F, BN, and C nanoshell elasticity from ab initio computations. Phys Rev B 64:235406–235410
28. Zhou X (2000) Strain energy and Young's modulus of single-wall carbon nanotubes calculated from electronic energy-band theory. Phys Rev B 62:13692–13696

29. Yu L, Ruzsinszky A, Perdew JP (2016) Bending two-dimensional materials to control charge localization and fermi-level shift. Nano Lett 16:2444–2449
30. Bertolazzi S, Brivio J, Kis A (2011) Stretching and breaking of ultrathin MoS_2. ACS Nano 5:9703–9709
31. Butler SZ, Hollen SM, Cao LY, Cui Y et al (2013) Progress, challenges, and opportunities in two-dimensional materials beyond graphene. ACS Nano 7:2898–2926
32. Castellanos-Gomez A, Poot M, Steele GA, van der Zant HSJN et al (2012) Elastic properties of freely suspended MoS_2 nanosheets. Adv Mater 24:772–775
33. Feldman L (1976) Elastic-constants of 2h-MoS_2 and 2h-$NbSe_2$ extracted from measured dispersion curves and linear compressibilities. J Phys Chem Solids 37:1141–1144
34. Liu KK, Zhang WJ, Lee YH, Lin YC et al (2012) Growth of large-area and highly crystalline MoS_2 thin layers on insulating substrates. Nano Lett 12:1538–1544
35. Najmaei S, Liu Z, Zhou W, Zou XL et al (2013) Vapour phase growth and grain boundary structure of molybdenum disulphide atomic layers. Nat Mater 12:754–759
36. van der Zande AM, Huang PY, Chenet DA, Berkelbach TC et al (2013) Grains and grain boundaries in highly crystalline monolayer molybdenum disulphide. Nat Mater 12:554–561
37. Kang K, Xie SE, Huang LJ, Han YM et al (2015) High-mobility three-atom-thick semiconducting films with wafer-scale homogeneity. Nature 520(7549):656–660
38. Watson AJ, LuW, Guimarães MHD, Stöhr M (2021) Transfer of large-scale two-dimensional semiconductors: challenges and developments. 2D Mater 8:032001
39. Liu K, Yan QM, Chen M, Fan W et al (2014) Elastic properties of chemical-vapor-deposited monolayer MoS_2, WS2, and their bilayer heterostructures. Nano Lett 14:5097–5103
40. Zhu T, Li J (2010) Ultra-strength materials. Prog Mater Sci 55:710–757
41. Cadelano E, Palla PL, Giordano S, Colombo L (2009) Nonlinear elasticity of monolayer graphene. Phys Rev Lett 102:235502–235504
42. Wei X, Fragneaud B, Marianetti CA, Kysar JW (2009) Nonlinear elastic behavior of graphene: Ab initio calculations to continuum description. Phys Rev B 80:205407
43. Defo RK, Fang S, Shirodkar SN, Tritsaris GA et al (2016) Strain dependence of band gaps and exciton energies in pure and mixed transition-metal dichalcogenides. Phys Rev B 94:155310
44. Duerloo KAN, Li Y, Reed EJ (2014) Structural phase transitions in twodimensional Mo- and W-dichalcogenide monolayers. Nat Commun 5:1–9
45. Hou W, Azizimanesh A, Sewaket A, Peña T et al (2019) Strain-based room-temperature non-volatile MoTe2 ferroelectric phase change transistor. Nat Nanotechnol 14:668–673
46. Wei N, Xu L, Wang HQ, Zheng JC (2011) Strain engineering of thermal conductivity in graphene sheets and nanoribbons: a demonstration of magic flexibility. Nanotechnology 22:105705
47. Han Y, Zhou J, Gao L (2021) Experimental nanomechanics of 2D materials for strain engineering. Appl Nanosci 11:1075–1091
48. Berry J, Zhou S, Han J, Srolovitz DJ et al (2017) Dynamic phase engineering of bendable transition metal dichalcogenide monolayers. Nano Lett 17:2473–2481
49. Song S, Keum DH, Cho S, Perello D et al (2016) Room temperature semiconductor-metal transition of $MoTe_2$ thin films engineered by strain. Nano Lett 16:188–193
50. Wang SW, Medina H, Hong KB, Wu CC et al (2017) Thermally strained band gap engineering of transition-metal dichalcogenide bilayers with enhanced light-matter interaction toward excellent photodetectors. ACS Nano 11:8768–8776
51. Zhao R, Wang Y, Deng D, Luo X (2017) Tuning phase transitions in 1T-TaS2 via the substrate. J Robinson Nano Lett 17:3471–3477
52. Kretschmer S, Komsa HP, Boggild P, Krasheninnikov AV (2017) Structural transformations in two-dimensional transition-metal dichalcogenide MoS_2 under an electron beam: insights from first-principles calculations. J Phys Chem Lett 8:3061–3067
53. Zhao X, Kotakoski J, Meyer JC, Sutter E et al (2017) Engineering and modifying two-dimensional materials by electron beams. MRS Bull 42:667–676
54. Zhu J, Wang Z, Yu H, Li N et al (2017) Argon plasma induced phase transition in monolayer MoS_2. J Am Chem Soc 139:10216–10219

55. Wang R, Yu Y, Zhou S, Li H et al (2018) Strategies on phase control in transition metal dichalcogenides. Adv Funct Mater 28:1802473
56. Wang Y, Xu N, Li D, Zhu J (2017) Thermal properties of two dimensional layered materials. Adv Funct Mater 27:1604134
57. Chen XK, Zeng YJ, Chen KQ (2020) Thermal transport in two-dimensional heterostructures. Front Mater 7:578791
58. Zhang LM, Jiao BB, Yun SC, Kong YM et al (2017) A CMOS compatible MEMS pirani vacuum gauge with monocrystal silicon heaters and heat sinks. Chin Phys Lett 34:025101
59. Ezheiyan M, Sadeghi H, Tavakoli MH (2016) Thermal analysis simulation of germanium zone refining process assuming a constant radio-frequency heating source. Chin Phys Lett 33:058102
60. Hu R, Hu JY, Wu RK, Xie B et al (2016) Examination of the thermal cloaking effectiveness with layered engineering materials. Chin Phys Lett 33:044401
61. Zhang G, Zhang YW (2017) Thermal properties of two-dimensional materials. Chin Phys B 26:034401
62. Lei JM, Peng XY (2016) Improved kernel gradient free-smoothed particle hydrodynamics and its applications to heat transfer problems. Chin Phys B 25:020202
63. Cheng Y, Wu X, Zhang Z, Sun Y (2021) Thermo-mechanical correlation in two-dimensional materials. Nanoscale 13:1425–1442
64. Kim SE, Mujid F, Rai A, Eriksson F (2021) Extremely anisotropic van der Waals thermal conductors. Nature 597:660–665
65. Balandin AA (2011) "Thermal properties of graphene and nanostructured carbon materials. Nat Mater 10:569–581
66. Pop E (2010) Energy dissipation and transport in nanoscale devices. Nano Res 3:147–169
67. Sahoo S, Gaur APS, Ahmadi M, Guinel MJF et al (2013) Temperature-dependent Raman studies and thermal conductivity of few-layer MoS_2. J Phys Chem C 117:9042–9047
68. Yan R, Simpson JR, Bertolazzi S, Brivio J et al (2014) Thermal conductivity of monolayer molybdenum disulfide obtained from temperature-dependent Raman spectroscopy. ACS Nano 8:986–993
69. Taube A, Judek J, Lapinska A, Zdrojek M (2015) Temperature-dependent thermal properties of supported MoS_2 monolayers. ACS Appl Mater Interfaces 7:5061–5065
70. Liu X, Zhang YW (2018) Thermal properties of transition-metal dichalcogenide. Chin Phys B 27:034402
71. Chen CC, Li Z, Shi L, Cronin SB (2014) Thermal interface conductance across a graphene/hexagonal boron nitride heterojunction. Appl Phys Lett 104:081908
72. Chen J, Walther JH, Koumoutsakos P (2015) Covalently bonded graphene-carbon nanotube hybrid for high-performance thermal interfaces. Adv Funct Mater 25:7539–7545
73. Liu X, Zhang G, Zhang YW (2015) Graphene-based thermal modulators. Nano Res 8:2755–2762
74. Peng B, Zhang H, Shao H, Xu Y et al (2016) Thermal conductivity of monolayer MoS2, MoSe2, and WS2: interplay of mass effect, interatomic bonding and anharmonicity. RSC Adv 6:5767–5773
75. Mingo N, Broido DA (2005) Carbon nanotube ballistic thermal conductance and its limits. Phys Rev Lett 95:096105
76. Nika DL, Pokatilov EP, Askerrov AS, Balandin AA (2009) Phonon thermal conduction in graphene: role of Umklapp and edge roughness scattering. Phys Rev B 79:155413–155512
77. Kumanek B, Janas D (2019) Thermal conductivity of carbon nanotube networks: a review. J Mater Sci 54:7397–7427
78. Zhao Y, Cai Y, Zhang L, Li B (2020) Thermal transport in 2D semiconductors—considerations for device applications. Adv Funct Mater 30:1903929
79. Cai Y, Lan J, Zhang G, Zhang YW (2014) Lattice vibrational modes and phonon thermal conductivity of monolayer MoS_2. Phys Rev B 89:035438–035448
80. Zhang Z, Xie Y, Ouyang Y, Chen Y (2017) A systematic investigation of thermal conductivities of transition metal dichalcogenides. Int J Heat Mass Transf 108:417–422

81. Pop E, Varshney V, Roy AK (2012) Thermal properties of graphene: fundamentals and applications. MRS Bull 37:1273–1281
82. Klemens PG (2000) Theory of the A-plane thermal conductivity of graphite. J Wide Bandgap Mater 7:332–339
83. Klemens PG (2001) Theory of thermal conduction in thin ceramic films. Int J Thermophys 22:265–275
84. Liu Y, He D (2021) Approach to phonon relaxation time and mean free path in nonlinear lattices. Chin Phys Lett 38:044401
85. Jeong C, Datta S, Lundstrom M (2011) Full dispersion versus Debye model evaluation of lattice thermal conductivity with a Landauer approach. J Appl Phys 109:073718
86. Kaasbjert K, Thygesen KS, Jacobsen W (2012) Phonon-limited mobility in n-type single-layer MoS_2 from first principles. Phys Rev B 85:115317
87. Gu X, Yang R (2014) "Phonon transport in single-layer transition metal dichalcogenides: a first-principles study. Appl Phys Lett 105:131903
88. Jiang JW, Zhuang XY, Rabczuk T (2013) Orientation dependent thermal conductance in single-layer MoS_2. Sci Rep 3:2209
89. Zhou WX, Chen KQ (2015) First-principles determination of ultralow thermal conductivity of monolayer WSe_2. Sci Rep 5:15070
90. Su J, Liu ZT, Feng LP, Li N (2015) Effect of temperature on thermal properties of monolayer MoS_2 sheet. J Alloys Compd 622:777–782
91. Zhang X, Sun D, Li Y, Lee GH et al (2015) Measurement of lateral and interfacial thermal conductivity of single- and bilayer MoS_2 and $MoSe_2$ using refined optothermal raman technique. ACS Appl Mater Interfaces 7:25923–25929
92. Lin Z, Liu W, Tian S, Zhu K et al (2021) Thermal expansion coefficient of few-layer MoS_2 studied by temperature-dependent Raman spectroscopy. Sci Rep 11:7037
93. Yu Y, Yu Y, Xu C, Su L et al (2016) Engineering substrate interactions for high luminescence efficiency of transition-metal dichalcogenide monolayers. Adv Funct Mater 26:4733–4739
94. Dai Z, Liu L, Zhang Z (2019) Strain engineering of 2D materials: issues and opportunities at the interface. Adv Mater 31:1805417
95. Yan Z, Liu G, Khan JM, Balandin AA (2012) "Graphene quilts for thermal management of high-power GaN transistors. Nat Commun 3:827
96. Gong F, Fang HH, Wang P (2017) Visible to near-infrared photodetectors based on MoS_2 vertical Schottky junctions. Nanotechnology 28:484002
97. Wang Y, Gaspera ED, Carey BJ (2016) Enhanced quantum efficiency from a mosaic of two dimensional MoS_2 formed onto aminosilane functionalized substrates. Nanoscale 8:12258–12266
98. Buscema M, Steele GA, Van ZHSJ (2014) The effect of the substrate on the Raman and photoluminescence emission of singlelayer MoS_2. Nano Res 7:561–571
99. Liu Z, Amani M, Najmaei S, Xu Q et al (2014) Strain and structure heterogeneity in MoS_2 atomic layers grown by chemical vapour deposition. Nat Commun 5:5264
100. Zhang X, Tan QH, Wu JB (2016) Review on the Raman spectroscopy of different types of layered materials. Nanoscale 8:6435–6450
101. Lu X, Luo X, Zhang J (2016) Lattice vibrations and Raman scattering in two-dimensional layered materials beyond graphene. Nano Res 9:3559–3597
102. Zhang X, Qiao XF, Shi W (2015) Phonon and Raman scattering of two-dimensional transition metal dichalcogenides from monolayer, multilayer to bulk material. Chem Soc Rev 44:2757–2785
103. Najmaei S, Ajayan PM, Lou J (2013) Quantitative analysis of the temperature dependency in Raman active vibrational modes ofmolybdenum disulfide atomic layers. Nanoscale 5:9758–9763
104. Late DJ, Shirodkar SN, Waghmare UV (2014) Thermal expansion, anharmonicity and temperature-dependent raman spectra of single- and few-layer $MoSe_2$ and WSe_2. ChemPhysChem 15:1592–1598

105. Huang X, Gao Y, Yang T (2016) Quantitative analysis of temperature dependence of Raman shift of monolayer WS_2. Sci Rep 6:32236
106. Chen Y, Wen W, Zhu Y (2016) Temperature-dependent photoluminescence emission and Raman scattering from $Mo_{1-x}W_xS_2$ monolayers. Nanotechnology 27:445705
107. Su L, Zhang Y, Yu Y (2014) Dependence of coupling of quasi 2-D MoS_2 with substrates on substrate types, probed by temperature dependent Raman scattering. Nanoscale 6:4920–4927
108. Su L, Yu Y, Cao L (2015) Effects of substrate type and material-substrate bonding on high-temperature behavior of monolayer WS_2. Nano Res 8:2686–2697
109. Su L, Yu Y, Cao L (2017) In situ monitoring of the thermal-annealing effect in a monolayer of MoS_2. Phys Rev Appl 7:034009
110. Jin T, Kang J, Su KE (2013) Suspended single-layer MoS_2 devices. J Appl Phys 114:164509
111. Zhang R, Koutsos V, Cheung R (2016) Elastic properties of suspended multilayer WSe_2. Appl Phys Lett 108:042104
112. Aiyiti A, Hu S, Wang C (2018) Thermal conductivity of suspended few-layer MoS_2. Nanoscale 10:2727–2734
113. Lee JU, Kim K, Cheong H (2015) Resonant Raman and photoluminescence spectra of suspended molybdenum disulphide. 2D Mater 2:044003
114. Tamulewicz MD, Kutrowska-Girzycka J, Gajewski KR (2019) Layer number dependence of the work function and optical properties of single and few layers MoS_2: effect of substrate. Nanotechnology 30:245708
115. Yang Y, Lin Z, Li R, Li Y et al (2021) Thermal expansion coefficient of monolayer MoS2 determined using temperature-dependent Raman spectroscopy combined with finite element simulations. Microstructures 1:2021002
116. Hu X, Yasaei P, Jokisaari J, Öğüt S et al (2018) Mapping thermal expansion coefficients in freestanding 2D materials at the nanometer scale. Phy Rev Lett 120:055902
117. Hu X, Hemmat Z, Majidi L, Cavin J et al (2019) Controlling nanoscale thermal expansion of monolayer transition metal dichalcogenides by alloy engineering. Small 1905892
118. Zhang D, Wu YC, Yang M, Liu X (2016) Probing thermal expansion coefficients of monolayers using surface enhanced Raman scattering. RSC Adv 6:99053–99059
119. Özbal G, Senger RT, Sevik C, Sevinçli H (2019) Ballistic thermoelectric properties of monolayer semiconducting transition metal dichalcogenides and oxides. Phys Rev B 100:085415
120. Poudel B, Hao Q, Ma Y, Lan Y et al (2008) High-thermoelectric performance of nanostructured bismuth antimony telluride bulk alloys. Science 320:634–638
121. Snyder GJ, Toberer ES (2008) Complex thermoelectric materials. Nat Mater 7:105–114
122. Bell LE (2008) Cooling, heating, generating power, and recovering waste heat with thermoelectric systems. Science 321:1457–1461
123. Zhang X, Zhao LD (2015) Thermoelectric materials: energy conversion between heat and electricity Auth. J Materiomics 1:92–105
124. Sothmann B, Sánchez R, Jordan AN (2015) Thermoelectric energy harvesting with quantum dots. Nanotechnology 26:032001
125. Guilmeau E, Maignan A, Wan C, Koumoto K (2015) On the effects of substitution, intercalation, non-stoichiometry and block layer concept in TiS2 based thermoelectrics. Phys Chem Chem Phys 17:24541–24555
126. Sztein A, Ohta H, Sonoda J, Ramu A et al (2009) GaN-based integrated lateral thermoelectric device for micro-power generation. Appl Phys Express 2:111003
127. Kaiwa N, Hoshino M, Yaginuma T, Izaki R et al (2007) Thermoelectric properties and thermoelectric devices of free-standing GaN and epitaxial GaN layer. Thin Solid Films 515:4501–4504
128. Pantha BN, Dahal R, Li J, Lin JY et al (2008) Thermoelectric properties of $In_xGa_{1-x}N$ alloys. Appl Phys Lett 92:042112
129. Yamaguchi S, Iwamura Y, Yamamoto A (2003) Nearly total spin polarization in La2/3Sr1/3MnO_3 from tunneling experiments. Appl Phys Lett 82:2065–2067

130. Sztein A, Haberstroh J, Bowers JE, DenBaars SP et al (2013) Calculated thermoelectric properties of InxGa1−xN, InxAl1−xN, and AlxGa1−xN. J Appl Phys 113:183707
131. Goyal V, Teweldebrhan D, Balandin AA (2010) Mechanically-exfoliated stacks of thin films of Bi2Te3 topological insulators with enhanced thermoelectric performance. Appl Phys Lett 97:133117
132. Huang W, Da H, Liang G (2013) Thermoelectric performance of MX2 (M = Mo,W; X = S,Se) monolayers. J Appl Phys 113:104304
133. Tao WL, Lan JQ, Hu CE, Cheng Y et al (2020) Thermoelectric properties of Janus MX_2(M = Pd, Pt; X, Y = S, Se, Te) transition-metal dichalcogenide monolayers from first principles. J Appl Phy 113:104304
134. Amani M, Chin ML, Birdwell AG, O'Regan TP et al (2013) Electrical performance of monolayer MoS2 field-effect transistors prepared by chemical vapor deposition. Appl Phys Lett 102:193107
135. Wang T, Zhu R, Zhuo J, Zhu Z et al (2014) Direct detection of DNA below ppb level based on thionin-functionalized layered MoS2 electrochemical sensors. Anal Chem 86:12064–12069
136. Pospischil A, Furchi MM, Mueller T (2014) Solar-energy conversion and light emission in an atomic monolayer p–n diode. Nat Nanotechnol 9:257–261
137. Kufer D, Konstantatos G (2015) Highly sensitive, encapsulated MoS2 photodetector with gate controllable gain and speed. Nano Lett 15:7307–7313
138. Zhuang HL, Hennig RG (2013) Computational search for single-layer transition-metal dichalcogenide photocatalysts. J Phys Chem C 117:20440–20445
139. Jin Z, Liao Q, Fang H, Liu Z et al (2015) A revisit to high thermoelectric performance of single-layer MoS_2. Sci Rep 5:18342
140. Gu X, Yang R (2014) Phonon transport in single-layer transition metal dichalcogenides: a first-principles study. Appl Phys Lett 105:131903
141. Guo SD, Dong J (2018) Biaxial strain tuned electronic structures and power factor in Janus transition metal dichalchogenide monolayers. Semicond Sci Technol 33:085003
142. Tao WL, Mu Y, Hu CE, Cheng Y et al (2019) Electronic structure, optical properties, and phonon transport in Janus monolayer PtSSe via first-principles study. Philos Mag 99:1025–1040
143. Lu AY, Zhu H, Xiao J, Chuu CP et al (2017) Janus monolayers of transition metal dichalcogenides. Nanotechnol 12:744–749
144. Peng R, Ma Y, Huang B, Dai Y (2019) Two-dimensional Janus PtSSe for photocatalytic water splitting under the visible or infrared light. J Mater Chem A 7:603–610
145. Li Y, Feng Z, Sun Q, Ma Y (2021) Electronic, thermoelectric, transport and optical properties of MoSe2/Bas van der Waals heterostructures. Results Phys 23:104010
146. Ramanathan AA, Khalifeh JM (2021) Thermoelectrics of MoS2(1-x)N2x compounds. Phys Sci Biophys J 5:000167
147. Pallecchi I, Manca N, Patil B, Pellegrino L et al (2020) Review on thermoelectric properties of transition metal dichalcogenides. 4:032008
148. Zhu ZY, Cheng YC, Schwingenschlogl U (2011) Giant spin-orbit-induced spin splitting in two-dimensional transition-metal dichalcogenide semiconductors. Phys Rev B 84:1–5
149. Yun WS, Han SW, Hong SC, Kim IG et al (2012) Thickness and strain effects on electronicstructures of transition metal dichalcogenides $2H^-$. Phys Rev B 85:033305
150. Bahmani M, Lorke M, Faghihnasiri M, Frauenheim T (2021) Reversibly tuning the optical properties of defective transition-metal dichalcogenide monolayers. Phys Status Solidi B 258:2000524
151. Mouri S, Miyauchi Y, Matsuda K (2013) Tunable photoluminescence of monolayer MoS2 via chemical doping. Nano Lett 12:5944–5948
152. Kim HJ, Yun YJ, Yi SN, Chang SK et al (2020) Changes in the photoluminescence of monolayer and bilayer molybdenum disulfide during laser irradiation. ACS Omega 5:7903–7909
153. Su H, Wu S, Yang Y, Leng Q et al (2020) Surface plasmon polariton–enhanced photoluminescence of monolayer MoS_2 on suspended periodic metallic structures. Nanophotonics 10:975–982

154. Bernardi M, Ataca C, Palummo M, Grossman JC (2015) Optical and electronic properties of two-dimensional layered materials. Nanophotonics 5:111–125
155. Tayyab M, Hussain A, Asif QA, Adil W et al (2021) A computational study of MoS2 for band gap engineering by substitutional doping of TMN (T = transition metal (Cu), M = metalloid (B) and N = non-metal (C)). Mater Res Express 8:046301
156. Timpel M, Ligorio G, Ghiami A, Gavioli L et al (2021) 2D-MoS2 goes 3D: transferring optoelectronic properties of 2D MoS2 to a large-area thin film. npj 2D Mater Appl 5:64
157. Castellanos-Gomez A, Quereda J, van der Meulen HP, Agraït N, Rubio-Bollinger G (2016) Spatially resolved optical absorption spectroscopy of single- and few-layer MoS2 by hyperspectral imaging. Nanotechnology 27:115705
158. Zhang Y, Chang TR, Zhou B, Cui YT (2014) Direct observation of the transition from indirect to direct bandgap in atomically thin epitaxial $MoSe_2$. Nat Nanotechnol 9:111–115
159. Zhao W, Ghorannevis Z, Chu L, Toh M et al (2013) Evolution of electronic structure in atomically thin sheets of WS_2 and WSe_2. ACS Nano 7:791–797
160. Bullen HJ, Vishwanath S, Nahm RK, Xing HG et al (2022) Nucleation, growth, and stability of WSe2 thin films deposited on HOPG examined using in situ, real-time synchrotron x-ray radiation. J Vacuu Sci Tech A 40:012201
161. Guerrero GA, Rodríguez AG, Moreno-García H (2019) Hydrazine-free chemical bath deposition of WSe2 thin films and bi-layers for photovoltaic applications. Mater Res Express 6:105906
162. Hotovy I, Spiess L, Mikolasek M, Kostic I (2021) Layered WS2 thin films prepared by sulfurization of sputtered W films. App Sur Sci 544:148719
163. Kim SJ, Kim D, Min BK, Yi Y (2021) Bandgap tuned WS2 thin-film photodetector by strain gradient in van der Waals effective homojunctions. Adv Mater 9:2101310
164. Lezama IG, Arora A, Ubaldini A, Barreteau C et al (2015) Indirect-to-direct band gap crossover in few-layer $MoTe_2$. Nano Lett 15:2336–2342
165. Krishna HM, Rao KR (2018) Realization of a novel hybrid super capacitor with stacked multi layered graphene with intervening bismuth telluride (Bi2Te3) semiconducting layer and 2-dimensional transition metal dichalcogenides (TMDs) acting as top and bottom layers for potential electrodes—possibilities and challenges. AIP Conf Proc 1992:040002
166. Afzal AM, Mumtaz S, Iqbal MZ, Waqas M (2021) Fast and high photoresponsivity gallium telluride/hafnium selenide van der Waals heterostructure photodiode. J Mater Chem C 9:7110–7118
167. Altaf S, Haider A, Naz S, Ul-Hamid A (2020) Comparative study of selenides and tellurides of transition metals (Nb and Ta) with respect to its catalytic, antimicrobial, and molecular docking performance. Nanoscale Res Lett 15:144
168. Lee CH, Silva EC, Calderin L, Nguyen MAT et al (2015) Tungsten ditelluride: a layered semimetal. Sci Rep 5:1–8
169. Di Sante D, Das PK, Bigi C, Ergönenc Z et al (2017) Three-dimensional electronic structure of the type-II Weyl semimetal WTe2. Phys Rev Lett 119:1–6
170. Lui CH, Frenzel AJ, Pilon DV, Lee YH et al (2014) Trion-induced negative photoconductivity in monolayer MoS_2. Phys Rev Lett 113:1–5
171. Koperski M, Molas MR, Arora A, Nogajewski K et al (2017) Optical properties of atomically thin transition metal dichalcogenides: observations and puzzles. Nanophotonics 6:1289–1308
172. Wang H, Zhang C, Chan W, Manolatou C (2016) Radiative lifetimes of excitons and trions in monolayers of the metal dichalcogenide MoS_2. Phys Rev B 93:1–11
173. Singh A, Moody G, Tran K, Scott ME (2016) Trion formation dynamics in monolayer transition metal dichalcogenides. Phys Rev B 93:1–5
174. Rezk AR, Carey B, Chrimes AF, Lau DWM et al (2016) Acousticallydriven trion and exciton modulation in piezoelectric two-dimensional MoS_2. Nano Lett 16:849–855
175. Christopher JW, Goldberg BB, Swan AK (2017) Long tailed trions in monolayer MoS_2: Temperature dependent asymmetry and resulting red-shift of trion photoluminescence spectra. Sci Rep 7:1–8

176. Lampert MA (1958) Mobile and immobile effective-mass-particle complexes in nonmetallic solids. Phys Rev Lett 1:450–453
177. Sanvitto D, Pulizzi F, Shields AJ, Christiansen PCM et al (2001) Observation of charge transport by negatively charged excitons. Science 294:837–839
178. Pulizzi F, Sanvitto D, Christianen PCM, Shields AJ et al (2003) Optical imaging of trion diffusion and drift in GaAs quantum wells. Phys Rev B 68:1–9
179. Policht VR, Russo M, Liu F, Trovatello C et al (2021) Dissecting interlayer hole and electron transfer in transition metal dichalcogenide heterostructures via two-dimensional electronic spectroscopy. Nano Lett 21:4738−4743
180. Conti S, Neilson D, Peeters FM, Perali A (2020) Transition metal dichalcogenides as strategy for high temperature electron-hole superfluidity. Condens Matter 5:22
181. Li Z, Wang T, Lu Z, Jin C et al (2018) Revealing the biexciton and trion-exciton complexes in BN encapsulated WSe2. Nat Commun 9:3719
182. Sie EJ, Frenzel AJ, Lee YH, Kong J et al (2015) Intervalley biexcitons and many-body effects in monolayer MoS_2
183. Chowdhury RK, Nandy S, Bhattacharya S, Karmakar M (2018) http://hdl.handle.net/123456789/2066
184. Li Z, Wang T, Miao S, Lian Z et al (2020) Fine structures of valley-polarized excitonic states in monolayer transitional metal dichalcogenides. Nanophotonics 9:1811–1829
185. Sie EJ, Lui CH, Lee YH, Kong J et al (2016) Observation of intervalley biexcitonic optical stark effect in monolayer WS2. Nano Lett 16:7421–7426
186. Lee HS, Kim MS, Kim H, Lee YH (2016) Identifying multiexcitons in MoS_2 monolayers at room temperature. Phys Rev B 93:1–6
187. He Z, Xu W, Zhou Y, Wang X et al (2016) Biexciton formation in bilayer tungsten disulfide. ACS Nano 10:2176–2183
188. Hao K, Specht JF, Nagler P, Xu L et al (2017) Neutral and charged inter-valley biexcitons in monolayer MoSe2. Nat Commun 8:1–7
189. Kidd DW, Zhang DK, Varga K (2016) Binding energies and structures of two-dimensional excitonic complexes in transition metal dichalcogenides. Phys Rev B 93:125423–125510
190. Yu H, Cui X, Xu X, Yao W (2015) Valley excitons in two-dimensional semiconductors. Natl Sci Rev 2:57–70
191. Scharf B, Van Tuan D, Žutić I, Dery H (2019) Dynamical screening in monolayer transition-metal dichalcogenides and its manifestations in the exciton spectrum. J Phys Condens Matter 22, 31:203001
192. Scharf B, Xu G, Matos-Abiague A, Žutić I (2017) Magnetic proximity effects in transition-metal dichalcogenides: converting excitons. Phys Rev Lett 119:127403–127406
193. Pospischil A, Mueller T (2016) Optoelectronic devices based on atomically thin transition metal dichalcogenides. Appl Sci 6:78
194. Pedersen TG, Latini S, Thygesen KS, Mera H (2016) Exciton ionization in multilayer transition-metal dichalcogenides. New J Phys 18:073043
195. Chernikov A, Berkelbach TC, Hill HM, Rigosi A et al (2014) Exciton binding energy and nonhydrogenic Rydberg series in monolayer WS2. Phys Rev Lett 113:076802–076805
196. He K, Kumar N, Zhao L, Wang Z et al (2014) Tightly bound excitons in monolayer WSe2. Phys Rev Lett 113:1–5
197. Hanbicki AT, Currie M, Kioseoglou G, Friedman AL (2015) Measurement of high exciton binding energy in the monolayer transition-metal dichalcogenides WS2 and WSe2. Solid State Commun 203:16–20
198. Ramasubramaniam A (2012) Large excitonic effects inmonolayers ofmolybdenum and tungsten dichalcogenides. Phys Rev B 86:1–6
199. Xiao K, Yan T, Liu Q, Yang S et al (2021) Many-body effect on optical properties of monolayer molybdenum diselenide. J Phys Chem Lett 12:2555
200. Tran Thoai DB (1990) Exciton binding energy in semiconductor quantum wells and in heterostructures. Phys B 164:295–299

201. Belov PA, Khramtsov ES (2017) The binding energy of excitons in narrow quantum wells. J Phys Conf Ser 816:012018
202. Ross JS, Klement P, Jones AM, Ghimire NJ et al (2014) Electrically tunable excitonic light-emitting diodes based on monolayer WSe2 p–n junctions. Nat Nanotechnol 9:268–272
203. Wen X, Gong Z, Li D (2019) Nonlinear optics of two-dimensional transition metaldichalcogenides. Info Mat 1:317–337
204. Janisch C, Wang Y, Ma D, Mehta N et al (2014) Extraordinary second harmonic generation in Tungsten disulfide monolayers. Sci Rep 4:1–5
205. Kilinc M, Cheney A, Neureuter C, Tarasek S (2021) Intracavity second harmonic generation from a WSe2 monolayer in a passively mode-locked picosecond fiber laser. Opt Mater Exp 11:1603–1613
206. Zhou KG, Zhao M, Chang MJ, Wang Q et al (2015) Size-dependent nonlinear optical properties of atomically thin transition metal dichalcogenide nanosheets. Small 11:694–701
207. Tang CY, Cheng PK,Wang XY, Ma S (2020) Size-dependent nonlinear optical properties of atomically thin PtS2 nanosheet. Opt Mater 101:109694
208. Zhang S, Dong N, McEvoy N, Obrien MC et al (2015) Direct observation of degenerate two-photon absorption and its saturation in WS2 and MoS_2 monolayer and fewlayer films. ACS Nano 9:7142–7150
209. Yang H, Wang Y, Tiu ZC, Tan SJ et al (2022) All-optical modulation technology based on 2D layered materials. Mater Micromach 13:92
210. Wang W, Wu Y, Wu Q, Hua J et al (2016) Coherent nonlinear optical response spatial self-phase modulation in MoSe2 nano-sheets. Sci Rep 6:1–6
211. Liu X, Guo Q, Qiu J (2017) Emerging low-dimensional materials for nonlinear optics and ultrafast photonics. Adv Mater 29:14
212. Yang J, Wang Z, Wang F, Xu R et al (2016) Atomically thin optical lenses and gratings. Light Sci Appl 5:16046
213. Hughes HP, Webb C, Williams PM (1980) Angle resolved photoemission from VSe2. J Phys C Solid State Phys 13:1125–1138
214. Jing Y, Zhou Z, Cabrera CR, Chen Z (2013) Metallic VS2 monolayer: a promising 2D anode material for lithium ion batteries. J Phys Chem C 117:25409–25413
215. Wang Z, Su Q, Yin GQ, Shi J et al (2014) Structure and electronic properties of transition metal dichalcogenide MX2(M = Mo,W, Nb; X = S, Se) monolayers with grain boundaries. Mater Chem Phys 147:1068–1073
216. Varma SJ, Kumar J, Liu Y, Layne K et al (2017) 2D TiS2 layers: a superior nonlinear optical limiting material. Adv Opt Mater 5:1700713
217. Huang P, Yuan L, Zhang K, Chen Q et al (2018) Room-temperature and aqueous solution-processed two-dimensional TiS2 as an electron transport layer for highly efficient and stable planar n-i-p perovskite solar cells. ACS Appl Mater Interfaces 10:14796–14802
218. Wang G, Chernikov A, Glazov MM, Heinz TF et al (2018) Excitons in atomically thin transition metal dichalcogenides. Rev Mod Phys 90:021001
219. Dyakonov M (2008) Springer series in solid-state science. Springer-Verlag, Berlin, p 157
220. Thilagam A (2014) Exciton complexes in lowdimensional transitionmetal dichalcogenides. J Appl Phys 116:053523
221. You Y, Zhang XX, Berkelbach TC, Hybertsen MS et al (2015) Observation of biexcitons in monolayer WSe2. Nat Phys 11:477
222. He XF (1991) Excitons in anisotropic solids: the model of fractional-dimensional space. Phys Rev B 43:2063–2069
223. Mitioglu A, Plochocka P, Granados del Aguila Á, Christianen P et al (2015) Optical investigation ofmonolayer and bulk tungsten diselenide (WSe2) in high magnetic fields. Nano Lett 15:4387–4392
224. Molas, M.R., Nogajewski, K., Slobodeniuk, A. O., J. Binder et al. 2017 Optical response of monolayer, few-layer and bulk tungsten disulfide"Nanoscale, 9, 13128-13141
225. Liu E, van Baren J, Lu Z, Taniguchi T (2021) Exciton-polaron Rydberg states in monolayer $MoSe_2$ and WSe_2. Nat Commun 12:6131

226. Kapuściński P, Delhomme A, Vaclavkova D, Slobodeniuk AO et al (2021) Rydberg series of dark excitons and the conduction band spin-orbit splitting in monolayer WSe2. Commun Phys 4:186–6
227. Berkelbach TC, Hybertsen MS, Reichman DR (2013) Theory of neutral and charged excitons in monolayer transition metal dichalcogenides. Phys Rev B 88:045318
228. Arora A, Drüppel M, Schmidt R, Deilmann T (2017) Interlayer excitons in a bulk van der Waals semiconductor. Nat Commun 8:639
229. Cudazzo P (2020) First-principles description of the exciton-phonon interaction: a cumulant approach. Phys Rev B 102:045136–045211
230. Qiu DY, da Jornada FH, Louie SG (2013) Optical spectrum of MoS2: many-body effects and diversity of exciton states. Phys Rev Lett 111:216805-5
231. Cheiwchanchamnangij T, Lambrecht WR (2012) Quasiparticle band structure calculation ofmonolayer, bilayer, and bulk MoS_2. Phys Rev B 85:205302-4
232. Molina-Sánchez A, Sangalli D, Hummer K, Marini A et al (2013) Effect of spin-orbit interaction on the optical spectra of single-layer, double-layer, and bulk MoS_2. Phys Rev B 88:045412
233. Brotons-Gisbert M, Proux R, Picard R, Andres-Penare D (2019) Out-of-plane orientation of luminescent excitons in two-dimensional indium selenide. Nat Commun 10:3913
234. Wang H, Liu W, He X, Zhang P et al (2020) An excitonic perspective on low-dimensional semiconductors for photocatalysis. J Am Chem Soc 142(33):14007–14022
235. Scholes GD, Rumbles G (2010) Excitons in nanoscale systems. Mater Sustain Energy 12–25
236. Scholes GD, Rumbles G (2006) Excitons in nanoscale systems. Nat Mater 5:683–696
237. Ernandes C, Khalil L, Almabrouk H, Pierucci D (2021) Indirect to direct band gap crossover in two-dimensional $WS_{2(1-x)}Se_{2x}$ alloys. npj 2D Mater Appl 5:7
238. Hybertsen MS, Louie SG (1986) Electron correlation in semiconductors and insulators: band gaps and quasiparticle energies. Phys Rev B 34:5390
239. Rohlfing M, Louie SG (2000) Electron-hole excitations and optical spectra from first principles. Phys Rev B 62:4927
240. Rasmussen FA, Thygesen KS (2015) Computational 2D materials database: electronic structure of transition-metal dichalcogenides and oxides. J Phys Chem C 119:13169–13183
241. Hooda MK, Yadav CS, Samal D (2020) Electronic and topological properties of group-10 transition metal dichalcogenides. J Phys Condens Matter 33:103001
242. Morales-Durán N, MacDonald AH, Potasz P (2021) Metal-insulator transition in transition metal dichalcogenide heterobilayer moiré superlattices. Phys Rev B 103:L241110
243. Maniyar A, Choudhary S (2020) Visible region absorption in TMDs/phosphorene heterostructures for use in solar energy conversion applications. RSC Adv 10:31730
244. Ermolaev GA, Stebunov YV, Vyshnevyy AA, Tatarkin DE (2020) Broadband optical properties of monolayer and bulk MoS_2. npj 2D Mater Appl 4:21
245. Vaquero D, Clericò V, Salvador-Sánchez J, Martín-Ramos A et al (2020) Excitons, trions and Rydberg states in monolayer MoS2 revealed by low-temperature photocurrent spectroscopy. Commun Phys 3:194
246. Berghäuser G, Malic E (2014) Analytical approach to excitonic properties of MoS_2. Phys Rev B 89:125309
247. Wu F, Qu F, MacDonald A (2015) Exciton band structure of monolayer MoS_2. Phys Rev B 91:075310
248. Fugallo G, Cudazzo P, Gatti M, Sottile F (2021) Exciton band structure of molybdenum disulfide: from monolayer to bulk. IOP Sci 3:014005
249. Li Y, Li G, Zhai X, Xiong S et al (2020) Spin splitting in a MoS2 monolayer induced by exciton interaction. Phys Rev B 101:245439
250. Merkl P, Mooshammer F, Brem S, Girnghuber A et al (2020) Twist-tailoring coulomb correlations in van der Waals homobilayers. Nat Commun 11:2167
251. Molina-Sanchez A, Wirtz L (2011) Phonons in single-layer and few-layer MoS2 and WS2. Phys Rev B 84:155413

252. Goldstein T, Wu YC, Chen SY, Taniguchi T et al (2020) Ground and excited state exciton polarons in monolayer $MoSe_2$. J Chem Phys 153:071101
253. Long C, Dai Y, Jin H (2021) Effect of point defects on electronic and excitonic properties in Janus-MoSSe monolayer. Phys Rev B 104:125306
254. Jiang Y, Chen S, Zheng W, Zheng B et al (2021) Interlayer exciton formation, relaxation, and transport in TMD van der Waals heterostructures. Sci Appl 10:72
255. Keldysh LV (1979) Coulomb interaction in thin semiconductor and semimetal films. JETP Lett 29:658
256. Kylänpää I, Komsa HP (2015) Binding energies of exciton complexes in transition metal dichalcogenide monolayers and effect of dielectric environment. Phys Rev B 92:205418
257. Ye Z, Cao T, O'Brien K, Zhu H et al (2014) Probing excitonic dark states in single-layer tungsten disulphide. Nature 513:214–218
258. Hill HM, Rigosi AF, Roquelet C, Chernikov A et al (2015) Observation of excitonic rydberg states in monolayer MoS2 and WS2 by photoluminescence excitation spectroscopy. Nano Lett 15:2992–2997
259. Wang G, Marie X, Gerber I, Amand T et al (2015) Giant enhancement of the optical second-harmonic emission of WSe2 monolayers by laser excitation at exciton resonances. Phys Rev Lett 114:09403–09406
260. Latini S, Olsen T, Thygesen KS (2015) Excitons in van der Waals heterostructures: The important role of dielectric screening. Phys Rev B 92:245123
261. Raja A, Chaves A, Yu J, Arefe G et al (2017) Coulomb engineering of the bandgap and excitons in two-dimensional materials. Nat Commun 8:15251
262. Stier AV, Wilson NP, Velizhanin KA, Kono J et al (2018) Magnetooptics of exciton rydberg states in a monolayer semiconductor. Phys Rev Lett 120:057405
263. Goryca M, Li J, Stier AV, Taniguchi T et al (2019) Revealing exciton masses and dielectric properties of monolayer semiconductors with high magnetic fields. Nat Commun 10:4172
264. Kormányos A, Burkard G, Gmitra M, Fabian J et al (2015) kp theory for two-dimensional transition metal dichalcogenide semiconductors. 2D Mater 2:022001
265. Zhang X, You Y, Zhao S, Heinz TF (2018) Experimental evidence for dark excitons in monolayer WSe_2. Phys Rev Lett 115:257403
266. Molas M, Faugeras C, Slobodeniuk AO, Nogajewski K et al (2017) Brightening of dark excitons in monolayers of semiconducting transition metal dichalcogenides. 2D Mater 4:21003
267. Zhang X, Cao T, Lu Z, Lin YC et al (2017) Magnetic brightening and control of dark excitons in monolayer WSe_2. Nat Nanotechnol 12:883–888
268. Zhou Y, Scuri G, Wild DS, High AA et al (2017) Probing dark excitons in atomically thin semiconductors via nearfield coupling to surface plasmon polaritons. Nat Nanotechnol 12:856–860
269. Zhang C, Chen Y, Johnson A, Li MY et al (2015) Probing critical point energies of transition metal dichalcogenides: surprising indirect gap of single layer WSe_2. Nano Lett 15:6494–6500
270. Hsu W, Lu LS, Wang D, Huang JK et al (2017) Evidence of indirect gap in monolayer WSe_2. Nat Commun 8:929
271. Selig M, Berghäuser G, Raja A, Nagler P et al (2016) Excitonic linewidth and coherence lifetime in monolayer transition metal dichalcogenides. Nat Commun 7:13279
272. Niehues I, Schmidt R, Drüppel M, Marauhn P et al (2018) Strain control of exciton-phonon coupling in atomically thin semiconductors. Nano Lett 18:1751–1757
273. Lindlau J, Robert C, Funk V, Förste J et al (2017) Identifying optical signatures of momentum-dark excitons in transition metal dichalcogenide monolayers. Preprint at https://arxiv.org/abs/1710.00988
274. Feierabend M, Berghäuser G, Knorr A, Malic E (2017) Proposal for dark exciton based chemical sensors. Nat Commun 8:14776
275. Berghäuser G, Steinleitner P, Merkl P, Huber R et al (2018) Mapping of the dark exciton landscape in transition metal dichalcogenides. Phys Rev B 98:020301R

276. Mueller T, Malic E (2018) Exciton physics and device application of two-dimensional transition metal dichalcogenide semiconductors. npj 2D Mater Appl 2:29
277. Molas MR, Slobodeniuk AO, Nogajewski K, Bartos M et al (2019) Energy spectrum of two-dimensional excitons in a nonuniform dielectric medium. Phys Rev Lett 123:136801–136806
278. Liu E, van Baren J, Taniguchi T, Watanabe K et al (2019) Magnetophotoluminescence of exciton Rydberg states inmonolayer WSe_2. Phys Rev B 99:205420–205428
279. Echeverry JP, Urbaszek B, Amand T, Marie X et al (2016) Splitting between bright and dark excitons in transition metal dichalcogenide monolayers. Phys Rev B 93:121107-5
280. Deilmann T, Thygesen KS (2017) Dark excitations in monolayer transition metal dichalcogenides. Phys Rev B 96:201113–201115
281. Bieniek M, Szulakowska L, Hawrylak P (2020) Band nesting and exciton spectrum in monolayer MoS_2. Phys Rev B 101:125423–125515
282. Chen Y, Huang Y, Lou W, Cai Y et al (2020) Electric-field-driven exciton vortices in transition metal dichalcogenide monolayers. Phys Rev B 102:165413–165512
283. Kirichenko EV, Stephanovich VA (2021) The influence of Coulomb interaction screening on the excitons in disordered two-dimensional insulators. Sci Rep 11:11956
284. Rodin A, Neto AC (2013) Excitonic collapse in semiconducting transition-metal dichalcogenides. Phys Rev B 88:195437
285. Stroucken T, Koch SW (2015) Optically bright p-excitons indicating strong Coulomb coupling in transition-metal dichalchogenides. J Phys Cond Matter 27:345003
286. Van Tuan D, Scharf B, Wang Z, Shan J et al (2019) Probing many-body interactionsin monolayer transition-metal dichalcogenides. Phys Rev B 99:085301
287. Van Tuan D, Scharf B, Žutić I, Dery H (2017) Marrying excitons and plasmons in monolayer transition-metal dichalcogenides. Phys Rev X 7:041040
288. Li Z, Wang T, Jin C et al (2019) Momentum-dark intervalley exciton in monolayer tungsten diselenide brightened via chiral phonon. ACS Nano 13:14107–14113
289. Poellmann C, Steinleitner P, Leierseder U, Nagler P et al (2015) Resonant internal quantum transitions and femtosecond radiative decay of excitons in monolayer WSe2. Nat Mater 14:889–893
290. Ceballos F, Cui QN, Bellus MZ, Zhao H (2016) Exciton formation in monolayer transition metal dichalcogenides. Nanoscale 8:11681–11688
291. Steinleitner P, Merkl P, Nagler P, Mornhinweg J et al (2017) Direct observation of ultrafast exciton formation in a monolayer of WSe2. Nano Lett 17:1455–1460
292. Guo L, Wu M, Cao T, Monahan DM et al (2019) Exchange-driven intravalley mixing of excitons in monolayer transition metal dichalcogenides. Nat Phys 15:228–232
293. Schmidt R, Berghäuser G, Schneider R, Selig M et al (2016) Ultrafast Coulomb-induced intervalley coupling in atomically thin WS2. Nano Lett 16:2945–2950
294. Plechinger G, Nagler P, Arora A, Schmidt R et al (2016) Trion fine structure and coupled spin-valley dynamics in monolayer tungsten disulfide. Nat Commun 7:12715
295. Mai C, Barrette A, Yu YF, Semenov YG et al (2014) Many-body effects in valleytronics: Direct measurement of valley lifetimes in single-layer MoS2. Nano Lett 14:202–206
296. Yan TF, Yang SY, Li D, Cui XD (2017) Long valley relaxation time of free carriers in monolayer WSe2. Phys Rev B 95:241406
297. Bertoni R, Nicholson CW, Waldecker L, Hübener H et al (2016) Generation and evolution of spin-, valley-, and layer-polarized excited carriers in inversion-symmetric WSe2. Phys Rev Lett 117:277201
298. Chernikov A, Ruppert C, Hill HM, Rigosi AF et al (2015) Population inversion and giant bandgap renormalization in atomically thin WS2 layers. Nat Photonics 9:466–470
299. Cunningham PD, Hanbicki AT, McCreary KM, Jonker BT (2017) Photoinduced bandgap renormalization and exciton binding energy reduction in WS2. ACS Nano 11:12601–12608
300. Hong XP, Kim J, Shi SF, Zhang Y et al (2014) Ultrafast charge transfer in atomically thin MoS2/WS2 heterostructures. Nat Nanotechnol 9:682–686
301. Ruppert C, Chernikov A, Hill HM, Rigosi AF et al (2017) The role of electronic and phononic excitation in the optical response of monolayer WS2 after ultrafast excitation. Nano Lett 17:644– 651

302. Sie EJ, Steinhoff A, Gies C, Luo CH et al (2017) Observation of exciton redshift-blueshift crossover in monolayer WS2. Nano Lett 17:4210–4216
303. Yuan L, Chung TF, Kuc A, Wan Y et al (2018) Photocarrier generation from interlayer chargetransfer transitions in WS2-graphene heterostructures. Sci Adv 4:700324
304. Wake DR, Yoon HW, Wolfe JP, Morkoç H (1992) Response of excitonic absorption spectra to photoexcited carriers in GaAs quantum wells. Phys Rev B 46:13452–13460
305. Manzke G, Henneberger K, May V (1987) Many-exciton theory formultiple quantum-well structures. Phys 139:233–239
306. Aivazian G, Yu HY, Wu SF, Yan JQ et al (2017) Many-body effects in nonlinear optical responses of 2D layered semiconductors. 2D Mater 4:025024
307. Cunningham PD, McCreary KM, Jonker BT (2016) Auger recombination in chemical vapor deposition-grown monolayer WS2. J Phys Chem Lett 7:5242–5246
308. Danovich M, Zólyomi V, Fal'ko VI, Aleiner IL (2016) Auger recombination of dark excitons in WS2 and WSe2 monolayers. 2D Mater 3:035011
309. Sun DZ, Rao Y, Reider GA, Chen GG et al (2014) Observation of rapid excitonexciton annihilation in monolayer molybdenum disulfide. Nano Lett 14:5625–5629
310. Mouri S, Miyauchi Y, Toh M, Zhao WJ et al (2014) Nonlinear photoluminescence in atomically thin layered WSe2arising from diffusion-assisted exciton-exciton annihilation. Phys Rev B 90:155449
311. Shi HY, Yan RS, Bertolazzi S, Brivio J et al (2013) Exciton dynamics in suspendedmonolayer and few-layer MoS2 2D crystals. ACS Nano 7:1072–1080
312. Zhao J, Zhao W, Du W, Su R et al (2020) Dynamics of exciton energy renormalization in monolayer transition metal disulfides. Nano Res 13:1399–1405
313. Thränhardt A, Kuckenburg S, Knorr A, Thomas P et al (2000) Interplay between coherent and incoherent scattering in quantum well secondary emission. Phys Rev B 62:16802
314. Kira M, Koch SW (2012) Semiconductor quantum optics, 1st edn. Cambridge University Press
315. Trovatello C, Katsch F, Borys NJ, Selig M et al (2020) The ultrafast onset of exciton formation in 2D semiconductors. Nat Commun 11:5277
316. Robert C, Lagarde D, Cadiz F, Wang G et al (2016) Exciton radiative lifetime in transition metal dichalcogenideMonolayers. Phys Rev B 93:205423
317. Ceballos F, Cui Q, Bellus MZ, Zhao H (2016) Exciton formation in monolayer transition metal dichalcogenides. Nanoscale 8:11681–11688
318. Yao K, Yan A, Kahn S, Suslu A et al (2017) Optically discriminating carrier-induced quasiparticle band gap and exciton energy renormalization in monolayer MoS2. Phys Rev Lett 119:087401
319. Borys NJ, Barnard ES, Gao S, Yao K et al (2017) Anomalous above-gap photoexcitations and optical signatures of localized charge puddles in monolayer molybdenum disulfide. ACS Nano 11:2115–2123
320. Brito-Cruz CH, Gordon JP, Becker PC, Fork RL et al (1988) Dynamics of spectral hole burning. IEEE J Quantum Electron 24:261–269
321. Hendry E, Hale PJ, Moger J, Savchenko AK et al (2010) Coherent nonlinear optical response of graphene. Phys Rev Lett 105:097401
322. Feierabend M, Brem S, Ekman A, Malic E (2021) Brightening of spin- and momentum-dark excitons in transition metal dichalcogenides. 2D Mater 8:015013
323. Han B, Robert C, Courtade E, Manca M et al (2018) Exciton states in monolayer MoSe2 and MoTe2 probed by upconversion spectroscopy. Phys Rev X 8:031073
324. Dean JJ, van Driel HM (2009) Second harmonic generation from graphene and graphitic films. Appl Phys Lett 95:261910
325. Autere A, Jussila H, Dai Y, Wang Y et al (2018) Nonlinear optics with 2D layered materials. Adv Mater 30:1705963
326. Malard LM, Alencar TV, Barboza APM, Mak KF et al (2013) Observation of intense second harmonic generation from MoS2 atomic crystals. Phys Rev B: Condens Matter Mater Phys 87:201401

327. Hsu WT, Zhao ZA, Li LJ, Chen CH et al (2014) Second harmonic generation from artificially stacked transition metal dichalcogenide twisted bilayers. ACS Nano 8:2951–2958
328. Shi J, Yu P, Liu F, He P et al (2017) 3R MoS2 with broken inversion symmetry: a promising ultrathin nonlinear optical device. Adv Mater 29:1701486
329. Säynätjoki A, Karvonen L, Rostami H, Autere A et al (2017) Ultra-strong nonlinear optical processes and trigonal warping in MoS2 layers. Nat Commun 8:1–8
330. Autere A, Jussila H, Marini A, Saavedra JRM et al (2018) Optical harmonic generation in monolayer group-VItransition metal dichalcogenides. Phys Rev B: Condens Matter Mater Phys 98:115426
331. Beams R, Cançado LG, Krylyuk S, Kalish I et al (2016) Characterization of few-layer 1T MoTe2 by polarization-resolved second harmonic generation and raman scattering. ACS Nano 10:9626–9636
332. Ribeiro-Soares J, Janisch C, Liu Z, Elias AL et al (2015) Second Harmonic generation in WSe2. 2D Mater 2:045015
333. Janisch C, Wang Y, Ma D, Mehta N et al (2015) Extraordinary second harmonic generation in tungsten disulfide monolayers. Sci Rep 4:5530
334. Cui Q, Muniz RA, Sipe JE, Zhao H (2017) Strong and anisotropic third-harmonic generation in monolayer and multilayer ReS2. Phys Rev B: Condens Matter Mater Phys 95:165406
335. Erkensten D, Brem S, Malic E (2021) Exciton-exciton interaction in transition metal dichalcogenide monolayers and van der Waals heterostructures. Phys Rev B 103:045426
336. Binder J, Howarth J, Withers F, Molas MR et al (2019) Upconverted electroluminescence via auger scattering of interlayer excitons in van der Waals heterostructures. Nat Commun 10:2335
337. Malic E, Selig M, Feierabend M, Brem S et al (2018) Dark excitons in transition metal dichalcogenides. Phys Rev Mater 2:014002
338. Moody G, Dass CK, Hao K, Chen CH et al (2015) Intrinsic homogeneous linewidth and broadening mechanisms of excitons in monolayer transition metal Dichalcogenides. Nat Commun 6:8315
339. Wang F, Dukovic G, Brus LE, Heinz TF (2005) The optical resonances in carbon nanotubes arise from excitons. Science 308:838–841
340. Lafrentz M, Brunne D, Rodina AV, Pavlov VV et al (2013) Second-harmonic generation spectroscopy of excitons in ZnO. Phys Rev B: Condens Matter Mater Phys 88:235207
341. Wen X, Gong Z, Li D (2019) Nonlinear optics of two-dimensional transition metal dichalcogenides. Info Mat 1:317–337
342. Rueda JH, Noordam ML, Komen I, Kuipers L (2021) Nonlinear optical response of a WS_2 monolayer at room temperature upon multicolor laser excitation. ACS Photon 8:550−556
343. Mertens J, Shi Y, Molina-Sánchez A, Wirtz L et al (2014) Excitons in a mirror: formation of optical bilayers using MoS2 monolayers on gold substrates. Appl Phys Lett 104:191105
344. Gillard DJ, Genco A, Ahn S, Lyons TP et al (2021) Strong exciton-photon coupling in large area MoSe2 and WSe2 heterostructures fabricated from two-dimensional materials grown by chemical vapor deposition. 2D Mater 8:011002
345. Wang Y, Wang Z, Yao W, Liu GB et al (2017) Interlayer coupling in commensurate and incommensurate bilayer structures of transition-metal dichalcogenides. Phys Rev B 95:115429
346. Tong Q, Yu H, Zhu Q, Wang Y et al (2017) Topological mosaics in moiré superlattices of van der Waals heterobilayers. Nat Phys 13:356–362
347. Wu F, Lovorn T, MacDonald AH (2017) Topological exciton bands in Moiré heterojunctions. Phys Rev Lett 118:147401
348. Zhang C, Chuu CP, Ren X, Li MY et al (2017) Interlayer couplings, Moiré patterns, and 2D electronic superlattices in MoS_2/WSe_2 hetero-bilayers. Sci Adv 3:e1601459
349. Zhou Y, Scuri G, Sung J, Gelly RJ et al (2020) Controlling excitons in an atomically thin membrane with a Mirror. Phys Rev Lett 124:027401
350. Scuri G, Zhou Y, High AA, Wild DS et al (2018) Large excitonic reflectivity of monolayer MoSe2 encapsulated in hexagonal boron nitride. Phys Rev Lett 120:037402

351. Wild DS, Shahmoon E, Yelin SF, Lukin MD (2018) Quantum nonlinear optics in atomically thin materials. Phys Rev Lett 121:123606; Hoi IC, Kockum A, Tornberg L, Pourkabirian A et al (2015) Probing the quantum vacuum with an artificial atom in front of a mirror. Nat Phys 11:1045
352. Sung J, Zhou Y, Scuri G, Zólyomi V et al (2020) Broken mirror symmetry in excitonic response of reconstructed domains in twisted MoSe2/MoSe2 bilayers. Nat Nanotech 15:750–754
353. Shimazaki Y, Schwartz I, Watanabe K, Taniguchi T et al (2020) Strongly correlated electrons and hybrid excitons in a moiré heterostructure. Nature 580:472–477
354. Horng J, Stroucken T, Zhang L, Paik EY et al (2018) Observation of interlayer excitons in MoSe2 single crystals. Phys Rev B 97:241404
355. Gerber IC, Courtade E, Shree S, Robert C et al (2019) Interlayer excitons in bilayer MoS2 with strong oscillator strength up to room temperature. Phys Rev B 99:035443
356. Deilmann T, Thygesen KS (2018) Interlayer excitons with large optical amplitudes in layered van der Waals materials. Nano Lett 18:2984–2989
357. Geim AK, Grigorieva IV (2013) Van der Waals heterostructures. Nature 499:419–425
358. Palummo M, Bernardi M, Grossman JC (2015) Exciton radiative lifetimes in two-dimensional transition metal dichalcogenides. Nano Lett 15:2794–2800
359. Rivera P, Schaibley JR, Jones AM, Ross JS et al (2015) Observation of long-lived interlayer excitons in monolayer MoSe2–WSe2 heterostructures. Nat Commun 6:6242
360. Fogler MM, Butov LV, Novoselov KS (2014) Hightemperature superfluidity with indirect excitons in van der Waals heterostructures. Nat Commun 5:4555
361. Calman EV, Fogler MM, Butov LV, Hu S et al (2018) Indirect excitons in van der Waals heterostructures at room temperature. Nat Commun 9:1895
362. Kang J, Tongay S, Zhou J, Li J et al (2013) Band offsets and heterostructures of two-dimensional semiconductors. Appl Phys Lett 102:012111
363. Kósmider K, Fernández-Rossier J (2013) Electronic properties of the MoS_2-WS_2 heterojunction. Phys Rev B 87:075451
364. Van der Donck M, Peeters FM (2018) Interlayer excitons in transition metal dichalcogenide heterostructures. Phys Rev B 98:115104
365. Xu K, Xu Y, Zhang H, Peng B et al (2018) The role of Anderson's rule in determining electronic, optical and transport properties of transition metal dichalcogenide heterostructures. Phys Chem Chem Phys 20:30351–30364
366. Yan J, Ma C, Huang Y, Yang G (2019) Tunable control of interlayer excitons in WS_2/MoS_2 heterostructures via strong coupling with enhanced mie resonances. Adv Sci 6:1802092
367. Tan Q, Rasmita A, Li S, Liu S et al (2021) Layer-engineered interlayer excitons. Sci Adv 7:eabh0863
368. Calman EV, Fowler-Gerace LH, Choksy DJ, Butov LV et al (2020) Indirect excitons and trions in MoSe2/WSe2 van der Waals heterostructures. Nano Lett 20:1869–1875
369. Chiu MH, Zhang C, Shiu HW, Chuu CP et al (2015) Determination of band alignment in the single-layer MoS_2/WSe_2 heterojunction. Nat Commun 6:7666
370. Quan CJ, Lu C, He C, Xu X et al (2019) Band alignment of MoTe2/MoS2 nanocomposite films for enhanced nonlinear optical performance. Adv Mater Interfaces 6:1801733
371. Wilson NR, Nguyen PV, Seyler K, Rivera P et al (2017) Determination of band offsets, hybridization, and exciton binding in 2D semiconductor heterostructures. Sci Adv 3:e1601832
372. Komsa HP, Krasheninnikov AV (2013) Electronic structures and optical properties of realistic transition metal dichalcogenide heterostructures from first principles. Phys Rev B 88:085318
373. Amin B, Singh N, Schwingenschlogl U (2015) Heterostructures of transition metal dichalcogenides. Phys Rev B 92:075439
374. Ruiz-Tijerina DA, Falko VI (2019) Interlayer hybridization and moire superlattice minibands for electrons and excitons in heterobilayers of transition-metal dichalcogenides. Phys Rev B 99:125424
375. Heo H, Sung JH, Cha S, Jang BG et al (2015) Interlayer orientation-dependent light absorption and emission in monolayer semiconductor stacks. Nat Commun 6:7372

376. Okada M, Kutana A, Kureishi Y, Kobayashi Y et al (2018) Direct and indirect interlayer excitons in a van der Waals heterostructure of hBN/WS2/MoS2/hBN. ACS Nano 12:2498–2505
377. Kunstmann J, Mooshammer F, Nagler P, Chaves A et al (2018) Momentum-space indirect interlayer excitons in transition-metal dichalcogenide van der Waals heterostructures. Nat Phys 14:801–805
378. Hanbicki AT, Chuang HJ, Rosenberger MR, Hellberg CS et al (2018) Double indirect interlayer exciton in a MoSe2/WSe2 van der Waals heterostructure. ACS Nano 12:4719–4726
379. Brem S, Lin KQ, Gillen R, Bauer JM et al (2020) Hybridized intervalley moire excitons and flat bands in twisted WSe2 bilayers. Nanoscale 12:11088–11094
380. Liu GB, Xiao D, Yao Y, Xu X et al (2015) Electronic structures and theoretical modelling of twodimensional group-VIB transition metal dichalcogenides. Chem Soc Rev 44:2643–2663
381. Tongay S, Fan W, Kang J, Park J et al (2014) Tuning Interlayer coupling in large-area heterostructures with CVD-Grown MoS2 and WS2 monolayers. Nano Lett 14:3185–3190
382. Liu KH, Zhang L, Cao T, Jin C et al (2014) Evolution of interlayer coupling in twisted molybdenum disulfide bilayers. Nat Commun 5:4966
383. Jin CH, Ma EY, Karni O, Regan EC et al (2018) Ultrafast dynamics in van der Waals heterostructures. Nat Nanotechnol 13:994–1003
384. Ceballos F, Bellus MZ, Chiu HY, Zhao H et al (2014) Ultrafast charge separation and indirect exciton Formation in a MoS2-MoSe2 van der Waals heterostructure. ACS Nano 8:12717–12724
385. Zhu HM, Wang J, Gong Z, Kim YD et al (2017) Interfacial charge transfer circumventing momentum mismatchat two-dimensional van der Waals heterojunctions. Nano Lett 17:3591–3598
386. Wang K, Huang B, Tian M, Ceballos F et al (2016) Interlayer coupling in twisted WSe2/WS2 bilayer heterostructures revealed by optical spectroscopy. ACS Nano 10:6612–6622
387. Zereshki P, Yao P, He D, Wang Y et al (2019) Interlayer charge transfer in ReS2/WS2 van der Waals heterostructures. Phys Rev B 99:195438
388. Ma EY, Guzelturk B, Li G, Cao L et al (2019) Recording interfacial currents on the subnanometer length and femtosecond time scale by terahertz emission. Sci Adv 5:eaau0073
389. Wang H, Bang J, Sun Y, Liang L et al (2016) The role of collective motion in the ultrafast charge transfer in van der Waals heterostructures. Nat Commun 7:11504
390. Liu F, Li QY, Zhu XY (2020) Direct determination of momentum-resolved electron transfer in the photoexcited van der Waals heterobilayer WS2/MoS2. Phys Rev B 101:201405(R)
391. Liu JY, Zhang X, Lu G (2020) Excitonic effect drives ultrafast dynamics in van der Waals heterostructures. Nano Lett 20:4631–4637
392. Komsa HP, Krasheninnikov AV (2012) Effects of confinement and environment on the electronic structure and exciton binding energy of MoS2 from first principles. Phys Rev B 86:241201(R)
393. Shi HL, Pan H, Zhang YW, Yakobson BI (2013) Quasiparticle band structures and optical properties of strained monolayer MoS2 and WS2. Phys Rev B 87:155304
394. Rivera P, Yu H, Seyler KL, Wilson NP et al (2018) Interlayer valley excitons in heterobilayers of transition metal dichalcogenides. Nat Nanotechnol 13:1004–1015
395. Radisavljevic B, Radenovic A, Brivio J, Giacometti V et al (2011) Single-layer MoS_2 transistors. Nat Nanotech 6:147–150
396. Van der Donck M, Zarenia M, Peeters FM (2017) Excitons and trions in monolayer transition metal dichalcogenides: a comparative study between the multiband model and the quadratic single-band model. Phys Rev B 96:035131
397. Mak KF, He K, Changgu Lee GH et al (2013) Tightly bound trions in monolayer MoS_2. Nat Mater 12:207–211
398. Ross JS, Wu S, Yu H, Ghimire NJ et al (2013) Electrical control of neutral and charged excitons in a monolayer semiconductor. Nat Commun 4:1474
399. Kormanyos A, Burkard G, Gmitra M, Fabian J et al (2015) kp theory for two-dimensional transition metal dichalcogenide semiconductors. 2D Mater 2:022001

400. Jones AM, Yu H, Ghimire NJ, Wu S et al (2013) Optical generation of excitonic valley coherence in monolayer WSe_2. Nat Nanotechnol 8:634–638
401. Singh A, Moody G, Wu S, Wu Y et al (2014) Coherent electronic coupling in atomically thin $MoSe_2$. Phys Rev Lett 112:216804
402. Plechinger G, Nagler P, Kraus J, Paradiso N et al (2015) Identification of excitons, trions and biexcitons in single-layer WS2 physica status solidi (RRL). Rapid Res Lett 9:457–461
403. Shang J, Shen X, Cong C, Peimyoo N et al (2015) Observation of excitonic fine structure in a 2D transition-metal dichalcogenide semiconductor. ACS Nano 9:647–655
404. Courtade E, Semina M, Manca M, Glazov MM et al (2017) Charged excitons in monolayer WSe2: experiment and theory. Phys Rev B 96:085302
405. Moody G, Schaibley J, Xu X (2016) Exciton dynamics in monolayer transition metal dichalcogenides. J Opt Soc Am B 33:C39
406. Dery H, Song Y (2015) Polarization analysis of excitons in monolayer and bilayer transition-metal dichalcogenides. Phys Rev B 92:125431
407. Ganchev B, Drummond N, Aleiner I, Fal'ko V (2015) Three-particle complexes in two-dimensional semiconductors. Phys Rev Lett 114:107401
408. Volmer F, Pissinger S, Ersfeld M, Kuhlen S et al (2017) Intervalley dark trion states with spin lifetimes of 150 ns in WSe2. Phys Rev B 95:235408
409. Wang Z, Zhao L, Mak KF, Shan J (2017) Probing the spin-polarized electronic band structure in monolayer transition metal dichalcogenides by optical spectroscopy. Nano Lett 17:740–746
410. Young PA (1968) Lattice parameter measurements on molybdenum disulphide. J Phys D 1:936
411. Kam KK, Parkinson BA (1982) Detailed photocurrent spectroscopy of the semiconducting group VIB transition metal dichalcogenides. J Phys Chem 86:463–467
412. Boker T, Severin R, Müller A, Janowitz C et al (2001) Band structure of MoS_2, $MoSe_2$, and $\alpha-MoTe_2$: angle-resolved photoelectron spectroscopy and ab initio calculations. Phys Rev B 64:235305
413. Scalise E, Houssa M, Stesmans A, Geoffrey P et al (2012) Strain-induced semiconductor to metal transition in the two-dimensional honeycomb structure of MoS2. Nano Res 5:43–48
414. Kumar A, Ahluwalia PK (2012) A first principle comparative study of electronic and optical properties of 1H–MoS_2 and 2H–MoS_2. Mater Chem Phys 135:755–761
415. Ataca C, Sahin H, Akturk E, Ciraci S (2011) Mechanical and electronic properties of MoS2 nanoribbons and their defects. J Phys Chem C 115:3934–3941
416. Bjorkman T, Gulans A, Krasheninnikov AV, Nieminen RM (2012) Are we van der Waals ready? J Phys Condens Matter 24: 424218
417. Scalise E (2016) Theoretical study of transition metal dichalcogenides. 2D materials for nanoelectronics. CRC Press, Boca Raton
418. Johari P, Shenoy VB (2012) Tuning the electronic properties of semiconducting transition metal dichalcogenides by applying mechanical strains. ACS Nano 6:5449–5456
419. Aghajanian M, Mostofi AA, Lischner J (2018) Tuning electronic properties of transition-metal dichalcogenides via defect charge. Sci Rep 8:13611
420. Gusakova J, Wang X, Shiau LL, Krivosheeva A (2017) Electronic properties of bulk and monolayer TMDs: theoretical study within DFT framework (GVJ-2e method). Phys Status Solidi A 2017:1700218
421. Bernardi M, Ataca C, Palummo M, Grossman JC (2017) Optical and electronic properties of two-dimensional layered materials. Nanophotonics 6:479–493
422. Villaos RAB, Crisostomo CP, Huang ZQ, Huang SM et al (2019) Thickness dependent electronic properties of Pt dichalcogenides. npj 2D Mater Appl 3:2
423. Huynh TMD, Nguyen DK, Nguyen TDH, Dien VK, Pham HD (2021) Geometric and electronic properties of monolayer HfX2 (X = S, Se, or Te): a first-principles calculation. Front Mater 7
424. Chhowalla M, Shin HS, Eda G, Li LJ et al (2013) The chemistryof two-dimensional layered transition metal dichalcogenide nanosheets. Nat Chem 5:263–75

425. Wilson JA, Yoffe AD (1969) Transition metal dichalcogenides discussion and interpretation of observed optical, electrical and structural properties. Adv Phys 18:193–335
426. Kobayashi K, Yamauchi J (1995) Electronic-structure and scanningtunneling-microscopy image of molybdenum dichalcogenide surfaces. Phys Rev B 51:17085–95
427. Terrones H, Terrones M (2014) Electronic and vibrational properties of defective transition metal dichalcogenide haeckelites: new 2D semi-metallic systems. 2D Mater 1:011003
428. Hinsche NF, Thygesen KS (2018) Electron-phonon interaction and transport properties of metallic bulk and monolayer transition metal dichalcogenide TaS2. 2D Mater 5:015009
429. Lopez-Sanchez O, Lembke D, Kayci M, Radenovic A et al (2013) Ultrasensitive photodetectors based on monolayer MoS2. Nat Nanotechnol 8:497–501
430. Britnell L, Ribeiro RM, Eckmann A et al (2013) Strong light–matter interactions in heterostructures of atomically thin films. Science 340:1311–1314
431. Sundaram RS, Engel M, Lombardo A et al (2013) Electroluminescence in single layer MoS2. Nano Lett 13:1416–1421
432. Klein A, Tiefenbacher S, Eyert V, Pettenkofer C et al (2001) Electronic band structure of single-crystal and single-layer WS2: influence of interlayer van der Waals interactions. Phys Rev B 64:205416
433. Ellis JK, Lucero MJ, Scuseria GE (2011) The indirect to direct band gap transition in multilayered MoS2 as predicted by screened hybrid density functional theory. Appl Phys Lett 99:261908
434. Kuc A, Zibouche N, Heine T (2011) Influence of quantum confinement on the electronic structure of the transition metal sulfide TS2. Phys Rev B 83:245213
435. Peelaers H, Van de Walle CG (2012) Effects of strain on band structure and effective masses in MoS2. Phys Rev B 86:241401
436. Komsa HP, Krasheninnikov AV (2012) Effects of confinement and environment on the electronic structure and exciton binding energy of MoS2 from first principles. Phys Rev B 86:241201
437. Mann J, Ma Q, Odenthal PM et al. 2-dimensional transition metal dichalcogenides with tunable direct band gaps: MoS2(1−x)Se2x monolayers. Adv Mater 26:1399–1404
438. Zhu ZY, Cheng YC, Schwingenschlogl U (2011) Giant spin-orbitinduced spin splitting in two-dimensional transition-metal dichalcogenide semiconductors. Phys Rev B 84:153402
439. Sun LF, Yan JX, Zhan D et al (2013) Spin-orbit splitting in single-layer MoS2 revealed by triply resonant Raman scattering. Phys Rev Lett 111:126801
440. Jin W, Yeh PC, Zaki N et al (2013) Direct measurement of the thickness-dependent electronic band structure of MoS2 using angle-resolved photoemission spectroscopy. Phys Rev Lett 111:106801
441. Li JB, Xiao S, Liang S et al (2017) Switching freely between superluminal and subluminal light propagation in a monolayer MoS2 nanoresonator. Opt Express 25:13567–76
442. Ifti IM, Hasan MM, Arif MAH, Zubair A (2020) Effect of vacancy on electronic properties of MX2 (M = Mo, W and X = S, Se) monolayers. In: 11th international conference on electrical and computer engineering (ICECE), pp 391–394
443. Jing Y, Liu B, Zhu X, Ouyang F et al (2020) Tunable electronic structure of two-dimensional transition metal chalcogenides for optoelectronic applications. Nanophotonics 9:1675–1694
444. Ma J, Li W, Luo X (2016) Ballistic thermal transport in monolayer transition-metal dichalcogenides: role of atomic mass. Appl Phys Lett 108:082102
445. Yu R, de Abajo FJG (2020) Chemical identification through two-dimensional electron energy-loss spectroscopy. Sci Adv 6:eabb4713
446. Chen G (1998) Thermal conductivity and ballistic-phonon transport in the cross-plane direction of superlattices. Phys Rev B 57:14958–14973
447. Markussen T, Jauho AP, Brandbyge M (2008) Heat conductance is strongly anisotropic for pristine silicon nanowires. Nano Lett 8:3771–3775
448. Lee J, Lim J, Yang P (2015) Ballistic phonon transport in holey silicon. Nano Lett 15:3273–3279

449. Li W, Carrete J, Mingo N (2013) Thermal conductivity and phonon linewidths of monolayer MoS2 from first principles. Appl Phys Lett 103:253103
450. Jiang JW (2014) Phonon bandgap engineering of strained monolayer MoS2. Nanoscale 6:8326–8333
451. Keyshar K, Berg M, Zhang X, Vajtai R et al (2017) Experimental determination of the ionization energies of MoSe2, WS2, and MoS2 on SiO2 using photoemission electron microscopy. ACS Nano 11:8223–8230
452. Wei X, Wang Y, Shen Y, Xie G et al (2014) Phonon thermal conductivity of monolayer MoS2: a comparison with single layer graphene. Appl Phys Lett 105:103902
453. Hong Y, Zhang J, Zeng XC (2016) Thermal conductivity of monolayer MoSe2 and MoS2. J Phys Chem C 120:26067–26075
454. Hemmat Z, Yasaei P, Schultz JF, Hong L et al (2019) Tuning thermal transport through atomically thin Ti3C2Tz MXene by current annealing in vacuum. Adv Funct Mater 29:1805693
455. Yasaei P, Hemmat Z, Foss CJ, Li SJ et al (2018) Enhanced thermal boundary conductance in few-layer Ti3C2 MXene with encapsulation. Adv Mater 30:1801629
456. Guo Z, Miao N, Zhou J, Pan Y et al (2018) Coincident modulation of lattice and electron thermal transport performance in MXenes via surface functionalization. Phys Chem Chem Phys 20:19689–19697
457. Gholivand H, Fuladi S, Hemmat Z, Salehi-Khojin A et al (2019) Effect of surface termination on the lattice thermal conductivity of monolayer Ti3C2Tz MXenes. J Appl Phys 126:065101
458. Sarikurt S, Çakır D, Keçeli M, Sevik C (2018) The influence of surface functionalization on thermal transport and thermoelectric properties of MXene monolayers. Nanoscale 10:8859–8868
459. Wang Y, Zhang K, Xie G (2016) Remarkable suppression of thermal conductivity by point defects in MoS2 nanoribbons. Appl Surf Sci 360:107–112
460. Jo I, Pettes MT, Ou E, Wu W et al (2014) Basal-plane thermal conductivity of few-layer molybdenum disulfide. Appl Phys Lett 104:201902
461. Peimyoo N, Shang J, Yang W, Wang Y et al (2015) Thermal conductivity determination of suspended mono-and bilayer WS2 by Raman spectroscopy. Nano Res 8:1210–1221
462. Zhang X, Sun D, Li Y, Lee G-H et al (2015) Measurement of lateral and interfacial thermal conductivity of single-and bilayer MoS2 and MoSe2 using refined optothermal Raman technique. ACS Appl Mater Interfaces 7:25923–25929
463. Liu R, Li W (2018) High-thermal-stability and high-thermal-conductivity Ti3C2Tx MXene/ poly (vinyl alcohol) (PVA) composites. ACS Omega 3:2609–2617
464. Singhal J, Jena D (2020) Unified ballistic transport relation for anisotropic dispersions and generalized dimensions. Phys Rev Res 2:043413
465. Fiore S, Klinkert C, Ducry F, Backman J et al (2022) Influence of the hBN dielectric layers on the quantum transport properties of MoS2 transistors. Materials 15:1062
466. Settino J, Citro R, Romeo F, Cataudella V et al (2021) Ballistic transport through quantum point contacts of multiorbital oxides. Phys Rev B 103:235120
467. Chen KX, Wang XM, Mo DC, Lyu SS (2015) Thermoelectric properties of transition metal dichalcogenides: from monolayers to nanotubes. J Phys Chem C 119:26706–26711
468. Ding G, Gao GY, Huang Z, Zhang W et al (2016) Thermoelectric properties of monolayer MSe2 (M = Zr, Hf): low lattice thermal conductivity and a promising figure of merit. Nanotechnology 27:375703
469. Qin D, Ge XJ, Ding G, Gao G et al (2017) Strain-induced thermoelectric performance enhancement of monolayer ZrSe2. RSC Adv 7:47243–47250
470. Sangwan VK, Arnold HN, Jariwala D, Marks TJ et al (2013) Low-frequency electronic noise in single-layer MoS2 transistors. Nano Lett 13:4351–4355
471. Tong X, Ashalley E, Lin F, Li H et al (2015) Advances in MoS2-based field effect transistors (FETs). Nano-Micro Lett 7:203–218
472. Samnakay R, Jiang C, Rumyantsev SL, Shur MS et al (2015) Selective chemical vapor sensing with few-layer MoS2 thin-film transistors: comparison with graphene devices. Appl Phys Lett 106:02311

473. Lin Y-F, Xu Y, Lin C-Y, Suen Y-W et al (2015) Origin of noise in layered MoTe2 transistors and its possible use for environmental sensors. Adv Mater 27:6612–6619
474. Na J, Lee YT, Lim JA, Hwang DK et al (2014) Few-layer black phosphorus field-effect transistors with reduced current fluctuation. ACS Nano 8:11753–11762
475. Cho I-T, Kim JI, Hong Y, Roh J et al (2015) Low frequency noise characteristics in multilayer WSe2 field effect transistor. Appl Phys Lett 106(2):023504
476. Ji H, Joo M-K, Yun Y, Park J-H et al (2016) Suppression of interfacial current fluctuation in MoTe2 transistors with different dielectrics. ACS Appl Mater Interfaces 8:19092–19099
477. Renteria J, Samnakay R, Rumyantsev SL, Jiang C et al (2014) Low-frequency 1/f noise in MoS2 transistors: low-frequency 1/f noise in MoS2 transistors: relative contributions of the channel and contacts. Appl Phys Lett 104:153104
478. Joo MK, Moon BH, Ji H, Han GH et al (2017) Understanding coulomb scattering mechanism in monolayer MoS2 channel in the presence of h-BN buffer layer. ACS Appl Mater Interfaces 9:5006–5013
479. Noh J-Y, Kim H, Kim Y-S (2014) Stability and electronic structures of native defects in single-layer MoS2. Phys Rev B 89:205417
480. Komsa H-P, Krasheninnikov AV (2015) Native defects in bulk and monolayer MoS2 from first principles. Phys Rev B 91:125304
481. Pandey M, Rasmussen FA, Kuhar K, Olsen T et al (2016) Defect-tolerant monolayer transition metal dichalcogenides. Nano Lett 16:2234
482. Khan MA, Erementchouk M, Hendrickson J, Leuenberger MN (2017) Electronic and optical properties of vacancy defects in single-layer transition metal dichalcogenides. Phys Rev B 95:245435
483. Qiu H, Xu T, Wang Z, Ren W et al (2013) Hopping transport through defect-induced localized states in molybdenum disulphide. Nat Commun 4:2642
484. Yu Z, Pan Y, Shen Y, Wang Z et al (2014) Towards intrinsic charge transport in monolayer molybdenum disulfide by defect and interface engineering. Nat Commun 5:5290
485. Haldar S, Vovusha H, Yadav MK, Eriksson O et al (2015) Systematic study of structural, electronic, and optical properties of atomic-scale defects in the two-dimensional transition metal dichalcogenides MX2 (M=Mo, W; X=S, Se, Te). Phys Rev B 92:235408
486. Li W-F, Fang C, van Huis MA (2016) Strong spin-orbit splitting and magnetism of point defect states in monolayer WS2. Phys Rev B 94:195425
487. Sarma SD, Hwang EH (2015) Screening and transport in 2D semiconductor systems at low temperatures. Sci Rep 5:16655
488. Kaasbjerg K, Thygesen KS, Jacobsen KW (2012) Phonon limited mobility in MoS2 from first principles. Phys Rev B 85:115317
489. Kaasbjerg K, Thygesen KS, Jauho A-P (2013) Acoustic phonon-limited mobility in two-dimensional MoS2: deformation potential and piezoelectric scattering from first principles. Phys Rev B 87:235312
490. Kaasbjerg K, Low T, Jauho AP (2019) Electron and hole transport in disordered monolayer MoS2: atomic vacancy induced short-range and Coulomb disorder scattering. Phys Rev B 100:115409
491. Wan X, Chen K, Xie W, Wen J et al (2016) Quantitative analysis of scattering mechanisms in highly crystalline CVD MoS2 through a self-limited growth strategy by interface engineering. Small 12:438–445
492. Brem S, Zipfel J, Selig M, Raja A et al (2019) Intrinsic lifetime of higher excitonic statesin tungsten diselenide monolayers. Nanoscale 11:12381
493. Carvalho BR, Wang Y, Mignuzzi S, Roy D (2017) Intervalley scattering by acoustic phonons in two-dimensional MoS2 revealed by double-resonance Raman spectroscopy. Nat Commun 8:14670
494. Puotinen D, Newnhan RE (1961) The crystal structure of MoTe2. Acta Crystallogr 14:691–692
495. Guguchia Z, Kerelsky A, Edelberg D, Banerjee S et al (2018) Magnetism in semiconducting molybdenum dichalcogenides. Sci Adv 4:eaat3672

496. Guguchia Z (2020) Unconventional magnetism in layered transition metal dichalcogenides. Condens Matter 5:42
497. Zhang Z, Liu X, Yu J, Hang Y et al (2016) Tunable electronic and magnetic properties of two-dimensional materials and their one-dimensional derivatives. WIREs Comput Mol Sci. https://doi.org/10.1002/wcms.1251
498. Deng Y, Yu Y, Song Y, Zhang J et al (2018) Gate-tunable room-temperature ferromagnetism in two-dimensional Fe3GeTe2. Nature 563:94–99
499. Chen W, Zhang J, Nie Y, Xia Q et al (2020) Electronic structure and magnetism of MTe2 (M = Ti, V, Cr, Mn, Fe, Co and Ni) monolayers. Mater 508:166878
500. Zhuang HL, Hennig RG (2016) Stability and magnetism of strongly correlated single-layer VS_2. Phys Rev B 93:054429
501. Jiang X, Liu Q, Xing J, Liu N (2021) Recent progress on 2D magnets: fundamental mechanism, structural design and modification. Appl Phys Rev 8:031305
502. Komsa HP, Krasheninnikov AV (2017) Engineering the electronic properties of two dimensional transition metal dichalcogenides by introducing mirror twin boundaries. Adv Electron Mater 3:1600468
503. Zhang Z, Zou X, Crespi VH, Yakobson BI (2013) Intrinsic magnetism of grain boundaries in two-dimensional metal dichalcogenides. ACS Nano 7:10475–10481
504. Gao D, Si M, Li J, Zhang J et al (2013) Ferromagnetism in freestanding MoS2 nanosheets. Nanoscale Res Lett 8:1–8
505. Tongay S, Varnoosfaderani SS, Appleton BR, Wu J et al (2012) Magnetic properties of MoS2: existence of ferromagnetism. Appl Phys Lett 101:123105
506. Le D, Rahman TS (2013) Joined edges in MoS2: metallic and half-metallic wires. J Phys Condens Matter 25:312201
507. Enyashin AN, Seifert G (2014) Electronic properties of MoS_2 monolayer and related structures. Nanosyst Phys Chem Math 5:517–539
508. Islam Z, Haque A (2021) Defects and grain boundary effects in MoS2: a molecular dynamics study. J Phys Chem Solids 148:109669
509. Zou X, Liu Y, Yakobson BI (2013) Predicting dislocations and grain boundaries in two-dimensional metal-disulfides from the first principles. Nano Lett 13:253–258
510. Zhou W, Zou X, Najmaei S, Liu Z et al (2013) Intrinsic structural defects in monolayer molybdenum disulfide. Nano Lett 13:2615–2622
511. Azizi A, Zou X, Ercius P, Zhang Z (2014) Dislocation motion and grain boundary migration in two-dimensional tungsten disulphide. Nat Commun 5:4867
512. Elibol K, Susi T, Argentero G, Reza M et al (2018) Atomic structure of intrinsic and electron-irradiation-induced defects in MoTe2. Chem Mater 30:1230–1238
513. Lin J, Pantelides ST, Zhou W (2015) Vacancy-induced formation and growth of inversion domains in transition-metal dichalcogenide monolayer. ACS Nano 9:5189–5197
514. Lin YC, Bjorkman T, Komsa HP, Teng PY et al (2015) Threefold rotational defects in two-dimensional transition metal dichalcogenides. Nat Commun 6:6736
515. Wang J, Xu X, Qiao R, Liang J et al (2018) Visualizing grain boundaries in monolayer MoSe2 using mild H2O vapor etching. Nano Res 11:4082–4089
516. Koós AA, Vancsó P, Szendrő M, Dobrik G (2019) Influence of native defects on the electronic and magnetic properties of CVD grown MoSe2 single layers. J Phys Chem C 123:24855–24864
517. Man P, Srolovitz D, Zhao J, Ly TH (2021) Functional grain boundaries in two-dimensional transition-metal dichalcogenides. Acc Chem Res 54:4191–4202
518. Chen X, Lei B, Zhu Y, Zhou J et al (2021) Diverse spin-polarized in-gap states at grain boundaries of rhenium dichalcogenides induced by unsaturated Re–Re bonding. ACS Mater Lett 3:1513–1520
519. He W, Kong L, Zhao W, Yu P (2022) Atomically thin 2D van der Waals magnetic materials: fabrications, structure, magnetic properties and applications. Coatings 12:122
520. Komsa HP, Kotakoski J, Kurasch S, Lehtinen O et al (2012) Two-dimensional transition metal dichalcogenides under electron irradiation: defect production and doping. Phys Rev Lett 109:035503

521. Huang B, Clark G, Navarro-Moratalla E, Klein DR et al (2017) Layer-dependent ferromagnetism in a van der Waals crystal down to the monolayer limit. Nature 546:270–273
522. Zhao J, Nam H, Ly TH, Yun SJ et al (2017) Chain vacancies in 2D crystals. Small 13:1601930
523. Susi T, Kepaptsoglou D, Lin YC, Ramasse QM et al (2017) Towards atomically precise manipulation of 2D nanostructures in the electron microscope. 2D Mater 4:042004
524. Wu H, Zhao X, Guan C, Zhao LD et al (2018) The atomic circus: small electron beams spotlight advanced materials down to the atomic scale. Adv Mater 30:1802402
525. Zhang W, Song XJ, Zhou N, Li H et al (2017) Magnetism and magnetocrystalline anisotropy in vacancy doped and (non)metal adsorbed single-layer PtSe2. Comp Mater Sci 129:171–177
526. Kuklin AV, Begunovich LV, Gao L, Zhang H et al (2021) Point and complex defects in monolayer PdSe2: evolution of electronic structure and emergence of magnetism. Phys Rev B 104:134109
527. Fu M, Liang L, Zou Q, Nguyen GD et al (2020) Defects in highly anisotropic transition-metal dichalcogenide PdSe2. J Phys Chem Lett 11:740
528. Cao D, Shu HB, Wu TQ, Jiang ZT et al (2016) First-principles study of the origin of magnetism induced by intrinsic defects in monolayer MoS2. Appl Surf Sci 361:199–205
529. Manchanda P, Skomski R (2017) Defect-induced magnetism in two-dimensional NbSe2. Superlatt Microstruct 101:349–353
530. Ramasubramaniam A, Naveh D (2013) Mn-doped monolayer MoS2: an atomically thin dilute magnetic semiconductor. Phys Rev B 87:195201
531. Andriotis AN, Menon M (2014) Tunable magnetic properties of transition metal doped MoS2. Phys Rev B 90:125304
532. Cai L, He J, Liu Q, Yao T et al (2015) Vacancy-induced ferromagnetism of MoS2 nanosheets. J Am Chem Soc 137:2622–2627
533. Han SW, Park Y, Hwang YH, Lee WG et al (2016) Investigation of electron irradiation-induced magnetism in layered MoS2 single crystals. Appl Phys Lett 109:252403
534. Kochat V, Apte A, Hachtel JA, Kumazoe H et al (2017) Re doping in 2D transition metal dichalcogenides as a new route to tailor structural phases and induced magnetism. Adv Mater 2017:1703754
535. Li B, Xing T, Zhong M, Huang L (2017) A two-dimensional Fe-doped SnS2 magnetic semiconductor. Nat Commun 8:1958
536. Zhang Q, Ren Z, Wu N, Wang W et al (2018) Nitrogen-doping induces tunable magnetism in ReS2. npj 2D Mater Appl 2:22
537. Coelho PM, Komsa HP, Lasek K, Kalappattil V et al (2019) Room-temperature ferromagnetism in MoTe2 by post-growth incorporation of vanadium impurities. Adv Electron Mater 5:1900044
538. Wang B, Xia Y, Zhang J, Komsa HP et al (2020) Niobium doping induced mirror twin boundaries in MBE grown WSe2 monolayers. Nano Res 13:1889–1896
539. Tiwari S, Van de Put ML, Sorée B, Van den Berghe WG (2021) Magnetic order and critical temperature of substitutionally doped transition metal dichalcogenide monolayers. npj 2D Mater Appl 5:54
540. Yu S, Tang J, Wang Y, Xu F et al (2022) Recent advances in two-dimensional ferromagnetism: strain-, doping-, structural- and electric field-engineering toward spintronic applications. Sci Technol Adv Mater 23:140–160
541. Ataca C, Ciraci S (2011) Functionalization of single-layer MoS2 honeycomb structures. J Phys Chem C 115:13303–13311
542. Zhang M, Huang Z, Wang X, Zhang H et al (2016) Magnetic MoS2 pizzas and sandwiches with Mnn (n = 1–4) cluster toppings and fillings: a first-principles investigation. Sci Rep 6:19504
543. Park CS, Chu D, Shon Y, Lee J et al (2017) Room temperature ferromagnetic and ambipolar behaviors of MoS2 doped by manganese oxide using an electrochemical method. Appl Phys Lett 110:222104
544. Cao S, Xiao Z, Kwan CP, Zhang K et al (2017) Moving towards the magnetoelectric graphene transistor. Appl Phys Lett 111:182402

545. Zhang K, Wang L, Wu X (2019) Spin polarization and tunable valley degeneracy in a MoS_2 monolayer via proximity coupling to a Cr_2O_3 substrate. Nanoscale 11:19536–19542
546. Yang Z, Gao D, Zhang J, Xu Q et al (2015) Realization of high Curie temperature ferromagnetism in atomically thin MoS2 and WS2 nanosheets with uniform and flower-like morphology. Nanoscale 7:650–658
547. Mishra R, Zhou W, Pennycook SJ, Pantelides ST et al (2013) Long-range ferromagnetic ordering in manganese-doped two-dimensional dichalcogenides. Phys Rev B 88:144409
548. Cheng YC, Zhu ZY, Mi WB, Guo ZB et al (2013) Prediction of two-dimensional diluted magnetic semiconductors: doped monolayer MoS2 systems. Phys Rev B 87:100401
549. Zhao X, Wang T, Xia C, Dai X et al (2017) Magnetic doping in two-dimensional transition-metal dichalcogenide zirconium diselenide. J Alloys Compd 698:611–616
550. Kosterlitz JM, Thouless DJ (1973) Ordering, metastability and phase transitions in two-dimensional systems. J Phys C 6:1181–1203
551. Saab M, Raybaud P (2016) Tuning the magnetic properties of MoS2 single nanolayers by 3d metals edge doping. J Phys Chem C 120:10691–10697
552. Zhang K, Pan Y, Wang L, Mei WN et al (2020) Extended 1D defect induced magnetism in 2D MoS2 crystal. J Phys Condens Matter 32:215302

Chapter 5
Spin-Valley Coupling in TMDs

5.1 Introduction

Since group VI transition metal dichalcogenides (TMDs) MX_2 (M = Mo, W; X = S, Se) are possessed X–M–X structure under a covalently bonded hexagonal quasi-2D network, in which weakly stacked weak Van der Waals forces exist. The MX_2 monolayer is the unit element of bulk crystal, in which each metal atom sits on a trigonal prismatic coordination center that is bound to six chalcogen ligands, whereas each chalcogen center is pyramidal to hold three metal atoms. By the way, trigonal prisms are interconnected by the layered structure, in which the metal atoms are sandwiched between layers of chalcogen atoms. Every TMD monolayer has a crystal symmetry D_{3h}^{1} and their inversion symmetry are explicitly broken. On the other hand, the bulk crystal has a *2H* stacking order with the space group D_{6h}^{4} and inversion symmetry. Materials in this family have similar band structures including physical properties. Usually, bulk TMDs are semiconductors with an indirect gap located between the top of the valence band and the bottom of the conduction band in the middle of the K and its Brillouin zone [1].

The atomically thin MX_2 films including monolayers and multilayers are considered to be chemically inert. This class of intrinsic 2D semiconductors is widely regarded as a platform for ultimate electronics. Thus the monolayer quantum confinement shapes the band structure from an indirect bandgap to a direct band located at the K points in their Brillouin zone [1–3]. This process formed direct bandgap in monolayers lies in the visible and near-infrared range, which makes them ideal candidates for optoelectronic applications. Because the conduction and valence band edges in the monolayers are positioned at the edge-junctions of the hexagonal Brillouin zone (BZ), thereby the K and –K points relates to time-reversal. This gives twofold degeneracy with the in-equivalent band maxima that establish a separate catalogue for lower energy electrons and holes, which is typically called valley pseudospin. Such analogous valley pseudospin to spintronics carriers are utilized for encoding formation in electronic devices. Therefore, it provides the notion of valley-dependent electronics

A. K. Singh, *2D Transition-Metal Dichalcogenides (TMDs): Fundamentals and Application*, Materials Horizons: From Nature to Nanomaterials,
https://doi.org/10.1007/978-981-96-0247-6_5

(or valleytronics) which has been extensively exposed for numerous materials by interpreting multivalley band structures [4–11].

5.2 Degree of Freedom and Electronic Structure

Quantum degrees of freedom of electrons are playing significant parts in several expanses of contemporary science and technology. Because they provides the base for the information storage and processing. It also provides a significant way to comprehend and handle a novel quantum degree of freedom that possibly unlocks an innovative model for statistics-based technology. The electron's degree of freedom in definite crystalline solids possesses a valley degree of freedom with the additional charge carriers and spin. Specifically, the valley degree of freedom defines the manifold degeneracy of energy maxima (or valleys) in the conduction or valence bands of the positioned electrons. Owing to valleys usually unglued by a great crystal momentum, therefore, the electron intervalley scattering time longer in spotless substance possessing in some atomic-scale imperfections. Thus the electronic valley provides carrier information for a new technology known as "valleytronics" [12].

Monolayer TMDs also provides a significant novel workroom for the investigation of valley physics and their valleytronic utility. The main use of transportation of interior quantum degree of freedom for information data transferring with the ability to discriminate among situations, that signify diverse fragments of information under the Hilbert space controlled dynamics by these states. This is an essential requirement for the internal quantum degree of freedom to associate with the quantifiable physical features because it allows one to couple the exterior perturbations of measurement to regulate.

The emerging 2D materials with hexagonal lattices, particularly TMDs have twofold valleys at the K and –K points in the Brillouin zone, it is typically called time-reversal that copy of each other [13–15]. Therefore, they are contrasted with valley Berry curvatures and magnetic moments that combine to exterior electric and magnetic fields, its gives upsurge to the valley Hall effect and valley Zeeman effect. Their overturn regularity contravention features property can be directly correlated to optical interband conversions at the time-reversal pair of valleys that may get to valley-dependence conversion selection rules. Such valley-dependence mechanisms allow it possibly manipulate the valley pseudospin via electric, magnetic, and optical properties. All such kinds of physical manipulations in valley pseudospin have been discovered for the monolayer TMDs because their lattice has a fundamental overturn irregularity. Furthermore, the robust spin–orbit connection to the transition metal elements also stretches upswing to an active attraction among the valley pseudospin and the spin that makes it possible to interaction amid the twofold degrees of freedom, by permitting spin modifications through the valley phenomena, as illustrated in Fig. 5.1 [16, 17].

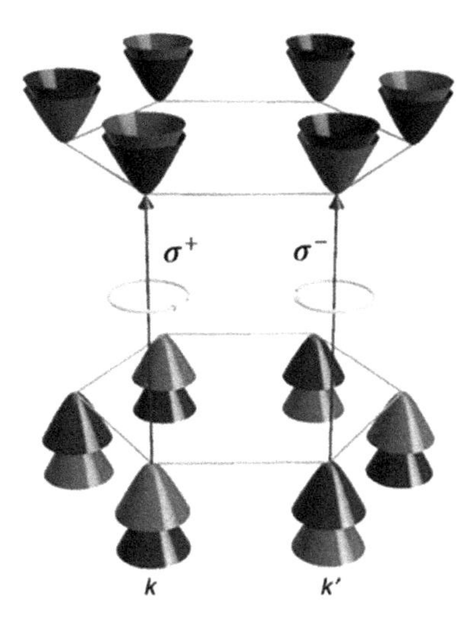

Fig. 5.1 Valley-dependent optical selection rules for the TMD monolayers: Schematic of the electronic band structure of TMD monolayer near the K and K′ points in the Brillouin zone. The dark and light black colors of the bands represent spin orientations spin-down and spin-up that reveal the optical selection rules. The optical transitions at the K and K′ valleys are excited by the right-handed (σ^+) and left-handed (σ^-) circularly polarized light, respectively (adopted from Cotrufo et al. [17])

In the electronic structure, most monolayer TMDs represents the quasi-2D hexagonal lattice, as illustrated in Fig. 5.2a–c. In which the M atom is synchronized by the six adjacent X atoms in a trigonal prismatic configuration. But their configurational reversal regularity is split in the monolayer due to the absence of inversion symmetry. The mirror regularity in the out-of-plane z-direction collectively with the time-reversal equilibrium due to an effective connection among the valley and the spin degrees of freedom.

According to first-principles approximation the band edge Bloch wave functions are nearby to conduction band minima (valence band maximum) at ± K, it mainly due to the $Md_{Z^2}\left(d_{x^2-y^2} \pm id_{xy}\right)$ orbital. It represents the eigenstate of the angular momentum operator L_z for the magnetic quantum number "m" $m = 0(m = \pm 2)$ [18, 19], including small contributions of X_{P_x} and X_{P_y} orbitals [16, 20, 21] (see Fig. 5.2d). Whereas the ± K points are the invariant high symmetry points under the C_3 process (2π/3 in-plane revolution). Therefore, Bloch wave functions at ± K belong to C3 eigenstates along with the eigenvalues $e^{i\frac{2}{3}l\pi}$, here l is an integer that depending to the selection of the revolution epicenter (see Fig. 5.2a) [20]. Corresponding typical orbital assistances and quantum numeral l for the conduction and valence band edge Bloch functions at the K point are listed in Table 4.1, in a similar way their complements at the − K point could be achieved by getting time-reversal.

Usually, TMDs have a robust spin–orbit coupler (SOC) creating through M-d orbitals [20–24]. The spin–orbit breaking at the band edges also noticed due to regularities. Monolayers having mirror regularity to the M-atom plane which usually connected to their Bloch states that are invariant under the mirror reflection

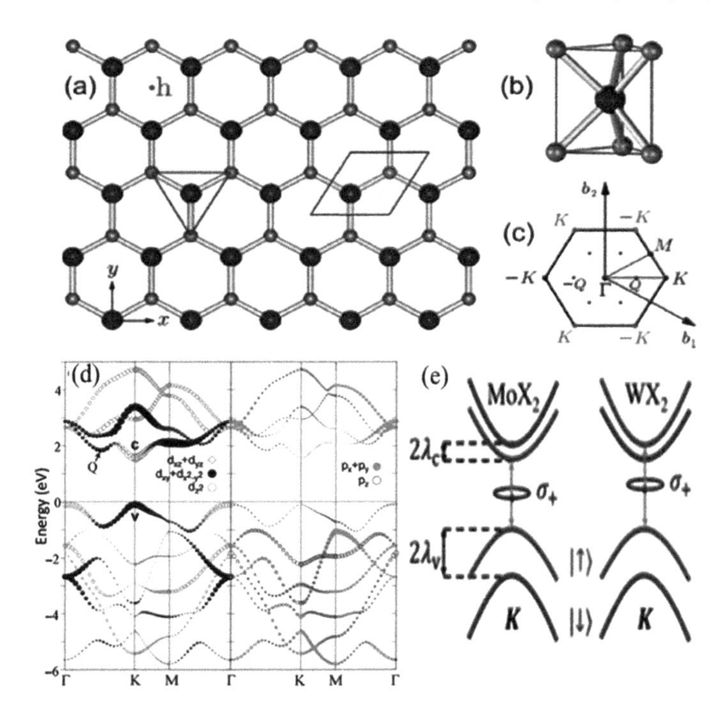

Fig. 5.2 **a** Top view of group-VIB TMD monolayer structure: The dark and light spheres represent M and X atoms respectively. The diamond region is the 2D unit cell with lattice constant **a**; **b** trigonal prismatic coordination geometry, corresponding to the blue triangle in top view; **c** the first Brillouin Zone, whereas the b_1 and b_2 are the reciprocal lattice vectors; **d** band structures of the MoS_2 monolayer from first-principles calculations without SOC, with the orbital composition illustrated. The metal d-orbitals and chalcogen p-orbitals are denoted by dots of different symbols, with the dot size proportional to the orbital weight in the corresponding state (**a**–**d** adopted from Yu et al. [16], pp. 279–294); **e** spin-valley coupling in ML TMDs: the conduction (valence) band spin-splits in the K valley by $2\lambda_{c(v)}$ due to strong spin–orbit coupling with the CB ordering reversed between MoX_2 and WX_2, the emitted/absorbed light has valleys elective helicity, σ_+ (adopted from Scharf et al. [199])

operation. This directly leads to the mirror reflection of the in-plane spin vector in its contradictory, whereas an out-of-plane spin vector Bloch states spin can be whichever parallel or antiparallel to the out-of-plane (z) direction. This gives the SOC breaking condition that should be toward the z-direction. Moreover, as per the time-reversal regularity requirements, the spin splits should have a reverse sign at an unsystematic pair of momentum space points. The neighborhood of the K and – K momentum space points, as per the SOC manifestation the effective coupling among the spin components $S_z = \pm 1/2$ with the corresponding valley pseudospin component τ_z(where $\tau_z = \pm 1$, for the $\pm$ K valley)[20–23]. The corresponding SOC Hamiltonian energy is generally obtained from the following relationship;

$$H_{\mathrm{SOC}} = \lambda \tau_z S_z \tag{5.1}$$

Based on first-principle calculations it has also demonstrated that the valence band at the K point SOC splits with λ_v ~ 0.15 eV for the MoX_2 monolayer and λ_v ~ 0.45 eV in WX_2 monolayer [21, 22, 25–27]. Moreover, the giant SOC splits also reveals the valence band edge of single-layer TMDs locked the spin index with the valley index, by means the K and –K valley are possessed lonely spin-up (down) lower energy holes, that is created by the removal of the spin-down (up) valence electrons, as depicted in Fig. 5.2e [23]. It has a determinate magnetic quantum number $m = \pm 2$ compared to the valence band edge from the $d_{x^2-y^2} \pm id_{xy}$ orbital, their conduction band edge from the d_{z^2} orbital having $m = 0$. On the site, the spin–orbit coupling may vanish for the leading order. Thus according to first-principles approximation the value of conduction band spin breaking λ_c is around some meV for MoS_2 and tens of meV for $MoSe_2$, WS_2, and WSe_2 [21, 23, 28–35], which initiates to slight X-p orbital alignments and second order influence from the remote conduction band that consists $M - d_{xz} \pm id_{yz}$ orbitals. So, in SOC two spin contributions have reverse signs and their opposition consequence gives us the sign variance of λ_c in case of the MoX_2 and WX_2 (see Fig. 5.2e) [21, 23, 31].

As per the simple theoretical approximation, a decent explanation of the topmost valence band and lowermost conduction bands is provided by the three band tight-binding estimation [12, 23]. According to this model the three M-d-orbitals namely d_{z^2} and $d_{x^2-y^2} \pm id_{xy}$ are the key orbital contributors in the formation of band edge Bloch functions of $\pm$ K valleys. Additionally, the M–M adjacent-neighbor hopping of the three-band tight-binding approximation explains the dispersals and Berry curvatures of the conduction and valence bands nearby to $\pm$ K points [12, 23]. Moreover, in addition to the third-nearest-neighbor hopping all three bands are in a good agreement with the first-principles approximation throughout BZ [12, 23]. Yet applicability of this model has been verified from the various studies by demonstrating the distinct kinds of edge states formation in TMDs, such as nanoribbons, quantum dots formation, intercellular orbital magnetic moment, magnetoelectronic and optical properties, magnetoluminescence [36–42].

Moreover, to get a good understanding of the larger neighborhood around the $\pm$ K points investigators have introduced another versatile model by expanding the tight-binding method, it is namely known as the *k.p* approximation. The *k.p* approximation describes the band boundary physics on the basis explanation of the overhead discussed three-band approximation by protecting the linear order of k, the minimal conduction and valence band edges in $\tau_z K$ valley have been defined in a rather simple form by following the relationship [22, 43, 44];

$$H_{k.p} = \mathrm{at}\left(\tau_z k_x \sigma_x + k_y \sigma_y\right) + \frac{\Delta}{2}\sigma_z - \lambda_v \tau_z S_z \tag{5.2}$$

where the σ_x, σ_y and σ_z are the Pauli matrix which represents the spanned of the conduction and valence states at the $\tau_z K$ point, S_z is the Pauli spin matrix, "*a*" represents the lattice constant, "*t*" belongs to the impressive hopping addition, "*k*" is the wave vector that measure by the $\tau_z K$ and λ_v to represent valence band SOC splits. The twofold band *k.p* approximation is considered to be a massive Dirac fermion model. This model allows to lower energy electronic constructions in the $\pm$ K valleys can be captured for the monolayer TMD, including band dispersal, Berry curvatures, and giant SOC splits of the valence band. Term τ_z is explicitly accounted to the valleys depending electronic constructions at the band boundaries. It is responsible for the valley Hall effect, valley magnetic moment, and the valley depending optical assessment criterion.

Due to the simplicity of *k.p* approximation, these two-band models have been extensively used to explore the various properties of TMD monolayers. Besides the simplicity of the *k.p* model, it is also not free from some unavoidable limitations during the applications. As an instance, this model cannot describe the electron–hole irregularity and the trigonal warping of band dispersal. Predominantly dual impacts are responsible for it include quadratic terms in k [21, 23, 43–46]. The amended approximations associating higher order terms have been utilized to define the optical conductivity, magneto-optical properties, plasmons, and spin relaxation [45–50]. Since the minor conduction band spin splits at $\pm$ K, disappears in the three-band tight-binding approximation, thereby, in the *k.p* extension also permits to a modification in Eq. 5.2 if desired.

More recently by adopting the correction idea in DFT study up to thirteen atomic orbitals per unit cell (4d orbitals of the molybdenum atom and the 3 s and 3p orbitals of the sulfur atom) of MoS_2 have been demonstrated. Considering the hopping and overlap couplings of each atom with its first nearest neighbors in each crystalline sublattice, their different values are allowing to intra-plane and interplane S–S hopping integrals. Therefore, a closed-form expression provides an effective mass tensor at stationary points. The key issues which have been dealt with in this study; the incorporation of the 3 s orbital for S, consideration of finite overlaps between the atomic orbitals of nearest neighbors in each crystalline sublattice, considering differences between intra-plane and interplane S–S hopping integrals. The adjustment of tight binding parameters overall is in good agreement with the other renowned methods to provide a closed-form expression for the effective mass tensor. It has been proven the valence and conduction bands are isotropic near the gap edges [44, 51].

5.3 Optical Valley Polarization

The urgency of 2D Dirac materials with hexagonal lattice appeals pronounced attention in valleytronics investigation, such as TMDs. It has been possessed two in-equivalent valleys in K and $-$K points in the Brillouin zone. They can couple and gives us far-field circularly polarized light to generate spin carriers in materials. These two valleys keep time-reversal symmetry with each other due to the existence of opposite Berry curvatures. It means, the polarized optical properties of TMDs are due to valley polarization which arises from the manipulation of the valley degree of freedom [52, 53]. Similar electrons spin due to an internal quantum degree of freedom TMDs can have band structures composed of two in-equivalent valley “states”. They offer capability (TMDs) to govern the valley degree of freedom that becomes due to the presence of Berry phase linked physical features and robust spin-orbital connection [54, 55]. So, TMDs are promising semiconductors for valleytronics, owing to their direct band gaps in the visible spectral range [56–58]. As MoS_2 monolayer has a hexagonal lattice structure with C_3 symmetry with a crystal unit 2H phase. The inversion symmetry broken property gives a honeycomb lattice with two sets of colors, where yellow and blue balls represent the S and Mo atoms, as depicted in Fig. 5.3a [3]. In which the energy extremum is located at the corners of hexagonal Brillouin zone (K and −K points). The corresponding optical transition between the conduction band and the valence band in momentum space is due to two valleys absorbing left and right-hand circularly polarized light separately [59]. Additionally, the involvement of C3 symmetry in hexagonal 2D materials allows in-equivalent A-B sublattices with the direct band gap at the K point, their pseudospin valleys are usually discussed [60].

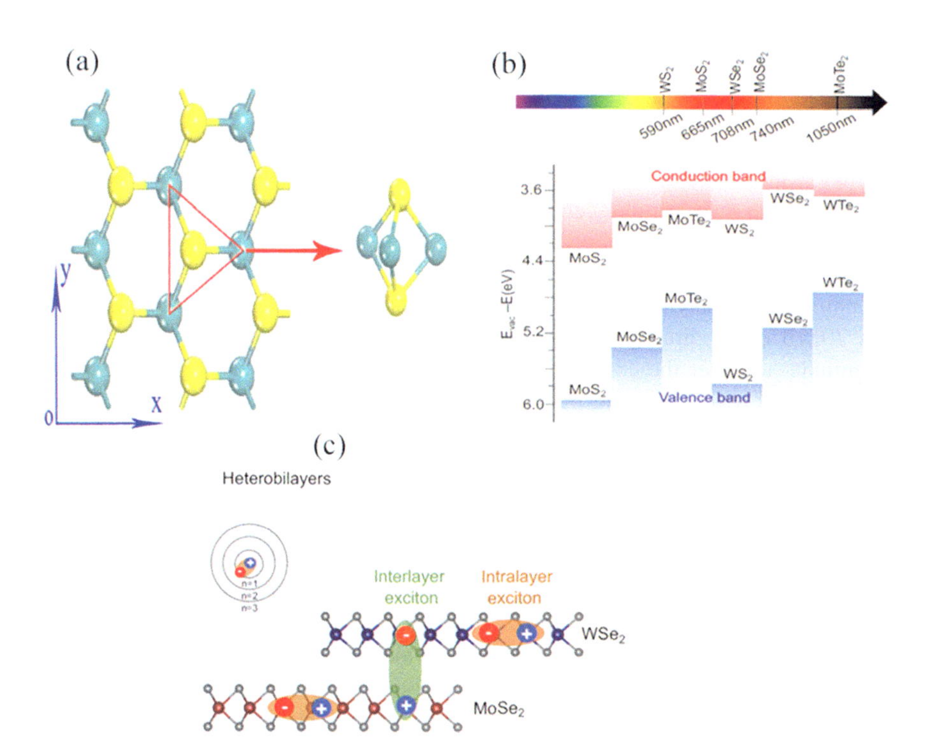

Fig. 5.3 **a** Structure of single-layer MoS_2, the top view of a hexagonal lattice Structure, the side view of the inset triangle. There are three molybdenum atoms as the first nearest neighbor of each sulfur atom. There are six sulfur atoms as the first nearest neighbor of each molybdenum atom (adopted from Jin, Z. et al. 2015 Sci Rep 5, 18342); **b** band alignment between 2D semiconductor monolayers for the different materials; **c** schematic of van der Waals hetero-bilayers (**b**, **c** adopted from Shree, S. et al. 2020 Nat. Rev. Phys., Nature, https://doi.org/10.1038/s42254-020-00259-1)

Fundamentally optical conversions in TMDs falls in the energy range $\approx$ 1.1 eV (single-layer $MoTe_2$) to $\approx$ 2.1 eV (single-layer WS_2), as depicted in band gaps in Fig. 5.3b. The ferromagnetic semiconductors (such as $CrBr_3$ and CrI_3) are also considered to possessed similar transition energies [61, 62]. The interlayer excitons in heterostructures like MoS_2/WSe_2 above the emission wavelengths 1100 nm (<1.1 eV) also belong to the telecommunication bands [63]. Moreover black phosphorous is a layered semiconducting material with a direct bandgap that intensely changing through the integer of layers and covered visible (single-layer) to mid-infrared (bulk) spectral regions [64]. A similar evolution of band gap change to the thickness has been recognized for the $PtSe_2$ [65], but it has an indirect band gap like Si. On the other hand, the spectrum of the layered hexagonal BN has a bandgap in the deep ultraviolet at 6 eV (200 nm) [66].

The valley polarization in TMDs is usually assessed from the measurements of the degree of circular polarization;

$$\rho^{\pm} = \frac{I_{\sigma_\pm}(\sigma^+) - I_{\sigma_\pm}(\sigma^-)}{I_{\sigma_\pm}(\sigma^+) + I_{\sigma_\pm}(\sigma^-)} \tag{5.3}$$

Here term $I_j(l)$ is expressed the measured PL spectrum for $j = (\sigma^+, \sigma^-)$ polarized excitation and $l = (\sigma^+, \sigma^-)$ polarized analysis [67]. A perfectly circularly polarized light for the, $\rho = 1$. Luminescence spectra of symmetric polarization for excitation with opposite helicities are recognized for the $\rho \approx \pm\, 0.32$ [68].

Since robust Coulombic interattraction is immensely crucial to dielectric screening accompanying with a three-dimensional non-homogeneous surroundings [69, 70]. It could directly correlate to exciton changeover energy with a greater magnitude; together the exciton binding energy and the free carrier band gap are usually tuned by engineering the local dielectric environment [71, 72]. Although, possible undesirable consequences are responsible for indigenous dielectric variations from randomness and defects which reflects in terms of greatly widened optical changeovers [73, 74]. Optical conversions from the dipole-momentum-spin are usually denoted to as 'bright' and their excitons recombination emitted a photon. Particularly in stacked materials excitons with diverse three-dimensional alignments of the optical dipole, either in-plane or out-of-the-layer plane, such as WSe_2 and $InSe_2$ [74–77]. The optical transitions rely upon due to phonon absorbance or emission or spin fraternization of the distinct electronic states resulting in a great numeral of probable optical transitions ('bright' and 'dark') for the definite material. This kind of emission and detection efficiency is generally optimized under the applied magnetic field by choosing a specific light polarization and proliferation direction [78, 79]. Specifically, when an electron and hole exists in the identical layer, then it is generally known as intralayer excitons, as depicted in Fig. 5.3c. Whereas the interlayer excitons forms the hetero-bilayers in TMD due to type-II (or staggered) band arrangement [80] with the photoexcited electrons and holes living in diverse layers (See Fig. 5.3b). Usually, these excitons are denoted as the indirect in actual space.

The distinction in the lattice coefficients among the single-layers forms the hetero-bilayer that influences the position of K-points, typically it determines whether the interlayer conversion is direct or indirect in reciprocal space. It means a phonon requires an addition to a photon in the emission process [81, 82] for the formation of moire effects/reconstructions [83, 84]. Therefore, a regular variation of the electrons and holes band structure depending on the distinction among the lattice coefficients and/or twist angle. Moreover, the depth and periodicity of the moire potential also creates localized emitters (separate excitons) or collective excitations (trapping of excitons) [85]. Using this analogy various processes (such as CVD, MBE) fabricated TMDs materials' optical properties have been also interpreted [86, 87].

The significant TMD monolayer energy of the A exciton transition is normally expressed in terms of single particle bandgap (of unbound electrons and holes) and exciton binding energy difference. The associated dielectric environment (and hence all energy scales linked to the Coulomb interaction) engineering reveals a significant change in the solitary atom bandgap and the exciton binding energy of the TMD single-layers [72]. However, the universal A exciton conversion energy shifts somewhat lesser while changing in a solitary atom band gap, and their exciton binding energies partly compensated for everyone to other. As the typical example compare to single-layer $MoSe_2$ conversion energies together and not-together to hBN encapsulation [88], as illustrated in Fig. 5.4a. In which the line width of absorption have expressively influenced by dielectric randomization [89]. Similarly, the bubbles, wrinkles, polymer residues, and hBN are also reflects distinct dielectric coefficients. Thus, a un-uniform dielectric environment influenced the energy of the exciton conversion and the inclusively profile of reflectivity spectra. The unchanging dielectric slabs like thick (tens to hundreds of nm) hBN layers is also examined to steer the absorption. So, the visibility of exciton resonances in absorption has impacted due to width of hBN and SiO_2 Fig. 5.4b. This is owing to thin-film interference effects, where the lowermost hBN width gives to how far the single-layer shaped the Si/SiO_2 interface that behaves as a mirror. This interpretation has also led to the choice of hBN thickness of the heterostructure which also optimized by using a transfer matrix approach to increase the visibility of the targeted transitions [90]. Thus, the specific energy of the excitonic resonance of the stacked semiconductor must be considered [90]. Other consequences on TMD materials located in front of mirrors have also revealed that the modulation of the absorbance forte can be up to 100% owing to interference/cavity influence [91, 92].

On the other hand multilayers absorption of layered semiconductors such as black phosphorous, $ReSe_2$ [65, 93, 94] also provides information on the crystal structure of the particular materials by showing the high anisotropy in their layer plane, it has considered to be direct evidence for the higher anisotropic lattice construction. In TMDs multiple-layers and bulk-robust excitonic characteristics have been also noticed even at room temperature [95]. Including these features, they also attributes to intralayer A and B exciton. Moreover, recognition of interlayer excitons (formation of carriers in 2 neighboring layers) has also realized for the bulk samples absorption [96, 97]. Absorbance of the interlayer excitons is also feasible from the homobilayers and homotrilayers of MoS_2 [98, 99]. This unique optical property of the system allows

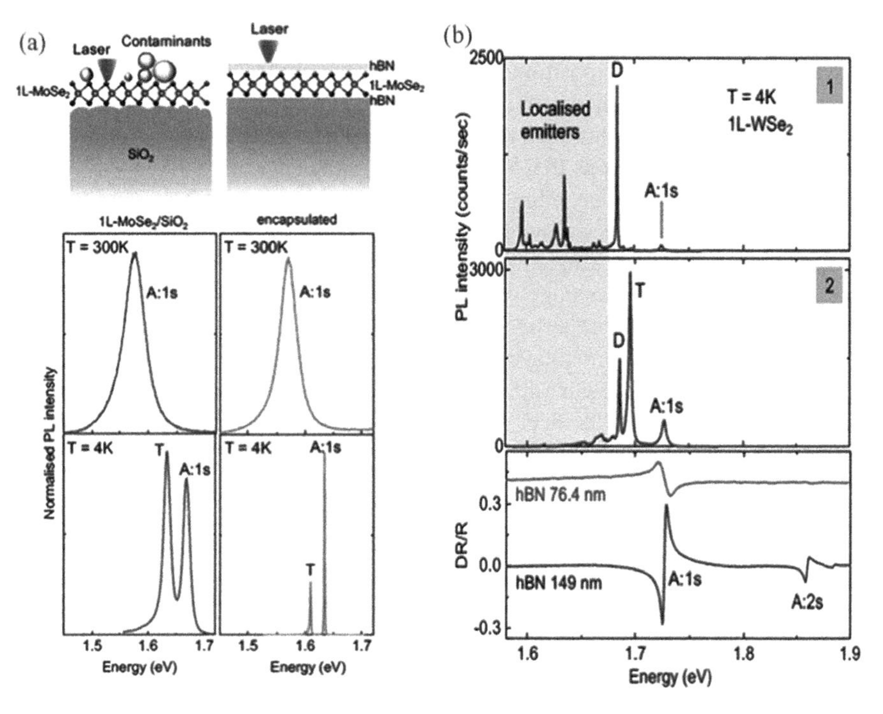

Fig. 5.4 **a** Schematic for the $MoSe_2$ monolayer structures in the diverse dielectric surroundings. The left panel is the structure of an un-encapsulated TMD single-layer on SiO_2 and the TMD single-layer encapsulated in hBN is on the right panel. Characteristic PL spectra noted at room temperature and cryogenic temperature for the encapsulated and un-encapsulated samples are depicted in the bottom panel. The line width decreases expressively for the encapsulated one at $T = 4$ K compared to the un-encapsulated; **b** characteristic PL spectra at $T = 4$ K noted at dual distinct positions 1 and 2 of the specimen as exposed in the figure. The key exciton changeovers (neutral excitons (A:1 s), charged excitons (T), and spin-forbidden dark exciton (D) are quenched and localized emitters appear when PL is documented on the bubbles or wrinkles. Though, robust PL emission analogous to the key excitons from WSe_2 single-layers is noted at a area of the specimen (adopted from Shree, S. et al. 2020 Nat. Rev. Phys., Nature, https://doi.org/10.1038/s42254-020-00259-1)

conversion energy of the absorbance can be tuned by the implication of an electric field vertical to the layers (Stark shift) above 120 meV and the interattraction among the interlayer and intralayer excitons [100, 101].

TMD materials' luminescence properties are also widely used to interpret the macroscopic optical properties and their microscopic electronic excitation, including evaluation of the crystalline quality. Precisely luminescence defines as an excess electromagnetic radiation (light) released to a solid, additionally, their symmetry radiation could be explained using Planck's law. The extra energy transforms in the form of luminescence radiation. In luminescence, the process electrons gets excited and attend high energy states, such as a light source utilized for the photoluminescence (PL). The subsequent charges relaxation energy via the phonon emission accompanied with the photon emission. Due to the involvement of successive relaxation and recombination process they consume comparatively a longer time, this is the main dissimilarity compared to distinct kinds of so-called secondary radiation that is generally expressed as reflected light and scattered light. Specifically, material excited to a light pulse luminescence and it remains upto deterioration certain time and finally it is observed from the time-resolved photoluminescence equipment. Therefore, PL emission spectrum transitions due to the dominance of population state rather than great oscillator forte and higher density of states.

Using the photoluminescence excitation spectroscopy (PLE) can get PL emission intensity of a selected energy by recording the distinct photon excitation energies under implication of a variable lasers or strong white light excitation source. By analyzing the line width and regulating step of the origin reveals the spectral resolution of the PLE experiment. Usually, the obtained PL intensity depends on two aspects, the first one is absorbance forte at the excitation energy, and the second one is the efficiency of energy relaxation followed by radiative recombination (in competition with non-radiative channels). So, it gives collectively together absorbance and energy release (frequently from the phonon emission) dependence spectroscopy. Thus the PLE is an interesting tool to investigate several materials including TMDs. Such as PL sign improvement of interlayer excitons for the resonant laser excitation energy to intralayer states confirms the interlayer excitons via carrier transferring procedure among the layers [82, 102–104].

As the typical PLE demonstration of MoS_2 monolayers, where B exciton states actively overlay to excited A exciton states (A:2 s, A:3 s…). The PLE spectroscopy offers to distinguished the excited states through collection of the emission intensity of the ground state A:1 s, as a function of the excitation laser energy, by scanning the throughout energies of A:2 s, A:3 s, etc. Not only s-symmetry for the states but also p-states evaluated. Since to assess the p-states dual-photon absorbance process is compulsory, therefore, the laser energy should be tuned to middle of the changeover energy [66]. For the recognition of the highly excited exciton states in one and dual-photon the PLE is generally considered an impressive tool to determined influence of the diverse dielectric surroundings on the energy of exciton states. Using this approach it is also feasible to get exciton binding energies and forecast the probable p exciton states splits [105, 106]. Here it should be keeping in view that the crystal regularity or randomization consequences are also provides a mix of s and p exciton states [107].

PLE is also useful technique for the identification of effective relaxation networks. As the PLE demonstration of $MoSe_2$ single-layers have an episodic fluctuation in energy throughout the energy range with unexpected exciton resonance (approximately steady absorbance) [108]. Maximum to every identically spaced energy due to longitudinal acoustic phonons at the M point of the Brillouin zone, whereas the LA (M) leads to efficient energy relaxation of excitons via emission of the LA (M) phonons [109]. The experimental observations are generally directly correlated to a spectrally narrow excitation source by resolving the fine separations between different peaks related to phonon emission [109].

Since electronic equilibrium states in single-layers and multiple-layers crystalline systems mainly directs the optical selection rules to the light polarization in emission

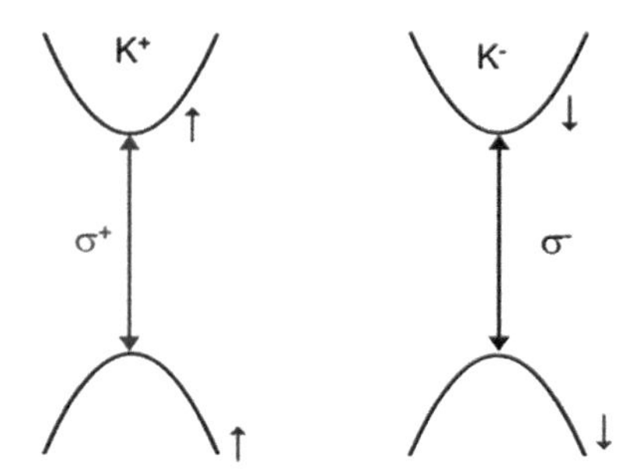

Fig. 5.5 Schematic illustrations for the valley-selective optical transition that obeys chiral selection rules to polarized light. Reflects the direct bandgap at the K-points of the Brillouin zone for TMD monolayers (adopted from Shree, S. et al. 2020 Nat. Rev. Phys., Nature, https://doi.org/10.1038/s42254-020-00259-1)

and absorption [22, 110–112], typically as depicted in Fig. 5.5. The time-integrated PL measurement emitting light provides us evidence about the spin and valley. The time-integrated experiment's circular polarization depends on the exact ratio of PL emission time τ_{PL} and depolarization time (τ_{depol}), the relationship is normally expressed as;

$$P_C = P_0 / \left(1 + \frac{\tau_{PL}}{\tau_{\mathrm{depol}}} \right) \tag{5.4}$$

here P_0 represents the primarily originated polarization depends on the excitation energy [113, 114].

Spin-valley polarization in TMD monolayers (typically MoS_2) leads to the inter-band transition due to chiral selection criterions, as optical conversions in the $K^+ (K^-)$ valley and $\sigma^+ (\sigma^-)$ polarization. Their neutral bright excitons intrinsic lifetime is an order of 1 ps. According to time-integrated PL experiment outcomes P_C values lie in the order of 50%, thereby, it is inferred as τ_{depol} order of extent. However, in actual exercise supplementary critical pump-probe experiments are frequently performed with the extremely small valley lifetimes to get neutral excitons [115]. Therefore, the ratio of τ_{PL} and τ_{depol} can be controlled or adjusted by incorporating single-layers in optical microcavities [116]. Longer valley lifetimes in monolayers have been also noticed for the resident carriers [117] not for the excitons that are measured by the pump-probe techniques, such as Kerr rotation employed to probe polarization for semiconducting or metallic nanostructures [75].

Whereas in TMD hetero-bilayers optical spectroscopy is usually used to probe the local atomic registry. This literal meaning provides information about in what way metal and chalcogen atoms are allied from the topmost layer to the bottom-most layer [88, 118]. This information is significant for the formation of nano-scale, periodic moiré potentials, as illustrated in Fig. 5.6a. TMD bilayer reconstruction of layers when they are brought in contact they can be visualized using imaging techniques such as transmission electron microscopy (TEM) or scanning electron microscopy (SEM). As the intense hBN-encapsulated twisted WSe_2 bilayers have revealed a three-dimensional variable rebuilding pattern owing to the interaction among the corresponding layers afterward stacking, as illustrated in Fig. 5.6b [15]. Moreover, polarization selection criterions probed in PL also provides evidences about the distinct layer-by-layer depositions (such as H-type or R-type for the 60° or 0° twist angle) [160]. So, polarization-resolved optical spectroscopy together with direct atomic-resolution imaging of the lattice is considered to be a very useful combination to analyze the formation of moire potentials [119]. In addition to these PL specimens above the spot diameter of 1 μm, at which moire potentials occurs with a periodicity of nanometers, lead varying effects (see Fig. 5.6a). The inherent lifetime of interlayer excitons are usually order of the ns at lower temperatures and not belonging to ps in case of the single-layers. That permits the imaging exciton and polarization three-dimensional dispersal in the PL maps [120, 121]. Thus physics of both intralayer and interlayer excitons can be described for the single-layer and bilayer areas by using the identical specimen, as illustrated in Fig. 5.6a, c.

Moreover, circular polarization also manipulates by applying external magnetic fields [24, 122]. Such as the hetero-bilayers in which a giant Zeeman splitting 26 meV at $B = 30$ T for interlayer excitons induced close to unity valley polarization have been also examined from the PL emission [123]. An applied filed 7 T on monolayer $MoSe_2$ has also revealed close to unity polarization of electrons probed in absorption and emission [124].

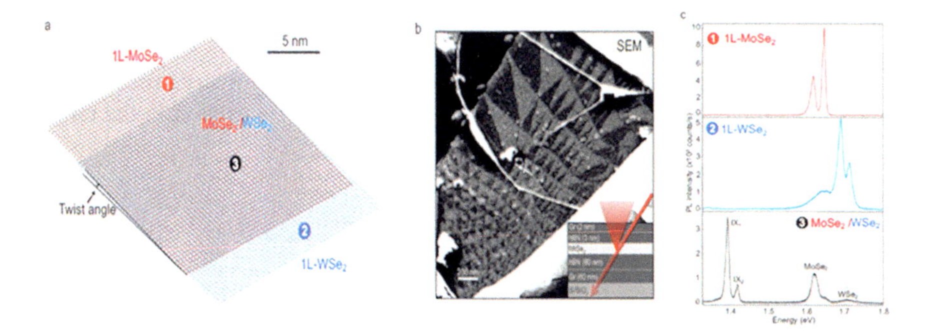

Fig. 5.6 Moire interlayer excitons in hetero-bilayers: **a** two distinct single-layer materials stacked vertically to show a moire pattern due to little lattice mismatch and twist angle. Various individual atomic arrangements in the heterostructure exhibited diverse optical features. Three spots labeled 1, 2, and 3 indicate about chosen for PL measurements on $MoSe_2$ single-layer, WSe_2 single-layer, and $MoSe_2/WSe_2$ heterostructure; **b** SEM image of hBN-encapsulated twisted WSe_2 bilayers together a spatially variable rebuilding design owing to the interaction among the corresponding layers afterward stacking; **c** the photoluminescence spectra to hBN-encapsulated $MoSe_2$ single-layer, $MoSe_2/WSe_2$ heterostructure (solid black curve) and WSe_2 single-layer (solid blue curve). Intralayer exciton emission is noticed in the $MoSe_2$ and WSe_2 single-layers. The interlayer exciton emission (IX_1 and IX_2) in energy underneath the intralayer resonances for the heterostructure (adopted from Shree, S. et al. 2020 Nat. Rev. Phys., Nature, https://doi.org/10.1038/s42254-020-00259-1)

5.4 Electrical Valley Polarization

Similar to charge and spin, valley information also gates from the transport of materials. Depending on carrier types of the valley pseudospin, such as free electrons and holes, doped electrons and holes, quasi-particles (like neutral), and charged and bound excitons. There are various distinct ways to control the transport of valley polarization [125].

Besides the several electrical valley polarization investigations [125–130], recently have been focused on electric field controlled layer polarization with a fixed hole density in the correct area of the moiré bilayer [131]. By interpreting twisted WSe_2 AB homobilayers as a prototype bilayer Hubbard system. Whereas the twist angle retain within 2–3° of AB stacking to uphold smooth bands for robust correlation effects, however, it can be extended up to a 58°-twisted bilayer. To form a triangular moiré lattice with a period of about 10 nm and a moiré density of $n_M \approx 1.1 \times 10^{12}$ cm^2. Samples have been encapsulated from the approximately dual symmetric gates that are made to the hexagonal boron nitride (hBN) gate dielectrics and graphite gate electrodes and grounded by a graphite contact, as depicted in Fig. 5.7. Use of two gates allows the self-governing regulation of the doping density (v) and vertical electric field (E) of the specimen. Moreover, a native dielectric sensor can also integrate into the right half of the device that contains WSe_2 single-layer to make electronically decoupled from the specimen through a ~ 1-nm-thick hBN spacer.

The contrast reflectance spectra of the right half of the device as a function of gate voltage (bottom axis) and doping density (top axis) in the moiré bilayer reveals the sensor layer carrier neutrality. Occurrences of resonance such as features nearby 1.725 and 1.85 eV saturates the color scale by showing 1 and 2 s excitons for the WSe_2 single-layer sensor. The corresponding red-shift of 1 s exciton in the range

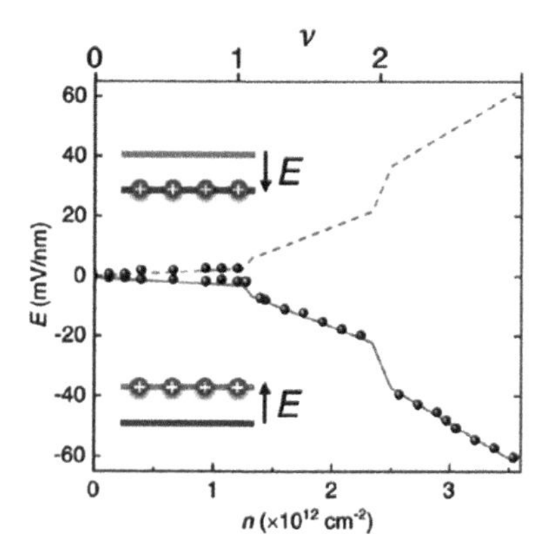

Fig. 5.7 Doping dependence of E_c (symbols) determination from the exciton sensor. The solid line guides the eye. The dashed line by inverting the solid line to $E = 0$. The insets illustrate the layer polarization beyond E_c (adopted from Xu, Y. et al. 2022, https://doi.org/10.48550/arXiv.2202.02055)

30–50 meV correlates to moiré exciton, moreover, their energy may lower due to interlayer connections of the WSe_2 moiré bilayer. A similar contrast reflectance spectra has been also noticed for the left half of the device with negative sensing features [131].

A sequence of non-conducting (or insulating) states have provided the sensor 2 s exciton in terms of sudden energy blue shift together an enhanced reflectance [132]. The strongest insulating state has been also demonstrated the charge neutrality situation ($\nu = 0$). Moreover, electron doping also induce extra insulating states such as $\nu = 1, 2, 3, 4$ with the various slight filling aspects. Whereas hole doping results in solitary $\nu = 1$ and 2 insulating states. So, the flatter moiré conduction bands due to several such states give rise to the correlation effect. Additionally, such insulating states of the electrons cannot consider sensitive in terms of out-of-plane electric field E, because it indicates a strong interlayer hybridization of the states [133]. On the other hand, the hole states are strongly dependent on E, this directly relates to their weak interlayer hybridization. Especially the weak interlayer coupling of the hole's doped correlation effects has attracted much attention for the investigation.

According to theoretical interpretation, the extraordinary gate regulation of the contending electronic phases in warped TMD AB bilayers unlocks a novel approach to experience a robust correlation effects and their applications. Specifically, due to their strong Mott gap and the flat Hubbard bands in every sturdily connected single-layer [134]. Whereas one layer can act as a Mott (or charge transfer) insulator and another layer contains its inherent electrons. These two layers have been recognized by regulating the electric field for $\nu > 1$. As depicted in Fig. 5.7. The sudden charge transfer near E_c has been attributed to the significance of a layer-selected Mott insulator for $1 < \nu < 2$. It also leads an electric field that switched the system at $\nu = 2$ from a unpolarized to a polarized layer insulator. Moreover, it also revealed a radical alteration in magnetization within a finite magnetic field, as illustrated in Fig. 5.8. Typically it is called the giant magnetoelectric effect (this means control magnetization through the electric field and vice versa) which is purely due to electronic origin contribution [135, 136].

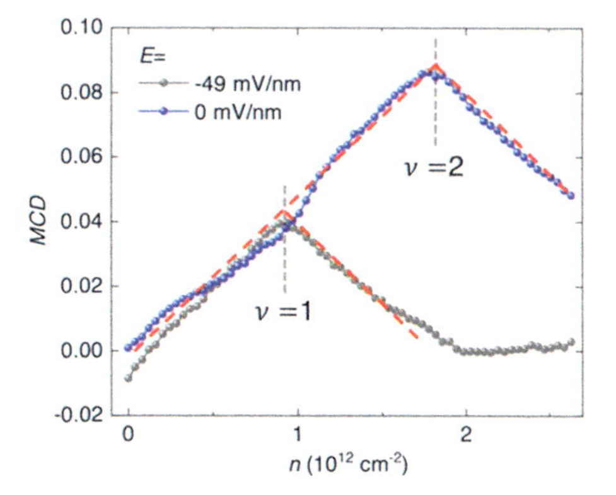

Fig. 5.8 Line cuts reflect the doping dependence of MCD with $P = 0$ (blue) and $P = 1$ (black). The red dashed lines denote the expected density dependence of the local moment density for both (adopted from Xu, Y. et al. 2022, https://doi.org/10.48550/arXiv.2202.02055)

Charge transfer between the two monolayers (WSe_2) in the twisted bilayer has also obtained by employing the parallel plate capacitor model. The critical electric field E_c to the layer polarization with the charge density νn_M is usually expressed as;

$$E_c = \frac{C_q + 2C_i + C_g}{C_q} \times \frac{\nu n_M e}{2\varepsilon_{BN}\varepsilon_0} \tag{5.5}$$

Here C_q represents the quantum capacitance of each monolayer, C_i is connected to the interlayer capacitance of the twisted bilayer, Cg is the gate capacitance, and e is the elementary charge. In the case of the non-interacting electrons $C_q \gg C_i \gg C_g$ and $E_c = \frac{\nu n_M e}{2\varepsilon_{BN}\varepsilon_0} \approx 28$ mV/nm for $\nu = 1$. This greater value to an order of magnitude than evaluated E_c leads to a significant electron–electron interaction in WSe_2. Such interactions significantly can also modify the value of C_q and C_i to favorable to the layer polarization state. However, the $\nu = 2$ inconsistency among the experiment ($E_c \approx 17$ mV/nm) and theoretical approaches ($E_c \approx 56$ mV/nm) has been also noted. Usually, such discrepancy is reduced E_c with the increasing doping density. This could be due to weaker interaction effects, and as a consequence, the Fermi level moves to higher-lying moiré bands, because moiré bandwidth normally upsurges to the band index [137].

Therefore, a jump in E_c at $\nu = 1$ and 2 due to the occurrence of the carrier gap in the insulating states. Thus an extra electric field is usually desired to overwhelm the carrier gap to completely transfer the carriers from one to another layer [138]. However, the higher jump in E_c at $\nu = 2$ than 1 has not well addressed. Although it is connected to enhance layer polarizability at lower doping densities, therefore interaction effect is significant. So, more theoretical and experimental explanations are desired to quantitatively adequately define density-dependent layer polarizability [131].

5.5 Valley Coherence Layer

Inversion symmetry breaking, strong spin–orbit coupling, and time-reversal equilibrium in single-layers of semiconducting transition metal dichalcogenides (TMDs) have revealed locking of the electronic spin orientation in a specific valley, K or K', at the edge of the Brillouin zone [68, 139, 140]. Due to this significant properties attracted a renewed interest in valleytronics of monolayers TMDs to get encoding information in their electronic valley degree of freedom and spins in spintronics. The formalism used for spins the evolution of the pseudospin could be described as illustrated in a Bloch sphere in Fig. 5.9. Here the poles respect to states $|K$ and $-K|$ are representing a recognized valley index and their equatorial plane respect to a linear overlapping of such states.

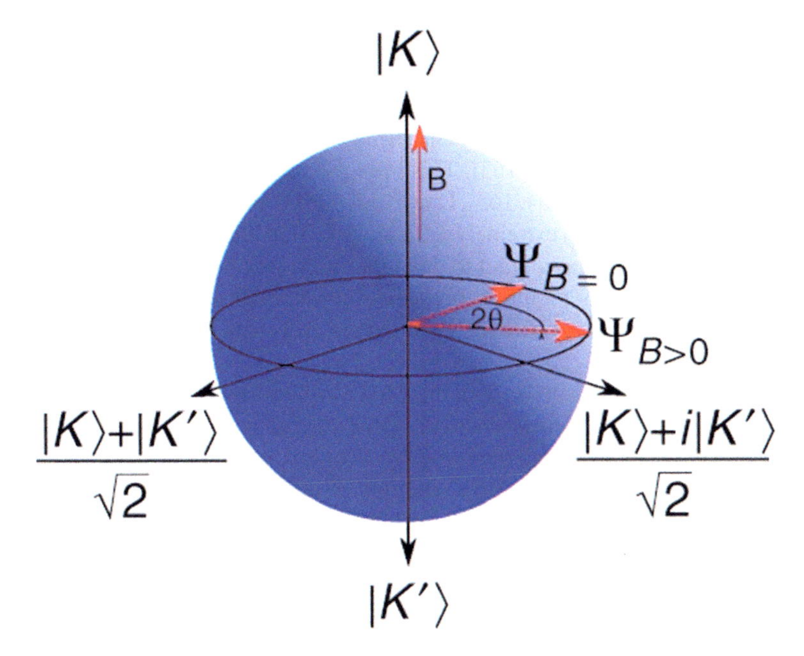

Fig. 5.9 Valley coherence in WSe_2 excitons: Bloch sphere representation of the valley pseudospin vector. Valley-polarized states ($|K$ and($|K'$)) lie on the poles of the sphere while valley coherence is represented by a Bloch vector oriented on the equator. The application of a perpendicular magnetic field leads to the precession of the pseudospin vector around the equator due to the valley Zeeman effect, evolving from a position $\Psi_{B=0}$ to $\Psi_{B>0}$ (adopted from Duerwiel, S. 2018 Nat Commun 9, 4797)

The occurrence of the bigger spin–orbit splatting's in TMDs makes the valley pseudospin robust against intervalley scattering [68, 139]. This led optical initialization of the valley states $|K$ and $|-K$ by the circularly polarized light that has been widely in TMDs excitons, like MoS_2, WS_2, and WSe_2 [68, 141, 142]. Additionally, it has also noticed the retention of linear polarization in WSe_2 and WS_2, due to the optical initialization of a superposition of valley states $\left|X = \frac{1}{\sqrt{2}}(|K + |-K)\right.$ [143–149]. Usually, the lifetime of such exciton valley coherence has projected up to a few hundred femtoseconds which limit the coherent employment of the valley pseudospin [143–149]. Moreover, the dephasing link to random momentum-scattering of excitons on disorder due to the occurrence of the electron–hole exchange interaction [150, 151], this process is recognized as Maialle–Silva–Sham (MSS) mechanism [152]. Additional factors also limit the capability to employ the valley coherence with the small exciton lifetime (1–2 ps) [143–149], owing to the valley index of electrons or holes, which reduces the optical govern [153].

Thus valley coherence is a bounded state of an electron and hole exciton due to the overlapping of the conduction and valence band states in the **K** and − **K** valleys [105, 154]. The **K** and − **K** valley excitons selectively are produced by circularly polarized light excitation of contradictory helicities [59, 68, 155, 156]. Similarly, the linearly polarized excitation also generates a hybrid **K** and − **K** exciton state in a perfect valley coherence situation [157, 158]. Although the valley coherence may degrade quickly owing to a collective impact of the fast scattering and intervalley interchange [159–161]. Significantly the values of valley coherence time lie in the range of 98−520 fs [116, 162, 163]. This life lime value is smaller than the exciton radiative lifetime of ~ 1 ps [164, 165]. Due to this TMDs are optical readouts of robust exciton valley coherence an extremely difficult task. That makes them enable to useful for the coherent excitons as a qubit to quantum information processing. A lengthier valley coherence time is required to execute any operation on it. Hence method that improves the valley coherence time expressively can have greater scientific importance [166].

As the typical mechanism has demonstrated to whole (100%) degree linear polarization (DOLP) in photoluminescence (PL) peak of $A1s$ exciton for the single-layer of MoS_2 which encapsulated from the some layers of graphene (FLG) at the uppermost and bottommost. The comprehensive retention of the produced valley coherence in steady-state PL suggests achievement a greater valley coherence time. This provides minimum ten times improvement in the valley coherence time [143, 151, 153, 162]. It depends on the linear polarization track of the exciting light. The excitons created at precise center-of-mass momentum (**Q**) points in the exciton band at time $t = 0$, as depicted in Fig. 5.10. In this lifetime the exciton may also undergo scattering and exchange interaction to couple together, as a result, degrade in valley coherence. However, the polarization of state is pseudospin through vector **S** in the Bloch sphere. Therefore, at $t = 0$, toward the **S** parallel to the interchange persuaded magnetic field (usually signified as the precession frequency Ω). If consider x—polarized excitation then it generates a pure state and represents from $S_x = 1, S_y = 0, S_z = 0$. The excitons scattering from the different **Q** points that are responsible for the pseudospin processes to feeling a determinate torque everywhere Ω (**Q**) owing to a mixed state, it is usually signified by a density matrix operator ρ. A useful **Q** space and the Bloch sphere mechanism representation are illustrated in Fig. 5.10. The decoupling of the density matrix as the numeral traces and the traceless matrix. σ (here σ signifies the Pauli matrices) dynamics for the valley pseudospin have been well expressed from the Maialle-Silva-Sham (MSS) mechanism [159, 166].

Another significant study has also demonstrated that the valley coherent exciton-polaritons in WSe_2 monolayers. According to investigation tunable zero-dimensional optical microcavity due to a top concave DBR into the optical path of piezo nanopositioners and two mirrors to a total optical cavity length of around 2.5 μm, as a schematic diagram with embedded monolayer is illustrated in Fig. 5.11a. The formed microcavity hemispherical shape is supported to zero-dimensional Laguerre Gaussian approaches. This kind of connection through the ground state longitudinal mode is significantly described by altering the mirror parting. Therefore, corresponding cavity resonance scanning is possible throughout the exciton resonance, this allows the visualization of features of the anti-crossing that indicates the construction of exciton–polariton twigs, as depicted in Fig. 5.11b. Corresponding fits of the peak locations and the coupled oscillator approximation yields a Rabi splits 26.2 ± 0.1 meV, as illustrated in Fig. 5.11c. The low intrinsic electron doping presence revealed that

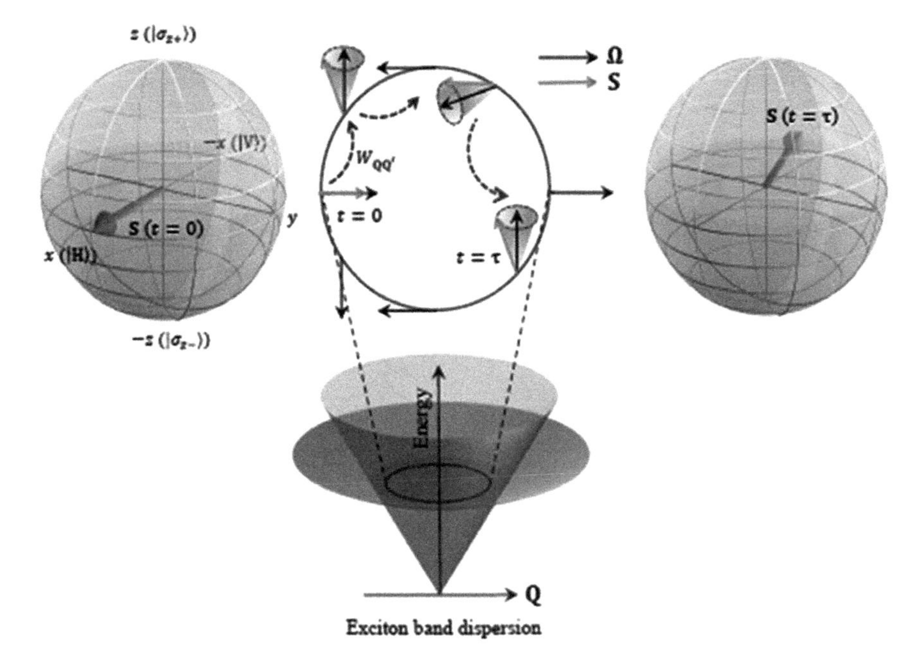

Fig. 5.10 Mechanism for the exciton valley decoherence: left panel for the linearly polarized light excitation (along x), an exciton pseudospin (S, solid arrow) points along the x direction in the Bloch sphere at $t = 0$, middle panel to top view of a ring inside the light cone of the exciton band showing the exciton decoherence dynamics due to scattering $W_{QQ'}$ within the light cone (dashed arrows) and subsequent precession because of intervalley exchange induced pseudo-magnetic field (dark black solid arrows), right panel to the emitting photon polarization depends on the direction of S when the exciton undergoes radiative recombination at a time at $t = \tau$ (adopted from Gupta, G. et al. 2021, arXiv:2106.03359v2)

weak coupling is possible between the trion and cavity [145, 166–169], as depicted in Fig. 5.11d.

5.6 Valley-Spin Concept for Monolayers

5.6.1 Selection Rules for Valley and Spin

Weak van der Waals bonds and close distance between the layers (or monolayer) allow inducing the proximity effect that also plays a significant role [170]. Such as one of the most striking proximity effects in the TMD heterostructures is the existence of Moiré superlattice [171–173] (see Fig. 5.12a). The Moiré superlattice electron of each layer experience additional Moiré potential on top of the layer-specific periodic potential. Due to this influence the Moiré superlattice in high symmetry points with

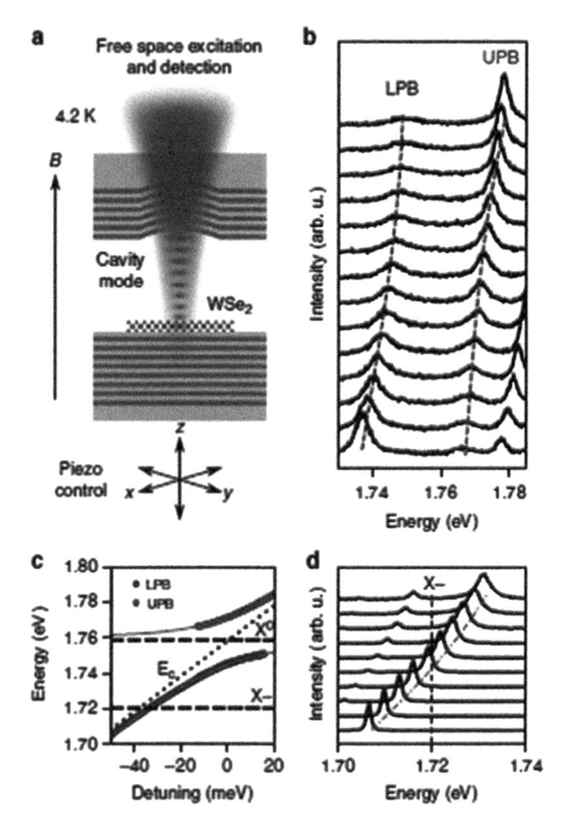

Fig. 5.11 Strong exciton-photon coupling in WSe_2 monolayers: **a** Schematic for the WSe_2 monolayer placed in an open hemispherical 0-dimensional microcavity, maintained at 4.2 K in an LHe bath cryostat with a superconducting magnet. Piezonanopositioners allow tuning of cavity length; **b** photoluminescence spectra for the cavity mode that scanned across the exciton resonance to show the formation of the lower and upper polariton branches, LPB and UPB, the dashed curves guide the approximate peak positions of the LPB (UPB); **c** polariton peak positions as a function of the exciton-photon detuning, where E_c (dotted line) is the energy of the tunable cavity mode, the dashed lines are the energies of the exciton and trion, the symbols show spectral positions of the UPB (LPB) peaks; **d** spectra corresponding to tuning the ground state mode through the trion energy to demonstrate the weak coupling between the cavity and trion and the dashed line indicates the energy of the trion (cavity mode) (adopted from Dufferwiel, S. et al. 2018 Nat Commun 9, 4797)

various local atomic registries (point A, B, and C in Fig. 5.12a). In which each registry has a different interlayer atomic configuration at their C_3 rotation center. Such a rotation process also affects in the electronic and exciton properties of TMD materials [171, 173–184]. As pointed-out the dependence of Moiré superlattice on the twist angle between the layers also opens up an additional degree of freedom to control the 2D heterostructure electronic and excitonic properties.

Moreover, the difference in interlayer atomic configuration at each local registry's C_3 rotation center reveals a unique exciton valley optical selection rule corresponding

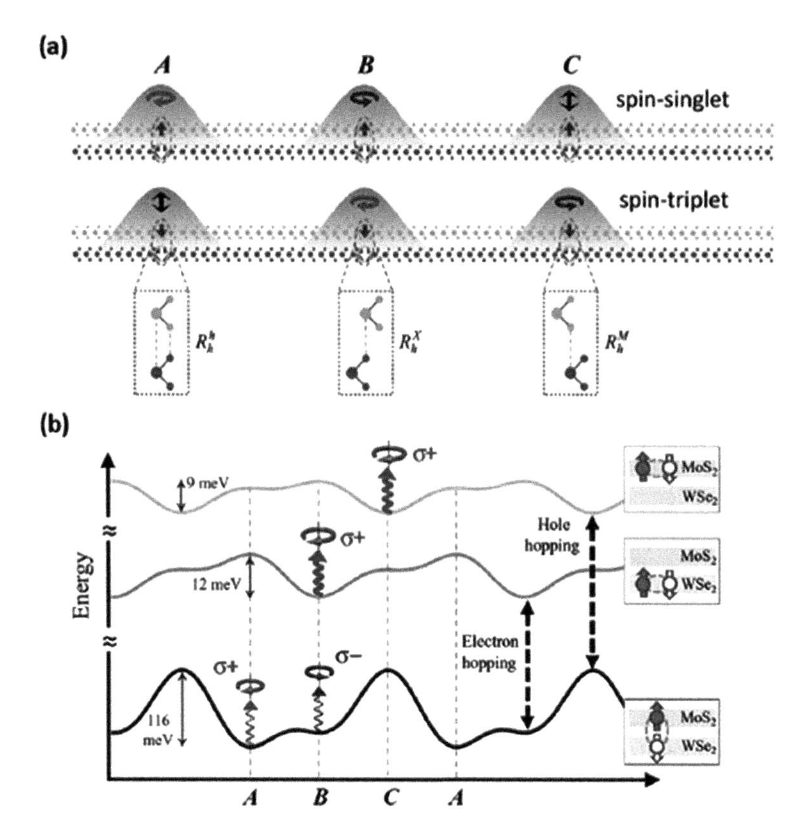

Fig. 5.12 Representation of moiré superlattice influence and optical selection rule in the 2D hetero-bilayer TMD: **a** interlayer exciton optical selection rule in K valley of AA-stacked hetero-bilayer TMD. The valley optical selection rule depends on the local high symmetry registry (A, B, or C) and the spin configuration (singlet or triplet) (adopted from Yu, H. Y. et al. 2018 2D Mater. 5, 035,021); **b** exciton energy spatial modulation due to Moiré potential in the K valley of AA-stacked MoS_2/ WSe_2 (adopted from Yu [85])

to each registry [53, 85, 97] (Fig. 5.12a). Including the triplet interlayer exciton emission in the out-of-plane direction having a different local optical selection rule compared to the singlet exciton [171, 184] (Fig. 5.12a). The existence of Moiré superlattice also offers the spatial modulation of the band gap for the individual monolayer and heterostructure band gap, which means the spatial modulation is possible for the both intralayer and interlayer exciton energy [173, 183] (Fig. 5.12b). This kind of band gap modulation allows the occurrence of the Umklapp process, therefore, the multiple peaks in both intralayer and interlayer exciton emission appears [179, 181, 183, 185]. As the experimental findings it has also demonstrated that the diverse peaks in the interlayer exciton emission may possess distinct valley optical selection rules [102, 184–186]. This condition is known as spatial modulation of exciton energy and optical selection rule [185] or the diverse selection rules for singlet and triplet exciton [102, 184, 186]. More conventionally it could say that, although the valley optical selection rule remains valid for the interlayer exciton, the selection rule depending on the emission location and emission energy. Moreover, in the inhomogeneity situation, the grouping of the non-homogeneous enlargement and the valley optical selection rule gives a lower degree of circular polarization in the PL emission. To sidestep such lengthening, the encapsulation of the TMD specimen using hexagonal Boron Nitride (h-BN) is commonly preferred [187].

Typically optical selection rules are interpreted by describing the total magnetic quantum number of the electronic states nearby the K and − K points. Whereas the wave function of the conduction band is mostly constituted from the transition metal atom d_{z^2} orbit with the magnetic quantum number 0, while the valence band belongs to $\left(d_{x^2-y^2} + i\tau d_{xy}\right)/\sqrt{2}$ orbit with the magnetic quantum number 2τ [59, 139, 188]. There are no interband optical transitions allows while alone atomic orbitals to be taken, because of the alteration in the magnetic quantum numeral –2τ, thereby, in this case, spin contributions are usually canceled. On the other hand, if also incorporate the alteration in valley orbital angular momentum (a valley pseudospin flip) [189], then the entire modification in the magnetic quantum numeral converts –τ. The alteration in the magnetic quantum numeral directly provides us the optical dipole selection rules, it is generally interpreted as the direct interband transitions at the K and − K points which are combined entirely to left and right circularly polarized light, as depicted in Fig. 5.13a. Corresponding estimations have also demonstrated that the selection rules are uphold good even to an important portion of the K and − K valleys too far from the band boundaries.

Since the photoexcited electron–hole (e–h) pairs are bounded through the robust Coulombic connections in TMDs to procedure excitons. So, according to Dirac dispersion of excitons under the consequence of the hydrogen model, it directs the

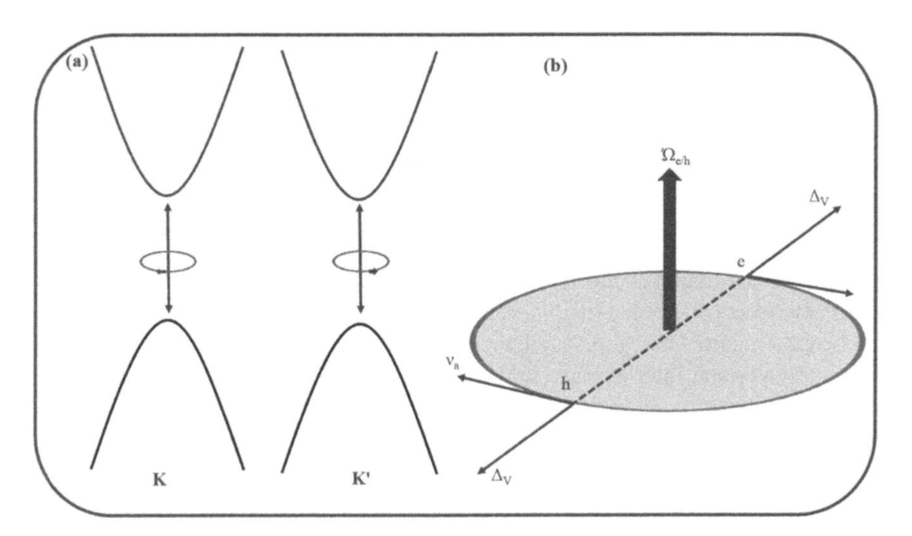

Fig. 5.13 **a** Single-layer TMDs electronic band dispersions near the *K* and *K*′ points of the Brillouin zone; **b** electron (**e**) and hole (**h**) in each valley due to an exciton under a central potential *V*

electron and hole together an effective mass govern to the band dispersals in the impact of the dominant Coulombic potential V. Their excitons states are labeled from the (n, m), where $n = 0, 1, 2, \ldots$ and $m = 0, \pm1, \pm2, \ldots$ that represents the radial quantum number and the angular momentum quantum number. The states $(n, \pm m)$ associating to the contradictory angular momentum degeneracy to usual semiconductors and respective band gap is generally positioned at the Γ point of the Brillouin zone. This is known as the universal significance of the time-reversal symmetry. The monolayer TMDs bright excitons (optical dipole conversion permissible) are having $|m| = 1 \pm \tau$ for right (+) and left (−) circularly polarized light [190, 191]. Herein the first integer term '1' is usually associated to phase winding numeral (= 1 for the monolayer TMDs), this is also known as the topological integer for the Bloch band which is generally absent in case of the common semiconductors for both s-like ($m = 0$) and d-like ($m = 2$) exciton states [191]. Whereas in the case of N-fold rotating symmetry more bright exciton states occurs having $m = 1 \pm \tau + jN$ (here j is a number) [189, 192]. Moreover, the p-type ($m = 1$) exciton states are considered to be optically dim. Specifically this state electron and hole gets a comparable speed in the normal direction owing to the occurrence of Berry curvatures, typically it illustrated in Fig. 5.13b. The occurrence of such an irregular term breaking the time-reversal symmetry in every valley, as the consequence, it provides an energy dissimilarity among the left ($m = 1$) and right ($m = -1$) rotating states, it is reflecting the condition of the time-reversal symmetry preservation in both valleys. Moreover, it also reveals in the energy splits in p levels directly proportional to Berry curvature flux penetrating exciton wrapping function in momentum space [193–195].

5.6.2 *Valley Zeeman Effect and Magnetic Moment*

Under an applied weak external magnetic field a periodic system is induced two different kinds of physical effects on the band extrema. Namely, the first one is recognized as the Bloch states in the extremely degenerate Landau levels (LLs), and the second one further shifts in LLs energy which is known as the Zeeman effect. Spin-valley combination due to braking overturn regularity in semiconducting single-layer (ML) H-phase transition metal dichalcogenides (TMDs) under the external magnetic fields has attracted much attention for the theoretical and experimental investigations [22, 24, 54, 71, 196–199]. Moreover, the inversion symmetry breaking and threefold rotational symmetry are also lead to valley-dependent optical selection rules [15, 22, 59, 200], which is helpful to quantify the valley-dependent Zeeman effect and Landau levels in the TMDs [24, 198]. Additionally, a description of the energy band boundaries at every valley response under the applied external magnetic field has been considered to be the basic mechanism.

The band edge Zeeman shift is usually described by the band and valley-dependent orbital magnetic moment m_{nk} (or Laudé g-factor g_{nk}^{orb}), whereas $m_{nk} = g_{nk}^{orb}\mu_B$ [201]. However, an alternative orbital magnetic moment approach has been extensively used to define Zeeman shift without LLs. In addition to this, two-band tight-binding (TB) models have been also discussed by considering the LL is asymmetric between the K and - K valleys [48, 202, 203]. Thus in a conclusive view based on experimental findings both valley Zeeman effect and LLs are important because LLs defines from the TB model can further shift due to an additional Zeeman term [198, 202, 204].

In general, the quantitative analysis of the lone band g-factor is described from the two methods [197]. The first one is depend on the phenomenological approach in which the orbital magnetic moment is separated into atomic and valley, therefore, the entire solitary band g-factor contains three quantities, namely, spin, atomic and valley [196, 205]. Meanwhile the valence band maxima (VBM) and conduction band minima (CBM) of TMDs are predominantly made of d-orbitals, their corresponding atomic participations are considered magnetic quantum numbers $\pm$ 2 and 0 [206]. The corresponding valley has been roughly contrariwise relative to the effective mass [15, 198, 207, 208]. Note that in these approaches with greater than double bands this kind of relationship does not uphold usually [201], including significant problem separating of the orbital magnetic moments into atomic and valley that cannot explain appropriately, precisely not clearly define whether they are additive in case of additional assistances or left. Moreover, another significant method is depending on Peierls substitution into an active multiband $\mathbf{k} \cdot \mathbf{p}$ Hamiltonian [201]. That is contradictory to the phenomenological model, for the argument has been given to atomic terms vanished, therefore, the impact of remote bands is considered to be significant, by providing the variety of different exciton g-factors from the Hamiltonians with the distinct numeral of bands [209]. Hence both approaches provides g-factors in justifiable settlement with the experimental findings, but their numeric depends on choose of the specific model Hamiltonian.

Moreover to interpret the valley Zeeman effects and LLs in 2D-TMDs proposed a model considering Hamiltonian incorporating spin–orbit coupling (SOC) of an electron in a periodic potential that is disconcerted by a homogenous external magnetic field. This investigation lead under the Luttinger-Kohn (LK) approach, the valley bandages splits in the LLs together a valley-dependent Zeeman shift [187, 210]. The outcome leads to the LL indices are symmetric for the K and $-$ K valleys. And their orbital magnetic moments described the Zeeman shift that is equivalent to a dense Berry curvature alike manifestation which governs from the semi-classical approximation [211]. Moreover, the LK approach of electrons in a nonlocal periodic potential for the condensed form of the magnetic moment could remain unaffected. This extension allows the inclusion of nonlocal exchange and correlation effects for the single-band g-factor. By implying this approach investigators have been evaluated the Berry curvature like expression for the g-factor without the summing above vacant bands by employing first-principles Hamiltonians [212].

Thus significantly overturn equilibrium breaking permits the occurrence of an orbital magnetic moment $m_V (or \mu_V)$. Spontaneously, that arose to the self-rotation of the electron wave packet [14]. However, the motion within the 2D plane is expressed as the orbital magnetic moment out of the plane. So, a system described from the 2D Dirac Hamiltonian approach orbital magnetic moment is simply expressed at K and $-$ K points, $m_V = \tau \frac{e\hbar}{2m^*} \hat{z}$ [14], here m * (> 0) represents the effective mass of the Bloch band and $\hat{z}$ corresponds to the out-of-plane unit vector. Compared to the Berry curvature, in this approach the orbital magnetic moment transmits possessing similar signs for both conduction and valence bands in the equivalent valley. Therefore, its behavior alike a spin magnetic moment associating an effective Bohr magneton $\pm\frac{e\hbar}{2m^*}$ along with the upper and lower sign to the conduction and valence band. Additionally, it also leads to the conduction and the valence states at the K point can carry a valley orbital angular momentum ħ/2 and − ħ/2. These signs could alternatively switch for the − K valley to satisfying the time-reversal symmetry. Such orbital magnetic moment connects under the applied magnetic field B (Zeeman-like interaction) which is typically defined as $-m_V.B$ (note that in different books it is also defined as $-\mu_V.B$) [213].

In a significant study, comparative solitary band and exciton g-factors calculations from the DFT and GW approaches for WSe_2 and $MoSe_2$-WSe_2 systems have been also interpreted [212]. The Perdew-Berke-Ernzerh of (PBE) parametrization and GW band structures for the ML WSe_2, with SOC effects, is illustrated in Fig. 5.14. This leads to GW calculations for the alike systems, the direct gap at K has improved by 0.9 eV owing to the GW self-energy modification [214]. This investigation leads to the LDA and PBE can have rather similar g-factors, while GW may change the g-factors significantly, due to the consequence of nonlocal exchange and correlation effects in the Hamiltonian and quasiparticle energies that play an important role in the prediction of the single-band g-factors.

Moreover the experimental exciton Laudé g-factor (g_{X0}) is usually obtained from the following relationship;

$$\Delta E_{X0} = E_{X0}(\sigma+) - E_{X0}(\sigma-) = g_{X0}\mu_B.B_Z \quad (5.6)$$

where E_{X0} is the Zeeman splitting of the first bright neutral exciton ($X0$) photoluminescence peak in H-phase TMDs, in the presence of an out-of-plane magnetic field Bz. According to first-principles calculations, in ML WSe_2 the conduction and valence bands K(− K) valley can be coupled to σ^+ (σ^-) circularly polarized light [214, 215]. Since $X0$ is usually extremely localized at the K and − K valleys inside

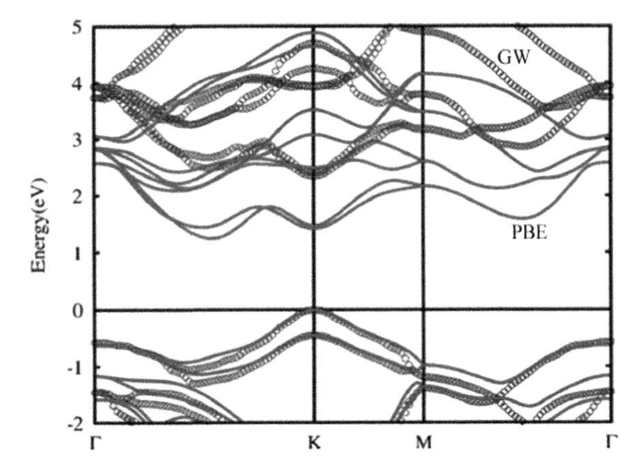

Fig. 5.14 PBE (lines) and GW (dots) band structure for the ML WSe_2, with SOC (adopted from Xuan et al. [212])

the area whereas the Bloch state band structure turn into quadratic. Therefore, it leads the energies of the states contributed in creation of the $X0$ exciton. Thus, the solitary-particle g-factors at K and – K have an identical extent with the contradictory sign. The bright exciton g-factor corresponds to $g_{X0} = 2\left(g_{v\uparrow}^{orb} - g_{c\uparrow}^{orb}\right)$ and dark exciton g-factor belongs to $g_{D0} = 2\left(g_{v\uparrow}^{orb} - g_{c\downarrow}^{orb}\right) - 4$, where – 4 arises due to spin involvements [212, 215]. These bright and dark exciton g-factors depending on the spin –permissible and spin-empty transitions as listed in Table 5.2. As the instance for the ML WSe_2 computed effective masses 0.32, 0.44 and – 0.39 m_e to c1, c2 and v1 bands is illustrated in Fig. 5.15. In which energy levels at the valleys associating LL and Zeeman effects, as depicted Fig. 5.15b–e, it revealed a good settlement with the experimental findings. Thus it reveals

LLs due to spin and valley polarization. Because these energy levels include both the LL and Zeeman effects (computed entirely from first-principles) and their corresponding g-factors include not only the spin and valley terms but also the atomic and cross terms. A similar results have been also obtained for the g-factor when

Table 5.1 Rotating regularity and orbital configurations of conduction and valence band Bloch function at K point in single-layer TMDs

	M_d	X_p	$l(M)$	$l(X)$	$l(h)$
CBM	d_{z^2}	$p_x - ip_y$	0	+ 1	– 1
VBM	$d_{x^2-y^2+id_{xy}}$	$p_x + ip_y$	+ 1	– 1	0

The Bloch functions are eigenstates of the C_3 revolution with eigenvalue $e^{i\frac{2}{3}l\pi}$, where l depending on the selection of revolution epicenter can be an M or X site, or the hollow center h of the M–X hexagon (as depicted in Fig. 5.2a [16])

Table 5.2 Single-band and exciton g-factors for the ML WSe_2 by the DFT versus GW at K

g^{orb}	PBE	GW	Different studies' experimental findings
$g_{c\uparrow}^{\text{orb}}$	– 2.81	– 4.15	
$g_{c\downarrow}^{\text{orb}}$	– 1.90	– 3.24	
$g_{v\uparrow}^{\text{orb}}$	– 4.86	– 6.40	
$g_{v\downarrow}^{orb}$	– 4.17	– 5.71	
g_{X0}	– 4.10	– 4.50	– 3.7, – 4.3, – 4.37, – 4.4
$(\tilde{g}_{X0})$	...	(– 4.26)	
g_{D0}	– 9.92	– 10.32	– 9.3, – 9.5, – 9.9
$(\tilde{g}_{D0})$	...	(– 9.76)	

For the single-band g-factors, c, and v refer to the frontier conduction and valence bands, respectively, while ↑ and ↓ refer to spin-up and spin-down bands at K. The superscript "orb" indicates only the orbital part of the g-factor. $X0$ and $D0$ refer to the lowest-energy bright (spin-allowed) and dark (spin-forbidden) excitons, respectively. The numbers in brackets include the effects of a frequency-dependent BSE kernel [212]

Table 5.3 C_{3x} eigenvalues of the first two bands of each valley computed for the large-scale DFT wave functions at high symmetry momentum points [289]

Band	Valley	K_+	K_-	γ
1	K	$e^{i\pi/3}$	$e^{i\pi/3}$	$e^{i\pi}$
1	K′	$e^{-i\pi/3}$	$e^{-i\pi/3}$	$e^{i\pi}$
2	K	$e^{-i\pi/3}$	$e^{-i\pi/3}$	$e^{i\pi/3}$
2	K′	$e^{i\pi/3}$	$e^{i\pi/3}$	$e^{-i\pi/3}$

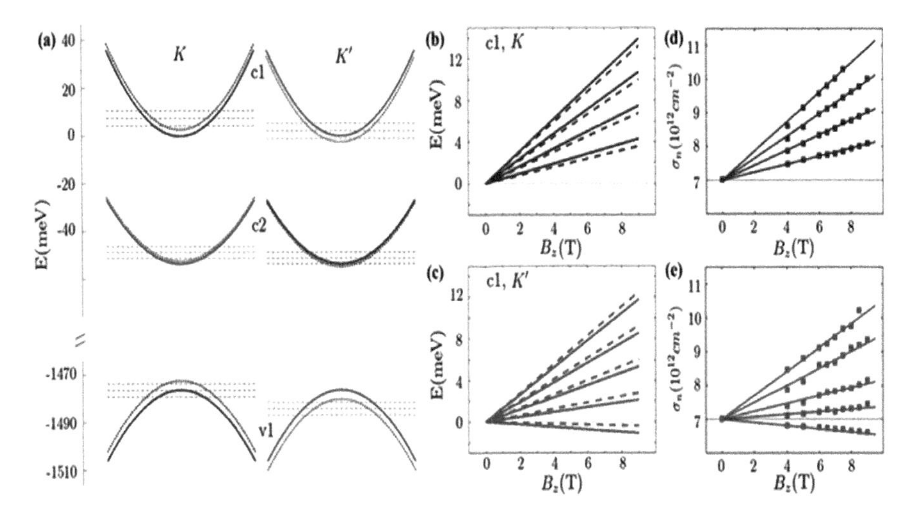

Fig. 5.15 Predicted energy levels at *K* and *K′* in ML WSe_2 in the presence of a uniform external magnetic field and comparison with the experiment: **a** horizontal dashed lines denote the energy levels (LLs) obtained using a magnetic field of 9T, the solid curves represent energy bands in the absence of a magnetic field, and dashed curves represent the Zeeman-shifted bands (ignoring LL effects), curves are representing spin-down and spin-up; **b, c** predicted energy of band $c1$ as a function of magnetic field strength at *K* and *K′*, solid lines and dashed lines are computed from the GW and DFT-PBE g-factors approaches; **d, e** magnetic field dispersion of the LLs of band $c1$ at the *K* and *K′* valleys. The vertical axes are the critical electron densities required to fill the LLs as probed from the onset of Pauli blocking of the corresponding inter-LL optical transitions (adopted from Xuan [212])

excitons of lowermost energy spin-permissible transitions in $MoSe_2$-WSe_2 hetero-bilayers for the AA and AA′ stacking [212].

5.6.3 Valley Hall Effect and Berry Phase Effect

Specifically, Berry phase and Berry curvature are usually associated to Bloch bands properties in 2D (or Dirac) materials, which includes various valley conflicting physical mechanisms [12]. The three-dimensional overturn equilibrium breaking in these kinds of systems induces valley-contrasting effective magnetic fields that arise in the momentum space; this is typically recognized as Berry curvature fields [216–219]. While applying the in-plane electric fields gets the Berry curvature driven by charges from contradictory valleys to flow in opposite perpendicular directions, ultimately it gives us valley Hall effects [216–219]. Moreover, among the Berry curvature fields breaking overturn equilibrium in single-layer TMDs also impacts the effective Zeeman fields in momentum space [59, 220], which are usually called Ising spin–orbit coupling (SOC) fields [221–224]. Moreover, the gated TMDs, or polar TMDs known as Rashba-type SOCs [225–227] that are raised naturally [221]. Thus co-occurrence of Ising and Rashba SOCs in gated/polar TMDs give us a new valley reduction in Berry curvatures with a special type of valley Hall effect, this is called spin–orbit coupling induced valley Hall effects (SVHEs). In this novel form the Berry curvatures the inversion symmetric hybridization from the diverse d-orbitals. This special kind of Berry curvature initiates owing to inversion-asymmetric spin–orbit interactions [59]. Hence two distinct type formations can be distinguished based on their physical origins. The Berry curvature persuaded by SOCs is also known as spin-kind Berry curvature, while the regular Berry curvatures/valley Hall effects are due to orbital degrees of freedom.

Spin-kind Berry curvature and spin–orbit coupler persuaded valley Hall effect in TMDs as the typical example gated monolayer MoS_2. In which electrostatic gating conduction band minimum nearby to the K-valleys with the partial filling [221, 228], whereas the electron bands created mainly from the d_{z^2} of Mo-atoms [23, 229]. Spins of d_{z^2} electrons is generally formed effective Hamiltonian near the K-valleys [221, 230, 231];

$$H_{spin}(k + \varepsilon K) = \xi_k^c \sigma_0 + \alpha_{SO}^c \left(k_y \sigma_x - k_x \sigma_y\right) + \varepsilon \beta_{SO}^c \sigma_z \tag{5.7}$$

where $\xi_k^c = \frac{|k|^2}{2m_c^*} - \mu$ represents the usual kinetic energy term, $2m_c^*$ is the effective mass of the electron band, μ is the chemical potential, $k = \left(k_x, k_y\right)$ is the momentum at K(−K)-valleys, $\varepsilon = \pm$ is the valley index. Term β_{SO}^c represents the Ising SOC that pinches the electron spins in out-of-plane directions, as depicted in Fig. 5.16a. Ising SOC due to the infringement of in-plane mirror regularity (mirror plane vertical to the 2D lattice plane) and atomic SOC from the transition metal atoms. Moreover, term α_{SO}^c represents the Rashba SOC which pinches the electron spins toward in-plane directions associating helical spin textures (see Fig. 5.16a). Rashba SOC arises due to out-of-plane mirror regularity (mirror plane parallel to the lattice plane) breaking by the gating or by lattice structure [232]. Thus, H_{spin} form due to a massive Dirac

Hamiltonian (if neglected the kinetic term there is no influence of the Berry curvatures) where the Ising SOC play a crucial role in the valley conflicting Dirac mass, that order of a few to tens of meVs [23].

Significantly Pauli matrices $\sigma = (\sigma_x, \sigma_y, \sigma_z)$ also impacts on the spin degrees of freedom. Therefore, in contrast to the massive Dirac Hamiltonian it is usually expressed as [22];

$$H_{\text{orb}}(k + \varepsilon K) = V_F(\varepsilon k_x \tau_x + k_y \tau_y) + \Delta \tau_z \tag{5.8}$$

where the Pauli matrices $\tau = (\tau_x, \tau_y, \tau_z)$ act on subspace that is constructed from the diverse d-orbitals, while the term V_F is incorporated due to electron hopping. The Dirac mass Δ generates due to a big band gap (~2Δ), typically in monolayer TMDs it is order of 1–2 eVs [22]. Since Ising and Rashba SOCs in nondegenerate

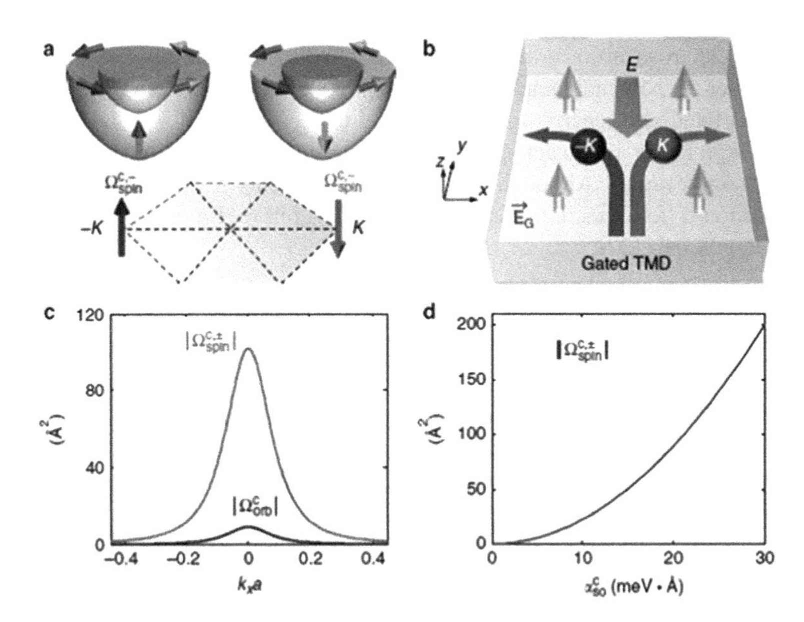

Fig. 5.16 Representation of spin–orbit coupling induced valley Hall effects: **a** schematics for the Ising spin–orbit coupling (SOC) (circular arrows), the Rashba SOC (vertical arrows), and the spin-type Berry curvatures $\Omega^{c,-}_{\text{spin}}$ (arrows at K and K′) in the lower spin bands represented by pockets above K/− K-points; **b** valley Hall effects due to $\Omega^{c,-}_{\text{spin}}$, the arrows indicate the out-of-plane gating fields/electric polarization labeled by E_G; **c** magnitudes of spin-type Berry curvature $\left|\Omega^{c,\pm}_{\text{orb}}\right|$ near the conduction band edge (larger solid curve) and orbital-type Berry curvature $\left|\Omega^{c}_{\text{orb}}\right|$ (small solid curve) near K-points. Rashba coupling strength is set to be $\alpha^{c}_{so} = 21.4$ meV Å according to $\alpha^{c}_{so}k_F$ = 3 meV, comparable to $2\left|\beta^{c}_{\text{so}}\right| = 3$ meV, parameters for Ω^{c}_{orb} are set to be: $\Delta = 0.83$ eV, $V_F =$ 3.51 eV Å. Shows that $\left|\Omega^{c,\pm}_{\text{spin}}\right|$ is nearly ten times of Ω^{c}_{spin}; **d** $\left|\Omega^{c,\pm}_{\text{spin}}\right|$ as a function of α^{c}_{so} at the K-points, therefore, $\left|\Omega^{c,\pm}_{\text{spin}}\right|$ scales quadratically with α^{c}_{so} (adapted from Zhou et al. [34])

spin subbands **close** the conduction band minimum. So, upper/lower spin-subbands of energy are generally described from the relationship;

$$E^{e}_{c,\pm}(k) = \xi^{c}_{k} \pm \sqrt{\left(\alpha^{k}_{SO}k\right)^{2} + \left(\varepsilon\beta^{c}_{SO}\right)^{2}} \tag{5.9}$$

The corresponding generated Berry curvatures by SOCs in lower spin bands having energy $E^{e}_{c,-}(k)$, which is defined as;

$$\Omega^{c,-}_{spin}(k+\varepsilon K) = \frac{(\alpha^{c}_{SO})^{2}\left(\beta^{c}_{SO}\right)^{2}}{2\left[\left(\alpha^{c}_{SO}k\right)^{2} + \left(\beta^{c}_{SO}\right)^{2}\right]^{3/2}} \tag{5.10}$$

Here note that $\Omega^{c,-}_{spin}$ reflects a valley-dependence signs owing to the valley conflicting Dirac mass and Ising SOCs. In case of an in-plane electric field, $\Omega^{c,-}_{spin}$ drives the electrons in lowermost spin bands for the contradictory valleys to flow in contradictory perpendicular directions, as illustrated in Fig. 5.16b. Whereas the upper spin band with energy $E^{e}_{c,+}(k)$ is equal to $\Omega^{c,+}_{spin} = -\Omega^{c,-}_{spin}$. Thus the valley currents from upper and lower spin bands are partially cancel out to each other and both of them occupied. Although non-zero valley currents are still generated owing to population differences in the spin-split bands, therefore, Eq. 5.10 is similar to its orbital counterpart [22].

$$\Omega^{c}_{orb}(k+\varepsilon K) = -\frac{\varepsilon V_{F}^{2}\Delta}{2\left[(V_{F}k)^{2} + \Delta^{2}\right]^{3/2}} \tag{5.11}$$

Term Ω^{c}_{orb} have important implications in valleytronic applications. Because TMDs magnitude of Ω^{c}_{orb} is usually an order of $10(\text{Å})^{2}$ owing to the big Dirac mass by the band gap 2Δ (~ 1–2 eV) [22]. However, the $\Omega^{c,\pm}_{spin}$ Dirac mass β^{c}_{SO} have a larger up to a few meVs near the conduction band edges [23]. The gated MoS_2 defined Rashba energy value $\alpha^{c}_{SO}k_{F} \approx 3\,\text{meV}$ is also close to the splitting of energy Ising SOCs, as depicted in Fig. 5.16c [23]. Therefore, the gated/polar TMDs also generates valley Hall signals.

Moreover, the strength of Ω^{c}_{orb} also determines by intrinsic parameters of the material, however, it is hard to tune it. On the other hand, $\Omega^{c,\pm}_{spin}$ represents a quadratic dependence for the Rashba coupler forte α^{c}_{SO}, therefore, it is usually govern by exterior gating fields, as depicted in Fig. 5.16d. Thus the value of $\left|\Omega^{c,\pm}_{spin}\right|$ strongly depends on the α^{c}_{SO}.

In actual (or real) the gated/polar TMDs, the spin-type Berry curvature Ω_{spin} that all times existing in association of the orbital-kind Berry curvature Ω_{orb}. So, it has an interplay between Ω_{spin} and Ω_{orb} near to K-valleys, as depicted in Fig. 5.17a, b. Generally single-layer MoS_2 is taken as an instance, to define the total Berry curvatures at the K-points by utilizing a realistic tight-binding (TB) model [23]. The simplest Berry curvatures at $K = (4\pi/3a, 0)$ and at –K and their time-reversal at

$\Omega(-K) = -\Omega(K)$. In the case of the conduction band whereas the Ising SOC strength β^c_{SO} so small. This means, in the absence of gating, the total Berry curvatures $\Omega^{c,\pm}_{spin}$ to the both spin-subbands at $K = (4\pi/3a, 0)$ contains only orbital contribution Ω^c_{orb} pointing in negative z-direction [8]. While changing in coupling strength, $\Omega^{c,\pm}_{spin}$ comes into in picture through a drastic change in $\Omega^{c,\pm}_{tot}$. In which the lower spin-subband ($\Omega^{c,-}_{spin}$) points to the negative z-direction (see Fig. 5.17a). Since, $\beta^c_{SO} < 0$ for the molybdenum (Mo) established TMDs [23, 233], this provides a negative magnitude of $\Omega^{c,-}_{spin}$. As the consequence $\Omega^{c,-}_{tot}$ in increasing order with α^c_{SO}, as depicted in Fig. 5.17c. However, the upper spin-subband $\Omega^{c,+}_{spin} = -\Omega^{c,-}_{spin}$ belongs to antiparallel its orbital counterpart $\Omega^{c,+}_{orb}$, so, it gives them to compete against each other with the increasing α^c_{SO}, thus, the $\Omega^{c,+}_{spin}$ become comparable to $\Omega^{c,+}_{orb}$, in this situation a zero total Berry curvature $\Omega^{c,+}_{tot} = 0$, at the certain α^c_{SO} that could be achieved. Furthermore, increase in α^c_{SO}, the $\Omega^{c,\pm}_{spin}$ dominates, this leads to distinct signs of total Berry curvatures for the higher ($\Omega^{c,+}_{tot} > 0$) and inferior ($\Omega^{c,+}_{tot} < 0$) spin bands. However, tungsten (W)-based TMDs have a contradictory situation for the molybdenum (Mo)-based materials because the conduction band different sign. As depicted in Fig. 5.17d, the total valence band of the orbital-type contribution $\Omega^{v,+}_{tot}$ may insensitive to α^v_{SO} and close to $\Omega^{v,+}_{orb}$ at $\alpha^v_{SO} = 0$.

Thus, the behavior of total Berry curvatures is generally understood by considering spin-type and orbital-type contributions separately. Because the total Berry curvature Ω^n_{tot} at the K-points for the under test band, n is the algebraic sum of Ω^n_{spin} and Ω^n_{orb}, so, it expressed as $\Omega^n_{tot}(\varepsilon K) = \Omega^n_{spin}(\varepsilon K) + \Omega^n_{orb}(\varepsilon K)$[34].

Moreover, the valley Hall Effect is also understood based on the existence of Berry curvature that expressively adjusts the electron dynamics and produce novel electrical transportation mechanism. Specifically, under the employed E, the electrons can improve an anomalous speed [200].

$$v_a = \frac{e}{\hbar} E \times \Omega \tag{5.12}$$

here e and ħ represents the elementary charge and the Planck constant. This expression is considered to be analogous the Lorentz force to electrons movement in a transverse magnetic field. Moreover, the Berry curvature is contradictory for the distinct valleys, as depicted in Fig. 5.17c. Therefore, the anomalous velocity gives us to the valley Hall effect [14, 15, 200, 234]. In which charge carriers of the distinct valleys flows in contradictory perpendicular directions under the applied electric field E without applying any magnetic field. This allows the probability to govern the electronic valleys through the electric field owing to facts that the K and − K electrons carry an orbital magnetic moment μ_V with the contradictory sign [14, 15, 200, 234, 235]. The transverse flow drives by the valley Hall effect to generate a true angular momentum current J_V (valley current) without a net charge in current, as depicted in Fig. 5.18. This kind of transverse valley current leads to an accumulation of excessive K (− K) electrons on the left (right) edges of the specimen, and a finite

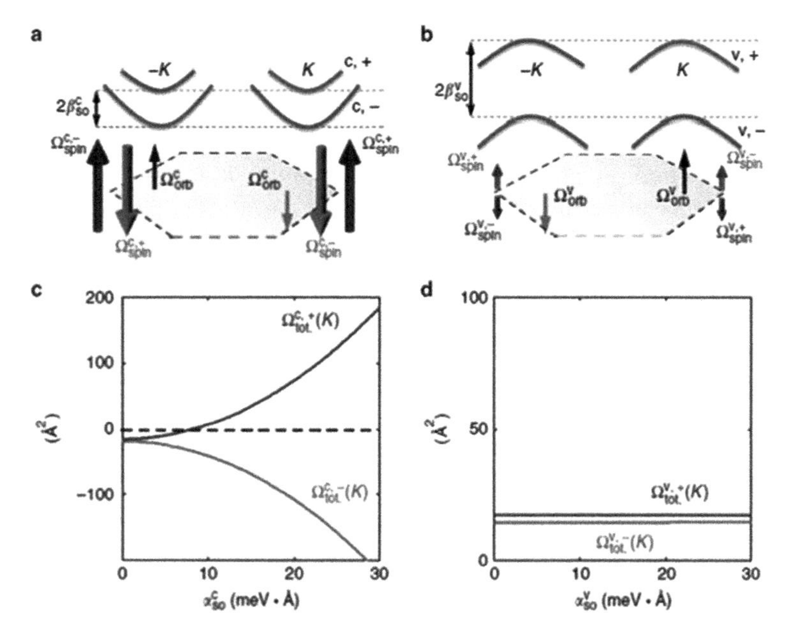

Fig. 5.17 The interplay between Ω_{spin} and Ω_{orb}: **a** the case near the conduction band edge; **b** the case near the valence band edge; **c** total Berry curvatures $\Omega_{\text{tot}}^{c,+}$ of upper (c, +)/lower(c, −) spin-subbands at + K-point versus α_{so}^{c} near the conduction band edge. As α_{so}^{c} increases, $\Omega_{\text{spin}}^{c,\pm}$ becomes dominant and changes (adopted from Zhou et al. [34]); $\Omega_{\text{tot}}^{c,+}$ dramatically; **d** total Berry curvatures $\Omega_{\text{tot}}^{v,+}$ of upper (v, +)/lower (v, −) spin-subbands at + K-point as a function of α_{so}^{v} at the valence band edge, obviously, $\Omega_{\text{tot}}^{v,\pm}$ are insensitive to α_{so}^{v} and remain close to $\Omega_{\text{orb}}^{v,\pm}$ at $\alpha_{\text{so}}^{v} = 0$

valley polarization (due to population imbalance) of the opposite sign arises within the valley, this means the free path from the edges [12].

The Valley Hall effect in single-layer TMDs initially has been demonstrated by detecting a Hall voltage when circularly polarized light is employed to create an imbalance between the two valleys in a MoS_2 Hall bar device [236]. Moreover, the effect is also directly imaged in MoS_2 transistors by magneto-optical Kerr rotation microscopy [237, 238]. This experimental arrangement is alike to that utilized to identify the spin Hall effect [239]. The boundary valley polarization is also captured by the polarization revolution or ellipticity of linearly polarized probe light nearby the important exciton resonance. The experimental demonstration relies on the significant combinations among the valley degree of freedom and formation of the circularly polarized light. Several studies have also contrasted such as the valley Hall effects and studied the bilayer 2H-MoS_2 [236, 240]. The pure 2H-bilayers have an overturn equilibrium compare to 3R-MoS_2 non-centrosymmetric to whichever layer thicknesses. However, the Berry curvature of the bilayer 2H-MoS_2 is zero when it associated to non-existing valley Hall effect. Their inversion symmetry is breaking through the implication of a transverse electric field from the gating [241–243]. Therefore, the

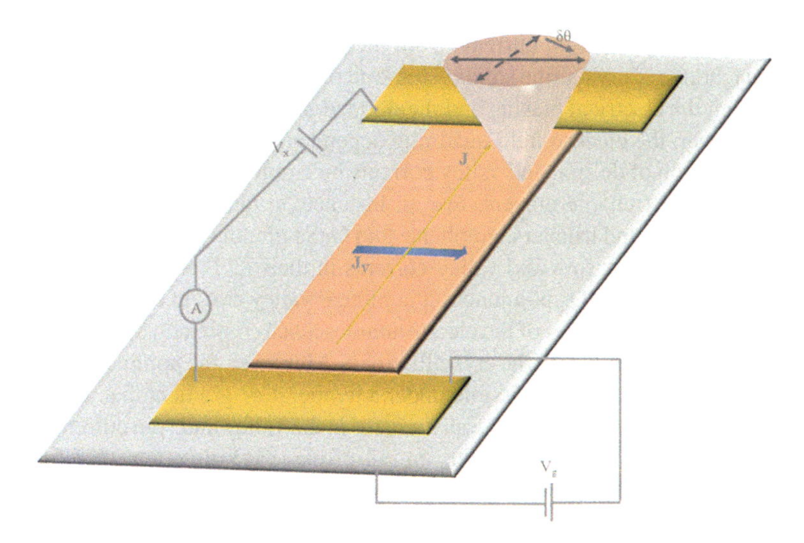

Fig. 5.18 Schematic for the magneto-optical Kerr rotation microscopy of a back-gated transistor. V_g, gate voltage. A longitudinal electron current J drives a transverse valley current JV by the VHE, giving rise to a Kerr rotation angle δθ on sample edges

Berry curvature and the valley Hall effect could uninterruptedly tune by the transverse electric field [241–243]. Specifically, the way of valley current flow can be exchanged in the opposite direction under the perpendicular electric field [237, 244].

5.6.4 Nonlinear Valley—Spin Currents Through Trigonal Twisting

The collective characteristic of 2D crystals is the occurrence of the conduction and valence band edges at homogenous extrema in momentum space which is frequently denote as valleys. Their Fermi surface made from the fine distinguishable pouches at the valleys that establish an impressive interior degree of freedom of the carrier. The manipulation of the valley pseudospin and spin in electronics are considerably significant for the device performances [6, 14, 147, 245, 246]. The creation and governance of spin and valley pseudospin currents are extremely valuable for spintronics and valleytronics, which is characterized by a variety of approaches such as spin injection or pumping, and optical injection [247–249]. In terms of time-reversal symmetric property, the spin Hall effect from spin–orbit coupling and valley Hall effect due to inversion symmetry breaking is also considered to be significant approaches [14, 22, 239, 250–252]. In general, the spin or valley Hall current continually associated through the longitudinal charge current, their typical values may greater. Therefore, decadence cannot be avoided when it has a similar rectilinear dependency on the field as the Hall currents.

Thus, the valley and spin currents from the anisotropy of Fermi pockets are the universal features of crystalline solids. This kind of valley and spin currents produced only through the electrical biasing and it exhibited in the second order to the electric field. Moreover, the electric field quadratic dependency allows current rectification for the generation of dc spin and valley currents by employing ac electric field under the absence of net charge current. For an instance, several 2D crystals including TMDs monolayers and trilayers, graphene, and GaSe monolayers are also possessed considerable nonlinear spin and valley currents in their K, Γ, and Λ valleys. Similarly, the monolayer TMDs p–n junction nonlinear valley current due to exceptional circular polarization patterns of the electroluminescence contingent on the alignment of the connection in respect to the crystalline axis. Moreover, the nonlinear valley and spin Seebeck effects are also possible under a temperature gradient that also plays a similar role to electric field and give an increase in the valley and spin currents [253].

Since 2D crystals mirror symmetry in out-of-plane (z) direction, therefore, the Bloch states should have their spin either parallel or antiparallel to the z-axis. If considering a spin-up Fermi pocket in valley A with dispersion $\varepsilon_{A\uparrow}(q)$, where q represents the wave vector measured from A. The corresponding in-plane electric field E, and function $f(q, E)$ is the steady-state distribution function of carriers. In case of $j_{A\uparrow}(E) \neq j_{A\uparrow}(-E)$ current is lacked due to 180° rotational symmetry, while in terms of a time-reversal equilibrium coordination, the current has a complement of $j_{\bar{A}\downarrow}$ due to a spin-down box at the valley $\bar{A}$ of the time-reversal A, with the corresponding energy $\varepsilon_{\bar{A}\downarrow}(q) = \varepsilon_{A\uparrow}(-q)$. So, the Boltzmann transportation expression from the relax time approach is written as $f_{\bar{A}\downarrow}(q, E) = f_{A\uparrow}(-q, -E)$[34]. As the values of $j_{\bar{A}\downarrow}(E) = -j_{A\uparrow}(-E)$ is illustrated in Fig. 5.19a-c, it directly correlates to under suitable conditions Fermi pocket anisotropy that probably exists. Whereas the contributed currents from the time-reversal pair of Fermi pockets have a finite difference $j_{A\uparrow}(E) - j_{\bar{A}\downarrow}(E) \neq 0$, this is typically called valley current and spin current.

The valley current and spin current are usually in the second order of the electric field. Likewise, by employing an electric field toward the x axis (without covering the Hall effect), the longitudinal and perpendicular constituents of $j_{A\uparrow}$, it is usually expressed as [253];

$$\begin{aligned} j^{x}_{A\uparrow}(E) &= \sigma^{xx}_{A\uparrow}E + \sigma^{xxx}_{A\uparrow}E^2 + O\left(E^3\right) \\ j^{y}_{A\uparrow}(E) &= \sigma^{yxx}_{A\uparrow}E^2 + O\left(E^3\right) \end{aligned} \tag{5.13}$$

In case of opposite current spins are vice versa, the charge current of odd function electric field can be obtained from the $j_{A\uparrow}(E) + j_{\bar{A}\downarrow}(E) = 2\hat{x}\sigma^{xx}_{A\uparrow}E^2 + O\left(E^3\right)$, while the even function of the field valley (spin) current corresponds to $j_{A\uparrow}(E) - j_{\bar{A}\downarrow}(E) = 2(\hat{x}\sigma^{xxx}_{A\uparrow} + \hat{y}\sigma^{yxx}_{A\uparrow})E^2$. Moreover, under the application of ac electric field $E_x = Ecos\omega t$, while the DC is equal to zero, and the corresponding valley(spin) current is equal to;

$$j_{A\uparrow}(E) - j_{\bar{A}\downarrow}(E) = (\hat{x}\sigma^{xxx}_{A\uparrow} + \hat{y}\sigma^{yxx}_{A\uparrow})E^2(1 + \cos 2\omega t) \tag{5.14}$$

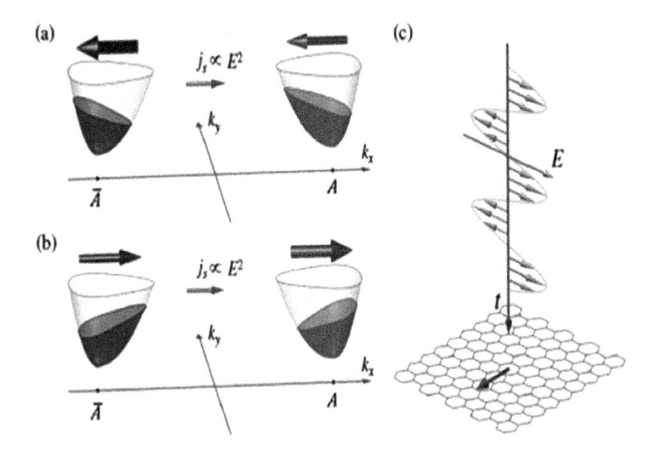

Fig. 5.19 **a, b** Charge carrier allocations of a spin-up Fermi pocket at valley A and a spin-down pocket at valley $\bar{A}$, in an electric field along + x (**a**) or – x (**b**) direction. The anisotropy of the Fermi pocket in a difference in the currents from A and $\bar{A}$, giving rise to a valley (spin) current quadratic in the field. The horizontal arrows correspond to the current from the Fermi pocket A ($\bar{A}$), with the arrow thickness denoting the magnitude; **c** quadratic dependence in the field makes possible the generation of dc valley and spin currents by ac electric field, in the absence of net charge current (adopted from Yu et al. [253])

Thus the addition of the second harmonic term leads to the valley (spin) current dc constituent. The Eq. (5.14) implicitly also leads the ω^{-1} greater to the momentum relax time τ, similar to the steady-state expression in Eq. (5.13), while the rectification effects are even beyond the $\omega\tau < 1$.

In the monolayer TMDs top valence band has a local maximum at Γ and K (– K) points, while the **lowermost** conduction band has two different kinds of native minimal, the first one corresponding to K (– K) point and the second one respect to lower regularity Λ (– Λ) points among the K (– K) and Γ, as illustrated in Fig. 5.20a–e. The Fermi pouches at K (– K) irregularity belong to trigonal warping [23, 45]. This type of anisotropy originated due to breaking the 180° rotational symmetry of the pouches. Owing to conduction and the valence bands spin splitting in the K valleys [22, 25]. When Fermi energy lies in-between bands splitting one can get only spin-up (-down) Fermi pouch at K (– K). Thus, the valley current is almost same as the spin current. Moreover, when the applied field is along a zigzag direction the valley (spin) current is either parallel or antiparallel owing to the reflection symmetry of the Fermi pocket (see Fig. 5.20a). While the application of electric field toward the armchair direction the valley (spin) current is transverse to the field as depicted in Figs. 5.20(d, e). Similarly, trigonal warping also occurs for the holes pouch at Γ point. Such warping may contradictory for the spin-up and spin-down pouches because of time-reversal symmetry (see Fig. 5.20b), which gives increase in nonlinear spin current. Whereas the six inferior symmetry Λ valleys in the conduction band (Fig. 5.20c) is responsible for the anisotropy of valley-dependent current and entire spin current involvement of total Λ and -Λ pouches. The direction of Γ or Λ pouches

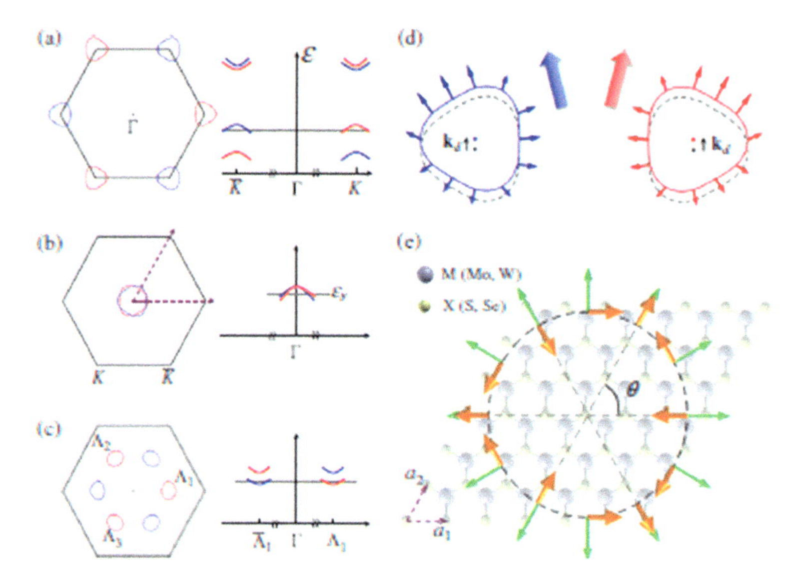

Fig. 5.20 **a** Hole pockets at K valleys; **b** at Γ point; **c** electron pockets at Λ valleys in monolayer TMDs, whereas red (blue) denotes spin-up (down); **d** displacement of K pockets by an electric field in the armchair direction, where the thick red (blue) arrow corresponds to the current from the Fermi pocket K ($\overline{K}$). The valley (spin) current flows perpendicular to the field. The thin red (blue) arrows illustrate the group velocity on the displaced K ($\overline{K}$) valley Fermi surface; **e** dependence of the spin-valley current direction (orange arrow) on the relative angle θ between the field (green arrow) and the crystalline axis (adopted from Yu et al. [253])

spin current as a function of field positioning is also almost alike for the K pouches (see Fig. 5.20e).

Nonlinear spin and valley currents magnitude are also depending on the dispersal of the Fermi pouches and their spreading functionality in the electric field. Therefore, the relaxation time approximation can be used for the K valleys of ML TMDs, by considering the dispersal of the electron or the hole nearby to Fermi surface, it is expressed from the following relationship;

$$\varepsilon_K(q) = \frac{\hbar^2 q^2}{2m^*}\left(1 + \beta q \cos 3\theta_q\right) \tag{5.15}$$

where β represents the weak dependence on the Fermi energy when neglecting the spin and valley relaxations, so, the spin and valley currents are obtained as;

$$J_s = J_v \frac{12\pi}{\hbar} \varepsilon_F \beta |K_d|^2 (\cos 2\theta - \sin 2\theta) + O\left(|K_d|^4\right) \tag{5.16}$$

where $K_d = e\tau E/\hbar$ and τ represents the momentum relaxation time.

The normalized spin and valley charge currents ratio is;

$$J_s/J_c = J_v/J_c = 3\beta e\tau E/\hbar \quad (5.17)$$

Moreover, the emerging monolayer and multilayer TMD p–n junction devices [254–258] are also useful for an ideal laboratory exploration of the nonlinear valley and spin currents. Like, forward bias electrons (holes) from the K valleys in the n (p) region reach up to the junction and produce electroluminescence (EL). When the junction is only toward to armchair direction then the valley (spin) current is collinear to the charge current, as the consequence, carriers are accumulated in the junction region due to the valley being polarized. Such as the valley-dependent optical selection rule in which EL is mostly due to circular polarization [253]. Whereas in case of junction does not along in the armchair direction, it has distinguished mechanism of EL polarization [22, 259]. Another case is when a p-n joint toward the zigzag direction in which the valley (spin) current is transverse to the charge current, then the charges could accrue associating contradictory valley polarizations at the dual distinct sides, therefore, EL has a conflicting circular polarization. Thus the three-dimensional dependency distinguishes the non-native valley transportation which become due to impact of the native carriers in the inhabitants of recombined electrons and holes through the electric field in the depletion area [259]. This effect is also considered to be valuable to differentiate the valley Hallcurrent [14, 236, 259]. Because in the p-n joint EL polarization on the either sides should alter the sign when it becomes due to the valley Hall effect, on the other hand, when it arises nonlinear in that case the valley current remains unchanged.

Similarly, a second-order nonlinear effect also exists with the temperature gradient ∇T by revealing a pure valley (spin) current. For the instance, in the case of $T/|\nabla T|$ much greater to mean free path, the ratio of the valley and carrier currents is generally obtained from the following expression;

$$j_v/j_c = \frac{6}{\hbar}\alpha\beta k_B|\nabla T|\tau \quad (5.18)$$

Here α is the dimensionless coefficient.

The valley and spin current's quadratic dependency on temperature gradient gives us the generation of the pure valley and spin flows. However, in the case of arbitrary inhomogeneous temperature distribution, temperatures at the two ends are expressed as the zero charge current, but a finite valley (spin) current.

Similarly, third harmonic generation is also possible in TMDs. In contrast to the second harmonic generation third harmonic generation does not require the absence of inversion symmetry and it can be observed both in odd- and even-layers. Usually in the third harmonic generation signal increase with the layer numbers, therefore, a thicker sample can have a longer light–matter interaction length [260]. It has a cubic dependence on the excitation power and its angle dependence is different from the second harmonic generation. The ML TMDs with D_{3h} space group third harmonic generation polarization signal is always the same with the excitation polarization and independent of the orientation of the crystal [261, 262]. Additionally, it could also vanish under circularly polarized light excitation [261, 262]. In contrast, in the second harmonic generation, the signal cannot vanish but decrease under circularly polarized light excitation compared to linearly polarized light excitation [261].

Moreover, four-wave mixing is also recognized in TMDs [263], to get it a pump and probe beam have been used by defining the difference between $\omega_{FWM} = \omega_{pump} - \omega_{prob}$. The four-wave mixing signal intensity usually increases with layer numbers for less than 13 layers, likely to third harmonic generation. However, four-wave mixing signal intensity is decreased exponentially when the number of layers is more than 13, due to higher reflection of incidence beam and higher reabsorption in a thick sample. The four-wave mixing polarization dependence is more complicated than the third harmonic generation owing to the two beams' involvement. Therefore, a dumb-bell shape response is possible while rotating the pump beam or the probe beam, thereby, a sixfold pattern has been achieved from the rotation of the pump and probe beam simultaneously. The four-wave mixing imaging has also combined with the second harmonic generation imaging to evaluate the crystal quality and the layer numbers. Additionally, the chemical reaction of thermal oxidation has been also monitored in situ by changing the four-wave mixing intensity.

Four-wave mixing microspectroscopy is also useful for the study of dephasing and exciton dynamics of MoS_2 in a resonant manner [264]. Specifically short pulses are arranged with a certain time delay to create a four-wave mixing pulse that serve as a probe beam to resolve the dynamics of the neutral exciton and trion. Four-wave mixing microspectroscopy are also capable to provide a high time resolution that's only limited by the pulse duration. Therefore, its resonant excitation provides more information regarding the relaxation process.

The high-harmonic generation is also possible in TMDs, which has a great importance in the strong-field and attosecond physics. As the fourth-order harmonics and high-harmonic generation up to 13th order in MoS_2 [260, 265]. In contrast to the perturbative limit where the nth order signal can have a relation with the incidence intensity of I^n order. The high-harmonic generation intensity is depending on the excitation intensity with $I^{3.3}$ for all high orders. The non-perturbative high-harmonic generation in MoS_2 is attributed due to intraband and interband electron transition [265–267]. The high-harmonic generation in ML TMDs has enhanced compared to bulk owing to stronger electron–hole interaction in monolayers. The high-harmonic generation selection rule of the ML TMD depends on their symmetry [265]. Moreover, the moiré patterns with strong electron correlation in TMD heterostructures also considered to be a promising feature for a high-harmonic generation [268–272].

5.7 Valley–Spin Concept for Bi-Layers

Since the AB-stacked bilayer, TMDs are composed of two layers rotated 180 degrees to each other. It enables to an additional degree of freedom that is typically called layer pseudospin. Their localization of charges on the top (bottom) layer is generally indexed by the layer pseudospin up (down). Moreover, the interlayer hopping and strong spin–orbit coupling causes the interplay between spin, therefore, the valley and layer pseudospin are possible in TMD materials [273, 274]. The valley-dependent spin–orbit coupling changes the sign in between two layers, so three degrees of freedom are coupled together, as a consequence several interesting optical and magnetoelectric effects appears [274–277]. Spin and valley correspond to magnetic moments and the layer pseudospin are usually correlated to an electric field [140, 274]. Thereby, the magnetoelectric transport with the different degrees of freedom over a possible direction toward potential device concepts. This kind of potential pseudospin of electrons usually noticed in pseudospintronics that is utilized in adjustable and considerable pseudospin polarization. Particularly the ability to feasible control the pseudospin polarization over a large energy range is considered to be a significant step in the development of pseudo-spintronic devices [140, 278, 279]. However, different degrees of freedom in bilayer TMDs are normally managed via dynamical approaches, such as magnetic and optical means [274–276, 280–282]. Including static means by gate electric field having advantages of compactness, power efficiency, and compatibility with the existing semiconductor technology [237, 283–287].

According to large-scale density functional theory estimations, the van der Waals density (vdW) intermediate-range interaction is due to the semilocal exchange term [288]. Like twisted bilayer of WSe_2 lattice relaxation have a significant impact on bands. The DFT calculation has also led at $\theta = 5.08^o$ around 762 atoms lie in each unit cell with a substantial alternation of the layer space d in distinct area, as depicted in Fig. 5.21a, b. As per representation, the $d = 6.7$ Å has a minimum stacking regions (MX and XM), in which metal atoms on the top layer are aligned and chalcogen atoms on the bottom layer and vice versa, whereas $d = 7.1$ Å is the largest in MM area in which metal atoms aligned on the layers. Like, an entirely stress-free structure at the lower energy moiré valence bands of twisted bilayer WSe_2 have been occurred at the $\pm$ K valley, as illustrated in Fig. 5.21c. Moreover, a continuum model depending on an effective mass explanation give us the construction of moiré bands in terms of three-dimensionally modulated interlayer tunneling $\Delta_T(r)$ and layer-dependent potential $\Delta_{1,2}(r)$. So, the Hamiltonian for the $\pm K$ valley bands can be expressed as [289];

$$H_\uparrow = \begin{pmatrix} -\frac{\hbar^2(\mathbf{k}-k_+)^2}{2m^*} + \Delta_1(r) \Delta_T(r) \\ \Delta_T^\dagger(r) \quad -\frac{\hbar^2(\mathbf{k}-k_-)^2}{2m^*} + \Delta_2(r) \end{pmatrix} \tag{5.19}$$

Similarly its time conjugate $H_\downarrow$ has also defined. Since the continuum approximation is holds only for little twist angles whereas the moiré wavelength is big sufficient. Therefore, the atom arrangement in any native area of a twisted bilayer can be similar to an untwisted bilayer in which single-layer laterally moved respect to the other consistent movement vector d_0. As the instance $d_0 = 0, -(a_1 + a_2)/3, (a_1 + a_2)/3$ has $a_{1,2}$ primitive lattice vector for the monolayer, in the corresponding regions MM, MX, and XM. So, the moiré potentials of twisted TMD bilayers $\Delta_T(r)$ and $\Delta_{1,2}(r)$ are generally verified to the valence band boundaries of the untwisted bilayer as a function of the respective shift vector [290]. According to the lowermost harmonic

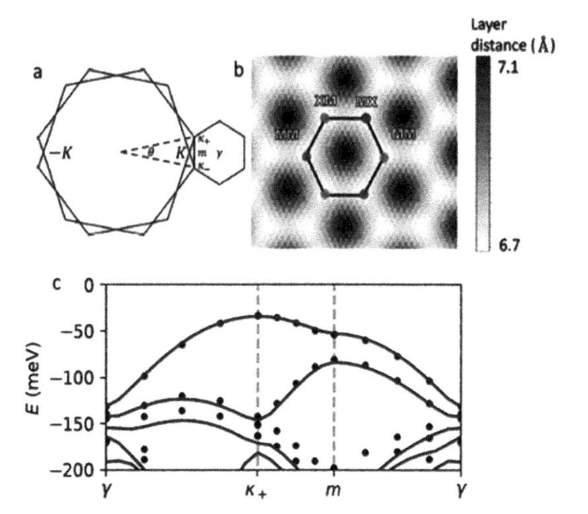

Fig. 5.21 **a** The κ± points of the moiré Brillouin zone that shaped to the K points of the monolayer Brillouin zones, which are rotated by ± θ/2; **b** the interlayer space of the twisted WSe_2 structure as per DFT to establish a big alternation among the MM and XM/MX areas; **c** the continuum band structure (blue lines) plots compared to large-scale DFT calculations (dots) at twist angle θ = 5.08°, with an outstanding settlement (adopted from Devakul et al. [289])

approach $\Delta_T(r)$ and $\Delta_{1,2}(r)$ sinusoids form that interpolates among the MM, MX, and XM areas are expressed as [175];

$$\Delta_{1,2}(r) = 2V \sum_{j=1,3,5} \cos(g_j.r \pm \psi) \tag{5.20}$$

$$\Delta_T(r) = w\left(1 + e^{-ig_2 r} + e^{-ig_3 r}\right) \tag{5.21}$$

Here g_j belongs to the $(j-1)\pi/3$ counter-clockwise rotations of the moire reciprocal lattice vector $g_1 = \left(4\pi\theta/\sqrt{3}a_0, 0\right)$, whereas a_0 is the lattice constant, thus, throughout the energy parameter of the continuum approach value depending solitary on their dimensionless factor $\alpha \equiv V\theta^2/\left(m^* a_0^2\right)$, w/V and ψ. The applicable DFT calculation based on this approach for the untwisted bilayers gives us the values of the quantities $V = 9.0$ meV, ψ = 128°, and $w = 18$ meV with the significant interlayer tunneling forte w value almost double, that is larger than the earlier noted one [175], as typical schematic is demonstrated in Fig. 5.21c.

Instead of several theoretical interpretations [289, 291–293], the significant large-scale DFT band is depending on similarities of the eigenvalues. The isolate bands of ±K valley have been computed in form of their C_{3z} eigenvalue for the highly symmetric momenta γ, κ±, ultimately it provides the Chern number [294]. Typically initial twice bands' C_{3z} eigenvalues are listed in Table 5.3, in terms of non-trivial valley Chern number $C_{K,1} = C_{K,2} = 1$.

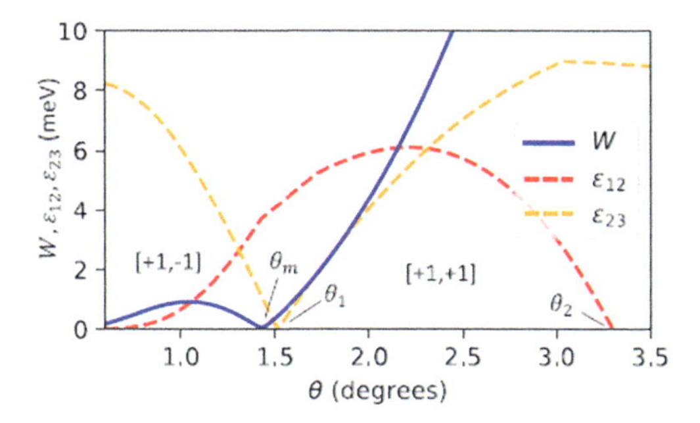

Fig. 5.22 The bandwidth of the first band W and the indirect band gap between the first two pairs of bands ε_{12} and ε_{23}. The bandwidth is minimized at θ_m while being well separated from the remaining bands. The Chern numbers of the first two bands, $\left[C_{K,1}, C_{K,2}\right]$, before and after the ε_{23} gap closing at $\theta = \theta_1$. For $\theta \geq \theta_2$, ε_{12} vanishes (adopted from Devakul et al. [289])

Using continuum approach theses parameters have been established by defining the lattice constant $a_0 = 3.317$ Å and effective mass $m^* = 0.43\ m_e$ [42–44], to calculate the band structure of the twisted bilayer of WSe_2. The various twist angles the $E_i(k)$ values have been defined in a study [289]. The corresponding first bandwidth has been achieved by $W = \max_k E_1(k) - \min_k E_1(k)$, and their direct or indirect band gaps ε_{ij} among the couples of the bands $(i, j) = (1, 2)$ and $(2, 3)$, $\varepsilon_{ij} = \min_k E_1(k) - \max_k E_1(k)$ as depicted in Fig. 5.22. Moreover, the topological characters of the initial two valence bands are also separated into three key regimes depending on the values $\theta_1 = 1.5°$ and $\theta_2 = 3.3°$.

In the case of $< \theta_1$, the topmost double bands are decently detached to remaining spectrum, and their contradictory Chern number is $\left[C_{K,1} C_{K,2}\right] = [+1, -1]$, but the corresponding first bandwidth "$W < meV$" may very small. So, corresponding twist angles behaviors of the topmost dual valence bands could understand with the help of the effective tight-binding approach of a moiré honeycomb lattice following the Kane–Mele model [175]. Moreover, at $\theta = \theta_1$, the band gap $\varepsilon_{2,3}$ could be close, the corresponding Chern number of the topmost dual bands changes $[+1, +1]$. Whereas, in the case of the $\theta_1 < \theta < \theta_2$, the dual topmost bands have the identical Chern number $[+1, +1]$, but they are separable from the sizable gap $\varepsilon_{12}, \varepsilon_{2,3} > 0$. Their first band gaps increases rapidly with the increasing θ values. Furthermore, for the $\theta_2 < \theta \lesssim 5.4°$, the indirect gap ε_{12} to be vanished, however the direct gap still sustain. The corresponding Chern number of the topmost dual bands respect to [+1, +1], moreover their bands overlaps in terms of energy that extremely dispersive. Hence, in the second and third orders $\varepsilon_{2,3} > 0$, the topmost dual band are formed a gapped $C = 2$ multifold. While the outside $\theta \gtrsim 5.4°$, the gap $\varepsilon_{2,3}$ has vanished and the corresponding topmost dual bands not lengthier remotely. As per the continuum approach topological value $\theta \approx 5^o$ is considered to be typical for the large-scale parameter evaluations [289].

In the case of the $\theta < \theta_2$, specifically near to θ_2 the primary band is supplementary dispersal together the spin Charn number $C = 1$ and $\varepsilon_{12} > 0$, it makes them favorable to quantum spin Hall insulators for the filling $n = 2$ holes each moiré unit cell. Additionally, the extensive variety of angles $\theta_1 < \theta \lesssim 5.4°$,$\varepsilon_{2,3} > 0$ and the topmost bands spin Chern number is possible around $C = 1$, it provide us to the possibility of a twofold quantum spin Hall state associating twin sets of counter-propagating spin-polarized edge at $n = 4$.

Similarly, parameter bandwidth W has also determined by interpreting the large and small range of the angles, using the analytical expression for the magic angle dispersion near γ. According to this approach, it has assumed that the bandwidth minimum is nearby the angle at which $E_1(\gamma)$ crossover from the minimum to maximum, so, the corresponding expression is expressed as;

$$E_1(\gamma + k) \approx E_1(\gamma) + \frac{k^2}{2m_y} + \mathcal{O}(k^3) \tag{5.22}$$

Thus the effective mass m_y should be diverged near the crossover. Further, suppose if θ_m be the angle at $m_y^{-1} = 0$, so we can consider only the six most relevant states at γ, which is expressed by the following relationship;

$$\theta_m^{-2} = \frac{8\pi^2}{9m^* a_0^2}\left(\frac{1}{\varepsilon_{n_0} - \varepsilon_{n_0+1}} + \frac{1}{\varepsilon_{n_0} - \varepsilon_{n_0-1}}\right) \tag{5.23}$$

where $\varepsilon_n = 2w\cos(\pi n/3) + 2V\cos(2\pi n/3 - \psi)$, and n_0 is the integer. Thus with help of Eq. (5.23), we can evaluate θ_m value.

Likely other parameters such as the Berry curvature $\mathcal{F}(k)$ of the bands (top band) and their first Chern number $\left(C_{K,1} = \frac{1}{2\pi}\int_{BZ} \mathcal{F}dk\right) = 1$, including energy distribution above, below, and near the critical angle has been also evaluated. In a similar fashion polarization under the broken inversion symmetry are also defined at k [289].

5.8 Inversion-Symmetry-Induced Spin Polarization

Under the non-occurrence of a magnetic field, the spin degeneracy of the band structure can be preserved in the bulk materials (TMDs), it is commonly known as inversion symmetry. Usually, it is assumed that a finite spin polarization only arises in non-centrosymmetric materials [295]. Since spin–orbit coupling (SOC) **exists in structures** associating satisfactorily lower crystalline equilibrium in spatial unpackaged sequential solids, typically it is known as the Dresselhaus effect (usually signified as D-1), which arises due to bulk inversion asymmetry [296]. Whereas the 2D quantum wells and heterostructures refer to the Rashba effect that arises due to 2D structural overturn irregularity [297]. Significant Rashba-like spin consistency in few 3D bulk crystals (usually signified as R-1) is considered to be important because it cannot play a role. Thus, both R-1 and D-1 are basic requirements for the nonexistence of crystal bulk overturn regularity, but their expectation too existence in the spin breaking of a 3D material that has bulk overturns regularities. Because SOC is an actual impact that is associated to specific nuclear sites in a solid [298]. Moreover, it is also symmetric of discrete atomic sites in the solid to form a decent preliminary point to explain the SOC persuaded spin polarization consequence, somewhat to the global symmetry of the unit cell. Therefore, the entire spin polarity of a crystal is the vector summation whole sites of the indigenous site (Rashba or Dresselhaus) spin polarities [296, 297, 299, 300].

But this explanation has missed dual previous procedures of spin polarities in nonmagnetic materials (usually called R-2 and D-2). Because the site point group of an atom in a 3D crystal missing the overturn regularity, the atomic sites whichever non-polar or polar that has whichever an overturn irregular indigenous surroundings or a native surroundings creates a site dipole field [301]. As the atomic SOC sites revealed a local Dresselhaus spin polarization and site dipole field lead an indigenous Rashba spin polarization, the incorporation of such sites regularities gives the global bulk space group lacked overturn regularity, as depicted various forms in Fig. 5.23 [296, 297]. Thus the R-1 and D-1 effects are generally categorized from the total spin polarizations associating features of the spin textures. Usually, R-1 has a helical texture in which spin positions are transverse to the dipole field axis, while D-1 belongs to a nonhelical texture [302–304]. So, in bulk R-1 and D-1 effects begins due to overlapping of the relevant site when the bulk space group of a solid has overturn regularity (centrosymmetric), this is also called hidden spin polarization (Fig. 5.23c). In which a layered crystal has separate layers ('sectors') possessing an indigenous regularity to produce a dipole field (a native Rashba spin polarization). Nonetheless, in the case of the entire crystal it has a space group consisting overturn regularity, therefore, Rashba spin polarization to every indigenous subdivision (α) are compensated by the supplementary subdivision of its overturn companion (β). It is essentially inattentive usually underneath the current paradigm, this is called R-2 spin polarization due to concealed by compensation.

To differentiate the existence or nonexistence of overturn regularity in the bulk space group (centrosymmetric and non-centrosymmetric), it should be clarified, because such inversion symmetry is due to the atomic site point group. The latter subgroups are construct by the subset of all symmetric processes of the crystal space group and leaves the atomic site invariant [305]. Since the total spin polarization of a bulk crystal is the summation of entire atomic sites of the indigenous spin polarities, so, it can be classified into three types of spin polarization considering diverse connection of bulk space group and site point group, as illustrated in Fig. 5.23. If all the site point groups belongs to overturn regularity (such as C_i, C_{2h}, D_{2h}, C_{4h}, D_{4h}, S_6, D_{3d}, C_{6h}, D_{6h}, T_h, and O_h) having no SOC persuaded indigenous spin polarity, hence the bulk spin polarization is vanish. However, their spin polarization can be governed by alteration the atomic organization [306].

Moreover, the **total** bulk spin polarization: R-1 and D-1 effects can be interpreted by considering their energy band's non-zero net spin polarization and combinations of bulk inversion asymmetry of the site point group's macroscopic spin polarization

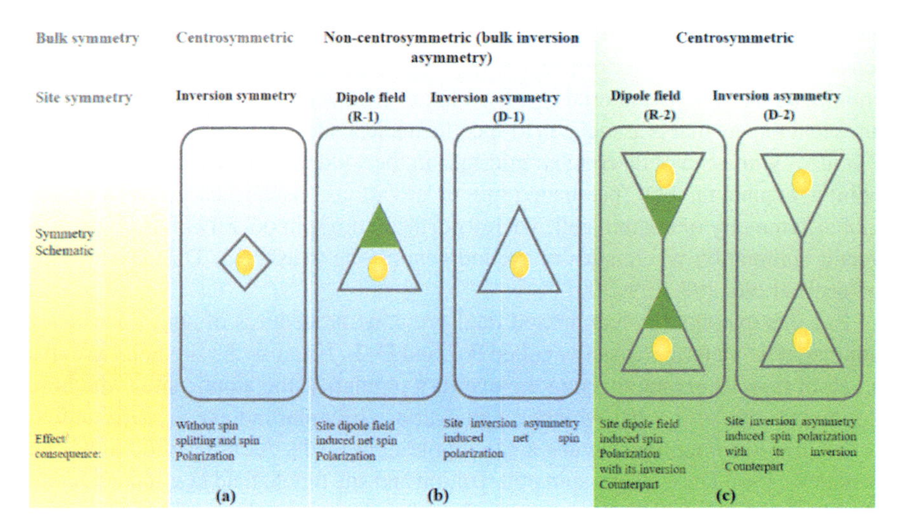

Fig. 5.23 **a** Representation of nonexistence of spin polarity in centrosymmetric crystals for all atomic sites are overturn symmetric. As the indigenous situation (crystal field) of centrosymmetric atomic sites does not create spin–orbit combination persuaded spin polarization, the total (bulk) spin polarity is lacking as well; **b** the total bulk spin polarizations (R-1 and D-1 effects): a native site dipole field or the site overturn irregularity gives to indigenous Rashba or indigenous Dresselhaus effects. The combined non-centrosymmetric space group, these indigenous effects developed bulk R-1 (Rashba) and D-1 (Dresselhaus) effects; **c** the compensated (hidden) bulk spin polarity (R-2 and D-2 effects): a native site dipole field or the site overturn irregularity provides the indigenous Rashba or indigenous Dresselhaus effects. In case together a centrosymmetric space group, these indigenous effects produce bulk R-2 (Rashba) and D-2 (Dresselhaus) effects. Here the spin polarity from every sector is hidden by compensation from their overturn associates but is willingly visible when the consequences from separate subdivisions are pragmatic

effects, as listed in Table 5.4. The D-1 effect is associated with a non-centrosymmetric space group (bulk inversion asymmetry), in which either all sites have non-polar point group symmetry (D_2, D_3, D_4, D_6, S_4, D_{2d}, C_{3h}, D_{3h}, T, T_d, and O), or few sites belong to polar point group symmetry (C_1, C_2, C_3, C_4, C_6, C_{1v}, C_{2v}, C_{3v}, C_{4v}, and C_{6v}), including all the dipoles site add up to zero, as the instance zinc-blende structure, InAs, GaSb, and GaAs, etc. [307]. Whereas the R-1 effect is always accompanied by the bulk crystals due to the D-1 effect. They originate due to a combination of a non-centrosymmetric space group (bulk inversion asymmetry) with the site dipole field. At least one site should have a polar site point group (C_1, C_2, C_3, C_4, C_6, C_{1v}, C_{2v}, C_{3v}, C_{4v}, and C_{6v}) and individual dipoles add to a non-zero value. Because a polar point group contains simultaneously a site dipole field and site inversion asymmetry. Since the R-1 effect in bulk crystals is always accompanied by the D-1 effect with a system-dependent relative magnitude. Therefore R-1 effect occurs in such types of bulk structure: ZnS-wurtzite ($P6_3mc$), GeTe (R3m), LiGaGe ($P6_3mc$), and BiTeI (P3m1). As the typical α-SnTe R-1 and D-1 effects at both the conduction bands (CB_1, CB_2) and valence bands (VB_1, VB_2), as illustrated in Fig. 5.24a–f. [306]. As per the representation of the basic crystal structure of α -SnTe (space group R3m), it has polar point groups C_{3v} corresponding to both Sn and Te sites. This phase is usually correlated to R-1 and D-1 classes. However, the optimized calculated band structure are revealed the R-1 and D-1 effect which lift the spin degeneracy among CB_1, and CB_2 and the VB_1 and VB_2, with the noteworthy (~ 300 meV) spin splits. The obtained spin surfaces from the entire atoms in the unit cells reflect a strong spin polarization. The 2D view of the CB_1 and VB_1 band's spin textures are found to be with helical pattern formation in-plane, which reflects the dominance of R-1 to D-1. Another class of compensated (hidden) spin polarization R-2 and D-2 effects belongs to crystal structures in which inversion symmetry exists for the bulk space group; nonetheless it is not for the site point groups. Because their sites bring whichever an indigenous dipole field (R-2) or a site overturn irregular crystal field (D-2). An adjoining of the bulk centrosymmetric space group together a site dipole field provides the bulk R-2 effect, while the adjoining of a global centrosymmetric space group together site overturns irregularity consequences in the bulk D-2 effect [306].

The experimental realization of inversion symmetry polarization in TMDs such as WSe_2, $MoSe_2$, ZrSiTe, and other 2 D materials has been also recognized in various studies, by interpreting different existing theoretical concepts based on inversion-induced polarization in these kinds of materials [295, 299, 300, 308–310].

Table 5.4 Cataloguing of spin polarization in nonmagnetic bulk materials **depending** on bulk space group and site point group [306]

Site point group Bulk space group	Non-centrosymmetric (at least one site)			Centrosymmetric (all sites) (C_i, C_{2h}, D_{2h}, C_{4h}, D_{4h}, S_6, D_{3d}, C_{6h}, D_{6h}, T_h, O_h)
	Non-polar (all sites) (D_2, D_3, D_4, D_6, S_4, D_{2d}, C_{3h}, D_{3h}, T, T_d, O)	Polar (at least one site) (C_1, C_2, C_3, C_4, C_6, C_{1V}, C_{2V}, C_{3V}, C_{4V}, C_{6V})		
		Dipoles add up to zero	Dipoles add up to non-zero	
Non-centrosymmetric (for example, F $\bar{4}$ 3 m)	**a** D-1 Example: GaAs, ZrCoBi	**b** D-1 Example: γ-LiAlO2	**c** R-1 & D-1 Example: BiTeI, α-SnTe	Not possible (Site point group cannot be centrosymmetric if space group is non-centrosymmetric)
Centrosymmetric (for example, R $\bar{3}$ m)	**d** D-2 Example: Si, NaCaBi	**e** R-2 & D-2 Example: MoS2, Bi2Se3, LaOBiS2		**f** Absence of spin polarization Example: β-SnTe

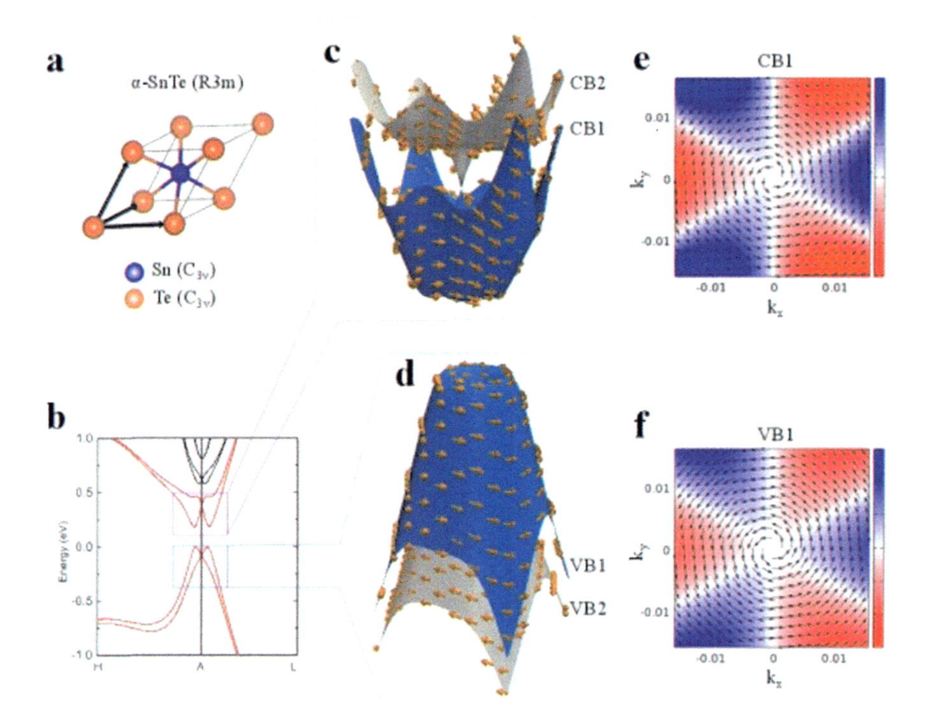

Fig. 5.24 Rhombohedral α-SnTe (*R3m*) with R-1 (dominant over D-1) spin textures: **a** crystal structure and site point groups of each atomic site; **b** band structure around the A point along the A–H and A–L directions in the Brillouin zone. The two lowest conduction bands (CB1 and CB2) and two highest valence bands (VB1 and VB2) represent in red; **c**, **d** spin polarization (contributed by the entire unit cell) represents from the brown arrows for the states of spin-split CB1 C CB2 and VB1 C VB2 bands in the k-plane containing the A–H and A–L directions near the A point that indicated by solid boxes in **b**; **e**, **f** corresponding 2D diagram of the spin polarizations of CB1 and VB1. The arrows indicate the in-plane spin direction and the color scheme for the out-of-plane spin component. The magnitude of each spin vector is renormalized to unity for simplicity unless otherwise noted (adopted from Zhang, X. et al. 2014, arXiv:1402.4446)

5.9 Optics-Induced Stark Effect

Since the broken inversion symmetry can give rise to the valley-selective electron populations and the valley pseudospin in monolayer transition metal dichalcogenides (TMDs) [22, 68]. The controlled valley degree of freedom can have implications for both spintronics and the emerging field of valleytronics. The circularly polarized light breaks the valley degeneracy and allow selective tuning of the exciton energy levels in each valley under the valley-selective optical Stark effect (OSE)[311, 312]. Otherwise, breaking the valley degeneracy requires a large magnetic field through the valley Zeeman effects [24, 313]. Thus, a circularly polarized light (or laser pulses) under the optical Stark effect (OSE) provides a large effective magnetic fields to break the spin degeneracy on sub-picosecond timescales and offer an all-optical alternative that enables more precise control over the quantum states[314, 315]. The OSE has been also considered a powerful tool to manipulate valley pseudospin in transition metal dichalcogenides (TMDs). As the pseudospin in monolayer TMDs arises from the K and K′ valleys in the band structure, however, usually they are degenerate in energy but have a separation in momentum space [140]. Under the optical transitions each valley governs by the polarization-sensitive selection rules, so the ultrafast optical Stark shifts are also used to break the valley degeneracy and to control the valley pseudospin in excitons [312, 316]. Hence the OSE is considered to be a coherent interaction between electronic energy levels in matter, and a pump photon field may results shift in their transition energies [317]. Such coherent interaction enhances by the strong light-matter couplings, such as in quantum wells, quantum dots, atomically thin semiconductors, and bulk organometallic perovskite films [318–320].

Normally, the OSE is estimated by a nonresonant photon field by avoiding the dissipation and incoherent influences related to excited state inhabitants. In cases faraway to the resonance the greater deceitful excited states are expressively insignificant and such material considered being a two-level system. Therefore, conventionally OSE has anticipated on the basis of the dressed-atom picture [321]. In this case, the coherent interattraction of the ground and excited states to the light field increases, typically it is known as the Floquet states. Such hybrid light-matter states are consisting electronic energy levels (ground and exciton states $|g\rangle$ and $|x\rangle$) that dressed from a photon $|g \pm h\upsilon\rangle$ and $|x \pm h\upsilon\rangle$. These short-lived Floquet states interrelated to the equilibrium matter states via the Coulombic potential and hybridized levels eventually it gives us the energy-level repulsion. This situation has been treated as the excited states of non-interrelating atoms that is significant to employ on the lower dimensional semiconductors for the reduction of the electrostatic screening and robust many-body interactions.

As in a typical example the valley-selective optical Stark effect for the excitation nearby A exciton resonance in single-layer WS_2 [322]. This interpretation leads together underneath and overhead resonance pumping associating positive helicity pulses (σ^+) have produced a blue shift of the K-valley in single-layer WS_2. Whereas no significant alteration has been remarked in the K-valley energy associating negative helicity pulses (σ^-) pumping. Thus, the photo persuaded absorption has also accompanied to the coherent formation of intervalley biexcitons. Additionally, it has been also remarked that a valley-selective optical Stark effect is feasible only the underneath resonance and it is qualitatively consisting to dressed-atom representation.

Typical photo persuaded variations of the K and − K-valleys in the single-layer WS_2 absorption spectrum is depicted in Fig. 5.25a. The chiral optical selection rules in WS_2 has been exploited by employing a circularly polarized pump excitation pulse to drive dynamics in a specific valley and a circularly polarized white-light probe pulse to achieve alternations persuaded in the valleys [1]. The advantage of this technique is to get time-dependent absorption spectra of every valley that allowing to straight measurement just after the photoexcitation. It also provides a direct observation of reduction in absorption associating to state-filling photo-induced absorption

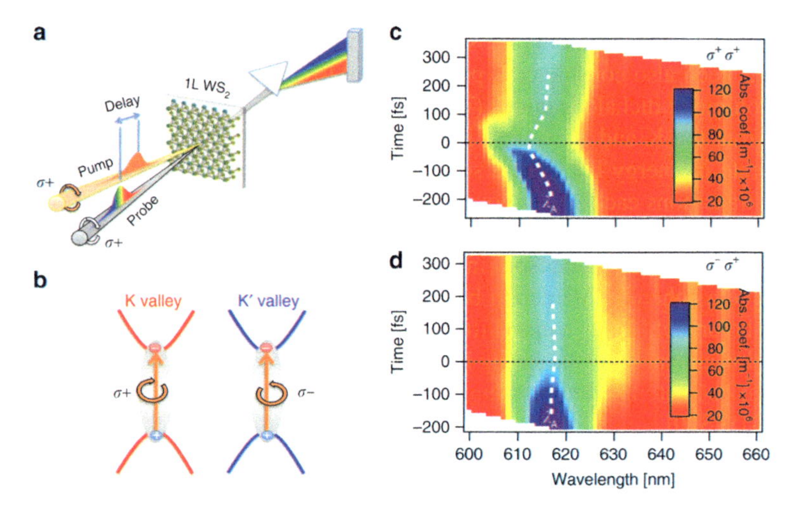

Fig. 5.25 Valley-selective optical Stark effect in monolayer WS_2: **a** schematic for the valley-resolved transient absorption experiment; **b** the K- and K′-valleys are populated by right-hand (σ^+) and left-hand (σ^-) circularly polarized light due to chiral selection rules; **c** summary of time-dependent, room temperature, absorption spectra of monolayer 1L WS_2 with a σ^+ white-light probe upon 610 nm excitation with co-($\sigma^+\sigma^+$); **d** the cross-circularly polarized ($\sigma^-\sigma^+$) light. The center wavelength of the A exciton resonance (λ_A) is represented by a dashed white light. Time zero indicates whiles the pump and probe pulse temporally overlapped (adopted from Cunningham et al. [322])

with the excited state conversions and variations in absorption line width including spectral alterations in energy levels with the OSE, dynamic of Coulombic screening and lattice heating [312, 323–326]. But the time-resolved Kerr rotation has considered an extra sensitive method because it has ability to resolve < 1 meV spectral shifts to nearby resonance excitation, therefore, large optical Stark shifts could be effortlessly dignified with the transient absorption [327].

Moreover, the room temperature valley-resolved spectrum of single-layer WS_2 with the photoexcitation wavelength of 610 nm is illustrated in Fig. 5.25 c, d. The obtained spectrums are just above the A exciton resonance positioned at 616 ± 1 nm. It demonstrates that the pump and probe pulses are the right-hand circularly polarized ($\sigma^-\sigma^+$). Their K-valley absorption blue is limited within the instrumental capability (see Fig. 5.25c). It has been also noticed that the blue shift exists only for the zero time while the pump and probe temporally overlay. This leads to the apparent shift due to coherent light-matter interconnections among the pump pulse and the remaining specimen somewhat incoherent procedures related to the photoexcited excitons or charge carriers. Whereas, there is not any important shift has been noticed for the cross-circular polarizations ($\sigma^-\sigma^+$) K-valley. This means a left-hand circularly polarized pump coherently governs by the − K-valley, as depicted in Fig. 5.25d. This is a reliable case of a valley-selective OSE in WS_2 [4]. In which the applied field $\sigma^+(\sigma^-)$ pulses are possibly coupled choosily to the K^- ($-$ K^-) valley. Whereas the driving the –K valley cannot have an impact on the K-valley energies. Such coherent light-matter interconnection is usually well lonely to a single valley, therefore, one can get only an optical Stark shift for circularly polarized light. However, a slight residual blue shift (~ 1 meV) may persist for the picoseconds afterward photoexcitation association with the bandgap renormalization to the photoexcited exciton inhabitants [328]. That is an incoherent procedure and it unconnected to OSE, results from a considerably leisurelier timescale.

The cross-circular polarizations of σ^- σ^+ can give a photo-induced absorption band of the A exciton resonance (see Fig. 5.25d), as the consequence, the coherent intervalley biexciton generation probably in the single-layer WS_2. Because in TMDs strong Coulomb many-body interactions gives the construction of steady biexcitons [329], due to degenerate K and –K valleys and minor spin–orbit splits of the conduction band. Therefore, many biexcitons are probable in TMDs between the intra and intervalley for the charged and neutral and combinations of bright and dark excitons. The neutral intervalley biexciton generation schematic is depicted in Fig. 5.26. The σ^- pumped photon allows an exciton in K valley and following σ^+ probed photon has been accessed from the |-xK⟩ → |xK, -xK⟩ transition. This is a general process for single-layer TMD semiconductors. The measured photo-induced absorption characteristics leads it position around 48 ± 7 meV underneath the A exciton [328]. Since this kind of absorption band is also **close** to the emission band in the case of the charged biexcitons [329–333], however, some converse situations are also possible in the different kinds of TMDs such as encapsulated in hBN and graphene [334].

Theoretically, OSE in TMDs describes by the different models, one of the basic models is the dressed-atom model of the optical Stark effect. In which exciton states are considered to be non-interacting atom-like states [311, 312]. The corresponding ground state, |g⟩, and A exciton, |x⟩, the hybridized Floquet (i.e., photon-dressed) states, |g + hυ⟩ and |x – hυ⟩ are the light-matter coupling. The frequency associated

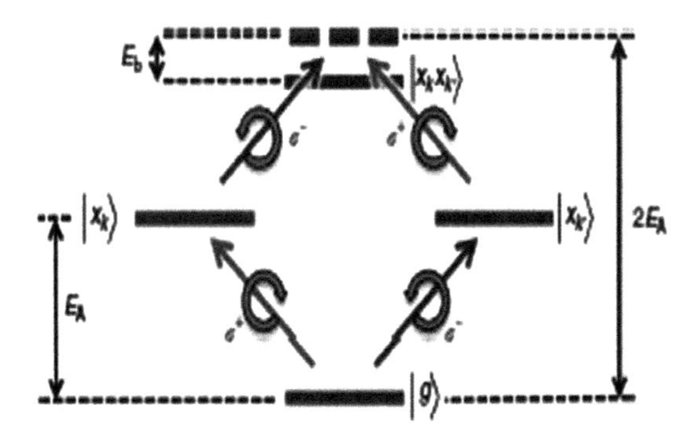

Fig. 5.26 Formation of intervalley biexciton from sequential photon absorption: a right-hand (σ^+) or left-hand (σ^-) circularly polarized pulse raises the system from the ground state, |g⟩, to populate either the K- or K′-valley with an exciton, $|x_K\rangle$ or $|x_{K'}\rangle$ respectively. A second photon of opposite chirality can then transition the system from the valley exciton to the intervalley biexciton state via the process $|x_K\rangle \rightarrow |x_{K'}x_K\rangle$ or $|x_K \rightarrow |x_K x_{K'}$, this bound intervalley biexciton is reduced by its binding energy, E_b, to below twice the exciton energy E_A (adopted from Cunningham et al. [322])

with the light-matter coupling $\nu_R = M_{gx}\mathcal{E}/h$ and their energy eigenstates are usually expressed from the following relationship [322];

$$E_{|x\,{}^{\text{upper}}_{\text{lower}}} = E_x - \frac{\Delta}{2} \pm \frac{1}{2}\sqrt{\Delta^2 + 4V^2} \tag{5.24}$$

$$E_{|g\,{}^{\text{upper}}_{\text{lower}}} = E_g - \frac{\Delta}{2} \pm \frac{1}{2}\sqrt{\Delta^2 + 4V^2} \tag{5.25}$$

Here Eqs. 5.24 and 5.25 reflects the quantum mechanical connection among the symmetry matter and photon-dressed states that reveals the energy-level repulsion by showing the construction of superior and inferior exciton-polariton bands. Whereas the energy shift of the exciton resonance has been described as;

$$\Delta E = -\Delta \pm \sqrt{\Delta^2 + 4V^2} \tag{5.26}$$

Here $\sqrt{\Delta^2 + 4V^2}$ is the generalized Rabi frequency. In the case of far resonance, the energy can be reduced as;

$$\Delta E \cong 2\frac{M_{gx}^2\mathcal{E}^2}{\Delta} \tag{5.27}$$

Significantly the two-level dressed-atom model has defended to enhancing photon energies, thereby the OSE energy shift changes the signs from the blue shift $h\upsilon < E_A$ to a red-shift for $h\upsilon > E_A$. But it could not explain the blue shift perseveres for the photon energies overhead the A exciton resonance.

To resolve the issue investigators have introduced another model based on the excitonic approach of the optical Stark effect including the many-body Coulomb interactions [335]. Assuming the many-body inter connection among virtual excitons by treating in a Hartree–Fock approach. Thus the optical Stark shift can be expressed from the following expression, for the low limit ($Na_0^2 \ll 1$) of the 2D materials;

$$\Delta E \cong 2\frac{M_{gx}^2\mathcal{E}^2}{\Delta}\left(\frac{|\emptyset_{1s}(r=0)|^2}{N_s^{\text{PSF}}}\right) + 8\pi N a_0^2\left(1 - \frac{315\pi^2}{4096}\right)E_b \tag{5.28}$$

here N represents the virtual exciton density and a_0 corresponds to the exciton Bohr radius (typically for the TMDs Bohr radius of 1 nm). This approximation is generally consisting decent for the virtual density of $1.1 \times 10^{-14}\,\text{cm}^{-1}$. The numerator, $|\emptyset_{1s}(r=0)|^2$ correlates to the 1S exciton wave function, which reveals the built excitons from linear combinations of atomic Bloch states. Whereas the denominator ($N_s^{\text{PSF}} \approx 1/2\pi a_0^2$) reflects the saturation density caused by phase-space filling, while the thermal energy is much less than the exciton binding energy (E_b).

The experimental and theoretical innovations on TMDs Stark effects are not limited to the above-discussed significant case but many more and continued by interpreting the novel concepts and findings [312, 316, 322, 327, 336, 337].

5.10 Conclusions

In the sequential order to provide various aspects of TMD materials, one has also paid attention to the spin-valley coupling. This work has provided an understating of the spin-valley coupling in TMD materials as well as their impact on physical properties. To build a better understating of spin-valley coupling in TMD materials a brief overview from before to current era information has been provided. Based on theoretical concepts and experimental findings descriptions of the degree of freedom and electronic structure, optical and electrical valley polarization, and valley coherence have been also discussed. Since the valleytronics properties of TMD materials are one of the significant factors for their applications, therefore, it is important to define the valleytronics properties of TMD materials and discussed the valley-spin concept for the monolayers and bilayers. Thus the selection rules for the valley and spin, valley Zeeman Effect and magnetic moment, valley Hall Effect and Berry phase effect, and nonlinear valley-spin currents through trigonal twisting for the monolayers and Bi-Layers descriptions have been discussed in detail. Moreover, inversion symmetry induced have also played a crucial role to define the physical properties of TMD materials, therefore, a separate discussion on inversion symmetry induced spin polarization and optics-induced Stark effect have been also incorporated.

References

1. Zhu B, Zeng H, Dai J, Cui X (2014) The study of spin-valley coupling in atomically thin Group VI transition metal dichalcogenides. Adv Mater 26:5504–5507
2. Mak KF, Lee C, Hone J, Shan J et al (2010)Atomically thin MoS_2: a new direct-gap semiconductor. Phys Rev Lett 105:136805
3. Splendiani A, Sun L, Zhang Y, Li T et al (2010) Emerging photoluminescence in monolayer MoS_2. Nano Lett 10:1271–1275
4. Karch J, Tarasenko SA, Ivchenko EL, Kamann J et al (2011) Photoexcitation of valley-orbit currents in (111)-oriented silicon metal-oxide-semiconductor field-effect transistors. Phys Rev B 83:121312
5. Isberg J, Gabrysch M, Hammersberg J, Majdi S et al (2013) Generation, transport and detection of valley-polarized electrons in diamond. Nat Mater 12:760–764
6. Zhu Z, Collaudin A, Fauqué B, Kang W et al (2012) Field-induced polarization of Dirac valleys in bismuth. Nat Phys 8:89–94
7. Ciccarino CJ, Christensen T, Sundararaman R, Narang P (2018) Dynamics and spin-valley locking effects in monolayer transition metal dichalcogenides. Nano Lett 18:5709–5715
8. Sun YY, Shang L, Ju W, An YP et al (2019) Tuning valley polarization in two-dimensional ferromagnetic heterostructures. J Mater Chem C 7:14932–14937

9. Hsu YT, Wu F, Sarma SD (2021) Spin-valley locked instabilities in moiré transition metal dichalcogenides with conventional and higher-order Van Hove singularities. Phys Rev B 104:195134
10. Wei Q, Chen D, Cai Y, Shen L et al (2022) Generation and enhancement of valley polarization in monolayer chromium dichalcogenides. J Supercond Nov Magn. https://doi.org/10.1007/s10948-021-06112-5
11. Husain S, Pal S, Chen X, Kumar P et al (2022) Large Dzyaloshinskii-Moriya interaction and atomic layer thickness dependence in a ferromagnet-WS_2 heterostructure. Phys Rev B 105:064422
12. Mak KF, Xiao D, Shan J (2018) Light–valley interactions in 2D semiconductors. Nat Photonics 12:451–460
13. Liu KK, Zhang W, Lee YH, Lin YC et al (2012) Growth of large-area and highly crystalline MoS2 thin layers on insulating substrates. Nano Lett 12:1538–1544
14. Xiao D, Yao W, Niu Q (2007) Valley-contrasting physics in graphene: magnetic moment and topological transport. Phys Rev Lett 99:236809
15. Yao W, Xiao D, Niu Q (2008) Valley-dependent optoelectronics from inversion symmetry breaking. Phys Rev B 77:235406
16. Yu H, Yao W (2017) Valley-spin physics in 2D semiconducting transition metal dichalcogenides. In: 2D materials properties and devices. Cambridge University Press, Cambridge
17. Cotrufo M, Sun L, Choi J, Alù A et al (2019) Enhancing functionalities of atomically thinsemiconductors with plasmonic nanostructures. Nanophotonics 8:577–598
18. Mattheiss LF (1973) Band structures of transition-metal-dichalcogenide layer compounds. Phys Rev B 8:3719
19. Pandey SK, Das R, Mahadevan P (2020) Layer-dependent electronic structure changes in transition metal dichalcogenides: the microscopic origin. ACS Omega 5:15169−15176
20. Liu GB, Xiao D, Yao Y, Xu X et al (2015) Electronic structures and theoretical modelling of two-dimensional group-VIB transition metal dichalcogenides. Chem Soc Rev 44:2643–2663
21. Kormányos A, Burkard G, Gmitra M, Fabian J et al (2015) k·p theory for two-dimensional transition metal dichalcogenide semiconductors. 2D Mater 2:022001
22. Xiao D, Liu GB, Feng W, Xu X et al (2012) Coupled spin and valley physics in monolayers of MoS_2 and other Group-VI dichalcogenides. Phys Rev Lett 108:196802
23. Liu GB, Shan WY, Yao Y, Yao W et al (2013) Three-band tight-binding model for monolayers of group-VIB transition metal dichalcogenides. Phys Rev B 88:085433
24. Srivastava A, Sidler M, Allain AV, Lembke DS (2015) Valley Zeeman effect in elementary optical excitations of monolayerWSe2. Nat Phys 11:141–147
25. Zhu ZY, Cheng YC, Schwingenschlögl U (2011) Giant spin-orbit-induced spin splitting in two-dimensional transition-metal dichalcogenide semiconductors. Phys Rev B 84:153402
26. Ramasubramaniam A (2012) Large excitonic effects in monolayers of molybdenum and tungsten dichalcogenides. Phys Rev B 86:115409
27. Li S, Wang H, Wang J, Chen H et al (2021) Control of light–valley interactions in 2D transition metal dichalcogenides with nanophotonic structures. Nanoscale 13:6357–6372
28. Kadantsev ES, Hawrylak P (2012) Electronic structure of a single MoS2 monolayer. Solid State Commun 152:909–912
29. Cheng Y, Zhu Z, Schwingenschlogl U (2012) Role of interlayer coupling in ultra thin MoS2. RSC Adv 2:7798–7802
30. Kośmider K, Fernández-Rossier J (2013) Electronic properties of the MoS_2-WS_2 heterojunction. Phys Rev B 87:075451
31. Kormányos A, Zólyomi V, Drummond ND, Burkard G (2014) Spin-orbit coupling, quantum dots, and qubits in monolayer transition metal dichalcogenides. Phys Rev X4:011034
32. Scharf B, Xu G, Abiague AM, Zuti I (2017) Magnetic proximity effects in transition-metal dichalcogenides: converting excitons
33. Berghäuser G, Bernal-Villamil I, Schmidt R, Schneider R et al (2018) Inverted valley polarization in optically excited transition metal dichalcogenides. Nat Commun 9:971

34. Zhou BT, Taguchi K, Kawaguchi Y, Tanaka Y et al (2019) Spin-orbit coupling induced valley Hall effects in transition-metal dichalcogenides. Commun Phys 2:26
35. Vitale V, Atalar K, Mostofi AA, Lischner J (2021) Flat band properties of twisted transition metal dichalcogenide homo- and heterobilayers of MoS_2, $MoSe_2$, WS_2 and WSe_2. 2D Mater 8:045010
36. Wang K, Hu G, Liu R, Zhang Y et al (2020) Polarization-driven edge-state transport in transition-metal dichalcogenides. Phys Rev Appl 13:054074
37. Luo G, Zhang ZZ, Li HO, Song XX et al (2017) Quantum dot behavior in transition metal dichalcogenides nanostructures. Front Phys 12:128502
38. Tang Y, Mak KF, Shan J (2019) Long valley lifetime of dark excitons in single-layer WSe2. Nat Commun 10:4047
39. Zheng W, Jiang Y, Hu X, Li H et al (2018) Light emission properties of 2D transition metal dichalcogenides: fundamentals and applications. Adv Opt Mater 6:1800420
40. Cortés N, Ávalos-Ovando O, Ulloa SE (2020) Reversible edge spin currents in antiferromagnetically proximitized dichalcogenides. Phys Rev B 101:201108
41. Jiang X, Kuklin AV, Baev A, Ge Y et al (2020) Two-dimensional MXenes: From morphological to optical, electric, and magnetic properties and applications. Phys Rep 848:1–58
42. Scrace T, Tsai Y, Barman B, Schweidenback L et al (2015) Magnetoluminescence and valley polarized state of a two-dimensional electron gas in WS2 monolayers. Nat Nanotech 10:603–607
43. Ruiz-Tijerina DA, Danovich M, Yelgel C, Zolyomi V et al (2018) Hybrid k .p-tight binding model for subbands and infrared intersubband optics in few-layer lms of transition metal dichalcogenides: MoS2, MoSe2, WS2, and WSe2. Phys Rev B 98:035411
44. Juniora LAM, Bruno-Alfonso A (2021) Thirteen-band tight-binding model for the MoS2 monolayer. Mater Res 24:e20210059
45. Kormányos A, Zólyomi V, Drummond ND, Rakyta P et al (2013) Monolayer MoS2: trigonal warping, the Γ valley, and spin-orbit coupling effects. Phys Rev B 88:045416
46. Rostami H, Moghaddam AG, Asgari R (2013) Effective lattice Hamiltonian for monolayer MoS2: tailoring electronic structure with perpendicular electric and magnetic fields. Phys Rev B 88:085440
47. Rostami H, Asgari R (2014) Intrinsic optical conductivity of modified Dirac fermion systems. Phys Rev B 89:115413
48. Rose F, Goerbig MO, Piéchon F (2013) Spin- and valley-dependent magneto-optical properties of MoS. Phys Rev B 88:125438
49. Scholz A, Stauber T, Schliemann J (2013) Plasmons and screening in a monolayer of MoS2. Phys Rev B 88:035135
50. Wang L, Wu MW (2014) Electron spin relaxation due to D'yakonov-Perel' and Elliot-Yafet mechanisms in monolayer MoS2: Role of intravalley and intervalley processes. Phys Rev B 89:115302
51. Abdi M, Astinchap B, Khoeini F (2022) Electronic and thermodynamic properties of zigzag MoS2/MoSe2 andMoS2/WSe2 hybrid nanoribbons: impacts of electric and exchange fields. Results Phys 34:105253
52. Hsu WT, Lu LS, Wu PH, Lee MH et al (2018) Negative circular polarization emissions from WSe2/MoSe2 commensurate heterobilayers. Nat Commun 9:1356
53. Hao K, Specht JF, Nagler P, Xu L et al (2017) Neutral and charged inter-valley biexcitons in monolayer MoSe2. Nat Commun 8:15552
54. Schaibley JR, Yu H, Clark G, Rivera P et al (2016) Valleytronics in 2D materials. Nat Rev Mater 1:16055
55. Abdullah R, Weibo G (2020) Opto-valleytronics in the 2D van der Waals heterostructure. Nano Res 14:1901–1911
56. Berghäuser G, Bernal-Villamil I, Schmidt R, Schneider R et al (2018) Inverted valley polarization in optically excited transition metal dichalcogenides. Nat Commun 9:971
57. Xia J, Wang X, Tay BK, Chen S et al (2011) Valley polarization in stacked MoS2 induced by circularly polarized light. Nano Res 10:1618–1626

58. Jiang Y, Low T, Chang K, Katsnelson MI et al (2013) Generation of pure bulk valley current in graphene. Phys Rev Lett 110:046601
59. Cao T, Wang G, Han W, Ye H et al (2012) Valley-selective circular dichroism of monolayer molybdenum disulphide. Nat Commun 3:1–5
60. Li Z, Xu B, Liang D (2020) Polarization-dependent optical properties and optoelectronic devices of 2D materials. 5464258:1–35
61. Bermudez VM, McClure DS (1979) Spectroscopic studies of the two dimensional magnetic insulators chromium trichloride and chromium tribromide. J Phys Chem Solids 40:129–147
62. Molina-Sanchez A, Catarina G, Sangalli D, Fernandez-Rossier J (2019) Magneto-optical response of chromium trihalide monolayers: chemical trends 1912.01888
63. Karni O, Barré E, Lau SC, Gillen R et al (2019) Infrared interlayer exciton emission in MoS_2/WSe_2 heterostructures. Phys Rev Lett 123:247402
64. Ling X, Wang H, Huang S, Xia F et al (2015) The renaissance of black phosphorus. Proc Natl Acad Sci 112:4523–4530
65. Ansari L, Zhang D, Yan XQ, Ren M et al (2019) Quantum connement-induced semimetal-to-semiconductor evolution in large-area ultra-thin $PtSe_2$ Films grown at 400°C. npj 2D Mater Appl 3:1–8
66. Cassabois G, Valvin P, Gil B (2016) Hexagonal boron nitride is an indirect bandgap semiconductor. Nat Photon 10:262–266
67. Lorchat E, Azzini S, Chervy T, Taniguchi T et al (2018) Room-temperature valley polarization and coherence in transition metal dichalcogenide–graphene van der Waals heterostructures. ACS Photon 5:5047–5054
68. Mak KF, He K, Shan J, Heinz TF (2012) Control of valley polarization in monolayer MoS_2 by optical helicity. Nat Nanotech 7:494–498
69. Rytova NS (2018) Screened potential of a point charge in a thin film. arXiv preprint arXiv: 1806.00976
70. Keldysh LV (1979) Coulomb interaction in thin semiconductor and semimetal films. Soviet J Exp Theor Phys Lett 29:658
71. Mueller T, Malic E (2018) Exciton physics and device application of two-dimensional transition metal dichalcogenide semiconductors. npj 2D Mater Appl 2:29
72. Waldecker L, Raja A, Rösner M, Steinke C et al (2019) Rigid band shifts in twodimensional semiconductors through external dielectric screening. Phys Rev Lett 123:206403
73. Rhodes D, Chae SH, Ribeiro-Palau R, Hone J (2019) Disorder in van der waals heterostructures of 2d materials. Nat Mater 18:541–549
74. Cadiz F, Courtade E, Robert C, Wang V et al (2017) Excitonic linewidth approaching the homogeneous limit in mos 2-based van der waals heterostructures. Phys Rev X 7:021026
75. Mak KF, Shan J, Ralph DC (2019) Probing and controlling magnetic states in 2d layered magnetic materials. Nat Rev Phys 1:646–661
76. Lo TW, Chen V, Zhang Z, Zhang Q et al (2022) Plasmonic nanocavity induced coupling and boost of Dark Excitons in monolayer WSe2 at room temperature. https://doi.org/10.1021/acs.nanolett.1c04360
77. Andres-Penares D, Habil MK, Molina-Sánchez A, Zapata-Rodríguez CJ et al (2021) Out-of-plane trion emission in monolayer WSe2 revealed by whispering gallery modes of dielectric microresonators. Commun Mater 2:52
78. Wang G, Robert C, Glazov MM, Cadiz F et al (2017) In-plane propagation of light in transition metal dichalcogenide monolayers: optical selection rules. Phys Rev Let 119:047401
79. Robert C, Han B, Kapuscinski P, Delhomme A et al (2020) Measurement of the spin-forbidden dark excitons in mos2 and mose2 monolayers. Nat Commun 11:4037
80. Liu Y, Stradins P, Wei SH (2016) Van der waals metalsemiconductor junction: Weak fermi level pinning enables active tuning of schottky barrier. Sci Adv 2:e1600069
81. Kang J, Tongay S, Zhou J, Li J et al (2013) Band off-sets and heterostructures of two-dimensional semiconductors. Appl Phys Lett 102:012111
82. Rivera P, Schaibley JR, Jones AM, Ross JS et al (2015) Observation of long-lived interlayer excitons in monolayer $MoSe_2$-WSe_2 heterostructures. Nat Commun6:1–6

83. Sushko A, Greve KD, Andersen TI, Scuri G et al (2019) High resolution imaging of reconstructed domains and moire patterns in functional van der Waals heterostructure devices. https://doi.org/10.48550/arXiv.1912.07446
84. van Der Zande AM, Kunstmann J, Chernikov A, Chene DA et al (2014) Tailoring the electronic structure in bilayer molybdenum disulde via interlayer twist. Nano Lett 14:3869–3875
85. Yu H, Liu GB, Tang J, Xu X et al (2017) Moire excitons: From programmable quantum emitter arrays to spin-orbit coupled articial lattices. Sci Adv 3:e1701696
86. Singh KJ, Ciou HH, Chang YH, Lin YS et al (2022) Optical mode tuning of monolayer tungsten diselenide (WSe2) by integrating with one-dimensional photonic crystalthrough exciton–photon coupling. Nanomaterials 12:425
87. Mandyam SV, Kim HM, Drndić M (2020) Large area few-layer TMD film growths and their applications. J Phys Mater 3:024008
88. Shree S, Paradisanos I, Marie X, Robert C et al (2020) Guide to optical spectroscopy of layered semiconductors. Nat Rev Phys Nat. https://doi.org/10.1038/s42254-020-00259-1
89. Raja A, Waldecker L, Zipfel J, Cho Y et al (2019) Dielectric disorder in two-dimensional materials. Nat Nanotechnol 14:832–837
90. Robert C, Semina MA, Cadiz F, Manca M et al (2018) Optical spectroscopy of excited exciton states in mos 2 monolayers in van der waals heterostructures. Phys Rev Mater 2:011001
91. Horng J, Martin EW, Chou YH, Courtade E et al (2020) Perfect absorption by an atomically thin crystal. Phys Rev Appl 14:024009
92. Back P, Zeytinoglu S, Ijaz A, Kroner M et al (2018) Realization of an electrically tunable narrow-bandwidth atomically thin mirror using monolayer $MoSe_2$. Phys Rev Lett 120:037401
93. Ho C, Huang Y, Tiong K, Liao P (1998) Absorptionedge anisotropy in res 2 and rese 2 layered semiconductors. Phys Rev B 58:16130
94. Zhang E, Wang P, Li Z, Wang H et al (2016) Tunable ambipolar polarizationsensitive photodetectors based on high-anisotropy $ReSe_2$ nanosheets. ACS Nano 10:8067–8077
95. Wilson JA, Yoe A (1969) The transition metal dichalcogenides discussion and interpretation of the observed optical, electrical and structural properties. Adv Phys 18:193–335
96. Liu Y, Elbanna A, Gao W, Pan J (2021) Interlayer excitons in transition metal dichalcogenide semiconductors for 2D optoelectronics. https://doi.org/10.1002/adma.202107138
97. Arora A, Drüppel M, Schmidt R, Deilmann T et al (2017) Interlayer excitons in a bulk van der waals semiconductor. Nat Commun 8:1–6
98. Gerber IC, Courtade E, Shree S, Rober C et al (2019) Interlayer excitons in bilayer mos 2 with strong oscillator strength up to room temperature. Phys Rev B 99:035443
99. Slobodeniuk A, Bala Ł, Koperski M, Molas MR et al (2019) Fine structure of k-excitons in multilayers of transition metal dichalcogenides. 2D Materials 6:025026
100. Leisgang N, Shree S, Paradisanos I, Sponfeldner L et al (2020) Giant Stark splitting of an exciton in bilayer MoS2. Nat Nanotechnol 15:901–907
101. Lorchat E, Selig M, Katsch F, Yumigeta K et al (2021) Dipolar and magnetic properties of strongly absorbing hybrid interlayer excitons in pristine bilayer MoS_2. Phys Rev Lett 126:037401
102. Ciarrocchi A, Unuchek D, Avsar A, Watanabe K et al (2019) Polarization switching and electrical control of interlayer excitons in two-dimensional van der Waals heterostructures. Nat Photon 13:131–136
103. Selig M, Katsch F, Schmidt R, de Vasconcellos SM et al (2019) Ultrafast dynamics in monolayer transition metal dichalcogenides: interplay of dark excitons, phonons, and intervalley exchange. Phys Rev Res 1:022007
104. Kim J, Jin C, Chen B, Cai H et al (2017) Observation of ultralong valley lifetime inWSe2/MoS2 heterostructures. Sci Adv 3:e1700518
105. Hill HM, Rigosi AF, Roquelet C, Chernikov A et al (2015) Observation of excitonic rydberg states in monolayer mos2 and ws2 by photoluminescence excitation spectroscopy. Nano Lett 15:2992–2997
106. Srivastava A, Imamoglu A (2015) Signatures of blochband geometry on excitons: nonhydrogenic spectra in transition-metal dichalcogenides. Phys Rev Lett 115:166802

107. Glazov M, Golub LE, Wang G, Marie X et al (2017) Intrinsic exciton-state mixing and nonlinear optical properties in transition metal dichalcogenide monolayers. Phys Rev B 95:035311
108. Wang G, Gerber IC, Bouet L, Lagarde D et al (2015) Exciton states in monolayer mose 2: impact on interband transitions. 2D Mater 2:045005
109. Shree S, Semina M, Robert C, Han B et al (2018) Observation of exciton-phonon coupling in mose 2 monolayers. Phys Rev B 98:035302
110. Dau MT, Gay M, Felice DD, Vergnaud C et al (2018) Beyond van der waals interaction: the case of mose2 epitaxially grown on few-layer graphene. ACS Nano 12:2319–2331
111. Li Z, Wang T, Miao S, Lian Z et al (2020) Fine structures of valley-polarized excitonic states in monolayer transitional metal dichalcogenides. Nanophotonics 9:1811–1829
112. Dyakonov MI (2017) Spin physics in semiconductors, vol 1. Springer
113. Kioseoglou G, Hanbicki AT, Currie M, Friedman AL et al (2012) Valley polarization and intervalley scattering in monolayer mos_2. Appl Phys Lett 101:221907
114. Tornatzky H, Kaulitz A-M, Maultzsch J (2018) Resonance proles of valley polarization in single-layer mos 2 and mose 2. Phys Rev Rett 121:167401
115. Zhang Q, Sun H, Tang J, Dai X et al (2021) Prolonging valley polarization lifetime through gate-controlled exciton-to-trion conversion in monolayer molybdenum ditelluride. https://doi.org/10.48550/arXiv.2111.05560
116. Duerwiel S, Lyons TP, Solnyshkov DD, Trichet AAP et al (2018) Valley coherent exciton-polaritons in a monolayer semiconductor. Nat Commun 9:1–7
117. Robert C, Park S, Cadiz F, Lombez L et al (2021) Spin/valley pumping of resident electrons in WSe2 and WS2 monolayers. Nat Commun 12:5455
118. Seyler KL, Rivera P, Yu H, Wilson NP et al (2019) Signatures of moire-trapped valley excitons in mose 2/wse 2 heterobilayers. Nature 567:66–70
119. Sushko A, De Greve K, Andersen TI, Scuri G et al (2019) High resolution imaging of reconstructed domains and moire patterns in functional van der Waals heterostructure devices. https://doi.org/10.48550/arXiv.1912.07446
120. Rivera P, Seyler KL, Yu H, Schaibley JR et al (2016) Valley-polarized exciton dynamics in a 2d semiconductor heterostructure. Science 351:688–691
121. Unuchek D, Ciarrocchi A, Avsar A, Sun Z et al (2019) Valley-polarized exciton currents in a van der Waals heterostructure. Nat Nanotechnol 14:1104–1109
122. Molas M, Slobodeniuk AO, Kazimierczuk T, Nogajewski K et al (2019) Probing and manipulating valley coherenceof dark excitons in monolayer wse 2. Phys Rev Lett 123:096803
123. Nagler P, Ballottin MV, Mitioglu AA, Mooshammer F et al (2017) Giant magnetic splitting inducing nearunityvalley polarization in van der Waals heterostructures. Nat Commun 8:1–6
124. Back P, Sidler M, Cotlet O, Srivastava A et al (2017) Giant paramagnetism-induced valley polarization of electrons in charge-tunable monolayer mose2. Phys Rev Lett 118:237404
125. Soni A, Pal SK (2021) Valley degree of freedom in two-dimensional van der Waals materials. arXiv:2109.01074
126. Ye T, Li Y, Li J, Shen H et al (2022) Nonvolatile electrical switching of optical and valleytronic properties of interlayer excitons
127. Khan I, Marfoua B, Hong J. Electric field induced giant valley polarization in two dimensional ferromagnetic WSe2/CrSnSe3 heterostructure. Light Sci Appl 11:23
128. Zhang A, Yang K, Zhang Y, Pan A et al (2021) Electrically switchable valley polarization, spin/valley lter, and valve effects in transition-metal dichalcogenide monolayers interfaced with two-dimensional ferromagnetic semiconductors. Phys Rev B 104:L201403
129. Hu H, Tong WY, Shen YH, Duan CG (2020) Electrical control of valley degree of freedom in 2D ferroelectric/antiferromagnetic heterostructures. J Mater Chem C 8:8098–8106
130. Ye Y, Xiao J, Wang H, Ye Z et al (2016) Electrical generation and control of the valley carriers in a monolayer transition metal dichalcogenide. Nat Nanotech 11:598–602
131. Xu Y, Kang K, Watanabe K, Taniguchi T et al (2022) Tunable bilayer Hubbard model physics in twisted WSe2. arXiv:2202.02055

132. Xu Y, Liu S, Rhodes DA, Watanabe K et al (2020) Correlated insulating states at fractional fillings of moiré superlattices. Nature 587:214–218
133. Liu GB, Xiao D, Yao Y, Xu X et al (2015) Electronic structures and theoretical modelling of two-dimensional group-VIB transition metal dichalcogenides. Chem Soc Rev 44:2643–2663
134. Dalal A, Ruhman J (2021) Orbitally selective Mott phase in electron-doped twisted transition metal-dichalcogenides: a possible realization of the Kondo lattice model. Phys Rev Res 3:043173
135. Sivadas N, Okamoto S, Xiao D (2016) Gate-controllable magneto-optic Kerr Effect in layered collinear antiferromagnets. Phys Rev Lett 117:267203
136. Fiebig M (2005) Revival of the magnetoelectric effect. J Phys D: Appl Phys 38:R123–R152
137. Li T, Zhu J, Tang Y, Watanabe K et al (2021) Charge-order-enhanced capacitance in semiconductor moiré superlattices. Nat Nanotechnol 16:1068–1072
138. Gu J, Ma L, Liu S, Watanabe K et al (2021) Dipolar excitonic insulator in a moire lattice. arXiv:2108.06588
139. Jiang M, Wu Z, Yang Q, Zhang Y et al (2022) Coherent spin dynamics of localized electrons in monolayer MoS2. J Phys Chem Lett 13:2661–2667
140. Xu X, Xiao D, Heinz TF, Yao W (2014) Spin and pseudospins in layered transition metal dichalcogenides. Nat Phys 10:343–350
141. Suzuki R, Sakano M, Zhang YJ, Akashi R et al (2014) Valley-dependent spin polarization in bulk MoS2 with broken inversion symmetry. Nat Nanotechnol 8:611–617
142. Surrente A, Dumcenco D, Yang Z, Kuc A et al (2017) Defect healing and charge transfer-mediated valley polarization in MoS2/MoSe2/MoS2 trilayer van der Waals heterostructures. Nano Lett 17:4130–4136
143. Wang G, Marie X, Liu BL, Amand T et al (2016) Control of exciton valley coherence in transition metal dichalcogenide mono layers. Phys Rev Lett 117:187401
144. Chen YJ, Cain JD, Stanev TK, Dravid VP et al (2017) Valley-polarized excitonpolaritons in a monolayer semiconductor. Nat Photonics 11:431–435
145. Duerwiel S, Lyons TP, Solnyshkov DD, Trichet AAP et al (2017) Valley-addressable polaritons in atomically thin semiconductors. Nat Photon 11:497–501
146. Solanas CA, Waldherr M, Klaas M, Suchomel H et al (2021) Bosonic condensation of exciton–polaritons in an atomically thin crystal. Nat Mater 20:1233–1239
147. Liu Y, Gao Y, Zhang S, He J et al (2019) Valleytronics in transition metal dichalcogenides materials. Nano Res 12:2695–2711
148. Lundt N, Nagler P, Nalitov A, Klembt S et al (2017) Valley polarized relaxation and upconversion luminescence from Tamm-plasmon trion-polaritons with a MoSe2 monolayer. 2D Mater 4:025096
149. Lei C, Ma Y, Zhang T, Xu X et al (2020) Valley polarization in monolayer CrX2 (X = S, Se) with magnetically doping and proximity coupling. New J Phys 22:033002
150. Wang G, Glazov MM, Robert C, Amand T et al (2015) Double resonant Raman scattering and valley coherence generation in monolayer WSe2. Phys Rev Lett 115:117401
151. Hao K, Moody G, Wu F, Dass CK et al (2016) Direct measurement of exciton valley coherence in monolayer WSe2. Nat Phys 12:677–682
152. Schmidt R, Arora A, Plechinger G, Nagler P et al (2016) Magnetic-eld-induced rotation of polarized light emission from monolayer WS2. Phy Rev Lett 117:077402
153. Ye Z, Sun D, Heinz TF (2017) Optical manipulation of valley pseudospin. Nat Phys 13:26–29
154. Chernikov A, Berkelbach TC, Hill HM, Rigosi A et al (2014) Exciton binding energy and nonhydrogenic Rydberg series in monolayer WS2. Phys Rev Lett 113:076802
155. Zeng H, Dai J, Yao W, Xiao D et al (2012) Valley polarization in MoS2 monolayers by optical pumping. Nat Nanotechnol 7:490–493
156. Lagarde D, Bouet L, Marie X, Zhu CR et al (2014) Carrier and polarization dynamics in monolayer MoS2. Phys Rev Lett 112:047401
157. Jones AM, Yu H, Ghimire NJ, Wu S et al (2013) Optical generation of excitonic valley coherence in monolayer WSe2. Nat Nanotechnol 8:634–638

158. Kallatt S, Umesh G, Majumdar K (2016) Valley-coherent hot carriers and thermal relaxation in monolayer transition metal dichalcogenides. J Phys Chem Lett 7:2032–2038
159. Maialle MZ, De Andrada E, Silva EA, Sham LJ (1993) Exciton spin dynamics in quantum wells. Phys Rev B 47:15776–15788
160. Yu T, Wu MW (2014) Valley depolarization due to intervalley and intravalley electronhole exchange interactions in monolayer MoS2. Phys Rev B Condens Matter Mater Phys 89:205303
161. Conway MA, Muir JB, Earl SK, Wurdack M et al (2022) Direct measurement of biexcitons in monolayer WS2. 2D Mater 9:021001
162. Hao K, Xu L, Wu F, Nagler P et al (2017) Trion valley coherence in monolayer semiconductors. 2D Mater 4:025105
163. Schmidt R, Arora A, Plechinger G, Nagler P et al (2016) Magnetic-field-induced rotation of polarized light emission from monolayer WS2. Phys Rev Lett 117:077402
164. Palummo M, Bernardi M, Grossman JC (2015) Exciton radiative lifetimes in twodimensional transition metal dichalcogenides. Nano Lett 15:2794–2800
165. Gupta G, Majumdar K (2019) Fundamental exciton linewidth broadening in monolayer transition metal dichalcogenides. Phys Rev B 99:085412
166. Gupta G, Watanabe K, Taniguchi T, Majumdar K (2021) Observation of perfect valley coherence in monolayer MoS2 through giant enhancement of exciton coherence time. arXiv:2106.03359
167. Sidler M, Back P, Cotlet O, Srivastava A et al (2017) Fermi polaron-polaritons in charge-tunable atomically thin semiconductors. Nat Phys 13:255–261
168. Dufferwiel S, Schwarz S, Withers F, Trichet AAP et al (2015) Exciton–polaritons in van der Waals heterostructures embedded in tunable microcavities. Nat Commun 6:8579
169. Mol PR, Barman PK, Sarma PV, Kumar AS et al (2021) Anomalously polarised emission from a MoS2/WS2 heterostructure. Nanoscale Adv 3:5676
170. Žutić I, Matos-Abiague A, Scharf B, Dery H et al (2019) Proximitized materials. Mater Today 22:85–107
171. Yu H, Liu GB, Yao W (2018) Brightened spin-triplet interlayer excitons and optical selection rules in van der Waals heterobilayers. 2D Mater 5:035021
172. Zhang C, Chuu CP, Ren X, Li MY et al (2017) Interlayer couplings, Moiré patterns, and 2D electronic superlattices in MoS2/WSe2 hetero-bilayers. Sci Adv 3:e1601459
173. Yu H, Liu GB, Tang J, Xu X et al (2017) Moiré excitons: from programmable quantum emitter arrays to spin-orbit–coupled artificial lattices. Sci Adv 3:e1701696
174. Wu F, Lovorn T, Tutuc E, MacDonald AH (2018) Hubbard model physics in transition metal dichalcogenide Moiré bands. Phys Rev Lett 121:026402
175. Wu F, Lovorn T, Tutuc E, Martin I et al (2019) Topological insulators in twisted transition metal dichalcogenide homobilayers. Phys Rev Lett 122:086402
176. Regan EC, Wang D, Jin C, Bakti Utama MI et al (2020) Mott and generalized Wigner crystal states in WSe2/WS2 Moiré superlattices. Nature 579:359–363
177. Tang Y, Li L, Li T, Xu Y et al (2020) Simulation of Hubbard model physics in WSe2/WS2 Moiré superlattices. Nature 579:353–358
178. Shimazaki Y, Schwartz I, Watanabe K, Taniguchi T et al (2020) Strongly correlated electrons and hybrid excitons in a Moiré heterostructure. Nature 580:472–477
179. Seyler KL, Rivera P, Yu H, Wilson NP et al (2019) Signatures of Moiré-trapped valley excitons in MoSe2/WSe2 heterobilayers. Nature 567:66–70
180. Alexeev EM, Ruiz-Tijerina DA, Danovich M, Hamer MJ et al (2019) Resonantly hybridized excitons in moiré superlattices in van der Waals heterostructures. Nature 567:81–86
181. Jin C, Regan EC, Yan A, Iqbal Bakti Utama M et al (2019) Observation of Moiré excitons in WSe2/WS2 heterostructure superlattices. Nature 567:76–80
182. Jin C, Regan EC, Wang D, Iqbal Bakti Utama M et al (2019) Identification of spin, valley and Moiré quasi-angular momentum of interlayer excitons. Nat Phys 15:1140–1144
183. Wu F, Lovorn T, MacDonald AH (2018) Theory of optical absorption by interlayer excitons in transition metal dichalcogenide heterobilayers. Phys Rev B 97:035306

184. Wang T, Miao S, Li Z, Meng Y et al (2020) Giant valley-Zeeman splitting from spin-singlet and spin-triplet interlayer excitons in WSe2/MoSe2 heterostructure. Nano Lett 20:694–700
185. Tran K, Moody G, Wu F, Lu XA et al (2019) Evidence for Moiré excitons in van der Waals heterostructures. Nature 567:71–75
186. Zhang L, Gogna R, Burg GW, Horng J et al (2019) Highly valley-polarized singlet and triplet interlayer excitons in van der Waals heterostructure. Phys Rev B 100:041402
187. Rasmita A, Gao W (2020) Opto-valleytronics in the 2D van der Waals heterostructure. Nano Res 14:1901–1911
188. Liu GB, Shan WY, Yao Y, Yao W et al (2013) Three-band tightbinding model for monolayers of group-VIB transition metal dichalcogenides. Phys Rev B 88:085433
189. Mecklenburg M, Regan BC (2011) Spin and the honeycomb lattice: lessons from graphene. Phys Rev Lett 106:116803
190. Zhang X, Shan WY, Xiao D (2018) Optical selection rule of excitons in gapped chiral fermion systems. Phys Rev Lett 120:077401
191. Gong P, Yu H, Wang Y, Yao W (2017) Optical selection rules for excitonic Rydberg series in the massive Dirac cones of hexagonal two-dimensional materials. Phys Rev B 95:125420
192. Cao T, Wu M, Louie SG (2018) Unifying optical selection rules for excitons in two dimensions: band topology and winding numbers. Phys Rev Lett 120:087402
193. Srivastava A, Imamoğlu A (2015) Signatures of Bloch-band geometry on excitons: nonhydrogenic spectra in transition-metal dichalcogenides. Phys Rev Lett 115:166802
194. Zhou J, Shan WY, Yao W, Xiao D (2015) Berry phase modification to the energy spectrum of excitons. Phys Rev Lett 115:166803
195. Mandal KK, Gupta Y, Sohoni M, Gopal AV et al (2022) A photonic integrated chip platform for interlayer exciton valley routing. https://arxiv.org/pdf/2203.03586.pdf
196. Koperski M, Molas MR, Aroro A, Nogajewski K et al (2017) Optical properties of atomically thin transition metal dichalcogenides: observations and puzzles. Nanophotonics 6:1289
197. Wang G, Chernikov A, Glazov MM, Heinz TF et al (2018) Excitons in atomically thin transition metal dichalcogenides. Rev Mod Phys 90:021001
198. Wang Z, Shan J, Mak KF (2017) Valley- and spin-polarized Landau levels in monolayer WSe2. Nat Nanotechnol 12:144–149
199. Scharf B, Xu G, Abiague AM, Zutic I (2017) Magnetic proximity effects in transition-metal dichalcogenides: converting excitons. Phys Rev Lett 119:127403
200. Xiao D, Chang MC, Niu Q (2010) Berry phase effects on electronic properties. Rev Mod Phys 82:1959
201. Wang G, Bouet L, Glazov MM, Amand T et al (2015) Magneto-optics in transition metal diselenide monolayers. 2D Mater 2:034002
202. Cai T, Yang SA, Li X, Zhang F et al (2013) Magnetic control of the valley degree of freedom of massive Dirac fermions with application to transition metal dichalcogenides. Phys Rev B 88:115140
203. Li X, Zhang F, Niu Q (2013) Unconventional quantum hall effect and tunable spin hall effect in dirac materials: application to an isolated MoS_2 trilayer. Phys Rev Lett 110:066803
204. Fallahazad B, Movva HCP, Kim K, Larentis S et al (2016) Shubnikov–de Haas oscillations of high-mobility holes in monolayer and bilayer WSe_2: landau level degeneracy, effective mass, and negative compressibility. Phys Rev Lett 116:086601
205. Koperski M, Molas MR, Arora A, Nogajewski K et al (2019) Orbital, spin and valley contributions to Zeeman splitting of excitonic resonances in MoSe2, WSe2 and WS2 monolayers. 2D Mater 6:015001
206. Bieniek M, Korkusinski M, Szulakowska L, Potasz P et al (2018) Band nesting, massive Dirac fermions, and valley Landé and Zeeman effects in transition metal dichalcogenides: a tight-binding model. Phys Rev B 97:085153
207. Xuan F, Quek SY (2021) Valley-filling instability and critical magnetic field for interaction-enhanced Zeeman response in doped WSe2 monolayers. npj Comput Mater 7:198
208. Jana K, Muralidharan B (2022) Robust all-electrical topological valley filtering using monolayer 2D-Xenes. npj 2D Mater Appl 6:19

209. Rybkovskiy DV, Gerber IC, Durnev MV (2017) Atomically inspired k·p approach and valley Zeeman effect in transition metal dichalcogenide monolayers. Phys Rev B 95:155406
210. Luttinger JM, Kohn W (1955) Motion of electrons and holes in perturbed periodic fields. Phys Rev 97:869
211. Chang MC, Niu Q (1996) Berry phase, hyperorbits, and the Hofstadter spectrum: Semiclassical dynamics in magnetic Bloch bands. Phys Rev B 53:7010
212. Xuan F, Quek SY (2020) Valley Zeeman effect and Landau levels in two-dimensional transition metal dichalcogenides. Phys Rev Res 2:033256
213. Nandy S, Taraphder A, Tewari S (2018) Berry phase theory of planar Hall effect in topological insulators. Sci Rep 8:14983
214. Qiu DY, da Jornada FH, Louie SG (2013) Optical spectrum of MoS_2 many-body effects and diversity of exciton states. Phys Rev Lett 111:216805
215. Lyons TP, Dufferwiel S, Brooks M, Withers F et al (2019) The valley Zeeman effect in inter- and intra-valley trions in monolayer WSe_2. Nat Commun 10:2330
216. Nagaosa N, Sinova J, Onoda S, MacDonald AH et al (2010) Anomalous Hall effect. Rev Mod Phys 82:1539
217. Zhou X, Zhang RW, Zhang Z, Feng W et al (2021) Sign-reversible valley-dependent Berry phase effects in 2D valley-half-semiconductors. npj Comput Mater 7:160
218. Yu H, Chen M, Yao W (2020) Giant magnetic field from Moire induced Berry phase in homobilayer semiconductors. Nation Sci Rev 7:12–20
219. Nakatsuji S, Arita R (2022) Topological magnets: functions based on Berry phase and multipoles. Rev Condens Matter Phys 13:119–142
220. Xie YM, Zhang CP, Hu JX, Mak KF et al (2022) Valley-polarized quantum anomalous hall state in Moiré MoTe2/WSe2 heterobilayers. Phys Rev Lett 128:026402
221. Lu JM, Zheliuk O, Leermakersn I, Yuanu FQ et al (2015) Evidence for two-dimensional Ising superconductivity in gated MoS_2. Science 350:1353–1357
222. Xi X et al (2016) Ising pairing in superconducting NbSe2 atomic layers. Nat Phys 12:139–143
223. Xi X, Wang Z, Zhao W, Park JH et al (2016) Ising pairing in superconducting NbSe2 atomic layers. Nat Phys 12:139–143
224. Zhou BT, Yuan N et al (2016) Ising superconductivity and majorana fermions in transition metal dichalcogenides. Phys Rev B 93:180501
225. Lu AY, Zhu H, Xiao J, Chuu CP et al (2017) Janus monolayers of transition metal dichalcogenides. Nat Nano-Tech 12:744–749
226. Rashba EI (1959) Symmetry of bands in wurzite-type crystals. 1. Symmetry of bands disregarding spin-orbit interaction. Sov Phys Solid State 1:368
227. Ochoa H, Roldan R (2013) Spin-orbit-mediated spin relaxation in monolayer MoS_2. Phys Rev B 87:245421
228. Saito Y, Nakamura Y, Bahramy MS, Kohama Y et al (2016) Superconductivity protected by spin-valley locking in ion-gated MoS_2. Nat Phys 12:144–149
229. Cappelluti E, Roldan R, Silva-Guillen JA, Ordejon P et al (2013) Tightbinding model and direct-gap/indirect-gap transition in single-layer and multilayer MoS_2. Phys Rev B 88:075409
230. Yuan NFQ, Mak KF, Law KT (2014) Possible topological superconducting phases of MoS_2. Phys Rev Lett 113:097001
231. Taguchi K, Zhou BT, Kawaguchi Y, Tanaka Y et al (2017) Valley Edelstein effect in monolayer transition metal dichalcogenides. Phys Rev B 98:035435
232. Zhang J, Jia S, Kholmanov I, Dong L et al (2017) Janus monolayer transition-metal dichalcogenides. ACS Nano 11:8192–8198
233. Kośmider K, González JW, Fernández-Rossier J (2013) Large spin splitting in the conduction band of transition metal dichalcogenide monolayers. Phys Rev B 88:245436
234. Yuan H, Liu Z, Xu G, Zhou B (2016) Evolution of the valley position in bulk transition-metal chalcogenides and their monolayer limit. Nano Lett 16:4738–4745
235. Lensky YD, Song JCW, Samutpraphoot P, Levitov LS (2015) Topological valley currents in gapped Dirac materials. Phys Rev Lett 114:256601

236. Mak KF, McGill KL, Park J, McEuen PL (2014) The valley hall effect in MoS2 transistors. Science 344:1489–1492
237. Lee J, Mak KF, Shan J (2016) Electrical control of the valley Hall effect in bilayer MoS2 transistors. Nat Nanotech 11:421–425
238. Lee J, Wang Z, Xie H, Mak KF, Shan J (2017) Valley magnetoelectricity in single-layer MoS_2. Nat Mater 16:887–891
239. Kato YK, Myers RC, Gossard AC, Awschalom DD (2004) Observation of the spin Hall effect in semiconductors. Science 306:1910–1913
240. Cysne TP, Bhowal S, Vignale G, Rappoport TG (2022) Orbital Hall effect in bilayer transition metal dichalcogenides: from the intra-atomic approximation to the orbital magnetic moment approach. arXiv:2201.03491
241. Du L, Hasan T, Castellanos-Gomez A, Liu GB et al (2021) Engineering symmetry breaking in 2D layered materials. Nat Rev Phys 3:193–206
242. Mehdi M, Moayed R, Bielewicz T, Zöllner MS et al (2017) Rashba spin-orbit interaction in lead sulphide nanosheets. Nat Commun 8:15721
243. Wu S, Ross JS, Liu GB, Aivazian G et al (2013) Electrical tuning of valley magnetic moment through symmetry control in bilayer MoS2. Nat Phys 9:149–153
244. Szałowski K (2021) Spin state switching in heptauthrene nanostructure by electric field: computational study. Int J Mol Sci 22:13364
245. Gunawan O, Habib B, De Poortere EP, Shayegan M (2006) Quantized conductance in an AlAs two-dimensional electron system quantum point contact. Phys Rev B 74:155436
246. Rycerz A, Tworzydło J, Beenakker CWJ (2007) Valley filter and valley valve in graphene. Nat Phys 3:172–175
247. Tombros N, Jozsa C, Popinciuc M, Jonkman HT et al (2007) Electronic spin transport and spin precession in single graphene layers at room temperature. Nature (London) 448:571–574
248. Xiao J, Bauer GEW, Uchida KC, Saitoh E et al (2010) Theory of magnon-driven spin Seebeck effect. Phys Rev B 81:214418
249. Pulizzi F (2012) Spintronics. Nat Mater 11:367
250. Murakami S, Nagaosa N, Zhang SC (2003) Dissipationless quantum spin current at room temperature. Science 301:1348
251. Sinova J, Culcer D, Niu Q, Sinitsyn NA et al (2004) Universal intrinsic spin hall effect. Phys Rev Lett 92:126603
252. Jungwirth T, Wunderlich J, Olejník K (2012) Spin Hall effect devices. Nat Mater 11:382–390
253. Yu H, Wu Y, Liu GB, Xu X et al (2014) Nonlinear valley and spin currents from Fermi pocket anisotropy in 2D crystals. Phys Rev Lett 113:156603
254. Pospischil A, Furchi MM, Mueller T (2014) Solar-energy conversion and light emission in an atomic monolayer p–n diode. Nat Nanotechnol 9:257–261
255. Baugher BWH, Churchill HOH, Yang Y, Jarillo-Herrero P (2014) Optoelectronic devices based on electrically tunable p–n diodes in a monolayer dichalcogenide. Nat Nanotechnol 9:262–267
256. Li MY, Chen CH, Shi Y, Li LJ (2016) Layered materials and their potential applications. Materialstoday 19:322–335
257. Nazif KN, Kumar A, Hong J, Lee N et al (2021) High-performance p–n junction transition metal dichalcogenide photovoltaic cells enabled by MoOx doping and passivation. Nano Lett 21:3443–3450
258. Zhu W, Low T, Wang H, Ye P et al (2019) Nanoscale electronic devices based on transition metal dichalcogenides. 2D Mater 6:032004
259. Zhang YJ, Oka T, Suzuki R, Ye JT et al (2014) Electrically switchable chiral light-emitting transistor. Science 344:725
260. Saynatjoki A, Karvonen L, Rostami H, Autere A et al (2017) Ultra-strong nonlinear optical processes and trigonal warping in MoS2 layers. Nat Commun 8:893
261. Rosa HG, Ho YW, Verzhbitskiy I, Rodrigues MJFL et al (2018) Characterization of the second- and third-harmonic optical susceptibilities of atomically thin tungsten diselenide. Sci Rep 8:10035

262. Woodward RI, Murray RT, Phelan CF, de Oliveira REP et al (2017) Characterization of the second- and third-order nonlinear optical susceptibilities of monolayer MoS2 using multiphoton microscopy. 2D Mater 4:011006
263. Li D, Xiong W, Jiang L, Xiao Z et al (2016) Multimodal nonlinear opticalimaging of MoS2 and MoS2-based van der Waals heterostructures. ACS Nano 10:3766–3775
264. Jakubczyk T, Delmonte V, Koperski M, Nogajewski K et al (2016) Radiatively limited dephasing and exciton dynamics in MoSe2 monolayers revealed with four-wave mixing microscopy. Nano Lett 16:5333–5339
265. Liu HZ, Li YL, You YS, Ghimire S et al (2017) Highharmonic generation from an atomically thin semiconductor. Nat Phys 13:262–265
266. Vampa G, McDonald CR, Orlando G, Corkum PB et al (2015) Semiclassical analysis of high harmonic generation in bulk crystals. Phys Rev B 91:064302
267. Ghimire S, DiChiara AD, Sistrunk E, Ndabashimiye G et al (2012) Generation and propagation of high-order harmonics in crystals. Phys Rev A 85:043836
268. Jin C, Regan EC, Yan A, Utama MIB et al (2019) Observation of moire excitons in WSe2/ WS2 heterostructure superlattices. Nature 567:76–80
269. Seyler KL, Rivera P, Yu H, Wilson NP et al (2019) Signatures of moire-trapped valley excitons in MoSe2/WSe2 heterobilayers. Nature 567:66–70
270. Alexeev EM, Ruiz-Tijerina DA, Danovich M, Hamer MJ et al (2019) Resonantly hybridized excitons in moire superlattices in van der Waals heterostructures. Nature 567:81–86
271. Wen X, Gong Z, Li D (2019) Nonlinear optics of two-dimensional transition metal Dichalcogenides. Info Mat 1:317–337
272. Shao C, Lu H, Zhang X, Yu C et al (2022) High-Harmonic generation approaching the quantum critical point of strongly correlated systems. Phys Rev Lett 128:047401
273. Ciccarino CJ, Chakraborty C, Englundc DR, Narang NC (2019) Dynamics and spin valley layer effects in bilayer transition metal dichalcogenides. Faraday Discuss 214:175–188
274. Gong Z, Liu GB, Yu H, Xiao D et al (2013) Magnetoelectric effects and valley-controlled spin quantum gates in transition metal dichalcogenide bilayers. Nat Commun 4:2053
275. Schneider LM, Kuhnert J, Schmitt S, Heimbrodt W et al (2019) Spin-layer and spin-valley locking in CVD-grown AA0- and AB-stacked Tungsten-Disulde Bilayers. J Phys Chem C 123:21813–21821
276. Jiang C, Liu F, Cuadra J, Huang Z et al (2017) Zeeman splitting via spin-valley-layer coupling in bilayer MoTe2. Nat Commun 8:802
277. Tang B, Che B, Xu M, Ang ZP et al (2021) Recent advances in synthesis and study of 2D twisted transition metal dichalcogenide bilayers. Small Struct 2:2000153
278. Liu Z, Feng W, Xin H, Gao Y et al (2019) Two-dimensional spinvalley-coupled Dirac semimetals in functionalized SbAs monolayers. Mater Horiz 6:781–787
279. Jung SW, Ryu SH, Shin WJ, Sohn Y et al (2020) Black phosphorus as a bipolar pseudospin semiconductor. Nat Mater 19:277–281
280. Zeng H, Liu GB, Dai J, Yan Y et al (2013) Optical signature of symmetry variations and spin-valley coupling in atomically thin tungsten dichalcogenides. Sci Rep 3:1608
281. Yuan H, Bahramy MS, Morimoto K, Wu S et al (2013) Zeeman-type spin splitting controlled by an electric field. Nat Phys 9:563569
282. Jones AM, Yu H, Ross JS, Klement P et al (2014) Spin-layer locking effects in optical orientation of exciton spin in bilayer WSe2. Nat Phys 10:130–134
283. Yu ZM, Guan S, Sheng XL, Gao W et al (2020) Valley-layer coupling: a new design principle for valleytronics. Phys Rev Lett 124:037701
284. Huang B, Clark G, Klein DR, Neill DM et al (2018) Electrical control of 2D magnetism in bilayer CrI3. Nat Nanotech 13:544–548
285. Zhang L, Chen J, Zheng X, Wang B et al (2019) Gate-tunable large spin polarization in a few-layer black phosphorus-based spintronic device. Nanoscale 11:11872–11878
286. Li X, Wu X, Li Z, Yang J et al (2012) Bipolar magnetic semiconductors: a new class of spintronics materials. Nanoscale 4:5680–5685

287. Khani H, Pishekloo SP (2020) Gate-controlled spin-valley-layer locking in bilayer transition-metal dichalcogenides. Nanoscale 12:22281–22288
288. Peng H, Yang ZH, Perdew JP, Sun J (2016) Versatile van der waals density functional based on a meta-generalized gradient approximation. Phys Rev X 6:041005
289. Devakul T, Crépel V, Zhang Y, Fu L (2021) Magic in twisted transition metal dichalcogenide bilayers. Nat Commun 12:6730
290. Zhang Y, Liu T, Fu L (2021) Electronic structures, charge transfer, and charge order in twisted transition metal dichalcogenide bilayers. Phys Rev B 103:155142
291. Magorrian SJ, Enaldiev VV, Zólyomi V, Ferreira F et al (2021) Multifaceted moiré superlattice physics in twisted wse2 bilayers. Phys Rev B 104:125440
292. Tang H, Carr S, Kaxiras E (2021) Geometric origins of topological insulation in twisted layered semiconductors. Phys Rev B 104:155415
293. Kundu S, Naik MH, Krishnamurthy HR, Jain M (2022) Moiré induced topology and flat bands in twisted bilayer WSe2: a first-principles study. Phys Rev B 105:L081108
294. Fang C, Gilbert MJ, Bernevig BA (2012) Bulk topological invariants in noninteracting point group symmetric insulators. Phys Rev B 86:115112
295. Gatti G, Gosálbez-Martínez D, Roth S, Fanciulli M et al (2021) Hidden bulk and surface effects in the spin polarization of the nodal-line semimetal ZrSiTe. Commun Phys 4:54
296. Dresselhaus G (1955) Spin-orbit coupling effects in zinc blende structures. Phys Rev 100:580–586
297. Rashba EI (1960) Properties of semiconductors with an extremum loop 1 cyclotron and combinational resonance in a magnetic field perpendicular to the plane of the loop. Sov Phys Solid State 2:1109–1122
298. Herman F, Kuglin CD, Cuff KF, Kortum RL (1963) Relativistic corrections to the band structure of tetrahedrally bonded semiconductors. Phys Rev Lett 11:541–545
299. Riley JM, Mazzola F, Dendzik M, Michiardi M et al (2014) Direct observation of spin-polarized bulk bands in an inversion-symmetric semiconductor. Nat Phys 10:835–839
300. Gilardoni CM, Hendriks F, van der Wal CH, Guimaraes MHD (2021) Symmetry and control of spin-scattering processes in two-dimensional transition metal dichalcogenides. Phys Rev B 103:115410
301. Tilley R (2006) Crystals and crystal structures. Wiley, pp 67–79
302. Ishizaka K, Bahramy MS, Murakawa H, Sakano M et al (2011) Giant Rashba-type spin splitting in bulk BiTeI. Nat Mater 10:521–526
303. Winkler R (2004) Spin orientation and spin precession in inversion-asymmetric quasi-two-dimensional electron systems. Phys Rev B 69:045317
304. Winkler R (2003) Spin-orbit coupling effects in two-dimensional electron and hole systems. Springer
305. Flurry RL (1972) Site symmetry in molecular point groups. Int J Quant Chem 6:455–458
306. Zhang X, Liu Q, Luo JW, Freeman AJ et al (2014) Hidden spin polarization in inversion-symmetric bulk crystals. Nat Phys 10:387–393
307. Luo JW, Bester G, Zunger A (2009) Full-zone spin splitting for electrons and holes in bulk GaAs and GaSb. Phys Rev Lett 102:056405
308. Guimarães MHD, Koopmans B (2018) Spin accumulation and dynamics in inversion-symmetric van der Waals crystals. Phys Rev Lett 120:266801
309. Kobayashi T, Nakata Y, Yaji K, Shishidou T et al (2020) Orbital angular momentum induced spin polarization of 2D metallic bands. Phys Rev Lett 125:176401
310. Zhang Y, Liu P, Sun H, Zhao S et al (2020) Symmetry-assisted protection and compensation of hidden spin polarization in centrosymmetric systems. Chin Phys Lett 37:087105
311. Kim J, Hong X, Jin C, Shi SF et al (2014) Ultrafast generation of psuedo-magnetic field for valley excitons in WSe2 monolayers. Science 346:1205–1208
312. Sie EJ, McIver JW, Lee YH, Fu L et al (2015) Valley-selective optical Stark effect in monolayer WS2. Nat Mater 14:290–294
313. Ballottin MV, Mitioglu AA, Mooshammer F (2017) Giant magnetic splitting inducing near-unity valley polarization in van der Waals heterostructures. Nat Commun 8:1551

314. Press D, Ladd TD, Zhang B, Yamamoto Y (2008) Complete quantum control of a single quantum dot spin using ultrafast optical pulses. Nature 456:218–221
315. Berezovsky J, Mikkelsen MH, Stoltz NG, Coldren LA et al (2008) Picosecond coherent optical manipulation of a single electron spin in a quantum dot. Science 320:349–352
316. Mountain TL, Nelson J, Lenferink EJ, Amsterdam SH et al (2022) Valley-selective optical Stark effect of excitonpolaritonsin a monolayer semiconductor. Nat Commun 12:4530
317. Autler SH, Townes CH (1955) Stark effect in rapidly varying fields. Phys Rev 100:703–722
318. Unold T, Mueller K, Lienau C, Elsaesser T et al (2004) Optical Stark effect in a quantum dot: ultrafast control of single exciton polarizations. Phys Rev Lett 92:157401
319. Sim S, Lee D, Noh M, Cha S et al (2016) Selectively tunable optical Stark effect of anisotropic excitons in atomically thin ReS2. Nat Commun 7:13569
320. Yang Y, Yang M, Zhu K, Johnson JC et al (2016) Large polarization-dependent exciton optical Stark effect in lead iodide perovskites. Nat Commun 7:12613
321. Cohen-Tannoudji C, Haroche S (1969) Absorption et diffusion de photons optiques par un atome en interaction avec des photons de radiofréquence. J Phys Fr 30:153–168
322. Cunningham PD, Hanbicki AT, Reinecke TL, McCreary KM et al (2019) Resonant optical Stark effect in monolayer WS2. Nat Commun 10:5539
323. Cunningham PD, McCreary KM, Jonker BT (2016) Auger recombination in chemical vapor deposition-grown monolayer WS2. J Phys Chem Lett 7:5242–5246
324. Cunningham PD, Hanbicki AT, McCreary KM, Jonker BT (2017) Photoinduced bandgap renormalization and exciton binding energy reduction in WS2. ACS Nano 11:12601–12608
325. Gao S, Liang Y, Spataruy CD, Yang L (2016) Dynamical excitonic effects in doped two-dimensional semiconductors. Nano Lett 16:5568–5573
326. Ruppert C, Chernikov A, Hill HM, Rigosi AF et al (2017) The role of electronic and phononic excitation in the optical response of monolayer WS2 after ultrafast excitation. Nano Lett 17:644–651
327. LaMountain T, Bergeron H, Balla I, Stanev TK et al (2018) Valley-selective optical Stark effect probed by Kerr rotation. Phys Rev B 97:045307
328. Nagler P, Ballottin MV, Mitioglu AA, Durnev MV et al (2018) Zeeman splitting and inverted polarization of biexcion emission in monolayer WS2. Phys Rev Lett 121:057402
329. Conway MA, Muir JB, Earl SK, Wurdack M (2022) Direct measurement of biexcitons in monolayer WS2. 2D Mater 9:021001
330. Schlesinger I, Powers-Riggs NE, Logsdon JL, Qi Y et al (2020) Charge-transfer biexciton annihilation in a donor–acceptor co-crystal yields high-energy long-lived charge carriers. Chem Sci 11:9532–9541
331. Ye Z, Waldecker L, Ma EY, Rhodes D et al (2018) Efficient generation of neutral and charged biexcitons in encapsulated WSe2 monolayers. Nat Commun 9:3718
332. Paur M, Molina-Mendoza AJ, Bratschitsch R, Watanabe K et al (2019) Electroluminescence from multi-particle exciton complexes in transition metal dichalcogenide semiconductors. Nat Commun 10:1709
333. Barbone M, Montblanch ARP, Kara DM, Palacios-Berraquero C et al (2018) Charge-tunable biexciton complexes in monolayer WSe2. Nat Commun 9:3721
334. Raja A, Chaves A, Yu J, Arefe G et al (2017) Coulomb engineering of the bandgap and excitons in twodimensional materials. Nat Commun 8:15251
335. Schmitt-Rink S, Chemla DS (1986) Collective excitations and the dynamical Stark effect in a coherently driven exciton system. Phys Rev Lett 57:2752–2755
336. Scolfaro D, Finamor M, Trinchão LO, Rosa BLT et al (2021) Acoustically driven stark effect in transition metal dichalcogenide monolayers. ACS Nano 15:15371–15380
337. Sie EJ, Lui CH, Lee YH, Kong J (2016) Observation of intervalley biexcitonic optical stark effect in monolayer WS2. Nano Lett 16:7421–7426

Chapter 6
Heterostructures

6.1 Introduction

Semiconductor heterostructures and their superlattices are considered to provide an essential forum for various significant device applications like lasers, light-emitting diodes, solar cells, high-electron-mobility transistors, etc. Specifically, heterostructures belong to IV, III–V, and II–VI group semiconductors having covalent bonding among atoms at the heterointerface that are commonly developed by epitaxial progression. Therefore, the selection of material constituents is strappingly restricted by lattice mismatch. Their atomic inter diffusion during the growth process developed atomic-scale interface roughness and compositional inclines at the boundary to compromise the functionalities of such heterostructures, exclusively when thicknesses reduce to a single atomic layer.

TMDs covalent bonded distinct layers are usually hold to each other by week vdW forces having no surface dangling bonds. It gives an adequate advantage to assemble the discrete layers in the operative multiple-layer heterostructures to atomically sharp interfaces without atom interdiffusion, which offer digital control of the layered components under the none lattice parameter constraints. This could be verified by the opportunity to utilize vdWepitaxy in the growth of better quality ultrathin layers on substrates possessing a huge lattice incompatibility and atomically smooth edges [1–5]. As the epitaxial progression of MoS_2 using vdW strengths has been established in the beginning of 1990s [5, 6]. Not only the epitaxial development of MoS_2 but also other TMDs monolayers have been also mechanically constructed by taking together separate layers [7]. Therefore, successive stacking of layers is possible that sometimes also denoted as the vertical heterostructure. That signifies the superficial material assembling in a chosen arrangement as construction of Lego blocks with atomic accuracy [8]. Taking advantage of the method numerous 2D materials can be associated including graphene and h-BN. These kinds of structural configurations have been expansively explored both theoretically and experimentally along

A. K. Singh, *2D Transition-Metal Dichalcogenides (TMDs): Fundamentals and Application*, Materials Horizons: From Nature to Nanomaterials,
https://doi.org/10.1007/978-981-96-0247-6_6

with the key difficulties of the constancy of monolayers to oxygen and surface (interfacial) impurity. Moreover, another alternative approach to fabricate heterostructure is horizontal heterostructure by altering the alloying configurations in a manageable way for the identical single layer. Thus heterostructures of TMDs including other 2D materials have been investigated by employing different approaches [9].

Since a single-layer TMD consists of a single or few-atom-thick covalently bonded lattice-free of the dangling bonds atomic sheets that often exhibit extraordinary electronic and optical properties. In contrast, typical nanostructures consisting dangling bonds and trap states at the surface. The 2D TMDs' fully saturated chemical bonds on surface interactions with the neighboring layers are usually characterized by van der Waals forces. Like without direct chemical bonding van der Waals interactions allows to integration of highly disparate materials possessing no constraints of crystal lattice matching. This unique property also provides considerable freedom for the integration of 2D TMDs for the various nanoscale materials to create diverse van der Waals heterostructures that are usually not possible with conventional materials [10].

The integration of van der Waals heterostructures to other nanomaterials usually in form of a combination of 2D TMDs with 0D (plasmonic nanoparticles and quantum dots) and 1D nanostructures (nanowires and nanoribbons), as illustrated various materials heterostructures in Fig. 6.1a–f [11–15]. The creation of 0D–2D and 1D–2D van der Waals heterostructures opens up a new pattern for nanoscale material integration that makes enable them for many extraordinary devices, including broadband photodetectors with ultrahigh speed or gain and a new generation of atomically thin transistors with unprecedented speed and flexibility [11–15]. Whereas the van der Waals integration of different 2D TMDs is vertically stacking 2D–2D heterostructures and their superlattices give rise to possibilities to control and manipulate the next generation confinement and transport of charge carriers, excitons, photons, and phonons within the atomic interfaces. Moreover, the integration of 2D TMDs with traditional 3D bulk materials are also form 2D–3D heterostructures with novel functions for well-developed bulk materials and conventional electronic technologies [10].

6.2 TMDs Heterostructures

Since transition metal dichalcogenides (TMDs) could accompaniment or even outshine graphene (zero band gap) in electronic and optoelectronic applications owing to their robust spin–orbit connection and advantageous electronic and mechanical features. Specifically, their decent d electron arrangements offers numerous electronic constructions of 2D TMDs contingent on their stoichiometry and numeral of layers [16]. TMDs electronic constructions can be with the semiconductor (e.g., MoS_2, WS_2), semimetal (WTe_2, $TiSe_2$), metal (NbS_2, VSe_2), or superconducting ($NbSe_2$, TaS_2) features [17, 18]. Additionally, with the decreasing layer thickness they exhibit direct-to-indirect band gap transition. Thus significantly their electronic

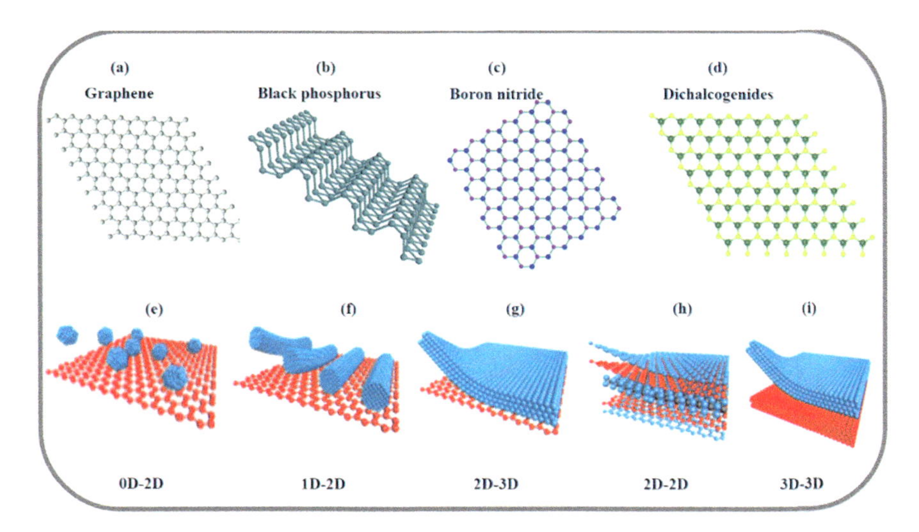

Fig. 6.1 Representation for the two-dimensional layered materials and van der Waals heterostructures: **a** monolayer graphene; **b** monolayer black phosphorus; **c** monolayer boron nitride; **d** monolayer dichalcogenides; **e–i** schematic illustrations of different dimensions layered materials integration (**e–i** adopted from Liu et al. 2019 Nature, 567, 324)

and optoelectronic properties regulates through the layer stacking order and composition, this could correlates to layer number/thickness without adjustment of lattice matching. For the instance, 2D semiconducting TMDs' proper band gaps are considered to be useful to convert solar energy more efficiently [19, 20]. Nonetheless, the fact is that the pristine 2D TMDs remain suffering from few difficulties associated to their structural or features due to informal restacking, sluggish charge dynamics, and/or poor visible-light harvesting or alteration. They also suffer from a few other factors, like light absorption range versus redox potentials and active site versus conductivity. So the amend or functionalities of pristine 2D TMD heterostructures are desired to get a variety of properties [11, 21].

Thus, the formation of heterostructures through the combination of dual materials considered to be a key approach to regulate the electrons in semiconductors has been extensively explored. The discoveries are based on possible functionalities such that integer, fractional, and quantum spin Hall effects including exciton–polariton condensations [22–27]. The functional materials based on heterostructures can have numerous utilizations covering semiconductor lasers and light-emitting diodes based on the work functions of the incorporated materials. In heterostructures predominately dual kinds of band arrangements are shaped. In the type-I arrangement both conduction band minima (CBM) and valence band maxima (VBM) are situated within the material having a slight band gap, as depicted in Fig. 6.2a, b. Where the electrons and holes excited in a wide-gap material transferring to a narrow-gap material, as arrows directed in schematics. The quantum confinement of the electrons and holes in an identical area is enabling the radiative regrouping, which is an essential requirement for the light-emitting utility [28]. Moreover, the carriers

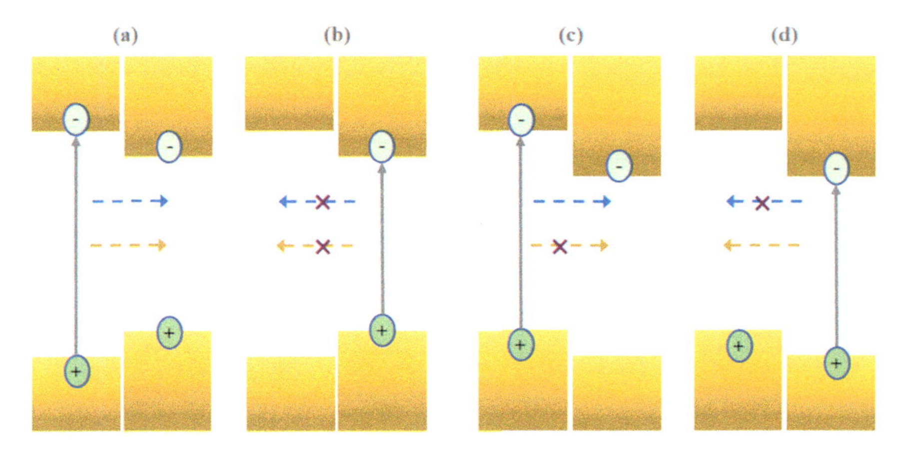

Fig. 6.2 Schematic for the band alignment between two semiconductors: **a, b** type-I; **c, d** type-II, alignments

excited in the narrow-gap material also inhibit mutual layer transferring owing to their inferior energies. On the other hand in type-II arrangement, both CBMand VBM can be positioned in diverse materials, as illustrated in Fig. 6.2c, d. In such circumstance, both CBM and VBM are situated in a narrow-gap with a wide-gap material. This situation leads to the excitation of the wide-gap material followed by the transfer of electrons but not from the holes. Whereas the opposite charge transfer occurs when the narrow-gap material is excited. Therefore, the separation of the electrons and holes of the different layers increases the lifetime, it is the basic need for photovoltaics and photodetection applications [29–32].

Moreover, studies of heterojunction physics on heterostructures have also suggested that it usually classified into three types, namely type-I, type-II, and type-III [33, 34]. By demonstrating the two semiconducting materials A and B could merge and their valence band maxima (VBM) and conduction band minima (CBM) should satisfy the relation $VBM_A < VBM_B < CBM_B < CBM_A$ for the type-I heterostructure and $VBM_A < VBM_B < CBM_A < CBM_B$ for type-II, while $VBM_A < CBM_A < VBM_B < CBM_B$ for type-III. According to interpretation type-I and II heterostructures are falls into the semiconducting category, while type-III heterostructures become metals. Stronger charge transfer in type-III heterostructures could enhance the magnetic proximity effect and the Coulomb interaction between the substrate, therefore, TMDs possibly enhanced the valley splitting. It also opens opportunities to get a large valley splitting in type-III vdW heterostructures [34].

6.2.1 Anderson's Rule Band Alignment

Significantly band alignment of two-dimensional heterostructures is also a great interest. To describe the band alignment from different approaches it has been considered. Such as, the most frequently used Anderson's rule, despite it being a macroscale approach directly or implicitly is used to determine the band alignments of TMDs [35–38]. So, Anderson's rule has been considered an appropriate approximation to describe band alignments of TMDs systems [39]. As per Anderson's rule approximation generally two type-II heterojunctions consists uncontaminated complexes, namely, A/C and B/C, in which A and B are homogeneously mixed and form $A_xB_{(1-x)}$. In case of the complex A in the A/C heterojunction belongs to a lower CBM and VBM than C and B/C heterojunction "B" possess a higher CBM and VBM compare to C, therefore, the band boundaries should bisect for the certain configuration $A_xB_{(1-x)}/C$ $(0 < x < 1)$, as diagrammatically illustrated in Fig. 6.3a, b. By explicitly assuming the alloys CBM and VBM can alter monotonically to the stoichiometric concentration of x. Hence in overall picture CBMs and VBMs of $A_xB_{(1-x)}$ and C crosses at $x = x_1$ and x_2 (usually $x_1 \neq x_2$). However, the type-I heterojunction belongs to the range $x_1 < x < x_2$, otherwise, it is type-II. Whereas in this case, CBM and VBM are not varying monotonically with stoichiometry and manifold bisecting points also exist. Like type-I heterojunctions they also form manifold distinct composition ranges. Hence, conclusively it is to say that the bisection of the CBM and VBM in uncontaminated complexes results in type-I/II transitions, therefore, the heterojunctions in TMD alloys must occur [40].

In the simple words, Anderson has suggested (like Schottky Mott) that the band alignment is purely dependent on the relative work function/affinities of the two constituents. However, relatively very few experimental systems have validated this approach. Beside of, this model has been considered the backbone to expand by including the effect of charge states at the interface [41]. Such charge states are due to the formation of new states through chemical bonding between the two systems and charge neutrality caused by the induced field creation, as the resultant of the two systems' electron affinities [42]. This kind of charge state also includes an interface (or metal) induced gap state, IIGS (or MIGS) [43, 44]. Moreover, energies of such states and the shape of the electrostatic field induced across the interface is characterized nontrivial because of the atomic bonding and system geometry. Additionally weakly interacting two-dimensional materials also offers to build an understanding of band alignments in these systems [35–37]. Likely in the 3D case, the dangling bonds are not present at the surface but newly formed bonds between the layers cannot have, however, substantially they do not exist in reconstruction [8]. Such TMDs heterostructures are usually close to real-world likenesses to Anderson's original thought experiment, in which the band alignments of two separate systems are close to infinity with minimal interaction [45].

6.2.2 *Strategies to Construct 2D TMDs Heterostructures*

Construction of proper heterogeneous for the targeted application is the critical task, to effectively regulate physical parameters. In overall, approaches incorporating structural defects (S vacancies), doping, adsorption of the molecules to sensitization,

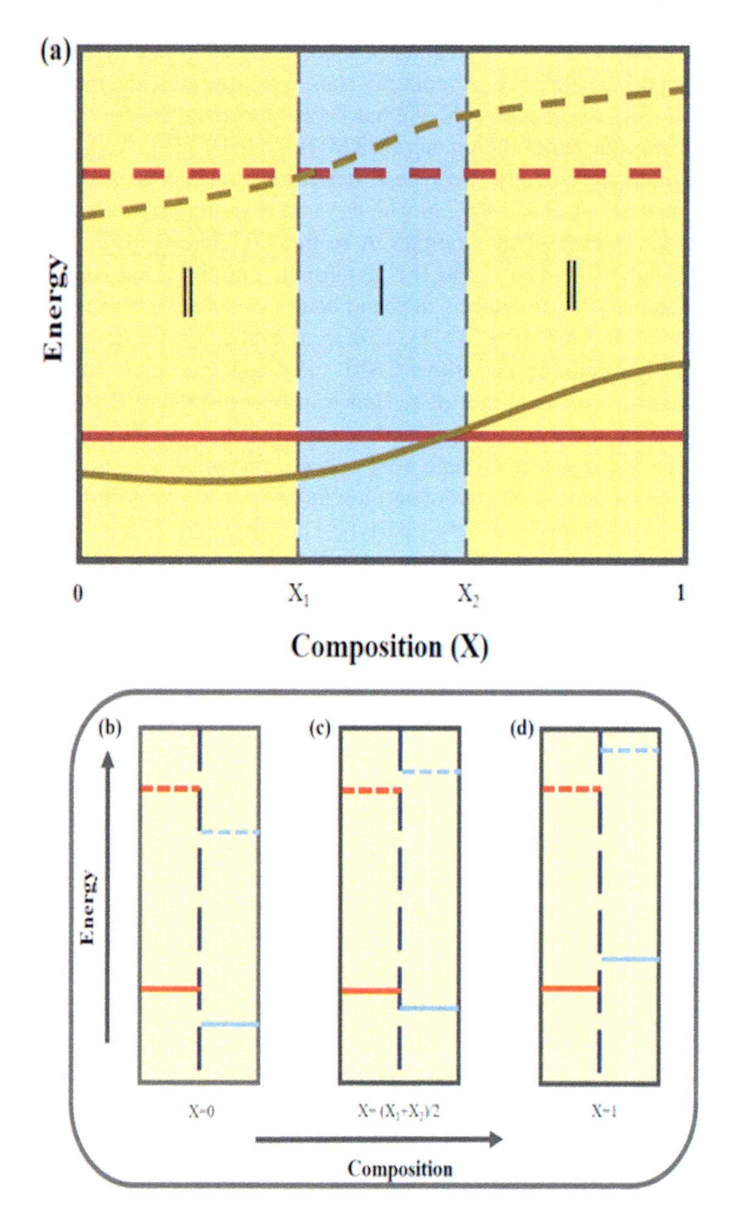

Fig. 6.3 **a** Diagram plot to the conduction band minima (CBM, dotted curved) and valence band maxima VBM (curved solid) in an $A_{(1-x)}B_x$ alloy (curved lines) as a function of compositional concentration of x and for a compound C (independent of x, flat lines) that are combined to procedure a heterojunction, in case of $x_1 < x < x_2$ the heterojunction is of type-I (i.e., the bandgap of C lies completely within the bandgap of the $A_{(1-x)}B_x$ alloy), while for $0 < x < x_1$ and $x_2 < x < 1$ the heterojunction of type-II (i.e., the bandgaps of compound C and the $A_{(1-x)}B_x$ alloy only partially overlap)—as illustrated in **b–d**

and hetero-association with other lower dimensional nanomaterials that involves energy band and surface-structure or interface engineering to the diverse situations, as depicted in Fig. 6.4a–r. Such strategies are helpful to get an improved equilibrium among associated aspects influencing the physical parameters. As the intense, the dynamic sites of 2D nanomaterials stem structural disorders create dislocations and vacancies, which leads to electron scattering and assembly among electrons and holes. While the normal surface adjustments scarcely lead to the building of ideal photocatalysts even they recover to certain extent the photoactivity of original 2D TMDs. As an instance, the introduced dopants (cobalt and nickel ions) into a MoS_2 structure also improves the catalytic action of basal planes with the cumulative density of the non-saturated atom places including additional organizational imperfections. By implications of these approaches, one can regulate the CB and band gap to improve the water reduction reaction and conductivity. Similarly, organic dye sensitizer light harvesting capability also improves by implication of a push–pull impact among the donor and acceptor, ultimately it improves the chemical strength of 2D TMDs [46].

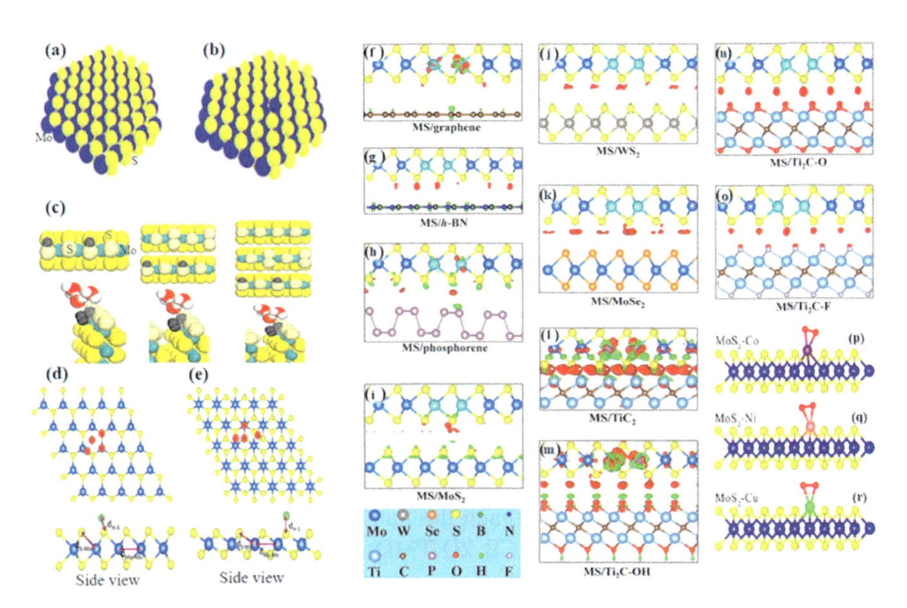

Fig. 6.4 Schematic illustrations of typical strategies for the photoactivity of 2D TMDs (as the typical example MoS2): **a** pristine; **b** defects to heteroatoms doping (**a** and **b** adopted from Pető et al. 2018, https://doi.org/10.1038/s41557-018-0136-2; oai:real.mtak.hu:88,048); **c** single, bilayer and trilayer heterostructures formation mechanisms (black spheres represent either adsorption or desorption center, while pink color corresponds to absorbing layer) (adopted from Bora Seo et al. 2015 ACS Nano 9, 3728–3739); **d, e** typical representation of possible substitutional and adsorbing positions in H-MoS_2 and T-MoS_2 (here letters A, B, C, D represents adsorption sites) (adopted from Xu et al. 2016 AIP ADVANCES 6, 075001); **f–r** different heterostructures of MoS_2 with other 2D structures nanomaterials by surface functionalization (**f–o** adopted from Ling et al. 2019 npj Comput Mater 5, 83 and **p–r** from Yoo et al. 2021 Nanomaterials 11, 832)

Therefore, to construct 2D TMD heterostructures strategies can be adopted based on the regulation of structural and features relations through elemental doping by the formation of faults (vacancies) or particular surfaces [47]. Moreover, considering their intrinsic limitations predominantly 2D TMDs-mixed-dimensional heterostructures have been classified into 0D–2D, 1D–2D, and 2D–2D heterostructures classes.

6.2.2.1 0D–2D Heterostructures

This type of heterostructures construction is usually achieved from the 0D nanomaterials to use as the co-catalysts. Such co-catalysts promotes the charge parting by accommodating electrons or holes in 2D TMDs. Additionally they also provides additional dynamic spots for the reduction or oxidation reactions, thereby, improves in overall constancy and reduction in activation energy to surface chemical reactions including suppressing lateral or backbone reactions [48, 49]. The incorporation of components and dimensionality/microstructure (core/shell) due to extent impacts from nanoparticles to single atoms are also useful for the flexible regulation of photoactivity [50, 51]. So, the electronic features, like organizational constancy and exciton generation of 0D semiconductors are predominantly depending on their size, shape, and composition. Like, if the extent is lesser than the Bohr radius (~ 18 nm for PbS, ~ 46 nm for PbSe, and ~ 5 nm for CdSe) allows light harvesting in visible and infrared areas for the solar light [52].

Moreover, 0D–2D hybrid heterostructures also creates synergy for photocatalytic activity. Specifically, the use of 0D metal nanoparticles as co-catalysts could enhance the light absorption of 2D TMDs due to the involvement of diverse mechanisms. Like a dipolar oscillation of electrons in the metallic nanostructure induces a robust absorbance of electromagnetic effect, therefore, 0D metallic nanostructures acts as plasmonic projections for proficient light harvesting and concentration [53]. Additionally, the surface plasmon resonance (SPR) of 0D metal nanomaterials is also impressively improve the indigenous optical field efficiently in 0D metal nanomaterials gap expanses [54]. The 0D–2D hybrid heterostructures combination also enhances the charge separation through the transformation of photo-excited electrons from metal nanoparticles to CB 2D TMDs and formed a Schottky junction [55, 56].

0D–2D hybrid heterostructures metal nanoparticles also behave as electron sinks and it work as conductive “wires” to accelerate the electron transmission or charge parting. Therefore, electrons injects in the CB of 2D TMDs. This is a significant process for the wide-band gap semiconductors, to increase the redox capabilities due to certain extent in the procedure may overcome the restriction of energy arrangement of 2D TMD that promotes the electron–hole separation, therefore, more active sites are available, and it also acts as a substrate for the suppressing [57]. Such heterostructures metal co-catalysts also catalyze the reorganization of H_2 and O_2 into the water and lead to a harm in total efficiency. However, to resolve this problem effective recombination-blocking materials (such as acid-tolerant SiO_2, MoO_x, or CrO_x) have been also used to prevent the facile reverse reaction [58, 59].

Hence, the 0D–2D hybrid heterostructures of 2D TMDs allow the variable regulation of charge transferring at their edges with the additional active sites to enhance the physical activity (such as photochemical reactions to decrease over potential). Additionally, such hybrid materials also offer an improved appreciation of energy level matching among donor and acceptor materials that are useful for various technical applications, such as solar energy is achieved more effectively [52].

6.2.2.2 1D–2D Heterostructures

Compared to 0D nanoparticles 1D or 2D nanostructures usually have greater mobility and lesser charge reintegration rates [60]. The 1D nanomaterials' large length-to-diameter ratios allow their more easy separation and recycling after completing the photoreactions or other physical activity. The 1D–2D heterostructures are entirely incorporate the qualities, but they have also mitigated the disadvantages of solitary units. Like the short surface region of 1D nanowires and the restacked of 2D nanosheets. However, 1D–2D heterostructures enable novel functions that cannot be achieved from their components [61]. The conductive of 1D nanomaterials (like carbon nanotubes and carbon nanofibers) is also useful to make electron transport mediators or outstanding electron acceptors, to get an improved electron transmission with the decreased reintegration rate of charge carriers [62]. Moreover, 1D nanomaterials possessing mesoporous constructions can have great surface regions to representation of suitable dynamic locations. Thus the combination of 1D–2D heterostructures are considered to be a good construction of hybrid systems for their potential technical applications [11].

6.2.2.3 2D–2D Heterostructures

2D–2D heterostructures have provided advantages when they are apprehended collected in form of a vertical stack through the vdW forces. These kinds of stacking allow a far bigger quantity of arrangements compare to any concessional growth approach. Therefore, the development of 2D–2D heterostructures is variable without additional necessities for lattice equivalency and processing compatibility. In such circumstances formed 2D–2D heterostructures vdW gaps are considered as 2D nanoscale gaps which changes the microenvironment of dynamic locations via quarantine impacts (arithmetical constraint and quarantine field). These systems' catalytic reactions in confined regions have quicker kinetics than in an exposed system [63, 64].

The interface is the utmost significant factor in the utilization of the devices [98]. Comparing the 0D–2D and 1D–2D, the 2D–2D hybrid heterostructures have an enhanced interconnected interface that also expressively improve the key photoactivity property. The 2D–2D hybrid heterostructure's intimate edge formation among dual kinds of 2D nanomaterials is also considered to be favorable for exciton dissociation. Because such formation built-in potential, as the consequence it overall improve the device quantum efficiency. However, due to the formation of intimate interfaces among 2D nanomaterials they face problem of lattice mismatch. Nevertheless, the great connected region in 2D–2D heterostructures is effectively promotes the relocation and parting of electron–hole pairs. For the instance, 2D–2D heterostructure's redox potentials are readily regulate to match. Thereby, the overall water-splitting reactions through the addition of a hydrogen evolution photocatalyst and/or an oxygen evolution photocatalyst. Moreover, the 2D–2D heterostructures are also beneficial to improve the stability of the systems [11].

6.2.3 Band Alignment in TMD Heterostructures

Band edge alignment in heterojunctions due to composed of pseudo-binary alloys. Usually, pure TMD compounds and alloys are generalized from the gradient approximation and Perdew-Burke-Ernzerhof (PBE) functions [65–70]. But these approximations also have certain drawbacks like local density and unreliable yield band gap predictions [71, 72]. PBE implies the band arrangement in $MoSe_2/WS_2$ heterojunctions that is type-I [65, 67], whereas the investigational findings have revealed that it is type-II [73, 74]. To overcome this problem various authors have presented different models to provide more accurate results. In this order, a significant approximation Strongly Constrained and Appropriately Normed semilocal density functional (SCAN) has been also introduced [75]. To consider this approximation can be provided more accurate properties of many materials compared to others [76]. As illustrated in Fig. 6.5, the type-II $MoSe_2/WS_2$ heterojunction together CBM WS_2 ~ 25 meV that is inferior to $MoSe_2$. Moreover, it also consists of the above interpretation by revealing the CBM and VBM monotonic variation to the composition (x) of the organizations. Nevertheless, CBM and VBM are generally expressed by a second-order polynomial $E(x) = E(1)x + E(0)(1-x) + wx(1-x)$, here $E(x)$ represents the band boundary energy (CBM or VBM) as a function of stoichiometry (x), $E(0)$ and $E(1)$ represents the band boundary energies of the uncontaminated TMD complexes, and w is the fitting parameter for the DFT data.

Using the approach of heterojunction band arrangement can also obtain the band boundary statistics, as typically defined for the six heterojunctions of dual diverse TMD pseudo-binary single layers. It reveals a probable fabrication of heterojunctions in which every side identical TMD forms the pseudo-binary alloy with distinct stoichiometry (such as $Mo_{0:3}W_{0:7}S_2 = Mo_{0:5}W_{0:5}S_2$), but such junctions are usually unavoidably type-II, where both VBM and CBM have identical monotonicity to stoichiometry (see Fig. 6.5). Their types of band alignments with the effective band gap dissimilarity among the lowermost CBM and topmost VBM of the two successive single-layers as a function of stoichiometry for all six TMD pseudo-binary heterojunctions, as depicted in Fig. 6.6a–f. It reveals that simply three pseudo-binary alloy couples have both type-I and type-II

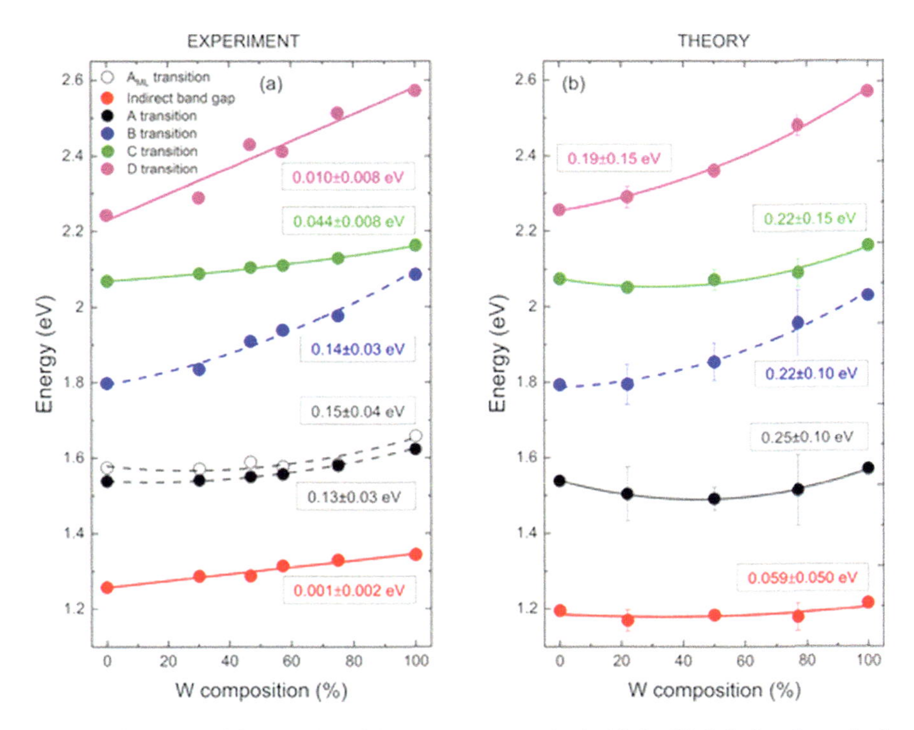

Fig. 6.5 Optical transition energies of the tungsten content for the Mo1−xWxSe2 alloy determined: **a** experimentally; **b** DFT calculations (adopted from Kopaczek et al. 2021 ACS Omega 6, 19893–19900)

heterojunctions, namely, $MoS_{2(1-y)}Se_{2y}/WS_{2(1-x)}Se_{2x}$, $WS_{2(1-y)}Se_{2y}/Mo_{(1-x)}W_xSe_2$, and $MoS_{2(1-y)}Se_{2y}/Mo_{(1-x)}W_xS_2$, whereas the remaining have solitary type-II heterojunctions [40].

As depicted in Fig. 6.6a, type-I/type-II alteration in $MoS_{2(1-y)}Se_{2y}/WS_{2(1-x)}Se_{2x}$ heterostructures could be described with these words; the type-I area lies nearby the upper left corner because the relatively small CBM gap among $MoSe_2$ and WS_2. This directly relates to comparatively slight anion doping into WS_2 or $MoSe_2$ that converted the $MoSe_2/WS_2$ heterojunction in type-I. Similar transition has also noticed in $MoS_{2(1-y)}Se_{2y}/WS_{2(1-x)}Se_{2x}$ pseudo-binary heterojunction transition. Thus, in such a transition MoS_2CBM (VBM) is underneath to the WS_2 and $MoSe_2$ overhead to the WS_2. It directly indicates to certain stoichiometry CBM (VBM) difference between these two successive layers of materials must be zero, as depicted in Fig. 6.6b. It also leads the compositions (y) type-I heterojunction throughout the composition range (x). Likewise, stoichiometry (x) nearly type-I heterojunction for the whole stoichiometric range of (y). Here one should keep in view that type-I stoichiometric range while psychiatrists to zero at the lowest left corner (see Fig. 6.6b), in this case it should be $x = 0$, $y = 0$, resulting a vertical heterojunction that is a simple homogeneous MoS_2 bilayer (it's a homo in-spite of heterojunction). An analogous description can also adopt for the $WS_{2(1-y)}Se_{2y}/Mo_{(1-x)}W_xSe_2$ pseudo-binary

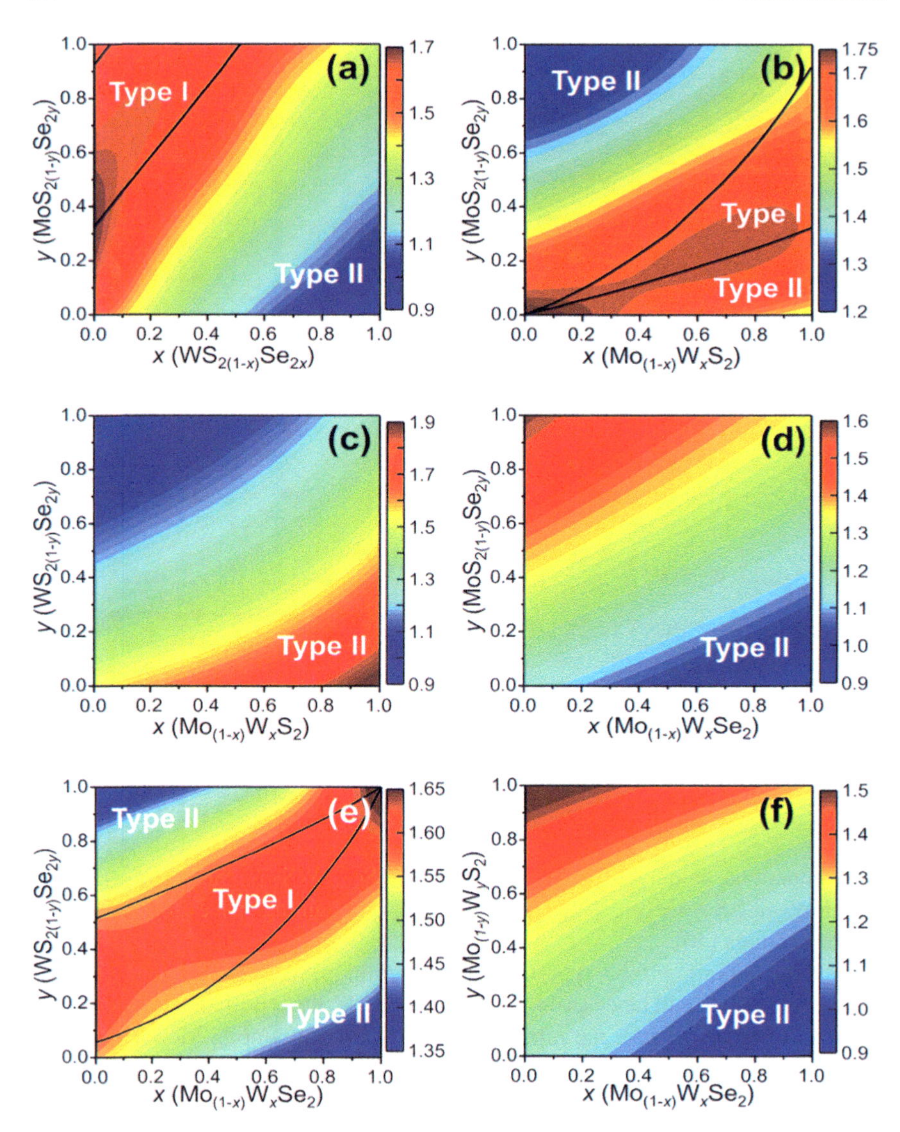

Fig. 6.6 Representation of pseudo-binary heterojunction type diagrams for all six possible pseudo-binary alloy heterojunctions. The contours show the effective bandgap as a function of composition (fit 25 compositions each). The black curves represent type-I/II transition boundaries, corresponding to the zero contour lines of the CBM difference or VBM difference (adopted from Zhou et al. 2019, https://www.osti.gov/servlets/purl/1767702)

vertical heterojunction except for the limits $x = 1$, $y = 1$ to the WSe_2 bilayers, as typical representation is given in Fig. 6.6e [40].

Thus adjusting the heterojunction since type-II to type-I by influencing the stoichiometric ratio of (MoW) $(SeS)_2$TMD pseudo-binary systems has provided an opportunity to regulate the CBM (VBM) difference of the active band gap for the heterojunction, as an intense control of the light emission wavelength in type I heterojunction. Typically type-I heterojunctions' effective band gaps categories are in the range 1.6–1.8 eV, whereas the impressive range of type-II heterojunctions band gaps can be altered in the range 0.9–2.0 eV [40].

6.2.4 Vertical Heterostructures

Besides, the significant utilization of 2D monolayers several prominent issues that restrict their applications in high-quality use in optoelectronics. Such as the extremely-thin thickness of 2D TMD crystals cannot absorb adequate light power, because it directly affects the exterior quantum efficiency and detection ability of optoelectronic devices [77]. Additionally, it is also problematic to adjust the intra-layer excitons from the discrete 2D TMD layer crystals due to a little lifetime that restricts their utilization as exciton devices [78]. To resolve such problems of 2D TMD monolayers the vertically (van der Waals) stacked heterostructures using twofold or additional kinds of 2D TMD crystals via van der Waal forces have been introduced that find a valuable attention in spectroscopy, electronics, photocatalysis, optoelectronics, and materials science. Such vertically stacked 2D TMD layers crystals originate new features which are diverse to get discrete 2D TMD, comprising distinct interlayer excitons, photovoltaic effect, tunneling-assist carrier recombination, ultrafast and high-efficiency charge transfer, and magnetic effect [79–84]. To get a desired output usually high-quality 2D TMDs-based vertical heterostructures are used, such as high-quality performance, lower power consumption, optoelectronics, spintronics, tunneling field effect transistors (FETs), photodetectors, photovoltaics, light emission devices, and catalysts.

The principal vertical stacking construction was demonstrated with the graphene/h-BN heterostructure by the process of exfoliation and transferring, in 2010 [85]. Later on, several diverse strategies have been employed to make vertically stacked construction, like mechanical exfoliation and transfer [86–88], molecular beam epitaxy (MBE) [89–91], pulsed laser deposition (PLD) [92, 93], metal–organic chemical vapor deposition (MOCVD) [94–96], chemical vapor deposition (CVD) [77, 97–100], and atomic layer deposition (ALD) [101, 102] to produce 2D TMD vertical heterostructures.

Though, by use of the conventional exfoliation and transferring processes fabricated 2D TMD vertical heterostructures controllability limits the surface structural construction, number of layers, low yield, time-consumption, 3D selectivity including crystal positioning. Additionally, throughout the transferring procedure hard to evade the incorporation of a few impurities owing to connection among the

TMD crystals and transferring mediators, therefore, ultimately deterioration may appear in the device's performance [103]. Similarly, the growth development with MBE, PLD, and MOCVD usually suffers to the thermodynamic symmetry and crystal excellence compared to as-grown 2D materials (normally lesser grain sizes). In the gas assist process, post-growth thermal procedure or annealing in a sulfur atmosphere is preferred to enhance the crystal excellence. While very few reports with LAD process fabricated 2D TMD vertical heterostructures are available. Thus, overall almost all approaches remain desire to optimize for their advances and universality to fabricate 2D TMDs-based vertical heterostructures [104]. Besides the advantages and disadvantages of the CVD method, it has been utilized most extensively to construct 2D TMDs vertical heterostructures with the controllable approach of high-quality scalable production. At a sought size, stacking sequence, chemical composition, physical properties, and contamination free to edges [105–108]. By using the CVD method various 2D TMDs vertical heterostructures have been fabricated and interpreted, like MoS_2/WS_2, WSe_2/MoS_2, $WSe_2/MoSe_2ReS_2/WS_2$, $MoSe_2/WSe_2$, p-MoS_2/n-MoS_2, MoS_2/WSe_2, $PtS_2/PtSe_2$, NbS_2/MoS_2, $MoSe_2/MoS_2$, $MoTe_2/MoS_2$, MoS_2–WS_2/WS_2, VSe_2/MX_2(M: Mo, W; X: S, Se), VS_2/WSe_2, $PtSe_2/MoSe_2$, MTe_2/WX_2 (M = V, Nb, Ta; X = S, Se), $WS_2/WSe_2(MoSe_2)$, WSe_2/WS_2, $NiTe_2/MoS_2$, MTe_2/WSe_2 (M = Ni, Co, Nb, V), VSe_2/WSe_2, WX_2–MoX_2/WX_2–MoX_2 (X = S, Se), $WSe_2/WS_{2(1-x)}Se_{2x}$, $Mo_6Te_6/MoS_2(1-x)Te_{2x}$, $CrSe_2/WSe_2$, $NbSe_2/MoSe_2$, $WS_2/Mo_{1-x}W_xS_2$, etc. [109].

6.2.4.1 Matching Lattice Heterostructures

Despite various theoretical works based on the lattice-matched heterostructures, a remarkable systematic study became by describing the dual distinct stacking ordering, as illustrated in Fig. 6.7 [110]. As per the demonstration, the categorization of the stacking order corresponds to the bulk phase while the transition metal of one layer is over to chalcogen atoms of the other layer (B). While another stacking order is in which the chalcogen atoms of the two planes across the gap over to each other (A). Moreover, the used notation "bilayer (MoS_2, WS_2, B)" was defined for the bilayer made by MoS_2 and WS_2 with the B stacking. In addition to these, the innovation was also done with the bulk crystals by making the alternating layers, like "crystal (MoS_2, $MoSe_2$, A)" whereas both MoS_2 and $MoSe_2$ crystals layers were made in an alternatively repeated manner with the A stacking.

Therefore, based on these investigations the kinds of heterostructures are usually divided into four types. Type-I corresponds to WS_2/MoS_2 hybrids that are characterized by an indirect gap for both A and B stacking, in which B stacking is considered to be energetically more favorable. However, the discrepancy among the direct and indirect gaps is slightly lesser, therefore, it is difficult to identify what magnitude affected functional in DFT simulations. On the other hand, the WS_2/WSe_2 and MoS_2/WSe_2 bilayers are correlated to type-2 that have a hybrid direct gap at K for the both A and B bilayer stacking. Surprisingly, type-2 systems bulk crystals shaped by an endless numeral of stacking layers with A stacking, such as crystals WS_2, WSe_2, and A,

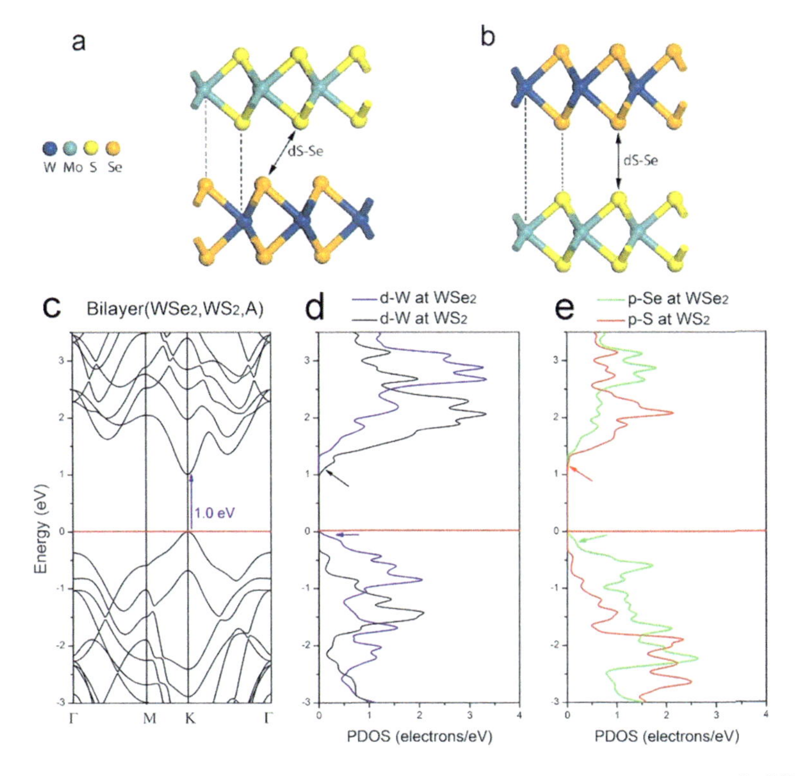

Fig. 6.7 Atomic models for the bilayer heterostructures of semiconducting TMDs with different stackings A and B (upper panel). The dotted lines indicate the alignment of atoms between layers, and the arrows correspond to the distance dS–Se between chalcogen atoms of different layers, the type-2 Bilayer (WSe_2, WS_2, A) (lower panel): **a** band structure for the direct band gap; **b** PDOS for the *d*-electrons of W at each of the layers. The blue arrow indicates the states at the top of the VB originating from W in the WSe_2 layer and the black arrow corresponds to the states at bottom of the CB due to W in the WS_2 layer; **c–e** PDOS shows the *p*-electrons of the chalcogen atoms. The green arrow exhibits the states at the top of the VB of Se in the WSe_2 layer and the red arrow to the states at the bottom of the CB of the S atoms in the WS2 layer (adopted from Terrones et al. [110])

with a direct band gap at K (ca. 1 eV). As per the claim, it is the principal event of a multiple-layer TMD organization associating a direct band gap [110]. Similarly, crystals are also formed by alternation of the layers with the type B stacking, such as crystals WS_2, WSe_2, and B, which have an indirect gap. The findings were revealed the states at the over to the VB for the WS_2/WSe_2, A is bilayer owing to W and Se atoms in the WSe_2 layer, while at the bottommost the CB created owing to W and S atoms of the WS_2 layer, thereby, both electrons and holes can be three-dimensionally discrete. This is also exciting to bilayer TMDs that have analogous manners in which inversion symmetry breaks in an employed exterior electric field [111, 112]. Where the indigenous Mulliken charges of the atoms in diverse layers have been intended, similar behavior has been noted for the WS_2/WSe_2, A bilayer, where the W atoms of the WS_2 layer possess a positive net charge (+ 0.16e) and the W atoms of theWSe_2

layer have a negative charge (− 0.09e). Moreover, S atoms of the WS_2 layer have a negative total charge and Se atoms possessed a positive overall charge. It should be noted that this manner can solitary appear while the distinct layers possess diverse chalcogen species. It has also established that the inherent electric field is due to electron–hole parting and the existence of a direct band gap, however, demonstration has not provided a perfect view of the respective reason and consequence.

Moreover, the hybrid containing $WS_2/MoSe_2$ layers and $MoS_2/MoSe_2$ layers are correlated to be type 3. As per interpretation both of them have a direct gap under A-type stacking, but they have an indirect gap under B-type stacking. The hybrids $WSe_2/MoSe_2$ with the indirect gap are further categorized as type 4. Moreover, the MoS_2/WS_2 heterostructures have been also explored by another group [113]. By comparing the heterostructures with bilayers of uncontaminated materials, it has remarked that the heterojunction interlayer connection contests to the energy discrepancy of the single-layer states. In this case, the VB at the Γ point nearly degenerates, and corresponding optically dynamic K-point states localized in distinct single-layers to make the lowermost energy electron–hole couples three-dimensionally detached. Here it should be renowned that the band structure also depends on the involved calculating particulars. As the typical band structure of MoS_2/WS_2 heterojunction is depicted in Fig. 6.8. The LDA and projector-augmented waves methods yield gives almost comparable energies of the VB at the K- and Γ-points [113]. However, the K-point converts expressively inferior if one uses the GW method [114].

Likewise, an interesting demonstration is also available for the heterostructures by demonstrating the electron transfer occurs only due to conjugative dual layers of the edge, while the remaining layers nearly un-impacted, such as $(MoS_2)_3$–$(MoSe_2)_3$ and $(MoSe_2)_3$–$(MoTe_2)_3$ interfaces [115]. These systems Mulliken population analysis lead that the S atom of the L_4 layer MoS_2 and six-fold building improvements solitary 0.01e, whereas the Se atom to L_3 $MoSe_2$ sufferers simply 0.03e, it directly

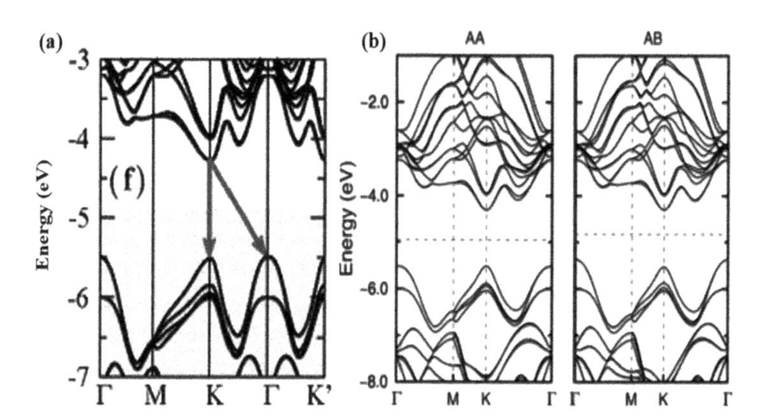

Fig. 6.8 Band structure of the MoS_2/WS_2 bilayer from DFT calculation using different functionals: **a** LDA and augmented waves (adopted from Kosmider et al. [113]); **b** from the GW approximation for the AA and AB stacking (adopted from Kim et al. [226])

correlates to close tendency of the electron affinity of chalcogen atoms replication. Moreover, it has been also noticed that the band gap of the heterostructures regularly adjusts underneath the exterior perpendicular electric field, which also helps to eventually achieve semiconductor–metal conversion at the acute fields. Therefore, the outcomes of the electronic features in an optimistic and deleterious field are dissimilar because of asymmetrical structures along the thickness direction and the intrinsic polarization. Moreover, it has been also addressed that the band gap of $(MoS_2)_3$–$(MoSe_2)_3$ [$(MoSe_2)_3$–$(MoTe_2)_3$] continuously decrease with the increasing positive applied electric field, therefore, the corresponding achieved semiconductor–metal conversion at $\sim$ 0.1 V/Å [0.05 V/Å]. In the visa versa case (when a negative field was applied) the band gap first increases until – 0.05 V/Å, and reached up to 0.45 eV [0.28 eV] for $(MoS_2)_3$–$(MoSe_2)_3$ [$(MoSe_2)_3$–$(MoTe_2)_3$], afterward it decreases linearly. It initially increases in the band gap due to the gradual neutralization of the intrinsic polarization of heterogeneous TMDs under a negative electric field [115]. Moreover, also noted that the critical electric field for the semiconductor–metal transition in $(MoSe_2)_3(MoTe_2)_{3\ is}$ lower than that in $(MoS_2)_3$–$(MoSe_2)_3$. All these initial investigations have established the fact, despite of the lesser band gap progressively diffusive behavior of the valence pz-orbitals from S to Te that enables extra charge transferring from the chalcogen to Mo atoms at the identical electric field.

Furthermore, another remarkable theoretical investigation demonstrated that the vertical heterostructures are made of nearly lattice-matched metallic (SL $NiTe_2$) and semiconducting (SL $MoTe_2$) TMDs. This result also leads to the effects of the size and dimensionality of constituents on the electronic structure of such heterostructures. Additionally, it has also defined that the vertical heterostructures can construct by the stacking of SL $MoTe_2$ and $NiTe_2$ layers in a specified manner V:$(MoTe_2)$p/ $(NiTe_2)q$ with $1 \leq p$ and $q \leq 5$, where p and q are the numbers of formula units for the primitive unit cell. Cause in vertically stacked TMDs interlayer coupling is rather weak. Besides the weak interlayer interaction, the electronic structures of vertically stacked heterostructures are very thin and their wide range alternation of constituents is significantly important. Like, the hexagonal and trigonal $MoTe_2$ and $NiTe_2$ TMDs vertically stacked heterostructures are stable [116]. Therefore, the stability of free-standing (SL) layers has been enhanced when they are grown on specific substrates. Similarly, the stability of bilayers and multilayer stacked TMDs are also enhances. Thus, in vertical heterostructures, the separation between the metal—semiconductor zones is in direct space and complete even for two layers of constituents. Due to minute charge transfer at the undoped junction and the band bending at the interface almost negligible. Whereas the fundamental band gap semiconducting side belongs to indirect and is smaller than the SL $MoTe_2$ constituent. This leads to the 2D metallic system being confined in a single layer when an SL $NiTe_2$ is inserted into the $MoTe_2$ vertical stacked heterostructure form.

Typical vertical heterostructures are also in layer-by-layer stacking of 2D SL constituents that interact weakly with adjacent layers. The vertical stack of p layers of one constituent is followed by a stack of q layers of the other constituent, as illustrated in Fig. 6.9. This kind of stacking sequence can be repeated periodically and continuously for a 3D hexagonal lattice. The 3d and 4 s, 4d and 5 s, and 5 s and 5p states to be treated as the valence states for Ni, Mo, and Te atoms. This has been employed in various theoretical simulation approaches to optimize the structures of vertical heterostructures, as the optimized parameters are the listed in Table 6.1 [117]. Moreover, the variation in electronic structures of $NiTe_2$ and $MoTe_2$ for the monolayer, bilayer, trilayer, and periodic multilayer bulk forms have been also explored. That leads to the bilayer, trilayer, and multilayer structures, interlayer distance, and stacking geometry easily optimized with the vdW interaction, while the situation is more critical for the semiconducting $MoTe_2$. Because the monolayer structure of $MoTe_2$ is the semiconductor with a direct band gap (1.15 eV) at the K-point of hexagonal BZ, while the bilayer band gap is slightly reduced and becomes indirect, as illustrated in Fig. 6.10. That occurs between the maximum of the valence band at the Γ- point and the minimum of the conduction band along the Γ–K direction. Stacking additional layers (such as trilayer) continuing also reduces the gap with indirect behavior. Therefore, the 3D layered $MoTe_2$ band gap falls into the indirect characteristic and it is saturated at 0.75 eV. However, the fundamental band gap of $MoTe_2$ slabs should reduce with an increasing number of layers and it changes direction into indirect. Though this theory consists of band gaps to deduce monolayers and multilayers of $MoTe_2$. That is significantly important for the context of band tenability to describe the electronic properties of vertical heterostructures.

In the case of a vertical heterostructure, V:$(MoTe_2)$p/$(NiTe_2)$q, p, and q are considered to be crucial parameters to control the electronic properties. Which leading the confinements of electronic phase separation for the large $p = q \geq 2$. Using this approach $MoTe_2$ electronic properties depending on p has been also addressed. Such as the weak vdW interaction between adjacent SL $MoTe_2$ and $NiTe_2$ layers cannot couple the states of the two layers [117]. Hence studies on matching lattice heterostructures continue to explore the novel properties of TMD materials [118–120].

6.2.4.2 Strain Minimize Approach

Usually, ultrathin TMDs are capable to sustain a considerably superior mechanical strain relative to their bulk complements [121–123], but ultrathin TMDs nanoscale devices are commonly underneath strain. Therefore, strain manufacturing is desired to get new and better physical features by altering their lattice and electronic construction. Strain could efficiently alter the atomic bond arrangement (bond length, bond angle, and bond strength) and their interconnection among electronic orbitals. As the consequence an innovative mechanism and features in ultrathin TMDs, such as thermal, electronic, optical, and magnetic properties. In a similar manner, the ability to alter the inherent features of ultrathin TMDs by the strain manufacturing may open marvelous prospects for their applicability in numerous arenas, such as sensing, electronic and photoelectric devices, and so on [123, 124].

There are various investigations based on DFT approximation have been presented by the investigators to explore unstrained and strained monolayer and bilayer TMDs

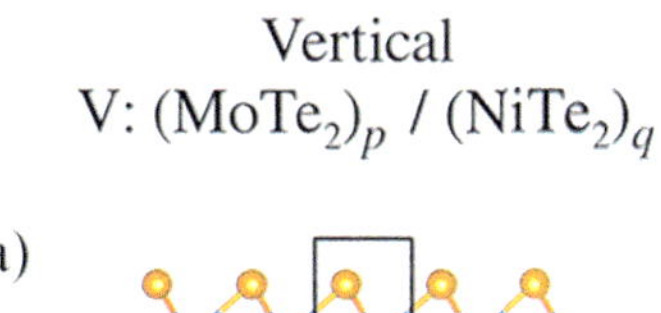

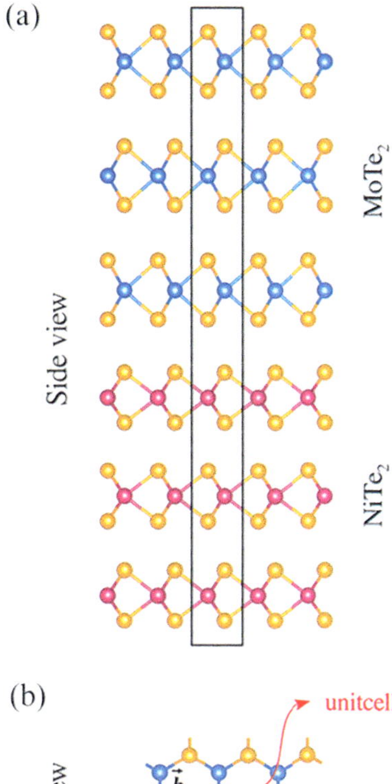

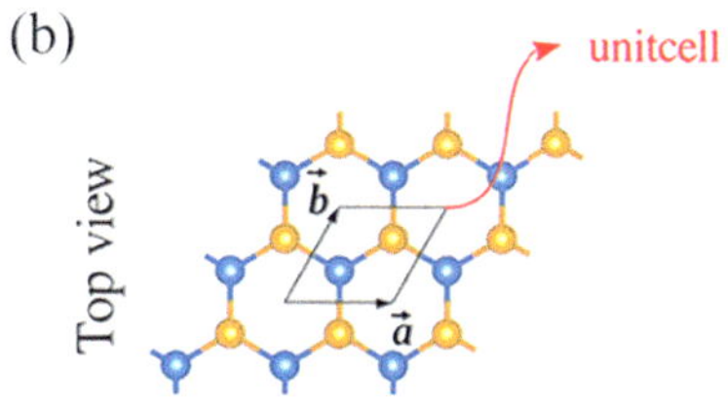

Fig. 6.9 **a** View of the vertical heterostructure V:($MoTe_2$)p/($NiTe_2$)q with $p = 3$ and $q = 3$; **b** top view. A three-dimensional hexagonal unit cell is delineated in side and top views (adopted from Araset et al. [66])

Table 6.1 Optimized values of vertical heterostructure, V: ($MoTe_2$)p/($NiTe_2$)q, and values of average cohesive energy, $\overline{E}_c$, average formation energy $\overline{E}_f$, superlattice constants $a = b$, c, interface spacing dint, and fundamental indirect band gap, E_g [117]

Composite (p/q)	$\overline{E}_f$ (eV)	$\overline{E}_c$ (eV)	$\overline{E}_f$ (eV)	$a = b$ (Å)	c (Å)	$d$int (Å)	E_g (eV)
V:(1/1)	– 0.075	11.988	– 0.075	3.589	12.797	3.108	
V:(2/2)	– 0.051	12.012	– 0.051	3.581	25.910	3.227	
V:(3/3)	– 0.044	12.019	– 0.044	3.579	38.887	3.271	0.641
V:(4/4)	– 0.040	12.023	– 0.040	3.583	51.667	3.203	0.657
V:(5/1)	– 0.032	12.734	– 0.032	3.528	41.192	3.306	0.722
V:(1/5)	– 0.033	11.328	– 0.033	3.651	36.239	3.105	

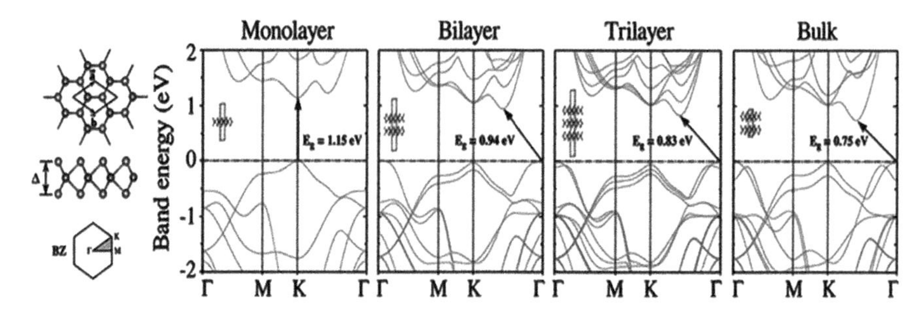

Fig. 6.10 Electronic energy band structure of monolayer (SL), bilayer, trilayer, and periodic bulk $MoTe_2$, whereas the zero of energy is set at the top of the valence band (adopted from Aras et al. [117])

[125–131]. To the similarity in their atomic structures and the electronic properties of MoS_2 and $MoTe_2$ monolayers, however, their bilayers have similarities but not completely equivalent trends. A typical study also performed the DFT simulation strain calculation for the six most emerging TMDs (MoS_2, $MoSe_2$, NbS_2, $NbSe_2$, ReS_2, $ReSe_2$), by changing the chemical constituents and topological atomic alignments, as atomic configurations of all six TMDs is illustrated in Fig. 6.11. The corresponding materials' true stress versus engineering strain responses is depicted in Fig. 6.12. This result has revealed a smaller strain (approximately 5%,) almost all TMDs followed a linear stress–strain relationship. But over the 5% of entire stress–strain curves convert into nonlinear. Thereby, TMDs like MoS_2 and $MoSe_2$, strain unstiffening levels off the stress–strain curve progressively till the until tensile strength (UTS) point, afterward mechanical unsteadiness exist. Though, compositions ReS_2, $ReSe_2$, NbS_2, and $NbSe_2$ may experience brittle breakage together an abrupt decline in stress values. Thus MoS_2 and $MoSe_2$ can have a greater UTS and superior failure strain compare to ReS_2 and $ReSe_2$. Moreover, they are also having greater asset and breakage strain [130].

TMDs liner mechanical features like elastic constants C_{11}, C_{22}, C_{12}, and C_{66} could be willingly estimated by the primary inclines of various stress–strain outcomes. In the case of C_{22} almost equal to C_{11}, then the layer modulus extent and signifies the resistance of a nanosheet in stretching, it is usually estimated as;

$$\gamma_{2D} = \frac{1}{2}(C_{11} + C_{12}) \tag{6.1}$$

Similarly, remaining linear mechanical features, such as Young's modulus (E), Poisson's ratio (n), and shear modulus (G), can be also described by given relationship;

$$E = \frac{\left(C_{11}^2 + C_{12}^2\right)}{C_{11}} \tag{6.2}$$

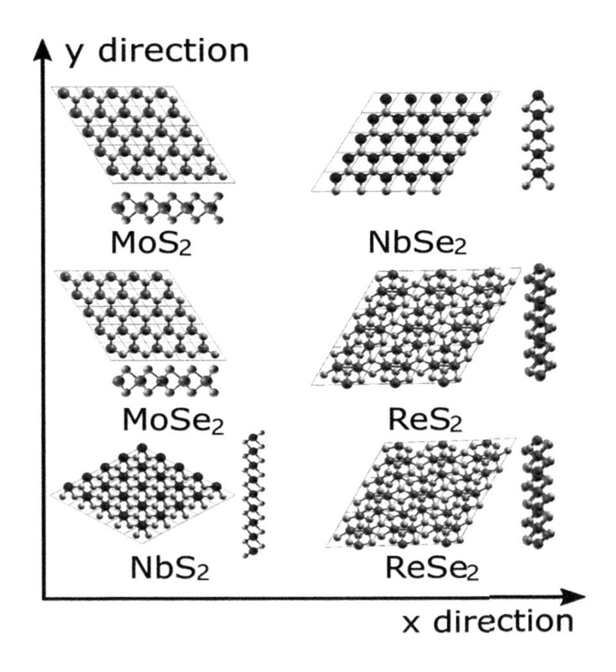

Fig. 6.11 Atomic topologies and the unit cells for the six TMDs, the two deformation directions (x and y) are represented by arrows (adopted from Sun [130])

and

$$\nu = \frac{C_{12}}{C_{11}}, \ G = C_{66} \tag{6.3}$$

The corresponding physical properties of all six compositions are listed in Table 6.2. It has revealed that the TMDs are belonging to smaller values of the elastic constants. Young's modulus, shear modulus, and 2D layer modulus have feebler covalent bonds in TMDs, though components in TMDs have sufficient amounts of protons in their nuclei. Thus, the bigger atomic radius and tougher protecting influence occur owing to the domination of extra internal electrons with the attractive furthest electrons and nuclei. Such flagging effect has been measured by the first ionization energies of diverse elements, like S (999 kJ mol^{-1}) > Se (941 kJ mol^{-1}) > Re (760 kJ mol^{-1}) > Mo (684 kJ mol^{-1}) > Nb(652 kJ mol^{-1}) [132]. Moreover, TMDs Young's modulus dependence on their chemical components has been also explored. Like the similar non-metallic elements Young's modulus is increased toward the transition metal when it drives from the V to VII group. As a typical example, for the same metallic element, TMD contains S atoms with larger Young's modulus than that possessing the Se atoms.

Similarly, the bending modulus (D) for a 2D nanosheet can be also obtained by using the following relationship [133];

$$D = \frac{Eh^2}{12(1 - \nu^2)} \tag{6.4}$$

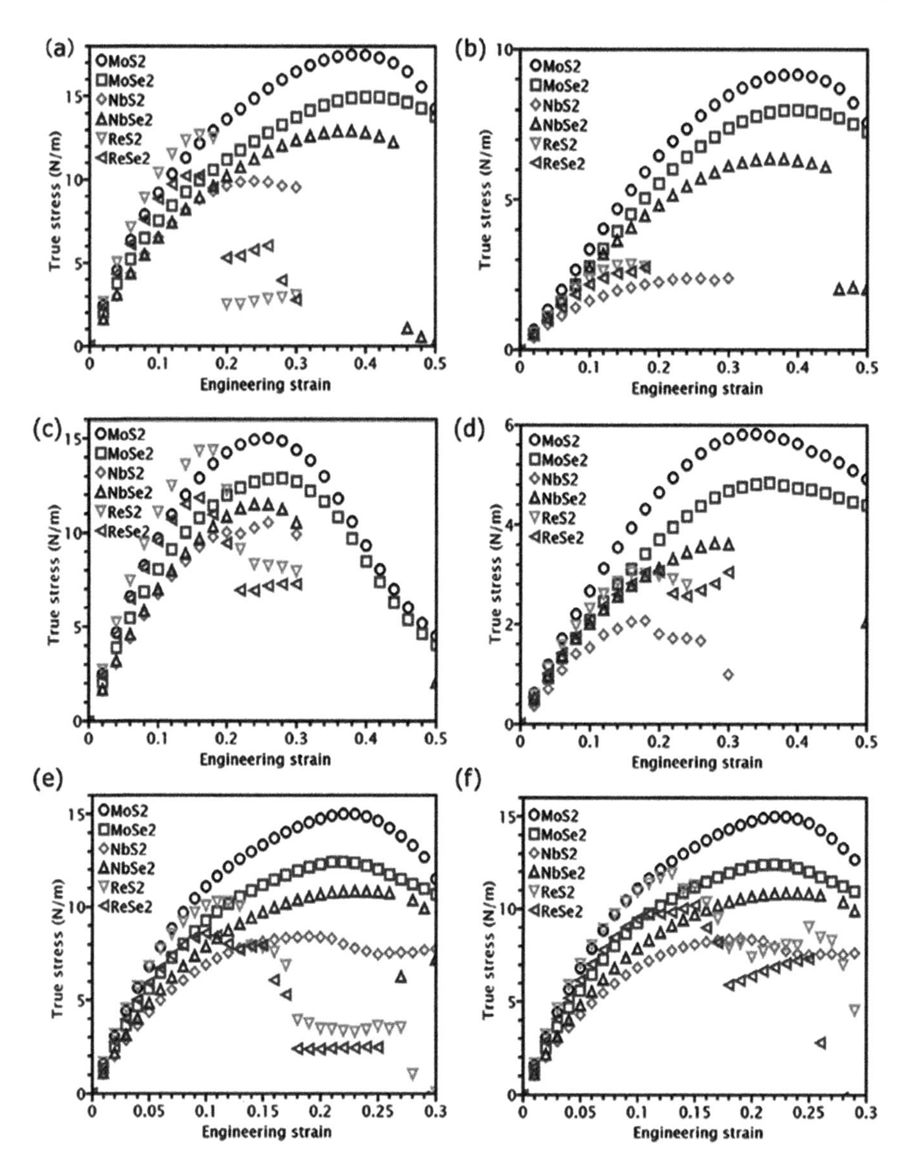

Fig. 6.12 Stress–strain curves for the six TMDs: **a**, **b** represents the stress–strain responses under x-uniaxial tension along the x and y directions; **c**, **d** stress–strain curves under y-uniaxial straining in the x and y directions; **e**, **f** the stress–strain responses in biaxial tension along the x and y directions (adopted from Sun [130])

Table 6.2 List of the DFT calculated mechanical properties for all TMDs, including elastic constants (C_{11}, C_{22}, and C_{12}), UTS, fracture strain (ε), Young's modulus (E), layer modulus (γ), bending modulus (D), critical bulking strain (ε_c), and Poisson's ratio (ν)

	MoS_2	$MoSe_2$	$NbSe_2$	NbS_2	$ReSe_2$	ReS_2	$NbSe_2$
C_{11}	133.36	114.56	86.58	90.09	120.55	140.77	86.58
C_{12}	37.05	31.89	30.74	25.61	26.55	31.12	30.74
UTS_x	17.48	14.97	12.95	9.91	10.28	12.7	12.95
ε_x	0.39	0.41	0.37	0.23	0.15	0.16	0.37
C_{22}	134.26	114.5	89.53	90.03	125.17	142.11	89.53
C_{21}	36.15	30.59	28.47	22.12	25.61	31.74	28.47
UTS_y	15.01	12.89	11.52	10.67	11.85	14.45	11.52
ε_y	0.26	0.27	0.25	0.23	0.16	0.17	0.25
UTS^x_{biax}	15.00	12.42	10.83	8.43	8.59	10.26	10.83
E^x_{biax}	160.57	135.04	114.71	108.74	144.39	167.48	114.71
UTS^{biax}_y	15	12.41	10.83	8.41	15.67	11.93	10.83
E^{biax}_y	160.54	134.98	116.11	111.79	151.37	170.58	116.11
ε_{biax}	0.23	0.22	0.25	0.19	0.09	0.12	0.25
G	48.6	41.13	28.08	18.65	53.53	59.42	28.08
γ_{2D}	85.21	73.23	58.66	57.85	73.55	85.95	58.66
E	123.07	105.68	75.67	82.81	114.70	133.89	75.67
ν	0.28	0.28	0.36	0.28	0.22	0.22	0.36
D (eV)	6.8	5.84	4.41	4.59	6.14	7.17	4.41
$\varepsilon_c L^2$	34.94	34.95	36.86	35.06	33.86	33.87	36.86

All moduli and strengths are taken in the N/m unit [130]

here h is the thickness of the TMD, although it is hard to evaluate precisely due to the electronic arrangement toward the perpendicular direction of changes while deformation. Nevertheless, the lowermost estimated D has been achieved by the absolute thickness of the nanosheet about ~ 3.13 A° for TMDs. Usually, TMDs have inferior modulus with higher bending moduli to graphene because of their much larger thicknesses. Thus the larger difference in thickness is responsible for the attributed three layer atomic construction in TMDs. That provides additional interaction terms preventive the bending motion.

Using values of D and E one can also know the buckling mechanism and calculate the precarious buckling strain (ε_c) by employing Euler's buckling theorem [133];

$$\varepsilon_c = -\frac{4\pi^2 D}{EL^2} \tag{6.5}$$

for the identical length (L). The critical buckling strain of the TMDs are upto ten times superior to graphene owing to a greater D and lesser E. Thus TMDs have extra vigorous to in-plane organizational deformations and additional resistant to buckling.

Similarly direction dependence Young modulus $E(\theta)$ and Poisson's ratio $\nu(\theta)$ toward the arbitrary in-plane direction θ (θ represents the angle associated to the x direction) has been also defined using the following elastic constants [134];

$$E(\theta) = \frac{C_{11}C_{22} - C_{12}^2}{C_{11}\sin^4(\theta) + C_{12}\cos^4(\theta) + \left(\frac{C_{11}C_{22} - C_{12}^2}{C_{66}} - 2C_{12}\right)\sin^2(\theta)\cos^2(\theta)} \tag{6.6}$$

$$\nu(\theta) = -\frac{\left(C_{11} + C_{22} - \frac{C_{11}C_{22} - C_{12}^2}{C_{66}}\right)\sin^2(\theta)\cos^2(\theta) - C_{12}\left(\sin^4(\theta) + \cos^4(\theta)\right)}{C_{11}\sin^4(\theta) + C_{22}\cos^4(\theta + \left(\frac{C_{11}C_{22} - C_{12}^2}{C_{66}} - 2C_{12}\right)\sin^2(\theta)\cos^2(\theta)} \tag{6.7}$$

The hexagonal structure of TMDs such as MoS_2, $MoSe_2$, and $NbSe_2$ are also considered to be approximately isotropic, therefore, their $E(\theta)$ and $\nu(\theta)$ should be independent of angular variation. While the ReS_2 and $ReSe_2$ have a slight anisotropic behavior, therefore, the corresponding $E(\theta)$ peaks make at 45° angle concerning the x-uniaxial loading direction, as a typical representation is illustrated in Fig. 6.13. The corresponding θ value to the biggest modulus has also shown the lowest Poisson ratio.

Moreover, NbS_2 have the strongest anisotropic behavior and their corresponding $E(\theta)$ approached at maxima in x and y directions, when the Poisson's ratio reached the minima.

Moreover, the invention has been also done with nonlinear mechanical features such as UTS and failure strain of all inspected TMDs in diverse loading approaches. It has been noted that the TMDs possessed inferior UTS, besides certain TMDs may have greater failure strain. Like the failure strain of MoS_2 and $MoSe_2$ are reaches

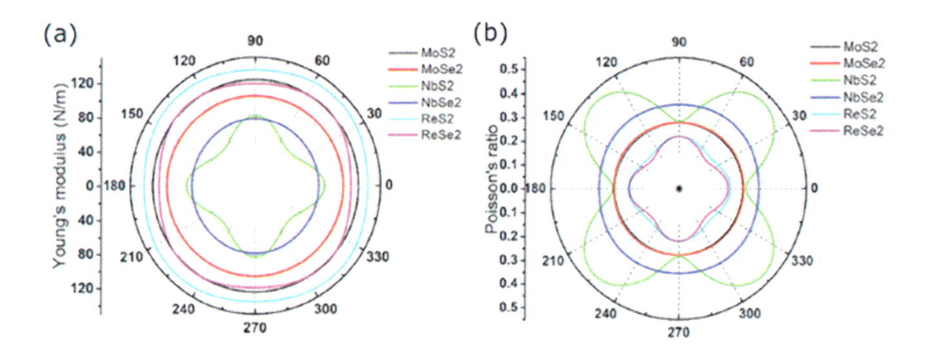

Fig. 6.13 **a** Angular dependence of Young's modulus; **b** Poisson's ratio, for six TMDs (adopted from Sun [130])

up to 0.4 in uniaxial tension. With the UTS approach alternatively one also finds the perfect asset of TMDs as *E*/10, here *E* is Young's modulus [135]. The UTS approach is also recognized as Griffth's strength limit [130].

Likewise, in another DFT simulation, the $MoTe_2$ and MoS_2 monolayer at zero strain has been studied. It revealed a direct band gap with VBM and CBM located at the six K points corners of the BZ, as depicted in Fig. 6.14. The obtained band gap values of the MoS_2 and $MoTe_2$ are 1:61 eV and 0:93 E_v, respectively. It has been noted that beyond the K points also Γ and Q critical points can be mentioned, because they are close to energetically K_v and K_c. So, under strain (or in multilayers) Q_c becomes the global CBM and Γ_υ the global VBM. Owing to facts, almost all group-VI TMDs edges conduction and valence bands are predominantly composed of d metal (M) and p chalcogen (X) orbitals [136]. However, types (symmetries) and contributions of the orbitals vary with chosen k point and the constituent atoms. At point K, the VB and CB are predominately described from the M-orbitals $\left(d_{xy}, d_{x^2-y^2}(d_{z^2})\right)$ with small X-orbitals (p_{xy}) contribution, as the typical orbital projection of band structure of $MoTe_2$monolayer p (Te) and d (Mo) orbitals, is depicted in Fig. 6.15. Whereas at the point Γ, the VB is mainly due to Md_{z^2} associated with a small Xp_z contribution.

Moreover, similar compositions Q_c analysis in terms of K_υ has revealed the involvement of minor contributions of the d_{z^2} and d_{xyz} orbitals [136]. Hence in these materials, the uniform tensile strain general reduction at the direct gap (at K) has been described from the simple tightly-bond approximation, because in TMDs minor overlaps among orbitals. However, it's also a real fact that the different orbitals composition applications to strain act differently at the various high-symmetry k-points.

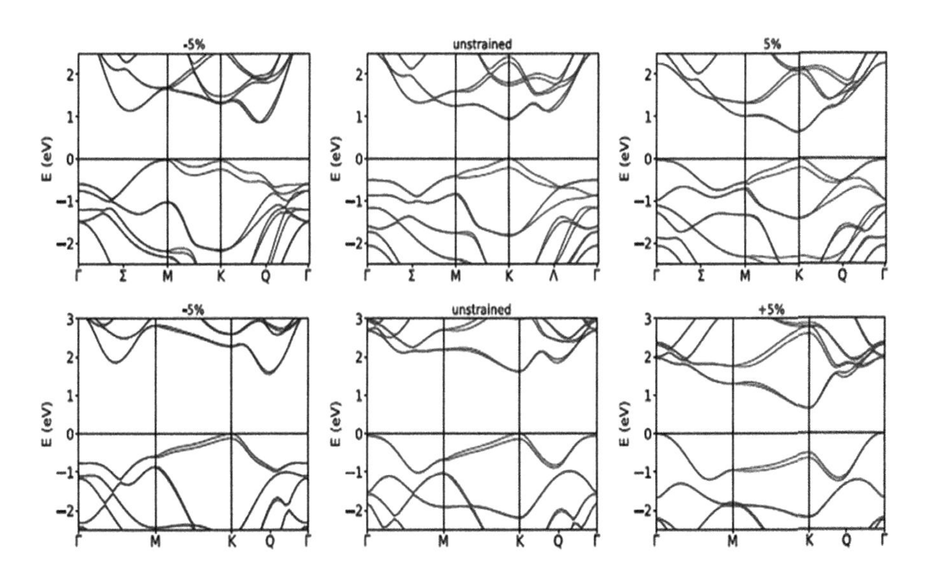

Fig. 6.14 Electronic band structure for monolayer $MoTe_2$ (top) and MoS_2 (bottom), under compressive strain (left), under tensile strain (right), and with zero strain (center) (adapted from Postorino et al. [131])

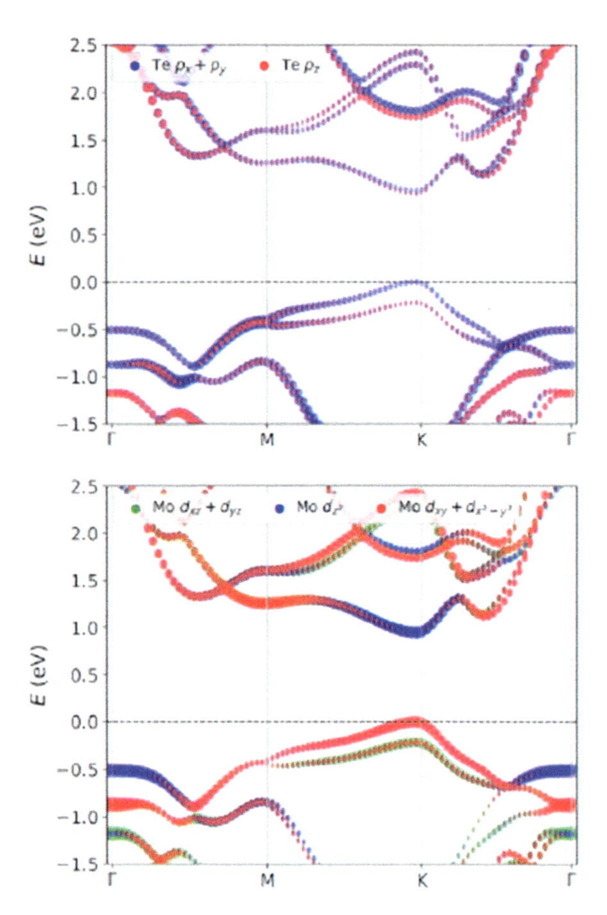

Fig. 6.15 Projected orbital band structure for the $MoTe_2$ ML: p orbitals of Te (top panel) and d-orbitals of Mo (bottom panel) (adopted from Postorino et al. [131])

Predominantly, the dominant contribution of d_{xy}, $d_{x^2-y^2}$ shifts the K_ν downward, while the d_{z^2} governs from Γ_ν that remains unaffected. This effect is responsible for the direct-to-indirect band gap crossover ($\Gamma_\nu - K_c$) under the applied tensile strain. Typically, in case of MoS_2 and $MoTe_2$ monolayer at $\varepsilon < 1$ (upto ~ 5)%. Such as the maximum delocalization of the Te, and p-orbitals to sulfur make the $MoTe_2$ less sensitive to strain, the corresponding situation is depicted in Fig. 6.16. Therefore, it leads to small amounts of tensile strain required to shift down the K_ν in MoS_2 [131].

Since the behavior of the MoS_2 and $MoTe_2$ are entirely different under the applied uniform compressive strain. Such as the CBM in MoS_2-ML moves from K to Q for the strain above $-$ 2%, as a result, the formation of an indirect gap ($K_\nu - Q_c$). While, in the case of the $MoTe_2$-ML, the compressive strain is increased and two CB minima along the L and S directions decrease their energies, whereas the valence band at M increase the energy. These contradictory facts have led to the first transition from a direct gap $K_c - K_\nu$ to an indirect gap $Q_c - K_\nu$ at about—2% of strain and a second transition due to another indirect gap $Q_c - M_\nu$, under the strain of about -5%. The contributions of the $d_{x^2-y^2}$, d_{z^2} orbitals become from the metal atoms, while the most

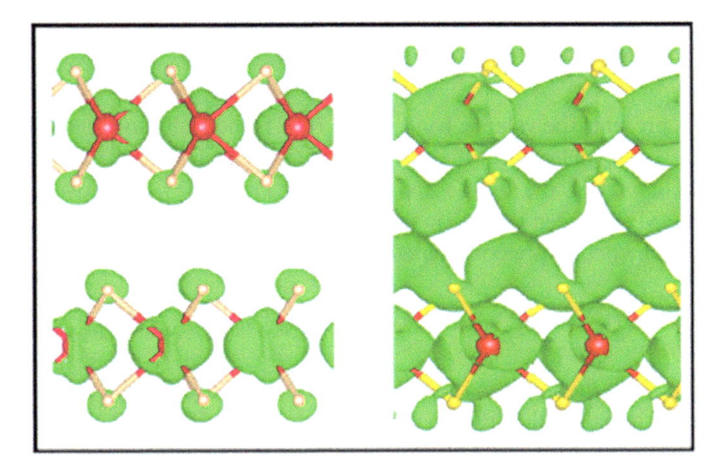

Fig. 6.16 Q_c of MoS_2 (left) and $MoTe_2$ (right) bilayers at zero strain green isosurface at 5%. Mo (red), S (pink), and Te (yellow) (adopted from Postorino et al. [131])

dominant contributions come from the p_{xyz} orbitals of the chalcogen atoms. The more delocalized behavior of the p orbitals of Te to S makes it more sensitive M_ν, that is more sensitive for the $MoTe_2$ than MoS_2 under a compressive strain. Moreover, the calculated physical parameters values and electronic gap trends with strain for the TMD monolayers have been also studied. It has been noted that the VBM, CBM, and vacuum potential curves are qualitatively similar to 2D nitrides, possessing quite a distinct slope behavior. Specifically, vacuum potential decreases more rapidly in $MoTe_2$ to the other cases [131].

A remarkable DFT simulation on strain dependence second harmonic generation (SHG) of the TMD materials (such as MoS_2, $MoSe_2$, WS_2, WSe_2, and $MoTe_2$) has been also interpreted by adopting both uniaxial and biaxial strain. The DFT calculations have been performed with the LDA functional and norm-conserving pseudopotentials [137–139]. By including the (4 s, 4p, 5 s, 4d), (5 s, 5p, 6 s, 5d), (3 s, 3p), (4 s, 4p), and (5 s, 5p) valence electrons for Mo, W, S, Se, and Te. It has been noted that the wave functions are expanded in a plane wave up to a kinetic energy cutoff of 1089 eV. The corresponding Brillouin zone belongs to the 12 × 12 × 1 Γ-centered k-point grid. The structures are also fully relaxed until the forces less than 0.01 eV/Å and the unit cell in the perpendicular direction, typically it set 20 Å to avoid spurious interactions between images. While the spin–orbit coupling includes only a relatively minor effect on the SHG spectra in TMDs that mostly consists of splitting of the lowest energy peak into the A and B exciton peaks. The SHG responses are due to the real-time time-dependent DFT calculations of the polarization through the interaction with a time-dependent electric field in the dipole approximation [140]. The nonlinear optical susceptibility is usually numerically calculated by the following equation;

$$i\hbar\frac{\partial}{\partial t}|v_{ik}\rangle = \left(H_k^{\text{sys}} + ie\varepsilon(t).\partial_k\right)|v_{ik}\rangle \tag{6.8}$$

where v_{ik} represents the valance band states, H_k^{sys} is the Hamiltonian of the system, and term $ie\varepsilon(t).\partial_k$ determines the coupling of the system with the time-dependent external field ε(t) in the dipole approximation [140–143]. Moreover, the time-dependent BSE has been used to include excitonic effects in the monolayer TMDs [141]. So, many bodies are contributed to Hamiltonian that can be expressed as;

$$H_k^{\text{sys}} = H_k^{\text{KS}} + \Delta H_{\text{scissor}} + V_h[\Delta\rho] + \sum\nolimits_{\text{SEX}}\left[\Delta\gamma\right] \tag{6.9}$$

where $\Delta H_{\text{scissor}}$ is the uniform scissor shift correction to the Kohn–Sham Hamiltonian, it is determined by the difference between the direct GW band gap and the direct LDA band gap [141, 143], $V_h[\Delta\rho]$ represents the time and field-dependent Hartreepotential [143, 144], and $\sum_{\text{SEX}}\left[\Delta\gamma\right]$ represents the screened-exchange self-energy. Calculations have been performed without excitonic effects under the independent particle approximation (IPA), for that only the Kohn–Sham term is considered.

To calculate IPA and BSE optical properties the equations of motion integration have been done with a 0.015-fs time step for a total simulation time of 45 fs with a 21 $\times$ 21 $\times$ 1 Γ-centered k-point grid and 0.2 eV damping 8, 3 valence and 12, 4 conduction bands. Additionally, high-resolution calculations have been also performed to identify the fine structure of the SHG spectra with the damping of 0.05 eV and simulation time of 160 fs. A total of 112 conduction bands have been included in the calculation of the screened Coulomb interaction BSE approach, as the typical SHG spectra of the BSE approach for the various TMDs are illustrated in Fig. 6.17.

Moreover, the LDA function cannot ignore in the calculation of the band gap of monolayer TMDs. Because it coincided with the exciton binding energy of the unstrained TMDs, which is approximately equal to the error of the computed band gap due to the large expense of GW calculations. Therefore, in G_1W_0 calculation a rigid scissor shift to the conduction bands considered to be proportional to the LDA band gap, to maintain consistency between different TMDs, it is commonly expressed as [145]:

$$\frac{E_{\text{shift}}(\text{MoS}_2)}{E_{\text{shift}}(\text{TMD})} = \frac{E_{\text{g}}^{\text{LDA}}(\text{MoS}_2)}{E_{\text{g}}^{\text{LDA}}(\text{TMD})} \tag{6.10}$$

where E_{shift} is the scissor shift and E_{g} stands for the band gap. Note that usually denoted exciton binding energies by E_{b} and quasiparticle band gaps represents by the E_g^{GW} [60]. Additionally, the degree of gap correction is considered insensitive to the application of strain in TMDs [146]. It has been also noted that the SHG intensity depends on supercell volume, which is less defined in 2D materials due to the inclusion of an arbitrary amount of vacuum in the unit cell volume, therefore, it is desired $\chi^{(2)}$ to normalize the multiplicative factor d_z/l_z, where d_z corresponds to the layer thickness and l_z represents the supercell length in the perpendicular

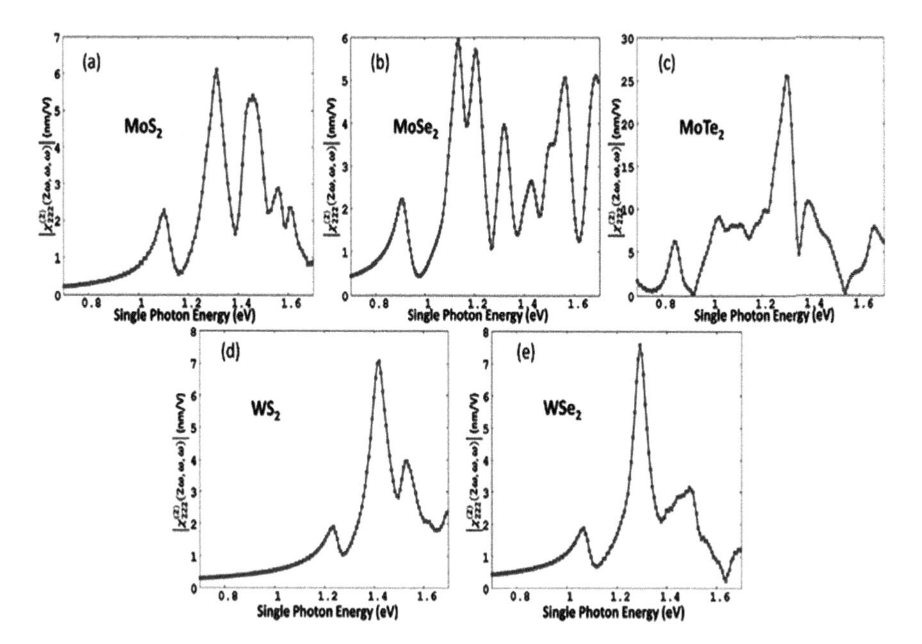

Fig. 6.17 SHG spectra calculated with the BSE method: **a** MoS_2; **b** $MoSe_2$; **c** $MoTe_2$; **d** WS_2; **e** WSe_2 with 0.05 eV damping (adopted from Beach et al. [148])

direction. In which only 222 (*yyy*) components of the second harmonic susceptibility $\chi^{(2)}(2\omega, \omega, \omega)$, where 2 ($y$) belongs to the armchair direction and 1 (x) corresponds to the zigzag axis. Thus, unstrained and biaxially strained TMDs have only one independent component of the $\chi^{(2)}$ tensor [147], however, in uniaxial strain reduces the symmetry from D_{3h} to C_{2v}, therefore, the different non-zero components of $\chi^{(2)}$ cannot be longer or equal in TMDs. Similarly, in the case of MoS_2 under 3% strain in armchair (y) direction (by employing the IPA method), therefore, the 222 components may prominent, including magnitude and other components that are non-zero in the unstrained case. In this instance, only one component appeared to be activated through the strain 212 (*yxy*) which belongs to only a small degree, typical second harmonic generation as a function of positive biaxial strain (expansion), spectrum for the WSe_2 is depicted in Fig. 6.18 [148].

Including above discussed significant properties of SHG the fine structure are also depending on the C exciton that lies in-between 2.4 and 3.3 eV for the TMDs. The changes in SHG peaks are directly express as a function of strain [148]. Their transitions at K and Γ can have opposite reactions to strain $E_K(C_2 - v_1)$ with a linear increase, while the tensile strain and $E_\Gamma(C_1 - v_1)$ to be decreased. Whereas the $E_Q(C_1 - v_1)$ also has a similar strain responsible for the $E_K(C_2 - v_1)$ respect to a roughly linear increase in transition energy together increasing tensile strain. Thus all three transitions of the biaxial strain reflect the largest slope with respect to strain [148]. As per DFT calculations, the transition between v_1 and c_2 at K involves only d orbitals of the transition metals and the corresponding move from in-plane

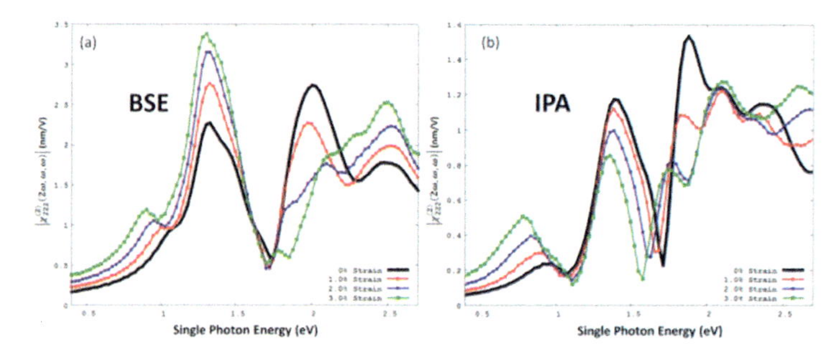

Fig. 6.18 Second harmonic generation as a function of positive biaxial strain (expansion): **a** from BSE; **b** from IPA for the WSe_2 (adopted from Beach et al. [148])

($d_{x^2-y^2}$ and d_{xy}) to out-of-plane ($d_{3z^2-r^2}$) orbital. Whereas the transition at Γ involves d to p-orbitals when moving from out-of-plane ($d_{3z^2-r^2}$ and p_z) to in-plane (p_x and p_y). Thus the application of tensile strain weakens the hybridization between the d and p orbitals [149], by disrupting the localization and modifying the SHG amplitude differently for each transition based on the orbital behavior of the transition. This interpretation could directly correlates to the concept of disparity behavior concerning strain between the two sub-peaks of the $MoSe_2$ C-exciton region. Moreover, the probing approach fine structure of the C-exciton region in SHG can also track the behavior of the various sub-peaks under a tunable strain degree of freedom, the dominant orbitals significant SHG spectral transitions in TMDs are summarized in Table 6.3 [148]. Thus the studies are not limited to above discussed significant DFT interpretations, but also many more theoretical and simulations investigations have been presented, such as a detailed theoretical description of strain-induced systems provided by Rostami et al. [150]. Similarly, a simulation study based on strain in van der Waals epitaxy and their collective macroscopic impact on a negligibly small perturbation has been also discussed [128]. Thus the story continues to explore the various dimensions of strain-induced TMDs by interpretation distinct properties [128, 129, 148–150].

Not only theoretical studies but also there are several experimental realizations on strain-induced ultrathin TMDs have been also interpreted to get novel thermal,

Table 6.3 Dominant orbital characters of the important transitions in the SHG spectra for the TMDs [148]

k-point	Band index	Characters of the valence band	Characters of the conduction band
K	$v_1 \rightarrow c_1$	$d_{x^2-y^2}$, d_{xy}	$d_{3z^2-r^2}$, p_x, p_y
K	$v_1 \rightarrow c_2$	$d_{x^2-y^2}$, d_{xy}	$d_{3z^2-r^2}$
Q	$v_1 \rightarrow c_1$	$d_{3z^2-r^2}$, $d_{x^2-y^2}$, d_{xy}	$d_{x^2-y^2}$, d_{xy}, p_x, p_y
Γ	$v_1 \rightarrow c_1$	$d_{3z^2-r^2}$, p_z	p_x, p_y
M	$v_1 \rightarrow c_1$	$d_{3z^2-r^2}$, $d_{x^2-y^2}$, d_{xy}, p_z	$d_{x^2-y^2}$, d_{xy}

electronic, optical, and magnetic properties [123, 125, 151]. Like strain engineering, it leads to a change in the bond length and angle due to lattice deformation [152–154], which altered the phonon characteristics of extremely-thin TMDs distinguished by Raman spectroscopy. As the example, a blue/red alteration occur in-plane vibrational E_{2g}^1 the mode in ultrathin group-VI TMDs underneath compressive/tensile strain [155, 156]. Additionally splitting of E_{2g}^1 mode directly correlates possibility of ultrathin TMDs formation under the uniaxial strain [126, 157–159], in which A_{1g} the mode is more inert to the in-plane strain, as depicted in Fig. 6.19a, b. It also leads the vibrational outcomes of Raman modes connect to the comparative positioning among the vibrational and strain directions. Moreover, Raman events of ultrathin TMDs may also endure a substantial alteration underneath tensile strain by revealing an increase and decrease activity of $E'(A_1)$ for the group-VI TMDs, whereas under the compressive strain opposite responses in Raman activity between E' and A_1 modes, as illustrated in Fig. 6.19c, d [160]. Similarly, the thermal conductivity of TMDs is also modify the semiconducting ultrathin films due to unstiffening/hardening of phonon modes over the tensile/compressive strain, as the typical illustration represents in Fig. 6.1e, f [123, 161]. Moreover, the induced strain effect experimental realizations have been also confirmed for the electronic, optical, and electrical transport, luminescence, and piezoelectricity alternation in TMDs [123, 151].

6.2.4.3 Individual (or Single) Layer Heterostructures

Since the diverse atomic features of 2D TMDs, the van der Waals (vdW) gaps among the atomic layers are considered to be a substratum for the exclusive electronic/optoelectronic characteristics of the materials and devices [89]. Therefore, significant vertical heterostructures have been investigated [162]. By interpreting the electronic structure and the lattice dynamics of $MX_1^{(1)}X_1^{(2)}$ monolayer heterostructures in which upper and lower planes formed due to the occurrence of different chalcogenides species $X^{(1)}$ and $X^{(2)}$. As per the plane-wave approach and DFT/PBE level of theory, the corresponding typical diagram is illustrated in Fig. 6.20 [162]. It has been noticed that in contrast to non-polar systems (with $X^{(1)} = X^{(2)}$), usually the polar systems have $X^{(1)} \neq X^{(2)}$, therefore the corresponding Rashba splitting occurs at the Γ-point for the uppermost valence band. Usually, it has been interpreted in terms of broken mirror symmetry. Theoretically, Rashba coupling parameters (αR) for the MoSSe, MoSTe, MoSeTe, WSSe, WSTe, and WSeTe structures have been determined at 2, 12, 4, 5, 14, and 10 meV·Å. It's also noted that the Rashba splitting enhances stronger SOC (M: from Mo to W and $X^{(1)}, X^{(2)}$,, from S to Te). Additionally, the Rashba effect is more sensitive to the distance difference of the chalcogen atoms from the transition metal planes than the atomic species.

Hence both theoretical and experimental efforts have been dedicated to the uninterrupted imagining of the vdW gaps and stacking arrangements for a broad spectrum of 2D TMD materials containing MoS_2, WS_2, $PtSe_2$, $PtTe_2$ layers, and their heterostructures/heterointerfaces, as schematic illustrations in Fig. 6.21a, b, for both horizontal

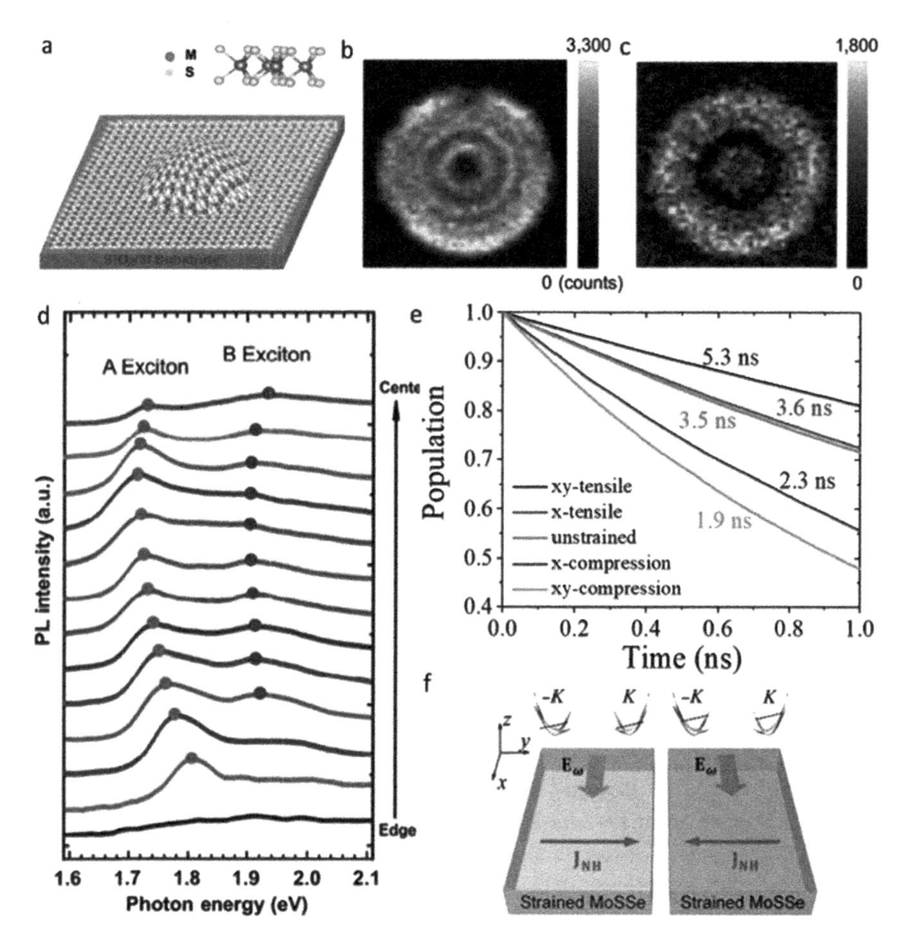

Fig. 6.19 **a** Diagramic illustration of a MoS_2 bubble with isotropic tensile strain; **b, c** PL intensity mapping pictures of strained MoS_2 bubble at the resonant energies for A excitons and B excitons; **d** PL spectrum of four-layer MoS_2 as a function of radial coordinate onto MoS_2 bubble, whose middle is described at $r = 0$; **e** nonradiative exciton recombination dynamics for the single-layer WSe_2 at diverse strains; **f** schematic diagram to show robust gate dependence of nonlinear Hall effects in strained MoSSe (adopted from Yan et al. [123])

and vertical aligned 2D layers [163–166]. The atomic-scale regularity with higher consistency horizontal stacking of 2D $PtTe_2$ layers has been achieved via HAADF-STEM (see Fig. 6.21c) [167]. That revealed a discrete crystalline construction associating a well-resolved vdW gaps (~ 0.52 nm). Whereas the vertically-arranged 2D MoS_2 layers are uniformly spaced vdW gaps (Fig. 6.21d) [168]. Moreover, the integral of the vertically-aligned 2D $PtSe_2$ layers has shown with the recognized vdW gaps of ~ 0.55 nm (Fig. 6.21e) [163]. Additionally, the distinct types of vertically-aligned 2D TMDs are also visualized for the few-layer single-crystalline 2D WS_2 (see

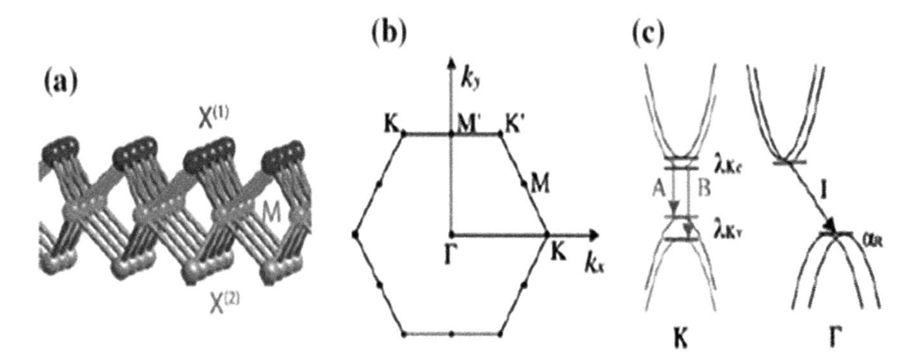

Fig. 6.20 **a** Schematic of an $MX^{(1)}$, $X^{(2)}$ monolayer heterostructure; **b** corresponding first Brillouin zone; **c** schematic view of the electronic band structure around the K and Γ-points, the band splittings at the K-point due to the SOC at the Γ-point, Rashba spin splitting of the uppermost VB (adopted from Cheng et al. [162])

Fig. 6.21f) with the well-discriminate stacking of WS_2 (002) layers and interlayer distant around ~ 0.62 nm [169].

Janus Monolayer in 2D TMDs

Demonstration of the essential mechanisms is depending on interaction among the atoms and electron beam. TEMis considered an appreciating tool for recognizing the atomic-scale features of 2D TMDs associating void and defects formations including dopants in the MoS_2 monolayer [170, 171], by distinguishing the asymmetric Janus structures of the MoSSe single-layer [172, 173]. Additionally, TMD single-layers degree of freedom is also helpful to display an inherent in-plane overturn irregularity [172]. Specifically, the replacement of the upper layer S to Se atoms in a single-layer of MoS_2. By employing an annular dark-field scanning TEM(ADF-STEM) a cross-section of the asymmetric Janus construction of the MoSSe single-layer can be also evidently distinguished from lowest S and topmost Se atoms, as depicted in Fig. 6.22a, b [172], where the picture disparity considered to be proportionate to square of the atomic numeral. Moreover, the Janus construction of the MoSSe single-layer are also shaped through governed sulfurization of monolayer $MoSe_2$ (top Se substituted by S) [173]. Similarly, the WSSe Janus monolayer is also possible from lower energy operation by employing tilted high-angle annular dark field Z-contrast scanning transmission electron microscopy (HAADF-Z-STEM). Their Janus construction with the highest S and lowest Se are normally discriminate from the picture intensities, as illustrated in Fig. 6.22c, d [174]. Moreover, atomic resolve low-voltage scanning TEM (STEM) imaging with atom-by-atom examination has been also utilized to inspect the atomic alloy nature of various molybdenum and tungsten ditelluride single-layers alloyed with sulfur or selenium ($MX_{2x}Te_{2(1-x)}$, M = Mo, W, and X = S, Se) for the 2H and 1T'phases. Thus the diverse alloying concentrations of

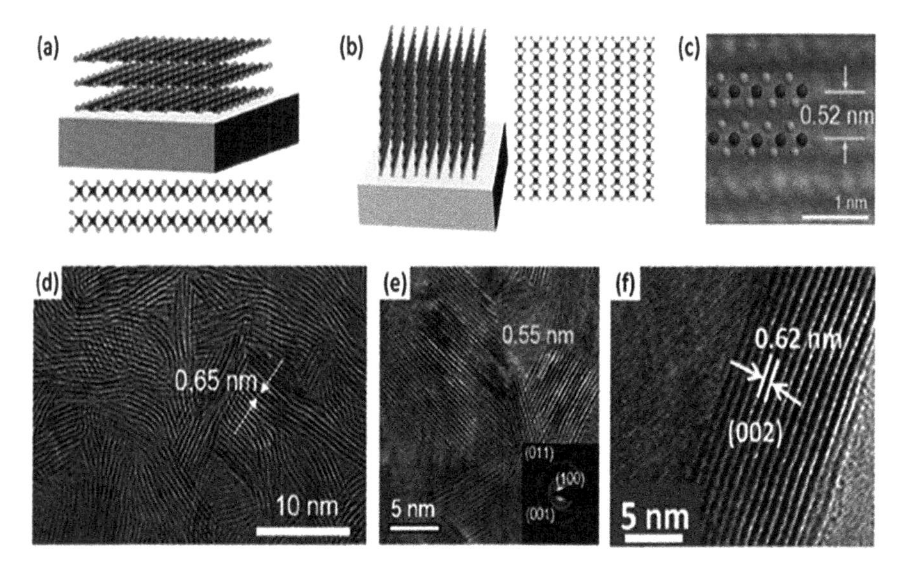

Fig. 6.21 **a** Diagram for the horizontally-arranged 2D TMD layers growth on a substrate; **b** diagram of vertically-arranged 2D TMD layers growth on a substrate; **c** cross-sectional HAADF-STEM picture of 2D $PtTe_2$ layers to display an interlayer distance of (B 0.52) nm with a layer-stacking order of (1T) Te–Pt–Te atomic chains (here blue and yellow dots indicate Pt and Te atoms); **d** HR-TEM picture of the as-grown vertically-arranged 2D MoS_2 layers, displaying surface-uncovered boundaries with uniform distance (B0.65 nm) vdW gaps; **e** HR-TEM picture of vertically- arranged 2D $PtSe_2$ layers. The inset displays the analogous SAED pattern indexing the $PtSe_2$ crystalline planes; **f** HR-TEM picture of 2D WS_2 (002) layers with an interlayer distance of B0.62 nm (adopted from Li et al. [89])

Z-contrast STEMimaging can be straightly correlated to the atomic numeral of the imaging species [175].

6.2.4.4 Layer Restacked Heterostructures

Almost all TMDs have two phases in nature, namely called 2H (semiconducting) and 1T (metallic phase) [176]. The bulk TMDs are also used as electrode material, but they have low surface limits to their applications. However, the thin 2D sheets of TMDs have advantages owing to high specific surface area and availability of abundant oxidation states, by providing facilitates of charge storage via faradaic and non-faradaic mechanisms [17, 177]. The electrochemical performance of such TMDs (Supper conductors) are usually very high with the anisotropic crystal structures [178]. Owing to the high surface area of 2D TMD nanosheets they are capable to provide more ion storage facilities. During charging and discharging (or reversible operations) such materials undesirably undergo a layer restacking process (usually known as a restacked process), as a consequence usually a decrease in charge storage

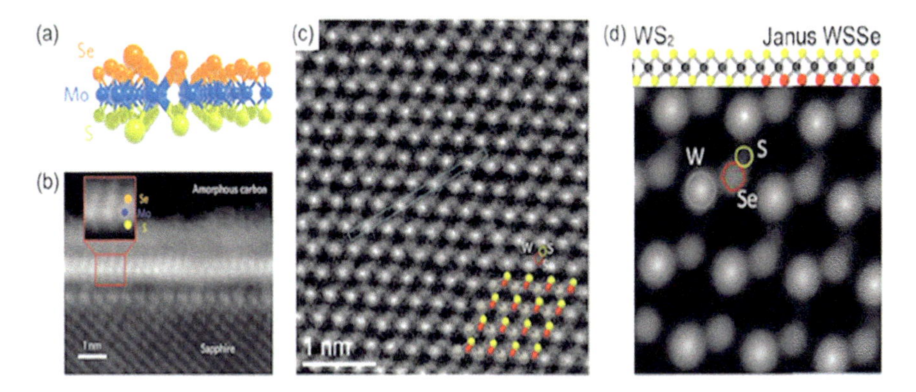

Fig. 6.22 **a** Molecular approach to fabricate Janus MoSSe single-layer; **b** ADF-STEM picture of a cross-section for the Janus MoSSe single-layer; **c** tilted HAADF-Z-STEM picture of ML Janus WSSe tilted at $x = +151$ along with the overlaid ball-and-stick approach exhibiting W atoms (gray), Se atoms (red), and S atoms (yellow); **d** molecular approach of Janus WSSe single-layer (top) and the respective simulated STEM picture for Janus WSSe tilted at $x = +151$, to the relevant intensity ratio defined experimentally (adopted from Li et al. [89])

due to restricted availability of the surface area by structural transformation of the TMDs [179].

Generally stacking ordering of 1H TMDs generates the 2H (AB ordering) or 3R (ABC ordering) arrangements. A variety of single-layer TMDs offers the possibility to restack them in desired sequences by making heterostructures, as illustrated in Fig. 6.23a, b [180–182]. The restacked (or folded) monolayer MoS_2 can have diverse features contrast to monolayer and bilayer 2H MoS_2. These kinds of inversion symmetry-breaking stacking sequences of bilayer TMDs also provide a supplementary pathway to tune the optical, electronic, and spintronics features of TMDs [183]. As the instance, the energy of the indirect conversion fluctuates from 1.49 eV to 1.62 eV to the twist angles 0 and 30 1C (Fig. 6.23c) [184] due to interlayer combination that depending on twist angle among the topmost and lowermost layers of MoS_2. The restacked bilayer MoS_2 has also a robust dependency on interlayer distances. Whereas the band structure of Mo–d states exists at K points, while a mixture of Mo–$d_z{}^2$ and S–p_z are in the valence band at G point. Thus the S–p_z states dependents on the interlayer stacking. Thereby, the energy state at the G point is considered to be more sensitive to the interlayer coupling than the K point. Moreover, the concentration of trions and excitons also depends on the twist angle [185], and as the consequence, similar results from the folded MoS_2 [186].

The mechanism behind this, as the TMDs MX_2 layered structure, X atoms are usually in trigonal prismatic coordination around the M atoms [187]. The atomic stacking sequence within a single XMX monolayer represents βAβ. In which 2H structure gives rise to metallic edge states that are associated with electrocatalytic activity [188]. The primitive unit cell of the 2H phase has a hexagonal structure (however, for the simplicity of DFT calculations some cases have considered the non-primitive rectangular unit cell depending on approximations). Typical 2H structures

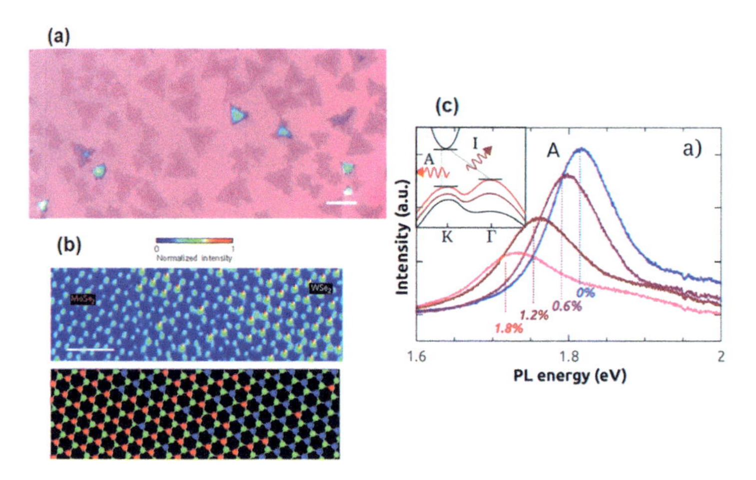

Fig. 6.23 **a** Optical picture for the triangular $MoSe_2$–WSe_2 heterostructure crystals developed by CVD, the core of the triangles contains $MoSe_2$ whereas WSe_2 epitaxially develops on the $MoSe_2$ boundaries; **b** top: the color map of the chemical stoichiometry of the $MoSe_2$–WSe_2 heterostructure recognized by STEM, the chemical stoichiometry is acquired by the scattered electron intensity. Bottom: restructured structures of the identical area, scale bar 1 nm; **c** amendment in the photoluminescence spectrum of monolayer MoS_2 by the implication of tensile strain. Inset: the band structure of monolayer MoS_2 for 0% (black), ~ 5% (maroon), and ~ 8% (red) of strain (adopted from Voiry et al. [186], https://doi.org/10.1039/c5cs00151j)

within a rectangular unit cell with the lattice constants a and b are illustrated in Fig. 6.24. Suppose one of the 2H structure's X layers shifts then the X atoms become octahedral coordination around the M atoms and the whole crystal transform to metallic, the corresponding phase is usually referred to as 1T, like IV and group V TMDs, TiS_2, and $TaSe_2$[187]. Moreover, group VI TMDs have a stable metallic structure under the octahedral M–X coordination. This lower-symmetry phase also refers to the 1T′ phase, which is a distorted version of the 1T structural phase, as depicted in Fig. 6.24 [187, 189, 190]. This kind of structural phase transformation of TMDs enhances the electrocatalytic activity, such as WS_2, WTe_2, $MoTe_2$ [187, 189, 191], and a metastable phase instanced chemically exfoliated and restacked MX_2 monolayers [192].

Despite several theoretical and experimental studies on restacked heterostructures for various applications [193–197]. Recently the combined 2D structures of MXenes and TMDs have attracted much attention. Because it can preserve the best features of both the MXenes and TMDs. While combining the MXenes and TMDs, restacking of the layers is hindered, as the consequence, an increase in the contact area between the electrolyte and electrode [198, 199]. Moreover, the 2D heterointerface between MXenes and TMDs consists of van der Waals interactions to prevent Fermi level pinning. Due to this effect, they can operate synergistically with the work function adjustability of metallic MXenes to afford Schottky barrier-free contact and reduce the contact resistance [200]. Moreover, MXene–TMD composites are considered to be excellent reversible specific capacitance and superior coulombic efficiency with

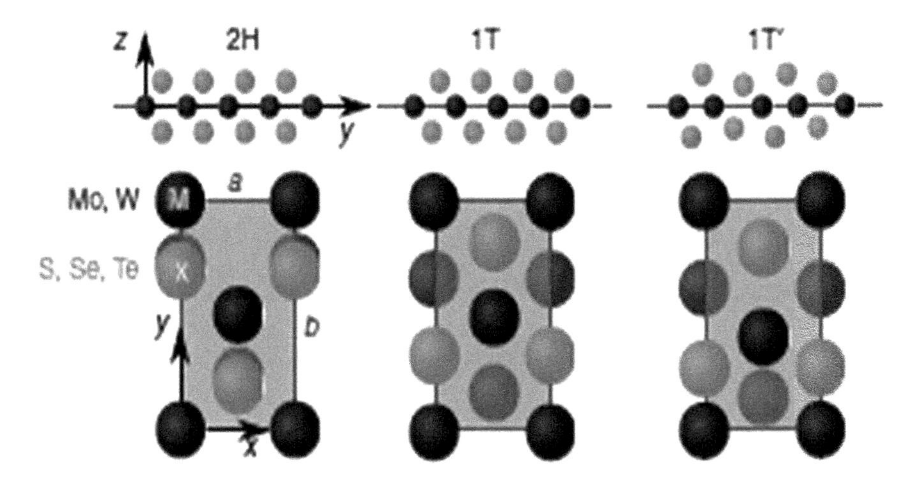

Fig. 6.24 Three crystalline phases of 2D group VI TMDs represented in a rectangular unit cell with dimensions a and b. All three phases consist of a metal (Mo/W) atom layer sandwiched between two chalcogenides (S/Se/Te) layers. The corresponding semiconducting 2H phase is often referred to as the trigonal prismatic structure, and the metallic 1T and 1T′ are defined as the octahedral and distorted octahedral, respectively (adopted from Karel-Alexander et al. [192])

improved cyclability with high rate performance in electrochemical applications, as the typical schematic illustrated in Fig. 6.25. For the instance, 2D MoS_2 as an anode material in rechargeable Li-ion batteries increases specific capacity (approximately 1131 mA h g^{-1}) with a higher coulombic efficiency than the pristine MoS_2 electrode [201]. Therefore, the $MoS_2/Ti_3C_2T_x$ composite delivers a stable reversible capacitance (614.4 mA h g^{-1} at 100 mA g^{-1}) with a low electrochemical impedance with better rate performance, as the typical results for the MoS2-MXene are depicted in Fig. 6.26a, b. The unique structure and synergistic effect hindered the agglomeration and volume expansion of the MoS_2 nanosheets because of the large conductive surface of the $Ti_3C_2T_x$ substrate (see Fig. 6.26c). That typically promotes rapid electron/Li + transfer kinetics. Similarly, assembling the MoS_2/Ti_3C_2–MXene@C nanohybrids as anodes lithium batteries has been considered to be an ideal mass loading of electrodes of approximately 0.8–1.0 mg cm^{-2}. However, studies are not limited to only a few discussed cases, a vast number of MXenes-TMDs batteries and other reversible properties based on modern devices have been also demonstrated [202].

6.2.4.5 Layers Vertically Aligned Heterostructures—Experimental Studies

Vertically aligned heterostructures experimentally have been extensively explored to fabricate various kinds of devices for specific uses by employing different synthetic methods [89, 203]. As an instance, the WSe_2 and MoS_2 are the semiconductors with

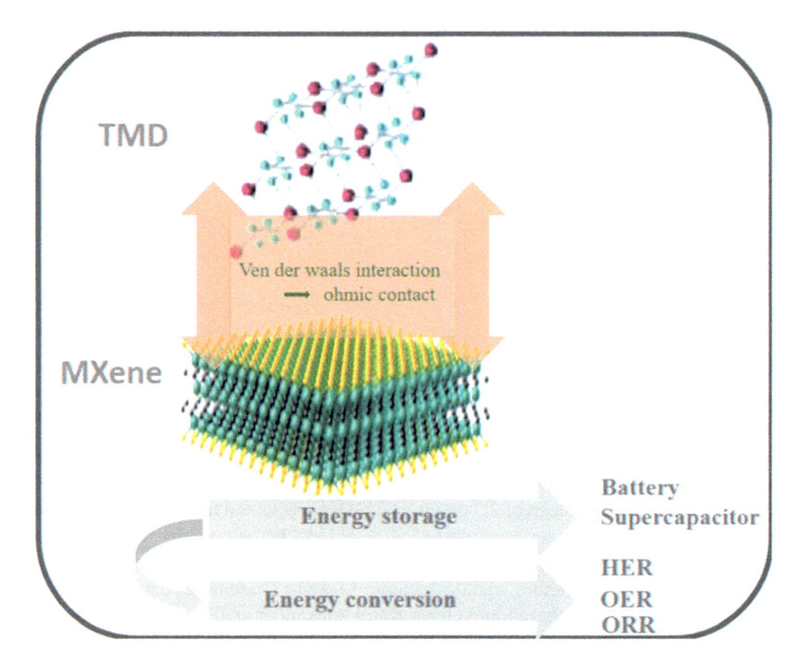

Fig. 6.25 Schematic for the 2D TMD/MXene hybrid structure toward energy storage and conversion (partially adopted from Lim et al. 2020, ACS Nano 14, 10834–10864)

p-type and n-type behaviors [204, 205]. Their combinations are particularly important due to high band offsets among transition metal dichalcogenide semiconductors by creating the pn junction diode [205]. Such vertical heterostructures of WSe_2 and MoS_2 with vertically aligned layers can be synthesized by the sequential growth of WSe_2 and MoS_2 via kinetically controlled rapid selenization and sulfurization process (typically for the 30 nm thick tungsten film and followed 100 nm thick edge-exposed WSe_2 film at 600 °C) at desired temperatures, as illustrated in Fig. 6.27a, b [206]. Successful fabrication of WSe_2 with vertically aligned layers around 15 nm thick molybdenum thin film by sputtering is depicted in Fig. 6.27 c. Subsequently, the achieved Mo/WSe_2 thin film has also rapidly sulfurized (at 600 °C) to produce MoS_2/WSe_2thick (around 140 nm) heterostructure (see Fig. 6.27d, e). The different steps formed WSe_2 and MoS_2/WSe_2heterostructures polycrystalline structure and alloying elemental amounts presence have been confirmed from the STEM, TEM, and mapping analysis. Additionally, the formation of MoS_2/WSe_2 heterostructures has been also verified from the Raman spectroscopy. In which expected dominant out-of-plane M–X vibration mode (A_{1g}) over to in-plane M–X vibration mode $\left(E^1_{2g}\right)$ has noticed. However, it is hard to resolve the Raman spectrum of WSe_2 due to the small energy difference between the A_{1g} mode (253 cm^{-1}) and E^1_{2g} (250 cm^{-1})[207]. On the other hand, Raman spectrum of MoS_2 is clearly distinguished with the stronger peak intensity of A_{1g} mode (408 cm^{-1}) than E^1_{2g} mode (383 cm^{-1})[207, 208]. Moreover, the analysis of Raman peak analysis has also confirmed the edge termination of

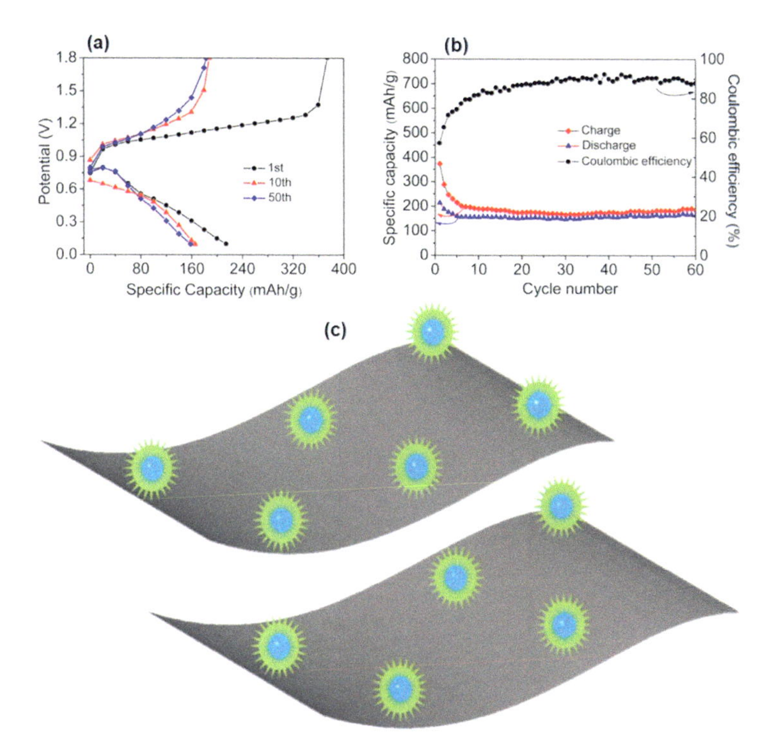

Fig. 6.26 MoS_2-MXene cathodes: **a** discharge/charge curves; **b** curves of the cycling performance; **c** schematic structure for the MoS_2-MXene composites to cycling performance (adopted from Tan et al. 2021, Energy Fuels 35, 12666–12670)

MoS_2. It is normally correlated with the in-plane M–X vibration mode (E_{1g} and E^1_{2g}) of MoS_2 that shifts to the lower frequency side, while the WSe_2 peak shifts to the higher frequency side. Such frequency shifts originated from the atomic weight difference of both constituent cations (Mo and S) and anions (S and Se). Whereas the A_{1g}vibration mode of MoS_2 is due to only anion atoms vibration in the out-of-plane direction, therefore, its peak move to a lower frequency side when alloying with the sulfur and selenium.

To provide the depth-resolved elemental profile of the MoS_2/WSe_2 heterostructure film has been also examined by scanning Auger electron spectroscopy. The depth-resolved elemental profile of up to only 13% of the film has been detected for all four component elements in the analyzed spectra. Therefore, the overlap zone has been estimated at 18 nm in thickness, where the surface roughness of the bottom WSe_2 layer and the depth resolution of Auger spectroscopy profiling is around 9 and $\sim$ 15 nm. This lead to the degree of inter diffusion of the component elements are only limited to the boundary between MoS_2 and WSe_2. The experimental spectral findings (both Raman spectra and Auger electron spectroscopy depth-profile) have revealed there is no significant alloying occurs in the MoS_2/WSe_2 heterostructures [206].

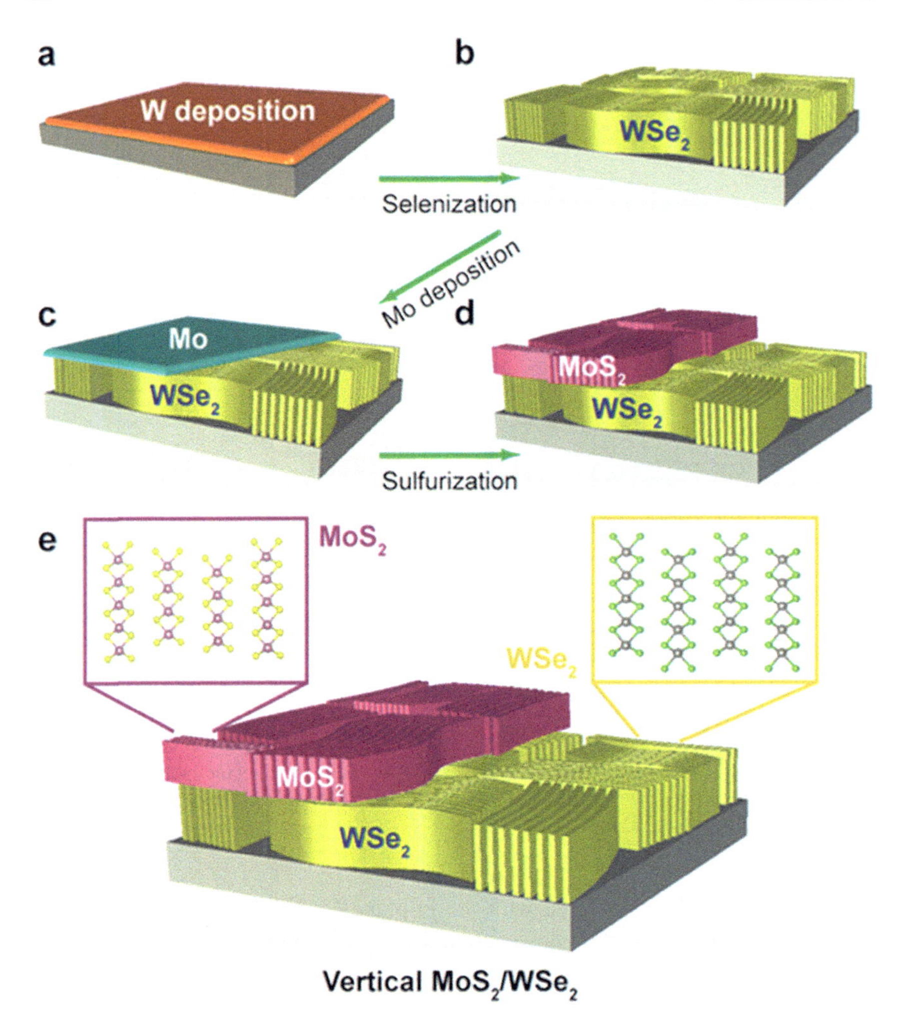

Fig. 6.27 Schematics for the MoS_2/WSe_2 vertical heterostructure synthesis: **a** tungsten (W) coated onto the substrate; **b** WSe_2 film with a vertically aligned layer formed via rapid selenization; **c** Molybdenum (Mo) coating on top of the synthesized WSe_2 film; **d** MoS_2/WSe_2 vertical heterostructure formation via rapid sulfurization; **e** the heterostructure consists of the MoS_2 and WSe_2, in which their van der Waals layers are aligned perpendicular to the substrate (adopted from Yu et al. [206])

Thus the large area MoS_2/WSe_2 vertical heterostructure are also possibly fabricate in a sandwich structure device in which Pd contacted with WSe_2 as a positive electrode and Ti/Au contacted MoS_2 as a negative electrode, as illustrated in Fig. 6.28a, b. The current–voltage characteristic curve of the device has a typical diode characteristic with an ON/OFF current ratio of $\sim$150 and an ideality factor of 1.5 at a lower voltage range below 1 V, as depicted in Fig. 6.28c, d. To know the actual origin of the device diode characteristic, the individual two-terminal devices Pd–WSe_2–Pd and

Ti/Au–MoS_2–Ti/Au have been analyzed. The current–voltage characteristic curves of both two-terminal devices have revealed nearly an ohmic characteristic. However, it is well known that MoS_2 and WSe_2 are intrinsically n-type and p-type semiconductors. Therefore, it has been concluded that the rectification behavior of the device originated from the pn junction between n-type MoS_2 and p-type WSe_2. Thus, the pn junction diode of MoS_2/WSe_2 with vertically aligned van der Waals layers ON/OFF current ratio are usually high enough. But several issues remain unresolved in the initial demonstration such as the the absolute value of the ON current density is small compared to other high-purity polycrystalline heterostructures, as depicted in Fig. 6.28a–f [206].

Experimental investigations with the vertically aligned van der Waals layers day by day improve to fabricate more robust and efficient devices for different purposes with the structural modifications of materials based on the initial findings. As an instance, the in-plane/vertical heterostructures of $Re_xMo_{1-x}S_2/MoS_2$ and ReS_2/MoS_2 have been demonstrated. To synthesize these materials one of the most powerful CVD methods have been used frequently [89, 118, 120, 203, 209], as the typical schematic is illustrated in Fig. 6.29 a. usually the high purity Mo and Re materials to be used to deposit on top of two Si/SiO_2 substrates with a 285-nm SiO_2, at 600 °C along with a ramping rate of 50 °C/min by maintaining desired temperature for 10 min. Additionally, the furnace also purged with 500 sccm argon to the top cover of the furnace elevated for fast cooling of the samples. In this way one can fabricate in-plane $Re_xMo_{1-x}S_2/MoS_2$ and vertical ReS_2/MoS_2 heterostructures on the Si/SiO2 substrates. Moreover, to fabricate the devices titanium/gold metal layer (Ti: 3 nm, Au: 35 nm) e-beam evaporation tool have been used to make source-drain electrodes. Typical heterostructures morphology optical microscopy schematic of in-plane $Re_xMo_{1-x}S_2/MoS_2$ heterostructure is illustrated in Fig. 6.29b, while its typical optical images are depicted in Fig. 6.29c, d. Here visualized the $Re_xMo_{1-x}S_2$ alloy at the center area surrounded by the MoS_2 at the outer edge. Moreover, a schematic of the ReS_2/MoS_2 vertical heterostructures with the corresponding optical morphology is illustrated in Fig. 6.29e–g which is composed of a bottom MoS_2 layer and an overlap to the ReS_2 top layer.

Herein one has paid attention only to vertical ReS_2/MoS_2heterostructure, therefore, excluded the discussion in detail on experimental findings of in-plane $Re_xMo_{1-x}S_2/MoS_2$ heterostructure. The successively two layers formed on Si/SiO_2 substrate have revealed the vertical heterostructure of ReS_2/MoS_2. Their Raman characterization demonstrated that the two prominent peaks appearance at E^1_{2g} (382.5 cm^{-1}) and A_{1g} (403.0 cm^{-1}) with a frequency distance 20.5 cm^{-1}. Similar results have been also noted for the Raman spectroscopic analysis with the central bilayer area, in which both the peaks for upper ReS_2 and covered MoS_2 appear with the E^1_{2g} (382.7 cm^{-1}) and A_{1g} (404.2 cm^{-1}), whereas the intensities of MoS_2 nearly unchanged. Moreover, this system's PL spectrum has been also examined for the edge monolayer with ~ 680 nm peak. This has directly correlated to bottom layer MoS_2 semiconducting behavior. Moreover, a slight peak acquired from the central points of the bilayer sample that correlated to the PL quenching effect of the metallic layer

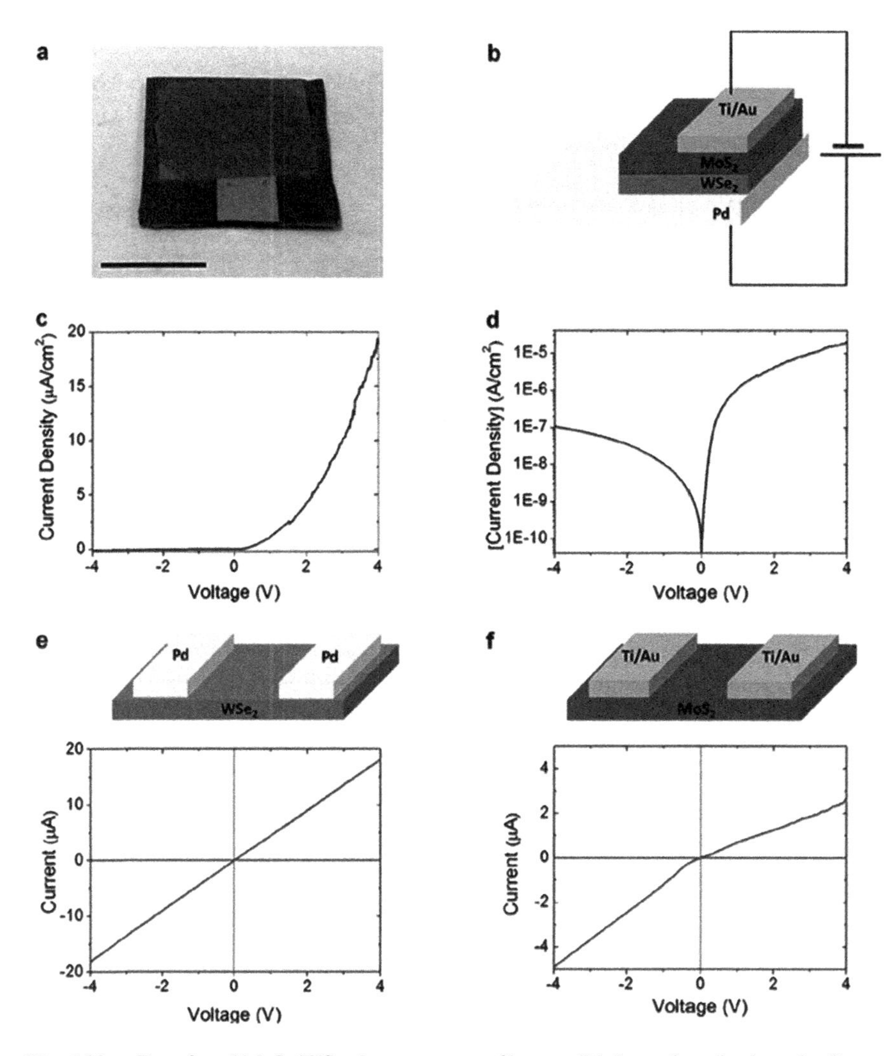

Fig. 6.28 **a** Transferred MoS_2/WSe_2 heterostructure film on a Pd electrode pad, where the dimension of the pad is 2 cm × 2 cm, scale bar 1 cm; **b** schematic for the MoS_2/WSe_2 heterostructure diode, in which heterostructure sandwiched between Pd and Ti/Au, to defined as positive and negative electrodes, respectively; **c** current–voltage characteristic for the MoS_2/WSe_2 heterostructure diode; **d** logarithmic plot of the current–voltage curve in panel **c**; **e** schematic and current–voltage characteristic for the WSe_2 thin film two-terminal device at which Pd electrodes placed; **f** schematic and current–voltage characteristics of MoS_2 thin film two-terminal device on which Ti/Au electrodes placed (adopted from Yu et al. [206])

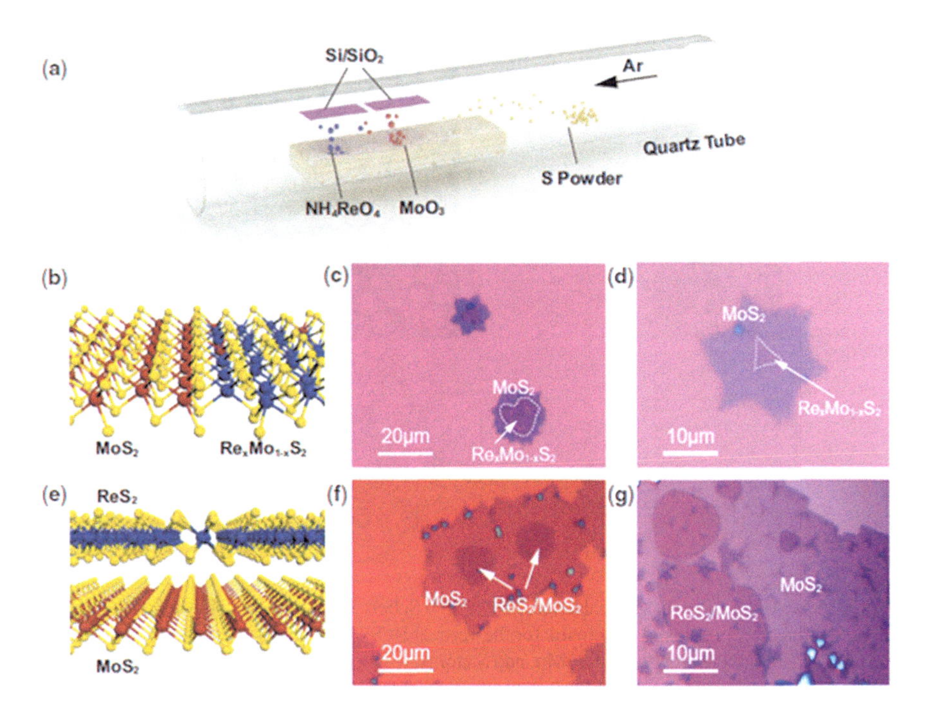

Fig. 6.29 **a** Schematic of the synthesis process for the in-plane/vertical heterostructures by CVD approach; **b** schematic for the in-plane $Re_xMo_{1-x}S_2/MoS_2$ heterostructures; **c, d** Typical optical images for the in-plane $Re_xMo_{1-x}S_2/MoS_2$ heterostructures; **e** schematic of vertical ReS_2/MoS_2 heterostructures; **f, g** Typical optical images for the vertical ReS_2/MoS_2 heterostructures. Here the blue, red, and yellow spheres in **b** and **e** represent Re, Mo, and S atoms, respectively (adopted from Ma et al. [118])

in the bilayer region. Additionally, the crystal structure of the vertical ReS_2/MoS_2 heterostructure of the as-grown sample has been also characterized using aberration-corrected STEM Z-contrast imaging, with the as-recorded annular dark field (ADF) images filtered by a Gaussian function (full-width half maximum = 0.12 nm) to remove high-frequency noise, as depicted in Fig. 6.30a. Since the intensity of the ADF image is directly related to the atomic number Z, therefore the heavier atoms like Re ($Z^2 = 5625$) should be brighter than lighter atoms Mo ($Z^2 = 1764$), and both of them much brighter than S atoms ($Z^2 = 256$). Using Fig. 6.30a it further divides into the part, in the bottom appears many dark atoms, while in the upper part, several bright atoms have appeared. Therefore, the Re and Mo atoms can be differentiated by their local intensity, as the atomic identification schematic is depicted in Fig. 6.30b. In which Re and Mo atoms represent the blue and red colors. That revealed the upper left domain of MoS_2 and the lower right domain of $Re_xMo_{1-x}S_2$ alloy. Their atomic arrangements are also identified for the both $Re_xMo_{1-x}S_2$ and ReS_2 regions that have a T′ phase structure, while the MoS_2 regions belong to the H phase structure. The intensities profile across the red/blue rectangles is depicted in

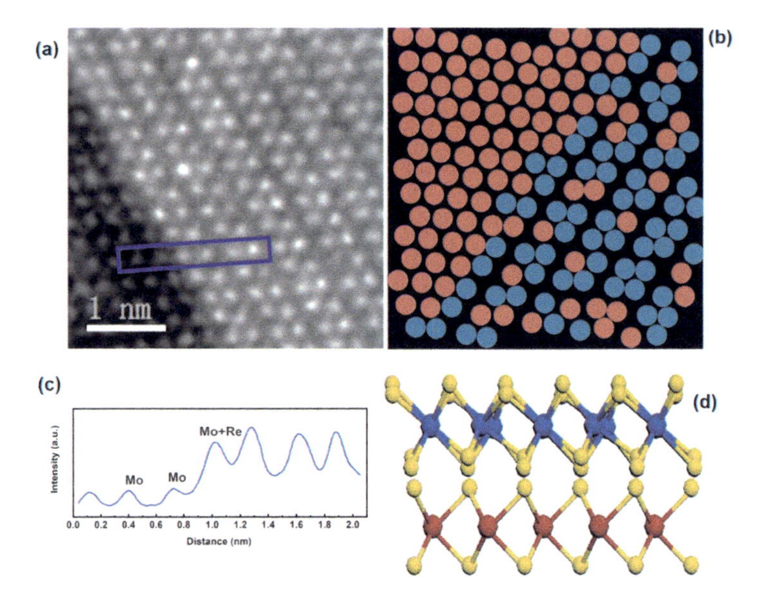

Fig. 6.30 **a** ADF image for vertical heterostructure; **b** Schematic for the Re and Mo atoms arrangement; **c** Intensities profiles; **d** Schematic for the vertical ReS_2/MoS_2 heterostructure. Note the blue, red, and yellow spheres represent Re, Mo, and S atoms, respectively (adopted from Ma et al. [118])

Fig. 6.30c. Their dimmer atom sites have also revealed that the bottom supporting layer belongs to MoS_2 and this layer partially overlaps with an upper ReS_2 layer. Thus the ReS_2/MoS_2 vertical heterostructure crystal structure can possibly deduce from the ADF images, as illustrated in Fig. 6.30d.

6.2.5 Lateral Heterostructures (LHs)

Since vertical heterostructures are generally realized either by mechanical exfoliation or via direct growth. But, atomically stitching materials in the lateral direction is only possible via direct growth. The first successful synthesis of LHs (also called planar or in-plane) of 1L $MoSe_2$-WSe_2 has been carried out from the single-step physical vapor transport method using a mixture of TMDs bulk powder counterparts as precursors [9, 210]. Adopting analogy differences in the volatility and vapor pressure of various TMDs precursors the growth of $MoSe_2$ followed by the edge-epitaxy of WSe_2 domains has been achieved. Though TMD domains are generally vulnerable to multiple exchanges of growth systems and chemical environments, therefore, it is a challenging task to get a sequential growth of multi-junction LHs.

To overcome the issue reverse flow strategy has been adopted for the conventional physical vapor transport system to a sequential growth of the heterogeneous TMDs LHs, while the superlattices via a multi-step process. The bulk TMD precursors are to be used directly depending on the types of 2D domains in the LHs. It makes it possible to realize different transition metal and chalcogenide components without the formation of alloys across interfaces, in a separate growth chamber of the synthesized constituents of TMD domains [211]. Similarly, the monolayer multi-junction TMDs LHs and lateral superlattices have been also synthesized under a bidirectional flow of carrier gases to mitigate the problem associated with uncontrolled nucleation during the temperature ramping stage and avoid cross-contamination during the exchange of precursors. Moreover, the individual domains with the LHs are usually controlled by varying the growth temperature and reaction time. Though, such bidirectional carrier gas-based strategies are complicated for the growth of multi-junction TMDs LHs as well as lateral superlattices. Moreover, this technique is also not free from the limitations such as lack of scalability on account of multiple exchanges of growth chambers to fabricate individual domains.

Therefore, the atomically thin TMD layers are prone and degrade during the multiple exchanges of growth processes leading to a change in a growth strategy that can allow continuous and direct growth of multiple 2D layers without breaking the growth condition or exchanging CVD systems. Such as the water-assisted one-pot CVD situ growth approach of multi-junction LHs with the precise number of 2D layers, the width of individual domains, the degree of alloys, and atomically sharp interfaces, simply by changing the carrier gases [212]. Whereas a mixture of TMD bulk powders at high temperature (~ 1060 °C) allows them to react with different carrier gases, while the substrates are placed in the temperature range of ~ 750–820 °C. Whereas the N_2 carrier gas through H_2O promotes the selective evaporation of Mo-related precursors that lead to the formation of the MoX_2 domain. Moreover, a switch to reducing gases like Ar with H_2 (5%) has stopped the growth of MoX_2 and promoted the nucleation of WX_2 domains. The key advantage of this process is the simplicity of vapor phase modulation induced in situ growth of multi-junction LHs, which does not require any source switching (or toxic precursors). Additionally, in this process number of layers of the LHs can be controlled by adjusting the kinetic coefficient of adatoms. Considering the remarkable advantages of this approach more recently continuous tuning of the band gap of each domain of monolayer (1L) LHs composed by the ternary alloys $MoS_{2(1-x)}Se_{2x}$–$WS_{2(1-x)}Se_{2x}$, as the typical schematic of growth mechanism and LHs formation are illustrated in Fig. 6.31(a, b) [213]. It also leads to the lateral side of each domain being controlled by increasing or decreasing the time of the corresponding growth step, by revealing the presence of different domains in heterostructures, like optical and electronic contrasts Fig. 6.31c, e. Here low-magnification optical image (see Fig. 6.31d) correlates to the abundance of 2D heterostructure islands on the substrate surface and the coexistence of islands for the different thicknesses of the same sample.

Another approach to the realization of edge, and ohmic contact via resistance reduction of hexagonal WSe_2 (1 and 2L as a core layer) and tetragonal CoSe LHs, has been also demonstrated by using a two-step vapor transport method. The bulk WSe_2 powder is useful to grow the WSe_2 layer when the $CoCl_2$ precursor is used for the growth of the metallic layer [214]. This innovation has revealed that the growth temperature may impact the lateral size of the metallic layer. Additionally, it has

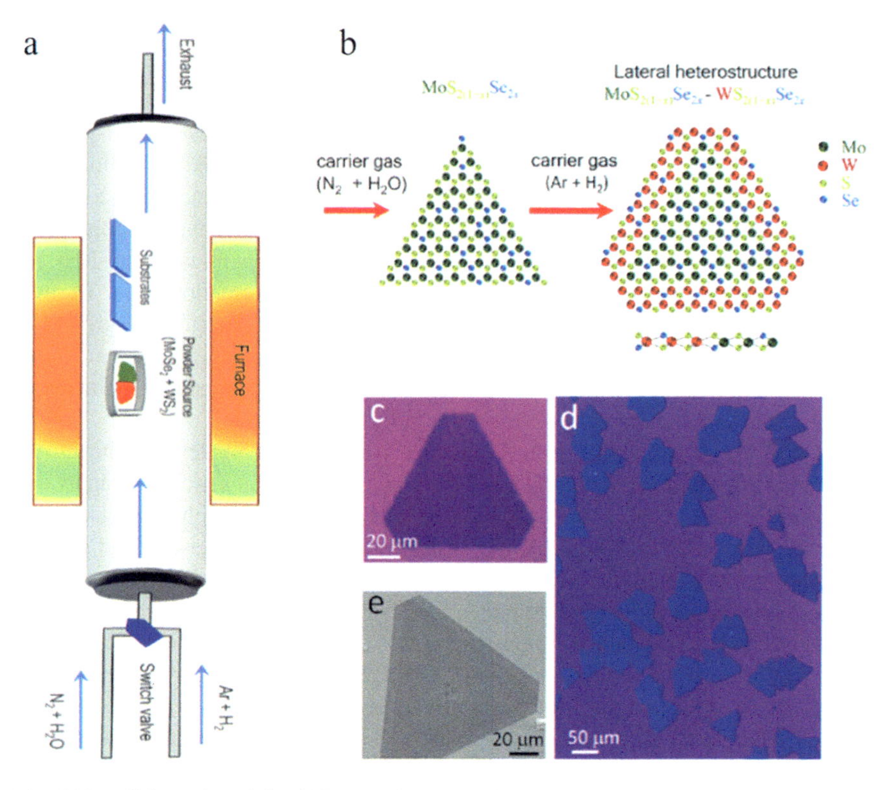

Fig. 6.31 **a** Schematics of the CVD growth process; **b** cartoons to show the atomic structure of the lateral heterostructures based on ternary alloys of $MoS_{2(1-x)}Se_{2x}$–$WS_{2(1-x)}Se_{2x}$; **c**, **d** High and low-magnification optical images for the as-grown heterostructures; **e** Scanning electron microscopy (SEM) image for a lateral alloy heterostructure (adopted from Nugera et al. [213], https://www.researchgate.net/publication/358199994)

been noted that with the rising growth temperature (in between 535 and 570 °C) the lateral width of the 1L CoSe layer significantly increased, whereas the thickness of the CoSe layer remains identical at the edge of 1L-WSe_2 with the increasing growth temperature, but the thickness of the layer increases on the edges of 2L-WSe_2. Therefore, the growth and diffusion rate of adatoms has been influenced by the increasing growth temperature even slightly. The lower growth temperature also promotes the adatom's effective diffusion toward the edges and contribute to lateral growth, on the other hand, high temperature favored the growth of high-energy sites and forms comparatively thick layers.

6.2.5.1 Edge Epitaxy

Edge epitaxy relies upon HSs owing to the growth of the TMD crystals on the active edge of a dissimilar TMD crystal. Because the relatively small lattice mismatch between various TMDs produced the unsaturated edge in a TMD film, it can serve as an active growth front in the lateral epitaxy of a dissimilar TMD film, as depicted in Fig. 6.32a. This approach is useful for the several variations including single-step [116, 182], two-step [215], and sequential [216] schemes. Like, in the single-step approach, two stoichiometric TMD crystals such as MX_2 and $M'X_2$ are simultaneously placed in a single boat at high temperatures. Their vapor phase is carried to the colder zone of the reaction chamber to condense side by side on a substrate and formed MX_2-$M'X_2$ lateral HSs, such as MoS_2-WS_2, $MoSe_2$-WSe_2, MoS_2-$MoSe_2$, etc. [116, 182, 217]. As the typical lateral HSs are depicted in Fig. 6.32b, in which the HSs are composed of a central TMD crystal surrounded by a dissimilar TMD crystal on their periphery. But the single-step edge epitaxy process encounters two major issues that are mostly routed in the coexistence of multiple precursors in the vapor phase. First, it often yields an alloyed junction with a finite width instead of an abrupt junction with an atomically sharp interface, as illustrated in Fig. 6.32c [116, 182]. The second one necessitates a growth condition that works for both TMD crystals that are intended to form the lateral junction.

However, these two issues can be addressed by the two-step process in which a TMD crystal is first grown and then transferred into another chamber for the lateral epitaxy of the second TMD crystal on the edge of the first one. This approach offers two major advantages; first, it eliminates the interference of the precursors that are needed for the growth of the two TMD films to make atomically sharp junctions, another one it enables to change of both transition metal and chalcogen elements on either side of the junction. It allows expanding the variety of lateral junctions via edge epitaxy, such as WSe_2-MoS_2 lateral HSs with atomically sharp interfaces and dissimilar chalcogens and transition metals, as illustrated in Fig. 6.32d [215].

The edge epitaxy is also achieved from the sequential approach to composed structures in terms of multiple junctions between two heterogenous TMDs, as depicted in Fig. 6.32e–g. In this process, cyclic exchange of active precursors can use to alternatively perform the epitaxy on the fresh edge of the last grown TMD film. In this process, solid MoX_2 and WX_2 (X: S or Se) crystals are simultaneously usually placed in a boat that is held at a moderate temperature. The subsequent flow of carrier gasses $N_2 + H_2O$ into the reaction chamber promotes only the growth of MoX_2. The subsequent switching of carrier gas $Ar + H_2$ (5%) terminates the growth of the MoX_2 and selectively promotes the epitaxy of WX_2 at the age of MoX_2. This approach offers a fast depletion of pre-existing precursors [120, 216].

6.2.5.2 Lithographic Pattering Approach

Lateral TMD LHs are also achieved from the lithographic approach, by the post-growth alloying process of lithographic patterned TMD. The lithographic step followed by the deposition of an impermeable mask to protect the designated parts of the MX_2 thin film from the elemental exchange during the post-growth alloying process, as illustrated in Fig. 6.33a. Therefore, in this process only exposed regions of the MX_2 undergo the compositional changes, while the protected regions remain

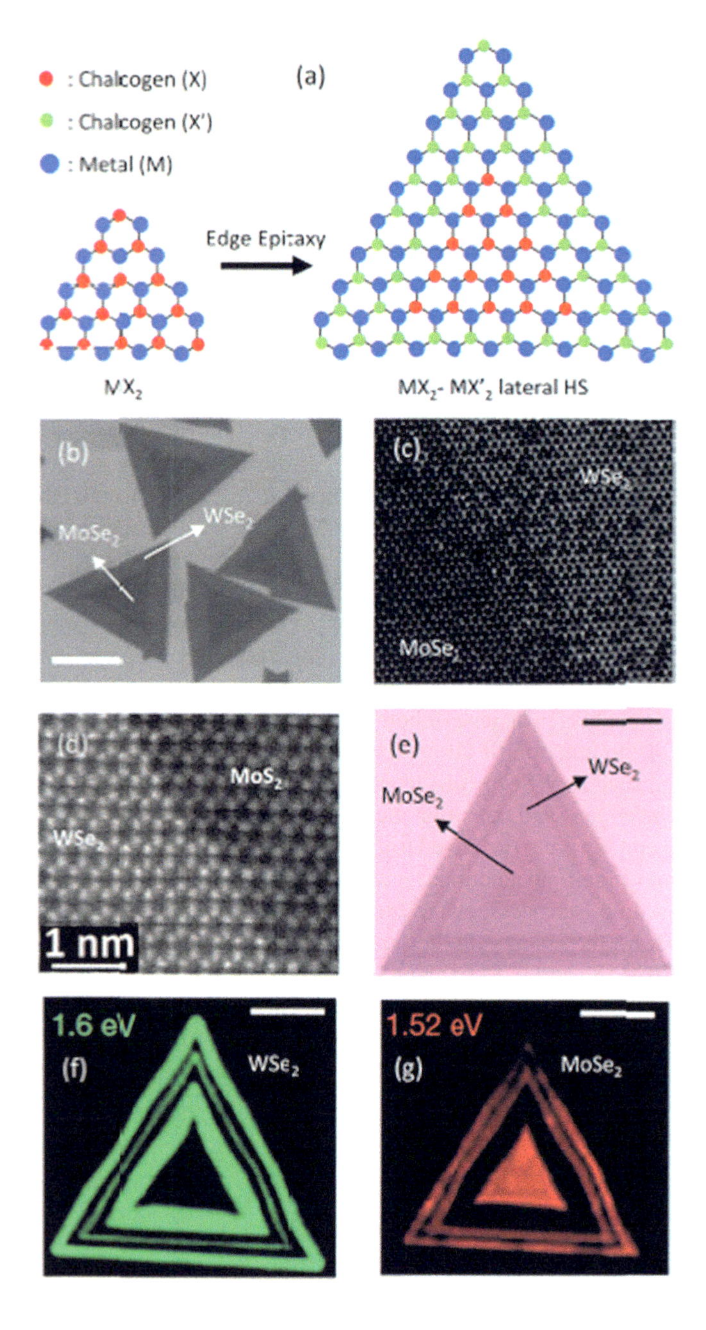
: Chalcogen (X)
: Chalcogen (X')
: Metal (M)
(a)
Edge Epitaxy
MX_2
MX_2- MX'_2 lateral HS
(b)
$MoSe_2$
WSe_2
(c)
WSe_2
$MoSe_2$
(d)
MoS_2
WSe_2
1 nm
(e)
WSe_2
$MoSe_2$
1.6 eV
(f)
WSe_2
1.52 eV
(g)
$MoSe_2$

◀**Fig. 6.32** **a** schematic illustration of the edge epitaxy for the synthesis of an MX_2-MX'_2 lateral HS; **b** STEM image for the $MoSe_2$-WSe_2 HSs synthesized through a single-step edge epitaxy reaction; **c** STEM images for the $MoSe_2$-WSe_2 junction, to show the formation of an alloyed interface; **d** STEM image to illustrates the formation of an atomically sharp interface via a two-step epitaxy of MoS_2 on the edge of a WSe_2 monolayer; **e** optical image of a multi-junction $MoSe_2$-WSe_2 Lateral HS synthesized via a sequential edge epitaxy technique; **f**, **g** PL maps with 1.6 eV (WSe_2) and 1.52 eV ($MoSe_2$) emission energies, to confirms the formation of a multi-junction structure (adopted from Taghinejad et al. [120])

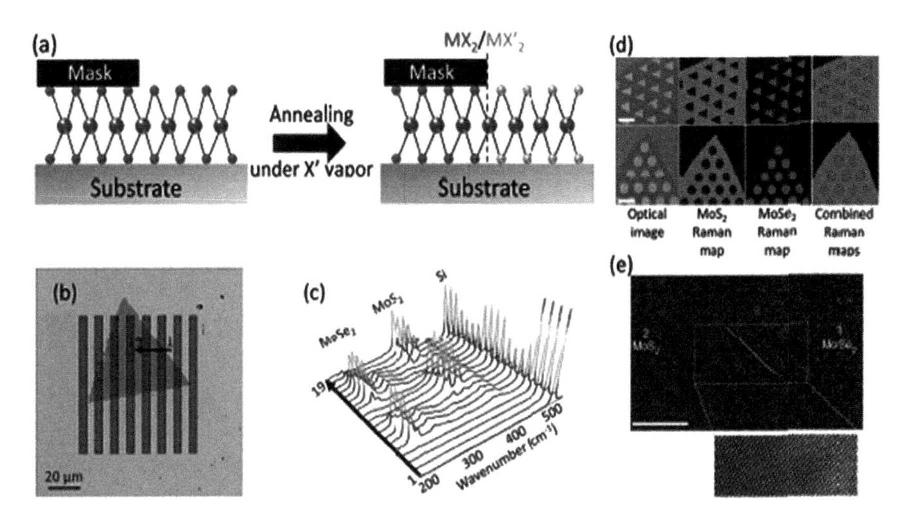

Fig. 6.33 **a** Schematic for the lithographic patterning used for the fabrication of lateral MX_2-MX'_2 HSs; **b** optical image for the $MoSe_2$-MoS_2 HS synthesized via the sulfurization of a patterned $MoSe_2$ monolayer; **c** Raman spectra, alternating appearance in $MoSe_2$ Raman modes (in protected regions) and those of MoS_2 (in exposed regions) confirms the formation of a lateral $MoSe_2$-MoS_2 junction; **d** optical images and Raman maps for the different HS geometries synthesized via the lithographic patterning approach; **e** STEM imaging of the $MoSe_2$-MoS_2 interface to show this approach produced an alloyed junction instead of an atomically sharp one. The inset shows a Z-contrast image intensity obtained from the highlighted box (adopted from Taghinejad et al. [120])

intact to form a lateral junction, as the intense typical $MoSe_2$- MoS_2 LHs are depicted in Fig. 6.33b, c [218, 219]. The key advantage of this technique is lateral LHs to be produced with arbitrary shapes and their lateral dimensions in pre-defined locations, therefore, its adequate features offer unprecedented flexibility for the fabrication of practices, typical examples depicted in Fig. 6.33d, e.

6.2.5.3 Bandgap Engineering in LHs

Band engineering technique offers to interpret different regions of the PL spectrum based on their alloying compositions. Such as PL spectra of 1L-LHs' different amounts of WS_2 and $MoSe_2$ peak positions as depicted in Fig. 6.34. The left to right

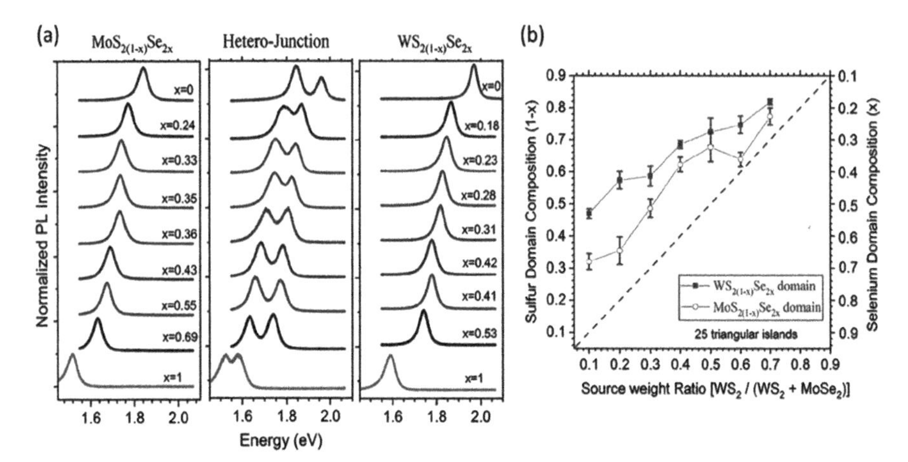

Fig. 6.34 **a** PL spectra of seven different samples grown from powder sources with different compositions. The spectra on the left, center, and right panels were collected from the Mo-rich junction, and W-rich domain; **b** correlation between the average composition in the Mo-rich and W-rich domains versus the WS_2 content in the powder source (WS_2:$MoSe_2$) (adopted from Nugera et al. [213], https://www.researchgate.net/publication/358199994)

PL at different points in the heterostructures reveals the Mo-rich domain (left panel) at the heterojunction (central panel), and the W-rich domain (right panel). The corresponding chalcogen concentration (x) at each domain is usually calculated from the measured photoluminescence peak positions by employing Vegard's law (for ternary alloy);

$$E_g MS_{2(1-x)}Se_{2x} = (1-x)E_g(MS_2) + xE_g(MSe_2) - bx(1-x) \tag{6.11}$$

where $M = [\text{Mo or W}]$ and b is the band parameter, for the Mo and W-based alloys $b = 0.05$ and 0.04 [65]. It has been noticed that the PL peaks of two domains in heterostructures consistently possess a blue shift when the percentage of WS_2 relative to $MoSe_2$ is increased. This blue shift correlates to a decrease in the selenium content ($2x$) at each domain due to the higher abundance of sulfur species in the carrier gas during the growth. It also allows to PL position and corresponding band gap to be continuously tuned between 1.63 and 1.77 eV for the $MoS_{2(1-x)}Se_{2x}$ domain and 1.74–1.87 eV for the $WS_{2(1-x)}Se_{2x}$ domain. Whereas at the interface (central panel in Fig. 6.34a) superposition of two PL peaks occurs due to simultaneous excitation of the PL from both domains, by the ≈ 0.4 μm laser spot. Usually, broad heterojunctions display in a single PL peak at an intermediate energy shift, and their position is related to a compositional gradient or alloying of the metallic element at the interface [216]. Hence, the relatively sharp interface also relates to local PL displays of two separated peaks corresponding to every individual domain [216]. However, others/all interfaces (including all relatively weak peaks) due to the possibility of a small alloying in the metal element to the region close to the interface, as depicted in Fig. 6.34b. Both domains are very similar with slightly higher sulfur content for the tungsten domain compared to the molybdenum domain. Whereas the dashed line represents the ideal case the composition of the source is stoichiometrically transferred to the sample. It has been also noted that usually, samples possess a higher sulfur content compared to sources that have a higher abundance of sulfur species in the carrier gas due to differences in volatility between sulfur and selenium [213].

Moreover, investigations have been also performed with the Raman spectra by interpreting two different monolayer-based heterostructures with distinct chalcogen contents. In the case of the 1L-LHs, the composition $MoS_{1.32}Se_{0.68}$–$WS_{1.53}Se_{0.47}$ (high sulfur content) has revealed a clear contrast between the Mo-rich (darker contrast) and the W-rich (lighter contrast) domains, whereas the growth times of the Mo-rich and W-rich domains are around 12 and 3 min. The average size of the islands is around 60 μm, while the lateral size of the W-rich domain is perpendicular to the interface ($\approx$ 6 μm). Raman spectrum of the Mo-rich domains are also correlated to the formation of the in-plane and out-of-plane MoS_2-like phonon modes at 376 cm^{-1} (E^1_{2g} (S-Mo)) and 409 cm^{-1} (A_{1g} (S-Mo)) [220–222]. The peak around 270 cm^{-1} generally relates to the A_{1g} phonon mode for both Se and S atoms around the Mo atom (Se-Mo-S) [36]. The peak at 351 cm^{-1} is considered to be composition-dependent E^1_{2g}, due to Se-Mo-S impurity [36]. Because such composition (sulfur-rich) Raman peaks (Se-Mo-S) may weaker compare to MoS_2 (S-Mo) modes. On the other hand, Raman spectra of the W-rich domain can have three optical phonon modes at 265 cm^{-1} (A_{1g} (Se-W)), 358 cm^{-1} (E^1_{2g} (S-W)), and 415 cm^{-1} (A_{1g} (S-W)) [223]. Their two additional peaks at 165 and 330 cm^{-1} are due to longitudinal acoustic (LA) and its second-order replica 2LA [224].

Similarly, in the case of 1L-LHs, the higher selenium concentration ($MoS_{0.48}Se_{1.52}$–$WS_{0.86}Se_{1.14}$) has been also addressed by a clear contrast between the Mo-rich (darker contrast) and the W-rich (lighter contrast) domains. Their PL intensity maps at 1.63 eV (Mo-rich domain) and 1.73 eV (W-rich domain) reflect a spatial distribution of the laterally connected domains with intradomain compositional homogeneity. So, the intensity variation of the PL signal has been noticed that is connected to the Mo domains of both monolayer heterostructures. However, this type of alloy with high selenium content Raman spectroscopic peak splits into two ($MoSe_2$ A_{1g}) peaks. That can further separate as the selenium (sulfur) content decrease or increases [220, 222]. These two peaks are usually called $MoSe_2$- like A_{1g} (Se-Mo-S) phonon modes and correspond to different Se/S configurations around the Mo atom, which appeared at 222 and 267 cm^{-1} [225]. Additionally, their peak shoulder at 353 cm^{-1} is generally related to a hybrid mode E^1_{2g} (Se-Mo-S). The MoS_2-like modes usually appear at 370 cm^{-1} (E^1_{2g} (S-Mo)) and 403 cm^{-1} (A_{1g} (S-Mo)). Thus the $WS_{0.86}Se_{1.14}$ domain correlates to A_{1g} (S-W) mode peaks at 408 cm^{-1} and A_{1g} (Se-W) at 261 cm^{-1}, which reflects a redshift concerning the sulfur-rich alloy [223]. Moreover, the E^1_{2g} (S-W) mode peak at 358 cm^{-1} and the LA and 2LA modes red shifts to 147 and 294 cm^{-1}.

6.2.5.4 Composition-Dependent Band Alignment

Based on various experimental findings (including AFM results) a comparative plot of composition-dependent work function has been demonstrated using the expected theory for the both ternary domains $MoS_{2(1-x)}Se_{2x}$ and $WS_{2(1-x)}Se_{2x}$, by assuming undoped samples (Fermi level at mid-band gap). This calculation is useful for the values of the valance and conduction band energies of the binary compounds employing the GW method and Vegard's law with band-bowing parameters [65, 226]. The expected work function differences according to their chalcogen compositions of the heterostructures CPD change across the junction (144, 132, and 112 meV) that may differ from the expected values for undoped samples (141, 228, and 203 meV) [213]. On the other hand, only heterostructure with lower selenium content has a good agreement between experiment and theory. However, such discrepancy between the expected and measured work function differences across the heterojunctions is very common according to literature [215, 227], because TMDs and other 2D materials unintentionally doped during the synthesis. The unintentional doping changes the position of the Fermi level, thereby, the difference in work functions measured experimentally. A typical band diagram schematic is illustrated in Fig. 6.35, considering the differences in the relative position of the Fermi levels for undoped (dashed lines) and doped (dashed line) samples. The shallower CPD profiles across the junctions can be also obtained experimentally [216].

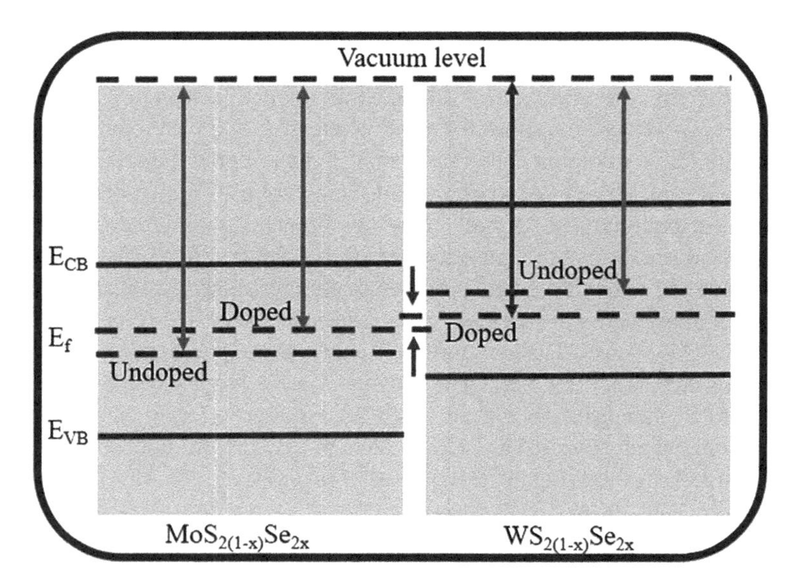

Fig. 6.35 Schematics to show the change in work function differences (arrows) due to electronic doping (dashed lines for E_f) compared to undoped samples (dashed line for E_f)

Not only experimental but also theoretical studies on band aliment of LHs have been done by the different research groups by employing distinct approaches to explore various dimensions of such kinds of materials [228–233].

6.2.6 Others TMDs Heterostructures—Mixed Lateral and Vertical Hybrid Structures

TMD bulk solid source materials are also sensitive to humidity while the reaction time is at high temperatures to produce complex oxides and hydroxides. Because of the presence of oxide species reduces the growth temperature. Such as gypsum [$CaSO_{4.2}H_2O$] acts as a solid water source to control the vapor pressure inside the CVD reactor. The controlled thermal decomposition of the water source inside the CVD reactor in one-step vapor transport-based growth of complex HSs with lateral, vertical, and hybrid MoS_2-WS_2 heterointerface has been fabricated [234]. In which initially the WS_2 layer has been deposited at a lower temperature of the water source (~85 °C), that results a low amount of vapor phase precursor to create a lower super saturation of WS_2 particulates. Additionally, it also promotes screw dislocation to form the spiral WS_2 nanoplates. While the use of the higher source temperature of the water source (~120 °C) invokes layer-by-layer deposition of WS_2 with a large lateral dimension. The subsequent growth of the MoS_2 layer along the edges of the pre-deposited WS_2 layer also give rise the formation of LHs composed of single or multiple layers depending on the size of the predeposited WS_2 layer. In the initial stage WS_2 layer has formed different sizes, however, the prolonged reaction duration creates another layer of MoS_2-WS_2 that fully covered the bottom MoS_2-WS_2 LHs to rise a hybrid-like structure.

The use of oxide source precursors allows producing a hybrid HS consisting lateral with vertical growth direction of WS_2/MoS_2- WS_2 from a single-step CVD method by adjusting the growth temperature and flow direction of carrier gas, as depicted in Fig. 6.36a [235]. In the beginning, 1L MoS_2 is formed due to low submission temperature, subsequent formation of WS_2 layer at the edges of the core MoS_2 layer that predominately possessed a hexagonal or triangular shape. Increase in temperature (> 850 °C) the adatoms gets gain enough energy to overcome the energy barriers, therefore, contribute to vertical growth, further it forms the second WS_2 layer that partially or fully cover the initially formed MoS_2-WS_2 LHs. On the other hand, at higher temperatures, the simultaneous growth along lateral and vertical directions becomes more kinetically favorable.

Similarly, the direct growth of 2D Metal–Semiconductor-based lattice-matched multilayer metallic IT-VSe_2 on 1L 1HMX_2 HS can also produce via a two-step CVD process (see Fig. 6.36b) [236]. In such a process initially, nucleation of the VSe_2 layer preferably occurred at the edge-elongated stack along the edges of 1H-MX_2, which eventually merged into the completion of the vertical layer. Therefore, a significant difference in the diffusion of VCl_3, a source VSe_2, and transition metal

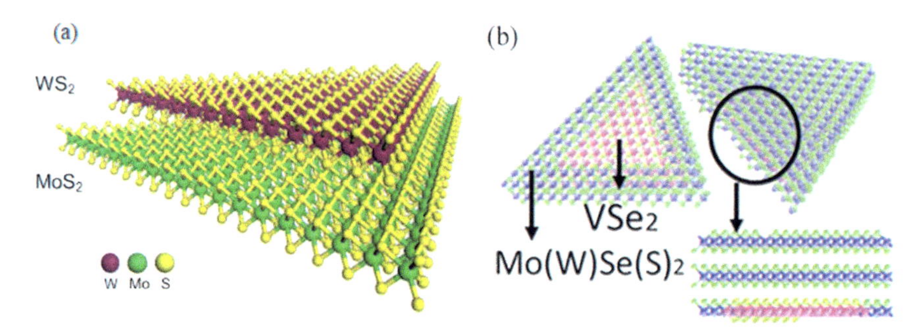

Fig. 6.36 **a** Hybrid structure of WS_2-MoS_2 (adopted from Pak et al. 2017, Nano Lett. 17, 5634–5640); **b** schematic for the VSe_2 and TMDs hybrid structure where multilayer VSe_2 as grown at the edges of 1L-TMDs followed by coverage of top layer via two-step CVD process (adopted from Zhang et al. [236])

oxide precursors to be distinguished. Moreover, to prevent the formation of the alloy during the second growth step a small lattice mismatching between different 2D layers benefited from the lattice misfits ~ 2.2 and ~ 1.9% in the formation of VSe_2/WSe_2 and $VSe_2/MoSe_2$ heterostructures. Specifically, it leads to the formation of commensurate lattice stacking along the vertical direction. This kind of growth behavior of the VeS_2 layer is possibly due to the choice of substrate and growth template. However, existing dangling bonds on the sapphire surface impede the outer lateral growth mode. Indeed the dangling bond-free top MX_2 surface (metallic VSe_2 layers) can grow easily.

Because the lattice structures of transition metal mono-chalcogenides in MX-type are different from MX_2-type TMDs. Therefore, heterostructures composed of MX/MX_2 structures forms incommensurate lattices, such as Moire patterns with unusual physical phenomena. Though, LHs between heterogeneous layered materials with large lattice misfits make it challenging to synthesize from the CVD method. Nonetheless, hybrid heterostructures like 2D GaSe and $MoSe_2$ (lattice mismatch ~ 13%) can be formed in both lateral and vertical directions by employing a two-step lower-pressure CVD method. In which MoO_3, Se, and a mixture of GaSe + $Ga2Se_3$ powder have been used as precursors [237]. A single-layer GaSe is to be directly achieved in the CVD-grown $MoSe_2$ layer by adjusting the carrier gas flow rate. It has been noted that at a low flow rate, the vapor phase deposition of GaSe is slower than the diffusion on the substrate. Additionally, the diffusion barrier of GaSe on the $MoSe_2$ layer is lower than the absorbance energy of the GaSe molecule on the SiO_2 substrate; this leads to the deposition of GaSe on the $MoSe_2$ terrace and slowly cover the entire $MoSe_2$ surface. While the increase in the flow rate, the deposition rate may dominate over the diffusion rate, therefore, increase in the number of nucleation sites to promote the growth along the lateral direction. The large lattice misfit also produces rough interfaces for the vertical and lateral directions.

Thus extensive studies are desirable to understand the role of other growth parameters, such as temperature, pressure, and flux ratios toward controlling the interface characteristics. Additionally, the vertical and lateral heterostructures creation and their integration into a hybrid structure for the exciting functionalities. Tuning the humidity environment, growth temperature, and flow rate during reaction time can also promote vertical or lateral growth direction to eventually form a hybrid structure.

6.3 Conclusions

In the conclusive remarks, one has discussed the various aspects of the TMD materials, by providing a deep understanding of TMD heterostructures. This work has provided an understanding of TMD heterostructures by covering the past to present significant developments in such materials, including the important Anderson's rule for the band alignment theoretical description. A significant description of TMDs materials heterostructures construction for their potential technical uses has been also provided with an understanding of the topic. A brief discussion on strategies to construct 2D TMD heterostructures as well as their possible 0D–2D, 1D–2D, and 2D–2D heterostructures combination description would be helpful for the specific utility of the material. Moreover, the general description of crucial band alignment in TMD heterostructures has been also provided which is helpful to build a better understanding of band formation behavior in these kinds of materials. A detailed segment has been provided on significantly important vertical heterostructures with the sub-sections matching lattice heterostructures, strain minimizes approach, individual (or single) layer heterostructures, Janus monolayer 2D TMDs, layer restacked heterostructures and vertically aligned heterostructures based on the experimental studies. This would be useful to understand the behavior of vertical TMD heterostructures and the determination of their potential uses for specific purposes. Moreover, a complete segment description has been also devoted on technically more demanding lateral TMD heterostructures including key significant topics, edge epitaxy, lithographic pattering approach, band gap engineering of LHs, and composition-dependent band alignment in LHs. Description of lateral TMD heterostructures with their key governing physical parameters would provide a sound knowledge to build a better understanding on their various aspects that may decisive in selection for the specific targeted potential applications. Furthermore, a useful discussion on mixed lateral and vertical hybrid structures has been also provided. That would be helpful to know the structural and other physical parameter changes in hybrid structures, therefore, the significant impact on overall technical performance based on such TMD materials devices. Thus one has provided depth and up-to-date knowledge on TMD heterostructures that would be useful for their potential technical utility in various areas.

References

1. Koma A, Sunouchi K, Miyajima T (1985) Fabrication of ultrathin heterostructures with van der Waals epitaxy. J Vac Sci Technol B3:724
2. Koma A, Yoshimura K (1986) Ultrasharp interfaces grown with van der Waals epitaxy. Surf Sci 174:556–560
3. Koma A (1992) Van der Waals epitaxy—a new epitaxial growth method for a highly lattice-mismatched system. Thin Solid Films 216:72–76
4. Koma A (1999) Van der Waals epitaxy for highly lattice-mismatched systems. J Cryst Growth 201:236–241
5. Ohuchi F, Shimada T, Parkinson B, Ueno K et al (1991) Growth of MoSe2 thin films with van der Waals epitaxy. J Cryst Growth 111:1033
6. Ueno K, Saiki K, Shimada T, Koma A (1990) Epitaxial growth of transition metal dichalcogenideson cleaved faces of mica. J Vac Sci Tech A 8:68
7. Lei X (2022) Optimization of mechanically assembled Van Der Waals heterostructure based on solution immersion and hot plate heating. J Phys Conf Ser 2152:012007
8. Geim A, Grigorieva I (2013) Van der Waals heterostructures. Nature 499:419–425
9. Chakraborty SK, Kundu B, Nayak B, Dash SP et al (2022) Challenges and opportunities in 2D heterostructures for electronic and optoelectronic devices. iScience 25:103942
10. Liu Y, Weiss NO, Duan X, Cheng HC et al (2016) Van der Waals heterostructures and devices. Nat Rev Mater 1:16042
11. Gan X, Lei D, Ye R, Zhao H et al (2021) Transition metal dichalcogenide-based mixed-dimensional heterostructuresfor visible-light-driven photocatalysis: dimensionality and interface engineering. Nano Res 14:2003–2022
12. Liu Y, Cheng R, Liao L, Zhou H et al (2011) Plasmon resonance enhanced multicolour photodetection by graphene. Nat Commun 2:579
13. Konstantatos G, Badioli M, Gaudreau L, Osmond J et al (2012) Hybrid graphene–quantum dot phototransistors with ultrahigh gain. Nat Nanotechnol 7:363–368
14. Liao L, Bai J, Qu Y, Lin Y et al (2010) High-k oxide nanoribbons as gate dielectrics for high mobility top-gated graphenetransistors. Proc Natl Acad Sci USA 107:6711–6715
15. Liao L, Lin YC, Bao M, Cheng R et al (2010) High-speed graphene transistors with a self-aligned nanowire gate. Nature 467:305–308
16. Sun X, Deng HT, Zhu WG, Yu Z et al (2016) Interface engineering in two dimensional heterostructures: towards an advanced catalyst for Ullmann couplings. Angew Chem Int Ed 55:1704–1709
17. Choi W, Choudhary N, Han GH, Park J et al (2017) Recent development of two-dimensional transition metal dichalcogenides and their applications. Mater Today 20:116–130
18. Lv RT, Robinson JA, Schaak RE, Sun D et al (2015) Transition metal dichalcogenides and beyond: synthesis, properties, and applications of single- and few-layer nanosheets. Acc Chem Res 48:56–64
19. Su TM, Shao Q, Qin ZZ, Guo ZH et al (2018) Role of interfaces in two-dimensional photocatalyst for water splitting. ACS Catal 8:2253–2276
20. Rahmanian E, Malekfar R, Pumera M (2018) Nanohybrids of two dimensional transition-metal dichalcogenides and titanium dioxide for photocatalytic applications. Chem Eur J 24:18–31
21. Fadojutimi PO, Gqoba SS, Tetana ZN, Moma J (2022) Transition metal dichalcogenides [MX2] in photocatalytic water splitting. Catalysts 12:468
22. von Klitzing K (1986) The quantized Hall effect. Rev Mod Phys 58:519–531
23. Tsui DC, Stormer HL, Gossard AC (1982) Two-dimensional magnetotransport in the extreme quantum limit. Phys Rev Lett 48:1559–1562
24. Bernevig BA, Zhang SC (2006) Quantum spin hall effect. Phys Rev Lett 96:106802
25. Konig M, Wiedmann S, Brune C, Roth A et al (2007) Quantum spin hall insulator state in HgTe quantum wells. Science 318:766–770

26. Kasprzak J, Richard M, Kundermann S, Baas A et al (2006) Bose–Einstein condensation of excitonpolaritons. Nature 443:409–414
27. Bellus MZ, Li M, Lane SD, Ceballos F et al (2017) Type-I van der Waals heterostructure formed by MoS_2 and ReS_2 monolayers. NanoscaleHoriz. 2:31–36
28. Lu ZH, Lockwood DJ, Baribeau JM (1995) Quantum confinement and light emission in SiO2/Si superlattices. Nature 378:258–260
29. Wang F, Wang ZX, Xu K, Wang FM et al (2015) Tunable GaTe-MoS2 van der Waals p–n junctions with novel optoelectronic performance. Nano Lett 15:7558–7566
30. Wang Z, Yin H, Jiang C, Safdar M et al (2012) ZnO/ZnS_xSe_{1-x}/ZnSe double-shelled coaxial heterostructure: enhanced photoelectrochemical performance and its optical properties study. Appl Phys Lett 101:253109
31. Lo SS, Mirkovic T, Chuang CH, Burda C et al (2011) Emergent properties resulting from type-II band alignment in semiconductor nanoheterostructures. Adv Mater 23:180–197
32. Peng P, Milliron DJ, Hughes SM, Johnson JC et al (2005) Femtosecond spectroscopy of carrier relaxation dynamics in Type II CdSe/CdTe tetrapod heteronanostructures. Nano Lett 5:1809–1813
33. Özçelik VO, Azadani JG, Yang C, Koester SJ et al (2016) Band alignment of two-dimensional semiconductors for designing heterostructures with momentum space matching. Phys Rev B 94:035125
34. Li Q, Chen KQ, Tang LM (2020) Large valley splitting in van der Waals heterostructures with Type-III band alignment. Phys Rev Appl 13:014064
35. Ahn J, Jeon PJ, Raza SRA, Pezeshki A et al (2016) Transition metal dichalcogenide heterojunction PN diode toward ultimate photovoltaic benefits. 2D Mater 3:04501
36. Choi W, Akhtar I, Rehman MA, Kim M et al (2019) Twist-angle-dependent optoelectronics in a few-layer transition-metal dichalcogenide heterostructure. ACS Appl Mater Interfaces 11:2470
37. Wang W, Li K, Wang Y, Jiang W et al (2019) Investigation of the band alignment at MoS2/PtSe2 heterojunctions. Appl Phys Lett 114:201601
38. Chaves A, Azadani JG, Alsalman H, da Costa DR et al (2020) Bandgap engineering of two-dimensional semiconductor materials. npj 2D Mater Appl 4:29
39. Chiu MH, Tseng WH, Tang HL, Chang YH et al (2017) Band alignment of 2D transition metal dichalcogenide heterojunctions. Adv Funct Mater 27:1603756
40. Zhou S, Ning J, Sun J, Srolovitz DJ (2020) Composition-induced type I and direct bandgap transition metal dichalcogenidesalloy vertical heterojunctions. Nanoscale 12:201–209
41. Heine V (1965) Theory of surface states. Phys Rev 138:A1689
42. Kroemer H (1975) Problems in the theory of heterojunction discontinuities. CRC Crit Rev Solid State Sci 5:555–564
43. Louis E, Yndurain F, Flores F (1976) Metal-semiconductor junction for (110) surfaces of zinc-blende compounds. Phys Rev B 13:4408
44. Tejedor C, Flores F (1977) A simple approach to heterojunctions. J Phys C: Solid State Phys 11:L19
45. Mönch W (1993) Semiconductor surfaces and interfaces. Springer, Berlin
46. Radisavljevic B, Radenovic A, Brivio J, Giacometti V et al (2011) Single-layer MoS2 transistors. Nat Nanotechnol 6:147–150
47. Gan XR, Lei DY, Wong KY (2018) Two-dimensional layered nanomaterials for visible-light-driven photocatalytic water splitting. Mater Today Energy 10:352–367
48. Ran JR, Jaroniec M, Qiao SZ (2018) Cocatalysts in semiconductor based photocatalytic CO_2 reduction: achievements, challenges, and opportunities. Adv Mater 30:1704649
49. Zhang J, Zhu ZP, Tang YP, Müllen K et al (2014) Titanianano sheet mediated construction of a two-dimensional titania/cadmium sulfide heterostructure for high hydrogen evolution activity. Adv Mater 26:734–738
50. Jin Y, Jiang DL, Li D, Xiao P et al (2017) $SrTiO_3$nanoparticle/$SnNb_2O_6$nanosheet 0D/2D heterojunctions with enhanced interfacial charge separation and photocatalytic hydrogen evolution activity. ACS Sustain Chem Eng 5:9749–9757

51. Bi WT, Li XG, Zhang L, Jin T et al (2015) Molecular co-catalyst accelerating hole transfer for enhanced photocatalytic H2 evolution. Nat Commun 6:8647
52. Tan L, Li PD, Sun BQ, Chaker M et al (2017) Stabilities related to near-infrared quantum dot-based solar cells: the role of surface engineering. ACS Energy Lett 2:1573–1585
53. Pincella F, Isozaki K, Miki K (2014) A visible light-driven plasmonic photocatalyst. Light Sci Appl 3:e133
54. Chen W, Zhang SP, Kang M, Liu WK et al (2018) Probing the limits of plasmonic enhancement using a two-dimensional atomic crystal probe. Light Sci Appl 7:56
55. Shan HY, Yu Y, Wang XL, Luo Y et al (2019) Direct observation of ultrafast plasmonic hot electron transfer in the strong coupling regime. Light Sci Appl 8:9
56. Tian Y, Tatsuma T (2005) Mechanisms and applications of plasmon-induced charge separation at TiO_2 films loaded with gold nanoparticles. J Am Chem Soc 127:7632–7637
57. Zhang F, Zhuang HQ, Song J, Men YL et al (2018) Coupling cobalt sulfide nanosheets with cadmium sulfide nanoparticles for highly efficient visible-light-driven photocatalysis. Appl Catal B Environ 226:103–110
58. Maeda K, Teramura K, Lu DL, Saito N et al (2006) Noble-metal/Cr_2O_3 core/shell nanoparticles as a cocatalyst for photocatalytic overall water splitting. Angew Chem Int Ed 45:7806–7809
59. Bau JA, Takanabe K (2017) Ultrathin microporous SiO_2 membranes photodeposited on hydrogen evolving catalysts enabling overall water splitting. ACS Catal 7:7931–7940
60. Meng FK, Li JT, Cushing SK, Bright J et al (2013) Photocatalytic water oxidation by hematite/reduced graphene oxide composites. ACS Catal 3:746–751
61. Xu B, He PL, Liu HL, Wang PP et al (2014) A 1D/2D helical CdS/$ZnIn_2S_4$ nano-heterostructure. Angew Chem Int Ed 53:2339–2343
62. Zhang X, Chen YJ, Xiao YT, Zhou W et al (2018) Enhanced charge transfer and separation of hierarchical hydrogenated TiO_2 nanothorns/carbon nanofibers composites decorated by NiS quantum dots for remarkable photocatalytic H_2 production activity. Nanoscale 10:4041–4050
63. Li HB, Xiao JP, Fu Q, Bao XH (2017) Confined catalysis under two-dimensional materials. Proc Natl Acad Sci USA 114:5930–5934
64. Fu Q, Bao XH (2017) Surface chemistry and catalysis confined under two-dimensional materials. Chem Soc Rev 46:1842–1874
65. Kang J, Tongay S, Li J, Wu J (2013) Monolayer semiconducting transition metal dichalcogenide alloys: stability and band bowing. J Appl Phy 113:143703
66. Yang JH, Yakobson BI (2018) Unusual negative formation enthalpies and atomic ordering in isovalent alloys of transition metal dichalcogenide monolayers. Chem Mater 30:1547–1555
67. Gan LY, Zhang Q, Zhao YJ, Cheng Y et al (2014) Order-disorder phase transitions in the two-dimensional semiconducting transition metal dichalcogenide alloys Mo1−xWxX2 (X = S, Se and Te). Sci Rep 4:6691
68. Perdew JP, Burke K, Ernzerhof M (1996) Generalized gradient approximation made simple. Phy Rev Lett 77:3865
69. Zhang C, Gong C, Nie Y, Min KA et al (2017) Systematic study of electronic structure and band alignment of monolayer transition metal dichalcogenides in Van der Waals heterostructures. 2D Mater 4:015026
70. Liu R, Wang F, Liu L, He X et al (2021) Band alignment engineering in two-dimensional transition metal dichalcogenide-based heterostructures for photodetectors. Small Struct 2:2000136
71. Perdew JP, Levy M (1983) Physical content of the exact Kohn-Sham orbital energies: band gaps and derivative discontinuities. Phy Rev Lett 51:1884
72. Mori-Sánchez P, Cohen AJ, Yang W (2008) Localization and delocalization errors in density functional theory and implications for band-gap prediction. Phy Rev Lett 100:146401
73. Kozawa D, Carvalho A, Verzhbitskiy I, Giustiniano F et al (2016) Evidence for fast interlayer energy transfer in MoSe2/WS2 heterostructures. Nano Lett 16:4087–4093
74. Keyshar K, Berg M, Zhang X, Vajtai R et al (2017) Experimental determination of the ionization energies of $MoSe_2$, WS_2, and MoS_2 on SiO_2 using photoemission electron microscopy. ACS Nano 11:8223–8230

75. Sun J, Ruzsinszky A, Perdew JP (2015) Strongly constrained and appropriately normed semilocal density functional. Phy Rev Lett 115:036402
76. Sun J, Remsing RC, Zhang Y, Sun Z et al (2016) Accurate first-principles structures and energies of diversely bonded systems from an efficient density functional. Nat Chem 8:831–836
77. Chen F, Su W, Ding S, Fu L (2020) The fabrication and tunable optical properties of 2D transition metal dichalcogenides heterostructures by adjusting the thickness of Mo/W films. Appl Surf Sci 505:144192
78. Liu Y, Zhang S, He J, Wang ZM et al (2019) Recent progress in the fabrication, properties, and devices of heterostructures based on 2D materials. Nano-micro Lett 11:1–24
79. Yu Y, Hu S, Su L, Huang L et al (2015) Equally efficient interlayer exciton relaxation and improved absorption in epitaxial and nonepitaxial MoS_2/WS_2 heterostructures. Nano Lett 15:486–491
80. Furchi MM, Pospischil A, Libisch F, Burgdörfer J et al (2014) Photovoltaic effect in an electrically tunable van der Waals heterojunction. Nano Lett 14:4785
81. Lee CH, Lee GH, Zande AMVD, Chen W et al (2014) Atomically thin p–n junctions with van der Waals heterointerfaces. Nat Nanotechnol 9:676–681
82. Hong X, Kim J, Shi SF, Zhang Y et al (2014) Ultrafast charge transfer in atomically thin MoS2/WS2 heterostructures. Nat Nanotechnol 9:682
83. Sierra JF, Fabian J, Kawakami RK, Roche S et al (2021) Van der Waals heterostructures for spintronics and opto-spintronics. Nat Nanotechnol 16:856–868
84. Ge M, Wang H, Wu J, Si C et al (2022) Enhanced valley splitting of WSe_2 in twisted van der Waals WSe2/CrI3 heterostructures. npj Comput Mater 8:32
85. Dean CR, Young AF, Meric I, Lee C et al (2010) Boron nitride substrates for high-quality graphene electronics. Nat Nanotechnol 5:722–726
86. Chiu MH, Zhang C, Shiu HW, Chuu CP et al (2015) Determination of band alignment in the single-layer MoS2/WSe2 heterojunction. Nat Commun 6:7666
87. Rigosi AF, Hill HM, Li Y, Chernikov A et al (2015) Probing interlayer interactions in transition metal dichalcogenide heterostructures by optical spectroscopy: MoS2/WS2 and MoSe2/WSe2. Nano Lett 15:5033–5038
88. Cao X, Jiang C, Tan D, Li Q et al (2021) Recent mechanical processing techniques of two-dimensional layered materials: a review. J Sci Adv Mater Devices 6:135–152
89. Li H, Yoo C, Ko TJ, Kim JH et al (2022) Atomic-scale characterization of structural heterogeny in 2D TMD layers. Mater Adv 3:1401
90. Ohtake A, Sakuma Y (2021) Two-dimensional WSe2/MoSe2 heterostructures grown by molecular-beam epitaxy. J Phys Chem C 125:11257–11261
91. Mortelmans W, Mehta AN, Balaji Y, Sergeant S et al (2020) On the van der Waals epitaxy of homo-/heterostructures of transition metal dichalcogenides. ACS Appl Mater Interfaces 12:27508–27517
92. Zatko V, Dubois SMM, Godel F, Carrétéro C et al (2021) Band-gap landscape engineering in large-scale 2D semiconductor van der Waals heterostructures. ACS Nano 15:7279–7289
93. Jie W, Yang Z, Zhang F, Bai G et al (2017) Observation of room-temperature magnetoresistance in monolayer MoS2 by ferromagnetic gating. ACS Nano 11:6950–6958
94. Cohen A, Patsha A, Mohapatra PK, Kazes M et al (2021) Growth-etch metal–organic chemical vapor deposition approach of WS2 atomic layers. ACS Nano 15(1):526–538
95. Lee DH, Sim Y, Wang J, Kwona SY (2020) Metal–organic chemical vapor deposition of 2D van der Waals materials—the challenges and the extensive future opportunities. APL Mater 8:030901
96. Jin G, Lee CS, Okello OF, Lee SH et al (2021) Heteroepitaxial van der Waals semiconductor superlattices. Nat Nanotechnol 16:1092–1098
97. Chen F, Wang Y, Su W, Ding S et al (2019) Position-selective growth of 2D WS2-based vertical heterostructures via a one-step CVD approach. J Phys Chem C 123:30519–30527
98. Chen F, Yao Y, Su W, Ding S et al (2020) Optical performance and growth mechanism of a 2D WS2–MoWS2 hybrid heterostructure fabricated by a one-step CVD strategy. Cryst Eng Commun 22:660–665

99. Bhowmik S, Rajan AG (2022) Chemical vapor deposition of 2D materials: a review of modeling, simulation, and machine learning studies. iScience 25:103832
100. Alahmadi M, Mahvash F, Szkopek T, Siaj M (2021) A two-step chemical vapor deposition process for the growth of continuous vertical heterostructure WSe2/h-BN and its optical properties. RSC Adv 11:16962–16969
101. Balasubramanyam V, Merkx MJM, Verheijen MA, Kessels WMM et al (2020) Area-selective atomic layer deposition of two-dimensional WS2 nanolayers. ACS Mater Lett 2(5):511–518
102. Chang SJ, Wang SY, Huang YC, Chih JH et al (2022) Van der Waals epitaxy of 2D h-AlN on TMDs by atomic layer deposition at 250C. Appl Phys Lett 120:162102
103. Zhang Y, Yao Y, Sendeku MG, Yin L et al (2019) Recent progress in CVD growth of 2D transition metal dichalcogenides and related heterostructures. Adv Mater 31:1901694
104. Pham PV, Bodepudi SC, Shehzad K, Liu Y et al (2022) 2D heterostructures for ubiquitous electronics and optoelectronics: principles, opportunities, and challenges. Chem Rev 122(6):6514–6613
105. Cai Z, Liu B, Zou X, Cheng HM (2018) Chemical vapor deposition growth and applications of two-dimensional materials and their heterostructures. Chem Rev 118:6091–6133
106. Ji Q, Zhang Y, Zhang Y, Liu Z (2015) Chemical vapour deposition of group-VIB metal dichalcogenide monolayers: engineered substrates from amorphous to single crystalline. Chem Soc Rev 44:2587–2602
107. Zhang T, Fu L (2018) Controllable chemical vapor deposition growth of two-dimensional heterostructures. Chem 4:671–689
108. Song L, Li H, Zhang Y, Shi JJ (2022) Recent progress of two-dimensional metallic transition metal dichalcogenides: syntheses, physical properties, and applications. J Appl Phys 131:060902
109. Jiang X, Chen F, Zhao S, Su W (2021) Recent progress in the CVD growth of 2D vertical heterostructures based on transition-metal dichalcogenides. Cryst Eng Comm 23:8239–8254
110. Terrones H, López-Urías F, Terrones M (2013) Novel hetero-layered materials with tunable direct band gaps by sandwiching different metal disulfides and diselenides. Sci Rep 3:1549
111. Ramasubramaniam A, Naveh D, Towe E (2011) Tunable band gaps in bilayer transition-metal dichalcogenides. Phys Rev B 84:205325
112. Xiao J, Long M, Li X, Zhang Q et al (2014) Effects of van der Waals interaction and electric field on the electronic structure of bilayer MoS_2. J Phys Cond Matter 26:405302
113. Kosmider K, Fernández-Rossier J (2013) Electronic properties of the MoS2-WS2 heterojunction. Phys Rev B 87:075451
114. Bernardi M, Palummo M, Grossman JC (2013) Extraordinary sunlight absorption and one nanometer thick photovoltaics using two-dimensional monolayer materials. Nano Lett 13:3664
115. Kou L, Frauenheim T, Chen C (2013) Nanoscale multilayer transition-metal dichalcogenide heterostructures: band gap modulation by interfacial strain and spontaneous polarization. J Phys Chem Lett 4:1730
116. Gong Y, Lin J, Wang X, Shi G et al (2014) Vertical and in-plane heterostructures from WS2/ MoS2 monolayers. Nat Mater 13:1135–1142
117. Aras M, Kılıç Ç, Ciraci S (2018) Lateral and vertical heterostructures of transition metal dichalcogenides. J Phys Chem C 122:1547−1555
118. Ma Y, Xu S, Wei J, Zhou B et al (2022) In-plane and vertical heterostructures from 1T0/2H transition-metal dichalcogenides. Oxford Open Mater Sci 1:itab016
119. Chowdhury T, Sadler EC, Kempa TJ (2020) Progress and prospects in transition-metal dichalcogenide research beyond 2D. Chem Rev 120:12563–12591
120. Taghinejad H, Eftekhar A, Adibi A (2019) Lateral and vertical heterostructures in twodimensionaltransition-metal dichalcogenides. Opt Mater Exp 9:1590
121. Pu J, Yomogida Y, Liu KK, Li LJ et al (2012) Highly flexible MoS2 thin-film transistors with ion gel dielectrics. Nano Lett 12:4013–4017
122. Castellanos-Gomez A, Poot M, Steele GA, van der Zant HS et al (2012) Elastic properties of freely suspended MoS2 nanosheets. Adv Mater 24:772

123. Yan Y, Ding S, Wu X, Zhu J et al. Tuning the physical properties of ultrathin transition-metal dichalcogenides via strain engineering. RSC Adv 10:39455
124. Peng Z, Chen X, Fan Y, Srolovitz DJ et al (2020) Strain engineering of 2D semiconductors and graphene: from strain fields to band-structure tuning and photonic applications. Light Sci Appl 9:190
125. Liao C, Zhao Y, Ouyang G (2018) Strain-modulated band engineering in two-dimensional black phosphorus/MoS2 van der Waals heterojunction. ACS Omega 3:14641–14649
126. Wang Y, Cong C, Yang W, Shang J et al (2015) Strain-induced direct–indirect bandgap transition and phonon modulation in monolayer WS2. Nano Res 8:2562–2572
127. Peng Z, Chen X, Fan Y, Srolovitz DJ et al (2020) Strain engineering of 2D semiconductors and graphene: from strain fields to band-structure tuning and photonic applications. Light: Sci Appl 9:190
128. Zhu Z, Cui P, Jia Y, Zhang S et al (2019) Strain in van der Waals epitaxy and evidence for a collective macroscopic effect of a negligibly small perturbation. Phys Rev B 100:035429
129. Feierabend M, Morlet A, Berghäuser G, Malic E (2017) Impact of strain on the optical fingerprint of monolayer transition-metal dichalcogenides. Phys Rev B 96:045425
130. Sun H, Agrawal P, Singh CV (2021) A first-principles study of the relationship between modulus and ideal strength of single-layer, transition metal dichalcogenides. Mater Adv 2:6631
131. Postorino S, Grassano D, D'Alessandro M, Pianetti A et al (2020) Strain-induced effects on the electronic properties of 2D materials. Nanomater Nanotechnol 10:1–11
132. Lang PF, Smith BC (2003) Ionization energies of atoms and atomic ions. J Chem Educ 80:938
133. Jiang JW (2015) Graphene versus MoS2: a short review. Front Phys 10:287–302
134. Cadelano E, Colombo L (2012) Effect of hydrogen coverage on the Young's modulus of graphene. Phys Rev B 85:245434
135. Hess P (2018) Predictive modeling of intrinsic strengths for several groups of chemically related monolayers by a reference model. Phys Chem Chem Phys 20:7604–7611
136. Liu GB, Xiao D, Yao Y, Xu X et al (2015) Electronic structures and theoretical modelling of two-dimensional group-vib transition metal dichalcogenides. ChemSoc Rev 44:2643–2663
137. Perdew JP, Wang Y (1992) Accurate and simple analytic representation of the electron-gas correlation energy. Phys Rev B 45:13244
138. Krack M (2005) Pseudopotentials for H to Kr optimized for gradient-corrected exchange-correlation functionals. Theor Chem Acc 114:145–152
139. Hartwigsen C, Goedecker S, Hutter J (1998) Relativistic separable dual-space Gaussian pseudopotentials from H to Rn. Phys Rev B 58:3641
140. Souza I, Íñiguez J, Vanderbilt D (2004) Dynamics of Berry-phase polarization in time-dependent electric fields. Phys Rev B 69:085106
141. Attaccalite C, Grüning M, Marini A (2011) Real-time approach to the optical properties of solids and nanostructures: time-dependent Bethe-Salpeter equation. Phys Rev B 84:245110
142. Grüning M, Sangalli D, Attaccalite C (2016) Dielectrics in a time-dependent electric field: a real-time approach based on density-polarization functional theory. Phys Rev B 94:035149
143. Attaccalite C, Grüning M (2013) Nonlinear optics from an ab initio approach by means of the dynamical Berry phase: application to second- and third-harmonic generation in semiconductors. Phys Rev B 88:235113
144. Attaccalite C, Nguer A, Cannuccia E, Grüning M (2015) Strong second harmonic generation in SiC, ZnO, GaN two-dimensional hexagonal crystals from first-principles many-body calculations. Phys Chem Chem Phys 17:9533
145. Qiu DY, da Jornada FH, Louie SG (2016) Screening and many-body effects in two-dimensional crystals: monolayer MoS_2. Phys Rev B 93:235435
146. Song W, Yang L (2017) Quasiparticle band gaps and optical spectra of strained monolayer transition-metal dichalcogenides. Phys Rev B 96:235441
147. Boyd RW (2008) Nonlinear optics, 3rd edn. Academic Press, Burlington
148. Beach K, Lucking MC, Terrone H (2020) Strain dependence of second harmonic generation in transition metal dichalcogenide monolayers and the fine structure of the C exciton. Phys Rev B 101:155431

149. Rasmussen FA, Thygesen KS (2015) Computational 2D materials database: electronic structure of transition-metal dichalcogenides and oxides. J Phys Chem C 119:13169
150. Rostami H, Roldan R, Cappelluti E, Asgari R et al (2015) Theory of strain in single-layer transition metal dichalcogenides. Phys Rev B 92:195402
151. De Palma AC, Cossio G, Jones K, Quan J et al (2020) Strain-dependent luminescence and piezoelectricity in monolayer transition metal dichalcogenides. J Vac Sci Technol B 38:042205
152. Ghorbani-Asl M, Borini S, Kuc A, Heine T (2013) Strain-dependent modulation of conductivity in single-layer transition-metal dichalcogenides. Phys Rev B: Condens Matter Mater Phys 87:235434
153. Guo SD, Wang Y (2018) Biaxial strain tuned electronic structures and power factor in Janus transition metal dichalchogenide monolayers. Semicond Sci Technol 33:085003
154. Khalatbari H, Vishkayi SI, Oskouian M, Soleimani HR (2021) Band structure engineering of NiS_2 monolayer by transition metal doping. Sci Rep 11:5779
155. Hui YY, Liu X, Jie W, Chan NY et al (2013) Exceptional tunability of band energy in a compressively strained trilayer MoS2 sheet. ACS Nano 7:7126
156. Zhang Q, Chang Z, Xu G, Wang Z et al (2016) Strain relaxation of monolayer WS2 on plastic substrate. Adv Funct Mater 26:8707
157. Zhu CR, Wang G, Liu BL, Marie X et al (2013) Strain tuning of optical emission energy and polarization in monolayer and bilayer MoS2. Phys Rev B: Condens Matter Mater Phys 88:121301
158. Scalise E, Houssa M, Pourtois G, Afanas'ev VV et al (2014) Phonon and magnon excitations in Raman spectra of an epitaxial bismuth ferrite film. Phys E 56:416
159. McCreary A, Ghosh R, Amani M, Wang J et al (2016) Effects of uniaxial and biaxial strain on few-layered terrace structures of MoS2 grown by vapor transport. ACS Nano 10:3186–3197
160. Yagmurcukardes M, Bacaksiz C, Unsal E, Akbali B et al (2018) Strain mapping in single-layer two-dimensional crystals via Raman activity. Phys Rev B 97:115427
161. Vargas-Bernal R (2019) Electrical properties of two-dimensional materials used in gas sensors. Sensors 19:1295
162. Cheng YC, Zhu ZY, Tahir M, Schwingenschlogl U (2013) Spin-orbit–induced spin splittings in polar transition metal dichalcogenide monolayers. EPL 102:57001
163. Han SS, Ko TJ, Yoo C, Shawkat MS et al (2020) Automated assembly of wafer-scale 2D TMD heterostructures of arbitrary layer orientation and stacking sequence using water dissoluble salt substrates. Nano Lett 20:3925–3934
164. Huang H, Fan X, Singh DJ, Zheng W (2020) Recent progress of TMD nanomaterials: phase transitions and applications. Nanoscale 12:1247–1268
165. Wang M, Li H, Ko TJ, Shawkat MS et al (2020) Manufacturing strategies for wafer-scale two-dimensional transition metal dichalcogenide heterolayers. J Mater Res 35:1350–1368
166. Ko TJ, Wang M, Yoo C, Okogbue E et al (2020) Large-area 2D TMD layers for mechanically reconfigurable electronic devices. J Phys D: Appl Phys 53:313002
167. Wang M, Ko TJ, Shawkat MS, Han SS et al (2020) Wafer-scale growth of 2D PtTe2 with layer orientation tunable high electrical conductivity and superior hydrophobicity. ACS Appl Mater Interfaces 12:10839–10851
168. Islam MA, Li H, Moon S, Han SS et al (2020) Vertically aligned 2D MoS2 layers with strain-engineered serpentine patterns for high-performance stretchable gas sensors: experimental and theoretical demonstration. ACS Appl Mater Interfaces 12:53174–53183
169. Rout CS, Joshi PD, Kashid RV, Joag DS et al (2013) Selective optical assembly of highly uniform nanoparticles by doughnut-shaped beams. Sci Rep 3:1–8
170. Chen Q, Li H, Zhou S, Xu W et al (2018) Ultralong 1D vacancy channels for rapid atomic migration during 2D void formation in monolayer MoS2. ACS Nano 12:7721–7730
171. Komsa HP, Kotakoski J, Kurasch S, Lehtinen O et al (2012) Two-dimensional transition metal dichalcogenides under electron irradiation: defect production and doping. Phys Rev Lett 109:035503
172. Lu AY, Zhu H, Xiao J, Chuu CP et al (2017) Janus monolayers of transition metal dichalcogenides. Nat Nanotechnol 12:744–749

173. Zhang J, Jia S, Kholmanov I, Dong L et al (2017) Janus monolayer transition-metal dichalcogenides. ACS Nano 11:8192–8198
174. Lin YC, Liu C, Yu Y, Zarkadoula E et al (2020) Low energy implantation into transition-metal dichalcogenide monolayers to form Janus structures. ACS Nano 14:3896–3906
175. Lin J, Zhou J, Zuluaga S, Yu P et al (2018) Anisotropic ordering in 1T′ molybdenum and tungsten ditelluride layers alloyed with sulfur and selenium. ACS Nano 12:894–901
176. Eftelthari A (2017) Molybdenum diselenide (MoSe2) for energy storage, catalysis, and optoelectronics. Appl Mater Today 8:1–17
177. Eftekhari A (2017) Tungsten dichalcogenides (WS2, WSe2, and WTe2): materials chemistry and applications. J Mater Chem A 5:18299–18325
178. Tanwar S, Arya A, Gaur A, Sharma AL (2021) Transition metal dichalcogenide (TMDs) electrodes for supercapacitors: a comprehensive review. J Phys: Condens Matter 33:303002
179. Hemanth NR, Kim T, Kim B, Jadhav AH et al (2021) Transition metal dichalcogenide-decorated MXenes: promising hybrid electrodes for energy storage and conversion applications. Mater Chem Front 5:3298
180. Lu N, Guo H, Li L, Dai J et al (2014) MoS2/MX2 heterobilayers: bandgap engineering via tensile strain or external electrical field. Nanoscale 6:2879–2886
181. Kou L, Frauenheim T, Chen C (2013) Nanoscale multilayer transition-metal dichalcogenide heterostructures: band gap modulation by interfacial strain and spontaneous polarization. J Phys Chem Lett 4:1730–1736
182. Duan X, Wang C, Shaw JC, Cheng R et al (2014) Lateral epitaxial growth of two-dimensional layered semiconductor heterojunctions. Nat Nanotechnol 9:1024–1030
183. Jiang T, Liu H, Huang D, Zhang S et al (2014) Valley and band structure engineering of folded MoS2 bilayers. Nat Nanotechnol 9:825–829
184. van der Zande AM, Kunstmann J, Chernikov A, Chenet DA et al (2014) Tailoring the electronic structure in bilayer molybdenum disulfide via interlayer twist. Nano Lett 14:3869–3875
185. Huang S, Ling X, Liang L, Kong J et al (2014) Probing the interlayer coupling of twisted bilayer MoS2 using photoluminescence spectroscopy. Nano Lett 14:5500–5508
186. Voiry D, Mohite A, Chhowalla M (2015) Phase engineering of transition metal dichalcogenides. Chem Soc Rev 44:2702–2712
187. Wilson JA, Yoffe AD (1969) The transition metal dichalcogenides discussion and interpretation of the observed optical, electrical and structural properties. Adv Phys 18:193–335
188. Jaramillo TF, Jørgensenjacob KP, Nielsen BH, Horch S et al (2007) Identification of active edge sites for electrochemical H_2 evolution from MoS2 nanocatalysts. Science 317:100–102
189. Voiry D, Yamaguchi H, Li J, Silva R et al (2013) Enhanced catalytic activity in strained chemically exfoliated WS2 nanosheets for hydrogen evolution. Nat Mater 12:850–855
190. Heising J, Kanatzidis MG (1999) Structure of restacked MoS2 and WS2 elucidated by electron crystallography. J Am Chem Soc 121:638–643
191. Brown BE (1966) The crystal structures of WTe2 and high-temperature MoTe2. Acta Crystallogr 20:268–274
192. Duerloo KAN, Li Y, Reed EJ (2014) Structural phase transitions in two-dimensional Mo- and W-dichalcogenide monolayers. Nat Commu 5:4214
193. Joensen P, Crozier E, Alberding N, Frindt R (1987) A study of single-layer and restacked MoS2 by x-ray diffraction and x-ray absorption spectroscopy. J Phys C 20:4043
194. Tsai HL, Heising J, Schindler JL, Kannewurf CR et al (1997) Exfoliated-restacked phase of WS2. Chem Mater 9:879–882
195. Goloveshkin AS, Bushmarinov IS, Lenenko ND, Buzin MI et al (2013) Structural properties and phase transition of exfoliated-restacked molybdenum disulfide. J Phys Chem C 117:8509–8515
196. Prouzet E, Heising J, Kanatzidis MG (2003) Structure of restacked and pillared WS2: an x-ray absorption study. Chem Mater 15:412–418
197. Gordon R, Yang D, Crozier E, Jiang D et al (2002) Structures of exfoliated single layers of WS2, MoS2, and MoSe2 in aqueous suspension. Phys Rev B 65:125407

198. Luan S, Han M, Xi Y, Wei K et al (2020) MoS_2-decorated 2D Ti_3C_2 (MXene): a high performance anode material for lithium-ion batteries. Ionics 26:51–59
199. Vyskocil J, Mayorga-Martinez CC, Szokolova K, Dash A et al (2019) 2D stacks of MXene Ti3C2 and 1T-phase WS2 with enhanced capacitive behavior. Chem Electro Chem 6:3982–3986
200. You J, Si C, Zhou J, Sun Z (2019) Contacting MoS2 to MXene: vanishing p-type Schottky barrier and enhanced hydrogen evolution catalysis. J Phys Chem C 123:3719–3726
201. Xiao J, Choi D, Cosimbescu L, Koech P et al (2010) Exfoliated MoS2 nanocomposite as an anode material for lithium ion batteries. Chem Mater 22:4522–4524
202. Hemanth NR, Kim T, Kim B, Jadhav AH et al (2021) Transition metal dichalcogenide-decorated MXenes: promising hybrid electrodes for energystorage and conversion applications. Mater Chem Front 5:3298
203. Sukanya R, da Silva Alves DC, Breslin CB (2022) Review—recent developments in the applications of 2D transition metal dichalcogenides as electrocatalysts in the generation of hydrogen for renewable energy conversion. J Electrochem Soc. https://doi.org/10.1149/1945-7111/ac7172
204. Kim H, Kim WY, O'Brien M, McEvoy N (2018) Optimized single-layer MoS_2 field-effect transistors by non-covalent functionalization. Nanoscale 10:17557–17566
205. Kang J, Tongay S, Zhou J, Li JB et al (2013) Band offsets and heterostructures of two-dimensional semiconductors. Appl Phys Lett 102:012111
206. Yu JH, Lee HR, Hong SS, Kong D et al (2015) Vertical heterostructure of two-dimensional MoS2 and WSe2 with vertically aligned layer. Nano Lett 15:1031–1035
207. Mead IDG, Can JC (1977) Long wavelength optic phonons in WSe2. J Phys 55:379−382
208. Kong D, Wang H, Cha JJ, Pasta M et al (2013) Synthesis of MoS_2 and $MoSe_2$ films with vertically aligned layers. Nano Lett 13:1341–1347
209. Choudhary N, Chung HS, Kim JH, Noh C et al (2018) Strain-driven and layer-number-dependent crossover of growth mode in van der Waals Heterostructures: 2D/2D layer-by-layer horizontal epitaxy to 2D/3D vertical reorientation. 5:1800382
210. Hong X, Kim J, Shi SF, Zhang Y et al (2014) Ultrafast charge transfer in atomically thin MoS_2/WS_2 heterostructures. Nat Nanotech 9:682–686
211. Zhang Z, Chen P, Duan X, Zang K et al (2017) Robust epitaxial growth of two-dimensional heterostructures, multiheterostructures, and superlattices. Science 357:788–792
212. Sahoo PK, Memaran S, Xin Y, Balicas L et al (2018) One-pot growth of twodimensional lateral heterostructures via sequential edge-epitaxy. Nature 553:63–67
213. Nugera FA, Sahoo PK, Xin Y, Ambardar S et al (2022) Bandgap engineering in 2D lateral heterostructures of transition metal dichalcogenides via controlled alloying. Small 18:2106600–2106611
214. Ma H, Huang K, Wu R, Zhang Z et al (2021) In-plane epitaxial growth of 2D CoSe-WSe2 metal-semiconductor lateral heterostructures with improved WSe_2 transistors performance. Info Mat 3:222–228
215. Li MY, Shi Y, Cheng CC, Lu LS et al (2015) Epitaxial growth of a monolayer WSe2-MoS2 lateral p-n junction with an atomically sharp interface. Science 349:524–528
216. Sahoo PK, Memaran S, Xin Y, Balicas L et al (2018) One-pot growth of two-dimensional lateral heterostructures via sequential edge-epitaxy. Nature 553:63–67
217. Bogaert K, Liu S, Chesin J, Titow D et al (2016) Diffusion-mediated synthesis of MoS2/WS2 lateral heterostructures. Nano Lett 16:5129–5134
218. Mahjouri-Samani M, Lin MW, Wang K, Lupini AR et al (2015) Patterned arrays of lateral heterojunctions within monolayer two-dimensional semiconductors. Nat Commun 6:7749
219. Taghinejad H, Eftekhar AA, Campbell PM, Beatty B et al (2018) Strain relaxation via formation of cracks in compositionally modulated two-dimensional semiconductor alloys. Npj 2D Mater Appl 2:1
220. Feng QL, Mao NN, Wu JX, Xu H (2015) Growth of MoS2(1–x)Se2x (x = 0.41–1.00) monolayer alloys with controlled morphology by physical vapor deposition. ACS Nano 9:7450–7455

221. Feng QL, Zhu YM, Hong JH, Zhang M et al (2014) Growth of large-area 2D MoS2(1–x)Se2x semiconductor alloys. Adv Mater 26:2648
222. Jadczak J, Dumcenco DO, Huang YS, Lin YC et al (2014) Composition dependent lattice dynamics in MoSxSe(2–x) alloys. J Appl Phys 116:193505
223. Duan XD, Wang C, Fan Z, Hao GL (2016) Synthesis of WS2xSe2–2x alloy nanosheets with composition-tunable electronic properties. Nano Lett 16:264–269
224. Berkdemir A, Gutierrez HR, Botello-Mendez AR, Perea-Lopez N et al (2013) Identification of individual and few layers of WS2 using Raman spectroscopy. Sci Rep 3:1755
225. Lee J, Pak S, Lee YW, Park Y et al (2019) Direct epitaxial synthesis of selective two-dimensional lateral heterostructures. ACS Nano 13:13047–13055
226. Kim HG, Choi HJ (2021) Thickness dependence of work function, ionization energy, and electron affinity of Mo and W dichalcogenides from DFT and GW calculations. Phys Rev B 103:085404
227. Zheng BY, Ma C, Li D, Lan JY et al (2018) Band alignment engineering in two-dimensional lateral heterostructures. J Am Chem Soc 140:11193
228. Ávalos-Ovando O, Mastrogiuseppe D, Ulloa SE (2019) Lateral heterostructures and one-dimensional interfaces in 2D transition metal dichalcogenides. J Phys Condens Matter 31:213001
229. Cheng K, Guo Y, Han N, Jiang X et al (2018) 2D lateral heterostructures of group-III monochalcogenide: potential photovoltaic applications. App Phys Lett 112:143902
230. Sun M, Wu T, Huang B (2020) Anion charge density disturbance induces in-plane instabilities within 2D lateral heterojunction of TMD: an atomic view. Nano Energy 70:104484
231. Dong R, Kuljanishvili I (2017) Review article: progress in fabrication of transition metal dichalcogenides heterostructure systems. J Vac Sci Technol B 35:030803
232. Zhang C, Chen Y, Huang JK, Wu X et al (2015) Visualizing band offsets and edge states in bilayer–monolayer transition metal dichalcogenides lateral heterojunction. Nat Commu 7:10349
233. Imani S, Yengejeh Wen W, Wang Y (2021) Mechanical properties of lateral transition metal dichalcogenide heterostructures. Front Phys 16:13502
234. Zhao B, Karpiak B, Khokhriakov D, Johansson A et al (2020) Unconventional charge–spin conversion in Weyl-semimetal WTe2. Adv Mater 32:2000818
235. Zhang X, Xiao S, Nan H, Mo H et al (2018) Controllable one-step growth of bilayer MoS2–WS2/WS2 heterostructures by chemical vapor deposition. Nanotechnology 29:455707
236. Zhang Z, Gong Y, Zou X, Liu P et al (2019) Epitaxial growth of two-dimensional metal–semiconductor transition-metal dichalcogenide vertical stacks (VSe2/MX2) and their band alignments. ACS Nano 13:885–893
237. Li X, Lin MW, Lin J, Huang B et al (2016) Two-dimensional GaSe/MoSe2 misfit bilayer heterojunctions by van der Waals epitaxy. Sci Adv 2:e1501882

Chapter 7
Topological Insulating Nature of TMDs

7.1 Background

This decade has countersigned the apparently unstoppable growth of topological concepts, concluding in the 2016 Nobel Prize in Physics. The word *topological materials* comprise a wide-range of structural constructions that display phases described by a topological invariant [1, 2] that still unaffected by distortions in the organization Hamiltonian. This is due to the invariance of the Chern number (or Z_2). In additional general terms the Berry curvature, the integral over the Brillouin zone yields the Chern number. Topological materials can be in form of topological insulators (TI), Weyl and Dirac semimetals (WSM, DSM), and topological superconductors (TSC). However, it is difficult to distinguish them because the distinction is not always clear-cut, since some categories overlap. For instance, certain transition metal dichalcogenides (TMDs) becomes topological insulators under appropriate circumstances, while others, like WTe_2 behaves Weyl semimetals [2].

Therefore, the TIs are the materials with a band structure comprising a bulk band gap crossed by surface states with linear dispersion [3–6]. Such surface states exhibits spin-momentum locking, by implying the spin of a particle in the surface states that can be determined from its momentum [7]. Ultimately it leads to the prohibition of backscattering, as a consequence, a reduction in scattering at other angles from non-magnetic impurities, vacancies, and other defects, to full fill the essential requirements of a change in momentum with the change in spin. On the other hand, surface electrons may remain show phonon scattering (although the asset of this interaction is remain not well-understood) and interactions with the bulk states [8, 9]. But such states' linear dispersion provides to fundamentally massless particles together a big Fermi velocity. Like, when the Fermi energy of the TI is positioned in the bulk band gap then entire conduction ideally takes place on the surface, which enables the construction of devices that reveal the uncommon features of such surfaces [10].

A. K. Singh, *2D Transition-Metal Dichalcogenides (TMDs): Fundamentals and Application*, Materials Horizons: From Nature to Nanomaterials,
https://doi.org/10.1007/978-981-96-0247-6_7

Thus 2D topological materials offer prospects that did not explore earlier. Such as in their constructions the electrons gas on the surfaces that are straightly obtainable, different to semiconductor heterostructures, in which it is hidden. This flexibility also provides outstanding electrostatic govern onto the conduction in the 2D network, which supports their transistor applicability as well as approachability of light for optical uses. Moreover, several topological materials display extremely robust spin-orbit combination underneath the suitable circumstances that may yield dissipation less edge states conduction in absence of huge magnetic fields, it makes them enables for the potential transistor utility following the quantum spin and anomalous Hall effects. Moreover, TIs spin-orbit torques are enormously, predominantly the lower mobilities of TIs, that not a concern for the long large currents, it could be obtained by enhancing the electron density, such as in Bi_2Te_3 [11]. Therefore, such manifold uses of topological materials are also possible from the TMD WSe_2 that exhibits an extraordinarily bigger spin-orbit combination, as a consequence a widespread direct band gap and a robust anisotropic rise of the valley homogeneity in a magnetic field, it makes them perfect for estimating the valley degree of freedom. Additionally, TMD materials are also useful to make an atomically thin and direct band gap which is the ideal platform for optical emitters and detectors. They also displays robust excitonic influences and piezoelectricity [12]. The topological superconducting state is impressive for the fabrication with stimulating progresses, to achieve innovative impacts such as chiral superconductivity, and Majorana edge modes [13], therefore, the fractional Josephson Effect and non-conventional Cooper pairing exist.

Hence, more specifically TMDs are a category of 2D materials associating a hexagonal honeycomb lattice. Their typical organizational principle is MX_2, here M and X sit at the alternative boundaries of the hexagon and signify as transition metal atoms, alike Mo or W, and chalcogen atoms like S, Se, or Te [14, 15]. Their topological configurations known as the geometric organizations altered from a parental material through the procedures, like stretching, twisting, crumpling, and bending. The himself-organized topological configurations also comprise wrinkles, folds, and scrolls. To stabilized unsteady 2D materials to originate various photoelectrical characteristics that usually not appear with their parental materials [16–19]. Like graphene, the gradient strain on a small wrinkle is induced a pseudo-magnetic field to produce pseudo-Landau levels (PLLs) together evenly spaced conductance peaks [20]. However, their Hofstadter's spectrum and van Hove singularities could be distinct in monolayer stackings [21–25]. Moreover, the topological TMD configurations have wrinkled structures that largely impact the electronic characteristics of the 2D material, counting the electronic configuration, dipole moment, planar mobility, carrier density, on/off ratios, indigenous charge distribution, higher photoresponsivity, energy gap [26, 27]. While the doubled TMD configurations convert into inherent single-layers to procedure multifaceted profiles and ultimately distorted the lattice configurations. Such exceptional stacking feature influences the interlayer combination and singularity of the materials [28, 29]. Thus, it provides an approach to adjust the second harmonic generation (SHG) intensity, additionally, a backbone for exploration of valley electronics established on the uncommon status of K and K' in the Brillouin zone owing to their lattice unevenness [30–33]. Furthermore, the scrolled configuration of TMDs hybridized together different loads, like graphene oxide, pentacene, copper (II) phthalocyanine, and C60. Such kinds of hybrid materials control the topological configurations of TMDs among the import of the supplementary features and functionality [34]. Their hybridization depending on scrolled constructions also steady with air owing to self-encapsulation, which reveals their excessive possibility to fabricate highly stable devices. Therefore, the existence of wrinkles, folds, and scrolls provides innovative opportunities for TMDs.

7.2 Bulk TMD Topological Properties

Since TIs retain their distinctive properties even when diluted with impurities owing to their versatility in remarkable quantum spin Hall (QSH) effect insulating property that makes them useful for numerous applicability, such as quantum computing, spintronics, optoelectronics, energy harvesting, etc. [35, 36]. The TIs lack a large band gap behind its hindering advancement. In the beginning, many three-dimensional (3D) compounds theoretically and experimentally have been explored, like Bi_2Se_3, Bi_2Te_3, and Sb_2Te_3, etc. [35]. That has been characterized with the help of four Z_2 indices, $[v_0, (v_1, v_2, v_3)]$. Such kind of 3D TIs secure to time-reversal equilibrium and further they separated in robust and feeble TIs materials, respect to $v_0 = 1$ and 0 [37]. Their experimental verification from ARPES experiment has played a conclusive part [38]. In the beginning, experimentally $Bi_{1-x}Sb_x$ was explored first that recognized as 3D TIs [39]. However, their electronic configuration is extremely complex because of the co-occurrence of manifold bands in the surface states. But $Bi_{1-x}Sb_x$ has been recognized not more suitable than the 3D TIs systems. To get more suitable 3D TIs devotion was rapidly changed toward the complexes of the tetradymite family, such as Bi_2Se_3, Bi_2Te_3, Sb_2Te_3, and their products. This family of 3D TIs has a key advantage the surface states properties represents from the single Dirac cone, associating a big bulk band gap, as illustrated in Fig. 7.1a [7, 40, 41]. Specifically, the twofold alloys (e.g., Bi_2Se_3) are willingly formed in bulk single crystals, like, thin films, nanobelts, or nanoplates. However, the issue with them, generally their tendency to possess a high density of defects, as the consequence, the bulk always conducting evens the liquid helium temperature [42, 43]. To overcome the issue it was preferred to make ternary and quaternary compounds with an impressive methodology to suppress the bulk conductivity. As the instance $(Bi_{1-x}Sb_x)_2$ Te_3 chemical potential can be regularly adjusted for the bulk conduction band to the valence band among the enhancing Sb quantity [44, 45]. Such kind of doping approach was depended on the consideration of undoped Bi_2Te_3 belongs to n-type while the Sb_2Te_3 to be a p-type. Likewise, $(Bi_{1-x}Sb_x)_2$ $(Te_{1-y}Se_y)_3$ govern of the composition to upsurge the bulk resistivity order of ohm·cm for a single crystal [46]. Thus, the numerous 3D TI's structural, electrical, optical, magnetic, superconducting, etc., properties have been extensively explored for both undoped and doped compositions [2], as the typical band diagrams of Bi_2Se_3 and Bi_2Te_3 are depicted in Fig. 7.1b, c.

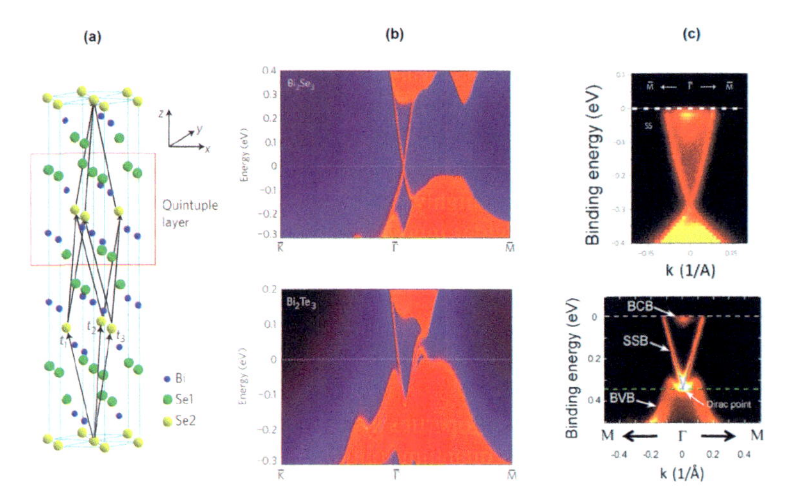

Fig. 7.1 **a** Crystalline configuration of Bi_2Se_3 to represents a layered configuration of the Van der Waals type, with a separation into quintuple layers (QLs) of Se–Bi–Se–Bi–Se. Every QL is around 1 nm thick; **b** band pictures of Bi_2Se_3 and Bi_2Te_3 according to first-principles approach; **c** the gapless surface states band structures for the Bi_2Se_3 and Bi_2Te_3 defined from ARPES (adopted from Culcer et al. [2])

Similar to bulk 3D TMDs, a small bulk band gap in two-dimensional 2D TMD TI materials also impedes the realistic application [35]. The first experimentally tested confirmed 2D TI materials were HgTe/CdTe and InAs/GaSb quantum wells at only extremely low temperatures [47–49]. Compared to 3D materials, 2D TIs can have more advantageous features, including better flexibility and easier integration into current electronics [50]. The 2D materials are readily integrate from microfabrication technologies to get high performances and high-density logic and memory devices. Moreover, compared to 2D surface of 3D TIs is not protected against backscattering in any direction other than 180°, whereas the 2D TIs have robust edge states that prevent backscattering [51]. Additionally, their (2D) structural chemical bonding allows flexibility to easily modify in post-synthesis processes, to tune the band gap and achieve desired properties. But only tiny band gaps in 2D TIs are impeding their progress. Therefore, an extensive effort has been made to investigate 2D TIs with a large band gap at room temperature [52–58]. To achieve this goal a few graphene-like 2D honeycomb structures materials have been proposed for the QSH insulators, including silicene, germanene, stanene, etc. [51, 59]. Although, it is a mirror fact that the development of 2D TIs with a large band gap is difficult and challenging to realize the QSH effects at room temperature [35]. Various innovations by employing several methods lead to an increase in the band gap in 2D TI, such as by placing the materials on a substrate and chemical functionalization. In the process of placing the structure on a substrate, the electronic structure can be modified through the interaction between the material and substrate, possibly destroying the topological order. Another approach by chemical functionalization of TI monolayers is considered a very effective approach to widen the band gap and improve the structural stability with preserving nontrivial topological order. Therefore, the bonds are possibly modify with surface engineering, thus these materials have been experimentally realized [51, 60–62]. Such as materials that have heavy atoms, like Pb and Bi, are also considered to be favorable for the TIs, because their heavier atoms generally have a stronger SOC with a larger band gap. The element Pb is the heaviest in group IV, thereby, it may drive a nontrivial topological order with a large band gap. Similarly, Bi is the largest element in group V having an unusually low toxicity. It is also known for its strong SOC and ability to drive a material as the nontrivial topological state [63]. Hence, such property materials are considered to be potential candidates for the 2D TIs.

In addition to these, a class of nontrivial topology also produces topologically prevented edge states at the zero-dimensional (0D) finishing points. These are also recognized as higher-order topological insulators (HOTI) materials that also generalized to the idea of TI [64]. As the intense, a two-dimensional (2D) second-order TI (SOTI) hosts topologies to prevent corner states at its 0D corners. The edge of 2D SOTI is similar to a traditional TI, but a very few 2D materials are considered to be as HOTI [64].

7.2.1 Berry Curvature and Chern Invariance

As one has discussed previously (in Chap. 5) the forces can impact the electron transportation of the topological organization that originates from the interaction among the spin and orbital degrees of freedom of the electron. The total energy (or Hamiltonian) of an electron in a nontrivial multiple band configuration depends on the moderate location of the electron and the wave momentum in the BZ. To resolve the problem Berry considered the influence of a slow change in Hamiltonian and explored the supplementary geometric accumulated phase (Berry) [65]. Moreover, the dynamical phase energy $\mathrm{e}^{-iEt/\hbar}$, and the geometric phase should be consider for the calculation of the velocity of the wave packet, by employing the following relationship;

$$\hbar\dot{x} = \nabla_k\varepsilon - \hbar\dot{k} \times \Omega(k) \tag{7.1}$$

Here term Ω is defined as Berry curvature. Further correction in Berry equation in terms of crystal momentum Bloch wave has also defined, it is usually expressed as;

$$\Omega = \nabla_{\mathrm{k}} \times R_{\mathrm{i}} \tag{7.2}$$

Hear R_{i} represents the Bloch wave part.

Moreover, at low temperatures electrons in a material occupy the lowest energy states according to the exclusion principle. Therefore, the physical quantity can be calculated to a specified energy and summation to all energies in the Fermi regime. The equations of motion for the small applied fields should comprise functions of the dispersal relation (this means the Berry curvature and their derivatives). In the case of the insulator, this should be integrated with the filled band over the BZ. This kind of simplest observable integral of the Berry curvature throughout the BZ is known the Chern number.

$$C = \frac{1}{2\pi} \oint_{BZ} \Omega . dS \tag{7.3}$$

This equation represents the condition of the topological invariant, the term invariance meaning is perturbations **in** the bands that cannot alter without the band arrangement reordering, however, the band crossover points either create or destroy, but changes should be as an integer [2].

7.2.2 Chern for the Insulator

In a 2D crystal due to spin-orbit connection interaction, the total Hamiltonian is commonly expressed as;

$$H_{2D} = \hbar \nu_F (\sigma_x k_x + \sigma_y k_y) + m\sigma_z \tag{7.4}$$

here ν_F represents the Fermi velocity and m related to the mass. However, in the real materials few extra non-linear terms are also incorporated, however for simplicity it is described by associating a few utmost fundamental constituents. This kind of building is similar to the relativistic 2D massive Dirac equation and their easiest explanation is expressed as the well-known Chern insulator. In which the Chern number is state-forward related to the Hall conductivity in the existence of chemical potential in the gap.

Under an applied electric field for the perturbed states transverse current is calculated by summation of the filled energy states;

$$\sigma_{xy} = -\frac{e^2}{h}\text{sgn}(m) = \frac{e^2}{h}C \tag{7.5}$$

It is known as anomalous Hall effect (AHE) owing to fact that the Hall current is depending on an external magnetic field and an interior equilibrium contravention term. Therefore, the Chern insulator constantly comprises a couple of fermions. In 3D realizations, the topological insulator pairs reside on dual contradictory surfaces of the material, while in the 2D realization, two sublattices contain the pairs. When fermions pairs attend contradictory signs of m, then AHE has vanished, otherwise, the AHE conductivity is an integer multiple of the quantum e^2/h [2].

7.2.3 Topological Protection of Transport Properties

The perturbation of the band configuration alters the Berry curvature, while their integral amount is known Chern number cannot alter until and unless the bands are rearranged. However their moderate disorder and interactions generally does not destroy this situation because of its filled states [2]. This leads to uninterrupted deformations of the bands with the unchanged throughout the transport constant that is directly connected to a topologically invariance extent, to count characteristics of the band configuration, like the number of nontrivial band intersections. Such a 2D system's transport properties are categorized as topologically protected.

7.2.4 Topological Insulator as Dirac System

Since in actual crystal exists many bands, when existing bands come into as completely degenerate pairs then both time-reversal (without magnetization all form) and inversion symmetry (center of overturn) can be visualized. Like, if suppose an electronic spin-up state belongs to k, then the corresponding time-reversal state at $-k$ associated a down spin. While another overturn equilibrium state associated spin down necessarily exist right on top of the one possessing spin-up at k, therefore, at least two electrons for each momentum. But in the perturbation condition it is not like that because a gap cannot open as long and it does not hold spin-dependence forces, therefore, the calculation should be in terms of the Weyl point. Moreover, in case of extremely robust spin-orbit connection the spin degeneracy lifted, that is also called twofold degeneracy. As per Kramers' theorem, a Weyl point at k^* in the BZ is also supplemented additional one possessing an identical chirality at $-k^*$. Whereas the inversion momentum desired a Weyl point associating a contradictory chirality at k once more, therefore, it is necessary to have dual contradictory chirality Weyl fermions that sit over each other. So, to define it usually desired to 5 functions in which 3 momenta parameters for the vanishing. This construction is not probable except the enforced by supplementary regularities or adjustment. Those systems that satisfy these conditions are known as Dirac semimetals because two chiral Weyl associates at the similar momentum point create a Dirac particle with zero mass. Besides the gap opens to the constructed massive Dirac fermion but probably it is not a trivial insulator.

Thus, in the case of linearized band configuration while the Weyl fermions of conflicting chirality overlapping at the equal momentum point. The perturbation opens the mass gap (m) with the both sign and as the consequence, the linearized Hamiltonian gives rise as;

$$H = \begin{pmatrix} -\hbar v_{\mathrm{F}} \sigma (k - k^*) & m \\ m & \hbar v_{\mathrm{F}} \sigma (k - k^*) \end{pmatrix} \tag{7.6}$$

This is a representation of the four components of the Dirac equation including mass "m". While applying the boundary condition on the material, the "m" changes the sign. As the intense, border is recognized by $z = 0$, with $m = |m|\mathrm{sgn}(z)$, then the localized state wave function;

$$|\psi\rangle = e^{-\int_0^z m(z')dz'} (|\psi_1\rangle,\ |\psi_2\rangle)^T \tag{7.7}$$

where ψ_1 and ψ_2 are representing two spins that satisfy the following term

$$|\psi_1\rangle = -i\sigma_z|\psi_2\rangle \tag{7.8}$$

To solve the Dirac expression, the two-spinor ψ_2 should be an eigenstate of the following equation;

$$H_{\mathrm{boundary}} = \hbar v_{\mathrm{F}}\left[\sigma_{\mathrm{x}}\left(k_{\mathrm{x}} - k_{\mathrm{x}}^*\right) + \sigma_{\mathrm{y}}\left(k_{\mathrm{y}} - k_{\mathrm{y}}^*\right)\right] \tag{7.9}$$

It directly gives us the gapless mode localization at the interface. Usually, a gapless Hamiltonian is relative to the product of $\sigma_{\|}.k_{\|}$ parallel to the surface. The normal insulators usually (including the vacuum) follow the Dirac expression possessing a constructive mass. When a gapless surface mode does not exist on the boundary of an insulator, it is typically called an ordinary, while, if it existing, then the crystal is recognized as topological insulator. This is the microscopic understanding. In which bands are nonentity nonetheless coalesced atomic orbitals. When a robust spin-orbit combination converses the energy ordering of atomic orbitals, then the band gap is called inverted, therefore, the sign of the Dirac mass "m" can be also changed. However, in an ordinary insulator at a boundary, the atomic orbitals should reverse to their normal order at which point they have to encounter at the similar energy, therefore, the gap should be nearby at the edge.

7.2.5 *Hamiltonians and Kinetics*

Generally the surface states of topological insulators are defined from the subsequent Dirac–Rashba Hamiltonian relationship;

$$H_{\mathrm{TI}} = A\left(s_{\mathrm{x}}k_{\mathrm{y}}\right) + \lambda s_{\mathrm{z}}\left(k_{\mathrm{x}}^3 - 3k_{\mathrm{x}}k_{\mathrm{y}}^2\right) \tag{7.10}$$

here A and λ are devoted to the material-specific constraints, s represents the vector of Pauli spin matrices, and k corresponds to 2D wave vector.

However, the Dirac and Weyl semimetals are also described in the different forms of Hamiltonian equation [66],

$$H_{DWSM} = \hbar v \mathrm{T_x} \sigma.\mathrm{k} + \mathrm{m}\mathrm{T_z} + \mathrm{b}\sigma_Z + \mathrm{b'}\mathrm{T_z}\sigma_Z \qquad (7.11)$$

where σ is the pseudospin degree of freedom, T the valleys or nodes, m is a mass parameter, and b and b' represent effective internal Zeeman fields. For the Dirac semimetals a single node, therefore, $m = b = b' = 0$. But according to the simplest model of a Weyl semimetal, it has two nodes for the case $|b| > |m|$ and $b' = 0$, that act as the source and sink of Berry curvature. However, in various known materials, many more pairs of nodes exist. Such as, when Weyl points become owing to fragmented overturn equilibrium concerning M_x, M_y mirror planes then four couples of Weyl nodes (±kx, ±ky, ±kz) exist accompanying their time-reversal associates. Similarly, eight couples in the TaAs group may also occur owing to the extra C_4 rotational regularity (±ky,±kx, ±kz) [67, 68].

Therefore, the distinction among a Weyl node and its associates containing contradictory chirality are determined with help of the electromagnetic outcome of the material which associates so-called axion term'. In term source of chiral magnetic effect (CME) and the quantum anomalous Hall effect (QAHE) in Weyl semimetals [69–71]. Therefore, transition metal dichalcogenides are defined from the following Hamiltonian relationship [28];

$$H_{TMD} = A\left(\mathrm{T}\sigma_x k_x + \sigma_y k_y\right) + \frac{\Delta}{2}\sigma_z - \lambda \mathrm{T}\frac{\sigma_z - \|}{2} s_z \qquad (7.12)$$

where s belongs to spin $\mathrm{T} = \pm$ is the valley, σ is the orbital pseudospin index, and $\|$ is the identity matrix in two dimensions. While 2 λ represents the spin splitting at the valence band top due to spin-orbit coupling. Moreover, the additional terms encapsulated the spin splitting of the conduction band electron-hole asymmetry and trigonal warping of the spectrum [72, 73]. In form of the trigonal warping equation can be written as;

$$H_{\mathrm{TW}} = \frac{k}{2}\left(\sigma_+ k_+^2 + \sigma_- k_-^2\right) \qquad (7.13)$$

where $\sigma_{\mp} = x \pm iy$. It reveals there is always a gap between the valence and conduction bands that manifested in terms of the mass appearing in each of the sets of the Dirac Hamiltonian. Moreover, to satisfy time-reversal invariance materials should have two valleys that are related by time-reversal [74].

Indeed, the presence of spin and pseudospin-charge combination in topological material Hamiltonians due to electromagnetic fields that persuaded the interband dynamics and considered a suitable interaction to the disorder. These kind impacts more appropriately could be visualized from the density matrix theory [75]. The single-particle density matrix follows the quantum Liouville equation;

$$\frac{d\rho}{dt} + \frac{i}{\hbar}[H, \rho] = 0 \qquad (7.14)$$

here H represents the total Hamiltonian. According to the explanation, the density matrix could be decomposed into two segments. The first portion is signified by $\langle \rho \rangle$ to reflect the overall moderate impurity throughout the structures, while the remains ultimately combined over and designated by g: $= \langle \rho \rangle + g$, together $\langle g \rangle = 0$. In the case of the linear outcome of an electric field E, the density matrix of the symmetric and non-symmetric constituents are comprised as $\langle \rho \rangle = \langle \rho_0 \rangle + \langle \rho_{\mathrm{E}} \rangle$, here $\langle \rho_E \rangle$ represents the correction term in the symmetric density matrix $\langle \rho_0 \rangle$, correspond to first order E. Therefore, the linearized kinetic equation to E is expressed as;

$$\frac{d\langle \rho_{\mathrm{E}} \rangle^{mm'}}{dt} + \frac{i}{\hbar}[H_0, \langle \rho_{\mathrm{E}} \rangle]^{mm'} + J(\langle \rho_{\mathrm{E}} \rangle)^{mm'} = \frac{eE}{\hbar} \cdot \left\{ \delta^{mm'} \frac{\partial f_0(\varepsilon_{\mathrm{k}}^m)}{\partial k} + iR_{\mathrm{k}}^{mm'} \left[f_0(\varepsilon_{\mathrm{k}}^m) - f_0\left(\varepsilon_{\mathrm{k}}^{m'}\right) \right] \right\} \tag{7.15}$$

here H_0 represents the band Hamiltonian, $J\langle \rho \rangle$ signifying the generalized scattering term, $f_0(\varepsilon_{\mathrm{k}}^m)$ represents the Fermi-Dirac distribution function and $iR_{\mathrm{k}}^{mm'} = \left\langle u_{\mathrm{k}}^m \middle| i \frac{\partial u_{\mathrm{k}}^{m'}}{\partial k} \right\rangle$ corresponds to the Berry connection. Thus the operative term give upsurge to the Berry curvature inherent involvement in the Hall conductivity together the fragmented time-reversal equilibrium, including additional outcomes characteristics, moreover, the difference factor of the Fermi occupation number allowing to term operates off-diagonal response, hence the interband coherence assistances in the electrical response. The advantages of this methodology one can instantaneously distinct the inherent impacts, extrinsic impacts, and effects of the combined interband coherence and disorder [2].

7.3 Monolayer TMD Topological Properties

Monolayer transition metal dichalcogenides (TMDs) are formed from a layer of transition metal atoms sandwiched between two layers of chalcogen atoms in the trigonal prismatic structure [76]. Due to their two-dimensional nature, the high electron mobility and the massive Dirac energy spectrum allow to being considered candidates for next-generation transistors [77–79]. However, the crystal structure of TMDs has no inversion symmetry but they have out-of-plane mirror symmetry. This also supports a strong atomic spin−orbit coupling (SOC) field in TMD materials [80–83]. Several TMD materials such as MoS_2, $MoSe_2$, WS_2, and $NbSe_2$ also have a superconducting state even down to atomic thickness [18–20]. More significantly monolayer superconducting of MoS_2 and $NbSe_2$ are also possessed extremely high in-plane upper critical field Hc_2 (more than six times higher than the Pauli limit) [80–82]. In which the SOC pins electron spins of the Cooper pairs to the out-of-plane directions, which prevent the electron spins being aligned with external in-plane magnetic fields, therefore, it protects the superconductivity [84].

As per initial predictions the monolayer TMDs 1T′ (or distorted octahedral) phase in the two-dimensional TIs [85], with this structure both bulk and few-layered synthesis are possible on a large scale retaining high quality [86]. However, understanding their topological nature is still a challenging task. Because their a spontaneous structural phase transition from 1T (or octahedral) to 1T′ phase, as illustrated in Fig. 7.2a, b causes the band inversion, thereby, it leads topological phase transition from trivial to nontrivial [85, 86]. Additionally, the energy dispersion of 1T′-MoS_2 can strongly resemble the common feature in topological materials, as depicted in Fig. 7.2c. Therefore, the band inversion in 1T′-TMDs could be interpreted based on the common speculation that the chalcogenide p_x band [85, 87], which usually formed the valence band, by moving above the metal d_{xz} band that reflects the d-p inversion mechanism. Later on, findings revealed that the common features are not the same, therefore, instead of counterintuitively p_x and d_{xz} bands they inverted even before the structural phase transition takes place. However, the band topology of Te-based TMDs, such as $MoTe_2$ has an entirely different feature compared to MoS_2 [85, 86]. Moreover, the additional phases of TMDs (such as S and H') are also gives a nontrivial band topology [87–89].

To resolve the issue a remarkable study with the topological phase transition in monolayer TMDs has been introduced by showing the band inversion mechanisms entirely different from the conventional speculation. According to interpretation, the 1T′-MX_2 can have two distinct types of band inversion. Such as, 1T′-MS_2 and 1T′-MSe_2 bands opposite to the conventional speculation, in which the X-p band pulls down below the M-d band, and it is referred to as the p-d-type band inversion. Similarly, 1T′-MTe_2 can also have two different M-d bands that are inverted during the topological phase transition, typically it is called d-d-type band inversion.

In the layered TMDs (MX_2, where M = Mo, W, X = S, Se, Te) the most stable polytype trigonal prismatic 2H phase possessed a moderate band gap (>1.0 eV) with a strong valley polarization due to inversion symmetry breaking, while an octahedral 1T phase is metallic with inversion symmetry. The 1T phase can be more stable than the 1T′ phase, it also has inversion symmetry under the Peierls-like distortions, as illustrated in Fig. 7.2b. While going from the 1T phase structural transformation to the 1T′ phase transformation in monolayer MX_2, the band topology is changed from trivial to nontrivial. Their intermediate structural changes referred to 40%-1T′, 60%-1T′, and 80%-1T′, which linearly interpolate to the atomic positions and lattice parameters between the 1T and the 1T′ phases. It is subsequently optimized only by the X atoms by fixing the M atoms at the interpolated positions. Usually, the topology of the time-reversal-invariant band structure is characterized by the Z_2 invariant ν in which ν = 0 and ν = 1, which correlates to the trivial and nontrivial states. The lattice inversion symmetry presence in all intermediate structures and the 1T and 1T′ phases allows that Z_2 invariant to be calculated in terms of the product of the parity eigenvalues at the time-reversal-invariant momenta (TRIM), Γ, X, Y, and S [37, 86, 91]. Such structures are illustrated by employing the Kohn-Sham wave function for the nth band at the wave vector k;

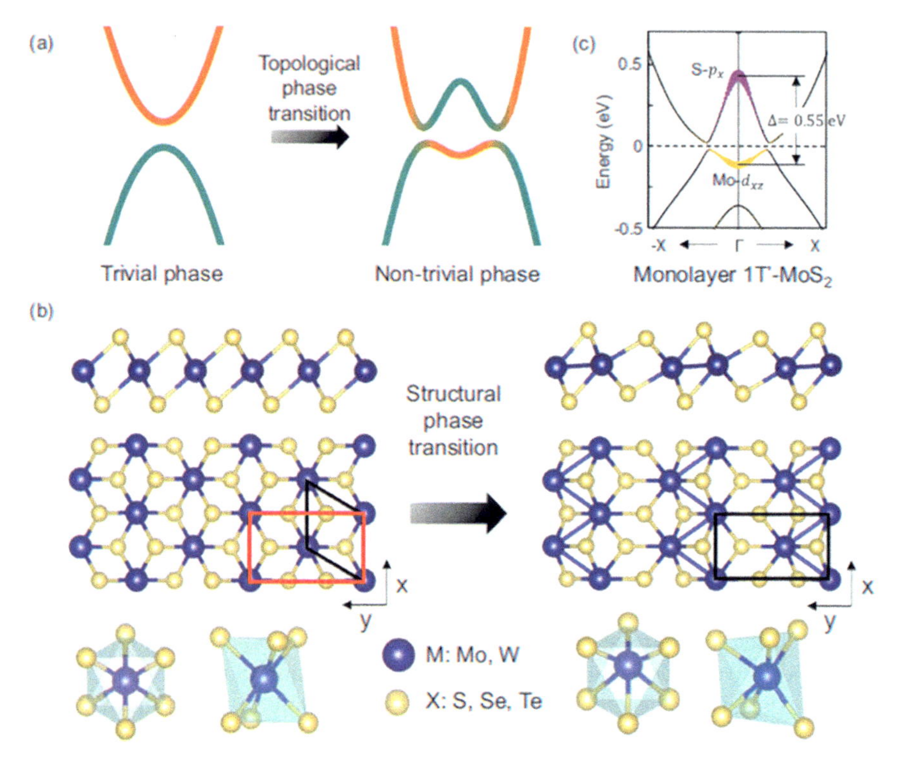

Fig. 7.2 **a** Schematic for the topological phase transition in TIs. The valence band rises above the conduction band, and an interband hybridization opens the nontrivial energy gap. The band inversion mechanism can be directly inferred from the band structure; **b** structural transformation from the 1T to the 1T′ phase in TMDs, the primitive cells are denoted as black solid lines, whereas red solid lines indicate the rectangular supercell for the 1T phase; **c** band structure for the monolayer 1T′-MoS_2 near the Γ point strongly resembles exhibiting an inverted gap of $\Delta = 0.55$ eV and thus misleading the band inversion mechanism in 1T'-TMDs (adopted from Duk, H. C. et al. 2016 Phys. Rev. B 93, 125109) [90]

$$\left|\psi_{\mathrm{k,n}}\right\rangle = \sum_{\mathrm{i,l,m}} c_{\mathrm{k,n}}^{i,l,m} |i, l, m\rangle \tag{7.16}$$

Here $|i, l, m\rangle$ represents the PAW projector for the ith atom, l is the angular momentum, and m is the magnetic quantum number. The term complex coefficient $c_{\mathrm{k,n}}^{i,l,m}$ can be decomposed as;

$$c_{\mathrm{k,n}}^{i,l,m} = \left|c_{\mathrm{k,n}}^{i,l,m}\right| e^{i\phi_{\mathrm{k,n}}^{i,l,m}} \tag{7.17}$$

Here $\left|c_{k,n}^{i,l,m}\right|$ represents the orbital weight and $e^{i\phi_{k,n}^{i,l,m}}$ is the phase factor.

If the inversion function is selected as the origin, then the phase factors at the Γ point should follow the relationship, owing to inversion symmetry;

$$e^{i\phi_{\Gamma,n}^{A,l,m}} = \pm e^{i\phi_{\Gamma,n}^{A',l,m}} \tag{7.18}$$

Here A and A' are denoted to an inversion pair of the M or X atoms in MX_2.

Similarly, the effective parity index also defines by using the following relationship;

$$P_{\mathrm{k,n}}^{A,l,m} = (-1)^l \cos(\phi_{\mathrm{k,n}}^{A,l,m} - \phi_{\mathrm{k,n}}^{A',l,m} \tag{7.19}$$

In the range of -1 to $+1$, the odd parity belongs to (-1) and even parity corresponds to (+1). By using the value of the parity index, the degree of symmetry of each wave function is usually defined, typically their values are equivalent to the parity eigenvalues at the Γ point. Additionally, this concept probably also expands to other k points near the Γ point where the parity eigenvalues are not well defined.

Moreover, the OP weight also is defined with the help of the following relationship;

$$w_{\mathrm{k,n}}^{A,l,m} = \left|c_{\mathrm{k,n}}^{A,l,m}\right|^2 P_{\mathrm{k,n}}^{A,l,m} \tag{7.20}$$

In case of the centrosymmetric materials $w_{\mathrm{k,n}}^{A,l,m}$ it gives us a clear identification of both orbital characters and parities of the wave functions in a band structure. Such as, both orbital and parity to be visualized for the 1T'-MoS_2 [90]. Furthermore, the OP projected band structure around the Γ point during the 1T-to-1T′ transition in MoS_2 has been also explored. It has been noted that the S-p_x orbital band with odd parity and the Mo-d_{xz} orbital band with even parity near the Fermi level, typically represents as $|S\_p_x^-\rangle$ and $|Mo - d_{xz}^+\rangle$. In the case of the 1T′ phase $|S\_p_x^-\rangle$ and $|Mo - d_{xz}^+\rangle$ bands represent the inverted-like energy dispersion by the d-p inversion having an inverted gap $\Delta = 0.55$ eV [90]. However, the band ordering between the $|S\_p_x^-\rangle$ and $|Mo - d_{xz}^+\rangle$ states may remain unchanged throughout the whole structural transformation. Similarly, in a more general condition $|X\_p_x^-\rangle$ the state always exists above the $|M - d_{xz}^+\rangle$ state for the both 1T and 1T′ phases of the MX_2. That directly indicates these two states are irrelevant to the topology change for all the group-VI TMDs, including the SOC to band gap open in 1T′-MX_2 [85, 86].

On the other hand, three orbital bands; the $M - d_{yz}$ orbital band with even parity $\left(\left|M - d_{yz}^+\right\rangle\right)$ play a crucial role in the topological phase transition. Additionally, the $X - p_y$ orbital band with odd parity $\left(\left|X - p_y^-\right\rangle\right)$ and $M - d_{z^2}$ orbital band with odd parity $\left(\left|M - d_{z^2}^-\right\rangle\right)$ can also play a crucial role in the topological phase transition. As the OP projected band structure of MoS_2 has revealed the p-d-type band inversion in which the state $\left|S - p_y^-\right\rangle$ moves down below the $Mo - d_{yz}^+$ state. Moreover, in case of 60% T′ MoS_2, it has noticed that the both $\left|S - p_y^-\right\rangle$ and $Mo - d_{yz}^+$ bands cross near the Fermi level. However, these two bands move apart from each other with a significant modification of their band shape and eventually form a large inverted gap of $\delta_{MoS_2} = 1.05$ eV. But a distinction has been also noticed in characteristic features of the $MoTe_2$. like $|Mo - d_{z^2}^-\rangle$ the band pulls down below the $|Mo - d_{yz}^+\rangle$ with an inverted gap of $\delta_{MoTe_2} = 0.94$ eV. Such d-d band inversion is considered to be a fundamental mechanism for the topology change in $MoTe_2$, in contrast to MoS_2.

Therefore, a band inversion mechanism in MX_2 to be interpreted where the p-d and d-d inversion processes occurs, as the schematic is illustrated in Fig. 7.3a. Schematic mechanism of $|M - d_{xz,yz}^+\rangle$, $|X - p_{x,y}^-\rangle$ and $|M - d_{z^2}^-\rangle$ states predominately contributed to the energy states around the Fermi level, while the structural transformation from 1T to 1T′. Particularly in the case of the 1T phase, the octahedral crystal fields with trigonal distortion (D_{3d} group symmetry) are degenerate $|M - d_{xz}^+\rangle$ and $|M - d_{yz}^+\rangle$ states with the associated $|X - p_x^-\rangle$ and $|X - p_y^-\rangle$ degenerate states. As the consequence, the metal M atoms deviate from their octahedral positions, therefore, degeneracies of the d- and p-orbital bands are fully lifted, as depicted in the energy splitting of the $|M - d_{xz,yz}^+\rangle$ states in Fig. 7.3a. Moreover, the energy differences of $|M - d_{xz,yz}^+\rangle$, $|X - p_{x,y}^-\rangle$ and $|M - d_{z^2}^-\rangle$ state for the 1T-MX_2 (where X = S, Se, Te, and Po) is depicted in Fig. 7.3b, whereas Fig. 7.3c represents the band ordering in 1T-MX_2. Hence findings have revealed while increasing the atomic number (X) the $|X - p_{x,y}^-\rangle$ levels are also increased owing to valence electrons loosely bound to the nucleus. Like, in 1T-MS_2 and 1T-MSe_2, $|X - p_{x,y}^-\rangle$ states are positioned below the $|M - d_{z^2}^-\rangle$, therefore, the energy splitting of the $|X - p_{x,y}^-\rangle$ states levels crossing between the $|X - p_y^-\rangle$ and $|M - d_{z^2}^-\rangle$ bands, it gives the p-d band inversion. However, in the case of 1T-$MoTe_2$ and 1T-MPo_2 the $|X - p_{x,y}^-\rangle$ states energy is higher, so, the splitting of this state cannot affect the band topology. Whereas, the $|M - d_{z^2}^-\rangle$ band pulling down below the Fermi level which leads to the level crossing between the $|M - d_{z^2}^-\rangle$ and $|M - d_{yz}^+\rangle$ bands, typically it is called d-d band inversion. Moreover, $|M - d_{z^2}^-\rangle$ and $|X - p_{x,y}^-\rangle$ bands can be close to each other (such as 1T-WTe_2 and 1T-WPo_2) then two orbital bands strongly hybridized during the structural transformation and resulting in a mixture of the p-d and d-d inversions.

7.3.1 Binary Materials

As we discussed TIs are a class of material that have a semiconducting bulk band gap and surface states with Dirac-like dispersion (typical schematic is illustrated in Fig. 7.4a), and their remarkable theoretical investigations have shown the existence of the state on the edges of certain 2D materials [92], as in the first experimental realization for the HgTe quantum wells was noticed [47]. Subsequently, theoretical investigations have also predicted that similar behavior could notice in

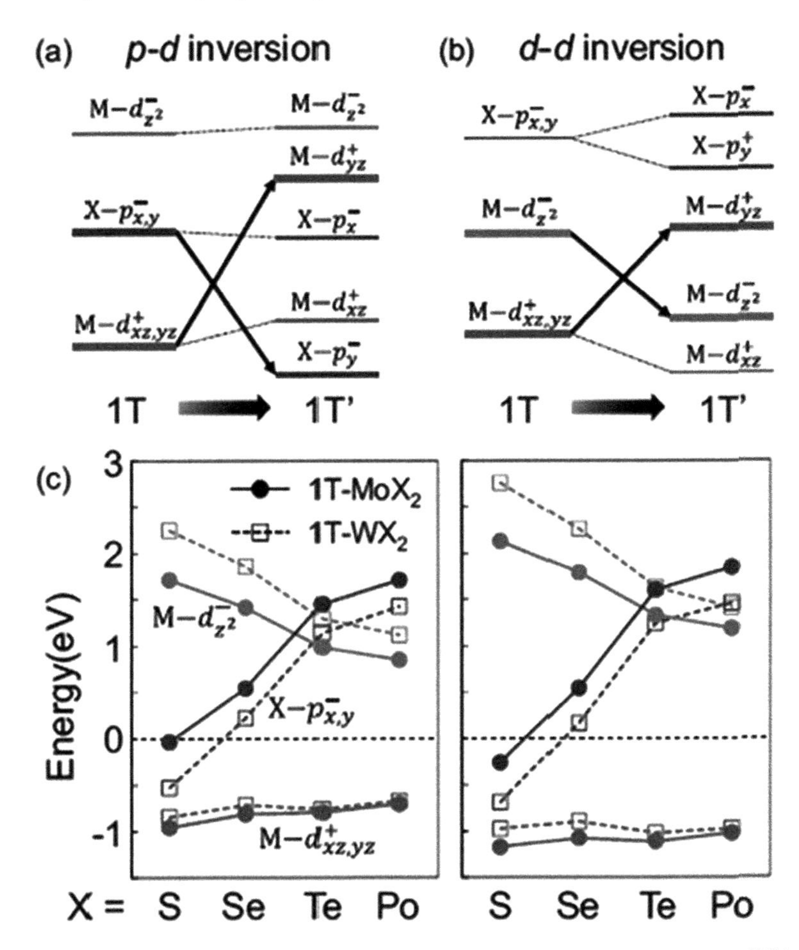

Fig. 7.3 **a, b** Schematics for the p-d and d-d band inversion processes in MX_2; **c** the GGA (left panel) and HSE06 (right panel) calculations for the band ordering in 1T-MX_2 (adopted from Duk, H. C. et al. 2016 Phys. Rev. B 93, 125109) [90]

three-dimensional materials with surface states [93]. Their surface states have spin-momentum locking to topologically protect against the defects, so, such properties make them enable as the TIs materials for a variety of logic and memory devices. The first experimental demonstration on a 3D TI became with the $Bi_{1-x}Sb_x$, in which the surface states directly have been observed by using angle-resolved photoelectron spectroscopy (ARPES) [84]. Afterward, various TMDs binary compositions such as Bi_2Se_3, Bi_2Te_3, Sb_2Se_3, Sb_2Te_3, and Bi_2Se_3 are also recognized as 3D TIs possessing a bulk bandgap of 0.3 eV. Their adequate 3D TIs properties make them enable the utilization of the surface states at room temperature, which is considered to be most suitable TIs for device applications [7]. Usually, such materials have a layered structure with each unit cell (known as a quintuple layer) composed of five alternating layers of X-M-X-M-X (M is metal and X is the chalcogen), as depicted in Fig. 7.4b for Bi_2Se_3 [94]. Their van der Waals gap between the successive quantum layers also makes them highly suitable for the van der Waals epitaxy (VDWE) [95].

Most of the growth studies on 3D TIs have been performed on the binary TIs such as Bi_2Se_3, Bi_2Te_3, Sb_2Se_3, Sb_2Te_3, MoS_2, XBi (X = B, Al, Ga, and In), etc. [95–98]. The lattice mismatch grown TIs unstrained lattice spacing can be recognized from RHEED that reflects the expected bulk lattice constant of the first quantum layer to be formed, as depicted in Fig. 7.4c. It indicates that the immediate unstrained growth in such ultrathin films consists of vdWE [95]. However, an abrupt layered growth have been also noted at the Bi_2Se_3/SiC interface in the TEM image, with a complete absence of any misfit dislocations (see Fig. 7.4c) [99].

Typically, such growths are also performed using a relatively high chalcogen-to-metal flux environment to get stoichiometric films. Like, Bi evaporated for the growth of Bi-based TIs using low Bi fluxes and substrate temperatures. In contrast to this, Se, Te, and Sb fluxes are predominantly composed of clustered molecular species, like Sb tetramers (Sb_4) and Se and Te dimers (Se_2 and Te_2) which are unlikely to participate in the growth process, so, it shows the importance of high fluxes to improve the growth quality [100, 101]. Moreover, the growth temperature could be limited to below 320 °C for the significant chalcogen loss that allows from grains above the temperature [101]. Because in the ideal case usually growth is performed between 200 and 300 °C with chalcogen-metal ratios >10:1. [101, 102]. Moreover, the use of a chalcogen cracker increases the percentage of atomic species in the chalcogen flux, which ultimately reduces the required chalcogen-metal ratio to 1.5:1 [103]. Though in this process, the required flux ratios are much smaller than those required for the TMDs which primarily has to do with the differences in the surface

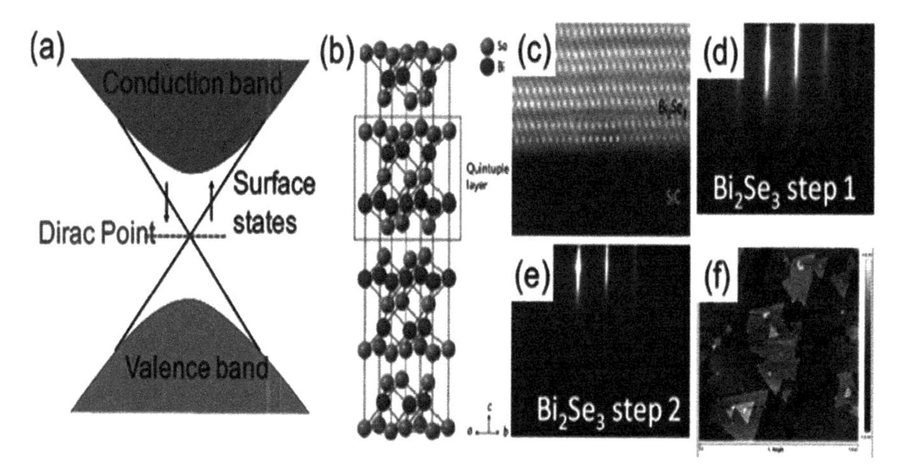

Fig. 7.4 **a** Schematic of the surface states of a topological insulator: **b** the layered structure of Bi_2Se_3; **c** cross-sectional transmission electron microscopy (TEM) image of the Bi_2Se_3/SiC interface; **d, e** Reflection high energy electron diffraction (RHEED) images of Bi_2Se_3 grown on sapphire after for the first and second growth steps; **f** A 5 × 5 μm AFM image of Bi_2Se_3 grown using the two-step method (adopted from Lee Walsh, A. et al 2017 Appl. Mater. today 9, 504–515) [95]

mobility of Bi compared to transition metals, like W. In addition to cooling of the samples post-growth in presence of a chalcogen flux also helps to improve the crystal quality [104].

TIs growth characterized from the AFM and SEM images have shown the triangular shape grains that reveal the hexagonal crystal structure with limited grain sizes (typically <250 nm) [95]. Such a shape's appearance are usually connected to the used two-step growth process of the III-V semiconductor that increases the grain sizes from 100 nm to >1 μm [105]. Typically in the first step, 1–2 quintuple layers quantum layers are deposited at a low temperature (110–130 °C), for the high sticking coefficient and low diffusion of adatom, to ensure good coverage across the substrate. But the low-temperature growth limits the grain sizes and lead a poor crystallization. As the initial 1–2 quantum layers are formed then usually growth is stopped, so that successively specimen annealed to give a higher temperature (~300 °C) to improve the crystallization of the formed thin layer. Therefore, the reaming film is deposited at a higher temperature to achieve large grain sizes and higher quality films. As consequence grain sizes are increased with increasing substrate temperatures in the range of 270–320 °C, afterward Se losses and degradation in material quality begins.

The typical RHEED images after the first and second growth steps are illustrated in Fig. 7.4d, e, which reveals an improvement in crystallinity. Using the same method further improvements can be achieved in the grain sizes up to (~1.5 μm to >2 μm), as depicted in Fig. 7.4f [101].

Likewise, investigations of large grain sizes and crystalline film's various electrical and spectroscopic measurements have been also demonstrated with the adequate topological properties of TIs surface states. Typically, Fig. 7.5a–e represents the ARPES measurements of MBE-grown Bi_2Se_3 films as a function of thickness [106]. It also consisted of the theoretical predictions; the Dirac cone can be associated with the surface states only becomes evident once the layer thickness reaches at least 5 quintuple layers. Below the 5 quintuple layers opposite surface wave functions could interact and result the hybridization to open a gap in the surface states. Moreover, Shubnikov de Haas oscillations have been also realized in the magnetic field-dependent resistivity measurements that signify the existence of the surface states on both the top and bottom surfaces of the TI film [107]. The quality growth has been also explored for the other binary TIs such as Bi_2Te_3 and Sb_2Te_3 [108].

More recently robust tunable large band gap quantum spin hall states have been investigated in binary Cu_2S monolayers and studied their electronic and topological properties [109]. As per the investigation, the unit cells of the Cu_2S monolayer have been formed by two Cu atoms and one S atom. A honeycomb structure has formed from the first Cu and one S atom, while the other Cu atoms lie in the center of the honeycomb, as the top view of the atomic structure is illustrated in Fig. 7.6a. Whereas each S atom is surrounded by six Cu atoms, in which each Cu atom is bonded with three Cu and three S atoms. Therefore, the Cu_2S system has formed three types of honeycomb-like structures, namely, planar, two-sided buckled, and one-sided buckled, as depicted in Fig. 7.6a–d, while the hexagonal lattice associated Brillouin zone (BZ) as exhibited in Fig. 7.6e. The planar structure possesses space group P6/mmm, which is similar to the typical planar structure of M_2X materials.

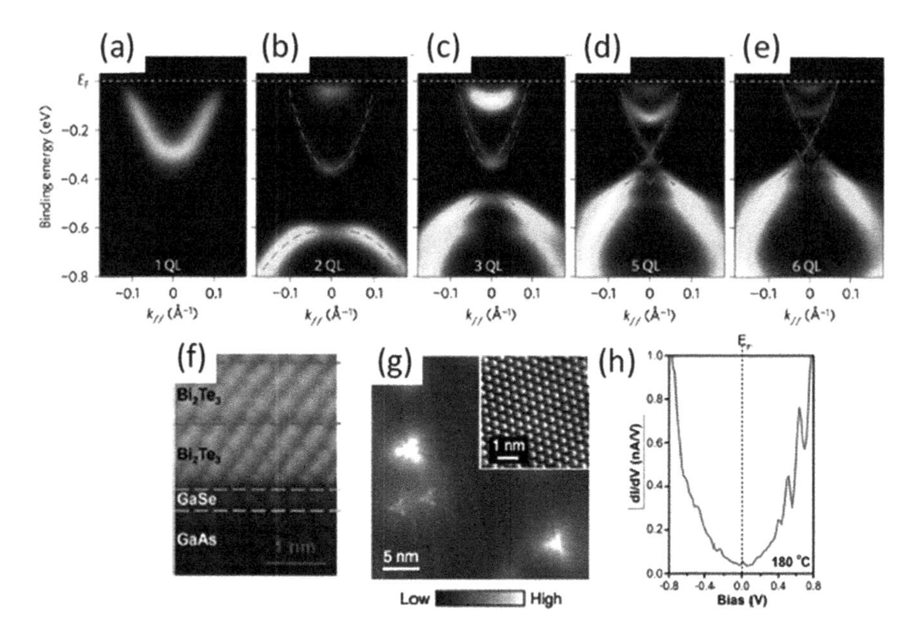

Fig. 7.5 **a–e** Thickness-dependent ARPES of Bi_2Se_3 to show the appearance of surface states after 5 QL; **f** TEM of Bi_2Te_3 grown on GaAs with a GaSe interlayer; **g** STM image of Bi_2Se_3 surface to show the common Se vacancy defects; **h** STS of Bi_2Se_3 grown on sapphire that showing E_f near midgap due to the reduction of Se vacancies (adopted from Lee Walsh, A. et al 2017 Appl. Mater. today 9, 504–515)

Whereas the one-sided and two-sided buckled structures can have space groups P6mm and P3m1. The high buckling can reduce the symmetry of the atomic structure. Moreover, these states total energies are relatively close. Among all three structures, the two-sided buckled structure has been recognized as the most stable structure that possessed lower energy (8.6 and 10 meV/ atom) then the one-sided buckled (see Fig. 7.6d) and planar structure (Fig. 7.6b). The associated structure has a gradual decrease in the buckling height until reaching a plateau at approximately 4.22 Å, as illustrated in Fig. 7.6g, It directly has been correlated to both one-sided and two-sided buckled structures transformation into the planar honeycomb structure with larger tensile strain. Furthermore, their band analysis has revealed that in the absence of spin−orbit coupling (SOC), the monolayer Cu_2S has gapless behavior with the conduction band minimum (CBM) and heavy-hole valence band maximum (VBM) degenerates at the Γ point, while with the SOC has transformed into a gapped insulating state with a large direct band gap of 220 meV, as depicted in Fig. 7.7a–d. A significant band gap opening in Cu_2S has been achieved with the heavier elements. Where for the monolayer origin Rashba splitting induces an effective electric field by breaking the inversion symmetry of the crystal structure. Thus in case of without SOC, the CBM and VBM mainly derived to s orbital of Cu and the p_x and p_y orbitals of S atoms, while the p orbitals of Cu and p_z orbital of S do not contribute to states near the Fermi level (see Fig. 7.7c). Whereas, in the case of including SOC, the

CBM, and VBM can be switched and their valence band edge comprised of the s orbital of Cu and the conduction band edge comprised of p_x and p_y orbitals of S, which is opposite to the band alignment in the case without SOC (see Fig. 7.7d). It leads the band structure of monolayer Cu_2S inverted and followed the characteristics of QSH insulators. Moreover, the topological property of monolayer Cu_2S could be also defined by calculating the Z_2 topological invariant parameter value ($Z_2 = 1$) [109].

Theoretically binary TIs are also extensively studied, such as various DFT simulations that have presented structural and electronic properties by using the Kohn-Sham formulation [110–116]. Specifically, in terms of TI materials like Bi_2Se_3 and Bi_2Te_3. DFT has also exploited for dual main things; the leading one is the explanation of the ground-state atomic locations (relaxation) and the subsequent one is the narrative of the converged charge density. Both of them can significantly contribute in quantifying the surface states and the selection of the suitable exchange-correlation functionals (XCFs). The TIs are mostly studied with XCFs including the local density approximation (LDA) and generalized gradient approximation (GGA) [4, 106, 117–124], as well as kinetic density functional (meta GGA). All three with the XCFs provide the rungs on a ladder with chemical accuracy when applied to the many-body

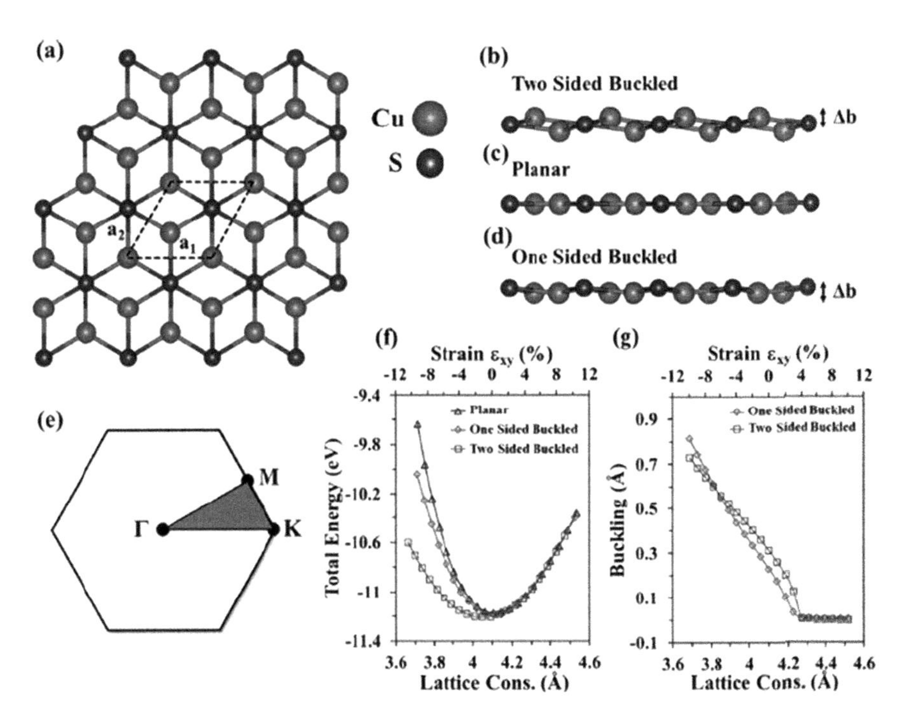

Fig. 7.6 **a** Top view of the crystal structure of monolayer Cu_2S honeycomb; **b** side views of two-sided buckled; **c** planar; **d** one-sided buckled structures; **e** first Brillouin zone with high symmetry points; **f** energies; **g** vertical buckling heights as a function of strain (adopted from Sufyan et al. [109])

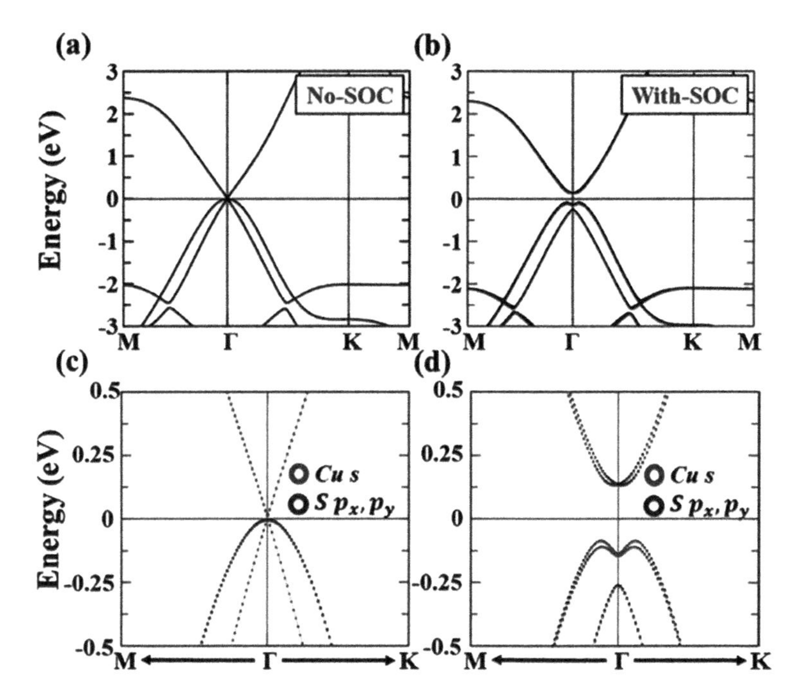

Fig. 7.7 **a–d** Hybrid-functional-calculated band structure of the one-sided buckled monolayer Cu_2S without (left column) and with SOC (right column), here dots represent s/p_{xy} orbitals (adopted from Sufyan et al. [109])

perturbation theory in the GW approximation (where G is the single-particle Green's function and W is the screened Coulomb interaction) to improve the description of the band structures [125, 126]. It is also significant for the realization of the atomic construction that is tortuously connected to electronic features. As the intense, the GW approximations or DFT-derived atomic structures are frequently apply to experimentally. Whereas, regarding layered TI materials like Bi_2Se_3 and Bi_2Te_3, treatment of the vdW forces have been considered more important [127]. Due to a slab (or film) fixed n quintuple layers have an association among the thickness of the slab or film (i.e., due to strain) and an opportunity to notice a topological surface state. In which the vdW forces facilitate the interactions among the quintuple layers across the inter-quintuple layer and their space—depending on the magnitude of influence, as a consequence they creates smaller or larger vdW gaps. Thus, the vdW-corrected DFT provides us with a different overall thickness that may different type of surface state [128, 129]. However, it should be noted that the majority of the thickness of a slab is made up of the intra-quintuple layer separations, therefore, one should evaluate the contributions of the intra-quintuple layer in inclusive thickness [127]. As the typical schematic of the bulk hexagonal unit cells and the slab models are illustrated in Fig. 7.8a–c. Where the slabs are centered toward the perpendicular direction and a vacuum layer of thickness ~8.8Å is supplemented to both sides of the surfaces to build whole vacuum thickness ~17.6Å because this amount of vacuum thickness warrants the insignificant dipole interaction among the surfaces under the vacuum [130–132]. Using this approach a systematic investigation with one quintuple layer to eight quintuple layers has been **executed** by employing the projector-augmented wave pseudo potentials at a 7 × 7 × 1 k-point grid span for the two-dimensional Brillouin zone. Moreover, the kinetic energy cutoff of the plane waves have been recorded at 500 eV and precision tag set accuracy generates for the 162 000, 225 000, 288 000, 352 800, 432 000, 486 000, 540 000, and 604 800 numbers of plane waves to 1, 2, 3, 4, 5, 6, 7, and 8 quintuple layers. With the additional support of tag adding a grid is almost twice in size as a regular grid to get high-quality geometrical optimizations, it has been noticed that each quintuple layer in a multi-quintuple layer structure is feebly connected to the remaining by the vdW force. Moreover, from this approach, the comparative alteration in the width of the slab can be also obtained by using the relationship $\Delta t/t = \left(t_{XCF} - t_{expt}\right)/t_{expt}$, where t_{XCF} and t_{expt} are the adjusted width and perfect bulk-cut width. Typical n quintuple layer function and XCF for the Bi_2Se_3, Bi_2Te_3 is illustrated in Fig. 7.9a,b. In which the parallel line passing from the zero for $\Delta t/t$ to reflect the width of the construction together investigational lattice by using the LDA functional, a negative value of $\Delta t/t$ has also achieved. Additionally, other structural parameters and electrical properties of these materials have been also interpreted [127].

7.3.2 *Ternary Materials*

Though MX_2 binary TIs have shown remarkable properties for their specific applications, nonetheless they are also suffering at some stage of performances, such as Bi_2Se_3 single and less-warping Dirac cone that influences the magneto transport measurements by showing the bulk conductance dominates even in low carrier samples, that raises the question of possible scattering channels responsible for the reduced surface mobility. Additionally, from the Dirac point of view, the surface state in Bi_2Se_3 has predicted to close the bulk valence band maximum. It means the electron scattering channel from surface states to bulk continuum states opens and finally topological transport regime collapse, as the typical example is $TlBiSe_2$ [133]. In contrast to layered binary chalcogenides that has inert surface due to the weak bonding between the layers, the ternary compounds possessed broken bonds that give rise to trivial surface states. The theoretical studies have revealed the presence of such surface states in addition to the topological ones [134, 135]. Thus a variety of group VIB MX_2 (M = Mo, W; X = S, Se, Te) compounds that can have a semiconducting H phase and a metallic T/T′ phase under the phase transition, typically the resistive random access memory (RRAM) device (such as $MoTe_2$-based devices) performance further improves by alloying the additional element, such as $Mo_{1-x}W_xTe$-based devices [133].

The typical crystal structure of ternary single-layer MSSe (M = Mo, W) is illustrated in Fig. 7.10a. Each unit cell consists of four M, four S, and four Se atoms,

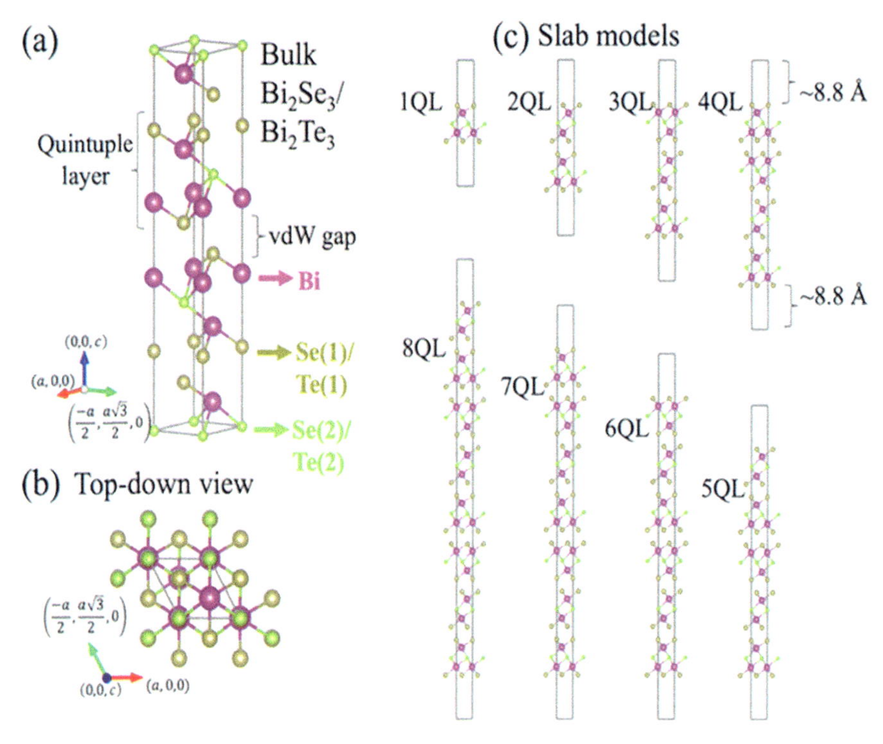

Fig. 7.8 **a, b** Schematic the bulk crystal lattice of Bi_2Se_3 and Bi_2Te_3 in a hexagonal system with space group R3m (166); **c** the slab models constructed from the bulk lattice with varied quintuple layers (adopted from Reid et al. [127])

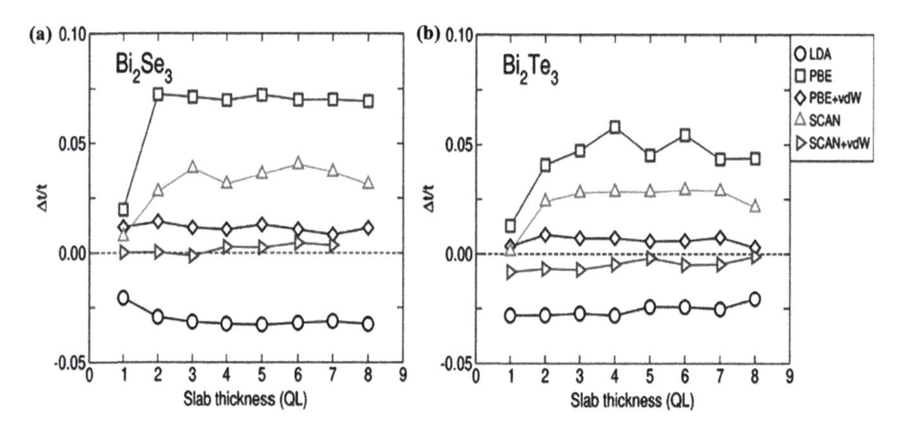

Fig. 7.9 **a, b** Represents the change in slab thickness in percent to bulk values for Bi_2Se_3 and Bi_2Te_3 (adopted from Reid et al. [127])

with the M layer sandwiched between the S and Se layers. This kind of stacking of the atomic layers may break their inversion symmetry. Through the charge in 1S′–MSSe it distributes unevenly in the top and bottom surfaces which gives rise to spontaneous polarization along the out-of-plane direction [136]. The calculated surface dipole moment value has been obtained 0.713 (0.754) Debye per unit cell for 1S′–WSSe (MoSSe). The top view of four- and eight-membered MSSe rings are also visualized, wherein basal planes seem to be structurally similar to most of the TMDs, as the corresponding bonding characters are illustrated in MSSe, in Fig. 7.10b. Whereas Fig. 7.10c represents the plot of the electrons localization function (ELF) map in a transverse plane comprising dual M, S, and Se atoms. This leads to the electron localizations being mostly dispersed at the M and chalcogen sites and surround the midpoints among them, to form a sharing behavior M-S/Se bonds. Whereas aside to the middles among M and chalcogen atoms a few electrons localization has also appears among two metal atoms which signify a chemical bond formation in adjacent two nearest metal atoms. The corresponding crystal orbital Hamilton population (COHPs) of the M-M bonds and their local densities of states at the metal centers (LDOS) is depicted in Fig. 7.10d. The relative energies of the four phases of monolayer MSSe are depicted in Fig. 7.10e. Moreover, typical electronic band structures of 1S′–MSSe and 1S′–WSSe are illustrated in Fig. 7.11a–d. Both ternary compounds' TIs electronic bands lies in the locality of the Fermi level predominantly made of the M-$d_z{}^2$ orbitals, associating other orbitals residing far to the Fermi level. In the absence of SOC, the valence and conduction bands degenerated at the Fermi level that overpass point lying nearby the Γ point (see Fig. 7.11a). Whereas including SOC the degenerated M-$d_z{}^2$ orbitals may splits in binary states, together the Fermi level that separates the valence and conduction bands (see Fig. 7.11b). This correlated to non-conducting states of 1S′–WSSe and MoSSe associating a global band gap 38 and 17 meV. It is also noted that the 1S′–MoSSe has a feebler SOC forte of Mo contrast to W leading to the lesser gap opening, however, their band dispersions are almost similar. Moreover, the calculated band of the para elastic phase-p-WSSe (because this phase is similar for both TIs) with and without SOC is illustrated in Fig. 7.11c, d. It leads to the p-WSSe being considered to be a gapless semiconductor for the absence of SOC, it altered in an non-conducting state by governing the SOC. Additionally, the bands of p-WSSe close to the Fermi level predominately constitution of the W-$d_z{}^2$ orbitals. While excluding the SOC these two bands are participated in the band intersection, therefore, a parabolic character in p-WSSe, however, it could be linear for the 1S′ –WSSe. Their intersection points exist precisely at or in its place nearby to Γ point. Such inconsistencies are straightly linking to the lattice misrepresentation in p-WSSe. Surprisingly in the transformation of ferroelastic into a para elastic state, their band crossing 1S′ –WSSe shifts to Γ point. Similarly, the value of the Z_2 invariant parameter has been also determined for these ternary systems [136].

Moreover, day-by-day innovations in the area of TIs have revealed a new class of ternary TMDs that reflects a smoother phase transition between Type-I and Type-II Weyl fermions, with the improved physical properties in terms of the technical devices applications, it is namely known as MM'X_4 family (M= V, Nb, or Ta; M' = Co, Rh, or Ir; and X = Se or Te) [137–139]. Using the band gap engineering has been

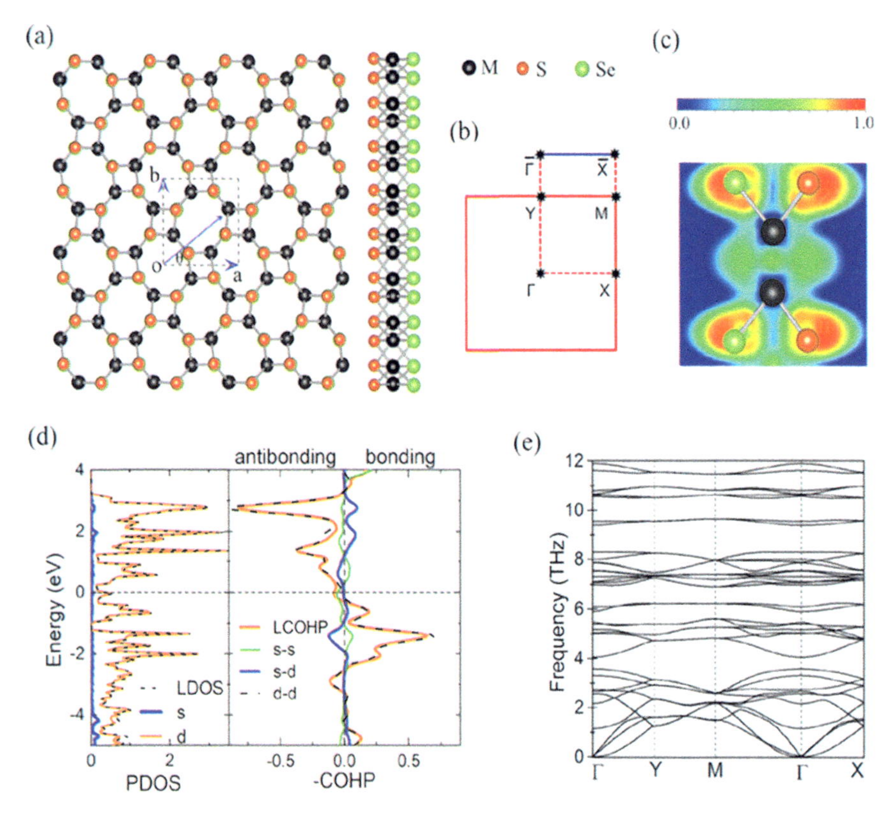

Fig. 7.10 **a** Crystal structures of $1S'$–MSSe from the top (left panel) and side (right panel) views, the black square marking the primitive cell; **b** 2D and projected one-dimensional Brillouin zones with high symmetry points; **c** ELF for $1S'$–WSSe; **d** PDOS and COHP analysis for the W-d orbital projections of the nearest-neighbor W-W interactions. LDOS and LCOHP stand for the total DOS and the total COHP at the respective metal atoms; **e** the phonon spectra for $1S'$–WSSe (adopted from Ma et al. [20])

achieved various ternary and other multielement systems such as $MoS_{2(1-x)}Se_{2x}$, $WS_{2x}Se_{2(1-x)}$, $Mo_{(1-x)}W_xS_2$, $ReS_{2x}Se_{2(1-x)}$, $MoW_{1-x}S_{2y}Se_{2(1-y)}$, etc. through adjusting the chemical composition of ionic or anionic elements [137, 139–144, 144, 144, 144–146]. Specifically, this family makes nine possible ternary combinations of group IV B transition metals (Ti, Zr, and Hf) with chalcogen elements (S, Se, and Te) in the monoclinic structure [147], however, other families of the MM'X_4 have been also interpreted in various studies, as listed in Table 7.1 [137].

The first-principles computations based on the density functional theory monolayer of group IVB transition metals (Ti, Zr, and Hf) with chalcogen elements (S, Se, and Te) are also have nine possible ternary combinations in monoclinic structure with a chemical formula ABX_4, where A/B = Zr, Hf, or Ti and X = S, Se, or Te, as depicted in Fig. 7.12a. The relaxed atomic structure of ABX_4 is similar to the original BAX_4, however, it can leave a total of nine distinct compounds.

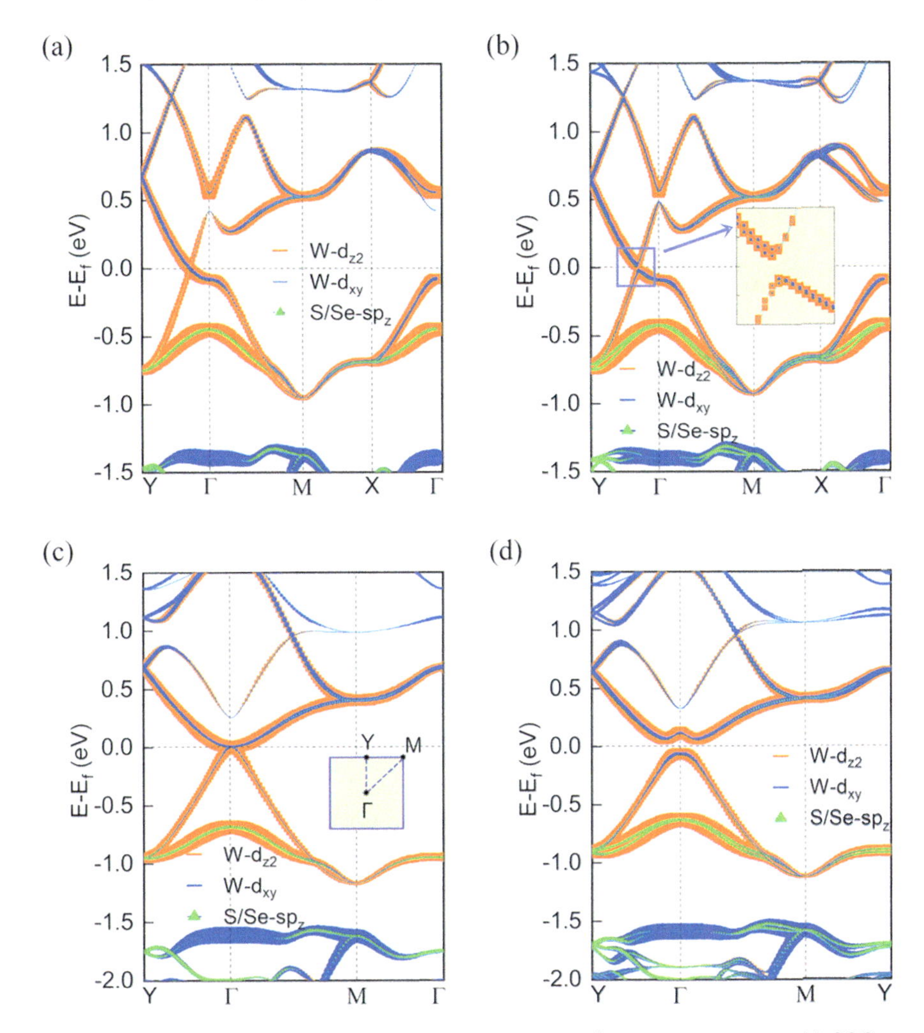

Fig. 7.11 **a, b** Representation of fat band structures for the 1S′–WSSe without and with SOC; **c, d** Fat band structures of SLWSSe in the para elastic state without and with SOC, to this the Fermi level set 0 eV (adopted from Ma et al. [20])

The corresponding lattice parameters and energy differences between the 1T and 2H structures are listed in Table 7.2. Moreover, the dynamical stability of monolayer ABX_4 is usually defined from the phonon dispersion calculations. Like all the ternary transition metal telluride monolayers ($HfTiTe_4$, $HfZrTe_4$, and $ZrTiTe_4$) have positive frequency modes that directly correlated to their dynamical stability, as depicted in Fig. 7.12b–d. Whereas the ternary transition metal sulfides ($HfTiS_4$, $HfZrS_4$, and $ZrTiS_4$) and ternary transition metal selenides ($HfTiSe_4$, $HfZrSe_4$, and $ZrTiSe_4$) are possessed small imaginary frequencies for the lowest acoustic phonons, the rest of the phonons may positive. To examine the electronic band structure of such ternary

Table 7.1 The magnetic states, equilibrium lattice parameters, band gaps, and topological invariants (Z_2 or Chern number) of MM'X_4 (M = Ta, Nb, or V; M' = Ir, Rh, or Co and X= Se or Te) compounds computed with GGA and HSE06 hybrid functionals. Here Chern numbers are indicated by 'C', however, it is Z2

Systems	Magnetic state	a (Å)	b (Å)	Band gap (eV)		Z_2/Chern		Net magnetic moment (μ_B)
				GGA	HSE	GGA	HSE	
$VCoSe_4$	NM	3.354	11.336	−0.058	0.019	1	0	0.000
$VIrSe_4$	NM	3.512	11.622	0.112	0.481	0	0	0.000
$VRhSe_4$	AFM	3.455	11.713	0.110	0.666	0	0	0.000
$NbCoSe_4$	NM	3.456	11.578	−0.034	−0.035	1	1	0.000
$NbIrSe_4$	NM	3.616	11.741	0.055	0.569	0	0	0.000
$TaIrSe_4$	NM	3.622	11.772	0.144	0.675	0	0	0.000
$NbRhSe_4$	NM	3.561	11.839	0.0004	0.201	0	0	0.000
$TaRhSe_4$	NM	3.597	11.874	0.087	0.544	0	0	0.000
$TaCoSe_4$	NM	3.446	11.533	−0.034	0.101	1	0	0.000
$TaCoTe_4$	NM	3.694	12.395	0.0817	0.135	1	1	0.000
$NbIrTe_4$	NM	3.795	12.543	0.040	0.048	1	1	0.000
$TaIrTe_4$	NM	3.807	12.543	0.027	0.281	1	0	0.000
$TaRhTe_4$	NM	3.781	12.624	0.068	0.072	1	1	0.000
$NbRhTe_4$	NM	3.763	12.606	0.054	0.161	1	1	0.000
$VIrTe_4$	FM	3.740	12.402	Metal	Metal	C = 0	C = 0	−2.79
	NM	3.743	12.465	−0.016	0.440	1	0	0.000
$VCoTe_4$	FM	3.626	12.224	Metal	Metal	C = 2	C = 0	2.46
	NM	3.615	12.202	0.066	0.051	1	1	0.000
$VRhTe_4$	FM	3.691	12.496	Metal	Metal	C = 1	C = 1	1.68
$NbCoTe_4$	NM	3.673	12.366	0.008	0.082	1	1	0.000

The non-magnetic and ferromagnetic states are indicated as NM and FM [137]

monolayers without and with the SOC were taken. Without SOC $HfTiS_4$, $HfZrS_4$, $ZrTiS_4$, and $HfZrSe_4$ have been recognized as normal semiconductors possessing reasonable band gaps; however, other compositions such as $ZrTiS_4$, $HfZrS_4$, and $HfZrSe_4$ are categorized as the indirect semiconductors.

On the other hand, the band structures of monolayers $HfTiSe_4$, $HfTiTe_4$, $ZrTiSe_4$, $ZrTiTe_4$, and $HfZrTe_4$ are also have nontrivial topological properties with semimetallic features without and with the SOC, without SOC their valence and conduction bands become degenerate at the Γ-point in all nontrivial compounds near to Fermi level. While in the case of including SOC lifts the degeneracy and opens a finite gap at the Γ point in which TRS is preserved, this indicates a clear signature of a band inversion. In addition to this Z_2 index has also achieved 1 which confirms their nontrivial topology. Moreover, the band structure calculation from HSE06 has

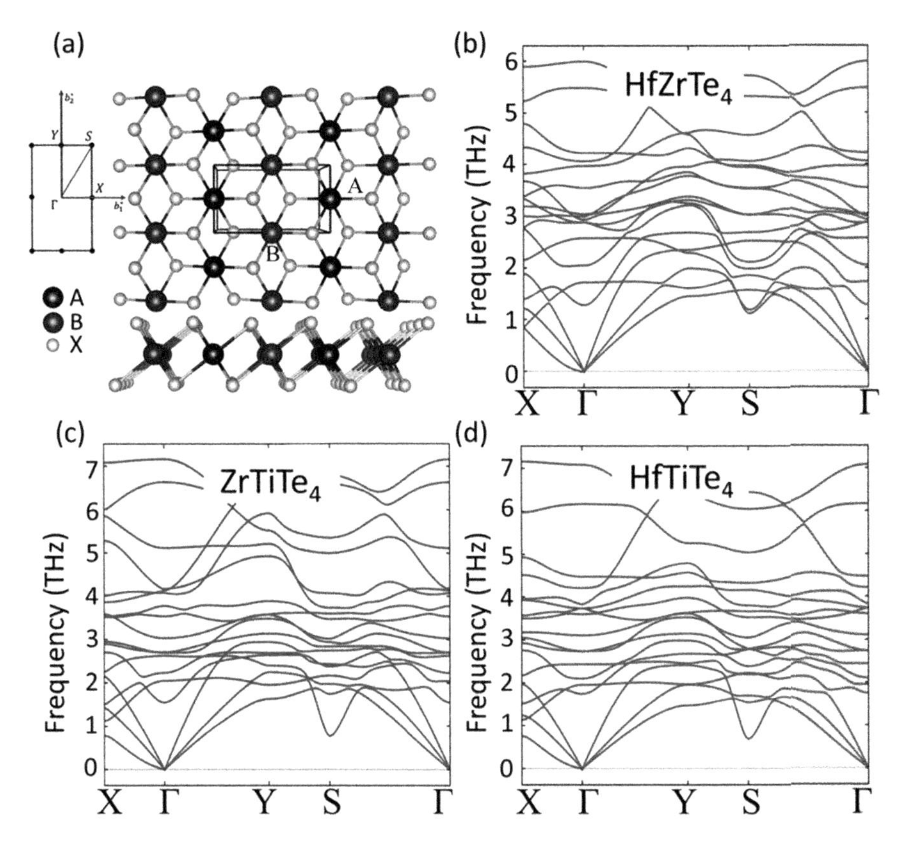

Fig. 7.12 **a** Top and side views of ABX_4 (A and B =Zr, Hf, or Ti and X = S, Se, and Te) compounds and the first Brillouin zone labeled with specific high symmetry points; **b–d** the phonon band structures for the $HfZrTe_4$, $ZrTiTe_4$, and $HfTiTe_4$ monolayers (adopted from Macam et al. [147])

Table 7.2 Calculated equilibrium lattice parameters and the energy differences between the 1T and 2H structures [147]

ABX_4	1T		2H		$\Delta E_{1T\text{-}2H}$ (eV)
	an (A°)	b(A°)	an (A°)	b (A°)	
$HfTiS_4$	6.100	3.523	5.835	3.378	−0.018
$HfZrS_4$	6.314	3.644	6.082	3.510	−0.012
$HfTiSe_4$	6.307	3.683	6.130	3.551	−0.009
$HfZrSe_4$	6.518	3.762	6.360	3.672	−0.009
$ZrTiS_4$	6.141	3.544	5.856	3.389	−0.012
$ZrTiSe_4$	6.365	3.672	6.151	3.564	−0.008
$ZrTiTe_4$	6.677	3.857	6.632	3.840	−0.005
$HfTiTe_4$	6.672	3.856	6.618	3.831	−0.007
$HfZrTe_4$	6.840	3.956	6.762	3.904	−0.006

revealed the $HfTiSe_4$ and $ZrTiSe_4$ do not maintain their nontrivial phase and become trivial insulators ($Z_2 = 0$) with the direct and indirect band gaps at 381meV and 338meV, whereas the $HfTiTe_4$, $HfZrTe_4$, and $ZrTiTe_4$ remains nontrivial and maintained their semimetallic features. The orbital analysis of the nontrivial compounds $HfTiTe_4$, $HfZrTe_4$, and $ZrTiTe_4$ have also revealed that non-SOC CBM and VBM lie near the Fermi level and mainly consist of Ti-$d_z{}^2$ and Te-p_x; p_y orbitals in $ZrTiTe_4$ and $HfTiTe_4$, while Zr-$d_z{}^2$ and Te-p_x; p_y orbitals in $ZrTiTe_4$, as depicted in Fig. 7.13a–c. The incorporation of SOC opens a local band gap around Γ and a p-d band inversion occurred [147].

Similarly, other nontrivial non-magnetic materials families are also demonstrated such as ($MM'X_4$, M = Ta, Nb, or V; M'= Ir, Rh, or Co; X = Se or Te), CaM_2X_2 (M = Zn or Cd, X = N, P, As, Sb, or Bi). Typically electronic band structures without and with the SOC of $CaCd_2Bi_2$ is

illustrated Fig. 7.14a, b. The structure of $BaCu_2S_2$ and high symmetry points and Dirac points for the $CaCd_2Bi_2$ and $CaCd_2SbBi$ are represented in Fig. 7.15a, b, in which along the high symmetry path it is categorized as a Dirac semimetal without SOC, it can turn out into a topological insulator after including SOC effects. Overall investigations with the ternary TIs are still challenging and continue to investigate for novel properties.

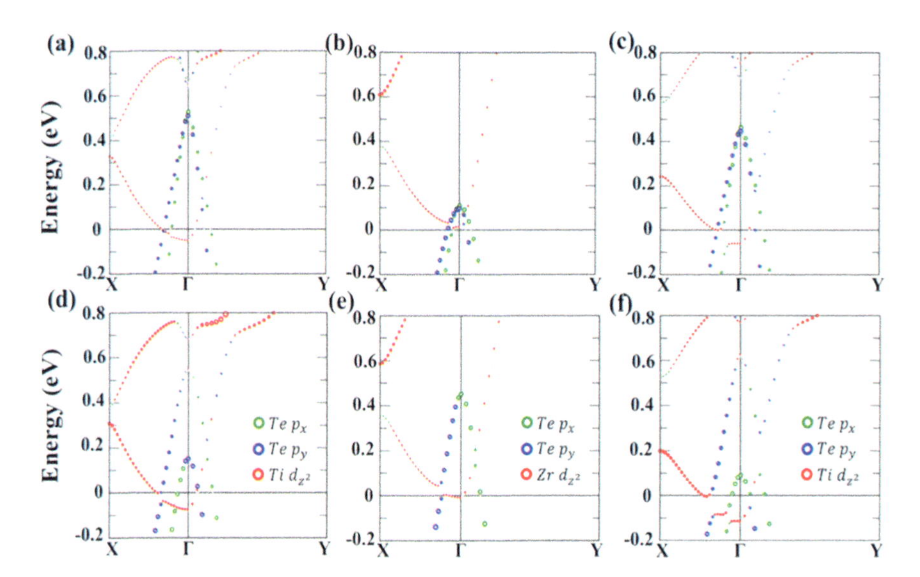

Fig. 7.13 Band structures with the various orbitals in $ZrTiTe_4$, $HfZrTe_4$, and $HfTiTe_4$ monolayers along X, U, and Y high symmetry points using HSE06 hybrid functional: **a, d** monolayer $ZrTiTe_4$ without and with SOC; **b, e** monolayer $HfZrTe_4$ without and with SOC; **c,f** monolayer $HfTiTe_4$ without and with SOC. The sizes of circles are proportional to the partial DOS from different orbitals that follow, the red circles for the $d_z{}^2$ orbital of B atoms, green circles for the p_x-orbital of X atoms, and blue circles for the pyorbital of X atoms (adopted from Macam [147])

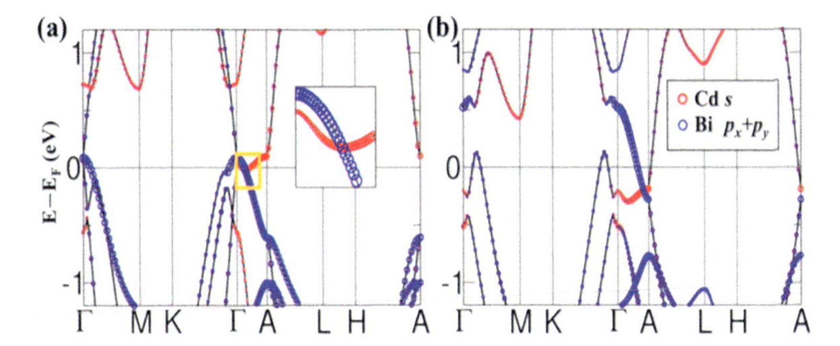

Fig. 7.14 **a, b** Band structures of bulk $CaCd_2Bi_2$ without and with SOC. The circle corresponds to the orbital contribution from Cd s orbital and Bi p_x and p_y orbitals, to identify the bands near Γ point (adopted from Feng et al. 2022 Sci. Rep. 12, 4582) [148]

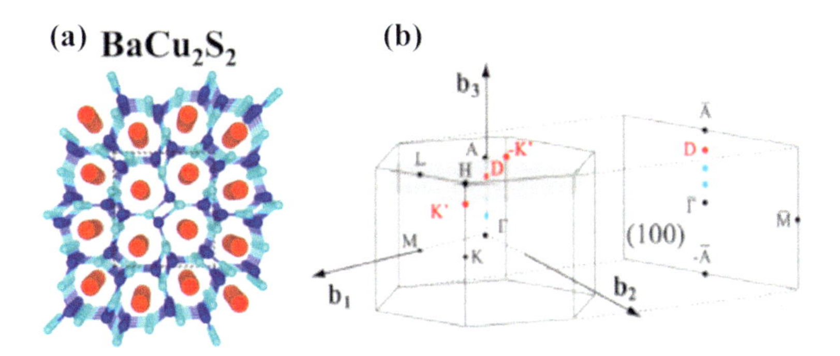

Fig. 7.15 **a** The possible bulk structure of AM_2X_2 for the $BaCu_2S_2$ (Pnma) Zintl phase; **b** The corresponding 3D first Brillouin zone (BZ) projected surface BZ at (100) plane. The high symmetry points and Dirac points (red for $CaCd_2Bi_2$ and blue for $CaCd_2SbBi$) are labeled (adopted from Feng et al. 2022 Sci. Rep. 12, 4582)

7.3.3 Crystal Structure

7.3.3.1 Monolayer TIs Crystal Structure—Theoretical Aspect

The monolayer TMDs is composed of an inner layer of metal M atoms ordered on a triangular lattice that is sandwiched between two layers of chalcogen X atoms located on the triangular lattice of alternating hollow sites in a triangular prismatic fashion. In which the X-M-X layers are bonded together by weak van der Waals forces, as the top view of MX_2 TMD is depicted in Fig. 7.16a. The corresponding triangular Bravais lattice is spanned by the basis vectors as;

$$\vec{R}_1 = (a, 0, 0), \quad \vec{R}_2 = \left(\frac{a}{2}, \frac{\sqrt{3}}{2}a, 0\right) \tag{7.21}$$

where a is lattice constant. The coordinates of the nearest-neighbors of a Mo atom are typically obtained as illustrated in Fig. 7.16b. Here θ represents the angle between the M-X bond and the M plane. The experimental values of both a and θ are summarized in Table 7.3. The reciprocal lattice, concerning the triangular Bravais lattice, is depicted in Fig. 7.16c, which can be defined with help of the following spanned vectors;

$$\vec{b}_1 = \frac{4\pi}{\sqrt{3}a}\left(\frac{\sqrt{3}}{2}, -\frac{1}{2}, 0\right), \quad \vec{b}_2 = \frac{4\pi}{\sqrt{3}a}(0, 1, 0) \tag{7.22}$$

The first Brillouin zone of TMD is hexagonal and corresponding high symmetry points at Γ, K, M are expressed as;

$$\Gamma = (0, 0), \quad K = \left(\frac{4\pi}{3a}, 0\right), M = \left(\frac{\pi}{a}, \frac{\sqrt{3}\pi}{3a}\right) \tag{7.23}$$

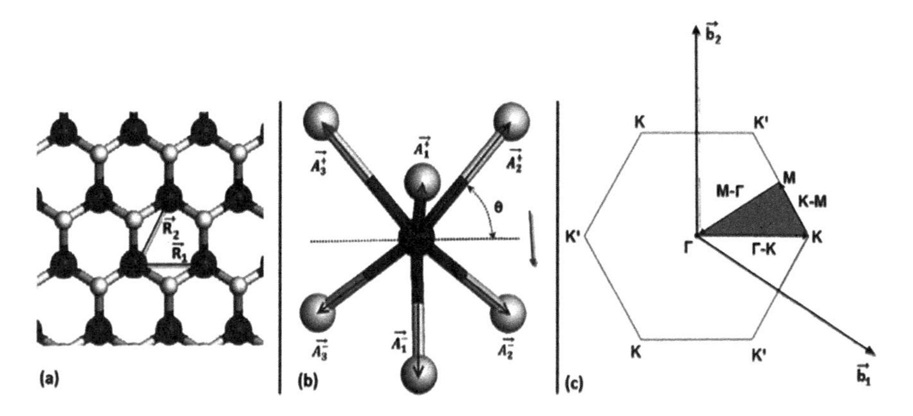

Fig. 7.16 **a** Top view of monolayer transition metal dichalcogenides MX_2; **b** Sketch of the atomic structure of the monolayer TMDs, in which six vectors $\vec{A}_i^{\pm}$ connected to nearest-neighbor *M* and *X* atoms with $i = 1, 2, 3$, separated by a distance $l = a/\sqrt{3}\cos\theta$ nm, *a* is lattice constant and *θ* is the angle between *M-X* bond and the *M* plane; **c** the first Brillouin zone and high symmetry points *Γ*, *K*, and *M* of TMDs in reciprocal space of the triangular lattice. Its primitive lattice vectors are $\vec{b}_1$ 1 and $\vec{b}_2$ (adapted from Dias et al. [149])

Table 7.3 Lattice constant and angle *θ* between the *M-X* bond and the *M* plane used in our calculations [149]

	$MoSe_2$	MoS_2	$MoTe_2$	WSe_2	WS_2
an (A°)	3.288	3.166	3.519	3.282	3.1532
θ (rad)	0.710	0.710	0.710	0.710	0.710

The outermost shells contain Mo and S atoms that belong to 4d and 3p orbitals. Whereas the subbands near the top of the valence bands and the bottom of the conduction bands predominantly construct by $d_z{}^2$, d_{xy}, $d_x{}^2{}_{-y}{}^2$. However, the d_{xz}, d_{yz} orbitals of molybdenum and p_x, p_y, p_z orbitals of sulfide including other inner orbitals usually occurred in subbands with higher energy. It directly revealed the contribution to the low-energy bands mainly due to $d_z{}^2$, d_{xy}, and $d_x{}^2{}_{-y}{}^2$ orbitals. The interactions between these d orbitals also plays an important role in the formation of low-energy subbands. Therefore, in contrast to tight-binding models the next-nearest-neighbor (NNN) interactions between M atoms theory could be considered more appropriate [149]. Aside from six nearest-neighbor (NN) S atoms (three on the top layer and the others on the bottom layer) $\vec{A}_i$ vectors ($i = 1, 2, 3$) of each Mo atom also interacts with six NN Mo atoms (see $\vec{S}_j$ and six NNN Mo-atoms, see $\vec{C}_j$, where $j = 1,...,6$), as illustrated in Fig. 7.17 [149]. It has also been noted that the two S atoms sit on top of each other, the corresponding coordinates of hopping vectors are summarized in Table 7.4.

7.3.3.2 Experimental Aspects

Experimentally topological structures of the TMDs are possibly generated by a variety of the synthesis process and transfer of the specimens and manipulation through polymer auxiliaries. Such as the significant structural formation of wrinkles, folds, and scrolls established by TMDs.

Wrinkled TMDs

Wrinkles are the utmost frequent topology in 2D materials that impulsively constructed by the thermodynamic and mechanical inconstancy in such materials [150, 151]. The TMDs wrinkles are also possible by following the synthesis and transfer steps [26, 27, 152–161]. Wrinkles are mostly explored from chemical vapor deposition (CVD) fabricated 2D materials/ or devices for different applications [162–164]. The single or few-layer 2D materials can be mechanically obtained to the unpackaged materials. In the process of micromechanical extraction, wrinkles are arbitrarily shaped due to the inhomogeneous contravention strengths in the layered materials. It has been also noted that when layered material's lengths and widths exceed 10 nm then prone wrinkles are produced [165, 166]. Similarly, releasing tape at a slower speed also gives us smaller flakes with fewer wrinkles [26]. However, the bending and wrinkling of MoS_2 have been attributed to its imperfect delamination to the substrates. The sizes of the wrinkles also depend on their interfacial adhesion. The typical wrinkle possessed height 43 nm and width 0.8 m as illustrated in Fig. 7.18a. The interfacial adhesion energies among MoS_2 and the substrates during transferring also generate wrinkles from the mechanical exfoliation of the SiO_2 and Si_3N_4 substrates, as depicted in Fig. 7.18b, c [158]. The interaction among the bending energy of the MoS_2 flake and interface adhesion energy also influence the creation

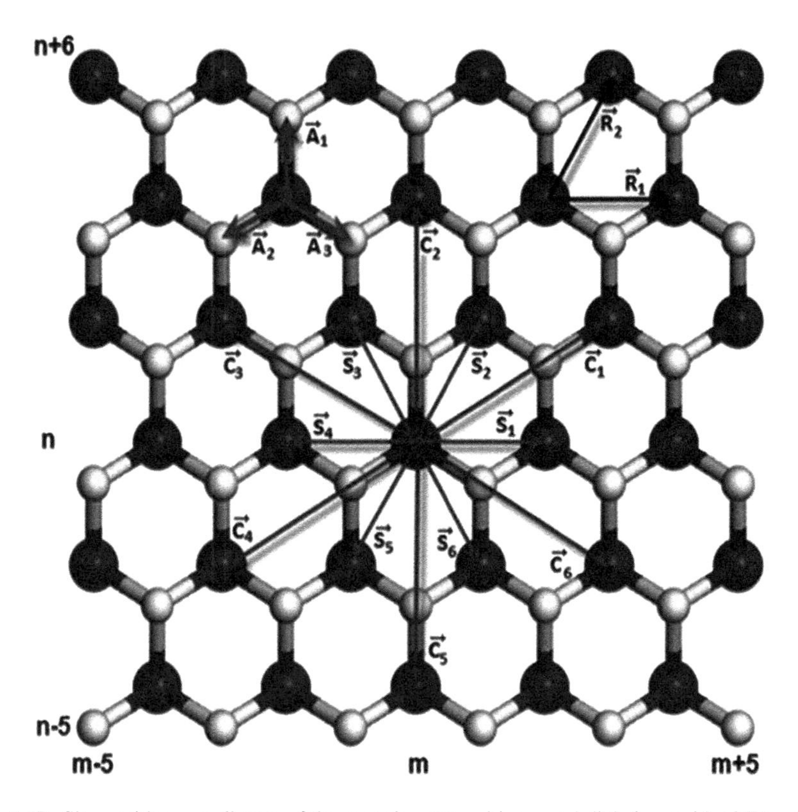

Fig. 7.17 Site-position coordinates of the monolayer transition metal dichalcogenides MX_2. dark and light spheres denote the metal (*M*) and chalcogenide (*X*) atoms, respectively. The arrows represent the nearest-neighbor ($\vec{S}_i$) and next-nearest-neighbor ($\vec{C}_i$) *M*-*M* hopping vectors, respectively, where $i = 1, \ldots, 6$. The arrows indicate the nearest-neighbor *M*-*X* hopping vectors ($\vec{A}_j$) with $j = 1, \ldots,$ 3.. Note that in actuality each of them is composed of two vectors $\vec{A}_j^{\pm}$, with + (−) corresponding to the *X* atoms in the upper (lower) plane, and $\vec{R}_1$ and $\vec{R}_2$ are basis vectors (adopted from Dias et al. [149])

of wrinkles. Typical adhesion energy (γ) has been achieved from the parameters width of the MoS_2 flake (t), amplitude (A), and wavelength (λ) of the wrinkles, by using the following relationship;

$$\gamma = \frac{\pi^4 E t^3 A^2}{6\lambda^4} \tag{7.24}$$

here E is Young's modulus of the MoS_2 flake. Moreover, the wrinkles are also generate as Gaussian-shaped nonetheless it bigger to ripples, as depicted in Fig. 4.18d [159, 167].

The CVD growth process is predominantly have two kinds of origins, the first one is wrinkles-rough growth on a substrate surface and the second one is the thermal

Table 7.4 Where the hopping vectors are used coordinates relative to the site $(m, m).\overrightarrow{A_i}$ with the nearest-neighbor *M-X* hopping vectors with $i = 1 \ldots, 3$, $\overrightarrow{S_i}$ and $\overrightarrow{C_i}$ corresponds to the nearest-neighbor and next-nearest-neighbor *M-M*/*X-X* hopping vectors [149]

Vector	Hopping	Coordinates
$\overrightarrow{A}_1^{\pm}$	$(m, n) \to (m, n+1)$	$l(0, \cos\theta, \pm\sin\theta)$
$\overrightarrow{A}_2^{\pm}$	$(m, n) \to (m-1, n-1)$	$l(-\frac{\sqrt{3}}{2}\cos\theta, -\frac{1}{2}\cos\theta, \pm\sin\theta)$
$\overrightarrow{A}_3^{\pm}$	$(m, n) \to (m+1, n-1)$	$l(\frac{\sqrt{3}}{2}\cos\theta, -\frac{1}{2}\cos\theta, \pm\sin\theta)$
$\overrightarrow{S}_1$	$(m, n) \to (m+2, n)$	$a(1,0, 0)$
$\overrightarrow{S}_2$	$(m, n) \to (m+1, n+2)$	$a(\frac{1}{2}, \frac{\sqrt{3}}{2}, 0)$
$\overrightarrow{S}_3$	$(m, n) \to (m-1, n+2)$	$a(-\frac{1}{2}, \frac{\sqrt{3}}{2}, 0)$
$\overrightarrow{S}_4$	$(m, n) \to (m-2, n)$	$a(-1,0, 0)$
$\overrightarrow{S}_5$	$(m, n) \to (m-1, n-2)$	$a(-\frac{1}{2}, -\frac{\sqrt{3}}{2}, 0)$
$\overrightarrow{S}_6$	$(m, n) \to (m+1, n-2)$	$a(\frac{1}{2}, -\frac{\sqrt{3}}{2}, 0)$
$\overrightarrow{C}_1$	$(m, n) \to (m+3, n+2)$	$d(\frac{\sqrt{3}}{2}, \frac{1}{2}, 0)$
$\overrightarrow{C}_2$	$(m, n) \to (m, n+4)$	$d(0, 1, 0)$
$\overrightarrow{C}_3$	$(m, n) \to (m-3, n+2)$	$d(-\frac{\sqrt{3}}{2}, \frac{1}{2}, 0)$
$\overrightarrow{C}_4$	$(m, n) \to (m-3, n-2)$	$d(-\frac{\sqrt{3}}{2}, -\frac{1}{2}, 0)$
$\overrightarrow{C}_5$	$(m, n) \to (m, n-4)$	$d(0, -1, 0)$
$\overrightarrow{C}_6$	$(m, n) \to (m+3, n-2)$	$d(\frac{\sqrt{3}}{2}, -\frac{1}{2}, 0)$

enlargement during the cooling owing to miss-match lattice coefficients among the growing substrates and TMDs. As the intense, the monolayer of MoS_2 over an Au foil and it transfers to a target substrate, such as SiO_2/Si [168]. This led that some wrinkles which have been noticed on the SiO_2/Si substrate, as depicted in Fig. 7.18e. Moreover, the wrinkle is also formed by altering the surface structures of the used grown substrates. Using this approach a large area wrinkled of MoS_2 has been synthesized on graphene substrates; it can also adjust by the graphene width, grain edge and nucleation density, as illustrated in Fig. 4.18f [154]. The wrinkled MoS_2 is also produce from the patterned sapphire substrates. Likely to CVD, the high vacuum co-evaporation technique produced wrinkled $MoSe_2$ thin films over spotted constructions, their lateral periodicity has been also tuned with the film thickness, as depicted in Fig. 4.18g [153]. Moreover, the self-assembled wrinkle TMDs could have a superior governability in the growth direction and periodicity through transfer on pre-strained substrates. As the intense, $ReSe_2$, MoS_2, and CdS thin films have developed from the chemical bath deposition (CBD) approach and it transferred over to distinct pre-stretched elastomeric substrates, as depicted in Fig. 4.18h–k [27, 152, 169]. Therefore, the wrinkled microstructures are produced a controlled direction,

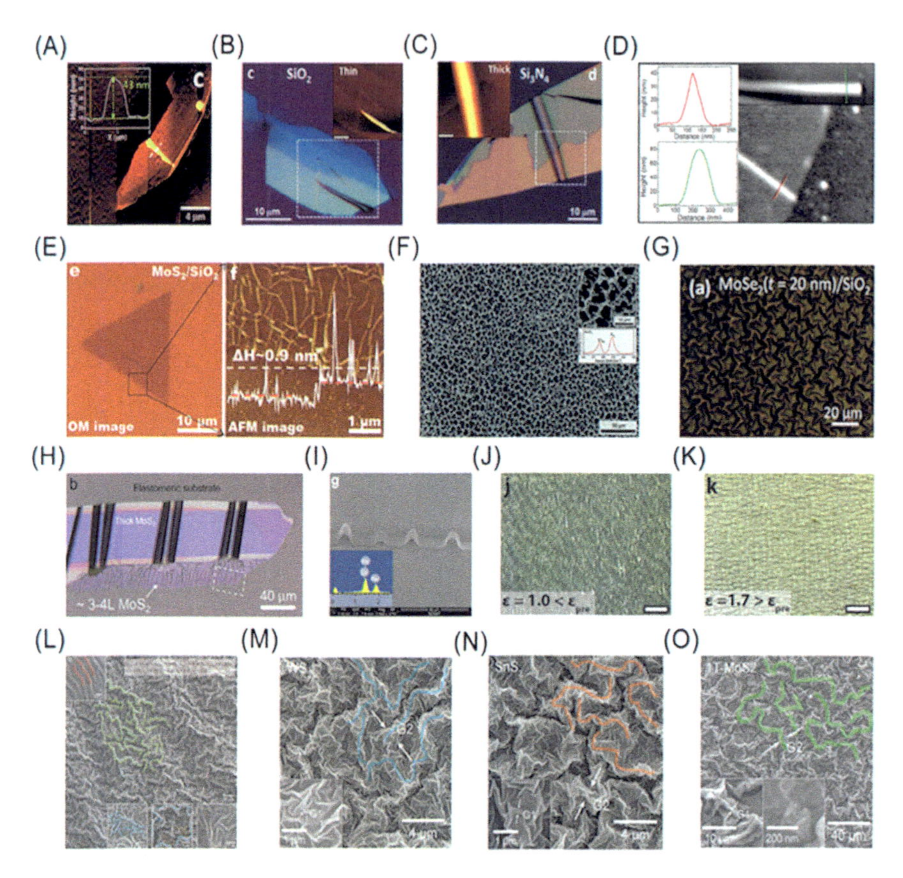

Fig. 7.18 **a** Wrinkled MoS_2 by mechanical exfoliation on a SiO_2 substrate; **b, c** optical images of wrinkled MoS_2 flakes on SiO_2 and Si_3N_4; **c** substrates, inset: respective atomic force microscopy (AFM) pictures; **d** AFM pictures of Gaussian-shaped wrinkles on exfoliated WSe_2 single-layers, inset: relevant profiles alongside the dashed lines; **e** Wrinkled MoS_2 on SiO_2/Si substrate and corresponding AFM image; **f** SEM pictures of wrinkled MoS_2 manufactured over graphene, inset: Raman spectra of the wrinkled MoS_2 films; **g** optical picture of wrinkled $MoSe_2$ developed over the substrate; **h, i** wrinkles on MoS_2 and $ReSe_2$, flakes produced by liberating pre-stretched substrates. **j, k** optical pictures of the mechanical govern CdS film microstructure by commanding pre-strain on polydimethylsiloxane substrates at $\varepsilon < \varepsilon_{pre}$, J, and $\varepsilon > \varepsilon_{pre}$; (**l–o**) scanning electron microscope pictures of the hierarchical wrinkles using sacrificial layers: wrinkled MoS_2, WS_2, SnS_2, and 1T-MoS_2, O flakes (adopted from Zhang et al. [14])

such as wave length and amplitude afterward liberating the strain. By employing this methodology, the GaSe stretchable devices have been constructed by transporting them over pre-strained elastomeric substrates for the creation of GaSe wrinkles [155]. The heating thermoplastic substrates associating TMDs could also consider alternative impassive approach to harvest wrinkled TMDs. The hierarchical 2H-MoS_2, 1T-MoS_2, WS_2, and SnS_2 wrinkled thin films on substrates are also achievable by the repetitive heating process, as illustrated in Fig. 7.18l–o. Thus different kinds of

Table 7.5 The fabrication process, the figure of merit, and controllability of wrinkled TMDs [14]

Fabrication process	Materials	Figures of merit	Controllability
Mechanical exfoliation	MoS_2	Adhesion energy of MoS_2 thin films affected by wrinkles	Uncontrollable
Mechanical exfoliation	MoS_2	Realignment of band structure by the wrinkle	Uncontrollable
CVD synthesis	MoS_2 on graphene	Wrinkle dimension influenced by underlying graphene	Controllable
CBD on pre-stretched soft surfaces	CDs	Reversible control of thin film microstructures	Controllable
Pre-strained elastomeric substrates	MoS_2	Local strain engineering and the funnel effect	Controllable
Mechanical exfoliation	WSe_2	Strain-tunable single photon sources	Uncontrollable
High-vacuum co-evaporation	$MoSe_2$	Formation of wrinkle pattern by etching or gold deposition	Controllable
Pre-stretched elastomeric substrates	$ReSe_2$	Tunable optical, magnetic, and electrical properties, and the funnel effect	Controllable
Pre-strained PDMS	GaSe	Strain sensors	Controllable
Thermally-induced shrinkable Substrates	MoS_2, WS_2, SnS_2, 1T-MoS_2	Universal method for creating hierarchical wrinkles	Controllable

wrinkled have been successfully formed by using different approaches, as listed in Table 7.5.

Folded TMDs

In recent year's generation of folded TMDs have attended great attention that can be developed from being random to controllable. For instance, the fold structures of the MoS_2 exfoliated samples [170]. While releasing the pre-stretched substrates of MoS_2 monolayer it can generate wrinkles, further, it collapses into folds owing to reduced bending rigidity, as depicted in Fig. 7.19a, b. TMDs growth from the CVD method also produces folds during the transferring procedure [85]. As the typically poly (methyl methacrylate) (PMMA) film has used as the charge carrier layer and it further liquefied in acetone. Their slight portion has been stripped from the substrate to get self-assembled folded TMDs, as illustrated in Fig. 7.19c. To get self-assembled folds in a controlled way have been also interpreted with help of the 3D imaging folded boundaries from (scanning) transmission electron microscopy (STEM) [171]. The high-resolution scanning transmission electron microscopy (HRSTEM) annular

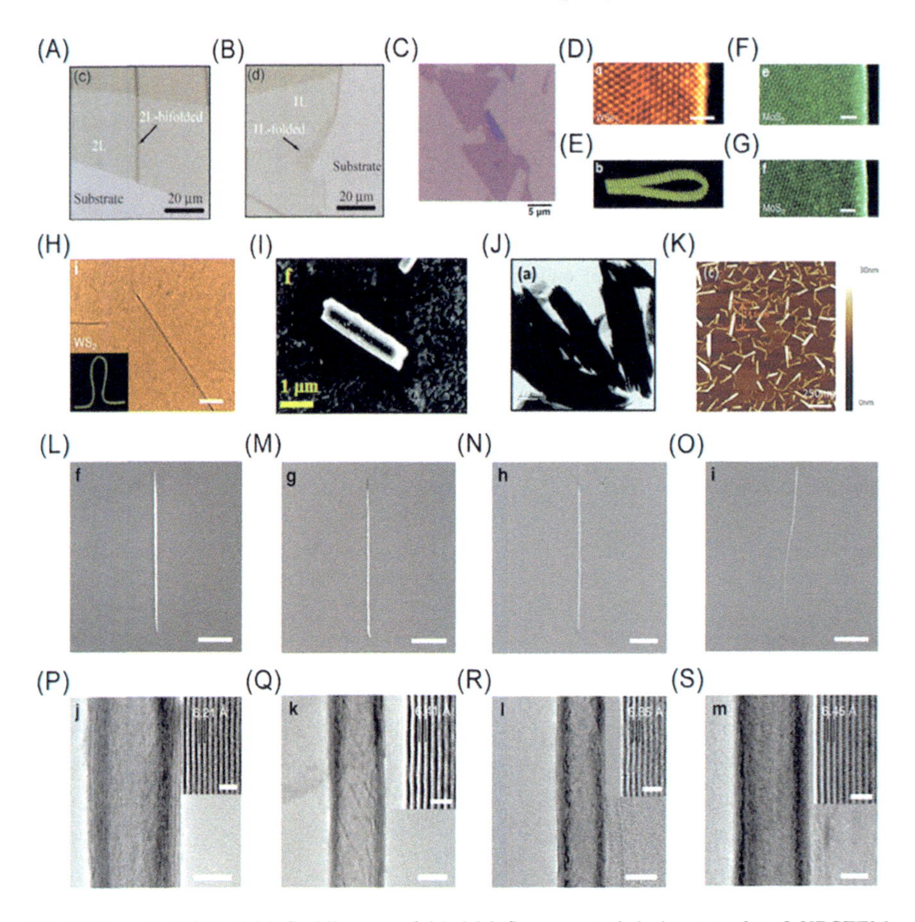

Fig. 7.19 **a, b** Bifolded MoS_2 bilayer; **c** folded MoS_2 generated during transfer; **d** HRSTEM annular dark-field (ADF) **picture** for an armchair WSe_2 fold; **e** atomic model for an armchair WSe_2 fold; **f** original ADF picture for a chiral MoS_2 fold; **g** detached ADF picture of a chiral MoS_2 fold; **h** transmission electron microscope (TEM) picture of WS_2 fold, inset is a model: **i** SEMpicture of exfoliated MoS_2 nanoscrolls (NSs); **j** high-resolution TEM(HR-TEM) pictures of 1T@2H MoS_2 NSs by adding certain amounts of LCA; **k** MoS_2 NSs formation along the unglued boundaries due to a feeble argon plasma treatment: (**l–s**) scanning electron microscope (SEM) and TEMpictures of transition metal dichalcogenides NSs shaped by one drop of ethanol, MoS_2 NSs, WS_2 NSs, $MoSe_2$ NSs, and WSe_2 NSs (adopted from Zhang et al. [14])

dark-field (ADF) pictures to the WSe_2, MoS_2, and WS_2 folds, including respective atomic models are illustrated in Fig. 4.19d–h. It noted that the bending behavior has also followed the elastic shell model. Various methods to produce folded TMDs are summarized in Table 7.6.

Table 7.6 Fabrication methods, the figure of merit, and controllability for the folded and scrolled TMDs [14]

Fabrication process	Materials	Figures of merit	Controllability
Folded TMDs			
Collapse of wrinkles	MoS_2	Reduced interlayer coupling	Uncontrollable
Folding exfoliated monolayers	MoS_2	SHG anisotropy	Uncontrollable
Transfer process	MoS_2	The blue shift of the A-exciton peak, enhanced PL quantum yield	Uncontrollable
HR (S)TEM to reconstruct the 3D structures	WSe_2, MoS_2, WS_2	Bending behavior via linear elastic shell model	Controllable
Scrolled TMDs			
Exfoliation of bulk MoS_2 by supercritical fluids	MoS_2	PL in the range of 420–600 nm	Controllable
Ar-plasma treatment	MoS_2, WS_{2-x}, WSe_{2-x}	Sulfur-deficient catalysts	Controllable
One drop of ethanol	MoS_2, WS_2, $MoSe_2$, WSe_2	Increased electron mobility and stability	Controllable
Scrolling MoS_2 nanosheets at the edge via LCA	2H MoS_2, 1T MoS_2	Highly thermal-stable paramagnetism and phase conversion	Controllable

Scrolled TMDs

Scrolled TMDs fabrication is quite difficult because the mechanical forte and chemical constancy restricted the multiple bending [172]. To get a better controllability of the formed TMD nanoscrolls (NSs) requires an exterior supplementary. Like, For MoS_2 NSs fabrication via one-pot exfoliation of layered bulk MoS_2 by supercritical fluids in dimethylformamide, as the typical structure is illustrated in Fig. 4.19i [173]. It noted that the fabricated 2H MoS_2 can be scrolled by the interaction among amine functional groups and boundaries of MoS_2 nanosheets, as depicted in Fig. 7.19j. Further, the 1T metallic phase is produced by the bending strain after scrolling. Under the thermal action, the glide rolling of the sulfur plane is also considered to be additionally clear for the 1T@2H MoS_2-NSs. This way production of a maximum of 1T phase content can be achieved [174]. The addition of successive multilayer MoS_2 NSs also produces at the ends or grain boundaries of MoS_2 sheets from the action of argon plasma, as depicted in Fig. 4.19 k [175]. Their topmost layer sulfur atoms can be choosily detached by the plasma bombardment, thus scrolls production is also possible under in-plane tensile stress. Moreover, a preparing process of the TMD NSs from one drop of ethanol possessing closely 100% yield in 5 seconds has been also recognized [176]. By employing this method obtained morphologies of the scrolled TMDs are represented in Fig. 4.191–s. In this process production of NSs are possible by releasing a portion of TMDs flakes to the substrate through the ethanol intercalation, afterward, it is scrolled under inherent strain. Moreover, their lengthy and periodic NS arrays have been also constructed in form of ribbon arrays together well-defined thicknesses and directions using focused ion beam microscopy. Thus the studies are not limited to the above-discussed few cases, but also extensive investigations are still ongoing to fabricate various TMDs scrolled structures with high precision for cutting-edge technical applications [177], as the various fabrication methods for the different TMDs are listed in Table 7.6.

7.3.4 Band Structure in Normal State

TMD materials, such as MoS_2, $MoSe_2$, WS_2, $NbSe_2$, WS_2, can have superconducting or insulating sate even down to atomic thickness [178–182]. Significantly monolayer superconducting state of the MoS_2 and $NbSe_2$ [80–82, 183] are possessed extremely high in-plane upper critical field Hc_2, which is more than six times higher than the Pauli limit. Enhancement in Hc_2 due to the presence of the strong Ising SOC [80–82, 183]. Owing to this fact, the Ising SOC pins electron spins of the Cooper pairs in out-of-plane directions that prevents the electron spins from being aligned with external in-plane magnetic fields and hence protects the superconductivity.

As the instance, the typical monolayer of $NbSe_2$ has been formed by a layer of Nb atoms with triangular lattice sandwiched by two layers of Se atoms and a hexagonal lattice structure in the out-of-plane direction and their broken A−B sub lattice symmetry, as depicted in Fig. 7.20a. Their in-plane mirror symmetry along the y-direction has broken and gives rise to the Ising SOC that pins electron spins into the out-of-plane directions and splits in the energy bands [184]. Therefore, the band structure of monolayer $NbSe_2$ is described with the help of first-principle calculations by taking into account SOC, as illustrated in Fig. 7.20b. The band structure of $NbSe_2$ has very much similar to other 2D TMD monolayers, such as MoS_2, $MoSe_2$, WSe_2, WTe_2 [178–182, 185]. While unlike Mo- and W-based materials that are insulating intrinsically, the Nb atom has one less d-electron in the outermost shell than Mo and W atoms, thereby, the chemical potential of $NbSe_2$ lies in the valence band. Moreover, also computed the band splitting at the K points due to Ising SOC around 150 meV. Their band splits significantly at the Fermi energy, the Fermi surface is far away from the K points. Therefore, the Ising SOC at the Fermi energy also plays a crucial role to protect the superconductivity from the paramagnetic effects of in-plane magnetic fields.

Thus, to describe the superconducting (or insulating) properties of such materials a six-band tight-binding model has been addressed that consists of d_{z^2}, d_{xy}, and $d_{x^2-y^2}$ orbitals of the Nb atoms [185, 186]. The third-nearest-neighbor hoppings is generally explained from the band structure DFT calculations. Moreover, the tight-binding model for the $NbSe_2$ has been also discussed based on the valance band of

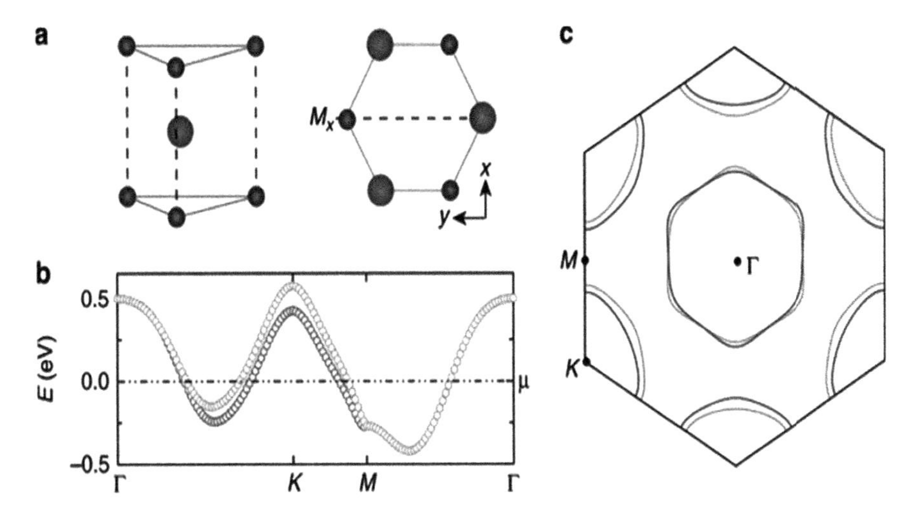

Fig. 7.20 **a** The lattice structure of monolayer $NbSe_2$ (left) and its top view (right), where M_x denotes the in-plane mirror symmetry; **b** band structure for the normal state of monolayer $NbSe_2$ as per DFT; **c** the energy contour at the Fermi level from tight-binding model (adopted from He et al. [84])

the monolayer and it is defined as;

$$c_k d_{z^2}, \uparrow, c_k d_{xy}, \uparrow, c_k d_{x^2-y^2}, \uparrow, c_k d_{z^2}, \downarrow, c_k d_{xy}, \downarrow, c_k d_{x^2-y^2}, \downarrow \tag{7.25}$$

The corresponding normal state tight-binding model up to the third-nearest-neighbor hopping is defined as;

$$H_N(k) = \sigma_0 \otimes H_{TNN}(k) + \sigma_z \otimes \frac{1}{2}\lambda L_z + \frac{1}{2}\mu BgH.\sigma I_3 \tag{7.26}$$

where

$$H_N(k) = \begin{pmatrix} V_0 & V_1 & V_2 \\ V_1^* & V_{11} & V_{12} \\ V_2^* & V_{12}^* & V_{22} \end{pmatrix}, L_z = \begin{pmatrix} 0 & 0 & 0 \\ 0 & 0 & -i \\ 0 & i & 0 \end{pmatrix} \tag{7.27}$$

where I_3 represents 3×3 identity matrix. The parameters V_0, V_1, V_2, V_{11}, V_{12}, and V_{22} are expressed in terms of $(\alpha, \beta) = \left(\frac{1}{2}k_x a, \frac{\sqrt{3}}{2}k_y a\right)$.

The specific parameters of the third-nearest-neighbor tight-binding model are usually fitted from the band structure by the first-principle, as summarized in Table 7.7. The band energy depicted in Fig. 7.20b, it reveals that the third-nearest-neighbor tight-binding model matches the first-principle calculation outcomes. Concerning monolayer $NbSe_2$, it has out-of-plane mirror symmetry, in which their S_z quantum number labels the two bands around the Fermi level. As a consequence,

Table 7.7 Fitting parameters for the Hamiltonian HTNN (k), the energy parameters ε_1 to λ are expressed in units of eV [84]

ε_1	ε_2	t_0	t_1	t_2	t_{11}	t_{12}	t_{22}	r_0	r_1
r_2	r_{11}	r_{12}	u_0	u_1	u_2	u_{11}	u_{12}	u_{22}	λ
1.4466	1.8496	−0.2308	0.3116	0.3459	0.2795	0.2787	−0.0539	0.0037	−0.0997
0.0385	0.0320	0.0986	0.0685	−0.0381	0.0535	0.0601	−0.0179	−0.0425	0.0784

the basis of $[c_{\mathrm{k}}, \uparrow, c_{\mathrm{k}}, \downarrow]$ the corresponding normal state Hamiltonian HN (k) is reduced to a 2 × 2 effective Hamiltonian $H_N^{eff} = \epsilon(k)\sigma_0 + \beta_{SOC}(k)\sigma_z$ with the energy dispersion $\epsilon(k)$ around the Fermi level and the Ising spin-orbit coupling $\beta_{SOC}(k)$. Moreover, at the K points, the Ising spin-orbit coupling $\beta_{SOC}(K) = \lambda$ reveals the band splitting around 150 meV. If the large Ising spin-orbit coupling, then in-plane upper critical field H_{c2}, it may strongly enhance and its theoretical value even goes higher than the experimental data.

Furthermore, the monolayer $NbSe_2$ that contains ripples, therefore, the Rashba spin-orbit coupling_ $\alpha_{\mathrm{R}}(k) = V_{\mathrm{r}}\left[\sqrt{3}\sin\beta\cos\alpha\sigma_x - (\sin2\alpha + \sin\alpha\cos\beta)\sigma_y\right]$ induces by the ripples [151] that are treated as a perturbation for the effective Hamiltonian $H_{\mathrm{N}}^{\mathrm{eff}}$ at the Fermi level. It affects the in-plane upper critical field H_{c2}. Generally, Rashba spin-orbit coupling competes with the Ising spin-orbit coupling; therefore, theoretical in-plane upper critical field H_{c2} is usually lowered to the experimental value. Moreover, the Fermi surfaces are constructed from the tight-binding models as depicted in Fig. 7.20c. At the Fermi level around the Fermi pockets Γ, K, and −K points, generally these bands split due to Ising SOC except to states lying along the Γ−M lines. The Fermi surfaces of $NbSe_2$ (see Fig. 7.20c) have been also recognized from the angle-resolved photoemission spectroscopy [83, 187]. Thus the normal band structure of topological insulators can be described with the help of tight-binding approximation.

7.3.5 Ising Superconductivity in Monolayer

Some of the 2D TMDs with the lattice structure 2H-MX_2 (M = transition metals; X = chalcogenides) have maintained a superconducting state up to a magnetic field of tens of teslas, it is the far beyond the Pauli limit [80, 164, 183, 188]. Therefore, to the absence of strong spin-orbital scattering an Ising pairing mechanism has been proposed based on the non-centrosymmetric structure which leads to an Ising-type SOC. The Ising SOC pins electron spins to the out-of-plane directions to make an in-plane upper critical magnetic field far beyond the Pauli limit. Considering this view many superconductors with huge critical magnetic fields have been investigated in those materials that have a broken inversion symmetry [189]. Typically it is classified into the following two categories.

7.3.5.1 2D Ising Superconductivity-Type-I

Since SOC is experienced by a moving electron with momentum k, that is proportional to $k \times E.\sigma$, where E is the electric field experienced by the electron and s denoted the Pauli matrices. In the case of 2D material, the motion of electrons are restricted to the plane. In the monolayer of 2H MX_2, their top view is due to a honeycomb lattice structure with a broken in-plane inversion symmetry to ensure electrons effectively should lie in-plane electric field, as depicted in Fig. 7.21a. Consequently, the electron spin is also strongly locked to the out of-plane orientation due to an effective Zeeman field and opposite directions of electrons with opposite momentum. Therefore, MX_2 monolayer electrons at K and K′ valleys experience the opposite effective Zeeman fields, as illustrated in Fig. 7.21b. This type of Zeeman field is commonly described as an Ising Zeeman field. However, in some superconducting 2H-MX_2, the Ising Zeeman field protects spins of electrons in the Cooper pairs from being realigned in-plane direction, leading to quite large in-plane upper critical fields that exceed several times of Pauli limit [189].

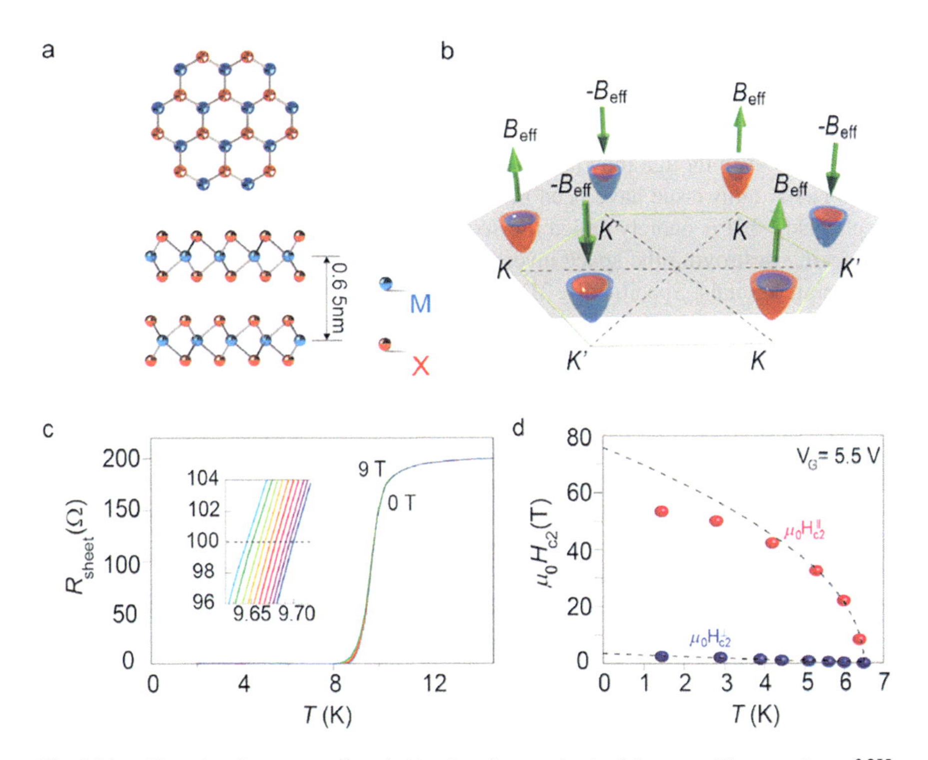

Fig. 7.21 **a** Top view (upper panel) and side view (lower view) of the crystalline structure of 2H-MX_2; **b** conduction band electron pockets near the K and K′ points in the hexagonal Brillouin zone in monolayer 2H-MX_2; **c** temperature-dependent resistance for the MoS_2-EDLT device under in-plane magnetic fields; **d** in-plane and out-of-plane upper critical fields as a function of temperature (adopted from Li et al. [36])

In general, the non-zero electronic density of states (DOS) at the Fermi level is the precondition for superconductivity. Therefore, charge doping or gating has been widely used to induce or enhance superconductivity in 2D materials [190]. As per the initial innovations, a few-layered 2H-MoS_2 created a high-density 2D electron system in an outermost layer [80, 191]. It has been noted that the in-plane upper critical fields several times higher than the Pauli limit, as depicted in Fig. 7.21c, d. Moreover, spin-orbit scattering can also play a role in the enhancement of the critical field. That is usually explained by the Ising SOC, by adopting the concept of pins of the electron spins to out-of-plane directions. So, the in-plane magnetic fields cannot effectively polarize electron spins to in-plane directions.

7.3.5.2 Ising Superconductivity-Type-II

Ising superconductivity is considered to exist only in materials that break the space inversion symmetry. But unconventional superconducting behavior was uncovered as identified in 1T′-MoS_2 that retains the inversion symmetry with the in-plane upper critical field even far exceeds the Pauli paramagnetic limit, as depicted in Fig. 7.22b, c [192]. Therefore, the spin valley locking and Rashba spin-orbit interaction cannot interpret as enhancement in H_{c2} owing to the centrosymmetric structure in 1T′-MoS_2 nanosheets, as depicted in Fig. 7.22a. Particularly, the H_{c2}/H_p in 1T′-MoS_2 is smaller than $MoTe_2$ even though they have a similar thickness.

To resolve this issue have been proposed a new type of Ising superconductivity mechanism, that is namely called the second type of Ising superconductivity based on lacking destroying the space inversion symmetry in 2D materials with multiple degenerate orbitals. Specifically, 2D materials have lattice inversion symmetry, and their energy degeneracy (such as in-plane PXY orbit) in the orbit of electrons opens the possibility to explore new physics of spin-orbit coupling. The spin-orbit coupling effect reveals the clockwise and counterclockwise in-plane electron orbit produce effective magnetic fields perpendicular to the plane in opposite directions. Thus due to the intrinsic magnetic field, the electron spin can be polarized, and their polarization direction out of the plane. About centrosymmetric materials despite the spin degeneracy of their electrons with opposite spin bound under the reverse orbit motion, it leads to a so-called spin-orbital locking. Additionally, the direction of the intrinsic magnetic field is felt by the electrons with different opposite orbits, as illustrated in Fig. 7.22d. So, materials with heavy elements SOC are produced ultrahigh intrinsic magnetic fields up to thousands of teslas that effectively resist the external in-plane magnetic field perpendicular to it. As a consequence superconductivity survives in a strong in-plane magnetic field within the material, this is typically called type-II Ising superconductor. As per the theoretical calculation, the upper critical field H_{c2} of type-II Ising superconductors can be much higher than usual superconductors, which significantly breaks the Pauli limit, as illustrated in Fig. 7.22e. A large number of Ising superconductor promising candidates have been predicted that further need to experimentally verify for practical applications, as depicted in Fig. 7.22f [84, 189].

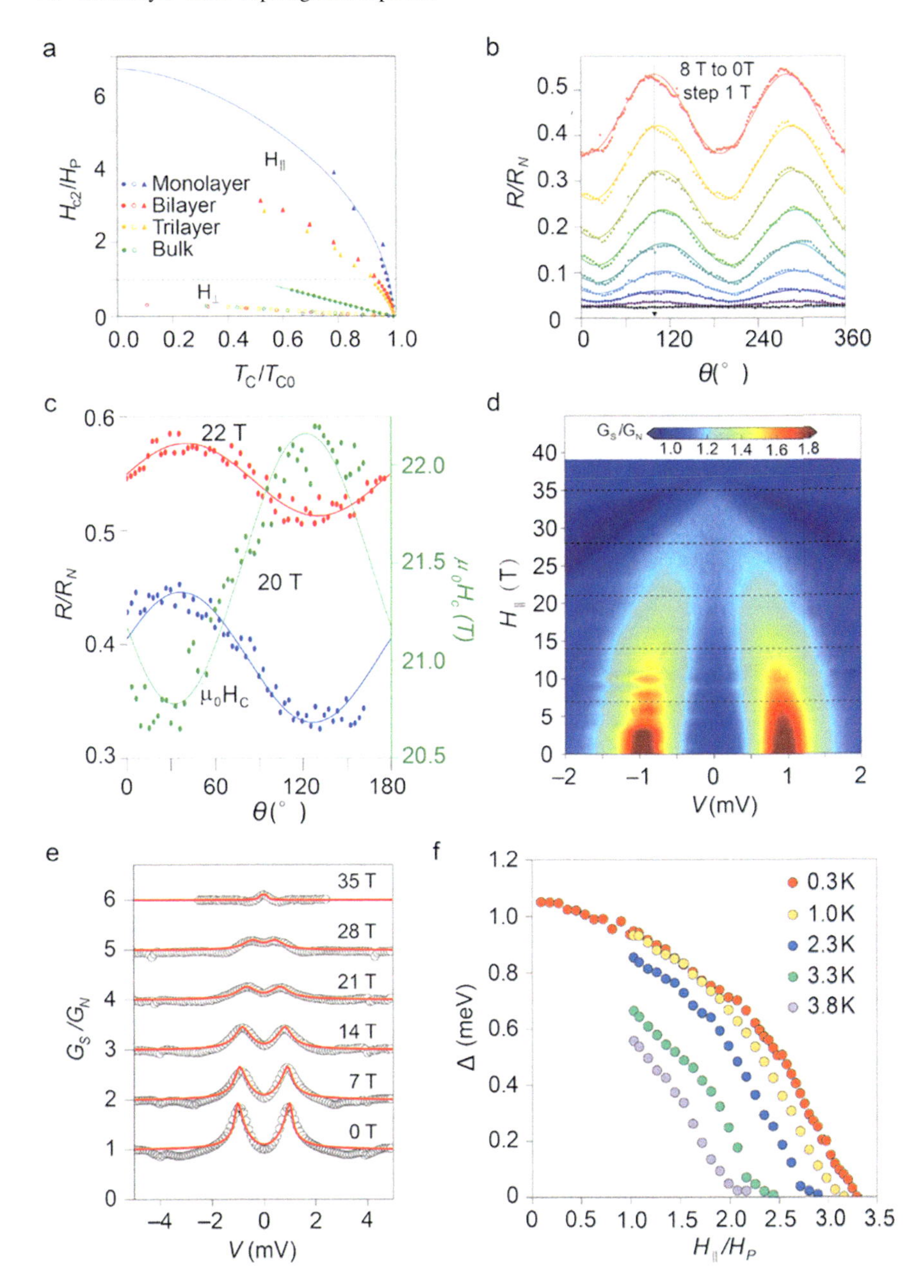

Fig. 7.22 **a** In-plane and out-of-plane critical fields HC_2/HP as a function of transition temperature T_C/T_{C0} for $NbSe_2$ with different thickness; **b** field-dependent magnetoresistance for the $NbSe_2$; **c** effective critical field (green, right axis) and magnetoresistance characterization (blue and red, left axis) for a $Pt/CrBr_3/NbSe_2$ magnetic tunnel junction; **d** contour plot for the differential conductance of trilayer $NbSe_2$ as a function of bias voltage and in-plane field at 0.3 K; **e** differential tunneling conductance spectra for the different fields (dashed lines in (**d**)); **f** superconducting gap as a function of H_k at various temperatures (adopted from Li et al. [36])

7.3.5.3 Theoretical Aspect

Most of the TMDs have been explained based on the tight-binding approximation, as the instance 2H-$NbSe_2$.The layered structure of 2H-$NbSe_2$ reveals the nodal topological phase that's useful for the device a minimum phenomenological model that faithfully displays the essential features of a real material. Their bands may nearby the Fermi energy that is mostly driven by Nb d orbitals. Such material band structures can be qualitatively reproduced by a model with one orbital per site on a triangular lattice. Therefore, this approximation is considered to deal with the system effectively as a quasi-2d structure, by neglecting the less significant Se-derived 3d bands near the Γ point. The tight-binding parameters and the band structure vary with the number of layers, thereby; the quasi-2d bands remain largely unchanged due to the weak interlayer coupling [193]. Thus, the Hamiltonian for a triangular lattice with the zigzag edge parallel to the x-direction is written as;

$$\hat{H} = \hat{H}_{\text{kin}} + \hat{H}_{\text{SOC}} + \hat{H}_{\text{M}} + \hat{H}_{\text{SC}} \tag{7.28}$$

where

$$\begin{aligned}
\hat{H}_{\text{kin}} &= -t\sum\nolimits_{\langle \text{i,j}\rangle.\sigma} c_{\text{i}\sigma}^{\dagger} c_{\text{j}\sigma} - \mu\sum\nolimits_{\text{i}\sigma} c_{\text{i}\sigma}^{\dagger} c_{\text{i}\sigma} - t_2\sum\nolimits_{\langle\langle \text{i,j}\rangle\rangle.\sigma} c_{\text{i}\sigma}^{\dagger} c_{\text{j}\sigma} \\
\hat{H}_{\text{SOC}} &= -i\lambda\sum_{\text{i,j},\sigma'} e^{3i\theta_{ij}}\sigma_{\text{z}}^{\sigma\sigma'} c_{\text{i}\sigma}^{\dagger} c_{\text{j}\sigma'} \\
\hat{H}_{\text{M}} &= -\sum_{\text{i},\sigma,\sigma'} M_{\text{y}}(i)\sigma_{\text{y}}^{\sigma,\sigma'} c_{\text{i}\sigma}^{\dagger} c_{\text{i}\sigma'} \\
\hat{H}_{\text{SC}} &= \sum_{\text{i}} \Delta\left(c_{\text{i}\uparrow}^{\dagger} c_{\text{i}\downarrow}^{\dagger} + H.c\right)
\end{aligned}$$

Here t and t_2 represent the nearest and next-nearest-neighbor hopping and μ is the chemical potential. Moreover, λ is parameterized by the out-of-plane Ising-type SOC, M_y is the magnetic splitting and Δ corresponds to superconducting pairing, σ is the Pauli matrices spin degrees of freedom. The symbols $\langle i, j\rangle$ and $\langle\langle i, j\rangle\rangle$ correspond to summation over nearest and next-nearest-neighbors respectively, θ_{ij} is the angle the vector connecting i and j sites makes with the positive x-axis (so that $e^{3i\theta_{ij}} = \pm 1$).

Such a system is more conveniently analyzed by passing to momentum space and working with the Nambu basis

$$\psi = \left(c_{k\uparrow}, c_{k\downarrow}, c_{k\uparrow}^{\dagger}, c_{k\downarrow}^{\dagger}\right)^{T} \tag{7.29}$$

Therefore, the Bogoliubov–de Gennes Hamiltonian has been defined as;

$$H = E_0(k)\tau_z + E_{SO}(k)\sigma_z + M\sigma_y + \tau_y\sigma_y \tag{7.30}$$

The additional set of Pauli matrices τ acts in the particle-hole space. The normal and spin-orbit hopping terms are expressed as;

$$E_0(k) = -2t\left[\cos(k_x a) + 2\cos\left(\frac{k_x a}{2}\right)\cos\left(\frac{k_y a\sqrt{3}}{2}\right)\right] - 2t_2\left[\cos\left(k_y a\sqrt{3}\right) + 2\cos\left(\frac{k_x 3a}{2}\right)\cos\left(\frac{k_y 3a}{2}\right)\right] \quad (7.31)$$

$$E_{SO}(k) = 2\lambda\left[\sin(k_x a) - \sin\left(\frac{k_x a}{2} - \frac{k_y 3a}{2}\right) - \sin\left(\frac{k_x a}{2} + \frac{k_y 3a}{2}\right)\right] \quad (7.32)$$

Therefore, the diagonalization of the Hamiltonian for the four energy bands is usually expressed [194];

$$E^2 = E_0^2 + E_{SO}^2 + M^2 + \Delta^2 \pm 2\sqrt{E_0^2\left(E_{SO}^2 + M^2\right) + M^2\Delta^2} \quad (7.33)$$

7.3.6 Nodal Topological Phase

Since superconductivity can survive in a regime even if the applied magnetic field is higher than the Pauli limit field. Therefore, the spectral function $A(E, R) = -\frac{1}{\pi}Tr[ImG(E, R)]$ of the a semi-infinite (such as $NbSe_2$) stripe can be calculated. Here, E represents energy, R is the point on the armchair edge and G is the Green's function of the BdG Hamiltonian. The parallel armchair edges to the y-direction are considered to be subject to periodic boundary conditions. The in-plane magnetic field applied is chosen to be $H = 3H_P$. Their bulk Brillouin zone has six pairs of nodal points lying in the $\Gamma-M$ line, as depicted in Fig. 7.23a. The projection onto the edge of the Brillouin zone along the k_y direction gives us four pairs of projected nodal points. In which two outer pairs of the projected nodal points are connected by two sections of Majorana flat bands. The inner two pairs of projected nodal points may lie in the middle four pairs of nodal points (see Fig. 7.23a). Hence, they are connected by two sections of doubly degenerated Majorana flat bands. In the edge of the Brillouin zone, there are four projected nodal points and four sections of Majorana flat bands that exist, as depicted in Fig. 7.23b. However, other finite energy modes emerges below the bulk continuum spectrum. They are called trivial edge modes that are removed by the local potentials.

Therefore, to describe the nodal topological phase an effective Hamiltonian approach is considered to be appropriate in terms of a symmetry point of view. In such systems, Ising SOC near the K points is very strong. It directly correlates to superconducting states near K points that are hardly affected by the in-plane magnetic

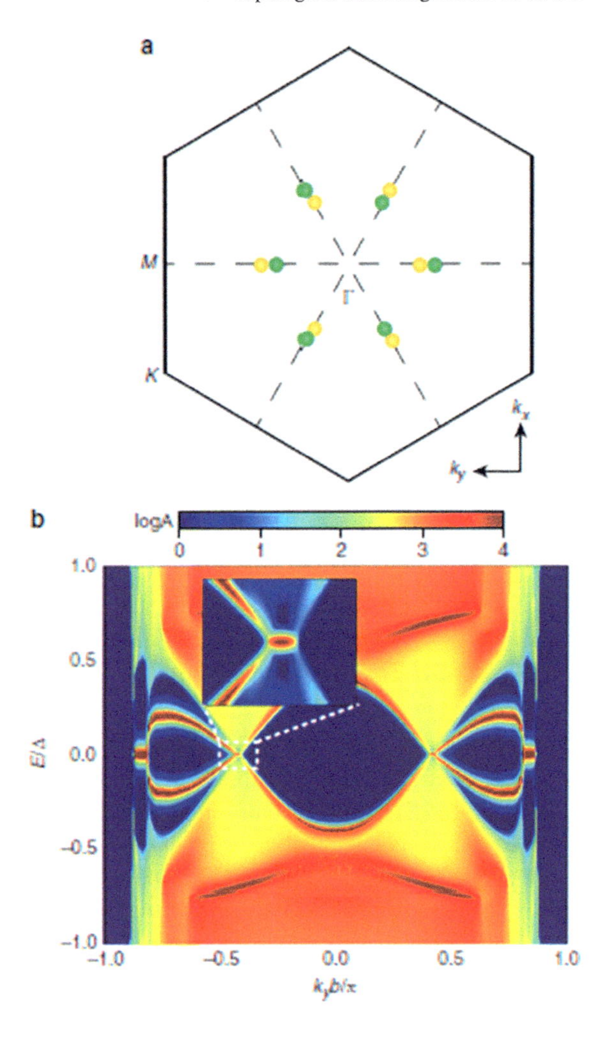

Fig. 7.23 The nodal topological superconducting monolayer $NbSe_2$ driven by an in-plane magnetic field with H = 3HP: **a** position for the six pairs of point nodes; **b** the project into four pairs of nodal points on the armchair edges are shown, the logarithmic plot of the spectral function A(E, R) on the armchair edge in a semi-infinite strip of $NbSe_2$ with open boundary condition in the x-direction and periodic boundary condition in the y-direction, the Majorana flat bands connecting the point nodes at the edge. Insert shows a pair of nodal points connected by a Majorana flat band (adopted from He et al. [84])

field at $H = 3H_p \ll H_{c2}$. Additionally, it also provides an understanding of the gap-closing effect of the in-plane magnetic field, therefore, attention is only required to pay to the states in the Γ pocket where the Ising SOC is weak.

Since around the Γ pocket, the $d_z{}^2$ orbital dominated [185, 186]. Therefore, for the basis $(c_{k\uparrow}, c_{k\downarrow})$ in the absence of external magnetic fields, Hamiltonian should satisfy the point group symmetries $M_z = i\sigma_z, M_x = i\sigma_x$ and $C_3 = e^{-i(\pi/3)\sigma_z}$ and time time-reversal symmetry $T = i\sigma_y K$. Moreover, existence of these symmetries restricts the effective Hamiltonian, up to the third order in k, thus one can express the effective Hamiltonian as;

$$\mathcal{H}_0 = \left[\frac{k_x^2 + k_y^2}{2m} - \mu\right]\sigma_0 + \lambda_{SOC}\left(k_+^3 + k_-^3\right)\sigma_z \tag{7.34}$$

Here, μ is the chemical potential, and $k_{\pm} = k_x \pm ik_y$. It is also important to note that term λ_{SOC} could vanish along the Γ −M lines. Furthermore, when includes the in-plane magnetic field, the spin-singlet pairing Δ within the basis $c_{k\uparrow}c_{k\downarrow}c^{\dagger}_{-k\uparrow}c^{\dagger}_{-k\downarrow}$, the Hamiltonian of the superconducting phase is expressed as;

$$\mathcal{H}_s = \left[\frac{k_x^2 + k_y^2}{2m} - \mu\right]\tau_z + \lambda_{SOC}\left(k_+^3 + k_-^3\right)\sigma_z + \frac{1}{2}\mu_B g H_x \sigma_x \tau_z + \frac{1}{2}\mu_B g H_y \sigma_y + \Delta\sigma_y\tau_y \tag{7.35}$$

Here, τ_z represents the Pauli matrices in the particle-hole basis, H_x and H_y are magnetic fields in the *x*- and *y*-directions, μ_B is the Bohr magneton and g is the electron's gyromagnetic ratio.

Along the Γ−M line, the Ising SOC vanishs, thereby, the electrons' spins are in-plane polarized by the in-plane magnetic field. As a consequence spin-singlet Cooper pair's time-reversal partners are strongly suppressed. Once Zeeman energy $\frac{1}{2}\mu_B g H$ exceeds Δ, the pairing gap becomes close to along the Γ−M line. Whereas, in case away from the Γ−M line, the Ising SOC pins the time-reversal partners, and their spin in the opposite direction along the z direction, this leads spin-singlet Cooper pair condensate, ultimately it makes the pairing gap remain finite. Additionally, also leads the six pairs of point nodes to split out along the Γ−M line when the in-plane magnetic field is gradually applied, as illustrated in Fig. 7.23a. These six pairs of point nodes to protected by the chiral symmetry even if time-reversal has been broken by the external magnetic field. $\mathcal{H}_s$ is due to a time-reversal like symmetry is normally expressed as;

$$U_T \mathcal{H}_s\left(k_x, k_y\right) U_T^{-1} = \mathcal{H}_s\left(-k_x, k_y\right) \tag{7.36}$$

where $U_T M_z T \tau_z$ and k_y are unchanged under the symmetry operation. Moreover, in the case of $\mathcal{H}_s$ to 1D particle-hole-like symmetry $U_P \mathcal{H}_s\left(k_x, k_y\right) U_P^{-1} = \mathcal{H}_s\left(-k_x, k_y\right)$ with $U_P = \tau_x K$. It gives us to $\mathcal{H}_s$ respect to chiral symmetry $C\mathcal{H}_s\left(k_x, k_y\right)C^{-1} = -\mathcal{H}_s\left(k_x, k_y\right)$, where $C = \tau_x \sigma_x$. Hence for any fixed k_y, $\mathcal{H}_s$ lies in the BDI class, therefore, the Hamiltonian is topologically nontrivial [195–198] in the range of k_y, where $\mathcal{H}_s$ is nontrivial to the Majorana zero-energy modes on the edge of the system, as depicted in Fig. 7.23b. With the tuning of parameter k_y, the system undergoes a topological phase transition from a trivial regime to a nontrivial regime by closing the bulk gap. Such topological phase transitions at k_y points are called the nodal points (see Fig. 7.23b) [84, 199].

7.3.7 Aging Effects

Electrical control of spins is one of the key objectives in the field of spintronics. Topological insulating materials have a strong spin-orbit coupling, wherein the host spin-momentum locked gapless modes are confined to the boundary of an insulating bulk. The existence of such helical boundary modes offers the possibility to produce spin polarization and spin currents together electrical means. Therefore, investigations of the topological insulators to spintronics view have been dedicated on 3D TIs, their 2D surface hosts have a massless helical Dirac fermion [200].

Though, the contamination scattering restricts the prospects of using the 3D TI surface layers for spintronics. Besides the direct backscattering $k \rightarrow -k$ of the Dirac electrons is forbidden due to time-reversal equilibrium, the scattering by any other angle is allowed, therefore, the loss of momentum and spin conservation at a scale set by the elastic mean free path [201]. Similarly, current-induced spin accumulation also limits from the mean free path [202]. Moreover, the impurity scattering is also much more restricted to 2D TIs because their boundary modes are confined in 1D. Such helical modes have only 2 momentum directions, left and right, and time-reversal symmetry (TRS) forbids elastic backscattering between the two states. In this situation, modes continue ballistic (or retain their spin) at distances underneath the inelastic mean free path [203, 204]. Similarly, the current persuaded out of symmetry spin polarization of a 2D TI edge cannot be restricted from the elastic non-magnetic imperfection scattering. Due to a bias voltage V (or charge current e^2V/h) a spin accumulation per density $\langle S_z/n \rangle = eV/4E_F$ at the age of a 2D TI, which is an independent scalar of disorder numerous spin rotation equilibrium contravention process on the 2D TI edge have been investigated and studied their charge transport [204–207]. By considering translational invariant edge, in which the spin rotational symmetry is broken due to bulk or structural inversion asymmetry that can lead to a momentum space spin rotation of the helical edge modes without breaking time-reversal symmetry [208, 209]. Likewise, the spin quantization axis rotates in real space in the presence of a random Rashba spin-orbit effect [32, 210]. Such a system's TRS mechanisms cannot provide elastic backscattering, however it can adjust the charge conductance at non-zero temperatures inelastically [204, 208, 211]. However, elastic backscattering is possible when TRS is under broken symmetry [204, 207, 212]. A detailed theoretical explanation of 2D TIs edges could be explained by the example of transport properties of WTe_2 based on the tight-binding model [200]. Typical straight-edge termination in WTe_2 band structure in a pristine sample, as the TI and metallic phases, is illustrated in Fig. 7.24. The Fermi level has placed within the valence band ($\mu = -400meV$) to allow an abundance conducting bulk mode, while when the Fermi level is placed near to center (56 meV) an extensive bulk gap (see E = 0 in Fig. 7.24) confirms solitary the edge modes in the pure sample, at the zero-temperature boundary. While every plot to be utilized for a horizontal straight-edge termination [200], to ensure a Dirac point suppressed inside the valence band (see Fig. 7.24). Moreover, similar results are also obtained belonging to zigzag termination that has a Dirac point in the bulk gap.

Thus, the topological insulators have time-reversal symmetry. Their edge/surface states consist of pairs of states with opposite spins and propagate in opposite directions. They are considered to be degenerate as per Kramers' theorem and pure spin currents of the Hall effect [213]. As the typical representations for the two dimensions and three dimensions are depicted in Fig. 7.25a, b. This leads that insulators

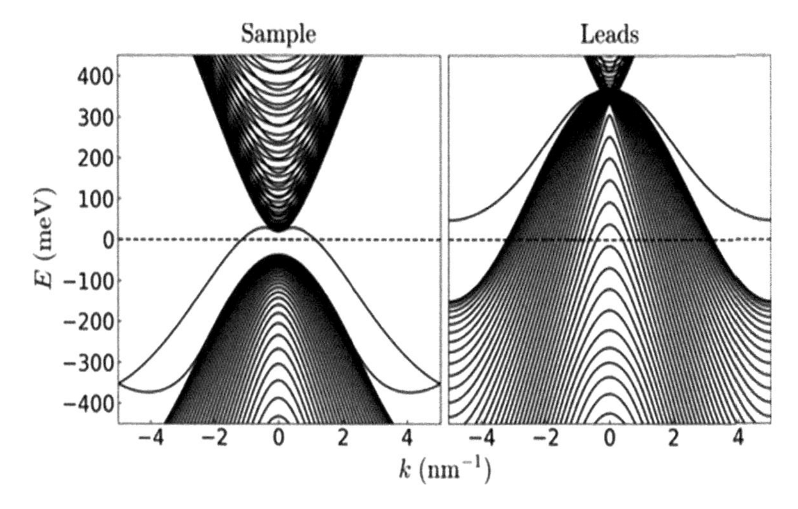

Fig. 7.24 Straight-edge termination of WTe_2 band structure in a pristine sample (left) and leads (right), corresponding to the TI and metallic phases. The dashed line shows the chemical potential; the lead bands are shifted by 400 meV relative to the sample (adopted from Copenhaver and Vayrynen [200])

can be classified into either topological insulators or ordinary insulators based on Z_2 topological numbers [214, 215]. Typically $\nu = 0$ is for an ordinary insulator, while $\nu = 1$ for the topological insulator (2D). The edge states differences between ordinary and topological insulators are defined by considering a semi-infinite plane of a 2D system. Their translational symmetry along the boundary of the semi-infinite plane (edge) to be assumed, to get a good number of the Bloch wave number k along the direction, as the schematics of the edge states of the ordinary insulators and topological insulators are illustrated in Fig. 7.26a, b. In which a clear difference is noticed where the edge states are connected to bulk valence or conduction bands. The ordinary insulators on both sides of the edge-state dispersion are connected to the same bulk bands, while the topological insulators' edge-state dispersion connects the bulk conduction and bulk valence bands. Due to spin-orbit coupling the edge states are spin-splits; typically it is called Rashba splitting. Due to time-reversal symmetry, the symmetric edge states to $k = 0$ are considered to be Kramers degenerate that have opposite spins. Moreover, 3D topological insulators are one of the typical surface state insulators with a single Dirac cone. As the instance (111) surface of Bi_2Se_3 or Bi_2Te_3 is considered to be a single Fermi surface that encircles the Γ point. Their surface-state dispersion has a linearity in the wave number, as the schematic is illustrated in Fig. 7.26d, typically it is called a Dirac cone. In the case of Bi_2Se_3 and Bi_2Te_3 the surface, states have been also noticed experimentally in terms of a single Fermi surface around the Γ point, these surface states due to a single Dirac cone (see Fig. 7.26d) [4, 5, 7, 216].

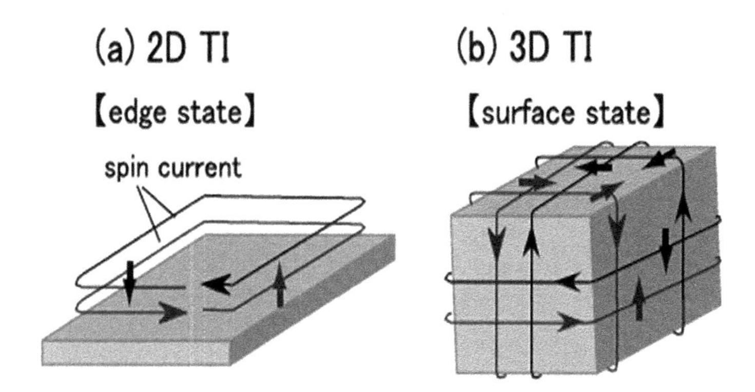

Fig. 7.25 **a** Schematic for a 2D topological insulator; **b** a 3D topological insulator (adopted from Murakami [213])

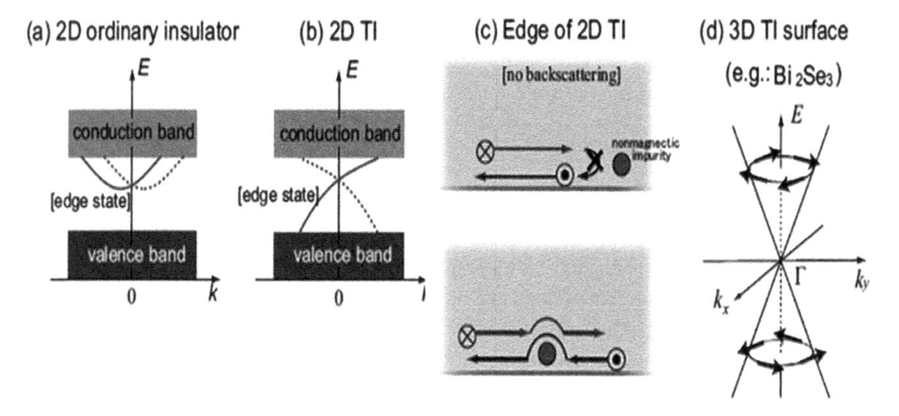

Fig. 7.26 Schematics for the edge states **a** 2D ordinary insulator; **b** topological insulator; **c** backscattering is prohibited for edge states of 2D topological insulators; **d** Dirac cone of the surface states of 3D topological insulators (adopted from Murakami [213])

7.3.7.1 Edge States Transport of 2D Topological Insulators

The topological insulators slope of edge-state dispersion corresponds to the electron velocity $v = \frac{1}{\hbar}\frac{\partial E}{\partial k}$ (see Fig. 7.26b). Therefore, the two edge states are propagated in opposite directions having an opposite spin due to the time-reversal symmetry. These edge states carry pure spin current with the fixed velocity directions for each spin [217, 218]. So that impurities cannot flip spins of the edge-state electrons, thereby, they do not exhibit backscatter edge states, as depicted in Fig. 7.26c [204, 212]. It gives edge states are form perfectly conducting channels. This kind of perfectly conducting channels of edge states have been also noticed in CdTe/HgTe/CdTe quantum well [47, 219, 220]. Their gap opens by making into a quantum well with CdTe and the system behave as a topological insulator when the good thickness d exceeds $d_c = 60$ A° [21], however, if the good thickness d *is* less than d_c it is considered to be an ordinary insulator. These predictions have been verified from the transport experiments. It also noted that the Fermi energy belongs to the bulk band gap and possessed no conducting channels corresponding to $d < d_c$ and two conducting channels for the $d > d_c$. Thus the conductance from $G = 2e^2/h$, when $d > d_c$ and $G = 0$ for the $d < d_{c,}$ it has also verified experimentally [47, 219]. The perfectly conducting channels have been also experimentally verified from the nonlocal transport measurements for the multi-terminal geometry [220].

Moreover, such perfectly conducting channels also contributes to thermoelectric transport due to their interesting overlaps between TI materials and efficient thermoelectric materials. Such overlaps due to thermoelectric transport calculation based on special boundary states, like as edge states of 2D topological insulators, surface states of 3D topological insulators, and helical states on the dislocations in case of 3D topological insulators [213, 221, 222]. Whereas the edge/ dislocation states of the 2D and 3D topological insulators electrons do not undergo elastic backscattering, this is one of the basic requirements for good thermoelectric transport [221, 222]. Due to the finite temperatures, the inelastic scattering causes the decoherence of the 1D helical states, thereby the thermoelectric transport is reduced. So, at a lower temperature, the helical 1D states become gradually dominant over the transport by the bulk carriers, therefore, the thermoelectric figure of merit is increased.

7.3.7.2 Impact of Topological Edge States on Optoelectronic Properties

The Z_2 topological edge state can be also visualized in monolayer 2D TMDs, by showing high PL intensity and wide PL emission wavelength range [223]. Such as the WSe_2 monolayer PL emission at different positions measurements with a 633 nm excitation laser and recorded their normalized spectra, as illustrated in Fig. 7.27a, b [224]. It has been noted that the overlapped band gap of shrunken and expanded (photonic crystal) PhCs (See Fig. 7.27b, as indicated in yellow shadow) band gap narrower than the theoretical prediction, it arises due to imperfection of the fabrication process. Moreover, the PL emission of the top WSe_2 monolayer is strongly affect from the underneath (PhC) structure, while the spontaneous emission rate of the WSe_2 monolayer modulated and positively correlate with the optical local density of state (LDOS) of PhC structure [225–227]. Both bulk states of expansion and shrunken PhCs reveals the PL resonances at the wavelength 778 nm and 802 nm. In the interface area between the shrunken and expansion PhCs, a PL emission around 785 nm has been recorded in the overlapped band gap range with the 5 times enhanced PL emission. It has directly correlated to the existence of a topological edge state (see Fig. 7.27b). To provide further confirmation of the bulk-edge bulk state transition the PL mappings with the wavelength of 778 nm, 788 nm, and 802 nm have been measured to directly visualized the distribution of resonant states, as depicted in Fig. 7.27c–e. It noted that the mapping wavelength of the edge state slightly shifts from the PL peak center to avoid the overlap of the expanded bulk state. Therefore, an obvious bulk-edge bulk state has been also recorded by changing the bright inside and dark outside pattern into the dark inside and bright outside one. Additionally, the inside bulk state decays from the center to the interface area have shown an attenuation pattern. The PL mapping of the edge state also demonstrated a bright area close to the interface and their intensity gradually decays to bulk, as depicted in Fig. 7.27d. Therefore, an unusual dark line of the Z_2 topological edge state has been noticed at the interface, which leads to a topological edge state exactly localized at the interface area. Thus the bright area intensity of WSe_2 PL emission is strongly suppressed at the position of the dark line, where the quantum emission efficiency is critically affected. Therefore, the physical insight of the dark line is considered to be a significant parameter for future topological devices as the optical diffraction limit [224].

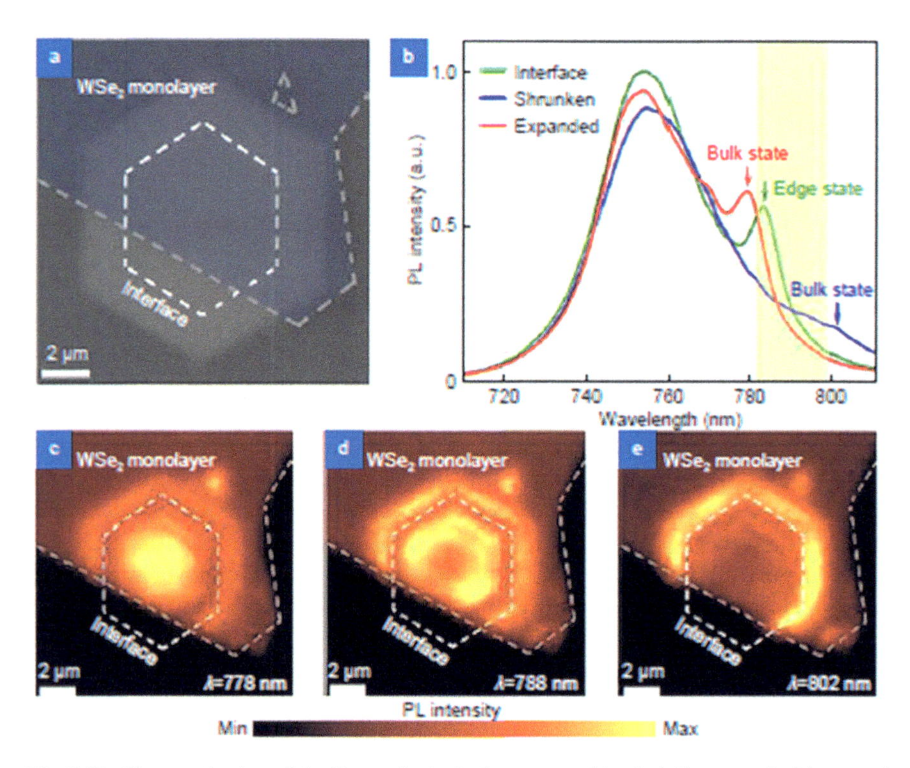

Fig. 7.27 Characterization of the Z_2 topological edge state and its dark line: **a** optical image of the Z_2 PTI with a WSe_2 monolayer transferred on the top, the interface between shrunken and expanded PhC regions, and the area of WSe_2 monolayer fringe are marked with white and gray dashed lines; **b** PL spectra of WSe_2 monolayer upon the shrunken PhC (blue), expanded PhC (red), and near interface (green). The yellow shadow indicates the overlapped bandgap of shrunken and expanded PhCs; **c–e** PL mappings for the bulk (expanded)-edge bulk (shrunken) state at the resonance wavelength of 778, 788, and 802 nm. The defect triangle region of the WSe_2 monolayer due to the transferring process (adopted from He et al. [224])

7.3.8 Cracking

Topological insulators (TIs) have a remarkable difference of existing electronic insulators whereas protecting metallic, conductive surfaces. Usually, in the normal materials, defects like cracks and deformations are considered to be barriers to electrical conduction, automatically it makes the material additional electrically resistive. While in the case of 3D TIs, they convert into superior conductors by incorporation of cracks due to cracks itself behave as conductive topological surfaces that can offer supplementary pathways for the electrical current. Expressively, a TI material every surface or prolonged defect could harbor this kind of conduction [228]. For the instance, 3D TI samarium hexaboride (SmB_6) increased the electrical conduction due to its unique single crystals that are wide sufficient to anchorage cracks that strangely not suffering to conduction by bulk defects [229, 230]. However, the SmB_6 topological nature is not relevant to all TIs with cracks, but also in their films with grain boundaries.

Moreover, in general 3D TIs suffers to the limited bulk conduction that polluted or overwhelm the surface in electrical characterizations. Therefore, the most of transportation studies have been performed with the surface-dominated extremely thin films of the TIs. The SmB_6 is an exceptional owing to its truthfully insulating bulk construction and strong surface conduction [229, 230]. Though SmB_6 has been recognized as the mixed-valent insulator to over the four decades, the theoretical and experimental evidence has revealed that it can host a topologically prevented surface with a completely insulating bulk. But its topological behavior of the surface is still debatable [231]. Nevertheless, the accurately insulating bulk character of SmB_6 can be connected to a strong surface conduction throughout its surfaces associating few uncommon characteristics of TIs in bulk [232].

Considering the discrepancy in surface and bulk electrical conductivities in SmB_6 3D TIs a model has been demonstrated based on cracks within the materials [233, 234].By describing the most appropriate Corbino disk structures for the electronic conduction in 3D TIs, as depicted in Fig. 7.28a, b [233, 234]. It offers the transport properties of different crystalline surfaces individually. The resistance of the Corbino sample is inversely proportional to the sheet conductivity that is expressed as;

$$R_{\text{Corbino}} = \frac{\ln(r_{\text{out}}/r_{\text{in}})}{2\pi}\frac{1}{\sigma_s(0)} \tag{7.37}$$

where $\sigma_s(0)$ (in $\frac{1}{\Omega}$) is the surface conductivity at zero magnetic fields, and $r_{\text{in}}(r_{\text{out}})$ represents the inner(outer) radius of the Corbino disk. In the case of an ideal 2D layer, the carrier density is independent of the magnetic field, while the Corbino resistance has magnetic field-dependent behavior therefore, it reduced the mobility that follows the Lorentz force factor:

$$R_{\text{Corbino}}(B) = \frac{ln(r_{\text{out}}/r_{\text{in}})}{2\pi}\frac{\left(1 + (\mu B\cos\theta)^2\right)}{ne\mu} \tag{7.38}$$

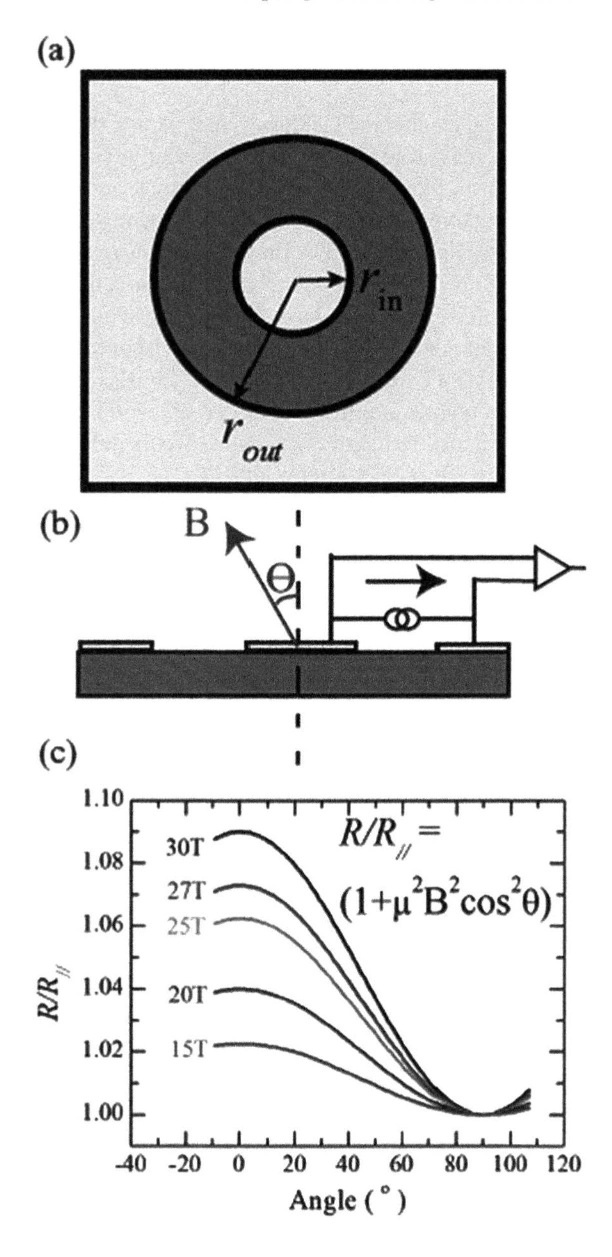

Fig. 7.28 **a** Setup of a Corbino disk on sample: **b** simulated resistance ratio versus magnetic field angle fixed at different magnetic field values (Eo et al. [233])

where B is the applied magnetic field, θ represents the angle between the field and the surface normal direction, n represents the surface carrier density, e is the electron charge and μ represents surface mobility. The change in resistance changes with an angle at a fixed magnetic field is illustrated in Fig. 7.28c. Moreover, the polishing and surface treatment by oxygen plasma-induced oxidization also give us a rougher polishing that possessed lower resistance plateau values than the ones finely polished. That is typically opposite to expectation because the surface roughness can contribute to extra scattering, therefore, reduced mobility is possible.

To describe the key discrepancy it has hypothesized the subsurface cracks can be produced by irregular polishing and it work as additional surface conduction pathways. As the typical SmB_6 is considered to be a 3D TI, in which subsurface cracks destruction of the bulk should be topologically prevented surfaces. This hypothetical concept has been also consisting to the noted resistance plateau assessment tendency for the various polishing substances. It is also well established that coarser polishing generates subsurface cracks at the bigger distance scales. The experimental verification of the subsurface cracks concept has also provided similar results by exploring the SmB_6 sample [233]. As the noticed subsurface cracks in ion-milled prepared SmB_6 crystal by coarse polishing (P1200 grit, which produced micron-level surface roughness), as illustrated in Fig. 7.29a–h. The cracks have been visualized (see Fig. 7.29h) up to 1 μm long in either transverse or vertical direction SEM images. However, cracks produced at finer polishing do have not a reason to hypothesize their existence and contribution to the total surface conduction due to the resolution limit of SEM. Additionally, using the length scales of subsurface cracks have been also determined the size of the grit particles. The thinning and polishing potentially also creates cracks, a single crystal that begins with a large enough surface. Moreover, the role of disorder on the surface of native oxide, as the most likely Sm_2O_3 forms on the surface of SmB_6 [235]. Thus various characteristic parameters of the 3D TIs (like SmB_6) have been defined by governing the induced cracks.

7.3.9 *Carrier Density*

Since most of the interesting applications of the TIs are demonstrated with thin film optoelectronic devices. Therefore, it is essential to their thin films should exhibit conductivity via only surface states along with few non-carriers and higher mobility of the topological carriers. Several outcomes of TIs thin films have shown that the ARPES data only, in which investigators extracted numerous evidences, like Fermi energy location to Dirac point, inclusive carrier density, and mobility. Most ARPES outcomes have been conducted under a vacuum instant subsequent development or on newly cleaved surfaces. Moreover, the ARPES technique can also consider a suitable technique for the characterized surface band structure of TIs, because ARPES is a surface sensitive approach that usually merely to leading some layers of the samples in every material. Therefore, it is fairly beneficial for consideration the features of the surfaces of TI films containing the dispersal of the topological states, the existence or nonexistence of a trivial band-bending 2DEG, and the location of the surface Fermi energy. The ARPES technique's small penetration depth cannot probe, therefore, it is difficult to distinguish the bulk character of a material.

However, understanding the electrical conductivity character of the bulk specimen is critical for their inclusive features. The convectional Hall experimentations can

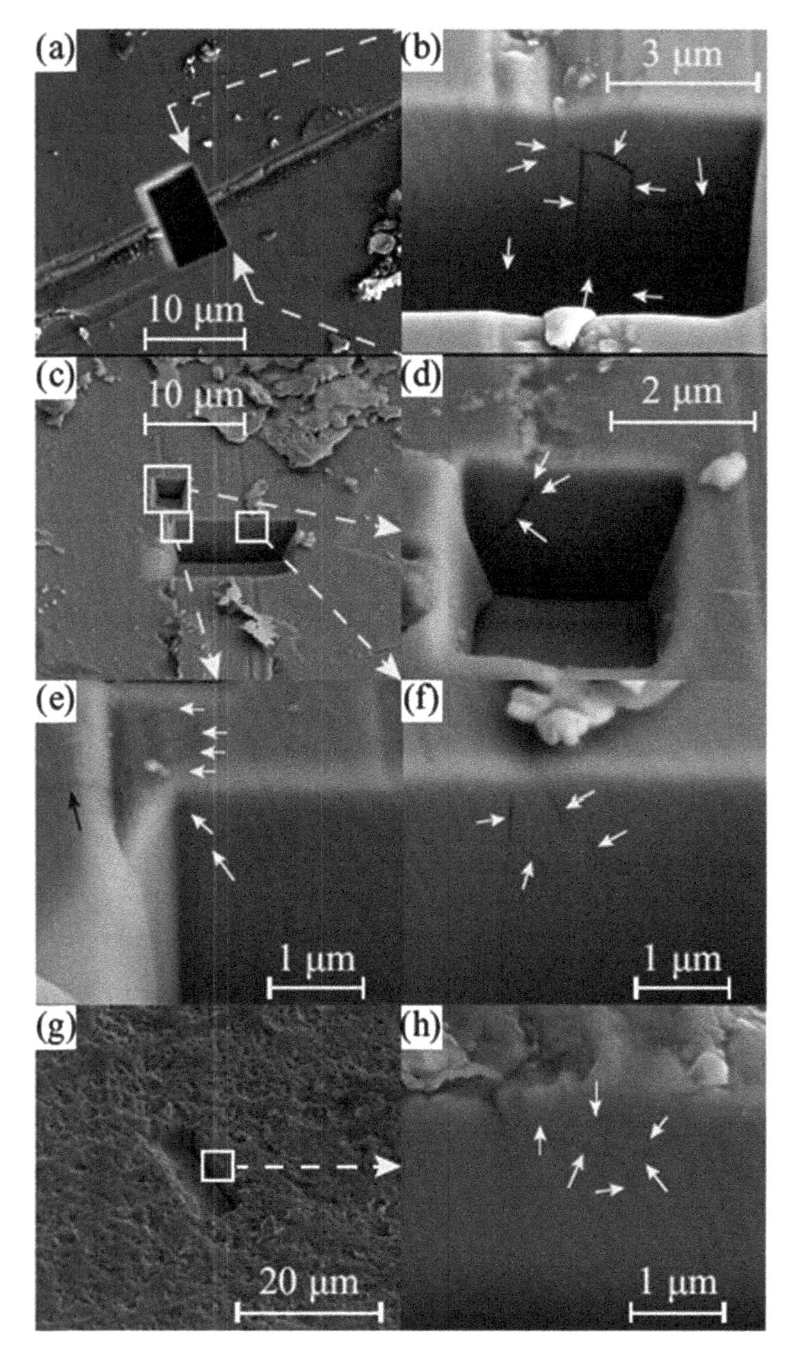

Fig. 7.29 **a** Scratch for ion-milled cross-section across; **b** subsurface cracks visualization below the scratch; **c** ion-milled cross-sections across another scratch; **d–f** subsurface cracks visualized below the scratch; **g** ion-milled cross-section on a rough-polished surface; **h** subsurface cracks below the rough surface (Eo et al. [233])

deliver more authentic view of the carrier density and mobility in a thin film. The high-field Schubnikov-de Haas measurement and low-field Hall experiment under a horizontal conduction approach offers a good understanding of distinguish electrical assistances to the different conducting networks in TI films and crystals [10, 38, 49, 236–238]. Moreover, the electrical extents tending to executed afterward disclosure in air which provides an additional appropriate representation to those willing to fabricate electronic devices depending on TIs. Thus the typical carrier density and mobility to be determined from the ARPES analyses, as summarized in Table 7.8 [10, 239].

Compared to the conventional 2H-semiconducting phase of TMD materials, the distinct polymorphic phases exist with intriguing states of matter in 2D topological insulators, superconductors, and Weyl semimetals [240–242]. Their metallic phase is usually characterized by V-shaped Dirac-Fermion-like carriers, carrier concentrations that are higher than 10^{13} cm^{-2}. Their weak gate modulation and environmental stability is usually resolved from the Schottky-limited transport in 2D-TMDs [243]. As the instance, there are seven monolayer 1T-metallic phase TMD/metal structures theoretically predicted, the zero Schottky barrier contacts due to efficient interface charge transfer based on the amounts of interface orbitals overlapping and lattice misfits [244]. Thus, the selective area semiconducting-to-metallic phase transformation proves a defect-free atomically sharp interface and matched band alignment between the work function and the conduction band energy, that leads seamlessly ohmic side contacts with RC as low as 200–300 $\Omega \mu m$ with an improved device characteristic, as depicted in Fig. 7.30 [243, 245–250]. The metallic-1T phases of MoS_2, $MoTe_2$, and WSe_2 are considered to be stable up to 100, 300, and 180 °C, however, further thermal annealing or laser irradiation may reverse the transition back to the 2H-semiconducting phase [198, 201, 202] Therefore, the experimental and theoretical calculations of sub-10 nm 1T/2H/1T MoS_2 structures have revealed an off-current density of 10 pA μm^{-1}, an on–off ratio of 10^4–10^7 and a SS of 69–120 mV dec^{-1}, which is only limited by intraband tunneling at 3.3 nm [250].

7.3.10 Structural Quality of Thin Films

TMDs thin films have been extensively studied to explore their structural and device fabrication possibilities for different purposes [251, 252]. In several TMDs thin films, it has been noticed that the decreasing thickness of the monolayer displays a transition from an indirect to a direct band gap [253–255]. Generally, their ordering of the atoms has been classified as trigonal prismatic (hexagonal, H), octahedral (tetragonal, T), and distorted octahedral (T′). Additionally, their structures are also accompanied by a different band structure and consequently, different electrical and optoelectrical properties [54, 86, 256]. But the major hurdle with 2D TMDs thin films is to get a reliable mono- or few-layer-scaled film with good layer controllability and large area uniformity for practical applications in electronic, optical, and magnetic devices [251].

Table 7.8 Electrical properties of binary TIs, here mobility (μ), sheet density (n_{2D}), bulk density (n_{3D}), film thickness (t), and temperature (T) at which electrical measurements are performed [10]

Material	Substrate	μ (cm^2/Vs)	n_{2D} (cm^{-2})	n_{3D} (cm^{-3})	t (nm)	T (K)
Bi_2Se_3	Mica	472	6.1×10^{13}		10	
Bi_2Se_3	Graphene		Intrinsic		<10	
Bi_2Se_3	$SrTiO_3(111)$		$1.9–3.1 \times 10^{13}$		5–50	
Bi_2Se_3	Graphene and $Al_2O_3(0001)$			$\sim 1 \times 10^{19}$	11–45	
Bi_2Se_3	InP(111)A	3500		$\sim 1 \times 10^{18}$		2
Bi_2Se_3	Si(111) Vicinal	1600		5×10^{18}	200	2
Bi_2Se_3	Si(111) Flat	1200		3×10^{19}	200	2
Bi_2Se_3	Si(111)		1.25×10^{13}		7	200
Bi_2Se_3	SiO_2		$\sim 4 \times 10^{13}$		20	2
Bi_2Se_3	Graphene	5000		$\sim 1.1 \times 10^{19}$	400	20
Bi_2Se_3	AlOx	623	$\sim 3 \times 10^{13}$		100	300
Bi_2Se_3	GaAs (111)B	100–1000		$8.06–40 \times 10^{18}$		4.2
Bi_2Se_3	SiO_2/Si		$\sim 2.2 \times 1013$		7	
Bi_2Se_3	$Al_2O_3(110)$	150		9.4×10^{18}	20	13
Bi_2Se_3	$Al_2O_3(0001)$			$0.5–3.5 \times 10^{19}$	2–40	
Bi_2Se_3	Rough InP(111)B	2000		9×10^{17}	100	4
Bi_2Se_3	CdS(0001)	~4000	$\sim 6 \times 10^{11}$		6	2
Bi_2Se_3	CdS(0001)	~400	$\sim 1 \times 10^{13}$		6	2
Bi_2Se_3	Si (100)	2880	8.4×10^{10}		8	
Bi_2Se_3	Si (100)	54	4.5×10^{12}		8	
Bi_2Te_3	SiO_2	2.8		1×10^{21}	~100	300
Sb_2Te_3	SiO_2	42		2.4×10^{19}	~100	300
Bi_2Te_3	SiO_2	80		2.7×10^{19}	~100	300
Sb_2Te_3	SiO_2	402		2.6×10^{19}	~100	300
Bi_2Te_3	$BaF_2(111)$				1.3×10^{19}	
Bi_2Te_3	Si(111)	1030		4.7×10^{19}		
Bi_2Te_3	$BaF_2(111)$	2000–4600	$2–4 \times 10^{12}$		10–50	
Bi_2Te_3	Prestructured Si(111) SOI	110	1.6×10^{14}			
Sb_2Te_3	Prestructured Si(111) SOI	636	6.7×10^{13}			
Bi_2Te_3	Si(111)		7×10^{12}	1.32×10^{19}	5.3	300
Bi_2Te_3	n-type Si(111)	35	1.2×10^{14}		4	2
Bi_2Te_3	$BaF_2(111)$	1500		$2–10 \times 10^{18}$	350	77
Bi_2Te_3	Mica	4800	2.45×10^{12}		4	

(continued)

Table 7.8 (continued)

Material	Substrate	μ (cm^2/Vs)	n_{2D} (cm^{-2})	n_{3D} (cm^{-3})	t (nm)	T (K)
Bi_2Te_3	GaAs(111)B		10.6×10^{12}			2
Sb_2Te_3	CdTe(111)B	279		7.48×10^{18}	~300	290
Sb_2Te_3	V-GaAs(111)B	781		9.13×10^{18}	400	1.8

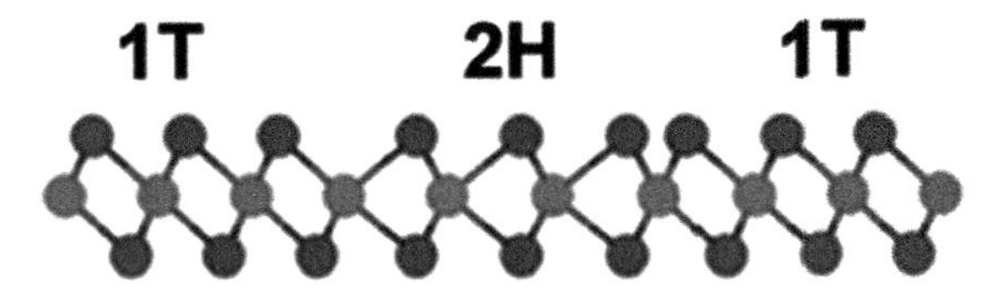

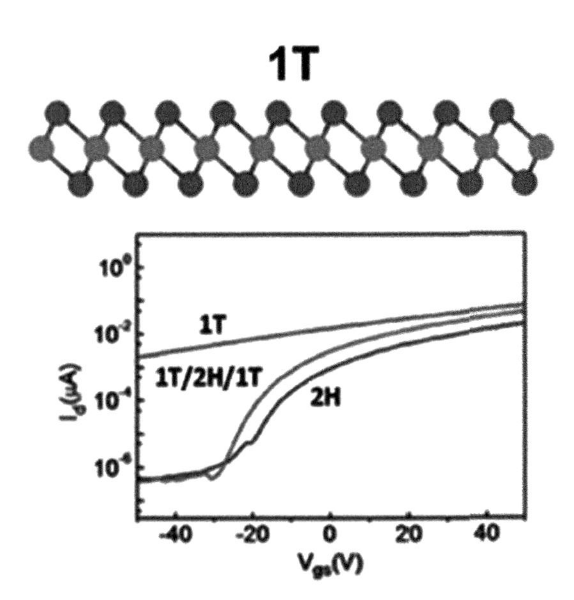

Fig. 7.30 Schematic for the oxygen plasma exposure of WSe_2 devices based on the 2H to 1T phase transformation of MoS_2 by argon plasma and the corresponding transfer characteristics (adopted from Retamal et al. [250])

Specifically, TI films are suffering to various kinds of structural defects, like mosaicity-tilt, mosaicity-twist, antiphase domain boundaries (APBs), stacking faults, and twin defects [257–260]. However, the topological surface state is strong and occurs even in imperfect films. Regarding electrical device fabrication, the key significant things are configurations of the film excellence that should have high carrier density. Therefore, the higher excellence TI thin films should possess a lower trivial charge density with a higher mobility to the electrons in the surface states. The trivial charge density trends are achieved to the point imperfections, like vacancies and antisite flaws, and surface band bending. In-plane threefold symmetry of the crystal structure of TI materials can have highly susceptible structural defects, as typically illustrated for the Bi_2Se_3 in Fig. 7.31a, b. That is achieved possibly from three orientations of a TI domain; A, B, and C. A faultless stacking arrangement in the growing direction reveals ABCABCABC together every quintuple layer alternated 120° to the preceding quintuple layer. A stacking imperfection may occur while a solitary quintuple layer is detached, therefore, a new order is achieved ABABCABC. Similarly, incorporation of a solitary quintuple layer is also change the ordering ABBCABCABC, as depicted in Fig. 7.32. Such stacking imperfections have an intense impact on the electrical characteristics of the film, covering the formation of topologically prevented states toward the imperfection or fluctuation of the Dirac point for the pure band structure [261]. However, such kinds of imperfections are challenging to inhibit.

Moreover, twin imperfections are also occurred while the dual domains possessed diverse stacking orders to encounter. Usually, twin flaws exist onto an abundant greater expanse than stacking imperfections. When vertical direction changes in stacking then it provides a lamellar twin, whereas the alteration in stacking in plane direction it gives us a rotational twin. The rotational twin defects have appeared as oppositely-oriented triangular domains in atomic force microscopy (AFM) scans. While these kinds of twin flaws are also recognized from transmission electron microscopy (TEM) [262, 263]. Usually, a better structural excellence film shows inferior and bigger triangular domains that connected to higher mobility (see Table 7.8). A larger domain size to be also achieved by growing the thin film over a lattice-consistent substrate (or by adjusting the substrate temperature), by controlling the growing rate and flux ratios [257, 260, 262, 264, 265].

Similarly, the antiphase domains also exist due to the distinction in the unit cell depth among the substrate and the TI film. In the case of a wider substrate, the

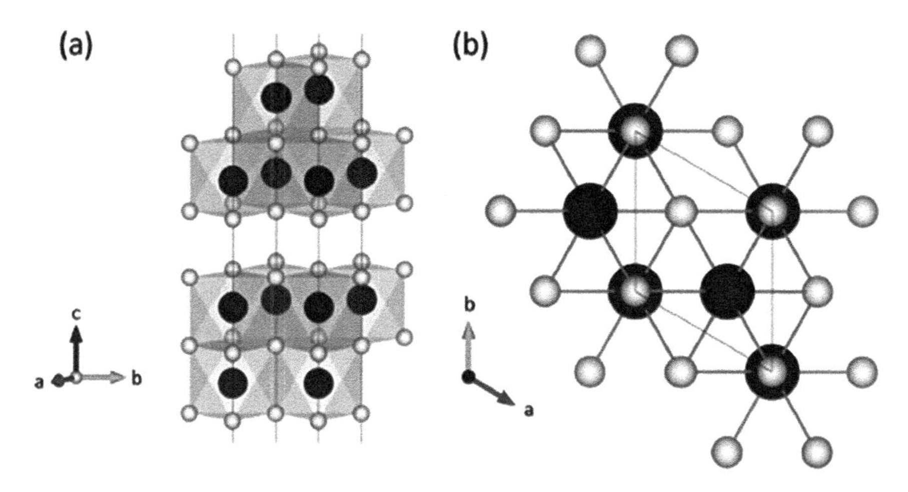

Fig. 7.31 Crystal structure of tetradymite Bi_2Se_3, Bi atoms represents from large and Se atoms smaller one; **a** crystal structure along the a-b plane showing two quintuple layers with the van der Waals gap between them: **b** crystal structure along the c-axis (adopted from Ginley et al. [10])

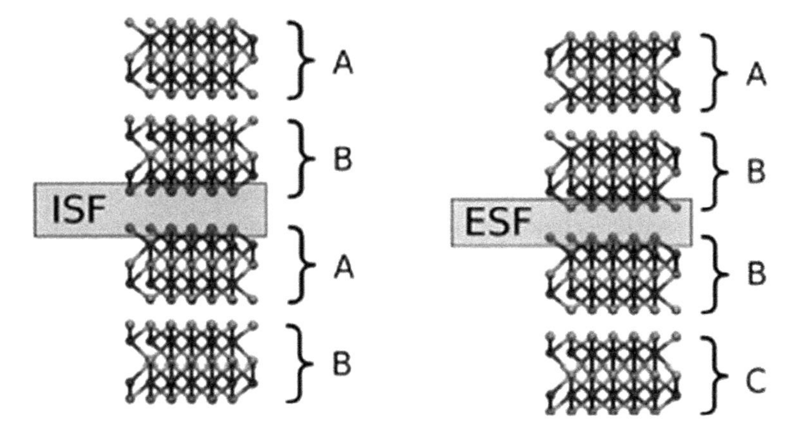

Fig. 7.32 Two types of stacking faults (ISF, or intrinsic stacking fault, and ESF, or extrinsic stacking fault) are possible in a TI film (adopted from Seixas et al. [261])

flat terraces together step depths equivalent to the substrate unit cell size. Owing to successive TI film domains are nucleate on a terrace that may superior to the terrace on which the next TI film domain nucleates. The combination of such two domains revealed an antiphase domain and formed a unit cell of the TI that shifts perpendicularly by the stature of one atomic step in the original substrate. Usually, this kind of flaw has also a deleterious impact on the electrical characteristics of the film [257], it is challenging to eliminate unless developing the isostructural constructions.

In addition to these, the mosaicity-twist and mosaicity-tilt also exist while the film nucleated from numerous minor domains. The mosaicity-twist is the result of the domain's slight rotating to each other in the plane of the film. While the domain's slightly misaligned out-of-plane gives the mosaicity-tilt. Using x-ray diffraction patterns interpretation their degree of mosaicity is defined. For instance, the mosaicity of the Bi_2Se_3 thin films is low, therefore, it is not reflecting a robust impact on the film transportation features, however, the roughness of the substrate to be enhanced by the strong effect [10]. Thus the configurable flaws in TI films have a great deleterious effect on the transportation features of the film. Even though the topological surface states remain prevents even in a imperfective film, the mobility of such states might be inferior compared to an ideal film, therefore, the topologically prevented states are supplementary problematic to differentiate from trivial states.

As the intense, typical Bi_2Se_3 films have a pyramid structure, as illustrated in Fig. 7.33 Appearance pyramids structures in Bi_2Se_3 thin film are usually correlated to the rotational equilibrium of the triangular in-plane lattice that allows to islands nucleate in diverse alignments, thereby, producing the twining and another organizational imperfections. The thin film growths on neighboring or uneven surfaces forces nucleated the islands to bring into line toward step edges and reduced the twin flaws [264]. Therefore, films are considered to be best for the electrical properties to have fewer and larger pyramids. It attributes to distinct development circumstances.

Such as the surface mobility while the development is high enough, in this circumstances bismuth and selenium adatoms possessed sufficient time to diffuse as the consequence gets utmost energetically promising locations. It also reduces the quantity of vacancy flaws (refining electrical features) by allowing discrete domains to grow superior (resulting in fewer but larger pyramids). While in case of low surface mobility due to a low substrate temperature or higher development rate, therefore, the adatoms possessed fewer time to attend satisfactory locations, as a consequence, additional vacancy flaws and pyramids to be formed. Moreover, in the case of extremely thin films (6–20 nm), material Bi_2Se_3 forms numerous minor triangular terraces, it indicates that the island creation is in the initial periods of development. This kind of island development (or Volmer-Weber growth) is owing to dominant adatom-adatom connections over adatom-substrate exchanges. With the increasing thickness of the films, the islands grow and merge, and as the consequence leads the fewer but larger pyramids. It directly correlates to switching in step flow growth toward the boundaries of the terraces. The change over in step flow development specifies that the adatoms are possessed high mobility on an accomplished Bi_2Se_3 quintuple layer due to most likely van der Waals bonding between quintuple layers.

Thus, to resolve major issues of the essential research fields of TMDs, to synthesize reliable mono- or few-layer-scaled 2D TMDs with good layer controllability and large area uniformity for practical applications in electronic, optical, and magnetic devices. During the preparation of thin TMD films, much attention is paid to controlling the growth habits of the substrate. In this order, many fabricating tools have been developed and adopted to optimize ultrathin TMD films, such as mechanical exfoliation (ME) from bulk materials, molecular beam epitaxy (MBE), chemical vapor deposition (CVD), and atomic layer deposition (ALD) [266–268]. Among all the CVD and ALD techniques considered to be the best methods to obtain superior conformity in terms of uniform surface homogeneity and enable comparably easy control of thin film thickness and defects [269–272]. Although MBE can also serve as a near defect-free crystal platform for intrinsic charge transport studies. Nonetheless, the synthetic strategies from the CVD and ALD also offers a high volume manufacturing of monolayer semiconductors, as the key component in industrial-scale devices which are highly desired. Some of the important methods of the TMDs thin film preparation are summarized in Table 7.9 [251].

7.3.11 Choice of Substrate

Topological insulators are also like other VDW materials that quite accommodate when it comes to choosing an appropriate substrate owing to their relaxed lattice matching criteria. It does not mean to substrate effects are considered completely negligible. As illustrated in ARPES, TEM, and STM images in Fig. 7.5a–g, the twinning defects are seen due to the threefold symmetry of the TI surface caused by different stacking sequences [95]. Moreover, various investigations with the ability to control such as twinning by grown on intentionally offcut vicinal Si substrates [262].

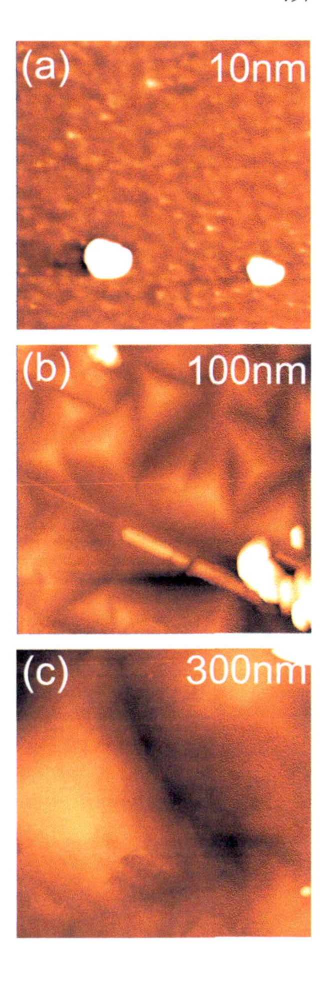

Fig. 7.33 The 2 × 2 μm atomic force microscopy pictures of Bi_2Se_3 films of thickness: **a** 10 nm; **b** 100 nm; **c** the triangular terraces grow superior and fuse as the film converts thicker (adopted from Ginley et al. [10])

The existence of a high density of step edges on the surface leads to the preferential growth of the TI film out from the step edges along a dominant crystal growth direction. The substrate effects have been also noticed in the rotational alignment between the TI and substrate [273]. Therefore, the growth is possible on lattice-mismatched substrates in which the TI nucleation sites are not located at integer multiples of the TI lattice spacing away from each other, thereby, while the grains coalesce an incomplete unit cell and small-angle grain boundaries exist [95].

The most commonly used substrate is c-sapphire (Al_2O_3) for the growth of TIs in the MBE technique that has in plane structure likely to TIs. Most of the work with Bi_2Se_3 growth is based on a sapphire substrate, however, the other materials can be also used on a different substrate. Owing to fact sapphire have numerous benefits, such as its low cost, easy availability, high surface quality, and common chemical manners. But it has been also noted that a huge lattice incongruity among sapphire and TIs (around 15% Bi_2Se_3) that may provide to distinct organizational

Table 7.9 Various synthesis methods used precursors and substrates for the 2D TMDs

Material	Synthesis approach	Temperature range (^{0})	Precursor		Substrate
			I	II	
Co_9S_8	ALD	80–200	bis(*N*,*N*′-diisopropylacetamidinato) cobalt (II)	H_2S	Porous nickel foam
CoS_2	PD CoOx, and Sulf CVD	550	S powder		FTO
Cu_2S	ALD	110–104	$[Cu(Bis\text{-}S\text{-}amd)]_2$ and $Cu_2(DBA)_2$	H_2S	Silicon and fused silica
FeS_2	CVD	230–550	$Cu_2(DBA)_2$, $FeCl_3$, Iron acetylacetonate, $Fe(CO)_5$ and S powder	CH_3CSNH_2, *tert*-Butyl disulfide, $[(CH_3)_3C]_2S_2$	Glass, silicon, silicon, moly/glass, and GC
$FeSe_2$	CVD	220–300, 250–350	$Fe(CO)_5$	H_2Se	GaAs
HfS_2	CVD	900–1000	$HfCl_4$	S powder	SiO_2/Si, $G/SiO_2/Si$, SiO_2, Mica, Sapphire
MnS	ALD	100–225	$Mn(EtCp)_2$	H_2S	Si, fused quartz plate
MoS_2	ALD	150–500	$Mn(EtCp)_2$, $Mo(CO)_6$, $MoCl_5$, $Mo(NMe_2)_4$	H_2S, $CH_3CH_2SS–CH_2CH_3$, H_2S, 1,2-ethanedithiol	3D nickel foam, Silicon, SiO_2, apphire, Silica NP
	CVD	550–1100	MoO_3, Mo(CO), Tetrakis(diethylaminodithiocarbomate) molybdate (IV)	S powder, $(C_2H_5)_2S$, Dimethyl sulfide, DMS	Sapphire, GaN/sapphire, SiO_2, Silicon, PET, PDMS, HOPG, $BiFeO_3$, h-BN, graphene, NP/Ti, NS/G, Au foil, Si_3N_4, $SrTiO_3$, Fused silica, GC, Ta foil
	Prepared Mo, Sulf CVD	200–850	H_2S, S powder		Quartz, SiO_2, GC, Mo foil
	PD MoO_x and Sulf CVD	500–1000	S powder		SiO_2, Sapphire, Silicon, FTO, Graphene

(continued)

Table 7.9 (continued)

Material	Synthesis approach	Temperature range (^{0})	Precursor		Substrate
			I	II	
	Solution on SiO_2 by spin coating, CVD	780	S powder		SiO_2
$MoSe_2$	CVD	450–900	$MoCl_5$, MoO_3	Diethyl selenide, di-*tert*-butylselenide, Se powder	Glass, Graphene network, SiO_2, Sapphire, mica, silicon
	PD Mo*Ox* Sele CVD	700–800	Se powder		Graphene
	Prepared Mo, Sele CVD	550–600	Se powder		GC, quartz, SiO_2, Silicon NW, Mo foil
$MoTe_2$	CVD	600	MoO_3	Te powder	SiO_2
	Prepared Mo, Tell CVD	700	Te powder		SiO_2, FTO/SiO_2
	PD MoO_x and Tell CVD	600	Te powder		SiO_2
NbS_2	CVD	650–1050	$NbCl_5$, Nb	S powder	BN, Sapphire, rGO,
	Prepared Nb, Sulf CVD	900	H_2S		SiO_2
$NbSe_2$	CVD	720	$NbCl_5$	Se powder	SiO_2
NiS_2	ALD	90–350	2,2,6,6-Tetramethylheptane-3,5-dionate nickel, N,N'-di-*tert*-butylacetamidinato nicke	H_2S	SiO_2

(continued)

Table 7.9 (continued)

Material	Synthesis approach	Temperature range (^{0})	Precursor		Substrate
			I	II	
	PD NiO_x and Sulf CVD	550	S powder		FTO
ReS_2	ALD	200–500	$ReCl_5$	H_2S	Al_2O_3
	CVD	450–900	$ReCl_5$, NH_4ReO_4, ReO_3,	H_2S, S powder	Mica, SiO_2, CNF, gold foil, sapphire
$ReSe_2$	CVD	625–760	ReO_3	Se powder, pellet	SiO_2
TaS_2	CVD	700–1000	Ta film, Ta_2O_5, TaC, TaN, TaO_x, Ta powder, $TaCl_5$	S pellets, powder	Sapphire, SiO_2, graphene
TiS_2	ALD	75–400	$TiCl_4$	H_2S	TiO_2, ZnS, rubbed glass
	CVD	150–450	$TiCl_4$	H_2S, $HS(CH_2)_2SH$, $HSC(CH_3)_3$, $S(Si(CH_3)_3)_2$	ZnS, rubbed glass, Cell ($Li/Li_{3.6}Si_{0.6}P_{0.4}O_4$), SiO_2, SS, CR
$TiSe_2$	CVD	820	Titanium oxide	Se powder	SiO_2
VS_2	CVD	550–600	VCl_3	S powder	SiO_2
WS_2	ALD	170–225	$W(CO)_6$	H_2S, NH_3	Stainless steel, SiO_2
	CVD	300–1060	WO_3, $W(CO)_6$, $WOCl_4$, WCl_6, $W(CO)_6$, WS_2 powder	S powder, $HS(CH_2)_2SH$, $HSC(CH_3)_3$, $(C_2H_5)_2S$	Al_2O_3, Au, SiO_2, PET, PDMS, HOPG, $BiFeO_3$, h-BN, graphene, Glass, Fused silica
	PD WO_x, and Sulf CVD	400–1000	S powder, H_2S		FTO, SiO_2, Graphene, Graphene

(continued)

Table 7.9 (continued)

Material	Synthesis approach	Temperature range (^{0})	Precursor		Substrate
			I	II	
	Prepared W, Sulf CVD	600	S powder		W film
	Prepared WO_3 NW and Sulf CVD	650–850	S powder		WO_3 nanowire
WSe_2	CVD	600–950	WO_3, $W(CO)_6$,, WSe_2	Se powder, H_2Se, $(CH_3)2Se$	Sapphire, Si_3N_4/SiO_2, SiO_2, BN
	PD W, and Sele CVD	600	Se powder		Si NW
	PD WO_x and Sele CVD	700–800	Se powder		Graphene
	Prepared WO_3 NR, Tell CVD	650	Se powder		WO_3 nanoribbon
WTe_2	PD W, Tell CVD Prepared WO_3 NR, Tell CVD	650–760	Te powder		SiO_2, WO_3 nanoribbon
ZrS_2	CVD	600–950	$ZrCl_4$	S powder	BN, SiO_2, sapphire

Here PD: predeposited, Sulf: sulfurization, Sele: selelnization, Tell: tellurization, HOPG: highly ordered pyrolytic graphite, rGO: reduced graphene oxide, SS: stainless steel, CNF: carbon nanofiber, GC: glassy carbon, FTO: fluorine-doped tin oxide, BN: boron nitride, NW: nanowire, CR: carbon ribbon, NR: nanoribbon, PDMS: polydimethylsiloxane [251] imperfections, such as mosaicity-tilt, mosaicity-twist, antiphase domain boundaries, and twin flaws [95, 257–260].

The major issue to use of various close lattice matching substrate existence of dangling bonds and surface oxide layers. As an instance, InP (111) lattice is narrowly matching with the Bi_2Se_3. Whereas InP has surface oxidation that is usually thermally desorbed at comparatively high temperatures, therefore, phosphorous to be out gassing to the specimen, it should be remunerated. Traditionally III/V chamber has been used for the easy phosphorous flux supply, but the use of this kind of chamber in the case of chalcogenide lead materials is difficult. However, a few successful have been made to supply chalcogen atoms in form of excess after absorbing surface oxides III/V and other materials substrates. Therefore, a large number of substrates are available with surface dangling bonds. A few successes are also made obtained in the passivation of such surfaces through the chalcogenide overpressure or by deposition of bismuth monolayer [10].

The evidence is also available on the growth of the Bi_2Se_3 by using the buffer layers or virtual substrates. The In_2Se_3 also have trivial band insulating properties with a similar layered crystal structure and lattice constant to the TIs, as per the theoretical predictions use of the In_2Se_3 as a buffer layer can improve the crystal excellence of Bi_2Se_3 thin films. Investigators have systematically optimized the MBE growth constraints of In_2Se_3 on Si(111) substrate [274], as the various substrate's lattice constants are compared for the binary TIs in Fig. 7.34. It reveals the In_2Se_3 layers considered to be greatest appropriate for the Bi_2Se_3 development under a three-step growth arrangement: Se passivation of the Si substrate at 100 °C, In_2Se_3 deposition in the temperature range of 100 to 400 °C and subsequently followed by Se annealing at 400 °C. Moreover, also successful growth of Bi_2Se_3 over the In_2Se_3 together a clear interface has been demonstrated [275] by adopting $(In_xBi_{1-x})_2Se_3$ as a buffer layer among Bi_2Se_3 and sapphire substrate. The reduction of flaws in Bi_2Se_3 crystal reveals an order of magnitude improvement in the mobility of the thin films developed with no buffer layer.

Furthermore, it has been also noticed that the growth on non-vdW substrates (such as Si and GaAs) is more difficult than on inert surfaces (such as HOPG) due to the dangling bonds at the interface that lead to chemical interaction between the TI and substrate [104]. Therefore, by using either H-passivation or deposition of an initial thin chalcogen layer on Si (111) an inert surface are also enables to vdWE [104]. In the case of the GaAs (111), the Se treatment has been used to form the GaSe surface terminated layer that enables a sharp TI/substrate interface, as depicted in the TEM image in Fig. 7.5f [276]. In this sequence, some studies are also available about the selective area growth of topological insulators, specifically Bi_2Te_3 and Sb_2Te_3, on pre-patterned Si (111) substrates [277]. A large number of used suitable substrates for the thin film growth of TMD materials are also summarized in Table 7.9.

Hence the robust innovations in the area of TIs by using various existing and novel techniques (such as CVD, ALD, etc) and theoretical predictions including experimental findings have indicated that the choice of substrates and their preparation are considered to be a crucial parameter for the high-quality growth of TIs [251, 278].

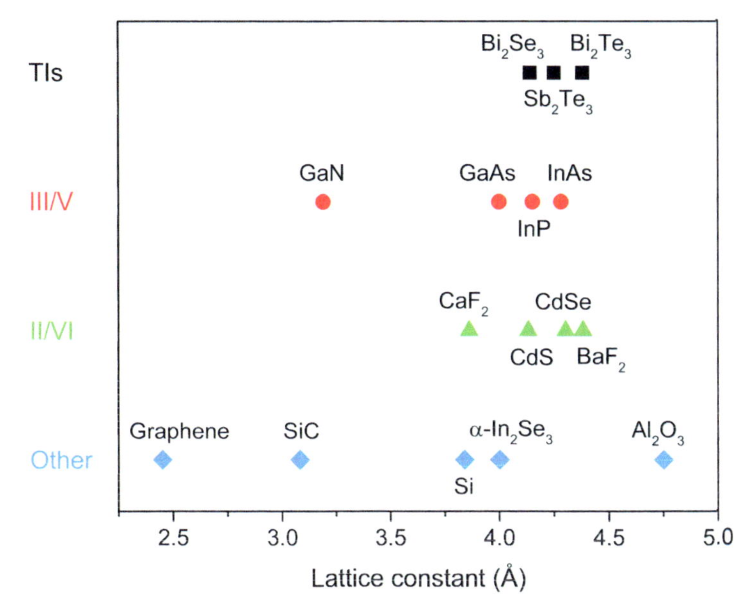

Fig. 7.34 Lattice constants for the different substrates contrast to lattice coefficients of twofold TIs. For zincblende and cubic structures, the lattice constant of the (111) face is illustrated. The wurtzite structures, graphene, In_2Se_3, and Al_2O_3 belong to the lattice constant of the hexagonal (001) face (adopted from Ginley, T. P. et al 2013 Crystals 6, 154)

7.4 Heterostructures and Superlattices

Heterostructures associate to various construction accompanying binary or supplementary inconsistent crystalline materials with precise boundaries among them, whereas the superlattices are periodic organizations of dual or additional material layers. That are controlled by the material stoichiometry, layer thickness, or doping concentrations of every layer. Such structures' electronic and optical characteristics could be modified. The heterostructures and superlattices containing TIs mechanisms have an exceptional surface state for the TIs layers that generates various fascinating events at the interfaces. Therefore, various theoretical estimations and experimental evidences of the heterostructures or superlattices have been related to the TIs for the different kinds of materials.

Since p-type and n-type binary topological insulators alloys gives us a reduced bulk conductivity. So, the addition of additional elements results the ternary alloys, and topological insulating p-n junctions are developed by taking typical binary alloys such as Sb_2Te_3/Bi_2Te_3. Wherein Sb_2Te_3 becoming to p-type doping owing to Sb/Te antisite flaws whereas Bi_2Te_3 makes n-type doped cause of Te vacancies. The development of Bi_2Te_3 and Sb_2Te_3 on GaAs (111) substrates has been achieved with the prospect to produce Sb_2Te_3/Bi_2Te_3 p-n junction [279]. Additionally, it also offers non-vincinal GaAs substrates that can construct high excellence Q_2Te_3 (Q = Sb, Bi) films. Moreover, the Sb_2Te_3/Bi_2Te_3 heterostructures on Si(111) substrates also grow from the alternation in thickness of the Sb_2Te_3 layer, the chemical potential of the underlayer Bi_2Te_3 can be tuned [280]. The corresponding ARPES data has also provided that Dirac cone surface states remain occur at the surface of Sb_2Te_3. It has been also renowned that the Sb_2Te_3/Bi_2Te_3 heterostructures developed on Si(111) also form the ternary complexes at the boundaries owing to Sb/Bi diffusion among the layers [281]. The controllability of the bulk carrier density of the Bi_2Te_3 layer has been achieved through the variation of Sb_2Te_3 layer thickness. Such variation trend is not limited only to discussed case but in recent years also have been obtained in other materials compositions like VSe_2/Bi_2Se_3, HgTe/CdTe, etc. [282, 283].

TI materials could also fabricate by combining to various kinds of semiconductors or band insulating materials. Such as In_2Se_3, which has a similar quintuple layer structure to Be_2Se_3 with relatively low lattice mismatch (~3.4%) possessing a large band gap of ~ 1.3eV. It has been noted that the MBE-grown In_2Se_3/ Be_2Se_3 superlattices TI film thickness can be tuned, therefore, the whole material is switched between the 2D and 3D TI states [284]. The superlattices growth of In_2Se_3/ Be_2Se_3 on Si(111) substrate also offers a good uniformity of thin film with sharp interfaces. Significantly it can also switch from In_2Se_3 to Bi_2Se_3 by exchanging In and Bi fluxes whereas the Se flux is remain unchanged. Moreover, this combination of superlattice also offers to sandwich a layer of In_2Se_3 among dual layers of Bi_2Se_3 [285]. The variation of the thickness of the interlayer leads to the coupling between the twofold TIs layers that can be altered from intensely combined to completely uncombined. A similar interlayer has been also achieved from the combination $(In_xBi_{1-x})_2Se_3$. It noted that the indium composition in the range of 1 to 0.3 combination forte among neighboring of TI layers is progressively enhanced at definite interlayer thickness. A theoretical detail tune-ability of the In_2Se_3 with Bi_2Se_3 for the various compositions has been also discussed by considering various significant parameters, such as phase diagram, Fermi energy level, in terms of the topological property of the materials [286].

In addition to these, the combination of II-VI semiconductor layers has been also grown as the TI materials. Specifically, such structures are interesting due to their superlattices containing manifold suppressed topological/trivial interfaces that allow the inherent transportation character even afterward the film is eliminate to the MBE chamber. This family also offers band structure manufacturing of the devices with different kinds of band gaps and band arrangements among the TI layers and the usual materials. This class material also permits to formation of numerous kinds of band arrangements inside the semiconductor layers that gives us an exceptional functionality in the TI heterostructures devices. Such as in TIII–VI heterostructures utilized the ZnSe as the semiconductor layer owing to its comparatively minor lattice incongruity to Bi_2Se_3(~3.2%) [287]. Moreover, the $Zn_xCd_{1-x}Se/Bi_2Se_3$ and $Zn_xCd_yMg_{1-x-y}Se/Bi_2Se_3$ superlattices on sapphire substrates have been grown with the enhanced physical parameters [288]. Additionally, it also noted that the tuning of the composition in the II-VI combination of the lattice of the layer coordinated to Bi_2Se_3 has reduced the strain at the interface. A combination of the $Zn_{0.49}Cd_{0.51}Se$ and $Zn_{0.23}Cd_{0.25}Mg_{0.52}Se$ layers also reveals the enhanced excellence comparative to

ZnSe. Similarly, the MBE-grown $Zn_{0.49}Cd_{0.51}Se/Bi_2Se_3$ layer has also shown the TI realization [289]. Moreover, a theoretical study of $Cr_{0:41}Sb_{1:59}Te_3/Dy_{0:6}2Bi_{1:38}Te_3$ exchange bias in terms of magnetic topological insulator superlattices has been also interpreted [290].

Additionally, the heterostructures also offers the TIs layered with magnetic insulators (MIs) compared to traditional insulators and semiconductors. In which the surface state of TIs shows gapped behavior owing to the contravention of time-reversal symmetry (TRS) via interchange connection to the MI. A chain of heterostructures of Bi_2Se_3 and magnetic or antiferromagnetic insulators have been explored [291]. It has been noted that the Bi_2Se_3/MnSe heterostructures are the supreme perfect for comparatively robust interchange connection with a humble surface band structure. The MBE-developed Bi_2Se_3/MnSe heterostructure could have an energy gap of 90 meV at the TI surface state owing to the contravention of TRS [292]. The supplementary complex materials such as $(Bi_xSb_{1-x})_2Te_3$/yttrium iron garnet (YIG) heterostructure has been also fabricated [293]. The key benefit of these kinds of combinations is controllability of the magnetic features at the inter-connecting boundaries [294].

Moreover, the band structure of Bi_2Se_3 can be also engineered by developing alternative 2D material over it [295]. The dual kinds of constructions Bi_2Se_3 or Sb_2Te_3 are usually developed on a graphene-terminated 6H-silicon carbide (0001) substrate enclosed by 1 QL of Bi_2Te_3 and Bi_2Se_3 protected by 1 bilayer (BL) of Bi(111). Their ARPES data have revealed that both Bi_2Te_3/Bi_2Se_3 and Bi_2Te_3/Sb_2Te_3 are possessing alike surface states to Bi_2Te_3. This proximity process is explained by realizing all the Bi_2Se_3, Sb_2Te_3, and Bi_2Te_3 have similar Z_2 number [214, 296]. Although, the band construction is altered considerably while the Bi_2Se_3 film is enclosed by the BL Bi(111). Their ARPES data exhibited that the Dirac points for the BL Bi(111) and Bi_2Se_3 film existed, but their band dispersal is expressively amended. It has been also noted that the MBE-developed BL Bi(111) over the Bi_2Te_3 heterostructures reflects a change in the Dirac point associating a band-bending impact at the Bi_2Te_3 and Bi BL interface [297, 298].

The possibility of finding the Majorana fermions in superconductivity in TIs through the proximity effect has been also predicted. As the intense, the atomically-flat Bi_2Se_3 films fabricated from the MBE technique over the 2H-$NbSe_2$(0001) that have been categorized as the s-wave superconductor substrate. Such heterostructures have a superconducting gap at the Bi_2Se_3 surface. The ARPES interpretation also revealed a nonconformity to the ideal theoretical exemplary for the $Bi_2Se_3/NbSe_2$ heterostructures because of the co-occurrence of manifold bands at the chemical potential [299, 300]. The $Bi_2Te_3/NbSe_2$ heterostructures also have superconducting ordering in TI surface states due to the proximity impact [301–303]. Therefore, evidence of a Majorana mode in the magnetic vortex in $Bi_2Se_3/NbSe_3$ heterostructures has been also recognized. Moreover, the epitaxial growth of Bi_2Se_3 over the cuprate-superconductor $Bi_2Sr_2CaCu_2O_8$ (BSCCO) substrate has been also demonstrated, in which the electronic construction **nearby** the Fermi level of the system interpreted [303]. By exploring the ARPES data confirmation to the co-occurrence of dual crystal phases in the film it has been visualized. Besides the crystalline nature of the film, it did not provide evidence of the superconducting proximity effect for the TI film [304, 305]. The nonappearance of proximity effect is due to a slight superconducting coherence distance and the existence of the topological surface state with the diverse regularities at the interface. Thus Bi_2Se_3 combination with a variety of materials such as $V_xSb_{2-x}Te_3$, $Cr_xSb_{2-x}Te_3$, etc has been explored as the TIs for the various potential uses [306, 307]. Such a material's topological phases are usually controlled by certain defined rules [308].

In recent years it has been considered rather than incorporating 3d transition metals into the crystal lattice, a more advantageous strategy is to place two ferromagnetic materials on the top and bottom surfaces of a 3D TI [309] because it can break the TRS for topological surface states (TSS) at the two surfaces through the magnetic proximity. Therefore, it opens an exchange gap and gives rise to a QAH insulator or axion insulator phase, depending on the thickness of the TI layer [310, 311]. Although the induction of sufficient magnetic order to open a sizeable exchange gap via proximity effects is a challenging thing owing to the undesired influence of the abrupt interface potential, this leads to a lattice mismatch between the magnetic material and TI [312]. Considering the minimization of the interface potential and above-discussed key factors has been demonstrated the epitaxial growth of a heterostructure that comprises 4 QL Bi_2Te_3 sandwiched between two single septuple layers (SL) of $MnBi_2Te_4$ ($[MnBi_2Te_4]_{1SL}$–$[Bi_2Te_3]_{4QL}$–$[MnBi_2Te_4]_{1SL}$). The ARPES spectra of the $MnBi_2Te_4]_{1SL}$–$[Bi_2Te_3]_{4QL}$–$[MnBi_2Te_4]_{1SL}$ have revealed that the band structure away from the Fermi energy (>1 eV) similar to 4QL (Bi_2Te_3), while the bands near the Dirac region is noticeably different. Whereas the Dirac cone on the surface of the heterostructure is elevated above the M-shaped valence band and the Fermi velocity of the Dirac electron band away from the Dirac region along ΓM increases the $v_F = 5.0 \pm 0.5 \times 10^5 ms^{-1}$ [313].

Moreover, DFT calculations of the Chern number revealed that the band structure near the Fermi level with spin-orbit coupling (SOC) along the ΓK and ΓM directions with the gapped region. The calculated energy window is ±0.2 eV, in which the 59 meV gap region between −0.029 and 0.03 eV the conductance has quantized at $-e^2/\hbar$. This quantization indicates a non-zero Chern number of −1 and corresponds to the entire 59 meV band gap region. Further, explore the magnetic coupling of the TI surface state and whether a gap is opened. The raw ARPES spectra of the $[MnBi_2Te_4]_{1SL}$–$[Bi_2Te_3]_{4QL}$–$[MnBi_2Te_4]_{1SL}$ thin film at hv = 40 eV has revealed that the band dispersions can be extracted from energy distribution curve (EDC) analysis (red circles) and MDC analysis (yellow and green circles), as illustrated in Fig. 7.35a. Whereas there is some spectral weight near Γ in the Dirac point region due to Te-orbital-related matrix elements effects (promptly in-between hv = 47–60 eV). This has been correlated to be gapped system. An accurate band gap can be obtained from analysis of the EDC around the Γ point, as depicted in Fig. 7.35b. The five distinct peak features are recognized in which two are located just below the Fermi level that corresponds to the bulk conduction bands (BCB, labeled in red). The third and fourth peaks exist in the Dirac point regime at 0.21 and 0.28 eV (labeled in blue) while the fifth peak has a broad feature that is attributed to the bulk valence band ≈0.40 eV (labeled in green). The clear peak splitting in the EDC spectra has confirmed a band gap and the size of the band gap obtained by a fit of the EDC profile with five individual Lorentzian line shapes. The fitting yields the bottom of the electron band to be 0.205 eV and the top of the hole band to be 0.280 eV, thereby a bandgap of around 75 $\pm$ 15 meV. The stacking plot (black) of the EDCs within the k range of $\pm$0.1 Å^{-1} around Γ is depicted in Fig. 7.35c, while the yellow points correspond fitting of the BCB. The plots of the DFT band structure along the Γ–M direction based on the spectral weight from atoms in the $[MnBi_2Te_4]_{1SL}$ layer on the top surface is illustrated in Fig. 7.35d. The predicted band gap is 59 meV. Moreover, the stacking plot of the MDCs at 0.2 eV up to E_F is depicted in Fig. 7.35e. At 0.2 eV only a single peak is located at Γ, moreover, moving toward the Fermi level this peak splits into two dominant peaks that disperse linearly outwards and represent the Dirac electron band, (blue arrows). Near to Fermi level the additional weak components on either side of the Dirac band $\approx$0.2 Å^{-1} away from Γ (black arrows). Based on the dispersion and weak spectral weight can assign to the components of the bulk conduction band, while an additional weak component has been noticed to be near Γ close to the Fermi level correlated to the bulk conduction band (red arrows). Additionally, the plots of the band gap as a function of photon energy are depicted in Fig. 7.35f, in the energy resolution $\pm$15 meV [313].

Moreover, this investigation has also demonstrated that the anisotropy and band warping in the $[MnBi_2Te_4]_{1SL}$–$[Bi_2Te_3]_{4QL}$–$[MnBi_2Te_4]_{1SL}$ heterostructures by fitting to a gapped Dirac band model. The plot of the ARPES constant energy contours and the DFT calculation for the iso-energy contours, highlight the constant energy contours at 0.13 and 0.4 eV, as illustrated in Fig. 7.36a–d. It has been noted that in both cases while approaching the massive Dirac gapped region from the Fermi level, it can be considered a clear evolution in the Fermi surface from hexagram to hexagon. It also shrinks in the gapped region to a point corresponding to the bottom of the electron band at $\approx$0.2 eV, therefore, at 0.25 eV in the hole band the surface texture evolves back into a hexagram upon moving further into the hole band. Here should be noted a few discrepancies between the experiment and theory, because the experimental Fermi surface has revealed a hexagon shape and it overlaps well with the center part of the contour in the DFT calculations. Therefore, the band anisotropy exists in the Dirac electron band, and the warping strength to be explained and modeled by adopting and modifying the Dirac Hamiltonian model of Bi_2Te_3 [314]. The exchange coupling terms have been added to define the Pauli matrix from the top and bottom surface [314];

$$\varepsilon(k) = D - \sqrt{\left(\Delta_c + \lambda k^3 \cos\left(3\theta^3\right) v_F^2 k^2\right)} \tag{7.39}$$

where D represents the doping level, Δ_e is the exchange gap, λ is the corresponding warping strength, v_F is the asymptotic Fermi velocity at large momenta away from the gapped region, k is the wave vector, and θ is the polar angle to the ΓK direction.

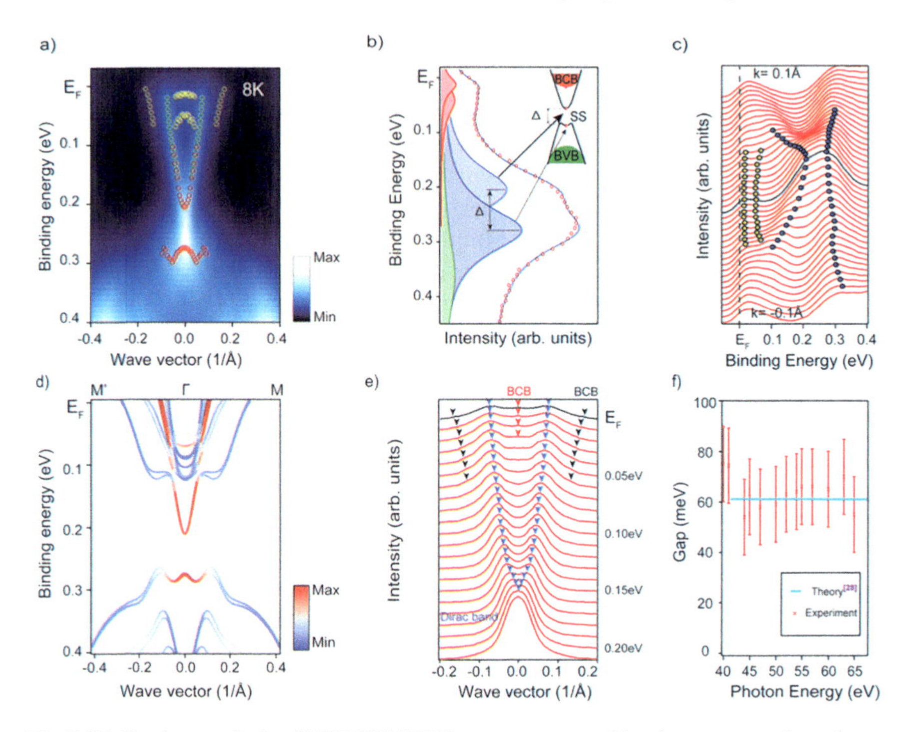

Fig. 7.35 Bandgap analysis of MBT/BT/MBT heterostructure and its photon energy dependence: **a** ARPES spectra of MBT/BT/MBT at $h\nu = 40$ eV at 8 K with p-polarized light, the red and yellow circles represent the dispersion of the Dirac bands and bulk conduction bands in EDC analysis. The green and yellow circles represent the Dirac electron band and bulk conduction band obtained from the MDC analysis: **b** EDC curve (red open circles) at the Γpoint, with the fitted curve plotted as blue solid line, the shaded blue peaks correspond to the Dirac electron band and hole band, the red and green peaks represent the bulk conduction band (BCB) and bulk valence band (BVB), separation of the hole band and Dirac electron band corresponds to $\Delta = 75 \pm 15$ meV; **c** stack plot of EDC curves taken within the wave vector range $\pm$ 0.1 A^{-1}, the EDC curve at Γ is highlighted in black and the blue points are the peak positions of the Dirac electron band and the yellow points are the positions of the bulk conduction bands; **d** Surface projected DFT bands along M′–Γ–M; **e** MDC stacking plot taken from 0.2 eV binding energy up to the Fermi level where the Dirac electron band marked by the blue arrows and the bulk conduction bands with the red and black arrows; **f** bandgap defined from EDC analysis as a function of photon energy (red markers), the DFT-predicted bandgap is plotted with a horizontal blue line, the average bandgap across all photon energies is 64 ± 15 meV (adopted from Li et al. [313])

7.5 Conclusions

Thus the author has discussed TMDs' topological insulating properties of different kinds of materials by providing a sound background discussion including bulk TMD TIs aspects and defined the Berry curvature including chern invariance, Chern of the insulator, topological protection of transport properties, TIs as a Dirac system, Hamiltonians and kinetics. Most of the topological properties have been provided in

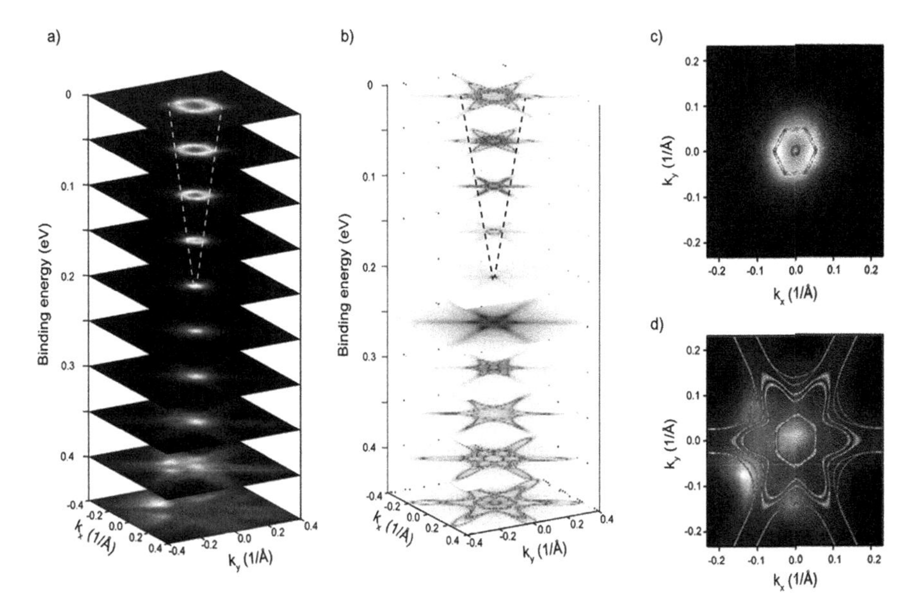

Fig. 7.36 Strong hexagonal warping in the MBT/BT/MBT heterostructure: **a** iso-energy scans taken on MBT/BT/MBT across the gap region between the Fermi level and 0.4 eV; **b** DFT-calculated iso-energy scans in the same energy range showing good agreement. The dashed lines of the Dirac electron band are a guide to the eye; **c** iso-energy cut at 0.13 eV; **d** at 0.4 eV binding energy overlaid with DFT-calculated contours. The angular intensity distribution shows threefold symmetry which agrees with the contour calculated by DFT (adopted from Li et al. [313])

terms of surface-to-bulk conductivity, therefore, a concrete discussion on monolayer TMD topological properties has been also discussed, including a description of binary and ternary compositions of topological insulating materials. Crystal structures of the topological insulating materials that are plying crucial role to define their topological phase, therefore, a discussion on the crystal of structures of TIs has been also provided including a theoretical aspect on monolayer TIs crystal structure and experimental pieces of evidence by providing the concepts of the wrinkled folded and scrolled TMDs. The band structures of topological insulating materials are quite different from the conventional materials, to provide a clear view of it a concrete general discussion on TIs band structure in a normal state has been also incorporated. In addition to this, to discriminate the different kinds of superconductivity in topological phase monolayers a discussion on Ising superconductivity in monolayer TIs has been also provided by discussing Type-I, and Type-II including a theoretical aspect. The different kinds of topological phases occur in TIs, therefore, the significant nodal topological phase and aging effects including edge states transport of 2D TIs and the Impact of topological edge states on optoelectronic properties have been also addressed. Including all these TIs properties, a general discussion of various significant properties of the TIs has been also discussed such as cracking, carrier density, structural quality of thin films, choice of substrate and heterostructures, and superlattices.

References

1. Ren Y, Qiao Z, Niu Q (2016) Topological phases in two-dimensional materials: a review. Rep Prog Phys 79:066501
2. Culcer D, Keser AC, Li Y, Tkachov G et al (2020) Transport in two-dimensional topological materials: recent developments in experiment and theory. 2D Mater 7:022007
3. Chen YL, Analytis JG, Chu JH, Liu ZK et al (2009) Experimental realization of a three-dimensional topological insulator, Bi2Te3. Science 325:178–181
4. Xia Y, Qian D, Hsieh D, Wray L et al (2009) Observation of a large-gap topological-insulator class with a single Dirac cone on the surface. Nat Phys 5:398–402
5. Hsieh D, Xia Y, Qian D, Wray L et al (2009) A tunable topological insulator in the spin helical Dirac transport regime. Nature 460:1101–1105
6. Mandal PS, Springholz G, Volobuev VV, Caha O et al (2017) Topological quantum phase transition from mirror to time reversal symmetry protected topological insulator. Nat Commun 8:968
7. Osterhage H, Gooth J, Hamdou B, Gwozdz P et al (2014) Thermoelectric properties of topological insulator Bi_2Te_3, Sb_2Te_3, and Bi_2Se_3 thin film quantum wells. Appl Phys Lett 105:123117
8. Zhu X, Santos L, Howard C, Sankar R et al (2012) Electron-phonon coupling on the surface of the topological insulator bi2se3 determined from surface-phonon dispersion measurements. Phys Rev Lett 108:185501
9. Pan ZH, Fedorov AV, Gardner D, Lee YS et al (2012) Measurement of an exceptionally weak electron-phonon coupling on the surface of the topological insulator Bi2Se3 using angle-resolved photoemission spectroscopy. Phys Rev Lett 108:187001
10. Ginley TP, Wang Y, Law S (2016) Topological insulator film growth by molecular beam epitaxy: a review. Crystals 6:154–226
11. Wang Y, Zhu D, Wu Y, Yang Y et al (2017) Room temperature magnetization switching in topological insulator-ferromagnet heterostructures by spin-orbit torques. Nat Commun 8:1364
12. Rostami H, Guinea F, Polini M, Roldan R (2018) Piezoelectricity and valley chern number in inhomogeneous hexagonal 2D crystals. npj 2D Mater Appl 2:15
13. Avila J, Pearanda F, Prada E, San-Jose P et al (2019) Non-hermitian topology as a unifying framework for the Andreev versus Majorana states controversy. Commun Phys 2:133
14. Zhang W, Zhang Y, Qiu J, Zhao Z et al (2021) Topological structures of transition metal dichalcogenides: a review on fabrication, effects, applications, and potential. InfoMat 3:133–154
15. Xu H, Xu S, Xu X, Zhuang J et al (2022) Recent advances in two-dimensional van der Waals magnets. Microstructures 2:2022011
16. Pan H, Feng Y, Lin J (2005) Ab initio study of electronic and optical properties of multiwall carbon nanotube structures made up of a single rolled-up graphite sheet. Phys Rev B 72:085415
17. Chen Y, Lu J, Gao Z (2007) Structural and electronic study of nanoscrolls rolled up by a single graphene sheet. J Phys Chem C 111:1625–1630
18. Chang C, Ortix C (2017) Theoretical prediction of a giant anisotropic magnetoresistance in carbon nanoscrolls. Nano Lett 17:3076–3080
19. Liu Z, Gao J, Zhang G, Cheng Y et al (2017) From twodimensional nano-sheets to roll-up structures: expanding the family of nanoscroll. Nanotechnology 28:385704
20. Ma C, Sun X, Du H, Wang J et al (2018) Landau quantization of a narrow doubly-folded wrinkle in monolayer graphene. Nano Lett 18:6710–6718

21. Li G, Luican A, Lopes dos Santos J, Castro Neto AH et al (2010) Observation of Van Hove singularities in twisted graphene layers. Nature Phys 6:109–113
22. Kim K, Coh S, Tan LZ, Regan W et al (2012) Raman spectroscopy study of rotated double-layer graphene:misorientation-angle dependence of electronic structure. Phys Rev Lett 108:246103
23. Dean CR, Wang L, Maher P, Forsythe C et al (2013) Hofstadter's butterfly and the fractal quantum hall effect in moiré superlattices. Nature 497:598–602
24. Ponomarenko LA, Gorbachev RV, Yu G, Elias DC et al (2013) Cloning of Dirac fermions in graphene superlattices. Nature 497:594–597
25. Hunt B, Sanchez-Yamagishi JD, Young AF, Yankowitz M et al (2013) Massive Dirac fermions and Hofstadter butterfly in a Van der Waals heterostructure. Science 340:1427–1430
26. Deng S, Che S, Debbarmaa R, Berry V (2019) Strain in a single wrinkle on an MoS_2 flake for in-plane realignment of band structure for enhanced photo-response. Nanoscale 11:504–511
27. Yang S, Wang C, Sahin H, Chen H et al (2015) Tuning the optical, magnetic, and electrical properties of $ReSe_2$ by Nanoscale strain engineering. Nano Lett 15:1660–1666
28. Xiao D, Liu GB, Feng W, Xu X et al (2012) Coupled spin and valley physics in monolayers of MoS_2 and other group-VI dichalcogenides. Phys Rev Lett 108:196802
29. Yao W, Xiao D, Niu Q (2008) Valley-dependent optoelectronics from inversion symmetry breaking. Phys Rev B 77:235406
30. Shinde S, Dhakal1 K, Chen X, Yun WS et al (2018) Stacking-controllable interlayer coupling and symmetric configuration of multilayered MoS2. NPG Asia Mater 10:e468
31. Kumar N, Najmaei S, Cui Q, Ceballos F et al (2013) Second harmonic microscopy of monolayer MoS2. Phys Rev B 87:161403
32. Malard LM, Alencar TV, Barboza APM, Mak KF et al (2013) Observation of intense second harmonic generation from MoS2 atomic crystals. Phys Rev B 87:201401
33. Zhu Z, Cheng Y, Schwingenschlog U (2011) Giant spin-orbitinduced spin splitting in two-dimensional transition-metal dichalcogenide semiconductors. Phys Rev B 84:153402
34. Gao L, Ni GX, Liu Y, Liu B et al (2014) Faceto- face transfer of wafer-scale graphene films. Nature 505:190–194
35. Wu L, Gub K, Li Q (2019) New families of large band gap 2D topological insulators in ethynylderivative functionalized compounds. Appl Surf Sci 484:1208–1213
36. Li M, Sinev I, Benimetskiy F, Ivanova T et al (2021) Experimental observation of topological Z2 excitonpolaritons in transition metal dichalcogenide monolayers. Nat Commun 12:4425
37. Hasan MZ, Kane CL (2010) Topological insulators. Rev Mod Phys 82:3045–3067
38. Qi XL, Zhang SC (2011) Topological insulators and superconductors. Rev Mod Phys 83:1057–1110
39. Hsieh D, Qian D, Wray L, Xia Y et al (2008) A topological dirac insulator in a quantum spin hall phase. Nature 452:970–974
40. Ren Z, Taskin AA, Sasaki S, Segawa K et al (2010) Large bulk resistivity and surface quantum oscillations in the topological insulator Bi2Te2Se. Phys Rev B 82:241306
41. Jia S, Ji H, Climent-Pascual E, Fuccillo MK et al (2011) Low-carrier-concentration crystals of the topological insulator Bi2Te2Se. Phys Rev B 84:235206
42. Ando Y (2013) Topological Insulator Materials. J Phys Soc Japan 82:102001
43. Yang WM, Lin CJ, Liao J, Li YQ (2013) Electrostatic field effects on three-dimensional topological insulators. China Phys B 22:097202
44. Zhang J, Chang CZ, Zhang V, Wen J et al (2011) Band structure engineering in (Bi1−xSbx)2Te3 ternary topological insulators. Nat Commun 2:574
45. Kong D, Chen Y, Cha JJ, Zhang Q et al (2011) Ambipolar field effect in the ternary topological insulator (BixSb1–x)2Te3 by composition tuning. Nat Nanotechnol 6:705–709
46. Ren Z, Taskin AA, Sasaki S, Segawa K et al (2011) Berry phase of nonideal Dirac fermions in topological insulators. Phys Rev B 84:035301
47. König M, Wiedmann S, Brüne B, Roth A et al (2007) Quantum spin Hall insulator state in HgTe quantum wells. Science 318:766–770

48. Knez I, Du RR, Sullivan G (2011) Evidence for helical edge modes in inverted InAs/ GaSb quantum wells. Phys Rev Lett 107:126603
49. Bernevig AB, Hughes TL, Zhang SC (2006) Quantum spin Hall effect and topological phase transition in HgTe quantum wells. Science 314:1757–1761
50. Ma Y, Kou L, Du A, Heine T (2015) Group 14 element-based non centrosymmetric quantum spin Hall insulators with large bulk gap. Nano Res 8:3412–3420
51. Xu Y, Yan B, Zhang HJ, Wang J et al (2013) Large-gap quantum spin hall insulators in tin films. Phys Rev Lett 111:136804
52. Chaves A, Azadani JG, Alsalman H, da Costa DR et al (2020) Bandgap engineering of two-dimensional semiconductor materials. npj 2D Mater Appl 4:29
53. Zhou W, Gong H, Jin X, Chen Y et al (2022) Recent progress of two-dimensional transition metal dichalcogenides for thermoelectric applications. Front Phys 10:842789
54. Choi W, Choudhary N, Han GH, Park J et al (2017) Recent development of two-dimensional transition metal dichalcogenides and their applications. Mater Today 20:116–130
55. Tian Y, Cheng Y, Huang J, Zhang S et al (2022) Epitaxial growth of large area ZrS2 2D semiconductor films on sapphire for optoelectronics. Nano Res https://doi.org/10.1007/s12274-022-4308-4
56. Yang WX, Zhou HL, Su D, Yang ZR et al (2022) Recent progress in 2D material van der Waals heterostructure-based luminescence devices towards the infrared wavelength range. J Mater Chem C 10:7352–7367
57. Ma Y, Kou L, Dai Y, Heine T (2016) Two-dimensional topological insulators in group-11 chalcogenide compounds: M 2 Te (M=Cu, Ag). Phys Rev B 93:235451
58. Pham A, Gil CJ, Smith SC, Li S (2015) Orbital engineering of two-dimensional materials with hydrogenation: a realization of giant gap and strongly correlated topological insulators. Phys Rev B 92:035427
59. Vogt P, Padova PD, Quaresima C, Avila J (2012) Silicene: compelling experimental evidence for graphenelike two-dimensional silicon. Phys Rev Lett 108:155501
60. Zhang RW, Zhang CW, Ji WX, Li SS et al (2016) Functionalized thallium antimony films as excellent candidates for large-gap quantum spin hall insulator. Sci Rep 6:21351
61. Sante D, Eck P, Bauernfeind M, Will M et al (2019) Towards topological quasifreestanding stanene via substrate engineering. Phys Rev B 99:035145
62. Reis F, Li G, Dudy L, Bauernfeind M et al (2017) Bismuthene on a SiC substrate: a candidate for a high-temperature quantum spin Hall material. Science 357:287–290
63. Song Z, Liu CC, Yang J, Han J et al (2014) Quantum spin Hall insulators and quantum valley Hall insulators of BiX/SbX (X=H, F, Cl and Br) monolayers with a record bulk band gap. NPG Asia Mater 6:e147
64. Jung J, Kim YH (2022) Hidden breathing kagome topology in hexagonal transition metal dichalcogenides. Phys Rev B 105:085138
65. Berry MV (1984) Quantal phase factors accompanying adiabatic changes. Proc R Soc A 392:45–57
66. Armitage N, Mele E, Vishwanath A (2018) Weyl and Dirac semimetals in three-dimensional solids. Rev Mod Phys 90:015001
67. Xu SY, Belopolski I, Alidoust N, Neupane M et al (2015) Discovery of a Weyl fermion semimetal and topological Fermi arcs. Science 349:613
68. Yan B, Felser C (2017) Topological Materials: Weyl semimetals. Ann Rev Condens Matter Phys 8:337–354
69. Zyuzin AA, Burkov AA (2012) Topological response in Weyl semimetals and the chiral anomaly. Phys Rev B 86:115133
70. Goswami P, Tewari S (2013) Axionic field theory of (3+1)-dimensional Weyl semimetals. Phys Rev B 88:245107
71. Chen Y, Wu S, Burkov AA (2013) Axion response in Weyl semimetals. Phys Rev B 88:125105
72. Kormanyos A, Zolyomi V, Drummond ND, Rakyta P et al (2013) Monolayer MoS2: trigonal warping, the Γvalley, and spin-orbit coupling effects. Phys Rev B 88:045416

73. Kormanyos A, Burkard GG, Gmitra M, Fabian J et al (2015) k·p theory for two-dimensional transition metal dichalcogenide semiconductors. 2D Mater 2:022001
74. Gong Z, Liu GB, Yu H, Xiao D et al (2013) Magnetoelectric effects and valley-controlled spin quantum gates in transition metal dichalcogenide bilayers. Nat Commun 4:2053
75. Culcer D, Sekine A, MacDonald AH (2017) Interband coherence response to electric fields in crystals: berry-phase contributions and disorder effects. Phys Rev B 96:035106
76. Mattheiss LF (1973) Band structures of transition-metal-dichalcogenide layer compounds. Phys Rev B 8:3719
77. Radisvljevic B, Radenovic A, Brivio J, Giacometti V et al (2011) Single-layer MoS2 transistors. Nat Nanotechnol 6:147–150
78. Wang QH, Kalantar-Zadeh K, Kis A, Coleman JN et al (2012) Electronics and optoelectronics of two-dimensional transition metal dichalcogenides. Nat Nanotechnol 7:699–712
79. Bao W, Cai X, Kim D, Sridhara K et al (2013) High mobility ambipolar MoS2 field-effect transistors: substrate and dielectric effects. Appl Phys Lett 102:042104
80. Lu JM, Zheliuki O, Leermakersn I, Yuanu NFQ et al (2015) Evidence for two-dimensional Ising superconductivity in gated MoS2. Science 350:1353–1357
81. Saito Y, Nakamura Y, Bahramy MS, Kohama Y et al (2016) Superconductivity protected by spin-valley locking in ion-gated MoS2. Nat Phys 12:144–149
82. Sohn E, Xi X, He WY, Jiang S et al (2018) An unusual continuous paramagnetic-limited superconducting phase transition in 2D NbSe2. Nat Mater 17:504–508
83. Bawden L, Cooil SP, Mazzola F, Riley JM et al (2016) Spin-valley locking in the normal state of a transition-metal dichalcogenide superconductor. Nat Commun 7:11711
84. He WY, Zhou BT, He JJ, Yuan NFQ et al (2018) Magnetic field driven nodal topological superconductivity in monolayer transition metal dichalcogenides. Commun Phys 1:40
85. Qian X, Liu J, Fu L, Li J (2014) Quantum spin Hall effect in two-dimensional transition metal dichalcogenides. Science 346:1344
86. Keum D, Cho S, Kim JH, Choe DH et al (2015) Bandgap opening in few-layered monoclinic MoTe2. Nat Phys 11:482
87. Ma Y, Kou L, Li X, Dai Y et al (2015) Quantum spin Hall effect and topological phase transition in two-dimensional square transition-metal dichalcogenides. Phys Rev B 92:085427
88. Nie SM, Song Z, Weng H, Fang Z (2015) Quantum spin Hall effect in two-dimensional transition-metal dichalcogenide haeckelites. Phys Rev B 91:235434
89. Ma Y, Kou L, Li X, Dai Y et al (2016) Two-dimensional transition metal dichalcogenides with a hexagonal lattice: room-temperature quantum spin Hall insulators. Phys Rev B 93:035442
90. Duk HC et al (2016) Understanding topological phase transition in monolayer transition metal dichalcogenides. Phys Rev B 93:125109
91. Fu L, Kane CL (2007) Topological insulators with inversion symmetry. Phys Rev B 76:045302
92. Kane CL, Mele EJ (2005) $z_2$topological order and the quantum spin hall effect. Phys Rev Lett 95:146802
93. Fu L, Kane CL, Mele EJ (2007) Topological insulators in three dimensions. Phys Rev Lett 98:106803
94. Peng H, Lai K, Kong D, Meister S et al (2010) Aharonov–Bohm interference in topological insulator nanoribbons. Nat Mater 9:225–229
95. Walsh LA, Hinkle CL (2017) Van der Waals epitaxy: 2D materials and topological insulators. Appl Mater Today 9:504–515
96. Gao W, Zheng Z, Wen P, Huo N (2020) Novel two-dimensional monoelemental and ternary materials: growth, physics and application. Nanophotonics 9:2147–2168
97. Li X, Zhang S, Wang Q (2017) Topological insulating states in 2D transition metal dichalcogenides induced by defects and strain. Nanoscale 9:562–569
98. Freitas RRQ, Rivelino R, de Brito Mota F, de Castilho CMC (2015) Topological insulating phases in two-dimensional bismuth-containing single layers preserved by hydrogenation. J Phys Chem C 119:23599–23606
99. Kepaptsoglou DM, Gilks D, Lari L, Ramasseet QM et al (2015) STEM and EELS study of the graphene/Bi2Se3 interface. Microsc Microanal 21:1151

100. Li HD, Wang ZY, Kan X, Guo X et al (2010) The van der Waals epitaxy of Bi2Se3 on the vicinal Si(111) surface: an approach for preparing high-quality thin films of a topological insulator. New J Phys 12:103038
101. Taskin AA, Sasaki S, Segawa K, Ando Y (2012) Achieving surface quantum oscillations in topological insulator thin films of Bi_2Se_3. Adv Mater 24:5581–5585
102. Tsipas P, Xenogiannopoulou E, Kassavetis S, Tsoutsou D et al (2014) Observation of surface dirac cone in high-quality ultrathin epitaxial Bi2Se3 topological insulator on AlN(0001) dielectric. ACS Nano 8:6614–6619
103. Ginley TP, Law S (2016) Growth of Bi2Se3 topological insulator films using a selenium cracker source. J Vac Sci Technol B 34:02L105
104. Bansal N, Kim YS, Edrey E, Brahlek M et al (2011) Epitaxial growth of topological insulator Bi2Se3 film on Si(111) with atomically sharp interface. Thin Solid Films 520:224–229
105. Akiyama M, Kawarada Y, Kaminishi K (1984) Growth of single domain GaAs layer on (100)-oriented si substrate by MOCVD. Jpn J Appl Phys 23:L843
106. Zhang Y, He K, Chang CZ, Song CL et al (2010) Crossover of the three-dimensional topological insulator Bi_2Se_3 to the two-dimensional limit. Nat Phys 6:584–588
107. Analytis JG, Chu JH, Chen Y, Corredor F et al (2010) Bulk Fermi surface coexistence with Dirac surface state in Bi_2Se_3: a comparison of photoemission and Shubnikov–de Haas measurements. Phys Rev B 81:205407
108. Roy A, Guchhait S, Sonde S, Deyet R et al (2013) Two-dimensional weak anti-localization in Bi2Te3 thin film grown on Si(111)-(7 $\times$ 7) surface by molecular beam epitaxy. Appl Phys Lett 102:163118
109. Sufyan A, Macam G, Huang ZQ, Hsu CH et al (2022) Robust tunable large-gap quantum spin hall states in monolayer Cu2S on insulating substrates. ACS Omega 7:15760–15768
110. Cohen AJ, Mori-Sánchez P, Yang W (2008) Insights into current limitations of density functional theory. Science 321:792–794; Cohen AJ, Mori-Sánchez P, Yang W (2012) Challenges for density functional theory. Chem Rev 112:289–320
111. Santra B, Perdew JP (2019) Perdew-Zunger self-interaction correction: how wrong for uniform densities and large-Z atoms? J Chem Phys 150:174106
112. Adeagbo WA, Ben Hamed H, Nayak SK, Böttcher R et al (2019) Theoretical investigation of iron incorporation in hexagonal barium titanate. Phys Rev B 100:184108
113. Jenkins GS, Sushkov AB, Schmadel DC, Butch NP et al (2010) Terahertz Kerr and reflectivity measurements on the topological insulator Bi_2Se_3. Phys Rev B 82:125120
114. Varsano D, Palummo M, Molinari E, Rontani M (2020) A monolayer transition-metal dichalcogenide as a topological excitonic insulator. Nat Nanotechnol 15:367–372
115. Lv MH, Li CM, Sun WF (2022) Spin-orbit coupling and spin-polarized electronic structures of janus vanadium-dichalcogenide monolayers: first-principles calculations. Nanomaterials 12:382
116. Zeng J, Liu H, Jiang H, Sun QF (2021) Multiorbital model reveals a second-order topological insulator in 1H transition metal dichalcogenides. Phys Rev B 104:L161108
117. Zhang H, Liu CX, Qi XL, Dai X et al (2009) Topological insulators in Bi_2Se_3, Bi_2Te_3 and Sb_2Te_3 with a single Dirac cone on the surface. Nat Phys 5:438–442
118. Zhang Y, He K, Chang CZ, Song CL et al (2012) Quasiparticle effects in the bulk and surface-state bands of Bi_2Se_3 and Bi_2Te_3 topological insulators. Phys Rev B 85:161101
119. Rusinov IP, Nechaev IA, Chulkov EV (2013) Theoretical study of influencing factors on the dispersion of bulk band-gap edges and the surface states in topological insulators Bi_2Te_3 and Bi_2Se_3. J Exp Theor Phys 116:1006–1017
120. Li W, Wei XY, Zhu JX, Ting CS et al (2014) Pressure-induced topological quantum phase transition in Sb2Se3. Phys Rev B 89:035101
121. Seixas L, West D, Fazzio A, Zhang SB et al (2015) Bruton's tyrosine kinase is essential for NLRP3 inflammasome activation and contributes to ischaemic brain injury. Nat Commun 6:7360
122. Seibel C, Braun J, Maa H, Bentmann H et al (2016) Photoelectron spin polarization in the Bi2Te3 (0001) topological insulator: initial- and final-state effects in the photoemission process. Phys Rev B 93:245150

123. Datzer C, Zumbülte A, Braun J, Förster T et al (2017) Unraveling the spin structure of unoccupied states in Bi_2Se_3. Phys Rev B 95:115401
124. Tao LL, Tsymbal EY (2018) Persistent spin texture enforced by symmetry. Nat Commun 9:2763
125. Förster T, Krüger P, Rohlfing M (2015) Two-dimensional topological phases and electronic spectrum of Bi_2Se_3 thin films from GW calculations. Phys Rev B 92:201404
126. Förster T, Krüger P, Rohlfing M (2016) GW calculations for Bi2Te3 and Sb2Te3 thin films: electronic and topological properties. Phys Rev B 93:205442
127. Reid TK, Alpay SP, Balatsky AV, Nayak SK et al (2020) First-principles modeling of binary layered topological insulators: structural optimization and exchange-correlation functional. Phys Rev B 101:085140
128. Li J, Li Y, Du S, Wang Z et al (2019) Intrinsic magnetic topological insulators in van der Waals layered MnBi 2 Te 4-family materials. Sci Adv 5:eaaw5685
129. Cao G, Liu H, Liang J, Cheng L et al (2018) Rhombohedral Sb2Se3 as an intrinsic topological insulator due to strong van der Waals interlayer coupling. Phys Rev B 97:075147
130. Sahoo S, Alpay SP, Hebert RJ (2018) Surface phase diagrams of titanium in oxygen, nitrogen and hydrogen environments: a first principles analysis. Surf Sci 677:18–25
131. Sahoo S, Khanna SN, Entel P (2015) Controlling the magnetic anisotropy of Ni cluster supported on graphene flakes with topological defects. Appl Phys Lett 107:043102
132. Sahoo S, Hucht A, Gruner ME, Rollmann G (2010) Magnetic properties of small Pt-capped Fe Co, and Ni clusters: a density functional theory study. Phys Rev B 82:054418
133. Kuroda K, Ye M, Kimura A, Eremeev SV et al (2010) Experimental realization of a three-dimensional topological insulator phase in ternary chalcogenide $TlBiSe_2$. PRL 105:146801
134. Lin H, Markiewicz RS, Wray LA, Fu L et al (2010) Single-dirac-cone topological surface states in the TlBiSe2 class of topological semiconductors. Phys Rev Lett 105:036404
135. Eremeev SV, Koroteev YM, Chulkov EV (2010) Ternary thallium-based semimetal chalcogenides Tl-V-VI2 as a new class of three-dimensional topological insulators. JETP Lett 91:594–598
136. Ma Y, Kou L, Huang B, Dai Y et al (2018) Two-dimensional ferroelastic topological insulators in single-layer Janus transition metal dichalcogenides MSSe(M = Mo, W). Phys Rev B 98:085420
137. Sufyan A, Macam G, Hsu CH, Huang ZQ et al (2021) Theoretical prediction of topological insulators in two-dimensional ternary transition metal chalcogenides (MM'X4, M = Ta, Nb, or V; M'= Ir, Rh, or Co; X = Se or Te). Chin J Phys 73:95–102
138. Liu J, Wang H, Fang C, Fu L et al (2017) Van der Waals stacking-induced topological phase transition in layered ternary transition metal chalcogenides. Nano Lett 17:467–475
139. Chen M, Zhu L, Chen Q, Miao N (2020) Quantifying the composition dependency of the ground-state structure, electronic property and phase-transition dynamics in ternary transition-metal-dichalcogenide monolayers. J Mater Chem C 8:721–733
140. Dolui K, Rungger I, Das Pemmaraju C, Sanvito S (2013) Possible doping strategies for MoS_2 monolayers: An ab initio study. Phys Rev B 88:075420
141. Chen W, Hou X, Shi X, Pan H (2018) Two-dimensional janus transition metal oxides and chalcogenides: multifunctional properties for photocatalysts, electronics, and energy conversion. ACS Appl Mater Interfaces 10:35289–35295
142. Cui F, Feng Q, Hong J, Wang R et al (2017) Synthesis of large-size 1T′ ReS2x Se2(1–x) alloy monolayer with tunable bandgap and carrier type. Adv Mater 29(46):1705015
143. Susarla S, Hachtel JA, Yang X, Kutana A et al (2018) Thermally induced 2d alloy-heterostructure transformation in quaternary alloys. Adv Mater 30:1804218
144. Park J, Kim MS, Park B, Oh SH et al (2018) Composition-tunable synthesis of large-scale Mo1–xWxS2 alloys with enhanced photoluminescence. ACS Nano 12:6301–6309
145. Gong Y, Liu Z, Lupini AR, Shi G et al (2014) Band gap engineering and layer-by-layer mapping of selenium-doped molybdenum disulfide. Nano Lett 14:442–449
146. Saito Y, Robertson J (2018) Direct transition of a HfGeTe4 ternary transition-metal chalcogenide monolayer with a zigzag van der Waals gap. APL Mater 6:046104

147. Macam G, Sufyan A, Huang ZQ, Hsu CH et al (2021) Tuning topological phases and electronic properties of monolayer ternary transition metal chalcogenides (ABX4, A/B = Zr, Hf, or Ti; X = S, Se, or Te). Appl Phys Lett 118:111901
148. Feng et al (2022) Prediction of topological Dirac semimetal in Ca-based Zintl layered compounds CaM2X2 (M = Zn or Cd; X = N, P, As, Sb, or Bi). Sci Rep 12:4582
149. Dias AC, Qu F, Azevedo DL, Fu J (2018) Band structure of monolayer transition-metal dichalcogenides and topological properties of their nanoribbons: next-nearest-neighbor hopping. Phys Rev B 98:075202
150. Duan X, Wang C, Fan Z, Hao G et al (2016) Synthesis of WS2xSe2-2x alloy nanosheets with composition-tunable electronic properties. Nano Lett 16:264–269
151. Zhang Y, Ye J, Matsuhashi Y, Iwasa Y (2012) Ambipolar MoS_2 thin flake transistors. Nano Lett 12:1136–1140
152. Castellanos-Gomez A, Roldán R, Cappelluti E, Buscema M et al (2013) Local strain engineering in atomically thin MoS_2. Nano Lett 13:5361–5366
153. Yang S, Lee Y, Kim JH, Kim HJ et al (2019) Single-step synthesis of wrinkled $MoSe_2$ thin films. Curr Appl Phys 19:273–278
154. Kim S, Kwona O, Kim D, Kim J et al (2018) Influence of graphene thickness and grain boundaries on MoS_2 wrinkle nanostructures. Phys Chem Chem Phys 20:17000–17008
155. Wang C, Yang SX, Zhang H, Du LN et al (2016) Synthesis of atomically thin GaSe wrinkles for strain sensors. Front Phys 11:116802
156. Jung W, Cho K, Lee W, Odom T et al (2018) Universal method for creating hierarchical wrinkles on thin-film surfaces. ACS Appl Mater Interfaces 10:1347–1355
157. Jung W, Yun G, Kim Y, Kim M et al (2019) Relationship between hydrogen evolution and wettability for multiscale hierarchical wrinkles. ACS Appl Mater Interfaces 11:7546–7552
158. Deng SK, Gao EL, Xu ZP et al (2017) Adhesion energy of MoS_2 thin films on silicon-based substrates determined via the attributes of a single MoS_2 wrinkle. ACS Appl Mater Interfaces 9:7812–7818
159. Brivio Jacopo TL, Alexander D, Kis A (2011) Ripples and layers in ultrathin MoS_2 membranes. Nano Lett 11:5148–5153
160. Shanmugam V, Mensah RA, Babu K, Gawusu S et al (2022) A review of the synthesis, properties, and applications of 2d materials Part. Part Syst Charact 39:2200031
161. Yana Y, Lianga S, Wangc X, Zhang M et al (2021) Robust wrinkled MoS2/N-C bifunctional electrocatalysts interfaced with single Fe atoms for wearable zinc-air batteries. PNAS 118:e2110036118
162. Lee J, Wang Z, He K, Shan J et al (2013) High frequency MoS_2 nanomechanical resonators. ACS Nano 7:6086–6609
163. Sun L, Ying Y, Huang H et al (2014) Ultrafast molecule separation through layered WS2 nanosheet membranes. ACS Nano 8:6304–6311
164. Xing Y, Zhao R, Shan P, Zheng F et al (2017) Ising superconductivity and quantum phase transition in macro-size monolayer $NbSe_2$. Nano Lett 17:6802–6807
165. Zhang J, Xiao J, Meng X, Monroeet C et al (2010) Free folding of suspended Graphene sheets by random mechanical stimulation. Phys RevLett 104:166805
166. Meng X, Li M, Kang Z, Zhang X et al (2013) Mechanics of self-folding of single-layer graphene. J Phys D Appl Phys 46:055308
167. Iff O, Tedeschi D, Martín-Sánchez J, Moczała-Dusanowska M et al (2019) Strain-tunable single photon sources in WSe_2 monolayers. Nano Lett 19:6931–6936
168. Shi J, Ma D, Han GF, Zhang Y et al (2014) Controllable growth and transfer of monolayer MoS_2 on au foils and its potential application in hydrogen evolution reaction. ACS Nano 8:10196–10204
169. Taylor J, Argyropoulos C, Morin S (2016) Soft surfaces for the reversible control of thin-film microstructure and optical reflectance. Adv Mater 28:2595–2600
170. Jiang T, Liu H, Huang D, Zhang S et al (2014) Valley and band structure engineering of folded MoS_2 bilayers. Nat Nanotechnol 9:825–829

171. Zhao J, Deng Q, Ly T, Han GH et al (2015) Two-dimensional membrane as elastic shell with proof on the folds revealed by threedimensional atomic mapping. Nat Commun 6:8935
172. Wang G, Dai Z, Xiao J, Feng SZ et al (2019) Bending of multilayer van der Waals materials. Appl Phys Lett 123:116101
173. Thangasamy P, Sathish M (2016) Rapid, one-pot synthesis of luminescent MoS_2 Nanoscrolls using supercritical fluids processing. J Mater Chem C 4:1165–1169
174. Hwang D, Choi K, Park J, Suh D (2017) Highly thermal-stable paramagnetism by rolling up MoS_2 nanosheets. Nanoscale 9:503–508
175. Meng J, Wang G, Li X, Lu X et al (2016) Rolling up a monolayer MoS_2 sheet. Small 12:3770–3774
176. Cui X, Kong Z, Gao E, Huang D et al (2018) Rolling up transition metal dichalcogenide nanoscrolls via one drop of ethanol. Nat Commun 9:1301
177. Alam S, Chowdhury MA, Shahid A, Alam R et al (2021) Synthesis of emerging two-dimensional (2D) materials—advances, challenges and prospects. FlatChem 30:100305
178. Ye JT, Zhangr YJ, Akashim R, Bahramy S et al (2012) Superconducting dome in gate-tuned band insulator. Science 338:1193–1196
179. Taniguchi K, Matsumoto A, Shimotani H, Takagi H (2012) Electric-fieldinduced superconductivity at 9.4?K in a layered transition metal disulphide MoS2. Appl Phys Lett 101:042603
180. Shi W, Ye J, Zhang Y, Suzuki R et al (2015) Superconductivity series in transition metal dichalcogenides by ionic gating. Sci Rep 5:12534
181. Li YW, Zheng HJ, Fang YQ, Zhang DQ et al (2021) Observation of topological superconductivity in a stoichiometric transition metal dichalcogenide 2M-WS2. Nat Commun 12:2874
182. Samarawickrama P, Dulal R, Fu Z, Erug U et al (2021) Two-dimensional 2M-WS2 nanolayers for superconductivity. ACS Omega 6:2966–2972
183. Xi X, Wang Z, Zhao W, Park JH et al (2016) Ising pairing in superconducting NbSe2 atomic layers. Nat Phys 12:139–143
184. Yuan NFQ, Mak KF, Law KT (2014) Possible topological superconducting phases of MoS2. Phys Rev Lett 113:097001
185. Liu GB, Shan WY, Yao Y, Yao W et al (2013) Three-band tight-binding model for monolayers of group-VIB transition metal dichalcogenides. Phys Rev B 88:085433
186. Lebegue S, Eriksson O (2009) Electronic structure of two-dimensional crystal from ab initio theory. Phys Rev B 79:115409
187. Ugeda MM, Bradley AJ, Zhang Y, Onishi S et al (2016) Characterization of collective ground states in single-layer NbSe2. Nat Phys 12:92–97
188. Sn AW, Hun YD, Yuan ZJ et al (2016) Nature of the quantum metal in a two-dimensional crystalline superconductor. Nat Phys 12:208–212
189. Li W, Huang J, Li X, Zhao S et al (2021) Recent progresses in two-dimensional Ising superconductivity. Mater Today Phys 21:100504
190. Chen J, Ge Y (2021) Emergence of intrinsic superconductivity in monolayer W_2N_3. Phys Rev B 103:064510
191. Saito Y, Nojima T, Iwasa Y (2016) Gate-induced superconductivity in two-dimensional atomic crystals. Supercond Sci Technol 29:093001
192. Peng J, Liu Y, Luo X, Wu J et al (2019) High Phase purity of large-sized 1T′-MoS2 monolayers with 2D superconductivity. Adv Mater 31:1900568
193. Ménard GC, Guissart S, Brun C, Pons S et al (2015) Coherent long-range magnetic bound states in a superconductor. Nat Phys 11:1013–1016
194. Głodzik S, Ojanen T (2020) Engineering nodal topological phases in Ising superconductors by magnetic superstructures. New J Phys 22:013022
195. Schnyder AP, Ryu S, Furusaki A, Ludwig AWW (2008) Classification of topological insulators and superconductors in three spatial dimensions. Phys Rev B 78:195125
196. Tewari S, Sau JD (2012) Topological invariants for spin−orbit coupled superconductor nanowires. Phys Rev Lett 109:150408

197. Wong CLM, Liu J, Law KT, Lee PA (2013) Majorana flat bands and unidirectional Majorana edge states in gapless topological superconductors. Phys Rev B 88:060504
198. Yuan NFQ, Lu Y, He JJ, Law KT (2017) Generating giant spin currents using nodal topological superconductors. Phys Rev B 95:195102
199. Guguchia Z, Gawryluk DJ, Brzezinska M, Tsirkin SS et al (2019) Nodeless superconductivity and its evolution with pressure in the layered dirac semimetal 2M-WS2. npj Quantum Mater 4:50
200. Copenhaver J, Väyrynen JI (2022) Edge spin transport in the disordered two-dimensional topological insulator WTe_2. Phys Rev B 105:115402
201. Pesin D, MacDonald AH (2012) Spintronics and pseudospintronics in graphene and topological insulators. Nat Mater 11:409–416
202. Culcer D, Hwang EH, Stanescu TD, Sarma SD (2010) Two-dimensional surface charge transport in topological insulators. Phys Rev B 82:155457
203. McGinley M, Cooper NR (2020) Fragility of timereversal symmetry protected topological phases. Nat Phys 16:1181–1183
204. Wu C, Bernevig BA, Zhang SC (2006) Helical liquid and the edge of quantum spin hall systems. Phys Rev Lett 96:106401
205. Kainaris N, Gornyi IV, Carr ST, Mirlin AD (2014) Conductivity of a generic helical liquid. Phys Rev B 90:075118
206. Tanaka Y, Furusaki A, Matveev KA (2011) Conductance of a helical edge liquid coupled to a magnetic impurity. Phys Rev Lett 106:236402
207. Novelli P, Taddei F, Geim AK, Polini M (2019) Failure of conductance quantization in two-dimensional topological insulators due to nonmagnetic impurities. Phys Rev Lett 122:016601
208. Schmidt TL, Rachel S, von Oppen F, Glazman LI (2012) Inelastic electron backscattering in a generic helical edge channel. Phys Rev Lett 108:156402
209. Rod A, Schmidt TL, Rachel S (2015) Spin texture of generic helical edge states. Phys Rev B 91:245112
210. Kim K, Lee Z, Malone BD, Chanet KT et al (2011) Multiply folded grapheme. Phys Rev B 83:245433
211. Lezmy N, Oreg Y, Berkooz M (2012) Single and multiparticle scattering in helical liquid with an impurity. Phys Rev B 85:235304
212. Xu C, Moore JE (2006) Stability of the quantum spin Hall effect: effects of interactions, disorder, and Z2 topology. Phys Rev B 73:045322
213. Murakami S (2011) Two-dimensional topological insulators and theiredge states. J Phys Conf Ser 302:012019
214. Kane CL, Mele EJ (2005) Z2 topological order and the quantum spin hall effect. Phys Rev Lett 95:146802
215. Fu L, Kane CL (2006) Time reversal polarization and a Z2 adiabatic spin pump. Phys Rev B 74:195312
216. Hsieh D, Xia Y, Qian D, Wray L et al (2009) Observation of time-reversal-protected single-dirac-cone topological-insulator states in Bi2Te3 and Sb2Te3. Phys Rev Lett 103:146401
217. Murakami S, Nagaosa N, Zhang SC (2003) Dissipationless quantum spin current at room temperature. Science 301:1348–1351
218. Sinova J, Culcer D, Niu Q, Sinitsyn NA et al (2004) Universal intrinsic spin hall effect. Phys Rev Lett 92:126603
219. Konig M, Buhmann H, Molenkamp LW, Hughes T et al (2008) The quantum spin hall effect: theory and experiment. J Phys Soc Jpn 77:031007
220. Roth A, Brune C, Buhmann H, Molenkamp LW (2009) Nonlocal transport in the quantum spin hall state. Science 325:294–297
221. Takahashi R, Murakami S (2010) Thermoelectric transport in perfectly conducting channels in quantum spin Hall systems. Phys Rev B 81:161302(R)
222. Tretiakov OA, Abanov Ar, Murakami S, Sinova J (2010) Large thermoelectric figure of merit for three-dimensional topological Anderson insulators via line dislocation engineering. Appl Phys Lett 97:073108

223. Huang CM, Wu SF, Sanchez AM, Peters JJP et al (2014) Lateral heterojunctions within monolayer MoSe2-WSe2 semiconductors. Nat Mater 13:1096–1101
224. He X, Liu D, Wang H, Zheng L et al (2022) Field distribution of the Z2 topological edge state revealed by cathodoluminescence nanoscopy. Opto-Electron Adv 5:210015
225. Zu S, Han TY, Jiang ML, Liu ZX et al (2019) Imaging of plasmonic chiral radiative local density of states with cathodoluminescence nanoscopy. Nano Lett 19:775–780
226. Gan XT, Gao YD, Fai Mak K, Yao XW et al (2013) Controlling the spontaneous emission rate of monolayer MoS2 in a photonic crystal nanocavity. Appl Phys Lett 103:181119
227. Carminati R, Cazé A, Cao D, Peragut F et al (2015) Electromagnetic density of states in complex plasmonic systems. Surf Sci Rep 70:1–41
228. Ran Y, Zhang Y, Vishwanath A (2009) One-dimensional topologically protected modes in topological insulators with lattice dislocations. Nat Phys 5:298
229. Wolgast S, Kurdak C, Sun K, Allen JW et al (2013) Low-temperature surface conduction in the Kondo insulator. Phys Rev B (R) 88:180405
230. Kim DJ, Thomas S, Grant T, Botimer J et al (2013) Surface "Hall Effect and Nonlocal Transport in SmB6." Sci Rep UK 3:3150
231. Hlawenka P, Siemensmeyer K, Weschke E, Varykhalov A et al (2015) Samarium hexaboride: a trivial surface conductor. arXiv:1502.01542 [condmat.str-el]
232. Wolgast S, Eo YS, Kurdak C (2015) Conduction through subsurface cracks in bulk topological. arXiv:1506.08233v1 [cond-mat.str-el]
233. Eo YS, Wolgast S, Rakoski A, Mihaliov D et al (2020) Comprehensive surface magnetotransport study of SmB6. Phys Rev B 101:155109
234. Li L, Sun K, Kurdak C, Allen JW (2020) Emergent mystery in the Kondo insulator samarium hexaboride. Nat Rev Phys 2:463–479
235. Sundermann M, Yavas H, Chen K, Kim DJ et al (2018) 4f crystal field ground state of the strongly correlated topological insulator SmB_{6}. Phys Rev Lett 120:016402
236. Hasan MZ, Kane CL (2010) Colloquium: topological insulators. Rev Mod Phys 82:3045–3067
237. Bernevig BA, Zhang S (2006) Quantum spin hall effect. Phys Rev Lett 96:106802
238. Jauregui LA, Joe AY, Pistunova K, Wild DS Electrical control of interlayer exciton dynamics in atomically thin heterostructures. 366:870–875
239. Brahlek M, Koirala N, Bansal N, Oh S (2015) Transport properties of topological insulators: band bending, bulk metal-to-insulator transition, and weak anti-localization. Solid State Commun 215:54–62
240. Yang H, Kim SW, Chhowalla M, Lee YH (2017) Structural and quantum-state phase transitions in van der Waals layered materials. Nat Phys 13:931–937
241. Zhao A, Gu Q, Haugan TJ, Bullard TJ et al (2022) Quantum spin Hall effect in two-dimensional metals without spin-orbit coupling. arXiv:2204.01949
242. Li P, Wen Y, He X, Zhang Q (2017) Eidence for topological type-II Weyl semimetal WTe2. Nat Commun 8:2150
243. Zhu J, Wang Z, Yu H, Li N (2017) Argon plasma induced phase transition in monolayer MoS2. J Am Chem Soc 139:10216–10219
244. Ouyang B, Xiong S, Jing Y (2018) Exciton physics and device application of two-dimensional transition metal dichalcogenide semiconductors. NPJ 2D Mater Appl 2:13
245. Katagiri Y, Nakamura T, Ishii A, Ohata C (2016) Gate-tunable atomically thin lateral mos2 schottky junction patterned by electron beam. Nano Lett 16:3788–3794
246. Nourbakhsh A, Zubair A, Sajjad RN, Tavakko"li AKG et al (2016) MoS2 field-effect transistor with sub-10 nm channel length. Nano Lett 16:7798–7806
247. Kappera R, Voiry D, Yalcin SE, Branch B (2014) Phase-engineered low-resistance contacts for ultrathin MoS2 transistors. Nat Mater 13:1128
248. Cho S, Kim S, Kim JH, Zhao J et al (2015) Phase patterning for ohmic homojunction contact in $MoTe_2$. Science 349:625–628
249. Lin YC, Dumcenco DO, Huang YS, Suenaga K (2014) Atomic mechanism of the semiconducting-to-metallic phase transition in single-layered MoS_2. Nat Nanotechnol 9:391

250. Retamal JRD, Periyanagounder D, Ke JJ, Tsai ML (2018) Charge carrier injection and transport engineering in two-dimensional transition metal dichalcogenides. Chem Sci 9:7727
251. Park GH, Nielsch K, Thomas A (2018) 2D transition metal dichalcogenide thin films obtained by chemical gas phase deposition techniques. Adv Mater Interfaces 6:1800688
252. Łapinska A, Kuzniewicz M, Gertych AP, Łosiewicz KC et al (2020) Study of structural and optoelectronic properties of thin films made of a few layered WS2 flakes. Materials 13:5315
253. Mak KF, Lee C, Hone J, Shan J et al (2010) Atomically thin MoS2: a new direct-gap semiconductor. Phys Rev Lett 105:136805
254. Jin W, Yeh PC, Zaki N, Zhang D et al (2013) Direct measurement of the thickness-dependent electronic band structure of MoS2 using angle-resolved photoemission spectroscopy. Phys Rev Lett 111:106801
255. Kuc A, Zibouche N, Heine T (2011) Influence of quantum confinement on the electronic structure of the transition metal sulfide TS2. Phys Rev B 83:245213
256. Empante TA, Zhou Y, Klee V, Nguyeon AE et al (2017) Chemical vapor deposition growth of few-layer MoTe2 in the 2H, 1T′, and 1T phases: tunable properties of MoTe2 films. ACS Nano 2017(11):900–905
257. Tarakina NV, Schreyeck S, Luysberg M, Grauer S et al (2014) Suppressing twin formation in Bi2Se3 thin films. Adv Mater Interfaces 1:1400134
258. Kampmeier J, Borisova S, Plucinski L, Luysberg M (2015) Suppressing twin domains in molecular beam epitaxy grown Bi2Te3 topological insulator thin films. Cryst Growth Des 15:390–394
259. Richardella A, Zhang DM, Lee JS, Koser A et al (2010) Coherent heteroepitaxy of Bi2Se3 on GaAs (111)B. Appl Phys Lett 97:262104
260. Schreyeck S, Tarakina NV, Karczewski G, Schumacher C et al (2013) Molecular beam epitaxy of high structural quality Bi2Se3 on lattice matched InP(111) substrates. Appl Phys Lett 102:041914
261. Seixas L, Abdalla LB, Schmidt TM, Fazzio A et al (2013) Topological states ruled by stacking faults in Bi2Se3 and Bi2Te3. J Appl Phys 113:023705
262. Wang ZY, Li HD, Guo X, Ho WK et al (2011) Growth characteristics of topological insulator Bi2Se3 films on different substrates. J Cryst Growth 334:96–102
263. Tarakina NV, Schreyeck S, Borzenko T, Schumacher C (2012) Comparative study of the microstructure of Bi2Se3 thin films grown on Si(111) and InP(111) substrates. Cryst Growth Des 12:1913–1918
264. Li HD, Wang ZY, Kan X, Guo X et al (2010) The van derWaals epitaxy of Bi2Se3 on the vicinal Si(111) surface: an approach for preparing high-quality thin films of a topological insulator. New J Phys 12:103038
265. Takagaki Y, Jenichen B (2012) Epitaxial growth of Bi2Se3 layers on InP substrates by hot wall epitaxy. Semicond Sci Technol 27:35015
266. Ye M, Zhang D, Yap YK (2017) Recent advances in electronic and optoelectronic devices based on two-dimensional transition metal dichalcogenides. Electronics 6:43
267. Chhowalla M, Shin HS, Eda G, Li LJ et al (2013) The chemistry of two-dimensional layered transition metal dichalcogenide nanosheets. Nat Chem 5:263
268. Samadi M, Sarikhani N, Zirak M, Zhang H (2018) Group 6 transition metal dichalcogenide nanomaterials: synthesis, applications and future perspectives. Nanoscale Horiz 2018(3):90–204
269. Keyshar K, Gong Y, Ye G, Brunetto G et al (2015) Chemical vapor deposition of monolayer rhenium disulfide (ReS2). Adv Mater 27:4640–4648
270. Cong C, Shang J, Wu X, Cao B et al (2014) Synthesis and optical properties of large-area single-crystalline 2D semiconductor WS2 monolayer from chemical vapor deposition. Adv Opt Mater 2:131
271. Tripathi TS, Karppinen M (2017) Atomic layer deposition of p-type semiconducting thin films: a review. Adv Mater Interfaces 4:1700300
272. Park J, Choudhary N, Smith J, Lee G et al (2015) Thickness modulated MoS2 grown by chemical vapor deposition for transparent and flexible electronic devices. Appl Phys Lett 106:012104

273. Tarakina NV, Cannon RJ, Sarjeant AN, Ok KM et al (2012) Directed synthesis of noncentrosymmetric molybdates. Cryst Growth Des 12:1913–1917
274. Rathi SJ, Smith DJ, Drucker J (2014) Optimization of In2Se3/Si(111) Heteroepitaxy To Enable Bi2Se3/In2Se3 bilayer. Growth Cryst Growth Des 14:4617–4623
275. Koirala N, Brahlek M, Salehi M, Wu L et al (2015) Record surface state mobility and quantum hall effect in topological insulator thin films via interface engineering. Nano Lett 15:8245–8249
276. He L, Kou X, Wang KL (2013) Review of 3D topological insulator thin-film growth by molecular beam epitaxy and potential applications. Phys Status Solidi (RRL) Rapid Res Lett 7:50–63
277. Kampmeier J, Christian W, Martin L, Melissa S et al (2016) Selective area growth of Bi_2Te_3 and Sb_2Te_3 topological insulator thin films. J Cryst Growth 443:38–42
278. Pandey A, Yadav R, Kaur M, Singh P (2021) High performing flexible optoelectronic devices using thin films of topological insulator. Sci Rep 11:832
279. Zeng Z, Morgan TA, Fan D, Li C et al (2013) Molecular beam epitaxial growth of Bi2Te3 and Sb2Te3 topological insulators on GaAs (111) substrates: A potential route to fabricate topological insulator p–n junction. AIP Adv 3:72112
280. Eschbach M, Młynczak E, Kellner J, Kampmeier J et al (2015) Realization of a vertical topological p–n junction in epitaxial Sb2Te3/Bi2Te3 heterostructures. Nat Commun 6:8816
281. Lanius M, Kampmeier J, Weyrich C, Kölling S et al (2016) P–n junctions in ultrathin topological insulator Sb2Te3/Bi2Te3heterostructures grown by molecular beam epitaxy. Cryst Growth Des 16:2057–2061
282. Bezryadina TV, Eremeev SV (2019) Heterostructures based on magnetic and topological insulators. Russian Phys J 61:1964–1970
283. Islam R, Ghosh B, Cuono G, Lau A et al (2022) Topological states in superlattices of HgTe class of materials for engineering three-dimensional flat bands. Phys Rev Res 4:023114
284. Zhao Y, Liu H, Guo X, Jiang Y et al (2014) Crossover from 3D to 2D quantum transport in Bi2Se3/In2Se3 superlattices. Nano Lett 14:5244–5249
285. Brahlek MJ, Koirala N, Liu J, Yusufaly TI et al (2016) Tunable inverse topological heterostructure utilizing $(Bi_{1-x}In_x)_2Se_3$ and multichannel weak-antilocalization effect. Phys Rev B 93:125416
286. Shibayev PP, König EJ, Salehi M, Moon J et al (2019) Engineering topological superlattices and phase diagrams. Nano Lett 19(2):716–721
287. Li HD, Wang ZY, Guo X, Wong TL et al (2011) Growth of multilayers of Bi2Se3/ZnSe: heteroepitaxial interface formation and strain. Appl Phys Lett 98:043104
288. Chen Z, Garcia TA, Hernandez-Mainet LC, Zhao L et al (2014) Molecular beam epitaxial growth and characterization of Bi2Se3/II-VI semiconductor heterostructures. Appl Phys Lett 105:242105
289. Chen Z, Zhao L, Park K, Garcia TA et al (2015) Robust topological interfaces and charge transfer in epitaxial Bi2Se3/II-VI semiconductor superlattices. Nano Lett 15:6365–6370
290. Liu J, Singh A, Liu YYF, Ionescu A et al (2020) Exchange bias in magnetic topological insulator superlattices. Nano Lett 20:5315–5322
291. Luo W, Qi XL (2013) Massive Dirac surface states in topological insulator/magnetic insulator heterostructures. Phys Rev B 87:085431
292. Matetskiy AV, Kibirev IA, Hirahara T, Hasegawa S et al (2015) Direct observation of a gap opening in topological interface states of MnSe/Bi2Se3 heterostructure. Appl Phys Lett 107:091604
293. Jiang Z, Chang CZ, Tang C, Zheng JG et al (2016) Structural and proximity-induced ferromagnetic properties of topological insulator-magnetic insulator heterostructures. AIP Adv 6:055809
294. Liu J, Hesjedal T (2021) Magnetic topological insulator heterostructures: a review. Adv Mater 2021:2102427
295. Chang CZ, Tang P, Feng X, Li K et al (2015) Band engineering of dirac surface states in topological-insulator-based van der waals heterostructures. Phys Rev Lett 115:136801

296. Sato T, Sugawara K, Kato T, Nakata Y et al (2021) Manipulation of dirac cone in topological insulator/topological insulator heterostructure. ACS Appl Electron Mater 3:1080–1085
297. Chen M, Peng JP, Zhang HM, Wang LL, He K, Ma XC, Xue QK (2012) Molecular beam epitaxy of bilayerBi(111) films on topological insulator Bi2Te3: a scanning tunneling microscopy study. Appl Phys Lett 101:081603
298. Lei T, Jin KH, Zhang N, Zhao JL et al (2016) Electronic structure evolution of single bilayer Bi(111) film on the 3D topological insulators Bi2SexTe3-x surfaces. J Phys: Condens Matter 28:255501
299. Xu SY, Alidoust N, Belopolski I, Richardella A et al (2014) Momentum-space imaging of cooper pairing in a half-Dirac-gas topological superconductor. Nat Phys 10:943–950
300. Zhu Z, Zheng H, Jia J (2021) Majorana zero mode in the vortex of artificial topological superconductor. J Appl Phys 129:151104
301. Dai W, Richardella A, Du R, Zhao W et al (2017) Proximity-efect-induced superconducting gap in topological surface states—a point contact spectroscopy study of NbSe2/ Bi2Se3 superconductor-topological insulator heterostructures. Sci Rep 7:7631
302. Xu JP, Wang MX, Liu ZL, Ge JF et al (2015) Experimental detection of a Majorana mode in the core of a magnetic vortex inside a topological insulator-superconductor Bi2Te3/NbSe2 heterostructure. Phys Rev Lett 114:017001
303. Xu SY, Liu C, Richardella A, Belopolski I et al (2014) Fermi-level electronic structure of a topological-insulator/cuprate-superconductor based heterostructure in the superconducting proximity effect regime. Phys Rev B 90:085128
304. Zhang M, Liu Q, Liu L, Zeng T (2022) Proximity-induced magnetism in a topological insulator/half-metallic ferromagnetic thin film heterostructure. Coatings 12:750
305. Yilmaz T, Pletikosic I, Weber AP, Sadowski JT et al (2014) Absence of a proximity effect for a thin-films of a Bi2Se3 topological insulator grown on top of a Bi2Sr2CaCu2O8+ cuprate superconductor. Phys Rev Lett 113:067003
306. Chi H, Moodera JS (2022) Progress and prospects in the quantum anomalous Hall effect. arXiv:2205.15226
307. Yaoa Q, Ji Y, Chen P, Hed QL, Kou QL (2020) Topological insulators-based magnetic heterostructures. Adv Phy. X 6:1870560
308. Costa M, Costa AT, Freitas WA, Schmidt TM (2018) Controlling topological states in topological/normal insulator heterostructures. ACS Omega 3:15900–15906
309. Bhattacharyya S, Akhgar G, Gebert M, Karel J et al (2021) Recent progress in proximity coupling of magnetism to topological insulators. Adv Mater 33:2007795
310. Chong SK, Han KB, Nagaoka A, Tsuchikawa R et al (2018) Topological insulator-based van der waals heterostructures for effective control of massless and massive dirac fermions. Nano Lett 18:8047–8053
311. Yao X, Gao B, Han MG, Jain D et al (2019) Record high-proximity-induced anomalous hall effect in (Bi x Sb 1–x) 2 Te 3 thin film grown on CrGeTe 3 substrate. Nano Lett 19:4567–4573
312. Katmis F, Lauter V, Nogueira FS, Assaf BA et al (2016) A high-temperature ferromagnetic topological insulating phase by proximity coupling. Nature 2016(533):513–516
313. Li Q, Trang CX, Wu W, Hwang J et al (2022) Large magnetic gap in a designer ferromagnet-topological insulator–ferromagnet heterostructure. Adv Mater 34:2107520
314. Fu L (2009) Hexagonal warping effects in the surface states of the topological insulator Bi_2Te_3. Phys Rev Lett 103:266801

Chapter 8
Applications of TMDs Materials

This chapter work is predominately devoted to potential novel applications of TMDs in diverse areas by describing two-dimensional, three-dimensional, and new-generation topological insulators (TIs). A detail description of the field effect transistors (FETs) is also provided by including flexible top-gated FETs, FETs as logic inverters, large-scale circuits, and analog and RF-applicable devices. TMDs are also used as light-emitting devices; therefore, a complete section devoted to this topic by describing different kinds of devices descriptions like light-emitting diodes (LEDs), LEDs basis on TMDs in vertical stacking, single photon emitter, organic light-emitting diodes (OLEDs), quantum dots (QDs) based light-emitting diodes. TMDs are also useful as lasers and photodetectors, therefore, two separate sections are devoted to these topics by providing a brief description of such devices. TMDs applications as photovoltaic devices are also discussed by providing descriptions of the traditional and transparent TMDs based solar cells. Moreover, TMDs solar water splitting applicability is also discussed by providing descriptions of electrocatalytic and photocatalytic types of water splitting with the usual water splitting into TMDs and their devices. TMDs also have a verity of biosensing applications, thereby; distinct kinds of biosensing devices' applicability are also discussed, including fluorophore-labeled, label-free, absorption, electrochemical, and FET biosensors based on TMDs. In addition to these descriptions on TMD–TMD heterostructures and their devices and Graphene/TMD heterostructure are also provided in two sections. Additionally, a section is also devoted to the TMDs based miscellaneous applications to describe some significant applicability, like thermoelectric devices, TMDs for drug delivery, use of TMDs in photothermal therapy, and biomedical imaging.

A. K. Singh, *2D Transition-Metal Dichalcogenides (TMDs): Fundamentals and Application*, Materials Horizons: From Nature to Nanomaterials,
https://doi.org/10.1007/978-981-96-0247-6_8

8.1 Overview

Present-day materials science is mainly driven by the pursuit of novel functional materials to allow for the ongoing development of high-processing electronic, quantum computing, energy-saving technologies, etc. These functional materials encompass multiferroics, high-temperature superconductors, and topologically non-trivial materials, i.e., materials exhibiting unconventional combinations of electronic conductivity and magnetism. Therefore, great attention is attracted to the topological materials that allow the dissipation of less spin transport on their surfaces that are undisturbed by structural imperfections or inclusions [1].

Owing to the majority of information processing being carried out by using complementary metal oxide semiconductors (CMOSs), their architecture is based on a series of interconnected semiconductor devices each performing a different function that aids in the overall goal of processing digital information. However, day-by-day increasing societal demand for smaller-sized electronic devices such as computers and cellular phones with improved functionality has forced not only the sizes of the constituent components of CMOS information processing to rapidly and inexorably shrink with an increase in operational frequencies. The requirements for an increase in device performance have led traditional semiconductor device technology to reach fundamental and perhaps ultimately insurmountable limits of several key quantities, such as size, weight, and power (SWaP). Particularly, in terms of information processing architectures, the most significant SWaP limitation the power consumption. Therefore, the quest continues toward ever smaller yet more efficient information processing systems, thereby, the architectures have undergone major renovations, like multiple logical core processing to circumvent the limitations in power consumption. Though, the imposition of multicore processing has been restricted to a slower rate of improvement than previously demonstrated by increasing the transistor scaling [2]. Therefore, it is ultimately desired to find a methodology to reduce the SWaP of constituent CMOS devices with an increased overall operational efficiency to fulfill an eminent international grand challenge [3].

After several decades (more than 70 years) extensive research in the area of materials breakthrough has come from the forefront of condense matter physics by demonstration of the remarkable theoretical approach by Landau and Lifshitz [4, 5]. According to them, the theory of phase transitions in condensed matter systems has required an extension to include the ideas of topological phase transitions and topological order to account for experimental observations of the integer and fractional quantum Hall regime [6–8]. Thus, the topological phase that is unique does not break any of the underlying symmetries of the system that cannot be described by a local order parameter [9]. In simple words, the inherent properties of the system cannot be changed due to adiabatic shifts in materials parameters unless the system passes a quantum critical point associated with a phase transition. The materials exhibited topological phases due to band structure effects and are usually referred to as topological insulators (TIs). TIs are a class of materials with an insulating time-reversal invariant (TRI) band structure to strong spin–orbit interactions that lead to an inversion of the band gap at an odd number of time-reversed points in the Brillouin zone [10–13]. The unique properties of the topological states have a significant impact on post-CMOS devices and information-processing technologies [3].

The journey of the TIs begins with the discovery of the quantum spin Hall Effect (QSH) in 1980. QSH was the first example of a quantum state with spontaneously broken symmetry, by means it comprises a symmetric probability distribution in which any pair of outcomes has the same probability of occurring. This has also noted that it depends on the topology, or physically-continuous geometric of a material and its geometry at quantum levels. Specifically, the QSH effect has occur when a strong magnetic field is applied to a two-dimensional gas of electrons in a semiconductor. Because at low temperatures and high magnetic fields, "s" the electrons travel in different regions along the edge of the semiconductor while at normal temperatures they remain scattered below and about the surface. The separation of these electrons into separate regions has also provided material with unique properties, such as the ability to simultaneously conduct and insulate. TIs are essentially a hybrid between an insulator and a conductor, where conduction is possible on the surface of the material while insulation occurs beneath the surface. Dr. Charles Kane and Dr. Eugene Mele of the University of Pennsylvania led the discover of the QSH state, which realizes in certain theoretical models by the spin–orbit coupling effect wherein electrons retain their angular momentum or rotational motion due to the geometric symmetry of the material. Moreover, such coupling causes electrons to move through a crystal by feeling a force without an external magnetic field that's sight contradictory to physical laws. Ultimately these models have predicted a class of materials that fit the description of the TIs [14].

Thus, TIs are the kinds of special materials that are categories into solid materials by dividing three forms conductor, insulator, and semiconductor. Conductor conductance is due to their electrons and holes in the bulk conduction band in which the unbounded electrons move frequently without restrictions to overpass the band gap. Insulators cannot conduct frequently owing to their extensive band gap among the valence band and conduction band, therefore, in this class electrons should acquire sufficient energy to overpass the band gap and move to the conduction band. While the empty gap of the semiconductor lies among the conductor and the insulator. But TIs cannot be straight-forward just categorized like as discussed classifications. Owing to fact that TI has a bulk electronic state together a slight band gap because of non existence free carriers within the bulk states. Nonetheless it is considered to be a topologically prevented metallic surface. This kind of non-gap surface state has a Dirac point which passes from the band gap, ultimately it makes the TI surface conducting [15]. The exceptional surface states formation owing to their inner robust spin–orbit combination effects and the time-reversal equilibrium. By mean the TIs reduces or evade the scattering impacts of the non-magnetic defects [16]. The kinds of Dirac points over diverse TIs surfaces are different therefore, differences in their topological insulating properties. Based on their properties or performances of devices they are further categorized [17].

8.1.1 Two-Dimensional TIs

Innovations with TIs began from the interpretation of two-dimensional (2D) topological state. Predominantly this kind of state has been explored theoretically by deriving the 2D materials like graphene and unchanging gradient in 2D semiconductors. The TIs achievements from the 2D materials are feasible because of the bulk band inversion. In 2007, B.A. Bernevig, T. Hughes, and S.C. Zhang positively authenticated this theoretical concept [18]. The QSH can be achieved through alternation the depth of the HgTe layer in CdTe/HgTe/CdTe semiconductor quantum well [19]. The occurrence of QSH predominantly due to the spin–orbit combination impact in CdTe, however, it was recognized comparatively feeble. In case of the increasing layer depth of the HgTe also equivalent enhancement in spin–orbit combination of HgCdTe system was recognized. Moreover, while the layer depth approached around 6.5 nm, a bulk band overturn occur, thereby, the p orbital moved onto the s orbital in the organization. Initially, the topological insulating state in the system $Hg_{1-x}Cd_xTe$ has been experimentally well approved as a 2D TIs.

8.1.2 Three-Dimensional TIs

Due to the current era, rapid technical demand boosted the growth procedure of three-dimensional (3D) TIs sharply. It is also categorized as a third-generation TIs and has been studied extensively focusing on its improved stability, whereas the leading generation was focused on the $Bi_{1-x}Sb_x$ binary alloy ($x = 0.07 \sim 0.22$). The difficulty noted with the first generation its proportion and unstability associating unpure chemical phase. Additionally, the first-generation surface state and Fermi level crosses nearly five times that has been considered unstable [20]. The surface construction of $Bi_{1-x}Sb_x$ was also complex together a shallow gap. Therefore, first-generation $Bi_{1-x}Sb_x$ has not considered being appropriate for the scientific research and technical applicability. Considering the major drawbacks of the first-generation TIs a quest was made to get a much more stable topological configuration for the technical operations by further optimization. Therefore, second-generation TIs became in the picture such as Bi_2Se_3, Bi_2Te_3, and Sb_2Te_3 [21, 22]. Usually, second-generation TIs have a hexagonal structure with narrow band gaps [23]. To contribute to this generation of materials a breakthrough became when (in 2009) S.C. Zhang forecast the occurrence of the next (second) generation of topological properties in various materials. Afterward, Xia utilized the angle-resolved photoemission spectroscopy (ARPES) to examine the surface state of a solitary Dirac cone and estimated the surface state by employing first-principles approach. This remarkable research has been provided the investigational backbone to the projection [24]. Second-generation TIs possessed a shallow bulk band gaps associating a normal structural configurations, therefore, easier to explore them by adopting simpler synthesis procedures. Many of them at present utilized as TIs. Typically, their surface state is lonely dimension inferior to the bulk structure, whereas the Dirac electrons transport to the surface. Specifically, in the 2D TIs, the electrons transport toward the periphery of the material. However, the Bi_2Se_3, Bi_2Te_3, and Sb_2Te_3 have individual Dirac point which passes via the band gap, thereby, the surface state contains a Dirac cone [25]. Significantly in the spectroscopic outcomes they appears as the couples that further can be divided from the diverse surfaces of the TIs based on topology. Like when the indigenous state density is estimated for the open boundary by creating the biggest native Wannier function, therefore, a solitary Dirac cone appears for the surface state. Hence, the second generation TIs offers numerous features superior compare to first generation TIs, such as the compositional proportion being informal to govern that allowing the easier preparation of their uncontaminated chemical phase. Therefore, the singular Dirac point makes them strong TIs and a big bulk band gap that encounters the necessity of experimentation and innovations. Existence of such characteristics additionally promises the Bi_2Se_3, Bi_2Te_3, and Sb_2Te_3 have a pronounced potential for the innovative research and their technical applicabilities.

Advances in research in the 2D TMDs as TIs introduced the third-generation topological crystalline insulators (TCIs). When (in 2013) Liu et al. have fabricated TCIs and interpreted the calculative band gap structure with analysis of the corresponding topological band [26]. This generation of materials has a bulk band gap, nonetheless they have a boundary state associating spin filter characters protected from the mirror equilibrium (001) over the edge. In general, they have symmetric numeral of Dirac cones that are protected to the mirror equilibrium of the lattice in the place of time-reversal regularity. Therefore, an innovative topological phase to be recognized in the SnTe and $Pb_xSn_{1-x}Se(Te)$ (001) thin films. However, the third generation TIs have more excellent features such as their band gap are controllable. Under the controlled electric field transverse on the film, therefore, the mirror equilibrium of the configuration may diminished, as the consequence a governable band gap over the edge state exist. Hence third generation TIs are quite different than second and first generations Tis, their precise surface alignment is also permitting a symmetric integer of Dirac cones associating a gapless surface states, while their bulk form has a shallow band gap. The third-generation TIs is also useful to fabricate many devices such as the photodetector [27, 28].

8.1.3 New Generation of TIs

The further development has also led to novel topological complexes such as $ZrTe_5$, $HfTe_5$, and Weyl semimetals (WSMs) that attended a great attention. This generation of single-layer of $ZrTe_5$ and $HfTe_5$ has noted with the wide gap 2D TIs [29]. Specifically, their direct and indirect bulk band gaps of 0.4 and 0.1 eV are significant. They also have topological characteristics that remain continuous throughout a long range of lattice strains (10% compression to 20% stretch), which makes them applicable to distinct substrates. The 3D $ZrTe_5$ and $HfTe_5$ phase alteration points among robust and feeble reflect the topological phase conversion in the minor adjustment of compression or temperature [30]. It has also established that the occurrence of Dirac fermions in $ZrTe_5$ [31, 32]. Their electronic structures are similar to Dirac semimetal (DSM) and the chiral magnetic effect due to magnetotransport is exist [33, 34]. Moreover, the optical and electrical characteristics of some layer (ZrTe5) have around 50% discrepancy to diverse in-plane directions, like hole mobility (3000 and 1500 cm^2/V.s toward the a-axis and c-axis) [35]. They also attributes superconductivity which induces by pressure [36, 37]. Moreover, they are also showing chiral irregularity and ultrahigh mobility ($1.82 \pm 0.01 \times 106$ cm^2/V.s for the $HfTe_5$ crystals) [38].

This class materials are also considered advanced and promising for the kinds of topological semimetals that have many excellent properties. They also have irregularity and topologically prevented surfaces Fermi curves and Weyl nodes in a bulk structure [39]. Such original states occur with no bulk gap at the positions of the Weyl points. Their Weyl semimetals phase lies among the normal and topological phases [40]. Significantly their effective band possessed a linear Dirac-type dispersal at the lower energy. Their chiral anomaly sensational feature provides a number of electromagnetic depending on chiral irregularity, such as $Y_2Ir_2O_7$, $HgCr_2Se_4$, TaAs, NbP, etc. Moreover, they are calcified as type-I and type-II and Weyl semimetals, like WTe_2 and $MoTe_2$ with the tilted Dirac cone for the innovative characteristics [41]. Under the applied magnetic field in certain definite directions, the substance has an insulator manner, while it behaves like a conductor for another directions employed field. This is distinct to the type-I Weyl semimetals, while the magnetoresistance type-II Weyl semimetals are usually connected to the crystal alignment. Hence the resistivity increased similar to a usual metal while the magnetic field and current flow toward the definite crystal directions, whereas it decreases in Weyl semimetals for the other orientations. Additionally, their negative longitudinal magnetoresistance relates to a chiral anomaly [17].

Thus, versatile TMD materials can have a variety of applications in different areas owing to their adequate structural, and physical–chemical features; it is not limited only to above discuss most significant features of the TMDs, but also many more. Herein, one has made effort to cover the important applications of TMDs in the coming sections.

8.2 Field Effect Transistors

The field-effect transistor (FET) is considered to be the main kind of modern transistor. Typical it is composed of source and drain regions to serve as contacts for the thin area connecting them, this is known as the channel. That is covered with a thin dielectric material; typically it is called gate dielectric which is capped with a metal electrode on top of the gate. That is used to electrostatically control the conductivity of the channel by changing the charge carrier concentration. The transistor is to be used as a switch, therefore, the conductivity should be changed between a value with high resistance corresponding to the open position of the switch, typically it is called the OFF state, while a highly conductive is called the ON state corresponding to the closed position of the switch. An ideal switch should be able to instantly switch between the ON and OFF states.

Though the electrical properties of bulk semiconducting TMDs have been investigated since the middle of the twentieth century [42, 43], their use in electronics was very limited. It has been extensively studied when first time used as FETs only after the landmark innovation with graphene by the Manchester group [44, 45]. It has been noted with the monolayers of semiconducting TMDs as the FETs [46, 47]. Initially, the field-effect mobility of single-layer (SL) MoS_2 is found to be at least 3 orders of magnitude lower than graphene [47]. Therefore, great attention has been paid to the monolayer semiconducting TMD FETs in 2011 when top-gated SL-MoS_2 FETs were introduced, as schematics are illustrated in Fig. 8.1 [48]. It noted that the FET devise (thickness 6.5 Å) can have moderate motilities (~60–70 cm^2/Vs), large on/off ratios (~10^8), and low sub-threshold swings (74 mV/dec) at room temperature [49–53]. Their two-dimensional channel coupled with a large band gap (> 1 eV) and ultrathin top gate dielectric allowed a superior gate control, this makes it enables small off currents and large switching ratios. The device carrier mobility has been obtained around 200 $cm^2\ V^{-1}\ s^{-1}$, while exfoliation of single layers onto SiO_2 could decrease in mobility 0.1–10 $cm^2\ V^{-1}\ s^{-1}$. This device has a current ON/OFF ratio exceeding 10^8 at room temperature. Despite a very high ON/OFF ratio, the device has revealed off-state current smaller than 100 fA (25 fA/µm). A high degree of electrostatic control has been also noted through the subthreshold slope $S = (d(\log I_{ds})/dV_{tg})^{-1}$, it is as low as 74 mV/dec.

Since encapsulation of MoS_2 using insulating HfO_2 is considered to be crucial to get high mobility. Therefore, additional approaches were introduced to achieve high mobility, such as vacuum annealing to remove adsorbates. By using this approach a

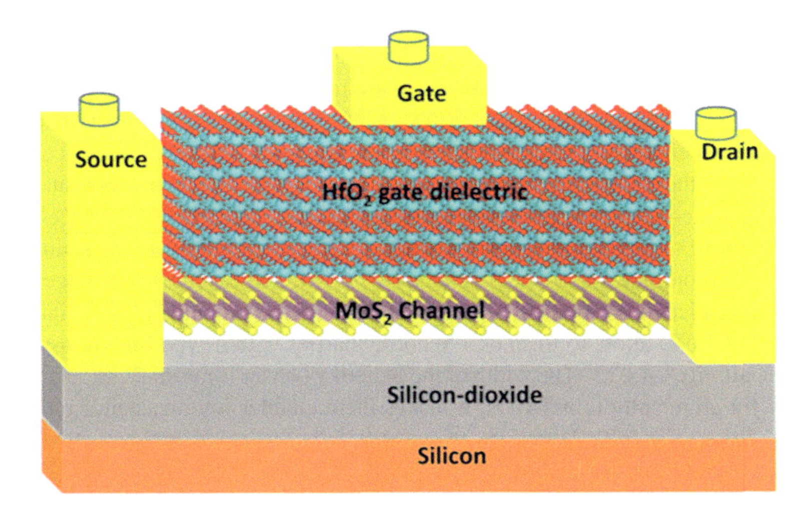

Fig. 8.1 Schematic of a typical device with a top gate dielectric (HfO_2) (adopted from [54])

higher intrinsic (field-effect) mobility of 1000 cm^2V^{-1} s^{-1} can be achieved at low temperatures in monolayer as well as bi-layers MoS_2 [55]. Moreover, the intrinsic mobility and conductivity of single-layer MoS_2 in single and dual-gated geometry has been also constructed [56]. It also noted that the charge carrier densities n_{2D} below $\sim 1 \times 10^{13}$ cm^{-2}, in single-layer MoS_2 have a decreasing conductance with the decreasing temperature. However, in the case of n_{2D} larger than 10^{13} cm^{-2}, an increment is notice in conductance with decreasing temperature, this result is considered to be a hallmark of a metal–insulator transition.

The high-performance single-layer MoS_2 transistors have been fabricated and the full-channel gating also addressed, as the typical parameters are illustrated in Fig. 8.2a–c. It has a maximum transconductance (g_m) 34 μS/μm at 4 V, the corresponding drain-source bias ($I_{ds} - V_{ds}$) characteristics are exhibited in Fig. 8.2d, where the back-gate voltage V_{bg} kept at 0 V. It noted that at a relatively low drain voltage of V_{ds} is 1 V, the current-carrying capacity of the MoS_2 charge-carrying channel has revealed a saturation. This was the first significant observation of drain current saturation in monolayer MoS_2 FET. As proposed a relatively high value of transconductance ($g_m = 20$ μS/μm for $V_{ds} = 1$ V and $g_m = 34$ μS/μm for $V_{ds} =$ 4 V) and a current saturation ($g_{ds} < 2$ μS/μm). Hence it was concluded that MoS_2 is also interesting for digital and analog applications by offering a gain > 10 [53]. The noticed saturation is also important to achieve the maximum possible operating speeds. Additionally, the breakdown current density in MoS_2 has also explored which is close to 5×10^7 A/cm^2. It is 50 times greater than the current carrying capacity of copper [53].

However, the Flicker noise or 1/f noise is considered to be a limiting factor for the TMDs based electronic device applications. Owing to the almost all-surface structures of ultrathin TMDs are sensitive to random perturbations in the environment. Like, the SL-MoS_2 offers a lowest Hooge parameter with high mobility of around 0.005 under a vacuum that is comparable to the noise amplitudes of carbon nanotubes and graphene [57]. The Hooge parameter increases as much as 3 orders of magnitude under ambient conditions, this noise interpretation has also confirmed the sensitivity of SL-MoS_2 due to adsorbates and trapped charges. However, its further encapsulation effectively reduced the noise amplitude in TMDs based devices [58].

Moreover, the contact resistance of metal electrodes also plays an important role to determine the conductance of the fabricated devices, specifically those devices fabricated by using reduced-dimension materials. Therefore, to fabricate ultrathin MoS_2 FET devices commonly used the Au contact to provide a linear current–voltage curve, this is also called ohmic contact [48, 59–61] (Fig. 8.3a). Particularly "Au" high work function (5.1 eV) allow to form of a Schottky barrier with n-type MoS_2 possessing electron affinity ~ 4.0 eV. The width of the Schottky barrier is normally exceptionally narrow for atomically thin TMDs; it makes them enables low-resistance tunneling through the barrier [62]. Therefore, the overall output is linear or electrically ohmic transport for ultrathin MoS_2 FETs even with high work function metal contacts [62]. A large number of theoretical and experimental studied have been also interpreted with the various metal contacts to ultrathin semiconducting TMDs transistors

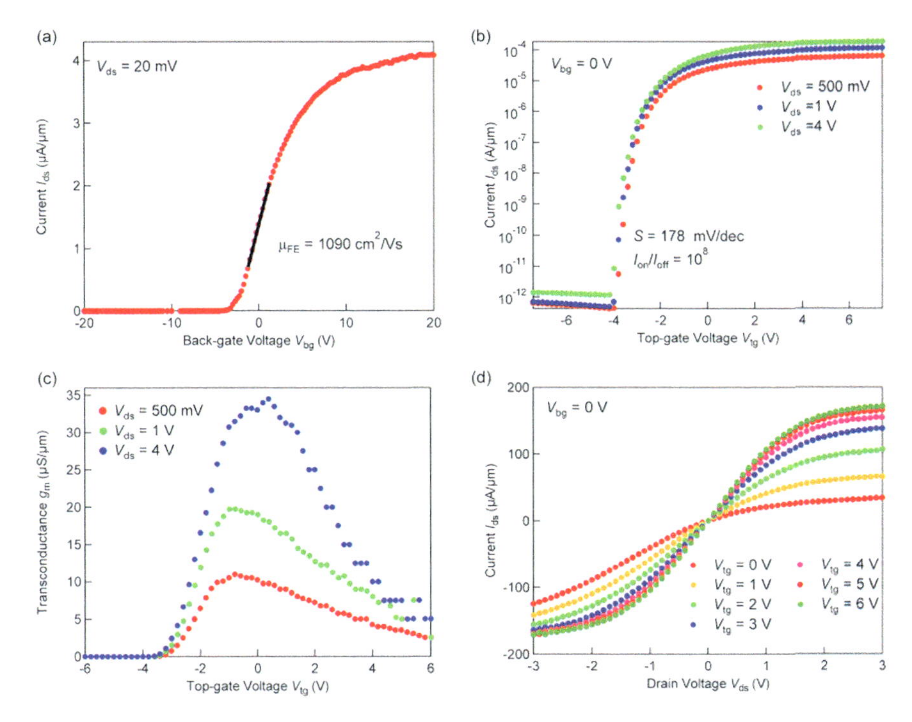

Fig. 8.2 Characteristics of a monolayer MoS_2 FET transistor: **a** Room-temperature back-gate transfer characteristic for the MoS_2 FET under the applied drain voltage $V_{ds} = 20$ mV. The estimated effective field-effect mobility $\mu_{fe} = 1090$ cm^2/V.s; **b** $I_{ds}-V_{tg}$ curve at three different drain voltages: $V_{ds} = 500$ mV, 1 V, and 4 V. The I_{on}/I_{off} ratio exceed 10^8. The sub-threshold swing S at $V_{ds} =$ 500 mV is 178 mV/dec; **c** Transconductance $g_m = dI_{ds}/dV_{tg}$ derived from $I_{ds}-V_{tg}$ characteristics. The peak transconductance for $V_{ds} = 4$ V is $g_{m,\,max} = 34$ μS/μm; **d** $I_{ds}-V_{ds}$ characteristics at the different top-gate voltages V_{tg} for drain voltages V_{ds} upto 3 V. This device, the on-current is $I_{on} =$ 344 μA (172 μA/μm) for $V_{tg} = 6$ V and $V_{ds} = 3$ V, corresponding to a current density of $J = 2.5 \times 10^7$ A/cm^2. This value is 25 times larger than the breakdown current density of copper. At typical bias voltages with V_{ds} exceeding~2 V, the drain current saturation is noted with the drain_source $g_{ds} = dI_{ds}/dV_{ds}$ close to 0 ($g_{ds} < 2$ μS/μm) at $V_{ds} = 3$ V. At the low bias voltages, the curves are linear that contacted to their ohmic behavior (adopted from Lembke et al. [53])

[62–67]. As per interpretation, the low work function metals (e.g., titanium and scandium) form lower resistance contact, as the consequence of the higher drain currents (Fig. 8.3b) [62]. Such as the ultrathin MgO and TiO_2 barrier between a ferromagnetic metal and MoS_2 which minimize Schottky barrier formation [68, 69]. Similarly, contact doping with nitrogen dioxide (NO_2), potassium, polyethyleneimine (PEI), etc. are also used for the semiconducting TMDs FETs [70–73].

Thus, the contact resistance ultimately dominates the charge transport with reduced channel dimensions [74]. This effect manifests itself as a reduction in drain current and mobility at channel length below 500 nm [74, 75], besides the quantum transport simulations predictions the excellent length may scale down to 10–15 nm

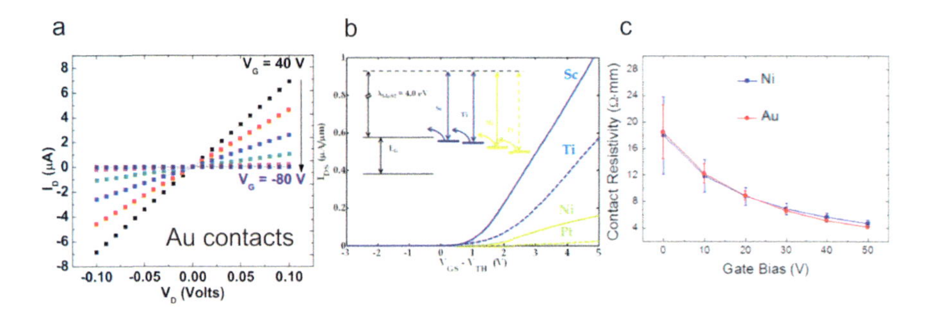

Fig. 8.3 Metal contacts: **a** linear $I_d - V_{ds}$ output curves in an unencapsulated, bottom-gated single-layer MoS_2 FET with Au contacts; **b** increase in drain current of a few-layer MoS_2 FET with decreasing work function of the contact metals; **c** gate voltage dependence of the contact resistance in a few-layer MoS_2 FET indicates a Schottky barrier at the contacts (adopted from Jariwala et al. [46]

[76–78]. It noted that the contact resistance of Au and Ni is considered to be a strong function for the gate bias, to tune the Schottky barriers (Fig. 8.3c) [75, 79].

Advances in this area left another key topic ionic-gating that have been extensively used to control and investigate the electronic properties of oxides [80, 81], nitrides [82], organic semiconductors [83–85], carbon-related materials [86], and III–V semiconductor nanowires [87–90]. The extreme tune-ability of charge carrier concentration can be achieved from this technique with the fundamental knowledge of the phase diagrams of novel materials, such as super conductivity in band insulating materials, $SiTiO_3$ (STO), ZrNCl, $KTaO_3$, etc. [82, 91, 92]. As the typical schematics of a TMD-based ILG transistor are illustrated in Fig. 8.4a–c. Wherein the channel is fabricated from the thin bilayer WSe_2 crystals by the mechanical ulterior transfer onto a SiO_2/Si substrate. The electrodes were fabricated by e-beam lithography and evaporation of titanium and gold (5/45 nm). The four electrodes were connected to the WSe_2 flake, two electrodes act as the gate (V_g), and the remaining refers as the (V_{ref}). The whole device was covered with ionic liquid (DEME-TFSI), therefore, it forms the conducting semiconductor channel as well as the reference and gate electrodes [93]. Moreover, the whole device was also covered from the polymethyl methacrylate (PMMA) and exposed, as shown in the rectangle window on top of the semiconducting channel to place the droplet (see Fig. 8.4c). Thus it reflects the behaviors of the device in different conditions [93].

Another key interesting topic is the tunneling field-effect transistors (TFETs) are also identified as a strong candidate for the next generation transistors, such as their potential ability to provide sub-60-mV/decade subthreshold slope (SS) and lower I_{OFF} [94]. The lower SS and I_{OFF} reduces the static and dynamic power consumption of TFET-based circuits. Therefore, the short-channel behavior of TFETs is improved by using 2D semiconductors as a channel material [95]. Such materials are allowing better gate control on the channel, thereby, an increase in the electric field at the source/channel interface. Additionally, the 2D materials also have negligible surface roughness and do not show any significant increase in band gap with the reduction in

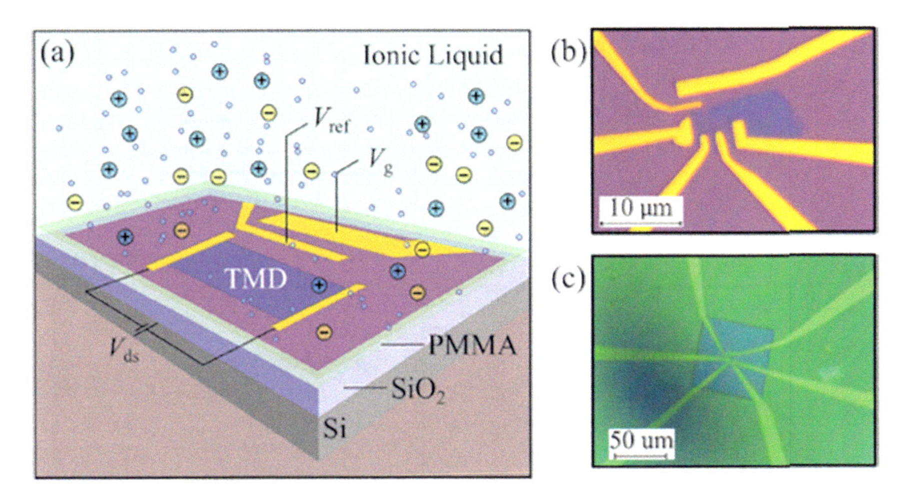

Fig. 8.4 **a** Schematics of an ionic-liquid-gated field-effect transistor (FET), by showing gate and reference electrodes, as well as the electrical circuit used to bias and measure the device; **b** optical microscope image of a bilayer of WSe_2 contacted in Hall bar configuration (the scale bar is 10 um); **c** optical microscope image of the device's polymethyl methacrylate (PMMA) windows (the scale bar is 50 um) (adopted from Vaquero et al. [93])

channel thickness [96, 97]. Significantly it is an important parameter for the TFETs to increase the channel band gap and reduce the I_{ON}. TMDs are considered to be an emerging 2D channel material for next-generation transistors. Because in contrast to graphene, TMD materials have a band gap [98]. Therefore, TMD-based transistors have a good I_{ON}/I_{OFF} ratio due to a higher band gap. There are several TMD materials have been utilized as the homo and heterojunction TFETs to improve the performance of the device [99, 100]. However, in the majority of cases, the conventional methods of doping have been used for source and drain formation. But conventional approaches are not very much fissile for the 2D materials [101–103]. Due to the fact, such devices have multiple supply voltages for source and drain doping, this is not an energy-efficient technique.

To overcome such problems the charge plasma-based TFET structures have been also interpreted [104]. The charge plasma-based monolayer TMD-TFET structures have been systematically investigated by utilizing a metal work function engineering to improve the performances of the devices, as the typical schematics of the dual-metal source (DMS)-TFET and single metal source (SMS)- TFET structures for both kinds of devices same gate and drain structure have been used [105]. By analyzing the several TMDs compositions it has been also noted that the WTe_2 would be one of the best compositions as a channel material for the devices. Moreover, such devices' various parameters have been also explored by interpreting the extensive simulations and experimental outcomes. Including a detail information about to design of sub-10-nm TFET devices, by presenting the different parameters calculations of the devices. The typical conversational and plasma-based devices schematics and their properties

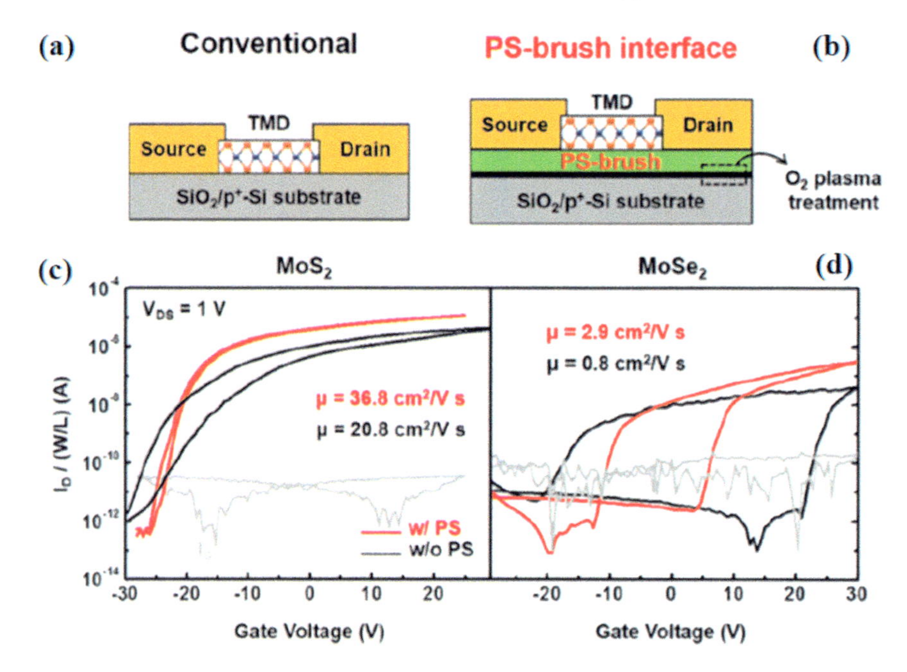

Fig. 8.5 **a**, **b** Schematic illustration of the conventional TMD FET and plasma-cleaned (PS-brush layer) structures; **c, d** $I_{DS} - V_{GS}$ curves of conventional and plasma-cleaned (PS-brushed) TMD transistors for n-MoS_2 and n-$MoSe_2$. The reduced hysteresis and higher mobility for the PS-brush treated FETs are noted (adopted from [106])

in the case of MoS_2 and $MoSe_2$ are depicted in Fig. 8.5a–d. A complete fabrication process schematic of the DMS TFET is depicted in Fig. 8.6a–o) [105].

8.2.1 Flexible Top-Gated Field-Effect Transistors

Since several devices are realized on rigid silicon, but in the modern era electronic demand is a non-planar form of devices that should be thin, light, and conformably attached to objects with unusual shapes, on the human skin, or even implanted into the human body [107, 108]. Therefore, to fulfill these objectives it is desired electronics should be realized on flexible substrates that should be robust to mechanical strain, easy to integrate, and capable of low-power consumption and high performance at the nanoscale [109, 110]. The 2D materials are considered to be good candidates for flexible substrates, due to their lack of dangling bonds, good carrier mobility in atomically thin (sub-1 nm) layers, reduced short-channel effects, and easy to transfer onto arbitrary substrates [75, 108, 110–113]. Specifically, monolayer TMD like MoS_2 has a lower power owing to their electronic band gap (~2 eV) that enables low off-current (~fA μm^{-1}) [114–117]. Moreover, the TMD transfer process with

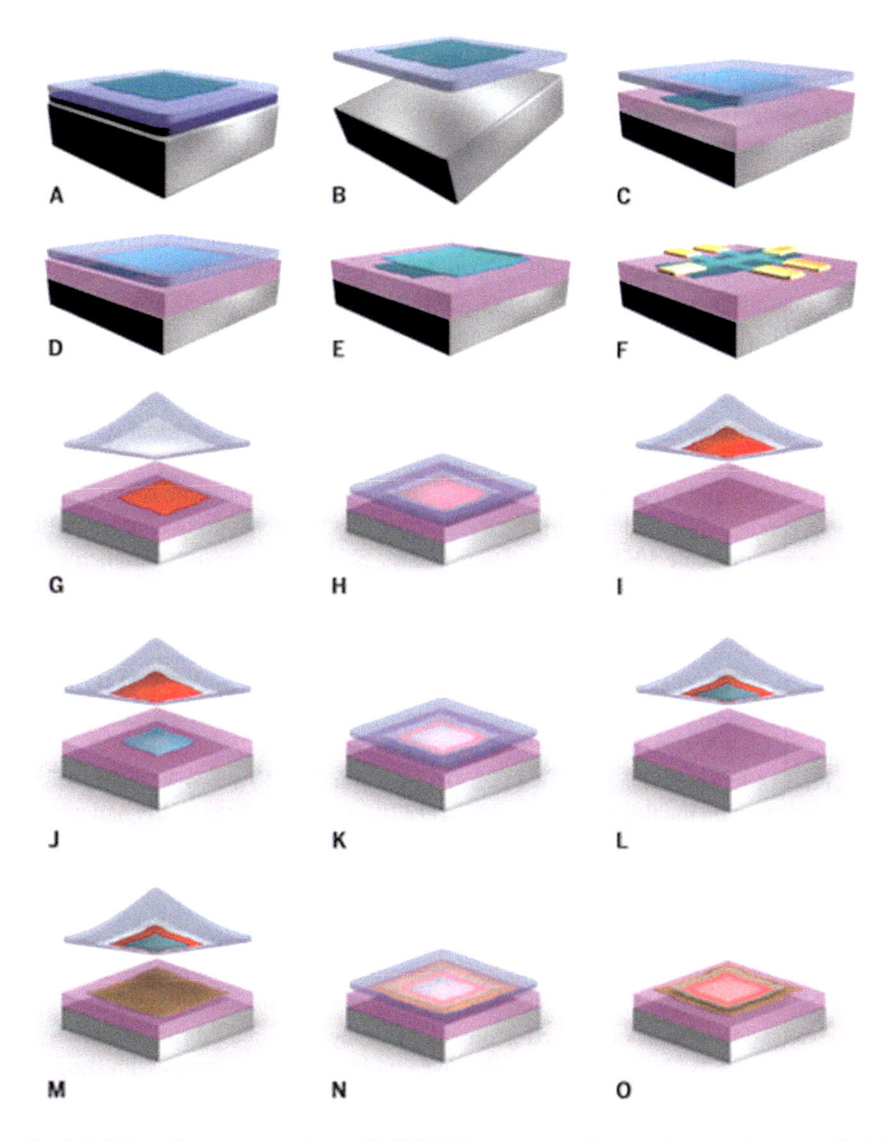

Fig. 8.6 Fabrication steps to realizing DMS TFET structure: **a** 2D crystal prepared on a sacrificial layer; **b** pick up the sacrificial layer with 2D material **c** the layer is then placed; **d** second layer is transferred the same way; **e** it is placed on top of the first layer; **f** deposition of metal on the proper location; **g** 2D material on a membrane is aligned; **h** then it placed on top of another 2D crystal; **i–l** the process is repeated to lift additional crystals; **m–o** the whole multilayer is placed on the substrate and the membraned is dissolved (adopted from Crystals 8, 8)

the embedded contact have been also achieved on the SiO_2/Si substrate from the CVD method [118–121]. Subsequently followed by lithographically patterned Au metal contact on the top. Then conformally cover the pre-patterned structures with ~ 5 μm thick flexible polyimide (PI) that holds both TMD and metal to SiO_2/Si growth substrate, through immersion and agitation in DI water.

Afterward transferring the source/drain contacts embedded to the PI substrate and on the top of the TMD semiconductor. To prevent contamination in exposed TMD surfaces usually deposited an Al_2O_3 gate dielectric immediately before any patterning steps. For instance, in the MoS_2 $MoSe_2$ and WSe_2 FETs in which **associates** are patterned earlier the transfer, their channel widths is usually larger to the electrode widths designated as "Type A" devices. While the FETs wherein the MoS_2 network is prior defined by responsive ion etching (RIE) previously transfer that facilitates precise channel width denoted to as the "Type B" devices. As the typical schematic of the device is illustrated in Fig. 8.7a, while the corresponding optical pictures of the WSe_2, $MoSe_2$, and MoS_2 FETs are depicted in Fig. 8.7b–d. Both Type A and Type B devices can be fabricated through the transfer process. Typical output characteristics for the micron-scale FETs are depicted in Fig. 8.7e–j. The device parameters of the various TMDs are listed in Table 8.1. It noted that almost all devices have n-type behavior. Additionally, it also noted that the threshold voltage V_T and extrinsic field- effect mobility $\mu FE,_{ext}$ at the maximum transconductance (g_m), while using the Al_2O_3 gate oxide capacitance ($C_{ox} = 0.21$ to $0.32\ \mu F\ cm^{-2}$) to TMD FETs accumulation. Moreover, the 2 μm long monolayer WSe_2 FET has a maximum on-current $I_D = 3.5 \pm 0.05\ \mu A\ \mu m^{-1}$ at a drain-source voltage $V_{DS} = 1$ V. Whereas 3 μm long $MoSe_2$ FET has $I_D = 4.2 \pm 0.34\ \mu A\ \mu m^{-1}$ at $V_{DS} = 4$ V, it was a first demonstration on the flexible $MoSe_2$ FETs [122]. Moreover, the mobility and width-normalized current of Type A devices have also demonstrated that is also itemized together inaccuracy inns. Since the channel width cannot pattern therefore they suffer from some current spreading effects. As the unpatterned hexagonal crystals of the selenide-based FETs are depicted in Fig. 8.7b, c. The measured in plain current units (μA) is depicted in Fig. 8.7e, f and h, i), although, the data presented in the table included the error bars. But it noted that such corrections are not required in "Type B" devices due to their optimized geometry with the modified fabrication process. Similarly "Type B" MoS_2 FETs are also have width-normalized units with superior excellence material that permits a greater $I_D \approx 67.3\ \mu A\ \mu m^{-1}$ for a 5 μm long FET at $V_{DS} = 5$ V. Hence compared to "Type A" MoS_2 FETs it has a greater sub-threshold fluctuate SS and off-current that reduced the on/off ratio. Furthermore, the device hysteresis has been also explored in the range of ~ 0.1 V (WSe_2) to ~ 1.6 V ($MoSe_2$) for all devices. It directly correlates to the supplementary patterning step of "Type B" devices that do not worsen TMD interfaces. Additionally, it also noted that the constancy of the flexible TMD FETs underneath tensile bending almost possessed negligible variations for a bending radius 4 mm [122].

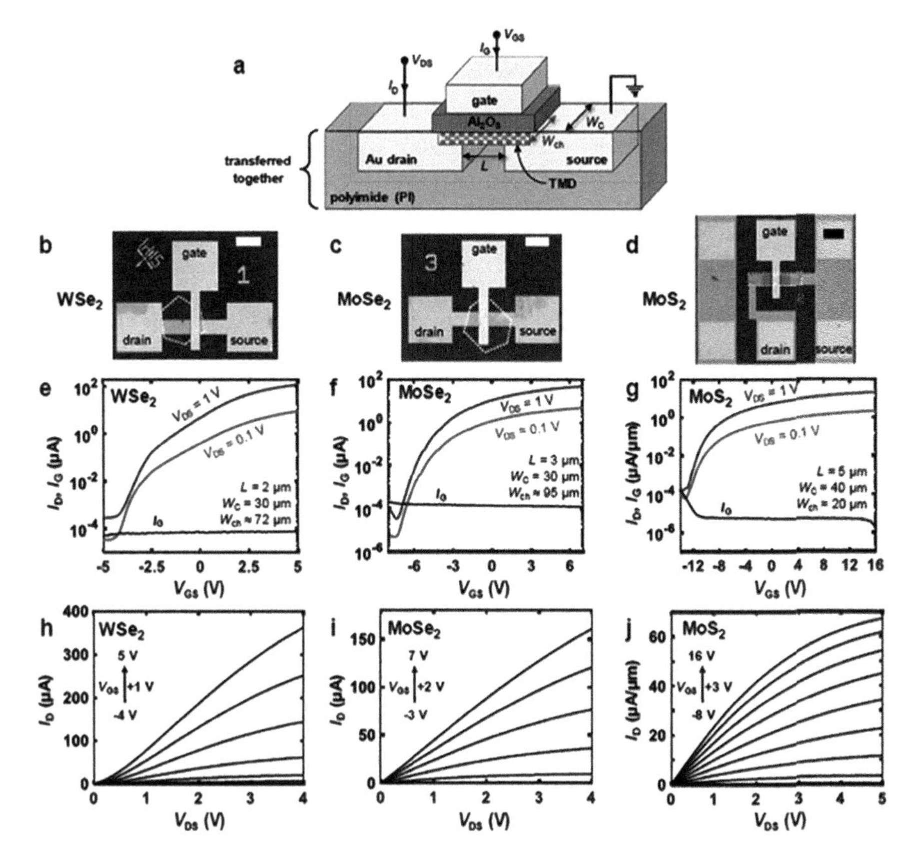

Fig. 8.7 Flexible field-effect transistors (FETs) with transition metal dichalcogenides (TMDs): **a** schematic cross-section optical microscope images of FETs; **b** WSe_2 (Type A); **c** $MoSe_2$ (Type A); **d** MoS_2 (Type B), scale bars: 50 μm; **e, h** corresponding transfer and output characteristics for WSe2 (Type A); **f, i** for $MoSe_2$ (Type A); **g, j** for MoS_2 (Type B). The gate current (I_G) is often negligible, although for some devices it can limit the on/off ratio (adopted from [123])

8.2.1.1 Nanoscale MoS_2 Flexible Transistors

When scaled down up to ~ 50 nm to typical MoS_2 FET by using the electron-beam lithography (EBL) on SiO_2/Si surface. It has noted that the cross-section of the Al_2O_3 gate dielectric covered the planar source/drain electrodes associating ~ 100 nm nanogap among them, it directly relates to the nonexistence of "steps" in surface topography that enables the manufacture method together connections surrounded in the varying substrate, as depicted in Fig. 8.8a. The electrical measurement has also revealed similar behavior for the Type B device for a 100 nm wide channel, as illustrated in Fig. 8.8b. A good on/off ratio (> 10^6) and high $I_D \approx 303$ μA μm^{-1} (at $V_{DS} = 1.4$ V) with μFE, ext ≈ 7.2 cm^2V^{-1} s^{-1} has been achieved for this device, as illustrated in Fig. 8.8b, c. It also recognized that the nanoscale FET device's mobility is lower than the micro-scale devices owing to the better influence of the

Table 8.1 Extrinsic field-effect mobility $\mu_{FE, ext}$ and threshold voltage V_T for the maximum g_m in linear FET operating regime at a drain-source voltage $V_{DS} = 100$ mV, the subthreshold swing (SS) value for the minimum [122]

Device	Channel length (nm)	$\mu_{FE, ext}$ ($cm^2V^{-1} s^{-1}$)	I_D at V_{DS} = 1 V (μA $μm^{-1}$)	V_T (V)	SS (mV $decade^{-1}$)	On/off ratio
WSe_2 (Type A)	2000	4.9 ± 0.07	3.5 ± 0.05	1.6	380	3×10^5
$MoSe_2$ (Type A)	3000	1.4 ± 0.09	1.3 ± 0.09	−2	430	10^6
MoS_2 (Type A)	5000	15 ± 2.2	5.4 ± 0.79	3.9	1700	3.6×10^3
MoS_2 (Type B)	5000	25.7	21	−5.2	850	10^5
MoS_2 (Type B)	100	7.2	229	0.6	730	2×10^6
MoS_2 (Type A)	70	20 ± 1.4	470 ± 45	6	1000	4×10^3

connection resistance (see Fig. 8.8c). This nano-scale device also has a signature about self-heating and velocity saturation due to onset current saturation at a lower V_{DS} with higher gate-source voltage V_{GS} [75, 124]. The estimated temperature for the MoS_2FET device is around 172 °C for the peak input power (see Fig. 8.8c). Additionally, it also noted that the Au contact is principally accountable to the lateral heat dispersion by the nanoscale device channel.

Moreover, also explored the different scale FET devices and provided supplementary understanding over the inherent device constraints by defining the parameters I_D (at an overdrive $V_{ov} = V_{GS} - V_T = 8$ V) and μFE, ext to the channel lengths 50 nm to 10 μm, as illustrated in Fig. 8.8d,e. By relating the I_D and $\mu_{FE, ext}$ to L, R_C, and the inherent field-effect mobility μ_{FE}, V_T, and $\mu_{FE,ext}$ different parameters are evaluates. Such as the I_D plateaus with a $\mu_{FE,ext}$ decreased value for the sub-1 μm channel lengths is visualized in Fig. 8.8d, e, it is directly correlated to devices limited from the R_C. Here dashed black lines correspond to Type B devices (red circles) that fitted with an average μ_{FE} (~12.9 $cm^2V^{-1} s^{-1}$) for the micron-scale devices that impacted to R_C small, by setting $R_C = 5$ kΩ μm that also follows the central distribution for the smaller L [122].

In the case of Type B devices, the solid black lines (see Fig. 8.8d, e) are keep at the greater μ_{FE} (~28 $cm^2 V^{-1} s^{-1}$) to get a fitting for $R_C = 2.4$ kΩ μm. On the other hand, a typical Type A device (blue symbols and inaccuracy inns, amended for current dispersal) fits with the $R_C \approx 310$ Ω μm for a device (at $L = 70$ nm) that possessed a slightly higher $\mu_{FE} = 29$ $cm^2V^{-1} s^{-1}$. Therefore, FET with the maximum on-current are achieved with an exciting $I_D = 470 \pm 45$ μA $μm^{-1}$ at $V_{DS} = 1$ V, as the corresponding electrical characteristics are illustrated in Fig. 8.8f.

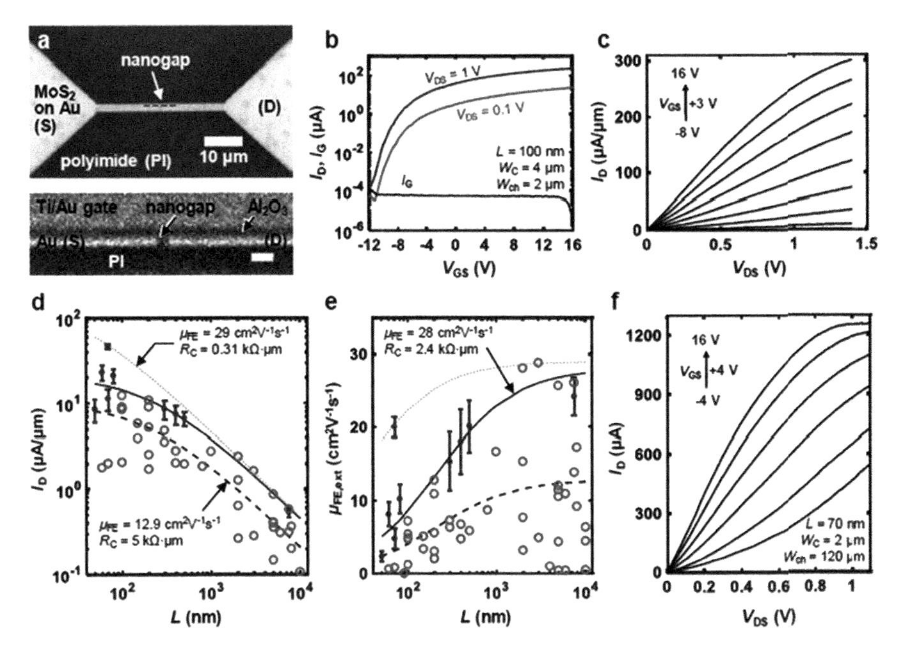

Fig. 8.8 Nanoscale MoS_2 field-effect transistors (FETs)**:** **a** Optical microscope picture of a nanoscale channel afterward the transfer (top) and cross-section scanning electron microscope (SEM) picture of a transistor having a channel length (L) = 100 nm (bottom, scale bar: 200 nm). The cross-section has taken from the dashed line in the upper picture; **b, c** Transfer and productivity individualities of a *Type B* MoS_2 FET to $L = 100$ nm; **d, e** drain current I_D and extrinsic field-effect mobility $\mu_{FE,\,ext}$ as a function of L for ~ 50 devices at a drain-source voltage $V_{DS} = 0.1$ V. I_D is exhibited at an overdrive voltage $V_{GS} - VT = 8$ V and $\mu_{FE,\,ext}$ take out at the extreme transconductance. Solid and dashed lines illustrate a fitting style for "best" and "typical" *Type B* devices (circles), whereas the scattered line associations to finest outcomes for all FETs, containing *Type A* devices; **f** Electrical physiognomies of a MoS_2 FET (*Type A*) with $L = 70$ nm with remarkable high $I_D > 1.2$ mA, which is ~ 470 μA μm^{-1} afterward current dispersal modification (adopted from Daus et al. [122])

As per estimated performance parameters in the most suitable case μ_{FE} and R_C, they are comparable to the high-performance single-layer MoS_2 on flexible substrates and SiO_2/Si rigid substrates [119, 124, 125]. Therefore, the maximum on-current I_D has achieved more than three times superior to the flexible MoS_2 FETs [126], compared to similar finest TMD FETs on inflexible substrates, even analogous to flexible FETs constructed on graphene and c-Si [127–129]. Thus, it noted that the manufacture procedure also empowers the scale of the flexible MoS_2 FETs for the shortest channel distances.

8.2.2 FET as Logic Inverter

Since the functionality of the distinct TMD-FETs should fulfills the essential necessities of working logic inverter, such as a current ON/OFF ratio of more than 1000 to voltage switch and moderate mobility for the higher frequency utilities. Additionally, also desired to design pull-up and pull-down FETs with matching width-to-length ratios to obtain suitable switching threshold voltage as well as noise margin for the inverter. These essential criteria of logic inverters and other logic gates fulfill only from n- or p-type FETs based on TMD materials.

Thus, semiconducting TMDs promise high-performance FETs to integrate them into more complex functional digital circuitry, including inverters and logic gates. Typically a logic inverter consists of two FETs in series that are designed to convert logical 0 (low input voltage) to logical 1 (high output voltage) and vice versa. The common electrode between the two FETs serves as the output, whereas the gates of the FETs serve as the input, as depicted in Fig. 8.9a. The significant metric of an inverter is the voltage gain and it characterized by the negative derivative of the output plot, as illustrated in Fig. 8.9a [130]. The single-layer MoS_2 inverters have been first introduced in 2011 [130]. Thereafter, various inverters based on a variety of semiconducting TMDs have been demonstrated with gains varying between 2 and 16 [131–133]. The required gain should be greater than the unity for the cascaded circuit.

inverters, such as ring oscillators. Typically a ring oscillator has an odd number of inverters in series to produce high-frequency clock signals, as illustrated in Fig. 8.9b [132]. Such ring oscillators have been successfully fabricated and interpreted with MoS_2, typically they are allowing to operate in a frequency range of up to 1.6 MHz with stage delays of ~ 60 ns, as depicted in Fig. 8.9c [132].

To deduce large-scale FETs the CVD-grown MoS_2 on flexible polyimide (PI) substrates have been also fabricated [134]. As interpreted that the high n-type doping of the MoS_2 allows to inverter integrate into a depletion-loaded circuit together a signal gain beyond 16 at a negative input voltage of − 15 V. Moreover, it has been also noticed that high-performance wafer-scale MoS_2 transistors introduced to multilayer MoS_2 islands on the film during CVD growth [135]. Therefore, a pseudo logic inverter has been apprehended in ML-MoS_2 FETs as pull-up and pull-down FETs, as depicted in Fig. 8.10a. Their electrical characterization has revealed an outstanding logic-level preservation and highest voltage gain above 23 at an input voltage of − 1.1 V. Moreover, construction on a wafer-scale extraordinary excellence MoS_2 film can also feasible from the ALD technique [136]. This kind of ALD-grown n-type inverter circuits usually have an exceptional level flip physiognomies comprising adjustment frequency up to 50 Hz. Thus, the experimental evidences are representing a methodological path for probable applicability of MoS_2 films under united circuit systems.

Typical MoS_2-based inverter integrated circuits based on the distinct two-dimensional layered materials (2DLMs) or associating 2DLMs to additional materials have been also fabricated. As an instance, an optical picture of the ReS_2 logic

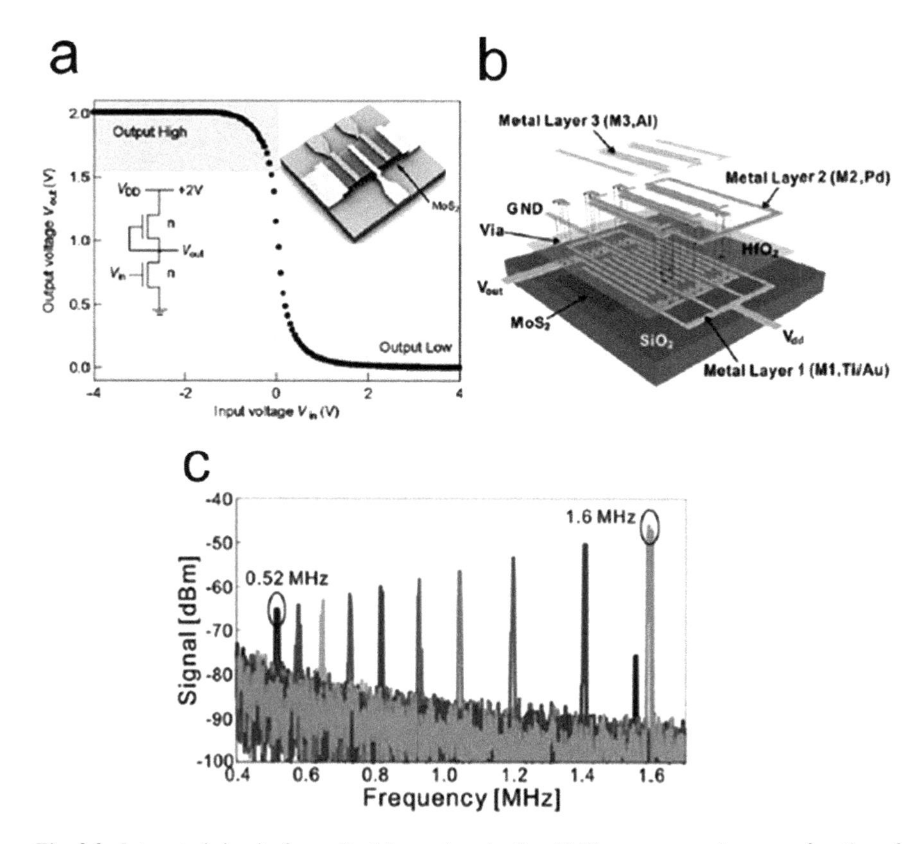

Fig. 8.9 Integrated circuits from ultrathin semiconducting TMDs. **a** output voltage as a function of input voltage of an inverter based on two single-layer MoS_2 FETs, inset shows the circuit diagram (left) and device schematic (right); **b** schematic for a ring oscillator circuit fabricated on ultrathin MoS_2; **c** power spectrum of the ring oscillator output signal as the supply voltage (V_{dd}) is increased from left to right (adopted from Jariwala et al. [46])

gates array (NOT, NAND, and NOR gates) is illustrated in Fig. 8.10b. The apprehension of NMOS logic devices constructed on a CVD developed ReS_2 film together graphene as electrodes have been also interpreted [138]. The NMOS inverter is also fabricated in depletion mode (D-mode) transistor arrangement by using the ReS_2 transistors. Such TMDs based FETs inverters offer to operate underneath a developing voltage of 1 V having a voltage advantage over 3.5. The high-performance device circuits established on wafer-scale MoS_2-graphene hetero construction have been also introduced [139]. This kind of integrated inverter can also fabricate by using two MoS_2-graphene FETs, in which MoS_2 and graphene assists as the transistor channel and connection electrode. The noticed signal inverter advantages nearby 12 in the operational voltage 3 V. Furthermore, the single-step logic gate from the phase-patterned development of extremely thin $MoTe_2$ also offers such construction. Using this process a PMOS inverter has been also achieved by taking two $MoTe_2$ FETs

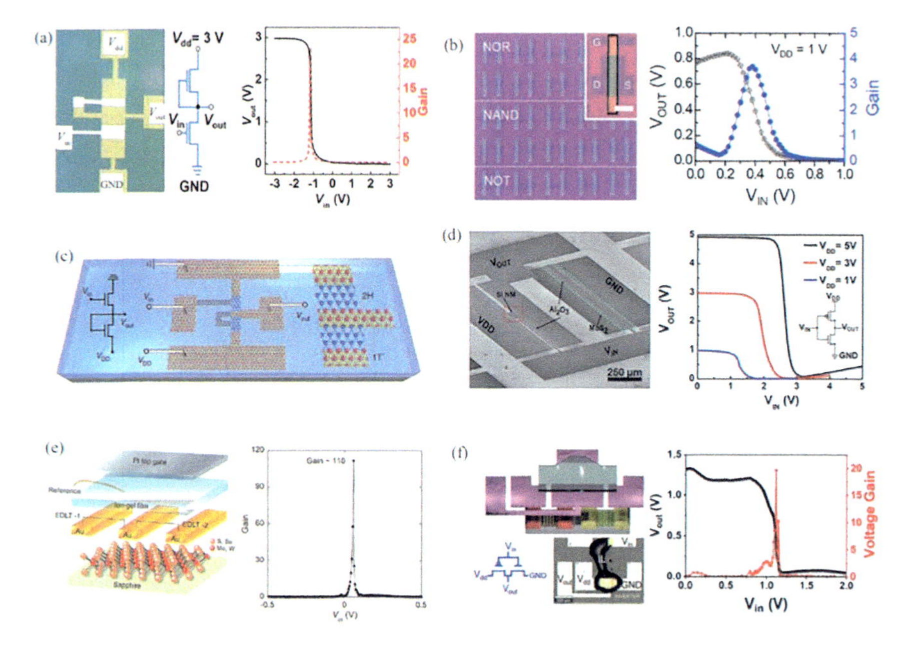

Fig. 8.10 **a** ML-MoS_2 FET and logic gate array on the wafer (left), the voltage transfer curve and gain of the inverter (right); **b** optical picture of the ReS_2 transistors and logic gates, such as NOR, NAND, and NOT gates (left), the voltage allocation physiognomies and signal gain of the NOT gate at VDD = 1 V (right); (c) diagram of the chemically manufactured $MoTe_2$ inverter, the left inset is the circuit diagram for the inverter; **d** diagramic design of a complementary inverter established on Si nanomembrane (NM) and MoS_2 FETs (left), the voltage transfer curves of the inverter at **diverse** VDD (right); **e** illustration of the monolayer MoS_2 and WSe_2 FET constructed on the sapphire substrate (left), the voltage gain plot of input voltage (right), the extreme gain surpasses 110 to lower input voltage; **f** diagramic drawing together the analogous optical microscopy picture of the CMOSinverter constructed on WSe_2 and $MoSe_2$, grown by MGSG (left), the output voltage and gain of the integrated inverter as a function of the input voltage (right) (adopted from [137])

Fig. 8.10c. Such an inverter is made of 2H $MoTe_2$ as channels and 1 T′ $MoTe_2$ as joints and interconnections. Additionally, this kinds of inverter also have a comprehensive rail-to-rail signal overturn and their voltage increase may exceed 35 in a procedural voltage − 6 V.

Similarly, the CMOS inverter is consisting together NMOS and PMOS that have been extensively utilized due to its advantages of better sound verge and lesser stationary power ingesting. To construct a CMOS inverter established on 2DLMs FETs, the transfer technique has frequently utilized to assimilate p-type material and n-type channels to fabricate the identical substrate [140]. The typical design of CMOS inverters is depicted in Fig. 8.10d, fabricated by combination a CVD developed MoS_2 n-type FET and a Si-NM p-type FET collectively [141]. Such logic inverters have an extreme DC voltage increase 16; they may integrate over the plastic substrate to get variable electrical characteristics. In addition to these, such construction of a balancing inverter by transporting p-type WSe_2 and n-type MoS_2 FETs over the sapphire substrate has been also explored (see Fig. 8.10e). This kind of inverter is gated by ion gel to get the uppermost voltage increase around 110. Moreover, different synthetic approaches have been proposed to permitting the concurrent increase in dual distinct materials over the identical wafer, as illustrated in Fig. 8.10f [142].

Due to the advantages of the metal directed choosy development (MGSG) method, it is considered to be useful for the production of extraordinary excellence p- and n-type TMDs for circuit applicabilities. Additionally, they also leads that a bottom-up CMOS inverter made of p-type WSe_2 and n-type $MoSe_2$ FETs, by showing a great voltage increase 23 free to the substrate location.

Hence respect to combination of dual distinct TMDs is additional encouraging for the real-world applications to obtain n- and p-type channels over the similar TMD film that is similar to the conventional CMOS technology. TMDs have a small band gap and show ambipolar behavior, therefore, to get the comprehension of n- and p-type FETs chemical doping into channel area or selecting diverse contact metals is also an impressive way [64, 143–145]. Like recognized behavior in both n- and p-types MoS_2 FETs adopting direct growth of MoS_2 film over to dissimilar kinds of source/drain silicon electrodes for a matching chip, where the p- and n-type MoS_2 FETs are associated as pull-up and pull-down transistors. Such a process achieved CMOS inverter has an extraordinary voltage increase above 20 [143]. However, a complementary logic inverter constructed over p- and n-doping $PtSe_2$ film fabricated on wafer-scale uninterrupted some-layer $PtSe_2$ films are governable for the p- or n-doping features. In this sequence by using p- and n-type $PtSe_2$ FETs a balancing inverter has also fabricated to get a voltage increase about 1 in the applied voltage 3 V. This innovative research effort unlocks the facile realization of huge- scale electronics centered on atomic thick $PtSe_2$ film. The various studies' findings on logic inverters based on TMDs are summarized in Table 8.2 [146].

8.2.3 Large-Scale Circuits

Despite ongoing enhancement in TMD material development and device fabrication methods, the precarious difficulties remains to enhance the mode (E-mode) FET that is significantly necessary to the understanding of multiple stage cascaded circuits. Usually, TMD established FETs have a noteworthy n-doping influence that is usually visualized and accompanied by a great deleterious threshold voltage (V_T) alteration in the transferring physiognomies. This kind of consequence is predominantly owing to the interfacial traps, dipoles, and charge impurities incorporation whiles the top dielectric ALD development procedure [147]. On the other hand, a lack of enhancement-mode TMD transistors together a positive V_T in the recognition of multiple-stage cascaded logic phases.

Considering the above-outlined problem effort has been also made to deposit gate metals together distinct work functions to comprehend E- and D-mode transistors constructed over extracted MoS_2 flakes [132]. However, to get wider region

Table 8.2 Significant large-scale continuous TMDs synthetic approaches [146]

Synthetic methods	Materials	Key preparation conditions	Doping type	Mobility (cm^2/Vs)	ON/OFF ratio	Domain size (μm)	Coverage (%)
	MoS_2	Independent carrier gas channels	n-type	40	~ 10^6	~ 2	100
	MoS_2	Aromatic molecules as seeding promotes	n-type	–	–	~ 60	60
	MoS_2	Low pressure to introduce multilayer dots	n-type	70	10^8	10–20	100
One-step direct deposition	WSe_2	Introduction of H_2 in reaction furnace	p-type	90	10^5	10–50	–
	WS_2	Substrate: cleaved graphite surface exceptionally high temperature at 1100 °C	Non-doped	–	–	15	–
	MoS_2	Face-to-face metal source supply substrate: soda-lime Glass	n-type	6.3–11.4	10^5–10^6	200	43–100
Two-step vapor chalcogenization	MoS_2	Mo metal evaporated by E-beam	n-type	4.1–8.7		–	100
	WS_2	WoO_3 deposited by ALD	p-type	3.9	–	0.01–0.02	100
	MoS_2	$(NH4)_2MoS_4$ decomposed into MoS_2 at 450 °C	n-type	14	5×10^2	–	–
MOCVD	MoS_2	MOCVD precisely control the concentration of precursors	n-type	30	10^6	1	100
MBE	$MoTe_2$	Modulating the source supply with mass flow	p-type	32	10^7	–	100

CVD grownup TMD films the gate-first method is normally adopted [148–150], as the gate electrodes and dielectric layer underneath in semiconductor layer are illustrated in Fig. 8.11a. Contrast to usual gate approaches the gate-first method improved the working ability of MoS_2 FET by reducing the static carriers in the gate dielectric. This useful technique also utilizes for the manufacturing of extremely even E-mode MoS_2 FETs and logic circuits [150]. Thereafter, the extraordinary functionality E-mode MoS_2 FETs have been utilized for the manufacturing of collective logic gates in consecutive circuits (AND, OR, NAND, NOR, XNOR, latch, and edge-trigger register). A typical optical image of the examining chip design is illustrated in Fig. 8.11b. The application of the gate-first technique also leads to the positive and contracted spreading of V_T, as depicted in Fig. 8.11c. It is considered to be critical for signal transmission and logic working in a digital circuit. Such device's test chip outcomes also showed a predictable logic working ability of an XNOR gate and a latch circuit (see Fig. 8.11d).

Additionally, efforts are also made toward integrating TMD basic logic gates into functional complicated circuits. The integration of MoS_2 FETs as a 1-bit implementation of a microprocessor is considered to be one of the most complex circuitry to make 2DLM FETs [148]. The circuit has been made from the NMOS inverters in which both pull-up and pull-down linkages are constructed by n-type E-mode FET. This device's voltage transfer characteristic as an inverter which reveals an excellent together a rail-to-rail logic signal alteration. Moreover, this circuit can also load according to user-defined programs that are stored in an external memory, to perform simple logical operations and communicate with its periphery. Moreover, the 1-bit microprocessor is also readily scalable to multiple bits operations.

It has also recognized as the extraordinary performing electronics established over the extremely unbroken solution-processable 2DTMD nanosheets [105]. Such devices' large-area arrays of locally back-gated MoS_2 transistors enable them for the practical logic gates and computational circuits, comprising an inverter, NAND, NOR, AND, and XOR gates, and a logic half-adder. These logic inverters are also showing a substantial voltage gain of about 20 with accurate functionality. Various devices have been designed and fabricated by using the XOR and AND gate the half-adder logic [146].

8.2.4 Analog and RF Applications

TMDs based FETs can also serve as an analog application, usually, FETs have been utilized as a mixture or amplifier of whichever current or voltage, where the frequencies of the input signals continuously belongs to RF range (3 kHz – 300 GHz) or greater [151]. The 2D materials constructed RF transistors have been activated after the innovation of graphene. The RF transistors perform finest in the overload command in which the conductance of drain keep as low as possible [152]. But ungap graphene cannot full fill the desired current saturation of the FETs, to get a voltage and power gain [153]. On the other hand, the lower mobilities and an appropriate

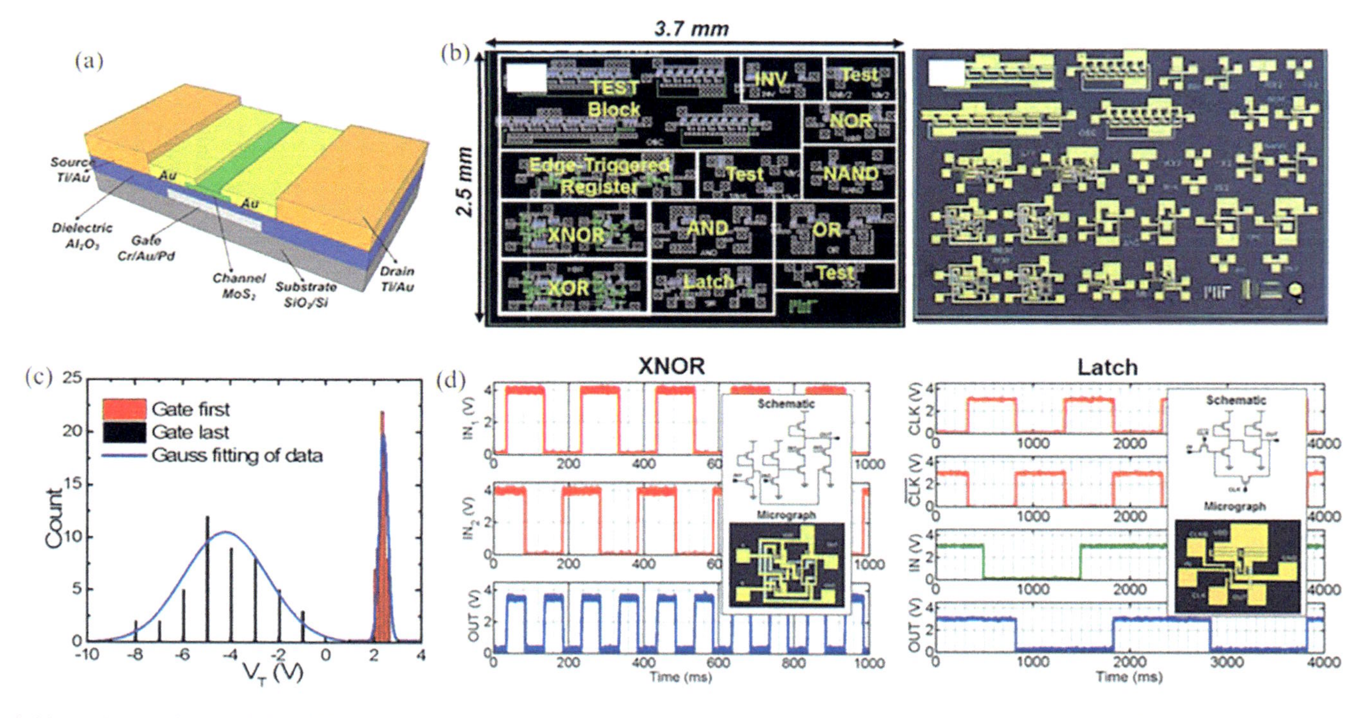

Fig. 8.11 **a** Diagram for the MoS_2 FET manufactured by the gate-first procedure; **b** design (left) and the optical picture (right) of the examination chip by the strategy flow; **c** figures of VT of MoS_2 FETs according to gate-last and gate-first manufacturing machineries; **d** diagramic micrograph, and waveform results of the XNOR gate (left) and latch circuit (right) (adopted from Tang [146])

band gap of TMDs are considered to be suitable which provides a suitable saturation behavior. Therefore, the superiority of the properties makes the TMDs desirable candidates for next-generation RF applications.

The first CVD-grown MoS_2 device performed at RF was based on the monolayer MoS_2 FETs [154]. It showed a short circuit current increase $|h_{21}|$ as a function of frequency with an extrinsic f_T of 2.8 GHz and an inherent f_T of 6.7 GHz. The MoS_2 FETs also have extrinsic f_{max} of around 3.6 GHz and intrinsic f_{max} of ~ 5.3 GHz. Moreover, their implementation as a common source (CS) amplifier together finest CVD MoS_2 FETs has been also explored by applying signal frequency 1.4 MHz with 100 mV peak-to-peak voltage. The achieved yield signal is around 500 mV, with the respective voltage increase $A_v = 14$ dB. The mixed dynamic frequency has been also apprehended from the MoS_2 FETs. Hence these devices showed entirely the predictable harmonics (2RF + LO, 2LO-RF,...) that have been accurately achieved.

Nevertheless, it is also a mirror fact that in almost all studies performed with the rigid substrate, however, recently effort has also made on a flexible substrate by introducing RF performance for the CVD-grown MoS_2 FET [125]. The devices based on the flexible substrates have revealed the intrinsic f_T (~5.6 GHz) and intrinsic f_{max} (~3.3 GHz), including CS amplifier property with 15 dB gain and RF mixing feature. Moreover, the applied frequencies may remain at 1.4 MHz.

Advances in research also explored the possibility to fabricate bilayer MoS_2 EFT, because the bilayer MoS_2 films are possessed advanced carrier mobility and greater density of states to the single-layer [155]. Similar research has been also performed with the bilayer CVD-grown MoS_2 films that revealed an exceptional arrangement of the gate to source and drain. The MoS_2 devices have possessed exterior cut-off frequency f_T and highest oscillation frequencies around 7.2 and 23 GHz. These devices have also shown the frequency mixing property for both rigid and flexible substrates that can operate at 1.5 GHz. Thus, these devices have mixing property in the range of gigahertz regime.

8.3 Light Emission Devices

When a semiconductor has a valence electron appropriate energy gain then it moves toward the conduction band lifting behind a hole in the valence band. It makes to an electron conductive that's unrestrictly moves throughout the crystal lattice. However, such an electron and the hole attracts too respectively due to the Coulombic strength of a bound state, this is typically known as exciton [156]. The exciton is an electrically impartial quasiparticle that exists in a semiconductor. Considering this concept the dual-atom arrangement in III-V semiconductors has been studied widely [157, 158]. Their exciton binding energy is generally defined as, the lowest energy necessary to discrete the electron and hole couples. As the intense, typical III-V semiconductors, like GaAs has an exciton binding energy of individually some meV, therefore, its exciton solitary recognized at the lower temperature. In compare to III-V semiconductor organizations, the 2D behavior single-layer TMDs have a robust Coulombic interattraction, as illustrated in Fig. 8.12a, b. Significantly a robust Coulombic interattraction allows the creation of tightly bounded excitons [159]. As the intense, the exciton binding energy to the MoS_2 has been assessed among 0.5 and 1 eV [160, 161], and its experimentally obtained worth is 240 meV [162]. Similarly, the exciton binding energy has defined 0.71 eV to the single-layer WS_2 [163]**,** 0.37 eV to the single-layer WSe_2 [164], **0.55 eV to the single-layer** $MoSe_2$ [165]**,** and 0.58 eV corresponding to single-layer $MoTe_2$ [64] that are enough superior to significant III-V semiconductors which solitary 3.71 meV to InP and 4.76 meV for the GaAs [166]. Thus the large exciton binding energies of the 2D materials consents us to explore the exciton dynamic at the normal room temperature. Additionally, TMDs systems also have a robust Coulombic interattraction and quantum confinement with a high-order quasiparticle exciton that delivers a separate basis to explore the many-body consequences in excitons. Specifically, while an exciton is bound to alternative electron or hole and ultimately it make a trion. Since exciton and trion are opposing to each other in single-layer TMDs, thereby, it allowing to tune through the electrostatic gating and chemical doping, as illustrated in Fig. 8.12c–h [167, 168]. In the case of dual impartial excitons bounded states make the four particles typically it is designated as the biexciton. The binding energy of biexciton has been found at ~ 20 meV for the single-layer $MoSe_2$ organization [169]. Moreover, the charged biexciton has been also interpreted by considering a five particles organization in h-BN covered single-layer WSe_2 that offers a straight way to deterministic govern in many-body quantum mechanism. Thus above discussed adequate different optical features of 2D TMDs have been prompted by a verity of optoelectronic device uses, like light emission diodes.

8.3.1 Light-Emitting Diodes (LEDs)

Electroluminescence property is owing to the emission of light to an electronic device in reply to an electric current, as the consequences the radiative recombination of charge carriers (electron–hole pairs or excitons) in a semiconductor. This process constructs the light-emitting optoelectronic devices, such as light-emitting diodes (LEDs) [170]. TMDs established LEDs have been constructed together various sophisticated arrangements, like metal–insulator–semiconductor junction, p-type semiconductor/n-type semiconductor (p–n) junction, and metal–semiconductor junction. Significantly the p–n junction of twofold TMDs layers or TMD associated to supplementary regular materials are also considered to be a key structural formation, such as the extremely luminescent WS_2 quantum dots/ZnO heterojunction devices to produce white light [171].

Since the extreme-thin behavior of TMDs is significantly sensitive to the neighboring dielectric situation which allows for the prospect of electrostatic gating govern of single-layer TMD doping. The construction of a p-n joint by employing dual split gates underneath the single-layer TMDs is the general method. Therefore, the simplest LED structure can have two local gates to define a p–n junction within the

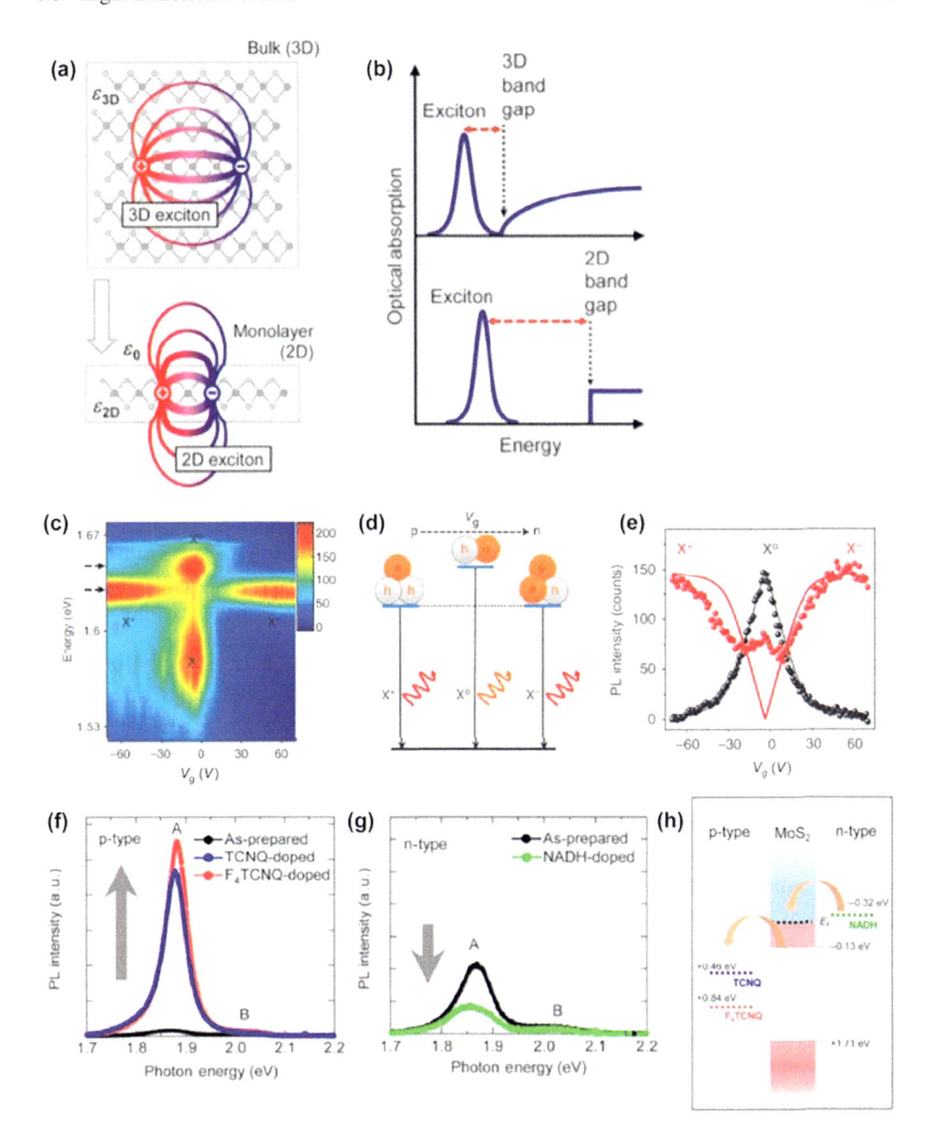

Fig. 8.12 Exciton possessions in 2D materials: **a** diagram of the confinement of charge carriers and the reduced dielectric screening owing to the non-appearance of contiguous layers in 2D materials compare to bulk 3D materials; **b** consequential upsurge in the band gap and exciton binding energy in semiconducting 2D materials (a and b adopted from Chernikov et al. [155]); **c** $MoSe_2$ PL plots as a function of back-gate voltage, close zero doping, mainly original and imperfection trapped excitons are remarked, whereas together a greater electron (hole) doping, negatively (positively) charged excitons leading the spectrum; **d** illustration of the gate-dependent trion and exciton quasi-particles and transitions; **e** trion and exciton peak intensity versus gate voltage, the solid lines are fits established on the mass action model (c-e adopted from Ross et al. [168]); **f** PL spectra of 1L-MoS_2 earlier and afterward being doped with p-type molecules (TCNQ and F4TCNQ);**(g)** PL spectra of 1L-MoS_2 earlier and afterward being doped with the n-type molecules (NADH); **h** diagram to the comparative capacities (vs. SHE) of 1L-MoS_2 and n- and p-type dopants (**f**–**h** adopted from Mouri et al. [167])

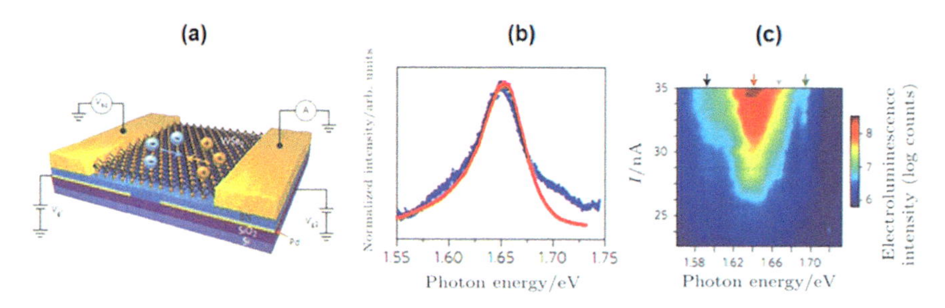

Fig. 8.13 LEDs founded on TMDs p–n junction: **a** diagramic plot of multiple single-layer WSe_2 p–n junction devices together palladium back gates (V_{g1} and V_{g2}) and source (S) and drain (D) contacts. The source-drain voltage (V_{sd}) is employed to one connection and the current (A) reading to another. While the electroluminescence in the WSe_2, electrons, and holes travel toward to each other (arrows) and recombine. The back gates are detached to the WSe_2 by h-BN. The device sits over a layer of silicon dioxide on a silicon substrate; **b** the Electroluminescence spectra originated by a current 5 nA narrowly resembles the photoluminescence spectra at 300 K; **c** the Electroluminescence intensity plot as a function of bias current and photon energy. Left to right, the arrows designate the contamination bounded exciton (X_I), the charged excitons (X_- then X_+), and the neutral exciton (X_0) (adopted from Yu et al. [175])

TMD sheet [172, 173]. As the p–n junction of the WSe_2 reveals the total photon emission rate up to ~ 16 million s^{-1} at an employed current of 35 nA, respect to one photon per 10^4 inserted electron–hole couples, as depicted in Fig. 8.13a–c) [174]. Where the right panel is constructed from left to right (see Fig. 8.13a) and their arrows indicated the electroluminescence features correspond to the impurity-bound exciton, two types of charged excitons, and the neutral exciton. However, this device entire assessed quantum efficiency is simply 0.01%. Hence the emission wavelength to be adjusted among establishments of contamination-bounded, charged, and impartial excitons, as illustrated in Fig. 8.13c [175].

This split gate notion is widely accepted to the TMDs established LEDs to fabricate electrically controlled single devices that have manifold functionalities. For instance, ambipolar single-layer WSe_2 devices together the splits of indigenous gates are also fabricated in both PN and NP arrangements possessing the diode ideal factors n = 1.9 and a rectification factor of 105. Their electroluminescence quantum efficiency is up to ~ 1% [172]. Therefore, the ambipolar field-effect transistor (FET) structure is considered to be suitable for the construction of p-n joint within 2D materials to light-emitting activity in single-layer TMDs [176]. Under the application of a gate voltage, the TMDs are efficiently doped to impartial atoms. It makes TMDs founded FET devices capable to work in the ambipolar injection establishment, in which both holes and electrons can be inject concurrently at the two opposite contacts and the light emission from the FET channel, as the typical illustration is given in Fig. 8.14a [177]. Their circularly polarized electroluminescence emits from p-i-n junctions that typically formed the electrostatic transistor in channels (see Fig. 8.14b). This approach is also considered to be useful to explain qualitative features of holes and electrons overlaps that are governed through the in-plane electric field. Such LED's

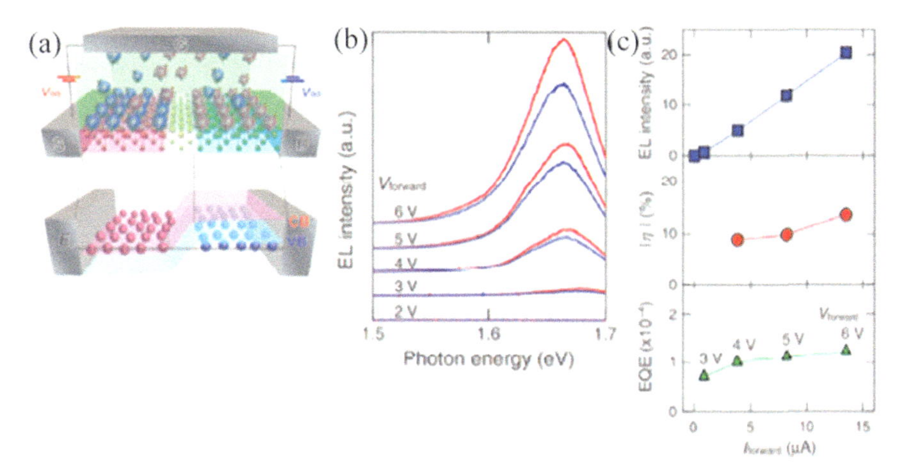

Fig. 8.14 **a** The upper panel is the device structure of the TMD electric-double-layer transistor (EDLT) under ambipolar charge accumulation, while the bottom panel shows the schematic band structure of EDLT induced p-i-n junction under equilibrium; **b** voltage dependence of EL spectra of the device; **c** the current dependence of total EL intensity (upper), current dependence of EL polarization (middle), current dependence of external quantum efficiency (bottom) (adopted from Zhang et al. [177] http://li.mit.edu/S/2d/Paper/Zhang15Oka.pdf)

external quantum efficiency (EQE) has been varied in-between 0.002% to 0.06% depending on the used TMD materials (see Fig. 8.15f).

Such kinds of basic LEDs are generally calcified as lateral p–n junction devices that are **restricted to** the **fine** one-dimensional joint interface, usually they are called low-efficiency LEDs.

8.3.2 LEDs Based on TMDs in Vertical Stacking

The modern era demands highly efficient and large area light-emitting LEDs for the desirable potential applications. Therefore, the vertically chip established LEDs have been explored thorough layer by layer stacking of 2D materials. This kind of configuration is normally consisting to graphene, h-BN, and TMDs. Such as, in the case of single optical active devices, the binary graphene layers can serve as translucent even electrode connections that are unglued from an h-BN/MoS_2/h-BN sandwich quantum well, as illustrated in Fig. 8.15a. The critical portion of such kind construction is the h-BN. Because some layers of h-BN can also serve as an adjustable obstacle that considerably reduce the leakage current while the device is operated underneath a current flow from top to bottom layers. Moreover, the width of the h-BN flake also plays a significant part in the LEDs action. Such as the width of the h-BN flake greater to single atomic layer also influence the lifetime of the carriers, as depicted in Fig. 8.15b. An increase in width of the h-BN permits current due to

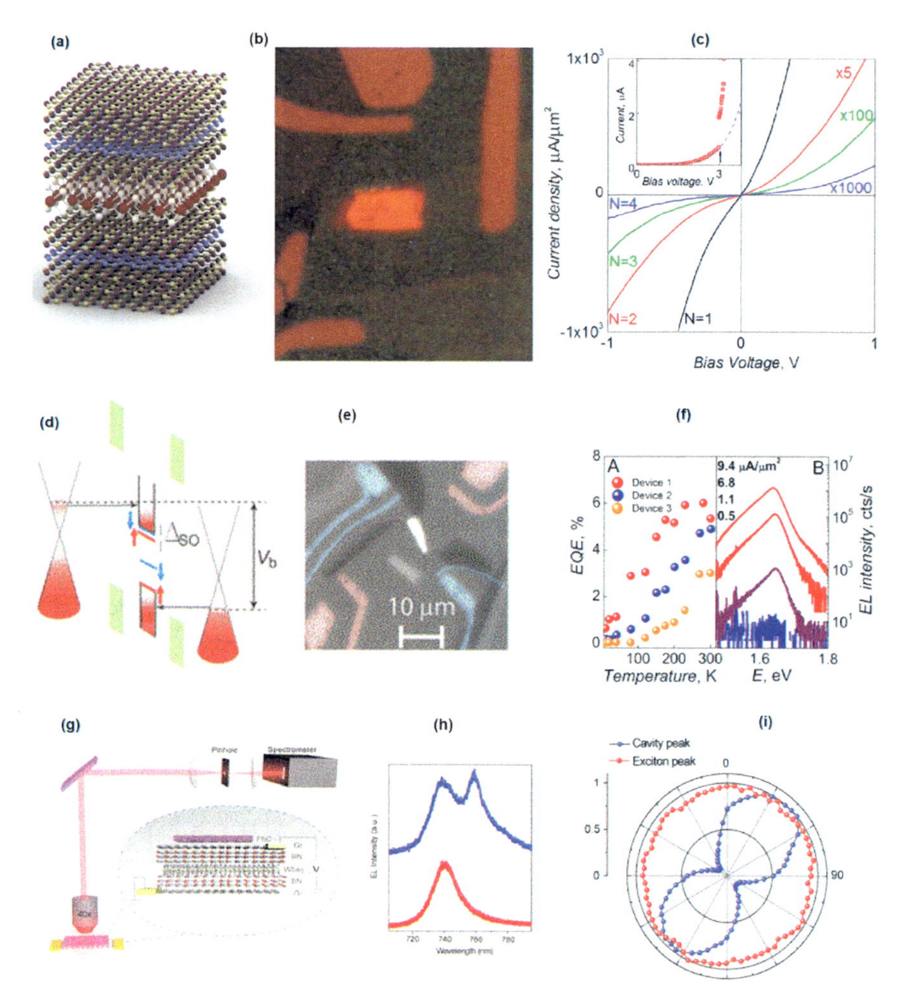

Fig. 8.15 LEDs constructed from the TMDs vertical stacking: **a** diagram for the SQW heterostructure h-BN/GrB/2 h-BN/WS_2/2 h-BN/GrT/h-BN; **b** optical picture of EL for the identical device, $V_b = 2.5$ V, $T = 300$ K (a and b adopted from Withers et al. [51]); **c** characteristic $I - V$ curves for the graphite/BN/graphite devices accompanied distinct widths of BN insulating layer, curve, a single-layer of BN, bilayer, triple-layer, and quadruple layer. (adopted from Britnell et al. [178]); **d** band arrangement at extraordinary bias of a WSe_2 LEQW; **e** magnification (50 ×) monochrome picture of a WSe_2 LEQW device under an employed bias of $V_b = 2$ V and current of 2 μA noted in suitable circumstances with feeble backlight illumination (central white region respect to robust electroluminescence), Au connections for the bottommost and topmost graphene; **f** left panel: temperature dependence of the quantum efficiency for three typical WSe_2 LED devices at bias voltages and injection currents of 2.8 V and $j = 0.15$ μA/μm^2 (device 1), 2.8 V and $j = 0.5$ μA/μm^2 (device 2), and 2.3 V and $j = 8.8$ μA/μm^2 (device 3), whereas the right panel shows the discrete electroluminescence spectra plots for four diverse inoculation current densities for device 3 (d-f adopted from Withers et al. [179]); **g** diagram of the EL measuremental set-up and device design; **h** EL obtained far (dots) and away (dots) from the cavity region with $V_b = 2$ V; **i** standardized cavity improved peak intensity (dark dots) and exciton peak intensity (light dots) as a function of polarization recognition angle (**g**–**i** adopted from Liu et al. [180])

tunneling by the h-BN and finally, it intensely decreased [178, 181], thereby very weak LED emissions resulted (see Fig. 8.15c). However, a well-designed solitary h-BN/MoS_2/h-BN quantum well LED device reveals ~ 1% quantum efficiency, further it may up to 8.4% by increasing the quantum well stacking up to four due to an increase in radiative recombination [117]. This kind of vertical stacking construction also normally increases the light-emitting quantum efficiency up to several orders of magnitude superior to the LEDs established on p–n junction configuration.

Moreover, the emission spectrum is also tuned by the combination of different 2D semiconductors. Such a device's quantum efficiency can be up to ~ 5%, which is almost comparable to organic and quantum dot LEDs. Thus, further modifications are desired to improve the efficiency of vertical LEDs. The substrate modulation and integration together photonic crystal cavity view adopted and fabricated a room temperature high-efficiency WSe_2 vertically assembled LEDs, as depicted in Fig. 8.15d, e. It has been noted that the EQE of WSe_2 devices increased among the temperature (see Fig. 8.15f), and its reaches upto 5% at room temperature [179], this is around 250 times greater **to** the finest MoS_2 quantum wells in suitable circumstances [117]. Moreover, the combined WSe_2 perpendicular stacking LEDs together photonic crystal cavities are illustrated in Fig. 8.15g, h. The overall emission efficiency has improved greater than 4 times contrast to LEDs on bare SiO_2/Si substrate [103]. These kinds of increments credited due to the robust combination among the photonic crystal mode and exciton EL in WSe_2, thereby, increasing the emission rate. However, the emission at the cavity resonance in single mode is highly linear polarized (84%) alone the cavity mode, as depicted in Fig. 8.15i.

Thus the photonic crystal cavity has been recognized by the electrically pumped single-mode light source, it is a necessary step toward on-chip optical information technologies. This is considered to be the benefit of the perpendicular stacked LEDs construction. However, vertical stacking LEDs are also not free from drawbacks one of the major challenges being the appropriate choice and grounding of the thickness of h-BN. So, many methods have been introduced to resolve the shortcomings of the LEDs to achieve higher quantum efficiency [156, 171, 182, 183].

8.3.3 Single Photon Emitter Based on TMDs

In recent, important development has been devoted to the realization of effective solitary photon emitters that are utilized for quantum information and other technological devices. Based on this view single-defect in monolayer TMDs are considered to be a novel kind of solitary-photon origin [184–187]. A solitary-defect emitter has been fabricated by utilizing the vertical and lateral van der Waals heterostructures of single-layer WSe_2 [188]. Moreover, the solitary-imperfection emitters are also facilitated through the incorporation of different kinds of optical constituents like photonic crystal cavities and waveguides (see Fig. 8.16a) [180]. The straightforward modulation of the solitary-mode EL at an extraordinary speed of ~ 1 MHz is depicted in Fig. 8.16b. The high-resolution PL spectrum of the 1L-WSe_2 flake has exposed a dual split in energy ($\Delta = 726$ meV) associating uneven intensities, as illustrated in Fig. 8.16c [187]. The robust antibunched photon release to the emitter has been also noticed. That further can also verify with the help of the fit by using the relation $g^2(\tau) = 1 - \rho^2 e^{-|\tau|/T}$ (See Fig. 8.16d) [189]. Moreover, the combination of solitary photons have been spotted to the distinct quantum emitters in WSe_2 and it further verifies via the noted antibunching in signal $g^2(0) = 0.42$ [190]. To date, various works have been done over the solitary-photon emitters of the layered TMDs by employing PL, though comparatively less research work has been done in terms of EL. In addition to these, compressed and crossbreed 2D semiconductor piezoelectric devices also allow adjusting the energy of solitary photons released and confined in wrinkled WSe_2 single-layers through the strain fields (see Fig. 8.16e) [191]. Their collective release owing to imperfection-assured excitonic complexes has been also noticed for the both 1L and 2L WSe_2, as depicted in Fig. 8.16f [189]. The nonappearance of proximity effect is due to a slight superconducting coherence distance and the existence of the topological surface state with the diverse regularities at the interface. Thus Bi_2Se_3 combination with a variety of mate atomically thin semiconductors at room temperature is considered to be a favorable step headed to achieve an inversion of less electrically motivated laser and ultrafast microcavity LEDs [112].

8.3.4 Organic Light-Emitting Diodes (OLEDs)

OLED technology has also attracted much attention and it is used widely in various high-processing technology devices such as smartphones, TV, wearable devices, etc. OLEDs devices provide a homogeneous emission over a large area having an internal charge-to-photon conversion efficiency of up to nearly 100% [192, 193], which makes them a promising candidate for new display applications. The OLEDs are typically comprised of organic semiconductors layered between carrier transfer layers with the electrodes. In OLED devices charges are usually injected from the opposite electrodes into the emission layer to emit the photons out of the multilayer structure. An efficient and long-lived OLED device should have characteristics like being energy level-aligned, transparent, good blocking, conductive, stable, and with an abundant charge transport layer (CTL).

A typical bottom-emitting OLED multilayer structure with a different charge transport layer is illustrated in Fig. 8.17. The multilayer structure of OLEDs have been also fabricated by vacuum thermal evaporation or spin coating techniques. The advantage of the vacuum evaporation technique is its non-destructive film deposition layers compared to the spin-coating technique that has physical damage under the layer. Therefore, most commercial OLEDs have been fabricated by the thermal evaporation process. OLED devices' most important parameters are the luminescence of the materials that reflects small molecule materials including polymer materials. The kinds of structures exist within OLEDs to form white light in terms of the number of emitting layers [194, 195]. Typically, single emission- layer OLEDs are fabricated by

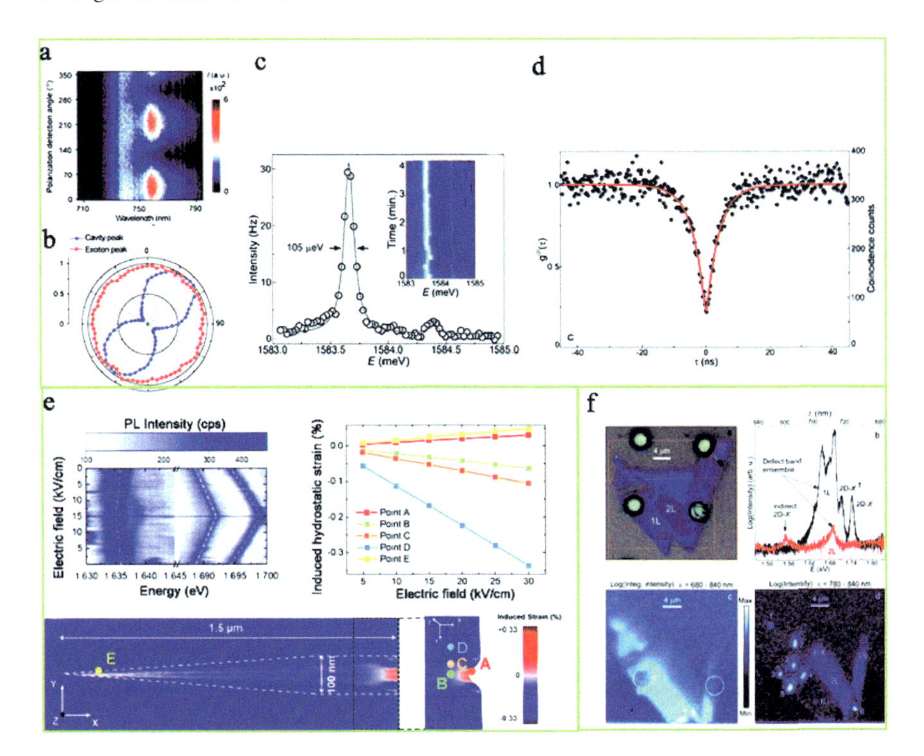

Fig. 8.16 The functionality of a solitary photon emitter constructed on layered TMDCs; **a** observed on-cavity EL as a function of polarization recognition angle; **b** standardized cavity improved peak intensity (blue dots) and exciton peak intensity (red dots) as a function of polarization recognition angle; **c** high-resolution PL spectrum of 0D-Xfrom the position of WSe_2 over the p-Si substrate; **d** standardized second-order relationship function $g_{(2)}(s)$ of the 0D-X lines for the WSe_2 over the p-Si substrate in the non-resonant cw excitation at $P = 0.5$ PN; **e** contour plot of the PL spectra of SPEs as a function of the electric field employed over a (001) piezoelectric device, the hydrostatic strain as a function of the electric field employed over a (001) piezoelectric device at five distinct points of the wrinkle; **f** Significant PL emission spectra for 1L (black) and 2L (red) WSe_2 for the even and unstrained positions of the flake; color-coded spatial maps of PL with (left) combined intensities in the spectral range of 680–840 nm, and (right) intensities in the spectral range of 780–840 nm (adopted from [171])

doping red, blue, and green phosphors into a single-emission layer. While the multi-emission-layer OLEDs are also consist of several emission layers of different color phosphors. Both kinds (OLEDs and pixelated OLEDs) of effective devices are generally fabricated in vertical stacking. Yet many emission layer structures of OLEDs have been fabricated [196–200]. Considering the white OLEDs first fabricated in the lab, nowadays the luminous efficacy of OLEDs exceeds 120 lm/W with the external quantum efficiency (EQE) greater than 63% under the backlight application. The color rendering index (CRI) values of white OLEDs for the lighting application are around 90 [201–204].

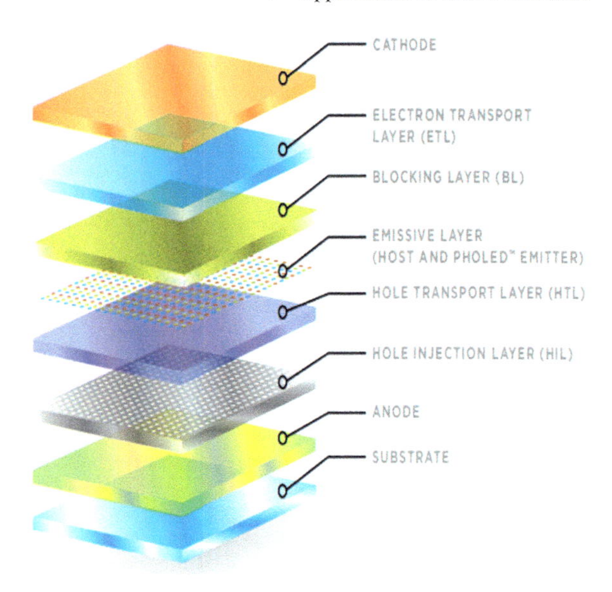

Fig. 8.17 Schematic representation of the organic materials for each layer of bottom-emitting OLEDs (adapted from https://oled.com/oleds/)

Besides the significant achievements of the OLEDs, they are also not free from certain shortcomings. Such as a broad and asymmetric emission spectrum of the OLEDs is a comprehensive issue. Communally monochromic OLEDs emit a light that is characterized by large full widths at half maximum (FWHM), typically > 60 nm [205–207]. Their broad emission spectrum contributes to a high CRI that diminish the color saturation of lighting devices. Therefore, color purity is a great obstacle for OLED technology to utilize for high-quality displays. Moreover, OLEDs have a complex device structure; thereby the higher power consumption, low-resolution pixel patterning, and limited stability are also restricts their widespread use as commercial products.

8.3.5 Quantum Dots (QDs) Based Light-Emitting Diodes

The optically and electrically excited QDs are usually generates photoluminescence (PL) and electroluminescence (EL) emissions. The optically excited QD-based LEDs are also utilized as nano phosphors in which light is generated by using a combination of QDs with different emission colors for LEDs backlight LCD's potential applications. Whereas, the electrically excited QD-based LEDs electrons and holes are directly injected into the QDs, resulted light being noticed through the radiative recombination of the injected carriers within QDs with different sizes and emission colors. The optically excited QD-based LEDs in backlight LCDs are significantly based on the three primary choices of the LCD design. The first one is the film configuration, where usually QDs placed over to light guide plate. So, the film is set away from the chip to generate a high-intensity flux and a lot of heat available to damage the QDs. The second one is the architecture of an edge configuration, where seldom is used in backlight displays due to the fragile QD tubes. The third one is the on-chip configuration, in which the QDs are sprayed directly on the blue GaN chip. Significantly, the on-chip configuration includes fewer consuming QDs but they need to withstand the high-intensity flux and heat. The most important reason to the use of QD-based LEDs for backlights is the color gamut that can be regulated by the NTSC. Such as perovskite QDs with narrowband emission are considered to be promising down converter candidates for LCD backlight applications [207]. Typically the all-inorganic $CsPbBr_3$ perovskite QDs are used in white-light LEDs for backlight applications, in which the color gamut overlap with the NTSC space of approximately 113%, which is higher than phosphor (86%) and Cd QDs (104%)-based white-light LEDs, as depicted in Fig. 8.18a [208]. But the use of QDs-based white-light LEDs in liquid crystal displays is consumed.

Particularly, the QLED technology based on electrically excited QDs has been widely used for charge transport in the fabrication technology of the materials. The EQE of RGB and white QLEDs may exceed 10% which makes them competitive rivals to OLEDs for future use in thin and flexible displays. Most significantly real-world applications of QLEDs can provide an efficient active-matrix QLED (AM-QLED) display. AM-QLED displays are self-emissive devices that are utilized for the current-driving mode, as the typical structure of an AM-QLED display is depicted in Fig. 8.18c [209]. The high performance of electrically driven QD-based red, green, blue, and white LEDs (EQE > 10%) is summarized in Table 8.3 [204]. The most efficient QLEDs have EQE over 20% with the ultra-thin layer in the tandem structure [210–214]. However, QLEDs are also facing several performance issues such as Auger recombination (AR), Förster resonance energy transfer (FRET), and

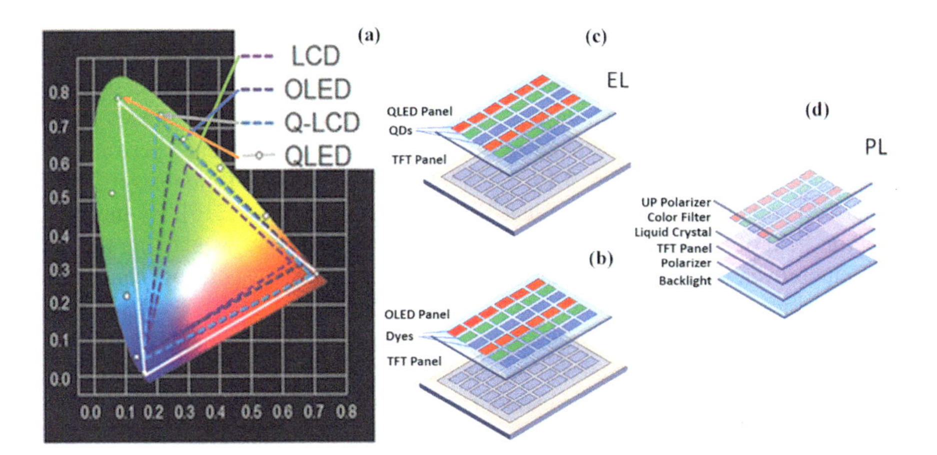

Fig. 8.18 **a** Color gamut of three different types of white-light LEDs; **b, c** schematic for the dyes and QDs displays; **d** liquid crystal display of the device (adapted from Wang et al. 2020 ACS Energy Lett. 5, 3374–3396) high energy and difficult for thin and flexible display applications, as the schematic of a typical liquid–crystal display structure is illustrated in Fig. 8.18b [209]

field-induced quenching (FIQ). These key issues should be addressed for the highly efficient and reliable performing QLEDs.

8.4 TMDs Based Lasers

Merged of two photons generates one photon with doubled frequency in nonlinear optical materials. Typically this process is known as second harmonic generation (SHG) which was first discovered in the 1960s, immediately after the invention of the laser. The development of SHG has led many applications in advanced technologies, such as on-chip light sources, imaging, sensing, and communications. For the instance, SHG-based imaging devices, which capture near-infrared (NIR) light and emit light in the visible range, this is also useful to develop novel all-optical NIR imaging technologies, such as night vision camera, etc.

The thinning down up to the monolayer of TMDs exhibits a direct band gap, strong luminescence, room-temperature stable excitons, and strong second-order nonlinearity. These unique optical properties make monolayers of TMDs an attractive choice to explore novel linear and nonlinear optical effects and their related applications. However, the atomic-scale interaction length with light led to a sole TMD monolayer that can emit an extremely low SHG signal which significantly hindered the development of practical nonlinear meta devices based on 2D materials. The high-refractive-index dielectric of nanoresonators is also considered a promising platform to enhance SHG. Despite of various features of dielectric nanoresonators the capability to exhibit strong confinement of light field is considered to be significant, it is a so-called bound state in the continuum (BIC) that induces a unique feature in dielectric nanoresonators [215]. The BIC's eigenfrequency lies in the continuum spectrum with a promising approach to enhance SHG in 2D materials. It has been recognized as second harmonic waves in resonance simultaneously to boost the conversion efficiency of SHG by TMD monolayer. Usually the pair of BICs cavity modes within a GaP grating slab are interpreted by transferring the TMDs monolayer onto the BICs slab, as depicted in Fig. 8.19. Thereby, the SHG signal of monolayer TMDs is largely amplified owing to the dual-BICs resonance process [216]. By mean the electric field of the fundamental light is significantly enhanced by exciting the first BIC, moreover, the excitation of the second BIC at harmonic wavelength can further boost the nonlinear emission.

But the key issue with the spatial mode matching of the TMD monolayer between the BIC-resonant fundamental and second harmonic wave. To resolve this issue has been proposed slightly tilting the incident angle of the fundamental wave that greatly improved the spatial mode matching within the TMD monolayer, therefore, it gives rise to four orders of magnitude enhancement of the SHG efficiency compared to that with a sole TMD monolayer, as depicted [see Fig.8.20a, b]. Moreover, it has also demonstrated that the patterning of the TMD monolayer can be optimized from spatial mode matching. It also further boosts the SHG process of the TMD monolayer

Table 8.3 High-performance QD-based red, green, blue, and white LEDs, PL, photoluminescence; EL, electroluminescence; FWHM, full widths at half maximum; QD, quantum dot; QY, quantum yield; EQE, external quantum efficiency; CE, current efficiency; PE, power efficiency; DHNRs, double heterojunction nanorod [204]

Device structure	Emitting layer	PL λ_{max} (nm)	EL λmax (nm)	EL FWHM (nm)	PL QY (%)	Peak EQE (%)	Peak CE (cd A^{-1})	Peak PE (lm W^{-1})	Von (V)	Max. L (cd m^{-2})	CIE coordinates	T50 @100 cd m^{-2}	Color of QD-LED
Conventional	CdZnSe/ZnS/OT	N/A	622	27	85	23.1	41.5	11.1	5.6	65,900	N/A	13,788	Red
Conventional	CdSe/CdS	N/A	640	28	> 90	20.5	N/A	N/A	1.7	42,000	(0.71, 0.29)	100,000	Red
Inverted	CdSe/CdS	610	620	N/A	N/A	18	19	25	1.5–1.7	50,000	N/A	4000	Red
Inverted	CdSe/CdS/ZnS	N/A	633	36	83	13.57	18.69	N/A	N/A	23,590	N/A	N/A	Red
Conventional	CdS/CdSe/ZnSe DHNR	N/A	612	N/A	~ 40	12.05	27.52	34.6	1.55	76,000	N/A	N/A	Red
Conventional	$Cd_{1-x}ZnxSe_{1-y}S_y$	N/A	625	25	N/A	12	15	18	1.5	21,000	(0.68, 0.31)	32,000	Red
Conventional	CdSe/CdS/ $Zn_xCd_{1-x}S$/ZnS	N/A	618	46	75	10.2	18.9	20.6	1.95	62,000	(0.626, 0.341)	84,000	Red
Conventional	CdZnSeS/ZnS/ oleic acid	N/A	534	29	85	27.6	124.5	31.8	6.1	115,500	N/A	53,808	Green
Inverted	CdZnSeS/ZnS	N/A	538	23	80	23.68	99.78	20.49	6.3	6.3 200,000	(0.2067, 0.7544)	N/A	Green
Inverted	CdSe@ZnS/ZnS	520	N/A	N/A	90	15.6	65.3	29.3	3.1	216,500	(0.157, 0.786)	N/A	Green
Conventional	$Zn_{1-x}CdxSe$/ZnS	N/A	517	45	91	15.4	53.4	39.1	2.35	121,000	(0.155, 0.736)	2400	Green

(continued)

Table 8.3 (continued)

Device structure	Emitting layer	PL $\lambda_{max\ (nm)}$	EL λmax (nm)	EL FWHM (nm)	PL QY (%)	Peak EQE (%)	Peak CE (cd A^{-1})	Peak PE (lm W^{-1}_)	Von (V)	Max. L (cd m_ $^{-2}$)	CIE coordinates	T50 @100 cd m^{-2}	Color of QD-LED
Conventional	$Cd_{1-x}ZnxSe_{1-y}S_y$	533	537	29	75	14.5	63	60	2	14,000	(0.27, 0.67)	> 90,000	Green
Conventional	CdSeZnS/ZnS	516	516	21	79–83	12.6	46.4	N/A	N/A	85,700	(0.101, 0.781)	N/A	Green
Conventional	CdZnSeS/ZnS	528	538	24	N/A	10.5	45.7	18	3	61,030	N/A	30,760	Green
Conventional	CdZnSeS/ZnS/ OT	N/A	474	28	85	21.4	17.9	4.2	8.8	26,800	N/A	107	Blue
Conventional	CdSe/ZnS	466	468	20	87	19.8	14.1	N/A	5.1	4890	(0.136, 0.078)	47.4	Blue
Conventional	$Zn_xCd_{1-x}S$/ZnS	442	445	25	93	15.6	2.94	1.71	5.8	4500	(0.143, 0.017)	4300	Blue
Conventional	ZnCdS/ZnS	441	443	21.5	90	12.2	2.1	1.6	2.6	7600	(0.14, 0.02)	N/A	Blue
Conventional	Cd_{1-x}ZnxS	N/A	455	20	N/A	10.7	4.4	2.7	2.6	4000	(0.16, 0.02)	1000	Blue
Conventional	CdZnS/ZnS CdZnSeS/ ZnSCdSeS/ZnS	447 519 605	N/A	20 22 25	82 78 75	10.9	21.8	N/A	N/A	23,352	(0.453, 0.333)	N/A	White

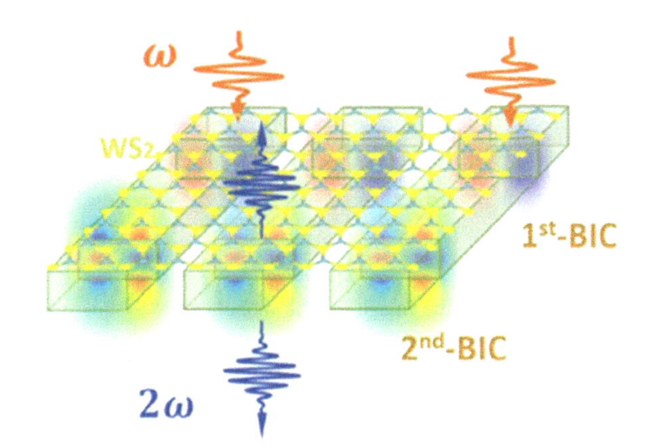

Fig. 8.19 The schematic diagram for the dual BICs scheme to boost the SHG with monolayer WS_2 on top of the photonic grating slab (adapted from https://phys.org/news/2022-07-boosting-harmonic-tmds-monolayer.html)

and amplifies SHG signal up to seven orders of magnitude, as illustrated in Fig. 8.20c, d [215, 216].

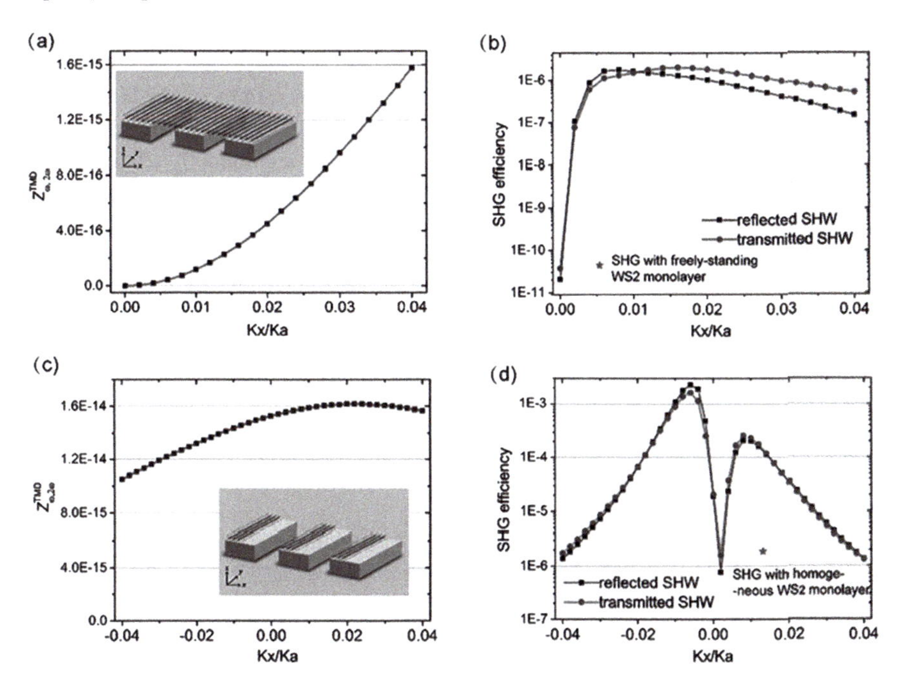

Fig. 8.20 **a**, **b** Spatial overlap coefficient and SHG efficiency with a homogeneous WS_2; **c, d** a patterned WS_2 on top of the photonic grating slab. **a, c** represents the K_x-dependent spatial overlap coefficient; **b, d** represents the K_x-dependent SHG efficiency at the reflected (top) and transmitted (bottom) side. The blue star in **b** and **d** is the reference point for the SHG efficiency with a free-standing WS_2 monolayer and that with homogeneous WS_2 on top of the grating. The fundamental wave is incident from the top side of the grating and the intensity is to be 0.1 GW/cm^2 (adopted from Hong et al. [216])

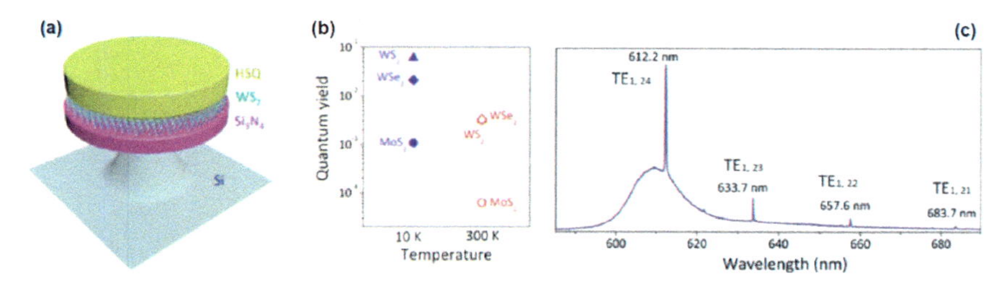

Fig. 8.21 Schematic image for the monolayer WS_2 microdisk laser, the sandwich structure of Si_3N_4/WS_2/HSQ ensures a higher confinement factor and leads to a larger modal gain; **b** the photoluminescence (PL) spectrum at 10 K when the pump intensity is above lasing threshold, for the whispering gallery modes at 612.2, 633.7, 657.6 and 683.7 nm; **c** PL quantum yield of monolayer MoS_2, WS_2, and WSe_2 at 300 and 10 K (adopted from Ye et al. [221])

Considering the above-discussed basic concepts of SHG, one can say that the spontaneous emission of TMDs monolayers enhances through the integration of the photonic crystals [217, 218] which distributed by the Bragg reflector micro-cavities [219, 220]. These two strategies initially have been used to develop monolayer MoS_2-based lasers. A typical schematic of the monolayer excitonic laser in a microdisk resonator is illustrated in Fig. 8.21a [221]. Wherein the embedded monolayer in between two dielectric layers in a Si_3N_4/WS_2/HSQ structure has strong optical confinement and leads to a larger modal gain. It is significantly crucial for the atomically thin monolayer gain medium. Because the resonant wavelength matches to the dominant peak of the lasing spectrum of the WS_2 monolayer that embedded 3.3 μm diameter microdisk at 612.2 nm (Fig. 8.21b), their quality factor $Q = \lambda/\Delta\lambda$ is around 2604 [221]. Additionally, it also noted that the lasing action has no decay even after several months. Thus, the Si_3N_4/WS_2/HSQ (hydrogen silsesquioxane) sandwich structure protects the monolayer from direct exposure to air. The typical comparative quantum yields of the various TMDs, such as MoS_2, WS_2, and WSe_2 monolayers, are illustrated in Fig. 8.21c.

In the second strategy, an atomically thin crystalline semiconductor (WSe_2) relates to a gain medium at the surface of a pre-fabricated photonic crystal cavity, as depicted in Fig. 8.22a, b [88].

It has been also noted that the continuous-wave nanolaser operate in the visible regime and achieve an optical pumping threshold as low as 27 nanowatts at 130 K. Additionally, the key lasing action lies in the monolayer of the gain medium that confined direct-gap excitons within one-nanometre cavity surface. The surface-gain geometry also allows the tailoring of gain properties through external controls like electrostatic gating and current injection to enable the electric pumping operation. This fabrication technique is considered to be scalable and compatible with integrated photonics for on-chip optical communication technologies [222].

Significantly QDs technology advances are also introduced the QD-based laser diode for the displays that enables to distinguish the features, such as conveniently tunable emission color, flexibility, low weight, and inexpensive processability. Especially, in the developments of QDs laser diodes are based on the intrinsic nature of

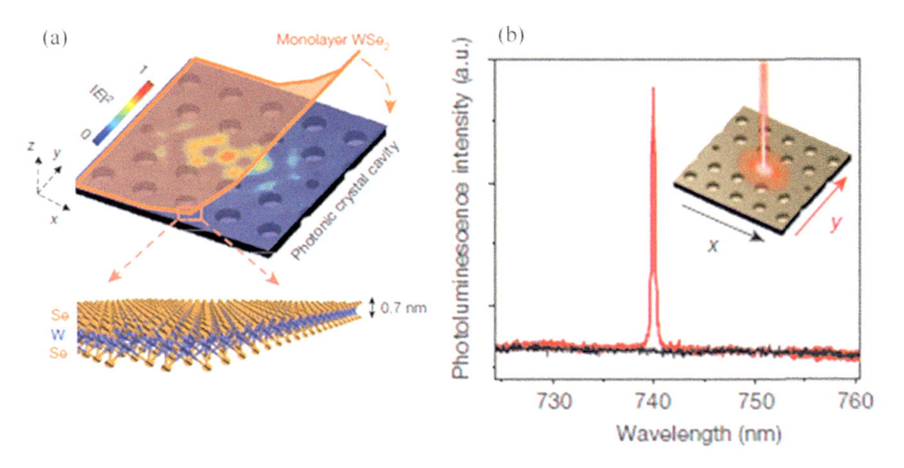

Fig. 8.22 **a** Cartoon of a laser architecture with a photonic crystal; **b** the polarization-resolved photoluminescence spectrum of a device at 80 K, to show a complete polarized narrow emission at ~740 nm. The black (red) line corresponds to the detected linear polarization in the *x*(*y*) direction (adopted from https://doi.org/10.48550/arXiv.2103.11064)

QDs, in which light amplification occurs only under multiexciton conditions owing to multifold degeneracy of the band-edge states and rapid decay of excitons through the nonradiative Auger recombination [223, 224]. It has directly related to an ultrashort lifetime of optical gain to complicate lasing action in QDs. Therefore, the state-of-the-art of QDs allows to substantially suppressed Auger recombination that opens up the potential path to achieve QD laser diodes [225, 226]. As the typical QLEDs with a tapered HTL have also provided the electrically excited cg-QDs with high current densities, as illustrated in Fig. 8.23a. Under high-current–density excitation ($J > 1$ A/cm^2) has also noticed a shoulder peak in the EL emission spectrum, but it is absent at lower current densities (see Fig. 8.23b). This leads to the primary condition of the population inversion of the band-edge state because emission from the higher state is feasible only when the lower-energy state is filled. It can be further verified by measurement of the optical gain directly (see Fig. 8.23c).

Moreover, as the QD laser diodes: the integration of an optical cavity also provides an optical-feedback path in an LED structure. Various types of optical cavíties, such as Fabry–Perot, whispering-gallery mode, and distributed feedback (DFB) have been fabricated. The integration of an optical cavity into a QLED structure has been also addressed [228]. It has been noted that a dual-functioning device can serve as both an optically pumped laser and an electrically operable LED, which is considered to be an intermediate step to achieving a laser diode. As a typically fabricated DFB optical cavity in the bottom indium tin oxide (ITO) electrode under the construction of a standard p-i-n-based LED-like device (see Fig. 8.23d). Hence by engineering the refractive index of ITO and mode confinement within the QD gain medium reveals a strong lasing emission from the LED-like multilayered device under optical excitation, even with an exceptionally thin (50 nm) gain medium, corresponds to ~

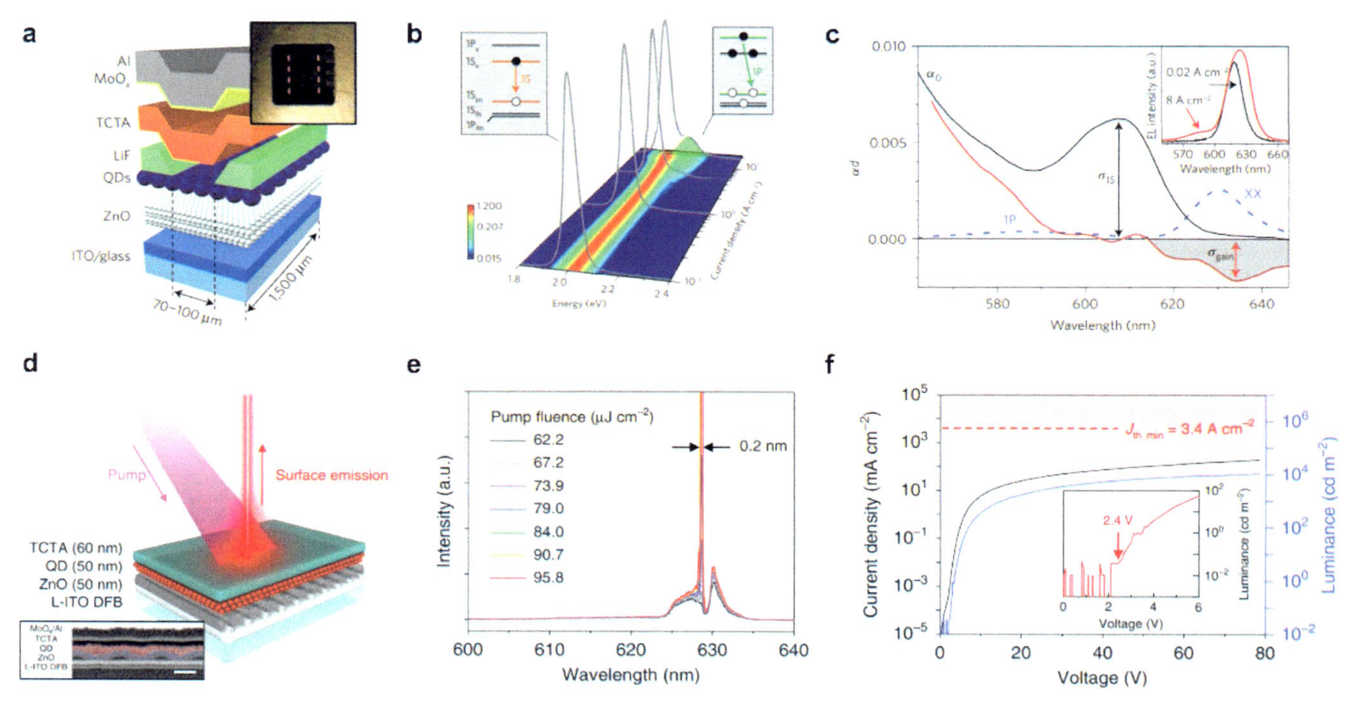

Fig. 8.23 Quantum-dot-based laser diodes: **a** QLED architecture to boost the current density up to 18 A/cm^2; **b** EL spectra as a function of current density, at high current density (>1 A/cm^2), the EL spectrum shows both 1S and 1P bands, indicating population inversion under electrical excitation; **c** optical-gain under electrical pumping that calculated by subtracting the normalized high- and low-current–density EL spectra; **d** schematic of an optically pumped LED-like structure with a DFB optical cavity; **e** surface-emission spectrum as a function of pump fluence, by showing single-mode lasing at 629 nm; **f** Current–density-voltage-luminance characteristics of QLEDs with an integrated DFB optical cavity, the red dashed line indicates the minimum current density required to achieve population inversion (adopted from Rhee et al. [227])

3 monolayers of QDs (see Fig. 8.23e). A thin gain medium is considered to be a favorable feature for electrical excitation; hereby device is also electrically operable like an LED with a decent EQE of 4.3% (see Fig. 8.23f). However, this interpretation has not revealed any indication of lasing under electrical excitation, because current densities are insufficient to achieve population inversion [227].

Among these, the TMDs based low-level diode lasers are also extensively used in biomedical applications for pain relief in the temporomandibular joint (TMJ) and muscles of mastication [229, 230], articular sounds and crepitus, joint locking, headache, earache [230, 231], mouth opening deviation, deflection, or restriction [229, 231], including in other therapy for the different diseases [232, 233].

8.5 Photodetectors

Photodetectors (PDs) generally convert the falling optical signal into charge–carrier flux. It established on the basis dual initial processes: (i) an exterior field-assisted transportation of photoexcited carriers, it means through the employed exterior field inspires a great amount of carrier transportation by detrapping of trapped photogenerated carriers; (ii) the photogenerated charge carriers are produced an electric field, this means with the decreasing inoculation blockade enlarged the charge inoculation. The dispersal and drift of photogenerated charge carriers are due to the rearrangement dynamics and the character of associates is explored by the strategy and establishing the device for a decent photodetector [234]. Characteristically, predominately dual kinds of constructions employed for the PDs: the first one is the lateral structure device and the second one is the perpendicular assembled device [235, 236]. In the lateral assembled device, the effective material is straightly associated to lateral electrodes, whereas in the perpendicular arrangement, the device contains distinct stacking layers. Usually, lateral configuration established PDs comprise photoconductors and phototransistors, while perpendicular assembled PDs comprise photodiodes, heterojunction PDs, and solar cell structure (p-i-n or n-i-p) devices. Generally, every kind of PDs belongs to dual-terminal devices except the phototransistors. In dual terminal PDs devices, single electrode serves as the cathode and second on as the anode. Whereas the phototransistors PDs together three-terminal also contains the gate, source, and drain electrodes [237, 238]. The typical simple schematic of a vertical and lateral structure PDs are illustrated in Fig. 8.24a, b.

A simplest lateral photoconductor organization has double electrodes that absorb light through the direct band gap semiconductors among the electrodes. Such kind of PDs requires an applied exterior bias to get light persuaded changes in conductivity. Moreover, such kinds of PDs generally have a higher improvement due to manifold carrier reexchanges. The metal–semiconductor Schottky joint is also participating significantly in their photodetection. The Schottky photodiode operates in diverse modes such as photoexcitation and carrier creation in semiconductors, and in the production of the carrier associated metal to semiconductor onto the Schottky barrier.

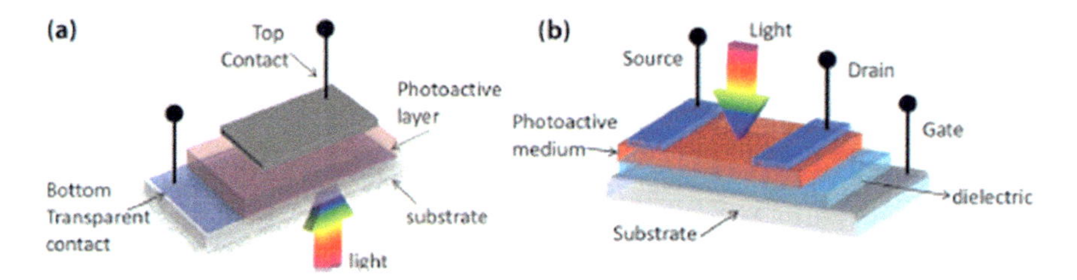

Fig. 8.24 Schematic for the PDs device configurations: **a** vertical structure; **b** lateral structure (adopted from Ghosh [239])

The Schottky PDs are also externally beneficial to the broadband photodetection [239].

Moreover, the photo field-effect transistors (photo FETs) have been also widely explored, as the typical schematic of the monolayer MoS_2 Field-Effect Transistors is illustrated in Fig. 8.25a [240]. The photo FET has comprised a photoconductive channel together drain and source electrodes over the double edges, along with a carrier transport channel that is governed from the employed gate bias. Although, the falling light is also produced by photocarriers to serve as a supplementary gate. Therefore, the electrical gate can be whichever open-circuited or outwardly biased to govern the photoresponse properties of the device. Moreover, a photo FETs channel materials should instantaneously show together extraordinary carrier mobility and better light harvesting ability for effective charge creation [241]. The noticed temporal photoresponses of the FET device under different incident light power densities and source-drain voltages (V_{ds}), when the full power density is 219.1 mW/cm^2, as depicted in Fig. 8.25b,c [240]. This led that when the light is switched on, the I_{ds} increased significantly immediately and gradually reached a saturation state. While in the case of the light switched off, the I_{ds} decreased promptly and then gradually return to their original level. The photoresponsivity and speed characteristics of the device under the light on and off conditions are illustrated in Fig. 8.25d, e [240].

Moreover, perovskites are also showing a similar behavior with exceptional photoelectric change efficiency and higher absorbance factor. Therefore, the perovskite/2D material hybrid arrangement has a significant channel material in photo FET. Underneath the enlightenment of light electrons and holes, couples are produced in the perovskite layer that promoted the transportations into the source and drain electrodes through the 2D material layer. Moreover, the applicable gate voltage efficiently increases the detachment of photocarriers. In this kind arrangement the photo FET usually solitary category of carrier is deliberately immobilized in the device, to make sure that another type of carrier to freely travel and contributes in photocurrent by appropriate selection of the resources and strategy of the device [242]. Typically this process is called photogating effect. It reveals, under the influence of the intrinsic enlargement the phototransistor has an extraordinary exterior quantum efficiency (EQE) that may exceed 100% with higher increase and better responsivity, and it could be controlled through the gate bias. However, practical photo FETs have

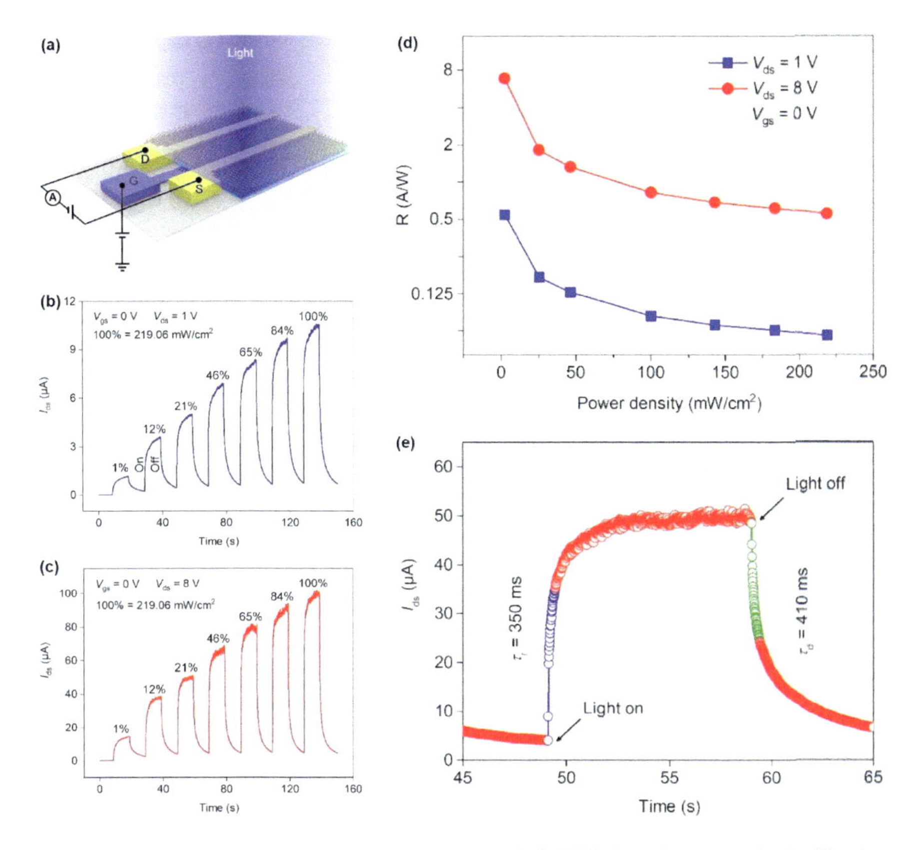

Fig. 8.25 Temporal photoresponse of the buried-gate MoS_2 FET photodetector under the illumination of different light power densities and source-drain voltages: **a** schematic for the experimental setup with LED illumination; **b, c** temporal photoresponses of the FET under various incident light power intensity, at source-drain voltages of 1 and 8 V; **d** the light power intensity dependent photoresponsivity of the device under different source-drain voltages; **e** rise (t_r) and decay times (t_f) of the photocurrent at $V_{ds} = 1$ V and $V_{gs} = 0$ V (adopted from Li et al. [240])

revealed a high responsivity, but their response speed is expressively lesser owing to the trapped carriers [239].

Thus, to fulfill an urgent demand novel semiconducting 2D materials (like TMDs and BP) have a greater optoelectronic properties with a wider band tunability compared to current semiconducting 2D materials [243, 244]. The 2D perovskite materials are also able to maintain a comparable optoelectronic performance with improved moisture resistance that is conducive to the long-term stability of optoelectronic devices [245]. The kinds of photodetectors based on thin 2D perovskite single crystals and their heterojunctions are depicted in Fig. 8.26, and the device performance parameters are summarized in Table 8.4 [246].

It has been also noticed that the device performance such as the detective, EQE, and IQE are greatly influenced by the layer thickness and quantum dots of 2D perovskite.

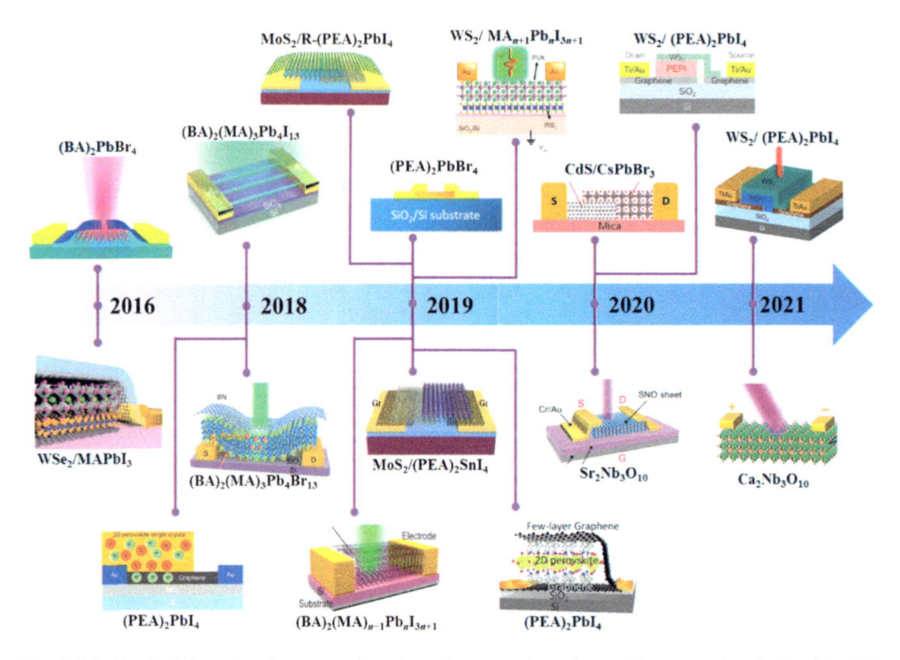

Fig. 8.26 Typical key developments in photodetectors based on thin or molecularly thin 2D perovskite single crystals or their heterojunctions (adopted from Li et al. [246])

The current – voltage curves of the optoelectronic devices are illustrated in Fig. 8.27a. It has revealed that with the increasing quantum well thickness, the open circuit voltage and short-circuit current are also increased and reached a maximum value for $n = 4$. The corresponding excitons are easily dissociated to free carriers owing to the layer-edge states enhancing the charges–transport and collection efficiency. A tunable EQE of photodetectors with a broad wavelength of vis – NIR and varying halide compositions and quantum good thicknesses in 2D perovskites devices are generally accessed from Fig. 8.27b [247]. Almost all fabricated devices have shown a single narrow peak that correlated to the filter less narrow band photodetection with FWHM below 60 nm. Specifically, the lowest value of 10 nm has a superior spectral resolution of around 420 nm. Moreover, the detectivity (D*) is provided by device sen-detective weak light, as illustrated in Fig. 8.27c. The calculated D* values of the monolayer are found much lower than the bulk perovskite flakes, whereas the IQE values of the monolayer are also lower than the bilayer and higher than to bulk [248]. The photoconductive gain of the device is around 1.7×10^5 due to the existence of shallow trap states and the self-doping effect in the hybrid perovskite. This is comparable to solution-processed 3D perovskite photodetectors with an interfacing hole-blocking layer. Overall, the 2D perovskite-based photodetectors with a flake thickness of about 20 nm reflect an ultrahigh responsivity of about 7.4×10^4 A W^{-1} with a superior detectivity of 1.2×10^{15} under the 532 nm illumination.

Table 8.4 Various device performances as photodetectors based on thin or molecularly thin 2D Perovskite single crystals or heterojunction [246]

Device configuration	Spectral region (nm)	Responsivity (A/W)	Detectivity (Jones)	t_{rise} (ms)	t_{decay} (ms)
Gr-$(BA)_2PbBr_4$-Gr	Vis-520	2100			
Au-$WSe_2/MAPbI_3$Au		950		22	37
Au-$(BA)_2(MA)_3Pb4I_{13}$Au		15,000	7×10^{15}	0.0276	0.0245
Au-$(BA)_2(MA)_3Pb4Br_{13}$Au		74,000	1.2×10^{15}		
Gr-$(PEA)_2PbI_4$-Gr		600			
Au-MoS_2/R-$(PEA)_2PbI_4$Au	Vis-600	0.45	2.2×10^{11}	100	100
Au-$(PEA)_2PbBr_4$Au	Vis-550	0.0054	1.07×10^{13}		
Gr-$MoS_2/(PEA)_2SnI_4$-Gr	600–700	0.121	8.09×10^9	34	40
Au-$(BA)_2(MA)_{n-1}PbnI_{3n+1}$-Au		10,000	4×10^{10}		
Gr-$(PEA)_2PbI_4$-Gr		730		80	
Au-WS_2/ $MA_{n+1}PbnI_{3n+1}$-Au		11,174.2		0.064	0.102
Au-$CsPbBr_3$/CdS-Au		0.49	3.67×10^{10}	0.084	0.023
Au$Sr_2Nb_3O_{10}$-Au	UV-400	1214	1.4×10^{14}	0.4	40
Gr-$WS_2/(PEA)_2PbI_4$-Gr		0.00013		200	200
Au$Ca_2Nb_3O_{10}$-Au	UV-350	14.94	8.7×10^{13}	0.08	5.6
Gr-$WS_2/(PEA)_2PbI_4$-Gr		0.000771			

The responsivity and response speed of photodetectors also depends on the layer thickness of the 2D perovskite. Therefore, thickness-dependent responsivity and response speed characteristics are depicted in Fig. 8.27e, f [249]. It leads to a thinner (typically 5.2 and 2.7) heterostructure-II band alignment that contributes to faster response speed and higher responsivity (as high as 10^4 A/W) compared to type-I band alignment. Such variations of optoelectronic device performance can be further verified from the transfer characteristic curves, as illustrated in Fig. 8.27d [249]. An n-type characteristic has been recognized without 2D perovskite, while slightly ambipolar behavior with both p-type and n-type conduction in the heterostructures with perovskite thickness 2.7 nm. However, a pure p-type behavior has been achieved with the perovskite thickness up to 36.2 nm. Therefore, the charge transport behaviors in 2D perovskite heterostructures, including pure p-type, ambipolar conduction, and pure n-type, to be tailored from the modifications in thickness. Moreover, open-circuit voltage variations have been also verified by employing illumination. The noted $I_{ds} - V_{ds}$ curve of the device under laser (10–40 mW) and dark conditions is depicted in Fig. 8.27g. This reflects an obvious photocurrent generation. The overall function between laser power and open-circuit voltage variation is summarized in Fig. 8.27h. This gives the maxima $|\Delta V_{oc}|$ reached at 0.37 V with the laser power of

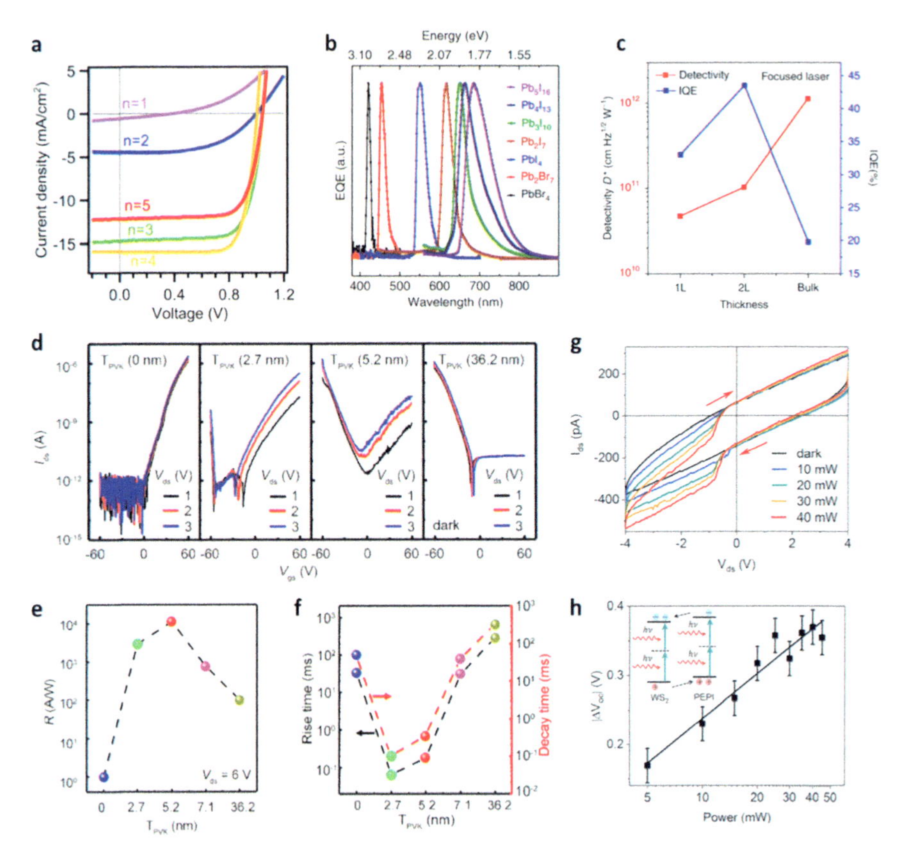

Fig. 8.27 Device performance of photodetectors based on 2D perovskites: **a** J – V curves of the device based on $(BA)_2(MA)_{n-1}Pb_nI_{3n+1}$ under illumination; **b** normalized EQE curves of vertical devices with varying layer numbers and halide compositions; **c** comparison of IQE and detectivity D* of $(BA)_2(MA)_3Pb_4Br_{13}$ devices based on bulk, bilayer, and monolayer flakes; **d** transfer characteristic curves; **e** thickness-dependent photoresponsivity; **f** thickness-dependent decay time and rise time of ultrathin perovskite/WS_2 vertical heterostructures with varying thickness and layer numbers; **g** $I_{ds} - V_{ds}$ curves; **h** V_{oc} values as a function of illumination power in $WS_2/(PEA)_2PbI_4$ heterostructure device (adopted from Li et al. [246])

40 mW. It infers that the upconversion-evoked photovoltaic effect associated with the open-circuit voltage variation through the two-photon absorption [246].

The PD quantitative performance characteristic parameters like photoresponsivity (R), increase, specific detectivity (D*), EQE, linear dynamic range (LDR), response speed, etc. are usually defined based on their merits. The photoresponsivity of a PD reflects in what way proficiently the detector replies to an event optical signal. Therefore, it is important to describe the event photon to the electrical signal changeover efficiency of the provided PD. The responsivity is usually expressed as;

$$R = \frac{I_{\text{ph}}}{P_{\text{in}}} \quad (8.1)$$

here I_{ph} represents the photocurrent that is expressing the change among the current of the device under illumination and dark condition ($I_{\text{ph}} = F_{\text{light}} - I_{\text{dark}}$), while, P_{in} reflects the incoming optical power over the inactive region of the PD. Generally, the responsivity is proportionate to EQE for a PD, therefore, a decent PD should have a higher exchange rate from photons to electron–hole couples. Moreover, the photoelectronic increase (G) is generally expressed in terms of the quantity of charge carriers flowing via an exterior circuit per incident photon.

$$G = R \times E_{\text{h}\upsilon} \quad (8.2)$$

where E_{hv} represents the energy of the incident photon. It has been noticed that the increasing photoconductive gain of a PD is also increased [250]. So, the gain can be obtained from the measurement of the transit time and carrier-recombination lifetime;

$$G = \frac{T_{\text{r}}}{t_{\text{tr}}} = \frac{T_{\text{r}}}{d^2/\mu V} \quad (8.3)$$

where d represents the device thickness, μ is the carrier mobility and V is the applied bias.

Similarly, the specific detectivity ($D*$) of a PD reflects the ability to detect lower-level light signals that have been also taken into account, because it influenced the photocurrent and dark current. The characteristic parameter D* is generally estimated as;

$$D^* = \frac{R}{(2qJ_{\text{d}})^{1/2}} \quad (8.4)$$

where R is the spectral responsivity, q represents the elementary electron charge and J_{d} ($J_{\text{d}} = I_{\text{dark}}$/area of the device) is the dark current density of the device. This is valid only under the approximation where the dark current of the PD is dominated by the shot noise. The high selectivity of a PD reflects its capability to notice extremely lower intensity light signals. The specific detectivity of a PD could also express in terms of the sound-comparable power and the electrical bandwidth of the device noise:

$$D^* = \frac{(AB)^{\frac{1}{2}}}{NEP} \quad (8.5)$$

here NEP corresponds to the sound comparable power, A represents the operational region of the PD, and B is the bandwidth. NEP of a PD defines the signal power that provides us a signal-to-noise ratio, therefore, a sensible PD should have a lesser NEP value.

Moreover, a better PD should notice falling light with a smaller responsivity covering an extensive range of light intensity. This typical feature of the PD is expressed as the LDR, it is normally written as;

$$\text{LDR} = 20\log\left(\frac{J_{\text{ph}}}{J_{\text{d}}}\right) \tag{8.6}$$

J_{ph} and J_{d} are the photocurrent and dark current densities of the device. The LDR within the limitation of a PD detects light. Additionally, another supplementary crucial performance parameter of a PD is the response speed. The response speed of a PD can be also obtained by calculating the upsurge and descent times of a transitory photocurrent [239, 251].

8.6 Photovoltaic Devices

8.6.1 Traditional TMDs Based Solar Cells

Photovoltaic cells are frequently utilized to convert solar energy into electricity. Solar cells that trust on inorganic or organic semiconductors have been industrialized since the preceding decades. Therefore photovoltaic is the technology that converts sunlight directly into electricity. Ultrathin or monolayer 2D nanomaterials, such as graphene and TMDs have shown great potential for solar cell application due to their intriguing electronic and optical properties [170]. Significantly TMDs have been also reassessed as an alternate for photovoltaics owing to their band gap lies in the visible region of the electromagnetic spectrum and robust optical absorbance [252]. TMDs are also exciting due to the technical reasons like chemically stability, ecologically durable and possibly to be manufactured at the lower price. Specifically, semiconducting 2D TMDs layers mono-to-few layer forms have attracted significant attention for PV applications [253]. The application of 2D TMDs for PV devices is mostly driven in two major directions. The first one focused on exploring the novel PV devices by employing the excellent mechanical flexibility and optical absorption inherent properties of 2D TMDs. This approach is mainly explored by the integration of multiple 2D TMDs layers with distinct electrical/ optical functionalities. The second one involved the incorporation of 2D TMDs into existing PV materials such as Si or conductive polymers. This approach focused on the best due to additional functionalities/merits of conventional PV devices, such as cost reduction. In the first case, the simplest form of 2D TMD-based PV hybrid devices has been constructed by the combination of two different 2D TMDs with distinct carrier types and/or appropriate band offsets. A large number of various 2D TMDs have been formed with the typical type-II band alignments by interfaced layers to each other that enable fast and efficient separation of photo-excited e^{-}-h^{+} via built-in potentials at their 2D/ 2D interfaces. Typically defined band alignments formed by the interface in various

2D TMD monolayers are depicted in Fig. 8.28a, which directly correlates to the suitability of 2D TMD/2D TMD hybrids for PV applications [254]. Moreover, a schematic diagram for the generation/separation of photo-excited e^--h^+ at the junction of 2D MoS_2/2D WS_2 for both 2D monolayers in presence of n-type transports is illustrated in Fig. 8.28b [255]. It has experimentally verified that the power conversion efficiencies (PCEs) of the 2D MoS_2/2D WS_2 hybrids remain below 2% [253]. The combination of 2D TMDs intrinsically distinct carrier types (i.e., p-type and n-type) has been extensively pursued in p–n junction PV applications. As the typical photo response from a vertically-stacked 2D WSe_2 (p-type)/2D MoS_2(n-type) p–n junction under the illumination of varying amplitudes of light for the PV effect is depicted in Fig. 8.28c [256, 257].

Thus 2D layered materials with a band gap in the visible region of the electromagnetic spectrum are useful as photosensitizers in photovoltaics and photodetectors. The monolayer 2D TMDs, such that MoS_2, $MoSe_2$, and WS_2, can absorb up to 5–10% incident sunlight within a thickness of less than 1 nm, thus TMDs can have 1 order of magnitude higher sunlight absorption than GaAs and Si [253]. The monolayers MoS_2 and WS_2 have been also successfully used in Schottky junction solar cells as the photoactive layers. It has been also noticed that the combination of monolayer TMDs with a lateral p–n junction architecture electrostatic doping lead to exciting functions including a photovoltaic effect in these atomically thin structures, as the intense, a p–n junction diode based on an electrostatically doped WSe_2 monolayer [173].

In which a split-gate electrode is coupled to two different regions of monolayer WSe_2 flakes for electrostatic doping, as schematically illustrated for the vertical device in Fig. 8.29a. A vertical photovoltaic solar cell have a maximum electrical output power of 9pW with a voltage of 0.64 V, a current of 14 pA, and a PCE of ~ 0.5%. Moreover, around ~95% transparency of the WSe_2 monolayer makes them attractive for semitransparent solar cells. However, a monolayer WSe_2 device with

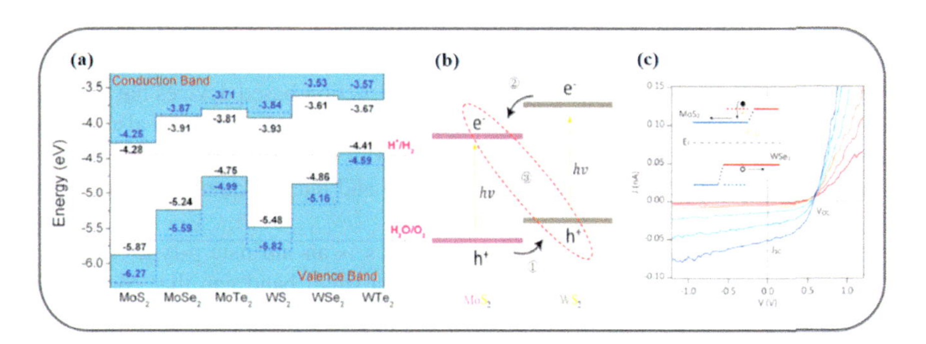

Fig. 8.28 **a** Band alignments for various 2D TMD monolayers (adopted from [258]); **b** band alignment of 2D MoS_2/WS_2 hybrids by showing a type II heterojunction; **c** photo response from a vertically-stacked 2D WSe_2/MoS_2 p–n junction under varying illumination (adopted from Furchi et al. [256])

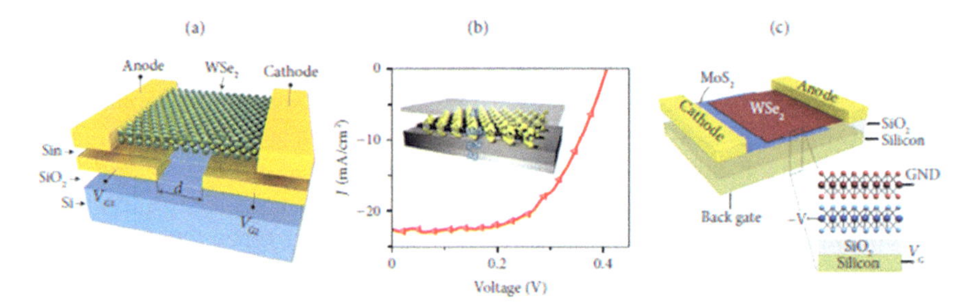

Fig. 8.29 **a** The p–n junction solar cell based on an electrostatically doped WSe_2 monolayer; **b** photovoltaic devices with a monolayer MoS_2 on p-type silicon substrates; **f** stacking van der Waals heterojunction using monolayers MoS_2 and WSe_2 (adopted from Fan et al. [261])

a similar structure has revealed photovoltaic power generation with a peak EQE of 0.2% [259]. In addition to these, photovoltaic devices with a monolayer MoS_2 on p-type silicon (p-Si) substrate have been also interpreted [260]. It has established the built an electric field near the interface between the MoS_2 and p-Si to promote the separation of photogenerated carriers. The associated current density–voltage has revealed a photocurrent density of 22.36 $mAcm^{-2}$ and an efficiency of 5.23%, as depicted in Fig. 8.29b. Hence the MoS_2 monolayer can be potentially integrated into the Si, moreover, it also demonstrates that 2D materials are also incorporated into a variety of Si-based electronic and optical devices.

In addition to these, it is also feasible to construct layer-stacked heterostructures by chosen monolayer materials to achieve enhanced efficiency of the photovoltaic devices. The layered materials are usually held together by van der Waals forces with atomically sharp interfaces. As the intense, van der Waals heterojunction formed by the monolayers MoS_2 and WSe_2 is depicted in Fig. 8.29c [256]. While the optical illumination the charge transfer occurs across the planar interface, and ultimately device shows a photovoltaic effect with a PCE of 0.2% and external quantum efficiency (EQE) of 1.5%. Similarly, graphene contacts of the monolayers MoS_2 and WSe_2 also improves the charge collection in the graphene-sandwiched 2-D heterojunctions [262]. Such kinds of monolayer heterojunction-based devices have a high EQE value of 2.4%. Likewise, lateral WS_2/MoS_2 heterostructures have also shown an open-loop voltage and closed-loop current of 0.12 V and 5.7 pA [263]. It noted that a lateral WS_2/WSe_2 photodiode has external and internal quantum efficiencies of 9.9% and 43% [264].

Moreover, the monolayer of black phosphorus is an intrinsic p-type semiconductor that possesses high carrier mobility and a small band gap that considered being suitable for broadband photodetectors [265, 266]. Phosphorene has been also recognized to serve as the light absorber and charge transfer layer in solar cells to enhance light absorption and electron recombination. It has been also noticed that an electrically tunable black phosphorus monolayer MoS_2 van der Waals heterojunction [267] has a maximum rectification ratio of $\sim$10 [268], a maximum responsivity of 418 mA/W and EQE 0.3% with a back gate voltage -30 V under 633 nm laser illumination. However, these devices substantially have a gain with a slower device response. Therefore, it is essential to optimize the gain and response time simultaneously for distinct applications. Hence, more efforts are needed to improve further the performance of 2D monolayer materials-based solar cells, including enhancement of light absorption, charge collection, and efficient exciton dissociation [261].

8.6.2 Transparent TMDs Based Solar Cells

More recently transparent solar cells (TSCs) have attracted much attention to overcome the limitations of traditional non-transparent solar cells [269]. By converting the diverse components, such as architectural windows, agricultural sheds, glass panels of smart devices, and even human skin into energy harvesting devices. Various TSCs with perovskites and organic semiconductors [270–272] have been examined with an average visible transparency (AVT) lower than 70%. To fabricate TSCs with very high AVT values (> 70%) near invisible solar cells (NISCs) technology has been also interpreted, but it is still challenging. The best NISCs have been realized only by combining several devices, such as transparent luminescent solar concentrators (TLSCs) absorb UV or IR and generate bright luminescence to propagate the edge of the TLSC through the internal reflection. The electrical power generation via the edge of the TLSCs using a conventional non-transparent Si solar cell to get a very high AVT (> 80%). Though a non-transparent part is still a question at the edge of the device, therefore, they are suffering from scalability issues [273, 274].

The two-dimensional (2D) transition metal dichalcogenides (TMDs), specifically, monolayer and few layered TMDs have been found most promising as the NISCs for TLSCs due to their adequate structural properties [275, 276]. Yet, many such TMD-based solar cells have been investigated by employing pn junctions [277, 278], heterojunctions [279, 280], Schottky junctions [281], etc. Although most of the TMDs based solar cells are also non-transparent owing to the use of opaque Si substrates and non-transparent metal electrodes [282]. However, the highest PCE of around 0.7% has been achieved using a triple-layer TMD with a Schottky junction structure [283]. This Schottky junction structure device can be expanded and applied even on a polyethylene naphthalate (PEN) substrate at cm-scales. It has been also noticed that the transparent electrodes and monolayer TMDs employed to realize TMD-based NISCs with a relatively high PT [284].

Moreover, NISC can also fabricate by using ITO and monolayer tungsten disulfide (WS_2) as transparent electrodes and a photoactive layer. The contact between ITO and WS_2 reveals a pure ITO–WS_2 junction having a relatively low Schottky barrier height (ΦB) (< 10 meV), likewise, a thin Cu-coated ITO electrode with insertion of WO_3(WO_3/Cu/ITO) to increase ΦB up to approximately 220 meV compared to WS_2. It is useful for carrier collection and generation within a Schottky-type solar cell. Moreover, such architectural design is vital to scale up the TMD-based solar cells to avoid an unexpected drop in the open circuit voltage (V_{OC}) and overcome the controlling aspect ratio of the unit device. It noted that the PT could reach up to 420 pW from a 1-cm^2 device with a high AVT of 79% [284].

A typical device structure with the ideal band structure of a Schottky-type solar cell is illustrated in Fig. 8.30a. The discrepancy between the work function of electrode A and the semiconductor leads to a built-in potential that is separated from the photogenerated electrons–holes pairs. The generated carriers travels to the opposite electrode, therefore, power generation is realized. A greater WF difference between the asymmetric electrodes gives a higher PCE. Thus, the WF should be controlled to ITO to get an optimal band structure for the Schottky-type solar cell. Their interface should be also better to avoid other interface impurities for the interface charge and energy distribution to avoid undesired surface recombination and reduction of carrier's separation efficiency at the interface barrier. Herein, the modulated WF for different types of thin metal films coated on ITO is depicted in Fig. 8.30b. It has been noticed that after an Mx coating (M is a metal and x is the thickness in nm) some of the Mx/ITOs maintained a high AVT greater than 80%, such as Ni1, Ni5, Fe1, Al1, Al5, Cu1, and Ag1. It reveals they are considered to be potential candidates to use as electrodes in TSCs. Moreover, the WF of the Mx/ITO has also revealed that the Mx/ITO is normally vary in the range of 4.2– 5.4 eV. Wherein most of them have transparency of more than 80%. The WS_2 monolayer grown by CVD naturally belongs to n-doped due to impurities and their WF is around 4.9 eV. Thus, the Mx/ ITOs with WF higher and lower than 4.9 eV are considered to be a promising feature for the transparent electrode with Schottky contacts (Ni1, Ni5, Cu1, and Fe1) and Ohmic-like contacts (Al5, Al1, Ag1, ITO), like WS_2 monolayer [284].

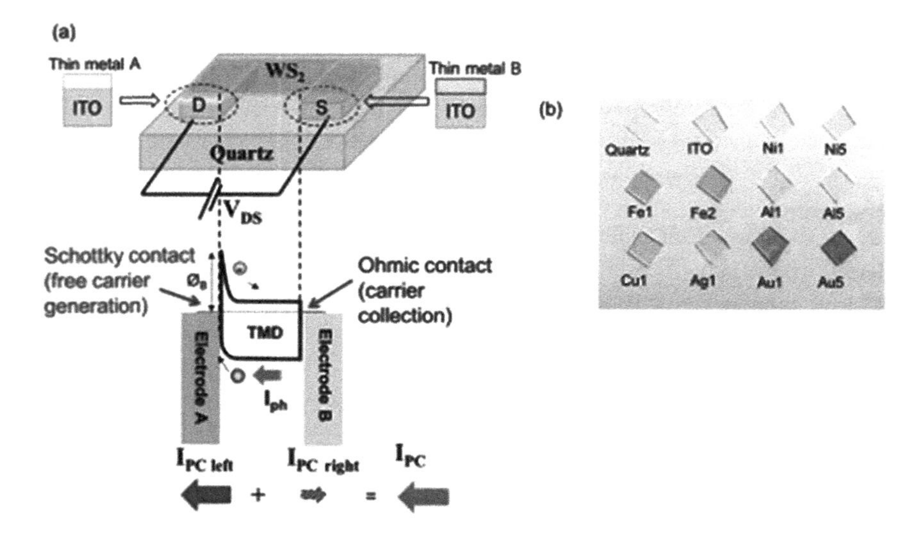

Fig. 8.30 **a** Schematic for the device structure and ideal optimal band structure for the transparent Schottky solar cell; **b** images of samples for WF and AVT measurement (adopted from He et al. [284])

To get a large scale (a small device at a μm-scale) fabrication should consider the important parameter P_T that is for TMDs relatively in a nascent phase. To scale up the devices it should be increased up to the practical level (more than 100 pW) [285]. Therefore, a suitable architectural design of the TMD-based solar cells is desired to scale up it. So, some important parameters should be addressed; (i) designing the structure of the unit device (UD), (ii) exploring the parallel connections, and (iii) investigating the series connections. (iv) Combining the parallel and series connections, (v) optimization of the architecture of the device. Considering these key parameters different widths (W) and channel lengths (L_{ch}) of solar cells have been fabricated and analyzed, as depicted in Fig. 8.31a, b. The performance of Schottky-type solar cells has been examined by defining values of parameters P_T, $V_{OC,}$ and short circuit current (I_{SC}). As the intensity, for the L_{ch} = 1 μm P_T is increased and W up to 33.5 μm, as depicted in Fig. 8.31c. It noted that, if W is larger than 33.5 μm then P_T is significantly decreased with W. It is considered as a threshold value of the critical width (W_{th}) to maintain a high P_T. The significant drop in P_T mainly appeared due to a drop in V_{OC}, as illustrated in Fig. 8.31d, e. Almost similar trends have been also noticed for the devices having L_{ch} = 2, 3, and 4 μm and W_{th} = 81.3, 116.3, and 142.0 μm. Significantly nearly a linear relationship between W_{th} and L_{ch} has been obtained, as depicted in Fig. 8.31f. This reveals that the aspect ratio of the device (W/L_{ch}) considered to be critical for the TMDs based large-scale solar cell designing, typically it should be lower than 36. As the intense, designing a unit device in a combination of several small channels connected in parallel and each channel should have a shunt resistance of R_{sh} (i) (i = 1, 2.., n). Subsequently, the total shunt resistance (R_{sh}-total) depends on the resistance of each channel (1/R_{sh}-total = Σ(1/R_{sh} (i)), like if the channel contains a metal-like pass then R_{sh} is very low, it leads to a low R_{sh}- total and low V_{OC}. The low R_{sh} is due to possible impurities, because of the 1 T phase of TMD or other impurities existing in TMD that is inevitably induced in chemical vapor deposition or mechanical exfoliation process [286]. Thus the existence of impurities leads to the formation of band tail localized states, as a consequence the percolation transport arises in 2D disordered materials with energy variations along the current-carrying path [287, 288]. However, with the increased W possibility an unexpected metal-like pass involved within the photoactive channel is also increased, as the results total a low R_{sh}-−, this is known as a well-known percolation model [289, 290]. Similarly using the other parameters such as an increase in P_T affects the number of series connections (N_{se}) the parallel and series combination connections are described. Thus based on the percolation model variations in physical characteristics parameters of the TMDs based transparent solar cells have been examined [284]. Moreover, an optimized device designed on a SiO_2/ Si substrate ((WO_3/Cu/ITO-ITO) and (UDM-C: UD × 3 parallel connections (UDM-A) × 4) in series connections (UDM-B) × 18,750 parallel connections)) on a quartz substrate architecture successfully has been fabricated [284].

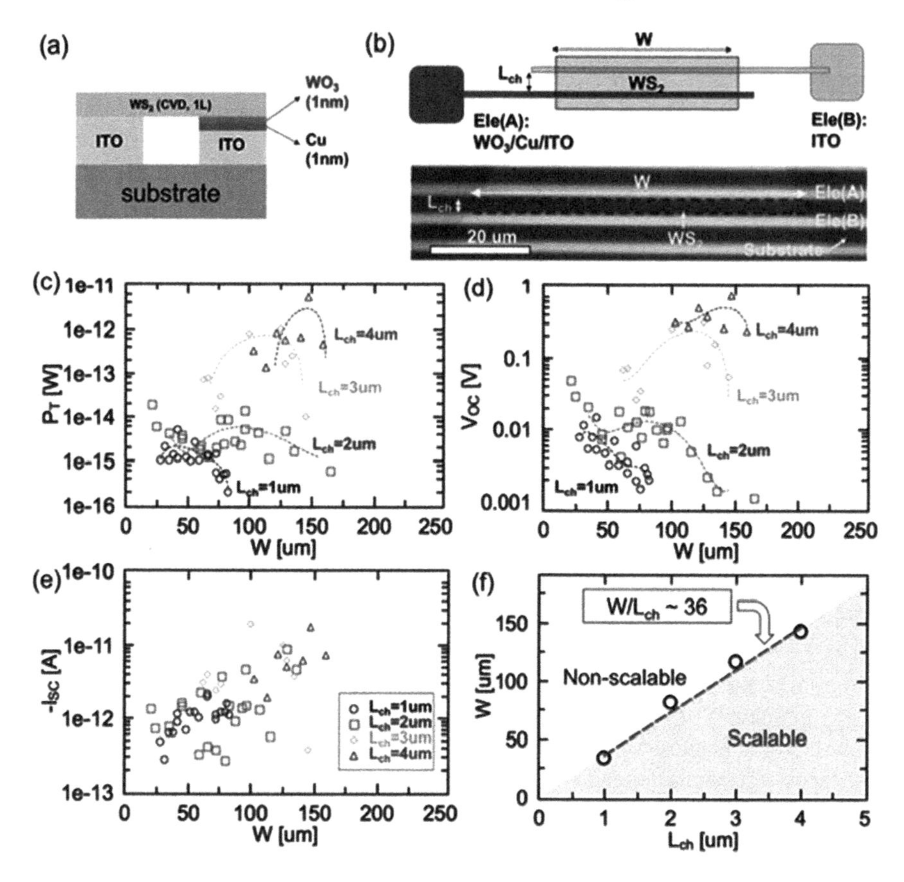

Fig. 8.31 **a** Structure of devices for architecture design of the UD; **b** schematic of UD structure; **c–e** P_T, V_{OC}, I_{SC} of unit devices with different L_{ch} and W; **f** critical width for a device with different L_{ch} (adopted from He et al. [284])

8.7 Solar Water Splitting

Hydrogen is considered to be the cleanest renewable source with a high energy density; it makes an ideal alternative to fossil fuels. Their easy storage and applications in fuel cells and combustion engines have already demonstrated the great potential of hydrogen as a promising energy carrier [291, 292]. In the modern era around 95% of hydrogen is generated from fossil fuel-based products, 4% from electricity, and 1% from biomass [293]. To pursue, a clean and carbon-free energy resource the water splitting from renewable energy resources for hydrogen evolution has gained much attention over the years [294]. Their practical performance on an industrial scale is mainly two approaches of water splitting, namely electrolysis and photolysis [294].

8.7.1 Electrocatalytic Water Splitting

The splitting of water in the electrocatalytic process mainly involves two half-reactions of the hydrogen evolution reaction (HER). The reaction process occurs at the cathode as per Eq. 8.7, while at the anode in the electrochemical cell takes place oxygen evolution reaction, as defined in Eq. 8.8.

$$2H^{+}(aq) + 2e^{-} \rightarrow H_2(g) \quad (8.7)$$

$$2H_2O(l) \rightarrow O_2(g) + 4H^{+}(aq) + 4e^{-} \quad (8.8)$$

It is well established that the HER can precede together the preliminary adsorption of a hydrogen atom, whichever via the Volmer–Heyrovsky or the Volmer–Tafel machines [295, 296]. That depends to the pH of the solution wherein the decrease in protons or water molecules ensues to left adsorbed hydrogen atoms over the surface of the metal or catalyst (*M–H*), here * corresponding to an efficient adsorption location onto the electrocatalyst (M), this adsorption process is well-recongnized as the Volmer reaction, it is usually demonstrated from the acidic media to Eq. 8.9 and their basic environments defined from the Eq. 8.10.

$$H^{+}(aq) + e^{-} + *M \rightarrow *M - H \quad (8.9)$$

$$H_2O(l) + e^{-} + *M \rightarrow *M - H + OH^{-}(aq) \quad (8.10)$$

Depends on the nature of the electrode and/or the electrocatalyst $H_2(g)$ that constructs by the combining double adsorbed hydrogen atoms (Tafel reaction), Eq. 8.11. Moreover, according to the Heyrovsky reaction, the adsorbed hydrogen atom combined to an electron and a proton (or water molecule) through the acidic and basic reactions (see Eqs. 8.12 and 8.13) to generate $H_2(g)$. In several cases, the Volmer reaction is considered to be a rate-determination step, therefore, the unvaluable metal established electrocatalysts, like dichalcogenides in which the reaction follows mostly together Volmer containing the Heyrovsky reaction to complete the Volmer–Heyrovsky process. The electrolyte pH similarly has an important role in the rate of HER reaction, as the Eq. 8.13. Therefore, water adsorption contributes a critical part in the Volmer reaction for the alkaline medium due to electrocatalysts provides the adsorption of water by the detachment and hydrogen construction. Usually, extraordinary overpotentials are desired for the alkaline solutions owing to the lethargic kinetics of the HER, however, it is still challenging for the growth of effective and steady HER electrocatalysts in alkaline medium [297]

$$*M - H + *M - H \rightarrow *M + *M + H_2(g) \quad (8.11)$$

$$*M - H + H^{+}(aq) + e^{-} \rightarrow *M + H_2(g) \quad (8.12)$$

$$*\mathrm{M}-H+\mathrm{H_2O}(l)+\mathrm{e}^- \rightarrow *\mathrm{M}+\mathrm{OH}^-(aq)+\mathrm{H_2}(g) \quad (8.13)$$

Similarly, OER is also similarly significant half-reaction **to** the electrolysis of water. A typical reaction process is commonly expressed from the Eqs. 8.14 and 8.15 for the acidic solutions and basic media.

$$2\mathrm{H_2O}(l) \rightarrow 4\mathrm{H}^+(aq)+\mathrm{O_2}(g)+4\mathrm{e}^- \quad (8.14)$$

$$4\mathrm{OH}^-(aq) \rightarrow 2\mathrm{H_2O}(l)+\mathrm{O_2}(g)+4\mathrm{e}^- E^0/V\,\mathrm{vs\ SHE} = 1.23-0.059(\mathrm{pH}) \quad (8.15)$$

These reaction processes involves the transfer of four protons coupled with four electrons including more complex bonds breaking and formations with a kinetically sluggish nature compared to HER. The overall efficiency of the electrocatalytic water splitting reaction in an electrolysis cell is limited by the OER process. The standard reduction potential E_o in OER is 1.23 V versus SHE. Though much higher applied potentials are required for the production of O_2(g), to get high overpotentials. Moreover, the mechanism of the OER is complex and extensively depends on the nature of the electrocatalyst, the most recognized OER mechanism is called the adsorbate evolution mechanism (AEM). In this process, the first step of alkaline electrolyte adsorption of OH^-(aq) at an adsorption site (*M) is to reveal the adsorbed *M–OH species, as expressed from the Eq. 8.16.

$$*\mathrm{M}+\mathrm{OH}^-(\mathrm{aq}) \rightarrow *\mathrm{M}-\mathrm{OH}+\mathrm{e}^- \quad (8.16)$$

The intermediate state can be further combined with OH^-(aq) and formed $H_2O(l)$ by adsorbing *M–O atoms, as per Eq. 8.17, it further follows the formation of adsorbed *M–OOH as expressed as Eq. 8.18. Moreover, the additional reaction with OH – (aq) to give $O_2(g)$ and $H_2O(l)$, OH^- (aq) to give $O_2(g)$ and $H_2O(l)$, as given in Eq. 8.19.

$$*\mathrm{M}-\mathrm{OH}+\mathrm{OH}^-(\mathrm{aq}) \rightarrow *\mathrm{MO}-\mathrm{H_2O}(l)+\mathrm{e}^- \quad (8.17)$$

$$*\mathrm{M}-\mathrm{O}+\mathrm{OH}^-(\mathrm{aq}) \rightarrow *\mathrm{MO}-\mathrm{OOH}+\mathrm{e}^- \quad (8.18)$$

$$*\mathrm{M}-\mathrm{OOH}+\mathrm{OH}^-(\mathrm{aq}) \rightarrow *\mathrm{MO}+\mathrm{O_2}(g)+\mathrm{H_2O}(l)+\mathrm{e}^- \quad (8.19)$$

Similarly, fresh water can also extract from sea-water, as an electrolyte in the electrolysis cell and it delivers limitless sustainable sources [298, 299].

Another typical approach, the light with semiconductors to generate electron (e^-) and hole (h^+) pairs that facilitate the reduction of water and give H_2(g) and the oxidation of OH^- ions or H_2O reveal the O_2(g) to generate photoelectrochemical (PEC) splits of water at a semiconductor photoanode and a metal cathode, as depicted in Fig. 8.32a, b, a semiconductor cathode and a counter electrode or a semiconductor anode and cathode, illustrated in Fig. 8.32c, d. As depicted in Fig. 8.32b, an electron promotes from the valence band to the conductance band by the absorbance of photons with the adequate matching energy possessing the band gap energy. So, it generates h^+ in the valence band and an e^- in the conduction band. The e^- can flow to the cathode where the reduction of water occurred and lead to the $H_2(g)$, and separation of charge is minimized from the e^-/h^+ recombination reaction. Therefore, the h^+ resides in a sufficiently low energy level for the oxidation of OH^- under a thermodynamically feasible reaction. Thus, the h^+ facilitates the oxidation event while in the process to generate $O_2(g)$. It also attended considerable interest in the fabrication of hybrid photovoltaic–photoelectrochemical (PV–PEC) [300, 301]. Where the production of $H_2(g)$ is due to the coupling of a photovoltaic solar cell with a water electrolysis system. In such systems, the properties of the semiconductors are important, such as the soil-plentiful and metal-less semiconductors together better constancy and strength make them proficient to harnessing light in the visible range at lower cost [302, 303].

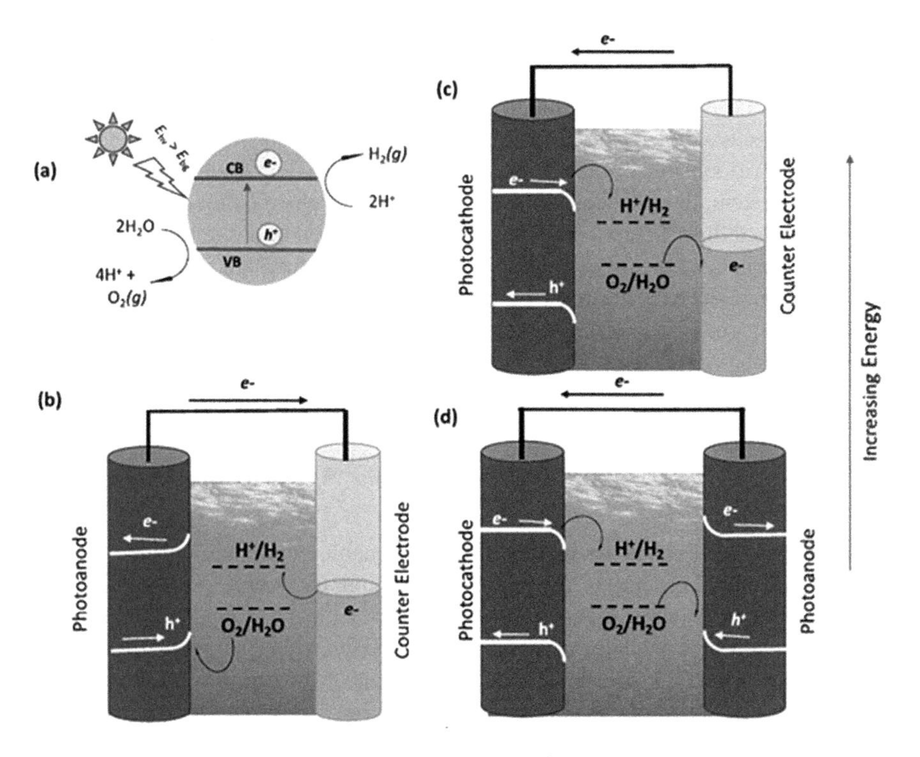

Fig. 8.32 Schematic representation: **a** generation of e^-/h^+ at a semiconductor on the absorption of a photon of light, sufficient to match the band gap energy; **b** the electron-transfer reactions on activation at a photoanode; **c** photocathode coupled with counter electrodes; **d** a photocathode coupled with a photoanode, where the E_o/SHE of $H+/H_2(g)$ is 0 V and E_o/SHE of $O_2(g)/H_2O$ is 1.23 V, under acidic conditions (pH = 0) (adopted from Sukanya et al. [304])

8.7.2 Photocatalytic Water Splitting

Photoelectrochemical (PEC) water splitting can also directly integrate water electrolysis to solar energy collection. It is a favorable machinery to alter recurrent solar energy into hydrogen. The electrocatalysts are not individually minimize the overpotential to HER but also provides the absorption of light at the lower cost for the effective PEC devices associating steady and extremely efficient HER photocathodes to combine a semiconductor photo absorber and pertinent cocatalysts.

To get additional competent novel materials or co-catalysts that coupled together to occurring semiconductors to improve strength and working ability as desired. A numerous innovative cost-effective and great surface space semiconductors have been also deliberated as photoelectrodes for the PEC devices. Specifically, 2D TMDs' variable bandgap energies together having ability to absorb photons in visible, ultraviolet and infrared regions of the electromagnetic spectrum make them interesting for PEC devices. Their 2D, 2H crystal phase (semiconductor) is usually absorb the light and generate e^{-}/h^{+} couples. While the 2D, 1 T metallic phase serves like an electron acceptor and suppress the charge recoupling procedure, whereas, together 2H and 1 T phases have applicability to the PEC cells. Moreover, 2D TMDs also have ability to work as the semiconductor photo-absorber and photocathode to produce hydrogen gas as a co-catalyst or as a semiconductor heterojunction. Particularly, the co-catalysts responsible to catch excited electrons in the CB and subsequently reduce the charge recoupling mechanism. Similar to hybrid materials that constructed by minimum dual diverse semiconductors for the formation of a p–n junction, associating an alternative layer of a p-type and n-type semiconductor, it reduces the charge reconnection. Hence the electrons accumulated in the n-type area and the holes exist in the p-type area [294, 304]. Subsequently, the photoelectrochemical functionality is also increase charge recoupling suppression [305].

8.7.3 Water Splitting in TMDs and Their Applications

Formation of a semiconductor interface in TMDs enables the creation of the advanced characteristics and properties that are usually lacking in a single 2D material [306–308] The heterostructures are typically formed by stacking 2D semiconducting materials via weak van der Waals (vdW) forces that widely investigated theoretically and experimentally [309–315] The vertically stacked heterostructures considered to be important because it not only preserves the intrinsic electronic prosperities of the single material but also enhances the electronic and optical properties [316]. For instance, the fabrication of $WSe_2/MoSe_2$ heterostructures are possible from the two-step method by chemical vapor deposition (CVD) [317, 318]. Similarly, WS_2/MoS_2 heterostructures have been also fabricated by employing the CVD method. Therefore, 2D semiconducting vdW heterostructures are typically classified into three main categories based on the relative positions of band edges of the two constituent monolayers [319], namely, type-I (symmetric), type-II (staggered), and type-III (broken), as depicted in Fig. 8.33. In the type-I heterostructure, the conduction-band minimum (CBM) of one layer is lower than that of the other layer, whereas its valence-band maximum (VBM) may higher than that of the other one. When both CBM and VBM of a layer are higher or lower than the other layer then VBM and CBM locate in different layers, this is called type-II heterostructures. Whereas in the type-III heterostructures, both CBM and VBM of one component are lower than the VBM of the other component.

Moreover, when in a type-I heterostructure the CBM and VBM are located in the same layer then it gives a higher combination rate of photogenerated carriers without a further opportunity for improvement in photocatalytic activity. Similarly, in type-III heterostructures, the photogenerated carriers cannot transfer across the interface, thereby, no advantage in photocatalytic activity. While the CBM and VBM are located in the different layers in a type-II heterostructure, they benefited from the relative positions of the band edges, therefore, the photogenerated electrons and holes are inherently separated. If a type-II heterostructure is used as a photocatalyst in water splitting, the water reduction and redox reactions is normally occurred in different layers, it is expressed as [320];

$$H_2O + H^+ \rightarrow 2H^+ + \frac{1}{2}O_2 \tag{8.20}$$

$$2H^+ + 2e^- \rightarrow H_2 \tag{8.21}$$

Thus, designing TMD-based type-II heterostructures are broaden the application of TMDs for photocatalysis activity.

Moreover, TMD-based vdW heterostructures are also useful as photocatalysts for water splitting due to their improved separation of photogenerated electrons – holes

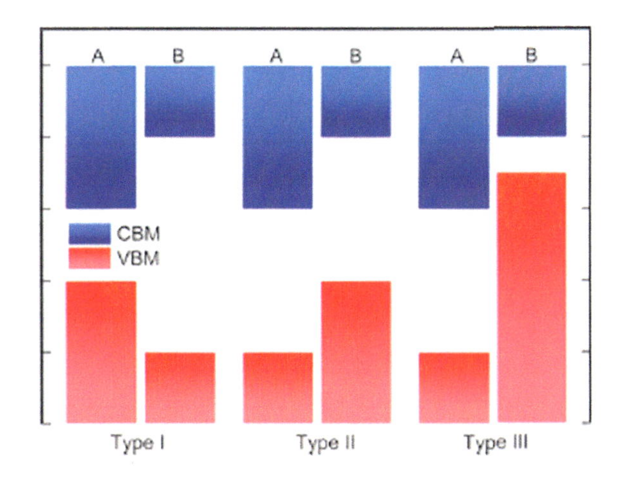

Fig. 8.33 Graphical illustrations for the type-I, type-II, and type-III van der Waals (vdW) heterostructures (adopted from Luo et al. [303])

pairs [321–324]. As a combination of the WS_2 and CN_2 forms a heterostructure in which the separation of the charges improved, this revealed their promises in water splitting [324]. Similarly, the designing of the TMD/ZnO heterostructures by the theoretical methods has also indicated outstanding characteristics and their display potential as photocatalysts [325].

On the other hand, the monolayered BSe is a single-layered semiconductor material that is thermally stable the possessing same hexagonal crystal structure as the TMDs having a small lattice mismatch [326, 327]. Therefore, very few BSe-based heterostructures have been designed that demonstrated, like the blue phosphorene/BSe vdW heterostructures considered to be a promising photocatalyst for water splitting. It is also beneficial to investigate the TMD/BSe bilayer from the theoretical calculations. A theoretical study also demonstrated that the stacked system to be fairly stable. Such as the MoS_2/BSe and WS2/BSe vdW heterostructures are possessed inherent type-II band alignment and their band edge positions suitable for water splitting. In addition to these, they also have excellent optical absorption properties. Hence, MoS_2/BSe and WS_2/BSe heterostructures are considered to be promising materials for photocatalytic and photovoltaic device applications.

The typical practical applications of TMDs for the 2D vertically aligned SnS_2 nanosheets are depicted in Fig. 8.34a. The photoanodes of the PEC water splitting can have a high photocurrent density of 1.92 mA cm^{-2} (100 mA cm^{-2} solar energy). While combining other materials with the TMDs are also enhanced the production of H_2(g). Such as, Si nanowire photocathode arrays **wrapped** in TMD layers **formation like** Si/MoS_2, Si/$MoSe_2$, Si/WS_2, and Si/WSe_2 photocathodes that have shown the excellent.

PEC performances with the photocurrents of 20–30 mA cm^{-2} (at 0 V vs RHE) in 0.5 M H_2SO_4 with negligible degradation of HER under 3 h solar irradiation (100 mW cm^{-2}) [329]. In another investigation, the vertically aligned MoS_2 nanoflakes have been formed at SiO_2/Si, it is usually used as a photoelectrode with a current density of 0.51 mA cm^{-2} (in 0.5 M H_2SO_4 at – 0.8 V (Ag|AgCl), with a 75 W xenon

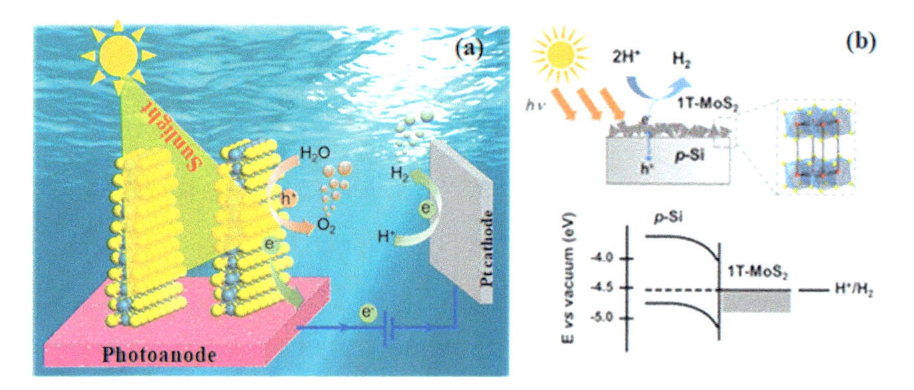

Fig. 8.34 **a** Schematic for the vertically aligned TMDs; **b** the interface heterojunction formed between the Si semiconductor and 1 T MoS_2 (b adopted from Ding et al. [328])

lamp) [330]. Similarly, the chemically exfoliated 1 T-MoS_2 has also combined with planar p-Si to get photocathodes for the PEC hydrogen generation with a high current density of 17.6 mA cm^{-2} possessing very good stability, as illustrated in Fig. 8.34b [328].

Similarly, modifications in TiO_2 with MoS_2 also lead enhance photocatalytic activity and production of hydrogen [331–333]. Their improved performance is due to the transfer of electrons from the CB of TiO_2 to the MoS_2 nanosheets assisted by the good contact between the 1 T-MoS_2 metallic phase and TiO_2 [331]. However other studies demonstrated that the MoS_2/TiO_2 is considered to be a heterojunction semiconductor, their high photocatalytic activity due to the transfer of the photogenerated electrons from MoS_2 to TiO_2 [334]. The difference in the mechanism of the enhanced PEC activity can connect to the ratio of the 1 T-MoS_2 and 2H-MoS_2 phases. Similarly, other semiconductors have also integrated with TMDs, such as MoS_2/g-C_3N_4 heterostructure fabricated by CVD and pulsed laser deposition technique. Thus the experimental findings have shown that the hybrid possessed low e^-/h^+ recombination rates when reached at values of 252 μmol h^{-1} to H_2(g) generation rates [335]. More examples includes like $Fe_2O_3/BiVO_4/MoS_2$ with a H_2(g) production rate of 46.5 μmol cm^{-2} over 2 h, $MoS_2/ZnIn_2S_4$ gives a maximum rate of 201 μmol h^{-1} for H_2(g), MoS2/ZnS with a maximum rate of 606 μmol h^{-1} g^{-1} of H_2(g) [304, 336].

Additionally, 2D TMDs are also finding numerous applications as cocatalysts in PEC devices to replace noble metal photocatalysts. For instance, MoS_2/rGO is useful as a co-catalyst with Cu_2O [337]. The ternary hybrid Cu_2O/MoS_2/rGO have a photoelectrochemical current density of 8.46 mA cm^{-2} which is about 26 times higher than noticed with Cu_2O. It connects to suppression of the charge recombination reaction and inhibition of the photo-corrosion of Cu_2O that ultimately gives higher photocurrents with greater stability. Moreover, MoS_2 has also employed as a protective layer for the stabilization of various Group III–V absorbers, such as GaInAsP/GaAs. In this case, a thin film of MoS_2 has modified the photo-absorber that is around > 5 times supplementary steady and fewer disposed to corrosion to the PtRu altered surface [338]. Remarkably, the 2H-MoS_2 also allows it to combine with the MXenes for the formation of a hybrid electrocatalyst to HER, it suppressing the oxidation of the MXene layers [339].

Another promising activity is the production of Janus TMD single-layers (XMY) in which solitary S layer usually replaces from the Se atoms to form SMoSe [340, 341]. It can achieve by exposing a monolayer of MoS_2 using CVD, an H_2(g) plasma to strip the top S layer, then Se powder is introduced to provide selenisation for the formation of SMoSe monolayers [340]. These kinds of emerging TMDs flack structural symmetry gives rise to an intrinsic dipole moment due to different electronegativities of two chalcogen atoms. Therefore, the layer with the lower electronegativity becomes positively charged, while the layer with the larger electronegativity adopts a negative charge [342]. Janus TMDs are considered to be heterojunction semiconductors that lead to the operative distinction of holes and electrons over the double diverse surfaces, thereby, photoelectrochemical action is improved.

8.8 TMDs Biosensors Application

TMDs are also considered emerging 2D materials due to their remarkable physicochemical properties and chemical versatility, including large specific surface areas, high mechanical properties, exceptional electronic performance, tunable chemical stability, high catalytic activity, photoluminescence, optical absorption, direct bandgap, and facile synthetic processes that have received tremendous attention, making them suitable for the construction of ultrasensitive biosensors [343–346]. The key properties of TMDs significantly compensate for the deficiency of many other 2D materials, especially graphene-based devices. However, the MoS_2 monolayer thickness is 6.5 Å. The layer thickness can significantly affect the properties of TMDs [347]. Therefore, a controlled synthesis of TMDs is essential for the investigation. Besides several synthesis approaches of 2D TMDs the "Bottom-up" process facilities to developed direct wet chemical and chemical vapor deposition (CVD) to achieve adequate control of layer structure. In the CVD process, a thin layer of metal or metal compound is usually first coated on a substrate and then exposed to a chalcogenide atmosphere to grow the 2D TMDs.

Specifically, the direct and tunable band gaps, excellent physicochemical properties, and the state-of-the-art structural design of TMDs getting much attraction for frontier research in the area of bioelectronic devices. A careful control reaction is desired, such as with the chalcogen content of the emerging 2D materials having unique shapes such as triangles, stars, and butterflies [348]. Adjusting the surface defects of low-dimensional materials also defines the atomic vacancies and step edges, thereby, the structure and function of TMDs are also play a crucial role [349].

Particularly cytotoxicity of TMDs is a great concern for biosensing in vitro and in vivo on a large scale while considering extensive labor work, such as screening for SARS-CoV-2 infection. However, graphene has shown the highest toxicity on human lung carcinoma epithelial cells (A549) than the WSe_2, MoS_2, and WS_2 [350]. Thus, the MoS_2 and WS_2 induce very low cytotoxicity in the cells compared to the other 2D materials. Moreover, the biocompatibility of S is better than the Se in terms of key chalcogen elements. While the Te showed latent toxicity (e.g., neurotoxicity and cytoskeleton disruption) to humans [351]. Hence, S is considered to be the best candidate as the chalcogen combination with the metals. Using these significant properties of the TMDs variety of biosensors can be designed for specific applications.

8.8.1 TMDs Based Optical Biosensors

The TMDs based photoluminescence biosensors are also considered to be excellent due to fluorescence quenchers via electron/charge transfer (ET) to "turn off" fluorophores at the initial state without the target [352]. As an instance, theoretically, TMDs provides a high 2D surface area for single-strand DNA (ssDNA) to adsorb on their surface via van der Waals force, thereby it can efficiently quench to adsorb fluorophores-bearing ssDNA. Typically, this mechanism permits the "turn on" of the fluorescence when ssDNA hybridized with the target nucleic acid sequence and forms double-stranded DNA that has a weak interaction with the TMDs and detached from the TMDs platforms.

8.8.1.1 TMDs Based Fluorophore-Labeled Biosensors

The biosensors are established on inherent fluorescence quenching characteristics of the 2D TMDs adopting probable fluorescence resonance energy (or electron) transformation (FRET). Such as sensing backbone built on different FRET couples (typically a fluorophore-2D MoS_2 couple), specifically considering dye-labeled DNA as the probe. Whereas the inflexible dual-stranded DNA (dsDNA) has a considerably feebler adsorption interattraction to 2D TMDs in contrast to solitary-stranded DNA (ssDNA), precisely the governable switching of ssDNA to dsDNA allows the fluorescence "turn on" and vice versa. A robust adsorption interattraction among the ssDNA and TMD NSs due to the van der Waals strength among the nucleobases and the basal plane TMD NSs [353, 354], subsequent hybridization of ssDNA together the objective, therefore the development of dsDNA, the nucleobases convert suppressed among the deleteriously charged helical phosphate supports in the greater density. That greatly deteriorates the interattraction among the developed dsDNA and TMD NSs. Moreover, it is also demonstrated that 1L MS_2 ($M = Mo, Ti, Ta$) has a greater fluorescence quenching efficiency and distinct affinities to ssDNA and dsDNA for realizing the sensitivity and choosy indentification of DNA and adenosine together a fluorescent dye-labeled aptamer as the probe and 1L MoS_2 as the substrate, as depicted in Fig. 8.35a [354, 355].

Similarly, the different DNA-TMDs NSs biosensors have been also fabricated for the finding of proteins using MoS_2 NSs as the substrate and a fluorescent dye-labeled DNA aptamer as the probe [353, 360, 361]. In which the fluorescence of the dye-labeled DNA aptamer is primarily quenched into the MoS_2 NSs, and subsequent hybridization together the balancing object molecules liberate the aptamer-probe far to the MoS_2, thereby, retention in fluorescence, as illustrated in Fig. 8.35b [353]. Afterward, a 2D WS_2 sensing platform is also employed to analyze the T_4 polynucleotide kinase (T_4 PNK) and T_4 PNK phosphatase (T_4 PNKP), as illustrated in Fig. 8.35c, d [360, 361]. Especially dual types of probes have been constructions made depending on DNA phosphorylation [360] and dephosphorylation of the $3'$ termini of nucleic acids [361]. In the case of the occurrence of T_4 PNK, a fluorescent dye-labeled dsDNA gives us phosphorylated afterward degradation of λ-exonuclease providing ssDNA that have a robust association to 2D WS_2, as the consequence, quenching of the fluorescence, as illustrated in Fig. 8.35c. In this compare, non-occurrence of T_4 PNK catalyzed by phosphorylation, so, the λ-exonuclease persuaded prohibited the reduction of dsDNA. Similarly, in the case of T_4 PNKP identification, it hydrolyzed the phosphorylated ssDNA probe in the ssDNA through a $3'$-hydroxyl end. Therefore, it is instantaneously extended to make dsDNA by Klenow fragment polymerase, as depicted in Fig. 8.35d.

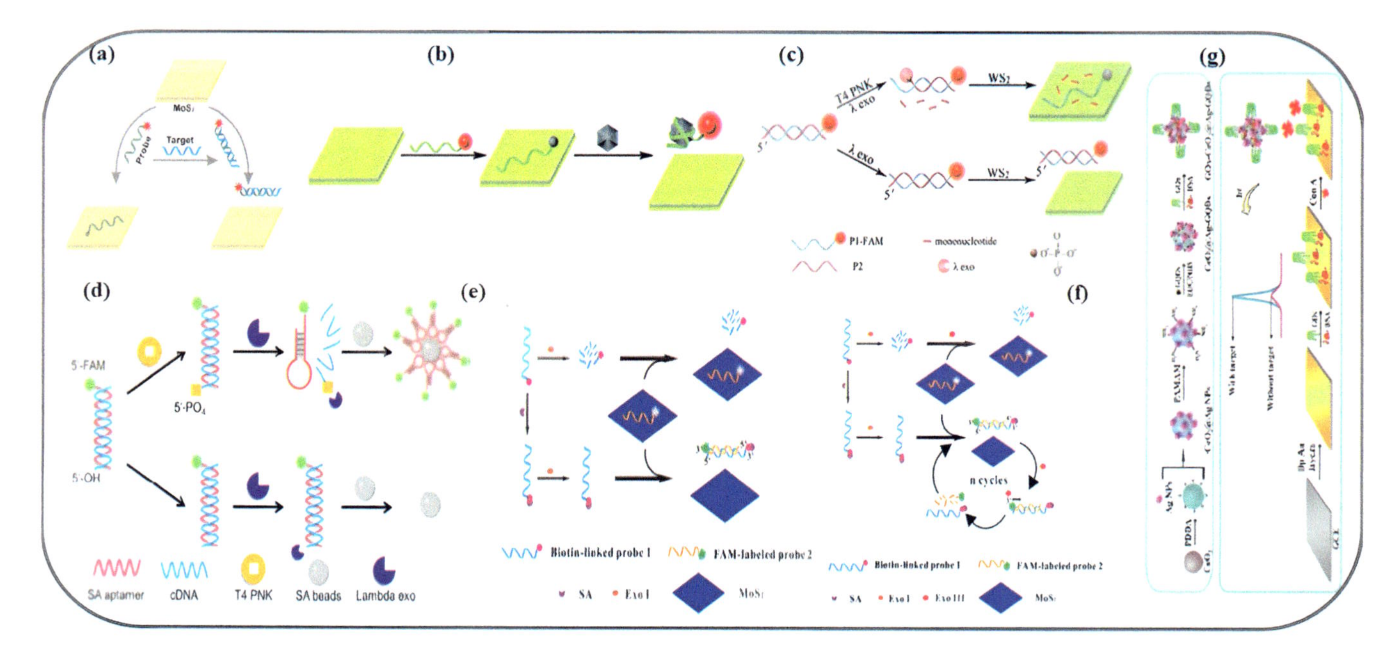

Fig. 8.35 **a** Scheme of a fluorimetric DNA assay using 2D MoS_2 as the sensing platform (adopted from (a) Zhu et al. [354]); **b** Scheme for the 2D MoS_2-based effective sensing system; **c** Scheme for the WS2 NS-based sensing platform for the detection of the T4 PNK activity and the inhibition analysis (b and c adopted from [356]); **d** Scheme for the 2D WS2-based platform for sensing the T4 PNKP activity and for analyzing the inhibition (adopted from [357]). **e** illustration of the mechanism of a 2D MoS_2-based biosensor for the detection of streptavidin via a combined process of the TPSLMD (adopted from [358]); **f** and Exo III-aided DNA recycling amplification (adopted from [356]); **g** PEDOT: PSS/Ti3C2/GQD modified SPCE for glucose detection (adopted from [359])

Moreover, a MoS_2-based biosensor can also use for streptavidin identification via a collective procedure of terminal safeguard of small-molecule-linked DNA (TPSMLD) and exonuclease III (Exo III) assisted DNA reprocessing enlargement [362]. In the TPSMLD process (see Fig. 8.35e) the streptavidin binds with biotin that is connected to probe 1, to avoid the enzymolysis the probe 1 by exonuclease I (Exo I). The subsisted probe 1 is consequently hybridized to probe 2 and releases probe 2 by the MoS_2 surface, this means a partial recovery of the fluorescence. However, the solitary TPSMLD progression has been unsuccessfully achieve sensitive streptavidin detection. Therefore, Exo III-aided DNA reprocessing enlargement could adapt to additional introduction into the TPSMLD procedure (see Fig. 8.35f). By demonstrating the presence of streptavidin with the bound biotin in probe 1 to protect the Exo I-catalyzed digestion. Upon development together Exo III and the probe 2-MoS_2 compound, probes 1 and 2 are hybridized and complemented through the slow reduction of probe 2 from 3'-to-5'-end from Exo III. This leads the residual probe 1 recurrently includes in fresh sequences of hybridization together probe 2, to release supplementary fluorescent dye-labeled probe 2 through the MoS_2 surface, thus the yield in a stronger fluorescence.

Additionally, it also noted that while the TMD NSs are used in quantum dots (QDs) the exotic PL features generates owing to the quantum confinement influence. Such deliberately created nano-scale point imperfections in the hexagonal lattice of pure 1L MS_2 (M = W, Mo) significantly affect the PL characteristics [363]. Therefore, the optical features of TMDs NSs can be modified through the size management and imperfection formation for optoelectronic utilizations. As the intense 10 ~ 20 nm-sized MoS_2 QDs have exhibited an intrinsic blue PL at 415 nm in the 300 nm UV illumination [364]. Subsequently, a FRET couple has been built through the attachment of the Alexa 430-labeled dsDNA together poly G tails over the MoS_2 QDs for the exploration of the acceptor and donor ability in the MoS_2 QDs [364]. It has been recognized the ssDNA attachment with the MoS_2 surface via the van der Waals strength which gives the hybridization of ssDNA and their subsequent creation and detachment of the dsDNA from the MoS_2 basal plane. Thus, the double characters of blue-luminescent MoS_2 QDs in the FRET process have been realized. As an instance, the typical fabrication mechanism of the PEDOT: PSS/Ti_3C_2/GQD modified SPCE for glucose detection is depicted in Fig. 8.35g.

8.8.1.2 TMDs-Based Label-Free Biosensors

Since fluorophore classify as the composite higher value and time taking method that opens the path for label-free strategies to provide numerous sensing ground to supplementary required real-world applicabilities [365–370]. The label-free fluorescence established identification of nuclease is usually obtained from the 2D WS_2 by utilizing a fluorescent combined polymer as the signal informer [371]. In-which polymer first forms a compound among the ssDNA and its fluorescence extinguished in the 2D WS_2 afterward the hydrolysis of ssDNA nuclease and consequent adsorption of the polymer in the 2D WS_2. Their chemiluminescence (CL) resonance energy transfer (CRET) among the donor and acceptor (2D TMDs) has been deduced from the label-free identification of biomolecules, like micro RNA [372]. Moreover, the fluorescence and CL also generates to different kinds of species rather than 2D TMDs, such as manipulation of the autofluorescence of TMDQDs is an efficient approach for the label-free identification of bioanalytics from the "turn on" and "turn off" fluorescence in a qualitative fashion [373, 374]. Additionally, the water soluble VS_2 QDs have been also produced as an autofluorescent basis and their fluorescence deeply extinguished through MnO_2 NSs. As the glutathione moved on the extinguished fluorescence of the VS_2QDs-MnO_2 NSs complex through the reduction of MnO_2 NSs over the Mn^{2+}, recognizing the label-free identification of glutathione, as depicted in Fig. 8.35g [373]. Thus, the optical fluorescence biosensor in the NIR excitation also reduces the contextual signals to avoid destruction to biological tissues that are usually triggered due to UV irradiation.

8.8.1.3 Absorption-Based TMD Sensors

The layered TMDs together semiconducting features also display visible absorption characteristics that is use for absorption established biosensors via a consistent alteration in their visible absorption. This kind of optical biosensors are generally lower value, usual and extremely sensible. Such as, water dispersal of MoS_2 NSs displays a quick aggregation results for the salts owing to a salt persuaded consequence [367] and a consistent spreading even in the occurrence of greater attentiveness salts while the 2D MoS_2 surface is secure to ssDNA layer. The exchange the ssDNA to dsDNA leads to harm in protecting impact, thereby the severe aggregation intiated. Therefore, label-free identification is realized with the discernment ability to ssDNA and dsDNA, and the size reliant optical absorption of MoS_2 NSs. In such optical biosensors, the salt persuaded aggregation of TMD NSs is also possible. Whereas a hybridization of chain reaction (HCR) is familiarized to supplementary improvement in DNA sensitivity identification, however, the LOD may lower to 1.54 nM through ssDNA hybridization upto 0.23 nM using HCR [375].

8.8.1.4 TMDs Based Colorimetric Biosensors

Fluorescence-based technique in which a dimensionally different MoS_2 has been utilized for the colorimetric assay by exploring its useful optical properties. Like MoS_2 possessed a strong optical absorbance over a wide range of wavelengths. The unique absorption of MoS_2 is also useful for the confinement of electronic movements without interlayer interference and confers monolayer band gaps. Both 0D and 2D MoS_2 have the potential to use as colorimetric-based biosensing applications. As the intense, 2D TMDs including MoS_2 exhibit peroxidase-like catalytic activities that are utilized to design various sensing platforms. Like detection of Fe II by using luminescent MoS_2-nanosheet-based peroxidase mimetics. The fabricated nanosheets catalyzes the oxidation of the peroxidase substrate phenylenediamine (OPD) in the presence of H_2O_2, therefore, it gives a yellow product. The Fe^{2+} can also enhance the catalytic activity of MoS_2 nanosheets largely, thereby, it influenced the OPD reaction which is followed by colorimetric measurements. This constitutes mainly based on the Fe II estimation in a sensitive manner [215]. Similarly, the MoS_2 nanosheets also catalyze the reaction 3, 3′,5, 5′-tetramethylbenzidine (TMB) with H_2O_2 to give a blue product. The formation of such blue products is followed calorimetric to provide a method for the sensitive and selective detection of H_2O_2 [216]. The same approach also follows in the determination of glucose. By combining MoS_2 nanosheets with GO_x it also catalyzes the oxidation of glucose to gluconic acid and oxygen of the solution converts into H_2O_2. The formation of H_2O_2 depends on the oxidation of glucose present in the system. Hence, the estimation of H_2O_2 using TMB provides a quantitative amount of glucose, as depicted in Fig. 8.36a [217]. Additionally, the peroxidase-like catalytic activities of MoS_2 have also possessed size-dependent optical absorption that utilizes in the detection of DNA molecules. Such as, in a salt solution 2D MoS_2 tend to agglomerate and often formed larger aggregates than those in an aqueous medium. As the consequence, the optical absorption capacity is decreased considerably. However, upon functionalization with ssDNA, this tendency of MoS_2 is hindered, thereby the original light absorption ability is restored. So, in the presence of target DNA, 2D MoS_2 again form aggregates and lead to a decrease in the optical absorbance. Thus, a diminished optical absorption of the MoS_2/ssDNA hybrid under the presence of target DNA is useful to utilize for the detection of DNA molecules, as depicted in Fig. 8.36b [218].

8.8.1.5 TMD Base Electrochemical Biosensors

Most of the electrochemical biosensors based on MoS_2 nanostructures employ potentiometric and amperometric techniques for the detection of bioanalytics. Although square wave voltammetry (SWV), differential-pulse voltammetry (DPV), and electron impedance spectroscopy (EIS) have been also frequently used techniques for the detection of bioanalytics with a high level of sensitivity [377–379].

A variety of analytes has been detected from the MoS_2- based electrochemical biosensors. Such as MoS_2-based hybrid materials fabricated sensor probes are used for the highly selective and sensitive detection of bioanalytics [376, 380–383]. The poly (xanthurenic acid) based on MoS_2 support is favorable to the fabrication of a highly electroactive biosensing matrix using an ultrasonic method, therefore, efficient detection of guanine and adenine [384]. Similarly, MoS_2 and PANI composite-based devices have been also fabricated for the in situ electrochemical detection of chloramphenicol [385]. Where the gold nanoparticle decorated 2D MoS_2 nanosheets have been used for the electrochemical sensing of glucose in the presence of commonly interfering species like ascorbic acid, dopamine, uric acid, and acetaminophen [386]. A myoglobin-immobilized 0D MoS_2 nanoparticles—GO hybrid has been also fabricated for the detection of H_2O_2, thus, an enhancement in electrochemical signal due

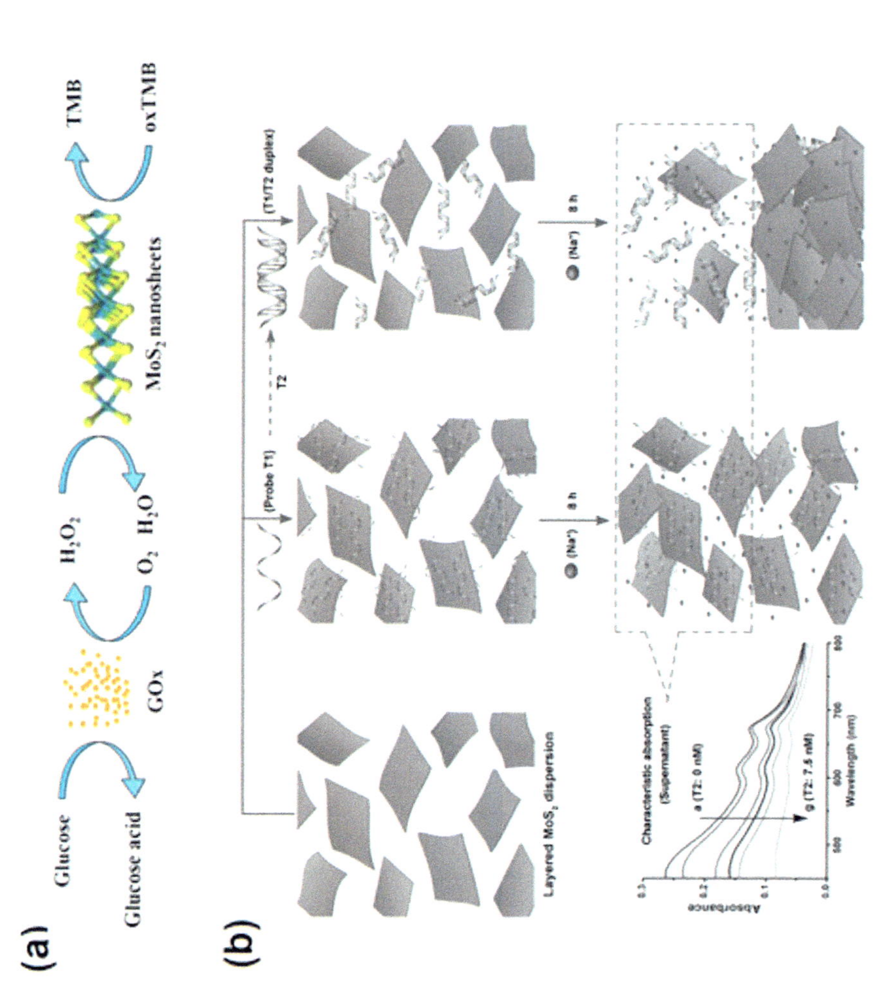

Fig. 8.36 **a** Colorimetric estimation of H_2O_2 and glucose; **b** Utilization of size-dependent optical absorption for the estimation of DNA (adopted from Barua et al. [376])

to the existence of GO that attributed to the high surface area available for the immobilization of myoglobin [387]. A similar approach can also employ in an extremely sensitive (2.5 nM) H_2O_2 biosensor based on MoS_2 quantum dots (~2 nm) [388]. In this order, a thionin-functionalized MoS_2-based electrochemical biosensor has been also fabricated for the direct detection of DNA with sensitivity up to the ppb level [389]. Moreover, the electrochemically reduced single-layer 2D MoS_2 nanostructures also exhibit a fast electron-transfer rate that helps in the detection of glucose with high sensitivity [390]. The micromolar-level electrochemical biosensing of glucose is also achieved by using a glucose oxidase (GOx)-immobilized gold/MoS_2 nanohybrid [387]. Thus fabrication of distinct types TMDs based electrochemical biosensors is still under extensive investigation to explore their various dimensions for detection based on the used material's characteristics. To get a high sensitivity, facile operation, low cost, reusability, prompt response, trace sampling, and fast analysis should be considered [391]. Typical schematics of electrochemical biosensors based on kinds of TMDs are illustrated in Fig. 8.37a–f which has been widely used for detection.

8.8.1.6 Field-Effect Transistor (FET) TMD-Based Biosensors

FET-based biosensors also attended great attention to getting highly desirable characteristics of label-free rapid electrical detection capabilities, low power consumption, mass production, and compactness with the possibility of on-chip integration on a chip for the measurement. Conventionally FET consists of two electrodes (drain and source) that are connected electrically via a semiconductor material (channel). The current flow between the drain and source through the channel has been modulated by an additional third electrode, namely, a gate. That is capacitively coupled through a dielectric layer by covering the channel. Capturing of biomolecules through the functionalized channel by producing an electrostatic effect that transduced into a readable signal in the form of a change in the electrical characteristics of the FET device [395]. However, such devices' performance characteristics depend on the biasing of the device [396]. A schematic of the TMD-based FET sensor is illustrated in Fig. 8.38a–f [368]. The device is based on typical reflecting sensing principles including detection strategy and chip design for MoS_2-based FET biosensors.

The direct band gap of MoS_2 allows better control between the conductive and insulated states that prevents leakage currents [347]. It also provides the flexibility to design FET to get higher sensitivity with accurate sensing at a very low concentration of the analytes. Moreover, the absence of out-of-plane dangling bonds in MoS_2 allows for reduced surface roughness scattering as well as interface traps. It also provides better electrostatic control owing to the lower density of the interface states on the semiconductor—dielectric interface. So, a reduction in low-frequency noise hinders the performance of FET-based biosensors [397].

Specifically, MoS_2-nanosheet-based FET biosensors are also able to detect proteins, pH, cancer biomarkers, etc., with high sensitivity and selectivity [398, 399]. While a few-layer MoS_2-film-based FET device is considered to be more stable with

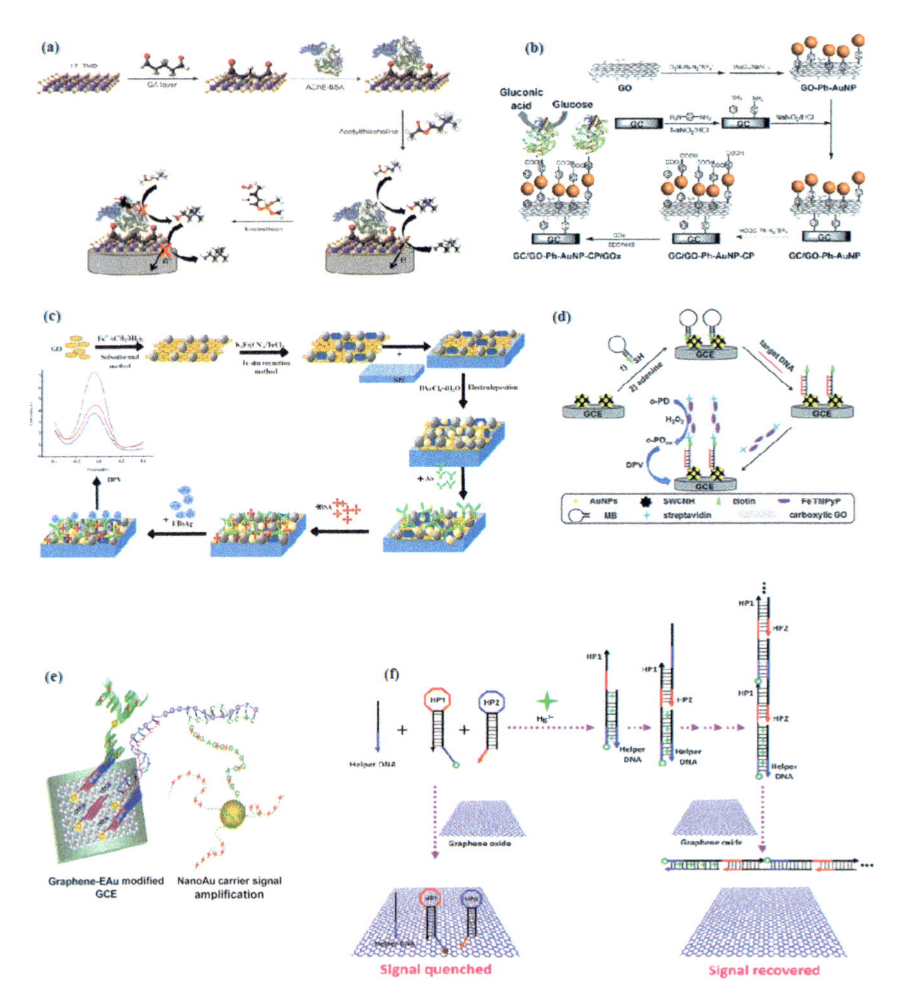

Fig. 8.37 **a** Design for the building of the pesticide biosensor by 1 T-phase TMD NSs (containg MoS_2, $MoSe_2$, WS_2, and WSe_2) as the basis (adopted from [392]); **b** Schematic of glucose biosensor by AuNP-decorated GO nanosheet. AuNPs has been decorated over GO nanosheet through a benzene bridge by employing aryldiazonium salt chemistry (GO-Ph-AuNPs) that afterward attached to a 4-aminophenyl altered GC electrode. The GC/GO-Ph-AuNPs supplementary functionalized together 4-carboxyphenyl (CP) earlier covalently attached GOx through amide bonds to form GC/GOPh-AuNPs-CP/GOx established glucose sensor; **c** Diagramic design for the construction of electrochemical immunosensor (adopted from [393]); **d** Diagram for the graphene/ferric porphyrin (FeTMPyP) established electrochemical sensor to DNA identification as horseradish peroxidase (HRP)-mimicking trace label; **e** Diagram for the sensing approach to identification of Hg_2^+ using a graphene/nano-Au compound for signal enlargement; **f** Enlarged fluorescent sensing process to identify Hg_2^+ through HCR. In the nonappearance of Hg_2^+, GO absorbs the DNA probes (helper DNA, HP1, and HP2) through noncovalent attachment and quenches the fluorescence of HP1. While in the occurrence of Hg_2^+, the assistant DNA unlocks HP1 owing to the construction of steady T–Hg_2^+–T arrangements and persuaded incessant HP1–HP2 hybridizations, that cannot adsorb over the GO, providing the creation of enlarged fluorescence (b, d, e, and f adopted from [394])

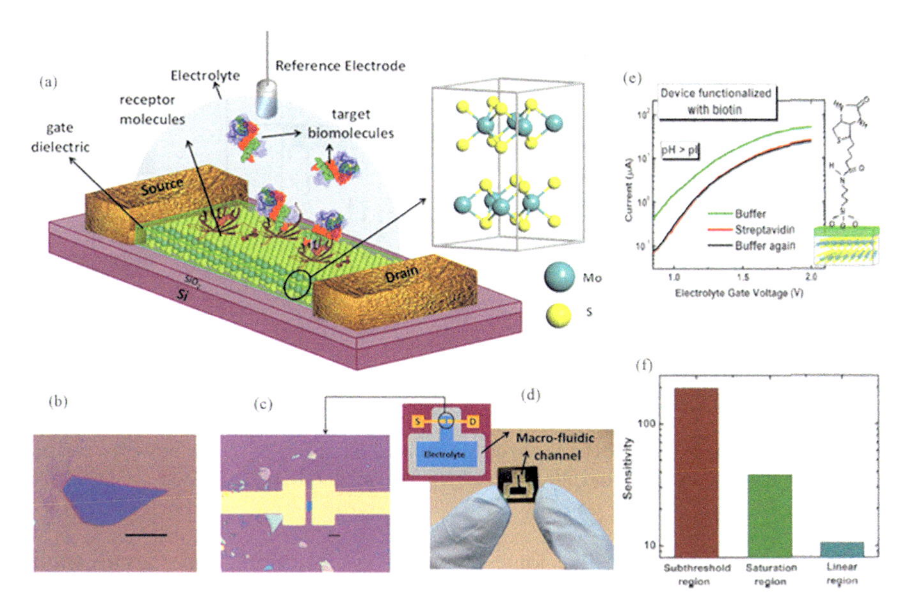

Fig. 8.38 MoS_2-based FET biosensor: **a** schematic for the sensing scheme by showing the MoS_2 channel functionalized with receptors for specifically capturing the target biomolecules and the drain and source contacts and the Ag/AgCl reference electrode for biasing the device; **b** optical image of a MoS_2 flake in a SiO_2/silicon substrate (scale bar = 10 μm); **c** optical image of the FET device; **d** image and schematic (inset) showing the microfluidic integrated FET chip; **e** transfer characteristics of a biotin-functionalized FET device measured in a pure buffer (0.01 × PBS), with the addition of a streptavidin solution (10 μM in 0.01 × PBS) and again with a pure buffer; **f** comparison of the sensitivities in the subthreshold, saturation, and linear regions of the transistor device (adopted from Sarkar et al. [368])

a highly sensitive response than the other monolayer-based devices [400]. Moreover, the chip integration with microfluidics also opens up huge prospects to realize compact sensor systems for clinically meaningful detection limits that are useful for their usage in point-of-care diagnosis applications [401, 402]. Additionally, the MoS_2 thin films-based FET devices also offer integrated logic circuits to provide the viability of realization, such as biosensors with low processing costs and multiplexed detection capabilities [376].

Moreover, not only MoS_2 FET biosensing devices but also other TMD materials-based devices such as WSe_2, WS_2, WSe_2, etc. have been fabricated and tested for their responses in terms of biomedical applications to diagnose distinct degases infected species [403–405]. In this sequence, the $MoTe_2$-based advanced DMFETs devices are also getting much attention due to their smaller bandgap, and lower thermal conductivity with a higher Seebeck coefficient compared to other 2D materials [406]. $MoTe_2$ also has unique and viable properties in terms of growth, bandgap engineering, carrier injection, etc. [406]. Additionally, a high ratio of on/off current with a low subthreshold swing of the $MoTe_2$ is also one of the most impressive

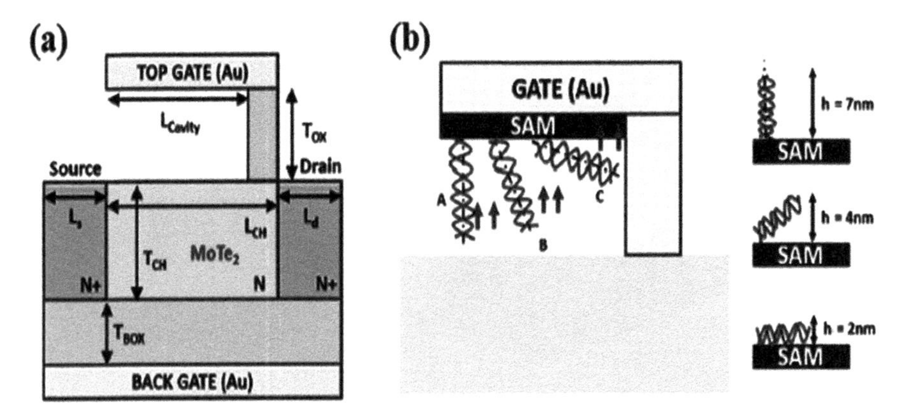

Fig. 8.39 **a** Schematic of the device based on $MoTe_2$ AMFET; **b** Device schematic for the concept of DNA orientation (adopted from https://doi.org/10.36227/techrxiv.14723343.v1)

features to make a device, like logic circuits, optoelectronics, and sensing applications. The cutting-edge day-by-day technologies are improving, therefore, high-level performing biosensing devices desired, like biosensing devices should have better biocompatibility performances, etc. A theoretical simulation on a typical $MoTe_2$-based "Accumulation Mode Field Effect Transistor (AMFET)" has been interpreted for the high-processing biosensing application, as schematics are illustrated in Fig. 8.39a. An established biosensor offers a superior sensitivity for neutral biomolecules, such as DNA biomolecules [405]. Therefore, the schematic device typically represents the DNA biomolecule orientation, as illustrated in Fig. 8.39b. Subsequently, the corresponding simulation has been interpreted the DNA detection with and without DNA-electrode interaction including the effect of DNA orientation and the impact of Back-Gate Bias on the device performance [405].

8.9 TMD–TMD Heterostructures and Devices

Semiconducting heterostructures and superlattices are also considered an influential material podium for the technical groundwork to every kinds of contemporary electronic and optoelectronic devices, like high-electron-mobility transistors, light-emitting diodes, lasers, sensors, solar cells, etc. [407]. The heterostructures are usually constructed from the atomically thin 2D materials that permit steep the modulation of electronic band structure by developing atomically piercing heterojunctions [408, 409]. The distinct 2D-TMD materials have formed numerous kinds of multifaceted TMD heterostructures to be architectured and developed by an approximately arbitrary combination of discrete 2D-TMD layers into multiple-layer heterostructures or superlattices with no restriction of lattice similarity. Such layered heterostructures have atomically piercing interfaces with zero inter-dispersal of atoms and their layered constituents are controlled digitally. Therefore, the creation of such TMD heterostructures opens up a novel aspect to the manufacturing the electronic including exploration of the optical features at the atomic level. It makes them enables the construction of a novel group of atomically thin electronic and optoelectronic devices [410–412].

The development of TMD–TMD heterostructures also allows the formation of different practical devices, like p–n diodes. The p–n diodes are the greatest essential construction chunks in several optoelectronic devices, such as photodiodes, solar cells, light-emitting diodes, etc. However, in the conventional semiconductor the p–n diodes are frequently achieved by selective doping of semiconducting materials in the p- and n-type areas. It is significant to make p–n diodes in 2D-TMDs owing to the problems in selective doping in 2D semiconductors and governing their doping profile. Therefore, 2D TMDs electrostatic doping are useful to generate planar p–n diodes, though a comparatively slow doping profile is restricted by the peripheral electrical field with the fairly lower optoelectronic efficiency [413]. To resolve this problem has been induced a perpendicularly stacked heterojunction among p-type WSe_2 and n-type MoS_2 flakes [256, 414, 415], as the typical heterostructures building block between monolayer WSe_2 and MoS_2 are formed with the unique structural and physical properties as the schematic illustrated in Fig. 8.40a, b [416]. The building of monolayer WSe_2 and multiple-layer MoS_2 (1L-WSe2/ML-MoS_2) heterojunctions have revealed a reliable configuration design in SEM images, as illustrated in Fig. 8.40c–e. The heterojunction current rectification behavior is gate-tunable **to** a p–n diode, as depicted in Fig. 8.40f. Commonly, the heterojunction p–n diode resistance is reduced exponentially together the enhancing bias voltage, whereas the series p-FET resistance is almost steady to the bias voltage. This means the heterojunction resistance is controlled through the p–n diode at the lower bias and it according to p-type 1L-WSe_2 FET in case of a higher forward bias. Moreover, the fits of the diode characteristics provides us ideality factor $n = 1.2$ that is already experimentally confirmed, it also has an exceptional diodic manner in the atomically piercing heterojunction p–n diode. Additionally, the photocurrent generated a spatially resolved map of the heterojunction p–n diode leading that the insight diode characteristics. The zero bias photocurrent mapping of automatically thin heterojunction p–n diode of WSe_2/MoS_2 has a broad overlapping region (see Fig. 8.40 g, h). In addition, an atomically thin p–n junction created in the 1L-MoS_2, likewise 1L-WSe_2 has also revealed that despite a superficial current rectification character their essential transportation machinery differs to the conventional p–n diode due to lack of depletion layer in the atomically thin heterostructures. Therefore, it supported the layer to layer tunneling and recoupling are contributed a significant part in their charge transportation physiognomies [414]. With the important note, in this kind of atomically thin p–n diodes a horizontal connection outdoor the junction overlying area that are useful for the photo-generated carriers to go through a comparatively longer lateral dispersal span, therefore, the struggle among interlayer recoupling and lateral dispersal compromises the charge gathering efficiency and hence restrict the inclusive EQE up to 34% in the ML-WSe_2/ ML-MoS_2 p–n diode [414].

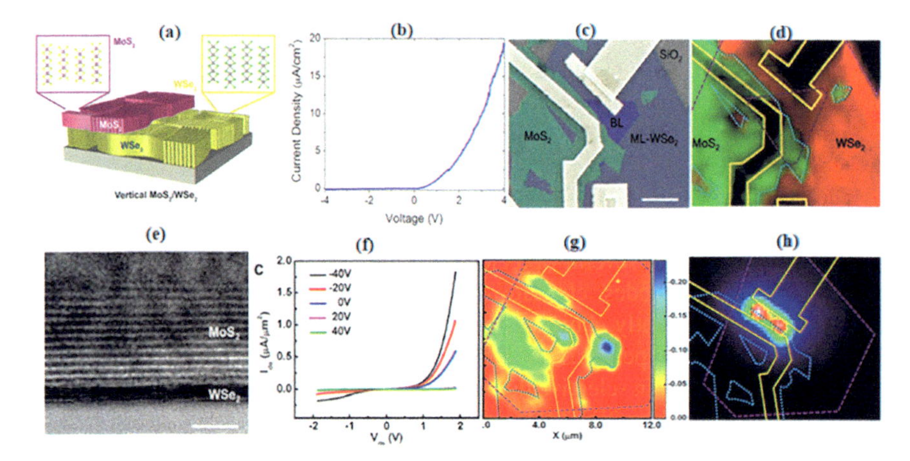

Fig. 8.40 **a** Represents the heterostructure consisting of the MoS_2 and WSe2, in which their van der Waals layers are aligned perpendicular to the substrate; **b** current − voltage characteristic of the MoS_2/WSe_2 heterostructure diode (a and b adopted from [417]); **c** The false color SEM image of the WSe_2/MoS_2 vertical heterojunction device, with ML-WSe_2 highlighted by blue color, BL-WSe_2 area by violet color, MoS_2 by green color, and metal electrodes by golden color, the scale bar is 3 μm; **d** PL mapping of the WSe_2/MoS_2 heterojunction device, with red color representing the PL from MoS_2 and the green color representing PL from WSe_2; **e** cross-sectional HRTEM image of the WSe_2/MoS_2 heterojunction interface, the scale bar is 5 nm.; **f** gate-tunable output characteristics of the WSe_2/MoS_2 heterojunction p − n diode; **g, h** False color scanning photocurrent micrograph of the WSe_2/MoS_2 heterojunction device acquired at Vds = 0 V and VBG = 0 V under irradiation 514 nm laser (5 μW). The purple square dotted line outlines the ML-WSe_2 and the dark purple square dotted line outlines the BL-WSe2. The blue circle dotted line outlines the MoS_2 and the golden solid line outlines the gold electrodes, the photocurrent noted for the entire overlapping junction area (c–h adopted from Cheng et al. [415])

In addition to these, the atomically sharp WSe_2/MoS_2 also exhibits electroluminescence (EL) properties. Specifically, under the forward bias EL is confined in the overlying region nearby the metal electrodes. This could be described through the electric field circulation in the heterojunction at the diverse biases. Usually, in photocurrent mapping at the zero biasing, the joint resistance of the p–n diode dominated the whole device, whereas the series resistance of the WSe_2 p-FET is almost comparatively irrelevant. So the photocurrent is due to complete overlying expanse while the p–n junction potential offsets. At an abundantly higher the forward bias exceeds the turn-on voltage of the p–n diode and the corresponding resistance of the 1L-WSe_2 p-FET is increased with a dominant constituent of the whole resistance. Thus, the mostly voltage declines crossways the heterojunction edge close to the electrodes owing to the great series resistance of the 1L-WSe_2 [413].

Innovation with TMD-TMD heterojunctions devices is not limited to above discussed initial observations, however, a vast number of TMD–TMD heterostructure-based devices with different TMDs compositions have been fabricated by various groups to explore their distinct dimensions for the high-performance optoelectronic, spintronics, energy conservation, water splitting, etc. devices [418–423].

8.10 TMD/ Graphene Heterostructure Devices

Although TMDs heterostructured have attracted considerable interest to fabricated kinds of atomically thin transistors with their optimized device geometry, such as a self-aligned gate, drain, and source high-performance few-layer devices [424]. Typically MoS_2 transistors can have an intrinsic cut-off frequency of 42 GHz and a maximum power-gain frequency of 50 GHz [126]. Besides the considerable progress, the performance of TMDs based heterostructured devices is still lagging behind the other counterpart technologies due to their excessive contact resistance at the source/drain interface including other significant parameters [407, 425–429]. Therefore, taking into account the key flexible property of van der Waals interaction other diverse heterostructures are fabricated with the 2D-TMDs atomic layered materials, typically with graphene, to enhance the overall performance efficiency of the devices [407, 425–427, 429]. The integration of graphene with 2D-TMD atomic layered materials can be modified the band gap of TMD/graphene heterostructure devices depending on dimensionality (such as zero, one, or two dimensions) by affecting the charge-transfer mechanism. As the integration of gapless graphene with the semimetallic TMDs has broadened the band light absorption terahertz (THz) to ultraviolet (UV) wavelength [430]. The combination of TMDs with graphene also forms a depletion area within the semiconductor owing to the distinction in their work functions; thereby its erected an electric field at the interface importantly facilitates the partition of the photogenerated electron–hole couples [431, 432]. Thus a combination of graphene with TMDs, like MoS_2, WS_2, WSe_2, and $MoTe_2$ [433–440] are possessed moderate band gaps with strong light-matter interaction compared to individual graphene. Therefore, a higher-performing photodetector established on graphene/TMD van der Waals (vdW) heterostructures among the significant Schottky barrier are possibly achieved, that makes a valuable influence in the photoexcited carrier partition owing to the existence of the in-built electric field [441]. In addition, the graphene/MoS_2 heterostructure is also unique appropriate selection to manufacture electronic/optoelectronic devices together aesthetic value and design flexibility. As an example, the high specific detection value 1.8×10^{10} Jones of a photodetector constructed by some layers MoS_2/glassy–graphene heterostructure underneath 532 nm light illumination [435]. Similar properties have been also obtained from the WS_2 and WSe_2 high-performance heterojunctions with graphene [436, 442]. However, the $MoTe_2$ has a band gap of 1.0 eV for its bulk form that is lesser to several TMDs, it is considered to be a better aspirant for the NIR photodetectors [443]. So, a perpendicular heterostructure fabricated associating a slight carrier travel pathway distance in $MoTe_2$ that provides a controllable Schottky blockade in $MoTe_2$/graphene configuration, whereas the $MoTe_2$ acts as a light-absorption layer, as schematic in Fig. 8.41a [438]. Similarly, the difference in the work function of Au and graphene in the Au–$MoTe_2$– graphene construction provides an in-built potential within the device, as depicted in Fig. 8.41b. Where the $MoTe_2$ bands energy drops after the connection and produced an interior electric field at the relevant points of $MoTe_2$ and graphene. Under light illumination, the electron–hole couples are produced in $MoTe_2$ and divided through the in-built electric field at the $MoTe_2$/graphene inter-junction. Specifically, holes move toward graphene, while the electrons stucked in $MoTe_2$ to act as an indigenous gateway, ultimately it appeals supplementary holes in graphene to decrease the resistance. As the typical electrical characteristic is illustrated in Fig. 8.41c, the device 45 nm-thick $MoTe_2$ generates a substantial photocurrent 2 μA at 150 μW input power with zero applicable voltage. It has been also noted that an effective partition of photoexcited electron–hole couples in the in-built electronic field by altering the $MoTe_2$ thickness and increase in the applicable bias, to get a higher bandwidth of 50 GHz. Additionally, a good design waveguide of the device also provides a higher responsivity up to 0.2 AW^{-1} at 1300 nm [438].

Moreover, the band energy also depends on the modulation of gate voltage, source-drain voltage, etc., that further improves the photoelectric performance. Like the asymmetry Schottky blockade at the Au–MoS_2 and MoS_2– graphene boundary that leads to a rectification ratio over 2×10^4 in the Au–MoS_2–graphene heterojunction, it is supplementary adjustable to the exterior gating [444]. Similarly, a device established over a hybrid vdW heterostructure of ReS_2 and graphene attracted much attention due to their layer dependence direct bandgap that connected to an anisotropic crystal construction of 1 T phase [445–448]. The identical work function of graphene and ReS_2 leads to diverse carrier transportation features among the gate voltage variation. In the case of $V_G < V_{Dirac}$, the holes injects in the graphene and the Fermi level of graphene moved below, therefore, a Schottky blockade generates. While the lighting, the electron–hole couples get excited in the ReS_2 layers and separate from the in-built potential. Whereas the holes moves to graphene and electrons stucked by the Schottky blockade that behave as a indigenous gate for the photogating consequence [446]. On the other hand, in the case of the $V_G > V_{Dirac}$, the photogenerated electrons toward graphene while the holes behave as a native gate. It benefits to the speedy charge transportation at the interface and higher carrier mobility in graphene with an extremely high photoresponsivity of $7 \times 10^5 AW^{-1}$, including a fast response time of less than 30 ms [446].

However, these devices usually suffered from undesired photocurrent dispersion surrounding zero-bias voltage. Therefore, efforts have been also made to incorporate asymmetry tunneling among the photocarriers and dark carriers via the h-BN layer. To get an improved current rectification with an abundant high current flow to the graphene/h-BN/ monolayer MoS_2 diode [449]. A highly sensitive photodetector has been also obtained with a MoS_2/h-BN/graphene heterostructure. Moreover, the great electron blockade of 2.7 eV at the graphene/h-BN boundary has suppressed the dark current with an ultrahigh detectivity of 2.6×10^{13} that is realized [450]. A machinery of loss of the zero-bias photocurrent has been also interpreted with the MoS_2/graphene photodiode and MoS_2/h-BN/graphene photodiode by employing the

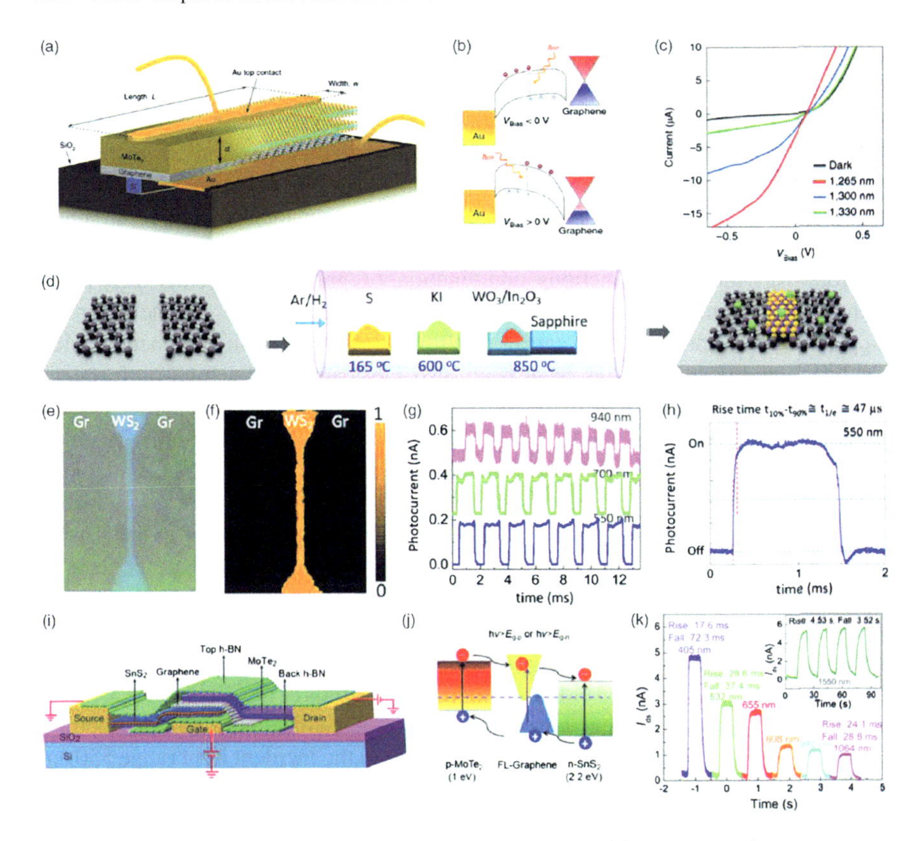

Fig. 8.41 Representation for the photodetectors based on TMD–graphene heterostructures; **a** Schematic illustration of a vertical $MoTe_2$–graphene heterostructure photodetector coupled to a silicon waveguide buried in SiO_2 claddings; **b** Corresponding band diagrams of the $MoTe_2$–graphene heterostructure under negative bias (upper) and positive bias (lower); **c** I–V curves with and without light for three different wavelengths; **d** Schematic to show the preparation of graphene–WS_2–graphene heterojunction by the in situ CVD growth method; **e** Optical image; **f** PL map of the graphene–WS_2–graphene junction; **g** Time-resolved photoresponse of the device under the illumination of different wavelengths; **h** Close-up look of a single ON/OFF switch of the device under 550 nm illumination; **i** schematic to show h-BN/$MoTe_2$/graphene/SnS_2/h-BN device configuration; **j** Band diagram and photoexcited carrier transport of the heterostructures; **k** Temporal photocurrent response under various wavelengths from the UV region to NIR, in the inset the temporal photocurrent response under 1550 nm wavelength (adopted from Liu et al. [421])

transferring attitude. Furthermore, the photoluminescence (PL) spectrum and photoelectronic physiognomies of the associated heterostructures have been also explored by considering the interlayer coupling of photocarriers at the MoS_2/graphene interface [434]. Such as a small band bending in MoS_2 leads to the depression of photocurrent near zero bias owing to photogenerated holes diffused to graphene that are consumed by the strong interlayer coupling with the photogenerated electrons in MoS_2, thereby, they cannot participate in the photocurrent. Thus, an extremely thin layer of h-BN over the heterostructure can inhibit the electron transportation among the layers while the photogenerated holes in MoS_2 tunnels through the ultrathin h-BN film under the influence of the internal electric field [434]. Therefore, the photocurrent of the self-powered tunneling photodiode **established over** the MoS_2/h-BN/graphene heterostructure is enlarged up to more than three orders of the value to an extremely high specific detectivity of 6.7×10^{10}. Moreover, in such vdW heterostructure phototransistors have also provided a near-ideal interface that makes enables low-voltage operation [451].

Moreover, the heterojunctions combination of graphene together 2D TMDs frequently designed the multiple-stage carrier transportation channels for the supplementary improvement in carrier partition effectiveness. The graphene–TMD–graphene sandwich buildings have been also fabricated to get high-performance photodetectors. As the instance, the utilization of indium adatoms could enclose the graphene–WS_2–graphene joints and give an extremely-high gain 6.3×10^3 electrons per photon with an extremely fast response time 45–60 μs [452], as depicted in Fig. 8.41d. A graphene sheet has moved over a sapphire substrate by following another chemical vapor deposition (CVD) technique and grown WS_2 in the channel area. The optical and PL pictures of the In/graphene–WS_2–graphene connection is illustrated in Fig. 8.41e, f. It reveals in this kind of device graphene serve as a translucent electrode and TMDs behave as a photoactive layer. In the case of the metal electrodes, their Fermi level pins down lower the conduction band minima of $WS_{2,}$, ultimately it provides a blockade height of hundreds of meV. However, contrast to metal connection, graphene connection prefers to reduce the blockade depth. A vibrant ON/OFF switching character of the device in the incident light range 550 and 700 nm has been visualized in Fig. 8.41 g, and a fast response time around 47 μs for the 550 nm light, as depicted in Fig. 8.41h [452]. The NIR photodetectors based on the graphene/ $MoTe_2$/graphene vertical vdW heterostructure via a site controllable layer-by-layer transfer approach [67], devices Schottky barrier height G_{top}/$MoTe_2$ and G_{bottom}/$MoTe_2$ remarkably influenced the photocurrent produce and transport that can be additionally modify by the back-gate voltage. Moreover, the larger scales graphene/WS_2/graphene vertical-stacked crossbar phototransistor device has been also invented to achieve layer-dependence photoconductivity of 1L WS_2 and 2LWS_2 [453].

Furthermore, the applicability of TMD–grapheme kind band arrangement is well known as p–g–n heterostructure device. Like 2D h-BN/p-$MoTe_2$/ graphene/n-SnS_2/h-BN p–g–n connection has revealed an exact detectivity up to $\approx 10^{13}$ Jones in the UV–vis–NIR range (see Fig. 8.41i) [454]. For the energy of photons greater to $MoTe_2$ or SnS_2 the photoactive layer are generally engross photons to create electron–hole couples, depicted in Fig. 8.41j. The electrons transport to SnS_2 in the contradictory direction in the in-built electron field whiles the holes transportation to $MoTe_2$. In the case of the energy of photons lesser than the band gap of $MoTe_{2,}$ 1 eV, whereas solitary graphene acts as a light absorbing layer and produced electrons transport to SnS_2 while the holes transport toward the contradictory sideways. The acquired broadband in the detection from the visible to middle infrared (MIR) wavelength is depicted in Fig. 8.41k [454].

Other p–g–n structures are also shaped from sandwiching the graphene into MoS_2 and WSe_2 to the transferring approach to reduce the dark current of the device and get an extended photoresponse range to the short-wavelength to infrared at the room temperature. Dual Schottky barricades and a robust in-built electric field confirm the partition of photoexcited electrons and holes [455]. However, more differences for the perpendicular stacked construction have been also interpreted with the lateral structure. Like a phototransistor composed laterally in the MoS_2/graphene/ MoS_2 heterostructure from the usual CVD method. That displayed broadband photoresponse for the visible to infrared range (532–1600 nm) [421]. The shaped dual Schottky barricades among graphene and MoS_2 can efficiently distinguish the photoexcited carriers in the graphene or MoS_2 area. Thus the construction of heterojunctions among TMDs and graphene is an operational way to enlarge the photoresponse range of the photodetectors to apprehend the efficient partition of carriers owing to the existence of the Schottky barricade and photogating consequence. Specifically, their in-built electric field expands the carrier transportation to accelerate the reduction in the recoupling possibility, ultimately it improve the photocurrent of devices [421]. Moreover, applications of TMD/graphene heterojunctions can be in numerous areas with their novel properties [456–460].

8.11 Miscellaneous

8.11.1 Thermoelectric Application

Due to growing energy demands thermoelectric (TE) materials-based devices have attracted much attention in recent years. TE devices alter the excess heat in the electrical power following the Seebeck effect principle and their efficiency depends on the dimensionless figure of merit (ZT) [461]. However, it also noted that dimensional materials are also advantageous for TE applicability owing to their piercing characteristics in the density of states (DOS) [462]. Specifically, the occurrence of piercing DOS produced by quantum confinement in twofold dimensional (2D, thin films), one-dimensional (1D, wires), and zero-dimensional (0D, quantum dots) structures enhance the Seebeck coefficient. Besides the other materials TMDs also considered to be prominent candidates for the TE application, as the first 2D Bi_2Te_3/Sb_2Te_3 superlattice has been introduced, their ultrahigh ZT value beyond 2.0 [463].

The key advantages of 2D TMDs their 2D crystal behavior provide piercing properties in DOS that offer interdependent electrical and thermal conductivity individual optimization with the maximizing energy conversion [464, 465]. TMDs layer stacking with the weakly bonded layered structures is also favorable for the TE application. Since intralayer atoms of every layer are connected by the robust chemical bonds, whereas the conjugative layers are linked by weak van der Waals (vdWs) interactions [424, 466]. Therefore, it is possible to separate the complex and matching different atomic layers to generate an extensive variety of vdWs heterostructures, lacking of the constraints of lattice similarity and procedural compatibility [424]. Moreover, it is also probable to adjust the band gap and carrier density of 2D-TMDs by varying the quantity of layers [467, 468], strain [469, 470], electric field [467, 471], and the physical configuration of the materials by hydrogenation [472, 473] or oxidation [472, 474]. Thus the crystal construction and chemical attachment of 2D TMDs themselves have a lower thermal conductivity in the out-of-plane direction and tremendously higher thermal conductivity in the in-plane direction. Such constructions have a great anisotropy and they are favorable to designing TE materials devices with the higher ZT. Specifically, 2D TMDs in the out-of-plane direction is considered to be decent for TE or thermionic uses, while 2D-TMDs in the in-plane direction is additionally appropriate for heat dissipation associated utilities. A typical schematic of TMD based TE device and their interdependent Seebeck coefficient (S) of the electrical conductivity (σ) and electronic (κel) and lattice (κlat) thermal conductivity to diverse charge carrier concentrations is illustrated in Fig. 8.42a, b [475].

TMD-based TE devices are also useful for the thermionic colors and power generators in solid-state [477, 478]. The solid-state thermionic power generators contains a semiconductive layer that is usually sandwiched among the metallic electrodes to provide a potential barricade for the carrier transportation, it also lowers the work function from some eVs to lesser than 1 eV [479]. Such TE devices work on principle a hot side electrons gains enough velocity while heated, thereby, they overwhelmed the energy barricade (material's work function) and travel toward the out-of-plane path. Ultimately, some electrons are composed at opposite side for the flowing in outer circuit [478–480]. This kind of current movement is normally denoted as the thermionic current. Usually, the effectiveness of a thermionic device is lesser than the effectiveness of a TE device having identical constraints. This is owing to the side impact and Kapitza resistances at the interface [481]. However, the major difficulty with such type devices is the backflow of the heat from the hot side to the cold side, cause of important thermal conductance of the semiconductor layer [480]. To shortout the problem a reduction in thermal conductivity up to a very lower value and a greater temperature decline should be recognized in the short distance range.

TMDs based heterostructure TE device's lower thermal conductivity in the out-of-plane direction originates due to weak vdWs bonds. They contribute significantly at a barrier; so, they are comfortably regulate through altering numeral of layers in 2D-TMDs. Using this property the barrier heights can be also optimized for thermionic transport. Moreover, analytical models are also used to investigate the performance of these devices by considering the perfect thermal conductive values necessity lower than 1 mW $m^{-2} \cdot K^{-1}$ [479]. Hence, the vdWs heterostructures meeting the requirements to designer for the higher performing solid-state thermionic power generators. Additionally, TMDs based high-performance TE devices are also fabricated by sandwiching the MoS_2 monolayer among twofold graphene single-layers. The out-of-plane ZT of graphene/MoS_2/graphene heterostructures has been achieved at around 2.8, that higher than monolayer MoS_2 (0.3) and monolayer graphene (negligible) [482]. Similarly, a solid-state thermionic power generator has also fabricated by sandwiching vdWs heterostructure between two graphene electrodes [483]. Such views heterostructures are composed of suitable multi-layer TMDs, like MoS_2, $MoSe_2$,

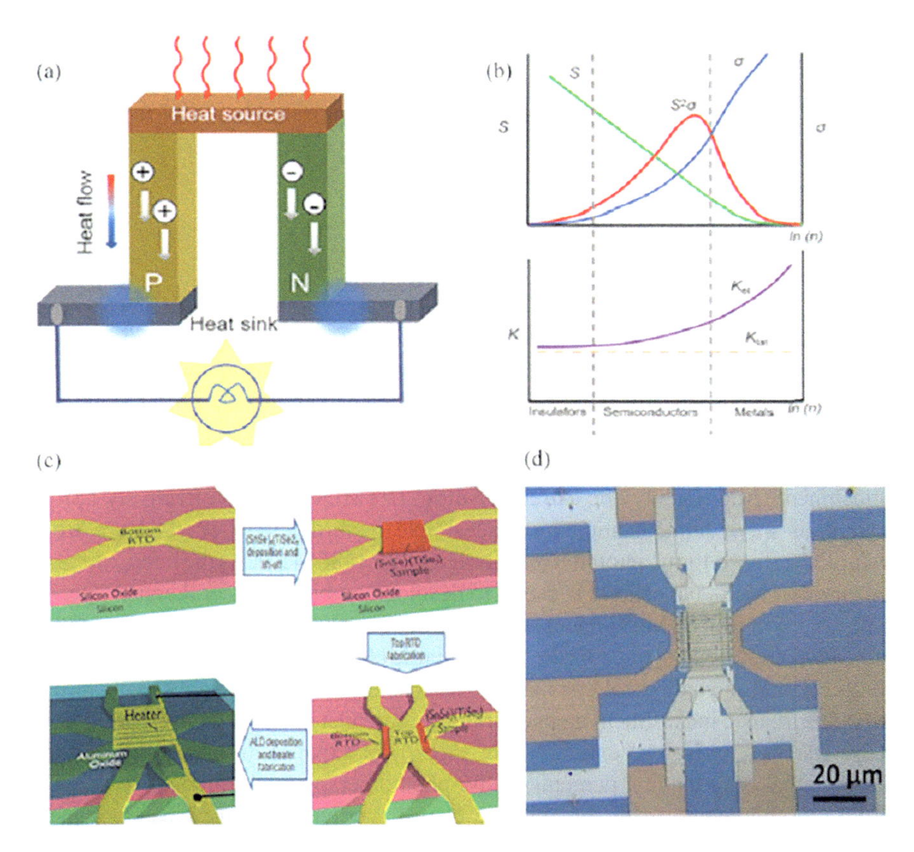

Fig. 8.42 **a** Schematic for the TE device base on the Seebeck effect; **b** Interdependent relation of Seebeck coefficient (S), electrical conductivity (σ), and the electronic (κ_{el}) and lattice (κ_{lat}) thermal conductivity for different charge carrier concentrations (a and b adopted from Zhou [475]); **c** Schematic diagrams for the device fabrication process; **d** Optical microscope image of an established device (c and d adopted Li et al. [476])

WS_2, and WSe_2. Such devices are also able to produce excess heat at 400 K with an effectiveness of approximately 7–8%.

The experimental and various theoretical interpretations have also emphasized the prospective of 2D TMDs for thermionic applicabilities with the prediction of large ZT values [479, 484]. Yet there is no successful experimental confirmation has been provided for the high TE working ability of the solid-state thermionic power generators with the proposed structures. Owing to difficulty in the measurements of thermal conductivity, Seebeck coefficient, and electrical conductivity of the monolayer and multiple layer 2D materials toward the out-of-plane direction. To shortout the technological challenges developed a specific platform [476], as a schematic diagram of devices is illustrated in Fig. 8.42c, d. Exhausting the device out-of-plane TE characteristics of $SnSe_2$ and $(SnSe)_n$ $(TiSe_2)_n$ (n = 1, 3, 4, 5) layered film it has been fabricated. The outcomes of this innovation has revealed the out-of-plane

Seebeck coefficient is reducing from −31 to -2.5 μV K^{-1}, whereas the out-of-plane effective thermal conductivity declining by the factor 2, in case of the n decreases from 5 to 1. Overall it enhances the interfacial phonon scattering [475, 485].

8.11.2 TMDs as Drug Delivery

2D TMDs are also useful in drug delivery, specifically, their carries reveals three basic functionalities; first, 2D TMDs compared with the liposomes and micelles to get stronger stability of the released and circumvent explosive drug release, the second one the wonderful extraordinary superficial expanse of 2D TMDs to provide a great quantity of anchor locations for upload molecules and third one, the superficial adornment of 2D TMDs that offers an easy physical adsorption or chemical attachment. Additionally, the extremely-thin 2D TMDs can also effectively upload numerous kind of medication molecules, like doxorubicin, 7-ethyl- 10-hydroxy camptothecin, chitosan, photodynamic reagent, etc. [486–492]. As the intense fictionalization of MoS_2 nanosheets is useful to develop combined cancer therapy. Specifically, the synthesized MoS_2 nanosheets have been modified from the lipoic acid PEG (LA-PEG) that is connected to the superficial MoS_2 nanosheets to expand biocompatibility and physiological constancy. The formed MoS_2-PEG nanosheet's strong NIR absorbance makes them an encouraging contender for photothermal treatment. This system's marvelous extraordinary superficial expanse to quantity proportion have a great drug loading percentages for chemotherapy medications, like doxorubicin, photodynamic mediator chlorine 6, and 7-ethyl-10-hydroxy camptothecin. The MoS_2-PEG doxorubicin uploaded has been also utilized for the chemotherapy and combined photothermal to in vitro cell culture tests and in vivo cancer treatment [486]. Moreover, MoS_2–PEG–CpG also provides a platform for immunotherapy. The synthesized MoS_2 nanosheet chemical exfoliation are also functionalized by cytosine–phosphate– guanine (CpG) and PEG to form the MoS_2–PEG–CpG nanoconjugates. Particularly MoS_2–PEG–CpG promotes the intracellular buildup of CpG and stimulates the manufacture of proinflammatory cytokines to elevate at the immune rejoinder level. Additionally, the co-cultured macrostage like cells, the MoS_2–PEG–CpG nanoconjugates are efficiently reduces the propagation action of cancer cells onto NIR radiation, this has considered a novel approach for cancer treatment [493].

2D TMDs based drug delivery is also use as an excellent gene delivery platform [494–496]. In which the polyethylenimine (PEI) and PEG are devoted to the superficial MoS_2 nanosheets through the disulfide bonds. Afterward, DNA interacts to the MoS_2–PEI– PEG complex nanocomposite through the electrostatic interaction and finally, it forms a compound together high constancy. Under the near-infrared light irradiation, the photothermally activated endosomal leakage is induced, as a consequence, the polymers are disconnected to the surface of MoS_2 nanosheets from the intracellular glutathione, thereby; the genes releases to the hybrid nanocomposite.

It is a successive procedure that considerably enhances the gene transport effectiveness with no severe cytotoxicity. Additionally, the MoS_2 nanocomposite has also providing a governable podium to supply genes for the cells [497].

Similarly, the amino termination of MoS_2-PEG-PEI nanosheets is assured to deleteriously charged siRNA, which works as a crucial controller to the cell cycle, like a well-established oncogene Polo-like kinase 1. The knockdown of Polo-like kinase 1 with siRNA are usually performed from the MoS_2-PEG-PEI nanosheets that interfere with the effectiveness, therefore, the transfection effects dignified concerning qPCR, western blot, and apoptosis assess. It reveals an innovative nanocarrier to MoS_2-PEG-PEI nanosheets with an extraordinary gene transportation capability and a better biocompatibility with decreased cytotoxicity [496]. The 2D TMDs also have a pronounced possibility to develop outstanding gene carriers, like graphene and other 2D nanomaterials. Since 2D TMDs have a sandwich construction in which the metal atoms lie as the intermediate layer in the organization together the surface chalcogen atoms. So, the medication molecules straightly uploaded through the interaction of the surface chalcogen atoms with a relatively simple reaction [98, 486]. As the typical schematic of drug delivery is illustrated in Fig. 8.43a–i. Thus compared to graphene, 2D TMDs are considered to be more suitable for drug loading [485, 498].

8.11.3 Use of TMDs in Photothermal Therapy

In photothermal therapy laser light is normally used to generate heat, thereby, the hyperthermia within tumor tissue is also heated due to the denaturation of proteins, disruption of the cell membrane, and irreversible damage to cancer cells [499]. The 2D TMDs' ability strong light absorption in near-infrared window areas makes them potential candidates to use as photothermal agents in photothermal therapy to remove cancer cells in vivo, as depicted in Fig. 8.44a-d [487, 493, 500–504]. As the figure extinction constant of extremely-thin MoS_2 nanosheets with a width nearby 1.54 nm around 29.2 L g^{-1} cm^{-1}, it is 7–8 times greater to the graphene. The use of MoS_2 concentration among 38 and 300 ppm gives us a rapid raise in solution temperature up to 40 due to the wave laser (800 nm) irradiation. Moreover, the in vitro human cervical cancer cell line it can also kill nearly entire cells afterward development together MoS_2 nanosheets, typically, 20 min underneath 800 nm near-infrared light. Likewise, 2D TMDs family like WS_2 nanosheets also has a high photothermal conversion coefficient. The figure extinction constant of WS_2-PEG nanosheets at 808 nm is up to 23.8 L g^{-1} cm^{-1}. The WS_2-PEG nanosheets can also inject into mice when irradiated from an 808 nm laser with an intensity of 0.8 W cm^{-2}. The tumor surface temperature of mice reached up to 65 within 5 min, thereby, the tumor is completely removed without any recurrence, which greatly improves the survival rate of mice [505]. Compared to the graphene figure extinction constant of MoS_2 nanosheets with a width of 1.54 nm, it is 7.8 folds for the near-infrared range, this leads the 2D TMDs

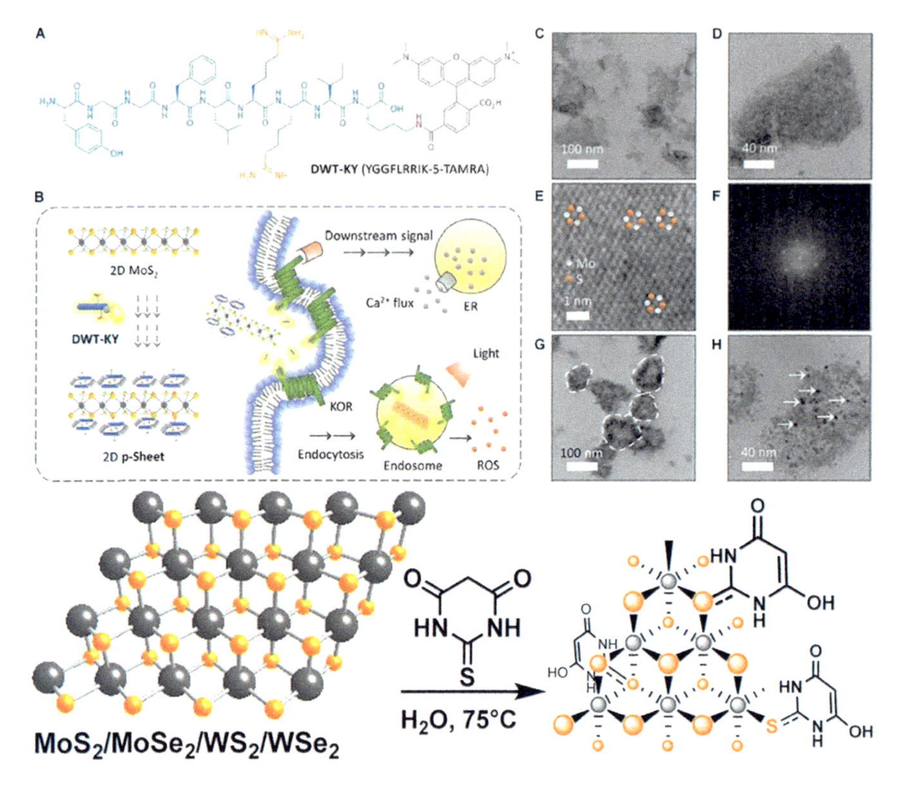

Fig. 8.43 Superficial alteration of 2D TMDs by the physical adsorption; **a** construction of the fluorescent agonist probe DWT-KY (YGGFLRRIK-5-TAMRA, where TAMRA is 5-carboxu etramethylrhodamine) for KOR binding; **b** diagram for the construction of a 2D p-Sheet among 2D MoS_2 and DWT-KY and the use of the material collaboration for the directed initiation of a KOR, that provides to (1) initiation of a downstream signaling direction to liberate Ca_2C flux from endoplasmic reticulum and (2) endocytosis of the material that released ROS intracellularly under the light irradiation; **c-e** high-resolution transmission electron microscopy (HRTEM) of 2D MoS_2; **(f)** fast Fourier transform pattern of a selected area from HRTEM of 2D MoS_2; **g, h** HRTEM picture of 2D p-Sheet (DWT-KY/2D MoS_2 = 1 mM/35 mg mL^{-1}); (i) superficial alteration of 2D TMDs via chemical attachment, typical production of thiobarbituric acid (TBA) altered MoS_2/$MoSe_2$/WS_2/WSe_2 together higher organic molecule treatment (adopted from Zhou et al. [346])

would be a superior selection to the graphene for the photothermal treatment [506, 507].

8.11.4 TMDs as Biomedical Imaging

Due to TMDs' exceptional chemical configuration, and the superior physical and chemical characteristics of the layered structures, they are efficiently utilized in

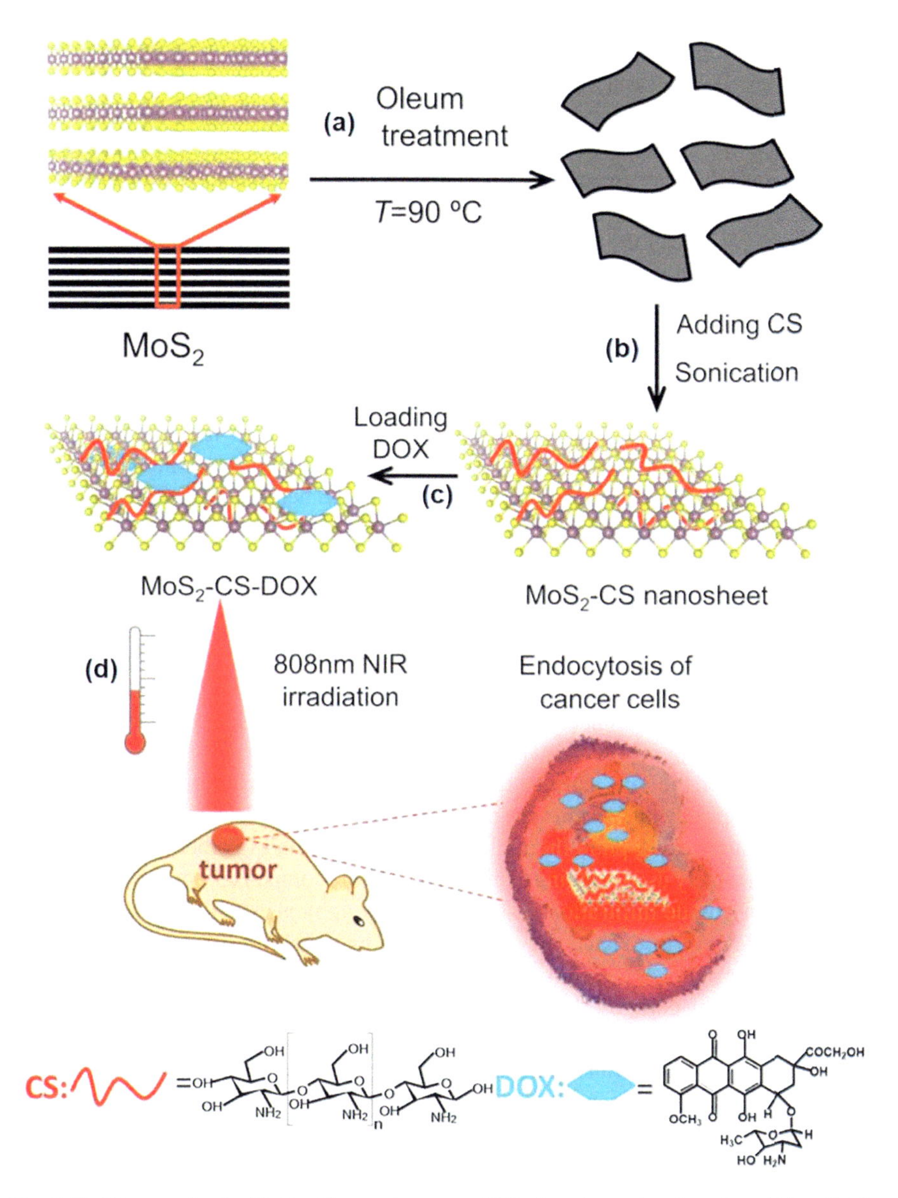

Fig. 8.44 Diagram for the MoS_2-CS nanosheets as a NIR photothermal activated medicine distribution organization for effective cancer treatment: **a, b** oleum treatment exfoliation procedure to create monolayer MoS_2 nanosheets and subsequent modification to CS; **c** DOX loading procedure; **d** NIR photothermal activated medicine distribution of the MoS_2 nanosheets to the tumor location (adopted from Zhou et al. [346])

numerous biological imaging. Specifically, 2D TMDs are used in the field of biological imaging that are further separated in three classes: fluorescence labeled imaging, photoacoustic imaging, and X-ray computed tomography (CT) imaging. If 2D TMDs are utilized for fluorescence-labeled imaging then it is compulsory to level 2D TMDs should be fluorescent molecules and the fluorescent tomography of cells or tissues is recognized by directing explicit cells or tissues together fluorescent-labeled nanosheets [508]. As the intense, the peptide ligand (TAMRA DN1K) links to the fluorophore assembly which is associated from MoS_2 nanosheets. Moreover, the nanosheets are recognized for the liver cancer cells or liver cancer tissues together extraordinary manifestation of CD47. Next in the same period the TAMRA DN1K over MoS_2 nanosheets are also coupled to the CD47 over the cell surface to comprehend the fluorescence tomography of the cancer cells or tissues together extraordinary manifestation of CD47 [509, 510]. Whereas, the photoacoustic imaging tomography (PAT) is the relatively novel biomedical tomography method. In which a material's light absorption capacity is produced a sound wave. The photoacoustic tomography principle is established on the photoacoustic impact of light captivators. Contrast to conversional in vivo optical tomography photoacoustic tomography expressively enhances the tomography depths and three-dimensional resolution owing to its robust absorbance in the near-infrared region [511]. As the intense, the intravenously inoculated among the WS_2-PEG nanosheets to make the mice attitude 4T1-tumors and visualized from the PAT tomography system associating a 700 nm laser as an excitation resource. So, it considerably enhances the PA signals in the region of the tumor by reflecting supplementary tumor accretion over the WS_2-PEG nanosheets [505]. Moreover, the TiS_2 nanosheets associating suitable surface modifications have been also efficiently utilized in vivo photoacoustic tomography with improved effectiveness of succeeding photothermal treatment [500]. On the other hand, X-ray CT imaging is a widely used imaging technology with the compensations of higher clarity and profound tissue infiltration. X-ray CT tomography is established on decreasing intensity of X-ray to expand the contrast imaging capability by the durable X-ray attenuation aptitude of the contrast negotiator [512]. Particularly manifold nanomaterials comprising components together a higher atomic number are considered to the good X-ray attenuators [513, 514]. Like the higher atomic numbers and robust X-ray attenuating aptitude of transition metal 2D TMDs offers outstanding contrast agent capability to X-ray CT tomography. So, the intravenous inoculation has modified the chitosan MoS_2 and WS_2-PEG nanosheets [487, 505, 515], thereby, producing an exceptional X-ray CT tomography aptitude in mice [516, 517]. Contrast to 2D TMDs, the X-ray attenuation capacity of carbon elements has lesser atomic numbers, which is feebler, so it is challenging to consider as a superior contrast negotiator for X-ray CT tomography [515].

8.12 Conclusions

The nonappearance of proximity effect is due to a slight superconducting coherence distance and the existence of the topological surface state with the diverse regularities at the interface. Thus Bi_2Se_3 combination with a variety of materials.

Overall in this work, one has discussed the possible applications of TMDs in distinct areas. Starting from the burring TMDs topological insulating topic for the high processing devices based on the two-dimensional, three-dimensional, and new generation topological insulating (TIs) materials. TMDs as field effect transistors (FETs) are one of the key device applications that can serve as superior to the usual transistor with novel properties like flexible top-gated FETs, logic inverters, large-scale circuits, and analog and RF devices. TMDs are most extensively used as light-emitting diodes for cutting-edge high-processing optoelectronics devices, therefore, one has discussed in detail by describing the light-emitting diodes (LEDs), LEDs based on TMDs in vertical stacking, single photon emitter, organic light-emitting diodes (OLEDs), quantum dots (QDs) based light-emitting diodes. TMDs are also offered lasing properties through the population invasion of charges due to their unique structural property therefore a brief discussion has been also devoted to TMDs based lasers. Moreover, TMDs based devices are used as kinds of photodetectors, so a brief description is also devoted to photo-detecting devices. TMDs are also considered to be the potential candidate for photovoltaic device applications that can provide high-efficiency renewal energy throughput at a cheaper cost. The TMDs mainly categorized into two categories based on their physical properties namely traditional and transparent TMDs based solar cells that have been also discussed in this work. Moreover, TMDs relatively novel properties that can predominately boost renewable energy-based research as the solar water splitting by mechanism of the electrocatalytic and photocatalytic types of water splitting, so, water splitting-based TMDs devices have been also discussed by providing descriptions of the distinct types of potential solar cells devices. The layered TMDs are also become hot materials due to their biocompatibility and applications in medical science/ technology as a variety of biosensors. So, various kinds of TMDs based biosensors' working functionality and their specific uses have been also accommodated, like fluorophore-labeled, label-free, absorption, electrochemical, and FET biosensors. Moreover, TMDs' adequate structural properties are also offered to integrate the TMD- TMD-based heterostructured devices by the layer stacking of either the same or distinct TMD materials. Such heterostructures-based devices are offered improved device performance than pure TMD-based devices, therefore, a discussion on this kind of device has been also provided. In this order graphene/TMD heterostructures have been also identified as the potential heterostructures to fabricate different kinds of high-processing devices, therefore, it is worth discussing these kinds of devices. Additionally, considering other significant utilities of the TMDs in the diverse area a miscellaneous section has been p also provided that is devoted to some important applications like thermo-electric devices, TMDs for drug delivery, use of TMDs in photothermal therapy, and biomedical imaging. Hence TMDs based devices can be also applied in various areas and are not limited to the above discuss few significant applications but many more to provide high-end performances with improved efficiencies. Thus TMDs materials are the burring materials that can have applications in diverse areas to fulfill the current/next-era high-processing technological demands for humans.

References

1. Folkers LC, Corredor LT, Lukas F, Sahoo M et al (2022) Occupancy disorder in the magnetic topological insulator candidate $Mn_{1-x}Sb_{2+x}Te_4$. Z Kristallogr 237:101–108
2. Institute of Electrical and Electronics Engineers (IEEE) (2018) The InternationalRoadmap for Devices and Systems (IRDS)
3. Gilbert MJ (2021) Topological electronics. Commu Phys 4:70
4. Lifshitz EM, Pitaevskii LP (1980) Statistical Physics. Pergamon Press
5. Nussinov Z, Ortiz GA (2009) A symmetry principle for topological order. Ann Phys 324:977–1057
6. Klitzing KV, Dorda G, Pepper M (1980) New method for high-accuracy determination of the fine-structure constant based on quantized Hall conductance. Phys Rev Lett 45:494
7. Thouless DJ, Kohmoto M, Nightingale MP, den Nijs M (1982) Quantized Hall conductance in a two-dimensional periodic potential. Phys Rev Lett 49:405
8. Stormer HL, Tsui DC, Gossard AC (1999) The fractional quantum Hall Effect. Rev Mod Phys 71:S298
9. Wen XG (2019) Choreographed entanglement dances: topological states of quantum matter. Science 363:eaal3099
10. Kane CL, Mele EJ (2005) Quantum spin Hall effect in grapheme. Phys Rev Lett 95:1757
11. Hasan MZ, Kane CL (2010) Colloquium: topological insulators. Rev Mod Phys 82:3045
12. Qi XL, Zhang SC (2011) Topological insulators and superconductors. Rev Mod Phys 83:1057
13. Hughes TL, Bernevig BA (2013) Topological Insulators and Topological Superconductors. Princeton University Press
14. Lowrance A (2019) Topological insulators and applications to quantum computing. JYI (Science news) 37
15. Beidenkopf H (2011) Spatial fluctuations of helical Dirac fermions on the surface of topological insulators. Nat Phys 7:939–943
16. Bonderson P, Nayak C, Qi XL (2013) A time-reversal invariant topological phase at the surface of a 3D topological insulator. J Stat Mech Theory Exp 2013:387–402
17. Tian W, Yu W, Shi J, Wang Y (2017) The property, preparation, and application of topological insulators: a Review. Materials 10:814
18. Zhang SC, Bernevig BA, Hughes T (2007) Quantum spin Hall effect and topological phase transition in HgTe quantum wells. Science 314:1757–1761
19. Büttner B, Liu CX, Tkachov G, Novik EG et al (2010) Single valley Dirac fermions in zero-gap HgTe quantum wells. Nat Phys 7:418–422
20. Lin YM, Rabin O, Cronin SB, Ying JY et al (2002) Semimetal-semiconductor transition in $Bi_{1-x}Sb_x$ alloy nanowires and their thermoelectric properties. Appl Phys Lett 81:2403–2405
21. Zhang H, Liu CX, Qi XL, Dai X et al (2009) Topological insulators in Bi_2Se_3, Bi_2Te_3, and Sb_2Te_3 with a single Dirac cone on the surface. Nat Phys 5:438–442
22. Zhang W, Yu R, Zhang HJ, Dai X et al (2010) First-principles studies of the three-dimensional strong topological insulators Bi_2Te_3, Bi_2Se_3 and Sb_2Te_3. New J Phys 12:065013
23. Augustine S, Mathai E (2001) Growth, morphology, and microindentation analysis of Bi_2Se_3, $Bi_{1.8}In_{0.2}Se_3$, and $Bi_2Se_{2.8}Te_{0.2}$ single crystals. Mater Res Bull 36:2251–2261
24. Xia Y, Qian D, Hsieh D, Wray L et al (2009) Observation of a large-gap topological-insulator class with a single Dirac cone on the surface. Nat Phys 5:398–402

25. Hsieh D, Xia Y, Qian D, Wray L et al (2009) Observation of time-reversal-protected single-Dirac-cone topological-insulator states in Bi_2Te_3 and Sb_2Te_3. Phys Rev Lett 103:146401
26. Liu J, Hsieh TH, Wei P, Duan W et al (2013) Spin-filtered edge states with an electrically tunable gap in a two-dimensional topological crystallin insulator. Nat Mater 13:178–183
27. Zhang H, Man B, Zhang Q (2017) Topological crystalline insulator SnTe/Si vertical heterostructure photodetectors for high-performance near-infrared detection. ACS Appl Mater Interface 9:14067–14077
28. Albright SD, Zou K, Walker FJ, Ahn CH (2021) Weak antilocalization in topological crystalline insulator SnTe films deposited using amorphous seeding on $SrTiO_3$. APL Materials 9:111106
29. Weng HM, Dai X, Fang Z (2014) Transition-metal pentatelluride ZrTe5 and HfTe5: A paradigm for large-gap quantum spin Hall insulators. Phys Rev X 4:339–345
30. Zhang Y, Wang CL, Yu L, Liu GD et al (2017) Electronic evidence of temperature-induced Lifshitz transition and topological nature in ZrTe5. Nat Commun 8:15512
31. Yuan X, Zhang C, Liu YW, Narayan A et al (2016) Observation of quasi-two-dimensional Dirac fermions in ZrTe5. NPG Asia Mater 8:e325
32. Wang J, Niu J, Yan B, Li X et al (2018) Vanishing quantum oscillations in Dirac semimetal ZrTe5. Proc Natl Acad Sci U S A 115:9145–9150
33. Li Q, Kharzeev DE, Zhang C, Huang Y (2016) Chiral magnetic effect in ZrTe5. Nat Phys 12:550–554
34. Ge J, Ma D, Liu Y, Wang H et al (2020) Unconventional Hall effect induced by Berry curvature. National Sci Rev 7:1879–1885
35. Qiu G, Du YC, Charnas A, Zhou H et al (2016) Observation of optical and electrical in-plane anisotropy in high-mobility few-layer ZrTe5. Nano Lett 16:7364–7369
36. Zhou YH, Wu JF, Ning W, Li NN et al (2016) "Pressure-induced superconductivity in a three-dimensional topological material ZrTe5. Proc Natl Acad Sci USA 113:2904–2909
37. Li Y, An C, Hua C, Chen X et al (2018) Pressure-induced superconductivity in topological semimetal $NbAs_2$. npj Quant Mater 3:58
38. Wang HC, Li CK, Liu HW, Yan JQ et al (2016) Chiral anomaly and ultrahigh mobility in crystalline HfTe5. Phys Rev B 93:165127
39. Burkov AA, Balents L (2011) Weyl Semimetal in a Topological Insulator Multilayer. Phys Rev Lett 107:127205
40. Halasz GB, Balents L (2012) Time-reversal invariant realization of the Weyl semimetal phase. Phys Rev B 85:035103
41. Soluyanov AA, Gresch D, Wang ZJ, Wu QS et al (2015) Type-IIWeyl semimetals. Nature 527:495–498
42. Fivaz R, Mooser E (1967) Mobility of charge carriers in semiconducting layer structures. Phys Rev 163:743
43. Grant A, Griffiths T, Pitt G, Yoffe A (1975) The electrical properties and the magnitude of the indirect gap in the semiconducting transition metal dichalcogenide layer crystals. J Phys C 8:L17
44. Lin J, Li H, Zhang H, Chen W (2013) Plasmonic enhancement of photocurrent in MoS_2 fieldeffect- transistor. Appl Phys Lett 102:203109
45. Carladous A, Coratger R, Ajustron F, Seine G et al (2002) Light emission from spectral analysis of Au/MoS_2 nanocontacts stimulated by scanning tunneling microscopy. Phys Rev B 66(4):045401
46. Jariwala D, Sangwan VK, Lauhon LJ, Marks TJ et al (2014) Emerging device applications for semiconducting two-dimensional transition metal dichalcogenides. ACS Nano 8:1102–1120
47. Chuang HJ, Tan X, Ghimire NJ, Perera MM et al (2014) High mobility WSe2 p-and n-type field-effect transistors contacted by highly doped graphene for low-resistance contacts. Nano Lett 14(6):3594
48. Radisavljevic B, Radenovic A, Brivio J, Giacometti V et al (2011) Single-layer MoS_2 transistors. Nat Nanotech 6:147–150

49. Wu W, De D, Chang SC, Wang Y et al (2013) High mobility and high on/off ratio field-effect transistors based on chemical vapor deposited single-crystal MoS_2 grains. Appl Phys Lett 102:142106
50. Sundaram R, Engel M, Lombardo A, Krupke R et al (2013) Electroluminescence in single layer MoS_2. Nano Lett 13:1416–1421
51. Withers F, Del Pozo-Zamudio O, Mishchenko A, Rooney A et al (2015) Light-emitting diodes by band-structure engineering in van der Waals heterostructures. Nat Mater 14:301–306
52. Ye Y, Ye Z, Gharghi M, Zhu H et al (2014) Exciton-dominant electroluminescence from a diode of monolayer MoS_2. Appl Phys Lett 104:193508
53. Lembke D, Kis A (2012) Breakdown of high-performance monolayer MoS_2 Transistors. ACS Nano 6:10070–10075
54. Santosh et al (2017) Computational study of MoS2/HfO2 defective interfaces for nanometer-scale electronics. ACS Omega 2:2827–2834
55. Baugher BW, Churchill HO, Yang Y, Jarillo-Herrero P (2013) Intrinsic electronic transport properties of high-quality monolayer and bilayer MoS_2. Nano Lett 13:4212–4216
56. Radisavljevic B, Kis A (2013) Mobility engineering and a metal-insulator transition in monolayer MoS_2. Nat Mater 12:815–820
57. Bao W, Cai X, Kim D, Sridhara K et al (2013) High mobility ambipolar MoS_2 field-effect transistors: substrate and dielectric effects. Appl Phys Lett 102:042104
58. Kim SK, Xuan Y, Ye PD, Mohammadi S et al (2007) Atomic layer deposited Al_2O_3 for gate dielectric and passivation layer of single-walled carbon nanotube transistors. Appl Phys Lett 90:163108
59. Jariwala D, Sangwan VK, Late DJ, Johns JE et al (2013) Band-Like transport in high mobility unencapsulated single-layer MoS_2 transistors. Appl Phys Lett 102:173107
60. Ghatak S, Pal AN, Ghosh A (2011) Nature of electronic states in atomically thin MoS_2 field-effect transistors. ACS Nano 5:7707–7712
61. Buscema M, Barkelid M, Zwiller V, van der Zant HS et al (2013) Large and tunable photothermoelectric effect in single-layer MoS_2. Nano Lett 13:358–363
62. Das S, Chen HY, Penumatcha AV, Appenzeller J (2012) High performance multilayer MoS_2 transistors with scandium contacts. Nano Lett 13:100–105
63. Popov I, Seifert G, Tománek D (2012) Designing electrical contacts to MoS_2 monolayers: a computational study. Phys Rev Lett 108:156802
64. Liu W, Kang J, Sarkar D, Khatami Y et al (2013) Role of metal contacts in designing high-performance monolayer n-type WSe_2 field effect transistors. Nano Lett 2013(13):1983–1990
65. Liu H, Si M, Deng Y, Neal AT et al (2014) Switching mechanism in single-layer molybdenum disulfide transistors: an insight into current flow across schottky barriers. ACS Nano 8:1031–1038
66. Das S, Sebastian A, Pop E, McClellan CJ et al (2021) Transistors based on two-dimensional materials for future integrated circuits. Nat Elect 4:786–799
67. Choi W, Choudhary N, Han GH, Park J et al (2017) Recent development of two-dimensional transition metal dichalcogenides and their applications. Mater Today 20:116–130
68. Chen J-R, Odenthal PM, Swartz A, Floyd GC et al (2013) Control of schottky barriers in single layer MoS_2 transistors with ferromagnetic contacts. Nano Lett 13:3106–3110
69. Dankert A, Langouche L, Kamalakar MV, Dash SP (2014) High-performance molybdenum disulfide field-effect transistors with spin tunnel contacts. ACS Nano 8:476–482
70. Fang H, Chuang S, Chang TC, Takei K, Takahashi T (2012) Javey, a high-performance single layered WSe_2 p-FETs with chemically doped contacts. Nano Lett 12:3788–3792
71. Fang H, Tosun M, Seol G, Chang TC et al (2013) Degenerate n-doping of few-layer transition metal dichalcogenides by potassium. Nano Lett 13:1991–1995
72. Du Y, Liu H, Neal AT, Si M et al (2013) Molecular doping of multilayer MoS_2 field-effect transistors: reduction in sheet and contact resistances. IEEE Electron Dev Lett 34:1328–1330
73. Pudasaini PR, Oyedele A, Zhang C, Stanford MG et al (2018) High performance multilayer WSe_2 field effect transistors with carrier type control. Nano Res 11:722–730

74. Liu H, Si M, Najmaei S, Neal AT et al (2013) Statistical study of deep submicron dual-gated field-effect transistors on monolayer chemical vapor deposition molybdenum disulfide films. Nano Lett 13:2640–2646
75. Liu H, Neal AT, Ye PD (2012) Channel length scaling of MoS_2 MOSFETs. ACS Nano 6:8563–8569
76. Yoon Y, Ganapathi K, Salahuddin S (2011) How good can monolayer MoS_2 transistors be? Nano Lett 11:3768–3773
77. Liu L, Lu Y, Guo J (2013) On monolayer MoS_2 field-effect transistors at the scaling limit IEEE Trans. Electron Dev 60:4133–4139
78. Rawat B, Vinaya MM, Paily R (2019) Transition metal dichalcogenide-based field-effect transistors for analog/mixed- signal applications. IEEE Trans Elect Devic. https://doi.org/10.1109/TED.2019.2906235
79. Baugher B, Churchill HO, Yang Y, Jarillo-Herrero P (2013) Intrinsic electronic transport properties of high quality monolayer and bilayer MoS_2. Nano Lett 13:4212–4216
80. Yuan H, Shimotani H, Tsukazaki A, Ohtomo A et al (2009) High-density carrier accumulation in ZnO field-effect transistors gated by electric double layers of ionic liquids. Adv Funct Mater 19:1046–1053
81. Yamada Y, Ueno K, Fukumura T, Yuan HT et al (2011) Electrically induced ferromagnetism at room temperature in cobalt-doped titanium dioxide. Science 332:1065–1067
82. Ye JT, Inoue S, Kobayashi K, Kasahara Y et al (2010) Liquid-gated interface superconductivity on an atomically flat film. Nat Mater 9:125–128
83. Xia BY, Cho JH, Lee J, Ruden PP et al (2009) Comparison of the mobility-carrier density relation in polymer and single-crystal organic transistors employing vacuum and liquid gate dielectrics. Adv Mater 21:2174–2179
84. Kim SH, Hong K, Xie W, Lee KH et al (2013) Electrolyte-gated transistors for organic and printed electronics. Adv Mater 25:1822–1846
85. Leger J, Berggren M, Carter S, Berggren M et al (2016) Iontronics: ionic carriers in organic electronic materials and devices CRC Press: Boca Raton. FL, USA
86. You A, Be MAY, In I (2007) Electrochemical carbon nanotube field-effect transistor. Appl Phys Lett 78:1291
87. Lieb J, Demontis V, Prete D, Ercolani D et al (2019) Ionic-liquid gating of inas nanowire-based field-effect transistors. Adv Funct Mater 29:1804378
88. Demontis V, Zannier V, Sorba L, Rossella F (2021) Surface nano-patterning for the bottom-up growth of III-V semiconductor nanowire ordered arrays. Nanomaterials 11
89. Ullah AR, Carrad DJ, Krogstrup P, Nygård J et al (2018) Near-thermal limit gating in heavily doped III-V semiconductor nanowires using polymer electrolytes. Phys Rev Mater 2:25601
90. Carrad DJ, Mostert AB, Ullah AR, Burke AM et al (2017) Hybrid nanowire ion-to-electron transducers for integrated bioelectronic circuitry. Nano Lett 17:827–833
91. Ueno K, Nakamura S, Shimotani H, Ohtomo A et al (2008) Electric-field induced superconductivity in an insulator. Nat Mater 7:855–858
92. Ueno K, Nakamura S, Shimotani H, Yuan HT et al (2011) Discovery of superconductivity in $KTaO_3$ by electrostatic carrier doping nat. Nanotechnol 6:408–412
93. Vaquero D, Clericò V, Salvador-Sánchez J, Quereda J (2021) Ionic-liquid gating in two-dimensional TMDs: the operation principles and spectroscopic capabilities. Micromachines 12:1576
94. Ionescu AM, Riel H (2011) Tunnel field-effect transistors as energy-efficient electronic switches. Nature 479:329–337
95. Jiang XW, Luo JW, Li SS, Wang LW (2015) How good is mono-layer transition-metal dichalcogenide tunnel field-effect transistors in sub-10 nm?—An ab initio simulation study. in IEDM Tech Dig 12.4.1–12.4.4
96. Ilatikhameneh H, Tan Y, Novakovic B, Klimeck G et al (2015) Tunnel field-effect transistors in 2-D transition metal dichalcogenide materials. IEEE J Explor Solid-State Comput Devices Circ 1:12–18

97. Cao W, Jiang Kang J, Sarkar D et al (2015) Designing band-to-band tunneling field-effect transistors with 2D semiconductors for next-generation low-power VLSI. IEDM Tech Dig 12–13
98. Chhowalla M, Shin HS, Eda G, Li LJ et al (2013) The chemistry of two-dimensional layered transition metal dichalcogenide nanosheets. Nature Chem 5:263–275
99. Ilatikhameneh H, Ameen TA, Klimeck G, Appenzeller J et al (2015) Dielectric engineered tunnel field-effect transistor. IEEE Electron Device Lett 36:1097–1100
100. Szabó Á, Koester SJ, Luisier M (2015) Ab-Initio simulation of van der Waals $MoTe_2$–SnS_2 heterotunneling FETs for low-power electronics. IEEE Electron Device Lett 36:514–516
101. Das S, Prakash A, Salazar R, Appenzeller J (2014) Toward lowpower electronics: tunneling phenomena in transition metal dichalcogenides. ACS Nano 8:1681–1689
102. Lu AKA, Houssa M, Radu IP, Pourtois G Toward an understanding of the electric field-induced electrostatic doping in van der Waals heterostructures: a first-principles study. ACS Appl Mater Interfaces 9:7725–7734
103. Ilatikhameneh H, Ameen TA, Chen CY, Klimeck G (2018) Sensitivity challenge of steep transistors. IEEE Trans Electron Devices 65:1633–1639
104. Anand S, Amin SI, Sarin RK (2016) Performance analysis of charge plasma based dual electrode tunnel FET. J Semicond 37:2016
105. Dubey PK, Kaushik BK (2019) A charge plasma-based monolayer transition metal dichalcogenide tunnel FET" IEEE Trans Elect Devices https://doi.org/10.1109/TED.2019.2909182
106. Cho et al (2019) Recent advances in interface engineering of transition-metal dichalcogenides with organic molecules and polymers. ACS Nano 13:9713–9734
107. He J, Nuzzo RG, Rogers JA (2015) Inorganic materials and assembly techniques for flexible and stretchable electronics. Proc IEEE 103:619–632
108. Akinwande D, Petrone N, Hone J (2014) Two-dimensional flexible nanoelectronics. Nat Commun 5:5678
109. Myny K (2018) The development of flexible integrated circuits based on thin-film transistors. Nat Electron 1:30–39
110. Chakraborty SK, Kundu B, Nayak B, Dash SP et al (2022) Challenges and opportunities in 2D heterostructures for electronic and optoelectronic devices. Science 25:103942
111. Salvatore GA, Münzenrieder N, Barraud C, Petti L et al (2013) Fabrication and transfer of flexible few-layers MoS_2 thin film transistors to any arbitrary substrate. ACS Nano 7:8809–8815
112. Gurarslan A, Yu Y, Su L, Yu Y et al (2014) Surface-energy-assisted perfect transfer of centimeter-scale monolayer and few-layer MoS_2 films onto arbitrary substrates. ACS Nano 8:11522–11528
113. Nourbakhsh A, Zubair A, Sajjad RN, Tavakkoli KG et al (2016) MoS_2 field-effect transistor with sub-10 nm channel length. Nano Lett 16:7798–7806
114. Gusakova J, Wang X, Shiau LL, Krivosheeva A et al (2017) "Electronic properties of bulk and monolayer TMDs: theoretical study within DFT framework (GVJ-2e method). Phys Status Solidi (a) 214:1700218
115. Ryou J, Kim YS, Santosh K, Cho K (2016) Monolayer MoS_2 bandgap modulation by dielectric environments and tunable bandgap transistors. Sci Rep 6:29184
116. Kshirsagar CU, Xu W, Su Y, Robbins MC et al (2016) Dynamic memory cells using MoS_2 field-effect transistors demonstrating femtoampere leakage currents. ACS Nano 10:8457–8464
117. Kang S, Eshete YA, Lee S, Won D et al (2022) Bandgap modulation in the two-dimensional coreshell-structured monolayers of WS_2. iScience 25:103563
118. Smithe KK, English CD, Suryavanshi SV, Pop E (2017) Intrinsic electrical transport and performance projections of synthetic monolayer MoS_2 devices. 2D Mater. 4:011009.
119. Smithe KK, Suryavanshi SV, Muñoz Rojo M, Tedjarati AD et al (2017) Low variability in synthetic monolayer MoS_2 devices. ACS Nano 11:8456–8463
120. Smithe KK, Krayev AV, Bailey CS, Lee HR et al (2018) Nanoscale heterogeneities in monolayer $MoSe_2$ revealed by correlated scanning probe microscopy and tip-enhanced Raman spectroscopy. ACS Appl Nano Mater 1:572–579

121. Chen J, Bailey CS, Hong Y, Wang L et al (2019) Plasmon-resonant enhancement of photocatalysis on monolayer WSe_2. ACS Photonics 6:787–792
122. Daus A, Vaziri S, Chen V, Koroglu C et al (2021) High-performance flexible nanoscale transistors based on transition metal dichalcogenides. Nat Elect 4:495–501
123. Daus et al (2020) High-performance flexible nanoscale field-effect transistors based on transition metal dichalcogenides. arXiv: 2009.04056
124. Smithe KK, English CD, Suryavanshi SV, Pop E (2018) High-field transport and velocity saturation in synthetic monolayer MoS_2. Nano Lett 18:4516–4522
125. Chang HY, Yogeesh MN, Ghosh R, Rai A et al (2016) Large-area monolayer MoS_2 for flexible low-power RF nanoelectronics in the GHz regime. Adv Mater 28:1818–1823
126. Cheng R, Jiang S, Chen Y, Liu Y et al (2014) Few-layer molybdenum disulfide transistors and circuits for high-speed flexible electronics. Nat Commun 5:5143
127. McClellan CJ, Suryavanshi SV, English CD, Smithe KKH et al (2020) 2D Device Trends http://2d.stanford.edu/2D_Trends.html
128. Lan Y, Xu Y, Wu Y, Cao Z et al (2018) Flexible graphene field-effect transistors with extrinsic fmax of 28 GHz. IEEE Electron Device Lett 39:1944–1947
129. Shahrjerdi D, Bedell SW, Khakifirooz A, Fogel K et al (2012) Advanced flexible CMOS integrated circuits on plastic enabled by controlled spalling technology. In Int Electron Devices Meeting (IEDM) 5.1. 1–5.1. 4 (IEEE, 2012) https://doi.org/10.1109/IEDM.2012.6478981
130. Radisavljevic B, Whitwick MB, Kis A (2011) Integrated circuits and logic operations based on single-layer MoS_2. ACS Nano 5:9934–9938
131. Song HS, Li SL, Gao L, Xu Y et al (2013) High-performance top-gated monolayer SnS_2 field-effect transistors and their integrated logic circuits. Nanoscale 5:9666–9670
132. Wang H, Yu L, Lee YH, Shi Y et al (2012) Integrated Circuits Based on Bilayer MoS_2 Transistors. Nano Lett 12:4674–4680
133. Huang JK, Pu J, Hsu CL, Chiu MH et al (2014) Large-area synthesis of highly crystalline WSe_2 monolayers and device applications. ACS Nano 8:923–930
134. Amani M, Burke RA, Proie RM, Dubey M (2015) Flexible integrated circuits and multifunctional electronics based on single atomic layers of MoS_2 and grapheme. Nanotechnology 26:115202
135. Xu H, Zhang HM, Guo ZX, Shan Y et al (2018) High-performance wafer-scale MoS_2 transistors toward practical application. Small 14:1803465
136. Zhang TB, Liu H, Wang Y, Liu J et al (2019) Fast-response inverter arrays built on wafer-scale MoS_2 by atomic layer deposition. Phys Status Solidi RRL 13:1900018
137. Tang et al (2019) Recent progress in devices and circuits based on wafer-scale transition metal dichalcogenides. Sci China Inf Sci 62:220401
138. Dathbun A, Kim Y, Kim S, Yoo Y et al (2017) Large-area CVD-grown sub-2 V ReS2 transistors and logic gates. Nano Lett 17:2999–3005
139. Yu LL, Lee YH, Ling X, Santos EJG et al (2014) Graphene/MoS_2 hybrid technology for large-scale two-dimensional electronics. Nano Lett 14:3055–3063
140. Zhang SM, Xu H, Liao FY, Sun Y et al (2019) Wafer-scale transferred multilayer MoS_2 for high performance field effect transistors. Nanotechnology 30:174002
141. Das T, Chen X, Jang H, Kwon Oh Il et al (2016) Highly flexible hybrid CMOS inverter based on Si nanomembrane and molybdenum disulfide Small 12:5720–5727
142. Chiu MH, Tang HL, Tseng CC, Han Y et al (2019) Metal-guided selective growth of 2D materials: demonstration of a bottom-up CMOS inverter. Adv Mater 31:1900861
143. Lan YW, Chen PC, Lin YY, Li MY et al (2019) Scalable fabrication of a complementary logic inverter based on MoS_2 fin-shaped field effect transistors. Nanoscale Horiz 4:683–688
144. Xu H, Zhang HM, Liu YW, Zhang S et al (2019) Controlled doping of wafer-scale PtSe2 films for device application. Adv Funct Mater 29:1805614
145. Tosun M, Chuang S, Fang H, Sachid AB et al (2014) High-gain inverters based on WSe_2 complementary field-effect transistors. ACS Nano 2014(8):4948–4953
146. Tang H, Zhang H, Chen X, Wang Y et al (2019) Recent progress in devices and circuits based on wafer-scale transition metal dichalcogenides. Sci China Inf Sci 62:220401–220419

147. Zou XM, Wang JL, Chiu CH, Wu Y et al (2014) Interface engineering for high-performance top-gated MoS_2 field-effect transistors. Adv Mater 26:6255–6261
148. Wachter S, Polyushkin DK, Bethge O, Mueller T et al (2017) A microprocessor based on a two-dimensional semiconductor. Nat Commun 8:14948
149. Lin ZY, Liu Y, Halim U, Ding M et al (2018) Solution-processable 2D semiconductors for high-performance large-area electronics. Nature 562:254–258
150. Yu L, El-Damak D, Ha S, Ling X et al (2015) Enhancement-mode single-layer CVD MoS_2 FET technology for digital electronics. IEEE Int Electron Dev Meeting (IEDM) 32.3.1–32.3.4. https://doi.org/10.1109/IEDM.2015.7409814
151. Agarwal A, Lang J (2005) Foundations of analog and digital electronic circuits. Elsevier, Amsterdam
152. Cheng R, Bai JW, Liao L, Zhou H et al (2012) High-frequency self-aligned graphene transistors with transferred gate stacks. Proc Natl Acad Sci USA 109:11588–11592
153. Schwierz F (2013) Graphene transistors: status, prospects, and problems. Proc IEEE 101:1567–1584
154. Sanne A, Ghosh R, Rai A, Yogeesh MN et al (2015) Radio frequency transistors and circuits based on CVD MoS_2. Nano Lett 15:5039–5045
155. Gao QG, Zhang ZF, Xu XL, Song J et al (2018) Scalable high performance radio frequency electronics based on large domain bilayer MoS_2. Nat Commun, 9:4778
156. Zhu Y, Sun X, Tang Y, Fu L et al (2021) Two-dimensional materials for light emitting applications: achievement, challenge and future perspectives. Nano Res 14:1912–1936
157. Miller RC, Kleinman DA, Tsang WT, Gossard AC (1981) Observation of the excited level of excitons in GaAs quantum wells. Phys Rev B 24:1134–1136
158. Miller RC, Kleinman DA (1985) Excitons in GaAs quantum wells. J Lumin 30:520–540
159. Chernikov A, Berkelbach TC, Hill HM, Rigosi A et al (2014) Exciton binding energy and nonhydrogenic Rydberg series in monolayer WS_2. Phys Rev Lett 113:076802
160. Cheiwchanchamnangij T, Lambrecht WRL (2012) Quasiparticle band structure calculation of monolayer, bilayer, and bulk MoS_2. Phys Rev B 85:205302
161. Ramasubramaniam A (2012) Large excitonic effects in monolayers of molybdenum and tungsten dichalcogenides. Phys Rev B 86:115409
162. Park S, Mutz N, Schultz T, Blumstengel S et al (2018) Direct determination of monolayer MoS_2 and WSe_2 exciton binding energies on insulating and metallic substrates 2D Mater. 5:025003
163. Zhu BR, Chen X, Cui XD (2015) Exciton binding energy of monolayer WS2. Sci Rep 5:9218
164. He KL, Kumar N, Zhao L, Wang ZF et al (2014) Tightly bound excitons in monolayer WSe_2. Phys Rev Lett 113:026803
165. Ugeda MM, Bradley AJ, Shi SF, da Jornada FH et al (2014) Giant bandgap renormalization and excitonic effects in a monolayer transition metal dichalcogenide semiconductor. Nat Mater 13:1091–1095
166. Kübler JK (1969) The exciton binding energy of III–V semiconductor compounds. Phys Status Solidi B 35:189–195
167. Mouri S, Miyauchi Y, Matsuda K (2013) Tunable photoluminescence of monolayer MoS_2 via chemical doping. Nano Lett 13:5944–5948
168. Ross JS, Wu SF, Yu HY, Ghimire NJ et al (2013) Electrical control of neutral and charged excitons in a monolayer semiconductor. Nat Commun 4:1474
169. Hao K, Specht JF, Nagler P, Xu LX et al (2017) Neutral and charged inter-valley biexcitons in monolayer $MoSe_2$. Nat Commun 8:15552
170. Pospischil A, Mueller T (2016) Optoelectronic devices based on atomically thin transition metal dichalcogenides. Appl Sci 6:78
171. Wang C, Yang F, Gao Y (2020) The highly-efficient light-emitting diodes based on transition metal dichalcogenides: from architecture to performance. Nanoscale Adv 2:4323
172. Thakar K, Lodha S (2020) Optoelectronic and photonic devices based on transition metal dichalcogenides. Mater Res Express 7:014002

173. Pospischil A, Furchi MM, Mueller T (2014) Solar-energy conversion and light emission in an atomic monolayer p–n diode. Nat Nanotechnol 9:257–261
174. Ross JS, Klement P, Jones AM, Ghimire NJ et al (2014) Electrically tunable excitonic light-emitting diodes based on monolayer WSe_2 p-n junctions. Nat Nanotechnol 9:268–272
175. Yu Y, Miao F, He J, Ni Z (2017) Photodetecting and light-emitting devices based on two-dimensional materials. Chinese Phys B 26:036801
176. Jo S, Ubrig N, Berger H, Kuzmenko AB et al (2014) Mono- and bilayer WS_2 light-emitting transistors. Nano Lett 14:2019–2025
177. Zhang YJ, Oka T, Suzuki R, Ye JT, Iwasa Y (2014) Electrically switchable chiral light-emitting transistor. Science 344:725–728
178. Britnell L, Gorbachev RV, Jalil R, Belle BD et al (2012) Electron tunneling through ultrathin boron nitride crystalline barriers. Nano Lett 12:1707–1710
179. Withers F, Del Pozo-Zamudio O, Schwarz S, Dufferwiel S et al (2015) WSe_2 light-emitting tunneling transistors with enhanced brightness at room temperature. Nano Lett 15:8223–8228
180. Liu CH, Clark G, Fryett T, Wu SF et al (2017) Nanocavity integrated van der waals heterostructure light-emitting tunneling diode. Nano Lett 17:200–205
181. Lee GH, Yu YJ, Lee C, Dean C et al (2011) Electron tunneling through atomically flat and ultrathin hexagonal boron nitride. Appl Phys Lett 99:243114
182. Guo Y, Gao F, Huang P, Wu R et al (2022) Light-emitting diodes based on two-dimensional nanoplatelets. Energy Mater Adv Article ID 9857943:1–24
183. Andrzejewski D, Myja H, Heuken M, Grundmann A et al (2019) Scalable large-area p–i–n light-emitting diodes based on WS_2 monolayers grown via MOCVD. ACS Photonics 6:1832–1839
184. Srivastava A, Sidler M, Allain AV, Lembke DS et al (2015) Valley Zeeman effect in elementary optical excitations of monolayer WSe_2. Nat Phys 11:141–147
185. Chakraborty C, Kinnischtzke L, Goodfellow KM, Beams R et al (2015) Voltage-controlled quantum light from an atomically thin semiconductor. Nat Nanotechnol 10:507–511
186. Koperski M, Nogajewski K, Arora A, Cherkez V et al (2015) Single photon emitters in exfoliated WSe_2 structures. Nat Nanotechnol 10:503–506
187. Tonndorf P, Schmidt R, Schneider R, Kern J et al (2015) Single-photon emission from localized excitons in an atomically thin semiconductor. Optica 2015(2):347–352
188. Capasso A, Matteocci F, Najafi L, Prato M et al (2016) Few-Layer MoS_2 flakes as active buffer layer for stable perovskite solar cells. Adv Energy Mater 6:1600920
189. Kumar S, Kaczmarczyk A, Gerardot BD (2015) Strain-induced spatial and spectral isolation of quantum emitters in mono- and Bilayer WSe_2. Nano Lett 15:7567–7573
190. Blauth M, Jurgensen M, Vest G, Hartwig O et al (2018) Coupling single photons from discrete quantum emitters in WSe_2 to lithographically defined plasmonic slot waveguides. Nano Lett 18:6812–6819
191. Iff O, Tedeschi D, Martın-Sanchez J, Moczała-Dusanowska M et al (2019) Strain-tunable single photon sources in WSe_2 Monolayers. Nano Lett 19:6931–6936
192. Goushi K, Yoshida K, Sato K, Adachi C (2012) Organic light-emitting diodes employing efficient reverse intersystem crossing for triplet-to-singlet state conversion. Nat Photonics 6:253–258
193. Li W, Pan Y, Xiao R, Peng Q et al (2014) Employing $\sim$100% excitons in OLEDs by utilizing a fluorescent molecule with hybridized local and charge-transfer excited state. Adv Funct Mater 24:1609
194. Shirota Y, Kageyama H (2007) Charge carrier transporting molecular materials and their applications in devices. Chem Rev 107:953–1010
195. Walzer K, Maennig B, Pfeiffer M, Leo K (2007) Highly efficient organic devices based on electrically doped transport layers. Chem Rev 107:1233–1271
196. Miao Y, Yin M (2022) Recent progress on organic light-emitting diodes with phosphorescent ultrathin (<1nm) light-emitting layers. iScience 25:103804
197. Patel KD, Juang FS, Wang HX, Jian CZ et al (2022) Quantum dot-based white organic light-emitting diodes excited by a blue OLED. Appl Sci 12:6365

198. Bauri J, Choudhary RB, Mandal G (2021) Recent advances in efficient emissive materials-based OLED applications: a review. J Mater Sci 56:18837–18866
199. Reineke S, Thomschke M, Lüssem B, Leo K (2013) White organic light-emitting diodes: status and perspective. Rev Mod Phys 85:1245–1293
200. Yook KS, Lee JY (2014) Small molecule host materials for solution processed phosphorescent organic light-emitting diodes. Adv Mater 26:4218–4233
201. Reineke S, Lindner F, Schwartz G, Seidler N et al (2009) White organic light-emitting diodes with fluorescent tube efficiency. Nature 459:234–238
202. Wang ZB, Helander MG, Qiu J, Puzzo DP et al (2011) Unlocking the full potential of organic light-emitting diodes on flexible plastic. Nat Photonics 5:753–757
203. Kuei CY, Tsai WL, Tong B, Jiao M et al (2016) Bis-Tridentate Ir(III) complexes with nearly unitary RGB phosphorescence and organic light-emitting diodes with external quantum efficiency exceeding 31%. Adv Mater 28:2795–2800
204. Yang Z, Gao M, Wu W, Yang X et al (2019) Recent advances in quantum dot-based light-emitting devices: challenges and possible solutions. Mater Today 24:69–93
205. Wan KB, Brovelli S, Klimov VI (2013) Spectroscopic insights into the performance of quantum dot light-emitting diodes. MRS Bull 38:721–730
206. Li Y, Wei Q, Cao L, Fries F et al (2019) Organic light-emitting diodes based on conjugation-induced thermally activated delayed fluorescence polymers: interplay between intra- and intermolecular charge transfer states. Front Chem 7:688
207. He J, Chen H, Chen H, Wang Y et al (2017) Hybrid downconverters with green perovskite-polymer composite films for wide color gamut displays. Opt Express 25:12915–12925
208. Wang HC, Lin SY, Tang AC, Singh BP et al (2016) Mesoporous silica particles integrated with all-inorganic CsPbBr 3 perovskite quantum-dot nanocomposites (MP-PQDs) with high stability and wide color gamut used for backlight display. Angew Chem Int Ed 55:7924–7929
209. Dai X, Deng Y, Peng X, Jin Y (2017) Quantum-dot light-emitting diodes for large-area displays: towards the dawn of commercialization. Adv Mater 29:1607022
210. Zhang H, Chen S, Sun XW (2018) Efficient red/green/blue tandem quantum-dot light-emitting diodes with external quantum efficiency exceeding 21%. ACS Nano 12:697–704
211. Dai X, Zhang Z, Jin Y, Niu Y et al (2014) Solution-processed, high-performance light-emitting diodes based on quantum dots. Nature 515:96–99
212. Zhang H, Sun X, Chen S (2017) Over 100 cd A^{-1} efficient quantum dot light-emitting diodes with inverted tandem structure. Adv Funct Mater 27:1700610
213. Song J, Wang O, Shen H, Lin Q et al (2019) Over 30% external quantum efficiency light-emitting diodes by engineering quantum dot-assisted energy level match for hole transport layer. Adv Funct Mater 29:1808377
214. Liu Z, Lin CH, Hyun BR, Sher CW et al (2020) Micro-light-emitting diodes with quantum dots in display technology. Light: Sci Appl 9:83
215. Boosting second harmonic generation with TMDs monolayer (2022, July 11) retrieved 18 July 2022 from "https://phys.org/news/2022-07-boosting-harmonic-tmds-monolayer.html"
216. Hong P, Xu L, Rahmani M (2022) Dual bound states in the continuum enhanced second harmonic generation with Transition Metal Dichalcogenides monolayer. Opto-Electron Adv 5: 200097
217. Wu S, Buckley S, Jones AM, Ross JS et al (2014) Control of two-dimensional excitonic light emission via photonic crystal. 2D Mater 1: 011001
218. Gan X, Gao Y, Mak KF, Yao X et al (2013) Controlling the spontaneous emission rate of monolayer MoS_2 in a photonic crystal nanocavity. App Phys Lett 103:181119
219. Liu X, Galfsky T, Sun Z, Xia F et al (2015) Strong light-matter coupling in two-dimensional atomic crystals. Nat Photon 9:30–34
220. Schwarz S, Dufferwiel S, Walker P, Withers F et al (2014) Two-dimensional metal-chalcogenide films in tunable optical microcavities. Nano Lett 14:7003
221. Ye Y, Wong ZJ, Lu X, Zhu H et al (2015) Monolayer excitonic laser. Nat Photon 9:733
222. Wu S, Buckley S, Schaibley JR, Feng L et al (2015) Monolayer semiconductor nanocavity lasers with ultralow thresholds. Nature 520(7545):69

223. Klimov VI, Mikhailovsky AA, Xu S, Malko A et al (2000) Optical gain and stimulated emission in nanocrystal quantum dots. Science 290:314–317
224. Fan F, Voznyy O, Sabatini RP, Bicanic KT et al (2017) Continuous wave lasing in colloidal quantum dot solids enabled by facet-selective epitaxy. Nature 544:75–79
225. Lim J, Park YS, Wu K, Yun HJ et al (2018) Droop-free colloidal quantum dot light-emitting diodes. Nano Lett 18:6645–6653
226. Lim J, Park YS, Klimov VI (2018) Optical gain in colloidal quantum dots achieved with direct-current electrical pumping. Nat Mater 17:42–49
227. Rhee S, Kim K, Roh J, Kwak J (2020) Recent progress in high-luminance quantum dot light-emitting diodes
228. Roh J, Park YS, Lim J, Klimov VI (2020) Optically pumped colloidal-quantum-dot lasing in LED-like devices with an integrated optical cavity. Nat Commun 11:271
229. Manfredin D, Guarda-Nardini L, Winocur E, Piccotti F (2011) Research diagnostic criteria for temporomandibular disorders: asystematic review of axis I epidemiologic findings. Oral Surg Oral Med Oral Pathol Oral Radiol Endod 112:453–462
230. Huang YF, Lin JC, Yang HW, Lee YH et al (2014) Clinical effectiveness of laser acupuncture in the treatment of temporomandibular joint disorder. J Formos Med Assoc 113:535–539
231. Khalighi HR, Mortazavi H, Mojahedi SM, Azari-Marhabi S et al (2022) The efficacy of low-level diode laser versus laser acupuncture for the treatment of myofascial pain dysfunction syndrome (MPDS). J Dent Anesth Pain Med 22:19–27
232. Zwiri A, Alrawashdeh MA, Khan M, Ahmad WMAW et al (2020) Effectiveness of the laser application in temporomandibular joint disorder: a systematic review of 1172 patients. Pain Res Manage 5971032:1–10
233. Rahimi A, Rabiei S, Mojahedi SM, Kosarieh E (2011) Application of low level laser in temporomandibular disorders. J Lasers Med Sci 2:165–170
234. Shi J, Xu X, Li D, Meng Q (2015) Interfaces in perovskite solar cells. Small 11:2472–2486
235. Wang H, Kim DH (2017) Perovskite-based photodetectors: materials and devices. Chem Soc Rev 46:5204–5236
236. Baeg KJ, Binda M, Natali D, Caironi M et al (2013) Organic light detectors: photodiodes and phototransistors. Adv Mater 25:4267–4295
237. Wang C, Zhang X, Hu W (2020) Organic photodiodes and phototransistors toward infrared detection: materials, devices, and applications. Chem Soc Rev 49:653–670
238. Obaidulla SM, Goswami DK, Giri PK (2014) Low bias stress and reduced operating voltage in SnCl2 Pc based n-type organic field-effect transistors. Appl Phys Lett 104:213302
239. Ghosh J, Giri PK (2021) Recent advances in perovskite/2D materials based hybrid Photodetectors. J Phys Mater 4:032008
240. Li Y, Li L, Li S, Sun J, Fang Y et al (2022) Highly sensitive photodetectors based on monolayer MoS_2 field-effect transistors. ACS Omega 7:13615–13621
241. Kufer D, Konstantatos G (2016) Photo-FETs: phototransistors enabled by 2D and 0D nanomaterials. ACS Photonics 3:2197–2210
242. Xie C, Mak C, Tao X, Yan F (2017) Photodetectors based on two-dimensional layered materials beyond grapheme. Adv Funct Mater 27:1603886
243. Qiu Q, Huang Z (2021) Photodetectors of 2D materials from ultraviolet to terahertz waves. Adv Mater 33:2008126
244. Nalwa HS (2020) A review of molybdenum disulfide (MoS_2) based photodetectors: from ultra-broadband, selfpowered to flexible devices RSC Adv 10:30529
245. Grancini G, Nazeeruddin MK (2019) Dimensional tailoring of hybrid perovskites for photovoltaics. Nat Rev Mater 4:4–22
246. Li Z, Hong E, Zhang X, Deng M et al (2022) Perovskite-type 2D materials for high-performance photodetectors. J Phys Chem Lett 13:1215–1225
247. Li J, Wang J, Ma J, Shen H, Li L et al (2019) Selftrapped state enabled filterless narrowband photodetections in 2D layered perovskite single crystals. Nat Commun 10:806
248. Leng K, Abdelwahab I, Verzhbitskiy I, Telychko M et al (2018) Molecularly thin two-dimensional hybrid perovskites with tunable optoelectronic properties due to reversible surface relaxation. Nat Mater 17:908–914

249. Yang T, Wang X, Zheng B, Qi Z et al (2019) Ultrahigh-performance optoelectronics demonstrated in ultrathin perovskite-based vertical semiconductor heterostructures. ACS Nano 13:7996–8003
250. McIntyre RJ (1985) Recent developments in silicon avalanche photodiodes. Measurement 3:146–152
251. Shao Y, Liu Y, Chen X, Chen C et al (2017) Stable graphene-two-dimensional multiphase perovskite heterostructure phototransistors with high gain. Nano Lett 17:7330–7338
252. Li Y, Chernikov A, Zhang X, Rigosi A et al (2014) Measurement of the optical dielectric function of monolayer transition-metal dichalcogenides: MoS_2, $MoSe_2$, WS_2 and WSe_2. Phys Rev B 90:205422
253. Bernardi M, Palummo M, Grossman JC (2013) Extraordinary sunlight absorption and one nanometer thick photovoltaics using two-dimensional monolayer materials. Nano Lett 13:3664–3670
254. Kang J, Tongay S, Zhou J, Li J et al (2013) Band offsets and heterostructures of two-dimensional semiconductors. Appl Phys Lett 102:012111
255. Hong X, Kim J, Shi SF, Zhang Y et al (2014) Ultrafast charge transfer in atomically thin MoS_2/WS2 heterostructures. Nat Nanotechnol 9:682–686
256. Furchi MM, Pospischil A, Libisch F, Burgdorfer J et al (2014) Photovoltaic effect in an electrically tunable van der waals heterojunction. Nano Lett 14:4785–4791
257. Choudhary N, Islama MA, Kima JH, Koa TJ et al (2018) Two-dimensional transition metal dichalcogenide hybrid materials for energy applications. Nano Today 19:16–40
258. Lee et al (2014) Nanoelectronic circuits based on two-dimensional atomic layer crystals. Nanoscale 6:13283
259. Baugher BWH, Churchill HOH, Yang YF, Jarillo-Herrero P (2014) Optoelectronic devices based on electrically tunable p-n diodes in a monolayer dichalcogenide. Nat Nanotech 9:262–267
260. Tsai ML, Su SH, Chang JK, Tsai DS et al (2014) Monolayer MoS_2 heterojunction solar cells. ACS Nano 8:8317–8322
261. Fan FR, Wu W (2019) Emerging devices based on two-dimensional monolayer materials for energy harvesting. Research 7367828:1–16
262. Lee CH, Lee GH, van der Zande AM, Chen W et al (2014) Atomically thin p-n junctions with van der Waals heterointerfaces. Nat Nanotech 9:676–681
263. Gong YJ, Lin JH, Wang XL, Shi G et al (2014) Vertical and in-plane heterostructures from WS_2/MoS_2 monolayers. Nat Mater 13:1135–1142
264. Duan XD, Wang C, Shaw JC, Cheng R et al (2014) Lateral epitaxial growth of two-dimensional layered semiconductor heterojunctions. Nat Nanotech 9:1024–1030
265. Li LK, Yu YJ, Ye GJ, Ge Q et al (2014) Black phosphorus field-effect transistors. Nat Nanotech 9:372–377
266. Liu H, Neal AT, Zhu Z, Luo Z et al (2014) Phosphorene: an unexplored 2D semiconductor with a high hole mobility. ACS Nano 8:4033–4041
267. Deng YX, Luo Z, Conrad NJ, Liu H et al (2014) Black phosphorusmonolayer MoS_2 van der Waals heterojunction p-n diode. ACS Nano 8:8292–8299
268. Bowen CR, Kim HA, Weaver PM, Dunn S (2014) Piezoelectric and ferroelectric materials and structures for energy harvesting applications. Energy Environ Sci 7:25–44
269. Traverse CJ, Pandey R, Barr MC, Lunt RR (2017) Emergence of highly transparent photovoltaics for distributed applications. Nat Energy 2:849–860
270. Liu G, Wu C, Zhang Z, Chen Z et al (2020) Ultraviolet-protective transparent photovoltaics based on lead-free double perovskites. Sol RRL 4:2000056
271. Chen S, Yao H, Hu B, Zhang G et al (2018) A nonfullerene semitransparent tandem organic solar cell with 10.5% power conversion efficiency. Adv Energy Mater 8:1800529
272. Song Y, Chang S, Gradecak S, Kong J (2016) Visibly-transparent organic solar cells on flexible substrates with all-graphene electrodes. Adv Energy Mater 6:1600847
273. Yang C, Sheng W, Moemeni M, Bates M et al (2021) Ultraviolet and near-Infrared dual-band selective-harvesting transparent luminescent solar concentrators. Adv Energy Mater 11:203581

274. Zhao Y, Meek GA, Levine BG, Lunt RR (2014) Near-infrared harvesting transparent luminescent solar concentrators. Adv Opt Mater 2:606–611
275. Coogan A, Gunko YK (2021) Solution-based '"bottom-up"' synthesis of group VI transition metal dichalcogenides and their applications. Mater Adv 2:146–164
276. Jariwala D, Davoyan AR, Wong J, Van der Atwater HA (2017) Waals materials for atomically-thin photovoltaics: promise and outlook. ACS Photonics 4:2962–2970
277. Nazif KN, Kumar A, Hong J, Lee N et al (2021) High-Performance p–n Junction Transition Metal Dichalcogenide Photovoltaic Cells Enabled by MoOx Doping and Passivation. Nano Lett 21:3443–3450
278. Frisenda R, Molina-Mendoza AJ, Mueller T, Castellanos-Gomez A et al (2018) Atomically thin p–n junctions based on two-dimensional materials. Chem Soc Rev 47:3339–3358
279. Wu K, Ma H, Gao Y, Hu W et al (2019) Highly-efficient heterojunction solar cells based on two-dimensional tellurene and transition metal dichalcogenides†. J Mater Chem A 7:7430–7436
280. Paul G, Pal AJ (2022) Impact of band-engineered polymer-interlayers on heterojunction solar cells based on 2D-layered perovskites (Cs3Sb2ClxI9−x). Sol Energy 232:196–203
281. Islam KM, Ismael T, Luthy C, Kizilkaya O et al (2022) Large-Area, High-Specific-Power Schottky-Junction Photovoltaics from CVD-Grown Monolayer MoS_2. ACS Appl Mater Interfaces 14:24281–24289
282. Pawar SA, Kim D, Kim A, Park JH et al (2018) Heterojunction solar cell based on n-MoS_2/p-InP. Opt Mater 86:576–581
283. Akama T, Okita W, Nagai R, Li C et al Schottky solar cell using few-layered transition metal dichalcogenides toward large-scale fabrication of semitransparent and flexible power generator. Sci Rep 7:11967
284. He X, Iwamoto Y, Kaneko T, Kato T (2022) Fabrication of near-invisible solarcell with monolayer WS_2. Sci Rep 12:11315
285. Aiello O, Crovetti P, Lin L, Alioto MA (2019) PW-power Hz-range oscillator operating with a 0.3–1.8-V unregulated supply. IEEE J Solid-State Circuits 54:1487–1496
286. Lin Z, Carvalho BR, Kahn E, Lv R et al (2016) Defect engineering of two-dimensional transition metal dichalcogenides. 2D Mater 3:022002
287. Wang L, Li Y, Gong X, Thean AVY, Liang G (2018) A physics-based compact model for transition-metal dichalcogenides transistors with the band-tail effect. IEEE Electron Device Lett 39:761–764
288. Wang L, Thean AVY, Liang G (2018) Percolation theory based statistical resistance model for resistive random access memory. Appl Phys Lett 112:253505
289. Stauffer D, Bunde A (1994) Introduction to percolation theory. Taylor and Francis, London, pp 93–120
290. Wang L, Thean AVY, Liang G (2019) A statistical Seebeck coefficient model based on percolation theory in two-dimensional disordered systems. J Appl Phys 125:224302
291. Hosseini SE, Wahid MA (2016) Hydrogen production from renewable and sustainable energy resources: promising green energy carrier for cean development. Renew Sust Energy Rev 57:850–866
292. Lucia U (2014) Overview on fuel cells. Renew Sust Energy Rev 30:164–169
293. Tee SY, Win KY, Teo WS, Koh LD et al (2017) Recent progress in energy-driven water splitting. Adv Sci 4:1600337
294. Tang J, Liu T, Miao S, Cho Y (2021) Emerging energy harvesting technology for electro/photo-catalyticwater splitting application. Catalysts 11:142–224
295. Wei J, Zhou M, Long A, Xue Y et al (2018) Heterostructured electrocatalysts for hydrogen evolution reaction under alkaline conditions. Nano-Micro Lett 10:75
296. Mete B, Peighambardoust NS, Aydin S, Sadeghi E et al (2021) Metalsubstituted zirconium diboride (Zr1-xTMxB2; TM = Ni, Co, and Fe) as low-cost and high-performance bifunctional electrocatalyst for water splitting. Electrochim Acta 389:138789
297. Faid AY, Barnett AO, Seland F, Sunde S (2021) NiCu mixed metal oxide catalyst for alkaline hydrogen evolution in anion exchange membrane water electrolysis. Electrochim Acta 371:137837

298. Kuang Y, Kenney MJ, Meng Y, Hung WH et al (2019) Solar-driven, highly sustained splitting of seawater into hydrogen and oxygen fuels. Proc Natl Acad Sci USA 116:6624
299. Wu L, Yu L, Zhang F, McElhenny B et al (2021) Heterogeneous bimetallic phosphide Ni2P-Fe2P as an efficient bifunctional catalyst for water/seawater splitting. Adv Funct Mater 31:2006484
300. Zhang H, Lu Y, Han W, Zhu J et al (2020) Solar energy conversion and utilization: towards the emerging photo-electrochemical devices based on perovskite photovoltaics. Chem Eng J 393:124766
301. Ahmet IY, Ma Y, Jang J-W, Henschel T et al (2019) Demonstration of a 50 cm_2 $BiVO_4$ tandem photoelectrochemical-photovoltaic water splitting device. Sustain Energy Fuels 3:2366
302. Li J, Chen Z, Yang H, Yi Z et al (2020) Tunable broadband solar energy absorber based on monolayer transition metal dichalcogenides materials using au nanocubes. Nanomaterials 10:257
303. Luo Y, Wang S, Ren K, Chou JP et al (2019) Transition-metal dichalcogenides/Mg(OH)2 van der Waals heterostructures as promising watersplitting photocatalysts: a first-principles study Phys. Chem Chem Phys 21:1791
304. Sukanya R, da Silva Alves DC, Breslin CB (2022) Review—recent developments in the applications of 2D transition metal dichalcogenides as electrocatalysts in the generation of hydrogen for renewable energy conversion. J Electroch Soc 169:064504
305. Seo S, Kim S, Choi H, Lee J et al (2019) Direct in situ growth of centimeters-cale multi-heterojunction $MoS_2/WS_2/WSe_2$ thin-film catalyst for photo-electrochemical hydrogen evolution. Adv Sci 6:1900301
306. Wu HY, Yang K, Si Y, Huang WQ et al (2019) Two-dimensional GaX/SnS2(X= S, Se) van der waals heterostructures for photovoltaic application: heteroatom doping strategyto boost power conversion efficiency. Phys Status Solidi RRL 13:1800565
307. Zhang X, Zhang Z, Wu D, Zhang X et al (2018) Computational screening of 2D materials and rational design of heterojunctions for water splitting photocatalysts. Small Methods 2:1700359
308. Xu L, Huang W-Q, Hu W, Yang K et al (2017) Two-dimensional MoS_2-graphene-based multilayer van der waals heterostructures: enhanced charge transfer and optical absorption, and electric-field tunable dirac point and band gap. Chem Mater 29:5504–5512
309. Zhang C, Nie Y, Liao T, Kou L et al (2019) Predicting ultrafast dirac transport channel at the one-dimensional interface of the two-dimensional coplanar ZnO/MoS_2 Heterostructure. Phys Rev B: Condens Matter Mater Phys 99:035424
310. You B, Wang X, Zheng Z, Mi W (2016) Black phosphorene/ monolayer transition-metal dichalcogenides as two-dimensional van der waals heterostructures: afirst-principles study. Phys Chem Chem Phys 18:7381–7388
311. Wang S, Ren C, Tian H, Yu J et al (2018) MoS_2/ZnO van der Waals heterostructure as a high-efficiency water splitting photocatalyst: a first-principles study. Phys Chem Chem Phys 20:13394–13399
312. Guan Z, Lian CS, Hu S, Ni S et al (2017) Tunable structural, electronic, and optical properties of layered two-dimensional C_2N and MoS_2 van der waals heterostructure as photovoltaic material. J Phys Chem C 121:3654–3660
313. Chuang H-J, Chamlagain B, Koehler M, Perera MM et al (2016) Low-Resistance 2D/2D ohmic contacts: a universal approach to high-performance WSe_2, MoS_2, and $MoSe_2$ Transistors. Nano Lett 16:1896–1902
314. Gong Y, Lin J, Wang X, Shi G et al (2014) Vertical and in-plane heterostructures from WS_2/MoS_2Monolayers. Nat Mater 13:1135–1142
315. Jin C, Regan EC, Yan A, Iqbal BaktiUtama M et al (2019) Observation of moiré excitons in WSe_2/WS_2 HeterostructureSuperlattices. Nature 567:76–80
316. Cai Z, Liu B, Zou X, Cheng HM (2018) Chemical vapor deposition growth and applications of two-dimensional materials and their heterostructures. Chem Rev 118:6091–6133
317. Gong Y, Lei S, Ye G, Li B et al (2015) Two-step growth of two- dimensional WSe_2/$MoSe_2$Heterostructures. Nano Lett 15:6135–6141

318. Tongay S, Fan W, Kang J, Park J et al (2014) Tuning interlayer coupling in large-area heterostructures with CVD-Grown MoS_2 and WS2Monolayers. Nano Lett 14:3185–3190
319. Özçelik VO, Azadani JG, Yang C, Koester SJ et al (2016) Band alignment of two-dimensional semiconductors for designing heterostructures with momentum space matching. Phys Rev B: Condens Matter Mater Phys 94:035125
320. Luo Y, Ren K, Wang S, Chou JP et al (2019) First-principles study on transition-metal dichalcogenide/bse van der waals heterostructures: a promising water-splitting photocatalyst. J Phys Chem C 123(37):22742–22751
321. Cui Z, Ren K, Zhao Y, Wang X et al (2019) Electronic and optical properties of van der waals heterostructures of g-GaN and transition metal dichalcogenides. Appl Surf Sci 492:513–519
322. Ren K, Sun M, Luo Y, Wang S et al (2019) First-principle study of electronic and optical properties of two- dimensional materials-based heterostructures based on transition metal dichalcogenides and boron phosphide. Appl Surf Sci 476:70–75
323. Luo Y, Wang S, Ren K, Chou JP et al (2019) Transition-metal dichalcogenides/Mg(OH)2 van der waals heterostructures as promising water-splitting photocatalysts: a first- principles study. Phys Chem Chem Phys 21:1791–1796
324. Kumar R, Das D, Singh AK (2018) C2N/WS2 van der Waals Type-II Heterostructure as APromising Water Splitting Photocatalyst. J Catal 359:143–150
325. Wang S, Tian H, Ren C, Yu J et al (2018) Electronic and optical properties of heterostructures based on transition metal dichalcogenides and graphene-like zinc oxide. Sci Rep 8:12009
326. Demirci S, Avazlı N, Durgun E, Cahangirov S (2017) Structural and electronic properties of monolayer group III monochalcogenides. Phys Rev B: Condens Matter Mater Phys 95:115409
327. Wang BJ, Li XH, Zhao R, Cai XL et al (2018) Electronic structures and enhanced photocatalytic properties of blue phosphorene/BSe van der Waals Heterostructures. J Mater Chem A 6:8923–8929
328. Ding Q, Meng F, English CR, Cabán-Acevedo M et al (2014) Efficient photoelectrochemical hydrogen generation using heterostructures of Si and chemically exfoliated metallic MoS_2. J Am Chem Soc 136:8504
329. Kwak IH, Kwon IS, Lee JH, Lim YR et al (2021) Chalcogen-vacancy group VI transition metal dichalcogenide nanosheets for electrochemical and photoelectrochemical hydrogen evolution. J Mater Chem C 9:101
330. Rosman NN, Mohamad Yunus R, Jeffery Minggu L, Arifin K et al (2021) Vertical MoS_2 on SiO_2/Si and graphene: effect of surface morphology on photoelectrochemical properties. Nanotechnology 32:035705
331. Liu Y, Li Y, Peng F, Lin Y et al (2019) 2H- and 1T- mixed phase few-layer MoS_2 as a superior to Pt co-catalyst coated on TiO_2 nanorod arrays for photocatalytic hydrogen evolution. Appl Catal B Environ 241:236
332. Paul KK, Sreekanth N, Biroju RK, Narayanan TN et al (2018) Solar light driven photoelectrocatalytic hydrogen evolution and dye degradation by metal-free few-layer MoS_2 nanoflower/ TiO_2(B) nanobelts heterostructure. Sol Energy Mater Sol Cells 185:364
333. Ren X, Qi X, Shen Y, Xiao S et al (2016) 2D co-catalytic MoS_2 nanosheets embedded with 1D TiO2 nanoparticles for enhancing photocatalytic activity. J Phys D: Appl Phys 49:315304
334. Liu X, Xing Z, Zhang Y, Li Z et al (2017) Fabrication of 3D flower-like black $N\text{-}TiO_{2-x}$@MoS_2 for unprecedented-high visible-light-driven photocatalytic performance. Appl Catal B Environ 201:119
335. Ye L, Zhang H, Xiong Y, Kong C et al (2021) Efficient photoelectrochemical overall water-splitting of MoS_2/g-C_3N_4 n-n type heterojunction film. J Chem Phys 154:214701
336. Andoshe DM, Jeon JM, Kim SY, Jang HW (2015) Two-dimensional transition metal dichalcogenide nanomaterials for solar water splitting Electron. Mater Lett 11:323–335
337. Tiwari S, Kumar S, Ganguli AK (2022) Role of MoS_2/rGO co-catalyst to enhance the activity and stability of Cu_2O as photocatalyst towards photoelectrochemical water splitting. J Photochem Photobiol A Chem 424:113622
338. Ben-Naim M, Britto RJ, Aldridge CW, Mow R et al (2020) Addressing the stability gap in photoelectrochemistry: molybdenum disulfide protective catalysts for tandem III-V unassisted solar water splitting. ACS Energy Lett 5:2631

339. Lim KRG, Handoko AD, Johnson LR, Meng X et al (2020) 2H-MoS_2 on Mo_2CTx MXene nanohybrid for efficient and durable electrocatalytic hydrogen evolution. ACS Nano 14:16140
340. Lu AY, Zhu H, Xiao J, Chuu CP et al (2017) Janus monolayers of transition metal dichalcogenides. Nat Nanotechnol 12:744
341. Zhang J, Jia S, Kholmanov I, Dong L et al (2017) Janus monolayer transition-metal dichalcogenides. ACS Nano 11:8192–8198
342. Ji Y, Yang M, Lin H, Hou T et al (2018) Janus structures of transition metal dichalcogenides as the heterojunction photocatalysts for water splitting. J Phys Chem C 122:3123–3129
343. Liu T, Liu Z (2018) 2D MoS_2 nanostructures for biomedical applications. Adv Healthc Mater 7:e1701158
344. Hu H, Zavabeti A, Quan H, Zhu W et al (2019) Recent advances in twodimensional transition metal dichalcogenides for biological sensing. Biosens Bioelectron 142:111573
345. Agarwal V, Chatterjee K (2018) Recent advances in the field of transition metal dichalcogenides for biomedical applications. Nanoscale 10:16365–16397
346. Zhou X, Sun H, Bai X (2020) Two-dimensional transition metal dichalcogenides: synthesis, biomedical applications and biosafety evaluation. Front Bioeng Biotechnol 8:236
347. Mak KF, Lee C, Hone J, Shan J et al (2010) Atomically thin MoS_2: a new direct-gap semiconductor. Phys Rev Lett 105:136805
348. Lin Z, Thee MT, Elias AL, Feng S et al (2014) Facile synthesis of MoS_2 and MoXW1-xS2 triangular monolayers. APL Mater 2:092514
349. Chen F, Tang Q, Ma T, Zhu B et al (2022) Structures, properties, and challenges of emerging 2D materials in bioelectronics and biosensors. InfoMat. 4:e12299
350. Teo WZ, Chng ELK, Sofer Z, Pumera M (2014) Cytotoxicity of exfoliated transition-metal dichalcogenides (MoS_2, WS_2, and WSe_2) is lower than that of graphene and its analogues. Chem A Eur J 20:9627–9632
351. Pessoa-Pureur R, Heimfarth L, Rocha JB (2014) Signaling mechanisms and disrupted cytoskeleton in the diphenyl ditelluride neurotoxicity. Oxidative Med Cell Longev 2014:1–21
352. Bhanu U, Islam MR, Tetard L, Khondaker SI (2014) Photoluminescence quenching in gold—MoS_2 hybrid nanoflakes. Sci Rep 4:5575
353. Ge J, Ou EC, Yu RQ, Chu X (2014) A novel aptameric nanobiosensor based on the self-assembled DNA–MoS_2 nanosheet architecture for biomolecule detection. J Mater Chem B 2:625–628
354. Zhu C, Zeng Z, Li H, Li F et al (2013) Single-layer MoS_2-based nanoprobes for homogeneous detection of biomolecules. J Am Chem Soc 135:5998–6001
355. Zhang Y, Zheng B, Zhu C, Zhang X et al (2015) Single-layer transition metal dichalcogenide nanosheet-based nanosensors for rapid, sensitive, and multiplexed detection of DNA. Adv Mater 27:935–939
356. Hu et al (2017) Two-dimensional transition metal dichalcogenide nanomaterials for biosensing applications. Mater Chem Front 1:24–36
357. Gao et al (2017) Detection of T4 polynucleotide kinase via allosteric aptamer probe platform. ACS Appl Mater Interfaces 9:38356–38363
358. Gong et al (2022) Molybdenum disulfide-based nanoprobes: Preparation and sensing application. Biosensors 12:87
359. Zhang et al (2022) Two-dimensional quantum dot-based electrochemical biosensors. Biosensors 12:254
360. Ge J, Tang LJ, Xi Q, Li XP et al (2014) A WS2 nanosheet based sensing platform for highly sensitive detection of T4 polynucleotide kinase and its inhibitors. Nanoscale 6:6866–6872
361. Liu X, Ge J, Wang X, Wu Z et al (2014) Development of a highly sensitive sensing platform for T4 polynucleotide kinase phosphatase and its inhibitors based on WS 2 nanosheets. Anal Methods 6:7212–7217
362. Xiang X, Shi J, Huang F, Zheng M et al (2015) MoS_2 nanosheet-based fluorescent biosensor for protein detection via terminal protection of small-molecule-linked DNA and exonuclease III-aided DNA recycling amplification. Biosens Bioelectron 74:227–232

363. Chow PK, Jacobs-Gedrim RB, Gao J, Lu TM et al (2015) Defect-induced photoluminescence in monolayer semiconducting transition metal dichalcogenides. ACS Nano 9:1520–1527
364. Feng Q, Zhu Y, Hong J, Zhang M et al (2014) Growth of large-area 2D $MoS_2(1-x)Se_2x$ semiconductor alloys. Adv Mater 26:2648–2653
365. Hu H, Xin J, Hu H, Wang X et al (2014) Organic liquids-responsive β-cyclodextrin-functionalized graphene-based fluorescence probe: label-free selective detection of tetrahydrofuran. Molecules 19(6):7459–7479
366. Lee DW, Lee J, Sohn IY, Kim BY et al (2015) Field-effect transistor with a chemically synthesized MoS_2 sensing channel for label-free and highly sensitive electrical detection of DNA hybridization. Nano Res 8:2340–2350
367. Li BL, Zou HL, Lu L, Yang Y et al (2015) Size-dependent optical absorption of layered MoS_2 and DNA oligonucleotides induced dispersion behavior for label-free detection of single-nucleotide polymorphism. Adv Funct Mater 25:3541–3550
368. Sarkar D, Liu W, Xie X, Anselmo AC et al (2014) MoS_2 field-effect transistor for next-generation label-free biosensors. ACS Nano 8:3992–4003
369. Wang L, Wang Y, Wong JI, Palacios T et al (2014) Functionalized MoS_2 nanosheet-based field-effect biosensor for label-free sensitive detection of cancer marker proteins in solution. Small 10:1101–1105
370. Yadav V, Roy S, Singh P, Khan Z et al (2019) 2D MoS_2-based nanomaterials for therapeutic, Bioimaging, and Biosensing Applications" Small 15:1803706
371. Li J, Zhao Q, Tang Y (2016) Label-free fluorescence assay of S1 nuclease and hydroxyl radicals based on water-soluble conjugated polymers and WS2 nanosheets. Sensors 16: 865
372. Zhao J, Jin X, Vdovenko M, Zhang L et al (2015) A WS_2 nanosheet based chemiluminescence resonance energy transfer platform for sensing biomolecules. Chem Commun 51:11092–11095
373. Du C, Shang A, Shang M, Ma X et al (2018) Water-soluble VS2 quantum dots with unusual fluorescence for biosensing. Sensors Actuators B: Chem. 255:926–934
374. Fahimi-Kashani N, Rashti A, Hormozi-Nezhad MR, Mahdavi V (2017) MoS_2 quantum-dots as a label-free fluorescent nanoprobe for the highly selective detection of methyl parathion pesticide. Anal Methods 9:716–723
375. Lan L, Yao Y, Ping J, Ying Y (2019) Ultrathin transition-metal dichalcogenide nanosheet-based colorimetric sensor for sensitive and label-free detection of DNA. Sensors Actuators B: Chem. 290:565–572
376. Barua S, Dutta HS, Gogoi S, Devi R et al (2018) Nanostructured MoS_2-based advanced biosensors: a review
377. Wang T, Zhu H, Zhuo J, Zhu Z et al (2013) Biosensor based on ultrasmall MoS_2 nanoparticles for electrochemical detection of H_2O_2 released by cells at the nanomolar level. Anal Chem 85:10289–10295
378. Wang T, Zhu R, Zhuo J, Zhu Z et al (2014) Direct detection of DNA below Ppb level based on thionin-functionalized layered MoS_2 electrochemical sensors. Anal Chem 86:12064–12069
379. Su S, Sun H, Cao W, Chao J et al (2016) Dual-target electrochemical biosensing based on DNA structural switching on gold nanoparticle-decorated MoS_2 nanosheets. ACS Appl Mater Interfaces 8:6826–6833
380. Cui Z, Li D, Yang W, Fan K et al (2022) An electrochemical biosensor based on few-layer MoS_2 nanosheets for highly sensitive detection of tumor marker ctDNA. Anal Methods 14:1956–1962
381. Lu K, Liu J, Dai X, Zhao L et al (2022) Construction of a Au@MoS_2 composite nanosheet biosensor for the ultrasensitive detection of a neurotransmitter and understanding of its mechanism based on DFT calculations. RSC Adv 12:798–809
382. Mphuthi N, Sikhwivhilu L, Ray SS (2022) Functionalization of 2D MoS_2 nanosheets with various metal and metal oxide nanostructures: their properties and application in electrochemical sensors. Biosensors 12:386
383. Li T, Shang D, Gao S, Wang B, Kong H et al (2022) Two-dimensional material-based electrochemical sensors/biosensors for food safety and biomolecular detection. Biosensors 12:314

384. Yang T, Chen M, Nan F, Chen L et al (2015) Enhanced electropolymerization of poly(xanthurenic acid)−MoS_2 film for specific electrocatalytic detection of guanine and adenine. J Mater Chem B 3:4884–4891
385. Yang R, Zhao J, Chen M, Yang T et al (2015) Electrocatalytic determination of chloramphenicol based on molybdenum disulfide nanosheets and self-doped polyaniline. Talanta 131:619–623
386. Zhu D, Liu W, Zhao D, Hao Q et al (2017) Label-free electrochemical sensing platform for microRNA-21 detection using thionine and gold nanoparticles co-functionalized MoS_2 nanosheet. ACS Appl Mater Interfaces 9:35597–35603
387. Yoon J, Lee T, Bapurao B, Jo J et al (2017) Electrochemical H_2O_2 biosensor composed of myoglobin on MoS_2 nanoparticle-graphene oxide hybrid structure. Biosens Bioelectron 93:14–20
388. Su S, Chao J, Pan D, Wang L et al (2015) Electrochemical sensors using two-dimensional layered nanomaterials. Electroanalysis 27:1062–1072
389. Kim H, Kim H, Ahn C, Kulkarni A et al (2015) In situ synthesis of Mos_2 on a polymer based gold electrode platform and its application in electrochemical biosensing. RSC Adv 5:10134–10138
390. Su S, Sun H, Xu F, Yuwen L et al (2013) Highly sensitive and selective determination of dopamine in the presence of ascorbic acid using gold nanoparticles-decorated Mos_2 nanosheets modified electrode. Electroanalysis 25:2523–2529
391. Hu H, Zavabeti A, Quan H, Zhu W et al (2019) Recent advances in two-dimensional transition metal dichalcogenides for biological sensing. Biosensors Bioelectronics 142:111573
392. Nasir et al (2017) Two-dimensional 1T-phase transition metal dichalcogenides as nanocarriers to enhance and stabilize enzyme activity for electrochemical pesticide detection. ACS Nano 11:5774–5784
393. Wei et al (2020) Recent advances and challenges in electrochemical biosensors for emerging and re-emerging infectious diseases. Biosensors 10:24
394. Krishnan et al (2019) A review on graphene-based nanocomposites for electrochemical and fluorescent biosensors. RSC Adv 9:8778
395. Matsumoto A, Miyahara Y (2013) Current and emerging challenges of field effect transistor based bio-sensing. Nanoscale 5:10702–10718
396. Nam H, Oh BR, Chen P, Yoon JS et al (2015) Two different device physics principles for operating MoS_2 transistor biosensors with femtomolar-level detection limits. Appl Phys Lett 107:012105
397. Deen MJ, Shinwari MW, Ranuárez J, Landheer D (2006) Noise considerations in field-effect biosensors. J Appl Phys 100:074703
398. Nam H, Oh BR, Chen P, Chen M et al (2015) Multiple MoS_2 transistors for sensing molecule interaction kinetics. Sci Rep 5:10546
399. Lee J, Dak P, Lee Y, Park H et al (2015) Two-dimensional layered MoS_2 biosensors enable highly sensitive detection of biomolecules. Sci Rep 4:7352
400. Li H, Zhang Q, Yap CCR, Tay BK et al (2012) From bulk to monolayer MoS_2: evolution of raman scattering. Adv Funct Mater 22:1385–1390
401. Yang Y, Yang X, Tan Y, Yuan Q (2017) Recent progress in flexible and wearable bio-electronics based on nanomaterials. Nano Res 10:1560–1583
402. Viswanathan S, Narayanan TN, Aran K, Fink D et al (2015) Graphene-protein field effect biosensors: glucose sensing. Mater Today 18:513–522
403. Lam CYK, Zhang Q, Yin B, Huang Y et al (2021) Recent advances in two-dimensional transition metal dichalcogenide nanocomposites biosensors for virus detection before and during COVID-19 outbreak. J Compos Sci 5:190
404. Nam H, Oh BR, Chen M, Wi S et al (2015) Fabrication and comparison of MoS_2 and WSe_2 field-effect transistor biosensors. J Vac Sci Technol B, 33:06FG01-7
405. De A, Shee S, Sarkar SK (2021) Investigation into $MoTe_2$ based dielectric modulated AMFET biosensor for label-free detection of DNA including electric variational effects. TechRxiv. Preprint. https://doi.org/10.36227/techrxiv.14723343.v1

406. Balendhran S, Walia S, Nili H, Ou JZ et al (2013) Two-dimensional molybdenum trioxide and dichalcogenides. Adv Funct Mater 23:3952–3970
407. Prabhu P, Jose V, Lee JM (2020) Design strategies for development of tmd-based heterostructures in electrochemical energy systems. Matter 2:526–553
408. Haigh SJ, Gholinia A, Jalil R, Romani S et al (2012) Cross-sectional imaging of individual layers and buried interfaces of graphene-based heterostructures and superlattices. Nat Mater 11:764–767
409. Liu Y, Zhang S, He J, Wang ZM et al (2019) Recent progress in the fabrication, properties, and devices of heterostructures based on 2D materials. Nano-Micro Lett 11:13
410. Wang T, Tan X, Wei Y, Jin H (2022) Unveiling the layer-dependent electronic properties in transition-metal dichalcogenide heterostructures assisted by machine learning. Nanoscale 14:2511–2520
411. Jia J, Choi JH (2022) Recent progress in 2D hybrid heterostructures from transition metal dichalcogenides and organic layers: properties and applications in energy and optoelectronics field. Nanoscale. https://doi.org/10.1039/D2NR01358D
412. Wu B, Zheng H, Ding J, Wang Y (2022) Observation of interlayer excitons in trilayer type-II transition metal dichalcogenide heterostructures. Nano Res. https://doi.org/10.1007/s12274-022-4580-3
413. Duan X, Wang C, Pan A, Yu R et al (2015) "Two-dimensional transition metal dichalcogenides as atomically thin semiconductors: opportunities and challenges" Chem. Soc Rev 44:8859–8876
414. Lin YC, Ghosh RK, Addou R, Lu N et al (2015) "Atomically thin resonant tunnel diodes built from synthetic van der Waals heterostructures" Nat. Commu 6:7311
415. Cheng R, Li D, Zhou H, Wang C et al (2014) Electroluminescence and photocurrent generation from atomically sharp WSe_2/MoS_2 heterojunction p–n diodes. Duan Nano Lett 14:5590–5597
416. Yu JH, Lee HR, Hong SS, Kong D et al (2014) Vertical heterostructure of two-dimensional MoS_2 and WSe_2 with vertically aligned layers
417. Yu et al (2015) Vertical heterostructure of two-dimensional MoS_2 and WSe_2 with vertically aligned layers. Nano Lett 15:1031–1035
418. Bian R, Li C, Liu Q, Cao G et al (2022) Recent progress in the synthesis of novel two-dimensional van der Waals materials. Natl Sci Rev 9:nwab164
419. Patil C, Dalir H, Kang JH, Davydov A et al (2022) Highly accurate, reliable, and non-contaminating two-dimensional material transfer system. Appl Phys Rev 9:011419
420. Pan Y, Fölsch S, Nie Y, Waters D et al (2018) Quantum-confined electronic states arising from the moiré pattern of MoS_2–WSe_2 heterobilayers. Nano Lett 18:1849–1855
421. Liu R, Wang F, Liu L, He X et al (2021) Band alignment engineering in two-dimensional transition metal dichalcogenide-based heterostructures for photodetectors. Small Struct 2:2000136
422. Bastonero L, Cicero G, Palummo M, Fiorentin MR (2021) Boosted solar light absorbance in PdS_2/PtS_2 vertical heterostructures for ultrathin photovoltaic devices. ACS Appl Mater Interfaces 13:43615–43621
423. Ma Y, Xu S, Wei J, Zhou B (2022) In-plane and vertical heterostructures from 1T′/2H transition-metal dichalcogenides. Oxford Open Materials Science, 1, itab016
424. Liu Y, Weiss NO, Duan X, Cheng HC et al (2016) Van der Waals heterostructures and devices. Nat Rev Mater 1:16042
425. Cheng R et al (2014) Few-layer molybdenum disulfide transistors and circuits for high-speed flexible electronics. Nat Commun 5:5143
426. Nieken R, Roche A, Mahdikhanysarvejahany F, Taniguchi T et al (2022) Direct STM measurements of R-type and H-type twisted MoSe2/WSe2. APL Mater 10:031107
427. Shim J, Jo SH, Kim M, Song YJ et al (2017) Light-triggered ternary device and inverter based on heterojunction of van der waals materials. ACS Nano 11:6319–6327
428. Patil C, Dalir H, Kang JH, Davydov A et al (2022) Highly accurate, reliable, and noncontaminating two-dimensional material transfer system Appl Phys Rev 9:011419

429. Chamlagain B, Withanage SS, Johnston AC, Khondaker SI et al (2020) Scalable lateral heterojunction by chemical doping of 2D TMD thin films. Sci Rep 10:12970
430. Wang M, Li Y, Fang J, Villa CJ et al (2020) Superior oxygen reduction reaction on phosphorus-doped carbon dot/graphene aerogel for all-solid-state flexible al–air batteries. Adv Energy Mater 10:1902736
431. Li C, Cao Q, Wang F, Xiao Y et al (2018) Engineering graphene and TMDs based van der Waals heterostructures for photovoltaic and photoelectrochemical solar energy conversion. Chem Soc Rev 47:4981–5037
432. Azadmanjiri J, Srivastava VK, Kumar P, Sofer Z et al (2020) Graphene-Supported 2D transition metal dichalcogenide van der waalsheterostructures. App Mater Today 19:100600
433. Zhou H, Chen Y, Zhu H (2021) Deciphering asymmetric charge transfer at transition metal dichalcogenide–graphene interface by helicity-resolved ultrafast spectroscopy. Sci Adv 7:eabg2999
434. Li H, Li X, Park JH, Tao L et al (2019) Restoring the photovoltaic effect in graphene-based van der Waals heterojunctions towards self-powered high-detectivity photodetectors. Nano Energy 57:214–221
435. Xu H, Han X, Dai X, Liu W et al (2018) High detectivity and transparent few-layer MoS_2/glassy-graphene heterostructure photodetectors. Adv Mater 30:1706561
436. Mehew JD, Unal S, Torres Alonso E, Jones GF et al (2017) Fast and highly sensitive ionic-polymer-gated WS_2–graphene photodetectors. Adv Mater 29:1700222
437. Yu W, Li S, Zhang Y, Ma W et al (2017) Near-infrared photodetectors based on $MoTe_2$/graphene heterostructure with high responsivity and flexibility. Small 13:1700268
438. Flory N, Ma P, Salamin Y, Emboras A et al (2020) Waveguide-integrated van der Waals heterostructure photodetector at telecom wavelengths with high speed and high responsivity. Nat Nanotechnol 15:118–124
439. Zhang K, Fang X, Wang Y, Wan Y et al (2017) Ultrasensitive infrared photodetectors based on a graphene–$MoTe_2$–graphene vertical van der waals heterostructure. ACS Appl Mater Interfaces 9:5392–5398
440. Kima G, Shin HS (2020) Spatially controlled lateral heterostructures of graphene and transition metal dichalcogenides toward atomically thin and multi-functional electronics. Nanoscale 12:5286–5292
441. Samuel WL, Prathamesh D, Kenji W, Takashi T et al (2019) Gate-tunable graphene–WSe_2 heterojunctions at the schottky-mott limit. Ung Adv Mater 31:1901392
442. Liu Y, Gao Z, Tan Y, Chen F (2018) Enhancement of out-of-plane charge transport in a vertically stacked two-dimensional heterostructure using point defects. ACS Nano 12:10529
443. Qingyuan H, Pengji L, Zhiheng W, Bin Y (2019) Molecular beam epitaxy scalable growth of wafer-scale continuous semiconducting monolayer $MoTe_2$ on inert amorphous dielectrics. Adv Mater 31:1901578
444. Huang HF, Xu WS, Chen TX, Chang RJ et al (2018) High-performance two-dimensional schottky diodes utilizing chemical vapour deposition-grown graphene–MoS_2 heterojunctions. ACS Appl Mater Interfaces 10:37258–37266
445. Corbet CM, McClellan C, Rai A, Sonde SS et al (2015) Field effect transistors with current saturation and voltage gain in ultrathin ReS_2. ACS Nano 9:363–370
446. Kang B, Kim Y, Yoo WJ, Lee C (2018) Ultrahigh photoresponsive device based on ReS2/graphene heterostructure. Small 14:1802593
447. Muhammad H, Lin G, Huiqiao L, Ying M et al (2016) Large-area bilayer ReS2 film/multilayer ReS_2 flakes synthesized by chemical vapor deposition for high performance photodetectors. Adv Funct Mater 26:4551
448. Li L, Han W, Pi LJ, Niu P et al (2019) Emerging in-plane anisotropic two-dimensional materials. InfoMat 2019(1):54–73
449. Jeong H, Oh HM, Bang S, Jeong HJ et al (2016) Van der Waals Heterostructures with high accuracy rotational alignment. Nano Lett 16:1858–1995
450. Vu QA, Lee JH, Nguyen VL, Shin YS et al (2017) Tuning carrier tunneling in van der waals heterostructures for ultrahigh detectivity. Nano Lett 17:453–459

451. Pak J, Lee I, Cho K, Kim JK et al (2019) Intrinsic optoelectronic characteristics of MoS_2 phototransistors via a fully transparent van der waals heterostructure. ACS Nano 13:9638–9646
452. Yeh CH, Chen HC, Lin HC, Lin YC et al (2019) Ultrafast monolayer In/Gr-WS2-Gr hybrid photodetectors with high gain. ACS Nano 13:3269–3279
453. Zhou YQ, Tan HJ, Sheng YW, Fan Y et al (2018) Utilizing interlayer excitons in bilayer ws2 for increased photovoltaic response in ultrathin graphene vertical cross-bar photodetecting tunneling transistors. ACS Nano 12:4669–4677
454. Li A, Chen Q, Wang P, Gan Y et al (2019) Ultrahigh-sensitive broadband photodetectors based on dielectric shielded $MoTe_2$/graphene/SnS_2 p–g–n junctions. Adv Mater 31:1805656
455. Long M, Liu E, Wang P, Gao A et al (2016) Broadband photovoltaic detectors based on an atomically thin heterostructure. Nano Lett 16:2254–2259
456. Ferrante C, Battista GD, López LEP, Batignani G et al (2022) Picosecond energy transfer in a transition metal dichalcogenide-graphene heterostructure revealed by transient Raman spectroscopy. Proc Natl Acad Sci U S A 12(119):e2119726119
457. Trovatello C, Piccinini G, Forti S, Fabbri F et al (2022) Ultrafast hot carrier transfer in WS_2/ graphene large area heterostructures. npj 2D Mater Appl 6, 24
458. Choi SH, Yun SJ, Won YS, Oh CS et al (2022) Large-scale synthesis of graphene and other 2D materials towards industrialization
459. Parappurath A, Mitra S, Singh G, Gill NK et al (2022) Interlayer charge transfer and photodetection efficiency of graphene–transition-metal-dichalcogenide heterostructures. Phys Rev Appl 17:064062
460. Sau JR, Calleja F, Aguilar PC, Ibarburu IM et al (2022) Phase control and lateral heterostructures of $MoTe_2$ epitaxially grown on graphene/Ir(111). Nanoscale. https://doi.org/10.1039/D2NR03074H
461. Shakouri A (2011) Recent developments in semiconductor thermoelectric physics and materials. Annu RevMater Res 41:399–431
462. Hicks LD, Dresselhaus MS (1993) Thermoelectric figure of merit of a one-dimensional conductor. Phys Rev B 47:16631–16634
463. Venkatasubramanian R, Siivola E, Colpitts T, O'Quinn B (2001) Thin-film thermoelectric devices with high room-temperature figures of merit. Nature 413:597–602
464. Hicks LD, Harman TC, Sun X, Dresselhaus MS (1996) Experimental study of the effect of quantum-well structures on the thermoelectric figure of merit. Phys Rev B 53:R10493–R10496
465. Dresselhaus MS, Chen G, Tang MY, Yang RG et al (2007) New directions for low-dimensional thermoelectric materials. Adv Mater 19:1043–1053
466. Geim AK, Grigorieva IV (2013) Van der Waals heterostructures. Nature 499:419–425
467. Hong J, Lee C, Park JS, Shim JH (2016) Control of valley degeneracy inMoS_2 by layer thickness and electric field and its effect on thermoelectric properties. Phys Rev B 93:35445
468. Hippalgaonkar K, Wang Y, Ye Y, Qiu DY et al (2017) High thermoelectric power factor in two-dimensional crystals of MoS_2. Phys Rev B 95:115407
469. Chen Y, Deng W, Chen X, Wu Y et al (2021) Carrier mobility tuning of MoS_2 by strain engineering in CVD growth process. Nano Res 14:2314–2320
470. Shi H, Pan H, Zhang YW, Yakobson BI (2013) Quasiparticle band structures and optical properties of strained monolayer MoS_2 and WS_2. Phys Rev B 87:155304
471. Dobusch L, Furchi MM, Pospischil A, Mueller T et al (2014) Electric field modulation of thermovoltage in single-layer MoS_2. Appl Phys Lett 105: 253103
472. Cho B, Hahm MG, Choi M, Yoon J et al (2015) Charge-transferbased gas sensing using atomic-layer MoS_2. Sci Rep 5:8052
473. Liu B, Chen L, Liu G, Abbas AN et al (2014) High-performance chemical sensingusing schottky-contacted chemical vapor deposition grown monolayer MoS_2 transistors. ACS Nano 8:5304–5314
474. Nan H, Wang Z, Wang W, Liang Z et al (2014) Strong photoluminescence enhancement of MoS_2 through defect engineering and oxygen bonding. ACS Nano 8:5738–5745

475. Zhou W, Gong H, Jin X, Chen Y et al (2022) Recent progress of two dimensional transition metal dichalcogenides for thermoelectric applications. Front Phys 10:842789
476. Li Z, Bauers SR, Poudel N, Hamann D et al (2017) Cross-plane seebeck coefficient measurement of misfit layered compounds $(SnSe)n(TiSe_2)n$ (N = 1,3,4,5). Nano Lett 17:1978–1986
477. Mahan GD, Woods LM (1998) Multilayer thermionic refrigeration. Phys Rev Lett 80:4016–4019
478. Shakouri A, Bowers JE (1997) Heterostructure integrated thermionic coolers. Appl Phys Lett 71:1234–1236
479. Zebarjadi M (2017) Solid-state thermionic power generators: an analytical analysis in the nonlinear regime. Phys Rev Appl 8:14008
480. Wang X, Zebarjadi M, Esfarjani K (2016) First principles calculations of solid-state thermionic transport in layered van der Waals heterostructures. Nanoscale 8:14695–14704
481. Vining CB, Mahan GD (1999) The B factor in multilayer thermionic refrigeration. J Appl Phys 86:6852–6853
482. Sadeghi H, Sangtarash S, Lambert CJ (2016) Cross-plane enhanced thermoelectricity and phonon suppression in graphene/MoS_2 van der Waals heterostructures. 2d Mater 4:015012
483. Liang SJ, Liu B, Hu W, Zhou K et al (2017) Thermionic energy conversion based on graphene van der waals heterostructures. Sci Rep 7:46211
484. Wang X, Zebarjadi M, Esfarjani K (2018) High-performance solid-state thermionic energy conversion based on 2D van der waals heterostructures a first-principles study. Sci Rep 8:9303
485. Derakhshi M, Daemi S, Shahini P, Habibzadeh A et al (2022) Two-dimensional nanomaterials beyond graphene for biomedical applications. J Funct Biomater 13:27
486. Liu T, Wang C, Gu X, Gong H et al (2014) Drug delivery with PEGylated MoS_2 nano-sheets for combined photothermal and chemotherapy of cancer. Adv Mater 26:3433–3440
487. Yin W, Yan L, Yu J, Tian G et al (2014) High-throughput synthesis of single-layer MoS_2 nanosheets as a near-infrared photothermaltriggered drug delivery for effective cancer therapy. ACS Nano 8:6922–6933
488. Han J, Xia H, Wu Y, Kong SN et al (2016) Singlelayer MoS_2 nanosheet grafted upconversion nanoparticles for near-infrared fluorescence imaging-guided deep tissue cancer phototherapy. Nanoscale 8:7861–7865
489. Yang B, Chen Y, Shi J (2018) Material chemistry of two-dimensional inorganic nanosheets in cancer theranostics. Chem 4:1284–1313
490. Wang F, Lv FY, Liu KL, Li ZH et al (2021) Synthesis of MoS_2 nanosheets drug delivery system and its drug release behaviors. Ferroelectrics 578:31–39
491. Yang Y, Wu J, Bremner DH, Niu S et al (2020) A multifunctional nanoplatform based on MoS_2-nanosheets for targeted drug delivery and chemo-photothermal therapy" Colloids and Surfaces B: Biointerfaces, 185:110585.
492. Zhang X, Wu J, Williams GR, Niu S et al (2019) Functionalized MoS_2-nanosheets for targeted drug delivery and chemo-photothermal therapy. Colloids Surf, B 173:101–108
493. Han Q, Wang X, Jia X, Cai S et al (2017) CpG loaded MoS_2 nanosheets as multifunctional agents for photothermal enhanced cancer immunotherapy. Nanoscale 9:5927–5934
494. Kim NY, Blake S, De D, Ouyang J et al (2020) Two dimensional nanosheet-based photonic nanomedicine for combined gene and photothermal therapy. Front Pharmacol 10:1573
495. Yin F, Gu B, Lin Y, Panwar N et al (2017) Functionalized 2D nanomaterials for gene delivery applications. Coord Chem Rev 347:77–97
496. Kou Z, Wang X, Yuan R, Chen H et al (2014) A promising gene delivery system developed from PEGylated MoS_2 nanosheets for gene therapy. Nanoscale Res Lett 9:587
497. Kim J, Kim H, Kim WJ (2016) Single-layered MoS_2–PEI–PEG nanocomposite-mediated gene delivery controlled by photo and redox stimuli. Small 12:1184–1192
498. An D, Fu J, Zhang B, Xie N et al (2021) NIR-II responsive inorganic 2D nanomaterials for cancer photothermal therapy: recent advances and future challenges. Adv Funct Mater 31:2101625

499. Shibu ES, Hamada M, Murase N, Biju V (2013) Nanomaterials formulations for photothermal and photodynamic therapy of cancer. J Photochem Photobiol C Photochem Rev 15:53–72
500. Qian X, Shen S, Liu T, Cheng L et al (2015) Two-dimensional TiS_2 nanosheets for in vivo photoacoustic imaging and photothermal cancer therapy. Nanoscale 7:6380–6387
501. Wang S, Chen Y, Li X, Gao W et al (2015) Injectable 2D MoS_2-integrated drug delivering implant for highly efficient NIR-triggered synergistic tumor hyperthermia. Adv Mater 27:7117–7122
502. Ranasinghe JC, Jain A, Wu W, Zhang K et al (2022) Engineered 2D materials for optical bioimaging and path toward therapy and tissue engineering. J Mater Res 37:1689–1713
503. Raja IS, Kang MS, Kim KS, Jung YJ et al (2020) Two-dimensional theranostic nanomaterials in cancer treatment: state of the art and perspectives. Cancers 12:1657
504. Molaei MJ (2021) Two-dimensional (2D) materials beyond graphene in cancer drug delivery, photothermal and photodynamic therapy, recent advances and challenges ahead: a review. J Drug Delivery Sci Tech 61(Complete). https://doi.org/10.1016/j.jddst.2020.101830
505. Cheng L, Liu J, Gu X, Gong H et al (2014) PEGylated WS_2 nanosheets as a multifunctional theranostic agent for in vivo dual-modal CT/photoacoustic imaging guided photothermal therapy. Adv Mater 26:1886–1893
506. Chou SS, Kaehr B, Kim J, Foley BM et al (2013) Chemically exfoliated MoS_2 as near-infrared photothermal agents. Angew Chem Int Ed 52:4160–4164
507. Salimi M, Shokrgozar MA, Hamid DH, Vossoughi M (2022) Photothermal properties of twodimensional molybdenum disulfide (MoS_2) with nanoflower and nanosheet morphology. Mater Res Bull 152:111837
508. Dou WT, Kong Y, He XP, Chen GR et al (2017) GPCR activation and endocytosis induced by a 2D material agonist. ACS Appl Mater Inter 9:14709–14715
509. Ma YH, Dou WT, Pan YF, Dong LW et al (2017) Fluorogenic 2D peptidosheet unravels CD47 as a potential biomarker for profiling hepatocellular carcinoma and cholangiocarcinoma tissues. Adv Mater 29:1604253
510. Lee YH, Won JH, Kim S, Auh QS et al (2022) Advantages of deep learning with convolutional neural network in detecting disc displacement of the temporomandibular joint in magnetic resonance imaging. Sci Rep 12:11352
511. Wang LV, Hu S (2012) Photoacoustic tomography: in vivo imaging from organelles to organs. Science 335:1458–1462
512. Shilo M, Reuveni T, Motiei M, Popovtzer R (2012) Nanoparticles as computed tomography contrast agents: current status and future perspectives. Nanomedicine 7:257–269
513. Liu Y, Ai K, Lu L (2012) Nanoparticulate X-ray computed tomography contrast agents: from design validation to in vivo applications. Acc Chem Res 45:1817–1827
514. Liu Z, Li Z, Liu J, Gu S et al (2012) Long-circulating Er3+- doped Yb2O3 up-conversion nanoparticle as an in vivo X-Ray CT imaging contrast agent. Biomaterials 33:6748–6757
515. Zhou X, Sun H, Bai X (2020) Two-dimensional transition metal dichalcogenides: synthesis, biomedical applications and biosafety evaluation. Front Bioeng. Biotechnol 8:236
516. Gharavi SM, Qiao Y, Faghihimehr A, Vossen J (2022) Imaging of the temporomandibular joint. Diagnostics 12:1006
517. Kimura K, Fujioka T, Mori M, Adachi T (2022) Dose reduction and diagnostic performance of tin filter-based spectral shaping CT in patients with colorectal cancer. Tomography 8:1079–1089

Chapter 9
TMDs Research with Atomic Layer Deposition (ALD) Technique

9.1 Introduction

Two-dimensional (2D) layered materials are either single-layer or multi-layers atomically stacked structures with the feature of van der Waals interaction between the adjacent layers but strong covalent bonding within the layer [1–4]. Contrast to their bulk complements, 2D materials have various unusual physical, chemical, and electronic features that offer extraordinary prospects for numerous emergent uses [5, 6]. Like optical and electronic devices, novel energy devices, catalysis, and bioengineering [7–13]. However, in the twentieth century, the scientific worker was ambiguous to the survival of 2D materials and normally supposed that films developed thermodynamically unsteady with reduced layers [14]. This situation was still not very abundantly clear until 2004 before a single-layer graphene sheet discovered [15]. This breakthrough effort promptly kindled marvelous investigation attention, and in the present time, various variety of 2D materials have been investigated, like graphene, silicene, and germanane [16], nitrides (e.g., h-BN and GaN) [17], oxides (e.g., MoO_3) [18], chalcogenides (e.g., WS_2, MoS_2, and SnS_x) [3, 6, 19] MXene (e.g., Ti_3C_2 and Ti_4N_3) [11, 20], layered organic materials [21, 22] and others.

Various approaches have been discovered to develop 2D layered materials, like exfoliation, physical and chemical vapor deposition (PVD and CVD), and solution established procedures [5, 23–25]. Though, a perfect technique proficient to fabricating larger amounts extraordinary-excellence 2D materials in a governable fashion is remains missing. Atomic layer deposition (ALD) has lately arisen as a novel production method, owing to its exclusive development machinery and extraordinary physiognomies. A large number of different kinds of 2D materials including TMDs have been explored upto the date by employing the ALD technique [26].

A. K. Singh, *2D Transition-Metal Dichalcogenides (TMDs): Fundamentals and Application*, Materials Horizons: From Nature to Nanomaterials,
https://doi.org/10.1007/978-981-96-0247-6_9

9.2 Atomic Layer Deposition (ALD)

As earlier one has discussed (in Chap. 2) atomic layer deposition (ALD) is a vapor-phase based method for the deposition thin-film materials over the different substrates under sequential and self-limiting superficial reactions. The ALD process can be achieved via dual distinct approaches: (i) atomic layer epitaxy (ALE), and (ii) molecular layering (ML) that was initially presented in the year 1970s [27]. ALD developed to fulfill the precise film thickness for small devices with high aspect ratios [27–29]. Usually, ALD thin films produced by the deposition of chemical gas or vapor phase species; typically it is known as precursors in a repeated manner [30]. ALD is also frequently known as the superior version of chemical vapor deposition (CVD). Contrasting CVD, ALD contains alternate pulses and subsequent purges of the precursors to get a deposition of the required film among an anticipated width and conformation [30, 31]. Moreover, ALD is frequently executed underneath the vacuum at different ranges of temperatures since room temperature to high temperatures, thereby; ALD has an extensive temperature window [32].

A general ALD process is illustrated in Fig. 9.1. Typically ALD process of each cycle is composed of two half-cycle. The ALD process consists of sequential alternating pulses of gaseous chemical precursors to react with the substrate. The individual gas-surface reactions are called 'half-reactions' that appropriately makeup only part of the materials. Thus in the first half-cycle, the precursor is transported through inactive gas and throbbed into the reactor underneath vacuum (< 1 Torr) for a deigned duration and reacts to the accessible active spots of the substrate and it chemisorbed over the surface [30]. The self-limiting process is not leaved supplementary any things than solitary single-layer at the surface. So, typically chamber is purged from an inactive carrier gas (typically N_2 or Ar) for the removal of excess unreacted precursor molecules. In the second half-cycle, the co-reactant pulses within the reactor react to the adsorbed precursor molecules over the substrate [33, 34]. The usual co-reactants for the ALD process are water vapor, O_2, O_3, and NH_3 [35]. In the end, the surplus co-reactant molecules and by-products of the reaction are purged out from the reactor. By repeating the double half cycles a desired film width and configuration are obtained [36].

Typically, ALD processes are conducted at modest temperatures (< 350 °C). The saturation range of temperature depends on the growth of a specific ALD process and it is referred to as the ALD temperature window. ALD process outside the temperatures window usually revealed a poor growth rate and non-ALD type deposition owing to slow reaction kinetics or precursor condensation (at low temperature) and thermal decomposition or rapid desorption of the precursor (at high temperature). To get benefit from the advantages of ALD, it should be operated within the designated ALD window for each deposition process [35].

The key advantages of ALD to the entire process are derived from the sequential manner, self-saturating, gas-surface reaction under the controlled deposition. The conformality of ALD-deposited films is considered to be a critical factor for the selection of ALD over contrast to deposition techniques such as CVD or sputtering.

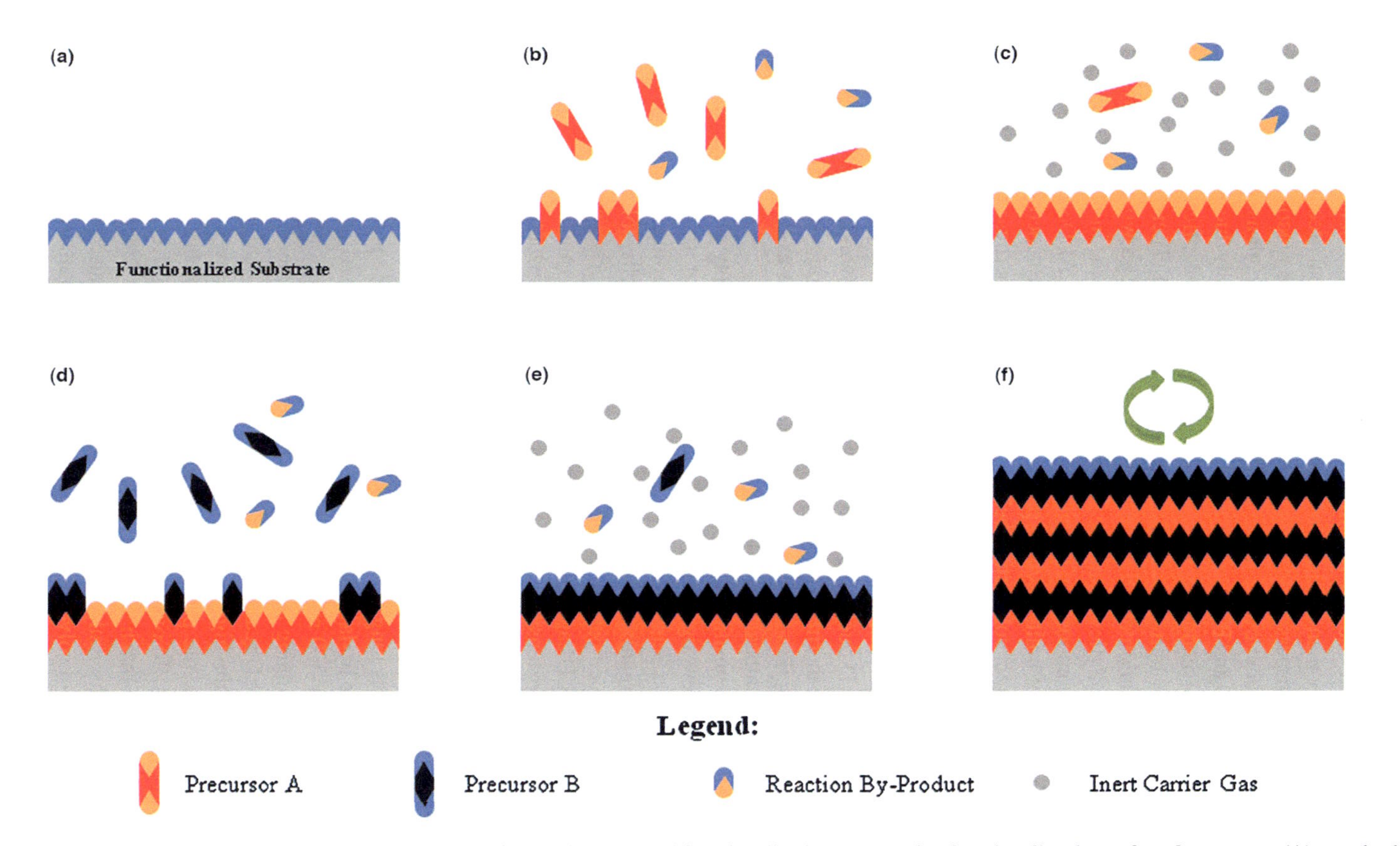

Fig. 9.1 Schematic for the ALD process: **a** substrate surface with the natural functionalization or treated to functionalize the surface; **b** precursor (A) to pulsed and reacts with the surface: **c** excess precursor and reaction by-products purged with the carrier gas; **d** precursor (B) pulsed and reacts with the surface; **e** excess precursor and reaction by-products purged with inert carrier gas; **f** steps 2–5 repeated until the desired material thickness achieved (adopted from Johnson et al. [35], p. 226)

With good conformality and a high aspect ratio three dimensionally-structured materials are possible from the self-limiting characteristic. That restricts the reaction at the surface to no more than one layer of precursor. The sufficient precursor pulse time allow to precursor disperse into deep trenches and complete reaction to the entire surface. The subsequent cycles allow uniform growth with high aspect ratio structures, while the CVD and PVD are suffered from non-uniformity due to faster surface reactions as well as shadowing effects. Moreover, ALD also offers the thickness control deposition of thin films by layer-by-layer deposition. Therefore, the thickness of a film can be tailored by the number of ALD cycles. The growth per cycle of the ALD films are typically less than one A/cycle, depending on the individual process. Additionally, ALD composition is controlled by tailoring the ALD super cycles that allow to composed of multiple ALD processes. Adjusting the supercycle ratios also allows to tailor different conduction behavior and optical properties of the films [35]. Though, it has been also noted that a nonlinear relationship between the cycle ratio and the film's atomic ratio is common in the ternary oxide processes which makes it less straightforward to deposit a film with a certain desired composition [35, 37, 38]. Moreover, complications are also in composition control to ternary and quaternary metal oxide ALD due to the requirement of the growth windows in individual ALD processes to be thermally compatible and a way of introducing dopants into the films as 'δ-doping' spikes that results in inhomogeneous films as-deposited, thereby typically it requires annealing [39].

ALD thin film deposition method has a widespread variety of applicability [3, 5, 9, 10]. Such as, few significant utilities based on ALD but it is not restricted to the subsequent areas: semiconductor manufacturing, lithium-ion batteries, microelectromechanical organizations (MEMS), capacitors, fuel cells, solar cells, transistors, drug delivery, medical and biomedical areas, dental materials and orthopedical implants [32, 40]. ALD capability of precise deposition conformal ultra-thin films is the interest for their extensive variety of applications in the production of microelectronics like gate oxides, semiconductors, and ferroelectrics [41–43]. This procedure has also attended great consideration in medical and biomedical utilities, especially when using organic substrates like polymers or biomaterials [44, 45].

Despite various encouraging characteristics ALD also suffered to slower deposition rates. Due to its long cycle times involvement while the pulsing and purging precursors and layer-by-layer deposition, as in most of the ALD deposition rates are the order of 100–300 nm/h [37, 46]. ALD rate of deposition strongly depends on their reactor designing and the aspect ratio of the substrate. If the surface area and volume of an ALD reactor increase then it takes more time for the pulsing and purging. Moreover, the high aspect ratio substrates also require a longer pulse and purge times to allow the precursor gas to disperse into trenches and other three-dimensional features. To overcome this, shortcoming spatial ALD has emerged as a promising technique that has significantly improved the throughput [47–49]. Spatial ALD has resolved the issue by the eliminating traditional pulse/purge chamber by replacing a spatially-resolved head, which exposed the substrate from a specific gas precursor depending on location. Such as the heat translates around the substrate to alter the exposed precursor, and finally get a film growth. Similarly, the spatial

ALD also works where the substrate moves past stationary precursor nozzles and it arranged to pass by them, so, the precursor cycling are finally achieved, the film grows. Overall, the spatial ALD techniques deposition rates can be improved up to around 3600 nm/h [50].

Thus spatial ALD (SALD) is a rather new tool that sentient for the comprehensive air if executed in the open air because extremely explosive precursors are used, however, many precursors cannot be useful for this process [51]. Therefore, before performing each ALD experiment, every condition should be determined and optimized. Hence the following parameters should be considered before performing the ALD experiment:

- The final applications should be very clear.
- Which kind of materials (forerunner, oxidizer and substrate) should be taken for the use.
- The exposer period pulse and purge should be estimated for an extraordinary-superior thin film.
- The growth rate depending on the determining circumstances should be specified.
- The reactor and precursor bubblers temperature and pressure should be predefined.
- Every material taken in use should be steady in employed circumstances.

By defining the above-described parameters or by adjusting the conditions (as desired) extraordinary-superior films together an adjusted growing rate can be achieved. It should be noted that to ensure experimental reproducibility each process repeats a few times for the samples. However, the frequent predecessors for ALD are in general expensive to perform various experiments under the adjustable circumstances makes it inefficient. Moreover, in certain cases, hazardous by-products are formed therefore; in such cases reducing the numeral of experimental cycles. Additionally, the theoretical modeling is another way to access the optimum circumstances with no actual experimentations, it significantly reduce the cost and time-consuming in the experiments.

So, a great attention has been also paid to ALD theoretical research based on computer simulations and theoretical approaches in combination to ALD use for the different determinations, like, thin film configuration and alloy and material choice [52]. Most of modeling procedure alters onto the multiple scales depending on the inter-attraction among atoms, molecules, particles, and groups of atoms (typically it is recognized as the functional groups). Moreover, the reverse engineering approach is adopted to define the various aspects of ALD. That would be useful for the experimental progression utilized in ALD growth to define in what way modeling and simulation provide a proper understanding to improve the ALD process [22].

9.2.1 ALD Precursors

ALD precursor is often a metal surrounded by organic functional groups, typically it is held in a vessel known as a bubbler. The bubbler temperature of the ALD system can be varied; it depends on the characteristics of the predecessor. Usually, the ALD predecessors are explosive, thermally steady and extremely reactive [37]. To select a specific precursor for a particular system ALD process a few physical properties should consider, like materials of attention, reactivity with supplementary co-reactant, ALD circumstances and ultimate applicability of the required film. Moreover, the dielectric constant, adsorption capability, gas impermeability, leakage current, electrical conductivity, photochemical activity, and antimicrobial activity should also pay attention to selecting a precursor [32]. In addition, these predecessors should be competent to react rapidly to the dynamic spots of the substrates and additional predecessor molecules. Therefore, predecessor chemistry is considered to be a crucial aspect in an ALD procedure because it also impacts the development machineries. In the advanced ALD process incorporation of fresh predecessors has also find abundant consideration. As the various theoretical approaches described, it could be appropriate to design innovative materials depending on available experimental data with the prediction of their features prior to manufacturing them [31].

The first-principles approach is also useful to investigate innovative materials with the expected characteristics which would be proven afterward experimentally [42, 53, 54]. Though, it is customary to theoretically predicted precursors should compare the various aspects because some of them are practically impossible from the experimentation. Hence theoretical approaches could be useful to define a precursor for the ALD, with their expected physical properties. For the instance, without experimental defined ALD conditions of silicon carbide (SiC) DFT calculations have been provided as the utmost favorable predecessors for the ALD method [55]. According to this interpretation combinations of disilane (Si_2H_6), silane (SiH_4), or monochlorosilane (SiH_3Cl) with ethyne (C_2H_2), carbon tetrachloride (CCl_4), or trichloromethane ($CHCl_3$) are the most promising materials for the ALD of SiC at 400 C. Thus the selection of materials and situations to the ALD possibly would propose without executing any experimentation that could save abutment quantity of time and efforts.

In this order also fabricated the zirconium precursor for the ZrO_2 ALD on silicon and their properties have been compared to the usually utilized zirconium predecessor [56]. The DFT calculations led to the novel predecessor would require an inferior progression rate owing to steric interruption with an additional separate segment of a single-layer in single cycle [56]. Similarly, the copper (I) carbene hydride DFT outcomes lead that it can act as a reducing agent and a predecessor to the Cu ALD [52]. The Cu established reducing agent has used in the case of a co-deposition where more copper deposition is desired [52]. Other materials DET calculations also defined the throughput with their expected physical properties, such as platinum [57], aluminum oxide [58, 59], hafnium oxide [60, 61], titanium oxide [62, 63], zirconium oxide [64, 65], etc. Similarly, aluminum ALD precursor can also achieve through replacement of solitary methyl in TMA and selective decoration of Pt nanoparticles among AlOx by ALD [58]. Moreover, dimethylaluminum isopropoxide (DMAI) is also utilized as an ALD predecessor with the decomposition process of DMAI [58]. In addition to these, the hafnium precursors are also used for ALD. Their DFT predictions have revealed the cyclic precursor possesses a lesser growing rate contrast to the alkyl amide predecessor resulted owing to the inferior possibility of the ending chemical adsorption of the bulky cyclic ligand over the surface [60].

Moreover, in this case of two similar precursors can be also used, to explain the DFT calculations are helpful by comparing various ALD predecessors without conducting actual investigation. As the intense, halide precursor's thermodynamics and kinetics are to be compared [62]. Similarly, the different aspects of TiI_4 and $TiCl_4$ can also compare [62]. The dissimilarity in bond forte among Ti-I and Ti-Cl are not allowing a substantial variation in the kinetics of their reactions with water. However, the diverse bond forces considerably affect the reaction thermodynamics. Additionally, the use of TiI_4 over $TiCl_4$ leads to a film containing fewer contamination and favorable to ALD at a lower temperature. Thus TiI_4 is considered to be a superior predecessor for TiO_2 ALD over the organic substrates contrast to $TiCl_4$ [62].

Moreover, some novel theoretical models have been also investigated to shape predecessor materials for ALD [66]. Like a GCM approach to estimate thermodynamic features (activity constants) of the operational groups which previously exist as the ALD predecessor materials, like tetrakis (dimethylamido) titanium (TDMAT), tetrakis (diethylamino) titanium (TDEAT), tetrakis (diethylamino) hafnium (TDEAH), tetrakis (ethylmethylamino) hafnium (TEMAH), etc. By utilizing assessed activity constants of these groups can formulate the CAMD outline for the optimal strategy to innovative predecessor materials for the ALD. They may also compare to saleable existing predecessors. As the highest optimum proposed predecessors have been projected with a ~ 40% upsurge in ALD growing rate [66]. The similar GCM approach could also employ for the figuring of the water contamination in an ALD reactor [67].

9.2.2 ALD Growth Characteristics

Usually, the ALD growth rate is also defined in terms of the width of the film by dividing the integer of cycles (nm/cycle). It is depending on various aspects, like predecessor flux reaching time to the substrate [30].

The growth rate of the ALD directly provides an understanding of the overall time taken in the designated thin film deposition. The ALD growth rate may different for the distinct system. Moreover, the growth rates are also impacted from the reactor temperature and pressure, pulse, and purge periods of the reactants, including the behavior of the substrate. In general, the ALD progression rate is considered to be continuous, in terms of the incline of the plot of film width to the numeral of cycles; though, it has also predicted that the development rate also changes with the numeral of cycles. As the example, a couple-lined plot of TiO_2 film width with the

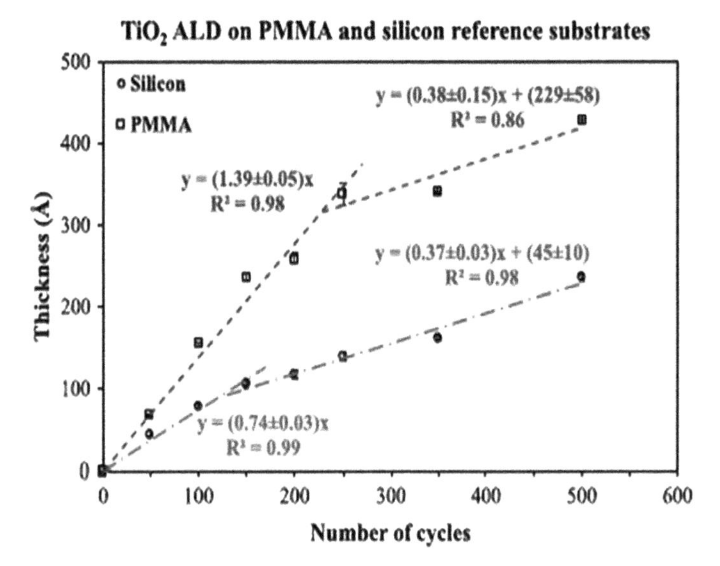

Fig. 9.2 Duo-linear plot of TiO_2 film thickness versus the number of cycles (adopted from Shahmohammadi et al. [32])

integer of ALD sequences. This leads the engrossed water molecules in pores of a polymethyl methacrylate (PMMA) substrate and liberated within the reactor while the preliminary cycles that affect the growing process, thereby, a greater growing rate contrast to later deposition progressions, as illustrated in Fig. 9.2 [31].

The significant ALD growth aspect is also to be considered, as how materials are settled over the surface of the substrate throughout the growing process [68]. Because ALD growth modes are usually attributed to one of the following modes: Volmer–Weber growth, Frank–van der Merwe growth, and Stranski–Krastanov growth [69, 70]. Volmer–Weber growth mode also known as island growth in which small clusters or islands nucleates first on the surface due to the interaction of the adsorbed atoms in themselves stronger compare to the interaction between the atoms and the substrate. Therefore, the cohesive force within the atoms is stronger than the surface adhesive force, so, atoms tend to accumulate. Such small clusters grow into larger three-dimensional ones and reach each other to cover the whole surface [69]. On the other hand, in the case of the Frank–van der Merwe growth mode (also known as bidimensional growth), the surface adhesive forces are stronger than the intra-atom cohesive force and produce layer-by-layer growth on the substrate. Therefore, the atoms covered the whole surface and produced a monolayer before the subsequent layer formed on top [71]. Moreover, when both Volmer–Weber and Frank–van der Merwe's growth modes are combined, the growth mode is called Stranski–Krastanov. The Stranski–Krastanov growth mode is more common because the film starts to form on the surface of the substrate as a whole layer. More successive layers formations lead to a pretty dense film; so, the growing style switched toward island development type [69]. Such transitions also impact the chemical and physical features of the film and substrate resources [69–71].

9.2.3 Surface Roughness

ALD process fabricated thin-films have been also widely explored to define the superficial unevenness. Owing to fact that, unevenness is portion of surface texture that defines like waviness or anomalies in the film surface. In this regard, the mapping technologies like AFM and optical or contact profilometry are used to define the surface roughness parameter. Moreover, the surface roughness is also significantly affects by the growth mode. For the instance, the Volmer–Weber growth mode starts with island nucleation, and the islands are further enlarged over time, therefore, their heights become larger than the thickness of one monolayer before converging to the whole superficial area. Similarly, in the case of the Volmer–Weber type growth mode, usually films are formed rough layers. However, smoother films can be achieved from the Frank–van der Merwe growth mode due to its ability to form layer-by-layer stacking [32]. Additionally, including all these the surface roughness also depends on other factors such as crystallinity, film thickness, and the nature of the substrate.

9.2.4 Step Coverage or Conformality

Conformality has defined as the deposition of a film having the same thickness with all topographic features, such as the top, sides, and bottom surfaces of a three-dimensional substrate. As the ALD process thin film deposition under the surface governed responses, therefore the formed thin films have an exceptional conformity. Thus, the step coverage offered through the ALD for multifaceted building surfaces is sophisticated to the traditional deposition methods like CVD and physical vapor deposition (PVD). Conformity is also considered to be a significant parameter and it greatly requires to depositing compounds of three-dimensional nanostructured surface or substrates with the higher amounts ratios [72, 73].

9.2.5 Deposition Temperature Impact

Especially in the case of the thermal ALD temperature is the crucial governing strength of the procedure. Therefore, in an ALD process, a temperature window should be established respect to the temperature range at that self-governed development could occur with a fixed rate. Usually within the temperature ALD window of the grow rate alter considerably together enhance in the physisorption/condensation of predecessors over the surface (or inferior reaction rates), so, the ALD process leads an uncontrolled growth. Whereas outside the temperature window in ALD operation, the precursor molecules are decomposed or desorbed and heat the substrate, and as a consequence uncontrolled growth results [74]. The characteristic temperature window for the normal thermal ALD processes is defined in the range of 150–350 °C [30], however, in some cases it may exceed this well-defined temperature window. For the instance, in the case of the organic and heat-sensing substrates the reactor temperature must be regulated to avoid distortion and degradation of the substrate. Moreover, room temperature ALD is also possible on halogen materials owing to biomaterial functionalization [75, 76].

9.2.6 *Mechanisms of the Thermal ALD*

Since a binary ALD process has consisted of a dose–purge–dose–purge sequence in reactant formation in each ALD cycle. In the case of the thermal ALD surface reactions occur at a comparatively higher temperature. Moreover, the self- restricting feature of the ALD process permits to steadiness of the surface development on a substrate throughout the dosage steps earlier to additional unreacted materials are purged out from the reactor. Therefore, after the chemisorbed species saturated the entire accessible active spots, there is no further chemisorption is possible, although in practice additional reactants can occur within the reactor. So, the precise machineries of the ALD reactions are still crucial and exciting. Nonetheless, few modulations have projected ALD reaction machineries. Typically ALD machineries are separated in the three classes depending on their preliminary surface reactions, reaction directions, and predecessor decomposition [77].

9.2.6.1 Initial Surface Reactions

In the ALD procedure initially a determinate quantity of active locations are existing throughout the surface of a substrate for every cycle. The primary surface reactions due to active spots those engaged from the reactants and depleted at the completion of every half-cycle, as the consequence supplementary active spots are generated in a successive half-cycle. Therefore, ALD per cycle initial growths depending to the amounts of nucleation locations that can be classified in three groups: linear, surface-improved, and surface-inhibited [78].

Reaction Pathway Process

The reaction passageways usually happens in an ALD reactor during the creation of aluminum oxide (Al_2O_3) from trimethylaluminum (TMA) and water (H_2O) are defined. Typically, TMA behaves as the aluminum predecessor within the organization and water serves as the oxygen basis. As the generalized versions of a few

Table 9.1 Well-recognized ALD reaction pathways [32]

Thin film	Precursor	Co-reactant	Reaction pathway
Al_2O_3	TMA (trimethylaluminum)	H_2O	$-OH + AlMe_3 \rightarrow -OAlMe_n + (3 - n)\ CH_4$ $-AlMe + H2O \rightarrow -AlOH + CH_4$
MO_2 (Ti, Hf, Zr)	MCl_4	H2O	$n\ (-OH) + MCl_4 \rightarrow (-O-)_n MCl_{4-n} + n\ HCl$ $(-O-)_n MCl_{4-n} + (4 - n)\ H_2O \rightarrow$ $(-O-)_n M(OH)_{4-n} + (4 - n)\ HCl$
MO_2	TDMAM (tetrakis(dimethyl amido) metal (Ti, Hf, or Zr))	H_2O	$M(NMe_2)_4 + 2\ H_2O \rightarrow MO_2$ (solid) $+ 4\ HNMe_2$

frequent and well-recognized reaction pathways for the ALD procedures are given in Table 9.1.

9.2.6.2 Precursor Chemisorptions

Generally, mechanisms of the precursor chemisorptions of an ALD process have been described that are predominately characterized in three groups; ligand interchange, dissociation, and association [77, 79]. In the ligand interchange chemisorptions process the splitting of predecessor take place over the surface and its ligand swapped together a surface assembly, finally, a gaseous product is produced [77]. Whereas in the dissociation procedure, single or extra ligands of the predecessors are detached to the molecule and attached from the surface clusters, as the consequence creating active spots over the substrate [77]. Their molecule constituents are frequently unglued owing to an exterior source like light or heat [79]. whereas, in the association progression, there is no ligand splitting that occurs to the predecessor, however, a coordinated bond are also generally shaped among the predecessor and the surface dynamic spots [80].

As the intense ligand exchange precursor chemisorption of the Al_2O_3 from TMA and H_2O for the ALD process, the typical pathway schematic is illustrated in Fig. 9.3a–e. In the principal half-cycle of the reaction TMA molecules to be chemisorbed over a hydroxylated surface and react to OH clusters, in which a ligand interchange happens and methane gas is liberate. Afterward the complete surface saturation to the TMA molecules, entirely unreacted predecessors and byproducts of the reactions are purged out. Moreover, the water fragments entered into the reactor for the next half-reaction that reacts to the CH_3 dismissed surface locations. In this process once more ligand interchange occurs among the CH_3 clusters of chemisorbed TMA molecules and OH assemblages of water vapor, and methane gas is liberate till entire dynamic positions are occupied. In the end, by-products and unreacted water

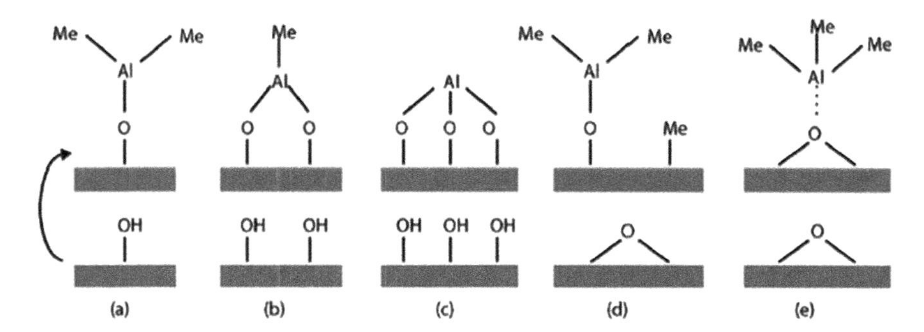

Fig. 9.3 Schematic of TMA reaction possibilities on oxidization of the surfaces: **a** ligand exchange with one; **b** two; **c** three OH groups; **d** dissociation; **e** association (adopted from Shahmohammadi et al. [32])

molecules are pumped out to the reactor [79]. Therefore, the surface is again hydroxylated to repeat the processing. Similarly, the ligand-exchange in ALD reactions also occurs in the deposition of metal oxides exhausting the alkoxide predecessors [81]. In the metals (such as copper, ruthenium, and platinum) ALD process frequently happens through the dissociation of chemisorption [79, 82]. However, the association machinery is difficult to recognize because it typically occurs afterward gas-phase dissociation or earlier to a ligand-interchange reaction, such as metal halide precursors [79].

9.3 TMDs ALD

Usually, 2D TMDs deposition on the TMDs is desired to fabricate various devices [83, 84], such as a hybrid material for energy uses together inorganic/organic materials [85, 86]. There are abundant deposition procedures, incorporating vapor deposition (physical or chemical), solution procedures etc. The ALD is also unique specified chemical vapor depositions consisting consecutive disclosure of chemical species. It provides supreme promising deposition circumstances for the nanoscale device manufacturing [87, 88]. Additionally, their self-restricting development features, predecessor molecules chemisorb over the surface reaction locations and successive reactant molecules reaction to the adsorbed predecessor species creating a subnanometer-thick film underneath a cycle makes ALD an impressive technique. ALD unique deposition can also offer numerous benefits, like atomic-scale width governability with great expanse consistency. The highly smooth coating even over the 3D composite configurations, such as nanoflakes fabricated from this technique [89, 90]. Moreover, the 2D material has a fairly diverse surface chemistry related to traditional 3D materials, such as the growing physiognomies of ALD over the 2D material have a considerably distinct feature. Contrasting to traditional 3D materials, ALD of 2D TMD materials has an absence of dangling bonds throughout the surface due to a chemically inactive surface over the basal plane [3]. This directly correlates to the absence of chemisorption chemical species availability over entire surface. However, the manufactured 2D TMDs can have intrinsic imperfections that are produced while the procedure, thus irregular circulation of reaction spots over the TMDs persuaded uneven deposition throughout the surface afterward the ALD action [91]. This is one of the critical limitations to employing ALD over the 2D materials. Therefore, most of the attention has been focused toward get a uniform deposition from ALD with no destruction or damage an original TMD structure.

9.3.1 Significant Surface Properties of 2D TMDs

As one has discussed in Chap. 2, the basic distinction among traditional materials and 2D TMDs is the occurrence of the reaction spots throughout the surface. A traditional material has a dangling bond over the surface that allows the reactions to diverse chemical species. Whereas an ideally in finite 2D material free of the dangling bond theoretically due to its strange crystal formation [26], this specifies that 2D materials are chemically inactive. Thus nonexistence of the dangling bond over the basal plane is considered to be key benefits of 2D TMDs. This provides the advantage to avoid electrical deterioration due to interface traps [92]. In the ALD process, the dangling bond-free condition allows for the suppression of material progression over the surface. For the example, if a predecessor or reactant is unprotected to the reactor over a dangling bond rich substrate, a chemical reaction happens on topmost of the surface through the reaction to dangling bonds (surface reaction sites), so, the film creation is started. But it should be noted that there is a nonexistence of a reaction spots on pure 2D TMDs which inhibits the film development on a basal plane [93].

Another side, for the un-ideal determinate 2D TMDs, the situation is fairly diverse because of distinct surface chemistry. Like utmost frequently used TMD flake by synthesized mechanically or chemically exfoliated [23, 94, 95] to get a single-crystal TMD. Moreover, the mechanical exfoliation process has been also used as an adhesive tape to peel out the ultrathin 2D TMD flake to the bulk TMD crystal and ensure the TMD flakes together solitary crystals with extraordinary clarity and impeccability. Particularly, a mechanically obtained 2D TMD flake has an extremely lower extent of imperfection concentration; therefore, it is useful to comprehend the inherent physiognomies of 2D TMD [96, 97]. It should be also noted that a few additional imperfections or edge spots may occur over the extracted flake and generates while the production or transporting procedure (point imperfections or dislocations) also significantly influence over the surface physiognomies [98]. The existing dangling bonds are chemically reactive [99–101] whereas the imperfections or edge locations of fabricated TMD acts as reaction spots to the gas molecules or chemicals [102, 103]. Therefore, such developed 2D TMDs show a better characteristic that is appropriate for laboratory-level device construction, but it cannot be considered too suitable for the larger amounts production with the controlled width and size of the flakes. To scale up the large-scale fabrication CVD-grown TMDs have been preferred owing to their ability to a comparatively greater level of imperfection concentrations owing to imperfections induced during the production procedure, it may also induce several types of organizational imperfections in the material. Thus the surface chemistry leads to a higher level of imperfection concentrations over the 2DTMD while a reaction of chemical species are considered to be more favorable to grown TMD from the CVD when once exfoliated.

Moreover, 2D TMDs consisting dual or additional diverse components including transition metal and chalcogen, therefore, it has many possibilities of various types of defects, such as point defects in 2D MoS_2 in terms of mono sulfur vacancy, disulfur vacancy, vacancy compounds containing Mo and three adjacent sulfur atoms, vacancy compounds associating Mo and three disulfur couples, antisite imperfections, and Mo atom, etc. [104]. Therefore, the adsorption of chemical species on 2D TMDs different to chemical species or a type of imperfection spot in other types of 2D materials. As the inspected adsorption of numerous molecules over the 2D MoS_2, thereby, the adsorption dynamisms among several molecules (CO_2, CO, O_2, H_2O, NO, N_2 and H_2) and basal plane or imperfections (S vacancy, S divacancy, Mo vacancy, single or dual Mo atoms over S solitary and divacancy, and single or dual S atoms over Mo solitary and divacancy) of 2D MoS_2 have been calculated from the density functional theory (DFT) approach. It demonstrated that every kinds of molecules have superior adsorption energy onto the imperfection locations contrast to the basal plane. Their adsorption energy can be altered contingent on the surface ingredients and the kind of chemical constituents. Thus, the imperfections over the 2D TMDs key contributions as the extremely reactive spots and trap locations to diverse molecules through the disclosure of chemical functional clusters [105].

9.3.2 *Impact of Physisorption on ALD*

According to traditional ALD concept, a film is constructed over the reaction locations of the surface through the chemical reaction among chemical constituents such as predecessor and reactant [28]. Such kind of chemical reaction is known as dissociative chemisorptions that happen via external energy and form a chemical bond among adsorbent (solid) and adsorbate (gas). It has believed while the ALD procedure, the chemical constituents could solitary react to the surface reaction spots of the substrate and remaining molecules cannot react therefore they purged out, such self-restricted development depending on the chemisorption and irretrievability in overall [106]. Compare to chemisorption established reaction the distinct adsorption machinery, physisorption is also considered to be significant in the contemporary ALD progression, specifically for the 2D materials. As per the physisorption procedure, the chemical constituents approach over the topmost of the 2D surface that is physically combined to the surface together a feeble intermolecular strength (van der Waals force) with no bond alteration among an adsorbent and adsorbate. However, some exceptions have been also recognized such as physisorption having fundamental bonding theories in carbon-based materials (carbon nanotube or graphene) [107, 108].

To resolve this problem several investigators made the effort with novel theories and concepts. As depicted in Fig. 9.4a, b, a plot of the Lennard–Jones potential approach is commonly utilized for the molecular adsorption exemplary. Here the deepness of the potential well relates to the attraction strength among the solid (substrate) and gas (precursor or reactant) and the x-axis reflects the span among them. According to this representation, the attraction force of physisorption is reasonably small (~ 40 meV) for the comparatively larger span (> 3 Å) contrast to the chemisorptions (~1 eV) together a lesser span (< 3 Å) [109, 110]. Due to its feeble bond forte and comparatively extended span the physically combined molecules can be effortlessly disconnected through the exterior energy sources, such as in thermal heating or long purging time [111, 112]. Moreover, the physisorption process does not allow the bond adjustment among the adsorbate and the adsorbent; therefore, it is considered to be reversible. Thus the ALD over the 2D substrate is advantageous for the imperfection or edge spots and it does not lie over the basal plane. Moreover, if specific chemical constituents are physically combined to the basal plane of the 2D material, in that case, they may behave as a reaction location for the predecessor, as the consequence nucleation over the basal plane. This leads, besides molecules physically combined over the whole basal plane an unbroken thin film fabrication on dangling bond-free 2D TMDs could be also possible from the ALD.

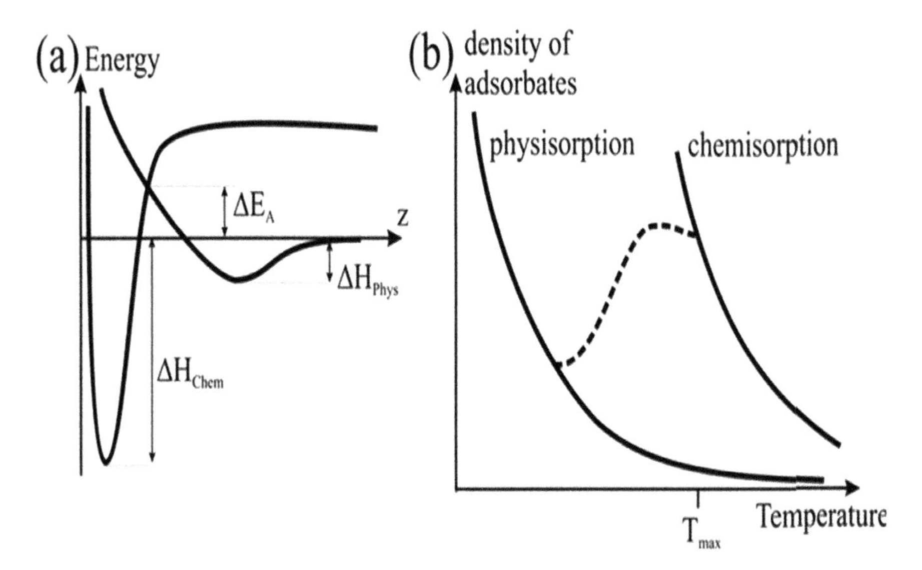

Fig. 9.4 **a** Lenard-Jones model for physisorption and chemisorption of molecules; **b** typical adsorption isobars whereas the solid lines are equilibrium physisorption and chemisorptions isobars, the dashed line represents irreversible chemisorption (adopted from Batzill, M. 2006 Sensors 6, 1345–1366)

9.3.3 Modern High-Quality Uniform Film Deposition from ALD

Nowadays ALD of TMDs offers highly unbroken films up to a sub-10 nm thickness level which is necessary for nanoscale device incorporation for the extraordinary working ability, as one has discussed in Chap. 2 [113, 114].

9.3.3.1 Significance of ALD Without Any Surface Treatment

The key significance of ALD without any surface treatment has been described from the experimental observations and theoretical interpretations. As it has been noticed that the combined mobility of 2D materials together the high-k dielectrics is enhanced due to dielectric screening [116–120]. This process fabricated FET device's electrical properties are also superior due to the screening effect. Although a direct ALD over a raw MoS_2 surface could show abundant nucleations, therefore insignificant changes in surface structure appeared earlier and afterward ALD HfO_2. Thus, it gives usual nucleation and growth of ALD films favorably occur on surfaces that have a high concentration of hydroxyl (–OH)-terminated groups, which are considered as the adsorption sites for the precursors [121]. On the other hand, the mechanically exfoliated TMD flake has no inherent –OH species, so relatively a flat thin film is achieved [122].

In the begging efforts have been made to get an unbroken thin film (< 10 nm) over the 2D TMDs by adjusting the procedure temperature and demonstrated the Al_2O_3 ALD over exfoliated MoS_2 then boron nitride (BN) [93]. Specifically deposition of ALD Al_2O_3 over mechanically exfoliated 2D MoS_2 and BN films from trimethylaluminum (TMA) and H_2O with different temperatures. As the atomic force microscopy (AFM) picture of MoS_2 and BN substrates afterward 111 cycles of ALD Al_2O_3 for the different procedure temperature (200, 300, and 400 °C) is illustrated in Fig. 9.5a. It noted a nominal thickness variation in Al_2O_3 with the growing rates of Al_2O_3 over the SiO_2 substrate for the 10 nm with different temperatures. This lead chemical adsorption of TMA over OH collections are almost unaffected by the temperature. On the other hand, the surface structure of ALD Al_2O_3 over the BN and MoS_2 is intensely impacted by the growth temperature. Such as at 200 °C, ALD Al_2O_3 deposition over the BN and MoS_2 has an outstanding consistency lacking of voids or pinholes. It also leads to the increasing procedure temperature of ALD Al_2O_3 also increasing several voids and pinhole formations, as revealed at 400 °C, ALD Al_2O_3 a numerous islands of Al_2O_3 on the surface. Moreover, the surface coverage of ALD Al_2O_3 over MoS_2 is marginally greater the BN due to the effect of the procedure temperature. Therefore, the feature of film development depends on the procedure temperature and it is directly correlated from the physisorption of the predecessor.

Moreover, to examine the inter-attraction among predecessors and the surface of 2D materials and comprehend the reaction machinery of early ALD cycles over 2D materials, adsorption energy among TMA and MoS_2 or BN has been also estimated

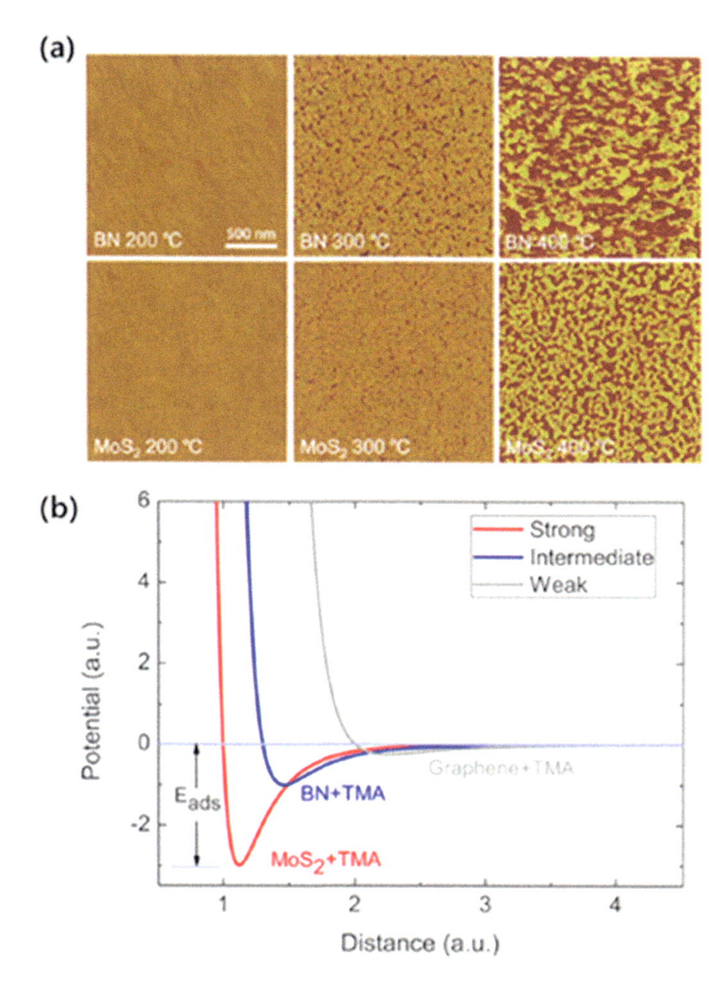

Fig. 9.5 **a** Atomic force microscope picture of BN and MoS_2 after 111 cycles of ALD Al2O3 at 200, 300, and 400 °C (scan size: 2 × 2 μm^2 with a scale bar of 500 nm). **b** illustration of Lennard–Jones potential model for physical adsorption at 2D crystal surfaces (adopted from Nam et al. [114])

by adopting the DFT approach. The physical adsorption energy of TMA over the BN superficial is around 8.7 kcal/mol that greater than the H_2O over BN. Similarly, the physisorption energy of TMA over MoS_2 is around 21.9 kcal/mol, it also superior than H_2O onto MoS_2. Thus TMA is more intensely physisorbed over the 2D materials compare to H_2O. Additionally, TMA also has the greatest inter-attraction energy among MoS_2 due to a few reasons. Like positively charged Al could extra positively interact to the negatively charged N, S and subsequently O while the polarizabilities of TMA and MoS_2 are superior to H_2O and BN, as the Lennard–Jones potential is depicted in Fig. 9.5b. Hence, the greater coverage of Al_2O_3 over MoS_2 compare to BN in the higher temperature range (300–400 °C) has been established with the note temperature-dependence progression of ALD Al_2O_3 over the 2D substrate is predominantly governed from the physisorption of the predecessors at the place of chemisorption.

9.3.3.2 Significance of Uniform Deposition by Surface Functionalization

Lowering the process temperature from the possible methods to get a uniform thin film on 2D TMDs also brings undesirable effects as well. Like, low-temperature ALD have a poor film density and much incorporation of impurities from precursor (e.g., C, N, O, etc.) due to insufficient activation energy for the reaction between precursor and surface reaction sites [63, 64]. So, it is extremely desired to get unbroken film from the ALD with no humiliation in the film excellence. To fulfill this requirement molecules attaching mechanism has been identified as a prominent technology.

Significance of Molecule Functionalization

To show the significance of molecular function effort has been with the ALD of HfO_2 on a mechanically exfoliated MoS_2 substrate [91]. By depositing 30 nm of HfO_2 over MoS_2 to cover the complete MoS_2 layer un-uniform and subsequently fabricated FET by the coating of organic materials shadowed by elimination from the solvents which is done afterward the transferring or lithography procedure [123, 124]. It has been remarked that the generated residues during processes acts as reaction sites [125]. But note that it should be defined whether deliberately combined molecules (organic or solvent contaminants) would encourage nucleation of ALD HfO_2 over the MoS_2 flakes by soaking in the acetone or N-methyl-2-pyrrolidone (NMP) or spin-coated from poly(methyl methacrylate) (PMMA) earlier ALD HfO_2. To explore this view, it has been also demonstrated that even no important alteration afterward ALD when MoS_2 immersed into acetone as depicted in Fig. 9.6a, however, a vibrant change in the phase picture of AFM appears afterward PMMA coating or the NMP action. Moreover, it also leads the organic remainder or solvent that remains over the MoS_2 substrate to act as a nucleation raise locations, thereby, improving the consistency of ALD HfO_2. Additionally also noted that the purge time of the ALD procedure affected the physisorption of chemical constituents. The exceptionally lengthy purge period among predecessor and reactant exposure (over 6 h) revealed none of HfO_2 associated peaks in the x-ray photoelectron spectroscopy (XPS) spectrum (see Fig. 9.6b). Whereas the vibrant Hf4f and O1s peaks for the predecessor and reactant purge period for the 5 and 10 s. This is direct evidence of physisorbed precursor molecules that detached during the longer purge period. Hence, the physisorption or molecule connection (functionalization) enhances the nucleation of ALD to the lower quantity of reaction spots over the TMD surface.

Furthermore, the successful deposition of Al_2O_3 over the exfoliated WSe_2 from organic functionalization via titanyl phthalocyanine (TiOPc) single-layer also

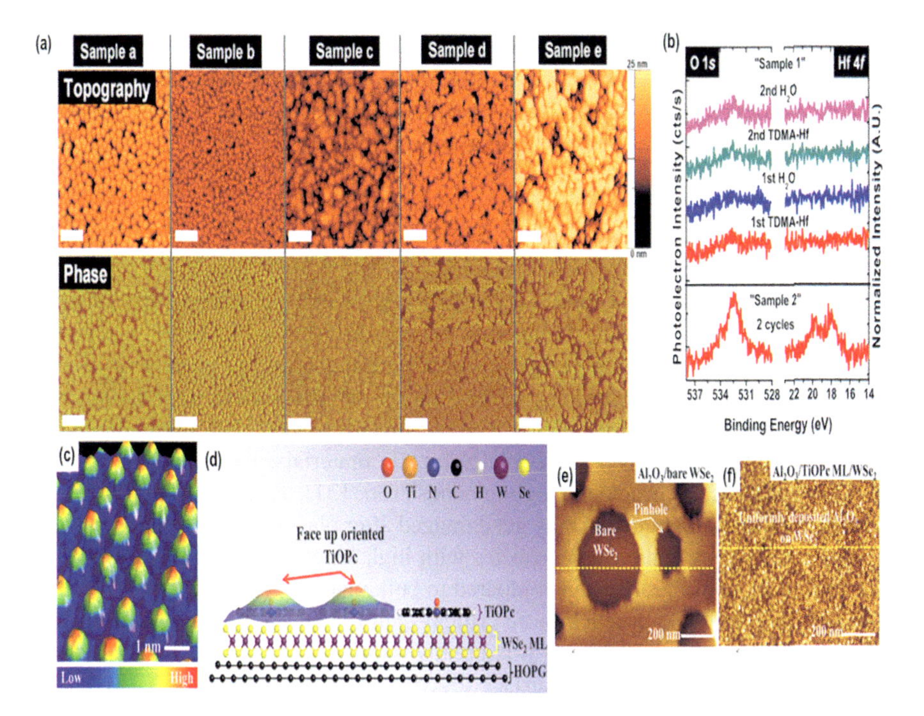

Fig. 9.6 AFM topography and phase pictures of the ALD HfO_2; **a** crude MoS_2 afterward mechanical exfoliation, MoS_2 immersed in acetone for 2.5 h (sample b), MoS_2 immersed in NMP for 2.5 h (sample c), MoS_2 coated from PMMA and immersed in acetone for 2.5 h (sample d), coated to PMMA and immersed in NMP for 2.5 h (sample e); **b** XPS spectra for the ALD HfO_2 over raw MoS_2 together longer purge period (> 6 h) and little purge period (5 and 10 s for predecessor and reactant purge period); **c** three-dimensional STM picture of the TiOPc single-layer together a molecular resolution; **d** diagram to the TiOPc single-layer, including the WSe_2 single-layer and the HOPG; **e, f** AFM pictures for the Al_2O_3/bulk WSe_2 without and with the TiOPc layer (adopted from Nam et al. [114])

followed the concept [68, 69]. Their first principle calculation also predicted that robust binding energy among dimethyl aluminum and their surface constituents generates by dissociation of TMA to the surface hydroxyl assembly and components of TiOPc (C, N, or O), for more than 1.5 eV. This also indicated that TMA is attached to the TiOPc layer positively through the reaction spots with the successive reactants. Moreover, the molecular beam epitaxy (MBE) deposited WSe_2 has been shown highly ordered pyrolytic graphite (HOPG), and a successive single-layer of TiOPc has been also fabricated employing the MBE. The scanning tunneling microscopy (STM) pictures afterward TiOPc deposition as illustrated in Fig. 9.6c. An extremely well-organized morphology having no imperfection specifies that an unbroken thin TiOPc single-layer is achievable from the MBE developed WSe_2. The corresponding energy level is depicted from the color profiles that reveal each solitary TiOPc molecule has a lone bright peak that reflects it negatively charged;

additionally it also gives the probable binding locations for the polar ALD predecessors (see Fig. 9.6d). The effective nucleation of the TiOPc ALD Al_2O_3 on bare and TiOPc-deposited WSe_2 has been also demonstrated from the AFM and RMS roughness parameters, as depicted in Fig. 9.6e, f.

Significance of Plasma Treatment

Plasma action is the utmost commonly utilize method in the contemporary construction procedure that willingly highpoints of any surface for the better acceptance [126–128]. Owing to highly energetic species are made during plasma ignition in which the functional groups are generated on the substrate surface by the reaction between reactive ion/electron species and the surface group [129]. Particularly, this kind of surface treatment is more useful for 2D materials, like carbon nanotubes (CNTs) graphene has a lack of dangling bonds [130, 131]. This technology is critical with the pristine substrate that may get damaged after plasma ignition because of ion bombardment from ion/electron species with high kinetic energy [132]. Regard to 2D TMDs are considerably thin compared to bulk materials, so the effect of plasma damage on 2D TMDs is much more critical than the conventional bulk materials. Therefore, to use this technology it is greatly required to functionalize the surface of the extremely thin 2D TMD surface together fewer destruction. 2D TMDs like MoS_2 and $MoSe_2$ are widely ALD using this approach to get a highly smooth thin film for the high-performance device. The plasma-treated ALD surface morphologies of MoS_2 and $MoSe_2$ are illustrated in Fig. 9.7a–f.

Significance of UV-O_3 Treatment

This is room temperature treatment using ultraviolet-ozone (UV-O_3) of the ALD-developed TMDs thin films. Like room temperature ultraviolet-ozone (UV-O_3) action on grow of ALD Al_2O_3 over 2D MoS2 [133]. The UV-O_3 action is one of the old-fashioned approaches in that removal of surface pollutants through the reaction to the extremely reactive extremists created due to the reaction between UV and O_3 [134]. Though, in the present time, it becomes a supplementary charming approach to construct a hydrophilic surface by formation of a hydroxyl ingredient over the surface [135]. Using this process, it can anticipate the ALD on 2D TMD would provide an even film holding its native structure. It has been also recognized as a nondestructive MoS_2 functionalization technique lacking of contravention sulfur-molybdenum bonds by utilizing UV-O_3 exposure to eliminate the topmost layers of bulk MoS_2. An additional small peak of Mo 3d may appear afterward the UV-O_3 action that related to S-O, though, the oxygen associated peak (Mo–O) has been not originate, as depicted in Fig. 9.8a. The supplementary doublet peak close to 164.8 eV has also appeared afterward UV-O_3 exposure, in S 2p spectra. It is overlapping together the O1s peak as illustrated in Fig. 9.8b. This is corresponding to the S–O feature which has noted in the Mo 3d. The newly generated S 2p peaks

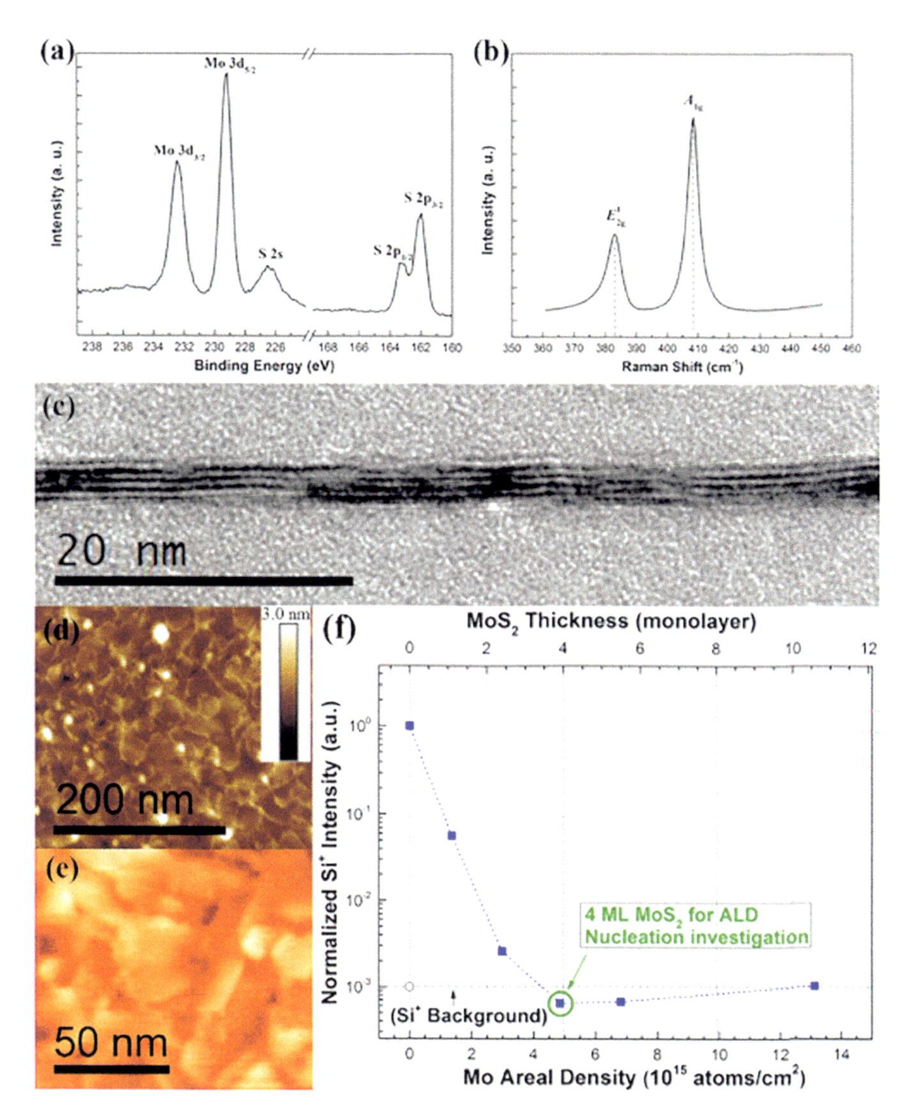

Fig. 9.7 Chemical state, 2D structure, surface morphology, and continuity investigation for the MoS_2: **a** XPS spectra of Mo 3d, S 2s, and S 2p core levels'; **b** Raman spectra of the as-grown MoS_2; **c** cross-sectional TEM image for the MoS_2; **d, e** AFM images for the MoS_2 surface in tapping mode and non-contact mode; **f** TOF–SIMS Si^+ signal decay as a function of Mo content (adopted from Nam et al. [114])

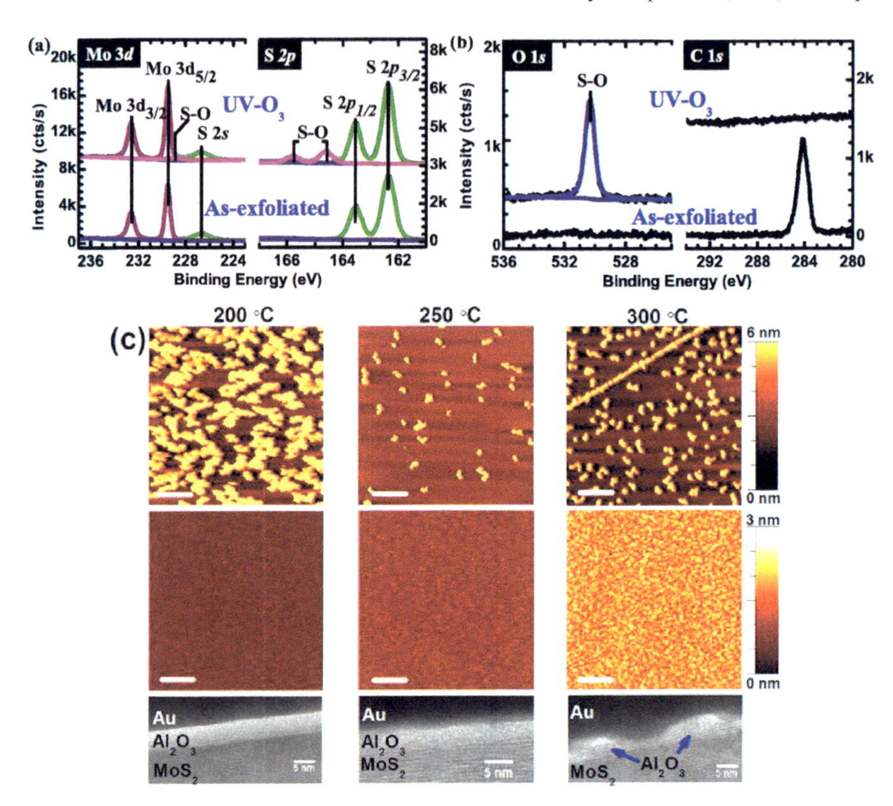

Fig. 9.8 **a** XPS spectra of Mo3d and S2p; **b** O1s and C1s of MoS_2 earlier and afterward the UV-O_3 action; **c** AFM pictures of ALD Al_2O_3 previously (top) and later (middle) the UV-O_3 treatment and cross-sectional TEM pictures of ALD Al_2O_3 over MoS_2 afterward the UV-O_3 action (bottom) (adopted from Azcatl et al. [133])

have also matched with S_4^+ that is corresponding to MoS_2 oxidation instead of S_2^- [136]. These spectral peak interpretations led to the spectral changes that have been deduced by the S–O formation on topmost of the MoS_2 surface with no Mo–S bond cleavage. Moreover, the calculation for the functionalized oxygen formation associating monolayer coverage has been also provided by estimation of the peak intensities of S 2p (IS-O/Is) that approves the functionalization from UV-O_3, it can also induce sufficient reaction spots for consequent predecessors over MoS_2. Using this technique carbon remainder from the surface are commonly removed after well UV-O_3 treatment (see Fig. 9.8b). Thus the changes occurred when the UV and O_3 have been simultaneously exposed because UV is significantly cumulative the reaction-ability of O_3, associating highly reactive O radicals generate O_2 dissociation by UV absorption [136]. The DFT approach has shown that the construction of S–O bonds could be dynamically promising with no cleaving Mo–S bonds. Additionally, it is hard to substitute the S atom in O because the S vacancy formation energy is quite large.

To explore growing physiognomies and consistency of film development of treated MoS_2, the ALD Al_2O_3 on UV-O_3 functionalization of MoS_2 has been also executed, as the surface structure is illustrated in Fig. 9.8c. Without UV-O_3 treatment surface morphology has revealed the nucleation happened choosily on the edges locations of MoS_2. At the 200 °C, the concentration of Al_2O_3 bunches over the MoS_2 appears to be greater compare to formation of sophisticated temperature circumstances owing to the thermal distortion of chemisorbed oxygen ingredients are suppressed at comparatively lower temperatures. Whereas the coverage is expressively enlarged over UV-O_3 functionalized MoS_2. The Al_2O_3 films exhibited very slight RMS roughness magnitude 0.14, 0.17, and 0.30 nm, together ALD procedure temperatures of 200, 250, and 300 °C. The HR-TEM cross-section pictures of the Al_2O_3 over UV-O_3 functionalized MoS_2 have revealed Al_2O_3 films' exceptional consistency at 200° and 250 °C. However, Al_2O_3 developed at 300 °C revealed island kind progression, this is due to S–O species generation afterward UV-O_3 action that it stands up to 250 °C, afterward ongoing to decompose or desorbed at 300 °C. The XPS examination afterward ALD Al_2O_3 over the functionalized MoS_2 revealed the S–O constituents vanished afterward ALD Al_2O_3. This indicates that superficial S–O constituents act as a sacrificial layer for ALD nucleation. However, there is no Mo–Al or S–Al peak due to a noncovalent bond of Al_2O_3 over MoS_2.

Similarly, the surface chemistry of WSe_2 and $MoSe_2$ associating UV-O_3 action has been interpreted [92]. It has been noted that the reactivities of TMDs dissimilar from their molecular constituents, and the diverse oxidation together the UV-O_3 action. Contrasting to MoS_2, both $MoSe_2$ and WSe_2 have also shown oxidation of transition metal (i.e., Mo and W) afterward 6 min of UV-O_3 disclosure. This could be owing to the reactivity dissimilarity in the direction of oxidation. Moreover, UV-O_3 treatment is also useful as a buffer layer nucleation promoter [137–139].

Significance of Ozone ALD (O_3-ALD) or Direct Plasma-Enhanced ALD (PE-ALD)

Since plasma action via O_2 or H_2O makes the morphology hydroxyl ingredients and empowers the reaction together the chemical constituents in the ALD procedure. Nonetheless it could also destruct the original material, therefore, device working abilities deteriorates [133]. The thickness of 2D flakes is usually utilized for the construction of FET. The incompletely oxidized or spoiled 2D flakes offer fairly diverse physiognomies to unspoiled materials. Therefore, a new approach is desired to resolve this problem to provide a unvarying dielectric deposition over 2D TMDs. The PE-ALD is fulfill to this requirement by offering highly reactive chemical species to generate during plasma reaction [140–142]. The involvement of highly reactivity radicals in this technique offers to process lower temperatures with a high-density film compared to normal thermal ALD [143, 144]. To further reduce the destruction of 2D TMDs to the extremely responsive plasma constituents the isolated plasma or comparatively mild reactant **basis** is generally used [117]. In this process, both TMA and TEMAHf are used as predecessors and H_2O and O_2 plasma as a reactant for ALD

Al_2O_3 and HfO_2 to get smooth surface thin films. So, using this approach one can get a uniform thin film up to around 3.5 nm together sub-nanometer scale unevenness [145, 146]. Moreover, the PE-ALD process fabricated FET devices are also offer a large shift in their band gap energies that directly correlates to the involvements of the species in the reaction process.

Significantly this process fabricated MoS_2-based FET device's electrical properties does not deteriorate but enhanced the performance. The advantage of this process is fabricated devices with thick flakes the effect of oxidation or damage is negligible. Thus this technique is useful for a higher number of layers-based devices, such as bi and trilayer of MoS_2 [147]. Overall, this technique offers to generate H-terminated vacancies within the material that is considered to be most stable to promote the nucleation and grow thin film on the top most 2D TMD layer [148–150].

Selective Area ALD for Nanopatterning

Nanopatterning is also an important parameter, especially for semiconductor manufacturing. The key patterning process relies on photolithography that makes enabled the downscaling of features through light sources from deep ultraviolet to extreme ultraviolet [151, 152]. However, from the conventional top-down methods, it is difficult owing to challenges in advanced nodes. Such as the edge placement error considered to be the primary problem in aligned manufacturing [153]. That may cause shorting between metal and neighboring metal parts.

The bottom-up area-selective atomic layer deposition (AS-ALD) offers film growth on the desired area and gets a self-aligned nanopatterning due to its addictive process, as illustrated in Fig. 9.9a–g. It also provides atomic-scale accuracy from streamlines processing steps for the semiconductor industry. The AS-ALD works based on the differences in nucleation behavior of precursors and co-reactants on different substrates. The nucleation discrepancy between the growth and non-growth areas is exploited to obtain AS-ALD. The maximum number of cycles before the growth initials on the non-growth areas could define from the selective window. Typically process selectivity is defined in two ways: first by comparing the selectivity for a specific thickness and second by comparing the thickness which is achieved after setting a specific selectivity. Several cycles of selectivity are reduced to a specific thickness that leads to unexpected deposition on the non-growth areas.

Yet, various approaches have been interpreted to extend the selective window, as depicted in Fig. 9.9a–g. Most of the approaches include templated and inherent AS-ALD. Other supplementary approaches have been also introduced [40].

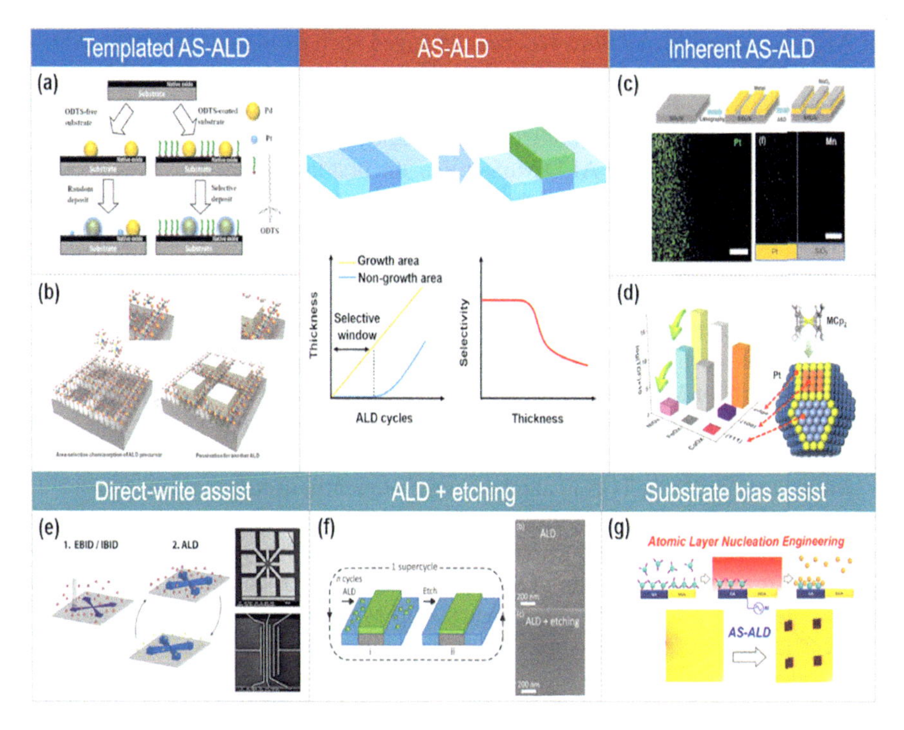

Fig. 9.9 **a** Templated for the AS-ALD achieved by SAMs; **b** templated AS-ALD achieved by SMIs; **c** inherent AS-ALD on metal/oxide substrates; **d** inherent adsorption difference of metal oxides on different metal facets; **e** direct-write-assisted AS-ALD; **f** ALD + etching process of AS-ALD; **g** substrate-bias-assisted AS-ALD (adopted from Zhang et al. [40], pp. 191–208)

9.3.4 Significant TMDs Fabrications from ALD

9.3.4.1 MoS_2

MoS_2 is one of the widely obtained TMD from the ALD processes by employing the different precursors, as summarized in Table 9.2 The precursors $MoCl_5$ and H_2S are used for the ALD processes ALD at 300 °C, corresponding surface chemistry is illustrated in Fig. 9.10a [154]. It revealed that the as-developed MoS_2 films are almost crystalline (inside in Fig. 9.10a), while the post-treatment in the sulfur vapor enhanced their crystals. It has been also noted that the annealing at 800 °C MoS_2 films is reconstructed with a stripper width and advanced crystallinity. Thus the annealing process has altered the film width, typically 1.7–0.8 nm for the ten-cycle ALD MoS_2 and from 3.2 to 1.3 nm in the case of 20-cycle ALD MoS_2. The ten-cycle ALD MoS_2 annealed films have 2-mm triangular crystals (see Fig. 9.10b), whereas thicker films are produced a continuous films after the annealing process. Moreover, this experimental sequence has also verified the development of MoS_2 films in the span of 350–475 °C [155–158]. However, at much higher growth temperature behavior of ALD-deposited MoS_2 films has been changed such as in the range 500–900 °C [159]. Additionally, the MoS_2 films deposition with greater to 120 ALD cycles at development temperatures of 500, 700, and 900 °C thickness can also achieve up to 2, 1.4, and 0.8 nm in the form of tri, bi and single layers [160].

Moreover, in the typical ALD MoS_2 films, the first layer growth of $MoCl_5$ chemically adsorbs the sensitive spots throughout the surface of the substrate and forms MoS_2 layer by reacting to the supplied H_2S. Initially, the MoS_2 layer is intrinsically inactive and lacking of appropriate adsorption locations over the surface. While in the fabrication process molecules forces proceed with the physical adsorption of $MoCl_5$ molecules over the MoS_2 layer. Therefore, the adsorption/desorption of predecessor molecules are optimized from the development temperatures. When a crucial numeral of MoS_2 layers are shaped at a designated growing temperature, the predecessor molecules cannot be additionally adsorbed over the MoS_2 basal plane, and no further layers are formed.

Moreover, $Mo(CO)_6/H_2S$ is another most frequently used precursor pair for the ALD process of MoS_2. With this precursor pair, it has been noted that the room temperature or a relatively low growth temperature (~ 170 °C) the amorphous MoS_2 films are formed [161]. As the in situ quartz crystal microbalance (QCM), ex-situ X-ray reflectivity (XRR), and ellipsometry ALD MoS_2 have realized a GPC of 2.5 Å/cycle. The in situ FTIR characterizations have authenticated the surface chemistry of ligand interchange among Mo (CO) and H_2S. It has also prove the confirmation of the combinations of the C=O elasticity afterward Mo $(CO)_6$ doses and its elimination subsequently the H_2S doses. Additionally, ALD MoS_2 can also perform in the range of 155–175 °C and their as-developed amorphous films promote to crystalline stage through the post-annealing in-between the 500–900 °C under an H_2S/Ar atmosphere, as depicted in Fig. 9.10c [162]. However, it has been also noted that the nanocrystalline 2H-MoS_2 films are achieved straightly at 170 °C through a PEALD using predecessors of $Mo(CO)_6$ and H_2S plasma (PH_2S), together a GPC of 0.5 Å/cycle [163].

However, in the place of H_2S the attached $Mo(CO)_6$ together dimethyldisulfide ($CH_3S_2CH_3$, DMDS) has been also used to grow the MoS_2 through the ALD [164]. In such an ALD procedure surface principally experiences decarbonylation of physisorbed $Mo(CO)_6$ and creates chemisorbed sub carbonyls ($Mo(CO)_n$, $n \leq 5$) over the surface. Consequently DMDs dissociated and produce a chemisorbed methyl-thiolate intermediary (CH_3S–) by S–S bond cleavage. Further, superficial methyl thiolates form sulfur adatoms through C–S bond cleavage, catalyzed from the chemisorbed $Mo(CO)_n$ surface ingredients, therefore it releases gaseous molecules like $(CH_3)_2S$ and C_2H_6, which is usually expressed as;

$$|-S + Mo(\mathrm{CO})_6(g) \rightarrow |-S - Mo(\mathrm{CO})_n + (6-n)\mathrm{CO}(g) \tag{9.1}$$

Table 9.2 Various MoS_2 film fabrications through ALD **procedures** [26]

Precursor A		Precursor B	Growth temperature (C) and mechanism	GPC (Å / cycle)	Crystallinity
Composition	Temperature (°C)				
$MoCl_5$	120	H_2S	300, R1	~ 1.6	2H (post-treatment in sulfur vapor)
	105		350–450, R1	0.87	2H
	70		375–475, R1	–	2H (post-treatment in sulfur vapor)
	200		430–470, R1	~ 4.0	2H
	200		450, R1	6.9	2H
	90		500–900, R2	0.07–0.17	2H
$Mo(CO)_6$	RT	H_2S	155–170, R1	~ 2.5	2H
			155–175, R1	0.74	2H (post-treatment in H_2S/Ar gas)
		PH_2S	175–225, R1	~ 0.5	2H
		$CH_3S_2CH_3$	100–120, R1	1.1	2H (post-treatment in Ar either)
		O_3	160, R3	~ 0.8	2H and 3R (sulfurization in sulfur vapor)
$Mo(NMe_2)_4$	RT	H_2S	60, R1	1.2	2H (post-treatment in sulfur vapor)
		$HS(CH_2)_2SH$	50, R1	–	2H (post-treatment in Ar either)
$Mo(the)_3$	115	H_2S	300, R1	0.25	2H
MoF_6	RT	H_2S	200, R1	0.46	2H (post-treatment in Ar either)
$(tabun)_2(NMe_2)_2Mo$	RT	$^PO_2$	150, R3	~ 2.0	2H (sulfurization in sulfur vapor)

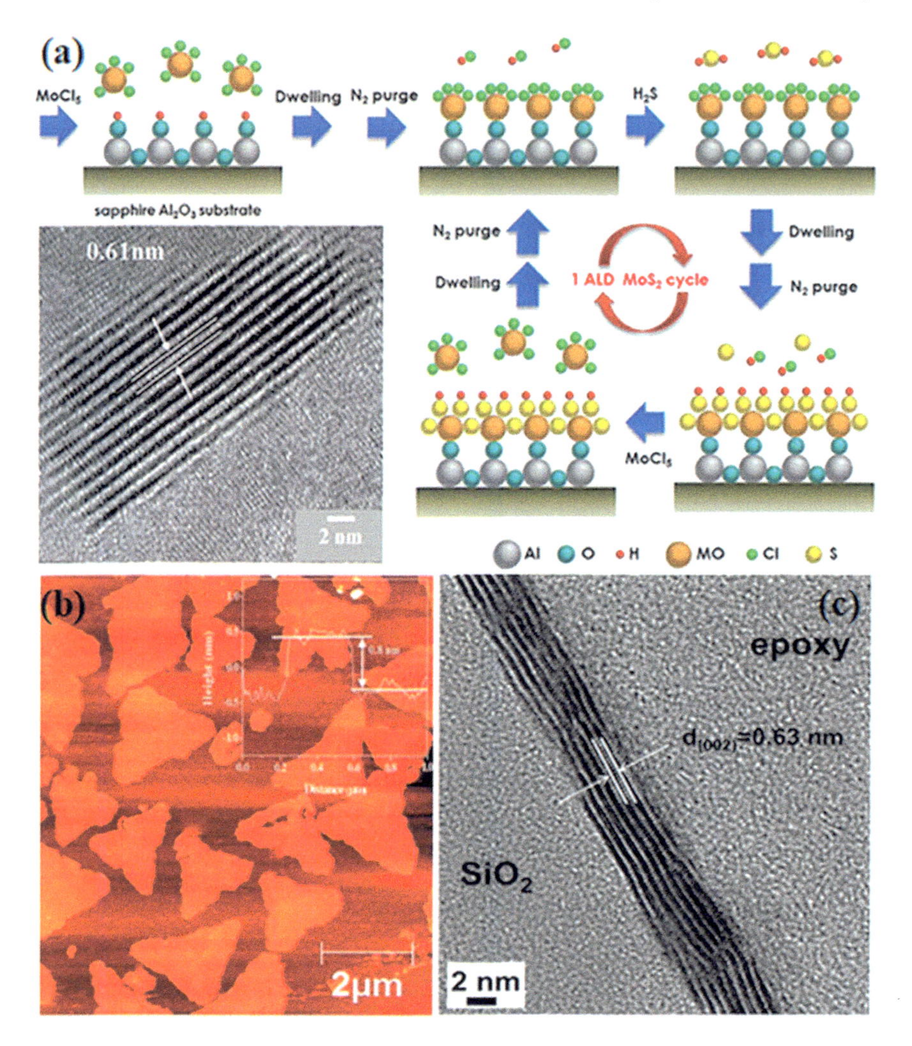

Fig. 9.10 **a** Schematic of the ALD MoS_2 procedure taking the predecessor couple of $MoCl_5$/ $H_2S_{.77,}$ the lattice spacing to a cross-sectional TEM interpretation (inset), **b** AFM pictures for the ALD MoS_2 single-layer annealed at 800 °C, inset depicts the depth profile of the triangular flakes; **c** cross-sectional interpretation of HR-TEM picture to the ALD MoS_2 films annealed at 900 °C for 5 min within a H_2S atmosphere (adopted from Cai et al. [26])

$$|{-S} - Mo(\mathrm{CO})_n + \mathrm{CH_3S_2CH_3}(g) \rightarrow |{-S} - Mo - S + \mathrm{biproducts}(g) \begin{pmatrix} a_1 & & 0 \\ & \ddots & \\ 0 & & a_n \end{pmatrix} \tag{9.2}$$

This interpretation also leads the GPC of the MoS_2 films is a parameter of the growth in the temperature range 100–120 °C [164]. A rapid increase in the GPC has been noticed above the temperature of 120 °C owing to uncontrolled thermal decomposition of the precursors. Additionally, in the temperature range 100C–120 °C the overall nature of films is amorphous, however, the post-annealing around 900 °C up to 5 min underneath an Ar atmosphere has considerably improved the crystals of the as-developed MoS_2 films.

To escape corrosion or noxious by-products (HCl to a $MoCl_5$ predecessor and sulfur impurities to a DMDS predecessor) and decrease the growing temperature [165] ALD process has been also developed for MoS_2 at 60 °C. By using the tetrakis (dimethylamido)-molybdenum(IV) (TDMA-Mo, $Mo(NMe_2)_4$) and H_2S as predecessors. The in situ QCM experimentation has provided the self-restrictive ALD development at 60 °C. This process as-constructed MoS_2 films have a GPC of 1.2 Å/cycle and their overall behavior amorphous. The post-operation at 1000 °C in the sulfur surrounding up to 5 h also improves the crystals of the MoS_2 films. Moreover, TDMA-Mo is also utilized for the coupling together 1,2-ethanedithiol ($HS(CH_2)_2SH$) to established a novel ALD procedure for the MoS_2 that enables developments at 50 °C [166]. Such process as-developed films are in compositional ratio nonetheless its overall behavior still amorphous. The amorphous MoS_2 films 2H phase can be also developed through the post-operation whichever to H_2 at 450 °C or for Ar at 800 °C.

Moreover, from the use of tris (2,2,6,6-tetramethylheptane-3,5-dionato) molybdenum ($Mo(thd)_3$) and H_2S can also develop crystals of MoS_2 films in the temperature between 250 and 400 °C via ALD. Below the 300 °C deposition, it is self-restrictive and even with an inferior GPC $\leq$ 0.028 Å/cycle. The inferior GPC is feasibly due to these aspects: the steric interruption and the little concentration of responsive spots. This process developed as-deposited MoS_2 films at ~ 300 °C has a good crystallinity, in which $Mo(thd)_3$ prone is decomposed at the temperature > 350 °C. Similarly, MoF_6 and H_2S are also used for the deposition of MoS_2 at 200 °C, their ALD surface reactions are generally expressed in the following sequence [167]:

$$|-(\mathrm{SH})_x + \mathrm{MoF_6}(g) \rightarrow |-(S)_x\,\mathrm{MoF}_{(6-x)} + x\mathrm{HF}(g) \tag{9.3}$$

$$|-(S)_x\mathrm{MoF}_{(6-x)} + 3\mathrm{H_2\,S}(g) \rightarrow |-\mathrm{S_2\,Mo(SH)}_x + S(s) + (6-x)\mathrm{HF}(g) \tag{9.4}$$

here "*s*" represents the solid stage. By the in situ QCM experimentations it ensure that the self-restrictive development together a GPC of 0.46 Å/cycle. While using the structural and spectroscopic techniques it demonstrates an amorphous phase of the as-developed MoS_2 films, the crystallinity, and stoichiometry are also improves through the post-operation at ~ 350 °C under the H_2 surrounding.

Additionally, the ALD process through R3 also has revealed the 2H-MoS_2 films, whereas the ALD-fabricated molybdenum oxides and sulfurized over the MoS_2. Typically molybdenum oxides are grown at 160 °C through an ALD adopting Mo

$(CO)_6$ and ozone (O_3), including 70% MoO_3 and 30% Mo_2O_5. Their sulfurization process has been carried out at ~ 850 °C in the sulfur vapor to transform the molybdenum oxides into a superiority of 2H-MoS_2 phase associating few minor 3R phase. Moreover, 2H-MoS_2 films have been also sulfurized by MoOx. The MO_3 films to fabricate through dual ALD procedures: the leading unique thermal ALD procedure uses the bis (tert-butylimido) bis (dimethylamido)-molybdenum ($(tBuN)_2(NMe_2)_2Mo$) and O_3 at 300 °C and the subsequent one is the PEALD by using $(tBuN)_2(NMe_2)_2Mo$ and O_2 plasma (PO_2) at 150 °C. Compared to thermal ALD, PEALD has increased the reactivity of the predecessors at the comparatively lesser temperatures and provide an advanced nucleation concentration. To get amorphous MoS_2 films the as-fabricated MoO_3 has been also sulfurized in the sulfur vapor at ~ 300 °C. The post-annealing in the sulfur vapor at 900 °C or greater temperature is responsible to transform the amorphous MoS_2 films in 2H-MoS_2 films containing extraordinary crystals and compositional ratio.

9.3.4.2 $MoSe_2$

Using the ALD process $MoSe_2$ films have been also obtained, as summarized in Table 9.3 [168]. The deposition of $MoSe_2$ films are also demonstrated from the two ALD processes adopting whichever $Mo(CO)_6$ or $MoCl_5$ coupled to the $((CH_3)_3Si)_2Se$. Typical deposition temperatures are in the range of 167 and 300 °C. The fabricated films to $MoCl_5$ have approximately compositional ratio and crystal 2H-$MoSe_2$ nano-flakes, while the films obtained to the $Mo(CO)_6$ have a noteworthy quantity of MoO_x owing to a complex of amorphous MoO_x and crystals of $MoSe_2$. Moreover, the dual-step method (R3 route) can also produce on the wafer-scale $MoSe_2$ films, by first deposition of MoO_3 films at 162 °C from the PEALD procedure adopting $Mo(CO)_6$ and PO_2 as predecessors and subsequent selenization at a higher temperature (~ 820 °C) in the H_2/Ar surroundings, as depicted in Fig. 9.11a [169]. The fabricated $MoSe_2$ film hexagonal crystal structure has been confirmed from the XRD analysis. The thickness of the films can be controlled by controlling the ALD-deposition of MoO_3. Typically $MoSe_2$ film thickness is 1.7 nm (2 layers), 2.5 nm (3 layers), and 6.2 nm (8 layers) after selenizing 60, 70, and 120 ALD MoO_3 cycles. Raman spectrum of the (see Fig. 9.11b) probed constructions of 3-layer $MoSe_2$ nanosheets have three characteristic peaks, whereas the HR-TEM (see Fig. 9.11c) characterization exposed a polycrystalline crystal together a honeycomb-like construction.

9.3.4.3 WS_2

WS_2 films are also achieved from the ALD process [170]. The ALD procedure of 2H-WS_2 thin films at 300 °C adopting WF_6 and H_2S as the predecessors gives us the WS_2 films [171]. It has been noted that WS_2 films grow only on certain substrates [172, 173]. Like ZnS, the sub-layer promotes the development of WS_2 films with the

Table 9.3 Different $MoSe_2$ and WSe_2 thin films synthesized by ALD processes [26]

Materials	Precursor A		Precursor B	Growth temperature (°C) and mechanism	GPC (Å/Cycle)	Crystallinity
	Composition	Temperature (°C)				
$MoSe_2$	$MoCl_5$	120	$((CH_3)_3Si)_2Se$	300, R1	–	2H
	$Mo(CO)_6$	55	$(CH_3)_3Si)_2Se$	167, R1	–	2H
		RT	$^{P}O_2$	162, R3	–	2H (MoO_3 selenization under H_2/Ar vapor)
WSe_2	WCl_6	85	DESe	600–800, R2	–	2H
		110	H_2Se	390, R1	5	2H

GPC ~ 1.1 Å/cycle (see Fig. 9.12a) [171, 173]. It's also noticed that the WS_2 growth gradually reduces with the increasing numeral of cycles up to it stabilized at ~ 0.1 Å/cycle. Moreover, ZnS also serve as a decreasing representative to WF_6 and enhance the surface adsorption, by following the reaction's mechanism [173]:

$$|{-}\mathrm{WSH} + \mathrm{WF_6}(g) \rightarrow |{-}\mathrm{WSWF}_x + \mathrm{HF}(g) \tag{9.5}$$

$$|{-}\mathrm{WSWF}_x + \mathrm{H_2S}(g) \rightarrow |{-}\mathrm{WSWSH}_y + \mathrm{HF}(g) \tag{9.6}$$

The fabricated film 2H crystalline phase has been confirmed from the characterizing tools like Raman, XRD, and HR-TEM [171, 173]. Common impurities Zn and F in the as-developed WS_2 thin films can be eliminated by the use of semiconductor -matching Si thin layers and H_2 plasma (PH2) as decreasing representative in the temperature range 300–450 °C, by using WF_6 and H_2S as the precursors [174]. Because Si thin layers respond to the WF_6 in the temperatures 250–450 °C, the steps are given below:

$$\mathrm{WF_6}(g) + 3\mathrm{Si}(s) \rightarrow \mathrm{W}(s) + 3\mathrm{SiF_2}(g) \tag{9.7}$$

Since metallic W and explosive SiFx ingredients are uninterruptedly manufactured through the reactions till the sacrificial Si layers entirely finished. Then typical WS_2 thin films are constructed through the sulfurization of metallic W utilizing H_2S, as followed the following reaction;

$$\mathrm{W}(s) + \mathrm{H_2S}(g) \rightarrow \mathrm{WS_2}(s) + 2\mathrm{H}(g) \tag{9.8}$$

It has been noted that simply six layers of WS_2 are completely sulfurized at 450 °C. Therefore, a thicker WS_2 is produced at rather higher temperatures above 450 °C. The H_2 plasma is used as a decreasing representative in the PEALD procedure under three

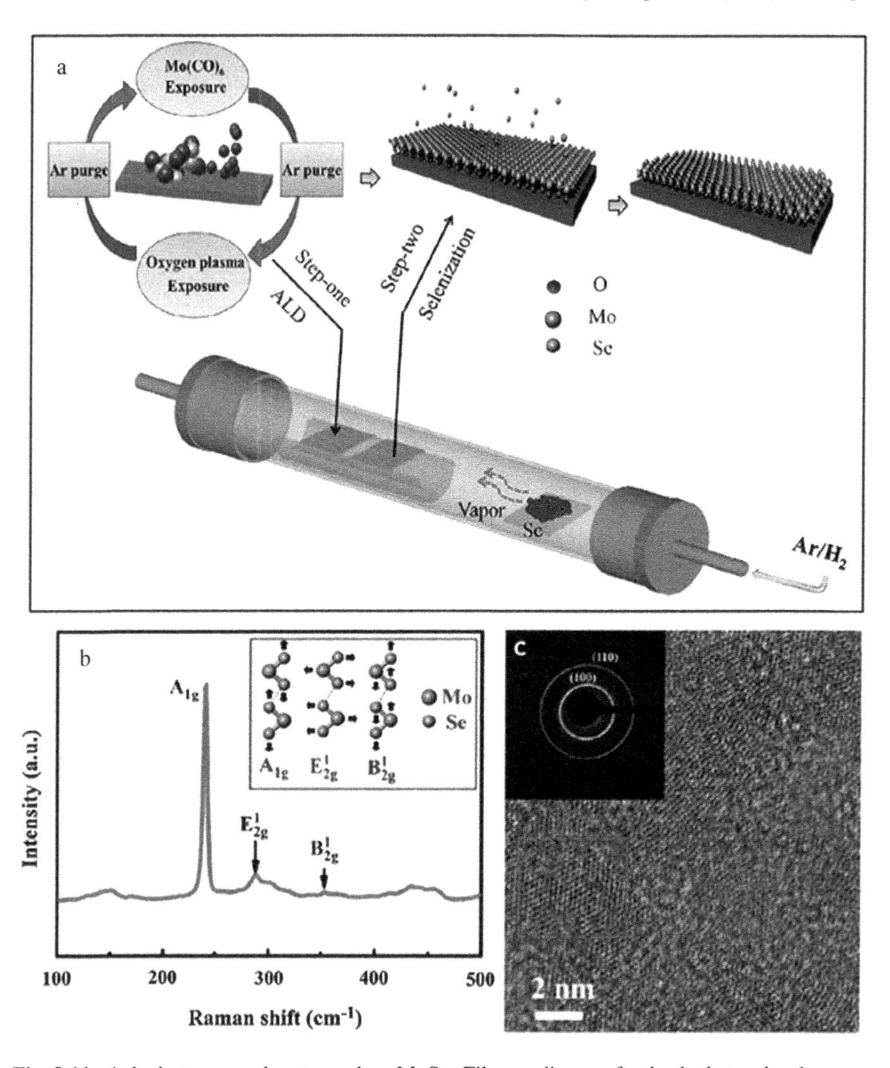

Fig. 9.11 A dual-step procedure to produce $MoSe_2$ Films: **a** diagram for the dual-step development procedure; **b** Raman spectrum to the as-developed three-layer $MoSe_2$ nanosheet (inset: schematic of Raman active vibration modes); **c** HR-TEM picture consisting inset of the SAED pattern for the three-layer$MoSe_2$ nanosheet (adopted from Cai et al. [26])

consecutive doses of WF_6, H_2 plasma, and H_2S. It has been also noticed that both plasma strength and disclosure period have influenced the growth and stoichiometry of the WS_2 thin films. Therefore, a 2H crystalline as-grown WS_2 film could be achieved using either of the above-said dual decreasing representatives. Moreover, the polycrystalline grains of WS_2 are also possible at about 5 and ~ 10 nm with the decreasing representatives of Si and H_2 plasma. The nucleation machinery of WS_2 films has been also described at 300 °C utilizing WF_6, H_2 plasma, and H_2S as

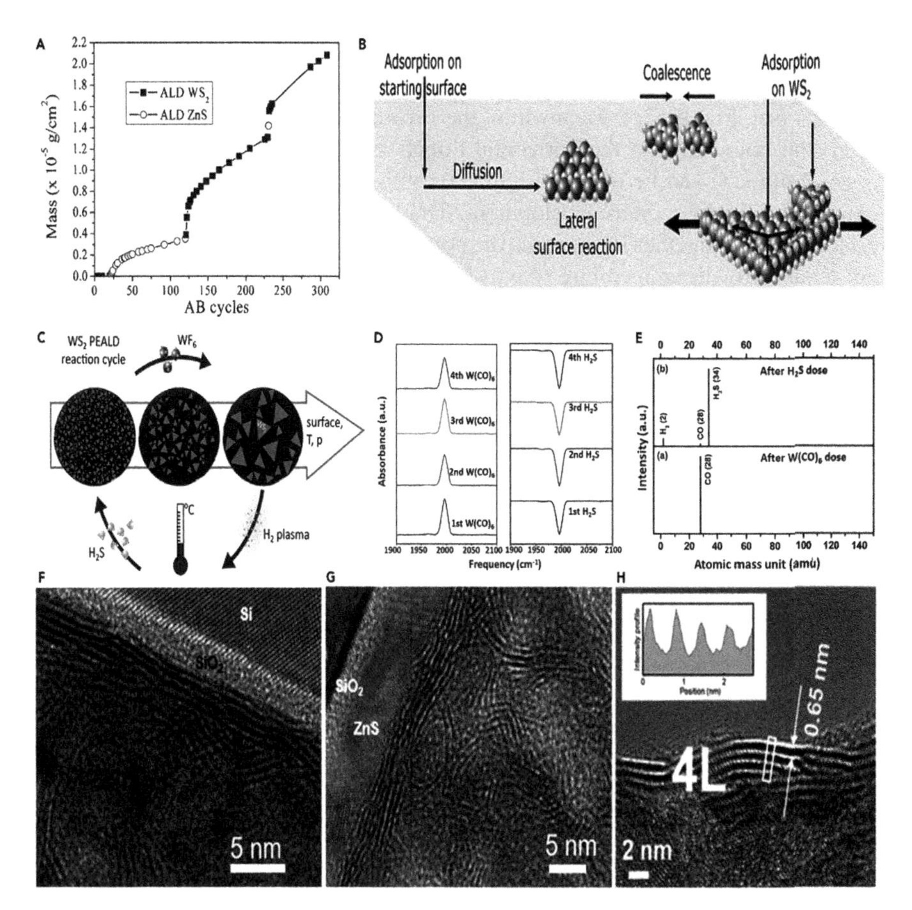

Fig. 9.12 **a** QCM mass gain for the ALD WS_2 and ZnS that pulsed consecutively over a SiO_2-coated crystal; **b** diagram for the ternary WS_2 PEALD reaction cycle, made of a WF_6 reaction, H_2 plasma exposure, and H_2S reaction. **c** diagram to qualitative exemplary to the nucleation nature of WS_2; **d** various FTIR spectrum to the numerous ALD cycles earlier and afterward dosing to $W(CO)_6$ in the ALD reactor; **e** RGA spectra just afterward the $W(CO)_6$ dose and H_2S dose; **f, g** cross-sectional HR-TEM picture to theWS_2 films developed over Si substrate and the ZnS-coated Si substrate (adopted from Cai et al. [26])

predecessors (see Fig. 9.12b) [175]. As per redox chemistry exemplary of the gas–solid reactions of the WF_6, H_2S, and H_2 plasma, whereas the WF_6 reacts to either –SH assemblies or surface S atoms to produce WF_x surface constituents that reduces late from the H_2 plasma in the metallic W which belongs to a lesser oxidation state. The successive H_2S disclosure is sulfurized by the metallic W to forms the intermediary oxidation state W over the WS_2. Although obtained S/W proportion is unsufficient ~ 1.9 ± 0.2 with the non-metallic W, this implies that S vacancies nearby the surface [175]. Therefore, the horizontal crystal development of WS_2 is revealed and three stages are recommended [176]: (i) the adsorption of predecessors on substrates, (ii) the dispersal of adsorbed predecessors near the lateral sensitive spots, (iii) the surface reaction of predecessors which control the development of 2D materials (see Fig. 9.12c). Meanwhile, the favored development of WS_2 is toward the crystal boundaries in the horizontal direction, therefore, it believes a superior crystal grain size can be obtained by decreasing the nucleation concentration and encouraging the horizontal development. Therefore it has been favored to initiate surface growing together fewer reactive spots (i.e., SiO_2), lesser H_2 plasma strength, and greater growth temperature (e.g., 450 °C) with the increased reactor pressure [176, 177].

To provide a more reliable ALD process for WS_2 film formation a new approach has been also introduced utilizing $W(CO)_6$ and H_2S as predecessors temperature in-between 155 and 230 °C. The in situ FTIR and remaining gas investigation (RGA) have been considered to obtain the surface constituents and the gaseous by-products in every half-reaction. The ALD half reactions to be expressed as:

$$|-W(\mathrm{CO})_x + \mathrm{H_2S}(g) \rightarrow |-W - \mathrm{SH} + x\mathrm{CO}(g) + 0.5\mathrm{H_2}(g) \tag{9.9}$$

$$|-W - \mathrm{SH} + W(\mathrm{CO})_6 \rightarrow |-W - S - W(\mathrm{CO})_x + (6 - x)\mathrm{CO}(g) + 0.5\mathrm{H_2}(g) \tag{9.10}$$

Using the FTIR characterization it has been confirmed the presence of $W(CO)_x$ surface clusters afterward the $W(CO)_6$ doses and their vanishing afterward H_2S doses (see Fig. 9.12d). Though, the significance of the SH surface assembly is still missing, this is possibly owing to the sensitivity limits of the FTIR. Moreover, using the RGA has established the by-products of CO and H_2 (see Fig. 9.12e). QCM has exposed a temperature-dependence development of WS_2 films. There is no development below 150 °C it might be owing to the thermodynamic limitation of $W(CO)_6$. But in the temperature between 175 and 205 °C the GPC is steady at 0.2 Å/cycle. However, overhead 205 °C, the GPC increases linearly, this is possibly owing to the thermal decomposition of $W(CO)_6$.

Though, by the use of $W(CO)_6$ and H_2S precursor at 350, 400 °C fruitful construction of crystalline 2H-WS_2 films through ALD has been demonstrated [178, 179]. It conflicts with the claim of W(CO)6 decomposed at the temperature more than 205 °C. Moreover, it has been also demonstrated that the development of WS_2 films is efficient over dual kinds of substrates: Si(100) wafers together a 4-nm original oxide SiO_2 layer (SiO_2/Si) and 8-nm ZnS-coated Si(100) wafers (ZnS/Si). It has been also noted that compared to SiO_2/Si substrate, the ZnS/Si substrate can have a slight impact over the GPC of ALD WS_2, nonetheless a small improvement in the crystals quality of WS_2 films (see Fig. 9.12f–h). Therefore the highest GPC of 3.6 Å/cycle has been achieved. Moreover, by replacing H_2S with H_2S plasma (PH_2S) has also coupled associating $W(CO)_6$ predecessor, thereby, the PEALD fabricated WS_2 films at 350 °C achieved with a GPC of ~ 1.0 Å /cycle. Additionally, the density functional theory (DFT) estimations also revealed the H_2S plasma creates many imperfections over the surface of WS_2, like S vacancy, W vacancy, and WS_2 anti-site imperfections. Inclusive such imperfections are promising for the adsorption of $W(CO)_6$ and $W(CO)_5$ ligands.

Thus, the precursor's developments to extensively studied WS_2 are not limited only to above discuss significant investigations but many more for different complex TMD compositions ALD depositions [180–182].

9.3.4.4 WSe_2

Using ALD can also fabricate the WSe_2 films by employing WCl_6 and diethyl selenide (DESe, $Se(C_2H_5)_2$) as the predecessors [183]. Usually, the WSe_2 ALD procedure is conducted at higher temperatures such as 600, 700, and 800 °C that are processed through R_2. The development of WSe_2 films depends exclusively to the temperature rather than the number of ALD cycles. The crucial widths of WSe_2 films developed at 600, 700, and 800 °C are 3.9 nm (5 layers), 2.5 nm (3 layers), and 0.8 nm (1 layer) (see Fig. 9.13a–c). Their AFM surface characterization has revealed no WSe_2 development afterward the initial 100 ALD cycles at each temperature (see Fig. 9.13d). By using the HR-TEM examination the ALD-fabricated WSe_2 films together a grain size of ~ 200 nm and their honeycomb-like construction are visualized, as illustrated in Fig. 9.13e. Moreover, their XPS characterizations revealed an unchanged compositional (Se/W) ratio of 2, without Cl ingredients. However, the ALD procedure of some layered WSe_2 films at 390 °C with the precursors WCl_5 and H_2Se Raman and XPS analysis has verified the development of WSe_2, whereas the HR-TEM exposed their layered construction together a GPC of 5 Å/cycle, as depicted in Fig. 9.13f and significant parameters are summarized in Table 9.3.

9.3.4.5 HfS_2

Similarly, other groups of TMD like HfS_2 films are also fabricated from the ALD by using three precursor pairs, $HfCl_4/H_2S$, HfI_4/H_2S, and $Hf(NEtMe)_4/H_2S$ [184]. It has been noticed that the $HfCl_4/H_2S$ pair precursors HfS_2 film deposition are also possible in the range of 200 and 500 °C. At the low temperature (~ 350 °C) HfS_2 films have amorphous nature with the chlorine remainders, whereas the crystalline 1T HfS_2 films have acquired at higher deposition temperatures among 350 and 500 °C. Their XRD analysis revealed the formation of monoclinic HfO_2 in HfS_2 films. It also noted that the HfS_2 films deposited at 400 °C have a better crystallinity with the purity. Moreover, the fabrication of HfI_4 has been also demonstrated by replacing the $HfCl_4$ precursor **to** ALD of HfS_2 [184]. This method fabricated films at 300 °C enabling them to form 1T HfS_2 with a rougher surface morphology. Furthermore, the $Hf(NEtMe)_4$ has also used to form films, nonetheless the as-constructed HfS_2 films have inferior crystallinity, un-uniform coverage and inconstant configuration.

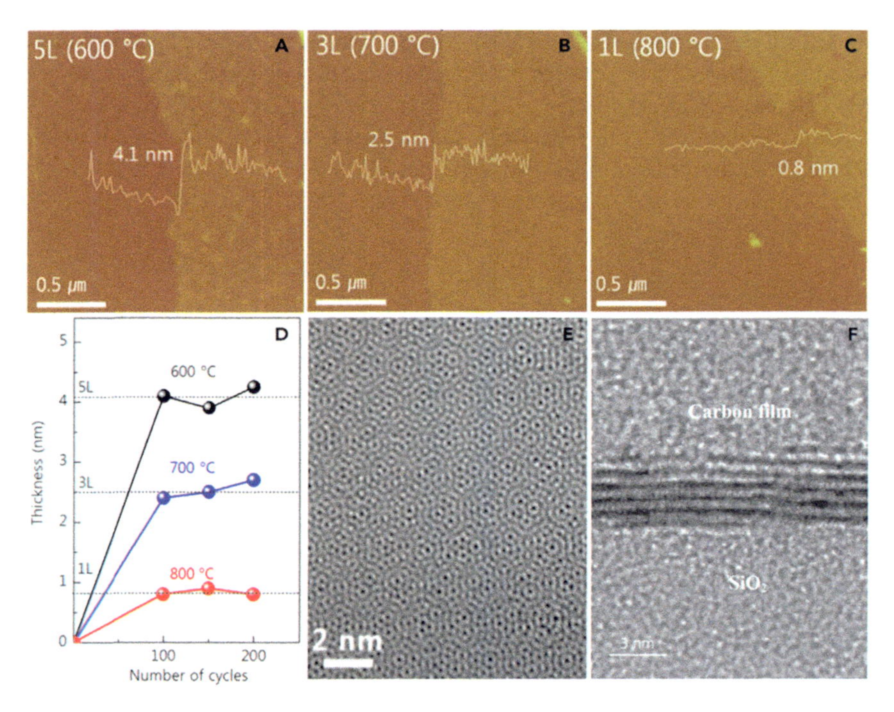

Fig. 9.13 2D Layered WSe_2 Materials from ALD procedure: AFM pictures and height profiles for the distinct layers, **a** 5, **b** 3, **c** 1, WSe_2 afterward transporting over a SiO_2 substrate; **d** thickness vs the numeral of cycles for WSe_2 at 600, 700 and 800 °C; **e** HR-TEM picture of a choosy expanse for the WSe_2 film (adopted from Cai et al. [26])

9.3.4.6 ReS_2

The significant ReS_2 films can also fabricate from the ALD technique with the 2D TMD misleading octahedral (1T′) single-layer configuration [185]. Significantly misleading 1T′ phase construction persuades the intrinsically decreased crystal regularity, providing to robust in-plane anisotropic optical and transportation features. The ReS_2 films also have amazing layer-independence features that making them impressive for various uses [185, 186]. ReS_2 films ALD are also possible by using the $ReCl_5$ and H_2S as precursors [187]. It noted that the development of ReS_2 could have a robust reliance on temperature. In the temperature between120°C–200 °C the GPC of ReS_2 increases ~ 0.3 Å/cycle to ~ 0.9 Å/cycle with the improved compositional ratio of the films from a S/Re ratio of 1–2 and the crystallinity changed from amorphous to crystals. Moreover, at the advanced temperatures the GPC is almost unmoved 0.8–0.9 Å/cycle at 200–300 °C but it gradually decreases up to ~ 0.2 Å/cycle at 450 °C. Thus ReS_2 films retain crystallinity and compositional ratio beyond 200 °C, whereas the crystals positioning is changed together the temperature. It also remarked in-between the temperatures range 200–300 °C the ReS_2 crystals oriented parallel to the substrate, whereas beyond the 350 °C the plate-like crystallites have supplementary arbitrary alignments to the substrate. Hence an extraordinary-excellent crystal with the compositional ratio ReS_2 film could be fabricated beyond 200 °C from the ALD procedure, and their crystal positioning is variable through the adjustment of growth temperature.

9.3.4.7 Miscellaneous

Since ALD is one of the most versatile techniques to fabricate sub-nano level thin films including TMDs. Innovation with the ALD process of TMD materials is not limited above discussed a few significant compositions of TMDs, but many more based on different group elements combinations by using novel precursor combinations under varying deposition conditions. Therefore, a vast number TMDs films have been fabricated from the ALD with the different precursors for their potential uses in a distinct area, such as TiS_2, ZrS_2, FeS, NiS, InSe, In_2Se_3, GeS, GeSe, SnS, SnS_2, Sb_2Te_3, Bi_2Se_3, Bi_2Te_3, MoO_3, WO_3, etc. [188–203].

9.4 ALD Deposited TMD Optoelectronics Devices

ALD-fabricated TMDs ultrathin optoelectronics devices are considered to be the most appropriate candidates for the semiconductor industries to downscaling of ICs [204]. As in modern technologies, billions of transistors are used instead of old techniques, so, the gate length of transistors and size have been evolved from 10 μm to 3 nm and even 1 nm in future [205]. The demands of nanodevices to a further reduction in their feature size down to atomic scale with the existing process and encountered difficulties. Thus, the rapid requirement for thin-film techniques in the semiconductor industry is driven by the development of ALD-fabricated TMDs devices [206]. Hence the high-efficiency TMDs optoelectronics devices can be also fabricated from the ALD technique, like ICs, photodetectors, transistors, FET, solar cells, biomedical applicable devices, etc. [40, 207].

9.4.1 Photodetector

Though, various TMD materials have been used for the device fabrication such as $MoSe_2$, ReS_2, TiS_2, WSe_2, etc., by the ALD process. The HfS_2 and ZrS_2 are also considered to fabricate at a large scale because they are possibly deposited over large substrate areas (also cost-effective) with a good control over uniformity and thickness. To get continuous films usually preferred a low-deposition temperature to prepare HfS_2 and ZrS_2 thin films using atomic layer deposition (ALD). In which ALD precursors typically one for each element supplied onto the substrate individually by isolated pulses to complete the gas phase reactions alimentation. To ensure the ALD film the growth should occur only via self-limiting surface reactions that uniformly coat both large areas and complex 3D shapes accurately with the reproducible control thickness [33, 35]. Moreover, ALD processes at relatively low temperatures below 500 °C can be scaled up to use in semiconductors and other industries [208, 209]. Since the HfS_2 and ZrS_2 film crystals are oxidized under ambient conditions [210–212]. Therefore, an ALD oxide film continuous HfS_2 and ZrS_2 films are possibly deposited up to a few monolayers to tens of nanometers in thickness using simple halide precursors $HfCl_4$ and $ZrCl_4$ together with H_2S, the crystalline films are usually deposited at relatively low temperatures of approximately 400 °C [213, 214]. The films are also possibly grown on oxide-terminated silicon substrates and different oxide, sulfide, and metal surfaces. The sensitivity of HfS_2 and ZrS_2 should be considered during deposition, storage, and applications, therefore, encapsulation of HfS_2/ZrS_2 films is desired under a continuous vacuum [184].

Typical HfS_2 and ZrS_2-based ALD have been fabricated for photodetectors and optoelectronic applications. The photodetector (photoresistor) devices are fabricated by evaporating patterned Au electrodes on glass substrates through a shadow mask followed by deposition of HfS_2 (~ 10 nm) or ZrS2 (~8 nm) with protective $Al_xSi_yO_z$ (~ 50 nm) layers by ALD, as illustrated in Fig. 9.14a, b. The devices 30 μm × 1 mm have been illuminated by the transparent $Al_xSi_yO_z$ layer using a laser source of different wavelengths. The fabricated photodetectors have been tested in photoresistor mode by using a modest bias of 1 V between the electrodes. While the illumination of 405 nm light, the conductivity (current) of both of the materials is increased due to an increase in photogenerated charge carriers, it can be separated by the electric field between the electrodes, as depicted in the current–voltage profile at the logarithmic scale in Fig. 9.14c, d. The photoresistor with no photoresponse has been also recognized in the absence of an external bias. On the other hand ZrS_2 device, the current increase is rather linear with increasing bias voltage both in dark and under illumination, while the HfS_2 device current increase more slowly at lower bias voltages, however, a faster increase has been noticed at larger biases. This means nearly ohmic contact to ZrS_2/Au in contrast to a Schottky junction for HfS_2/Au. As many metal/TMD pairs form Schottky junctions [215]. Additionally, it has been also noted that the current flow in the HfS_2 device is one to two orders of magnitude lower compared to ZrS_2 in both dark and under illumination, it is usually correlated to their higher electrical conductivity.

However, both HfS_2 and ZrS_2 devices have shown relatively fast photoresponse under pulsed 405 nm illuminations, as illustrated in Fig. 9.14e. The protective $Al_xSi_yO_z$ layer has shown no decline in performance under normal air conditions. In a 1 V bias and a 22 mW/cm^2 incident power density, the photoresponsivity of the ZrS_2 device has achieved at ~ 50 mA/W with rise and decay times 35 and 230 ms. Whereas under identical conditions, the responsivity of the HfS_2 device has much lower approximately 0.18 mA/W, with rise and decay times < 100 and 300 ms. The photoresponsivity (R) is also calculated by establishing a relationship for the photocurrent (I_{photo}) under illumination (I_{light}) minus the dark current (I_{dark}).

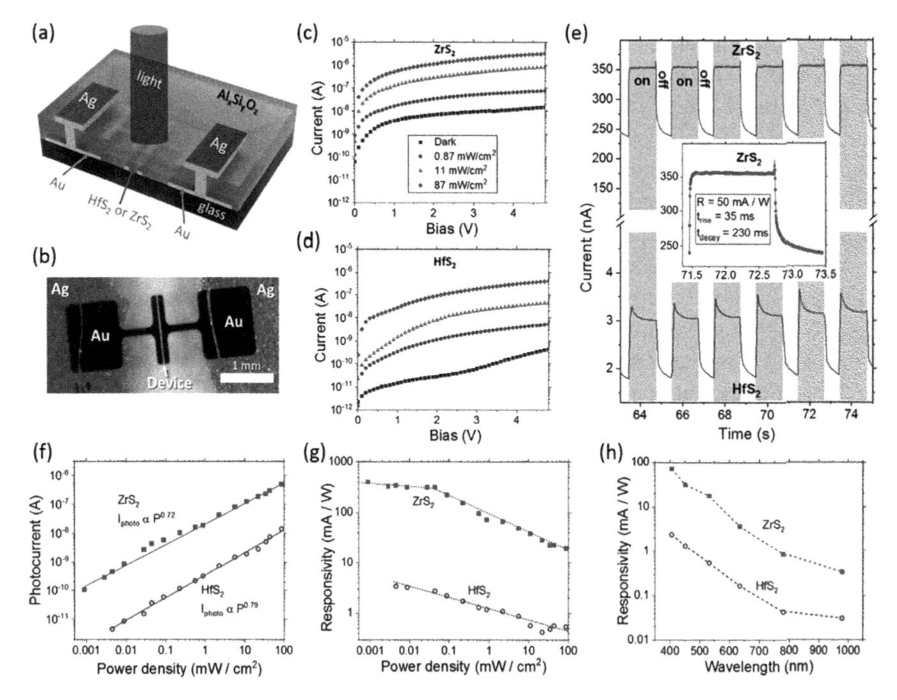

Fig. 9.14 **a** Schematic structure; **b** photograph (taken from the side of the glass substrate) for the $Al_xSi_yO_z/ZrS_2$/glass photodetector structure with Au electrodes; **c, d** I–V curves for the ZrS_2 and HfS_2 photodetectors in dark and under illumination with a 405 nm laser using different power densities; **e** time-dependent response (I–t) of HfS_2 and ZrS_2 photodetectors under pulsed illumination with a 405 nm laser at a power density of 22 mW/cm^2, the inset shows one on/off cycle for ZrS_2 in more detail along with the extracted responsivity as well as rise and decay times, the lines in the inset are two-phase exponential functions fitted to the data; **f** photocurrent as a function of illumination power density under 405 nm illumination, the lines represent power law curves fitted into the data; **g** responsivity as a function of illumination power density under 405 nm illumination, the dashed lines to guide the eye; **h** responsivity as a function of illumination wavelength under a power density of approximately 1 mW/cm^2, the dashed lines to guide the eye (adopted from Mattinen et al. [184])

$$R = \frac{I_{\text{photo}}}{P_{\text{light}}} = \frac{I_{\text{light}} - I_{\text{dark}}}{Pd.A} \tag{9.11}$$

where the P_{light} represents the incident power, which means the power density of the light source (Pd) times the area of the device (A). The rise (decay) time is defined as the time to reach 90% (10%) of the photocurrent when the laser is turned on (off).

Moreover, the photoresponse behavior of both materials has fast and slow components in different processes. Therefore a two-phase exponential function is desired to fit the photoresponse of the ZrS_2 device as depicted in Fig. 9.14e, it leads that the presence of two processes with approximately an order of magnitude difference in time constants 5/28 and 16/230 ms to the current rise and decay. Additionally, the third component can have slower processes thereby a slow decay current back to the dark current level of the time scale minutes. The slow response and high responsivity typical of 2D photodetectors have been attributed to the photogating effect, whereas the photogenerated electrons (holes) are trapped near the conduction (valence) band. Their trapped charge carriers mostly end up at the film/substrate interface that revealed an additional electric field perpendicular to the film to regulate the conductance along the film similar to an external gate [216]. The effect of photogating can also reduce by employing an external gate voltage to get a shorter response time at the expense of decreased responsivity [217].

Moreover, the photocurrent also increases with increasing power density in a sublinear fashion (common in 2D photodetectors). Their characteristic is usually described by a power law I_{photo} α P^x, the exponent x has defined between 0.79 and 0.72 for the HfS_2 and ZrS_2, as depicted in Fig. 9.14f. The highest responsivity values have been achieved at the lowest power densities, as illustrated in Fig. 9.14g. It has been also noticed the responsivity reached 3.4 mA/W for HfS_2 (0.0044 mW/ cm^2) and 500 mA/ W for ZrS_2 (0.00087 mW/cm^2). Significantly, the lowest power density ($\leq$ 0.044 mW/cm^2) has only a weak dependency between power density and responsivity in the case of the ZrS_2. The sublinear characteristic at larger power densities predominately govern by defects and charged impurities either within the 2D film or at the interfaces [216, 218]. Thus the trap states with a long lifetime are likely to be present as evidenced by the slow photoresponse component that also contributes. The response times may longer at smaller power densities which are also supported by the occurrence of the trap states with a long lifetime. Usually, the relatively low bias voltage of 1 V is preferred to minimize the dark current and consequently power consumption of the photodetectors, however, the higher bias voltages increase the responsivity at the cost of increasing the dark current. As demonstrated by the 5 V bias with a low power density of 0.087 mW/ cm^2, the corresponding responsivity is 1.25 A/W for ZrS_2.

Additionally, the spectral responses of the HfS_2 and ZrS_2 photodetectors of the used different lasers wavelengths 405–980 nm is illustrated in Fig. 9.14h. It leads the broadband photodetection, however, a quite small photocurrent are also recognized even in the infrared wavelengths range 780 and 980 nm. These device's responsivity rapidly increases toward smaller wavelengths for both of the materials. The largest responsivity has obtained under 405 nm illumination. Moreover, the indirect band gaps of the HfS_2 and ZrS_2 films have been accessed at 1.87 eV (660 nm) and 1.82 eV (680 nm). Nonetheless, a small but measurable photoresponse at wavelengths above the band gap due to the presence of defect states within the band gap. Thus ALD process fabricated TMD materials-based photodetection devices have been also recognized [184].

9.4.2 *Field Effect Transistor (FET)*

9.4.2.1 HfS_2 FETs

ALD process fabricated TMDs based EFT devices are also attended much attention, such as the design and fabrication of the top gated HfS_2 FETs devices with high- k HfO 2 dielectric. Top-gate HfS_2 FETs can be fabricated with ultrathin HfO_2 (5 nm) as dielectric and Al and Y buffer layers. Their on/off ratio are as high as 10^5 with the low subthreshold swing (SS) ~ 95 mV dec^{-1}. Significantly, even without any functionalization the top-gate HfS_2 FETs with 5 nm HfO_2 dielectric have excellent properties, this is key to the difference in TMD FETs from others. The uniform and ultrathin HfO_2 films are also almost free of pinhole-like defects, thereby it leads a self-functionalization of HfS_2. Moreover, the integration of HfS_2 as channel materials and HfO_2 as high- k dielectrics largely contributes to FET application.

Typical the top-gate HfS_2 FETs with high-k HfO_2 dielectric ultrathin HfS_2 flakes are often mechanically exfoliated from commercially available crystals by using scotch-tape technique and transferred to Si substrate with 300 nm thick SiO_2. The thickness of HfS_2 flakes is generally examined by an optical microscope and atomic force microscope (AFM). The source and drain regions have been recognized by covering copper grid shadow masks with typical gaps. The subsequent Au (50 nm) metal have deposited by the thermal evaporation machine. The direct deposition of Al_2O_3 films by ALD has formed island-like clusters with large pinholes [133, 219]. Because of the absence of out-of-plane dangling bonds in the initiated ALD reaction. Subsequently, devices have put into the chamber of ALD with tetrakis (dimethylamino) hafnium and water as precursors at 100 °C. The top-gate electrode has been constructed by the electron-beam lithography followed by thermal deposition of Au (20 nm) to conventional lift-off. The morphology and lattice vibration of bare HfS_2 and HfS_2 after metals and HfO_2 deposition has been normally characterized by a variety of techniques such as AFM and Raman spectra. The formed Al buffer layer has been visualized from the AFM image to verify the growth of HfO_2 film completely uniform and continuous on HfS_2, rather than an island-like cluster with pinhole formations. Raman spectra of HfS_2 before and after metal deposition have been also recognized with the two most prominent peaks at 260 and 336 cm^{-1}, which reflects the in-plane and out-of-plane modes, therefore, no obvious shift among in overall structure, by means there is no obvious damage or bond-disorder into HfS_2 materials [220].

Moreover, the thicker top-gated HfS_2 FETs have been also fabricated and interpreted with the remakes the thinner devices more susceptible to charge impurities at the interface. However, the thicker device's gate electric field are influenced by the charge screening effect. The thinner devices in the range of 7–12 nm HfS_2 flakes have excellent electronic properties, as the typical transfer characteristics of top-gate HfS_2 FETs with Al buffer layer (Al-HfS_2 FETs). Such fabricated devices have been designated as n-type conduction and transistor behavior [221]. Their transfer characteristic can also modulate the channel conductivity of HfS_2 under the applied top-gate voltage. Their on/off ratio of Al-HfS_2 FETs are at least 10^4 which fulfills the basic application demands [222]. Moreover, the extremely thin HfO_2 dielectric is only up to 5 nm thickness, such a device has excellent gate control on a channel. Their SS characterization to be also define in term rate of increase in current below the threshold voltage, it is expressed from the following relationship:

$$SS = dV_g/d\left(I_g I_d\right) \tag{9.12}$$

here V_g represents the top-gate voltage and I_d reflects the drain current. These kinds of devices can have SS around 182 mV dec^{-1}. The gate current of the typical Al-HfS_2 FETs are up to 250 pA μm^{-1}. The reliability and universality of such devices are usually accessed by introducing the yttrium metal as a buffer layer in top-gated HfS_2 FETs (Y-HfS_2 FETs). The fabricated device's transfer characteristics on/off ratio are around 10^5, which is usually greater than the HfS_2 FETs with an Al buffer layer. Meanwhile, this kind of devices SS may reduce up to 95 mV dec^{-1}. Moreover, significantly the gate current of Y-HfS_2 FETs dramatically decreases up to 250–4 pA μm^{-1} for the Al-HfS_2 FETs. This is due to Y's higher melting point that makes it less sensitive to the coalescence at the ALD process temperature, thereby, the Y-HfO_2 film has less amount of pinholes that is useful to make a more compact and uniform film. Hence all the characteristic parameters of the performance of Y-HfS_2 FETs are superior to Al-HfS_2 FETs. However, the on-current level of the HfS_2FETs has been found to be lower than the MoS_2 FETs, and their field-effect mobility of 0.1 cm^2 V^{-1} s^{-1} that is usually defined from the following relationship

$$\mu = Lg_m/W(\varepsilon_0\varepsilon_r/d)V_d \tag{9.13}$$

here L corresponds to the channel length, W represents the channel width, g_m is the transconductance which extracted from the transfer characteristics, ε_0 is 8.85 $\times$ 10^{-12} F m^{-1}, ε_r is 19 for the HfO_2, d represents the thickness of HfO_2 and V_d reflects the drain voltage of 0.1 V. The relatively low mobility due to metal contact, surface scattering, and the low conductivity of HfS_2 itself [115, 223, 224]. The room temperature electrical conductivity of HfS_2 has been obtained around $\approx 3.0 \times 10^{-3}$ Ω^{-1} cm^{-1}, however, it is lower than the conductivity $\approx$of 7.6×10^{-1} Ω^{-1} cm^{-1} the ZrS_2 [225]. The output characteristics of HfS_2 FET has also no obvious current saturation, it is due to various factors, such as HfS_2/HfO_2 interface and contact metal. Nonetheless, the thin dielectric especially (5 nm) thickness that suppressed the short channel effect to realizes the high-performance and low-power devices [213].

Moreover, not only the top gate but also the back gate and dual-gate sub 10 nm HfS_2 FET have been fabricated and analyzed. Considering the merit and demerits of top and back gates the dual gate features the HfS_2 FET has been recognized as more impassive device for the high-k dielectric in terms of mobility. Therefore, the kinds of configurations have been used to get desired output characteristics of the HfS_2 FET. Like a typical back gate, the HfS_2 FET device structure and its characteristics are illustrated in Fig. 9.15a–d.

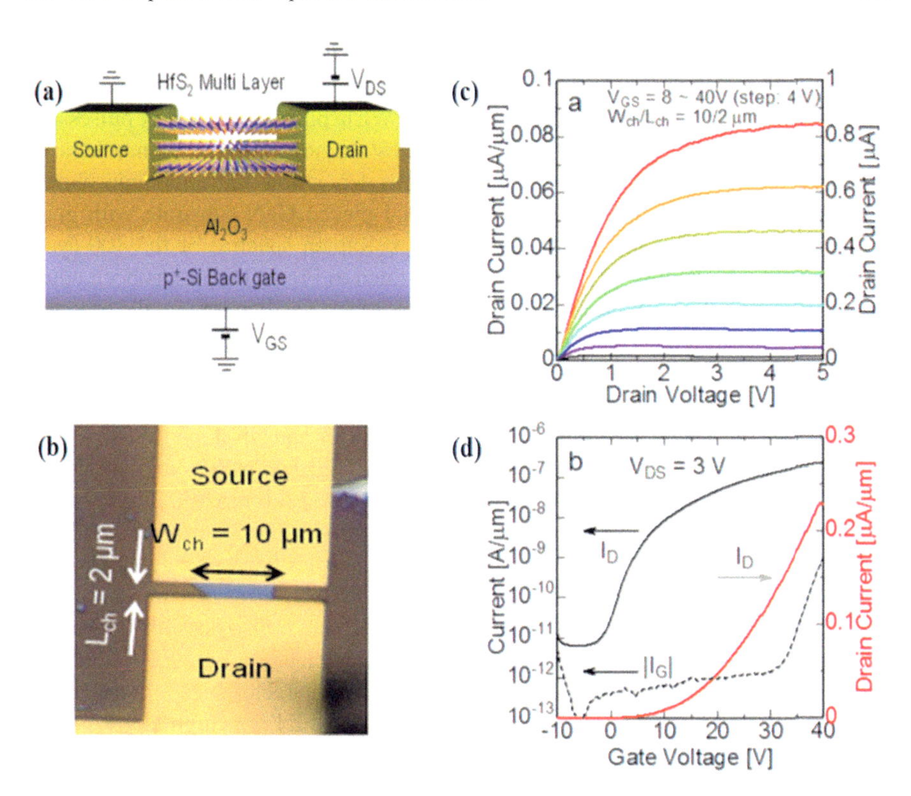

Fig. 9.15 Schematic and characterizations of back-gate HfS_2 FET: **a** structural schematic of back-gate HfS_2 FET; **b** the optical image of the back-gate HfS_2 FET to represent channel length and width at the source edge; **c** Output characteristics at RT which shows the current modulation property and robust saturation behavior; **d** Transfer characteristics with $V_{DS} = 3$ V, the maximum drain current is noticed around 0.2 μA/μm. The on/off current ratio under the applied voltage condition is over 10^4 (adopted from Kanazawa, T. et al. 2016 Sci. Rep. 6, 22277)

9.4.2.2 WS_2 FETs

Recently successfully has been also demonstrated that the wafer-scale fabrication of WS_2 films by ALD with controllable in situ p-type doping, on 8-inch α-Al_2O_3/Si wafer, 2-inch sapphire wafers and 1cm^2 GaN substrate pieces [226]. The growth mechanisms of ALD WS_2 and in situ Nb doping have been also demonstrated with the remarks the doping concentration can be controllable by altering Nb cycle numbers and analyzing their physical and electrical properties. Typical mechanisms of the ALD process for WS_2 growth and in situ Nb doping is illustrated in Fig. 9.16a. Whereas the reactor temperature can be up to 400 °C, while the WCl_6, $NbCl_5$, and HMDST kept at 93, 60 °C, and room temperature. The first complete cycle of WS_2 deposition includes 1 s WCl_6 pulse, followed by 8 s purge (Argon) and 1 s HMDST pulse for the 5 s purge sequentially performed. Similarly, Nb can also use to dope by $NbCl_5$ and HMDST precursors. One cycle of NbS_2 deposition includes a 1 s

$NbCl_5$ pulse, followed 8 s purge (Argon), and a 1 s HMDST pulse, followed by a 5 s purge. The growth rate of WS_2 film has been calibrated at about 0.036 nm/ cycle for controllable in situ doping. By replacing the WCl_6 pulses with $NbCl_5$ pulses and the doping concentration are normally adjusted by varying $NbCl_5$ pulse numbers, as photographs of wafer-scale 400-cycle WS_2 films deposition on 8-inch amorphous-Al_2O_3/Si wafer and 2-inch sapphire wafer, and pieced GaN substrate with good uniformity is depicted in Fig. 9.16b. Moreover, Raman spectra for the 400-cycle annealed WS_2 films at 950 °C for 2 h are illustrated in Fig. 9.16c. It has been noted that in the initial stage, WCl_6 and HMDST vapor are exposed directly to the sapphire substrates to form WS_2 layers laterally on sapphire substrates. Thereafter, subsequent layers have deposited onto the initial WS_2 layer to connect the isolated flakes and form films. A post-annealing process also improves the film quality.

Moreover, also have been fabricated the WS_2 n-FETs and Nb-doped WS_2 p-FETs CMOS-compatible processes from as-prepared ALD grown n-WS_2 and Nb-doped p-WS_2 films, as illustrated in Fig. 9.17a. The fabricated 4.6 nm WS_2 n-FETs and Nb-doped WS_2 p-FETs, as the top-gate transistors with the 2 μm gate width on the sapphire substrate. These devices electrical properties have revealed a CMOS-compatible process flow. Likewise the typical ALD Al_2O_3 films (20 nm) is used as high-k dielectrics, the equivalent oxide thickness of 13 nm is also possibly deposited. The transfer characteristic of 8-layer WS_2 n-FET is depicted in Fig. 9.17b, with the varying V_d from 0.1 V to 0.5 V, whereas the output characteristics Vg are varied in-between 1 and 5 V. The transfer on-current of WS_2 n-FET is as high as 0.4 μA/μm at $V_d = 0{:}5$ V with the on–off ratio up to 10^5. Moreover, the mobility (30 tested) of WS_2 n-FETs plots is depicted in Fig. 9.17b. The maximum and minimum mobilities of n-FETs have been achieved at 6.85 and 0.32 $cm^2\ V^{-1}\ s^{-1}$, whereas the median mobility is around 3.58 $cm^2\ V^{-1}\ s^{-1}$. The obtained mobility is over 70% of WS_2 n-FETs in the range of 1–5 $cm^2V^{-1}\ s^{-1}$.

Similarly, the transfer characteristic of a 4.6 nm Nb-doped WS_2 p-FET with 15-cycle varying V_d from 0.1 to 0.5 V and their output characteristics in terms of varying V_g in the range -2 V to -6 V have been also interpreted, as illustrated in Fig. 9.17c. In contrast to WS_2 n-FET, their charge carrier types change from electron to hole, it proves the Nb substituted at W atom sites in the WS_2 lattice. However, this device on and off-current only up to 5×10^{-2} at $V_d = 0{:}5$ V, which is far less than the WS_2 n-FET. Though hole mobility of Nb-doped WS_2 p-FET can be up to 0.016 $cm^2V^{-1}\ s^{-1}$ with the on/off ratio 10^1. Moreover, the Hall effect resistivity measurement of 15-cycle Nb-doped WS_2 up to 5 orders higher than the undoped WS_2, however, the mobility of 15-cycle Nb-doped WS_2 is far less than the undoped WS_2 at 300 K. It also noted that the field-effect mobility of WS_2 FETs smaller than the Hall effect of WS_2 due to the influence of transistors' electrical contacts on the underestimation of field-effect mobility. Since Hall mobility has been roughly estimated through field-effect mobility owing to the nonlinear dependence of carrier concentration on gate voltage [227]. Therefore, the stability should be accessed by the measurement on-current of Nb-doped WS_2 p-FET with gate length variation (for this device 5–50 μm), as depicted in Fig. 9.17c. As the distribution of $I_{d\ sat}$ (at $V_g = -4$ V, $V_d = 0{:}5$ V) for the 132 Nb-doped WS_2 p-FET with 20-cycle Nb doping. The $I_{d,\ sat}$ decreases with the

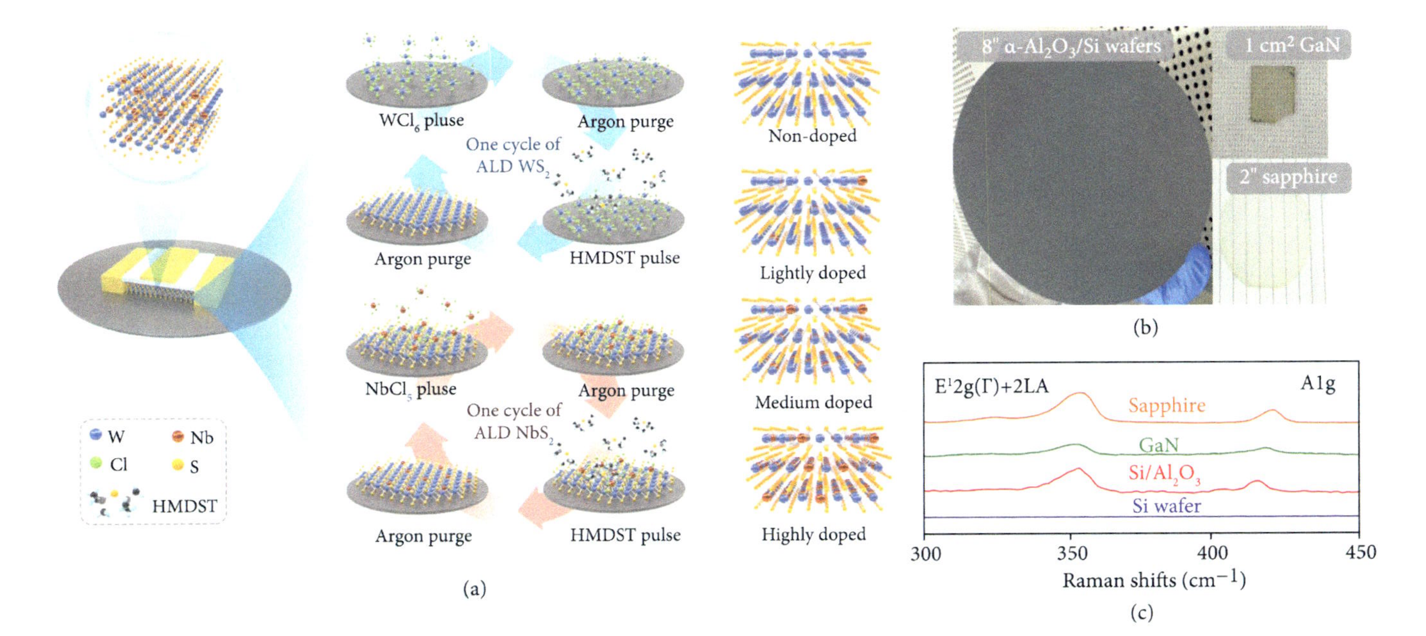

Fig. 9.16 ALD growth mechanisms and characterizations: **a** schematic for the idealized mechanisms of ALD process and WS_2 growth under in situ Nb doping, the doping concentration controlled by adjusting NbS_2 cycle numbers; **b** photographs of 400-cycle WS_2 films deposited on 8-inch α-Al_2O_3/Si wafer, 2-inch sapphire wafer, and pieced GaN substrates; **c** Raman spectra of annealed WS_2 on Si/Al_2O_3, GaN, and sapphire substrate surface (adopted from Yang et al. [226])

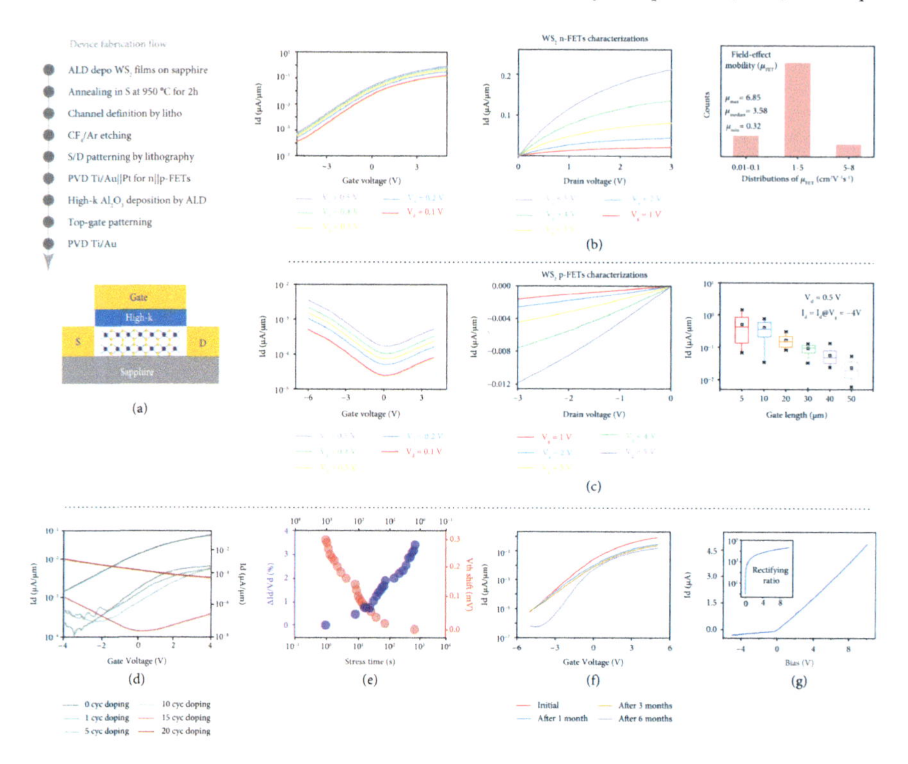

Fig. 9.17 Electrical properties of WS_2 n-FETs and Nb-doped WS_2 p-FETs; **a** CMOS-compatible process flow of FETs and schematic of device structures; **b** transfer and output characteristics of WS_2 n-FET with 2 μm gate width and the mobility distribution of 30 WS_2 n-FETs, the on-current reached 0.4 μA/μm, and the on/off ratio was up to 10^5; **c** transfer and output characteristics of 15-cycle Nb doped WS_2 p-FET with 2 μm gate width and the distribution of I_d at $V_d = 0.5$ V and $V_g = -4$ V for 132 Nb-doped WS_2 p-FETs with 25-cycle Nb doping, the carrier type changed from electron to hole, and the on-current was 5×10^{-2} μA/μm; **d** doping effects on WS_2 FETs, where the Nb dopants varied from 1 to 20 cycles, the Nb-doped WS_2 FET did not show p-type behavior but with a decreased on- and off-current until reaching 15 cycles, after 20-cycle Nb doping, the device shows heavily p-type behavior, indicating the controllable doping; **e** the PBTI of WS_2 n-FET at RT, the stress set to 5.5 V, after 1000 s stress, the on-current degraded only for 3.5%, while the ΔVth up to 300 mV; **f** the air stability of WS_2 n-FET in ambient for 1, 3, and 6 months, the on-current degraded slightly within one order, while the degradation of off-current less obvious after 3 months than that of 6 months; **g** the I-V curve for the WS_2/Nb-doped WS_2 p-n structures with the rectifying ratio of over 10^4, the inset figure shows the rectifying ratio (adopted from Yang et al. [226])

increasing gate length, therefore, a recessive fabrication and well-controlled uniform process are desired. Moreover, the Nb doping controllability has been also defined from the transfer characteristics of Nb-doped WS_2 FETs under Nb doping range 1 cycle to 20 cycles, as illustrated in Fig. 9.17d. Initially, the Nb-doped WS_2 FET is not exhibited p-type behavior until a decrease in on- and off-current up to 15 cycles. Further increasing in Nb concentrations the current of p-FET also increases and the on/off ratio decreases, therefore the resistivity and mobility of Nb doped WS_2 film are decreased, that has noticed identical to the Hall Effect measurement outputs. The heavily p-doped WS_2 FET after 20-cycle Nb doping leads to good controllability on in-situ Nb doping by ALD.

Additionally, the lack of dangling bonds at the surface of WS_2 makes it difficult to deposit high-quality high-k dielectrics. Therefore, the PBTI of WS_2 n-FET has been carried out to analyze the reliability of Al_2O_3 high-k dielectric. To get it, a stress has applied to the gate and biased at 5.5 V. The DC transfer characteristics at V_d = 0.5 V has been measured after the removal of PBTI stress at room temperature. It has also noted that after 1000 s stress application a degradation in on-current up to 3.5% (as depicted in Fig. 9.17e). However, the V_{th} shifts only 300 mV which is 6% of the max-applied gate voltage. It reveals the instability of high-k films has affected the electrical properties of WS_2 n- FET. So, higher-quality high-k dielectrics are desired to improve the electrical property of WS_2 n-FET [228]. Moreover, the air stability of WS_2 film has been also defined by placing the WS_2 n-FET in an ambient atmosphere and their transfer characteristics testing, the typical testing of these devices at V_d = 0.5 V after 1, 3, and 6 months are depicted in Fig. 9.17f. It leads the on-current of WS_2 n-FET degraded slightly over several months. However, deterioration in off current could hardly notice after 3-month exposure, its one order of magnitude after 6-month exposure. The on/off ratio has decayed from 10^5 to 10^4 after 6 months. Similarly, the vertical p-n structure based on WS_2 and Nb-doped WS_2 films have been also fabricated. Such FET device's electrical property p-n structure with rectifying ratio of 10^4 is illustrated in Fig. 9.17g. Their ideal factor 2.3 revealed conspicuous recombination of electron–hole [226].

Hence using the ALD fabrication process TMD based optoelectronics devices can be efficiently fabricated, not only limited to the above few examples but also many more for distinct uses in various areas [229].

9.5 Conclusions

In the concluding remarks, one has discussed the Atomic Layer Deposition (ALD) technique and its possible use in the fabrication of TMD-based optoelectronic devices. By discussing brief historical developments in the ALD technique including some significant prior fabricated TMD-based devices. By describing types of ALD precursors, growth characteristics, surface roughness, conformality, the effect of deposition temperature, and mechanics of thermal ALD with the initial surface reactions, reaction pathway process, and precursor chemisorptions. In addition to these one has also discussed ALD application on TMD materials to understand the basic surface properties of 2D TMDs and the physisorption impact on ALD. Moreover to understand the formation of a uniform film on 2D TMDs from ALD, the ALD on 2D TMDs without any surface treatment and uniform deposition on 2D TMD by surface functionalization have been also discussed. Moreover, to realize the actual fabrication application of the ALD process for the TMD materials some significant examples have been also discussed by providing descriptions of the MoS_2, $MoSe_2$, WS_2,

WSe_2, HfS_2, and ReS_2. Additionally, also provided a brief miscellaneous section to cover those materials that have been previously synthesized by the ALD technique for various uses. Moreover, to realize the actual device fabrication and their performances some important TMD-based ALD process device fabrications have been also discussed, as their key physical characteristics including significant advancements in devices in terms of existing and future applicability. So, ALD-fabricated TMD-based photodetector and Field Effect Transistor (FET) (HfS_2 FETs and WS_2 FETs) devices and their critical working performances including significant physical characteristics have been interpreted.

Thus the applicability of the ALD technique for TMD-based materials has been widely accepted due to the formation of high-quality thin film devices for modern and future optoelectronics device fabrications with reduced sizes and cost-effectiveness. Although several challenges with the ALD process fabrications of TMDs-based materials devices remain, that should be resolved in future from both theoretical and experiential innovations. Therefore, an intensive quest is desired on TMDs-based device fabrication from the ALD technique to provide more experimental data to make a conclusive view of such devices. Thus ALD technique could be considered a potential technique for the fabrication of future high-performing devices based on TMD materials.

References

1. Xu M, Liang T, Shi M, Chen H (2013) Graphene-like two-dimensional materials. Chem Rev 113:3766–3798
2. Tan C, Zhang H (2015) Two dimensional transition metal dichalcogenide nanosheet-based composites. Chem Soc Rev 44:2713–2731
3. Chhowalla M, Shin HS, Eda G, Li LJ et al (2013) The chemistry of two-dimensional layered transition metal dichalcogenide nanosheets. Nat Chem 5:263–275
4. Naguib M, Mochalin VN, Barsoum MW, Gogotsi Y (2014) 25th anniversary article: MXenes: a new family of two dimensional materials. Adv Mater 26:992–1005
5. Bhimanapati GR, Lin Z, Meunier V, Jung Y et al (2015) Recent advances in two-dimensional materials beyond graphene. ACS Nano 9:11509–11539
6. Butler SZ, Hollen SM, Cao L, Cui Y et al (2013) Progress, challenges, and opportunities in two-dimensional materials beyond graphene. ACS Nano 7:2898–2926
7. Lin YM, Dimitrakopoulos C, Jenkins KA, Farmer DB et al (2010) 100-GHz transistors from wafer-scale epitaxial graphene. Science 327:662
8. Liu M, Yin X, Ulin-Avila E, Geng B et al (2011) A graphene-based broadband optical modulator. Nature 474:64–67
9. Kim KS, Zhao Y, Jang H, Lee SY et al (2009) Large-scale pattern growth of graphene films for stretchable transparent electrodes. Nature 457:706–710
10. Zhu Y, Murali S, Stoller MD, Ganesh KJ et al (2011) Carbon-based supercapacitors produced by activation of graphene. Science 332:1537–1541
11. Anasori B, Lukatskaya MR, Gogotsi Y (2017) 2D metal carbides and nitrides (MXenes) for energy storage. Nat Rev Mater 2:16098
12. Deng D, Novoselov KS, Fu Q, Zheng N (2016) Catalysis with twodimensional materials and their heterostructures. Nat Nanotechnol 11:218–230
13. Xu M, Fujita D, Hanagata N (2009) Perspectives and challenges of emerging single-molecule DNA sequencing technologies. Small 5:2638–2649

14. Miro P, Audiffred M, Heine T (2014) An atlas of two-dimensional materials. Chem Soc Rev 43:6537–6554
15. Novoselov KS, Geim AK, Morozov SV, Jiang D (2004) Electric field effect in atomically thin carbon films. Science 306:666–669
16. Dávila ME, Xian L, Cahangirov S, Rubio A et al (2014) Germanene: a novel two-dimensional germanium allotrope akin to graphene and silicone. New J Phys 16:095002
17. Wang W, Jiang H, Li L, Li G (2021) Two-dimensional group-III nitrides and devices: a critical review. Rep Prog Phys 84:086501
18. Xie H, Li Z, Cheng L, Haidry AA et al (2022) Recent advances in the fabrication of 2D metal oxides. iScience 25:103598
19. Mohl M, Rautio AR, Asres GA, Wasala M et al (2020) 2D Tungsten chalcogenides: synthesis, properties and applications. Adv Mater Interface 7:2000002
20. Firouzjaei MD, Karimiziarani M, Moradkhani H, Elliott M et al (2022) MXenes: the two-dimensional influencers. Mater Today Adv 13:100202
21. Wang GE, Luo SZ, Di T, Fu ZH et al (2022) Layered organic metal chalcogenides (OMCs): from bulk to two-dimensional materials. Angew Chem Int Ed 61:e202203151
22. Cai SL, Zhang WG, Zuckermann RN, Li ZT et al (2015) The organic flatland—recent advances in synthetic 2D organic layers. Adv Mater 27:5762–5770
23. Nicolosi V, Chhowalla M, Anatids MG, Strano MS et al (2013) Liquid exfoliation of layered materials. Science 340:1226419
24. Bonaccorso F, Lombardo A, Hasan T, Sun Z, Colombo L et al (2012) Production and processing of graphene and 2d crystals. Mater Today 15:564–589
25. Zhong YL, Tian Z, Simon GP, Li D (2015) Scalable production of graphene via wet chemistry: progress and challenges. Mater Today 18:73–78
26. Cai J, Han X, Wang X, Meng X (2020) Atomic layer deposition of two-dimensional layered materials: processes, growth mechanisms, and characteristics. Matter 2:587–630
27. Puurunen RL (2014) A short history of atomic layer deposition: Tuomo Suntola's atomic layer epitaxy. Chem Vap Depos 20:332–344
28. George SM (2010) Atomic layer deposition: an overview. Chem Rev 110:111–131
29. Leskelä M, Ritala M (2003) Atomic layer deposition chemistry: recent developments and future challenges. Angew Chemie Int Ed 42:5548–5554
30. Oviroh PO, Akbarzadeh R, Pan D, Coetzee RAM (2019) New development of atomic layer deposition: processes, methods and applications. Sci Technol Adv Mater 20:465–496
31. Shahmohammadi M, Pensa E, Bhatia H, Yang B (2020) Enhancing the surface properties and functionalization of polymethyl methacrylate with atomic layer-deposited titanium(IV) oxide. J Mater Sci 55:17151–17169
32. Shahmohammadi M, Mukherjee R, Sukotjo C, Diwekar UM et al (2022) Recent advances in theoretical development of thermal atomic layer deposition: a review. Nanomaterials 12:831
33. Knoops HCM, Potts SE, Bol AA, Kessels WMM (2015) Atomic layer deposition. In: Handbook of crystal growth: thin films and epitaxy. In: Kuech T (ed). Amsterdam, The Netherlands, vol 3, pp 1101–1134
34. Hagen DJ, Pemble ME, Karppinen M (2019) Atomic layer deposition of metals: precursors and film growth. Appl Phys Rev 6:41309
35. Johnson RW, Hultqvist A, Bent SF (2014) A brief review of atomic layer deposition: from fundamentals to applications. Mater Today 17:236–246
36. Shahmohammadi M, Yang B, Takoudis CG (2020) Applications of Titania atomic layer deposition in the biomedical field and recent updates. Am J Biomed Sci Res 8:465–468
37. Leskela M, Ritala M (2002) Atomic layer deposition (ALD): from precursors to thin film structures. Thin Solid Films 409(1):13–146
38. Weber AZ, Borup RL, Darling RM, Das PK et al (2014) A critical review of modeling transport phenomena in polymer-electrolyte fuel cells. J Electrochem Soc 161:F1254
39. Gorkum AV, Nakagawa K, Shiraki Y (1989) Electroluminescence of doped organic thin films. J Appl Phys 65:2485–2492

40. Zhang J, Li Y, Cao K, Chen R (2022) Advances in atomic layer deposition. Nanomanuf Metrol. https://doi.org/10.1007/s41871-022-00136-8
41. Sheng J, Han KL, Hong T, Choi WH et al (2018) Review of recent progresses on flexible oxide semiconductor thin film transistors based on atomic layer deposition processes. J Semicond 39:11008
42. Chang S, Selvaraj SK, Choi YY, Hong S et al (2016) Atomic layer deposition of environmentally benign SnTiOx as a potential ferroelectric material. J Vac Sci Technol A Vacuum Surfaces Film 34:01A119
43. Koveshnikov S, Goel N, Majhi P, Wen H et al (2008) In0.53 Ga 0.47 As based metal oxide semiconductor capacitors with atomic layer deposition ZrO_2 gate oxide demonstrating low gate leakage current and equivalent oxide thickness less than 1 nm. ApPhL 92:222904
44. Bishal AK, Sukotjo C, Jokisaari JR, Klie RF et al (2018) Bioactivity of collagen fiber functionalized with room temperature atomic layer deposited titania. ACS Appl Mater Interfaces 10:34443–34454
45. Darwish G, Huang S, Knoernschild K, Sukotjo C et al (2019) Improving polymethyl methacrylate resin using a novel titanium dioxide coating. J Prosthodont 28:1011–1017
46. Ritala M, Niinisto J (2009) Chemical vapour deposition: precursors, processes and applications. Roy Soc Chem. https://doi.org/10.1039/9781847558794
47. Werner F, Veith B, Tiba V, Poodt P et al (2010) Very low surface recombination velocities on *p*- and *n*-type c-Si by ultrafast spatial atomic layer deposition of aluminum oxide. Appl Phys Lett 97:162103
48. Yota J, Shen H, Ramanathann R (2013) Characterization of atomic layer deposition HfO_2, Al_2O_3, and plasma-enhanced chemical vapor deposition Si_3N_4 as metal–insulator–metal capacitor dielectric for GaAs HBT technology. J Vac Sci Technol A 31:01A134
49. Poodt P, Lankhorst A, Roozeboom F, Speeet K et al (2010) High-speed spatial atomic-layer deposition of aluminum oxide layers for solar cell passivation. Adv Mater 22:3564–3567
50. Poodt P, Tiba V, Werner F, Schmidt J et al (2011) Ultrafast atomic layer deposition of alumina layers for solar cell passivation. J Electrochem Soc 158:H937–H940
51. Muñoz-Rojas D, Nguyen VH, de la Huerta CM, Aghazadehchors S et al (2017) Spatial Atomic Layer Deposition (SALD), an emerging tool for energy materials. Application to new-generation photovoltaic devices and transparent conductive materials. Comptes Rendus Phys 2017(18):391–400
52. Dey G, Elliott SD (2014) Copper(I) carbene hydride complexes acting both as reducing agent and precursor for Cu ALD: a study through density functional theory. Theor Chem Acc 133:1–7
53. Parker WD, Rondinelli JM, Nakhmanson SM (2011) First-principles study of misfit strain-stabilized ferroelectric $SnTiO_3$. Phys Rev B 84:245126
54. Agarwal R, Sharma Y, Chang S, Pitike KC et al (2018) Room-temperature relaxor ferroelectricity and photovoltaic effects in tin titanate directly deposited on a silicon substrate. Phys Rev B 97:54109
55. Filatova EA, Hausmann D, Elliott SD (2017) Investigating routes toward atomic layer deposition of silicon carbide: Ab initio screening of potential silicon and carbon precursors. J Vac Sci Technol A Vacuum Surfaces Film 35:01B103
56. Jung JSS, Lee SKK, Hong CSS, Shin JHH et al (2015) Atomic layer deposition of ZrO_2 thin film on Si (100) using η5: η1-$Cp(CH_2)_3NMeZr$ $(NMe_2)_2/O_3$ as precursors. Thin Solid Films 589:831–837
57. Karasulu B, Vervuurt RHJ, Kessels WMM, Bol AA (2016) Continuous and ultrathin platinum films on graphene using atomic layer deposition: a combined computational and experimental study. Nanoscale 8:19829–19845
58. Yang J, Cao K, Hu Q, Wen Y et al (2020) Unravelling the selective growth mechanism of AlOX with dimethylaluminum isopropoxide as a precursor in atomic layer deposition: a combined theoretical and experimental study. J Mater Chem A 8:4308–4317
59. Seo S, Yeo BC, Han SS, Yoon CM et al (2017) Reaction mechanism of area-selective atomic layer deposition for Al Al_2O_3 nanopatterns. ACS Appl Mater Interfaces 9:41607–41617

60. Park S, Park BE, Yoon H, Lee S et al (2020) Comparative study on atomic layer deposition of HfO_2: via substitution of ligand structure with cyclopentadiene. J Mater Chem C 8:1344–1352
61. Mukhopadhyay AB, Musgrave CB, Sanz JF (2008) Atomic layer deposition of hafnium oxide from hafnium chloride and water. J Am Chem Soc 130:11996–12006
62. Hu Z, Turner CH (2007) Atomic layer deposition of TiO_2 from TiI_4 and H_2O onto SiO_2 surfaces: ab initio calculations of the initial reaction mechanisms. J Am Chem Soc 129:3863–3878
63. Hu Z, Turner CH (2006) Initial surface reactions of TiO_2 atomic layer deposition onto SiO_2 surfaces: density functional theory calculations. J Phys Chem B 110:8337–8347
64. Cui C, Ren J (2012) A density functional theory study on the reactions of chlorine loss in ZrO_2 thin films by atomic-layer deposition. Comput Theor Chem 979:38–43
65. Han JH, Gao G, Widjaja Y, Garfunkel E (2004) A quantum chemical study of ZrO_2 atomic layer deposition growth reactions on the SiO_2 surface. Surf Sci 550:199–212
66. Shahmohammadi M, Mukherjee R, Takoudis CG, Diwekar UM (2021) Optimal design of novel precursor materials for the atomic layer deposition using computer-aided molecular design. Chem Eng Sci 234:116416
67. Shahmohammadi M, Mukherjee R, Takoudis CG, Diwekar UM (2021) Quantification of water impurity in an atomic layer deposition reactor using group contribution method. Res Dev Mater Sci 15:1703–1706
68. Puurunen RL (2004) Random deposition as a growth mode in atomic layer deposition. Chem Vap Depos 10:159–170
69. Fornari CI, Fornari G, Paulo HdO, Abramof E et al (2018) Monte Carlo simulation of epitaxial growth. In: Epitaxy, Zhong M (ed). BoD–Books on Demand, Norderstedt
70. Venables J (2000) Surface processes in epitaxial growth. In: Introduction to surface and thin film processes. Cambridge University Press, Cambridge, pp 144–151
71. Itagaki N, Nakamura Y, Narishige R, Takeda K et al (2020) Growth of single crystalline films on lattice-mismatched substrates through 3D to 2D mode transition. Sci Rep 10:4669
72. Vandalon V, Kessels WMM (2017) Revisiting the growth mechanism of atomic layer deposition of Al_2O_3: a vibrational sum-frequency generation study. J Vac Sci Technol A Vacuum Surfaces Film 35:05C313
73. Cremers V, Puurunen RL, Dendooven J (2019) Conformality in atomic layer deposition: Current status overview of analysis and modeling. Appl Phys Rev 6:21302
74. Dezelah CL (2012) Atomic layer deposition BT-encyclopedia of nanotechnology. In: encyclopedia of nanotechnology. In: Bhushan B (ed). Springer Netherlands, Dordrecht, pp 161–171
75. Bishal AK, Sukotjo C, Jokisaari JR, Klie RF et al (2018) Enhanced bioactivity of collagen fiber functionalized with room temperature atomic layer deposited titania. ACS Appl Mater Interfaces 10:34443–34454
76. Bishal AK, Sukotjo C, Takoudis CG (2017) Room temperature TiO_2 atomic layer deposition on collagen membrane from a titanium alkylamide precursor. J Vac Sci Technol A Vacuum Surfaces Film 35:01B134
77. Puurunen RL (2003) Growth per cycle in atomic layer deposition: a theoretical model. Chem Vap Depos 9:249–257
78. Kim J, Kim TW (2009) Initial surface reactions of atomic layer deposition. Jom 61:17–22
79. Richey NE, De Paula C, Bent SF (2020) Understanding chemical and physical mechanisms in atomic layer deposition. J Chem Phys 152:1–17
80. Puurunen RL (2005) Correlation between the growth-per-cycle and the surface hydroxyl group concentration in the atomic layer deposition of aluminum oxide from trimethylaluminum and water. Appl Surf Sci 245:6–10
81. Mui C, Musgrave CB (2004) Atomic layer deposition of HfO_2 using alkoxides as precursors. J Phys Chem B 108:15150–15164
82. Hu X, Schuster J, Schulz SE, Gessner T (2015) Surface chemistry of copper metal and copper oxide atomic layer deposition from copper(ii) acetylacetonate: a combined first-principles and reactive molecular dynamics study. Phys Chem Chem Phys 17:26892–26902

83. Choi W, Choudhary N, Han GH, Park J et al (2017) Recent development of two-dimensional transition metal dichalcogenides and their applications. Mater Today 20:116–130
84. Ahn EC (2020) 2D materials for spintronic devices. npj 2D Mater Appl 4:17
85. Huang X, Tan C, Yin Z, Zhang H (2014) 25th anniversary article: hybrid nanostructures based on two-dimensional nanomaterials. Adv Mater 26:2185
86. Koppens FH, Mueller T, Avouris P, Ferrari AC et al (2014) Photodetectors based on graphene, other two-dimensional materials and hybrid systems. Nat Nanotechnol 9:780–793
87. Zaidi SJA, Basit MA, Park TJ (2022) Advances in atomic layer deposition of metal sulfides: from a precursors perspective. Chem Mater. https://doi.org/10.1021/acs.chemmater.2c00954
88. Schilirò E, Lo Nigro R, Roccaforte F, Giannazzo F (2021) Substrate-driven atomic layer deposition of high-κ dielectrics on 2D materials. Appl Sci 11:11052
89. Zhu C, Xia X, Liu J, Fan ZD et al (2014) TiO_2 nanotube @ SnO_2 nanoflake core–branch arrays for lithium-ion battery anode. Nano Energy 4:105–112
90. Ansari MZ, Parveen N, Nandi DK, Ramesh R et al (2019) Enhanced activity of highly conformal and layered tin sulfide (SnSx) prepared by atomic layer deposition (ALD) on 3D metal scaffold towards high performance supercapacitor electrode. Sci Rep 9:10225
91. McDonnell S, Brennan B, Azcatl A, Lu N et al (2013) HfO_2 on MoS_2 by atomic layer deposition: adsorption mechanisms and thickness scalability. ACS Nano 7:10354–10361
92. Velický M, Toth PS (2017) From two-dimensional materials to their heterostructures: an electrochemist's perspective. Appl Mater Today 8:68–103
93. Liu H, Xu K, Zhang X, Ye PD (2012) The integration of high-k dielectric on two-dimensional crystals by atomic layer deposition. Appl Phys Lett 100:152115
94. Magda GZ, Petõ J, Dobrik G, Hwang C (2015) Highly responsive MoS_2 photodetectors enhanced by graphene quantum dots. Sci Rep 5:3
95. Coleman JN, Lotya M, O'Neill A, Bergin SD et al (2011) Two-dimensional nanosheets produced by liquid exfoliation of layered materials. Science 331:568–571
96. Bolotin KI (2014) Graphene: properties, preparation, characterisation and devices. Woodhead Publishing, Sawston Cambridge, p 199
97. Addou R, Colombo L, Wallace RM (2015) Surface defects on natural MoS_2. ACS Appl Mater Interfaces 7:11921–11929
98. Verhagen T, Guerra VLP, Haider G, Kalbac M et al (2020) Towards the evaluation of defects in MoS_2 using cryogenic photoluminescence spectroscopy. Nanoscale 12:3019–3028
99. Liao F, Yu J, Gu Z, Yang Z et al (2019) Enhancing monolayer photoluminescence on optical micro/nanofibers for low-threshold lasing. Sci Adv 5:eaax7398
100. Hess P (2021) Bonding, structure, and mechanical stability of 2D materials: the predictive power of the periodic table. Nanoscale Horiz 6:856–892
101. Marinov D, de Marneffe JF, Smets Q, Arutchelvan G et al (2021) Reactive plasma cleaning and restoration of transition metal dichalcogenide monolayers. npj 2D Mater Appl 5:17
102. Wang X, Tabakman SM, Dai H (2008) Atomic layer deposition of metal oxides on pristine and functionalized graphene. J Am Chem Soc 130:8152–8153
103. Cheng L, Qin X, Lucero AT, Azcatl A et al (2014) Atomic layer deposition of a high-k dielectric on MoS2 using trimethylaluminum and ozone. ACS Appl Mater Interfaces 6:11834–11838
104. Zhou W, Zou X, Najmaei S, Liu Z et al (2013) Intrinsic structural defects in monolayer molybdenum disulfide. Nano Lett 13:2615–2622
105. Ding X, Peng F, Zhou J, Gong W et al (2019) Defect engineered bioactive transition metals dichalcogenides quantum dots. Nat Commun 10:41
106. Puurunen RL (2005) Surface chemistry of atomic layer deposition: a case study for the trimethylaluminum/water process. J Appl Phys 97:121301
107. Farmer DB, Gordon RG (2006) Atomic layer deposition on suspended single-walled carbon nanotubes via gas-phase noncovalent functionalization. Nano Lett 6:699–703
108. Rammula R, Aarik L, Kasikov A, Kozlova J et al (2013) Atomic layer deposition of aluminum oxide films on graphene. 2013 IOP Conf Ser Mater Sci Eng 49:012014
109. Lazic P (2016) Physics of surface, interface and cluster catalysis. IOP, Bristol, pp 2.1–2.25
110. Lippert E (1960) The strengths of chemical bonds. Angew Chemie 72:602

111. Jeong SJ, Kim HW, Heo J, Lee MH et al (2016) Physisorbed-precursor-assisted atomic layer deposition of reliable ultrathin dielectric films on inert graphene surfaces for low-power electronics. 2D Mater 3:035027
112. Muneshwar T, Cadien K (2018) Surface reaction kinetics in atomic layer deposition: an analytical model and experiments. J Appl Phys 124:095302
113. Mattinen M, Gity F, Coleman E, Vonk JFA et al (2022) Atomic layer deposition of large-area polycrystalline transition metal dichalcogenides from 100 °C through control of plasma chemistry. Chem Mater. https://doi.org/10.1021/acs.chemmater.2c01154
114. Nam T, Seo S, Kim H (2020) Atomic layer deposition of a uniform thin film on two-dimensional transition metal dichalcogenides. J Vac Sci Technol A 38:030803
115. Radisavljevic B, Radenovic A, Brivio J, Giacometti V et al (2011) Single-layer MoS2 transistors. Nat Nanotechnol 6:147–150
116. Jena D, Konar A (2007) Enhancement of carrier mobility in semiconductor nanostructures by dielectric engineering. Phys Rev Lett 98:136805
117. Konar A, Fang T, Jena D (2010) Effect of high-K gate dielectrics on charge transport in graphene-based field effect transistors. Phys Rev B 82:115452
118. Chen F, Xia J, Ferry DK, Tao N (2009) Dielectric screening enhanced performance in graphene FET. Nano Lett 9:2571–2574
119. Newaz AKM, Puzyrev YS, Wang B, Pantelides ST et al (2012) Probing charge scattering mechanisms in suspended graphene by varying its dielectric environment. Nat Commun 3:734
120. Robertson J (2004) High dielectric constant oxides. Eur Phys J Appl Phys 28:265–291
121. Kobayashi NP, Donley CL, Wang SY, Williams RS (2007) Atomic layer deposition of aluminum oxide on hydrophobic and hydrophilic surfaces. J Cryst Growth 299:218–222
122. Son S, Yu S, Choi M, Kim D et al (2015) Improved high temperature integration of $Al2O_3$ on MoS2 by using a metal oxide buffer layer. Appl Phys Lett 106:021601
123. Fang H, Chuang S, Chang TC, Takei K et al (2012) High-performance single layered WSe2 p-FETs with chemically doped contacts. Nano Lett 12:3788–3792
124. Qiu H, Pan L, Yao Z, Li J et al (2012) Electrical characterization of back-gated bi-layer MoS2 field-effect transistors and the effect of ambient on their performances. Appl Phys Lett 100:123104
125. Lin YC, Lu CC, Yeh CH, Jin C et al (2012) Graphene annealing: how clean can it be? Nano Lett 12:414–419
126. Chandra L, Goyal M, Srivastava D (2021) Minimally invasive intraarticular platelet rich plasma injection for refractory temporomandibular joint dysfunction syndrome in comparison to arthrocentesis. J Family Med Prim Care 10:254–258
127. Zotti F, Albanese M, Rodella LF, Nocini PF (2019) Platelet-rich plasma in treatment of temporomandibular joint dysfunctions. Narrative Rev Int J Mol Sci 20:277
128. Nitecka-Buchta A, Walczynska-Dragon K, Kempa WM, Baron S (2019) Platelet-rich plasma intramuscular injections—antinociceptive therapy in myofascial pain within masseter muscles in temporomandibular disorders patients: a pilot study. Front Neurol 10:250
129. Kumar N, Yanguas-Gil A, Daly SR, Girolami GS et al (2009) Remote plasma treatment of Si surfaces: Enhanced nucleation in low-temperature chemical vapor deposition. Appl Phys Lett 95:144107
130. Hsueh YC, Wang CC, Liu C, Kei CC et al (2012) Deposition of platinum on oxygen plasma treated carbon nanotubes by atomic layer deposition. Nanotechnology 23:405603
131. Vervuurt RHJ, Karasulu B, Verheijen MA, Kessels WMM et al (2017) Uniform atomic layer deposition of Al_2O_3 on graphene by reversible hydrogen plasma functionalization. Chem Mater 29:2090–2100
132. Nan H, Zhou R, Gu X, Xiao S et al (2019) Recent advances in plasma modification of 2D transition metal dichalcogenides. Nanoscale 11:19202–19213
133. Azcatl A, McDonnell S, Santosh KC, Peng X et al (2014) MoS2 functionalization for ultra-thin atomic layer deposited dielectrics. Appl Phys Lett 104:111601
134. Vig JR (1985) UV/ozone cleaning of surfaces. J Vac Sci Technol A 3:1027

135. Clark T, Ruiz JD, Fan H, Brinker CJ et al (2000) A new application of UV−ozone treatment in the preparation of substrate-supported, mesoporous thin films. Chem Mater 12:3879–3884
136. Brown NMD, Cui N, McKinley A (1998) An XPS study of the surface modification of natural MoS_2 following treatment in an RF-oxygen plasma. Appl Surf Sci 134:11–21
137. Zhang H, Chiappe D, Meersschaut J, Conard T et al (2017) Nucleation and growth mechanisms of Al_2O_3 atomic layer deposition on synthetic polycrystalline MoS_2. J Chem Phys 146:052810
138. Wang CT, Ting CC, Kao PC, Li SR et al (2018) Improvement of OLED performance by tuning of silver oxide buffer layer composition on silver grid surface using UV-ozone treatment. Appl Phys Lett 113:051602
139. Su Z, Wang L, Li Y, Zhao H et al (2012) Ultraviolet-ozone-treated PEDOT:PSS as anode buffer layer for organic solar cells. Nanoscale Res Lett 7:465
140. Kim H (2011) Characteristics and applications of plasma enhanced-atomic layer deposition. Thin Solid Films 519:6639–6644
141. Rossnagel SM, Sherman A, Turner F (2000) Plasma-enhanced atomic layer deposition of Ta and Ti for interconnect diffusion barriers. J Vac Sci Technol B 18:2016
142. Kim H, Rossnagel SM (2002) Plasma–surface interactions. J Vac Sci Technol A 20:S145
143. Potts SE, Keuning W, Langereis E, Dingemans G et al (2010) Low temperature plasma-enhanced atomic layer deposition of metal oxide thin films. J Electrochem Soc 157:P66
144. Kim H, Chung I, Kim S, Shin S et al (2015) Improved film quality of plasma enhanced atomic layer deposition SiO_2 using plasma treatment cycle. J Vac Sci Technol A 33:01A146
145. Azcatl A, Qin X, Prakash A, Zhang C et al (2016) Covalent nitrogen doping and compressive strain in MoS_2 by remote N_2 plasma exposure. Nano Lett 16:5437–5443
146. McDonnell S, Addou R, Buie C, Wallace RM et al (2014) Defect-dominated doping and contact resistance in MoS_2. ACS Nano 8:2880–2888
147. Price KM, Najmaei S, Ekuma CE, Burke RA et al (2019) Plasma-enhanced atomic layer deposition of HfO_2 on monolayer, bilayer, and trilayer MoS_2 for the integration of high-κ dielectrics in two-dimensional devices. ACS Appl Nano Mater 2:4085–4094
148. Balasubramanyam B, van Ommeren F et al (2020) Probing the origin and suppression of vertically oriented nanostructures of 2D WS2 layers. ACS Appl Mater Interfaces 12:3873–3885
149. Kozhakhmetov A, Torsi R, Chen CY, Robinson JA (2020) Scalable low-temperature synthesis of two-dimensional materials beyond graphene. J Phys Mater 4:012001
150. Walter TN, Lee S, Zhang X, Chubarov M et al (2019) Atomic layer deposition of ZnO on MoS_2 and WSe_2. Appl Sur Sci 480:43–51
151. Bratton D, Yang D, Dai J, Ober CK (2006) Recent progress in high-resolution lithography. Polym Adv Technol 17:94–103
152. Seisyan RP (2011) Nanolithography in microelectronics: a review. Tech Phys 56:1061–1073
153. Mackus AJM, Merkx MJM, Kessels WMM (2019) From the bottom-up: toward area-selective atomic layer deposition with high selectivity. Chem Mater 31:2–12
154. Tan LK, Liu B, Teng JH, Guo S et al (2014) Atomic layer deposition of a MoS2 film. Nanoscale 6:10584–10588
155. Browning R, Padigi P, Solanki R, Tweet DJ et al (2015) Atomic layer deposition of MoS_2 thin films Mater. Res Express 2:035006
156. Valdivia A, Tweet DJ Jr, Conley JF (2016) Atomic layer deposition of two dimensional MoS2 on 150 m006D substrates. J Vac Sci Technol A 34:021515
157. Huang Y, Liu L, Zhao W, Chen Y (2017) Preparation and characterization of molybdenum disulfide films obtained by onestep atomic layer deposition method. Thin Solid Films 624:101–105
158. Liu L, Huang Y, Sha J, Chen Y (2017) Layer-controlled precise fabrication of ultrathin MoS_2 films by atomic layer deposition. Nanotechnology 28:195605
159. Kim Y, Song JG, Park YJ, Ryu GH et al (2016) Self-limiting layer synthesis of transition metal dichalcogenides. Sci Rep 6:18754
160. Yu Y, Li C, Liu Y, Su L et al (2013) Controlled scalable synthesis of uniform, high-quality monolayer and few layer MoS_2 films. Sci Rep 3:1866

161. Nandi DK, Sen UK, Choudhury D, Mitra S et al (2014) Atomic layer deposited MoS2 as a carbon and binder free anode in Li-ion battery. Electrochim Acta 146:706–713
162. Pyeon JJ, Kim SH, Jeong DS, Baek SH et al (2016) Wafer-scale growth of MoS_2 thin films by atomic layer deposition. Nanoscale 8:10792–10798
163. Jang Y, Yeo S, Lee HBR, Kim H et al (2016) Wafer-scale, conformal and direct growth of MoS_2 thin films by atomic layer deposition. Appl Surf Sci 365:160–165
164. Jin Z, Shin S, Kwon DH, Han SJ et al (2014) Novel chemical route for atomic layer deposition of MoS_2 thin film on SiO_2/Si substrate. Nanoscale 6:14453–14458
165. Jurca T, Moody MJ, Henning A, Emery JD et al (2017) Low-temperature atomic layer deposition of MoS_2 films. Angew Chem Int Ed 56:4991–4995
166. Cadot S, Renault O, Fregnaux M, Rouchon D et al (2017) A novel 2-step ALD route to ultra-thin MoS2 films on SiO_2 through a surface organometallic intermediate. Nanoscale 9:538–546
167. Mane AU, Letourneau S, Mandia DJ, Liu J (2018) Atomic layer deposition of molybdenum disulfide films using MoF_6 and H_2S. J Vac Sci Technol A 36:01A125
168. Krbal M, Prikryl J, Zazpe R, Dvorak F et al (2018) 2D MoSe2 structures prepared by atomic layer deposition. Phys Status Solidi RRL 12:1800023
169. Dai TJ, Fan XD, Ren YX, Hou S et al (2018) Layer-controlled synthesis of waferscale $MoSe_2$ nanosheets for photodetector arrays. J Mater Sci 53:8436–8444
170. Kim DH, Ramesh R, Nandi DK, Bae JS et al (2021) Atomic layer deposition of tungsten sulfide using a new metal-organic precursor and H 2 S: thin film catalyst for water splitting. Nanotechnology 32:075405
171. Scharf T, Prasad SV, Mayer T, Goeke R et al (2004) Atomic layer deposition of tungsten disulphide solid lubricant thin films. J Mater Res 19:3443–3446
172. Scharf TW, Diercks DR, Gorman BP, Prasad SV et al (2009) Atomic layer deposition of tungsten disulphide solid lubricant nanocomposite coatings on rolling element bearings. Tribol Trans 52:284–292
173. Scharf TW, Prasad SV, Dugger MT, Kotula PG et al (2006) Growth, structure, and tribological behavior of atomic layer-deposited tungsten disulphide solid lubricant coatings with applications to MEMS. Acta Mater 54:4731–4743
174. Delabie A, Caymax M, Groven B, Heyne M et al (2015) Low temperature deposition of 2D WS2 layers from WF6 and H2S precursors: impact of reducing agents. Chem Commun 51:15692–15695
175. Groven B, Heyne M, Nalin Mehta A, Bender H et al (2017) Plasmaenhanced atomic layer deposition of twodimensional WS2 from WF6, H2 plasma, and H2S. Chem Mater 29:2927–2938
176. Groven B, Nalin Mehta A, Bender H, Meersschaut J et al (2018) "Two dimensional crystal grain size tuning in WS2 atomic layer deposition: an insight in the nucleation mechanism. Chem Mater 30:7648–7663
177. Groven B, Mehta AN, Bender H, Smets Q et al (2018) "Nucleation mechanism during WS2 plasma enhanced atomic layer deposition on amorphous $Al2O_3$ and sapphire substrates. J Vac Sci Technol A 36:01A105
178. Sun Y, Chai Z, Lu X, He D (2017) A direct atomic layer deposition method for growth of ultra-thin lubricant tungsten disulfide films. Sci China Technol Sci 60:51–57
179. Yeo S, Nandi DK, Rahul R, Kim TH et al (2018) Low-temperature direct synthesis of high quality WS2 thin films by plasma-enhanced atomic layer deposition for energy related applications. Appl Surf Sci 459:596–605
180. Kunene TJ, Tartibu LK, Ukoba K, Jen TC (2022) Review of atomic layer deposition process, application and modeling tools. Mater Today Proc 62:S95–S109
181. Mattinen M, Leskelä M, Ritala M (2021) Atomic layer deposition of 2D metal dichalcogenides for electronics, catalysis, energy storage, and beyond. Adv Mater Interfaces 8:2001677
182. Zhukush M, Camp C, Galipaud J, Perathoner S et al (2022) Surface organometallic chemistry for ALD growth of ultra-thin films of WS2 and their photo(electro)catalytic performances. In: ISGC 2022: international symposium on green chemistry, La Rochelle, France. ⟨hal-03670014⟩

183. Park K, Kim Y, Song JG, Jin Kim S et al (2016) Uniform, large-area self-limiting layer synthesis of tungsten diselenide. 2D Mater 3:014004
184. Mattinen M, Popov G, Vehkamaki M, King PJ et al (2019) Atomic layer deposition of emerging 2D semiconductors, HfS2 and ZrS2, for optoelectronics. Chem Mater 31:5713–5724
185. Rahman M, Davey K, Qiao SZ (2017) Advent of 2D rhenium disulfide (ReS2): fundamentals to applications. Adv Funct Mater 27:1606129
186. Tongay S, Sahin H, Ko C, Luce A et al (2014) Monolayer behaviour in bulk ReS2 due to electronic and vibrational decoupling. Nat Commun 5:3252
187. Hama lainen J, Mattinen M, Mizohata K, Meinander K et al (2018) Atomic layer deposition of rhenium disulphide. Adv Mater 30:1703622
188. Basuvalingam SB, Zhang Y, Bloodgood MA, Godiksen RH et al (2019) Low-temperature phase-controlled synthesis of titanium Di- and Tri-sulfide by atomic layer deposition. Chem Mater 31:9354–9362
189. Mahuli N (2015) Atomic layer deposition of titanium sulfide and its application in extremely thin absorber solar cells. J Vacu Sci Tech A 33:01A150
190. Zhang Q, Ren K, Zheng R, Huang Z et al (2022) First-principles calculations of two-dimensional CdO/HfS2 Van der Waals heterostructure: direct Z-scheme photocatalytic water splitting. Front Chem 10:879402
191. Wang X (2021) Atomic layer deposition of iron, cobalt, and nickel chalcogenides: progress and outlook. Chem Mater 33:6251–6268
192. Guo Z, Wang X (2018) Atomic layer deposition of the metal pyrites FeS_2, CoS_2, and NiS_2. Angew Chem Int Ed 57:5898
193. Wells SA, Henning A, Gish JT, Sangwan VK et al (2018) Suppressing ambient degradation of exfoliated InSe nanosheet devices via seeded atomic layer deposition encapsulation. Nano Lett 18:7876–7882
194. Mughal MA, Engelken R, Sharma R (2015) Progress in indium (III) sulfide (In2S3) buffer layer deposition techniques for CIS, CIGS, and CdTe-based thin film solar cells. Sol Energy 120:131–146
195. Kim SB, Sinsermsuksakul P, Hock AS, Pike RD et al (2014) Synthesis of N-heterocyclic Stannylene (Sn(II)) and Germylene (Ge(II)) and a Sn(II) amidinate and their application as precursors for atomic layer deposition. Chem Mater 26:3065–3073
196. Kwon DS, Jeon W, Kim DG, Kim TK et al (2021) Improved properties of the atomic layer deposited Ru electrode for dynamic random-access memory capacitor using discrete feeding method. ACS Appl Mater Interfaces 13:23915–23927
197. Sinsermsuksakul P, Heo J, Noh W, Hock AS et al (2011) Atomic layer deposition of tin monosulfide thin films. Adv Energy Mater 1:1116–1125
198. Lee N, Lee G, Choi H, Park H et al (2019) Atomic layer deposition growth of SnS2 films on diluted buffered oxide etchant solution-treated substrate. Appl Sur Sci 496:143689
199. Zhang K, Nminibapiel D, Tangirala M, Baumgart H et al (2013) Fabrication of Sb2Te3 and Bi2Te3 multilayer composite films by atomic layer deposition. ECS Trans 50:3
200. Baitimirova M, Andzane J, Viter R, Fraisse B et al (2019) Improved crystalline structure and enhanced photoluminescence of ZnO nanolayers in Bi2Se3/ZnO heterostructures. J Phys Chem C 123:31156–31166
201. Lim SS, Kim KC, Jeon H, Kim JY et al (2020) Enhanced thermal stability of Bi2Te3-based alloys via interface engineering with atomic layer deposition. J Euro Ceramic Soc 40:3592–3599
202. Wei Z, Hai Z, Akbari MK, Qi D et al (2018) Atomic layer deposition-developed two-dimensional α-MoO_3 windows excellent hydrogen peroxide electrochemical sensing capabilities. Sens Actuat B: Chemica 262:334–344
203. Wei Z, Hai Z, Akbari MK, Hu J et al (2018) Ultrasensitive, sustainable, and selective electrochemical hydrazine detection by ALD-developed two-dimensional WO_3. Chem Electro Chem 5:266
204. Salahuddin S, Ni K, Datta S (2018) The era of hyper-scaling in electronics. Nat Electron 1:442–450

205. Zhang X, Li S, Yu H (2018) Research of science and technology strategic base on the international technology roadmap for semiconductors. In: Proceedings of the international conference on electronics and electrical engineering technology. ACM, New York, pp 69–72
206. Cavin RK, Lugli P, Zhirnov VV (2012) Prolog to the section on science and engineering beyond Moore's law. In: Proceedings of the IEEE, pp 1718–1719
207. Liao W, Zhao S, Li F, Wang C (2020) Interface engineering of two-dimensional transition metal dichalcogenides towards next-generation electronic devices: recent advances and challenges. Nanoscale Horiz 5:787–807
208. Ritala M, Niinistö J (2009) Industrial applications of atomic layer deposition. ECS Trans 25:641–652
209. Raaijmakers IJ (2011) Current and future applications of ALD in micro-electronics. ECS Trans 41:3–17
210. Mirabelli G, McGeough C, Schmidt M, McCarthy EK et al (2016) Air sensitivity of MoS2, MoSe2, MoTe2, HfS2, and HfSe2. J Appl Phys 120:125102
211. Chae SH, Jin Y, Kim TS, Chung DS et al (2016) Oxidation effect in octahedral hafnium disulfide thin film. ACS Nano 10:1309–1316
212. Mañas-Valero S, García-López V, Cantarero A, Galbiati M (2016) Raman spectra of ZrS2 and ZrSe2 from bulk to atomically thin layers. Appl Sci 6:264
213. Xu K, Huang Y, Chen B, Xia Y et al (2016) Toward high-performance top-gate ultrathin HfS2 field-effect transistors by interface engineering. Small 12:3106–3111
214. Kanazawa T, Amemiya T, Upadhyaya V, Ishikawa A et al (2017) Performance improvement of HfS2 transistors by atomic layer deposition of HfO2. IEEE Trans Nanotechnol 16:582–587
215. Xu Y, Cheng C, Du S, Yang J et al (2016) Contacts between two- and three- dimensional materials: Ohmic, Schottky, and p-n heterojunctions. ACS Nano 10:4895–4919
216. Fang H, Hu W (2017) Photogating in low dimensional photodetectors. Adv Sci 4:1700323
217. Thakar K, Mukherjee B, Grover S, Kaushik N et al (2018) Multilayer ReS2 photodetectors with gate tunability for high responsivity and high speed applications. ACS Appl Mater Interfaces 10:36512–36522
218. Yan C, Gan L, Zhou X, Guo J et al (2017) Space-confined chemical vapor deposition synthesis of ultrathin HfS2 flakes for optoelectronic application. Adv Funct Mater 27:1702918
219. Yang J, Kim S, Choi W, Park SH et al (2013) Improved growth behavior of atomic-layer-deposited high-k dielectrics on multilayer MoS2 by oxygen plasma pretreatment. ACS Appl Mater Interfaces 5:4739–4744
220. Roubi L, Carlone C (1988) Resonance Raman spectrum of HfS_2 and ZrS2. Phys Rev B 37:6808
221. Xu K, Wang ZX, Wang F, Huang Y et al (2015) Ultrasensitive phototransistors based on few-layered HfS2. Adv Mater 27:7881–7887
222. Cao W, Kang JH, Sarkar D, Liu W et al (2015) 2D semiconductor FETs—projections and design for sub-10 nm VLSI. IEEE Trans Electron Devices 62:3459
223. Allain A, Kang JH, Banerjee K, Kis A (2015) Electrical contacts to two-dimensional semiconductors. Nat Mater 14:1195–1205
224. Wang X, Wang P, Wang J, Hu W et al (2015) Ultrasensitive and broadband MoS2 photodetector driven by ferroelectrics. Adv Mater 27:6575–6581
225. Conroy LE, Park KC (1968) Electrical properties of the Group IV disulfides, titanium disulfide, zirconium disulfide, hafnium disulfide and tin disulfide. Inorg Chem 7:459–463
226. Yang H, Wang Y, Zou X, Bai R et al (2021) Wafer-scale synthesis of WS_2 films with in situ controllable p-type doping by atomic layer deposition. Research 9862483:1–9
227. Nazir G, Khan MF, Iermolenko VM et al (2016) Two and four-probe field-effect and Hall mobilities in transition metal dichalcogenide field-effect transistors. RSC Adv 6:60787–60793
228. Park T, Kim H, Leem M, Ahn W et al (2017) Atomic layer deposition of Al_2O_3 on MoS_2, WS_2, WSe_2, and h-BN: surface coverage and adsorption energy. RSC Adv 7:884–889
229. Aspiotis N, Morgan K, März, Müller-Caspary K et al (2022) Scalable, highly crystalline, 2D semiconductor atomic layer deposition process for high performance electronic applications. https://arxiv.org/ftp/arxiv/papers/2203/2203.10309.pdf

9789819602469